GEOMAGNETISM

Treatise on Geophysics

GEOMAGNETISM

Editor-in-Chief
Professor Gerald Schubert
Department of Earth and Space Sciences and Institute of Geophysics and Planetary Physics,
University of California Los Angeles, Los Angeles, CA, USA

Volume Editor
Dr. Masaru Kono
Tokyo Institute of Technology, Tokyo, Japan

ELSEVIER

AMSTERDAM • BOSTON • HEIDELBERG • LONDON • NEW YORK • OXFORD
PARIS • SAN DIEGO • SAN FRANCISCO • SINGAPORE • SYDNEY • TOKYO

Elsevier B.V.
Radarweg 29, 1043 NX Amsterdam, the Netherlands

First edition 2009

British Library Cataloguing in Publication Data
A catalogue record for this book is available from the British Library

Library of Congress Control Number: 2009929983

ISBN: 978-0-444-53461-3

For information on all Elsevier publications
visit our website at elsevierdirect.com

Contents

Preface

Geophysics is the physics of the Earth, the science that studies the Earth by measuring the physical consequences of its presence and activity. It is a science of extraordinary breadth, requiring 10 volumes of this treatise for its description. Only a treatise can present a science with the breadth of geophysics if, in addition to completeness of the subject matter, it is intended to discuss the material in great depth. Thus, while there are many books on geophysics dealing with its many subdivisions, a single book cannot give more than an introductory flavor of each topic. At the other extreme, a single book can cover one aspect of geophysics in great detail, as is done in each of the volumes of this treatise, but the treatise has the unique advantage of having been designed as an integrated series, an important feature of an interdisciplinary science such as geophysics. From the outset, the treatise was planned to cover each area of geophysics from the basics to the cutting edge so that the beginning student could learn the subject and the advanced researcher could have an up-to-date and thorough exposition of the state of the field. The planning of the contents of each volume was carried out with the active participation of the editors of all the volumes to insure that each subject area of the treatise benefited from the multitude of connections to other areas.

Geophysics includes the study of the Earth's fluid envelope and its near-space environment. However, in this treatise, the subject has been narrowed to the solid Earth. The *Treatise on Geophysics* discusses the atmosphere, ocean, and plasmasphere of the Earth only in connection with how these parts of the Earth affect the solid planet. While the realm of geophysics has here been narrowed to the solid Earth, it is broadened to include other planets of our solar system and the planets of other stars. Accordingly, the treatise includes a volume on the planets, although that volume deals mostly with the terrestrial planets of our own solar system. The gas and ice giant planets of the outer solar system and similar extra-solar planets are discussed in only one chapter of the treatise. Even the *Treatise on Geophysics* must be circumscribed to some extent. One could envision a future treatise on Planetary and Space Physics or a treatise on Atmospheric and Oceanic Physics.

Geophysics is fundamentally an interdisciplinary endeavor, built on the foundations of physics, mathematics, geology, astronomy, and other disciplines. Its roots therefore go far back in history, but the science has blossomed only in the last century with the explosive increase in our ability to measure the properties of the Earth and the processes going on inside the Earth and on and above its surface. The technological advances of the last century in laboratory and field instrumentation, computing, and satellite-based remote sensing are largely responsible for the explosive growth of geophysics. In addition to the enhanced ability to make crucial measurements and collect and analyze enormous amounts of data, progress in geophysics was facilitated by the acceptance of the paradigm of plate tectonics and mantle convection in the 1960s. This new view of how the Earth works enabled an understanding of earthquakes, volcanoes, mountain building, indeed all of geology, at a fundamental level. The exploration of the planets and moons of our solar system, beginning with the Apollo missions to the Moon, has invigorated geophysics and further extended its purview beyond the Earth. Today geophysics is a vital and thriving enterprise involving many thousands of scientists throughout the world. The interdisciplinarity and global nature of geophysics identifies it as one of the great unifying endeavors of humanity.

The keys to the success of an enterprise such as the *Treatise on Geophysics* are the editors of the individual volumes and the authors who have contributed chapters. The editors are leaders in their fields of expertise, as distinguished a group of geophysicists as could be assembled on the planet. They know well the topics that had to be covered to achieve the breadth and depth required by the treatise, and they know who were the best of their colleagues to write on each subject. The list of chapter authors is an impressive one, consisting of geophysicists who have made major contributions to their fields of study. The quality and coverage achieved by this group of editors and authors has insured that the treatise will be the definitive major reference work and textbook in geophysics.

Each volume of the treatise begins with an 'Overview' chapter by the volume editor. The Overviews provide the editors' perspectives of their fields, views of the past, present, and future. They also summarize the contents of their volumes and discuss important topics not addressed elsewhere in the chapters. The Overview chapters are excellent introductions to their volumes and should not be missed in the rush to read a particular chapter. The title and editors of the 10 volumes of the treatise are:

Volume 1: Seismology and Structure of the Earth

> Barbara Romanowicz
> University of California, Berkeley, CA, USA
>
> Adam Dziewonski
> Harvard University, Cambridge, MA, USA

Volume 2: Mineral Physics

> G. David Price
> University College London, UK

Volume 3: Geodesy

> Thomas Herring
> Massachusetts Institute of Technology, Cambridge, MA, USA

Volume 4: Earthquake Seismology

> Hiroo Kanamori
> California Institute of Technology, Pasadena, CA, USA

Volume 5: Geomagnetism

> Masaru Kono
> Okayama University, Misasa, Japan

Volume 6: Crust and Lithosphere Dynamics

> Anthony B. Watts
> University of Oxford, Oxford, UK

Volume 7: Mantle Dynamics

> David Bercovici
> Yale University, New Haven, CT, USA

Volume 8: Core Dynamics

> Peter Olson
> Johns Hopkins University, Baltimore, MD, USA

Volume 9: Evolution of the Earth

> David Stevenson
> California Institute of Technology, Pasadena, CA, USA

Volume 10: Planets and Moons

> Tilman Spohn
> Deutsches Zentrum für Luft-und Raumfahrt, GER

In addition, an eleventh volume of the treatise provides a comprehensive index.

The *Treatise on Geophysics* has the advantage of a role model to emulate, the highly successful *Treatise on Geochemistry*. Indeed, the name *Treatise on Geophysics* was decided on by the editors in analogy with the geochemistry compendium. The *Concise Oxford English Dictionary* defines treatise as "a written work dealing formally and systematically with a subject." Treatise aptly describes both the geochemistry and geophysics collections.

The *Treatise on Geophysics* was initially promoted by Casper van Dijk (Publisher at Elsevier) who persuaded the Editor-in-Chief to take on the project. Initial meetings between the two defined the scope of the treatise and led to invitations to the editors of the individual volumes to participate. Once the editors were on board, the details of the volume contents were decided and the invitations to individual chapter authors were issued. There followed a period of hard work by the editors and authors to bring the treatise to completion. Thanks are due to a number of members of the Elsevier team, Brian Ronan (Developmental Editor), Tirza Van Daalen (Books Publisher), Zoe Kruze (Senior Development Editor), Gareth Steed (Production Project Manager), and Kate Newell (Editorial Assistant).

G. Schubert
Editor-in-Chief

Contributors

W. Baumjohann
Space Research Institute, Austrian Academy of Sciences, Graz, Austria

J. E. T. Channell
University of Florida, Gainesville, FL, USA

C. Constable
University of California at San Diego, La Jolla, CA, USA

S. Constable
University of California, San Diego, La Jolla, CA, USA

D. J. Dunlop
University of Toronto, Toronto, ON, Canada

D. A. D. Evans
Yale University, New Haven, CT, USA

C. C. Finlay
ETH Zürich, Zurich, Switzerland

J. S. Gee
University of California, San Diego, La Jolla, CA, USA

G. Hulot
Institut de Physique du Globe de Paris/CNRS and Université Paris 7, Paris, France

A. Jackson
ETH Zürich, Zurich, Switzerland

C. L. Johnson
University of British Columbia, Vancouver, BC, Canada

D. V. Kent
Rutgers University, Piscataway, NJ, USA

J. L. Kirschvink
California Institute of Technology, Pasadena, CA, USA

M. Kono
Okayama University, Misasa, Japan

C. Laj
Laboratoire des Sciences du Climat et de l'Environment, Unité Mixte CEA-CNRS-UVSQ, Gif-sur-Yvette, France

P. McFadden
Geoscience Australia, Canberra, ACT, Australia

R. Nakamura
Space Research Institute, Austrian Academy of Sciences, Graz, Austria

N. Olsen
Danish National Space Center/DTU and Niels Bohr Institute, University of Copenhagen, Copenhagen, Denmark

Ö. Özdemir
University of Toronto, Toronto, ON, Canada

M. E. Purucker
Goddard Space Flight Center/NASA, Greenbelt, MD, USA

J. L. Rasson
Institut Royal Meteorologique, Dourbes, Belgium

T. D. Raub
Yale University, New Haven, CT, USA

C. V. Reeves
Earthworks BV, Delft, The Netherlands

T. J. Sabaka
Raytheon at NASA Goddard Space Flight Center, Greenbelt, MD, USA

L. Tauxe
University of California San Diego, La Jolla, CA, USA

G. M. Turner
Victoria University of Wellington, Wellington, New Zealand

K. A. Whaler
University of Edinburgh, Edinburgh, UK

T. Yamazaki
Geological Survey of Japan, Tsukuba, Ibaraki, Japan

1 Geomagnetism in Perspective

M. Kono, Okayama University, Misasa, Japan

1.1 Early History

The Earth has its own magnetic field (the geomagnetic field), which is confined by the action of the solar wind into a volume called the magnetosphere (*see* Chapter 3). This field is not steady, but varies with time due partly to the interaction with the solar wind, but more importantly by its own physical processes. Direct observation of such changes has been carried out only in the last few centuries, but with indirect measurements we can understand the field behavior millions of years back in time. In this extended time frame, there is evidence that the polarity of the magnetic field reversed frequently, and that the magnetic dipole axis in very ancient times was significantly displaced from the present rotational axis (the North and South geographic Poles).

It is of considerable interest how such knowledge was acquired over several centuries. We will take a brief tour of the historical events that provided important steps in formulating our understanding of the geomagnetic field. In doing so, we have to rely solely on the written records, which is the reason why only the European and Chinese histories are referred. There are many works on this topic; among them, the important ones are Mitchell (1932–46), Harradon (1943–45), Needham (1962), and Yamamoto (2003). The English translations of Chinese literature below were taken from Needham (1962). Chinese sentences given together with English were taken from the Japanese translation of this book (Hashimoto *et al.*, 1977).

When we talk about the earliest recognition of the magnetism of the Earth, we should be careful to discriminate two separate issues; that is, the attractive force exerted by a magnet on iron, and the north- (or south-) seeking property of the magnet. The former can be taken as the forerunner to the science of magnetism, while the latter is the basis for appreciation of the magnetic field associated with the Earth. Our main interest is in the geomagnetic field, but it is necessary to look into magnets first.

1.1.1 Attractive Force of the Magnets

The earliest observation of the natural magnets (lodestone or loadstone) is attributed to the Greek philosopher Thales of Miletos (624–546 BC). Thales did not leave any writings of his own, but Aristoteles (384–332 BC) wrote about him in *De Anima* ('On the soul') about two centuries later. According to this, Thales taught that the lodestone has a soul, because it could set another body (iron) in motion. Diogenes Laertius also wrote that Thales admitted that souls exist even in nonliving matters based on the observation that the magnets and ambers can attract things. This suggests that Thales knew not only about the attractive force of magnets, but also that due to the static electricity of ambers, that can be seen when they are rubbed by clothes (Mitchell, 1937).

References to the attractive force of magnets appear quite often in Greek manuscripts (e.g., Platon, Aristoteles, Democritus, Lucretius), and there is no doubt that this force was well known to the ancient Greeks. This may be because the attractive force appeared to them as a very remarkable phenomenon since it can act on materials which are not in contact. In these, the magnets were referred to mostly as the rock of Magnesia ($\lambda\iota\theta o\varsigma\ \mu\alpha\gamma\nu\eta\sigma\iota\eta$). Magnesia is the name of a place either in Macedonia, Crete, or Asia Minor. The names of magnetism as well as magnetite (Fe_3O_4) were derived from this Greek word.

Ancient Chinese people made similar observations, but the records are somewhat later than the corresponding Greek ones. In *Lü Shih Chhun Chhiu* 呂氏春秋 (Master Lü's Spring and Autumn Annals), written in the late third century BC, it is said that "the lodestone calls the iron to itself, or attracts it" 慈石召鐵、或引之也. After that, reference to magnets appear abundantly in the Chinese literature (e.g., *Huai Nan Tzu* 淮南子 (The Book of Huai Nan) in the first century BC, and *Lun Hêng* 論衡 (Discourses Weighed in the Balance) in 83 AD).

The attractive force that magnets exert on iron was a wonder in ancient times, and it was often attributed to magical power. Its full understanding had to wait until the nineteenth century when the magnetic force was explained by physical theorems such as Ampère's and Gauss's laws in the framework of electromagnetic theory.

1.1.2 Early Chinese Compasses

The fact that magnets have the property to align in the north–south direction was discovered by the ancient Chinese. From about the second century AD, there are many Chinese texts referring to "south-pointing carriage" which, in many instances, were described as guiding the soldiers in thick fogs to the right direction to beat the enemies. Many people thought that this was a device that used the property of magnets. However, it is now considered to be some mechanical device made up of gears and axles rather than an instrument similar to a magnetic compass (Needham, 1962). A more interesting sentence appears in the above-mentioned *Lun Hêng* (AD 83), which means that "when the south-controlling ladle is thrown upon the ground, it comes to rest pointing at the south" 司南之杓。投之於地。其柢指南. Wang Chen-To (1948) suggested that the first two letters (south-controlling) were changed in the process of hand copying from the original 'south-pointing', the fourth letter (ladle) means a spoon worked out from a lodestone into that shape, and the eighth letter (ground) actually indicates a diviner's board. Now, a diviner's board was used in ancient China for the purpose of telling fortunes, and it is inscribed with the constellation of Great Bear in the center, and the names of 24 directions on the circle around it. With these interpretations, the sentence can be taken to describe an instrument for seeking south using a magnet! Note that the Great Bear is the symbol of the pole and the spoon also has a shape reminiscent of its form.

Wang (1948) went further to show the credibility of his interpretation, by making a model of this instrument, with a bronze earth-plate and a spoon cut from the lodestone (see **Figure 1**). A photo of the actual instrument is shown in Needham (1962) and reproduced in Merrill *et al.* (1996). When Needham visited China, he was shown by Wang Chen-To himself the experiment in which the lodestone spoon gradually rotated to the southward direction and settled there. Although this effort is very impressive, it is rather doubtful if Chinese at this early age really used an instrument which can be identified as the ancestor of the magnetic compass. The interpretation, as suggested by Wang, is not completely convincing. Moreover, there is a conspicuous absence of the references to compass-like instruments for about a thousand years afterwords.

A well-known early record about the magnetic compass is *Měng Chhi Pi Than* 夢溪筆談 (Dream Pool

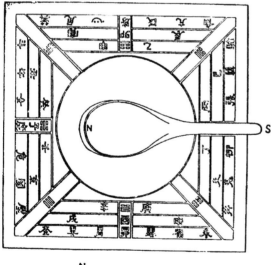

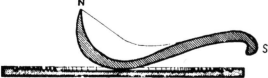

Figure 1 An ancient Chinese compass with the lodestone cut in the shape of a spoon, restored by Wang (1948).

Essays) written by Shen Kua 沈括 at about 1088 AD. In this book, it is said that "Magicians rub the point of a needle with the lodestone; then it is able to point to the south. But it always inclines slightly to the east, and does not point directly at the south" 方家以磁石磨針鋒。則能指南。然常微偏東。不全南也。。 The text explains how to make a magnetic needle, its south-seeking property, and moreover the fact that there is slight difference between the true south and its pointing direction (i.e., the first mention of the declination). Shen Kua further says that "It is best to suspend it by a single cocoon fibre of new silk attached to the centre of the needle by a piece of wax the size of a mustard-seed – then, hanging in a windless place, it will always point to the north" 以芥子許蝋綴於針腰。無風處懸之。則針常指南. This is the earliest written record about the magnetic compass using a magnetic (magnetized) needle.

Regarding the oldest compass, existence of an even earlier record was pointed out also by Wang (1948). The text was found in *Wu Ching Tsung Yao* 武經總要 (Collection of the Important Military Techniques) which is a compendium of military technology edited by Tsêng Kung-Liang 曾公亮 and completed in 1044. In this, it is said that "When troops encountered gloomy weather or dark nights, and the directions of space could not be distinguished, they let an old horse go on before to lead them, or else they made use of the south-pointing carriage, or the south-pointing fish to identify the directions." 若遇天景噎霾夜色瞑黑。又不能辨方向。則當縱老馬前行令識道路。或出指南及指南魚。以辨所向。 After that, how to make this fish is described. "Now the carriage method has not been handed down, but in the fish method a thin leaf of iron is cut into the shape of a fish two inches long and half an inch broad, having a pointed head and tail. This is then heated in a charcoal fire, and when it has become thoroughly red-hot, it is taken out by the head with iron tongs and placed so that its tail points due north. In this position, it is quenched with water in a basin, so that its tail is submerged several tenth of an inch. It is then kept in a tightly closed box. To use it, a small bowl filled with water is set up in a windless place, and the fish is laid as flat as possible upon the water-surface so that it floats, whereupon its head will point south".

Apparently, this magnetic pointer (fish) is given a thermoremanent magnetization (TRM) by quenching from high temperature and keeping it in the north–south direction. The record is very convincing as the description is detailed as well as correct. This communication can be taken as the first description

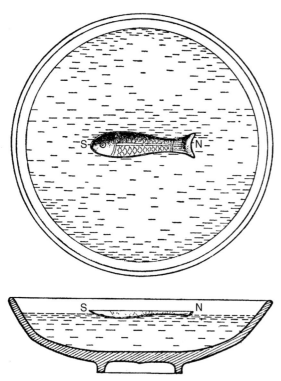

Figure 2   Illustration of a Chinese fish compass by Wang (1948).

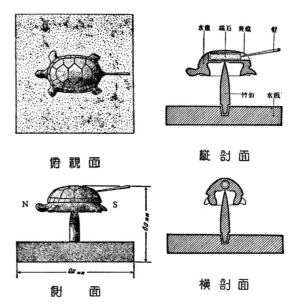

Figure 3   Illustration of a Chinese turtle compass by Wang (1948). Clockwise from top left, plan view, length-wise section, transverse section, and side view.

of the magnetic compass (wet type). **Figure 2** shows Wang's reconstruction of this compass.

The dry pivoted compass was described in *Shih Lin Kuang Chi* 事林廣記 (Guide through the Forest of Affairs), an encyclopedia compiled around 1150 by Chhen Yuan-Ching 陳元青見. In this compass, an wooden turtle has a tail made of a magnetic needle. A thin bamboo stick stands from the baseboard and holds the turtle at the hole made in its belly. The turtle rotates and points to the north because of the magnetic needle (**Figure 3**).

1.1.3 Magnetic Compass in European Documents

It is not clear when the knowledge of the magnetic compass reached Europe, and when it was first used in navigation. Gilbert wrote that it was brought to Europe by the Venetian Marco Polo, but there is evidence that the compass was used well before his return to Europe in 1295. It is often thought that the knowledge of the compass came from China through the intermediary of the Islam civilization. There is no written evidence, however, and the appearance of the

compass is earlier in European documents than in Islamic ones (Mitchell, 1932; Needham, 1962).

The earliest record of the north–south-seeking property of the compass in Europe appears to be that of Alexander Neckham (1157–1217) of St. Albans, England. In two treatises, *De Utensibibus* and *De Naturis Rerum* written about 1190, he described the use of the magnetic needle in navigation to indicate north, and that the needle is put on a pivot which may be the form of a primitive compass (Mitchell, 1932). Guyot de Provins of France (1184–1210) wrote a poem called *La Bible* around 1205, in which he described a floating compass. Jacques de Vitory of Kingdom of Jerusalem (1165–1240) left a similar document (*c.* 1218). These people were all monks or priests, and they only referred to the compass as having the noble property (to point always to the same direction). It is therefore natural to think that the properties of the compass were known to mariners well before it became popular so that the priests could use it for allegory in these writings (Mitchell, 1932).

1.1.4 *Epistola* of Petrus Peregrinus

Petrus Peregrinus (Roman name for Pierre Pélerin) wrote *Epistola* (*Epistola Petri Peregrini de Maricourt ad Sygerum de Foucaucourt militem: De magnete*, Letter of

Pierre Pélerin of Maricourt to Sygerus of Foucaucourt, soldier, concerning the magnet) in 1269, while he took part in the siege of the southern Italian town of Lucera, by the army of Charles the Count of Anjou. Maricourt and Foucaucourt are names of towns in Picardy, France. It has the form of a letter (*Epistola*) to a soldier called Sygerus of Foucaucourt. Although it was not published in printed form until 1558, many hand copies were circulated widely in western Europe in medieval times. In this booklet (**Figure 4**), Peregrinus explained various properties of the magnet (lodestone) based on experimental observations. In fact, this can be regarded as the first scientific treatise describing observations and experiments carried out for the purpose of clarifying natural phenomenon. The conclusions were derived logically based on observations and experiments.

The *Epistola* is composed of two parts. The first part is made up of 10 chapters, in which properties of magnets are discussed. The second part covers the technical use of the magnets, such as the construction

of a magnetic compass. The second part also contains a discussion of a perpetual motion machine using magnets. This is certainly invalid from the present-day knowledge of physics, but it cannot be blamed as an error, because it was written long before the concepts of thermodynamics or energy conservation were formed.

All of the material written in *Epistola* may not have been discovered by Peregrinus himself (Harradon, 1943), but this does not decrease the importance of *Epistola* as the first scientific paper in the human history. The most important properties of magnets described by Peregrinus were as follows.

1. *Finding the two magnetic poles.* To show that a magnet has two poles, grind and polish the magnet into a spherical shape. Next, put a small needle-shaped iron on this sphere, and write a line along its direction dividing the sphere into two equal parts (this defines a magnetic meridian). Move the needle to a different position and write the second division line. Peregrinus noted that, even if the above process is repeated at many places, all the lines meet at the same point on the sphere. The intersection of great circles defines two poles, the north and south magnetic poles.

2. *Determination of the polarity of the poles.* To determine which is the north and which is the south pole, place a magnet on a wooden plate which floats on the water surface in a large-enough container. After some time, the magnet settles into a north–south direction, and thus the north and south magnetic poles can be determined. This method is quite similar to the method of using a natural magnet as a compass (to put on water a wooden fish containing a natural magnet) in Chinese documents (see **Figure 2**). Peregrinus concludes that the magnet rotates so that the two poles are in the same direction as the celestial poles.

3. *Forces between two magnetic poles.* Using two magnets with poles marked as above, one magnet floating on the water and another held by a hand, it can be shown that the two poles attract each other if the S pole of the second is brought near the N of the first, or vice versa. On the other hand, they repel each other if the two poles are of the same polarity (N to N, or S to S).

4. *A magnetic pole cannot be isolated.* To show this, Peregrinus describes an experiment of cutting the magnet into two halves. Then, new poles appear at the cut end. The polarity of these new poles are opposite to the one at the other end of the cut pieces,

Figure 4 The title page of *Epistola*.

which can be seen by repeating the third experiment with halved magnets. This is the first realization of the dipoles and that monopoles either do not exist or are very rare. (One is hypothesized to exist in the Higgs field).

Figure 5 is the illustration of an instrument to determine the azimuth of the Sun and others shown by Peregrinus. A floating magnet (*Magnes*) determines the north direction. This part is a magnetic compass using a natural magnet (instead of a magnetized needle). In addition, the instrument has a ruler with a pin at each end. To determine the direction of the Sun or the Moon, the ruler is rotated until the shadow of a pin falls along its length. Peregrinus also showed another instrument in which an iron needle magnetized by a magnet is used in place of the magnet in this figure.

Many of the above findings have usually been attributed to Gilbert. For instance, the points 1, 2, and 4 are described very similarly to the description in *Epistola* but in more detail in Part I, Chapters 3–5 of *De Magnete* (Gilbert, 1600). As an example, **Figure 6** shows an illustration from *De Magnete* that describes the above point 4; cut a magnet in half, and poles of opposite polarities appear at the new edges. A rather curious fact is that Gilbert did not refer to Peregrinus in these descriptions. The neglect of the contribution of Peregrinus in the later years is perhaps caused by the popularity of Gilbert's work and the lack of proper citation in it.

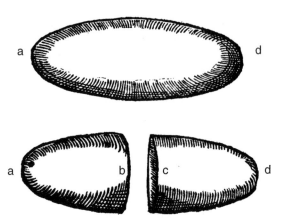

Figure 6 Experiment of cutting lodestone to show the appearance of magnetic pole at the new edges as described in *De Magnete*.

1.1.5 European Recognition of Declination

Before Needham (1962) showed the precedence of Chinese discovery, it was widely believed that Christopher Columbus discovered the declination on his first journey to the West Indies in 1492. According to the journal of the ship, quoted in a book by Las Casas and in the biography of Columbus by his son Fernando, magnetic direction changed "from northwest to northeast" on September 13. However, Mitchell (1937) argued convincingly that the declination must have been known in Europe at the time of the voyages of Columbus. In the biography written by Fernando, it is said that on the second voyage (1493), Columbus carried compasses from Genoa as well as from Flanders. In the Genoan type, the magnetic needle was fixed in the N–S direction of the compass card, but in Flemish type the needle was fixed one point (11.25°) to the east of true north, which shows that the declination in west-central Europe at that time was compensated in the latter compass. From the speed of information transfer in those times, it is deemed not probable that this knowledge is the outcome of the first journey of Columbus. The artisans of Flanders must have known the difference between true and magnetic north for some time.

It remains true, however, that the description by Columbus is the first scientific record of declination measurement, and that important discoveries were made of (1) the variability of declination from place to place, and (2) the position of the agonic line (the line connecting the place where $D = 0$) at the end of fifteenth century.

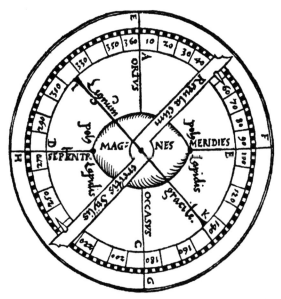

Figure 5 The compass using a magnet shown in *Epistola*.

1.1.6 Inclination

The fact that the magnetic needle deflects not only from the horizontal north–south direction but also from the horizontal plane was first recognized by Georg Hartmann and Robert Norman. Hartmann's findings were written only briefly in a letter (1544) which was mostly unknown until it was rediscovered in the nineteenth century, and his value of the inclination (9°) is too shallow for Europe at that time. On the other hand, Norman published his measurement of the inclination and discussion of the results in 1581 as *The Newe Attractive*. (*The Newe Attractive, Containyng a short discourse of the Magnes or Lodestone: and amongst other his vertues of a newe discovered secret and subtill propertie, concernyng the Declinyng of the Needle, touched therewith under the plaine of the Horizon. Now first founde out by Robert Norman Hydrographer.*)

Norman knew about the magnetic devices quite well since he was a sailor for 20 years, and his later job was the construction and sales of navigational instruments. He was aware of the tendency of the magnetic needle to make some angle with the horizontal plane from his vast experience in making magnetic compasses. He usually corrected this by putting a small weight on the needle, but at some time, he decided to study this phenomenon in more detail. He constructed a special instrument to measure the deflection from the horizontal plane by preparing a magnetic needle from carefully balanced iron, and making it freely rotate around the horizontal pivot (**Figure 7**). This is the first dip circle, and similar instruments were used in observatories until very recent times. By repeating experiments, he found out that this angle (the inclination) was 71° 50′ in London. As the title of the book indicates, Norman not only carried out the measurement but gave discussions about the nature of this force. In it he denied the old idea that there are attractive points in heaven (such as the celestial poles of Petrus Peregrinus) or in the Earth (the magnetic mountain in the Arctic, etc.), which is important in the development of the notion of remotely acting forces (Yamamoto, 2003).

1.1.7 *De Magnete* of William Gilbert

William Gilbert (1544–1603) was a well-established medical doctor and was the President of the Royal College of Physicians. He was famous in his time because he became the physician to Queen Elizabeth for the period 1601–03, but he is now considered as one of the founders of the physical sciences by the

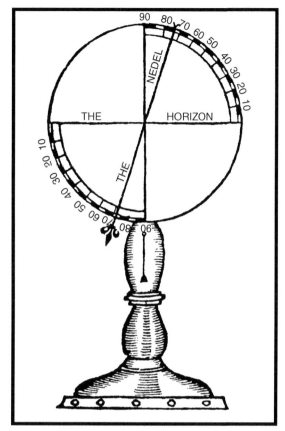

Figure 7 The dip circle of Robert Norman.

publication of *De Magnete* (*De Magnete, magneticisque corporibus, et de magno magnete tellure: Physiologia nova, plurimis & argumentis, & experrimentis demonstrata*; On magnets, magnetic materials, and the Earth as a large magnet: New natural philosophy proved by arguments and experiments). This book (**Figure 8**) had a profound influence in Europe and prepared for Newton's theory of gravity by establishing the fact that the magnetic force is a remotely working interaction (Yamamoto, 2003). Gilbert also founded the science of electricity by the experiments described in *De Magnete*. Here, we will only discuss what was made clear by him about magnetism. His most important contributions may be summarized as follows.

1. *Magnets and iron are the same.* This may sound a bit queer, but lodestone and iron were considered to belong to different kinds of matter ('rock' and 'metal') from the time of Aristoteles. Gilbert concluded that they are the same because most of the best magnets were found in an iron mine, magnets heated in the furnace produced good-quality iron, etc.

GVILIELMI GIL
BERTI COLCESTREN-
SIS, MEDICI LONDI-
NENSIS,

DE MAGNETE, MAGNETI-
CISQVE CORPORIBVS, ET DE MAG-
no magnete tellure; Phyfiologia noüa,
plurimis & argumentis, & expe-
rimentis demonftrata,

LONDINI

EXCVDEBAT Petrvs Short ANNO
M D C.

Figure 8   The title page of *De Magnete*.

Figure 9   Illustration of *Versorium*, a pivoted iron needle used by Gilbert to measure the direction of the field arround *Terella*.

2. *Similarity of the spherical magnet and the Earth.* Gilbert formed a lodestone into a sphere and called it *terella*. He used a small magnetic needle suspended at the pivot (*versorium*, see **Figure 9**) to find the direction of magnetic force on this sphere. Like Peregrinus, he identified poles, meridians, and the equator on the sphere. From the results of this model experiment, he concluded that the Earth is a magnetic body and thus a magnet (Globus terrae est magneticus & magnes).

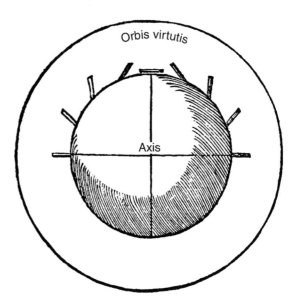

Figure 10   Gilbert's *Terella*, or the spherical magnet.

3. *Inclination is determined by the (magnetic) latitude.* Gilbert argued that if *terellas* are made from lodestones of different strength, the inclination might be larger in one because of the stronger attraction of the magnet, yet the actual inclinations were the same. This observation strengthened his conclusion that the Earth is a large spherical magnet (**Figure 10**).

His experiment using a spherical magnet was very similar to the one described by Peregrinus. However, the conclusion was quite different: Peregrinus found correspondence between the spherical magnet and the celestial globe. Gilbert, on the other hand, concluded that it is the same as the Earth, or the Earth is a spherical magnet. The way to reach this idea based on experiments is not quite perfect in the standards of the present-day science, but Gilbert reached the correct conclusion for the first time that the Earth has a dipole field which is aligned roughly parallel to the rotational axis.

1.2   Further Developments in Observations

Isaac Newton published his *Philosophiae Naturalis Principia Mathematica* (Mathematical Principles of Natural Philosophy) in 1668, which firmly established modern physics. It is clear that Newton's work received a big influence from Gilbert's *De Magnete*, especially the idea of the forces between objects that are not in contact (Yamamoto, 2003). Geomagnetism as a physical science took shape at about the same time, and it is remarkable that progress was made through observations even from these old times.

1.2.1 Secular Variation

In October of 1580, William Borough, who later served as a naval officer under Captain Drake, measured the declination at Limehouse in London, and recorded that it was $11\frac{1}{2}°$ E. Edmund Gunter, Professor of astronomy at Gresham College, measured the declination at Deptford and Limehouse in June of 1622, and obtained the values of $6° 15'$ and $5° 56'$ E (Limehouse value was the mean of eight observations). Gunter pointed out that temporal change was very likely, but he was cautious to admit the possibility of large error in Borough's measurement. Henry Gellibrand, Gunter's successor at Gresham College, went back in 1635 to Deptford, repeated the measurement, and found a value of about $4°$ E. This marks the discovery of secular variation.

1.2.2 Short-Term Variations

George Graham, a London clock maker, observed declination very frequently in 1722–23 using a very sensitive (12-inch) compass. He discovered that the declination changed even in 1 day, and sometimes the magnitude of change reached up to $30'$. Following this, Andreas Celcius and Olof Hjorter performed the same experiments in Uppsala in 1740–47. They not only confirmed Graham's results, but found out that the activity of the northern lights (aurora) was accompanied a large change in declination. Later, they cooperated with Graham in London to make simultaneous measurements, and found out that large disturbances occurred at the same time in two places on 5 April 1741. These are the discoveries of magnetic storms and solar quiet-day variations.

1.2.3 Magnetic Charts

After the voyage of Columbus, the practical utility of magnetic measurements was recognized because of their importance in navigation. A precise knowlede of the position of the ship, and the direction it is heading are very important for navigation. The latitude part of the position and direction are relatively easy. The angle between the horizon and the pole star (which is the astronomical latitude if the diurnal motion of the pole star is adjusted) can be measured with enough accuracy using sextants. The pole star shows the north direction at night, and in the daytime, the mean of the two measurements of the sun's azimuth gives the south direction, if they are obtained at times when the sun's height is the same (e.g., at the sunrise and sunset).

On the other hand, the determination of longitude is quite difficult. If a clock can keep very accurate time after the ship has left a port, the difference between the clock time and the local time as determined from the position of the sun (e.g., 12:00 noon for southing) gives the longitude difference between the current ship's position and the port. But the needed accuracy could not be achieved by the chronometers of those times. As it was found by the voyages of Columbus that the declination changed from NE to NW as the ship crossed the Atlantic from east to west, it was considered that the longitude may have a simple relation with the magnetic declination.

At the end of the seventeenth century, Edmund Halley organized an expedition devoted to the determination of the declination in various parts of the Atlantic. He thus published in 1701 the first isogonic map of the Atlantic (*General Chart of the Variation of the Compass (1701)*). This was the first magnetic chart, and its importance in the study of geomagnetism cannot be overstated. Ironically, however, this map did not solve the problem of the longitude, because of the unlucky coincidence that the isogonic lines (the traces on which the declination is equal) ran mostly east–west in the northern Atlantic at that time (**Figure 11**).

For further discussion of magnetic data obtained by the measurements at sea and their use in the study of geomagnetism.

1.2.4 Measurement of Intensity

The determination of the intensity of the magnetic field started much later than that of the direction. First came relative measurements. It was noticed that if the needle of a dip circle was displaced from its rest position, the needle oscillated about the equilibrium position with a period which decreased as the strength of magnetic field increased. This property was used by the explorer Alexander von Humboldt, who measured the oscillation period of a dip circle at many places while he traveled in South America with Aime Bonpland in 1798. He could show that the intensity systematically increased as he went further south from the magnetic equator which was near Cajamarca in Peru. Actually, the period of oscillation

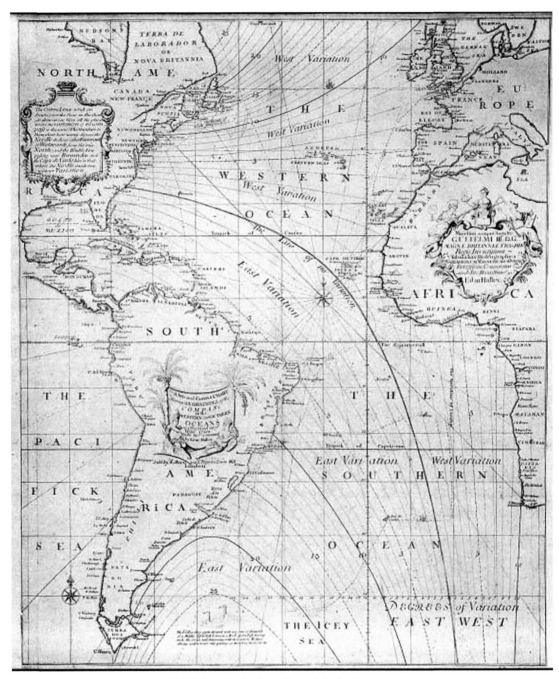

Figure 11   Halley's Atlantic chart of 1701 showing the lines of equal declination.

(T) depends not only on the strength of the magnetic field (H), but on the moment of inertia (I) and magnetic moment (M) of the magnetic needle.

$$T = 2\pi\sqrt{I/MH} \qquad [1]$$

Thus it is difficult to assume that the change in the period is entirely due to the change in the field

strength, and not due to changes in the property of the magnet.

Carl Friedlich Gauss devised a method of absolute intensity determinaton in 1832. His method was to use a suspended magnet (A in **Figure 12(a)**) and make it oscillate in the horizontal plane on both sides of the magnetic meridian. As the moment of inertia of the magnet is not known, he added weights

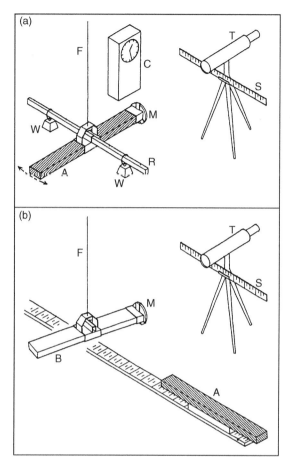

Figure 12 (a,b) Principle of Gauss' method of intensity determination. Modified from Malin SRC (1987). Historical introduction to geomagnetism. In: Jacobs JA (ed.) *Geomagnetism*, vol. 1, pp. 1–49. London: Academic Press. See text for explanation.

magnetic moment M. Observing the angular deflection for different values of d and again utilizing the method of least squares, it is possible to obtain M/H. With the values of $M H$ and M/H, Gauss could determine the horizontal field intensity in Göttingen as 17.8 µT. Gauss's method of intensity determination was quickly taken up by other people, such as Hansteen and Sabine, for measurements in other parts of the world. Present knowledge of the intensity of the magnetic field is discussed in Chapter 13.

1.2.5 Magnetic Observatories

More or less continuous observations of D started in London about 1652 and in Paris in 1663. By the time of Gauss, there were already a number of such observatories, and it was necessary to coordinate them to exploit the magnetic data. In the 1830s, Gauss and Humboldt organized the Göttingen Geomagnetism Union, which can be regarded as the first international geophysical organization. About 50 observatories (among them 15 were outside Europe) took part in this collaboration to carry out standardized observations; e.g., daily observation at the assigned time, and intensive observations of D on the six selected days per year.

1.3 Deciphering the Past Using Remanent Magnetization

Geomagnetism has a unique position in all the Earth sciences because it can go back in time and treat changes that occurred over millions of years or an even longer timescales. In other branches of geophysics, there are some efforts to go back in history, such as paleoseismology based on estimation of past movements of active faults uncovered by trench excavation. Geology is in a sense a study of all the processes operating on the Earth since its formation. However, in all of these studies, the quality of data are much inferior to those which are obtained for the present observations. There are many reasons for this. For instance, the amount of information used in the analysis of earthquakes today (e.g., wave forms recorded on seismograms at the observatories) is many orders of magnitude larger than the information from active fault traces. Geological processes themselves modify the original information into quite different forms (e.g., the fossils are different from the living organisms, irreversible processes such as metasomatism changes the properties of sediments, etc.).

(W) to the magnet so that the moment of inertia can be changed by a precisely known amount. By observing the period T for various increments in the moment of inertia, it is possible to determine the product $M H$ (through least-squares method, which is another invention of Gauss).

Next, near another suspended magnet (B in **Figure 12(b)**), he placed the magnet (A) in the direction perpendicular to the magnetic meridian. The magnetic field at the suspended magnet is rotated by an angle H'/H, where H' is the magnetic field due to the magnet

$$H' = \frac{M}{d^3}\left[1 - \frac{3}{8}\left(\frac{l}{d}\right)^2 + \cdots\right]$$
[2]

where d is the distance between the centers of the two magnets, and l is the length of the magnet with the

Magnetic field itself cannot be preserved over time, which is similar to the case of seismic waves. Only the remanent magnetization in rocks and other materials may survive, and they are also subject to geological processes that can destroy or change them. Nevertheless, the relation between the magnetic field (B) and the remanent magnetization (M) acquired in that field is very straightforward; their directions are usually parallel and their intensities are more or less proportional to each other with the proportionality constant determined by the magnetic properties of rocks and minerals. Remanent magnetizations can survive some of the changes that affect rocks; e.g., compaction, consolidation, and sometimes even metasomatism in sediments. The direct relation between B and M as well as the stability of magnetization over millions of years make it possible to study the magnetic field in the remote past with some confidence. The properties of rocks and minerals that are the basis of magnetic memories. Here, we will discuss the major developments in paleomagnetism.

1.3.1 Early Observation of the Remanence of Rocks

The fact that some rocks carry strong magnetization was known for a few hundred years prior to the actual measurements of their remanent magnetization because a magnetic compass needle is sometimes deflected near such rocks. Humboldt attributed these effects to lightening strikes. First measurements of the natural remanent magnetization (NRM) carried by rocks were done by people like Delesse and Melloni around 1850. These workers noted that some of the rocks carry magnetization which is parallel to the Earth's magnetic field. In 1899, Folgerheither (1899) studied the magnetizations in bricks and potteries, and suggested that the record of direction changes of the Earth's magnetic field may be obtained by measurements of such objects, knowing the place and time of firing at a kiln. This opened up the possibility of archeomagnetism. The current knowledge in centennial and millenial field changes is discussed in Chapter 7.

1.3.2 Polarity Reversals

David (1904) and Brunhes (1906) reported the discovery of NRM which was roughly antiparallel to the present field direction at the sampled locality. They found that the directions of NRMs in baked clays and in overlying lava flows which heated them were

essentially the same. These observations strongly suggested that the Earth's magnetic field had reversed in the past. Matuyama (1929) studied Quaternary and late Tertiary volcanic rocks in Japan, Korea, and Manchuria (NE China). He found both normally and reversely magnetized rocks, but noted that the reversely magnetized rocks were older than normally magnetized rocks, in all the localities (**Figure 13**). Thus he concluded that the Earth's magnetic field was reversed in the early Quaternary. Similar comparison was attempted earlier by Mercanton (1926) on rocks from widely separated areas such as Greenland, Iceland, Scotland, and others, but he could not draw a firm conclusion because the inferred ages of the rocks were too widely scattered.

After World War II, the study of paleomagnetism was greatly intensified in many countries. In a few places, measurements of NRMs were done in an effort to understand changes in a sequence of lava flows or sediments. Roche (1951) concluded that the last reversal occurred in the early Pleistocene based on his study of lavas of Massif Centrale in France. Hospers (1953) suggested that it is possible to use the changes in the NRM polarity for stratigraphic correlation in Icelandic lavas. Khramov (1958) studied sedimentary sequences in Turkmenia, and proposed the possibility of building a reversal timescale that could be applied to the whole world.

Even with these efforts, it was not quite obvious at that time if the Earth's magnetic field really reversed its polarity in the past. There were speculations that magnetization itself could orient in the direction

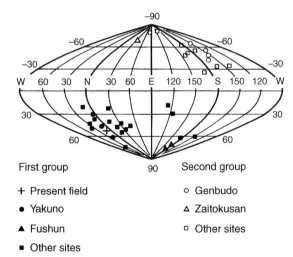

Figure 13 The magnetic directions of basalts in Japan, Korea, and NE China observed by Matuyama (1929). Vertical and horizontal directions correspond to the inclination and declination.

opposite to the magnetic field. Néel (1951) showed how self-reversal in the acquisition of TRM could occur. Soon afterwards, self-reversal of artificially induced TRM was actually discovered in a dacite lava from Mount Haruna, Japan (Nagata *et al.*, 1951). The controversy about field reversal versus self-reversal continued for several years, because it was pointed out that some self-reversal mechanisms cannot be reproduced in the laboratory because of the long length of time involved (Verhoogen, 1956). Experimentally, a strong correlation between the oxidation state of magnetic minerals and the NRM polarity was found in Scottish and other lava flows (Ade-Hall *et al.*, 1968).

The reality of field reversal was finally established by showing the synchronicity of reversals over the world. Two groups, one at US Geological Survey, Menlo Park, and the other at Australian National University combined K–Ar dating with paleomagnetism and showed that the reversals indeed occurred at the same time at widely separated places, such as Alaska, California, Iceland, and Hawaii (Cox *et al.*, 1963; McDougal and Tarling, 1963). Allan Cox and his colleagues developed a convention of calling the long and short intervals of the same polarity, 'polarity epochs' and 'polarity events' respectively (these were replaced by chrons and subchrons as internationally defined nomenclature later). With the discovery of the Jaramillo event in the Brunhes epoch (Doell and Dalrymple, 1966), the polarity timescale for the last 4 million years was essentially completed (**Figure 14**). There is no doubt that before the present normal polarity interval (Brunhes Chron, 0–0.8 Ma), the field was reversed for more than 1 My (Matuyama Chron, 0.8–2.6 Ma) with occasional short episodes of normal polarity, and before that another period dominated by normal polarity (Gauss chron), etc. The study of polarity

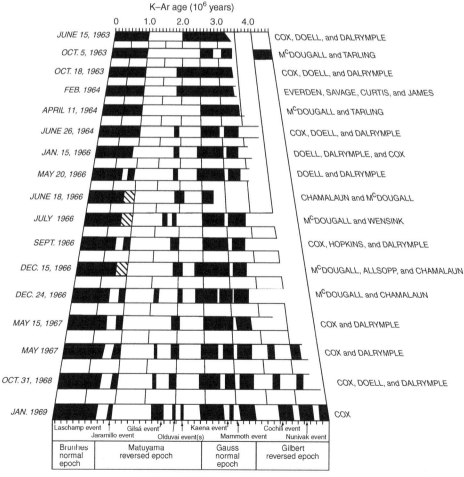

Figure 14 Early development of the geomagnetic polarity timescale (Dalrymple, 1972).

reversals were vigorously continued to the present time, and an increasingly complex detailes are uncovered as described in chapter 10.

1.3.3   Sea-Floor Spreading

In the 1950s, the biggest topic in paleomagnetism was continental drift rather than polarity reversals. Two groups from England led by Keith Runcorn and Patrick Blackett showed that paleomagnetic data from different continents indicated that large-scale movement of landmasses occurred in the past (Runcorn 1962; Blackett *et al.*, 1960). This revived the once-rejected theory of continental drift by Alfred Wagener (1929). However, acceptance of continental drift was not very smooth in many parts of the geophysics community, because of scant evidence. The reality and the importance of geomagnetic polarity reversals was not enthusiastically received until later.

The situation changed drastically in the few years between 1963 and 1966. During this time, the combined magnetic–geochronological data of Cox and others became compelling. However, the decisive step was put forward by Fred Vine and Drummond Matthews (1963) by their theory of tape-recorder-like acquisition of magnetization at the mid-ocean ridges. Based on the measurement of magnetic anomalies across the Mid-Atlantic and Indian Ocean ridges and the sea-floor spreading hypothesis of Hess (1962) and Dietz (1961), they proposed that the oceanic crust acquires normal or reversed remanence depending on the polarity of the magnetic field at that time, and that the crust moves away from the ridges, causing the alternating normal-reverse magnetic anomalies. Vine (1966) further gave very convincing evidence of the reality of the Vine–Matthews theory (**Figure 15**).

Subsequently, the plate tectonics model was pioneered by Jason Morgan (1968), Dan McKenzie and Robert Parker (1967), and Xavier Le Pichon (1968). The direction of the current plate movement was inferred from the source mechanism of earthquakes occurring at the plate boundaries, and the trend of the transform faults separating two plates. Also, the past plate motion was strongly constrained by the magnetic anomalies through the Vine–Matthews mechanism. It was this definitive property of the magnetic anomaly data which was one decisive factor in the quick establishment of the plate tectonics (compared with, for instance, continental drift theory). For the interplay between ancient tectonic motions and the magnetic records *see* Chapter 14..

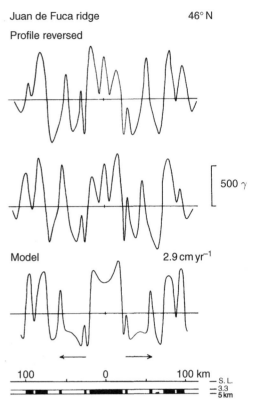

Figure 15   The magnetic anomaly across the Juan de Fuca ridge as interpreted by normal-reverse magnetization in the oceanic crust. Reproduced with permission from Vine FJ (1966) Spreading of the ocean floor: New evidence. *Science* 154: 1405–1415. Copyright 1966 AAAS.

1.3.4   Long-Term Behaviors of the Geodynamo

Geomagnetism provided a very useful tool for recovering past plate motions through the analysis of oceanic magnetic anomalies. There was a great reward for this service; the resulting intense study of sea-floor spreading in the world oceans led to the establishment of the polarity reversal timescale.

When plate tectonics was initiated, the polarity timescale was available only for the last 3.5 My from the paleomagnetic and geochronological studies of the volcanic rocks on land. Following Vine's first attempt to extend this timescale to tens of millions years in the past (Vine, 1966), Heirtzler *et al.* (1968) produced a timescale that covered all the Cenozoic era, based on comparison of the magnetic anomaly patterns in the Atlantic, Pacific, and Indian Oceans. This was a tremendous step for exploring the evolution of oceanic plates, because it provided a way to date the ocean floor, relying mostly on the pattern of magnetic anomalies.

The reversal timescale ended abruptly at about 80 Ma due to the 'Magnetic Quiet Zone' where magnetic anomaly was absent. Later, Larson and Pitman (1972) succeeded in establishing the polarity timescale for the parts of older oceans beyond the Quiet Zone. With Cenozoic and Mesozoic timescales, it is possible to determine the age of the all ocean floors (**Figure 16**).

The polarity timescale has been refined and updated many times. The current standard timescale is the one presented by Cande and Kent (1992, 1995). Further development is discussed in Chapter 12. Polarity timescales are the source of very important information about the nature of the geomagnetic field. The Quiet Zone is the part of the ocean where magnetic anomalies are absent because polarity reversals did not occur. The long period without reversals is called the Cretaceous Long Normal (CLN), and is speculated that the core dynamo state was significantly different at that time from both before (Jurassic and early Cretaceous) and after that time (Tertiary and Quaternary).

From the analysis of the polarity timescale, it was concluded that the reversals occur without memories of the past (Poisson process). The reversal process is not steady; the probability of a reversal increases monotonically from the end of the CLN to the present (McFadden, 1984). The typical timescale of the dynamo process in the core is considered to be about 15 000 years (e.g., Kono and Roberts, 2002). The geomagnetic polarity timescale provides the record of field behavior for time intervals much longer than this time. Thus there is a strong possibility that the change in the dynamo behavior over such long times represents the changes in the environment in which the dynamo process operates (McFadden and Merrill, 1984).

1.4 Descriptions of the Earth's Magnetic Field

1.4.1 Spherical Harmonic Analysis

Perhaps Gauss's most important legacy to geomagnetism is the invention of spherical harmonic analysis (SHA) and its application to the Earth's magnetic field. Because of Gauss's law of magnetism and the fact that the atmosphere is a very poor conductor of electricity, the magnetic field B is both divergence-free ($\nabla \cdot \boldsymbol{B} = 0$) and curl-free ($\nabla \times \boldsymbol{B} = 0$) at the surface of the Earth. With these conditions, it can be shown that the magnetic field can be expressed as the gradient of a scalar potential W, and this potential satisfies the Laplace equation

$$\nabla^2 W = 0 \tag{3}$$

For the spherical geometry of the Earth, the geomagnetic potential can be expressed as

$$W = a \sum_{l=1}^{l_{max}} \sum_{m=0}^{l} \left(\frac{a}{r}\right)^{l+1} P_l^m(\cos\theta)\left(g_l^m \cos m\phi + b_l^m \sin m\phi\right) \tag{4}$$

where a is the mean radius of the Earth (6371 km), r, θ, ϕ are the spherical coordinates (radius, colatitude, and longitude), P_l^m is the Schmidt-normalized associated Legendre functions of degree l and order m, and g_l^m and b_l^m are the Gauss coefficients. The components of the magnetic field on the surface of a sphere of radius r are, by the relation $\boldsymbol{B} = -\nabla W$,

$$X = \sum_{l=1}^{l_{max}} \sum_{m=0}^{l} \left(\frac{a}{r}\right)^{l+2}\left(g_l^m \cos m\phi + b_l^m \sin m\phi\right)\frac{\partial P_l^m}{\partial\theta} \tag{5}$$

$$Y = \sum_{l=1}^{l_{max}} \sum_{m=0}^{l} \left(\frac{a}{r}\right)^{l+2}\left(g_l^m \sin m\phi - b_l^m \cos m\phi\right) \times \frac{m P_l^m(\cos\theta)}{\sin\theta} \tag{6}$$

$$Z = -\sum_{l=1}^{l_{max}} \sum_{m=0}^{l} (l+1)\left(\frac{a}{r}\right)^{l+2}\left(g_l^m \cos m\phi + b_l^m \sin m\phi\right) \times P_l^m(\cos\theta) \tag{7}$$

where X, Y, Z are positive northward, eastward, and downward components. If the flattening of the Earth can be neglected, the values at the surface of the Earth are obtained by setting $r = a$ in the above expressions. In the above, we assumed that the magnetic field is produced entirely by sources within the Earth. If there are external sources, the potential and the forces also contain terms of r^l and r^{l-1}, respectively.

To determine the geomagnetic potential for 1835, Gauss read the field values at 84 regularly spaced points (at every 30° longitude and 0°, 20°, 40°, and 60° N and S latitudes) from the magnetic charts available to him (Barlow's for declination, Horner's for inclination, and Sabine's for total intensity). The coefficients of the potential up to degree 4 (24 coefficients) were calculated in such a way that the sum of the squares of differences between the observed and calculated values of the field components takes the minimum value of all the possible cases (the least-squares method). Through this analysis, Gauss could show that (1) the Earth's magnetic field can be divided into various modes represented by the spherical harmonics $P_l^m(\cos\theta)(\cos, \sin)m\phi$, (2) the source of this magnetic field lies within the Earth,

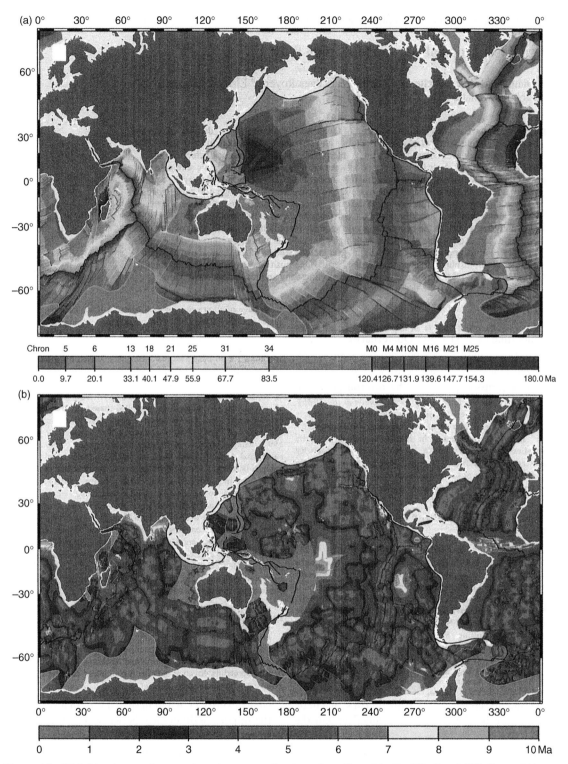

Figure 16   Digital isochrons of oceans based on magnetic chronology. From Mueller RD, Roest WR, Royer J-Y, Gahagan LM, and Sclater JG Digital isochrons of the world's ocean floor. *Journal of Geophysical Research* 102: 3211–3214.

(3) the dipole term is dominant in the geomagnetic field, and (4) the geomagnetic field can be well represented by low-degree field components. These are the fundamental properties of the geomagnetic field, which requires no modification even today.

Since Gauss's time, the SHA became the standard method of analysis of the geomagnetic field. Other methods exist which can also uniquely describe the magnetic field, such as the current loops or distributed dipole sources on the surface of the core. However, the SHA is in most cases preferred because the components obtained by this analysis (dipole, quadrupole, etc.) have definite physical meaning and because these can be conveniently continued to the surface of the core in which the dynamo process generates the magnetic field. After Gauss, SHA was repeatedly applied to the magnetic observations of later periods by other authors. In these, degree 6 (with 48 coefficients) was about the maximum the modelers could go to until about 1960. This limit was imposed by the accuracy of data as well as the computational loads to derive coefficients.

1.4.2 International Geomagnetic Reference Field

With the advent of electronic computers, larger and larger computations became possible. If enough data with good accuracy are available, there is practically no limit in the truncation level of the models with the computer power of today. As an example, the gravitational potential has a form quite similar to the geomagnetic potential (except for the $l = 0$ term, which is absent in the latter), and the gravitational potential deduced from satellite measurements goes up to 200 or more in degrees (>40 000 coefficients). The limit for the SHA of the geomagnetic potential comes from other reasons (see later sections).

As more and more models were produced by different researchers, it became necessary to integrate all of these efforts to obtain the best model for that time. An internationally coordinated effort was carried out under the aegis of the International Association of Geomagnetism and Aeronomy (IAGA), and produced the International Geomagnetic Reference Field (IGRF) for the year 1965. This was a model based on the results of a dozen or so research groups around the world and the truncation level was set to degree 10. Since then, the IGRF has evolved into a series of models at 5-year intervals covering the period of 1900–2005. By constant check with newly acquired data and with the combined effort of research groups,

the IGRF is considered to be the best model to describe the geomagnetic field in this period.

1.4.3 Satellite Measurements

Good global data coverage was difficult in the past because of various problems on land, but more so because of the oceanic area which occupies two-thirds of the surface area. The use of a magnetic satellite in a near-polar orbit solved this problem, and provided data with a uniform quality. The earlier satellites (such as *POGO*) measured only field intensity, which was not ideal for the purpose of the main-field analysis because of the nonuniqeness problem called Backus effect (Backus, 1970). *Magsat* and later satellites were equipped with vector as well as scalar magnetometers, with good internal consistency between these measurements.

The severest problem in satellite magnetic measurements of the vector field is the determination of the absolute orientation. The controlling factor of the accuracy of the data is not the magnetometer error but error in the orientation. Errors in the angle of 1 min and 1 s correspond to 0.03 % (15 nT) and 15 ppm (0.25 nT), respectively. Instrumental accuracy of this level is easily reached by most magnetometers. *Magsat* data (1980) had an absolute accuracy of about 6 nT, while the more recent *Ørsted* satellite (2000) achieved 1–2 nT accuracy. The reduction of errors for the *Ørsted* satellite reflects the improvement in orientation techniques (use of star sensor, etc.). With these satellites, models with degrees and orders up to about 13 were constructed (Langel *et al.*, 1980; Olsen *et al.*, 2000). This is a great improvement compared to the ground-based models, which can go up to about degree 10 (e.g., the IGRF models). Global observation and description of the magnetic field is further detailed in Chapter 2.

Figure 17 compares the vertical component of the magnetic field between the 1835 model of Gauss and the 2000 model based on *Ørsted* measurements. It is remarkable that the two models show essentially the same features discussed in the last section. The primary features of both figures are the dominance of the dipole field. The similarity is also quite apparent even in the secondary features. Some examples are (1) the magnetic equator moves northward in Southeast Asia, and southward near Peru in South America, (2) the two positive lobes near the North Pole, and (3) the single negative patch between Australia and Antarctica. These essential properties of the geomagnetic field have not changed in the last two centuries.

(a)

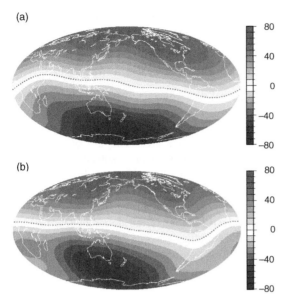

(b)

Figure 17 Vertical component of the magnetic field at the surface of the Earth in (a) 1835 (Gauss's model) and (b) 2000 (Ørsted model).

(a)

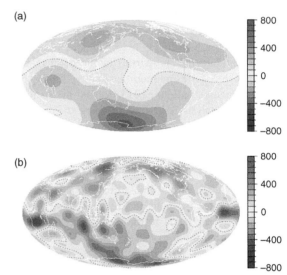

(b)

Figure 18 Vertical component of the magnetic field at the core surface (CMB) based on (a) 1835 (Gauss model) and (b) 2000 (Ørsted model) models.

Actually, the biggest difference in the two models lies in their resolution. Whereas the model of Gauss could express the spatial variation with the shortest wavelength of about 10 000 km ($l=4$), the most recent models can go down to about 3000 km ($l \approx 13$). However, because the short-wavelength features have very small amplitude at the surface of the Earth, their effect is quite small if we compare two models there. If we continue the potential field down to the core–mantle boundary (CMB), the higher-degree terms are greatly enhanced by the $(a/r)^{l+2}$ factor appearing in the expression for the force. The vertical component at the CMB would show that the contribution form the shorter-wavelength components are quite significant (**Figure 18**).

1.4.4 Geomagnetic Spectrum

The mean square of the field value over some area or volume ($\langle B^2 \rangle$ in the case of the magnetic field) is called the power. The power can be defined as a single number, or by parts if the field can be decomposed by some appropriate method. The power of the magnetic field on a spherical surface (Lowes, 1974) for the spatial degree l can be written as

$$R_l = \left(\frac{a}{r}\right)^{2l+4} \sum_{m=0}^{l} \left[(g_l^m)^2 + (b_l^m)^2 \right] \qquad [8]$$

which shows that the power decreases as r^{-2l-4} with change in the radius of the sphere. In a process that has a large amount of freedom, it is often found that the powers due to various degrees l are not much different from each other. One well-known example is the classic problem of black body radiation. In essence, there are a very large number of independent modes in the system, which interact with each other, accompanied by efficient energy transfer between various modes. The consequence is that the equipartition of energy is achieved among different modes. The power spectrum for such a process will be nearly white. Of course, the spectrum cannot continue to be flat to infinity (remember the quantum effect of Planck). In fluid motion, it is expected that the spectrum decays rapidly, once the scale becomes small enough for diffusion process to be effective.

This principle appears to apply to the dynamo process in the core. The resolution of the earlier models was not enough to tackle this problem, but the vector field data from satellites has enough resolution. The analysis of *Magsat* data (Langel and Estes, 1982) showed that the spatial (Lowes) spectrum is nearly flat to about degree 13 if it is continued down to the CMB (**Figure 19**). At the higher degrees, the field has undoubtedly sources in the crust.

Since full-scale dynamo simulations became possible (Glatzmaier and Roberts, 1995; Kageyama *et al.*, 1995), the power spectra of many dynamo models were analyzed. It was found that most of the dynamo results also show the nearly flat spectrum at the surface of the fluid

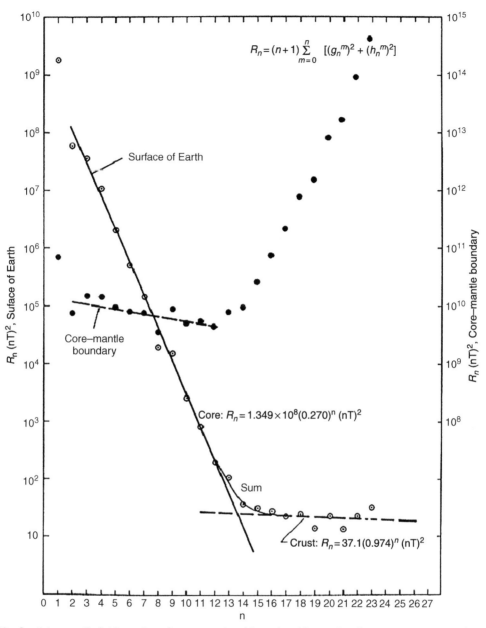

$$R_n = (n+1) \sum_{m=0}^{n} [(g_n^m)^2 + (h_n^m)^2]$$

Surface of Earth

Core–mantle boundary

Core: $R_n = 1.349 \times 10^8 (0.270)^n$ (nT)^2

Sum

Crust: $R_n = 37.1(0.974)^n$ (nT)^2

R_n (nT)^2, Suface of Earth

R_n (nT)^2, Core–mantle boundary

n

Figure 19 Spatial magnetic field spectrum (Lowes spectrum) based on *Magsat* data (Langel and Estes, 1982).

sphere (e.g., Kono and Roberts, 2002). The nearly flat spectrum continues to very high degrees such as the truncation level ($l = 40 \sim 100$ for most models). In a very intensive calculation (with no hyperdiffusivity in the model), Roberts and Glatzmaier (2000) found that the trend continues to about $l = 200$ and then a steep decay with l seems to start. This corresponds to the (core surface) wavelength of about 200 km.

This conclusion does not exclude the possibility that the power of a particular degree l is different from the general trend by a factor of $O(1)$. We may take note of the fact that Uranus and Neptune have magnetic fields in which the dipoles are much smaller than the quadrupoles. In the present geomagnetic field, the dipole term ($l = 1$) is about 2–3 times above and the quadrupole term ($l = 2$) is by a similar factor below the general trend. Similar departures have been observed in the simulation of various dynamo models. It is speculated that the departure of the $l = 1$ term above the trend indicates that the dipole is somehow preferred in the dynamo process in the Earth. The behaviors of the geodynamo in a long time span is discussed in Chapter 11.

1.4.5 Inverse Problem

Geophysical problems are often characterized by a combination of three elements; that is, (a) a source for the process, (b) a physical mechanism to induce some observable phenomena, and (c) measurements of these quantities. An example is the occurrence of an earthquake which has (a) geometry of the fault plane (width, height, azimuth, dip) and source function (a time-space description of the fault movement), (b) excitation and propagation of the seismic waves (which can be described by a theory), and (c) wave forms recorded at seismic observatories. When the source is known, it is relatively easy to predict what occurs and what will be observed. This is called the forward problem. Most geophysical problems have a different nature. It is more common that we have to guess the source when a set of observations is available. This is called the inverse problem. The determination of seismic source (e.g., centroid moment tensor solution) is an example of the inverse problem.

In the case of the magnetic field, there are (c) observations (field elements such as X, Y, Z or I, D, F at many points over the surface of the Earth), and we know (b) that the field is expressed by a potential and how the field components can be derived from it. We seek (a) the source (a scalar potential given by Gauss coefficients) which cause the observed magnetic field. It is thus a typical inverse problem.

Gauss's analysis of the 1835 field is the first of such inversion efforts. Suppose that a field element (one of the linear elements X, Y, Z) at a point i is designated by y_i, Gauss coefficients (g_l^m, h_l^m) by x_j, and the factors for each x_j depending on the position (θ_i, ϕ_i) by A_{ij}. The difference between the observation and the value derived from the potential is

$$e_i = y_i - \sum_j A_{ij} x_j \qquad [9]$$

or $e = y - Ax$ in matrix form, gives an estimate of error for this observation. The method of Gauss was to minimize the square of errors $|e|^2$. To obtain the minimum, the standard approach is to seek the local minimum for each x_j by equating the derivative to zero. This leads to the ordinary least-squares solution

$$x = (A^T A)^{-1} A^T y \qquad [10]$$

where A^T and A^{-1} indicate the transpose and the inverse of A, respectively. For the analysis of the present field, this simple method usually works (with the addition of the weight factor to take care of the reliability of individual data). In the case of data that are nonlinear field elements (I, D, F), the simple method is not sufficient any more. One reason is that the local minimum is no longer the global minimum. Another reason is that there is a fundamental non-uniqueness in some of the inverse mapping of these components (Backus, 1970; Kono, 1976; Proctor and Gubbins, 1990; Hulot et al., 1997).

As an easy example, we will discuss the case of declination. To define the declination, we need to know the horizontal components X and Y given by [5]. Because they are obtained by differentiation with respect to θ and ϕ, the factor $(a/r)^{l+1}$ in the potential [4] may be equated to 1. Consider two potentials

$$W_1 = a(b \cos\theta + c \sin\theta \sin\phi) \qquad [11]$$

$$W_n = a(W_1/a)^n \qquad [12]$$

Note that W_1 represents a geocentric dipole inclined by $\tan^{-1}(c/b)$ in the $90°$ meridian, that is, $(g_1^0, g_1^1, h_1^1) = (b, 0, c)$. For this dipole field, the horizontal components are

$$X_1 = \frac{1}{a}\frac{\partial W_1}{\partial \theta} = -b \sin\theta + c \sin\theta \sin\phi \qquad [13]$$

$$Y_1 = \frac{1}{a \sin\theta}\frac{\partial W_1}{\partial \phi} = c \cos\phi \qquad [14]$$

Similarly, the potential W_n gives the horizontal components

$$X_n = n(-b \sin\theta + c \cos\theta \sin\phi) \times (b \cos\theta + c \sin\theta \sin\phi)^{n-1} \qquad [15]$$

$$Y_n = nc \cos\phi (b \cos\theta + c \sin\theta \sin\phi)^{n-1} \qquad [16]$$

It can be seen from these that $Y_n/X_n = Y_1/X_1$, showing that $\tan D$ is the same for the two potential fields. By using D ($|D| \leq \pi$) instead of $\tan D$, the nonuniqueness may be removed in some cases, since the factor $(b \cos\theta + c \sin\theta \sin\phi)^{n-1}$ can take both positive and negative values. However, for odd n, even the signs of horizontal components are the same. As an example of odd n, the case of $n = 3$ gives

$$W_3/a = \frac{6}{5}\cos\theta - \frac{11}{58}(3\cos\theta + 5\cos 3\theta)$$
$$+ \frac{6}{5}\sin\theta\sin\phi + \frac{3}{108}\frac{3}{}(\sin\theta + 5\sin 3\theta)\sin\phi$$
$$- \frac{1}{10}\frac{15}{4}(\cos\theta - \cos 3\theta) - \frac{1}{60}\frac{15}{4}(3\sin\theta - \sin 3\theta) \qquad [17]$$

Thus, W_3 is a combination of dipole and octupole fields with nonzero coefficients

$$\left(g_1^0,\ b_1^1,\ g_3^0,\ b_3^1,\ g_3^2,\ b_3^3\right)$$
$$= \left(\frac{6}{5},\ \frac{6}{5},\ -\frac{1}{5},\ \frac{3\sqrt{6}}{10},\ -\frac{\sqrt{15}}{5},\ -\frac{\sqrt{10}}{10}\right) \quad [18]$$

Figure 20 shows that the two potential fields, W_1 and W_3, are quite different, and yet the declination is the same everywhere on the surface of a sphere. The arbitrary linear combination of W_n has the same property. It can be concluded that the inverse mapping of the declination to the potential is severely nonunique (Kono, 1976).

Serious treatments of geophysical inversion were initiated by the seminal paper of Backus and Gilbert (1967). The idea was to minimize the sum of errors and some auxiliary function. In their original treatment, the function was chosen so that the obtained solution had the best resolution (the 'deltaness'). This would be optimal for problems such as obtaining the mantle structure from seismic observations. However, it may not be appropriate for the geomagnetic inversion because the parameters to determine

are discrete numbers and not a continuous function of radius (say). Gubbins (1983) introduced the use of stochastic inversion for geomagnetic analysis. In this case, the function to be minimized is the sum of misfit (M) and a 'penalty function' (P)

$$S = M + \lambda P \quad [19]$$

where λ is the hyperparameter indicating the relative importance of the two terms. The ordinary least-squares method is the limit of small λ. When λ becomes large, the inversion chooses models for which P is small, that is, which satisfy some physically plausible constraints imposed by the function (e.g., minimum energy requirement, etc.). Although the penalty function as well as the hyperparameter are chosen arbitrarily, the application of the stochastic inversion over many years has established that this method gives reasonable models as solutions.

Stochastic inversion can treat both linear and nonlinear data elegantly. Gubbins, Bloxham, and various colleagues continued to collect inclination and declination data back in time, and inverted them to describe the secular variation before the time of Gauss (Bloxham *et al.*, 1989). This endeavor

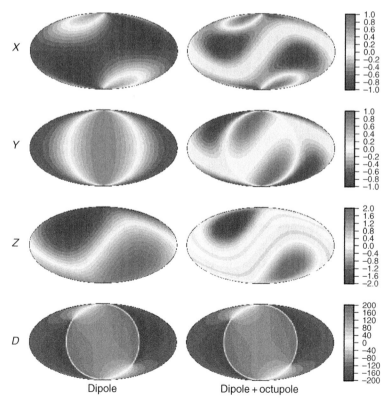

Figure 20 An example of an inclined dipole field (left) and a dipole+octupole field (right) which are quite different in X, Y, Z components, and yet, the values of declination D are the same everywhere.

culminated in the creation of the *gufm* 1 model of Jackson *et al.*, (2000), which is based on a large data set composed of field directions compiled by Jonkers *et al.* (2003) and covers the interval AD 1590–1995. These models are very important achievements that expand the time span for which we know some details of the magnetic field behavior.

1.5 Observational Constraints

In the long history of geomagnetism, and in the last one and half centuries in particular, the methods of measurements and analyses of the global field have undergone tremendous progress. Improvement in the sensitivity and accuracy of measurements have been matched by more and more elaborate modeling of the magnetic field. It is doubtful, however, if this trend of improvement will continue into the future. The reason for this lies in the nature of the Earth's magnetic field itself.

1.5.1 Geometrical Attenuation

As seen above, the geomagnetic spectrum at the surface of the Earth decays in power with the degree l roughly as $(R_c/a)^{2l+1}$, where R_c is the radius of the core. As the power is the mean square value of the magnetic field by definition, the field components can be seen to decay as $(R_c/a)^{l+(1/2)}$. Using this relation, the dipole field is attenuated by a factor of 0.407, while the quadrupole and octupole fields are reduced to 0.224 and 0.123, respectively, compared with the value at the CMB. This effect damps the shorter-wavelength signal quite drastically at the surface of the Earth. Assume that the spectrum is completely flat with the amplitude $R_l = (100\,\mu T)^2$ at the CMB. The field at the surface contains about $41\,\mu T$ dipole and $22\,\mu T$ quadrupole component, etc. The contributions from $l = 15$ and from $l = 25$ fields are only about 9.3 and 0.02 nT, respectively. Considering the fact that the intensity of the field due to the crustal source is roughly 10 nT, the measurement of the field generated in the core having short wavelength is practically impossible (**Figure 21**).

1.5.2 Accuracy of Data

The magnetic field is measured either by scalar or vector instruments. A typical example of scalar magnetometers is the proton-precession magnetometer. In this magnetometer, a strong field is applied to protons (which are in the molecules of water contained in the cylinder) and afterwards removed. The magnetic spins are aligned in the direction of the applied field, and upon the removal of the strong field, they precess around the direction of the (remaining) ambient field with a frequency proportional to the field strength. The relation between the field strength B and the signal frequency f can be expressed as $2\pi f = \gamma_p B$. The factor γ_p is called the gyromagnetic ratio, and it is the constant $\gamma_p = \mu/\hbar I$, where μ and $\hbar I$ are the magnetic moment and angular momentum of the proton. Measurement of the frequency of the signal therefore gives the absolute magnitude of the magnetic field. Optical pumping magnetometers, such as the rubidium vapor magnetometer, are similar.

On the other hand, the fluxgate magnetometer is widely used to obtain vector values of the field. It uses the nonlinear hysteresis property of a material such as mu-metal (a Ni-rich alloy). Because of its very high magnetic permeability, such sensors can easily be saturated in a weak magnetic field such as that of the Earth. In this magnetometer, an alternating current is applied to a coil wound around the mu-metal core. The signal induced by the hysteretic property of the core has twice the original frequency which is related to the strength of the ambient field. Through a negative feedback mechanism, electric current is supplied to oppose the ambient field, until the sum of the field becomes negligibly small. The compensation current is thus proportional to the magnetic field, but the proportionality depends on the core material as well as the shape of the coil. Thus, a fluxgate magnetometer measures the relative field intensity in a direction in which the core is aligned. In most cases, there are three sensors placed orthogonal to each other in a fluxgate magnetometer.

For both types of instruments, it is not very difficult to attain the sensitivity and accuracy of 0.1 nT, with proper facility for calibration in the case of the fluxgate magnetometer. Errors in the vector field accuracy of 6 nT (*Magsat*) and 1–2 nT (*Ørsted*) are primarily caused by inaccuracy in orientation determination. More details of the modern magnetic instruments can be found in Chapter 4.

1.5.3 Signal and Noise in Magnetic Field Measurements

The magnetic field we observe over the surface of the Earth is composed of three distinctly different components. The first is the internal magnetic field produced by the dynamo process in the core. This is also called the main field. The main field can be

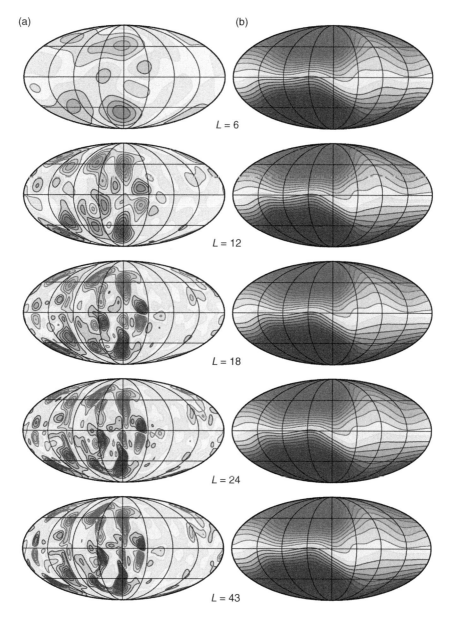

(a)

(b)

L = 6

L = 12

L = 18

L = 24

L = 43

Figure 21 The vertical magnetic component due to a dynamo model (a) at the CMB and (b) at the surface of the Earth when the spherical harmonics are truncated at various degrees. Modified from Kono M and Roberts PH (2002). Recent geodynamo simulations and observations of the geomagnetic field. *Reviews of Geophysics* 40: RG1013 (doi:10.1029/2000RG00102).

described by Gauss coefficients with degrees up to about 13. There must be shorter-wavelength components in the field generated in the core, but they cannot be determined with any confidence because of the reasons discussed below.

The second is the short-term fluctuations of the field. There are various types of such fluctuations. The largest of these is that accompanied with magnetic storms, which can sometimes exceed 1000 nT.

However, these are rather rare events, and can be excluded from the data set to extract the field that originates within the Earth. Variations of the field on the timescale of a day when solar activity is quiet (Sq) are small (about a few tens of nT), and their effect can be made even smaller by averaging many observations, or by taking values at local midnight.

The third is the field produced by the magnetization in the crust. The most famous example in this category is the oceanic magnetic anomalies, which are due to the alternatingly normally and reversely magnetized layers produced by sea-floor spreading from the mid-oceanic ridges. The average rate of reversals of the geomagnetic dipole moment is about five times per million years in the past few million years, and it is lower before. The present spreading rates are $1-10$ cm yr^{-1} at various ridges. Thus, the width of oceanic crust with either normal or reverse magnetization is about a few to tens of kilometers. This is much shorter than the wavelength of the main-field fluctuations.

There are also anomalies with longer wavelengths, as shown by the spectrum produced from *Magsat* observations (**Figure 19**). The surface spectrum is nearly flat above about $l = 13$, which suggests that this part of spectrum is produced by sources that are close to the observation locations.

The bulk of the mantle cannot be the source because the ambient temperature is much higher than the Curie points of magnetic minerals. Thus, the crust (and the uppermost mantle) appears to be responsible for this part of the spectrum. For more details about the magnetic properties of the mantle and crust, *see* Chapter 7. The intensity for many of the $l > 13$ terms is about 10 nT, and it is likely that they are mostly due to the sources in the crust (this is the subject of Chapter 6). It is therefore not possible to measure the core contribution for degrees $l > 13$, as the level of signal is significantly lower than that.

1.6 Magnetic Fields of the Planets and Satellites

The Earth is not unique in having an intrinsic magnetic field. The magnetic fields of the sun and Jupiter were detected by spectroscopic methods (Zeeman effect and kilometric radiation). With the advent of the space age, direct measurements by spacecrafts showed that many planets and a few satellites also have their own magnetic fields. These fields can originate either from active dynamos, induced fields, or ancient dynamos. Spacecraft observations have been carried out at various levels of sophistication. The first step is the 'flyby', in which a spacecraft approaches a planet (satellite) and observes the magnetic signature while it is close enough. In the next step, a spacecraft becomes an artificial satellite of the target body with low orbital inclination. This allows observations of the magnetic field at low latitudes. Finally, a spacecraft will be inserted into a near-polar

orbit, in which case most of the surface of the body can be observed.

The distinction between planets and satellites is not crucial in consideration of their magnetic fields. We therefore ignore the division, and call all these bodies 'planets' in the following. Until now, only the Moon (*Clementine* and *Lunar Prospector*), Venus (*Magellan*), and Mars (many spacecrafts in polar orbits) have been observed by polar orbit spacecrafts. Magnetic fields of other planets are known mostly by flyby observations, except Jupiter for which Galileo flew in a near-equatorial orbit encountering Jupiter's moons in a number of flybys.

1.6.1 Example of a Flyby Observation

As a typical flyby observation by a spacecraft, **Figure 22** shows the data from Mercury. Mariner 10 flew to Venus and used its gravity field to accelerate itself to reach Mercury orbit. On its first flyby (29 March 1974), a coherent field structure was discovered on the antisunward side of Mercury, with the maximum field strength of about 100 nT at the closest approach of 723 km (Ness *et al.*, 1975). From this observation, it was concluded that Mercury has an intrinsic magnetic field. The magnetic disturbance that occurred in the magnetosphere of Mercury after Mariner 10's closest approach resembles a substorm on the Earth.

Mercury is the only other terrestrial planet in the solar system in which an instrinsic magnetic field has been observed. The source mechanism for this magnetic field has not been settled yet; besides an ordinary dynamo, it is possible that remanent field or thermoelectric currents are responsible. Until present, Mercury has been visited only by Mariner 10. But now NASA's *Messenger* (launched 8 March 2004) is on its way to make a few flybys as well as enter into an elliptical orbit around Mercury (scheduled for 18 March 2011). Another spacecraft, *BepiColombo*, will also be sent to Mercury around 2012 in a joint effort of European countries and Japan. These orbiting spacecrafts can give us a picture of Mercury magnetic field as a whole. Undoubtedly, our understanding of planetary dynamos will be given a strong boost from these observations as Mercury's operating condition is quite different from that of the Earth.

1.6.2 Magnetic Mapping of Planets

Up to now, the magnetic field of an entire planet has been observed only for the Moon and Mars. A few spacecrafts circulated in polar orbit around Venus,

(a)

(b)

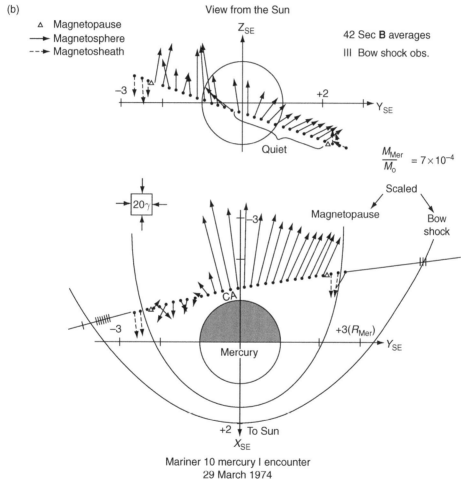

View from the Sun

△ Magnetopause
→ Magnetosphere
--→ Magnetosheath

42 Sec **B** averages

III Bow shock obs.

Z_{SE}

-3 $+2$ Y_{SE}

Quiet

$$\frac{M_{Mer}}{M_0} = 7 \times 10^{-4}$$

Scaled

Magnetopause

Bow shock

20γ

-3

CA

-3 $+3(R_{Mer})$ Y_{SE}

Mercury

$+2$ To Sun

X_{SE}

Mariner 10 mercury I encounter
29 March 1974

Figure 22 (a) Mercury (courtesy of NASA, photo PIA03103.jpg) and (b) its magnetic field observed by *Mariner 10* flyby. Modified from Ness NF, Behannon KW, Lepping RP, and Whang YC (1975). The magnetic field of Mercury. I. *Journal of Geophysical Research* 80: 2708–2716.

but there is no strong magnetic signal from this planet. The Moon and Mars lack an active dynamo, but very conspicuous magnetic anomalies have been observed on both of them. On the Moon, many magnetic signatures are correlated with surface features such as craters, indicating their possible relation with the impact events. Meteorite impacts are associated with a very high temperature, enough not only to vaporize rocks but even to ionize them. As remanences may be acquired in a magnetic field induced by impact events, whether the Moon had an ancient dynamo or not is still under debate.

Recent observations of the magnetic field on Mars revealed an astonishing feature. Very strong magnetic anomalies reaching $\pm 1500\,\mathrm{nT}$ (at 200 km altitude) were observed in a rather wide area in the Southern Hemisphere (Connerney *et al.*, 1999). Moreover, the anomalies appear as zones of alternating polarity. They are elongated in the east–west direction (**Figure 23**), reminiscent of the anomalies associated with sea-floor spreading on the Earth. However, they are quite different: the anomaly bands are much wider (typically 200 km) and their remanence intensity much higher ($20\,\mathrm{A\,m^{-1}}$ assuming a layer thickness of 30 km) than the typical oceanic anomalies on the Earth (5–50 km and 1–$10\,\mathrm{A\,m^{-1}}$). In addition, there is no other evidence of sea-floor spreading on Mars. If the remanent magnetization in the Martian crust was acquired in a process similar to plate tectonics on the Earth, it is required that some or all of the following conditions be satisfied on Mars: (1) the ancient magnetic field was very strong, (2) the polarity reversals occurred

more infrequently than on the Earth, (3) a thick magnetic layer was formed at the 'ridges', and/or (d) the magnetic minerals typical of the Martian crust acquired very strong remanences (e.g., due to the grain size being close to the single domain).

1.6.3 Planetary Dynamos: A Comparative study

The planetary bodies in the solar system (both planets and satellites) can conveniently be categorized by their primary composition. Terrestrial planets such as the Earth are mainly composed of rock (silicates and oxide with some metallic iron). Gas giants (Jupiter and Saturn) have rock cores as much as 10 times the mass of the Earth, but their main constituents are hydrogen and helium and the rock part is not important in their dynamo processes. The two other giants (Uranus and Neptune) also have a thick hydrogen envelope, but the main constituents are probably 'ice' (meaning nongaseous states) of H_2O, CH_4, and NH_3 and the ice layer may extend to about 80% of their radii. The distribution of these bodies indicate the condition under which the planets accumulated in the early solar system; in the inner part,

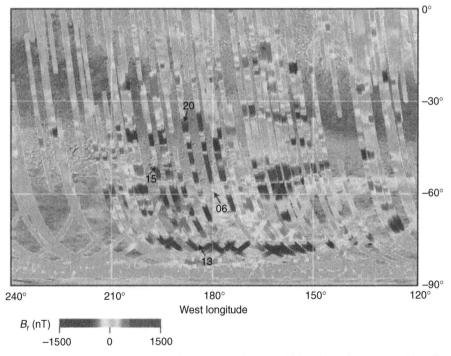

Figure 23 Magnetic field map of one-third of the Southern Hemisphere of Mars. Note the east–west trending bands of strong anomalies. Reprinted with permission from Connerney JEP, Acuna MH, Wasilewski PJ *et al.* (1999). Magnetic lineations in the ancient crust of Mars. *Science* 284: 794–798. Copyright 1999 AAAS.

large amounts of volatiles could not accrete due to high temperature, while gas and ice were the most abundant where temperature was low enough for them to accrete. Many of the satellites of Jupiter and other giants also contain large amounts of ice.

There are three possibilities for the origin of planetary bodies with magnetic fields: (1) dynamo, (2) remanence, and (3) induction. Exploration of the solar system with spacecrafts showed us all three of these are indeed operative in some of the planets (including satellites). We shall give a brief discussion of these occurrences.

A convective region with sufficiently large electrical conductivity is necessary for maintaining a dynamo. In the gas giants, hydrogen becomes metallic under the very high pressure, and has a high conductivity. There seems to be no problem of supplying energy sufficient to maintain the convection. In the icy giants, (dirty) water under very high pressure may provide the high conductivity necessary for a dynamo. Dynamo process may be confined to the thin outer shell, which may be responsible for the large tilt of dipole axis observed for Uranus and Neptune (Hubbard *et al.*, 1995).

In the terrestrial planets, only the metallic core has the possibility of providing dynamo generation. These cores are mainly composed of iron, but contain small but significant amounts of light elements such as Si, O, H, or S. The melting temperature of iron is considerably lower than that of silicate rocks (which is a primary reason why cores form to begin with). Inclusion of light elements (especially sulpher) lowers the melting temperature, so it is possible that the cores of not-too-small planets can remain largely molten for 4.6 billion years since their formation, even with the secular cooling of the silicate mantle overlying them. The molten state cannot guarantee, however, that convection in the core is vigorous enough to sustain a dynamo. In the Moon and Mars, the liquid part, if it still exists, is perhaps too small to generate a significant dynamo.

Venus, on the other hand, must have a liquid core similar to that of the Earth. The lack of an intrinsic magnetic field in Venus is perhaps not due to its core but more because of its mantle. The surface temperature is quite high on Venus, so that the temperature gradient within the mantle is not as large as in the Earth. Also, the volatile content of the upper mantle is quite low, inhibiting the formation of an asthenosphere, which lubricates the mantle and makes plate tectonics possible on the Earth. As a result, mantle convection on Venus is of the type with a thick lid on top, and heat transfer is not sufficiently high. The mantle does not draw high-enough heat from the core, so the core may be inactive with a subadiabatic temperature gradient. The state of Venus may have been different in the remote past; it may have had times when plate tectonics was in operation. Then the heat transfer was more efficient, which resulted in convection in the core needed for dynamo action. Unfortunately, the very high surface temperature prohibits a crustal remanence field to remain, and so there can be no evidence of dynamo action in Venus' past, unlike the cases of the Moon and Mars.

Recent observation of Jupiter's moons by the Galileo spacecraft brought us some shocks about their magnetic fields. All of the Galileo satellites show magnetic signatures in one way or another. Among them, the fields at Europa and Callisto seem to be of induction origin. Although they are farther away from Jupiter compared with the Moon from the Earth, the Jovian magnetic field is still strong enough to produce a clear signal of the induced magnetic field. The inductive region must be the 'dirty ocean' that exists below the surface ice cover. Io has a complex magnetic field, but it is perhaps also induced by the strong Jovian field. Ganymede, on the other hand, seems to have an intrinsic field and consequently an active dynamo. Although it is larger than the Moon (with the radius of 2638 km compared with 1738 km of the latter), it is not understood how the cores in such small bodies as Ganymede and Mars can continue to be convectively active until present.

1.6.4 The Magnetic Field of the Sun

In the previous section, we discussed the observed facts about the planetary magnetic fields. Before closing this topic, it is informative to consider the magnetic field of the sun. Direct observation of the sun's magnetic field became possible around 1950 when the solar magnetograph was constructed. Measurements are done by observing the shift of spectral lines due to the Zeeman effect. At first, the strong magnetic field associated with sunspots ($\sim$0.2 T) was detected. With an increase in sensitivity, it became possible to observe the global field of about 10^{-4} T, which is more or less of dipole type.

These observations provided us with a rich view of the solar magnetic field. The sun has a dipole field roughly aligned with its rotational axis, which reverses polarity every 11 years or so (the solar cycle). In one solar cycle, the sunspots first appear at middle latitudes (about $\pm 30°$). The area where sunspots are abundant migrates with time toward the equator. When the activity is confined to the equatorial region, a new group of sunspots appears in middle latitudes, and a new solar cycle begins (**Figure 24**). It is now known that the sunspots are an expression of the toroidal magnetic field floating up from the surface of the sun; the sunspots often form pairs that show the places where the field exits and enters the Sun's photosphere. The polarity of the dipole can be seen not only from the weak background field, but also in the polarity of the radial field in the sunspot pair (**Figure 25**). For instance, if the radial component of the background field is positive (negative) in the Northern Hemispere, which is similar to the reverse (normal) polarity on the Earth, the radial component accompanied with a sunspot pair will have $+-$ polarities from the west to east in the Northern Hemisphere ($-+$ in the Southern Hemisphere). When the background field is reversed, the polarity of the sunspot pairs also reverses.

There are two particularly wonderful things about the solar magnetic field. First, we can directly see what occurs in the dynamo region, at least near its surface. This is not possible for all the planetary dynamos, including that of the Earth. What we see from radiation of various wave-lengths is quite complex. The convection is chaotic and has characteristic length scales of 10^6 m (granules), 3×10^7 km (supergranules), and it even occurs on the global scale (7×10^8 km). The magnetic field generation is also chaotic and shows every complexity (Weiss, 2001). It is considered that the solar magnetic field is produced by the so-called $\alpha\omega$ dynamo process. This is consistent with the strong toroidal field evidenced in the sunspots.

Second, the sunspots have almost continuously been observed since the time of Galileo, and the solar cycle can be traced in the past. Because of the correspondence between the sunspots and the magnetic field, these records can be interpreted to represent the magnetic activity of the sun. The level of activity (which can be measured by the number of sunspots) between the dipole reversals (the sunspot or solar cycles) shows some changes over the observed period, but the most notable aspect of the record is that there was some period (1647–1715) when the sunspots were absent (the Maunder minimum, see **Figure 26**). This coincides with the time called the 'Little Ice Age' when the temperature was exceptionally low in Europe. It is still controversial but some people speculate that the Sun's activity was low when sunspots were few or none, and that the diminished amount of solar radiation caused the change in the climate.

In summary, the dynamo process in the sun is quite different from that of the Earth. It is in the turbulent

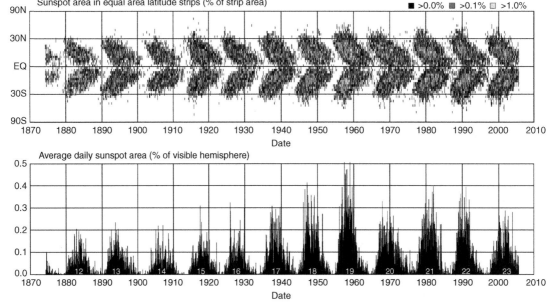

Daily sunspot area averaged over individual solar rotations

Figure 24   The butterfly diagram of the sunspots. Top: Time change in latitudinal distribution of sunspots. Bottom: Area of the solar surface occupied by sunspots. Reproduced from NASA, with permission.

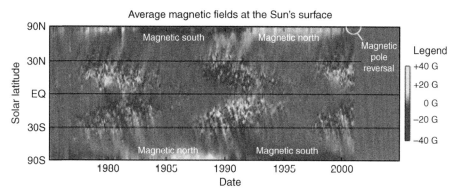

Figure 25 The magnetic butterfly diagram. Radial component of the magnetic field is longitudinally averaged. Note that the character of the figure is very similar to the ordinary butterfly diagram. Reproduced from NASA, with permission.

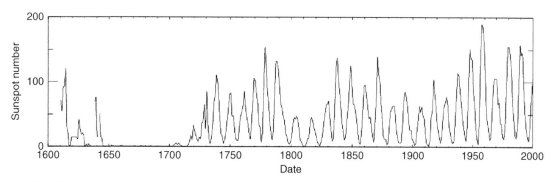

Figure 26 The yearly sunspot number for the period 1600–2000. Note the conspicuous absence of sunspots for half a century starting from about 1645 (the Maunder minimum). Reproduced from NASA, with permission.

regime, and can better be described by an $\alpha\omega$ process (the Earth's dynamo is thought to be closer to an α^2 process; e.g., Christensen *et al.*, 1999). Still, the magnetic field of the Sun gives important clues for considering the magnetic field of the Earth, clues that are otherwise unattainable. The surface of the solar dynamo region can be directly observed. The process is so rapid that 400 years's observation is enough to cover tens of polarity reversals.

1.7 Discussion and Conclusions

Magnetism seems to have a peculiar property that it is both very exotic and very practical, at different times or at the same time. It is sometimes just a target of curiosity, but at other times it is quite useful for daily life. Magnets have remained a source of wonder for thousands of years in human history. Today's life without magnets is difficult to conceive (just look around and you will see motors, tape recorders, hard disks, etc.).

The situation is very similar with the Earth's magnetic field. The north-seeking property of the

magnetic compass fascinated people in medieval times. It then became a very practical tool to measure direction (and sometimes position). All the ships in the world used magnetic compasses for navigational aid. Now, such days are gone (although most ships still carry compasses for contingency purposes), and it is the Global Positioning System (GPS) which gives you precise position as well as bearing. As to the target of curiosity, Albert Einstein counted the origin of the Earth's magnetism as one of the important problems of twentieth-century physics.

From the time of Gauss, the study of geomagnetism has steadily progressed and many of the wonders are now explained in terms of physical processes. The observational constraints, such as the geometrical decay and the crustal field hiding the short-wavelength part of the main field, are now well understood. Since 1995, when dynamo models based on first-principles calculations appeared, our understanding of field generation has been rapidly growing.

Is our curiosity going to be satisfied soon?

Far from it. The present is just a small window in the entire spectrum of what a magnetic field can do.

Paleomagnetists used to be called 'paleomagicians' because they tended to speak of unwieldly things such as 'continental drift' and 'polarity reversals'. Discovery of the Vine–Matthews mechanism, which combined magnetic reversals with sea-floor spreading, changed Earth sciences forever. Magnetism became the best tool to examine plate tectonic evolution since the Jurassic. Dynamo models can produce polarity reversals that are in many ways very similar to paleomagnetic observations (Glatzmaier *et al.*, 1999; Kono and Roberts, 2002). But these computer models include parameters that are quite different from those in the core of the Earth (particularly the viscosity). Moreover, the data from paleomagnetism are far from satisfactory; for instance, we are not at all sure of the field configurations during reversals. We need more and better observations. Likewise, theoretical advance is needed to ascertain that the computer simulation results apply to the Earth's dynamo.

Another consideration is that the mode of dynamo operation may have also changed with time. Existence of magnetic anomalies on the Moon and Mars indicates that such smaller bodies may have had self-sustaining dynamos early in their histories. Observations of Mercury's and Ganymede's magnetic fields suggest that there may be a mechanism by which the magnetic field can be actively sustained in these small bodies in the solar system. Venus does not show any intrinsic magnetic field, despite the fact that it has an internal structure quite similar to that of the Earth. Unfortunately, we cannot learn about the past of the Venutian dynamo, because the remanence field cannot exist there due to the very high surface temperature. In the present Venus, mantle convection is of thick lid type and quite different from plate tectonics, in which heat transfer is very effective in the mantle and can draw high heat flow from the core. It is speculated, however, that Venus may have had a time in the past when the mantle convected in the plate tectonics regime, and the core dynamo was active. A comparison of the magnetic fields of the planets with that of the Earth, together with the knowledge of other physical properties, will provide a better understanding of the origin and evolution of the geomagnetic field.

References

Ade-Hall JM, Khan MA, Dagley P, and Wilson RL (1968) A detailed opaque petrological and magnetic investigation of a single Tertiary lava flow from Skye, Scotland. 1. Iron–titanium oxide petrology. *Geophysical Journal of the Royal Astronomical Society* 16: 375–388.

Backus GE (1970) Non-uniqueness of the external geomagnetic field determined by surface intensity measurements. *Journal of Geophysical Research* 75: 6339–6341.

Backus GE and Gilbert JF (1967) Numerical applications of a formalism for geophysical inverse problems. *Geophysical Journal of the Royal Astronomical Society* 13: 247–276.

Blackett PMS, Clegg JA, and Stubbs PHS (1960) An analysis of rock magnetic data. *Proceedings of the Royal Society of London A* 256: 291–322.

Bloxham J, Gubbins D, and Jackson A (1989) Geomagnetic secular variation. *Proceedings of the Royal Society of London A* 329: 415–502.

Brunhes (1906) Recherches sur la direction de l'aimantation de roches volcanique. *Journal de Physique* 4^{e}sér., 5: 705–724.

Cande SC and Kent DV (1992) A new geomagnetic polarity timescale for the Late Cretaceous and Cenozoic. *Journal of Geophysical Research* 97: 13917–13951.

Cande SC and Kent DV (1995) Revised calibration of the geomagnetic polarity timescale for the Late Cretaceous and Cenozic. *Journal of Geophysical Research* 100: 6093–6095.

Christensen UR, Aubert J, Cardin P, *et al.* (2001) A numerical dynamo benchmark. *PEPI* 128: 25–34.

Christensen U, Olson P, and Glatzmaier GA (1998) A dynamo model interpretation of geomagnetic field structures. *Geophysical Research Letters* 25: 1565–1568.

Connerney JEP, Acuna MH, Wasilewski PJ, *et al.* (1999) Magnetic lineations in the ancient crust of Mars. *Science* 284: 794–798.

Cox A, Doell RR, and Dalrymple GB (1963) Geomagnetic polarity epochs and Pleistocene geochronometry. *Nature* 198: 1049–1051.

Dalrymple GB (1972) Potassium-argon dating of geomagnetic reversals and North American glaciations. In: Bishop WW and Miller JA (eds.) *Calibration of Hominid Evolution*, pp. 107–134. Edinburg: Scottish Academic Press.

David P (1904) Sur la stabilité de la direction d'aimantation dans quelque roches volcanique. *Comptes Rendus de l'Académie des Sciences* 138: 41–42.

Dietz RS (1961) Continent and ocean basin evolution by spreading of the sea floor. *Nature* 190: 854–857.

Doell RR and Dalrymple GB (1966) Geomagnetic polarity epochs – a new polarity event and the age of the Brunhes–Matuyama boundary. *Science* 152: 1060–1061.

Folgerhaiter G (1899) Sur les variations séculair de l'inclinaison magnétique dans l'antiquité. *Journal de Physique* 8: 5–16.

Gauss CF (1838) *Allgemeine Theorie des Erdmagnetismus* in *Resultate aus den Beobachtunger des magnetischen Vereins im Jahre 1836–1841*, 6 Hefte, Gottiugen und Leipzig. (Reprinted in Werke, Bd. 5, 121–193).

Gilbert W (1600) *De Magnete*.(translated by Mottelay PF (1958)). New York: Dover.

Glatzmaier GA and Roberts PH (1995) A three-dimensional self-consistent computer simulation of a geomagnetic field reversal. *Nature* 377: 203–209.

Gubbins D (1983) Geomagnetic field analysis - I. Stochastic inversion. *Geophysical Journal of the Royal Astronomical Society* 73: 641–652.

Harradon, Some early contributions to the history of geomagnetism. *Terrestrial Magnetism and Atmospheric Electricity* 48: 3–17, 79–91, 127–130, 197–199, 200–202, 1943; *ibid.* 49: 185–198, 1944; *ibid.* 50: 63–68.

Hashimoto M, Noya H, Omori M, and Miyajima K (1977) Japanese translation of Needham (1962), 515pp. Tokyo: Shisaku-sha.

Hess HH (1962) History of ocean basins. In: Engel AEJ, James HL, and Leonard BL (eds.) *Petrologic Studies*, pp. 599–620. Boulder, CO: Geological Society of America.

Heirtzler JR, Dickson GO, Herron EM, Pitman WC, III, and Le Pichon X (1968) Marine magnetic anomalies, geomagnetic field reversals, and motions of the ocean floor and continents. *Journal of Geophysical Research* 73: 2119–2136.

Hospers J (1953) Reversals of the main geomagnetic field, I–II. *Proceedings of the Koninklijke Nederlands Academic van Wetenschappen* B56: 467–491.

Hubbard WB, Podolak M, and Stevenson DJ (1995) The interior of Neptune. In: Cruickshank B (ed.) *Neptune*, pp. 109–138. Tucson, AZ: University of Arizona Press.

Hulot G, Khokhlov A, and Le Mouël J-L (1997) Uniqueness of mainly dipolar magnetic fields recovered from directional data. *Geophysical Journal International* 129: 347–354.

Jackson A, Jonkers ART, and Walker MR (2000) Four centuries of geomagnetic secular variation from historical records. *Philosophical Transaction of the Royal Society of London A* 358: 957–990.

Jonkers ART, Jackson A, and Murray A (2003) Four centuries of geomagnetic data from historical records. *Reviews of Geophysics* 41(2): 1006 (doi:10.1029/2002RG000115, 2003).

Kageyama A, Sato T, and Complexity Simulation Group (1995) Computer simulation of a magnetohydrodynamic dynamo, II. *Physics of Plasmas* 2: 1421–1431.

Khramov AN (1958) *Paleomagnetic Correlation of Sedimentary Strata*, 218pp. Gostoptekhizdat: Leningrad.

Kono M (1976) Uniqueness problems in the spherical harmonic analysis of the geomagnetic field direction data. *Journal of Geomagnetism and Geoelectricity* 28: 11–29.

Kono M and Roberts PH (2002) Recent geodynamo simulations and observations of the geomagnetic field. *Reviews of Geophysics* 40: RG1013 (doi:10.1029/2000RG00102).

Langel RA and Estes RH (1982) A geomagnetic field spectrum. *Geophysical Research Letters* 9: 250–253.

Langel RA, Estes RH, Mead GD, Fabiano EB, and Lancaster ER (1980) Initial geomagnetic field model from Magsat vector data. *Geophysical Research Letters* 7: 793–796.

Larson RL and Pitman WC, III (1972) World wide correlation of Mesozoic magnetic anomalies and its implications. *Geological Society of America Bulletin* 83: 3645–3662.

Le Pichon X (1968) Sea-floor spreading and continental drift. *Journal of Geophysical Research* 73: 3661–3697.

Lowes FJ (1955) Secular variation and the non-dipole field. *Annales de Geophysique* 11: 91–94.

Malin SRC (1987) Historical introduction to geomagnetism. In: Jacobs JA (ed.) *Geomagnetism*, vol. 1, pp. 1–49. London: Academic Press.

Matuyama M (1929) On the direction of magnetization of basalt in Japan, Tyosen and Manchuria. *Proceedings of the Imperial Academy Japan* 5: 203–205.

McDougall I and Tarling DH (1963) Dating of polarity zones in the Hawaiian Islands. *Nature* 200: 54–56.

McFadden PL (1984) Statistical tools for the analysis of geomagnetic reversal sequences. *Journal of Geophysical Research* 89: 3363–3372.

McFadden PL and Merrill RT (1984) Lower mantle convection and geomagnetism. *Journal of Geophysical Research* 89: 3354–3362.

McKenzie DP and Parker RL (1967) The North Pacific: An example of tectonics on a sphere. *Nature* 216: 1276–1280.

Mercanton PL (1926) Inversion de l'inclinaison magnétique terrestre aux âges géologiques. *Terrestrial Magnetism and Atmospheric Electricity* 31: 187–190.

Merrill RT, McElhinny MW, and McFadden PL (1996) *Magnetic Field of the Earth: Paleomagnetism, the Core, and the Deep Mantle*, 527pp. London: Academic Press.

Mitchell AC (1946) Chapters in the history of terrestrial magnetism. *Terrestrial Magnetism and Atmospheric Electricity* 37: 105–146, 1932; *ibid*. 42: 241–280, 1937; *ibid*. 44: 77–80, 1939; *ibid*. 51: 323–351.

Morgan WJ (1968) Rises, trenches, great faults, and crustal blocks. *Journal of Geophysical Research* 73: 1959–1982.

Mueller RD, Roest WR, Royer J-Y, Gahagan LM, and Sclater JG Digital isochrons of the world's ocean floor. *Journal of Geophysical Research* 102: 3211–3214.

Nagata T, Akimoto S, and Uyeda S (1951) Reverse thermo-remanent magnetism. *Proceedings of the Japan Academy* 27: 643–645.

Needham J (1962) *Science and Civilisation in China,* Vol. 4, *Physics and Physical Technology*, Part 1 Physics, 434pp. Cambridge: Cambridge University Press.

Néel L (1951) L'inversion de l'aimantation permanente des roches. *Annales de Geophysique* 7: 90–102.

Ness NF, Behannon KW, Lepping RP, and Whang YC (1975) The magnetic field of Mercury, I. *Journal of Geophysical Research* 80: 2708–2716.

Olsen N, *et al.* (2000) Ørsted initial field model. *Geophysical Research Letters* 27: 3607–3610.

Proctor MRE and Gubbins D (1990) Analysis of geomagnetic directional data. *Geophysical Journal International* 100: 69–77.

Roche A (1951) Sur les inversion des l'aimantation remanente des roches volcaniques dans les monts d'Auvergne. *Comptes Rendus de l'Académie des Sciences* 233: 1132–1134.

Runcorn SK (1962) Palaeomagnetic evidence for continental drift and its geophysical cause. In: Runcorn SK (ed.) *Continental Drift*, pp. 1–40. New York: Academic Press.

Verhoogen J (1956) Ionic ordering and self-reversal of magnetization in impure magnetites. *Journal of Geophysical Research* 61: 201–209.

Vine FJ (1966) Spreading of the ocean floor: New evidence. *Science* 154: 1405–1415.

Vine FJ and Matthews DH (1963) Magnetic anomalies over oceanic ridges. *Nature* 199: 947–949.

Wang CT (1948) Discovery and application of magnetic phenomena in China. 1. The lodestone spoon of the Han. *Chinese Journal of Archaeology* 3: 119.

Wegener A (1929) Die Entstehung der Kontinente und Ozeane. *The Origin of Continents and Oceans, 4th edn. Branschweig: Vieweg und Sohn (trans. from German by J. Biram), pp. 246 New York: Dover 1966.*

Weiss N (2001) Turbulent magnetic fields in the Sun. *Astronomy and Geophysics* 42: 3.10–3.17.

Yamamoto Y (2003) *Discovery of Magnetic and Gravitational Forces (in Japanese)*, vols. 1–3, 947 pp. Tokyo: Misuzu Publishers.

2 The Present Field

G. Hulot, Institut de Physique du Globe de Paris/CNRS and Université Paris 7, Paris, France

T. J. Sabaka, Raytheon at NASA Goddard Space Flight Center, Greenbelt, MD, USA

N. Olsen, Danish National Space Center/DTU and Niels Bohr Institute, University of Copenhagen, Copenhagen, Denmark

2.1 Introduction

The Earth's magnetic field (also known as the geomagnetic field) is usually defined as the magnetic field produced by all sources within and outside the solid Earth up to the so-called 'magnetopause', a boundary within which the Earth's magnetic field remains confined. Beyond this boundary is indeed the realm of the 'interplanetary magnetic field' (IMF) produced by the Sun, and transported by a plasma known as the 'solar wind'. In addition, and at least as far as the present chapter is concerned, it is also generally understood that the term Earth's magnetic field only encompasses the magnetic field produced by natural phenomena in the 0 (permanent) to typically a few hertz frequency range. This rather arbitrary limit amounts to dismiss phenomena such as electromagnetic waves (and many other high-frequency phenomena which can indeed be observed with appropriate instruments), and is largely dictated by the bandpass of the magnetometers traditionally used to investigate the permanent and slowly changing components of the Earth's magnetic field.

Bearing this definition in mind, the sources of the Earth's magnetic field can then be understood as falling into two main categories: magnetized media and electric currents. Obviously, magnetized sources can only be found inside the solid Earth. They occur in the form of rocks which have been magnetized in the past (permanent magnetization), but which also bear an additional magnetization proportional to the present ambient magnetic field (induced magnetization). Clearly also, such rocks can only be found in regions of the solid Earth, where the temperature is less than the Curie temperature of the minerals ultimately carrying the magnetization. This restricts magnetized rocks to lie in the uppermost layers of the Earth. All other sources of the Earth's magnetic field are electric currents. Those can be found in most regions of the Earth: inside the metallic core, in the mantle and crust, in the oceans, and finally above the neutral atmosphere, in the ionosphere and magnetosphere (cf. **Figure 1**).

Sources of the Earth's magnetic field thus differ in nature and location. The field they produce also widely differ in magnitude and spatiotemporal behavior. By far the most intense field (on the order of 30.000 nT at the equator, 60.000 nT at the poles) is the one produced within the core through a self-sustaining dynamo process. This field is known as the 'main field'. It changes on secular timescales, and its time derivative is therefore referred to as the 'secular variation.' It otherwise is mainly dipolar, but also has significant multipolar terms, decreasing in strength as higher orders (and therefore smaller spatial scales) are considered. The so-called 'crustal field,' produced by the magnetized rocks (which indeed mainly, but not exclusively, lie in the crust), is weaker on average. But it can vary substantially as a function of location, from fractions of a nanotesla to thousands of nanoteslas at the Earth's surface. This strong spatial variability of the crustal field results

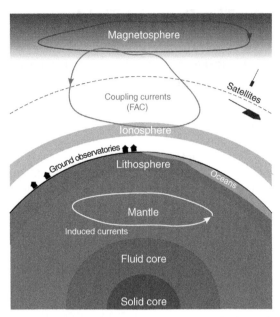

Figure 1   Sketch of the various sources contributing to the near-Earth magnetic field.

from the complex history of its magnetized sources, and from the various abilities of rocks to carry magnetization. As a result, the crustal field is characterized by a spatial spectrum with fairly comparable contributions at all length scales at the Earth's surface. The sum of the main field and of the crustal field then makes what is usually referred to as the 'internal field' (or 'field of internal origin'), since both have their sources inside the solid Earth. The internal field is dominated by the main field at large scales, and by the crustal field when scales on the order of less than typically 2000 km are considered at the Earth's surface.

As already stated, the Earth's magnetic field also has sources above the neutral atmosphere (some electrical currents are to be found in the neutral atmosphere; but they are very weak, except at times of thunderstorms, the magnetic signature of which is however traditionally not considered to be within the direct scope of Earth's magnetism). The field produced by such external sources is known as the 'external field' (or 'field of external origin'). It is produced by electrical current systems in the ionosphere and the magnetosphere, the physical origin of which can crudely be described in terms of currents driven by dynamo processes in the plasma of the ionosphere, which moves within the main field, and currents produced by the movement of charged particles traveling within the magnetosphere. Those two current systems are very different in physical nature, but are coupled to each other. The field produced by ionospheric currents is known as the 'ionospheric field' and that produced by magneto-spheric currents, as the 'magnetospheric field'. Their magnitude can strongly vary with time. A weak large-scale magnetospheric field (dominated by an external axial dipole) on the order of 20 nT at the Earth's surface is found at all time. During so-called 'magnetically quiet times', a weak ionospheric field can be found, which varies with local time. Its magnitude is minimum at night, peaks during day time, and is on the order of a few tens of nanoteslas at ground level. But at other times, known as 'disturbed magnetic times', both the magnetospheric and iono-spheric fields can become much more dynamic. The resulting external field can then occasionally reach magnitudes on the order of thousands of nanoteslas, and vary on timescales from a fraction of a second to several days.

Electrical currents can also be found in the crust and mantle of the Earth. Those arise because of the weak electrical conductivity of rocks, which reacts to the time-varying main and external fields described above. The resulting induced currents produce what is usually referred to as 'induced fields'. Currents induced by the main field are difficult to directly assess, and are thought to screen out high-frequency signals produced by the core. By contrast, currents induced by the external field produce so-called 'externally induced fields' that can easily be observed (and reach magnitudes of a fraction of the external inducing field). Similar induced currents are of course also to be found within the salty (and there-fore conducting) waters of the oceans. In that case, however, additional currents can also be found, which are generated by yet another dynamo process linked to the tidal and oceanic flows of the oceanic waters within the main field. Those however again produce very weak fields, on the order of a few nanoteslas at ground level.

The Earth's magnetic field is thus the result of the superposition of many fields produced by a great variety of sources. Unfortunately, magnetometers can only measure the resulting field, and a single measure-ment cannot distinguish the contributions from each source. However, provided enough measurements are being acquired and processed in appropriate ways, a separation of the contributions from each type of field can be achieved and global models of the Earth's magnetic field constructed. The purpose of the present chapter is to provide an up-to-date review of such

global models of the present field. Section 2.2 first describes the various types of magnetic data currently used to investigate the present magnetic field. Section 2.3 next reviews the models. Finally, special attention is being paid in Section 2.4 to the current behavior of the main field.

2.2 Magnetic Field Data Used in Modeling

2.2.1 Magnetic Elements, Coordinates, and Time

Magnetic measurements taken at the Earth's surface, including shipborne and airborne data, are typically given in a local 'topocentric (geodetic)' coordinate system (i.e., relative to a reference ellipsoid as an approximation for the geoid), shown in **Figure 2**. The magnetic elements X, Y, Z are the components of the field vector **B** in an orthogonal right-handed coordinate system, the axes of which are pointing toward geographic north, geographic east, and verti-cally down. Derived magnetic elements are: the angle between geographic north and the (horizontal) direc-tion in which a compass needle is pointing, denoted as declination $D = \arctan Y/X$; the angle between the local horizontal plane and the field vector, denoted as inclination $I = \arctan Z/H$; hori-zontal intensity $H = \sqrt{X^2 + Y^2}$; and total intensity $F = \sqrt{X^2 + Y^2 + Z^2}$. For the numerical calculation

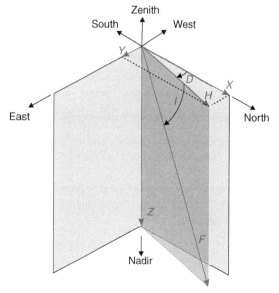

Figure 2 The magnetic elements in the local topocentric coordinate system, seen from northeast.

of declination D, it is recommended to use the 'four-quadrant inverse tangent' atan2, that is, $D = \text{atan2}(Y, X)$, to avoid the π-ambiguity of the arctan function. Note that the vertical component Z is given with respect to the local plumb line (i.e., with respect to the geoid), but interpreted as if it was given with respect to the reference ellipsoid. The difference between the two directions can be up to tens of arcseconds.

In contrast to magnetic observations taken at or near ground, satellite data are typically provided in the 'geocentric coordinate' system as spherical components B_r, B_θ, B_ϕ.

2.2.1.1   Geodetic and geocentric coordinates and components

Since spherical components (and coordinates) are used for geomagnetic field modeling, it is necessary to transform the locations of surface data, and the data themselves, from the geodetic to the geocentric coordinate system. This can be done in the following way:

The 'geodetic coordinates' (h, α, ϕ), where h is altitude, α is geodetic latitude, and ϕ is longitude, are transformed into the 'spherical geocentric coordinates' (r, θ, ϕ) by means of

$$\cot\theta = \frac{h\sqrt{a^2\cos^2\alpha + b^2\sin^2\alpha} + b^2}{h\sqrt{a^2\cos^2\alpha + b^2\sin^2\alpha} + a^2}\tan\alpha \quad [1]$$

$$r^2 = h^2 + 2h\sqrt{a^2\cos^2\alpha + b^2\sin^2\alpha}$$
$$+ \frac{a^4\cos^2\alpha + b^4\sin^2\alpha}{a^2\cos^2\alpha + b^2\sin^2\alpha} \quad [2]$$

$$\approx \frac{r_0^2}{1 + e^2\sin^2\alpha} \quad [3]$$

with $a = 6378.137$ km and $b = 6356.752$ km as the equatorial, respectively polar, radius of the Earth ellipsoid as defined by the World Geodetic System 1984 (WGS84), $e^2 = (a^2 - b^2)/a^2 = 6.69438 \cdot 10^{-3}$ as the eccentricity of the ellipsoid, and r_0 as the equatorial radius of the ellipsoid at height h ($r_0 = a$ at the surface).

The transformation of the local 'topocentric components' X, Y, Z to 'spherical components' B_r, B_θ, B_ϕ is done by means of

$$\begin{pmatrix} B_r \\ B_\theta \\ B_\phi \end{pmatrix} = \begin{pmatrix} -\sin\psi & 0 & -\cos\psi \\ -\cos\psi & 0 & +\sin\psi \\ 0 & 1 & 0 \end{pmatrix} \begin{pmatrix} X \\ Y \\ Z \end{pmatrix} \quad [4]$$

with $\psi = \theta - 90° + \alpha$.

2.2.1.2   Dipole coordinates and components

The first three Gauss coefficients, g_1^0, g_1^1, h_1^1, of a spherical harmonic expansion of the internal scalar magnetic potential (cf. Section 2.3.1) represent the field of a dipole with origin at the Earth's center, but inclined with respect to the axis of rotation. The dipole axis intersects the Earth's surface at the dipole pole; co-latitude θ_0 and longitude ϕ_0 of the dipole North Pole is given by

$$\theta_0 = 180° - \arccos\left(g_1^0/m_0\right) \quad [5]$$

$$\phi_0 = -180° + \text{atan2}\left(h_1^1, g_1^1\right) \quad [6]$$

where $m_0 = \sqrt{(g_1^0)^2 + (g_1^1)^2 + (h_1^1)^2}$ is the dipole strength and the output of the functions arcos and atan2 are given in degrees. Using the Gauss coefficients of DGRF 2000 gives as coordinates of the dipole North Pole $\theta_0 = 10.46°$, $\phi_0 = -71.57°$. A coordinate system with pole at (θ_0, ϕ_0) is called dipole or geomagnetic coordinate system (cf. **Figure 3**). Dipole (or geomagnetic) co-latitude and longitude are defined as

$$\theta_d = \arccos(\cos\theta\cos\theta_0 + \sin\theta\sin\theta_0\cos(\phi-\phi_0)) \quad [7]$$

$$\phi_d = \text{atan2}(\sin\theta\sin(\phi-\phi_0),$$
$$\cos\theta_0\sin\theta\cos(\phi-\phi_0) - \sin\theta_0\cos\theta) \quad [8]$$

The transformation of the spherical magnetic field components in the geographic frame, B_r, B_θ, B_ϕ, to

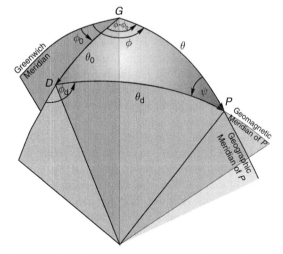

Figure 3   Relationship between geographic coordinates θ, ϕ, dipole (geomagnetic) coordinates θ_d, ϕ_d, and the (geographic) coordinates of the dipole pole, θ_0, ϕ_0. G is the geographic pole, D is the dipole pole, and P is the location of the point in consideration.

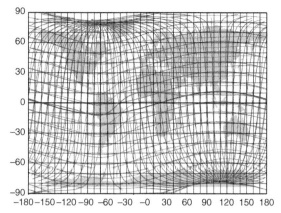

Figure 4 Map showing equidistant grids in latitude–longitude for various coordinate systems. Geographic coordinates are shown in black, dipole coordinates in blue, and QD coordinates in red. The respective equators (for which the respective co-latitude is equal to 90°) are indicated with thicker lines. The blue dots indicate the location of the dipole poles (for which $\theta_d = 0°$, respectively 180°), whereas the red dots indicate the dip poles (for which $\theta_q = 0°$, respectively 180°).

those in the dipole frame, B'_r, B'_θ, B'_ϕ, is done by means of

$$\begin{pmatrix} B'_r \\ B'_\theta \\ B'_\phi \end{pmatrix} = \begin{pmatrix} 1 & 0 & 0 \\ 0 & +\cos\psi & +\sin\psi \\ 0 & -\sin\psi & +\cos\psi \end{pmatrix} \begin{pmatrix} B_r \\ B_\theta \\ B_\phi \end{pmatrix} \qquad [9]$$

where

$$\psi = \text{atan2}\,(\sin\theta_0 \sin(\phi - \phi_0),\ \cos\theta_0 \sin\theta - \sin\theta_0 \cos\theta \\ \cos(\phi - \phi_0)) \qquad [10]$$

is the angle between the geographic and the dipole meridian at the location in consideration. The dipole equator (geomagnetic equator, not to be confused with the dip equator (magnetic equator), the line where inclination $I = 0°$) is defined as the line for which $\theta_d = 90°$; the dipole poles are then the points for which $\theta_d = 0°$, respectively 180°. A grid of constant dipole co-latitude and longitude is shown by the blue curves of **Figure 4**.

2.2.1.3 Quasi-dipole (QD) coordinates

Many ionospheric phenomena are naturally organized with respect to the geometry of the Earth's magnetic field due to its influence on the motion of charged particles. As a consequence, ionospheric conductivity is highly anisotropic, resulting in values that are so high parallel to the field that the magnetic

field lines are nearly equipotential lines. Therefore, it is often convenient to work in a coordinate system that follows the morphology of the magnetic main field, such as the quasi-dipole (QD) coordinates (θ_q, ϕ_q) proposed by Richmond (1995). The basic idea is to trace a field line (given by International Geomagnetic Reference Field (IGRF) of a specific epoch) from the point in consideration outward to the field line apex (i.e., its highest point above surface). The longitude of that point defines QD longitude ϕ_q. QD co-latitude θ_q is found by following an axial dipole field line from the apex downward to the ionospheric E-layer (typically assumed to be at an altitude of 115 km); the co-latitude of the intersection of the dipole field line and that altitude defines QD co-latitude θ_q. The dip-equator (magnetic equator) – the line where the magnetic field lines are horizontal, that is inclination $I = 0$ – is given by $\theta_q = 90°$; the dipole poles (for which the magnetic field lines are vertical) are defined by the points for which $\theta_q = 0°$, respectively $\theta_q = 180°$.

The QD coordinate system is not orthogonal and therefore mathematical differential operations (like curl, grad, and div) are nontrivial in QD coordinates. Richmond (1995) presents equations for these differential operators in the QD system.

2.2.1.4 Time

Local Time (LT), T, is defined as the difference in (geographic) longitude (in hours) between the subsolar point and the position in consideration:

$$T = (180° + \phi - \phi_s)/15 \qquad [11]$$

where ϕ and ϕ_s are the (geographic) longitudes (in degrees) of the location in consideration and of the subsolar point, respectively. Universal Time (UT), t, is defined as the LT of the Greenwich meridian:

$$t = (180° - \phi_s)/15 \qquad [12]$$

The connection between UT, t, and LT, T, follows as

$$T = t + \phi/15 \qquad [13]$$

Dipole (geomagnetic) local time (DLT) is defined relative to the dipole system.

$$T_d = (180° + \phi_d - \phi_{d,s})/15 \qquad [14]$$

where T_d is DLT in hours and ϕ_d and $\phi_{d,s}$ are dipole longitudes (in degrees) of the location in consideration and of the subsolar point, respectively. (Note that this equation is given incorrectly in Langel (1987) and Langel and Hinze (1998)). We use the notation DLT rather than Magnetic Local Time (MLT) to distinguish the above definition from that used in space physics, where MLT is defined with respect to a nonorthogonal coordinate system related to the IGRF. DLT is defined using only the first three Gauss coefficients of a field model.

Dipole Universal Time (DUT), t_d, is defined as the DLT of the dipole prime meridian, that is, the meridian from which dipole longitude is rendered. Hence t_d (in hours) is given by

$$t_d = \left(180° - \phi_{d,s}\right)/15 \qquad [15]$$

2.2.1.5   Coordinate systems for describing magnetospheric sources

It is advantageous to use different coordinate systems for describing the various current systems contributing to the magnetospheric field. The geocentric solar magnetospheric (GSM) frame is a Cartesian coordinate system which is preferably used to describe the interaction of the solar wind with the magnetosphere, especially magnetopause and -tail currents (which are located at distances greater than 8 Earth radii). The origin of the GSM frame is the center of the Earth. The x-axis is pointing toward the sun; the z-axis lies in the plane formed by the x-axis and the geomagnetic pole and points northward; the y-axis completes a right-handed coordinate system.

The solar magnetic (SM) frame is especially useful for describing the electrodynamics in the near-Earth space, for instance the magnetic effect of the ring-current. SM is a Cartesian coordinate system with origin in the Earth's center. Its z-axis is parallel with the dipole axis, x is in the plane defined by the subsolar point and the dipole axis, and y completes a right-handed system. The spherical coordinates of the SM system is identical to a dipole-colatitude/DLT coordinate system.

A detailed definition of SM and GSM is given in appendix 3 of Kivelson and Russell (1995).

2.2.2   Ground Data

The Earth's magnetic field is continuously monitored by a number (presently about 150) of geomagnetic observatories. Ground observations were the only data source for modeling the geomagnetic field for periods without satellite measurements. The distribution of the present observatory network is, however, very uneven, as can be seen from the dots in the left part of **Figure 5**, which shows observatories that provided data (hourly mean values) for (some of) the years 1995–2004. This considerably hampers determination of the global pattern of the magnetic field using observatory data. Satellite data provide excellent global coverage, but the spacecraft movement makes direct comparisons of satellite and observatory data difficult. Although geomagnetic observatory measurements are important also for periods for which satellite measurements are available, their role has changed: from being the only source for describing the static and time varying

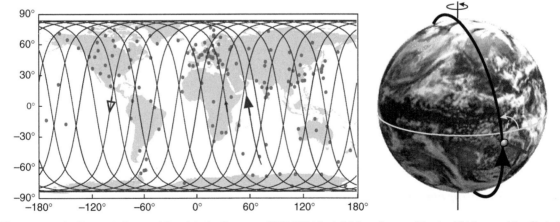

Figure 5   Left: Observatories providing data for the years 1995–2004 (red dots), and ground track of 24 hours of the Ørsted satellite on January 2, 2001 (blue curve). The satellite starts at 00 UT at 57° S and 72° E, moves northward on the morning side of the Earth, and crosses the Equator at 58° E (large black arrow). It continues crossing the polar cap, and moves southward on the evening side; 50 min after the first equator crossing, the satellite crosses again the equator, at 226° E (white arrow), on the dusk side of the orbit. The next equator crossing (after additional 50 min) is at 33° E (small black arrow), 24° westward of the first crossing 100 min earlier, while moving again northward. Right: The path of a satellite at inclination i in orbit around the Earth.

part of Earth's magnetic field, to providing information on its temporal change (secular variation as well as external field contributions and their induced counterparts) at a fixed location. The joint analysis of ground and satellite data combines the strengths of the two data sources: good global coverage (satellite data) and good temporal coverage (observatory data).

Although geomagnetic observatories aim at measuring the magnetic field components with an accuracy of 1 nT in the geodetic reference frame, it is presently not possible to take advantage of that (absolute) accuracy in field modeling, since the observatory data contain (unknown) contributions from the local crustal field of scale length shorter than those which present field models can resolve. Following Langel *et al.* (1982), this local field contribution is accounted for by 'observatory biases' for each observatory and each component. Joint analysis of observatory and satellite data allows to estimate these biases. As a consequence of this procedure, only the time-varying part of observatory data are used, and not their absolute (static) level. The need for an absolute baseline calibration of observatory data is therefore less important compared to an accurate determination of its time changes (for instance, drift). Recognizing this would simplify measurements, especially for ocean-bottom magnetometers, for which the exact determination of true north is an extremely difficult (and expensive) task.

Historically, observatory data are provided as annual mean values. The availability of hourly mean values, or even 1 min values, has, however, increased significantly recently, as can be seen from **Figure 6**, which shows the number of geomagnetic observatories providing annual means, hourly means, and 1 min values, respectively, through time. This allows for novel use of observatory data in combination with satellite data, for advanced modeling of the time changes of external field variations.

Observatory data are provided through the INTERMAGNET network and through the World Data Center (WDC) system.

Magnetic 'repeat stations' are permanently marked sites where high-quality magnetic observations are made every few years for a couple hours or even days (Newitt *et al.*, 1996). Their main purpose is to measure secular variation; repeat station data offer better spatial resolution than observatory data. However, unlike geomagnetic observatories, they do not measure the field continuously.

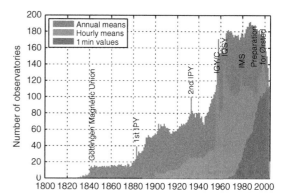

Figure 6 Number of geomagnetic observatories providing annual means, hourly means, and 1 min values, as a function of time. International campaigns, like the Göttingen Magnetic Union, the 1st and 2nd International Polar Year (IPY), the International Geophysical Year (IGY/C), the International Quiet Solar Year (IQSY), the International Magnetospheric Study (IMS), and the preparation of the Ørsted satellite mission have stimulated observatory data processing and encouraged the opening of new observatories.

2.2.3 Satellite Data

Ground-based magnetic measurements were the only data source before the first space-borne magnetic measurements were taken by the Sputnik 3 satellite in 1958 (Dolginov *et al.*, 1962). Satellite data for global field modeling were first taken by the POGO satellites (data from the earlier satellites Cosmos 26 and Cosmos 49 in 1964 were of much poorer quality).

Use of magnetic satellite data for modeling Earth's magnetic field is different from using ground data in several aspects. First, satellites sample the magnetic field over the entire Earth (apart from the 'polar gap,' a region around the geographic poles that is left unsampled if the inclination of the satellite orbit is different from 90°). Second, the data are obtained over different regions with the same instrumentation. Third, measuring the magnetic field from an altitude of 400 km or so corresponds roughly to averaging over an area of this dimension. Thus the effect of local magnetic heterogeneities (for instance due to magnetization of rocks) is reduced. And finally, since the satellite moves (with a velocity of about 8 km s^{-1} at an altitude of 400 km), it is not directly possible to decide whether an observed magnetic field variation is due to a temporal or spatial change.

Table 1 lists key parameters of satellites that have been used for modeling the geomagnetic field. In the following, we briefly describe these satellites.

Table 1   High-precision magnetic satellites

Satellite	Operation	Inclination	Altitude	Data
OGO-2	Oct. 1965–Sep. 1967	87°	410–1510 km	Scalar only
OGO-4	Jul. 1967–Jan. 1969	86°	410–910 km	Scalar only
OGO-6	Jun. 1969–Jun. 1971	82°	400–1100 km	Scalar only
Magsat	Nov. 1979–May 1980	97°	325–550 km	Scalar and vector
Ørsted	Feb. 1999–	97°	650–850 km	Scalar and vector
CHAMP	Jul. 2000–	87°	350–450 km	Scalar and vector
SAC-C/Ørsted-2	Jan. 2001–Dec. 2004	97°	698–705 km	Scalar only
Swarm	2010–2014	88°/87°	530/<450 km	Scalar and vector

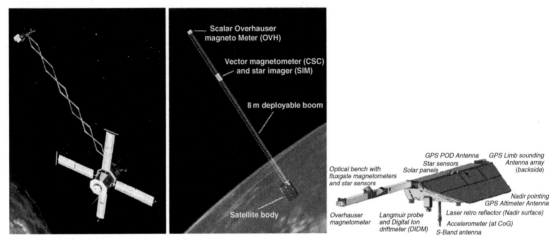

Figure 7   The Magsat (left, see Purucker (2007)), Ørsted (middle, see Olsen (2007)), and CHAMP (right, see Maus (2007)) satellites.

2.2.3.1 *POGO (OGO-2, OGO-4, OGO-6)*

The Polar Orbiting Geophysical Observatories (POGO) were the first satellites to measure the field intensity of Earth's magnetic field globally. They were equipped with optically pumped rubidium vapor absolute magnetometers. Three satellites of the POGO series are of interest for field modeling: OGO-2 measured the field between October 1965 and September 1967; OGO-4 between July 1967 and January 1969, with a few weeks of data overlap with OGO-2; and OGO-6 operated between June 1969 and June 1971. Intrinsic measurement error of all three satellites is believed to be below 1 nT, but contribution due to position uncertainty results in an effective magnetic error of about 7 nT. See Cain (2007) for more information on the POGO satellites. Data are available at Danish National Space Center's webpage.

2.2.3.2 *Magsat*

The US satellite Magsat (October 1979 to June 1980) made the first precise, globally distributed measurements of the vector magnetic field near Earth. The satellite flew at an altitude of 300–550 km, in a near-polar dawn–dusk orbit of inclination of 97°. Magsat carried a scalar (cesium 133 vapor optically pumped) magnetometer, and a triaxial fluxgate magnetometer, which sampled the ambient field at 16 Hz with a resolution of ±0.5 nT (**Figure 7**). Attitude was measured using two star-trackers on the spacecraft; transformation of attitude determined by these star trackers to the vector magnetometer on the end of the boom was done using a complicated optical system. Attitude errors limit the vector data accuracy to about 4 nT rms. See Purucker (2007) for more information on the satellite; data are available at Danish National Space Center's webpage.

2.2.3.3 *Ørsted*

The Danish Ørsted satellite is the first satellite mission after Magsat for high-precision mapping of the Earth's magnetic field. It was launched on 23 February 1999 into a near polar orbit. Being the first satellite of the International Decade of Geopotential Research, the satellite and its instrumentation has been a model for other present and forthcoming missions like CHAMP and Swarm.

Ørsted weighs 62 kg, measures $34 \times 45 \times 72$ cm and contains an 8-m-long boom, deployed shortly after launch, carrying the magnetic field instruments (**Figure 7**). The satellite is gravity gradient stabilized; attitude maneuvers are performed using magnetic torquers. The Ørsted orbit has an inclination of $96.5°$, a period of 100.0 min, a perigee at 650 km, and an apogee at 860 km. The orbit plane is slowly drifting, the local time of the equator crossing decreases by 0.91 min/day, starting from an initial local time of 02:26 on 23 February 1999 for the south-going track. Nominal lifetime of the mission was 14 months, but after more than 7 years in space the satellite is still healthy and provides high-precision magnetic data.

A proton precession Overhauser magnetometer (OVH), measuring the magnetic field intensity with a sampling rate of 1 Hz and an accuracy better than 0.5 nT, is mounted at the top of the deployable 8-m-long boom. At a distance of 6 m from the satellite body is the optical bench with the compact spherical coil (CSC) fluxgate vector magnetometer mounted closely together with the Star Imager (SIM). The CSC samples the magnetic field at 100 Hz (burst mode, at polar latitudes) or 25 Hz (normal mode) with a resolution better than 0.1 nT and is calibrated using the field intensity measured by the OVH. After calibration, the agreement between the two magneto-meters is better than 0.33 nT rms. Due to attitude errors, the accuracy of the vector components (B_r, B_θ, B_ϕ) is limited to 2–8 nT (4 nT rms), depending on the component. See Neubert *et al.* (2001) and Olsen (2007) for more information on the satellite; data are available at Danish National Space Center's webpage.

A copy of the Ørsted boom and payload (but with a scalar helium magnetometer instead of the Overhauser magnetometer) was launched in November 2000 on-board the Argentinean satellite SAC-C. However, due to a broken connection in a coaxial cable, no high-precision attitude data (and hence no reliable vector data) are available.

2.2.3.4   CHAMP

The German CHAMP satellite was launched on July 15, 2000 onto an almost-circular, near-polar (inclination $87.3°$) orbit with an initial altitude of 454 km, which has decreased to about 350 km by end of 2006. The satellite advances 1 h in LT within 11 days. Instrumentation is very similar to that of Ørsted;

however, attitude is obtained by combining measurements taken by two SIM heads, to minimize attitude error anisotropy. Accuracy of the scalar measurements is similar to that of Ørsted (better than 0.5 nT), while that of vector components is better than 2 nT when attitude is measured by both star imager heads (which happens for more than 60% of the time), otherwise the same accuracy as for Ørsted is achieved (**Figure 7**). See Maus (2007) for more information on the satellite; data are available at the ISDC (GFZ) page.

2.2.3.5   Swarm

The Swarm constellation mission, selected by the European Space Agency (ESA) in 2004, will provide the best-ever survey of the geomagnetic field and its temporal evolution. Scheduled for launch in 2010, the mission comprises a constellation of three satellites, with two spacecrafts flying side by side at lower altitude (450 km initial altitude) with a separation in longitude of about $1.4°$ (about 150 km at the equator), thereby measuring the east–west gradient of the magnetic field. The third satellite will be at higher altitude (530 km) and in a different LT compared to the lower satellite pair. High-precision and high-resolution measurements of the strength, direction, and variation of the magnetic field, complemented by precise navigation, accelerometer, and electric field measurements, will provide the necessary observations that are required to separate and model the various sources of the geomagnetic field. For more details, see Friis-Christensen *et al.* (2006).

2.2.3.6   Calibration and alignment of satellite magnetometers

Vector magnetometers, like those flown on Magsat, Ørsted, and CHAMP, are not absolute instruments, and hence have to be calibrated on a routine basis (e.g., Olsen *et al.*, 2003). This is done in orbit by comparing instrument output with the magnetic intensity measurements of a second, absolute scalar, magnetometer (Overhauser magnetometer in the case of Ørsted and CHAMP). In addition to this, 'calibration' of the vector magnetometer, the relative rotation between the coordinate system of the vector magnetometer and that of the attitude determination instrument (SIM in case of Ørsted and CHAMP) has to be determined with an accuracy of a few arcseconds, a procedure known as the 'alignment' of the magnetometer. The three 'Euler angles' (alignment parameters) describing this rotation are estimated in orbit by comparing the magnetic components measured by the calibrated vector magnetometer, combined with the attitude information provided by the SIM, with the magnetic components predicted from a geomagnetic field model, usually directly co-estimated with the Euler angles (e.g., Olsen *et al.*, 2006). It is important to recognize that alignment of space-borne magnetometers is rather different from alignment of observatory magnetometers. The latter is possible without any field model, by turning the magnetometer by respectively 180° around the three magnetometer axes and taking additional measurements. Proper combination of the magnetometer readings taken during this procedure, which is performed at the same position (no spatial change of the ambient magnetic field) and almost instantaneously (within a few minutes, to minimize the influence of temporal changes of the ambient field; remaining temporal field changes are corrected for by subtracting the field changes measured by a nearby variometer), removes the ambient magnetic field, and hence no knowledge of the field strength and direction of the ambient field is required.

Such a procedure is, however, not possible for satellite magnetometers in-orbit, due to movement of the satellite, and therefore satellite magnetometer alignment requires a model of the ambient field. See Olsen *et al.* (2003, 2006) for a description of satellite magnetometer alignment.

2.2.4 Indices

Several indices have been developed for characterizing the level of magnetic field contributions from ionospheric and magnetosphere sources. See Mayaud (1980) and Siebert and Meyer (1996) for a detailed description of the various indices. In the following, we will concentrate on the two indices Kp and Dst that are typically used for data selection and external field parametrization when deriving models of the geomagnetic field.

2.2.4.1 *Kp, an index of global geomagnetic activity*

Kp is a worldwide 3 h index that aims at describing the level of "all irregular disturbances of the geomagnetic field caused by solar particle radiation within the 3-hour interval concerned" (Siebert and Meyer, 1996). The index was introduced by Julius Bartels in 1938 and adopted by the IATME (International Association of Terrestrial Magnetism and Electricity, now International Association of Geomagnetism and Aeronomy (IAGA)) in 1951.

Kp is derived from (currently) 13 subauroral stations. For each of these observatories, the local disturbance level are determined by measuring the range (difference between the highest and lowest values) during three-hourly time intervals for the most disturbed horizontal magnetic field component. First, however, the regular daily variation, S_R, has to be removed from the magnetic data (the regular daily variation being defined as the mean daily variation of the 5 quietest days of each month). The range is then converted to a K value, taking values between 0 (quietest) and 9 (most disturbed) on a quasi-logarithmic scale. Using conversion tables based on statistical properties, the local K values are then converted to a standardized number, denoted as Ks. This number is given in a scale of thirds, ranging through 28 grades in the order 0° 0^+, 1^-, 1°, 1^+, 2^-, 2°, 2^+,...,8°, 8^+, 9^-, 9°. The planetary activity index Kp is the mean of the Ks value of a number of 'Kp stations' (originally 11, presently 13).

The 3 h index ap is a linearized version of Kp. For a station at about $\pm 50^\circ$ dipole latitude, ap may be regarded as the range of the most disturbed of the two horizontal field components, expressed in the unit of 2 nT. The daily average of the six 3 h values of ap is denoted as Ap.

Kp, ap, and Ap are calculated bi-weekly and are accessible through the webpage of the International Service of Geomagnetic Indices.

Only geomagnetic data taken during geomagnetic quiet conditions are typically used for geomagnetic field modeling. An often used criteria is $Kp \leq 2^0$, which is fulfilled for about 50% of all data. However, this is the average percentage of quietness; during years of solar maximum, considerable less data satisfy the condition $Kp \leq 2^0$. As an example, only about 30% of the solar maximum year 2003 have $Kp \leq 2^0$, while the percentage is about 60% for the solar minimum year 1996.

2.2.4.2 *Dst, an index of magnetospheric ring-current strength*

Dst is the north component of the axially symmetric part of the equatorial disturbance field at the Earth's surface caused by the magnetospheric ring current and its induced counterpart. It is determined on an hourly basis from the H component of the four low-latitude observatories Honolulu, San Juan, Hermanus, and Kakioka, after removal of secular variation (defined as the value during magnetically quiet conditions) and the regular daily variation S_R. This procedure is somewhat subjective and ignores a possible offset of *Dst* (i.e., the contribution of the magnetospheric ring-current for $Dst = 0$), which was determined by Langel and Estes (1985a) to be of about 20 nT. This offset, and – probably more important – baseline instabilities of the *Dst* index (e.g., Olsen *et al.*, 2005) make data selection based on (the absolute value of) *Dst* questionable. Instead, the use of the first time derivative (which is much less influenced by baseline instabilities compared to *Dst* itself) has been used recently. Olsen *et al.* (2006), for instance, recommend data selection according to $|dDst/dt| \leq 2\,\mathrm{nT\,h}^{-1}$. Similar to *Kp*, the distribution of d*Dst*/dt depends also on the level of solar activity, although to a smaller extend. The influence of various data selection criteria on the quality of field models has been investigated by Thomson and Lesur (2006).

Note that *Dst* does not only contain the magnetic signal of the magnetospheric ring current, but also contributions due to induction in the Earth's interior. Hence it contains both external (magnetospheric) and Earth-induced contributions; they can, however, be separated by means of a model of electrical conductivity of the Earth's mantle (Maus and Weidelt, 2004; Olsen *et al.*, 2005).

2.3 Global Models of the Earth's Magnetic Field

Most global models of the Earth's magnetic field rely on a spherical harmonic representation of the field. Those are the models we will describe in some detail. But other global representations can be and have been used, which we also ought to briefly mention.

Those first include the so-called 'natural orthogonal components' method extensively used by the Russian School in the 1970s (a detailed description of which can be found in Langel (1987)), and methods based on either dipole or monopole equivalent sources distributed over a source surface (see, for example, Mayhew and Estes (1983), Langel (1987), O'Brien and Parker (1994)). The latter have however found more applications in the context of crustal field modeling, and the interested reader is therefore referred to Chapter 6 for more information.

Modeling efforts have also been made to address the relatively poor adequacy of spherical harmonic representations with respect to uneven geographical data distributions. When a high concentration of data is to be accounted for in a specific geographical area, high-degree spherical harmonics are required, which

will represent the field very well in that area, but will unfortunately generate 'ringing effects' (i.e., oscillations at the corresponding small length scales) wherever else the data density is lower. Such ringing can be avoided by the use of appropriate strategies, such as the one developed by Shure *et al.* (1982) and Parker and Shure (1982), which consists in relying on so-called 'harmonic splines'. Those splines, which are linear combinations of spherical harmonics, make it possible to fit the data at the right level, while keeping the model to the minimum complexity required by the local distribution of the data (by using a Lagrange multiplier type of approach). A description of how this method can be implemented is provided by Langel (1987) to which the reader is referred for more details.

Introducing localized constraints in a more general sense but also via the use of linear combinations of spherical harmonics, as recently proposed by Lesur (2006), is another interesting possibility. However, when the geographical data distribution is particularly uneven, and/or very localized sources need to be accounted for (as is the case for some of the local ionospheric sources when dealing with satellite data), a strategy based on wavelets could be more appropriate. Because wavelets have a local geographical support, they indeed make it possible to keep the number of model parameters low by introducing smallscale wavelets only where needed. Interesting 'wavelet frames' have for instance been introduced in the context of global magnetic field modeling by Holschneider *et al.* (2003) and Chambodut *et al.* (2005), and Mayer and Maier (2006) make a very convincing case of the potential of such methods for modeling and analyzing the small-scale crustal field. Those promising methods are however still in their development stage.

An alternative approach based on the use of combined regional field representations has also recently been proposed by Thébault (2006). The idea consists in stitching together a dense coverage of regional models, the resolution of which can be adjusted to the regional density of data. This philosophy can in principle be applied to any classical type of regional representations, such as the 'rectangular harmonic analysis' introduced by Alldredge (1981) (*see* Chapter 6 for a review of those regional representations). Thébault (2006) however focuses on the use of so-called spherical cap harmonic analysis (SCHA). This regional representation was first introduced by Haines (1985) and subsequently improved by De Santis (1991), Korte and Holme (2003), and Thébault *et al.* (2005) who successfully overcame the major drawback of the original SCHA approach, which relied on an incomplete set of non-orthogonal basis functions and failed to properly describe the altitude (radial) dependency of the field. Thébault *et al.* (2005) exhibited the correct and complete set of orthogonal basis functions. This so-called revised spherical cap harmonic analysis (*R-SCHA*) now has all the appropriate mathematical properties for efficiently building a global field model from a combination of regional field models.

Spherical harmonic representations however keep many advantages. In particular, they are perfectly suited when the data to be modeled is reasonably well distributed over the globe, as is the case of the satellite data that make the bulk of the contemporary date. Also, they make it easy to accurately separate the field of internal origin from the field of external origin. For all those reasons, we will now review in more detail important examples of recent models based on this representation (for a review of the many models that have been published in the more distant past, see the well-named *Parade of Models* section 4 of Langel (1987), and Langel and Hinze (1998)).

2.3.1   Spherical Harmonic Representation of the Field

All those models rely on the following unique decomposition of a magnetic field **B** on a sphere of radius $r = R$ surrounded by sources (Backus, 1986; see also Backus *et al.* (1996)):

$$\mathbf{B}(R, \theta, \phi) = \mathbf{B}_{iR}(R, \theta, \phi) + \mathbf{B}_{eR}(R, \theta, \phi) + \mathbf{B}_{tor}(R, \theta, \phi) \qquad [16]$$

Here $\mathbf{B}_{iR}(R, \theta, \phi)$ and $\mathbf{B}_{eR}(R, \theta, \phi)$ are the potential fields produced on the sphere $r = R$ by all sources respectively below and above $r = R$. Those are in fact the values taken for $r = R$ by the potential fields $\mathbf{B}_{iR}(r, \theta, \phi)$ and $\mathbf{B}_{eR}(r, \theta, \phi)$ which may be defined more generally for respectively $r \geq R$ and $r \leq R$, with the help of $\mathbf{B}_{iR} = -\nabla V_{iR}$ and $\mathbf{B}_{eR} = -\nabla V_{eR}$, with

$$V_{iR}(r, \theta, \phi) = a \sum_{n=1}^{\infty} \left(\frac{a}{r}\right)^{n+1} \sum_{m=0}^{n} \left(g_{n,R}^m Y_n^{mc}(\theta, \phi) + h_{n,R}^m Y_n^{ms}(\theta, \phi)\right) \qquad [17]$$

$$V_{eR}(r, \theta, \phi) = a \sum_{n=1}^{\infty} \left(\frac{r}{a}\right)^{n} \sum_{m=0}^{n} \left(q_{n,R}^m Y_n^{mc}(\theta, \phi) + s_{n,R}^m Y_n^{ms}(\theta, \phi)\right) \qquad [18]$$

where the index R of the Gauss coefficients is here to recall that those coefficients describe the fields produced by all sources respectively below ($g_{n,R}^m$, $h_{n,R}^m$) and above ($q_{n,R}^m$, $s_{n,R}^m$) $r = R$. The additional field $\mathbf{B}_{tor}(R, \theta, \phi)$ field is then the toroidal field produced on the sphere $r = R$ by the local poloidal currents. It is defined by

$$\mathbf{B}_{tor}(r, \theta, \phi) = \sum_{n=1}^{\infty} \sum_{m=0}^{n} (T_n^{m,c}(r)\mathbf{C}_n^{m,c}(\theta, \phi) + T_n^{m,s}(r)\mathbf{C}_n^{m,s}(\theta, \phi)) \qquad [19]$$

where

$$\mathbf{C}_n^{m,(c,s)}(\theta, \phi) = -\hat{\mathbf{r}} \times \nabla_S Y_n^{m,(c,s)}(\theta, \phi) = \nabla \times Y_n^{m,(c,s)}(\theta, \phi)\mathbf{r} \qquad [20]$$

and the $Y_n^{m,c}(\theta, \phi)$ and $Y_n^{m,s}(\theta, \phi)$ are the Schmidt quasi-normalized real surface spherical harmonics (e.g., Langel, 1987) ($\hat{\mathbf{r}}$ being the unit radial vector). In all those equations, the Earth's radius a is taken as the reference radius.

When $R = a$, eqn [16] reduces to the sum of the two potential fields $\mathbf{B}_{ia}(a, \theta, \phi)$ and $\mathbf{B}_{ea}(a, \theta, \phi)$, since the neutral atmosphere has no local currents. Then the ($g_{n,a}^m$, $h_{n,a}^m$), are the sum of the Gauss coefficients describing the field of Internal Origin (simply denoted (g_n^m, h_n^m)), and of the analogous Gauss coefficients describing the Externally Induced Fields.

When R corresponds to a sphere within the F-region ionosphere where satellites evolve (typically between 400 km and 1000 km), local currents are to be found and $\mathbf{B}_{tor}(R, \theta, \phi)$ is non-zero. Then also the ($g_{n,R}^m$, $h_{n,R}^m$) describe the sum of the field of Internal Origin, of the Externally Induced Fields, and of the potential Field produced by all ionospheric sources located below $r = R$ (the main source of which is the E-region, located at about 110 km). The ($q_{n,R}^m$, $s_{n,R}^m$) then essentially describe the potential Magnetospheric Field and are simply denoted (q_n^m, s_n^m).

In discussing the various components of the Earth's magnetic field, and in particular the way each component contributes on average to the observed magnetic field, it will prove useful to deal with the concept of 'spatial power spectra'. This concept was introduced by Mauersberger (1956) and popularized by Lowes (1966, 1974), both in the case of potential fields. However, it is quite straightforward to introduce those also in the case of nonpotential fields.

Indeed, consider the sphere $S(R)$ of radius $r = R$ and assume the most general case when this sphere is surrounded by sources. Then, the field can be written in the form of eqn [16], and its average squared magnitude over $S(R)$ can be written in the form:

$$\left\langle \mathbf{B}^2(R, \theta, \phi) \right\rangle_{S(R)} = \frac{1}{4\pi} \int_0^{2\pi} \int_0^{\pi} \mathbf{B}(R, \theta, \phi) \cdot \mathbf{B}(R, \theta, \phi)$$
$$\sin\theta \, \mathrm{d}\theta \, \mathrm{d}\phi = W^i(R) + W^e(R)$$
$$+ W^T(R) \qquad [21]$$

where

$$W^i(R) = \sum_{n=1}^{\infty} W_n^i(R)$$

$$W^e(R) = \sum_{n=1}^{\infty} W_n^e(R) \qquad [22]$$

$$W^T(R) = \sum_{n=1}^{\infty} W_n^T(R)$$

with

$$W_n^i(R) = (n+1)\left(\frac{a}{R}\right)^{2n+4} \sum_{m=0}^{n} \left[\left(g_{n,R}^m\right)^2 + \left(b_{n,R}^m\right)^2 \right] \quad [23]$$

$$W_n^e(R) = n\left(\frac{R}{a}\right)^{2n-2} \sum_{m=0}^{n} \left[\left(q_{n,R}^m\right)^2 + \left(s_{n,R}^m\right)^2 \right] \quad [24]$$

$$W_n^T(R) = \frac{n(n+1)}{2n+1} \sum_{m=0}^{n} \left[\left(T_n^{m,c}(R)\right)^2 + \left(T_n^{m,s}(R)\right)^2 \right] \quad [25]$$

Equations [21] and [22] then show that each type of field – the potential field produced by all sources above $r = R$, the potential field produced by all sources below $r = R$, and the nonpotential (toroidal) field produced by the local (poloidal) sources on $r = R$ – and within each type of field, each degree n (in fact, each elementary field of degree n and order m, as is further shown by eqns [23]–[25]) contributes independently to the average squared magnitude $\left\langle \mathbf{B}^2(R, \theta, \phi) \right\rangle_{S(R)}$ on the sphere $r = R$. Hence, plotting $W_n^i(R)$ (resp. W_n^e (R), W_n^T (R)) as a function of n provides a very convenient mean of identifying which sources, and within each type of source, which degrees n, most contribute on average to the magnetic field $\mathbf{B}$ (R, θ, ϕ) on the sphere $r = R$. Such plots are known as 'spatial power spectra'.

In the special case, one is only interested in the contribution of the field of internal origin; the corresponding spectrum, then known as the Lowes–Mauersberger spectrum, becomes

$$W_n^i(R) = (n+1)\left(\frac{a}{R}\right)^{2n+4} \sum_{m=0}^{n} \left[\left(g_n^m\right)^2 + \left(b_n^m\right)^2 \right] \quad [26]$$

and can be computed for any value of $R \geq a$ (above the sources). In the even more special case when one is only interested in the core field, exactly the same formula

can be used for $R \geq b$ (where b is the core radius), provided one uses only core field Gauss coefficients.

2.3.2 Models of the Field of Internal Origin

A first series of models aims at describing the field of internal origin only. Such models assume the field to be defined as the gradient of a scalar potential of internal origin. Each model is then defined by a set of Gauss coefficients $(g_n^m(t), b_n^m(t))$, where some time dependency is introduced to account for the fact that the field of internal origin varies with time. Those models do not intend to describe the field of external origin, their associated externally induced fields, and other nonmodeled fields possibly contributing to the data used for the modeling. Yet, those need to be taken into account. This is done by using selecting and correcting procedures, by allowing for a simple parametrization of the remaining external fields, and by using appropriate damping and error covariance matrices, all in various combinations. Models thus differ because of the combination of tricks used to deal with those non-modelled fields, because of the type of data used, because of the maximum degree N of the involved Gauss coefficients, and because of the way the temporal behaviour of those Gauss coefficients are described.

2.3.2.1 The International Geomagnetic Reference Field

Probably the most used models are those from the IGRF/DGRF series, regularly published by IAGA. Those are (mostly empirical) representations of the slowly changing large-scale part of the field of internal origin. Their maximum degree used to be $N = 10$, but this was recently changed to $N = 13$. Each model is defined for a specific epoch, and therefore gives an instantaneous picture of the field every 5 years, starting in 1900.0 (i.e., 1 January 1900). Estimates of the field between those epochs can then be derived by simple linear interpolation.

Every time a new IGRF model is computed (the most recent version is IGRF-10, computed for epoch 2005.0, see Maus *et al.* (2005a)), all the data covering the past few years are taken into account, including geomagnetic measurements from satellites (when available), geomagnetic observatories, and repeat stations. Surveys taken by ships and aircrafts have also sometimes been used. 'Candidate models' in the form of Gauss coefficients (g_n^m, b_n^m) are then produced by various independent teams, usually using simple classical techniques. However, when satellite data are available, as is the present case, such models are often preliminary versions of more advanced models of the type described in the next sections. Those candidate models are next assessed (see, for example, Maus *et al.* (2005b)) and the IGRF model is finally derived as a (weighted) mean of those candidate models. This procedure is specific to the IGRF series of models and amounts to some sort of 'best guess' approach, based on the empirical experience of the IAGA working group in charge of producing the model.

The main purpose of the IGRF models is practical (see Meyers and Davis (1990) for a review of the many, sometimes unexpected, applications of the IGRF models). They are meant to predict the field of internal origin (in fact, the main field, given the low degrees modeled) not only at the epoch of validity (say 2005.0), but also for the 5 years to come. This is possible thanks to the secular variation model (i.e., the $(\dot{g}_n^m, \dot{h}_n^m)$ coefficients), published together with the fixed-epoch Gauss coefficients. This secular variation model is also a mean of 'candidate secular variation models', usually co-estimated with the field candidate models (because the data used usually cover a significant time span, and some temporal corrections are anyway needed). This model, of maximum degree $N=8$, is thus an estimate of the average temporal variation of the field in the few years preceding the release of the model, next extrapolated to predict the field over the next 5 years, that is, until the next IGRF models is released.

IGRF models are bound to be temporary. Once the next model has been released (some five years later), the previous IGRF is reconsidered, data that have become available in the mean time are taken into account, and a new Definitive Geomagnetic Reference Field (DGRF) is published. This new model is no longer an extrapolation, but an interpolation of the field. It is again computed for the same epoch and supersedes the old IGRF (DGRF 2000 recently superseded IGRF 2000). The predictive secular variation model that had been used in the previous five years is then also abandoned, and simply replaced by an interpolation between this DGRF model (DGRF 2000) and the latest IGRF model (IGRF 2005), for the period between those two models (while, of course, the new secular variation model published with the new IGRF is used to predict the future evolution of the field).

The most recent IGRF release (IGRF-10; see Maus *et al.* (2005a)) hence consists in 20 DGRF models for epochs 1900.0 to 1995.0 up to degree $N=10$ (one every five years), one DGRF model for epoch 2000.0 up to degree $N=13$, and one IGRF for epoch 2005.0 up to degree $N=13$, with a predictive secular variation model up to degree $N=8$. The next revision will occur in 2010.

The great advantage of the IGRF/DGRF series is their simplicity and predictive nature. But the chosen temporal and spatial parametrization is limited and one needs to be aware of those limitations. Maus *et al.* (2005b) note that IGRF for 2005.0 "is estimated to have a formal root mean square error over the Earth's surface of only 5 nT, though it is likely that the actual error is somewhat larger than this", while "the corresponding errors of the adopted secular variation model for 2005.0–2010.0 is estimated at 20 nT yr". In practice, the predicted field can thus easily be in error of more than 100 nT at the Earth's surface, all the more that those predictions are only for the slowly changing main field, and do not take into account other sources (such as crustal sources, which can again easily amount to several 100 nT; see, for example, Cohen *et al.* (1997)).

Model coefficients and software to synthesize field values at a given time and position are available at the WDC system. More details about IGRF/DGRF models can be found in Langel (1987).

2.3.2.2 *More advanced models*

Taking advantage of data obtained by the recent magnetic satellites Ørsted, CHAMP, and SAC-C, various geomagnetic field models of increasing complexity and accuracy have been derived.

The Ørsted initial field model (OIFM) of Olsen *et al.* (2000) was one of the first to be published. Similar to the IGRF model series, it represents a snapshot of the magnetic field at a given epoch t_0 (in this case $t_0 = 2000.0$), which was estimated from a few weeks of quiet nighttime Ørsted data around 1 January 2000. Data were selected according to the following criteria (recall Section 2.2.4 for definitions and purposes of indices): LT at about 22:00 to minimize ionospheric contributions; Kp, the index of global geomagnetic activity, and Dst, the ring-current index, have to satisfy $Kp \leq 1+$ for the time of observation, $Kp \leq 2^\circ$ for the previous 3 h interval, $|Dst| < 10$ nT and $|d(DST)/dt| < 3$ nT/hr, to ensure the data are collected at magnetic quiet times. An additional criteria based on the state of the interplanetary magnetic field (IMF, as provided by Advanced Composition Explorer (ACE) NASA/NOAA spacecraft), was used to further reduce contributions from polar cap ionospheric currents. Finally, only intensity data have been used at high magnetic latitudes ($|90^\circ - \theta_d| \geq 50^\circ$), to minimize the nonmodeled signature of field-aligned currents (FACs).

Spatial resolution of the OIFM model goes beyond that of the IGRF, since spherical harmonics up to degree $N=19$ are included. Some simple parametrization of the magnetospheric field is also included, which is modeled as a permanent potential field of external origin up to degree 2, and as a fast time changing degree 1 potential field of external origin with induced degree 1 potential field, the temporal variability of which is imposed as an independent information via the observatory-based Dst index (as a measure of the strength of the magnetospheric ring-current), following an earlier suggestion of Langel and Estes (1985b). Finally, covariance matrices taking into account correlated attitude errors have been used (based on Holme and Bloxham (1996)). The inverse method is a simple iterative least-squares fit (iterative, since scalar data are included), and does not involve any damping. Note also that this model does not involve any secular variation (the data used covering only a very short period of a few weeks and having been propagated to model epoch $t_0 = 2000.0$ by means of an *a priori* model of linear secular variation).

Similar models were subsequently published by Langlais *et al.* (2003) and Olsen (2002), but using much more (nighttime) data. Langlais *et al.* (2003) used 14 months of Ørsted data (between March 1999 and May 2000) and computed a model up to degree $N=29$ in very much the same way as the OIFM was computed, except for the fact that a secular variation model was then also co-estimated (up to degree $N=13$). Olsen (2002) used even more data (between March 1999 and September 2001). This model is also computed up to degree $N=29$, with a secular variation up to degree $N=13$. But a slightly different modeling approach was used, involving an improved statistical treatment of data errors – relying on an iteratively reweighted least squares (IRLS) inverse scheme with Huber weights (Holland and Welsch, 1977; Constable, 1988) – and a more advanced way of dealing with external field contributions. Permanent large-scale magnetospheric contributions were still estimated up to degree 2, but additional annual and semi-annual periodicity were included in their zonal terms. Also, fast-changing degree-1 coefficients (with induced counterparts), previously modulated by the Dst index, were modulated with a new, more appropriate $\tilde{R}C$ index, also estimated from globally distributed geomagnetic observatories. Exactly the same procedure was later

used to compute the first global model of the field of internal origin based on CHAMP data (Holme *et al.*, 2003). This *CO2* model used CHAMP and Ørsted vector and scalar data, between August 2000 and December 2001, and SAC-C scalar data throughout 2001.

The models discussed so far (and others which we did not describe in detail, such as that of Lesur *et al.* (2005)) either do not model the temporal variation of the field of internal origin (such as OIFM), or only model a constant secular variation. But as modern high-quality satellite data became available over longer and longer periods of time (they now have been available for more than 7 years), it became clear that this time variation had to be modeled in a more sophisticated way. This is possible using either a Taylor expansion around a specific epoch t_0, that is, assuming

$$g_n^m(t) = g_n^m|_{t_0} + \dot{g}_n^m|_{t_0} \cdot (t-t_0) + \frac{1}{2}\ddot{g}_n^m|_{t_0} \cdot (t-t_0)^2 + \cdots \quad [27]$$

(and similar for h_n^m), or a spline representation, that is assuming

$$g_n^m(t) = \sum_{l=1}^{L} g_{n,l}^m \cdot M_l(t) \quad [28]$$

with $M_l(t)$ as basic B-spline functions (Schumaker, 1981, De Boor, 2001) and L as number of basis functions.

While a Taylor expansion including linear and quadratic terms may be adequate for describing field changes over short time spans of a few years, a spline representation is probably preferable for periods longer than, say, 5 years, since the inclusion of higher (quadratic, cubic,...) temporal terms may lead to unwanted behaviors near the edges of the time interval. Cubic B-splines have been in use for quite some time now in the context of historical field modeling (*see* Chapter 5). In that case, they are used to represent the field itself (e.g., Bloxham and Jackson, 1992; Jackson *et al.*, 2000). But as we shall soon see, some models of the present field (such as the *Comprehensive Model* series (Sabaka *et al.*, 2002, 2004), see Section 2.3.5) rather use them for describing its first time derivative (i.e., the secular variation).

Recent models using Taylor expansion (eqn [27]), up to quadratic terms, include the POMME model series (Maus *et al.*, 2005c, 2006b). In its latest version (POMME-3.0 (Maus *et al.*, 2006b)), this model uses nighttime CHAMP data between July 2000 and August 2005. It goes formally up to degree $N=60$

but only Gauss coefficients up to degree $N = 16$ include nonzero first ($\dot{g}_n^m$, $\dot{h}_n^m$) and second ($\ddot{g}_n^m$, $\ddot{h}_n^m$) time derivatives. Also, it should be mentioned that contrary to the models discussed so far, some damping was introduced in the inverse procedure to keep those time derivatives well behaved. This damping was imposed on degrees 14 to 16 for the first time derivative, on degrees 10 to 16 for the second derivatives. This model is more sophisticated than previous models in many other respects. In addition to data selection criteria fairly similar to those used so far, Maus *et al.* (2006b) also implemented a selection criteria proposed by Stolle *et al.* (2006) to detect and exclude post-sunset low-latitude ionospheric plasma instabilities which produce local poloidal fields (which are not modeled). Corrections for predictable fields were also implemented, namely, the magnetic effect of the eight major oceanic tidal components (Kuvshinov and Olsen, 2005), the diamagnetic effect (Lühr *et al.*, 2003), and the signature of gravity-driven currents in the ionospheric *F*-layer (Maus and Lühr, 2006). Finally, magnetospheric sources have been taken into account (following the quiet-time magnetospheric field model of (Maus and Lühr ,2005), which uses both the *Dst* index and the IMF) in their intrinsic coordinate systems (i.e., solar magnetic (SM) coordinates for describing near-magnetospheric currents like the ring current, and geocentric solar magnetospheric (GSM) for describing far magnetospheric current systems like the tail currents, see Section 2.2.1.5). This treatment and the co-estimation of daily base line corrections for the *Dst* index (along the lines of Lesur *et al.* (2005), Olsen *et al.* (2005)), leads to a more accurate determination, especially of the secular variation.

The CHAOS model of Olsen *et al.* (2006) is an example of a model based on the alternative spline approach (eqn [28]). It was derived from essentially the same data set as POMME-3.0, namely, more than 6.5 years of high-precision geomagnetic measurements from the three satellites Ørsted, CHAMP, and SAC-C taken between March 1999 and December 2005. It goes formally up to degree $N = 50$, but all degrees above degree 18 are assumed constant. The time change of the low-degree ($n \leq 14$) coefficients is described by cubic B-splines (with 1 year knot separation), while that of coefficients with degree $n = 15–18$ is described by a linear Taylor expansion. In that case also, some damping is used to keep time derivatives well behaved, in a way that only affects the second (and higher) time derivatives. The modeling approach

also goes in several aspects beyond that used for other recent models. In particular, the magnetometer vector data are taken in the instrument frame, and the Euler angles describing the transformation from the magnetometer frame to the SIM frame (i.e., the so-called 'alignment parameters', recall Section 2.2.3) are co-estimated, avoiding the inconsistency of using vector data that have been aligned using a different (pre-existing) field model. The bending of the CHAMP optical bench connecting magnetometer and SIM is also accounted for by estimating CHAMP Euler angles in 10 day segments. Finally, in addition to describing magnetospheric fields using the SM and GSM coordinate systems, degree-1 external fields are co-estimated separately for every 12 h interval. The same improved statistical treatment of data errors, involving an IRLS inverse approach with Huber weights as in Olsen (2002), is otherwise used.

2.3.3   Models of the Core Field

All models discussed so far are models of the field of internal origin. But this field is the sum of the low-frequency core field, which reaches the Earth's surface (and makes the main field), and the crustal field. Fortunately, we know that the crustal sources can only lie very near the Earth's surface (above the Curie isotherm), in layers with reasonably well-known magnetic rock properties. This makes it possible to predict the likely contribution of the crustal field to the observed field of internal origin. Such predictions can be made either in a statistical way (e.g., Jackson, 1994) or in a deterministic way (e.g., Purucker *et al.*, 2002). In both cases, the conclusion is the same: the crustal field can easily explain the small scales of the field of internal origin but cannot explain its largest scales. The transition occurs around the degree 14 of the spherical harmonic representation of the field, and shows up nicely as a strong change of trend in the spatial Mauersberger–Lowes spectrum (recall eqn [26]) of the field of internal origin at the Earth's surface (see **Figure 14(a)** in Section 2.4.1, where this spectrum is further discussed; see also Chapter 6, where predictions of the crustal field contributions are illustrated). This then shows that the crustal field is dominating the small scales (say degree $N = 15$ and above) of the field of internal origin, to which it can be identified, while the core field is dominating the large scales (say degree $N = 13$ and below), to which it can also be identified. This of course also means that the smallest scales of the core field are permanently screened by a constant unknown crustal field and that the largest scales of the crustal field cannot be observed. Altogether it thus appears that at any given epoch, only the low temporal frequencies (because of the slightly conducting mantle) and the large spatial scales (because of the magnetized crust) of the core field can potentially be identified at the Earth's surface where it makes the Main Field. However, the screening of the core field by the quasi-permanent crustal field will not affect the first (and higher) time derivative of the small scales of the core field, which can unambiguously be identified to the observed secular variation up to the highest recoverable degrees, even beyond degree 13 (which can now be achieved, as will be shown in Section 2.4.1).

Because the transition at degree 14 in the spatial Mauersberger–Lowes spectrum is so sudden, most authors consider that identifying this core field by simply truncating the field of internal origin at degree 13 is a reasonable thing to do. But this has some potential drawback. When plotting maps, this will distort core field features which are described by Gauss coefficients with degrees across $N = 13$. This is not so much an issue when plotting such maps at the Earth's surface, because the core field is strongly dominated by its largest scales (because of its fast decreasing spectrum; recall **Figure 14(a)**). But the issue is potentially more serious when the core field is plotted at the core surface, where it originates from. Its spectrum is then much flatter (see **Figure 15(a)** in Section 2.4.1) and significant small-scale features are thus expected. These considerations led several authors to try and derive core field models directly from the data.

Such an approach is in fact common practice in the context of historical field modeling, the distribution of historical data being anyway not good enough to accurately recover the field of internal origin beyond degree 13. Examples of historical core field models produced in this way can be found in Chapter 5. Two interesting models of the recent and present core field have however also recently been published, which we ought to briefly describe.

The first model is the one of Wardinsky and Holme (2006), which covers the 1980–2000 period of time. This model only has Gauss coefficients up to degree $N = 15$, the temporal variations of which are described with the help of B-splines (recall eqn [28]). Its main characteristic is that magnetic observatory monthly means, annual means, and repeat station measurements are used in such a way that only field changes in time are taken into account. Satellite data also are used but only in an indirect way, by requesting the field model to be as close as possible to truncated (to degree 15) and next tapered versions of a MAGSAT model (Cain *et al.*, 1989) in 1980, and of the Ørsted model of Olsen (2002) in 2000. The underlying idea is that in doing so, the crustal signal is entirely removed from the data used between 1980 and 2000, while it already is minimized in the degrees 1–15 of the two MAGSAT and Ørsted models (because it has already been properly dealt with when constructing those models, which go to respectively $N = 63$ and $N = 29$). The tapering of the two MAGSAT and Ørsted models is introduced to avoid the possible spurious effects of a shear truncation. The main source of error linked to nonmodeled field sources then remains the field of external origin with its induced counterpart. Contrary to the models discussed above, none of those fields are either corrected for or even crudely modeled. It is dealt with as a source of error via an original two-step procedure. An 'interim' model, assuming this error is un-correlated (among the components of each vector data), is first computed. This model is then used to study residuals and infer a more sophisticated error covariance matrix, taking the so-inferred correlations among components into account. This new covariance matrix is then used to infer the final model. Finally, the inverse scheme also uses some regularization procedure (analogous to those used in the context of historical field modeling; *see* Chapter 5) based on some *a priori* information with respect to the ohmic dissipation in the core and to the temporal behavior of the field. We will comment on this model in Section 2.3.5.

The second recent core field model of interest is that of Jackson (2003), which is in fact a pair of models for epochs 1980 and 2000. Data used are a few months of MAGSAT data for epoch 1980, and December 1999 to January 2000 Ørsted data for epoch 2000, which are short enough periods for the secular variation to be ignored. The detailed data selection process used by the author is not specified in the original paper, but is likely similar to the one used by the same author to produce the Jackson *et al.* (2000) historical model. No correction nor any modeling procedure is used to account for the external and crustal fields, which are dealt with as sources of noise. Contributions from the crustal field in particular are kept minimum by using a coarse enough density of satellite data (1600 Z components from MAGSAT, 3684 vector data from Ørsted).

The way those models are computed is otherwise quite different from what we have seen so far. Although the models are eventually expressed in the form of Gauss coefficients, they are directly computed at the core surface, which is tesselated into 1442 almost equally spaced nodes and 2880 spherical triangles. The purpose of such a procedure is to make it possible to implement integral constraints over the core surface, imposing that $\Phi^+(1980)$ and $\Phi^+(2000)$ (resp. $\Phi^-(1980)$ and $\Phi^-(2000)$), the total flux of outgoing (resp. ingoing) field in 1980 and 2000, all be equal (in absolute values). Those constraints result from the combination of the fact that the field has no monopole source (which requires $\Phi^+(t)$ and $\Phi^-(t)$ to be equal (in absolute value) at any given epoch t), and that on short periods of time, magnetic diffusion (i.e., ohmic dissipation) can be assumed to play a negligible role within the core, so that the so-called 'frozen-flux approximation' applies (which implies $\Phi^+(t_1) = \Phi^+(t_2)$ and $\Phi^-(t_1) = \Phi^-(t_2)$ at any two given epochs (t_1, t_2); see Chapter 5). The two models are then solved for simultaneously using a so-called 'maximum entropy method' (Buck and Macaulay, 1991) which is well suited to solve such image reconstruction type of problems. No regularization is otherwise involved.

The specifics of such an approach is that, rather than looking for a core field model crudely truncated at a given degree with no regularization (which would distort partly resolved small-scale features), or for a model produced with the help of some regularization (which would smooth out those small-scale features), it looks for as many high-degree Gauss coefficients as tolerated by the data, to sharpen those features. Such a method is thus appropriate if one feels that core field features that are only partly resolved because of the overlying crustal signal (those which would need degrees larger than, say $N = 14$, to be properly described), are intrinsically sharp. That may well be the case, because dynamo simulations do show such sharp features (e.g., Christensen et al., 1999). One should however keep in mind that such a method may very well sharpen features that are either not so sharp in reality, or even more complex than reconstructed by the method. In other words, all methods have their drawbacks. Regularizing procedures tend to remove some information, but both maximum entropy methods, and simple truncation of a field of internal origin at degree 13, can introduce misleading features. They simply provide alternative views.

2.3.4 Models of the Crustal (Lithospheric) Field

Spherical harmonic models of the crustal field (also referred to as the 'lithospheric field' when global models are considered, such as here) can also straightforwardly be recovered from models of the field of internal origin. This only requires removing core field-dominated Gauss coefficients, up to, say, degree 14. This procedure removes the large-scale crustal field which we anyway cannot observe. It also unfortunately distorts crustal field features of intermediate length scales (which would need to be described by Gauss coefficients around degree 14). This is a well-known issue in the community involved in the study of crustal magnetism (see Chapter 6). Obviously, small-scale crustal field features will not be affected by those issues. The concern in that case has more to do with the limited spatial resolution of those models. The models discussed so far only go to maximum degrees $N = 50$ (CHAOS model of Olsen et al. (2006)) and $N = 60$ (POMME-3.0 model of Maus et al. (2006b)). In the quest of even better spatial resolutions, spherical harmonic models that only attempt to describe the crustal field beyond degree 15 have also recently been derived.

An interesting series of such models has been derived by Maus and co-workers, mainly using CHAMP data. The latest version of that series, *MF5* (Maus et al., 2007), was produced in a very similar way to the previous model *MF4* (Maus et al., 2006a), but using only the latest three years of CHAMP scalar and vector measurements (to take advantage of a lower flight altitude). It goes up to degree $N = 100$, but degrees above 80 have been damped.

Just like the POMME-3.0 model previously described (and which was produced by the same team), considerable care has been taken to avoid unmodeled field contributions, via a comparable procedure of selection and correction. CHAMP vector and scalar data at nonpolar latitudes (equatorward of $\pm 55°$ dipole latitude) were selected for $K_p \leq 2°$ and LT between 20:00 and 05:00. Effects of post-sunset low-latitude ionospheric plasma instabilities were identified and excluded by an automatic detection approach (Stolle et al., 2006). Data at polar latitudes have been used if the merging electric field at the magnetopause satisfies $E_m < 0.8\,\text{mV m}^{-1}$ and the IMF satisfies $|B_y| < 8\,\text{nT}$ and $-2\,\text{nT}\,B_z < 6\,\text{nT}$.

As had also been done for the POMME-3.0 model, the data have been corrected for the magnetic effect of the eight major oceanic tidal components (Kuvshinov and Olsen, 2005), the diamagnetic effect (Lühr *et al.*, 2003), and gravity-driven currents in the ionospheric *F*-layer (Maus and Lühr, 2006). Nonpolar data are also corrected for the effect of the polar electrojets. Polar latitude data have been 'super-selected' by discarding about 98% of the data as disturbed, based on the rms signal strength relative to neighboring tracks.

Since the model only intends to describe the crustal field starting from degree 16, contributions from lower degrees of the field of internal origin had to be removed. This was done by removing predictions from the POMME-3.0 model up to degree 15. Magnetospheric contributions (with their induced counter-parts), as parametrized in POMME-3.0, were also removed.

All tracks have next been high-pass-filtered by removing a degree-1 (external plus internal) field on an orbit-by-orbit basis. In addition, a set of three Euler angles, accounting for a remaining uncertainty in the satellite attitude, has been estimated for each orbit of CHAMP vector data. Finally, a line-leveling algorithm, which minimizes the distance-weighted misfit of the residuals against a predecessor of MF5, between all nearest pairs of measurements for all pairs of tracks, has been applied.

This selected and corrected data set has then been used to compute Gauss coefficients up to degree $N = 100$, using a standard least-squares approach. Coefficients above degree $n = 80$ were however regularized by 'selectively damping clusters of coefficients with increased power'. Details of this rather complicated processing scheme can be found in Maus *et al.* (2006a, 2007).

MF5 is arguably better than its predecessors, as is demonstrated in **Figure 8**. It probably provides the best picture so far of small-scale crustal (lithospheric) features resolvable from satellite data. However, there is considerable risk that some north–south trending features of the crustal field could have been removed by the processing scheme (high-pass filtering on a track-by-track basis). A comparison of the long-wavelength ($n \leq 40$) crustal field of CHAOS, CM4 (the comprehensive model of Sabaka *et al.* (2004); we will introduce this in some detail in the next section), and MF4 shows closer agreement between CHAOS and CM4, compared to MF4, despite the fact that CHAOS and MF4 are derived from rather similar data sets, compared to CM4 (which is based on pre-2002 data, and does not include any CHAMP vector data) (cf. Olsen *et al.*, 2006). Both differences CHAOS-MF4 and CM4-MF4 contain north–south trending stripes

and residuals of order of a few nanoteslas at middle latitudes. Similar effects can be seen when using MF5 in place of MF4. It is likely that these are caused by the orbit-by-orbit along-track filtering of the data used for MF4 and MF5.

2.3.5 Comprehensive Models

Over the past decade, a new approach to geomagnetic field modeling has been developed jointly by the magnetics groups at Goddard Space Flight Center (NASA/GSFC) and the Danish National Space Center (DNSC). This approach, known as the comprehensive modeling (CM) approach, takes advantage of the spatiotemporal properties of the various components of the near-Earth magnetic field, to make the best of observatory and satellite data, and co-estimate models of the internal, magnetospheric, and ionospheric fields (with their secondary, Earth-induced counterparts), in one huge inversion process. As a result, the CM approach differs in many ways from what we have seen so far. Indeed, all previous models relying on satellite data only used observatory data via the K_p and *Dst* indices, and ignored the much richer information provided by the raw observatory data. None took advantage of the theoretical possibility of modeling the ionospheric field by 'sandwiching' it between ground-based and space-borne observations. The CM approach precisely intends to take advantage of this possibility.

2.3.5.1 Modelling philosophy

In principle, if one assumes the ionospheric sources to essentially lie in the form of a current sheet in the E-region (at about 110 km altitude), and in the form of poloidal sources in the F-region (up to 1000 km), where they then only produce local toroidal fields, and if the field can be simultaneously determined everywhere at the Earth's surface and on a sphere of radius $r = R$ within the *F*-region, then the potential field produced by all sources below the Earth's surface, the potential field produced by the *E*-region ionospheric sources, the toroidal field locally produced by the *F*-region ionospheric sources at altitude $r = R$, and the potential field produced by all sources above $r = R$ can all be recovered.

But can observatory and satellite data provide enough information at the Earth's surface and on a sphere $r = R$, to achieve such a separation of the various fields in practice? Yes, provided a fairly

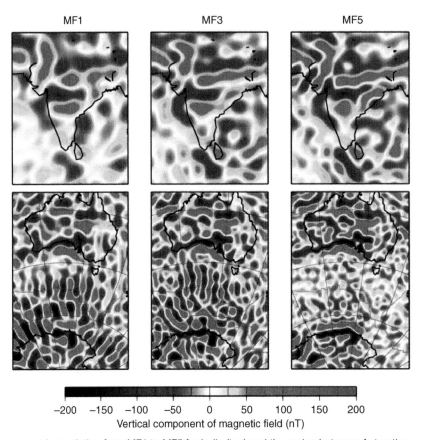

MF1   MF3   MF5

−200 −150 −100 −50   0   50   100   150   200
Vertical component of magnetic field (nT)

Figure 8 Improvement in resolution from MF1 to MF5 for India (top) and the region between Antarctica and Australia (bottom). Courtesy, S. Maus.

elaborate approach is being used, because of a number of practical limitations.

One is that observatories do not measure the field everywhere at the Earth's surface. Their actual geographical distribution is far from perfect (recall Section 2.2.2), and this limits the spatial resolution with which the Earth's surface is sampled.

Another is that satellites need time to complete a decent coverage of the Earth. A single orbit is completed in about 90 min and a few days are typically needed to recover data with a spatial distribution commensurate with that of the observatories (recall **Figure 5**). This is a problem because both the ionospheric and the magnetospheric fields can evolve significantly in the mean time.

Yet another difficulty is that some toroidal currents do flow in the F-region ionosphere, between the satellite's orbit and the E-region ionosphere (see, for example, Lühr *et al.* (2002), Maus and Lühr (2006), and Stolle *et al.* (2006)).

Finally, the potential field ultimately identified as being produced by sources below the Earth's surface is still anyway the sum of the 'real' field of internal origin (i.e., the sum of the core field and the crustal field), and of externally induced fields.

Fortunately, all those limitations can be overcome, thanks to a number of favorable circumstances the CM approach judiciously takes advantage of.

First is the fact that the most limiting fast changes in the field of external origin arise during magnetic disturbed times. Like all models we have seen so far, the comprehensive models therefore currently rely on some indexes (in particular, Kp and Dst) to select quiet-time data, and minimize those fast-changing signals. Note that, as a result, the present comprehensive models are quiet-time magnetic field models.

This selection procedure has two additional advantages. One, already recognized in previous modeling approaches, is that the remaining fast-changing field will essentially take the form of a dipolar contribution to the magnetospheric field, the variability of which can be parametrized by the *Dst* index. The other is that the remaining magneto-spheric, and all of the ionospheric, quiet-time fields are then essentially fixed with respect to the Earth–Sun line. They therefore produce regular temporal variations with restricted known temporal periodicities.

More advantage can then also be taken of the spatial properties of the quiet-time ionospheric and magnetospheric fields. First, the E-region sources can indeed reasonably be assumed to take the form of a sheet current at fixed (110 km) altitude. Next, those fields can be expressed with the help of relatively few parameters, provided appropriate coordinate systems are being used (QD for the ionospheric field, dipole for the magnetospheric field). The smallest modeled scales are then essentially those associated with the strong day-side E-region ionospheric current produced by the equatorial electrojet (*EEJ*) flowing along the magnetic equator, and F-region currents, all of which are low-order features in QD coordinates. Although the associated fields then need to be represented with the help of high-degree (n) spherical harmonics, only low-order (m) terms are required, which keeps the number of parameters low. Also, and this of course is important, those features have a geometry the distribution of observatories and a few days of satellite data are able to resolve.

A smart parametrization can thus be used to describe the field of external origin as the sum of a fast-changing essentially dipolar magnetospheric field modulated by the *Dst* index, and of almost Sun-synchronous, fixed-geometry quiet-time iono-spheric (both for E- and F-regions) and magnetospheric modes. Those modes are not exactly Sun-synchronous because at any instant, they are best represented in Earth-fixed dipole and QD coordinates. When converted into the standard spherical harmonic Earth-related geographic representation (i.e., the solid Earth's reference frame), they appear as oscillating modes with diurnal and a few higher harmonic periods. An annual and semi-annual periodicity also has to be included to account for seasonal effects.

This temporal behavior strongly contrasts with that of the field of internal origin which changes on much longer timescales within the Earth's reference frame. The core field changes on secular timescales, while the crustal field can reasonably be assumed constant. They can simply be represented using the same type of potential field spherical harmonic representations as in any of the previous models, together with a B-spline temporal expansion (eqn [28]) for the lowest degrees.

What about externally induced fields? We noted that those fields cannot be distinguished from the field of internal origin, both being produced by sources below the Earth's surface; that is, however, if we assume all fields to only be known at a given instant. But the purpose of a comprehensive model is not to describe the field only at such a given instant, but throughout a certain amount of time. Advantage can then be taken of the very different temporal behaviors of the field of internal origin and of the externally induced fields, which can be represented with similar temporal modes as the inducing fields of external origin.

The only truly limiting issue that then remains is that of the toroidal currents possibly flowing in the F-region between the satellite and the E-region iono-sphere. This issue could partly be dealt with using a selection procedure of the type proposed by Stolle *et al.* (2006) (and already used by Maus *et al.* (2006b) for the construction of the POMME-3.0 model; recall above). However, none of the published comprehensive models have yet implemented such procedures, and toroidal currents in the F-region have thus so far been ignored. Their contribution, together with contributions from all other nonmodeled field sources and intrinsic measurement errors, are dealt with as noise.

2.3.5.2 Early comprehensive models

Sabaka and Baldwin (1993) and Langel *et al.* (1996) described the first two versions (CM1 and CM2) of the comprehensive model (CM) series. In those early versions, not all fields were modeled. CM2, slightly more elaborate than CM1, only modeled the field of internal origin up to degree $N = 13$, with B splines. The crustal field was not solved for, except for so-called 'crustal biases' at observatories (i.e., for offsets, to account for the fact that taken together, all crustal sources produce significant constant fields at each observatory location). In the same way, not all external fields were solved for. F-region currents were ignored altogether, and the classical trick, consisting in only retaining intensity satellite data at high magnetic latitudes to avoid field-aligned currents, was used. Also, both the magnetospheric and the ionospheric E-region fields were modeled in simpler ways than later versions. One interesting specificity of those early models we should also mention is that, except for the component induced by the *Dst* variations, externally induced fields were modeled independently of the inducing external fields (i.e., without making any assumption with respect to the electrical structure of the Earth).

Perhaps the biggest merit of those early models is that they actually established the practical validity of the CM approach. In particular, it was not obvious at the onset that such an approach would work, given the awkward availability and properties of the various data sets. Both models relied on essentially the same data, which consisted in September 1965–August 1971 POGO scalar data, November 1979–May 1980 MAGSAT scalar and vector data, and quiet-day observatory hourly means (OHMs) during the same periods of time. Annual observatory means were also added, for the entire 1960–85 period covered by the models. The key is that, because of their *a priori* known temporal variability, the geometry of the external and induced fields could be constrained by observatory and satellite data when jointly available, while observatory data in the intermediate period provided the information needed to model the slow variation of the field of internal origin.

2.3.5.3 Recent comprehensive models

The third model of the series, CM3 (Sabaka *et al.*, 2002), still covers the same 1960–85 time period but is much more sophisticated and relies on more data. In particular, OHMs have been used for the quietest day of each month, as determined by *Kp*, and at two sampling rates: the OHM values closest to 01:00 LT, over the entire 1960–85 span of the model (allowing for the determination of broad-scale secular variation), and OHMs every two hours, during the POGO and Magsat missions envelopes (which is sufficient for analyzing external and induced contributions with periods down to a 6 h diurnal period). More POGO data were also used, especially quiet-time passes.

The scope of CM3 was much wider and its attention to detail much higher than its predecessors. The internal spherical harmonic expansion was extended to degree $N = 65$ in order to account for the crustal field as measured at satellite altitude, and crustal biases used only to account for smaller-scale crustal field contributions at observatory locations. Temporal changes of coefficients up to $N = 13$ were still described by cubic B-splines, but with some regularization to ensure a smooth secular variation. The

ionospheric and magnetospheric fields were better described using respectively QD and dipole coordinates, and including four (24 h, 12 h, 8 h, and 6 h) diurnal periods, with annual and semi-annual modulations. Additional solar activity influence was also taken into account through an amplification factor for the E-region ionospheric field (based on a 3 month moving mean of absolute $F_{10.7}$ solar radio flux values (Olsen, 1993)). The sources of this E-region ionospheric field was otherwise still considered as a sheet-current density (at 110 km altitude). As another improvement to CM2, F-region toroidal fields were also taken into account and modeled (assuming radial currents). Their temporal variability (diurnal and annual) was however kept simple to account for the fact that Magsat flew at relatively fixed local times (dawn and dusk; recall that only vector satellite data sense those fields).

Another significant change with respect to CM2 is that externally induced fields were no longer modeled independently of their inducing external fields. Rather, it was recognized that induced fields must be proportionate to the inducing fields in some way governed by the electrical properties of the solid Earth. An *a priori* four-layer 1-D radially varying conductivity model derived independently from European observatory data was thus introduced for that purpose (Olsen, 1998). This simplification allowed the externally induced fields to be formally coupled to their inducing fields, thereby reducing the number of free parameters in the model.

This model was much improved. But it still had one major drawback. It was only constrained by relatively few satellite vector data.

The most recent version of the model series, CM4 (Sabaka *et al.*, 2004), precisely aimed at correcting for this. It is a natural extension of CM3, spanning the years 1960–2002 and taking advantage of additional CHAMP (scalar) and Ørsted (vector and scalar) data, along with additional OHM values closest to 01:00 LT on the quietest day of each month, up to 2000. Additional Magsat data were also considered. This prompted some slight modifications in the parametrization of the F-region ionospheric toroidal fields, to account for the fact that Magsat and Ørsted do not sample the F-region at the same altitude and at the same local times (contrary to Magsat, Ørsted covers all LTs, and not only dawn and dusk). This parametrization was also improved to no longer restrict the corresponding poloidal currents to only be radial (they must more generally lie in a QD meridian).

In total, CM4 consists of 25 243 model parameters, of which 8840 are used to model the core and lithospheric field, 5520 for describing the ionospheric field, 800 for the magnetospheric field, 8448 for the F-region coupling currents, and 1635 parameters for observatory biases. As a final improvement, these model parameters were estimated with the help of an IRLS approach with Huber weights.

2.3.5.4 Predictions from CM4

Perhaps the best way to show the success and limits of the CM4 model is to illustrate its ability to account for observations. This can be done in a great variety of ways, and many such illustrations can be found in the original paper of Sabaka *et al.* (2004). Here, only a few examples will be provided, mainly to show how each type of fields contribute to the temporal variations of the geomagnetic field observed at a fixed location, as testified by observatories. Interestingly, and as we shall see shortly, the comprehensive model not only succeeds at explaining most of the geomagnetic field variations at quiet times, as would be expected (recall that only quiet-time data were used to build the model), it also is able to make remarkable predictions of the field at quite disturbed magnetic times.

Let us first consider time series covering the entire time span (1960–2002) the model is supposed to account for. **Figure 9** shows observed monthly mean values of the X (North) component at the Niemegk (NGK) observatory in Germany (black line) together with various predictions from the model. The red curve shows the sum of the predicted contributions from the field of internal origin (core field and crustal field). Adding the predicted contribution from the magnetospheric field with its induced counterpart leads to the green curve. Also adding the predicted contribution from the ionospheric field with its induced counterpart leads to the blue curve. The final prediction of the CM4 model would then require that we also add a constant value, corresponding to the local 'crustal bias', which is the local contribution from the small-scale crustal field not described by the spherical harmonic Gauss coefficients of the model (which only goes to $N = 65$). This crustal bias shows up here as an offset, and is on the order of 160 nT, a fairly typical value. This offset is one parameter of the CM4 model, because data from Niemegk have been used to infer CM4. But note that in general, like all global field models, CM4 can only predict the field to within

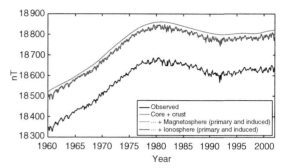

Figure 9 Observed and predicted (from CM4) monthly mean values of the X component at Niemeg (Germany).

such a constant value at the Earth's surface (unless the local crustal bias is known, of course). Note also that by contrast, no such macroscopic offset will be found between the field predicted by CM4 and the observed field at satellite altitude, because biases are produced by the very small-scale crustal field, the magnitude of which decreases fast with altitude.

Figure 9 illustrates the ability of the CM4 model to predict most of the observed time variations on these time scales. Of particular interest is the fact that both magnetospheric and ionospheric field contributions are important to achieve this success. They account not only for the strong annual (and semi-annual) variations, but also for significant irregularities, which have been captured by the Dst and $F_{10.7}$ parametrization.

Figure 10 provides a complementary way of understanding the contribution of each of the types of fields to the observed geomagnetic field on the same timescale and over the same time period. It shows time series of monthly-mean first differences (smoothed with a 12 month running average) again at Niemegk (NGK) in Germany, but also at Hermanus (HER) in South Africa. In each case, and for each field component (X, Y, Z; recall Section 2.2.1 for definitions) the observed signal is shown (black dots), together with predictions from just the core field (red line), from the core field plus the externally induced fields (green line), or from all fields (blue line). Note that this time, no more offset is to be seen. This illustrates the fact that the crustal bias is indeed a constant value, which has been removed by taking the time derivatives. Those curves otherwise illustrate the contribution of externally induced fields, and show that the $\dot{X}$, and especially $\dot{Y}$, horizontal components are least affected by the external

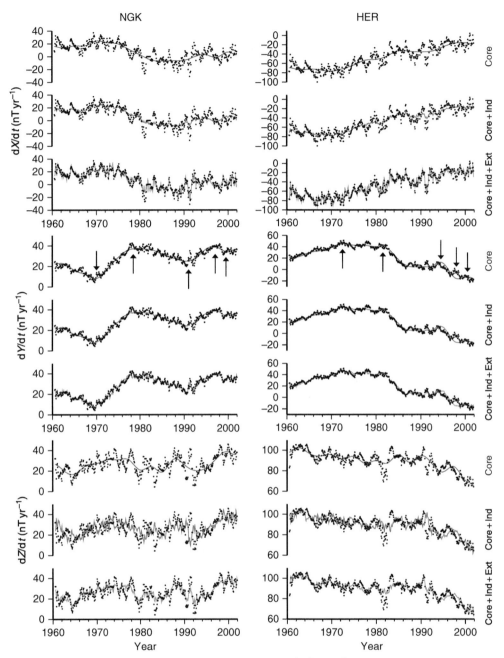

Figure 10   Comparison of smoothed monthly mean first differences, $\dot{X}$, $\dot{Y}$, and $\dot{Z}$, from observations (black symbols) and from predictions using a progression of sources from CM4 (colored lines) at NGK and HER from 1960 to 2002. Red, green, and blue lines present core, core plus induced, and core plus induced plus external sources, respectively. Arrows indicate approximate jerk locations from visual inspection of the core contribution to $\dot{Y}$. Four jerks previously identified, in 1969, 1979, 1992, 2000, are shown. The presence of an additional event in 1997 is also suggested. Modified from Sabaka, TJ Olsen, N and Purucker M (2004) Extending comprehensive models of the Earth's magnetic field with Ørsted and CHAMP data. *Geophysical Journal International* 159: 521–547 (doi:10.1111/j.1365–246X.2004.02421.x).

fields and their induced counterparts. This well-known feature is the reason why $\dot{X}$, and especially $\dot{Y}$, series have been extensively used in the past to investigate changes of trends in the field of internal

origin. As first noted by Courtillot *et al.* (1978), those series indeed reveal an intriguing piecewise pattern, with periods of stable trends, interrupted by sudden changes in those trends (as indicated by the arrows in

Figure 10). Those sudden changes are known as 'geomagnetic jerks'. When and how exactly they occur has been the subject of many recent studies, some of which rely on quite elaborate techniques (see, for example, Alexandrescu *et al.* (1996, 1999) and De Michelis and Tozzi (2005)). All those studies have however always relied on direct analysis of the observatory series, which do not guarantee that the suddenness and timing inferred for those jerks are not partly contaminated by the field of external origin and its induced counterpart. CM4 very clearly illustrates this issue. With induced and external contamination removed, the core field contribution from CM4 indeed reveals the well-known jerks in the $\dot{Y}$ components at NGK and HER. But local extrema due to the contribution of other sources exist in the data record near the times of each event. Those could lead to incorrectly date jerks by perhaps as much as 1 or 2 years. By better separating the core field from other fields, comprehensive models such as CM4 can provide an interesting alternative way of investigating geomagnetic jerks (see, for example, Sabaka *et al.* (2004) and Chambodut and Mandea (2005)).

It is interesting to also illustrate how the CM4 compares to the core field model of Wardinsky and Holme (2006) described earlier, which covers the years 1980–2000 (**Figure 11**). The reader will recall that this model was build by avoiding crustal signals and treating the field of external origin and its induced counterpart as a source of error in a two-step process. A first 'interim' core field model was built, simply treating those errors (i.e., the variations of the observed field about the green 'interim' model prediction of the $\dot{X}$, $\dot{Y}$, and $\dot{Z}$ components in **Figure 11**) as independent errors. But as can be seen in both **Figures 10** and **11**, this is incorrect. Fast changes due to the field of external origin and its induced counterpart are correlated among the $\dot{X}$, $\dot{Y}$, and $\dot{Z}$, components. The second (and final) model of Wardinsky and Holme (2006) precisely accounts for this, by using a covariance model with nondiagonal terms empirically determined from the correlation among the observed $\dot{X}$, $\dot{Y}$, and $\dot{Z}$ residuals with respect to the 'interim' model. As can be seen in **Figure 11**, this removes quite a few oscillations in the model prediction (now in red), which were undoubtedly due to external field contamination, and brings the model prediction much closer to that of the core field prediction of CM4 (in blue). Whether the weak oscillations this final model still predicts (compared to CM4) are real core field features or the result of remaining external field contamination (which we are inclined to believe) is left to the reader to decide. In any case, it is clear that treating external sources by means of a statistical method as suggested by Wardinsky and Holme (2006) brings very substantial improvement to core field modeling when, as opposed to the CM approach, other field sources are not explicitly modeled.

What about short-term geomagnetic field variations? **Figure 12** shows one month (April 1990) series of observed hourly mean values of X, Y, and Z components at the same NGK and HER observatories as before, together with CM4 predictions. This figure is particularly interesting because only a single point of each series was actually used to build CM4. April 1990 was indeed a month no satellite data was available, and as stated earlier, only the OHM value closest to 01:00 local time for the quietest day of the month was then used in CM4 (to assess the

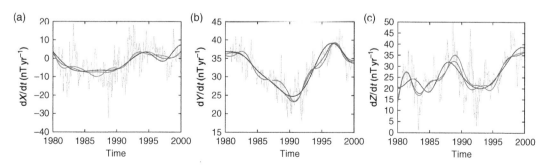

Figure 11 Comparison between smoothed monthly mean first differences for NGK (dashed line), and model predictions from CM4-core (blue) and two models by Wardinsky and Holme (2006) with (red) and without (green) error analysis. Modified from Wardinsky I and Holme R (2006) A method to refine main field modeling. In: *Proceedings of the First International Swarm Science Meeting,* volume WPP-261. ESA 2006.

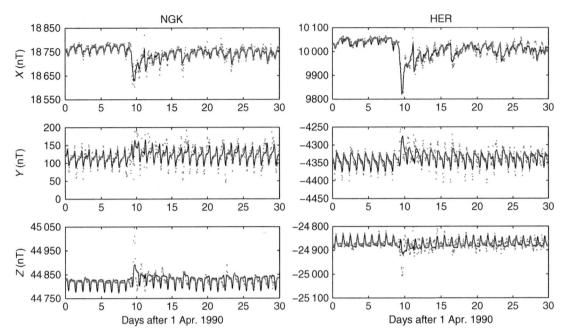

Figure 12   Observed (red dots) and CM4 predicted (black curves) field values at the observatories Niemegk (left) and Hermanus (right), for April 1990.

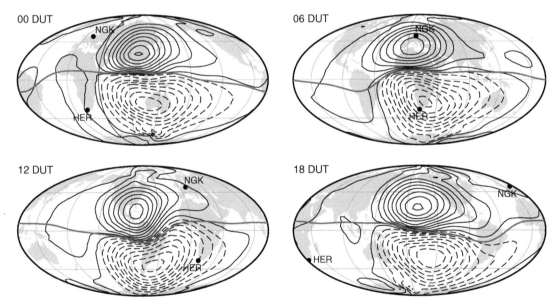

Figure 13   Global maps of ionospheric E-region equivalent current function Ψ as a function of DUT, t_d (cf. eqn [15]), for 21 Mar. 1980 and a value of $F_{10.7} = 140.0 \times 10^{-22}\,\mathrm{W\,m^{-2}\,Hz^{-1}}$, which is an average over the time span of the model. Contour interval is 20 kA; positive values are shown with solid, negative with dashed lines. The red curve represents the dip-equator. Locations of the Niemegk (NGK) and Hermanus (HER) observatories are also shown.

contribution of the Core and Crustal Fields). The fast variations predicted here are thus true predictions of the CM4 model, CM4 having its external field parameters independently constrained by simultaneous ground and satellite data acquired during the POGO and Magsat missions at other times.

Figure 13 precisely shows maps of the equivalent sheet current function Ψ (which defines the surface currents $\mathbf{J}_\mathrm{s} = -\hat{\mathbf{r}} \times \nabla \Psi$) of the regular E-region ionospheric field sources at 110 km as inferred from CM4, on a typical quiet day (21 March 1980) when Magsat data was available.

Those currents are essentially fixed with respect to the Earth–Sun line (maxima are always located near the central meridian line, below which the Earth rotates, when the DUT changes), but their detailed morphology follows important features of the main field. Note in particular that the current is amplified at the dip-equator (indicated by the red curve), which represents the EEJ, and that the latitudes of the Sq focuses in the Northern and Southern Hemispheres follow that of the dip-equator. Such main field related features were not captured by previous models of the regular daily variation (see, for example, Winch (1981), Campbell *et al.* (1989), and Chapter 3).

Similar maps of the regular E-region ionospheric field sources can be predicted from CM4 for April 1990 (which CM4 does by adjusting for seasonal effects and using the relevant $F_{10.7}$ index). The rotation of the Earth below those sources is then the cause of the diurnal periodicity of the quiet-time ionospheric field (and of its associated externally induced field, which arises precisely because of this periodic excitation) observed in observatories. Similar periodic signals are also produced by the essentially Sun-synchronous quiet-time magnetospheric field and its induced counterpart, which CM4 can also predict. All those fields and up to make for the periodic component of the predictions in **Figure 12**. The last nonperiodic contribution arises from the highly variable ring-current signal (and its induced counterpart), the variability of which CM4 predicts using the *Dst* index. As can be seen, the resulting prediction is quite successful. Note in particular that CM4 still manages to capture most of the field variations at NGK and HER, even when a storm starts on the 10th day and the magnetic conditions become severe (as testified by the *Dst* index, which goes way above the limit of 20 nT used to define quiet times in the modeling procedure). It should however be mentioned that less success is achieved at such magnetically disturbed times, when predicting field variations at higher latitudes, or closer to the equator (see Sabaka *et al.* (2004) or Chapter 5, where CM4 is also briefly discussed).

2.4   The Present Main Field

As the previous section amply illustrated, recent satellite missions boosted our ability to model and identify all components of the Earth's magnetic field. Many breakthroughs have been made, which we briefly described or alluded to: considerable improvement of our knowledge of the crustal field, better understanding of the ionospheric and magnetospheric fields and of their associated induced fields, and also detection of previously undetected or poorly known field components, such as the fields produced by oceanic currents at satellite altitude, or those locally produced in the F-region of the ionosphere. Those breakthroughs are not only important for the information they provide. They also contributed to our improved ability to investigate the main field, and most importantly, its temporal variation, the so-called 'secular variation'. It is the purpose of this final section to report on those recent progress and to provide an overview and brief discussion of the most important spatiotemporal characteristics of the present main field, and of the core field it testifies for. We focus on the 1980 to present time period, but will also occasionally refer to the known behavior of the main field over historical and archeological time periods, which are otherwise much more extensively addressed in Chapters 5 and 9.

2.4.1   Spatial Power Spectra and Timescales

A useful way of characterizing the present global spatiotemporal behavior of the Earth's main field, beyond the well-known fact that it is mainly dipolar and evolving on secular timescales, is to make use of the concept of spatial power spectra introduced in Section 2.3.1.

Applying this concept to the field of internal origin in 2000, as predicted by, for example, the CHAOS (Olsen *et al.*, 2006) and CM4 (Sabaka *et al.*, 2004) models, and plotting the Lowes–Mauersberger spectrum at the Earth's surface (eqn [26] for $R = a$), leads to **Figure 14(a)**. This figure (plotted in semi-log scale, beware) shows that the field of internal origin is indeed dominated by its dipole ($n = 1$) component. But it also shows that the spectrum is made of two almost linear branches, with a transition around degree 14. This dichotomy (first noted by Langel and Estes (1982) based on a 1980 Magsat-derived spectrum) is the signature of the fact that the field of internal origin is dominated by the crustal field for degrees above 14, and by the core field for degrees below 14, both fields probably contributing by similar amounts at degree 14 (recall Section 2.3.3). As a result, the main field

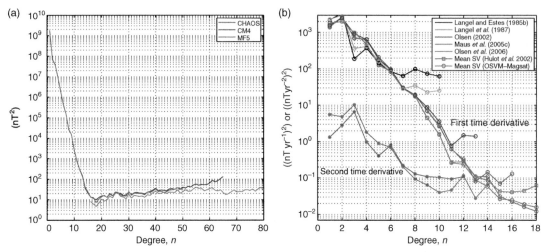

Figure 14   (a) Spatial power spectrum of the geomagnetic field at Earth's surface. (b) Spatial power spectrum of secular variation and secular acceleration at Earth's surface.

and therefore the core field, at any given epoch such as here in 2000, can only be recovered with some precision up to degree 13.

Figure 14(a) also illustrates the overall excellent agreement between various models of the field of internal origin, especially as far as the core field is concerned. Most disagreements are indeed to be found in the crustal field branch. Those likely trace back to the different way those two models deal with ionospheric field contributions, CHAOS relying on a selection process (recall Section 2.3.2), and CM4 modeling most of it (recall Section 2.3.5). It is unavoidable that at least a fraction of the ionospheric and unmodeled fields will be misinterpreted as possible crustal field contributions in both models, but in different ways, resulting in such disagreements. Taking extreme steps to get rid of those can also have some drawbacks, as is illustrated by the spectrum of the crustal field model MF5 of Maus *et al.* (2007) already discussed in Section 2.3.4. This spectrum (also shown in **Figure 14(a)**) shows that the estimated crustal field is then substantially weaker. By emphasizing the need to remove the external field, the modeling approach used by Maus *et al.* (2007) simply 'erased' part of the crustal field signal. Note however that this type of approach mainly affects the overall magnitude of the recovered crustal field and not so much its morphology (as discussed in more detail in Olsen *et al.* (2006)).

Plotting the analogous Lowes–Mauersberger spectrum for the first time derivative of the field of internal origin at the Earth's surface provides a

very interesting complementary picture. Because the crustal field can only very slowly change on planetary scales, we know that this time derivative, the secular variation, testifies for the first time derivative of the core field. **Figure 14(b)**, which shows such spectra predicted by several models (a number of which have been discussed in Section 2.3), is also an opportunity to illustrate the spectacular improvements the recent satellite missions have brought. Whereas Langel and Estes (1985b) could only recover the 1980 secular variation up to degree 6 or so, using data from 6 months of Magsat satellite data and 91 observatories (the 'flat' behavior above degree 6 being indicative of a noise of about 70 $(nT\,yr^{-1})^2$), a resolution Langel *et al.* (1988) could only slightly increase by the additional use of data from the DE-2 satellite between September 1981 and January 1983 (reducing the noise level to about 30 $(nT\,yr^{-1})^2$), the advent of Ørsted and CHAMP dramatically improved the situation. With just two years of Ørsted observations, the noise level was reduced to about 1 $(nT\,yr^{-1})^2$, making it possible to resolve the 2000 secular variation up to degree 11 or so (Olsen (2002)). Yet another order of magnitude in noise reduction was achieved by the model of Maus *et al.* (2005c) (derived by a combined analysis of Ørsted and CHAMP data spanning more than 5 years), resulting in a determination of the secular variation up to about degree 13. Finally, a noise level as low as 0.02 $(nT\,yr^{-1})^2$ could be achieved by the CHAOS model of Olsen *et al.*

(2006), allowing a determination of the secular variation likely up to degree 15. In fact, and as was already pointed out in Section 2.3.3, Ørsted and CHAMP observations now even make it possible to try and simultaneously infer the second time derivative of the field of internal origin, known as the 'secular acceleration', the spectrum of which is also shown in **Figure 14(b)** for the model of Maus *et al.* (2005c) (average acceleration over the 2000–2005 time period) and the CHAOS model of Olsen *et al.* (2006) (instantaneous acceleration for the 2002.5 epoch).

All those spectra typically reflect average estimates of the time derivatives over only a few years, that is, almost 'instantaneous' estimates of the present time secular variation and acceleration. But Magsat flew long enough to recover high-degree models of the field of internal origin in 1980, which can be compared to present-time high-degree models derived from Ørsted and CHAMP data, to produce 'average' secular variations models. The spectra of two such models are also plotted in **Figure 14(b)** (as inferred by Hulot *et al.* (2002) from Magsat (1980) and Ørsted (2000) models computed by Langlais *et al.* (2003), and as inferred here from the model of Cain *et al.* (1989) in 1980, and the OSVM model of Olsen (2002) in 2000). This shows that the recent improvements have brought the spectra of the instantaneous secular variation closer and closer to the average secular variation spectrum, further suggesting that those improvements are real.

In fact, **Figure 14(b)** would even suggest that the instantaneous secular variation model is better determined than the average secular variation from Hulot *et al.* (2002), the behavior of which suggests a noise contamination on the order of 0.06 $(nT\ yr^{-1})^2$, affecting degrees above 15. Note however that a perfect match between those spectra should not be expected since we also know that the secular variation is not stationary (as is clear from the fact that both the CHAOS and POMME models call for a secular acceleration over the past 6 years, and even clearer from the direct observation of geomagnetic jerks at observatories – recall **Figure 10**).

Plotting the analogous spectra for the core field and its secular variation at the core–mantle boundary (CMB) is also important, since this is where the core field originates from.

Figure 15(a) shows the core field spectrum at the CMB for epoch 2000 computed from the CHAOS model (all other models would predict virtually the same spectrum). The model was truncated at degree 13 to avoid the crustal field (recall Section 2.3.3) and eqn [26] was used with $R = b$, where b is the core radius, which amounts to downcontinue the core field to the CMB as a potential field (which we may do, since the core sources lie below the CMB, and the mantle can be assumed to be a source-free medium; recall Section 2.3.3). This spectrum reveals that the core field is still dominated by its dipole component at the CMB. But it now also shows that higher-degree

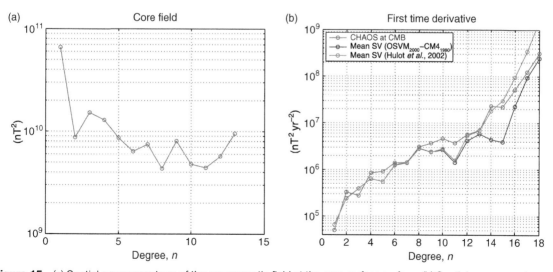

Figure 15 (a) Spatial power spectrum of the geomagnetic field at the core surface surface. (b) Spatial power spectrum of secular variation at the core surface.

components (and therefore smaller scales) play a much more prominent role at the CMB than at the Earth's surface. Their contributions only very weakly decrease on average as the degree increases.

The important role played by small scales at the CMB shows up in an even more striking way when plotting the secular variation spectrum at the CMB. **Figure 15(b)** shows three such spectra: the 'instantaneous' 2000 secular variation spectrum as inferred from the CHAOS model, and the two 'average' 1980–2000 secular variation spectra already shown at the Earth's surface in **Figure 14(b)**. All spectra have been boldly plotted up to degree 18, which we are theoretically entitled to do, since the secular variation is the first-time derivative of the core field, and thus also has all its sources below the CMB. In doing so, we also make the assumption that all degrees up to 18 are properly recovered. This however is not so obvious as the divergence between the two average secular variation spectra above degree 15 shows. As already noted, this divergence likely reflects noise contamination in the Hulot *et al.* (2002) average secular variation. But similar (weaker) contamination could also affect the other spectra, as the steeper trend beyond degree 15 would suggest. Those spectra show that contrary to the core field itself, the secular variation is generally not dominated by its largest scales at the CMB. In contrast, a clear increasing trend is now indeed to be seen, revealing a very strong small-scale dynamics, which could not be witnessed in such detail before the advent of the Ørsted and CHAMP missions.

A final very useful spectral quantity one can next easily derive is the 'correlation time' associated with a given degree n, and defined as

$$\tau(n) = \sqrt{\frac{W_n(r)}{W'_n(r)}} = \sqrt{\frac{\sum_{m=0}^{n}\left[\left(g_n^m\right)^2 + \left(h_n^m\right)^2\right]}{\sum_{m=0}^{n}\left[\left(\dot{g}_n^m\right)^2 + \left(\dot{h}_n^m\right)^2\right]}} \quad [29]$$

where $W_n(r)$ and $W'_n(r)$ are the degree n contributions to the Lowes–Mauersberger spectra of respectively the core field and its secular variation at radius r (as defined by eqn [26]). This quantity, which is independent of r, is best understood in statistical terms as a measure of the time it would take for the degree n core field to significantly evolve, were the core field to behave as if produced by a random statistical process (Hulot and Le Mouël, 1994). Although the core field is of course produced by a deterministic dynamo process within the core, it turns out that such a statistical interpretation is quite consistent with both the

known temporal evolution of the large scales of the core field recovered from historical and archeomagnetic data (see Hulot and Le Mouël, 1994; Hongre *et al.*, 1998), and the behavior of the field produced by numerical dynamo simulations (Bouligand *et al.*, 2005). **Figure 16** shows such estimates as inferred from the CHAOS model by simply making use of the two CHAOS spectra shown in **Figures 14(a)** and **14(b)**. Of course, any other choice of models for the same epoch would essentially lead to the same estimates. Of particular interest is the decreasing trend showing that the correlation time gets all the smaller that the degree considered increases, reaching values on the order of 20–30 years for the largest degrees such correlation times can be estimated. Small scales of the core field can change very quickly.

Several interpretations of the core field and secular variation spectra, as well as of the resulting correlation times, have been proposed (see, for example, McLeod (1996), Voorhies (2004), Holme and Olsen (2006) and references therein). Those interpretations usually start from the observation that a uniform random potential field or distribution of random sources (be it monopoles, dipoles, or current loops) at the CMB would lead to a slightly increasing core field spectrum at the CMB, in contradiction with the observed spectrum (recall **Figure 15(a)**). They thus conclude that such distributions would need to be slightly biased toward weaker contributions from small scales, or relocated at some distance below the CMB (implying a 'deeper' source). Such simple interpretations are however difficult to reconcile with the equations governing the core dynamo,

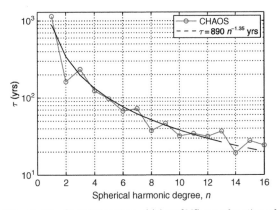

Figure 16 Correlation time $\tau(n)$ (eqn [29]), as a function of n, estimated from the CHAOS model. Modified from Olsen N, Lühr H, Sabaka TJ, *et al.* (2006) CHAOS - a model of Earth's magnetic field derived from CHAMP, Ørsted, and SAC-C magnetic satellite data. *Geophysical Journal International* 166: 67–75.

and most authors have therefore also developed more advanced interpretations, relying on dimensional analysis of those equations, to simultaneously account for the core field and secular variation spectra at the CMB (and therefore also for the correlation times). All those interpretations predict quite similar spectral behaviors, and it is difficult to tell which the observed spectra would favor. **Figure 16** shows a typical attempt to fit the correlation times with the help of a power law (based on a heuristic reasoning of Holme and Olsen (2006), which predicts such a power law with an exponent between −1 and −2). The issue is further complicated by the fact that the observed core field and secular variation spectra can be expected to fluctuate in time. In the case of **Figure 16**, such fluctuations could for instance account for the departures of the observed correlation times from the 'predicted' values. But how can we tell whether this is the case or not?

The statistical formalism introduced by Hulot and Le Mouël (1994) provides a mean to address such questions. By formally linking the core field and secular variation spectra (together with the associated correlation times) at a given epoch with the observed core field evolution, it indeed makes it possible to assess whether an observed present time remarkable property can possibly qualify as long-standing. Using this approach, Hongre *et al.* (1998) precisely showed that the low degree 2 correlation time seen in **Figure 16** is consistent with the core field behavior over the past 2000 years (as inferred from archeomagnetic data), that is over more than 10 times its value, strongly suggesting that this in fact is a robust long-standing feature, and not just the result of a fluctuation. In fact, Hongre *et al.* (1998) further showed that the present degree 2 core field is also remarkable because of the low value of its contribution to the field spectrum, showing that the distinct V shape made by the first three degrees in the core field spectrum (**Figure 15(a)**), which has been around for at least the past 2000 years, is also a robust long-standing feature. (Note incidentally that the long-term dominance of the dipole term, which evolves on longer timescales, requires more than 2000 years of observations to be confirmed, but is indeed well established thanks to paleomagnetic observation; see, for example, Merrill *et al.* (1998), Chapters 11 and 13.) Significant features in all those spectra can thus be found, which simple dimensional analysis of the dynamo equations fail to predict. However, direct comparison with results from numerical dynamo simulations is also possible

and quite straightforward. Interestingly, virtually all dipole dominated numerical dynamo simulations do predict such a V shape core field spectrum (see, for example, Dormy *et al.* (2000), Christensen and Olson (2003) of the present Treatise).

2.4.2 The Field at the Earth's Surface

Let us now consider the present main magnetic field at the Earth's surface. To also illustrate the way this field has been changing since it was first observed in detail by a high-precision satellite (MAGSAT), we in fact focus on two epochs, 2005.0 and 1980.0. For epoch 2005.0, we rely on the CHAOS model of Olsen *et al.* (2006), which is one of the only two recent models capable of describing that epoch without having to rely on some temporal extrapolation (the other being the POMME model of Maus *et al.* (2006b), which will anyway lead to extremely similar maps). We do not use the official IGRF-10 of Maus *et al.* (2005a) because, as we saw in Section 2.3.2, this model is only predictive for epoch 2005.0. For epoch 1980.0, we otherwise choose the CM4 model of Sabaka *et al.* (2004). In all cases, only degrees up to $N = 13$ have been used, since, as we saw, higher degrees correspond to crustal field contributions. In fact, it is worth noting that had we used the full resolution of each model (up to degree $N = 50$ for CHAOS, and $N = 65$ for CM4) and plotted the predicted field of internal origin, the resulting maps would have anyway looked very much the same, since crustal field contributions up to such degrees only amount to a few tens of nanoteslas (i.e., to less than 1% of the main field contribution). Note however also that nonmodeled smaller scales can contribute much more (as illustrated by the 'crustal bias' responsible for the shift between the observed and predicted field in **Figure 9**). The reader should therefore be warned not to make use of any of those maps to infer local values of magnetic field components without checking the local magnetic conditions (watching out not only for strongly magnetized local rocks, but also for anthropic activity and, of course, any possible ongoing magnetic storm – recall Section 2.1).

Figures 17 and **18** show maps of the declination (*D*) for epochs 2005 and 1980, while **Figures 19** and **20** show maps of the inclination (*I*) for the same two epochs. Those maps, which are all plotted using the same Hammer and polar (north and south) projections, illustrate the current morphology of the main field at the Earth's surface. A pure axial dipole field would lead to zero declination everywhere, and to

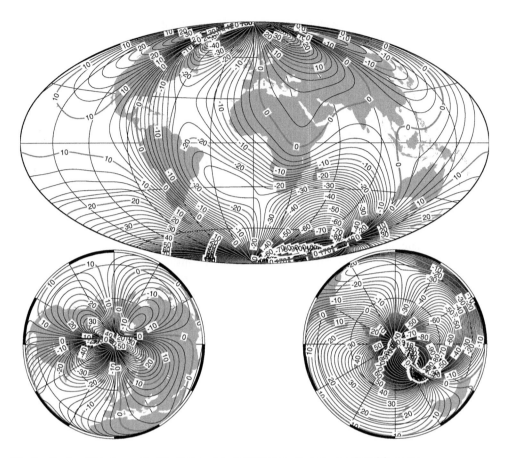

Figure 17   Declination *D* (in degrees) at Earth's surface in 2005.0 as given by the CHAOS model.

axisymmetric isoclines (lines of equal inclination), with the dip-equator (defined by $I = 0°$) lying on the geographic equator, and the north and south dip poles (also known as the north and south magnetic poles, and defined as the locations where $I = 90°$ and $I = -90°$) located at the north and south geographic Poles. But the Earth's main field is not an axial dipole, and this is the reason for both the distortion of the isoclines and the complex pattern of the 'isogones' (lines of equal declination). Much of this complexity is due to the tilt of the dipole component of the main field, but not all. This is best illustrated by the fact that the dip poles are not antipodal (which they would be, were the main field to be an inclined dipole). Indeed, using CHAOS and CM4, one can easily estimate the locations of those dip poles: 82.9° N, 117.3° W versus 64.2° S, 137.75° E for the north and south dip poles in 2005, and 76.45° N, 101.2° W versus 65.05° S, 139.15° E for the north and south dip poles in 1980. Those locations are also interesting to compare with the strictly antipodal

'geomagnetic poles' (the location of which are given by eqn [5] and [6]), the purely dipolar components of the main field would predict: 79.75° N, 71.8° W versus 79.75° S, 108.2° E for the north and south geomagnetic poles in 2005, and 78.8° N, 70.8° W versus 78.8° S, 109.2° E for the north and south geomagnetic poles in 1980. The very significant departure of the dip pole locations from those of the matching geomagnetic poles is a direct consequence of the non-dipolar components of the main field.

Figures 17–20 also illustrate the impact of secular variation. Although those maps look similar in 1980 and 2005, obvious changes can be found. Those are not trivial. In places such as in Europe, Africa, and over the Atlantic Ocean, those crudely amount to a 'westward drift' of the main field (as is best seen by focusing on, for example, the 'agonic' $D = 0°$ line), a drift that is also seen on historical time scales (*see* Chapter 5). But this westward drift is not universal. Isogones in the northern Pacific East of the United States clearly moved eastward. Also, whereas the

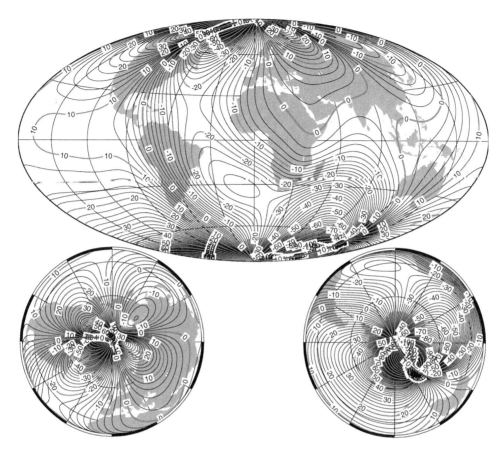

Figure 18 Declination D (in degrees) at Earth's surface in 1980.0 as given by the CM4 model.

south dip pole hardly moved (as is also confirmed by a series of direct local observations in 1986 and 2000 (Barton, 2002)), the north dip pole very quickly moved northward, at the impressive average pace of $20 \, km \, yr^{-1}$ (as is again confirmed by a series of direct local observations, in 1984, 1994, and 2001 (Newitt *et al.*, 2002)). All those changes reflect changes in the core field produced at the CMB, the morphology and evolution of which are discussed in the next section.

Maps of the field intensity at the Earth's surface reveal interesting additional features. **Figure 21** shows such maps for epoch 2005. Those maps again differ markedly from those of an axial dipole field (which would imply axisymmetric lines of equal intensity, or 'isodynamic lines', with one maxima at each geographic pole, and a minimum isodynamic line at the geographic equator). In particular, two remarkable features show up that a dipole, even inclined, cannot account for. One is the contrast between the relatively weak (about 60 000 nT) almost double northern maximum and its strong (66 700 nT) southern counterpart. The other is the very weak (less than 23 000 nT) minimum localized in the South Atlantic, known as the South-Atlantic Anomaly (SAA). This SAA is in fact mainly (but not only) due to the additional contribution of the quadrupole (degree-2) component of the main field. Adding this contribution to the dipole field results in a field which can be approximated by a so-called 'geomagnetic eccentric dipole', that is, a dipole with the same magnitude as the dipole component of the main field, but mathematically located slightly away from the center of the Earth (presently about 540 km, in the direction opposite to the location of the SAA). As far as the main field itself is concerned, this location does not have any special meaning (for a definition and a discussion of the concept of geomagnetic eccentric dipole, see, for example, Langel (1987), Lowes (1994)). But it provides a convenient way of understanding the consequence of the SAA for the near-Earth magnetospheric environment

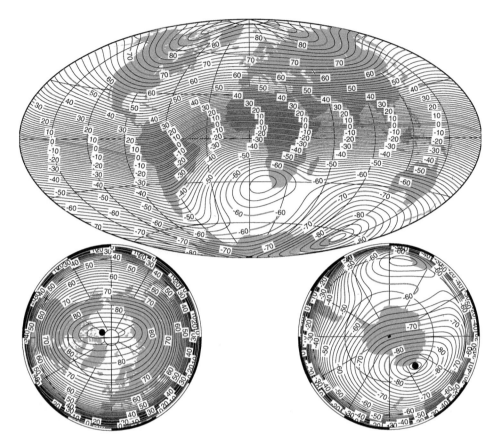

Figure 19   Inclination *I* (in degrees) at Earth's surface in 2005.0 as given by the CHAOS model. Also shown is the location of the magnetic poles.

within which low-earth orbit (LEO) satellites (and the space shuttle) evolve.

This near-Earth magnetospheric environment is a region above the ionosphere, where trapped particles are to be found which oscillate back and forth along the main field lines, slowly drifting westward if positively charged (mainly protons), eastward otherwise (electrons) (*see* Chapter 3). Those particles can have vary different energies. The most numerous but less energetic make for the magnetospheric ring current the reader must now be familiar with, and which is most intense at several Earth radii close to the equatorial plane. But much more energetic particles (protons and electrons typically above 1 MeV) are also to be found which contribute little to the electrical currents (and to the magnetic field) but have strong penetrating properties. Those make for what is then referred to as the radiation belts or Van Allen belts (see, for example, Kivelson and Russell (1995) for details). Electrons, which are most dangerous (because most penetrating) are in fact to be found in

two belts: an 'outer belt', which corresponds to electrons oscillating along main field lines between typically 60° latitude north and south, reaching 4–6 Earth radii at the equator; and an 'inner belt', with electrons traveling along low latitude (up to only 30°) field lines about the equator. Those field lines are near horizontal and force this axisymmetric inner belt to lie very close to the Earth. For a purely dipolar field of the present strength, its lower boundary would typically be only 1000 km away from the Earth's surface. But because the main field is better described by an excentric dipole than by a simple dipole, and because the radiation belts follow the geometry imposed by the main field, this lower boundary is brought even closer to the Earth's surface in the direction opposite to the shift, that is, above the SAA where the main field intensity is the weakest. As a result, the inner radiation belt is presently only a few hundred kilometers away from the Earth's surface above the SAA. This has many unpleasant consequences for the more and more numerous

Figure 20 Inclination I (in degrees) at Earth's surface in 1980.0 as given by the CM4 model. Also shown is the location of the magnetic poles.

LEO satellites, the instruments and electronics of which can severely suffer from radiation. Indeed, as reported by, for example, Heirtzler *et al.* (2002), nearly all spacecraft crossing this area at altitudes of 100 km to 1000 km have been damaged or degraded in performance to some extent (this has even been the cause of some problems with the electronic devices on Ørsted).

Looking at the way this SAA, and more generally the intensity of the main field, has been evolving in the recent years is of course also of substantial interest. **Figure 22** shows maps of the changes experienced by the main field intensity between 1980 (from CM4) and 2005 (from CHAOS). As can be seen, the biggest changes occurred over northern America, where the field intensity decreased by about 3000 nT, and over the Indian Ocean, where the field increased by about 1600 nT. Clearly also the average change corresponds to a global decrease, most of which is related to the average decrease (by 430 nT) of the axial dipole component of the main field over that period of time. This decrease, how-ever, is not the dominant signal in those maps, which reveal quite some structure. Of particular interest is the fact that relatively little changes occurred in the Pacific, and that most of the decrease occurred in the vicinity of the SAA, which has been slightly growing in size (a measure of this growth can also be given in terms of the shift of the equivalent eccentric dipole with respect to the center of the Earth, which was of 490 km in 1980, and is of 540 km in 2005). It is quite likely that this SAA will go on growing for some time, bringing the inner radiation belt closer to the ionosphere and increasing the hazard to LEO satellites.

Inclination, declination, and intensity maps such as those shown in **Figures 17–22** provide interesting information about the current morphology of the present main field at the Earth's surface. But unless considered simultaneously, each of these maps only provides partial information about the field. Plotting the radial component of the field B_r offers an

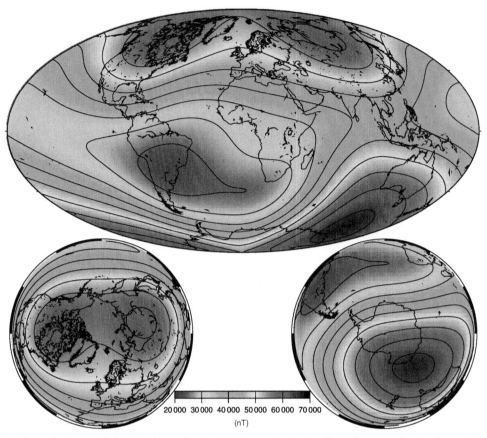

Figure 21 Scalar field intensity F at Earth's surface in 2005.0 as given by the CHAOS model. Contour interval 5000 nT.

opportunity to get a more synthetic image of the main field. Indeed, because the main field is entirely of internal origin, maps of this component at the Earth's surface entirely define the main field (see, for example, Backus *et al.* (1996)). **Figure 23** shows such maps for the main field in 2005 (from CHAOS). Interestingly, and somewhat paradoxically, those maps seem to contain less information than the corresponding maps of inclination, declination, and intensity. This, however, is only because of the overwhelming contribution of the dipole component of the field at the Earth's surface. The information is present, but only in the details. Indeed, the signature of all the basic features of the main field we already discussed can still be found. Note in particular the two northern high latitude lobes, the single southern latitude polar maximum, and the signature of the SAA. These maps will make it possible to identify the field lines at the CMB that cause those features, as we shall soon see.

Precisely for the same reasons, maps of the first time derivative of the radial component of the main

field are also provided (**Figure 24**). Note that those maps do not show changes between 1980 and 2005 as in **Figure 22**, but the radial component of the current 'instantaneous' secular variation (in 2005, from the CHAOS model). Comparing **Figure 24** to **Figure 22** is nevertheless interesting. It shows that the present secular variation essentially keeps with the trend of the past 25 years, producing fields of polarity opposite to that of the main field over the Americas (locally resulting in a decrease in intensity), and producing fields of correct polarity over India and the Indian Ocean (locally resulting in a weak increase in intensity). Note also the relatively weak secular variation in the Pacific Ocean, where indeed the intensity has only weakly changed over the past 25 years.

2.4.3 The Field at the Core Surface

Let us finally consider the present field at the core surface. **Figure 25** shows maps of the radial component of this field in 2005.0. Those maps

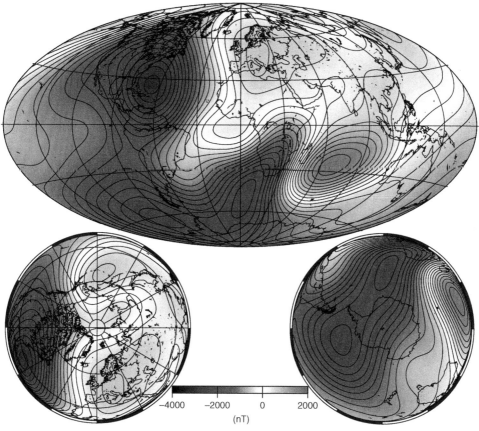

Figure 22 Difference of scalar field intensity F at Earth's surface between 2005 and 1980. Contour interval 200 nT.

correspond to exactly the same field as the one which led to **Figures 17**, **19**, **21** and **23** at the Earth's surface. They have been computed using the CHAOS model of Olsen *et al.* (2006), up to degree $N = 13$, and assuming no sources of any kind are to be found between the Earth's surface and the CMB. This amounts to neglect the arguably small contribution of the crustal field to degrees less than 13 (recall Section 2.3.3), and to ignore electrical currents possibly induced in the mantle by the core field (not an unreasonable assumption, but see, for example, Backus (1983), Benton and Whaler (1983), Alexandrescu *et al.* (1999)). **Figure 25** thus shows the radial component of the fraction of the field, which escapes the core at the CMB, and is responsible for the main field we can observe at the Earth's surface.

Not surprisingly, those maps reveal a lot more details than the analogous maps shown in **Figure 23**. This is because of the geometrical attenuation which affects small scales (large degrees n of the spherical harmonic representation, eqn [17]) much more

strongly than large scales (small degrees n), when moving away from the sources of the field. In particular, and as could be anticipated from the spectrum we already plotted (recall **Figure 15(a)**), the field is far less dipolar than at the Earth's surface.

Several striking features show up. Most remarkable is the large reverse patch that extends below the southern edges of southern America and Africa. Also remarkable are the two high intensity lobes to be found under Arctic Canada and Siberia, and the somewhat less intense, partly merged lobes to be found under Antarctica. Those features are the cause of the features we had already identified in the main field at the Earth's surface. In particular, the reverse patch at the core surface is the cause of the SAA, the growth of which is directly related to the growth of this reverse patch.

Maps of the first time derivative of the radial component of the field can of course also be plotted at the core surface. **Figure 26** shows such maps plotted from the CHAOS model for epoch 2005. In that case, for the reasons already outlined in Section

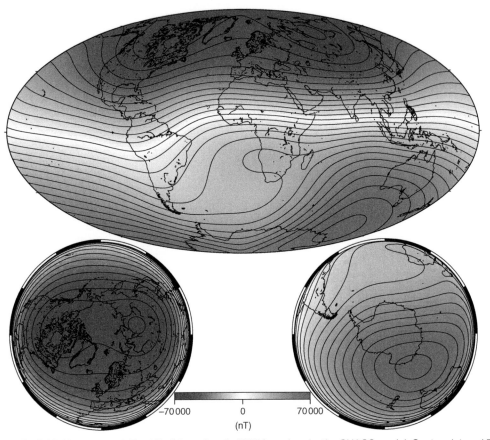

Figure 23   Radial field component, B_r, at Earth's surface in 2005.0 as given by the CHAOS model. Contour interval 5000 nT.

2.4.1, the full resolution of the model can be used, and those maps have therefore been plotted up to degree $N = 15$. Again, and not surprisingly, those maps reveal a lot more details than the analogous maps plotted at the Earth's surface (**Figure 24**). The important role played by the small scales, which dominate the signal, is obvious. Note in particular how the relatively large-scale regions of strong secular variation to be found over the Americas and the Indian Ocean at the Earth's surface are in fact produced by regions of equally intense but contrasted small-scale secular variation at the core surface. Note also how the relatively 'quiet' Pacific hemisphere at the Earth's surface is also associated with a similar quiet hemisphere at the core surface, even when small scales are taken into account.

Many of those striking features can also be identified (though with much less detail, especially in the far past) when analyzing the historical field, as is being done in Chapter 5. This then reveals that the growth of

the reverse patch responsible for the SAA probably started several centuries ago, that the Pacific hemisphere has essentially remained 'quiet', while the high-latitude lobes seem to have been fairly stationary over the same period of time.

Understanding the origin of those remarkable features (and other dynamical features, which are best revealed when analyzing the temporal evolution of the historical field; see again Chapter 5) is of course a problem of prime interest. But it also is a formidable task, as it requires some advanced understanding of the magnetohydrodynamics that governs the dynamo process responsible for the field generation within the core. Although very substantial progress has already been made in the numerical simulation of this process, direct simulation of the current behavior of the Earth's dynamo is not yet possible. Interesting studies can nevertheless be carried out by jointly analyzing the core field

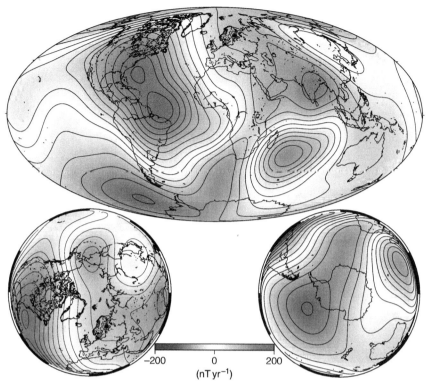

Figure 24 Secular variation of the radial field component, dB_r/dt, at Earth's surface in 2005.0 as given by the CHAOS model. Contour interval 200 nT yr^{-1}.

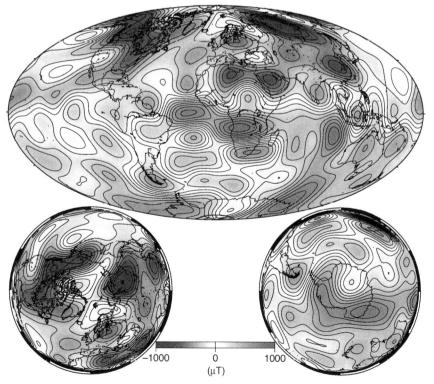

Figure 25 Radial field component, B_r, at the core surface in 2005.0 as given by the CHAOS model. Contour interval 100 μT.

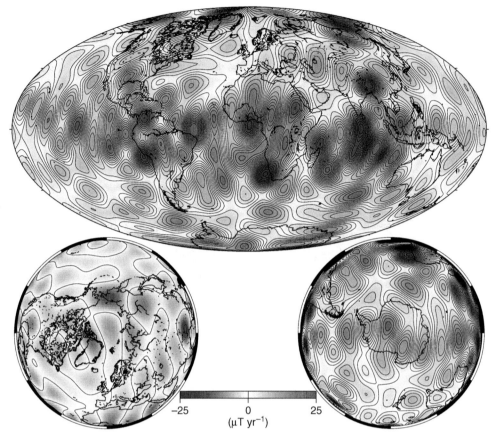

Figure 26 Secular variation of the radial field component, dB_r/dt, at the core surface in 2005.0 as given by the CHAOS model. Contour interval $2\,\mu T\,yr^{-1}$.

and its first time derivative, to also infer the core surface flows accounting for the secular variation. Although such an exercise requires quite a few important assumptions (*see* Chapter 5, where the rational for such an approach is exposed), it can indeed provide some insight into the processes possibly accounting for the current short-term behavior of the Earth's magnetic field, in particular its hemispheric asymmetry, the current growth of the SAA, and the current overall decline of the dipole field (see, for example, Hulot *et al.* (2002)). With such insight, the current fast progress in the field of dynamo numerical simulation, and the prospect of even better high-precision magnetic field satellite observations (thanks to the ESA Swarm mission which will soon take over the current Ørsted and CHAMP missions; see Friis-Christensen *et al.*, 2006), there is little doubt that more about those processes should soon be learned.

Acknowledgments

The authors thank Mike Purucker, Susan Macmillan, and Stefan Maus for kindly providing the Magsat Picture in Figure 7 and the material of Figures 6 and 8. Joseph Ribaudo is warmly acknowledged for pointing out mistakes in Figure 3, and eqns [7] and [10] in the first edition of this chapter. This is IPGP contribution 2213.

References

Alexandrescu M, Gibert D, Hulot G, Le Mouël J, and Saracco G (1996) Worldwide wavelet analysis of geomagnetic jerks. *Journal of Geophysical Research* 101: 21975–21994.

Alexandrescu MM, Gibert D, Le Mouël JL, Hulot G, and Saracco G (1999) An estimate of average lower mantle conductivity by wavelet analysis of geomagnetic jerks. *Journal of Geophysical Research* 104: 17735–17745.

Alldredge L (1981) Rectangular harmonic analysis applied to the geomagnetic field. *Journal of Geophysical Research* 86: 3021–3026.

Backus GE (1983) Application of mantle filter theory to the magnetic jerk of 1969. *Geophysical Journal of the Royal Astronomical Society* 74: 713–746.

Backus G (1986) Poloidal and toroidal fields in geomagnetic field modeling. *Reviews of Geophysics* 24: 75–109.

Backus G, Parker R, and Constable C (1996) *Foundations of Geomagnetism*. New York: Cambridge University Press.

Barton C (2002) Survey tracks current position of South magnetic pole. *EOS, Transactions American Geophysical Union* 83(27): 291.

Benton ER and Whaler KA (1983) Rapid diffusion of poloidal geomagnetic field through the weakly conducting mantle: A perturbation solution. *Geophysical Journal of the Royal Astronomical Society* 75: 77–100.

Bloxham J and Jackson A (1992) Time-dependent mapping of the magnetic field at the core–mantle boundary. *Journal of Geophysical Research* 97: 19537–19563.

Bouligand C, Hulot G, Khokhlov A, and Glatzmaier GA (2005) Statistical palaeomagnetic field modelling and dynamo numerical dynamo simulations. *Geophysical Journal International* 161: 603–626.

Buck B and Macaulay VA (eds.) (1991) *Maximum Entropy in Action*. Oxford, UK: Oxford University Press.

Cain JC (2007) POGO (OGO-2, -4 and -6 spacecraft). In: Gubbins D and Herrero-Bervera E (eds.) *Encyclopedia of Geomagnetism and Paleomagnetism*, pp. 828–829. Heidelberg: Springer.

Cain JC, Wang Z, Kluth C, and Schmitz DR (1989) Derivation of a geomagnetic model to n = 63. *Geophysical Journal* 97: 431–441.

Campbell WH, Schiffmacher ER, and Kroehl HW (1989) Global quiet day field variation model WDCA/SQ1. *EOS, Transactions American Geophysical Union* 70(5): 66.

Chambodut A and Mandea M (2005) Evidence for geomagnetic jerks in comprehensive models. *Earth Planets Space* 57: 139–149.

Chambodut A, Panet I, Mandea M, Diament M, Holschneider M, and Jamet O (2005) Wavelet frames: An alternative to spherical harmonic representation of potential fields. *Geophysical Journal International* 163(8): 875–899.

Christensen UR and Olson P (2003) Secular variation in numerical geodynamo models with lateral variations of boundary heat flow. *Physics of the Earth and Planetary Interiors* 138: 39–54.

Christensen U, Olson P, and Glatzmaier G (1999) Numerical modelling of the geodynamo: A systematic parameter study. *Geophysical Journal International* 138: 393–409.

Cohen Y, Alexandrescu M, Hulot G, and Le Mouël J-L (1997) Candidate models for the 1995 revision of IGRF, a worldwide evaluation based on observatory monthly means. *Journal of Geomagnetism and Geoelectricity* 49: 279–290.

Constable CG (1988) Parameter estimation in non-gaussian noise. *Geophysical Journal* 94: 131–142.

Courtillot V, Ducruix J, and Le Mouël J-L (1978) Sur une accélération récente de la variation séculaire du champ magnétique terrestre. *Comptes rendus Hebdomadaires des Séances de l'Académie des Sciences D* 287: 1095–1098.

De Boor C (2001) *Applied Mathematical Sciences, 27: A Practical Guide to Splines*. Berlin: Springer-Verlag.

De Michelis P and Tozzi R (2005) A local Intermittency Measure (LIM) approach to the detection of geomagnetic jerks. *Earth and Planetary Science Letters* 235: 261–272.

De Santis A (1991) Translated origin spherical cap harmonic analysis. *Geophysical Journal International* 106: 253–263.

Dolginov S, Zhuzgov LN, Pushkov NV, Tyurmina LO, and Fryazinov IV (1962) Some results of measurements of the constant geomagnetic field above the USSR from the third artificial Earth satellite. *Geomagnetism and Aeronomy* 2: 877–889.

Dormy E, Valet J, and Courtillot V (2000) Numerical models of the geodynamo and observational constraints. 1:paper number 2000GC000062, 2000.

Friis-Christensen E, Lühr H, and Hulot G (2006) Swarm: A constellation to study the Earth's magnetic field. *Earth, Planets and Space* 58: 351–358.

Haines GV (1985) Spherical cap analysis. *Journal of Geophysical Research* 90: 2583–2591.

Heirtzler J, Allen H, and Wilkinson D (2002) Ever-present South Atlantic Anomaly damages spacecraft. *EOS, Transactions American Geophysical Union* 83(15): 165–172.

Holland P and Welsch R (1977) Robust regression using iteratively reweighted least-squares. *Communications in Statistics-Theory and Methods* 6: 813–827.

Holme R and Bloxham J (1996) The treatment of attitude errors in satellite geomagnetic data. *Physics of the Earth and Planetary Interiors* 98: 221–233.

Holme R and Olsen N (2006) Core–surface flow modelling from high resolution secular variation. *Geophysical Journal International* 166: 518–528.

Holme R, Olsen N, Rother M, and Lühr H (2003) CO2: A CHAMP magnetic field model. In: Reigber C, Lühr H, and Schwintzer P (eds.) *First CHAMP Mission results for Gravity, Magnetic and Atmospheric Studies*, pp. 220–225. Berlin: Springer Verlag.

Holschneider M, Chambodut A, and Mandea M (2003) From global to regional analysis of the magnetic field on the sphere using wavelet frames. *Physics of the Earth and Planetary Interiors* 135(2): 107–124.

Hongre L, Hulot G, and Khokhlov A (1998) An analysis of the geomagnetic field over the past 2000 years. *Physics of the Earth and Planetary Interiors* 106: 311–335.

Hulot G, Eymin C, Langlais B, Mandea M, and Olsen N (2002) Small-scale structure of the geodynamo inferred from Ørsted and Magsat satellite data. *Nature* 416: 620–623.

Hulot G and Le Mouël JL (1994) A statistical approach to the Earth's main magnetic field. *Physics of the Earth and Planetary Interiors* 82: 167–183.

Jackson A (1994) Statistical treatment of crustal magnetization. *Geophysical Journal International* 119: 991–998.

Jackson A (2003) Intense equatorial flux spots on the surface of the Earth's core. *Nature* 424: 760–763.

Jackson A, Jonkers ART, and Walker MR (2000) Four centuries of geomagnetic secular variation from historical records. *Philosophical Transactions of the Royal Society of London A* 358: 957–990.

Kivelson MG and Russell CT (1995) *Introduction to Space Physics*. Cambridge, UK: Cambridge University Press.

Korte M and Holme R (2003) Regularization of spherical cap harmonics. *Geophysical Journal International* 153(1): 253–262.

Kuvshinov AV and Olsen N (2005) 3D modelling of the magnetic field due to ocean flow. In: Reigber C, Lühr H, Schwintzer P, and Wickert J (eds.) *Earth Observation with CHAMP, Results from Three Years in Orbit*. Berlin: Springer-Verlag.

Langel RA (1987) The main field. In: Jacobs JA (ed.) *Geomagnetism*, vol. 1, pp. 249–512. London: Academic Press.

Langel RA and Estes RH (1982) A geomagnetic field spectrum. *Geophysical Research Letters* 9: 250–253.

Langel RA and Estes RH (1985a) Large-scale, near-Earth magnetic fields from external sources and the corresponding induced internal field. *Journal of Geophysical Research* 90: 2487–2494.

Langel RA and Estes RH (1985b) The near-Earth magnetic field at 1980 determined from MAGSAT data. *Journal of Geophysical Research* 90: 2495–2509.

Langel RA, Estes RH, and Mead GD (1982) Some new methods in geomagnetic field modelling applied to the 1960–1980 epoch. *Journal of Geomagnetism and Geoelectricity* 34: 327–349.

Langel RA and Hinze WJ (1998) *The Magnetic Field of the Earth's Lithosphere: The Satellite Perspective*. Cambridge, UK: Cambridge University Press.

Langel RA, Ridgway JR, Sugiura M, and Maezawa K (1988) The geomagnetic field at 1982 from DE-2 and other magnetic field data. *Journal of Geomagnetism and Geoelectricity* 40: 1103–1127.

Langel RA, Sabaka TJ, Baldwin RT, and Conrad JA (1996) The near-Earth magnetic field from magnetospheric and quiet-day ionospheric sources and how it is modeled. *Physics of the Earth and Planetary Interiors* 98: 235–267.

Langlais B, Mandea M, and Ultré-Guérard P (2003) High-resolution magnetic field modeling: Application to MAGSAT and Ørsted data. *Physics of the Earth and Planetary Interiors* 135: 77–91.

Lesur V (2006) Introducing localized constraints in global geomagnetic field modelling. *Earth, Planets and Space* 58: 477–483.

Lesur V, Macmillan S, and Thomson A (2005) A magnetic field model with daily variations of the magnetospheric field and its induced counterpart in 2001. *Geophysical Journal International* 160(1): 79–88 (URL http://www.blackwell-synergy.com/doi/abs/10.1111/j.1365-246X.2004.02479.x.).

Lowes F (1974) Spatial power spectrum of the main geomagnetic field, and extrapolation to the core. *Geophysical Journal of the Royal Astronomical Society* 36: 717–730.

Lowes FJ (1966) Mean-square values on sphere of spherical harmonic vector fields. *Journal of Geophysical Research* 71: 2179.

Lowes FJ (1994) The geomagnetic eccentric dipole: Facts and fallacies. *Geophysical Journal International* 118: 671–679.

Lühr H, Maus S, and Rother M (2002) First *in-situ* observation of night-time *F* region currents with the CHAMP satellite. *Geophysical Research Letters* 29(10): 1489 (doi:10.1029/2001GL013845).

Lühr H, Rother M, Maus S, Mai W, and Cooke D (2003) The diamagnetic effect of the equatorial appleton anomaly: Its characteristics and impact on geomagnetic field modelling. *Geophysical Research Letters* 30(17): 1906 (doi:10.1029/2003GL017407).

Mauersberger P (1956) Das Mittel der Energiedichte des geomagnetischen Hauptfeldes an der Erdoberfläche und seine säkulare Änderung. *Gerlands Beitrage Zur Geophysik* 65: 207–215.

Maus S (2007) CHAMP magnetic mission. In: Gubbins D and Herrero-Bervera E (eds.) *Encyclopedia of Geomagnetism and Paleomagnetism*. Heidelberg: Springer.

Maus S and Lühr H (2005) Signature of the quiet-time magnetospheric magnetic field and its electromagnetic induction in the rotating Earth. *Geophysical Journal International* 162: 755–763.

Maus S and Lühr H (2006) A gravity-driven electric current in the earth's ionosphere identified in champ satellite magnetic measurements. *Geophysical Research Letters* 33: L02812 (doi:10.1029/2005GL024436).

Maus S, Lühr H, Rother M, et al. (2007) Fifth generation lithospheric magnetic field model from CHAMP satellite measurements. *Geochemistry Geophysics Geosystems*, 8, Q05013, (doi:10.1029/2006GC001521).

Maus S, Macmillan S, Chernova T, et al. (2005a) The 10th-Generation International Geomagnetic Reference Field. *Geophysical Journal International* 161: 561–565.

Maus S, Macmillan S, Lowes F, and Bondar T (2005b) Evaluation of candidate geomagnetic field models for the 10-th generation of IGRF. *Earth, Planets, and Space* 57(12): 1173–1181.

Maus S, McLean S, Dater D, et al. (2005c) NGDC/GFZ candidate models for the 10th generation International Geomagnetic Reference Field. *Earth, Planets and Space* 57: 1151–1156.

Maus S, Rother M, Hemant K, et al. (2006a) Earth's crustal magnetic field determined to spherical harmonic degree 90 from CHAMP satellite measurements. *Geophysical Journal International* 164: 319–330 (doi:10.1111/j1365-246x.2005.02833.x.).

Maus S, Rother M, Stolle C, et al. (2006b) Third generation of the Potsdam Magnetic Model of the Earth (POMME). *Geochemistry Geophysics Geosystems* 7: Q07008.

Maus S and Weidelt P (2004) Separating the magnetospheric disturbance magnetic field into external and transient internal contributions using a 1D conductivity model of the Earth. *Geophysical Research Letters* 31: L12614 (doi:10.1029/2004GL020232).

Mayaud PN (1980) *Geophysical Monograph 22: Derivation, Meaning, and Use of Geomagnetic Indices*. Washington, DC: American Geophysical Union.

Mayer C and Maier T (2006) Separating inner and outer Earth's magnetic field from CHAMP satellite measurements by means of vector scaling functions and wavelets. *Geophysical Journal International* 167: 1188–1203.

Mayhew M and Estes R (1983) Equivalent source modeling of the core magnetic field using Magsat data. *Journal of Geomagnetism and Geoelectricity* 35: 119–130.

McLeod MG (1996) Spatial and temporal power spectra of the geomagnetic field. *Journal of Geophysical Research* 101(B2): 2745–2763.

Merrill R, McFadden P, and McElhinny M (1998) *The Magnetic Field of the Earth: Paleomagnetism, the Core, and the Deep Mantle*. San Diego, CA: Academic Press.

Meyers H and Davis WM (1990) A profile of the geomagnetic model users and abusers. *Journal of Geomagnetism and Geoelectricity* 42: 1079–1085.

Neubert T, Mandea M, Hulot G, et al. (2001) Ørsted satellite captures high-precision geomagnetic field data. *EOS, Transactions of the American Geophysical Union* 82(7): 81–88.

Newitt LR, Barton CE, and Bitterly J (1996) Guide for magnetic repeat station surveys. International Association of Geomagnetism and Aeronomy.

Newitt L, Mandea M, McKee L, and Orgeval J (2002) Recent acceleration of the North magnetic pole linked to magnetic jerks. *EOS, Transactions of the American Geophysical Union* 83: 381–388.

O'Brien MS and Parker RL (1994) Regularized geomagnetic field modelling using monopoles. *Geophysical Journal International* 118: 566–578.

Olsen N (1993) The solar cycle variability of lunar and solar daily geomagnetic variations. *Annales Geophysicae* 11: 254–262.

Olsen N (1998) Estimation of *C*-responses (3 h to 720 h) and the electrical conductivity of the mantle beneath Europe. *Geophysical Journal International* 133: 298–308.

Olsen N (2002) A model of the geomagnetic field and its secular variation for epoch 2000 estimated from Ørsted data. *Geophysical Journal International* 149(2): 454–462.

Olsen N (2007) Ørsted. In: Gubbins D and Herrero-Bervera E (eds.) *Encyclopedia of Geomagnetism and Paleomagnetism*, pp. 743–746. Heidelberg: Springer.

Olsen N, Holme R, Hulot G, et al. (2002) Ørsted initial field model. *Geophysical Research Letters* 27: 3607–3610.

Olsen N, Lühr H, Sabaka TJ, et al. (2006) CHAOS – a model of Earth's magnetic field derived from CHAMP, Ørsted, and SAC-C magnetic satellite data. *Geophysical Journal International* 166: 67–75.

Olsen N, Sabaka TJ, and Lowes F (2005) New parameterization of external and induced fields in geomagnetic field modeling, and a candidate model for IGRF 2005. *Earth, Planets and Space* 57: 1141–1149.

Olsen N, Tøffner-Clausen L, Sabaka TJ, *et al.* (2003) Calibration of the Ørsted vector magnetometer. *Earth, Planets and Space* 55: 11–18.

Parker RL and Shure L (1982) Efficient modeling of the Earth's magnetic field with harmonic splines. *Geophysical Research Letters* 9: 812–815.

Purucker ME (2007) Magsat. In: Gubbins D and Herrero-Bervera E (eds.) *Encyclopedia of Geomagnetism and Paleomagnetism.* pp. 673–674. Heidelberg: Springer.

Purucker M, Langlais B, Olsen N, Hulot G, and Mandea M (2002) The southern edge of cratonic North America: Evidence from new satellite magnetometer observations. *Geophysical Research Letters* 29(15): 8000.

Richmond AD (1995) Ionospheric electrodynamics using magnetic Apex coordinates. *Journal of Geomagnetism and Geoelectricity* 47: 191–212.

Sabaka TJ and Baldwin RT (1993) Modeling the Sq magnetic field from POGO and Magsat satellite and contemporaneous hourly observatory data: Phase I. *Contract Report HSTX/G&G9302.*

Sabaka TJ, Olsen N, and Langel RA (2002) A comprehensive model of the quiet-time near-Earth magnetic field: Phase 3. *Geophysical Journal International* 151: 32–68.

Sabaka TJ, Olsen N, and Purucker M (2004) Extending comprehensive models of the Earth's magnetic field with Ørsted and CHAMP data. *Geophysical Journal International* 159: 521–547 (doi:10.1111/j.1365–246X.2004.02421.x.).

Schumaker LL (1981) *Spline Functions: Basic Theory.* New York: John Wiley & Sons.

Shure L, Parker RL, and Backus GE (1982) Harmonic splines for geomagnetic modelling. *Physics of the Earth and Planetary Interiors* 28: 215–229.

Siebert M and Meyer J (1996) Geomagnetic activity indices. In: Dieminger W, Hartman GK, and Leitinger R (eds.) *The Upper Atmosphere*, pp. 887–911. Berlin: Springer.

Stolle C, Lühr H, Rother M, and Balasis G (2006) Magnetic signatures of equatorial spread *F*, as observed by the CHAMP satellite. *Journal of Geophysical Research* 111: A02304 (doi:10.1029/2005JA011184).

Thébault E (2006) Global lithospheric magnetic field modelling by successive regional analysis. *Earth, Planets and Space* 58: 485–495.

Thébault E, Schott JJ, and Mandea M (2005) Revised Spherical Cap Harmonic Analysis (R-SCHA): Validation and Properties. *Journal of Geophysical Research* 111: B01102.

Thomson A and Lesur V (2006) An improved geomagnetic data selection algorithm for global geomagnetic field modelling. *Geophysical Journal International.* 169: 951–963 (doi:10.1111/j.1365-246X.2007.03354.x).

Voorhies CV (2004) Narrow-scale flow and a weak field by the top of Earth's core: Evidence from ørsted, magsat, and secular variation. *Journal of Geophysical Research* 109: B03106 (doi:10.1029/2003JB002833).

Wardinsky I and Holme R (2006) A method to refine main field modeling. In: *Proceedings of the First International Swarm Science Meeting*, volume WPP-261. ESA 2006.

Winch DE (1981) Spherical harmonic analysis of geomagnetic tides. *Philosophical Transactions of the Royal Society of London A* 303: 1–104.

Relevant Websites

http://isdc.gfz-potsdam.de – CHAMP Project, Information System and Data Center (ISDC), GFZ Potsdam.

http://www.intermagnet.org – INTERMAGNET Home Page.

http://isgi.cetp.ipsl.fr – ISGI Home page.

http://www.ngdc.noaa.gov – NOAA's Geophysical Data Center; Working Group V-MOD, IAGA.

http://www.spacecenter.dk – Scientific Data and Models, Danish National Space Center.

http://web.dmi.dk – WDC C1 for Geomagnetism, Solar–Terrestrial Physics Division, Danish Meteorological Institute.

3 Magnetospheric Contributions to the Terrestrial Magnetic Field

W. Baumjohann and R. Nakamura, Space Research Institute, Austrian Academy of Sciences, Graz, Austria

3.1 Introduction

The Earth's magnetic field is created and governed by processes and material in the Earth's interior. This field is not restricted to the inside, the surface, or the atmosphere of the Earth, but reaches far above the Earth into space. If that space were empty or only populated with neutral gases, there would be no consequences. However, that space is not a vacuum but, starting at a height of about 100 km, is filled with ionized gas.

The constituents of this ionized gas, a plasma of positively charged ions and negatively charged electrons, are not immobile but rather move around under the influence of externally applied and internally generated forces. This motion of charge carriers often results in a significant current flow and thus in the generation of magnetic field disturbances which can be of the same magnitude as the field generated inside the Earth at that altitude. If the spatial scale of such a current system is large enough, or if it flows close to the Earth's surface, as do for ionospheric currents, it can generate significant magnetic variations at the Earth surface.

In this chapter, we will describe the main sources of external magnetic field contributions. Nearly all of them result from an interplay between the magnetic field of the Earth and that of the Sun. They are highly variable, some changing on the scale of seconds, others on the scale of days, and typical disturbance amplitudes on the ground range between a few and some hundred nanotesla.

3.2 Geophysical Plasmas

A plasma is a gas of charged particles, which consists of equal numbers of free positive and negative charge carriers. Having roughly the same number of charges with different signs in the same volume element guarantees that the plasma behaves quasi-neutral. On average a plasma looks electrically neutral to the outside, since the randomly distributed particle electric charge fields mutually cancel. However, because of its sensitivity to electric and magnetic fields and its ability to carry electric currents and thus to generate magnetic fields, this fourth state of matter behaves quite different from a neutral gas.

Similar to a gaseous medium, the charged plasma particles are essentially free particles. Since the particles in a plasma have to overcome the Coulomb coupling with their neighbors, they must have thermal energies above some 10^5 K. Thus, a typical plasma is a hot and highly ionized gas. While only a few natural plasmas, such as flames or lightning strokes, can be found near the Earth's surface or below the ionosphere, plasmas are abundant in the universe. More than 99% of all normal matter (baryonic matter, not including dark matter) is in the plasma state.

Extraterrestrial plasmas have a wide spread in their characteristic parameters like density, temperature, and magnetic field. Even in the Earth's neighborhood, there are quite a number of different geophysical plasmas.

3.2.1 Solar Wind

The Sun emits a highly conducting plasma into interplanetary space as a result of the supersonic expansion of the solar corona. This plasma is called the solar wind. It flows with supersonic speed of about $500 \, \mathrm{km \, s^{-1}}$ and consists mainly of electrons and protons, with an admixture of 5% helium ions. Because of the high conductivity, the solar magnetic field is 'frozen' in the plasma (as in a superconductor, see below) and drawn outward by the expanding solar wind. Typical values for electron density and temperature in the solar wind near the Earth are $5 \, \mathrm{cm^{-3}}$ and 10^5 K, respectively. The interplanetary magnetic field strength is of the order of $5-10 \, \mathrm{nT}$ near the Earth's orbit.

When the solar wind impinges on the Earth's dipolar magnetic field, it cannot simply penetrate it,

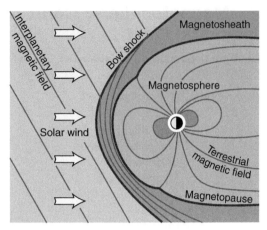

Figure 1 Solar wind interaction with the terrestrial magnetic field. Adapted from Baumjohann W and Treumann RA (1996) *Basic Space Plasma Physics*. London: Imperial College Press, with permission from Imperial College Press.

but is slowed down and, to a large extent, deflected around it. Since the solar wind hits the obstacle with supersonic speed, a bow shock wave is generated (see **Figure 1**), where the plasma is slowed down and a substantial fraction of the particles' kinetic energy is converted into thermal energy. The region of thermalized subsonic plasma behind the bow shock is called the magnetosheath. Its plasma is denser and hotter than the solar wind plasma and the magnetic field strength has higher values in this region.

3.2.2 Magnetosphere

The shocked solar wind plasma in the magnetosheath cannot easily penetrate the terrestrial magnetic field but is mostly deflected around it. This is a consequence of the fact that the interplanetary magnetic field lines cannot penetrate the terrestrial field lines and that the solar wind particles cannot leave the interplanetary field lines due to the aforementioned frozen-in characteristic of a highly conducting plasma.

The boundary separating the two different regions is called magnetopause and the cavity generated by the terrestrial field has been named magnetosphere (see **Figures 1** and **2**). The kinetic pressure of the solar wind plasma distorts the outer part of the terrestrial dipolar field. On the dayside it compresses the field, while the nightside magnetic field is stretched out into a long magnetotail which reaches far beyond lunar orbit.

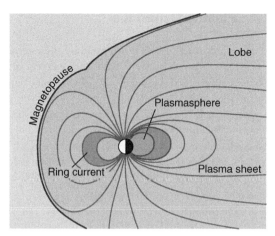

Figure 2 Plasma regions in the Earth's magnetosphere. Note that 'ring current' and 'plasmasphere' partially overlap in reality. Adapted from Baumjohann W and Treumann RA (1996) *Basic Space Plasma Physics*. London: Imperial College Press, with permission from Imperial College Press.

The plasma in the magnetosphere consists mainly of electrons and protons. The sources of these particles are the solar wind and the terrestrial ionosphere. In addition, there are minor fractions of He^+ and O^+ ions of ionospheric origin (more prominent at lower altitudes) and some He^{2+} ions originating from the Sun. However, the plasma inside the magnetosphere is not evenly distributed, but is grouped into different regions with quite different densities and temperatures. **Figure 2** depicts the topography of some of these regions.

The ring current lies on dipolar field lines between about 4 and $6 R_E$ (1 Earth radius $(R_E) = 6371$ km). It consists of energetic electrons and ions which move along the field lines and oscillate back and forth between the two hemispheres (see below). Typical electron densities and temperatures in the ring current are 1 cm^{-3} and 5×10^7 K. The magnetic field strength in this region is a few hundred nanotesla.

Most of the magnetotail plasma is concentrated around the tail midplane in an about $5-10 R_E$ thick plasma sheet. Near the Earth, it reaches down to the high-latitude auroral ionosphere along the field lines. Average electron densities and temperatures in the plasma sheet are 0.5 cm^{-3} and 5×10^6 K, with magnetic fields of $10-20$ nT.

The outer part of the magnetotail is called the lobe. It is threaded by magnetic field lines originating in the polar caps and contains a highly rarefied plasma. Typical values for the electron density, temperature, and the magnetic field strength are 10^{-2} cm^{-3}, 5×10^5 K, and 30 nT, respectively.

3.2.3 Ionosphere

The solar ultraviolet light impinging on the Earth's atmosphere ionizes a fraction of the neutral atmosphere. At altitudes above 80 km, collisions are too infrequent to result in rapid recombination and a permanent ionized population called the ionosphere is formed. Typical electron densities and temperatures in the mid-latitude ionosphere are 10^5 cm^{-3} and 10^3 K. The magnetic field strength is of the order of 10^4 nT.

The ionosphere extends to altitudes of about a thousand kilometers and, at low and mid-latitudes, gradually merges into the plasmasphere. As depicted in **Figure 2**, the plasmasphere is a torus-shaped volume inside the ring current. It contains a cool but dense plasma of ionospheric origin, which corotates with the Earth. In the equatorial plane, the plasmasphere extends out to about $4 R_E$, where the density drops down sharply to about 1 cm^{-3}. This boundary is called the plasmapause.

At high latitudes, plasma sheet electrons can precipitate along magnetic field lines down to ionospheric altitudes, where they collide with and ionize neutral atmosphere particles. As a byproduct, photons emitted by this process create the polar light, the aurora. These auroras are typically observed inside the auroral oval, which is a $5-10°$ wide belt around $70°$ northern or southern magnetic latitude, containing the 'footprints' of those field lines which thread the plasma sheet.

3.2.4 Currents

The plasmas discussed in the last section are usually not stationary but move under the influence of external forces. Sometimes ions and electrons move together, like in the solar wind. But in other plasma regions, ions and electrons often move in different directions, creating an electric current. Such currents create magnetic fields, which distort the Earth's internal field, most intensely at higher altitudes.

Actually, the distortion of the internal dipole field into the typical shape of the magnetosphere is accompanied by electrical currents. As schematically shown in **Figure 3**, the compression of the internal magnetic field on the dayside is associated with current flow across the magnetopause surface, the magnetopause

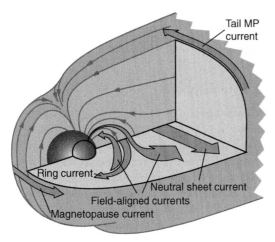

Figure 3 Magnetospheric current systems. MP current, magnetopause current. Adapted from Baumjohann W and Treumann RA (1996) *Basic Space Plasma Physics*. London: Imperial College Press, with permission from Imperial College Press.

current. The tail-like field of the nightside magnetosphere is accompanied by the current flowing on the tail magnetopause surface and the cross-tail neutral sheet current in the central plasma sheet, both of which are connected and form a Θ-like current system, if seen from along the Earth–Sun line.

Another large-scale current system, which mainly influences the configuration of the inner magnetosphere, is the ring current. The ring current flows around the Earth in a westward direction at radial distances of several Earth radii and is carried by trapped energetic particles, which oscillate back and forth along the field lines. In addition to their bouncing motion, these particles drift around the Earth. Since the protons drift westward while the electrons move in the eastward direction, this constitutes a net charge transport and thus a current.

A number of current systems exist in the conducting layers of the Earth's ionosphere, at altitudes of 100–150 km. Most notable are the auroral electrojets inside the auroral oval, the Sq currents in the dayside mid-latitude ionosphere, and the equatorial electrojet near the magnetic equator.

In addition to the currents that flow across the magnetic field lines, currents also flow along magnetic field lines. As shown in **Figure 3**, the field-aligned currents connect the magnetospheric currents to those flowing in the polar ionosphere. The field-aligned currents are mainly carried by electrons and are essential for the exchange of energy and momentum between these regions.

3.3 Plasma Dynamics

The dynamics of a plasma is governed by the interaction of the charge carriers with the electric and magnetic fields. If all the fields were of external origin, the physics would be relatively simple. However, as the particles move around, they may create local space charge concentrations and thus electric fields. Moreover, their motion can also generate electric currents and thus magnetic fields. These internal fields and their feedback onto the motion of the plasma particles make plasma physics complex.

In general, the dynamics of a plasma can be described by solving the equation of motion for each individual particle. Since the electric and magnetic fields appearing in each equation include the internal fields generated by every other moving particle, all equations are coupled and have to be solved simultaneously. Such a full solution is not only too difficult to obtain, but also of no practical use, since most of the time one is interested in knowing average quantities like density and temperature rather than the individual velocity of each particle. Therefore, one usually makes certain approximations suitable to the problem studied. For studying the macroscopic interaction between the solar wind and the Earth's magnetosphere, two approaches are most useful (the most developed theoretical approach, the so-called kinetic theory of plasmas is typically needed for microphysical aspects of space plasma physics).

The simpler approach is the single-particle motion or guiding-center description. It describes the motion of a particle under the influence of external electric and magnetic fields. This approach neglects the collective behavior of a plasma, but is useful when studying a very low-density plasma, threaded by strong magnetic fields, like that found in the ring current.

The magnetohydrodynamic approach, on the other hand, neglects all single particle aspects, but includes collective effects. The plasma is treated as a single conducting fluid with macroscopic variables, like average density, velocity, and temperature. The approach assumes that the plasma is able to maintain local equilibria and is suitable to study low-frequency wave phenomena in highly conducting fluids immersed in magnetic fields.

3.3.1 Single-Particle Motion

In a situation where the charged particles do not directly interact with each other and where they do not affect the external magnetic field significantly, the motion of each individual particle can be treated independently. This single-particle approach is only valid in very rarefied plasmas where collective effects are negligible. Furthermore, the external magnetic field must be much greater than the magnetic field produced by the electric current due to the charged-particle motion.

The equation of motion for a particle of charge q under the action of the Coulomb and Lorentz forces can be written as

$$m\frac{d\mathbf{v}}{dt} = q(\mathbf{E} + \mathbf{v} \times \mathbf{B}) \qquad [1]$$

where m represents the particle mass and $\mathbf{v}$ the particle velocity. Under the absence of an electric field and a homogeneous magnetic field, eqn [1] describes a circular orbit of the particle around the magnetic field, with the sense of rotation depending on the sign of the charge. The center of the orbit is called the guiding center. The gyroradius of the particle orbit increases with the particle's momentum and decreases for stronger magnetic fields. A possible constant velocity of the particle parallel to the magnetic field will make the actual trajectory of the particle three dimensional and look like a helix (see **Figure 4**).

Taking the electric field into consideration will result in a drift of the particle superimposed onto its gyratory motion. Since, due to the high mobility of electrons, parallel electric fields can typically not be maintained in geophysical plasmas. Solving eqn [1] yields

$$\mathbf{v}_E = \frac{\mathbf{E} \times \mathbf{B}}{B^2} \qquad [2]$$

The $\mathbf{E} \times \mathbf{B}$ drift is independent of the sign of the charge and thus electrons and ions move together with the same speed in the same direction.

Figure 5 shows the acceleration and deceleration effect of a perpendicular electric field and explains the $\mathbf{E} \times \mathbf{B}$ drift in an intuitive way. An ion is accelerated in the direction of the electric field, thereby increasing its gyroradius. But it is decelerated during

the second half of its gyratory orbit, now with a decreasing gyroradius. The changing gyroradii shift the position of the guiding center in the $\mathbf{E} \times \mathbf{B}$ direction. The electrons are accelerated when moving antiparallel to the electric field and decelerated when moving parallel. But since their sense of gyration is also opposite, their guiding centers drift into the same direction.

Up to now we have assumed that the magnetic field is homogeneous. This is definitely not the case in the magnetosphere, where the magnetic field has gradients and field lines are curved. This inhomogeneity of the magnetic field leads to a 'magnetic' drift of charged particles. As visualized in **Figure 6**, in a magnetic field configuration with a gradient in field strength, the gyroradius of a particle decreases in the upward direction and thus the gyroradius of a particle will be larger at the bottom of the orbit than during the top half. As a result, ions and electrons drift into opposite directions, perpendicular to both $\mathbf{B}$ and ∇B. Since ions and electrons gyrate in the opposite sense, ions and electrons also drift in opposite directions. The gradient drift

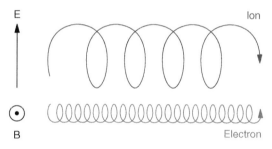

Figure 5   Particle drifts due to an electric field. Adapted from Baumjohann W and Treumann RA (1996) *Basic Space Plasma Physics*. London: Imperial College Press, with permission from Imperial College Press.

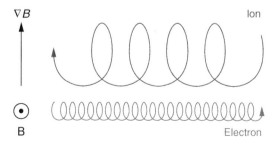

Figure 6   Particle drifts due to a magnetic field gradient. Adapted from Baumjohann W and Treumann RA (1996) *Basic Space Plasma Physics*. London: Imperial College Press, with permission from Imperial College Press.

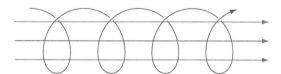

Figure 4   Ion orbit in a uniform magnetic field. Adapted from Baumjohann W and Treumann RA (1996) *Basic Space Plasma Physics*. London: Imperial College Press, with permission from Imperial College Press.

velocity is proportional to the perpendicular gyratory energy of the particle, $W_\perp = (1/2)mv_\perp^2$: more energetic particles drift faster, since they have a larger gyroradius and experience more inhomogeneity of the field. The opposite drift directions of electrons and ions lead to a transverse current.

The 'gradient' drift is only one component of the particle drift in an inhomogeneous magnetic field. When the field lines are curved, a 'curvature' drift appears. Due to their parallel velocity, the particles experience a centrifugal force. The curvature drift velocity is proportional to the parallel particle energy and perpendicular to the magnetic field and its curvature. It again creates a transverse current since ion and electron drifts have opposite signs.

In a cylindrically symmetric field, like in a dipole field, gradient and curvature drifts can be combined to

$$\mathbf{v}_B = \mathbf{v}_\nabla + \mathbf{v}_R = \left(v_\parallel^2 + \frac{1}{2}v_\perp^2\right)\frac{\mathbf{B} \times \nabla B}{\omega_g B^2} \qquad [3]$$

where $\mathbf{v}_\nabla$ and $\mathbf{v}_R$ are the gradient and curvature drift velocity, ω_g gives the gyrofrequency, and the subscripts $\perp$ and $\parallel$ denote components perpendicular and parallel to the ambient background field, respectively. The transverse current associated with this full magnetic drift creates the magnetospheric ring current, mentioned above and further detailed below.

3.3.2 Trapped Particles

In slowly varying magnetic and electric fields, not only the total energy of a particle, $W = W_\parallel + W_\perp$, is constant, but also the magnetic moment of a particle, $\mu = W_\perp/B$, does not change with time. Quantities like μ are called adiabatic invariants and are not absolute constants like total energy or total momentum, but change only very slowly.

Since the magnetic moment is invariant and the total energy is a constant of motion, only the ratio between perpendicular and parallel energy increases or decreases along the guiding center trajectory. In a converging magnetic field geometry, a particle moving into regions of stronger fields will have its transverse energy increasing at the expense of its parallel energy. If there is a point along the field line where all of the particle energy is in $W_\perp$, the particle cannot penetrate any further along the field line into the stronger field region. Actually, it does not stop, but is pushed back by the parallel

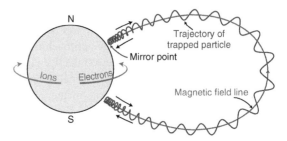

Figure 7 Trajectories of particles trapped on closed dipolar field lines. Adapted from Baumjohann W and Treumann RA (1996) *Basic Space Plasma Physics*. London: Imperial College Press, with permission from Imperial College Press.

component of the gradient force, the so-called mirror force, $-\mu\nabla_\parallel B$, from this mirror point.

A dipole magnetic field has a field strength minimum at the equator and converging field lines in both hemispheres. In such a configuration, particles will be trapped and bounce back and forth between their mirror points in the Northern and Southern Hemispheres (see **Figure 7**). The particles do not only gyrate and bounce, but undergo a slow azimuthal drift due to the combined effect of the gradient and curvature of the dipole magnetic field as described in eqn [3]. The ions drift westward while the electrons move eastward around the Earth. It is the current associated with this drift that constitutes the ring current.

3.3.3 Collisions and Conductivity

So far only the motion of single particles in external and slowly variable electromagnetic fields has been considered, but any interaction between the particles has been neglected. Interaction in plasmas is, however, unavoidable and collective effects constitute the very nature of plasma physics. The simplest kind of interaction between particles is a direct collision. The partially ionized plasma of the terrestrial ionosphere is a good example for such interactions. Here collisions between charged and neutral particles create electrical resistivity and current flow.

In the presence of collisions a collisional term has to be to added to the equation of motion [1] for a charged particle under the action of the Coulomb and Lorentz forces. Assuming all collision partners are at rest, then

$$m\frac{d\mathbf{v}}{dt} = q\left(\mathbf{E} + \mathbf{v} \times \mathbf{B}\right) - m\nu_c\mathbf{v} \qquad [4]$$

The collisional term on the right-hand side describes the momentum lost through collisions occurring at a frequency ν_c. It is often called frictional term since it impedes motion.

The friction term introduces a differential motion between electrons and ions and thus a current, even in homogeneous magnetic fields. In fact, the above equation reduces to a generalized Ohm's law

$$\mathbf{j} = \sigma(\mathbf{E} + \mathbf{v} \times \mathbf{B}) \qquad [5]$$

which is valid in all geophysical plasmas where the typical collision frequencies are low, and σ is the plasma conductivity.

While treating the plasma conductivity as a scalar is warranted in the dilute magnetospheric plasma, there is one place where we have to take the anisotropy introduced by the presence of a strong magnetic field into account. This is the lower part, the so-called E-region, of the partially ionized terrestrial ionosphere, at about 100–130 km altitude, where abundant collisions between the ionized and the neutral part of the upper atmosphere might even interrupt the cyclotron motion of electrons and/or ions, leading to an anisotropic conductivity tensor and a different form of Ohm's law:

$$\mathbf{j} = \sigma_{\|}\mathbf{E}_{\|} + \sigma_P\mathbf{E}_{\perp} - \sigma_H(\mathbf{E}_{\perp} \times \mathbf{B})/B \qquad [6]$$

The Hall conductivity, σ_H, determines the Hall current in the direction perpendicular to both the electric and magnetic field. The Hall conductivity maximizes near 100 km altitude, where the ions collide so frequently with the neutrals that they are essentially at rest, while the electrons already undergo a somewhat impeded $\mathbf{E} \times \mathbf{B}$-drift. The Pedersen conductivity, σ_P, governs the Pedersen current in the direction of that part of the electric field, $\mathbf{E}_{\perp}$, which is transverse to the magnetic field. The Pedersen conductivity maximizes near a height of 125 km, since here the ions are scattered in the direction of the electric field before they can start to gyrate about the magnetic field. The element $\sigma_{\|}$ is called parallel conductivity since it governs the magnetic field-aligned current driven by the parallel electric field component, $E_{\|}$.

3.3.4 Convection and Merging

While collisions play an important role in the ionosphere, most space plasmas are essentially collisionless. Hence, the conductivity in the magnetospheric or in the solar wind plasma is near infinite.

As in a superconductor, magnetic field lines are frozen in the plasma and both move together under the action of external forces. In particular, under the influence of an external electric field, the so-called flux tubes, bundles of field lines filled with plasma, simply drift following eqn [2]. On the other hand, if forces are exerted on the magnetic field lines leading to a motion of the flux tubes, an electric field will be generated. The latter is often called convection electric field.

However, there is an exception. Under certain conditions, especially in the thin and intense current sheets of the magnetopause and the magnetotail neutral sheet, strong plasma waves or inertial effects may substitute collisions and lower the conductivity to a finite value. In this case, the magnetic field lines can diffuse through the plasma. This rarely has major consequences, except for a situation as depicted in **Figure 8**.

Consider a magnetic topology with antiparallel field lines frozen into the plasma, as depicted in the left-hand diagram of **Figure 8**. If the flux tubes are stagnant and do not move, nothing will happen. However, when plasma and field lines on both sides move toward each other, the situation may change. When the conductivity becomes finite in a small volume of space, the magnetic field can vanish due to diffusion at a particular point. This results in the X-type configuration shown in the middle panel of **Figure 8**, with the magnetic field being zero at the center of the X, the magnetic neutral point. The result will be the situation depicted on the right-hand side of **Figure 8**. Plasma and field lines are being transported toward the neutral point from either side. At the neutral point the antiparallel field lines are cut into halves and the field line halves from one side are reconnected with those from the other side. The merged field lines are then expelled from the neutral point. The merged field lines will be populated by a mixture of plasma from both sides.

Figure 8 Magnetic field line reconnection.

3.4   Low- and Mid-Latitude Currents

The ions and, to a lesser degree, also the electrons in the ionospheric E-region are coupled by collisions to the neutral components of the upper atmosphere and follow their dynamics. Atmospheric winds and tidal oscillations of the atmosphere force the E-region ion component to move across the magnetic field lines, while the electrons move much slower at right angles to both the field and the neutral wind. The relative movement constitutes an electric current and the separation of charge produces an electric field, which in turn affects the current. Because of this, the E-region bears the name dynamo layer, the generator of which is the atmospheric wind motion. This wind-driven dynamo causes two current systems in the equatorial and mid-latitude ionosphere whose 'external' magnetic variations alter the geomagnetic field measured on the Earth's surface. A third current system results from electric and magnetic drifts of magnetospheric particles, the ring current. It is concentrated in the equatorial region of the Earth's magnetosphere.

3.4.1   Sq Current

The relation between current, conductivity, electric field, and neutral winds can be seen by replacing $\mathbf{E}_\perp$ with $\mathbf{E}_\perp + \mathbf{v}_n \times \mathbf{B}$ in the Ohm's law given above. For mid- and low-latitude dynamo currents, the dominant driving force for the current is actually the $\mathbf{E} \times \mathbf{B}$ field induced by the motion of ions, which are coupled to the neutral atmosphere via collisions and thus move with the neutral wind, across the magnetic field. (For auroral oval current systems discussed later, the neutral wind term is usually much smaller than the electric field term and can be neglected.)

The most important dynamo effect at mid-latitudes is the daily variation of the atmospheric motion caused by the tides of the atmosphere, that is, the diurnal and semi-diurnal oscillations, which are excited by the heating of the atmosphere due to solar radiation. The current system created by this tidal motion of the atmosphere is called solar quiet or Sq current. This current system creates daily magnetic variations, records of which are obtained at many different magnetic observatories distributed across the globe. These recordings can be used to construct the Sq current system. More sophisticated methods use measured wind patterns, conductivities, and disturbance magnetic fields and calculate electric

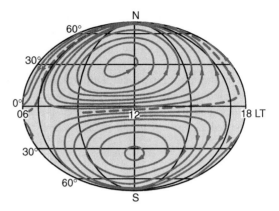

Figure 9   Dayside view of the Sq current system. Adapted from Baumjohann W and Treumann RA (1996) *Basic Space Plasma Physics*. London: Imperial College Press, with permission from Imperial College Press.

fields and currents based on Ohm's and Biot–Savart's laws.

Figure 9 presents a global view of the average Sq current system from above the terrestrial ionosphere: the lines give the direction of the current while the distance between the lines is inversely proportional to the height-integrated current density. The Sq currents form two vortices, one in the Northern and the other in the Southern Hemisphere, which touch each other at the geomagnetic equator. In accordance with the day–night contrast in the low- and mid-latitude E-region conductivities, the Sq currents are concentrated on the dayside.

3.4.2   Equatorial Electrojet

At the geomagnetic equator, the Sq current vortices of the Southern and Northern Hemispheres touch each other and form an extended nearly jet-like current in the ionosphere, the equatorial electrojet. However, the electrojet would not be so strong if it were formed only by the concentration of the Sq current. The special geometry of the magnetic field at the equator together with the nearly perpendicular incidence of solar radiation cause an equatorial enhancement in the effective conductivity which leads to an amplification of the jet current.

Since the magnetic field lines in the equatorial ionosphere are directed northward and parallel to the Earth's surface, the eastward ionospheric electric field drives an eastward Sq Pedersen current and a Sq Hall current, which flows vertically downward at the equator. As shown in **Figure 10**, the latter causes a charge separation in the equatorial ionosphere with negative

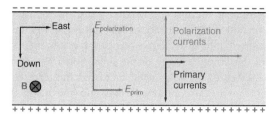

Figure 10 Eastward current enhancement at the magnetic equator.

charges accumulating on the top boundary and positive charges accumulating at the bottom of the highly conducting layer. This space charge distribution creates a secondary polarization electric field, directed vertically upward. The polarization electric field drives a vertical Pedersen current opposing the Hall current until it compensates it. Since the Hall conductivity is typically about 4 times higher than the Pedersen conductivity, the polarization field must also be 4 times stronger than the primary electric field. Moreover, the polarization electric field generates a secondary Hall current component flowing eastward, about 16 times stronger than the primary eastward Pedersen current, thus explaining the amplification of the equatorial electrojet current above the equator.

The strong horizontal jet current causes a magnetic field disturbance which weakens the horizontal component of the terrestrial magnetic field at the Earth's surface over a distance of about 600 km across the equator (similar to the effect of the ring current field; see below). Typical disturbance fields near the noon magnetic equator are of the order of 50–100 nT.

3.4.3 Ring Current

The westward drift of trapped ions and the eastward drift of trapped electrons around the Earth, depicted in **Figure 7**, represent a giant current loop of 1–10 MA, that can significantly alter the terrestrial field even at the Earth's surface.

Applying the magnetic drift velocity given in eqn [3] to the Earth's dipole field, one can calculate the current density caused by n particles with energy W circulating around the Earth at a certain radial distance or particular L-shell

$$j_d = nev_B = \frac{3L^2 nW}{B_E R_E} \qquad [7]$$

where L is measured in R_E but is dimensionless, B_E is the equatorial magnetic field on the Earth's surface,

and j_d results as an azimuthal current flowing in the westward direction.

Integrating over all energies, applying Biot–Savart's law, and then integrating over all L-shells, several symmetries in the equations lead to the simple expression

$$\Delta B_d = -\frac{\mu_0}{4\pi} \frac{3 U_R}{B_E R_E^3} \qquad [8]$$

field disturbance at the Earth's center, where U_R is the total energy of all ring current particles. The minus sign accounts for the fact that the disturbance field of the westward ring current is directed opposite the terrestrial dipole magnetic field.

The total magnetic field perturbation caused by the ring current must also include the diamagnetic contribution due to the cyclotron motion of the ring current particles. Again, symmetries result in a simple expression

$$\Delta B_\mu = \frac{\mu_0}{4\pi} \frac{U_R}{B_E R_E^3} \qquad [9]$$

This disturbance adds to the terrestrial dipole field, since the Earth's dipole moment and the magnetic moments of the ring current particles are co-aligned. The total magnetic field depression caused by the ring current, $\Delta B_R = \Delta B_d + \Delta B_\mu$, at the Earth's center is

$$\Delta B_R = -\frac{\mu_0}{2\pi} \frac{U_R}{B_E R_E^3} \qquad [10]$$

This is the famous Dessler–Sckopke–Parker relation, which directly relates the total energy contained in the ring current to the magnetic variation measured on the Earth's surface.

3.4.4 Storms and Sudden Commencements

The ring current and its associated disturbance field is not a stationary feature. At times more particles than usual are injected from the magnetotail into the ring current, mainly by an enhanced duskward solar wind electric field induced into the magnetotail. This way the total energy of the ring current is increased and the additional depression of the surface magnetic field can clearly be seen in near-equatorial magnetograms, as shown in **Figure 11**. For about 1 day, the equatorial terrestrial field was depressed by more than 150 nT. Strong depressions of the terrestrial field, up to 2–3% of the total surface field in extreme cases, have been noticed in magnetograms long

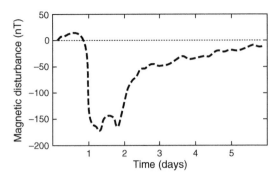

Figure 11   Magnetic field variation during a magnetic storm. Adapted from Baumjohann W and Treumann RA (1996) *Basic Space Plasma Physics*. London: Imperial College Press, with permission from Imperial College Press.

before one knew about the ring current and have been called magnetic storms.

A magnetic storm has two distinct phases. For some hours or days, an enhanced electric field injects more and more particles into the inner magnetosphere, building up the strong storm-time ring current and the associated magnetic disturbance field. After a day or two, the electric field amplitude and the rate of injection return to the normal level. The disturbance field starts to recover, since the ring current loses more and more storm-time particles. This recovery phase typically lasts several days.

The depression of the terrestrial dipole field given in eqn [10] is reflected in the Dst index. This index represents the average disturbance field at the Earth's equator and is calculated on the basis of hourly averages of the northward horizontal component recorded at four low-latitude observatories – Honolulu, San Juan, Hermanus, and Kakioka. All four observatories are 20–30° away from the dipole equator to minimize equatorial electrojet effects and are about evenly distributed in local time (longitude).

At each observatory, a magnetic perturbation amplitude is calculated by subtracting from the hourly averages a quiet time reference level and the Sq field, both of which vary with local time. All four magnetic disturbances are then averaged to further reduce local time effects and multiplied with the averages of the cosines of the observatories' dipole latitudes, to obtain the value of the ring current field at the dipole equator.

Magnetograms like in **Figure 11** often also show a positive excursion of the horizontal field magnitude, right at the beginning of the storm. This excursion is the magnetic signature of the solar wind impinging faster than usual onto the magnetopause. The

position of the dayside magnetopause is essentially determined as the surface of equilibrium between the magnetic pressure of the terrestrial magnetic field and the kinetic energy of the solar wind. Whenever the speed of the solar wind increases, the terrestrial field has to be compressed and thus the magnetopause has to recede to a new equilibrium position. If such a sudden compression of the dayside magnetospheric field occurs at the beginning of a magnetic storm, it is called storm sudden commencement (SSC), whereas when it is not followed by a storm, it is called sudden impulse (SI).

3.5   High-Latitude Currents

Intense ionospheric current systems are also flowing in the high-latitude ionosphere. However, in this region, the magnetic field lines are oriented approximately perpendicular to the ionospheric layers and so-called field-aligned currents connect the ionospheric currents to those flowing in the magnetosphere. Hence, the electrodynamics in the high-latitude E-region is coupled and even governed by the dynamics of the magnetosphere at large.

3.5.1   Magnetospheric Convection

The concurrent drift of plasma and field lines as one entity is called convection. Due to the infinite conductivity, the electric field is zero in the frame of reference moving with the plasma at a velocity $\mathbf{v}_c$. However, according to eqn [2], an observer in the Earth's fixed frame of reference will measure a convection electric field

$$\mathbf{E}_c = -\mathbf{v}_c \times \mathbf{B} \qquad [11]$$

Hence, the flow of the magnetized solar wind around the magnetosphere represents an electric field in the Earth's frame of reference. Since the solar wind cannot penetrate the magnetopause, this electric field cannot directly penetrate into the magnetosphere. However, when the interplanetary magnetic field has a southward component, the northward-directed terrestrial field lines at the dayside magnetopause can merge with the interplanetary magnetic field.

As depicted in **Figure 12**, when a southward-directed interplanetary field line encounters the magnetopause, it can merge with a closed terrestrial field line, which has both 'foot points' on the Earth. The merged field lines will split into two open field

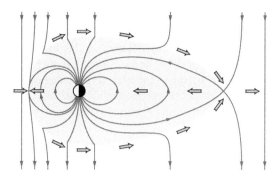

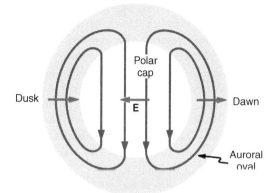

Figure 12 Reconnection and convection cycle in the magnetosphere. Adapted from Baumjohann W and Treumann RA (1996) *Basic Space Plasma Physics*. London: Imperial College Press, with permission from Imperial College Press.

Figure 13 Convection and electric field in the high-latitude ionosphere. Adapted from Baumjohann W and Treumann RA (1996) *Basic Space Plasma Physics*. London: Imperial College Press, with permission from Imperial College Press.

lines, each of which has one end connected to the Earth and the other stretching out into the solar wind. Subsequently, the solar wind will transport this field line across the polar cap down the tail and due to the stiffness of the field line, the magnetic tension, the magnetospheric part of the field line (inside the shaded region), will also be transported tailward. At the nightside end of the magnetosphere the two open field line halves will meet again and reconnect, leaving a closed but stretched terrestrial field line in the magnetotail and an open solar wind field line downtail of the magnetosphere. Due to magnetic tension, the stretched tail field line will relax and shorten in the earthward direction. During this relaxation it transports the plasma, to which it is frozen, toward the Earth.

For an observer on the Earth, the sunward transport of plasma in the magnetosphere caused by magnetic merging at the Earth's magnetopause is equivalent to an electric field. The total potential difference between the dawn and dusk magnetopause or, equivalently, across the polar cap corresponds to about 50–100 kV. For a cross section of the magnetosphere of about $30\,R_{\mathrm{E}}$, this amounts to a dawn-to-dusk directed field of some $0.2\text{–}0.5\,\mathrm{mV\,m}^{-1}$.

3.5.2   Ionospheric Convection

The motion of the flux tubes across the polar cap due to magnetic merging depicted in **Figure 12** also moves the ionospheric foot point of the flux tube and the plasma tied to it across the polar cap to the nightside. Similarly, the sunward convection of magnetospheric flux tubes leads to a sunward convection of the foot points of these flux tubes in the dawn- and

dusk-side high-latitude ionosphere, inside the auroral oval. This leads to a two-cell convection pattern in the polar ionosphere, depicted in **Figure 13**.

The convection pattern is equivalent to an electric potential pattern. Cold particles will drift along these contours: drawing equipotential contours and drawing $\mathbf{E} \times \mathbf{B}$ drift trajectories of the plasma is equivalent. Hence, we can take the two-cell convection pattern as a two-cell pattern of equipotential contours, which is equivalent to an ionospheric electric field that is directed toward dusk in the northern polar cap. Inside the Northern Hemisphere auroral oval the electric field is directed toward the pole on the dusk side, while it has a southward direction in the morning hours.

Since the ionospheric conductivity is a tensor with three different components (see previous section), three types of currents will be generated by the convection electric field. The first type is the field-aligned currents flowing parallel to the magnetic field into and out of the ionosphere. Second, there are the Pedersen currents which flow perpendicular to the magnetic field lines and parallel to the ionospheric convection field. Finally, Hall currents will flow perpendicular to both the magnetic and the electric field.

3.5.3   Auroral Electrojets

Since particles precipitating into the auroral oval cause significant ionization, its conductivity is much higher than that of the polar cap. As a result, the high-latitude current flow is concentrated inside the auroral oval, where it forms the auroral electrojets.

The auroral electrojets are the most prominent currents at auroral latitudes. They carry a total current of some million amperes. This is the same order of magnitude as the total current carried by the ring current, discussed in the previous section, but since the auroral electrojets flow only 100 km above the Earth's surface, they create the largest ground magnetic disturbance of all current systems in the Earth's environment. The disturbance fields have typical magnitudes of 100–1000 nT, but may reach 3000 nT during the largest magnetic storms.

It is important to distinguish between the convection auroral electrojets, shown in the left-hand panel of **Figure 14**, and the substorm electrojet on the right-hand side. The convection electrojets consist of eastward and westward electrojets. These are primarily Hall currents which originate around noon where they are fed by downward field-aligned currents. Typical sheet current densities range between 0.5 and $1 \, \text{A m}^{-1}$. The eastward electrojet flows in the afternoon sector and terminates in the pre-midnight region where it partially flows up magnetic field lines and partially rotates northward, joining the westward electrojet. The westward electrojet flows through the morning and midnight sector and typically extends into the evening sector along the poleward border of the auroral oval where it also diverges as upward field-aligned currents.

Similar to the ring current, which is 'measured' by the Dst index (see previous section), the auroral electrojet indices AE, AU, and AL were introduced as a measure of global auroral electrojet activity. The indices are based on 1 min samples of the northward component trace from auroral zone observatories located at 65–70° magnetic latitude with a longitudinal spacing of 10–40°. Referenced to a quiet-day level, the data of all observatories are plotted as a function of universal time. The upper and lower envelopes are defined as AU and AL, while AE is defined as the separation between the upper and lower envelopes. The upper and lower envelopes are thought to represent the maximum eastward and westward electrojet current, respectively, while AE represents the total maximum electrojet current.

3.5.4 Substorms

Convection is not a stationary process: magnetic merging between interplanetary and terrestrial field lines at the dayside magnetopause does not occur all the time, but only for southward-oriented interplanetary field lines, and is typically not in equilibrium with reconnection in the magnetotail. Only part of the flux transported into the tail is reconnected instantaneously in the deep tail and convected back to the dayside. The remaining field lines are added to the tail lobes, where they increase the magnetic flux density and, hence, enhance the cross-tail current in the neutral sheet. After some tens of minutes these intermediately stored field lines are suddenly reconnected at tail distances of 20–25 Earth radii and their magnetic energy is explosively released. The sudden reconnection of previously stored flux tubes has rather dramatic effects on the magnetospheric plasma and associated phenomena like aurora and magnetospheric and ionospheric currents. These effects, which last for 1–2 h, are summarized as magnetospheric substorm.

A substorm starts when the dayside merging rate is distinctively enhanced, typically due to a southward turning of the interplanetary magnetic field. The flux eroded on the dayside magnetopause is transported into the tail. Part of the flux is reconnected and convected back to the dayside magnetosphere. The enhanced convection causes enhanced current flow in the convection electrojets and an associated growth of the AE index.

The other part of the flux is added to the tail lobes. After 30–60 min, too much magnetic flux and thus magnetic energy has been accumulated in the tail. The tail becomes unstable and must release the surplus energy. This is the time of substorm onset and the beginning of the substorm expansion phase. At substorm onset, the aurora suddenly brightens and fills the whole sky. During the following 30–60 min, rather dramatic changes are seen in the auroral zone currents.

The sharp AE index seen in **Figure 15** to values of about 500 nT indicates that the ionospheric current flow is strongly enhanced. The unloading of

Convection electrojets   Substorm electrojet

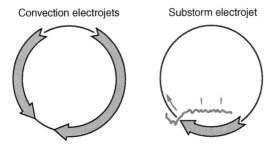

Figure 14 Auroral electrojets. The green line and arrows in the righthand panel indicate the boundary of the west- and northward expanding substorm auroral bulge. Adapted from Baumjohann W and Treumann RA (1996) *Basic Space Plasma Physics*. London: Imperial College Press, with permission from Imperial College Press.

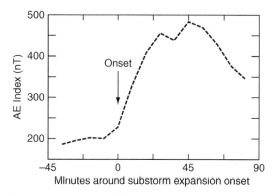

Figure 15 Variation of AE index during a substorm.

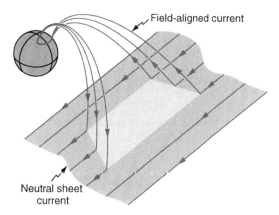

Figure 16 Substorm current wedge. Adapted from Baumjohann W and Treumann RA (1996) *Basic Space Plasma Physics*. London: Imperial College Press, with permission from Imperial College Press.

magnetic flux previously stored in the magnetotail leads to the formation of a substorm electrojet with strongly enhanced westward current flow in the midnight sector. The substorm electrojet is concentrated in the region of active aurora and expands westward during the course of the expansion phase. In contrast to the convection electrojets, where any increase is caused mainly by an increasing convection electric field, the strength of the substorm electrojet current is mainly determined by a strong increase in ionospheric conductance due to strong particle precipitation in the bright substorm aurora.

Since the substorm electrojet is governed by the strong increase of the conductivities inside the region of bright aurora, the situation is similar (except for directions) to that in the equatorial electrojet described in the previous section. However, in the present case, the polarization electric field and thus the enhancement of the westward current is not so strong, since field-aligned currents will remove part of the space charge deposited at the boundaries of the highly conducting channel.

Another difference between the convection and substorm electrojets is that in the case of the convection electrojets, the field-aligned currents are distributed over a wide local time range. In the case of the substorm electrojet, the jet itself and its field-aligned currents are much more concentrated in the midnight sector, forming a current wedge as depicted in **Figure 16**. The effects of this current wedge, in particular the magnetic disturbance associated with the field-aligned currents, can be seen also at mid-latitudes.

3.6 Geomagnetic Pulsations

As with any medium, a plasma carries waves in many different frequency ranges, from as low as millihertz to

as high as several tens of kilohertz. Typically, the higher-frequency waves can only be observed in the plasma itself, but the ultralow-frequency (ULF) waves, in particular, generate fast fluctuations of the Earth's surface magnetic field in the frequency range from a few millihertz up to a few hertz, corresponding to oscillation periods from several hundred seconds to a fraction of a second. These are the so-called geomagnetic pulsations, known of for about a century.

In most cases, the pulsating disturbance fields observed are associated with shear Alfvén waves. These waves constitute the simplest wave solutions of the magnetohydrodynamic equations and represent simple string-like oscillations of mass-loaded magnetic field lines. Shear Alfvén waves are purely transverse waves, that is, all variations have only components that are perpendicular to the ambient magnetic field. The magnetic component of this type of wave is parallel to the plasma velocity variation while the wave electric field points perpendicular to the magnetic and velocity variations.

An Alfvén wave may propagate parallel to the ambient field with the Alfvén velocity, v_A, which is essentially a 'magnetic sound' velocity, given by

$$v_A^2 = \frac{B^2}{\mu_0 nm} \qquad [12]$$

In the Earth's magnetosphere, typical Alfvén velocities range from some hundreds to several thousands of kilometers per second.

The ULF range, and hence the pulsations, are conventionally divided into five intervals, Pc1–Pc5,

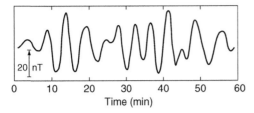

Figure 17 Ground magnetic disturbance of a Pc5 pulsation.

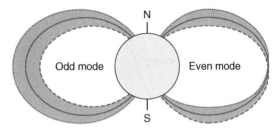

Figure 18 Fundamental poloidal field line resonances. Adapted from Baumjohann W and Treumann RA (1996) *Basic Space Plasma Physics*. London: Imperial College Press, with permission from Imperial College Press.

for continuous pulsations, and into two intervals, Pi1 and Pi2, for irregular pulsations. The class of continuous pulsations covers quasi-sinusoidal oscillations of narrow spectral bandwidth, as shown in **Figure 17**. They may have a comparably long duration from several minutes up to hours. Pc pulsations can generally be observed over a wide latitudinal and longitudinal range on the Earth's surface and in the magnetosphere. The irregular pulsations, in contrast, are shorter-lived, sometimes comprising only a few oscillations decaying in time.

3.6.1 Pc5 Pulsations

The Pc5 pulsations shown in **Figure 17** are caused by standing oscillations of magnetospheric field lines. For a standing wave, the length of the field line between the two reflection points in the ionosphere must be a multiple of half the parallel wavelength. Hence, each particular field line has a number of distinct eigenfrequencies or Alfvénic resonances. Since the length of the field lines increases with latitude, the resonance or eigenfrequency decreases with latitude. For an average Alfvén velocity of $1000 \, \mathrm{km \, s^{-1}}$, the fundamental resonance frequency on closed field lines ranges between 1 and 100 mHz and falls into the Pc3–Pc5 range.

Figure 18 schematically shows how the dipolar field configuration changes for two fundamental types of field line resonances. The foot points of the field lines are fixed in the ionosphere, but the field lines may either perform a 'breathing' motion (fundamental odd mode) or a 'wobbling' motion (even mode). In addition to these poloidal modes, the field lines can also exert toroidal oscillations, in which case the oscillation of the field line and the plasma bulk flow are purely azimuthal. Pc pulsations often are a mixture of poloidal and toroidal oscillations.

Pc5 pulsations are resonant oscillations of field lines. They are excited and driven via surface waves traveling along the magnetopause. These surface waves are caused by a Kelvin–Helmholtz instability excited by the flow of the solar wind around the magnetosphere and represent an evanescent wave mode. Being surface waves, their amplitude is strongly damped away from the magnetopause, yet they can still set the field line with a matching eigen- or resonance frequency into oscillation. All other field lines, whose resonance frequencies do not match, are only marginally excited and do not contribute to the pulsation.

Pc4–Pc5 pulsations might also be excited by packets of trapped particles bouncing up and down a field line, as long as the bounce period of these particles, which depends mainly on their energy, matches the eigenperiod of the field line.

3.6.2 Pi2 Pulsations

The short-period irregular Pi2 pulsations are associated with the development of the substorm current wedge described in the previous section. Whenever field-aligned currents are suddenly switched on somewhere in the magnetosphere, they must be transported to the ionosphere via Alfvén waves. Only this transverse magnetohydrodynamic wave mode can carry field-aligned current. Launched in the magnetosphere, the Alfvén waves are reflected back and forth between the ionosphere and the current generator in the tail until a stationary equilibrium is reached.

Figure 19 shows qualitatively the development of the magnetic disturbance field and thus the field-aligned current flow after switch-on of a current generator in the magnetotail. At $t = 0$, an Alfvén wave is launched which carries a current corresponding to the generator current and thus a particular magnetic disturbance field. This wave reaches the ionosphere at $t = t_A$, the Alfvén wave traveltime between magnetosphere and ionosphere, that is,

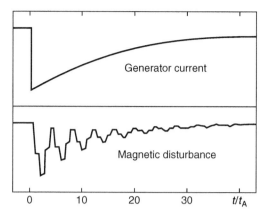

Figure 19 Magnetic disturbance due to switch-on of a current wedge. Adapted from Baumjohann W and Treumann RA (1996) *Basic Space Plasma Physics*. London: Imperial College Press, with permission from Imperial College Press.

some 30 s. Here, about 80% of its amplitude is reflected with the magnetic field of the reflected wave adding to the primary disturbance field.

At $t = 2t_A$, the reflected wave comes back to the generator and launches a third wave, whose magnetic disturbance must be of opposite polarity to decrease the total wave magnetic field such that the total disturbance matches that caused by the generator current at this time. Multiple bounces of the wave lead to magnetic field disturbances which oscillate with a period of $4t_A$ until they finally converge to match the generator current. The magnetic oscillations with periods of some 100 s are readily observable as Pi2 pulsations. They are often used as a good indicator for the onset of substorms.

3.7 Conclusions

This chapter gave a brief introduction into the fundamentals of magnetospheric physics. The topics and phenomena discussed were selected by their influence on the terrestrial field as measured on the Earth's surface. There are many more interesting features that can be observed in magnetospheric plasmas, let alone in other space plasmas, like solar or astrophysical plasmas.

However, all of them follow the same basic principles outlined above. If one includes the Earth's bow shock (which was not discussed here, since its physics has little influence on the terrestrial magnetic field), detailed *in situ* studies of the space plasmas in the Earth's neighborhood are an effective means to

understand many solar and astrophysical phenomena, from which we have only sparse observational information transmitted by electromagnetic radiation.

Hence, geomagnetism is not only the root of magnetospheric physics (ground-based observations of magnetic variations done by Humboldt and Gauss provided the first window into what was later called magnetosphere), but in a broader sense, that is, by measuring magnetic fields in the Earth's neighborhood, it is still essential to understand the plasma universe.

A more exhaustive description of all possible external sources of geomagnetic field variations (and on their use in diagnosing magnetospheric dynamics) is presented in Nishida (1978). A full description of theory and observations of space plasmas in the Earth's neighborhood can be found in Baumjohann and Treumann (1996). For those readers who want to know more about the guiding center approach, we recommend reading the monograph by Alfvén and Fälthammar (1963) or that by Northrop (1963). More about the physics of trapped particles, the ring current, and magnetic storms can be found in Lyons and Williams (1984), Kamide *et al.* (1998), and Daglis *et al.* (1999). The physics of the ionosphere and ionospheric currents is detailed in Hargreaves (1992) and Rishbeth and Garriot (1969). A good description of solar wind–magnetosphere coupling is given in a review article by Cowley (1982). Additional material on high-latitude current systems and magnetosphere–ionosphere coupling is found in a monograph by Kamide and Baumjohann (1993) and in a review article by Untiedt and Baumjohann (1993). For a detailed account on substorms see Baker *et al.* (1996). An exhaustive elementary description of fluid plasma waves is given in Bittencourt (1986) while further information about pulsations is contained in Glassmeier (1995) and McPherron (2005).

References

Alfvén H and Fälthammar CG (1963) *Cosmical Electrodynamics, Fundamental Principles*. Oxford: Clarendon Press.

Baker DN, Pulkkinen TI, Angelopoulos V, Baumjohann W, and McPherron RL (1996) Neutral line model of substorms: Past results and present view. *Journal of Geophysical Research* 101: 12975–13010.

Baumjohann W and Treumann RA (1996) *Basic Space Plasma Physics*. London: Imperial College Presse.

Bittencourt JA (1986) *Fundamentals of Plasma Physics*. Oxford: Pergamon Press.

Cowley SWH (1982) The causes of convection in the Earth's magnetosphere: A review of developments during the IMS. *Reviews of Geophysics and Space Physics* 30: 531.

Daglis IA, Thorne RM, Baumjohann W, and Orsini S (1999) The terrestrial ring current: Origin, formation, evolution, and decay. *Reviews of Geophysics* 37: 407–438.

Glassmeier KH (1995) ULF pulsations. In: Volland H (ed.) *Handbook of Atmospheric Electrodynamics*, vol. 2, pp. 463. Boca Raton: CRC Press.

Hargreaves JK (1992) *The Solar–Terrestrial Environment*. Cambridge: Cambridge University Press.

Kamide Y and Baumjohann W (1993) *Magnetosphere–Ionosphere Coupling*. Heidelberg: Springer.

Kamide Y, Baumjohann W, Daglis IA, *et al.* (1998) Current understanding of magnetic storms: Storm–substorm relationship. *Journal of Geophysical Research* 103: 17705–17728.

Lyons LR and Williams DJ (1984) *Quantitative Aspects of Magnetospheric Physics*. Dordrecht: D. Reidel Publication Company.

McPherron R (2005) Magnetic pulsations: Their sources and relation to solar wind and geomagnetic activity. *Survey of Geophysics* 26: 545–592.

Nishida A (1978) *Geomagnetic Diagnosis of the Magnetosphere*. Heidelberg: Springer.

Northrop TG (1963) *The Adiabatic Motion of Charged Particles*. New York: Interscience Publishers.

Rishbeth JA and Garriot OK (1969) *Introduction to Ionospheric Physics*. New York: Academic Press.

Untiedt J and Baumjohann W (1993) Studies of polar current systems using the IMS Scandinavian magnetometer array. *Space Science Reviews* 63: 245.

4 Observation and Measurement Techniques

G. M. Turner, Victoria University of Wellington, Wellington, New Zealand

J. L. Rasson, Institut Royal Meteorologique, Dourbes, Belgium

C. V. Reeves, Earthworks BV, Delft, The Netherlands

Nomenclature

f frequency	(Hz)	Q Koenigsberger ratio	(dimensionless)
m magnetic moment	(A m^2)	R resistance	(Ω)
B magnetic induction (field)	(T)	T, T_C, T_B temperature, curie temp., blocking temp.	(°C or K)
C capacitance	(F)	V voltage (signal or emf)	(V)
D, I or Dec, Inc declination, inclination	(degrees)	X, Y, Z N, E, Down components of geomagnetic field (induction)	(T)
E energy	(J)		
F total geomagnetic field	(T)	γ gyromagnetic ratio	(Hz T^{-1})
H magnetic field intensity	(A m^{-1})	λ, ϕ latitude, longitude	(degrees)
H horizontal component of induction	(T)	μ magnetic permeability	
		τ torque	(N m)
I moment of inertia	(kg m^2)	τ relaxation time	(sec, yr)
K mass (specific) susceptibility	(m^3 kg^{-1})	ϕ magnetic flux	(T m^2 = Wb)
		χ volume susceptibility	(dimensionless)
L angular momentum (spin)	(kg m^2 s^{-1})	ω angular frequency	(rad s^{-1})
L inductance	(H)		
M magnetization	(A m^{-1})		

4.1 Introduction

The geomagnetic field is a complex function of space and time, with contributions of both internal and external origin. Historically, knowledge of the morphology and variability of the field has been important in navigation, in understanding atmospheric and ionospheric processes, including radio transmission, as well as in geophysical studies of Earth's interior (the core, mantle, and the tectonic plates of that make up the lithosphere). The purpose of this chapter is to describe the instruments and practices used to observe and measure the full range of features of the geomagnetic field.

The significance of the predominantly dipole nature of the main field was recognized by Gilbert over 400 years ago; secular variation was documented at about the same time. The nondipole part of the field that originates from the core manifests itself at the surface as features that typically extend over thousands of kilometers. The International Geomagnetic Reference Field (IGRF, website) is a spherical harmonic model designed to accurately describe the spatial features of the field originating from the core. Much of the data incorporated into the IGRF comes from the global network of permanent geomagnetic observatories (INTERMAGNET, website).

By contrast, the remanent magnetization of crustal bodies typically results in magnetic anomalies of much smaller extent (up to tens or hundreds of kilometers). These are generally charted and studied through regional or local magnetic surveys.

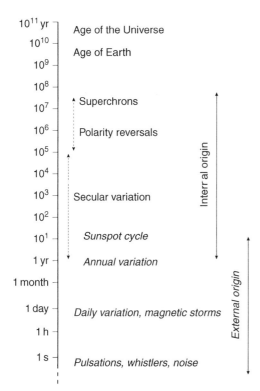

Figure 1   Time variations of the geomagnetic field in relation to the age of Earth and the Universe, indicating which are of internal and which are of external origin.

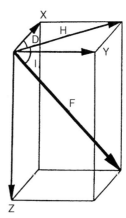

Figure 2   Decomposition of the geomagnetic vector, showing components used when describing the vector in Cartesian (*x, y, z*), cylindrical (*D, H, Z*) and spherical (*D, I, Z*) coordinate systems.

In order to describe the (instantaneous) geomagnetic field vector completely, three independent components must be specified. The three coordinate systems commonly used are as follows:

- Cartesian: X (North component), Y (East component), Z (vertical component, positive downwards);
- Cylindrical: D (magnetic declination angle, positive east of north), H (horizontal component), Z (as above);
- Spherical: D (as above), I (magnetic inclination angle, positive below horizontal), F (intensity of magnetic field).

The relations between these different systems are illustrated in **Figure 2**.

4.1.1   Instrumentation

The earliest known compass, made in China around the first century AD, was a spoon-shaped piece of lodestone, balanced on a horizontal plate so that it aligned with Earth's magnetic field, the handle pointing south. Although originally designed for the purpose of divining favorable orientations (*feng shui*), these early instruments were the forerunners of an extensive range of measuring instruments and navigational compasses based on pivoted or suspended magnets.

The magnetic compass is simply a needle balanced to rotate in a horizontal plane until it comes to rest in the plane of the magnetic meridian (with north, rather than south, now being the

In addition, the field has been observed to undergo a very broad spectrum of time variations, extending well beyond the limit of real time and historical observations (**Figure 1**). Perhaps the most astounding of these is the phenomenon of polarity reversals found only in the paleomagnetic record. Time-varying signals reaching Earth's surface from the core are attenuated due to the conductivity and permeability of the mantle, so that periodicities less than about an year cannot be detected. The shorter period magnetic field variations that are observed originate externally – from variations in the solar wind and the interplanetary magnetic field, and their interactions with the magnetosphere.

Although the vast majority of measurements of the geomagnetic field are made on Earth's surface, magnetometers are routinely flown in space probes and satellites, offering hugely superior spatial coverage. Modern models of the main field incorporate satellite measurements, while aeromagnetic measurements make a huge contribution to modern survey work.

principal sense). It was the first scientific instrument to employ a needle moving over a graduated scale, and though most moving needle instruments have by now been superseded by digital meters, the compass remains the simplest and easiest means of finding directions in remote locations, since it has no electronics and needs no power supply.

Very early compass-makers appear to have been unaware of the vertical component of the geomagnetic field. However, by the mid-sixteenth century, Hartmann and Norman had both noted and measured the angle of inclination, using a magnetized needle pivoted to rotate in the vertical north–south plane: a dip needle.

The dip needle was also used to make the first measurements of the intensity of the magnetic field. Small oscillations of the needle about its equilibrium position have a period given by $T = 2\pi\sqrt{I/(mB)}$, where I and m are the moment of inertia and the magnetic moment of the needle, and B is the magnetic field. The first recorded intensity measurements were made by de Rossel on the d'Entrecasteaux expedition of 1791–94 (Lilley and Day, 1993). Both the d'Entrecasteaux measurements, and those of Alexander von Humboldt, made between 1798 and 1803, clearly show the increase of intensity with latitude which is expected of a predominantly geocentric axial dipole (GAD) field.

Many of the first-generation magnetic field-measuring instruments were based on pivoted or suspended magnet systems. These include the quartz horizontal magnetometer (QHM) and the balance magnétique zero (BMZ) instruments used extensively in observatories, the astatic and parastatic rock magnetometers used in many of the pioneering paleomagnetic studies, variometer-type instruments used in early magnetic surveys and exploration work, and the Gough—Reitzel magnetometer commonly used until recently in magnetometer array studies.

From the 1950s onwards these elegant but delicate instruments have gradually been replaced by more robust electrical and electronic instruments such as the induction coil, and the fluxgate, which capitalizes on the high-permeability materials that were developed at the time. Electrical feedback systems have played a major role in stabilizing the response of these instruments. Still later, computer interfacing and data logging have led to partial or complete automation of many operations.

The most recent generation of magnetometers has grown out of research on magnetic properties at the atomic level, and the development of superconducting materials and devices, which are, in principle, capable of counting individual quanta of magnetic flux, and thus reaching the theoretical limit of resolution. These include the proton precession magnetometers (PPMs) and Overhauser effect magnetometers, used extensively in observatories, and the Superconducting quantum interference device (SQUID) which has revolutionized paleomagnetic measurements.

In this section we describe the principles of operation of the main types of sensors used in geomagnetism and paleomagnetism. Their incorporation into practical magnetometers is covered in the following sections on geomagnetic observatories, magnetic surveys, and paleomagnetism.

The induction coil, fluxgate, and SQUID are vector sensors: their orientation in space determines which component of the ambient field is measured. The proton precession, Overhauser and optically pumped magnetometers (OPM), on the other hand, are insensitive to the direction of the ambient magnetic field, and measure only its magnitude. This property stems from the fundamental physical principles put to work in their operation: Larmor precession, electron spin resonance, and Zeeman splitting.

4.1.2 Vector Magnetometer Sensors

4.1.2.1 The induction coil

The induction coil, or search-coil magnetometer is one of the simplest modern magnetometers: in principle, when the magnetic flux, ϕ, threading through a conducting circuit varies, an emf or voltage signal is induced in the circuit in accordance with Faraday's Law of electromagnetic induction. The induced voltage, V_i, is proportional to the rate of change of the magnetic flux.

In practice the flux linkage is enhanced by winding many turns of conducting wire on a core of high-permeability ferromagnetic material. For a solenoid of N turns, each of cross-sectional area A, and a core of relative permeability μ, $\phi = \mu NAB$, where B is the component of the magnetic field parallel to the axis of the solenoid.

For magnetic field variations of frequency ω, along the axis, $B = B_0 e^{-i\omega t}$, and the theoretical induced voltage is

$$V_i = -\frac{d\phi}{dt} = NA\mu(i\omega)B_0 e^{-i\omega t}$$

This signal must be amplified, and the actual measured voltage depends also on the resistance, inductance, and capacitance of the coil as well as the size of any damping resistor and the gain of the amplifier.

Due to their simplicity, induction coils are widely used for many applications, for instance in geophysical prospecting and metrology. In magnetic observatories they are used to monitor rapid variations of the geomagnetic field, such as pulsations. With suitable circuitry, induction coils can be designed to measure variations with frequencies between 10^{-4} and 10^{+7} Hz and fields from fractions of femto-tesla to tens of teslas (Korepanov *et al.*, 2001).

4.1.2.2 *The fluxgate*

The operation of a fluxgate sensor depends on the nonlinear relation between induced magnetization and magnetizing field in high-permeability easily saturated materials such as ferrites, permalloy, metallic glasses, and mu-metal.

Various different designs exist: most have two cores, either as two separate linear cores, or as two halves of a ring fluxgate.

The general principle of operation is described below for the case of the Vacquier two-core design shown in **Figure 3**. Identical primary coils are wound around the two high-permeability cores. The drive current (of frequency 50–1000 Hz) is of sufficient amplitude to saturate the cores during most of each cycle. The primary windings are connected in series opposition, so when the drive current is applied, the axial magnetic fields are out of phase (**Figure 4(a)**). If there is a steady ambient magnetic field, B_{amb}, parallel to the core axis, as shown in **Figure 3**, then when it augments the field in one coil, it reduces the total field in the other and vice versa (**Figure 4(b)**). The effect of the ambient field is thus that, during any given half-cycle, the core in which the field is augmented will be saturated for more of the half-cycle, while the core with the reduced field will be saturated for less of the half-cycle. The situation

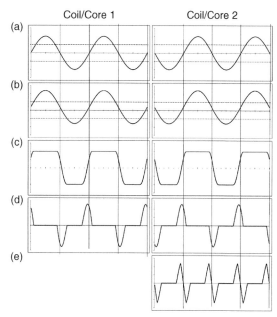

Figure 4 Signal processing in a two-core fluxgate sensor. (a) Sinusoidal drive current, frequency ω. In the absence of the iron cores, the drive signal results in antiphase magnetic fields in coils 1 and 2; (b) Addition of the ambient field B_{amb}, as shown in **Figure 3**, results in an asymmetry; (c) So, core 1 is saturated for more of the first half-cycle than core 2, while core 2 is saturated for more of the second half-cycle than core 1; (d) the magnetic flux changes in secondary coils 1 and 2 are therefore not synchronous, resulting in induced emf pulses as shown; (e) the output signal is the sum of the emfs in secondary coils 1 and 2, which has a dominant frequency of 2ω.

reverses in the next half-cycle of the driving signal (**Figure 4(c)**). This asymmetry between the magnetization of the two cores leads directly to an asymmetry between the rates of change of flux in the secondary coils (**Figure 4(d)**). The secondary coils are oppositely wound and series connected as shown (**Figure 3**), so the asymmetry results in a nonzero output twice per cycle, that is, at twice the frequency of the driving signal (**Figure 4(e)**). For small fields the amplitude of this second harmonic is proportional to the component of the ambient field parallel to the fluxgate axis. A phase-sensitive detector, with its reference signal set to twice the frequency of the driving signal is generally used to produce a DC output that is proportional to the ambient field.

In practical applications a feedback arrangement is almost always employed to cancel the field being measured, and hence linearize the response. Fluxgate sensors are highly directional, measuring only the

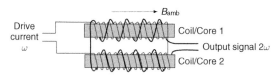

Figure 3 A two-core fluxgate magnetometer (Vacquier design).

field component parallel to the fluxgate axis. An arrangement of three orthogonal fluxgates may therefore be used to obtain the intensity and direction of the total field vector. If three independent units are used in close proximity; for example, in a spacecraft, problems can arise due to interference of the separate feedback systems. This has been overcome in the Compact Spherical Coil system, as used in the Oersted and CHAMP satellites, where all three sensors see the whole feedback field (Nielsen *et al.*, 1995).

The first fluxgate-type device was built and described by Aschenbrenner and Goubau (1936). Airborne fluxgates were used extensively during the Second World War for submarine detection. Subsequent developments mean that modern fluxgates have many applications: they are used for navigation on land and sea, as vector magnetometers in almost all space flights, in observatories, and in many medium-resolution rock magnetometers.

4.1.2.3   *The SQUID*

The development of the SQUID or superconducting quantum interference device, as a magnetic field sensor heralded a new era in sensitivity, particularly in paleomagnetic rock magnetometers.

The superconductivity of mercury was discovered by Kammerlingh Onnes in 1911, when he found that, below about 4 K, its electrical resistivity effectively vanished. Other metal superconductors, more suitable for the construction of practical devices, include niobium ($T_C = 9.2$ K) and lead ($T_C = 7.2$ K). More recently superconducting ceramic materials, for example, oxides of lanthanum or yttrium, with barium, strontium or calcium and copper, have been discovered with transition temperatures of 90 K and higher. However, practical sensors utilizing these 'high-temperature' superconductors have not yet been developed.

Superconductivity may be explained by the Bardeen, Cooper, Schrieffer (BCS) theory (Bardeen *et al.*, 1957). At sufficiently low temperatures a weak interaction causes electrons to form pairs. Such 'Cooper pairs' have boson-like properties; in particular, in the absence of an exclusion principle, they occupy a single quantum state, and may be described by a single coherent wave function. The pairing of electrons is critical to the theory. In the normal state, electrical resistivity results from collisions, or scattering of conduction electrons by the ions of the crystal lattice. In the superconducting state, one electron of

the Cooper pair effectively distorts the lattice, allowing the other free passage.

Another property of the superconducting state is the exclusion of magnetic flux from the bulk of the superconducting material (Meissner effect). If an external magnetic field exists through a ring of superconductor then, as it is cooled through the critical temperature, superconducting currents will be generated to cancel it. However, if the current exceeds a certain value, I_c, the superconductivity is lost, and the material reverts to its normal resistivity. The moment this happens the current drops and flux enters the ring. With the lowering of the current, superconductivity is restored. However, to maintain the coherency of the electron-pair wave function, the current can drop only by quantized amounts, and correspondingly only quantized amounts of flux may enter the ring. This flux quantum, ϕ_o, is equal to $h/2e = 2.09 \times 10^{-15}$ Wb. In principle, the number of flux jumps may be counted to give a measure of the strength of the external magnetic field. This effect is utilized in the SQUID.

Josephson (1962) predicted that a supercurrent should be able to pass through a thin layer of dielectric material sandwiched between two superconductors, if the wave functions of the electrons extend into the junction and join up coherently. A similar effect is achieved by creating a very narrow bridge or weak link between two pieces of superconductor.

The term SQUID has been applied to two types of devices that have evolved from the basic Josephson junction. The simpler concept is that of the DC SQUID. It consists of two Josephson junctions or weak links arranged symmetrically in a ring of superconductor. Magnetic flux through the ring will result in equal and opposite phase shifts of the wave functions across the two junctions. Interference therefore occurs between the currents in the two branches of the device, in a manner analogous to the interference of light waves in a double-slit experiment. The current is maximum if the flux is equal to an integral number of flux quanta, but is reduced to zero for an odd number of half-quanta.

The first superconducting rock magnetometers used radio frequency (RF)-driven SQUIDS (Goree and Fuller, 1976). These are not strictly interference devices, but rather single weak links contained in a ring or cylinder of superconducting material. The operation of a device similar to the Dayhem bridge used in the first superconducting rock magnetometers built by Superconducting Technologies Inc. (SCT) is illustrated in **Figure 5**.

The sensor is driven by an RF signal of frequency 20–30 MHz, through a coil wound around the outside. The amplitude of the RF signal is sufficient to exceed the critical current of the weak link. When this occurs the loop momentarily becomes resistive and a flux quantum is admitted before superconductivity is regained, as described above. This produces a pulse in the sensor coil. **Figure 5**(a) illustrates the situation when the critical current corresponds to 0.75 flux quantum; with no bias field, the output consists of two positive pulses and two negative pulses per cycle. The addition of an external magnetic field biases the RF signal, and changes the positions in the cycle at which the critical current is exceeded. **Figure 5**(b) shows that the positions of the pulses have been shifted, and from these shifts the DC field can be obtained. A common arrangement to avoid problems of nonlinearity and drift is to apply negative feedback, so the sensor is always at its most sensitive state, and to monitor the feedback signal needed to achieve this.

4.1.3 Scalar Magnetometers

4.1.3.1 Proton precession magnetometers

Thanks to groundbreaking work on nuclear magnetic induction by Bloch (1946) and its application to the measurement of weak magnetic fields by Packard and Varian (1954), the PPM became available from the late 1940s. It allows the absolute measurement of the scalar magnetic field at the push of a button. In the PPM the magnetic field is found by measuring the frequency of the Larmor precession of the proton magnetic moment about the ambient field, the protons being provided by a sample of water or other proton-rich fluid.

The spin angular momentum and magnetic moment vectors of the proton, $\mathbf{L}$ and $\mathbf{m_p}$ are parallel. Their ratio is a scalar constant known as the proton gyromagnetic ratio, γ_p.

$$\gamma_p = \frac{m_p}{L} = 2.67522205(23)$$
$$\times 10^8 \, \text{Hz} \, \text{T}^{-1} \, (\text{CODATA 2002, NIST website})$$

In an external magnetic field, $\mathbf{B}$, the magnetic moment experiences a torque and this results in a change in the angular momentum vector:

$$\boldsymbol{\tau} = \mathbf{m}_p \times \mathbf{B} = \frac{d\mathbf{L}}{dt}$$

$\mathbf{L}$ is parallel to $\mathbf{m}$, while the torque is perpendicular to $\mathbf{m}$, so $d\mathbf{L}$ is perpendicular to $\mathbf{L}$, and the effect is to make $\mathbf{L}$ and $\mathbf{m}$ precess about the direction of $\mathbf{B}$ (**Figure 6**). This is Larmor precession, and its frequency is given by

$$\omega_{\text{Larmor}} = 2\pi f = \gamma_p B$$

Hence a measurement of the Larmor frequency and division by the proton gyromagnetic ratio leads directly to the intensity of the ambient magnetic field. In practice the 'shielded' proton gyromagnetic ratio, γ', is used instead of γ. This takes account of the diamagnetism of water (or other proton-rich

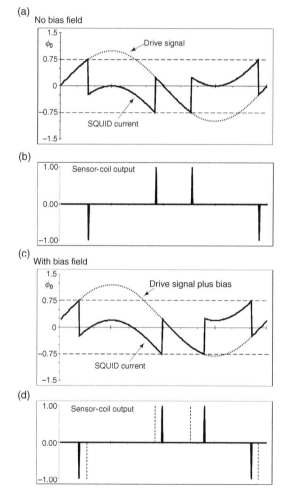

Figure 5 Signal processing by a weak link SQUID sensor, with and without a direct biasing field. (a) Sinusoidal drive signal that causes the super-current to exceed the critical value at 0.75 ϕ_0, at which point the flux through the SQUID jumps by ϕ_0. The bold curve shows the resulting SQUID current. (b) At each flux jump a pulse is induced in the sensor coil. (c) and (d) are the corresponding curves for the case of a direct bias field and show that the pulses in the sensor coil are offset by the direct field.

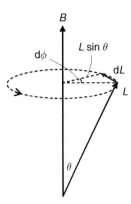

Figure 6 Larmor precession of the proton angular momentum vector, **L**, about the ambient magnetic field vector, **B**, in response to the torque, **m** x **B**, on the magnetic moment vector of the proton, **m**, which is parallel to **L**. See text for more details.

fluid) and the shape of the sample (usually taken to be spherical). For a spherical sample of water at 25°C

$$\gamma' = 2.67515333(23)$$
$$\times 10^8 \, \text{Hz} \, \text{T}^{-1} \, (\text{CODATA 2002, NIST website})$$

A PPM sensor is therefore ideally a spherical container filled with water, or other proton-rich fluid. A surrounding coil supplied with a direct current produces a direct field several orders of magnitude larger than the Earth's field, and aligns the magnetic moments of the protons. This is to partially remove the random orientation due to the thermal agitation of the protons. After the few seconds, necessary to obtain sufficient alignment, this DC field is removed and the proton magnetic moments precess about the ambient (geomagnetic) field vector. An emf induced by the precessing protons is then observed for a few seconds by a pick-up coil, until thermal agitation drowns the signal again (Primdahl, 2002). The observed signal is an exponentially decaying sinusoid at the Larmor frequency f. The emf signal in the geomagnetic field ranges from 850 to 3400 Hz and its amplitude is typically of a few microvolts. The signal-to-noise ratio is dependent upon the strength of the field; therefore, a PPM works better in strong fields. Often the same coil is used for the polarizing and pick-up operations. For maximum signal, the polarizing DC field should be applied at right angles to the field to be measured, in order to maximize the rotating component and the emf pick-up by the coil. Cylindrical coils will therefore have dead zones for certain orientations of the sensor, where the emf signals disappear in the noise. Toroidal coils have been used to eliminate this spatial dependency on the signal-to-noise ratio.

Efforts to improve the PPM have also addressed the way in which the frequency measurement is performed. In early PPMs, the frequency, f, of the output signal was multiplied by an integer M using a phase-locked loop. The resulting signal at frequency Mf was then counted by a digital counter for a time duration D. D was set by a custom-cut quartz crystal, serving as a frequency reference, and whose frequency F was adjusted to provide the correct scaling according to the value of γ'_p. For instance, the Geometrics G816 PPM used $M = 64$ and $F = 2.857252 \, \text{MHz}$. The interval D spanned 2^{20} cycles of F, about 0.37 s. The final count C was then equal to the magnetic field, B, in units of nT:

$$C = \frac{64 \times 2^{20}}{2857252}f = 23.4872f = \frac{2\pi}{\gamma'_p}f \cdot 10^9 \, \text{nT}$$

Nowadays, PPM frequency-measuring schemes are assisted by microcomputers and digital signal processors. This computing power allows use of every part of the proton precession signal, to adopt noise-canceling techniques and to extract qualitative information on the measurement (Sapunov et al., 2001; Jankowski and Sucksdorff, 1996; Primdahl, 2002). Elaborate statistical analysis on the digitized time series of the decaying precession signal and Fourier techniques are used in proprietary algorithms by PPM manufacturers in order to increase the true resolution of their PPMs.

Factors affecting the accuracy of the PPM originate in the following

1. Erroneous frequency measurement. This is the most easily avoided error as frequency standards are readily available for control either as precision oscillators or by broadcasting (time signals or global positioning system (GPS)). The reference oscillator in the PPM's electronics should regularly be checked against a frequency standard.
2. Magnetically unclean sensors, which distort the field to be measured. This contamination will result in 'heading errors' where the magnetometer readings will depend upon the sensor orientation.
3. Failure to adhere to the conditions defining γ'_p: nonspherical sample (Primdahl et al., 2005), use of another proton-rich fluid than water (Hrvoic, 2001), etc.
4. Mechanical rotation. Clockwise or anticlockwise sensor rotation around the magnetic field direction will add or subtract to the precession frequency. This effect is readily noticed in PPMs on board rotating platforms such as rockets or satellites (Alexandrov and Primdahl, 1993).

A drawback of the PPM is the low sampling rate, and the dead times corresponding to the polarization phases, when no data can be obtained. The low signal-to-noise ratio of the standard proton precession signal makes it difficult to perform geomagnetic field measurements with repeatability better than 0.1 nT. The latter limitations result in an instrument with an overall accuracy of 0.2 nT in a field of about 50 000 nT.

PPMs are used extensively in ground, marine, and airborne magnetic surveys; they are routinely carried on space probes and satellites, and are almost ubiquitous in geomagnetic observatories. Provided its frequency reference is checked and adjusted regularly, the PPM is an excellent realization of the International Magnetic Standard (IMS) (see Section 4.2).

4.1.3.2 *Overhauser effect proton magnetometers*

In 1953, Overhauser predicted an effect now known as dynamic nuclear polarization (Overhauser, 1953a, 1953b), which enhances the initial alignment of the proton magnetic moments considerably. His idea was to impose an appropriate RF signal on the atom and thereby excite the electronic spins to higher, nonthermal equilibrium states, a process now known as electron spin resonance or ESR. Because of a quantum mechanical coupling between the electron and the nuclear spins, as the excited electron spins try to equilibrate to their lower states they would reorient the nuclear spins. The nuclear-spin polarization achieved in this way would be increased by a factor of about 1000, the ratio of the electronic to the nuclear magnetic moments. The idea was met with much skepticism: it was even deemed by some to contravene the second law of thermodynamics until late 1953 when it was verified experimentally by Carver (Carver and Slichter, 1953, 1956).

Dynamic nuclear polarization of the proton magnetic moments is employed in the Overhauser PPM, where it increases the signal-to-noise ratio and lowers the power requirements, since little or no direct current is required for continuous polarization. An Overhauser PPM sensor should consist of a container filled with a proton-rich fluid having at the same time free electrons available for RF ESR. This is obtained by dissolving a substance containing free radicals such as Tempone, Proxyl, or Trityl into the fluid (Primdahl, 2002). A cavity resonator, supplied by an RF generator via a coaxial cable, surrounds the fluid container. A coil also surrounds the container, which plays a similar pick-up and polarization role as in the standard PPM. Three different polarization schemes are known and in use at present, each requiring specific free radical substances and RF excitation characteristics (Sapunov *et al.*, 2001).

The increased complexity in the chemistry of the fluid and the electronics of the sensor results in a much improved signal-to-noise ratio for the Overhauser instrument compared to the standard PPM. Noise levels and a repeatability approaching 1 pT are achieved, but the absolute accuracy is no better than the 0.2 nT of the standard PPMs, mainly because no testing and certification procedures exist to improve this figure (see below).

By avoiding the power-hungry direct field polarization of standard PPMs, the Overhauser magnetometer can achieve relatively low-power operation, which is an advantage for an instrument intended to operate continuously in an unattended mode. A single reading uses about 1 W s (1 J) and the standby power supply can be as low as 50 mW.

A drawback of the Overhauser PPM stems from the unknown long-term reliability of the device, compared with the tried and tested standard PPM. The single electron of the free radical makes it somewhat unstable chemically. The useful lifespan of the dissolved free radical is often specified as being between 5 and 10 years, and signal degradation in the best devices has been observed for continuous operation much longer than this.

4.1.3.3 *Optically pumped magnetometers*

Like proton magnetometers, OPMs are scalar instruments but, unlike the standard PPM, they deliver a continuous stream of data, in the form of a frequency that depends on the magnitude of the magnetic field. OPMs are based on the Zeeman splitting of the electron energy levels of some alkali metal and helium atoms in a magnetic field. The optical pumping scheme allows measurement of this energy splitting – and therefore the magnetic field – with very high resolution (Alexandrov and Bonch-Bruevich, 1992).

The state of an alkali metal (or metastable ^{4}He) atom is determined primarily by its outermost (valence) electrons. In the presence of a magnetic field, atomic energy levels are split by an amount that is proportional to the magnitude of the field: this is the Zeeman effect. Optical 'pumping' (Kastler, 1950) refers to the populating of one of these Zeeman sublevels at the expense of another.

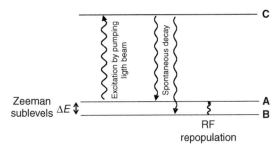

Figure 7 Schematic representation of the energy states of an atom in an optically pumped magnetometer. A and B are Zeeman sublevels, split by an amount ΔE in the presence of a magnetic field. C is a substantially higher energy state. See text for details of the optical pumping and RF repopulation processes.

The basic principle of optically pumped magnetometers is illustrated in **Figures 7** and **8**. A monochromatic light beam of suitable polarization is shone into a cell containing gaseous atoms of the alkali metal or helium. The wavelength and polarization are such that atoms from only one of the Zeeman sublevels, A, are excited to some higher state, C. After a very short time in state C, these excited atoms relax spontaneously to A or B with equal probability. Eventually, this process (excitation from only A, but relaxation to either A or B) leads to a situation with all the atoms in state B, and none in A. The light beam will no longer be absorbed – the cell becomes transparent to it.

Transmission of the light beam through the cell (to a photo-diode) serves as a switch to a coil around the cell that transmits an RF signal. The frequency of this RF signal is adjusted (via suitable feedback) until it corresponds to the Zeeman splitting between A and B. When this occurs A is repopulated and the light beam is once again able to excite atoms to state C, that is, light is again absorbed and the cell loses its transparency. This cycle therefore provides the basis of the feedback mechanism, while the RF frequency, f, provides the means to calculate B, the magnitude of the ambient magnetic field.

The Zeeman splitting, $\Delta E = \mathbf{\Delta\mu \cdot B} = hf$, where $\Delta\mu$ is the difference between the components of the atomic magnetic moment parallel to the magnetic field between the two sublevels. This is calculated from the quantum mechanics of the particular atom, using the Breit Rabi polynomial formula. The linear term is roughly equal to $28/(2I-1)$ Hz nT^{-1}, where I is the nuclear spin number, and determines the basic sensitivity of the magnetometer (**Table 1**). In some cases (e.g., potassium, helium) the Zeeman spectrum is 'resolved', and the polynomial coefficients can be calculated directly from fundamental physical constants: then the OPM has absolute accuracy (Alexandrov and Bonch-Bruevich, 1992; Gilles *et al.*, 2001).

The change in optical transmission upon application of the resonating RF signal depends in detail on the direction of the incident light beam relative to the

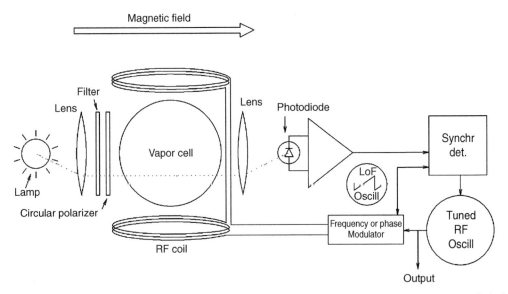

Figure 8 The operation principle of an M_z-mode optically pumped magnetometer. The RF frequency is usually in the hundreds of kilohertz while the low frequency, LoF, scans the spectral line at 10~100 Hz by modulating the frequency of the RF signal.

Table 1 OPM sensitivities and vaporization temperatures (depending on cell size)

Substance	Basic sensitivity ($Hz\,nT^{-1}$)	Vaporization temperature ($^\circ$C)
^{23}Na	7.00	100–130
^{39}K	7.00	40–60
^{41}K	7.00	40–60
^{87}Rb	7.00	25–35
^{85}Rb	4.66	25–35
^{135}Cs	3.50	20–30
^{4}He	28.0	–

ambient magnetic field. This has led to two categories of OPMs: M_z-mode OPM's, where the quasistationary change of the transmitted light intensity is monitored, and M_x-mode OPMs in which the modulation of an auxiliary light beam at the RF frequency is detected.

OPM's have a high sensitivity – some approaching noise levels of $0.1\,pT\,Hz^{-1/2}$. This makes them attractive for magnetic prospecting and aeromagnetic surveying, as well as for space-based observations. Their high cost, due largely to the short lifetime of the gas discharge lamp, has limited their use in observatories.

Rb and Cs are preferred by magnetometer designers, as their low vaporization temperatures are easy to achieve, and result in longer lifetimes; some have been measuring continuously for more than 10 years. ^{135}Cs, ^{85}Rb, and ^{87}Rb can all be used, but only ^{87}Rb has a well-resolved spectrum in the geomagnetic field range, making it suitable for an absolute observatory magnetometer. The first digital magnetic observatory was based on a rubidium OPM (Alldredge and Saldukas, 1964). Caesium OPMs are used extensively in aeromagnetic survey work. ^{39}K and ^{41}K also have resolved spectra in the geomagnetic field range and produce instruments of high sensitivity. However, the chemical reactivity of potassium, together with its higher vaporization temperature reduce the life of the vapour cell considerably. Furthermore, to achieve a high sensitivity requires a cell 150 mm or more in diameter, making a potassium OPM rather bulky. The fundamental physics of ^{4}He is rather simpler; however, there are technical problems to be overcome in its application (Gravrand et al., 2001; Blinov et al., 1984). Helium OPMs have been flown on several satellites and space missions.

4.2 Magnetic Observatories

A global network of geomagnetic observatories was proposed and initiated by Gauss and Weber in 1834 – the original set of observations providing the data for Gauss' first spherical harmonic analysis of the geomagnetic field. Nowadays geomagnetic observatories are charged with the task of maintaining continuous permanent records of all three components of the field. Measurements are made at intervals of between 1 h and 1 s and are referred to universal time. A resolution of $1 – 0.01\,nT$ is required. Vector measurements using, for example, fluxgate instruments are often complemented by scalar measurements of intensity to improve absolute accuracy. Geomagnetic observatory data is collected, collated, and disseminated by the World Data Centers (WDC). Nowadays near-real-time data is available through the INTERMAGNET website (see Section 4.2.3). Recently, the Virtual Global Magnetic Observatory Network has made geomagnetic data sets available online via search engines and object-building internet applications (Papitashvili et al., 2006).

4.2.1 Special Magnetic Conditions in an Observatory

A magnetic observatory should be constructed so that only the natural magnetic field is present. Therefore, all buildings intended to house magnetic instruments are made from nonmagnetic material (**Figure 9**). Additionally, the location is selected so that it is not situated on a local magnetic anomaly, be it from geological or artificial origin. Consequently, the magnetic field is very homogeneous: the magnetic field lines are parallel. Magnetic field differences within the observatory are very small and spatial gradients are low ($<1\,nT\,m^{-1}$). For the observation procedures to be valid, temporal magnetic field variations must be identical (within the observational error) inside the entire observation space.

As extremely high degrees of precision and accuracy are required in both angular and component measurements ($\sim$1 arcsec and 0.1 nT), recording instruments must be installed on specially constructed, highly stable 'pillars'. Horizontal directions are referenced to True North; therefore, a target should be available whose azimuth is known to the same accuracy and precision. Over time, these finely tuned conditions may degrade through secular changes in the environment or unauthorized introduction of

Figure 9 The geomagnetic observatory of Dourbes, showing the construction of the pavilions housing the observing instruments. To the left is the underground vault for the variometers: a high thermal inertia coupled with thick insulation ensures a low daily temperature variation in the vault. In the center is the absolute measurement pavilion, housing the reference pillar for the observatory. To the right, a technical pavilion for mains appliances and offices.

magnetic material to the observatory premises. Therefore, observatory conditions should be checked regularly and corrected if and as necessary.

4.2.2   Geomagnetic Observations Using the Baseline Concept

A complete description of the geomagnetic field requires the independent measurement of three angular or field components (see Section 4.1). The geomagnetic field varies with time, so observatory instrumentation must be of the recording type. At present, state-of-the-art component and orientation magnetometers, while potentially very sensitive and fast (up to the pT resolution at 100 Hz sampling rate), are not yet capable of measuring and recording all the field components continuously, with the required absolute accuracy. Therefore, a two-step observatory measurement procedure is generally employed:

1. A variometer is used to measure the variation of the field components about baseline values, in a continuous and unattended way, at the required sampling rate, say 1/minute.

2. Absolute measurements are performed manually (say 1/week) by an observer with adequate instrumentation (DIflux, proton magnetometer) to establish the values of the baselines mentioned in (1). Here the expression 'absolute measurement' means that the process of observation of the geomagnetic field must be fully traceable to metrological SI standards, and that the orientation of the geomagnetic vector is measured with respect to the local vertical and to geographic North.

If one can show that (1) the baselines remain stable between absolute measurements, and (2) the gradient between the variometer and the absolute measurement site is constant, then post-processing can be carried out to merge the two data sets and produce a final record having the accuracy of the absolute instrument, at the sampling rate of the variometer. It is expected that in the future fully automatic observatories will be available (see Section 4.2.4 on observatory automation) where the absolute measurement will also be made unattended.

4.2.2.1   *Absolute measurements of field components*
4.2.2.1.(i)   The International Magnetic Standard (IMS) Before the advent of truly absolute instruments such as the PPM (Section 4.1.3), based on fundamental principles of physics, absolute measurement of the geomagnetic field intensity was complicated and prone to many sources of error. The concept of an IMS was therefore introduced by Wienert (1970). The IMS is an ideal, hypothetical instrument with no systematic error, against which observatories' quasi-absolute instruments, such as quartz horizontal magnetometers (QHMs) and balances magnétique zero (BMZs) (see below) could be checked.

4.2.2.1.(ii)   Absolute measurement of the horizontal component, H, by the Gauss method Although rarely used nowadays, Gauss' method, devised in 1832, is of historical interest and importance. It enables measurement of both H and the magnetic moment, m, of a magnet, using only the magnet and a compass, and making measurements of only length, mass, time, and angle. In the past, this procedure was the cornerstone of the absolute determination of magnetic field.

1. The magnet is first suspended from its centre of mass and allowed to swing in the horizontal plane about its equilibrium position in the magnetic meridian. The period of oscillation, T, is measured. T is given by

$$T^2 = 4\pi^2 \frac{I}{mH}$$

where I is the moment of inertia of the magnet (about an axis through its center and perpendicular to its length), and can be calculated from measurements of its length and mass. The value of the product mH can therefore be found in terms of T and I.

2. The magnet is next laid in an East–West orientation and the compass placed a distance r from its center on the extrapolation of its axis. The compass needle experiences the resultant of the horizontal component of Earth's magnetic field, H, in a northerly direction, and the axial field of the dipole, H_r^* (* indicates the field on the axis of a dipole, moment m, a (large) distance r from its center is $H_r = 2\mu_0 m/(4\pi r^3)$), in an easterly direction.

The angle, α, of the compass needle from north is therefore given by

$$\tan\alpha = \frac{H_r}{H}$$

from which m/H can be found. Once both the product and the quotient of m and H are known, each can be found separately and the absolute measurement task is complete. For the highest accuracy it is necessary to consider further parameters (Laursen and Olsen, 1971) such as torsion of the suspension fiber, time changes in H, more accurate field patterns, and interactions of the axial magnet and compass needle, etc.

4.2.2.1.(iii) Quasi-Absolute measurement of H and Z using the QHM and BMZ The QHM was invented by Lacour in 1934, it was modified and developed by Lamont, and a Soviet version was developed at IZMIRAN. Although not strictly fulfilling the requirements of an IMS, the QHM provided valuable measurements in many observatories, particularly before the advent of the proton magnetometer. Its robustness, compactness, and practical design ensured its use until the

1980s, and the QHM is still used in some parts of the world.

An axial magnet, of moment m, suspended by a quartz fiber, is able to swing in the horizontal plane. The upper suspension point of the fiber is rotated until the mechanical torsion of the fiber equals exactly 360°, using the magnetic moment of the horizontal field on the magnet for detaining it. A total rotation of $360 + \theta$, where θ is the angle through which the magnet itself has turned is read from a graduated horizontal disk. If the torsion constant, τ, of the fiber is known, the horizontal component of the field, H, can be found by equating the torques on the magnet due to the fiber and due to H (Laursen and Olsen, 1971):

$$H = \frac{360\,\tau}{m\sin\theta}$$

Measurements made with a QHM are only quasi-absolute because they depend on recalibration of the torsion constant at 1–2 year intervals. The elastic properties of the quartz fiber and the magnetic moment of the magnet are temperature dependent, necessitating a temperature correction to each measurement made with the QHM.

The key part of the BMZ is a magnet balanced on knife edges so it is free to pivot about a horizontal axis. The vertical component of the ambient field is cancelled by a large magnet above the housing and fine adjustment of the position of a smaller magnet below the housing. The null position of the pivoted magnet is achieved by monitoring, through a telescope, the reflection of a light beam from a mirror mounted on the magnet.

Like the QHM, the BMZ requires careful calibration, and measurements made with it must be corrected for temperature, particularly of the upper magnet. It is therefore also a 'quasi-absolute' instrument.

4.2.2.1.(iv) Absolute measurements with the proton vector magnetometer (PVM) Although the proton precision magnetometer (PPM) is a scalar instrument, meaning it measures only the magnitude of the magnetic field vector, irrespective of direction (Section 4.1.3), its accuracy and simplicity of use have motivated researchers to devise ways in which it can be used for field component measurements as well. The PPM is placed inside a system of coils, by means of which known, auxiliary horizontal or vertical components can be added to the ambient geomagnetic field (**Figure 10**). Knowing the direction of the

Figure 10 A proton vector magnetometer, able to make absolute measurements of the horizontal and vertical component of the geomagnetic field. Note the special adjustments to individually level and orient the two sets of Helmholtz coils.

auxiliary field means that, in principle, information on the direction of the ambient field can be obtained from a comparison of PPM measurements made with and without the auxiliary field. A current generator able to set the coil current to about 1 part in 10^6 is required for results aiming at the 0.1 nT level of accuracy.

Considering the triangle defined by the ambient geomagnetic field vector, **F**, and an applied horizontal auxiliary field, **F**$_a$(**Figure 11**), the magnitude of the resultant vector, **F**$_r$ which will be measured by the proton magnetometer is given by

$$F_r^2 = F^2 + F_a^2 + 2FF_a \cos \alpha = F^2 + F_a^2 + 2F_a H$$

where $H = F \cos \alpha$ is the horizontal component of **F**, that is, the component in the direction of the auxiliary field. Direct use of this equation to find H requires knowledge of F and F_a. F is obtained simply from a PPM measurement with no auxiliary field. F_a however, requires accurate calibration of the coil system which is difficult to achieve to the same accuracy. Several methods of obtaining the horizontal H and vertical Z component of the geomagnetic field that avoid the need for this have been described. These generally involve switching the direction of an accurately constant auxiliary field, increasing F_a in multiples (de Vuyst and Hus, 1966), or varying F_a to minimize F_r (Hurwitz and Nelson, 1960).

The crucial aspect of the proton vector magnetometer (PVM) set-up is the need for accurate spatial orientation of the current carrying coils – which can be quite bulky. The magnetic axes should be oriented exactly along the local vertical or horizontally towards the magnetic cardinal directions (**Figure 10**). Orienting the mechanical symmetry axis of the coils often is not enough, and special manipulations must be performed to compensate for the non-coincidence of the magnetic and mechanical axes. Ultimately, a leveling precision of 3 arcsec must be achieved for H accuracy measurements of 0.7 nT in a typical midlatitude geomagnetic field.

Another drawback of the PVM method is that it needs quiet field conditions – the measurements involve several steps, during which the field should remain steady. It is therefore difficult to implement in high-latitude observatories, subject to frequent and large geomagnetic variations.

The PVM was being superseded by the DIflux (see below) at the end of the twentieth century, but it may well be revived again in the future; with smaller and faster proton sensors and compact coil systems available, it can be installed on the telescope of a nonmagnetic theodolite, alleviating the orientation and leveling problem (Sapunov *et al.*, 2006).

4.2.2.2   *Measuring the angular orientation of the geomagnetic field*
4.2.2.2.(i)   Declinometer The magnetic declination is the angle between the geodetic and magnetic meridian planes. The magnetic meridian plane can be defined as the vertical plane containing the magnetic axis of a magnet suspended so as to move freely in the horizontal plane.

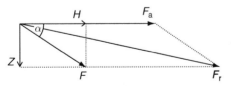

Figure 11 Diagram illustrating the principles of the PVM absolute component measurement. F is the ambient magnetic field vector, with horizontal and vertical components H and Z, respectively; F_a is the additional field (in this case horizontal) supplied by the auxiliary coil system. F_r is the total resultant field. See text for discussion.

These definitions lead to a straightforward principle for the absolute measurement of the magnetic declination – measuring the azimuth of the magnetic axis of a freely horizontally suspended magnet.

This conceptually simple procedure is not easy to realize, especially if accuracies of the order of the arc second are to be attained. Technical solutions (Laursen and Olsen, 1971) have to be found in order to suspend the magnet freely in a horizontal plane, to observe the orientation of its axis, and to measure angles between the meridians with the required accuracy, and to eliminate rheological effects, such as hysteresis and anelasticity, in the suspension fiber.

As a result the declinometer, or magnetic theodolite, was developed. It consists of a torsion head suspending the magnet in the required position. This head is mounted together with a telescope on a vertical-axis rotation table, indexed by a graduated circle. The telescope performs the dual task of observing a distant target with known azimuth, and to collimate on the magnet-end mirror. This mirror is mounted on the magnet with its optical plane normal to the magnet's magnetic axis. Angles can typically be obtained to an accuracy of 0.1 arcmin or even 1 arcsec.

The declinometer has been superseded nowadays by the DIflux, which offers operational simplification and an increased accuracy (see below).

4.2.2.2.(ii) Inclinometer

Historically the first inclinometer – the 'dip circle' – was built as a needle magnet free to move around a horizontal E–W axis. A vertical graduated circle, centered on this axis, allowed measurement of the angular position of the magnet in the magnetic meridian, and this gave the magnetic inclination. Despite clever procedures to eliminate the effect of gravity and to minimize collimation errors, the dip circle never achieved the high accuracy required in a magnetic observatory.

Another inclinometer, known as the 'earth inductor' with better metrological characteristics was introduced at the beginning of the twentieth century. This device was the first in geomagnetic observatories to use an 'electronic' sensor as a null indicator. This apparatus uses the emf induced by the geomagnetic field in a coil rapidly spinning about one of its diameters, as an indicator of its orientation relative to the field vector. When the coil's spin axis is collinear with the field, no emf is generated. Small deviations from colinearity can be amplified with high gain to give a restoring signal limited only by system noise. The early amplifiers were galvanometers deflecting light spots, and the rectification of the small alternating emf's by mechanical rotating contacts was not always satisfactory. Parasitic DC offsets were eliminated by observation with the coil oriented in symmetrical positions. Play in the bearings of the spin axis was probably the main limiting factor of accuracy. A modern version of the earth inductor, the Turbomag (Schnegg and Fischer, 1991), uses hydrodynamic bearings and contactless transfer of the emf to the observer.

In principle, the earth inductor could have been used for declination measurement also but, apart from the Turbomag, this did not occur probably because the addition of a telescope would have made it too bulky. Nevertheless, the earth inductor paved the way for the development of the DIflux which will be examined below.

4.2.2.2.(iii) DIflux (DIM)

For angular measurements, state-of-the-art instrumentation is now provided by a device called the 'DIflux' or declination/inclination magnetometer (DIM), which is assembled from a nonmagnetic theodolite and a fluxgate sensor mounted on a telescope. The magnetic axis of the fluxgate should be parallel to the optical axis of the telescope (**Figure 12**). The DIflux was first described by Tenani (1941). By the 1970s it had reached an advanced level of development (Meyer and Voppel, 1954; Serson and Hannaford, 1956; Trigg, 1970). Further developments by Daniel Gilbert, Jacques Bitterly, and Jean-Michel Cantin at IPG Paris have resulted in levels of precision,

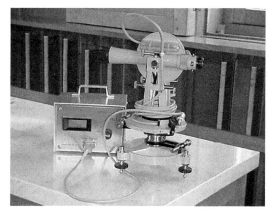

Figure 12 A DIflux with its electronic console. This instrument is able to measure the geomagnetic inclination and declination. A single-axis fluxgate sensor is located on top of the theodolite's telescope so that the magnetic axis is parallel to the telescope optical axis.

accuracy, resolution, and ease of use that make it currently the preferred instrument for angular measurements both in the observatory and in the field (Bitterly *et al.*, 1984).

The measurement principle takes advantages of the directional properties of the fluxgate; only the projection of the geomagnetic vector on the fluxgate axis is measured. Using the theodolite, an observer can orient the fluxgate in any direction while monitoring its electronic output and keeping track of its orientation in space via the theodolite's graduated horizontal and vertical circles. Usually, the preferred orientations of the fluxgate correspond to its magnetic axis being perpendicular to the geomagnetic field vector; then, the electronic output is close to zero and a high signal amplifier gain can be applied. The declination measurement involves setting-up the telescope axis into the horizontal plane, with the fluxgate axis normal to the magnetic meridian, so as to obtain a null from the fluxgate electronics. For inclination, the horizontal axis of the theodolite is set normal to the geomagnetic meridian and a fluxgate null is again sought. The essence of the measurement is therefore to determine the orientation of a plane perpendicular to the geomagnetic field vector. The plane's orientation is defined by two orthogonal lines in it, one horizontal and one in the geomagnetic meridian.

In practice a series of four measurements for declination and two for inclination (Lauridsen, 1985) is usually made, to minimize errors. The accuracy of a measurement with a DIflux depends on the accuracy and the magnetic cleanliness of the theodolite. The accuracy of a theodolite can be assessed by appropriate measurement techniques (Deumlich, 1980) and the magnetic cleanliness can also be measured and sometimes corrected to be below a given limit. We can therefore classify the 'clean' DIflux as an absolute instrument. In that sense it belongs to the family of the IMSs as defined in Section 4.2.2.

Instrumental errors originate in imperfect parallelism of optical and magnetic axes of telescope and fluxgate sensor, and in residual magnetization of the fluxgate core. Since the geomagnetic field may vary during a series of measurements, an external D and I variometer must keep track of this change if we want to have a truly absolute spot measurement of D or I. The proven best way to do that is by measuring the baseline of the variometer with the four D and two I measurements of the DIflux.

The DIflux performs well in equatorial and mid-latitude zones. The DIflux can also be operated in polar regions where the geomagnetic field vector is close to vertical. However, because the horizontal components are small ($<2000 \, nT$), horizontal angular measurements are not convenient there. Direct measurements of the small X and Y components using the fluxgate, properly oriented along geodetic North–South and East–West and with a scale factor directly determined with a proton magnetometer, are preferred (Gilbert *et al.*, 1988).

A problem facing the DIflux nowadays is in the supply of nonmagnetic theodolites. Therefore, the future is uncertain for the DIflux, and the development and marketing of a fully automatic DIflux would bring a solution (Section 4.2.4).

4.2.2.3 Variation measurements: variometers

In the present context, a variometer is a magnetometer designed to monitor the time changes of a magnetic field component relative to a fixed baseline. The variometers installed in magnetic observatories measure a variety of vectorial components depending upon the reference system used (see above). The most popular orientations are D, H, Z and X, Y, Z. The latter ensures that the variometer orientation will not have to be modified over time due to secular variation of the geomagnetic field.

The essential quality of a variometer is its ability to maintain a stable baseline between two absolute measurements. This means that it should have a very low drift over time and a very small dependence on temperature, pressure, humidity, etc. A variometer should respond to variations in the selected component only: contamination from other components must be eliminated. This often reduces to ensuring the correct orientation of the sensors. In the case of multiaxis sensors, the quality may be measured by the orthogonality of the set-up being close to perfection. The quality of the installation of the variometer must also to be considered: tilts or rotation of the pillar on which the variometer is installed will appear in the variometer's recording as a component variation.

4.2.2.3.(i) Classical magnet-based instruments Recording variometers originally used suspended magnets and photographic recording. They provided, and still do provide, much of the geomagnetic data going to the magnetic observatories databanks. Being suspended they are not

affected by pillar tilt and being mechanical they do not suffer from electrical noise; however, their sensitivity and dynamic range are limited, and they are now costly to maintain.

4.2.2.3.(ii) Digital variometers Many observatories now have electronic variometers with a digital data acquisition system that may be uploaded directly to data centers. Introduced from the late 1960s, electronic variometers were expected to obviate the disadvantages of photographic variometers, namely the lack of dynamic range and sensitivity. It was also felt that they would be more economical, being easier to install and offering the possibility of unattended operation. However, the introduction of these new instruments has come with some drawbacks, as the electronic components are less robust against electromagnetic/electrical disturbances than the mechanical configuration of suspended magnets. Many digital variometer installations have been lost as a result of a lightning strike on the observatory. Also, as the sensors are no longer suspended, the new instruments are more sensitive to tilt. The acquisition system samples an analog signal at discrete intervals, so low-pass filtering at at least twice the Nyquist frequency is necessary to avoid aliasing. This proves to be a problem with sensors such as proton magnetometers.

The sensors of digital variometers presently in use in magnetic observatories are mainly fluxgates (Rasmussen, 1990) or scalar magnetometers surrounded by backing-off coils (Alldredge and Saldukas, 1964; GEM website). Suspended (hanging or taut-fiber) magnets are still used in some designs, notably Bobrov, with attitude-restoring feedback coils (Jankowski *et al.*, 1984). As the necessary filtering cannot be applied to PPMs and as they are too slow to measure the high-frequency portion of the geomagnetic spectrum, OPMs should be used for variometric multiaxis recording (Gravrand *et al.*, 2001; Alexandrov *et al.*, 2004).

Perhaps the biggest problem with currently used variometers is their vulnerability to temperature variations. An instrument with a temperature coefficient of $0.1 - 1.0 \, \text{nT} \, \text{K}^{-1}$, in an environment that undergoes temperature variations of up to $10 \, \text{K}$ may result in errors of $10 \, \text{nT}$, or 30% of the typical daily variation in a component of the field. Clearly it is desirable to use variometers with low temperature coefficients to minimize temperature changes and to correct for temperature variation wherever possible.

4.2.2.4 Instrument certification and calibration

The community of geomagnetic observatories has always been concerned to set standards in order to maintain the quality of geomagnetic data produced and collated.

Not only should metrological standards be examined, but also other parameters that affect instrument performance, for example, temperature coefficients of variometers should also be monitored.

4.2.2.4.(i) Variometer certification The factors that affect a variometer's certification can in principle be deduced from an inspection of their baselines, which are obtained from measurements with absolute instruments. This is true for instrumental parameters like long-term stability, orthogonality, and scale factor accuracy. However, except for scalar measurements, the diurnal effects of temperature cannot be detected unless half a dozen or so absolute measurements are taken per day. This is not yet routinely possible. Therefore, the certification of a variometer is best obtained by intercomparison with a master variometer. This has been often realized at the 'Workshops on Geomagnetic Observatory Instruments, Data Acquisition and Processing' organized by International Association of Geomagnetism and Aeronomy (IAGA) every other year since 1987. Even better is the continuous running of two variometers set-up in slightly different conditions at the same observatory.

Some variometers have sufficient dynamic range for measuring the full field in all three axes. It is then possible to perform an experiment where the measuring variometer is oriented in various directions with respect to the field while a scalar magnetometer is recording close by. One may then equate the calculated modulus obtained from the variometer measurements to a reading taken from the scalar magnetometer. This will lead to a redundant set of equations from which a rotation matrix describing the errors of the variometer (orthogonality and scale factors) can be extracted. Subsequent correction of the variometer with the matrix operator will then provide a variometer reading with quasi-PPM accuracy (Merayo *et al.*, 2000; Gravrand *et al.*, 2001) for whatever orientation in space.

4.2.2.4.(ii) Calibration of scalar magnetometers Most scalar magnetometers do not in principle need to be calibrated, since their operation is traceable to fundamental physical constants. This is particularly true for the PPM, where, in addition to an accurate, precise value of the proton gyromagnetic ratio, only calibration of the frequency reference is required. The situation is more complicated for the absolute OPMs where several physical constants are required. Some, like the Landé factors for K, have not been determined to a high degree of precision, and these dominate the error budget of the OPM. Instrumental effects like light shifts and phase errors in electronic signals also need to be taken into account. The result of this is that scalar OPMs are quoted with varying and sometimes conflicting accuracies. A consensus exists, however, that the agreement between PPMs and OPMs is at the 0.1 nT level in a 50 000 nT field. This corresponds to 2 ppm. Two techniques exist for inter-comparing scalar magnetometers. The magnetometer exchange procedure (Rasson, 2005) is carried out in the ambient field and will therefore only give a difference for that value. The other technique uses artificial fields in a field stabilizer (Shifrin *et al.*, 2000) and will provide inter-comparison over an extended range of fields.

4.2.2.4.(iii)   Certification of DIflux (DIM)

Since the DIflux performs angle measurements, we cannot strictly speak of absolute measurements, as there need not exist a standard for angles. Nevertheless, a critical assessment of the accuracy can be made by investigating the angular accuracy (Deumlich, 1980) and the magnetic hygiene of the fluxgate-bearing theodolite.

Provided those two checks give results that comply with specified levels of accuracy, the DIflux can be declared certified. Note, however, that the two are not independent; an uncertainty (of, say 0.1 nT) in the fluxgate readings results in a corresponding uncertainty in declination or inclination (∼1 arc-second in declination, in a horizontal field of 20 000 nT).

Another frequently used procedure for certifying DIfluxes is through participation in the 'Workshops on Geomagnetic Observatory Instruments, Data Acquisition and Processing' already mentioned. One of the main activities in this kind of international workshop is the measurement of the baseline of a stable onsite variometer by all the participating DIfluxes. As all devices should measure the same constant baselines, any deviation of an instrument indicates a fault either in the angle reading or in the magnetic hygiene of the DIflux.

4.2.2.5   *Obtaining definitive absolute data*

As mentioned above, definitive geomagnetic field data at a magnetic observatory are obtained from two streams of data: variometric measurements and absolute measurements. Merging the two data sets is not a trivial or unique procedure and it requires all the skill of the observatory staff to result in an accurate time-series, giving the absolute value of the vector at each variometric sample. Important factors that impact on the procedure are

1. baseline stability of the variometer,
2. accuracy of the absolute measurements,
3. frequency and regularity of absolute measurements, and
4. data gaps in the variometer time-series.

Items (1) and (2) have been dealt with in earlier sections. Items (3) and (4) lead directly to the main problem facing definitive data production – the adoption of baselines.

In the simplest case of spherical coordinates for both absolute and variometric data, we have the fundamental magnetic observatory relationships between absolute (D, I, F) and variometric measurements (dD, dI, dF):

$$D = D_0 + dD$$

$$I = I_0 + dI$$

$$F = F_0 + dF$$

The subscript 0 indicates the baseline. In an ideal world with perfect instruments, the baselines would be merely constants, but instrumental drift due to temperature effects, mechanical creep, and pillar instabilities make the baselines wander in an unpredictable way. Values of the baselines D_0, I_0, and F_0, which are known every time an absolute measurement is taken (say one per week), vary slightly with time. But a value of the baselines must be available for every sample taken by the variometer. Adopting a baseline therefore entails the interpolation of the baseline from a sampling at one per week to the variometer sampling typically at one per minute. This can be conveniently done by fitting a mathematical expression like a low-order polynomial or a spline function to the baseline data sampled at one per week (**Figure 13**).

A final quality control should be made when the definitive data series has been obtained, in order to ensure that no errors resulting from faulty data processing software have occurred. A simple way of

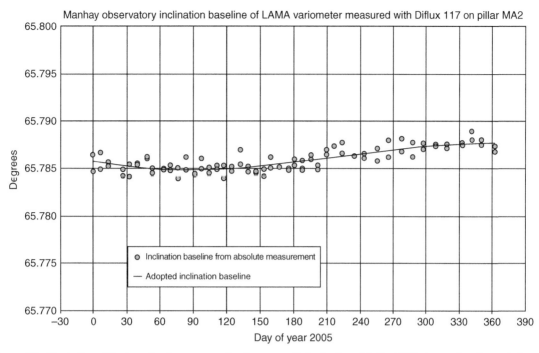

Figure 13 Illustration of the baseline (here measuring the magnetic inclination, *I*) in the MAB observatory for a whole year. Note the sensitive vertical scale of 0.005°/division, corresponding to about 5 nT/div. in *Z*. In this case, a third-order polynomial, fit by a least-squares procedure, has been adopted as a smooth baseline. The differences between measured and adopted baseline are of the order of 0.001°.

doing this is to go back to the absolute measurements performed during the data time series under consideration: the absolute spot measurements and the definitive data should agree. The adopted baselines should also agree with the absolute data baseline determinations.

4.2.3 The INTERMAGNET Magnetic Observatory Network

The International Real-time Magnetic Observatory Network (INTERMAGNET) was created in order to establish a worldwide cooperation of digital magnetic observatories. The networked observatories agreed to adopt modern standard specifications for measuring and data-logging equipment and to make data available in close to real time. Moreover, INTERMAGNET extends technical support for maintaining and upgrading existing magnetic observatories as well as for establishing new ones.

INTERMAGNET defines standards for the measurement and recording of the geomagnetic field, considering the state of the art. INTERMAGNET is constituted from existing groups who undertake geomagnetic observatory measurement. The acronym IMO is used to indicate an INTERMAGNET magnetic observatory (Green *et al.*, 1998).

Presently, INTERMAGNET data consists of time series of the geomagnetic vector, sampled at the round minute and carefully filtered to avoid aliasing effects. This data, collected at the IMOs represented on **Figure 14** (full squares), is continuously available from the geomagnetic information nodes (GINs, crossed circles) within 72 h. The data come in different accuracies and delays: reported (as recorded – near real-time), adjusted (corrected for artificial spikes and jumps), and definitive (reduced to baseline so that they have absolute accuracy). The latter are made available a few months after the end of each year at the earliest and finally with the production of a yearly CD-ROM.

People needing real-time data use the reported or adjusted data, available from the GINs in daily ASCII files either in INTERMAGNET imfv1.22 or IAGA 2002 format (e-mail request or ftp).

Access to recent definitive data is through the website (INTERMAGNET website). Older data are available on the CD-ROMs and at the website. Definitive data come in monthly files in a dedicated binary code. Multiplatform Java-based browsers are

INTERMAGNET observatories and GINs (2005)

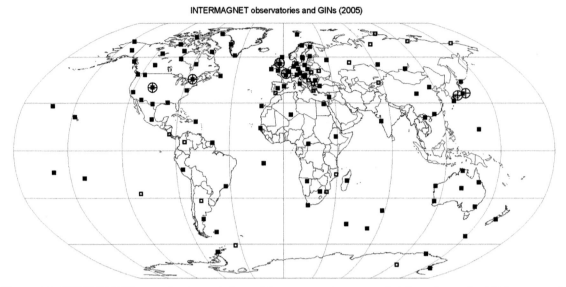

Figure 14   INTERMAGNET magnetic observatories as full squares and geomagnetic information nodes (GIN) as crossed circles representing the global coverage of the network in 2005. Open squares indicate some possible future IMOs (Robinson–Sterling pseudo-cylindrical projection).

available for easy perusal, inspection, and format conversion.

The data are available at no cost for *bona fide* scientific users. If any commercial aspects are involved, the user should contact the IMO directly for a financial arrangement.

All practical details about the data and their access are available from the website where the Technical Manual is also available for download.

4.2.4   Fully Automatic Magnetic Observatories

If the globe is one day to be covered by a uniform distribution of magnetic observatories separated by no more than say 2000 km, it will be necessary to install some in very remote and hard-to-access places such as on the deep sea-bottom (Chave *et al.*, 1995), on remote islands, and in inhospitable deserts. This is unlikely to be accomplished with classical observatories running the traditional way. Therefore, there is a need for fully automatic magnetic observatories.

Technology already exists for some geomagnetic measurements to be executed automatically.

1. Variometers are able to work for long time spans in an automatic mode, provided their baselines are regularly checked by absolute measurements. Only severe (>1 degree of arc) rotational motions

need to be monitored, but this task can also be performed automatically.

2. Provided electrical power is available, the magnitude of the geomagnetic field can be measured easily and absolutely by a scalar PPM in an unattended way for long periods. This is because a PPM does not require orientation or mechanical or electronic adjustments in order to produce absolute data.

Two approaches have been followed for obtaining automation of observatories:

1. The first is to set up a very precise (in terms of scale factor and orthogonality), stable, and temperature-insensitive variometer in a fixed orientation, in a clean geomagnetic environment. Then, baseline checks as far apart as 1 year may be acceptable (Gravrand *et al.*, 2001).

2. The second approach is to automate the full absolute measurement protocol, which comes down to automating the DIflux measurements, since the modulus can be measured automatically by a PPM (Rasson, 1996; van Loo and Rasson, 2006).

Some magnetometer set-ups flown on satellites have already reached a level of quality in baseline stability and precise orientation with respect to the stars such that they can truly be considered as automatic magnetic observatories in orbit (Nielsen *et al.*, 1995).

4.2.5 Magnetic Repeat Station Surveys

From the point of view of secular variation measurement, a magnetic repeat station plays the role of a miniature magnetic observatory, with elementary infrastructure and an observation schedule reduced to one measurement session every few years. This section on repeat surveys is therefore included with magnetic observatories, although, we are strictly dealing with a survey operation.

Instrumentation used during repeat surveys is nowadays similar to that used in an observatory. As pillars are seldom possible in the field, a nonmagnetic tripod is often used for setting up the DIflux and PPM.

For declination measurement, a device for measuring the geodetic azimuth of a distant target is necessary, at least when the repeat station is first installed. Provided there is sunshine, a theodolite with a solar filter is useful for this since then the astronomically known position of the sun can provide the target's azimuth (Rasson, 2005). Often the DIflux used for the magnetic measurements can be used for the sun sightings.

If the sun is not visible, a gyrotheodolite can be used. The true North direction is then obtained by sensing the direction of Earth's rotation axis by way of a suspended gyroscope. Differential GPS is also a possibility, but needs occupation of both the repeat station and target by the measuring team. A good topographic map may also help obtain the azimuth of the target.

When a measurement of the magnetic field is performed at a repeat station, it is a spot value measurement which can be notably different from the mean value at the site. Mean values, normally taken over one full year, are usually required in a repeat survey.

The difficulty then is to obtain this mean from a single measurement session at the station. Several procedures have been designed, where the spot measurement is linked in some way to a nearby observatory, where the annual mean is known. All methods try to eliminate or mitigate the main source of error in this linkage – differential field variations, mainly diurnal, between station and observatory.

The magnetic field variations at the repeat station site are recorded for a few days with a dedicated variometer. The spot value measurements are then used to establish the baselines of the onsite variometer. Several daily means can then be computed with absolute accuracy. Subtracting the corresponding daily means at the nearby observatory will then produce the difference in field elements Station – Observatory. The annual mean can then be obtained by simple addition.

The onsite spot value measurements are used to compute the baselines of a nearby magnetic observatory variometer. These baselines are subtracted from the baselines obtained from absolute measurements made at the observatory and give the difference in field elements Station – Observatory. The annual mean can then be obtained by simple addition. This method supposes that the (diurnal) variations are similar at both the station and observatory, at the time of the spot measurements. This method hence works well for a close-by observatory or for measurements taken when the diurnal variation at both places is small: dusk, sunrise, or sunset.

Onsite spot value measurements are performed at night, when the diurnal variation is negligible. With such a measurement, it is considered that an undisturbed field is obtained and that it reflects truly the secular variation. Direct comparison with a nearby observatory, also situated in the dusk zone, is allowed.

Obviously, quiet magnetic conditions are required for accurate results. IAGA has published a detailed guide for magnetic repeat survey practice (Newitt *et al.*, 1996), where the reader can find all the necessary practical information to set-up a magnetic repeat station network.

4.2.6 Products and Services Magnetic Observatories Can Provide

As mentioned in the introduction, magnetic observatories have a very practical purpose, as well as contributing to our knowledge of magnetic field of internal and external origin, and our understanding of Earth's deep interior. Navigation is a notable practical application, but there are also others like space weather, metrology, and magnetic signature determination. **Table 2** is an attempt to list, as completely as possible, the products and services a magnetic observatory is able to provide.

4.3 Magnetic Surveys for Geological Exploration

The general aim of a magnetic survey is to improve the resolution of the IGRF through a greater sampling density and/or frequency. This might be at a

Table 2   Products and services provided by geomagnetic observatories

Products/Services	Users
Value of the magnetic declination for various epochs and/or locations	Topographic and cartographic services, oil and gas companies, harbours, air traffic services, airports, military
Value of the secular variation of the magnetic declination	Topographic and cartographic services
International Geomagnetic Reference Field generation, magnetic charts	Topography, aeronautical mapping and safety services, maritime coastal and fluvial mapping, military
Time series of geomagnetic field	Aeromagnetic and marine surveyors, high-precision directional drilling for gas and oil, Earth orbiting satellite safety, space weather, military
Geomagnetically induced currents: nowcasting and forecasting	National and international electricity mains providers and operators, other cable network operators
High precision *in situ* measurements of the magnetic declination or other components	Airports, military, topography
Set-up and certification of compass roses (aircraft compass calibration and compensation facility)	Airports
Measurement of magnetic orientation of aircraft runways	Airports
Measurement of magnetic signatures of equipment, mechanical or electronic components	Manufacturers of magnetic resonance, military
Calibration and check of magnetic compasses	Leisure equipment industry, geometers, surveyors, antenna orientation for satellite TV
EM radiation standards of safety and security, Electrosmog	Government public safety agencies
Keeping of the magnetic induction metrological standards in the Earth's field range	Metrological community

regional level to better depict high-order features of the main field. Such low-density regional surveys are often repeated (reoccupying the same sites as far as possible) at 5–10 year intervals to monitor secular variation, and are discussed in Section 4.2.5.

At a more local level, surveys may be designed to resolve crustal magnetic anomalies (e.g., at a scale of 1:250 000), ore deposits or buried volcanic features (e.g., 1:20 000), or may be even smaller-scale investigations of sites of archaeological or environmental interest (e.g., 1:1000). Such surveys may be conducted over land, sea, or from the air. Usually only scalar (intensity) measurements are made but at fractional nT sensitivity. Absolute accuracy is of relatively little importance so such surveys are seldom rigorously linked to observatories, but meticulous care must be taken with the correction of temporal variations in the field. PPMs are ideally suited to local survey work on the ground, being inexpensive, quick and easy to use, and readily portable. Airborne surveys, of which many millions of line-km are flown each year, now usually employ cesium vapor magnetometers. At sea, a proton magnetometer may be towed behind a survey ship.

The main problem with making three-dimensional (3-D) vector measurements at sea and from

the air is the lack of a stable reference frame. Accurate navigation and determination of the vertical are essential and these factors ultimately limit the accuracy of results. A trade-off must usually be made between accuracy, which is improved by averaging over a number of measurements, and spatial resolution, which is compromised by such averaging. Most applied magnetic surveying therefore measures only the scalar magnitude of the total geomagnetic field, including the contribution of local anomalies.

4.3.1 Mapping Magnetic Anomalies

Local magnetic disturbances attributable to certain types of rock have long been recognized. Where these rock types are of economic importance, their magnetic effect may be used as a means of detecting them, even where the rocks are at depth or concealed below soil cover or overburden. Instrumentation to measure such local variations or 'anomalies' in the geomagnetic field in exploration mode – that is, carrying out a survey by making observations repeatedly in a pattern of observation points distributed systematically over a given search area – evolved rapidly during the second half of the twentieth century and eventually became the basis of an air-borne geophysical survey industry that now sets the pace in geological reconnaissance worldwide. By comparison, even with the aid of aerial photographs and satellite imagery, traditional geological mapping on the ground is prohibitively slow and expensive in many large areas of the world remote from modern infrastructure. Aeromagnetic surveys enable the groundwork to be done more selectively and efficiently and therefore play a significant role in mineral and petroleum exploration (Reeves, 2007).

Ground-based magnetometer surveys are still used to 'follow-up' selected anomalies mapped in airborne surveys in more detail, prior to drilling, etc., and have developed into specialized systems for particular localized applications such as engineering site investigations, detection of unexploded ordnance (UXO), and archeology. At sea, magnetometers are often towed behind marine research vessels. While this is slow and expensive on its own, cruises carried out primarily for other purposes can simultaneously conduct magnetometer traverses at little additional cost. Hence, applications of surveying with magnetometers are to be found in the air, on the ground, and at sea.

In recent decades essentially all magnetometers employed for work of this type have been electronic magnetometers that measure only the scalar magnitude of the magnetic field, regardless of its direction. This eliminates any need for precise orientational reference and so simplifies survey practice. After correction for temporal variations and subtraction of an appropriate IGRF, the anomaly recorded is then the component of the local-source anomaly in the direction of the main geomagnetic field (see Section 4.3.2). This is universally understood as the 'total field anomaly' in exploration magnetometry. Some increase in surveys that measure magnetic field gradients (either vertically or in a specified horizontal direction) has occurred in recent years. Advantages include improved resolution of near-surface sources. Usually gradients are measured by subtraction of the readings from two scalar sensors separated by a small distance.

Magnetometers have also been mounted in earth-orbiting satellites, most successfully in the OERSTED and CHAMP satellites of recent years. However, at satellite altitudes, the amplitudes of anomalies attributable to lithospheric sources are no more than about 30 nT and the resolution of detail no better than the altitude of the satellite, setting a rather coarse limit of 300–400 km for the scale of geological detail that is resolvable. (Chapter 6).

The production of magnetic anomaly maps and images in general has evolved rapidly in the last 50 years as a result of (1) improved, now exclusively electronic, magnetometer and ancillary instrumentation; (2) computer software and hardware to gather and process large volumes of data; and (3) image-processing techniques to display results in image format that lead to a largely intuitive appreciation of the geological information content.

4.3.2 The Origin of Magnetic Anomalies

While most rocks are only very weakly magnetic, any rock containing a quantity of a ferrimagnetic mineral such as magnetite distributed through its mineral fabric may display magnetic properties. The magnetization may be a long-lived, virtually permanent feature of the rock, having its origins in the geological history of the rock itself – for example, when it cooled through its Curie temperature (see Section 4.4) – and described as a remanent magnetization M_r with a direction that is, in general, unknown. On the other hand, the rock may have a certain magnetic (volume) susceptibility χ, in which case it will acquire an induced magnetization simply due to its presence in the present-day geomagnetic field. Such induced magnetization, M_i, is proportional to the product of the magnetic susceptibility and the inducing field and will be in the direction of the inducing field:

$$M_i = \chi F$$

Both types of magnetization may be present and their relative importance is quantified by the Koenigsberger ratio, Q:

$$Q = M_r / M_i$$

Considering, for simplicity, a single body of magnetized rock isolated within a large volume of nonmagnetic country rock, the magnetic rock body will possess a magnetic field that will be evident within its immediate vicinity and over a certain volume of space thereabout. This will be the case whether the magnetization of the rock is induced, remanent, or a combination of both. A scalar magnetometer measuring the total field strength within the vicinity of this body will experience the magnitude of the vector sum of the geomagnetic field and the local field due to the magnetic rock body (**Figures 15** and **16**). In the vast majority of cases, the latter is found to

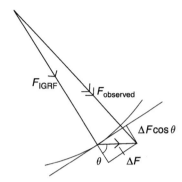

Figure 15 By convention, the 'total field anomaly' is the difference between the observed total field value, $|\mathbf{F}_{observed}|$, and the geomagnetic field value predicted by the IGRF, $|\mathbf{F}_{IGRF}|$. As long as the local anomaly is much smaller than the geomagnetic field, the total field anomaly so defined is a close approximation to the component of the magnetic field of the local source in the direction of the geomagnetic field, $\Delta F \cos \theta$.

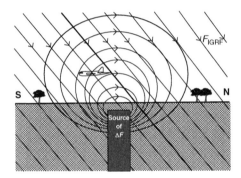

Figure 16 In the vicinity of a magnetized rock body, the passing magnetometer experiences the influence of both the ambient geomagnetic field, modeled by the IGRF, $\mathbf{F}_{IGRF}$, and a local-source field, $\Delta\mathbf{F}$.

be at least two orders of magnitude smaller than the former. The former is predictable at any given location from models of the geomagnetic field – the model provided by the IGRF (IGRF-10, website) being the one used almost universally in magnetic surveying.

What is measured by the surveying magnetometer is the vector sum of the global and local contributions and, by convention, the so-called 'total field magnetic anomaly' is what is left when the magnitude of the predicted field (IGRF) is subtracted from the observed field magnitude, once temporal variations have been eliminated. **Figure 15** illustrates that, to a close approximation, the anomaly defined as above is the component of the local-source field in the direction of the geomagnetic field. As a result of the

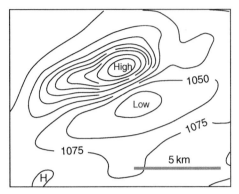

Figure 17 A typical aeromagnetic anomaly pattern in the vicinity of a single, compact and isolated geological source, showing both positive and negative parts (southern hemisphere). Contour interval 25 nT.

variations in the direction of the local magnetic field around the rock body, there will, in general, be areas where the local field tends to reinforce the geomagnetic field and others where it tends to oppose it. It follows that, when mapped over a given area around the body or 'source', some areas will show a positive magnetic anomaly value while others will show a negative one. A typical anomaly over a simple compact body is shown in **Figure 17**.

In general, the shape of any magnetic anomaly is a function of many factors, including the geomagnetic inclination and declination at the survey locality, the resultant direction of total (induced + remanent) magnetization within the body, the geometry of the body itself, its attitude in the ground (dip and strike), and the strength of its magnetization.

The shapes of anomalies over bodies of simple geometrical shape approximating to common geological bodies may be calculated directly, a process known as forward modeling. Such forward models help in interpreting the sources of observed anomalies in terms of the possible depth, dip, and geometrical form of the source body. While the inverse problem is, in principle, under-determined, reasonable geological assumptions or other *a priori* information can usually be brought to bear to add practical constraints in the inversion process that is often used in anomaly interpretation.

In practice, the flat-topped dipping dyke, extending down-dip to infinity and with parallel sides, is one of the more useful models for the geological sources of many anomalies. The anomaly to be expected over such a body with a vertical dip for a range of magnetic inclinations is illustrated in **Figure 18**.

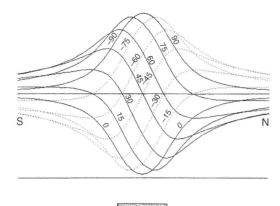

Figure 18 The magnetic anomaly over a vertical dyke-like body striking east–west and magnetized solely by induction varies with the angle of inclination *I* of the geomagnetic field at the locality of the body. The figure shows the curve family as I goes from +90° to 0° (northern hemisphere) as solid lines and from 0 degrees to −90° (southern hemisphere) as dotted lines. Note that the positive lobe of the anomaly is displaced from above the body in the direction of the equator in both hemispheres as the negative lobe grows on the side towards the nearest pole.

Approximations of magnetic sources by isolated poles or dipoles are found to be of little value in serious interpretation, particularly now that computer algorithms are readily available to compute accurately the effects of realistically dimensioned geological bodies in two and three dimensions. In specific applications, such as understanding the structure and stratigraphy in a potential petroleum province, the interpretation of gravity and magnetic data in terms of geological bodies and horizons of specified densities and magnetizations can be a helpful adjunct to understanding the stratigraphic and structural information contained in seismic sections. Since most sedimentary rocks are essentially nonmagnetic, the anomalies in such an area are usually of deep-seated origin, arising from the igneous and metamorphic basement below the sediments. Hence a rather small number of deep-seated anomalies are seen and their interpretation can help understand the structure of these older rocks, the downfaulting and subsidence of which led to the development of the sedimentary basin of interest for petroleum resources.

In normal survey practice, particularly in areas of Precambrian basement typical of solid mineral exploration environments, many hundreds or even thousands of anomalies are often detected

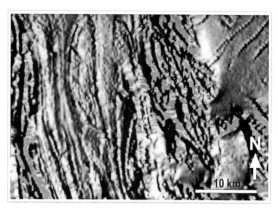

Figure 19 The magnetic anomalies in a certain area are displayed here as a gray-scale shaded-relief image. In many areas the geology gives rise to a plethora of anomalies that enable, in this case, the folds in the rock formations of the western half of the area to be traced quite clearly. Some subtle dyke anomalies appear in the NE corner of the area. Courtesy of Geoscience Australia.

(Figure 19). The extent to which any one of them may be quantitatively interpreted as suggested above depends to some extent on the degree of interference from neighboring anomalies. In any event, the trailing off to zero at infinite distance, predicted mathematically for all anomalies, is seldom observable to any great distance from the source. Often the interference from nearby anomalies is quite severe and the emphasis of the interpretation must then be mostly on qualitatively tracing patterns of anomalies from one place to another and reconciling these patterns with the (often isolated) exposures of certain suites of rock in the survey area. Even so, aeromagnetic anomaly mapping has proven to be a powerful tool in extending geological mapping beyond areas of relatively well-exposed geology in old (particularly Precambrian) terrains where the topography has become subdued and overlain by younger weathering products (see later sections).

It should be noted that it is rarely the case that eye-catching magnetic anomalies arise directly from important ore bodies. While this may have been the initial hope in developing the aeromagnetic survey method, experience has shown that the real value of aeromagnetic surveying is its role in building up a picture of the regional geology. From this picture, and knowledge of the types of geological environments in which certain types of economic minerals occur, selections can be made of limited areas in which more detailed and expensive investigations (such as ground geochemistry and drilling) may be worthwhile in order to isolate ore bodies of potential economic value. Even so, at a detailed scale of magnetic surveying on the ground surface, it is often the case that magnetic minerals occur in association with economic minerals even where the latter are themselves nonmagnetic. In such cases, magnetic anomalies may serve as pathfinders to economic targets and, in any case, provide an inexpensive way of tracing known economic occurrences along strike, for example.

4.3.3 Instrumentation Applied to Magnetic Surveying

4.3.3.1 Ground surveys

The earliest ground magnetometer surveys used sensitive dip needles, the deflections of which were sufficient to detect large deposits of magnetic iron ore. More sensitive mechanical instruments evolved, of which the Schmidt-type magnetic field balance became almost ubiquitous in mineral and petroleum exploration in the 1940s and 1950s, prior to the prevalence of electronic magnetometers. They were designed to measure the vertical (more rarely the horizontal) component of the geomagnetic field. The necessity to set up a tripod and level the instrument meant that progress was slow and surveys consequently costly, despite a poor sensitivity of only a few tens of nT.

The invention of the fluxgate magnetometer (Section 4.1.2) in the 1940s marked the arrival of the electronic age in magnetometry. Initially designed for airborne application, it was some time before the electronics were sufficiently lightweight and compact for hand-carried operations. The fluxgate element, if gimbaled to hang vertically in a damping medium, such as oil, could measure the vertical component of the magnetic field with a sensitivity sufficient to detect variations of a few tens of nT. Instruments such as the Jalander weighed only a few kilos and could be read in seconds so that long traverses (several kilometers) with frequent readings (every 5–50 m) could be accomplished each working day, even when combined with suitable survey practice to monitor and correct for temporal variations in the main geomagnetic field. As is often the case with exploration surveying, the quantity of observations and the uniformity of coverage of an area is more important than the absolute accuracy of each individual reading, the idea being to cover ground and detect interesting anomalies.

Improved performance in ground surveys was soon achieved by the PPM (see Section 4.1.3) in the 1950s and 1960s with the added advantage that it was the magnitude of the total field that was being measured, directly and absolutely to 1 nT accuracy, eliminating the need for any careful adjustment, orientation, or calibration of the instrument. This also brought ground surveys into line with airborne surveys where only the scalar magnitude of the total field has ever been measured from the early post-war days of the airborne fluxgate. Only relatively recently has the addition of a second or third magnetometer sensor been exploited to obtain magnetic gradient estimates either vertically, along track, or across track. Any possible theoretical advantages of making vector (as opposed to scalar) measurements in exploration practice have, even until the present, been outweighed by the additional technical complexity that would be entailed when working from a moving platform.

In the 1980s, with the advent of inexpensive and portable computer systems, ground magnetometry in which the operator had little to do except walk, push a button, and note down a reading became automated into integrated systems in which the x and y coordinates of each observation point and the time of the observation (for temporal variation corrections) were recorded directly in computer memory for later retrieval, correction with time-synchronized base-station readings, and plotted as contour maps or gray-scale images. Eventually, GPS readings became accurate enough that position could also be recorded directly (rather than the observer following pre-surveyed lines) and, in open terrain and small local areas, the use of multiple sensors could be employed to collect more than one line of observations with each passage of the observer or ground survey vehicle. Quad-bikes, mountain bikes, and many other types of transport suited to the terrain being surveyed have been pressed into service for this purpose. For archeological, unexploded ordinance, and engineering site investigation purposes, the readings may be as little as 50 cm apart giving resolution at a sub-meter scale where the sources are buried within the uppermost layers of the subsurface. Apart from ferrous metal objects that are magnetic, bricks and the clay in the immediate vicinity of open fireplaces will have acquired a magnetization that is detectable in such surveys and the method therefore offers considerable economy of effort in the excavation of sites of potential archeological interest. **Figure 20** shows, by way of example, the magnetic anomaly image over a settlement dated at 800–600 BC where ancient excavations of the subsoil to build

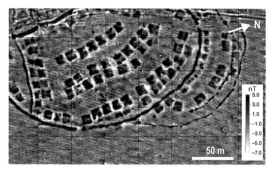

Figure 20 A carefully executed ground magnetometer survey over an archeological site in Siberia. The magnetic image reveals the locations of dwellings and ditches dated at 800–600 BC. The grid-squares on the figure are of dimension 40 m, and magnetic anomaly readings were made on a grid of 50 × 25 cm. Reproduced from Becker H and Fassbinder JWE (1999) Magnetometry of a Scythian settlement in Siberia near Cicah in the Baraba Steppe. In: Fassbinder JWE and Irlinger WE (eds.) *Archaeological Prospection, Arbeitsh. Bayerisches Landesamtf. Denkmalpflege*, Vol. 108, pp. 168–172. Munchen: Verlag Karl M. Lipp.

primitive habitations have left their mark in the details of the local magnetic field.

4.3.3.2 Airborne survey techniques

Airborne magnetic surveying has its origins in the Second World War in more ways than one. The necessary fluxgate magnetometer technology was originally developed for military applications such as submarine detection. But the war's legacy of aircraft and flying skills also contributed to a favorable environment for the application of the airborne magnetometer to exploration in peacetime. Early case histories were documented during the late 1940s (Reford and Sumner, 1964) as early fluxgate magnetometers underwent post-war refinement to improve their resolution and efficiency.

PPMs (see Section 4.1.3) made their appearance in the late 1950s and early adaptations to airborne use and to digital recording of the magnetic field readings are evident. Even so, the fluxgate magnetometer proved capable of further refinement and continued in service for many years. Optically pumped magnetometers (see Section 4.1.3) first came into airborne service in the early 1960s but they did not seriously replace earlier types until the late 1980s when expiry of the original patent led to their more general application. Since then they have become almost ubiquitous on account of their high sensitivity and fast reading rate. Helium, rubidium, cesium, and potassium types have all been used, but cesium vapor types seem now almost ubiquitous in the

industry. SQUID-type magnetometers (see Section 4.1.2) with vector capability are under development but so far only approach the overall sensitivity of standard systems based on the cesium vapor instrument.

4.3.4 Design and Execution of Surveys

4.3.4.1 Practical airborne magnetometry

The development of suitable magnetometers is only part of the story of aeromagnetic surveying. Many survey practicalities need to be addressed before the profiles of readings can be combined into a contour map – a hand-drafted map being the end product of early surveys (pre-1975 approximately) after several manually intensive intermediate steps. These steps included the planning and execution of an appropriately regular grid of parallel flight lines and their recovery in map form. In many frontier areas this was sufficiently challenging (due to poor or nonexistent topographic mapping) that early satellite imagery (1970s) was able to play a significant role as base maps in many surveys. Offshore, the conventional application of a downward-looking 35 mm camera to record the flight path was of little use and necessitated pressing various electronic navigation systems such as inertial navigation and Doppler navigators into service. A summary of these techniques, written at the time when they were about to be replaced by GPS, is given by Bullock and Barritt (1989).

To achieve good resolution of closely spaced near-surface geological sources, the airborne magnetometer needs to be flown at a low ground clearance, typically 60 or 80 m these days, and often only 150 m, even in the early years. For a similar reason, close spacing of the parallel survey lines in the grid of observations is essential if small features of geological importance are not to be missed. (Kimberlite pipes containing economic concentrations of diamonds, e.g., may be no more than 100 m in diameter). In the days of early reconnaissance work, the line spacing was often 1 km or more, but in modern surveys, particularly over the hard rock terrain typical of mineral exploration work, this is now commonly reduced to 400 or even 200 m. Worldwide, many millions of kilometers have been flown already in the attempt to achieve appropriate coverage of the geology of all continents. The national coverage of Australia, for example, now includes almost 20 million line-km of data gathered over more than 50 years (Geoscience Australia, website). Typical survey aircraft fly at about 250 km h^{-1} and operate for periods of 4–6 h each day, typically recording 20 000 line-km of data per month of operation.

Magnetometer readings are gathered at appropriately timed intervals – 1 s for the early proton magnetometer (equivalent to about 70 meters on the ground) and more recently ten or even one hundred times per second for modern cesium vapor instruments. Clearly there is a need for the electronic data gathering that made its appearance in the 1970s, first with dedicated custom-built systems and more recently with standard laptop computers and commercially available software.

Output of survey results in computer-generated anomaly contours, as opposed to contours handdrafted from analog paper-chart records generated during each flight, gradually became the norm in the late 1970s. Almost simultaneously, new computer software, originally developed for use with digital satellite imagery, was adapted for the display of aeromagnetic anomaly data sets as grayscale and color maps using, for example, shaded-relief effects to emphasize the geological content of the anomaly data sets (e.g., **Figure 19**) (O'Sullivan, 1991). Individual aeromagnetic survey projects with more than 1 million line-km of data have been carried out in recent years with the latest technology in Africa and Arabia. In Europe (outside of Scandinavia) and North America, generally older coverage to a lower standard is the norm, except where upgraded for oil exploration or, quite rarely now in Europe at least, for mineral exploration purposes. Images of outstanding quality and detail at scales compatible with regional geological maps (say 1:250 000) are now published routinely from these surveys.

Gamma-ray spectrometers are usually flown simultaneously to add information streams on the abundances of thorium, uranium, and potassium to the graphical output of many surveys. Airborne gravity anomaly mapping is sometimes carried out simultaneously with aeromagnetic readings, but this is much less common on account of the added cost of the gravimeter technology, making it attractive only in limited commercial situations such as frontier oil exploration areas with difficult ground access (Reeves and Bullock, 2006).

Arguably the biggest single innovation in more than 50 years of airborne magnetometry was the introduction of the GPS in the early 1990s. So many of the tedious and labor-intensive aspects of previous survey practice centered around navigation

of the survey aircraft as closely as possible along predetermined lines and, subsequently, the recovery of the flight path actually flown. Even after the arrival of the digital era in the 1970s, this part of the process remained primarily manual, up to the point where the flight path – recovered manually using the 35 mm strip films exposed during flight – was digitized and merged with the digitally acquired geophysical data. GPS provided quicker, better, and cheaper solutions to all these problems – in real time to an accuracy of about 5 m after the introduction of differential GPS. The regularity of the pattern of flight-lines so achieved was an added bonus to the quality of the survey data, quite apart from the accuracy with which individual features could be recovered on the ground. In addition, GPS facilitated the production of digital elevation models of the surveyed area, thanks to the combination of GPS (geocentric) heights for the aircraft and radar altimeter values for the aircraft's ground clearance.

4.3.4.2 Elimination of nongeological influences (aircraft, temporal changes)

A reading of a magnetometer on board an aircraft must clearly be corrected for any effect of the aircraft itself – primarily the engines and electrical currents, since most other parts of aircraft are now of nonferrous materials. The first-line approach is to mount the magnetometer sensor as far as possible away from such effects, usually in a fairing attached to the tail (**Figure 21**) and backing out as much as possible of any permanent magnetic effects.

Magnetic effects due to the varying attitude of the aircraft in the geomagnetic field (so-called maneuver noise) have been greatly reduced by the development of so-called automatic compensators. In calibration mode, such equipment records the variations in

Figure 21 An Aero-Commander Shrike aircraft fitted with a magnetometer in a tail stinger for aeromagnetic surveying. Courtesy of Geoscience Australia.

magnetometer reading as the aircraft executes a pre-determined series of maneuvers in pitch, roll, and yaw at high altitude (distant from geological sources) while flying on each of the cardinal compass directions. When switched to survey mode, the instrument then automatically applies in real time the appropriate correction for the actual attitude of the aircraft at any moment. A 'figure-of-merit' – the sum of the 12 terms, three maneuvers in each of four cardinal compass directions – is an assessment of the performance of the airborne system and commonly now is as low as about 1 nT. From the point of view of revealing geological detail, a major achievement in airborne survey practice has been the reduction of noise from all sources on the magnetic anomaly profile to as little as 0.1 nT. Since the interest for geological mapping is not confined only to the most magnetic rocks, even the most subtle anomalies are of potential value. So reducing the noise level lowers the threshold of what is detectable in anomaly mapping generally.

Noise arises from various sources. Maneuver noise is discussed above. The elimination of time variations in the geomagnetic field itself is achieved largely by monitoring temporal variations at a fixed base station during survey flying. The limit to this is set by the fact that geomagnetic micropulsations vary detectably from place to place, even over a few tens of kilometers, so the records of a single fixed base station do not perfectly represent the temporal variations experienced by the magnetometer in the aircraft throughout a survey area, parts of which are typically hundreds of kilometers apart. Nevertheless, the time-synchronized subtraction of total field readings at the fixed base-station magnetometer from the readings made in the surveying aircraft eliminates the bulk of the temporal variations. Corrections of several tens of nT are made typically in this way. The base station magnetometer should be sited in a magnetically quiet locality and its output carefully filtered to avoid the addition of (negative) noise when subtracted from the airborne profiles. On days when geomagnetic activity is high (magnetic storms), survey operations must be suspended. The magnetic base stations are rarely tied to a geomagnetic observatory. While this would be ideal, the remoteness of many survey areas dictates that the additional logistic effort required and the low commercial value of the absolute background level in any single survey (as opposed to the anomalies) usually makes this prohibitive. As a result, tying together a patchwork of hundreds of independently acquired surveys into a consistent national system

with a credible datum requires some innovative approaches (Tarlowski *et al.*, 1996).

The imperfections in the base-station subtraction process are almost always reduced further by flying surveys with a number of control lines or tie-lines at right-angles to the main survey direction. These are spaced such that an aircraft flying a survey line crosses a control line every few minutes – sufficiently often that departures from linearity in temporal variations are not severe. Once the aircraft magnetometer and compensator are properly calibrated, any difference between the two magnetometer readings (flight-line and tie-line) at each intersection point should be due only to temporal geomagnetic variations or, to be more precise, the imperfections of the base-station subtraction procedure in removing them. An iterative adjustment of each of the survey lines with low-order polynomial corrections to minimize these differences at all the hundreds or thousands of intersections in a survey usually makes, on average a few nT change to the near-DC level of each of the profiles.

The prevalent use of image-processing procedures for presentation of survey data as images (see next section) means that any remaining line-related noise may still be visible in certain image presentations. This residual can be reduced by applying a type of directional filtering procedure often called micro-leveling (Minty, 1992). Adjustments at this stage are at the level of a few tenths of a nT, but still as large as, or larger than, the noise level on each of the profiles. This is essentially a cosmetic procedure but it does significantly improve the signal-to-noise ratio in the final data when viewed as an image.

Exposing geological detail requires a minimum separation between airborne magnetometer and magnetic sources commensurate with operational safety of the aircraft. This leads to surveys with a terrain clearance of typically 60 or 80 m. This is achievable with safety in areas of subdued topography but needs to be adjusted in more rugged areas or around isolated hillocks due to the limited climbing capacity of any fixed-wing aircraft. Flying to a predetermined 'drape' on the topography means that similar altitudes are achieved at all the intersections of flight-lines with tie-lines in the survey, adding further precision to the data-reduction process in exchange for a small loss of geological resolution where terrain clearance needs to be greater than the minimum.

The free-air gradient of the undisturbed geomagnetic field with height is generally small (0.01–0.03 nT per meter), meaning that height differences at intersections need to be 10 m or more for differences in magnetic values due to this gradient to exceed the noise envelope in the profiles. This is comparable with the height discipline normally achieved in survey operations.

Positional accuracy for the aircraft of ±5 m is achievable routinely with differential GPS, meaning that the location of cross-over points is well determined, but it may still be prudent to exclude some cross-overs from the error analysis where exceptionally high magnetic gradients are noted and errors of a few nT may arise simply through a slight mispositioning of the crossover.

With all these precautions, flight-lines executed repeatedly over terrain with only very subtle anomalies show repeatability to about ±0.1 nT. This is comparable with the noise level on the profiles in general in modern data.

At the end of the data-reduction and processing cycle, maps and images revealing geological detail on the scale of a 50 m pixel are produced for interpretation. Commonly these maps and digital data sets are published by national government agencies as part of a nation's geological mapping program and compilations of such maps over whole continents will, eventually, provide valuable new insights into the, often hidden, geology of all the continental areas of the world. Thanks to modern communications, internet, and software technology, these data sets may be exchanged between organizations and clients worldwide, ultimately meaning that exploration decisions may be based on access to a maximum of factual *x-y* registered information from which intelligent interpretations of exploration priorities may be made.

Over the world's oceans, the magnetic anomaly data is mostly of shipborne origin and lacks the quality and resolution of airborne surveys since marine research vessels travel comparatively slowly, remotely from good monitoring of temporal variations, and seldom execute a regular pattern of closely spaced lines. A wealth of shipborne magnetometer data still awaits inclusion in rational worldwide databases (Reeves *et al.*, 1998). Nevertheless, the mapping of oceanic magnetic anomalies has played a key role in establishing our understanding of the way in which oceans grow at mid-ocean ridges with consequent separation of the once-adjacent continents and the improved appreciation of geological evolution that came with global tectonics. Over continental shelf areas critical for petroleum

exploration, airborne surveys will often be executed to the same critical standard as described above for land areas.

4.3.5 Data Presentation, Enhancement and Interpretation Methods

Since the data-capture and processing stream of modern magnetic anomaly surveys is entirely digital, the final product is also a digital data file or database in which all the collected data is preserved, organized sequentially by flight-lines, and where as many as necessary intermediate corrections are recorded against each reading. Assuming the data reduction and processing has been done correctly, a final corrected anomaly value will exist in this database for each of many millions of observation points. For a large survey this can involve gigabytes of data. Accessing small windows and individual profiles from such a database requires dedicated software if it is to be done conveniently and routinely. Commercial packages have been developed to enable users to do this on laptop PCs (e.g., Geosoft, website; Intrepid Geophysics, website).

For the human interpreter, the first priority is to visualize the data. This may be done on a profile-by-profile basis to inspect individual anomalies, but synoptic overviews of an entire survey are also necessary and best obtained by means of graphical displays of maps and images. The original method of display of magnetic anomaly data was contour maps. With the advent of personal computing power in the 1980s, contour displays have been largely replaced by images in which the anomaly values are converted to a gray scale or a color scale (Milligan *et al.*, 1991; Reeves *et al.*, 1997). The first step in this process is to interpolate the line-based data onto a regular raster or grid of values over the survey area. Typically the size of the grid element (pixel) is one-quarter to one-fifth of the flight-line spacing, so commonly 50–100 m for most surveys. Such a grid cannot represent the full information content of the original profiles along line (with samples at 7 m for a 0.1 s sampling interval) and anti-alias filtering is needed to avoid the smallest anomalies being represented as broader ones in the grid. Across-lines, there is often an under-sampling problem since the line spacing, for cost reasons, often does not meet the strictest sampling criteria. This can be solved in most cases by interpolation along the predominant strike direction of the geology where anomalies change most slowly from one flight-line to the next.

Once available in grid format, many methods of visualization of the gridded data are available (Milligan *et al.*, 1991). A gray-scale representation may be the simplest of these, with the highest magnetic values portrayed as white and the lowest as black (or vice versa) with a suitably stretched gray-scale between. Often more satisfactory is a gray-scale shaded-relief image in which the 'topography' of the anomaly field is displayed as though illuminated by a light source in a given direction (azimuth and altitude). Diffuse reflection from the surface of the magnetic topography is calculated such that slopes facing the illumination source appear brighter than those sloping away from it (**Figure 19**). This gives the human eye a clear impression of the undulations in the anomaly field and has the added benefit that the average magnetic anomaly background level (which may be ill-determined and of little value in studying the near-surface geology) is effectively lost among the average gray levels of the spectrum.

Color may also be used for display purposes, usually with a natural color spectrum and with the highest magnetic values in red and the lowest in blue. Such a presentation may even be combined with the gray-scale shaded-relief effect, or the color saturations may be adjusted to give the eye an optimum impression of the pseudo-topography (Milligan *et al.*, 1991).

All these techniques evolved quickly in the early 1990s and have become universally accessible to users working with commercial software packages. Display of the aeromagnetic data geographically registered with, for example, the pre-existing geological map (or any other geophysical or geochemical data set) is then a powerful tool for the interpreter to use and this is readily available in the geographical information systems (GIS) that are in common use with individuals and groups working with x,y-referenced geoscience data.

Once the data is in a digital gridded format, it is also readily amenable to digital filtering processes that may be operated in the space domain. These may be useful to smooth the data or to enhance certain types of feature, with a directional bias, for example. Calculating gradients across-strike will clearly enhance features striking in one direction, while the same features may be largely subdued by calculating the gradient along-strike. This is analogous to choosing a direction for illumination in the shaded-relief map. Clearly many different options are available.

A further suite of opportunities for data processing arises from the Laplacian nature of the magnetic anomaly field (Spector and Grant, 1970). Measurements of any potential field on one plane (if it is done thoroughly at all wavelengths, as is attempted in aeromagnetic surveys) are sufficient to calculate the field that would be observed on any different plane, permitting what is called upward and downward continuation of the data. Conveniently, this is done by way of a Fourier transform of the data into the wave number domain where upward and downward continuation operators are relatively simple filter functions that may be applied to the wave number and phase spectra. Inverse Fourier transformation restores the filtered result into a new grid in the space domain for display using any of the above gray-scale or color raster techniques.

Similar operations may be performed for calculating gradients in the data (vertical or in any chosen horizontal direction) and, by calculating suitable low-pass and high-pass filters based on the analysis of the wave number spectrum, regional and residual components of the anomaly field may be separated out.

The shape of a typical anomaly depends in part on the magnetic inclination at the survey locality (**Figure 18**). Through adjustments to the phase spectrum in the Fourier domain, anomalies may be transformed into the simple positive shape typical of bodies at the geomagnetic poles, bringing the positive anomaly above its source body. This process is called migration (or reduction) to the pole and is often considered an advantage in interpretation, though it is not without some problems at low magnetic inclinations where north–south trending bodies have very small amplitude anomalies in theory. The desired result assumes that source bodies are vertical and that remanent magnetization plays an insignificant role. Generally speaking, the results are nevertheless useful. The persistence of occasional negative anomalies in a map processed in this way is not disastrous and simply signals that one or both of these two assumptions is invalid locally.

The approach may be taken a step further by assuming the geology to be made up of a raster of vertical prisms of size equal to the grid-cell size of the data and calculating the equivalent magnetic susceptibility necessary for each prism to produce the observed magnetic anomalies. This, again, is seldom perfect in its application but is reasonably valid in many metamorphic terrains such as Precambrian shields. It can form an important step in the interpretation process that may be seen as attempting to go from the continuous Laplacian magnetic anomaly field to the discontinuous geology where changes in lithology and, for example, faults bring different rock types with different magnetic susceptibilities into contact over very short distances.

Other techniques used routinely to process data and produce outputs that help the extraction of geological information include the calculation and plotting of Euler solutions that indicate the depth of source bodies based on the curvature of their anomalies (Reid *et al.*, 1990) and the calculation of analytic signal expressions (Roest *et al.*, 1992). The latter are the positive envelope to the curve family shown in **Figure 18** and so are independent of magnetic inclination. Positive analytic signal values are also to be found over faults and contacts; however, they do not necessarily highlight only the main magnetic bodies in an area, though in experienced hands they do have considerable value.

Despite all these modes of assistance, the translation of the aeromagnetic anomalies into a geological map – interpretation *sensu stricto* – remains a largely intellect-driven endeavor at present. This is based on the physical constraints surrounding the origin of anomalies and the integration of quantitative interpretation results with more qualitative approaches as well as the constraints provided by other data such as the pre-existing geological map and other layers of geophysical (gamma-ray spectrometry, gravity) or geochemical information. An illustration of magnetic anomalies over an area of Precambrian shield in western Australia and their interpretation at the hands of a skilled geologist is given in **Figure 22**. An understanding of the magnetic properties of rocks in general (next section) undoubtedly plays an important part in this process, though information in any given area on this latter point is almost always incomplete. While a geological map shows the formations that outcrop, or at least sub-outcrop below soils and overburden, the magnetic anomaly pattern also contains information from more deeply buried geological sources. An interpretation map often, therefore, has to include elements of this third dimension of the geology that does not appear on conventional geological maps.

4.3.6   The Link between Magnetic Properties of Rocks and Regional Geology

The fact that magnetic properties are retained at all temperatures below the Curie point means that virtually the whole of the Earth's crust contributes to

magnetic anomalies and that it is the bulk of these igneous and metamorphic rocks that is represented in the anomaly patterns, though in general the effects of the shallower rocks will predominate over the deeper ones. The fact that the wavelength of anomalies increases with increasing vertical distance between source and magnetometer means that processes of wavelength-based filtering (e.g., after spectral analysis of anomaly patterns, see previous section) can be used to emphasize sources at different depths. Often, for geological mapping, it is the shallowest sources that are of most interest. These can be enhanced (and the deeper sources suppressed) by high-pass filtering or the calculation of a residual or a vertical derivative field from the observed field data (see previous section). Vertical gradient magnetometers similarly are more sensitive to the effects of shallow sources, so vertical gradient surveys tend to reveal the effects of near-surface rocks more clearly than total field surveys. Even quite magnetic rocks disposed in thin, flat-lying layers at or near the Earth's surface, such as lava flows, have relatively little effect in comparison to the 'basement' when it comes to geological mapping, however. So the ability to map underlying crystalline rocks below any surface formations is well-entrenched in the aeromagnetic method.

In the interpretation of anomalies, some understanding of the magnetic properties of rocks in general is an essential part of deriving useful solutions to the mathematically underdetermined problem of inverting anomaly patterns in two dimensions into models – even simplistic ones – of the subsurface in three dimensions. While the geometrical parameters of buried rock bodies in a given area may be unknown, the magnetic properties of representative rock-types may be determined independently in the field or laboratory. Experience gained from such measurements on common rock-types around the world may be helpful, even in an area where the local geology is essentially unknown.

Magnetic properties reside mainly, but not exclusively, in the magnetite grains that rocks contain. Magnetite content is, unfortunately, largely independent of the abundance of the main minerals that determine rock lithology as described by the geologist and petrologist. These basement rocks display a wide range of magnetic properties that are not readily summarized by simple rules that can be used to deduce rock types from their anomalies (Grant, 1985).

Apart from magnetite and the mixed oxides of iron and titanium, only the sulphide mineral

(a)

(b)

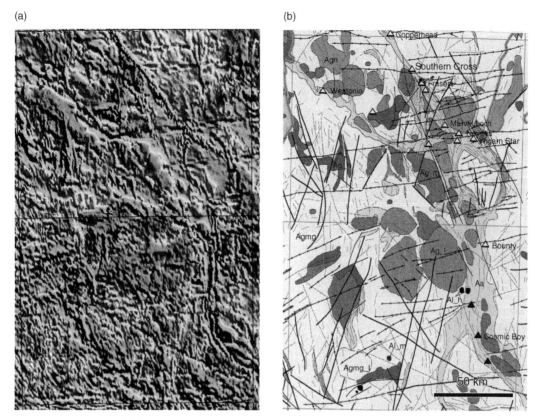

Figure 22   (a) An area of the composite aeromagnetic anomaly map of Western Australia and (b) its geological interpretation. Courtesy of Geoscience Australia and the Geological Survey of Western Australia.

pyrrhotite contributes to magnetic anomalies. At low concentrations, a fairly simple, monotonic relationship is found between magnetite (or pyrrhotite) content and the magnetic susceptibility of a rock. However, there is little relationship between magnetite content and the gross mineralogy of rocks and hence their more conventional petrological classification. It is more useful to relate the anomaly patterns to geological units known in the area, or known from their typical expressions elsewhere. For example, long rather straight anomalies cutting through all other patterns are typical of dykes intruded through cracks or fissures long after the bulk of the country-rock had solidified (**Figure 23**). Similarly, an oval area of anomalies with a texture distinctly different from the surrounding rocks is typical of a granite diapir emplaced long after the surrounding rock solidified.

While the direct relation of magnetic anomalies to specific igneous and metamorphic rock types is not possible in interpretational exercises (except where independent evidence is available from ground

studies), the fact that rock types have a bimodal distribution of susceptibility with the two populations separated by at least two orders of magnitude (**Figure 24**) means that many lithological boundaries in such geological terrains are coincident with magnetic property boundaries. Hence, even where two adjacent rock types are both unknown, their contact can often still be traced through aeromagnetic mapping. Magnetic anomaly patterns therefore provide a wealth of information on the general disposition of rock types in an area – their fold patterns, the brittle faults separating terranes with different tectonic histories, and the intrusion of geologically younger features (such as dykes, for example) that cross-cut the older rocks . They have therefore become an essential part of systematic and efficient geological mapping and mineral exploration in most parts of the world.

These evident difficulties notwithstanding, there appears to be no let up in the application of aeromagnetic anomaly mapping as a method of extending geological understanding of our continents and their

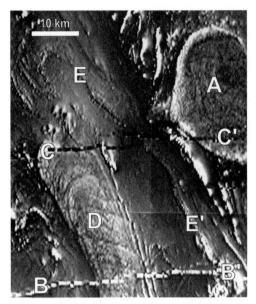

Figure 23 A gray-scale shaded-relief image of a small area of Western Australia showing the magnetic expression of several typical geological units. (A) Granite pluton with highly magnetized rim; (B-B') dyke with normal magnetization; (C-C') dyke with reversed magnetization; (D) granite; (E-E'): greenstones. Courtesy of Geoscience Australia.

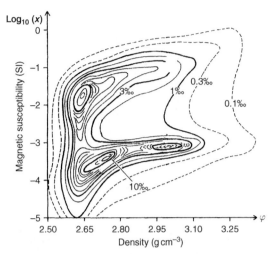

Figure 24 A frequency distribution plot of magnetic susceptibility against density for almost 30 000 rock specimens from northern Scandinavia shows a bimodal distribution between two populations of rocks with a difference in magnetic susceptibility of almost two orders of magnitude – effectively 'magnetic' and 'nonmagnetic' in terms of magnetic anomaly mapping. Note that, while some increase in susceptibility with density (and the higher proportion of mafic minerals) is evident, many highly magnetic rocks are of low density and must, therefore, contain few of the dark minerals found in mafic rocks. Modified from Henkel H (1991) Petrophysical properties (density and magnetisation) of rocks from the northern part of the Baltic shield. *Tectonophysics* 192: 1–19.

resources towards a more universal coverage of magnetic anomaly maps at global scale. Efforts are currently underway to compile the first global map of magnetic anomalies that, it is hoped, will be published in 2007.

4.3.7   Satellites and Space Probes

The first satellite to carry a magnetometer into Earth orbit was the Soviet Sputnik 3 in 1958. Since then most satellites and space probes have carried magnetometers of some sort. Early US missions were restricted by low payload requirements, and used a single search-coil design. Later on, single fluxgates were employed, and by varying the fluxgate orientation, three-component vector measurements were possible. The astronauts of the Apollo 12, 15, and 16 missions made vector measurements on the lunar surface, Apollo 16 using the newly developed ring-core fluxgate.

Relatively few satellites have been put into orbit specifically to carry out high-resolution magnetic surveys from space. MAGSAT (Langel *et al.*, 1982) carried a purpose-built fluxgate, together with a cesium vapor scalar instrument to calibrate the

intensity while in space. It took measurements during 1979 and 1980. Twenty years later, the Danish satellite OERSTED, launched in 1999, carried a three-component fluxgate mounted alongside a star camera for absolute orientation, as well as an Overhauser effect proton magnetometer. However, a better resolution, spatial and temporal coverage are achieved by Challenging Minisatellite Payload (CHAMP), launched in July 2000 and managed by GFZ, Potsdam (CHAMP, website). CHAMP is in a low-altitude, near polar orbit, which makes it ideal for high-resolution gravity and magnetic measurements. The magnetometers carried by CHAMP are similar to OERSTED's, but the low Earth orbit and improved GPS technology result in much better imaging of crustal anomalies. CHAMP's lifetime was originally predicted to be 5 years, but at the time of writing it has been reliably transmitting data for nearly 7 years. This duration means it is possible to begin to investigate short-period features of the geomagnetic secular variation: sudden changes such as jerks, westward drift, etc.

4.4 Paleomagnetic Methods

Direct measurements of the direction of the geomagnetic field exist only for the past 400 years. The record of intensity measurements is even shorter. However, these records indicate that many of the interesting features of the geomagnetic field operate on much longer timescales. So, to gain a complete understanding of the geomagnetic field, or even to be able to say whether Earth has always possessed a magnetic field, much longer records, extending back into the geological past, are required.

Paleomagnetism provides such records through the remanent magnetization of natural materials and artifacts for which the age of the magnetization can be reliably estimated by independent means.

Most rocks and sediments contain small concentrations of ferrimagnetic minerals, for example, magnetite, hematite. The two main primary processes by which these can attain a stable natural remanent magnetization parallel to the ambient geomagnetic field are thermoremanent magnetization (TRM) and detrital remanent magnetization (DRM). A TRM is acquired as a lava, volcanic or igneous rock, or a fired pot or kiln cools; as magnetic grains pass through their Curie temperatures they acquire magnetic moments, and these stabilize on further cooling through an, often well-defined, blocking temperature. At ambient temperatures thermal relaxation times can be of geological duration. Theory and experiment show that the intensity of a TRM or a partial TRM is directly proportional to the intensity of the magnetizing field. Hence it is, in principle, possible to retrieve information on the paleointensity of the geomagnetic field by normalizing a TRM. Delesse (1849) and Melloni (1853) were the first to study TRM in volcanic rocks. Reversely magnetized lavas and baked contacts were reported by David (1904) and Brunhes (1906), sparking the suggestion of polarity reversals, while Chevallier (1925) published remanent directions from lava flows on Mt. Etna, that showed evidence of secular variation.

By contrast, in DRM the grains already carry stable magnetic moments, and, on the average, they align with the ambient magnetic field during depositional and post-depositional processes. DRM occurs in sediments, sedimentary rocks, and sometimes in wind-blown deposits or loess, and is generally, though not always, weaker than TRM. Retrieval of the paleointensity of the geomagnetic field from a DRM is complicated by the many sedimentological factors that affect the magnetization process; at present, it is possible only to estimate relative paleointensities from sequences of sediments that are extremely uniform in character.

The accuracy and age control of paleomagnetic records are very variable. In favorable situations an accuracy of a few degrees can be attained in direction, but 10–20% is usually regarded as good for absolute paleointensity determinations. Whereas archaeological artifacts can often be dated to within decades or even years, radiogenic isotope methods are most commonly used for natural materials, and uncertainties of a few percent of the age are common. The sequential ordering of sedimentary deposits and sedimentary features such as climate-induced cyclic structures that can be tied to astronomical cycles can also provide invaluable constraints in dating sedimentary records of the past few million years. The fact that the main geomagnetic field has reversed its polarity at irregular intervals throughout geological time more generally means that it can be used as the basis of a unique and global timescale of immense value in dating geological events. The development and dating of the geomagnetic polarity timescale (GPTS) is described in Chapter 12.

Natural materials may also be magnetized or remagnetized through reheating, recrystallization, diagenetic and other secondary physical or chemical changes. Detailed descriptions of natural magnetization processes are given in Chapter 8.

4.4.1 Sampling

4.4.1.1 Introduction

High-resolution rock magnetometers (see below) are precision instruments, housed and operated in modern sophisticated laboratories. Natural and archaeological materials must therefore be carefully sampled and brought to the paleomagnetic laboratory for measurement. A crucial part of the sampling procedure is the recording of the orientation of the sample while *in situ*, that is, before removal from the outcrop, exposure, or site. Without such orientation the magnetization vector cannot be referred back to a geographic reference frame. It is also important to establish whether the material is still in the orientation in which it was magnetized, or, if not, whether a correction can be made for any disturbance that has occurred since that time. In rocks this usually means looking for an indicator of the paleohorizontal, for example, a bedding plane in sediments or possibly flow marks in lavas or ignimbrite flows, and careful measurement of its attitude.

4.4.1.2 Consolidated rocks

The most popular method of sampling consolidated rocks is by use of a high-speed drill with a diamond matrix-tipped drill stem, usually 2.5 cm in internal diameter, and usually water-cooled. Various models have been designed: modified chain-saw motors, petrol-fuelled; modified electric drills, used either with rechargeable batteries or a portable generator. A petrol-fuelled drill, with pressurized water-cooling is shown in **Figure 25(a)**. Outcrops are often remote and access is frequently difficult or awkward, so portability is important, as is the availability of water.

Drilled samples are usually between 5 and 15 cm in length, depending on the hardness and structural integrity of the rock. Before removal from the outcrop, the sample must be oriented with respect to north and the vertical. This is usually achieved by measuring the plunge below the vertical, and the azimuth of the core axis (or a pair of angles simply related to these), with an inclinometer and a compass (**Figure 25(b)**). A fiducial line is drawn lengthwise along the upper edge of the sample, the front (outside) face marked and the sample clearly labeled following laboratory conventions (**Figure 25(c)**). Every laboratory has its own set of conventions for the orientation and labeling of samples, which have usually been carefully worked out over many field seasons. Weakly magnetized rocks can usually be oriented with a geological magnetic compass;

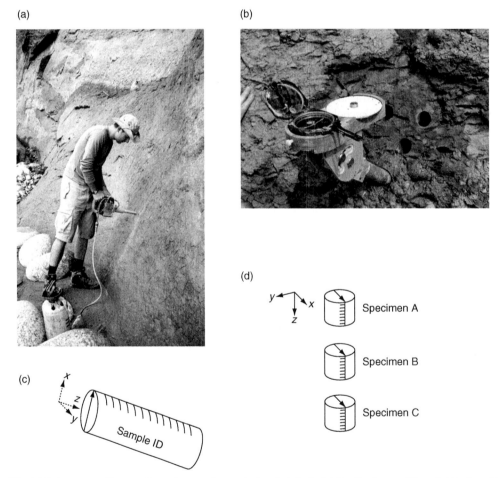

(a)

(b)

(d)

(c)

Figure 25   (a) Palaeomagnetic sampling using a water-cooled, petrol-fuelled drill with diamond tipped drill-stem. (b) Before the sample is broken loose, a line is marked on its exposed face and an orientation tool placed over it. A compass is used for azimuthal orientation of weakly magnetized rocks; the inclinometer level and scale are obscured beneath the compass. (c) A method of marking a drill sample once removed from the exposure; sample coordinate system. (d) Each sample is normally cut into a number of specimens. (a) Photo by S. Hüsing. (b) Photo by D. Michalk.

however, some basalts and other strongly magnetized rocks can influence the compass needle, and in such cases a sun compass is preferable. If a magnetic compass is used, the local magnetic variation (declination) must be known, so a correction to geographic north can be made. When a sun compass is used the location, date, and time of each measurement must be recorded, so that the azimuth can subsequently be calculated from tabulated or programmed positions of the sun.

Each drilled sample is normally cut into several 2.2 cm-long specimens for natural remanent magnetization (NRM) measurement (**Figure 25(d)**).

An alternative sampling method is to remove oriented blocks of rock by hand, or by breaking from the exposure with a hammer, and to subsequently drill samples from this block in the laboratory or workshop. The block is oriented in the field by marking strike and dip lines on a flat surface and orienting these lines with an inclinometer and compass. One advantage of this method is that the outcrop is not left with an unsightly array of holes. However, it is often difficult to find suitable outcrops as it is difficult to remove block-samples from rock that is completely firm, intact, and devoid of joints or cracks. Further, accurate orientation is much harder to achieve than with a drilled core.

Many unconsolidated sediments and soft materials such as ignimbrites and loess are too fragile to remain intact and keep their shape without being encased in some way. A popular practice with such materials is to carve a cube-shaped sample and to cover it with a small plastic box (one standard type is approx. 8 cm^3), which is then oriented in a manner modified from that described above for drill samples.

Most modern rock magnetometers are designed to accept both 2.5 × 2.2 cm cylindrical samples and one of the standard sized cubes.

4.4.1.3 Unconsolidated (lake and deep sea) sediments

Sedimentary sequences that represent periods of unbroken deposition often provide continuous records of geomagnetic changes spanning thousands, hundreds of thousands or even millions of years. Continuous cores of lacustrine or deep-sea sediments may thus yield invaluable information. Numerous coring devices have been designed for soft sediments. To obtain material suitable for paleomagnetic work, core samplers should not twist, shear, or vibrate the sediment: a simple push mechanism is preferable. Any twisting or shearing of material will compromise the determination of declination, while vibrating may

cause reorientation of magnetic grains within the sediment. Corers should contain an absolute minimum of ferrous materials: steel core barrels are known to induce steep components of magnetization in drill cores. There are very few devices that incorporate absolute orientation of the core sample, or even relative orientation between consecutive drives or sections. Individual drives should therefore be as long as possible, to minimize azimuthal matching between sections and the possibility of lost material. The most appropriate method depends on the field environment and the resolution required.

In relatively calm environments and up to about 100 m water depths, the Mackereth corer (**Figure 26**) has proved invaluable. Originally developed for geochemical studies (Mackereth, 1958, 1969), versions of this corer have been designed to collect single-drive cores between 1 and 18 m in length, the most common being the 6-m corer. It is built entirely from nonmagnetic materials and is operated pneumatically from a small boat anchored near the coring site, using cylinders of compressed air. Air is expanded above a piston (MP), forcing the core tube into the sediment in a single smooth drive; a second, fixed piston (KP) inside the core tube sits on top of the sediment throughout, preventing disturbance. At the completion of the coring stroke, air is diverted into a large anchor chamber (AC), which, until this moment, has anchored the corer, but which now becomes buoyant, lifting the corer and core to the surface. When properly deployed, a buoy at the water surface and/or a float chamber near its upper end supports the corer, and the core tube penetrates the sediment vertically.

In deeper water or less clement conditions gravity-driven piston corers or mechanical drills are more commonly used.

In shallow water corers pushed into the sediment on the end of a system of connected rods are popular. This method necessitates a stable platform: a well-anchored raft, or a winter platform of solid ice, are two possible options. Such systems and most drilling systems retrieve sections that are rarely more than 1.5 m long necessitating matching to obtain declination records.

The rotary corers traditionally used to core hard sediments and igneous basement on the sea floor are not suitable for the soft sediments that carry high-resolution paleomagnetic and climatic records as they cause too much disturbance and shearing of the material. The Ocean Drilling Program, which operates the ship JOIDES Resolution, and has drilled throughout the world's oceans for several decades,

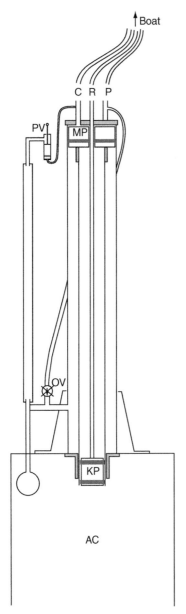

Figure 26 The Mackereth lake sediment corer is operated pneumatically from a boat. Passing air down line P, through a one-way valve OV, and the open valve PV creates a suction effect that pulls the anchor chamber AC into the sediment. Passing air down line C then closes PV and pushes the main piston MP and core tube into the sediment. The Kullenberg piston, KP, remains fixed with respect to the corer, minimizing disturbance to the uppermost sediment. At the completion of the coring stroke, air is diverted into the anchor chamber AC, which becomes buoyant lifting the corer to the lake surface for retrieval. Air-line R allows retraction of the core tube and emergency retrieval.

has developed an Advanced Piston Corer to overcome the problem (Ocean Drilling Program, website). The corer is dropped to the bottom of the coring wireline and then operates in a manner very similar to that of the Mackereth corer described above. Once the corer is in place, safety pins are broken by a burst of hydraulic pressure, which then drives the 9.5 m inner core barrel into the sediment in a single smooth stroke. The core section is brought to the surface for recovery, the drill bit and bottom hole assembly driven a further 9.5 m down and the whole process repeated to recover the next 9.5 m of sediment. The Advanced Piston Corer also carries an orientation tool so each section can be oriented for absolute paleomagnetic work. The Ocean Drilling Program quote a recovery rate of 1–4 sections per hour (9.5–$38.0 \, \mathrm{m \, h^{-1}}$), depending on water depth and the hardness of the material cored.

4.4.2   Rock Magnetometers

4.4.2.1   Introduction

Since its infancy in the 1950s paleomagnetism has seen three generations of rock magnetometers. The first astatic system was possibly that used by Melloni (1853) in his pioneering investigation of the TRM of Vesuvian lava flows. Astatic systems continued to be developed and used extensively for over 100 years, until post second world war developments in electronics led to the fluxgate, spinner magnetometers with online computer control and signal processing. This new generation of magnetometers proved rugged, reliable, and versatile, and most laboratories still employ spinner magnetometers for certain types of work. The ultimate in sensitivity is, in principle, provided by the cryogenic magnetometers, which first appeared in the early 1970s following Josephson's discoveries in superconductivity, including quantum properties of the Josephson junction and its close cousins, the weak link, and the SQUID. Modern cryogenic systems offer unprecedented sensitivity, fast response, no spinning or vibrating, full automation, including some demagnetization procedures if desired, data logging and presentation.

4.4.2.2   Astatic systems

A single suspended magnet suffers from time variations of the torque provided by the time-varying ambient magnetic field, and is really only suitable for rough measurements on samples with magnetizations strong enough to exceed this variability.

The principle of an astatic system is to compensate for time variations in a uniform horizontal field by using two magnets of equal moment, mounted antiparallel, one above the other, from the same

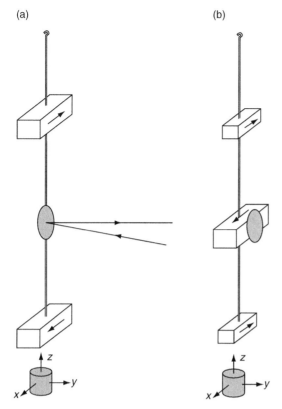

Figure 27 (a) Principle of the astatic magnetometer: the two magnets have equal and opposite moments. The angular deflection is measured for a number of different orientations of the specimen, one of which is shown. (b) Parastatic magnetometer system: the central magnet has a moment equal to twice that of the upper and lower magnets, and is oppositely oriented; this compensates for gradients in the horizontal magnetic field.

systems gave reliable results on standard-sized samples with magnetizations above about $0.2\,\mathrm{A\,m^{-1}}$: (most TRM-bearing volcanic and igneous rocks). However, their sensitivity was limited by vibration, field gradients, and variability caused by moving objects such as lifts, vehicles, etc.

These factors were particularly problematic in noisy city laboratories, and led to the establishment, in the 1960s, of many out-of-city laboratories, many of which are still in operation at the beginning of the twenty-first century. High sensitivity astatic magnetometers housed in these quiet environments, usually mounted on stable concrete slabs, and set in the center of large sets of Helmholtz coils, were capable, with patience, of accurate measurements to $0.001\,\mathrm{A\,m^{-1}}$: (weak sediments and limestones).

An alternative in magnetically noisy environments is a parastatic system, which employs three suspended magnets, as shown in **Figure 27(b)**. The upper and lower magnets are antiparallel to and each have half the moment of the center magnet. This arrangement is less sensitive to changes in the gradient of a horizontal field. Parastatic systems were first described by Thellier (1933), and were further developed in the 1960s (Pozzi and Thellier, 1963). Larochelle and Christie (1967) quote an accuracy of $1°$ in the measurement of a sample of $10^{-3}\,\mathrm{A\,m^{-1}}$, in a measurement time of 6 min.

4.4.2.3 Spinner magnetometers

The basic principle of all spinner magnetometers is that of electromagnetic induction: when the magnetic moment of a sample is rotated in or near either a coil or a fluxgate system, an alternating voltage signal is induced in the coil or fluxgate circuitry. For this reason they have also been called 'rock-generator' magnetometers (e.g., Nagata, 1961). In general, the amplitude of the induced signal is proportional to the component of the magnetic moment perpendicular to the rotation axis, and the phase can be used to determine the direction of this component with respect to some reference direction in the sample. The sample must be spun about at least two axes to obtain three orthogonal components of the magnetic moment and so compute the total vector. Measurements are often made in 4, 6, or 12 different orientations, creating a redundancy of information, but also enabling the calculation of a statistical measure of confidence in the result. This can provide valuable information on the homogeneity of magnetization in a sample and its short timescale viscosity.

suspension (**Figure 27(a)**). The total moment of the system is weak, and a weak torsional suspension can be used. When a sample is brought up symmetrically beneath the lower magnet, a torque is exerted on the system that is proportional to the magnetic moment of the sample (and depends also on a number of geometric factors), this results in an angular deflection that is measured by means of an optical lever. By measuring the deflection with the sample in a number of different positions with respect to the magnets, it is possible to obtain all three components of its magnetic moment, and hence calculate its magnitude and direction. Blackett (1952) detailed the design features, construction, and use of a highly sensitive astatic magnetometer, and this was then used in many of the pioneering paleomagnetic studies of the 1950s.

Moderate-sensitivity astatic systems were relatively easy to set up in most laboratories. Such

In induction coil-based systems, the induced voltage is actually a function of rotation frequency and the number of turns and dimensions of the coils, as well as the component of the magnetic moment of the sample perpendicular to the rotation axis. High rotation speeds (100–400 Hz) are required, and these have sometimes been achieved by means of an air turbine, to avoid magnetic field interference. However, rapid rotation is not always ideal for delicate samples.

Fluxgate sensors avoid the need for rapid rotation, since, in addition to the sample moment, the induced signal depends only on the fluxgate sensitivity (see Section 4.1.2) and the distance of the sample from it. Frequencies of less than 10 Hz are usual.

Several different designs of spinner magnetometers have been developed commercially. The Princeton Applied Research design includes a compensated (astatic) system of two coils connected in series opposition to eliminate field variations, while the Schonstedt magnetometer contains a single fluxgate housed in a mu-metal shield to eliminate external magnetic fields. The Institute of Applied Geophysics, Prague also manufactures a high-speed (85 Hz) spinner with a pair of uncompensated pickup coils, housed in a three-layer mu-metal shield.

The Digico instrument (Molyneux, 1971; Molspin, website) incorporated an innovative ring-shaped fluxgate, within which the sample was rotated at about 4 Hz. It was also the first magnetometer to be controlled from a computer terminal and to employ online signal processing – features which, a decade later, had become virtually taken for granted, but which made the Digico the most advanced magnetometer of its day. The basic operating principles, which were later miniaturized in the portable Minispin magnetometer, are illustrated in **Figure 28**. The sample rotates in the circular fluxgate. The signal is sampled 128 times per revolution – this is controlled by a photo-gate arrangement mounted over a slotted disk that rotates with the sample, at the lower end of the drive shaft. The signals pass through an analog-to-digital converter to the computer, where they are averaged over a preset number of revolutions. When the measurement is complete the output signal is Fourier-analyzed and the first harmonic gives the magnetic moment in the plane of the fluxgate: its components are calculated with respect to a reference direction on the sample holder, printed out, and saved. Once the sample has been spun in four or six different orientations, the direction and intensity of the magnetic moment are calculated, together with a measure of uncertainty. Further calculations can be programmed to perform field and/or bedding corrections, and so output declination and inclination in various reference frames. By increasing the number of revolutions over which each measurement is averaged, the user can increase the precision of the result. Collinson (1983) quotes that the practical lower limit of sensitivity of the Digico magnetometer is about $10^{-4}\,A\,m^{-1}$, with a measurement time of between 10 and 30 min, and a directional accuracy of about $5°$.

4.4.2.4 Cryogenic magnetometers

Rock magnetometers incorporating superconducting (SQUID) sensors (see Section 4.1.2) were developed in the early 1970s and began to appear in paleomagnetic laboratories around the world later that decade. Cryogenic magnetometers offer enormous improvements in both the sensitivity and speed of measurement, while causing minimum disturbance to delicate samples. Sensor design and signal processing are not the only technical problems in the development of a cryogenic magnetometer: maintaining the extremely low temperatures at which the superconducting components operate is an equally important challenge.

The first cryogenic rock magnetometers were built by SCT in the USA and by Cryogenic Consultants Ltd. (CCL) in the UK, and incorporated RF-driven weak-link SQUID sensors. The practical design of the sensors is dictated by constraints on the size of the superconducting loop, which, in order to mitigate the effects of thermal noise must have a very small inductance, and have a diameter of 1–2 mm (Fuller, 1987). This is an order of magnitude smaller than the standard paleomagnetic specimen, so the most convenient way to link the specimen's magnetic flux to the sensor is indirectly, via a transformer-like coupling. The specimen is inserted into a sensor coil, which is connected by to a more tightly wound field transfer coil, and the flux of this coil is measured by the SQUID detector. Both coils and the connections are superconducting: the arrangement therefore acts somewhat like a DC transformer, the coupling depending on the actual flux linkage rather than its rate of change. **Figure 29** illustrates the arrangement of pick-up coils and SQUID sensors in the original SCT machine, which measured two components of magnetic moment, vertical and horizontal. The Helmholtz configuration of the sensor coils ensures optimum sampling of a homogeneous specimen placed at their center: the horizontal axis coils are

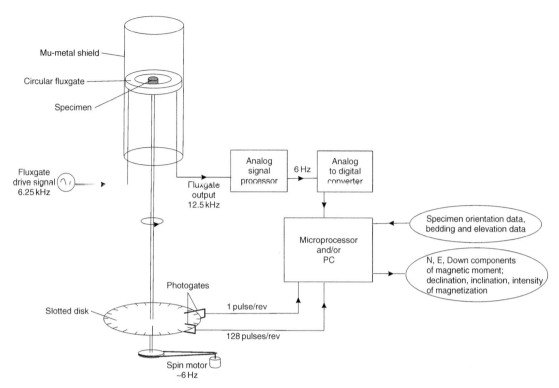

Figure 28 Principle of the Digico fluxgate magnetometer (Molyneux, 1971). The specimen is spun at the center of a single-axis ring-core fluxgate, signal processing is carried out by a microprocessor or computer, which calculates and outputs results as components of magnetic moment, or as required by the user.

less sensitive to sample position than the vertical axis set. Typically, a specimen is inserted into the sensor region, and measurements made at 90-degree steps as it is rotated about a vertical axis. This gives four estimates of the z component and two of each of the x and y components of magnetic moment. The specimen may then be inverted and the process repeated to give four more estimates of z and two more of x and y. Analyzing these, together with background and holder measurements allows calculation of the magnetic moment vector and estimation of the internal consistency of the overall measurement. The specimen holder is usually made of lightweight plastic or mylar, and the sensor region of the magnetometer is deep within a magnetically shielded, heavily insulated liquid helium cryostat.

Over the past 15 years several modifications and improvements have been made to commercially available cryogenic magnetometers. These include (1) replacement of RF-driven SQUIDS with DC SQUIDS, which effectively count individual flux quanta: improving sensitivity to ca. $10^{-12} \text{Am}^2 \text{Hz}^{-1/2}$ for a standard diameter access (2G Enterprises, website); and (2) design of a horizontal-axis machine, with

room temperature access at both ends of the cryostat, allowing long samples to be passed right through the sensor region. Whole, or half-cores of sediment can thus be measured intact, by stepping them through the sensors. Whole sediment cores typically have diameters of 5 cm or more, so this necessitates a wider access, which lowers the sensitivity for smaller samples. A further development, the U-channel, solves this problem. A long rectangular plastic trough up to 1.5 m long and 20 mm by 20 mm cross-section, the 'U-channel', is pressed into the flat surface of a half-core, the resulting sample removed, and the open face of the U-channel sealed with a lid. The U-channel sample can be passed continuously through a standard small-access magnetometer. At other times the same magnetometer can be used for discrete sample measurement without loss of precision. Both whole-core and U-channel measurements inevitably result in a convolution of the magnetization signal with the sampling function(s) of the sensor(s), which has a smoothing effect on the output. Methods of deconvolution are available, but not commonly used.

Until very recently the only practical means of maintaining the very low temperatures necessary to

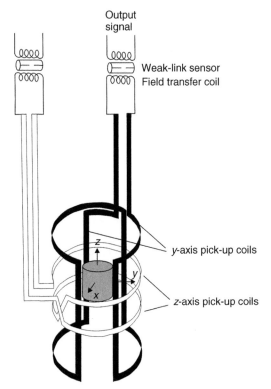

Output
signal

Weak-link sensor
Field transfer coil

y-axis pick-up coils

z-axis pick-up coils

Figure 29 Arrangement of horizontal axis (black) and vertical axis (white) superconducting pickup coils, field transfer coils, and weak-link (SQUID) sensors in the early SCT cryogenic magnetometer. Modified from Superconducting Technologies User Manual.

operate superconducting sensors was by means of liquid helium. Helium 4 has a boiling point of 4.2 K. It has a very low heat of vaporization (0.082 kJ mol^{-1}, or 2.56 kJ l^{-1} of liquid He), but, due to its low atomic mass, a relatively high specific heat capacity. This means that, without extremely efficient insulation, a dewar of liquid helium evaporates very rapidly. The first cryogenic magnetometers were insulated by means of high vacuums, super-insulation (layers of reflective aluminized mylar separated by fiberglass sheets), heat shields cooled by contact with the evaporated helium gas, and/or an outer dewar filled with liquid nitrogen. It is crucial to ensure that helium gas is excluded from the vacuum spaces, as even small amounts can result in conduction of heat into the sensor region and loss of superconductivity. By these means the rate of heat flow into a 30 l dewar can be kept to about 75 mW, resulting in a boil-off of about 2.5 l of liquid helium per day.

Huge improvements in liquid helium retention may be achieved by incorporating an active cooling device or 'cryocooler' in the magnetometer.

Unfortunately, the mechanical vibration and magnetic noise of such devices mean they must be physically separated from the magnetometer. A common arrangement is to place the compressor outside the building, or outside the magnetically shielded space or room which houses the magnetometer. The cryocooler cools a thermal shield around the helium reservoir, thus inhibiting evaporation. A modern magnetometer with a dewar of 50–90 l, and a well-maintained cryocooler, will typically run for between 500 and 900 days between fills.

An exciting new development in cryogenics is the 'pulse-tube cryocooler', which completely circumvents the need for liquid helium in order to reach temperatures of 4.2 K. The pulse tube cryocooler first appeared in 1999, and has very recently been incorporated in a rock magnetometer by 2G-Enterprises (2G Enterprises, website).

4.4.3 Progressive Demagnetization Techniques

4.4.3.1 Introduction

The NRM of a rock sample generally comprises several components carried by grains in different parts of the blocking temperature spectrum. The paleomagnetist usually (though not always) wants to isolate and determine the primary or characteristic component, acquired at the time of formation of the rock. In a straightforward situation involving only thermal activation/relaxation, but not chemical alteration of the magnetic minerals, the oldest component of magnetization will reside in the most stable grains, that is, those with the highest blocking temperatures and longest relaxation times. This will be the primary component unless the whole spectrum of grains has been remagnetized, for example, by a lightning strike or by heating to above the Curie temperature. Secondary partial thermoremanent magnetizations (pTRMs) and viscous components (VRMs) will be carried by less stable grains.

Such a situation is illustrated by the hypothetical example in **Figure 30**. Only those grains with blocking temperatures above T_2 retain a record of the primary magnetization. At some time in the rock's history, all the grains with blocking temperatures below T_2 have been remagnetized by a heating event and have gained a secondary pTRM in a different direction. Further, the least stable grains, those with $T_B < T_1$, have gained a VRM, probably parallel to the present-day field at the site. A single

(a)

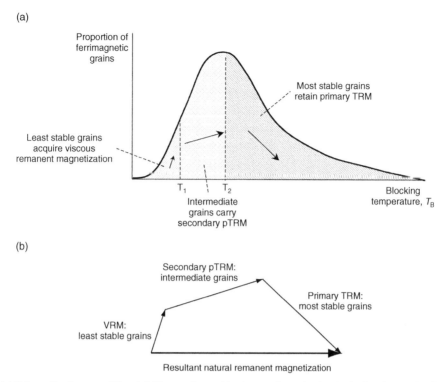

(b)

Figure 30 (a) Schematic diagram of the stability spectrum of ferrimagnetic grains in a typical rock sample, showing portions of the spectrum carrying the primary (TRM), secondary (pTRM), and viscous (VRM) components of remanent magnetization. (b) Vector diagram showing the addition of the various components of magnetization to yield the resultant natural remanent magnetization (NRM).

measurement of the NRM will yield the vector sum of these components, as shown in **Figure 30(b)**. In real situations there is almost always some overlap between the grains carrying each of the components of magnetization.

The association of the oldest component of magnetization with the most stable grains breaks down in more complex situations, for example, when chemical alteration has produced new, secondary ferrimagnetic minerals. In such cases it is possible to find a secondary chemical remanent magnetization carried by authigenic grains that have grown to such a size that they have the highest blocking temperatures in the sample, and which have formed at the expense of primary grains with high T_B's.

Partial or progressive demagnetization procedures are designed to incrementally remove the magnetization of a sample, working systematically through the blocking temperature or coercivity spectrum. This enables the separation and determination of components carried by grains in different intervals of the spectrum. Further information will be required to identify which is the primary or characteristic

component and to piece together the magnetization history of the sample completely.

4.4.3.2 Thermal demagnetization

This method appeals directly to the blocking temperature spectrum of the ferrimagnetic grains in a rock sample.

After measurement of NRM, a specimen is heated to a predetermined temperature, T_H, and then cooled back to room temperature in a carefully controlled, zero magnetic field environment. This procedure randomizes the magnetic moments of all grains whose blocking temperatures are exceeded during the heating. A subsequent measurement of remanence will now give the vector sum of the moments of only those grains with blocking temperatures above T_H. Vector subtraction of this from the original NRM will give the removed magnetization, that carried by grains with $T_B < T_H$. In principle a series of incremental heating steps, alternated with measurements of the magnetization remaining after each cooling will enable a complete analysis of the components of magnetization carried by a specimen.

A number of thermal demagnetizers are available commercially. In addition many paleomagnetic laboratories have purpose-built equipment. The basic requirements of a thermal demagnetizer are a magnetic field-free furnace and cooling chamber. Most furnaces employ noninductive electrical windings, though gas-fired furnaces are also used. Various designs take between 12 and 50 specimens at a time: either arranged around the axis of a cylindrical furnace inside multiple, nested mu-metal shields, or in a layered arrangement at the center of a large Rubens coil set-up. It is during the cooling stage that it is most important to maintain as low a magnetic field as possible. Some designs incorporate a separate chamber into which the samples are passed for cooling, while a second set of samples may be heated in the already hot oven. In other designs the samples are cooled in the same space, without moving them.

It is important to allow sufficient time for the specimens to equilibrate both at the maximum temperature and back to room temperature before measurement: typically 20–30 min for each stage.

4.4.3.3 Alternating field (AF) demagnetization

Working progressively through the coercivity spectrum of a specimen provides an alternative way of separating different components of magnetization. In AF demagnetization, the specimen is placed inside a solenoid, which is connected to a variable AC supply, while being housed in an environment of zero direct field. The amplitude of the alternating current is increased to a preset maximum corresponding to a desired maximum magnetic field, it is then smoothly and slowly ramped back down to zero. When the current is at maximum amplitude, the magnetic moments of all grains with coercivities below the corresponding magnetic field follow the oscillations back and forth along the axis of the solenoid. As the peak field falls below the coercivity of each grain, its magnetic moment becomes blocked: ideally equal numbers of grains become blocked in each opposing direction. All directions in the specimen should be exposed equally to the demagnetizing field of the solenoid. This may be achieved by successively aligning each of the three axes of the specimen with the axis of the solenoid while the alternating magnetic field is cycled up and down (static method). Alternatively, the specimen may be tumbled simultaneously about two perpendicular axes, by means of a small motor and belt-driven turntable, while the field is cycled just once. There are advantages and

disadvantages associated with each method. Certain specimens are susceptible to acquisition of a spurious magnetization during the tumbling procedure: a rotational remanent magnetization (RRM) antiparallel to the inner rotation axis was first documented by Wilson and Lomax (1972). RRM seems to be related to the more general phenomenon of gyroremanent magnetization (GRM), which may be acquired by single-domain grain-bearing specimens when they are rotated in steady or alternating magnetic fields (Stephenson 1980, 1981). Methods to correct for RRM and GRM have been proposed by Hillhouse (1977) and Giddings *et al.* (1997). However, these corrections are of an empirical nature, and to avoid any suspicion of doubt in results, many authors prefer to use thermal demagnetization in specimens that are likely to develop RRM or GRM.

4.4.4 Data Analysis and Statistics

Most rock magnetometers return the components of the magnetic moment of a specimen along the $x, y,$ and z axes of the specimen. The operating software usually offers the provision to enter the field orientation of the specimen and this enables transformation of the result to the present-day geographic reference frame: northerly, easterly, and downward components of magnetic moment; or declination, inclination, and total magnetic moment. The intensity of magnetization is further obtained by dividing the total magnetic moment by the specimen volume. Finally, if the bedding attitude is known, this may be entered and the magnetization vector referred to the pre-tilting coordinate frame.

4.4.4.1 Progressive demagnetization data

Progressive demagnetization of a specimen yields a series of vectors, describing the magnetization left after each step of demagnetization. Three different plots are useful in interpreting such data. Examples are shown in **Figure 31**.

4.4.4.1.(i) Intensity of magnetization against demagnetization level If the direction of remanence changes little during demagnetization, then this plot gives an approximation to the blocking temperature or coercivity spectrum of the specimen. For multicomponent magnetizations its interpretation is less straightforward (**Figure 31(c)**).

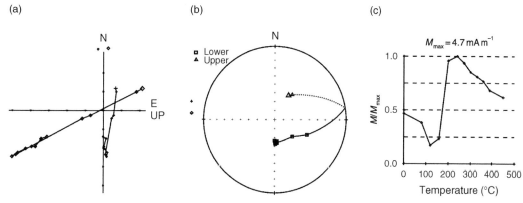

Figure 31 An example of progressive thermal demagnetization data for a (southern hemisphere) specimen carrying a primary magnetization of reversed polarity, overprinted by a normal polarity thermoviscous secondary component. (a) Vector component plot, showing the projection onto the horizontal plane (N vs E) and the projection onto the vertical NS plane (N vs UP), as crosses and diamonds respectively; (b) An equal angle stereographic projection of the sequence of remanent directions, (c) The intensity of magnetization plotted against demagnetization level (temperature).

4.4.4.1.(ii) **Stereographic projection** Stereo-graphic projection of the directions of the remanence remaining at each step contains no information regarding the intensity of magnetization. For a single-component magnetization the direction does not change during demagnetization and all the data points overlie each other. For a two-component magnetization, the initial NRM is the vector sum of the two components. As the lower blocking temperature component is progressively removed, the tip of the remanent unit vector moves along a great circle path towards the direction of the underlying more stable component. The direction of the secondary component lies on the backward extrapolation of this great circle. This process is best seen in **Figure 31(a)**, where the primary magnetization is of reversed polarity (to the south and downward, since the specimen is from a southern hemisphere site), and the secondary magnetization is close to the present-day field at the site ($D = 20°$, $I = -66°$).

4.4.4.1.(iii) **Vector component plots** Vector component plots combine both intensity and direction data, by projecting the successive remanent vectors onto two different planes and presenting them in a single diagram (Zijderveld, 1967). The actual planes chosen vary; in **Figure 31(b)**, the horizontal plane (N vs E) and the vertical north–south plane (N vs Up) are shown, with the northerly component plotted on the same (y) axis in both cases. Linear segments correspond to the removal of single components of magnetization, and the direction of such a component can be measured directly from the diagram. The

component residing in the grains with highest blocking temperatures or coercivities (often the primary component) should yield a linear segment terminating at the origin.

4.4.4.1.(iv) **Principal component analysis** In theory, the direction of a component of magnetization can be measured directly from a linear section of the vector component plot, as described above. In practice, however, the data are invariably scattered and estimation of the best straight line is often difficult and subjective. In such cases, principal component analysis, or PCA, (Kirschvink, 1980) provides a rigorous, objective way of estimating components of magnetization. In PCA the best-fit line is calculated to a pre-selected sequence of demagnetization data (vectors), which may be 'anchored' to the origin if desired, by analysis of a 3×3 matrix made up from the vector components: the 'orientation tensor'. The principal eigenvector gives the best-fit line, while a statistic of the quality of the fit, the 'maximum angular deviation' or MAD, is derived from the eigenvalues.

4.4.4.1.(v) **Remagnetization circles** In some situations involving multicomponent magnetization it is impossible to isolate the ChRM by progressive demagnetization. In such cases techniques involving extrapolation of trends in the demagnetization data (remagnetization circles) from a number of different samples until they intersect can be used to infer a common underlying component of magnetization or hypothetical endpoint (Halls, 1978; McFadden and

McElhinny, 1988). As with any extrapolation technique, extreme care is required in the interpretation of directions obtained from remagnetization circle analysis.

4.4.4.2 *The statistical treatment of paleomagnetic data*

When analyzing paleomagnetic data it is usual to treat the direction and intensity of the vectors separately. This is because the retrieval of the paleointensity of the geomagnetic field from magnetization intensity is a complex task and frequently cannot be achieved with the same degree of confidence as the paleodirection.

4.4.4.2.(i) Intensity of remanence and geomagnetic paleointensity The intensity of remanence (and also magnetic susceptibility and other laboratory-induced magnetizations) depends on the composition and concentration of the magnetic minerals and their granulometry, that is, grain size, shape distributions. Most such natural distributions do not follow a normal or Gaussian form, which is symmetrical about the mean value, but are highly skewed, with a long tail towards high values. It is often found that the logarithms of the values are normally distributed, however. The 'log-normal' distribution is found to give a good representation of natural remanent intensities and magnetic susceptibilities in rocks (Irving *et al.*, 1966; Tarling, 1966).

Methods for the determination of the intensity of the geomagnetic field from paleomagnetic measurements are covered in detail in Chapter 13, and only the rudimentary principles are given here. In general, the intensity of magnetization (TRM, DRM, CRM, etc.) of a sample will depend on many factors in addition to the strength of the prevailing field. These include intrinsic magnetic properties of the material such as the concentration of the remanence-bearing mineral(s), their saturation magnetization, and granulometry. In seeking the paleointensity one must firstly identify the portion of the ferrimagnetic grain spectrum that carries the primary or characteristic component of magnetization, and then normalize the intensity of magnetization of this portion to compensate for the intrinsic magnetic composition of the sample as described above.

Néel's theory of single-domain TRM (Néel, 1949, 1955) gives an analytical equation for magnetization intensity that is proportional to the external magnetic field (B_{ext}) in the weak-field approximation. Hence,

in principle absolute determinations of paleointensity can be made from TRM-bearing materials. The relationship between magnetization and paleointensity is more complicated in the case of a DRM, where the degree of alignment of ferrimagnetic grains depends on external factors that are much more difficult to quantify. The intensity of DRM is nevertheless still predicted to be proportional to the prevailing magnetic field, and methods to retrieve logs of relative paleointensity from uniform sequences of sediments are often successful. Such methods usually employ normalization of a portion of the NRM with a laboratory-produced remanence, often anhysteretic or isothermal remanent magnetization.

4.4.4.2.(ii) Paleomagnetic directions: the Fisher distribution The analysis of remanent directions usually reduces to dealing with a set of unit vectors obtained from a set of N samples. Each unit vector **M** can be expressed in terms of its declination and inclination, or as direction cosines l, m, and n, which are the northerly, easterly, and downward components of the unit vector (**Figure 32(a)**).

$$l_i = \cos(D_i)\cos(I_i) \quad D_i = \tan^{-1}\left(\frac{m_i}{l_i}\right)$$
$$m_i = \sin(D_i)\cos(I_i)$$
$$n_i = \sin(I_i) \qquad I_i = \sin^{-1}n_i$$

The mean direction is obtained via the vector sum **R** (**Figure 32(b)**).

$$\mathbf{R} = \left(\sum_{i=1}^{N} l_i, \sum_{i=1}^{N} m_i, \sum_{i=1}^{N} n_i\right) = (R_x, R_y, R_z)$$

$$\text{Mean } D = \tan^{-1}\left(\frac{R_y}{R_x}\right)$$

$$\text{Mean } I = \tan^{-1}\left(\frac{R_z}{\left(R_x^2 + R_y^2\right)^{1/2}}\right)$$

The length $R = \left(R_x^2 + R_y^2 + R_z^2\right)^{1/2}$ of the vector **R** forms the basis of the statistical analysis described below. If all N unit vectors are perfectly aligned, then $R = N$; the greater the scatter in the directions of the unit vectors, the smaller is the value of R.

The two-dimensional (2-D) probability density distribution function generally assumed to apply to paleomagnetic unit vectors is the Fisher distribution (Fisher, 1953). It is, in many ways, equivalent to a 2-D Gaussian or normal distribution, and is described mathematically by the function

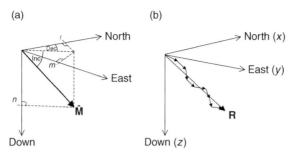

Figure 32 (a) The direction cosines l, m, and n of the unit magnetization vector $\hat{M}$ of a sample, and their relationship to declination and inclination. (b) Vector addition of unit magnetization vectors from a number of samples, $\hat{M}_1$ $\hat{M}_2$ $\hat{M}_3$ $\cdots$, to yield the vector R (see text).

$$P(\varphi) = \frac{\kappa}{4\pi \sinh \kappa} \exp(\kappa \cos \varphi)$$

where φ is the angle between a direction and the true mean direction of the whole population. κ is known as the precision parameter, and reflects the dispersion of the distribution: it is analogous to the invariance, or the reciprocal of the variance, of a Gaussian distribution. κ ranges between zero for a random distribution (maximum scatter) and infinity for perfect alignment of the entire population.

A palaeomagnetic data set is a subset of N samples taken from the whole population. For $N \gtrsim 6$ and $\kappa \gtrsim 3$, the best (minimum variance, unbiassed) estimate of κ is given by $k = (N-1)/(N-R)$ (Fisher, 1953; McFadden, 1980). Naturally, as more of the population is sampled, that is, as N increases, k becomes a better estimate of κ.

The angular standard deviation, or angular dispersion, S, is a useful measure of scatter, for example, in studies where secular variation should be adequately sampled. For a Fisher distribution S is the semi-angle of the cone that includes 63% of the population. For low scatter a good approximation is given by

$$S = \theta_{63} \approx \frac{81}{\sqrt{k}}$$

The closeness of the mean of a given data set to the true mean of the population is clearly of crucial importance in palaeomagnetism. The ideal situation is obviously a large data set that samples the population randomly.

For $k \gtrsim 10$ the confidence that can be placed in the sample mean is quantified by the relation

$$\alpha_{(1-P)} = \cos^{-1}\left\{ 1 - \frac{N-R}{R}\left(\left(\frac{1}{P}\right)^{1/(N-1)} - 1 \right) \right\}$$

where α is the semi-angle of the cone around the sample mean within which there is a probability $(1-P)$ of the true mean of the population lying. The levels of confidence most commonly quoted are 95% and 63%. The best estimates of α_{95} and α_{63} are given by

$$\alpha_{95} \cong \frac{140}{\sqrt{kN}} \quad \text{and} \quad \alpha_{63} \cong \frac{81}{\sqrt{kN}}$$

α_{63} is the direct analog of the standard error in the mean of a Gaussian distribution.

4.4.4.2.(iii)   Comparison of directions   In order to assess whether a paleomagnetically determined direction differs significantly from a direction that by comparison is completely specified, for example, the direction of the present-day field, α_{95} may be used directly. If the known direction lies outside the cone of 95% confidence of the paleomagnetic direction, then there is a 95% probability that the true directions differ.

It is frequently necessary to compare two paleomagnetically determined directions. It is often stated that the two directions are distinct (at the 95% level of confidence) if their cones of confidence do not overlap and are indistinguishable if they do. While the first statement is true, the second is not strictly correct. The cones may overlap by a small amount while the means are still distinct. More rigorous statistical tests have been devised and are discussed by McFadden and Lowes (1981) and Watson (1983).

4.4.4.3   Field tests

Progressive demagnetization experiments enable the analysis of natural remanent magnetization in terms of components carried by grains in different parts of the blocking temperature or coercivity spectrum. Statistical methods allow a meaningful comparison to be made between determined paleomagnetic directions. However, neither gives any information regarding the ages of the various components of magnetization, or can prove that the characteristic magnetization was acquired at the time the rock formed. Application of some or all of the field tests discussed below can help order events in the geological and magnetization histories of a rock and so help date components of magnetization.

4.4.4.3.(i)   *Consistency test*   Demonstration of the reproducibility of results is an important aspect of all branches of science. In paleomagnetism it is usually necessary to show that the characteristic magnetization is of geomagnetic origin and not produced by some spurious or local effect such as a lightning strike or physical disturbance. Replication of a signal in rocks or sediments from different provenances or of different lithologies is a strong argument in favor of a regional or global magnetization process. Multiple records also provide data for more sophisticated levels of statistical analysis.

4.4.4.3.(ii)   *Fold test*   Paleomagnetists often sample beds or sediments that are no longer flat lying. It is usual to assume that such sediments originally accumulated on a horizontal surface, and have later been tilted or folded by tectonic processes of some sort. If a stable characteristic magnetization predates the tilting or folding, then the magnetization vector will have been tilted with the beds (**Figure 33(a)**), and a correction to restore the beds to the horizontal will also restore the magnetization to its original orientation. To this end, it is routine to measure the attitude of all sedimentary strata sampled during fieldwork. In cases of simple bedding tilt, a single rotation about the horizontal strike of the beds by an angle equal to the dip is required. In cases where folding is evident two steps are necessary: restoration of the fold axis to horizontal followed by untilting of the beds in the limbs of the fold. This requires the additional field measurement of the plunge of the fold axis. A fold test is successful if, after these field corrections, the characteristic magnetization vectors from individual sites move closer together or site-average vectors become indistinguishable within confidence levels. It is then possible to conclude that the magnetization predates the folding or tilting episode. An unsuccessful or partially successful fold test can also yield information about the magnetization history of a unit.

4.4.4.3.(iii)   *Reversals test*   When a study involves sampling through a number of geomagnetic polarity reversals, for example, for the calculation of paleomagnetic pole positions (see below) or for magnetostratigraphy, the resulting data set contains records of both normal and reversed polarity. According to the geocentric axial dipole (GAD) hypothesis (explained below), if secondary components of magnetization have been completely removed so that the records are accurate reflections of the geomagnetic field, and if sufficient time intervals have been sampled to average over secular variation (10^4–10^5 years), then the averages of the normal polarity directions and the reversed polarity directions should be antipodal. **Figure 33(b)** illustrates data from a magnetostratigraphic study in NZ. After progressive thermal demagnetization and PCA to estimate characteristic magnetization directions, the mean normal and reversed directions are antipodal to better than 95% confidence. In this case, the difference between the overall mean direction and the GAD direction for the site is a combination of an anomaly due to the movement of the Australian plate over the past 3 My, and a local vertical axis rotation associated with the proximity of the site to the Australian/Pacific plate boundary (Turner *et al.*, 2005)

4.4.4.3.(iv)   *Baked contact test*   When an igneous magma or volcanic lava is emplaced adjacent to a country rock, a temperature gradient develops in the country rock as heat is conducted away and the magma or lava cools and solidifies. As the temperature falls below the blocking temperatures of its constituent ferrimagnetic grains, the igneous or volcanic rock gains a TRM, which (ideally) is parallel to the ambient geomagnetic field. The maximum temperature reached in the country rock will be a decreasing function of distance from the magma or lava, and the magnetic moments of ferrimagnetic grains with blocking temperatures up to this maximum will be reset, that is, the baked contact will acquire a secondary pTRM. This will be parallel to the field at the time of emplacement of the lava or magma, and hence parallel to the TRM of the lava or magma. In those parts of the country rock closest to the magma the entire blocking temperature spectrum might be affected, but further away only the grains with lower blocking temperatures will become reset, while grains with higher blocking temperatures will retain the primary magnetization of the country rock. The description given above constitutes a positive baked contact test.

(a)

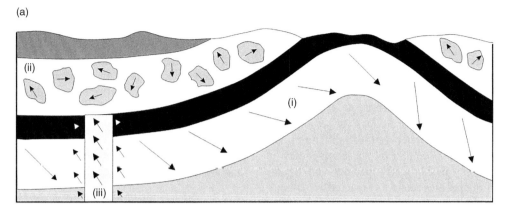

(b)

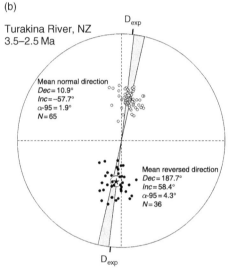

Figure 33 Paleomagnetic field tests: (a) Block diagram showing (i) the rotation of primary magnetization vectors during the folding of a rock unit (fold test); (ii) the randomization of the primary magnetization vectors of clasts incorporated in a conglomerate unit (conglomerate test); (iii) the remagnetization of country rocks immediately adjacent to an intruded dyke (baked contact test). (b) Example of a positive reversals test: equal area stereographic projection showing paleomagnetic directions from 101 sites sampled in a magnetostratigraphic study of the Turakina River Valley, New Zealand. The mean normal and reversed polarity directions are shown by stars, surrounded by an oval showing the cone of 95% confidence in the mean (alpha-95). (a) Adapted from Graham JW (1949) The stability and significance of magnetism in sedimentary rocks. *Journal of Geophysical Research* 54: 131–167. (b) From Turner GM, Kamp PJJ, McIntyre AP, Hayton S, McGuire DM, and Wilson GS (2005) A coherent middle Pliocene magnetostratigraphy, Wanganui Basin, New Zealand. *Journal of the Royal Society of New Zealand* 35: 197–227.

In the first part of the twentieth century baked contact tests provided invaluable evidence for the theory of polarity reversals. The opposing theory held that rocks found with reversed magnetizations had somehow become magnetized in the opposite direction to the ambient field (self-reversal). A worldwide compilation made by Wilson (1962a) listed 48 positive and three negative baked contact tests: 11 years later, McElhinny (1973) listed 154 positive results and still only 3 negative ones. Documented self-reversal mechanisms invariably involve particular, uncommon mineralogies, which are unlikely to be found simultaneously in large numbers of lavas and adjacent country rocks. As the number of positive baked contact tests accumulated, the likelihood of widespread self-reversal rapidly diminished and the idea of polarity reversals became generally accepted by the geomagnetism community. An unusual example of a doubly baked country rock was described by Wilson, (1962b). A lava flow and a dyke were successively emplaced in a laterite, which consequently carries secondary components of magnetization from both heating episodes. Both are of reversed polarity, but they can be separated by progressive demagnetization since their directions differ significantly. It is virtually impossible to conceive of this situation occurring through self-reversal.

4.4.4.3.(v) Conglomerate test
A conglomerate test may be applied when the rock (or sediment) of interest contains pebbles or clasts of a different lithology. It involves comparing the components of magnetization carried by the pebbles and by the matrix. If the pebbles retain a characteristic component of magnetization dating back to their previous history, then these vectors should have been randomized in the process of deposition, and no coherency should be observable in the data (**Figure 33(a)**). Any coherent component of magnetization found in the pebbles or between the pebbles and matrix raises the possibility of post-depositional remagnetization.

The conglomerate test is infrequently used, as it is not common to find suitable outcrops.

4.4.4.4   Poles
4.4.4.4.(i) Geomagnetic poles
The present-day geomagnetic field can, to a first approximation, be modeled as a geocentric dipole. In 2005, the axis of the best-fitting geocentric dipole intersected Earth's surface at $79.7°$ N, $288.2°$ E and $79.7°$ S, $108.2°$ E (IGRF-10, 2005). These locations are called the 'geomagnetic poles'. This tilted geocentric dipole model (**Figure 34(b)**) describes about 80–90% of the present field at Earth's surface – the remaining 'non-dipole' part may be described by the higher-order components of a spherical harmonic representation, that is, quadrupole, octupole, etc.

4.4.4.4.(ii) Virtual geomagnetic poles
If Earth's magnetic field could be modeled completely by a geocentric dipole aligned with the rotation axis (GAD) then at all points on the globe its horizontal component would be exactly northwards, that is, the declination would be zero, and the inclination would be a simple monotonic function of latitude (λ): $\tan(I) = 2\tan(\lambda)$. A corollary of this is that from any observed magnetic field direction (or paleomagnetically recorded direction), one can calculate the orientation of the geocentric dipole that would produce it, and the corresponding 'virtual geomagnetic poles' (VGPs). Imagine a globe: from the site of the observation draw a great circle in the direction of the horizontal component of the magnetic field or magnetization vector, the virtual pole must fall on this

great circle. The angle (at the center of Earth) between the site and the pole (or the virtual geomagnetic latitude) is calculated from the recorded inclination according to the relation noted above, and corresponds to a distance measured along the great circle. This process is illustrated in **Figure 34(c)**.

Alternatively the latitude and longitude of the VGP (λ_p, φ_p) may be computed from the observed paleomagnetic direction (D, I) and the site latitude and longitude, (λ_s, φ_s) using the following equations

$$\sin\lambda_p = \sin\lambda_s\sin\lambda + \cos\lambda_s\cos\lambda\cos D$$

$$\phi_p = \phi_s + \beta \quad \text{if } \sin\lambda \geq \sin\lambda_s\sin\lambda_p$$

$$\phi_p = \phi_s + \beta + 180 \quad \text{if } \sin\lambda < \sin\lambda_s\sin\lambda_p$$

$$\text{where} \quad \tan\lambda = (1/2)\tan I$$

$$\text{and} \quad \sin\beta = \frac{\cos\lambda\sin D}{\cos\lambda_p}$$

A pole calculated in this way is virtual in the sense that the geocentric dipole model employed in the calculation cannot be justified for an observation from a single location and a single point of time: it has no real physical meaning unless the effect of the non-dipole field just happened to be zero at the time and place of the observation.

VGPs, however, remain a favorite way of presenting paleomagnetic directions and are used extensively, for example, in mapping sequences of transitional directions from polarity reversals when the field was almost certainly not dipolar.

4.4.4.4.(iii) The GAD hypothesis
There is considerable evidence to indicate that, when the geomagnetic field is averaged over sufficiently long intervals of time (omitting intermediate directions recorded when the field was actually in the process of reversal), the mean positions of the geomagnetic poles coincide with the North and South poles of the rotation axis. For example, when the VGPs of a global distribution of rocks dated between 0 and 5 Ma are averaged the mean coincides with the rotation axis to within about $2°$ (McElhinny *et al.*, 1996). This leads to the GAD hypothesis: that the time-averaged geomagnetic field is a geocentric axial dipole field. There is some debate as to the minimum time period required to average out the secular variation (of both non-dipole and dipole fields): most sources quote between 10 000 and 100 000 years. The GAD hypothesis is fundamental to many applications

(a)   (b)   (c)

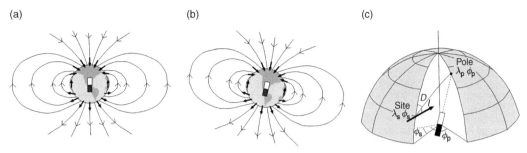

Figure 34 (a) The magnetic field due to a geocentric axial dipole. (b) The magnetic field due to a tilted geocentric dipole. (c) The principle of calculation of a virtual geomagnetic pole from an observed or recorded magnetic field direction at site (λ_S, ϕ_S).

of paleomagnetism, including providing crucial support for the theory plate tectonics and continental reconstructions.

4.4.4.4.(iv) Paleomagnetic poles and apparent polar wander paths
The term 'paleomagnetic pole' is applied to a pole calculated by averaging a paleomagnetic data set that spans a time interval long enough to average secular variation. The GAD hypothesis implies that such a pole gives the position of the rotation axis with respect to the sampling location at the time the magnetization was acquired. Paleomagnetic poles for the past 5 million years are coincident with the rotation axis. However, older poles differ significantly – the discrepancy generally increasing with age. The time-sequence of paleomagnetic poles from a given continent or stable cratonic block is called an apparent polar wander path (APWP). The concept of an APWP was introduced by Creer *et al.* (1954). Originally it was thought that, while the orientation of the rotation axis remains fixed in space, APWPs were produced by rotation of the whole Earth with respect to its rotation axis: this is now termed 'true polar wander'. The steady accumulation of data over the years soon showed that APWPs from different continents differ significantly. However, it was also found that sometimes, the application of a finite rotation brought a segment of one APWPs into coincidence with the contemporaneous section of another. If one imagines an APWP tied to the continent from which it is derived, the implication is that there have been periods of time when some continents have moved together, producing similar APWPs, and other periods when they have moved separately over Earth's surface, causing separation and new segments of their APWPs. This is the basis of the theory of plate tectonics, whereby

the continents move on thin lithospheric plates in response to mantle covection processes, constructive and destructive processes at plate boundaries.

References

Alexandrov EB, Balabas MV, Kulyasov VN, *et al.* (2004) Three-component variometer based on a scalar potassium sensor. *Measurement Science and Technology* 15: 918–922.

Alexandrov EB and Bonch-Bruevich VA (1992) Optically pumped magnetometers after three decades. *Optical Engineering* 31: 711–717.

Alexandrov EB and Primdahl F (1993) On gyro-errors of the proton magnetometer. *Measurement Science and Technology* 4: 737–739.

Alldredge LR and Saldukas I (1964) An automatic standard magnetic observatory. *Journal of Geophysical Research* 69: 1963–1970.

Aschenbrenner H and Goubau G (1936) Eine Anordnung Registrieung rascher magnetischer Storungen. *Hochfreg Tech. Elektroakust* 47: 178–181.

Bardeen J, Cooper LN, and Schrieffer JR (1957) Theory of superconductivity. *Physical Review* 108: 1175–1204.

Becker H and Fassbinder JWE (1999) Magnetometry of a Scythian settlement in Siberia near Cicah in the Baraba Steppe. In: Fassbinder JWE and Irlinger WE (eds.) *Archaeological Prospection, Arbeitsh. Bayerisches Landesamtf. Denkmlpflege*, Vol. 108, pp. 168–172. Munchen: Verlag Karl M. Lipp.

Bitterly J, Cantin JM, Schlich R, Folques J, and et Gilbert D (1984) Portable magnetometer theodolite with fluxgate sensor for earth's magnetic field component measurements. *Geophysical Surveys* 6: 233–239.

Blackett PMS (1952) A negative experiment relating to magnetism and the Earth's rotation. *Philosophical Transactions of the Royal Society Series A* 245: 309–370.

Blinov VE, Ginzburg BI, Zhitnikov RA, and Kuleshov PP (1984) Alkali-helium magnetometer with optically oriented potassium atoms. *Soviet Physics – Technical Physics* 29(2): 168–171. See also (in Russian): *Zhurnal Tekhnicheskoi Fiziki* 54: 287–292.

Bloch F (1946) Nuclear induction. *Physical Review* 70: 460–474.

Brunhes B (1906) Récherches sur les directions d'aimentation des roches volcaniques. *Journal of Physics* 5: 705–724.

Bullock SJ and Barritt SD (1989) Real-time navigation and flight-path recovery of aerial geophysical surveys: A review. In: Garland GD (ed.) *Proceedings of Exploration 87, Third*

Decennial International Conference on Geophysical and Geochemical Exploration for Minerals and Groundwater., Special Volume 3, pp. 960. Sudbury, Canada: Ontario Geological Survey.

Carver TR and Slichter CP (1953) Polarization of nuclear spins in metals. *Physical Review* 92: 212–213.

Carver TR and Slichter CP (1956) Experimenatl verification of the Overhauser nuclear polarization effect. *Physical Review* 102: 975–980.

Chave AD, Green AW, Evans RL, et al. (1995) *Report of a Workshop on Technical Approaches to Construction of a Seafloor Geomagnetic Observatory.* Technical Report WHOI-95-12. Woods Hole, USA: Woods Hole Oceanographic Institution.

Chevallier R (1925) L'aimentation des laves de l'Etna et l'orientation du champ terrestre en Sicile du XIIᵉ au XVIIᵉ siècle. *Annales de Physique* 4: 5–162.

Collinson DW (1983) *Methods in Rock Magnetism and Palaeomagnetism: Techniques and Instrumentation.* New York: Chapman and Hall.

Creer KM, Irving E, and Runcorn SK (1954) The direction of the geomagnetic field in remote epochs in Great Britain. *Journal of Geomagnetism and Geoelectricity* 6: 163–168.

David P (1904) Sur la stabilité de la direction d'aimentation dans quelques roches volcaniques. *Comptes Rendus Academie Sciences de Paris* 138: 41–42.

Delesse A (1849) Sur le magnétisme polaire dans les minéraux et dans les roches. *Annales Chimie et de Physique, Ser.* 325: 194–209.

Deumlich F (1980) *Instrumentenkunde der Vermessungstechnik.* Berlin (Ost): VEB Verlag für Bauwesen.

De Vuyst AP and Hus J (1966) Généralisation de la Mesure de l'Intensité des Composantes Verticale et Horizontale du Champ Magnétique Terrestre avec le Magnétomètre à Protons. *Annales De Geophysique* 22: 119–127.

Fisher RA (1953) Dispersion on a sphere. *Proceedings of the Royal Society of London, Series A* 217: 295–305.

Fuller M (1987) Experimental methods in rock magnetism and palaeomagnetism. In: Sammis CG and Henyey TL (eds.) *Methods of Experimental Physics*, pp. 303–471. Orlando: Academic Press.

Giddings JW, Klootwijk CT, Rees J, and Groenewoud A (1997) Automated AF-demagnetization on the 2G-Enterprises cryogenic magnetometer. *Geologie en Mijnbouw* 76: 35–44.

Gilbert G, Cantin JM, Bitterly J, Schlich R, and Folques J (1988) Absolute measurements of the Earth's magnetic field in French Observatories: Results obtained with the portable theodolite fluxgate magnetometer for the period 1979–86. *Proceedings of the International Workshop on Magnetic Observatory Instruments, Geomagnetism Series 32*, paper 88–17. Ottawa, Canada: Geological Survey of Canada.

Gilles H, Hamel J, and Chéron B (2001) Laser pumped ⁴He magnetometer. *Review of Scientific Instruments* 72: 2253–2260.

Goree WS and Fuller MD (1976) Magnetometers using RF-driven SQUIDs and their application in rock magnetism and palaeomagnetism. *Reviews of Geophysics and Space Physics* 14: 591–608.

Graham JW (1949) The stability and significance of magnetism in sedimentary rocks. *Journal of Geophysical Research* 54: 131–167.

Grant FS (1985) Aeromagnetics, geology and ore environments, I. Magnetite in igneous, sedimentary and metamorphic rocks: An overview. *Geoexploration* 23: 303–333.

Gravrand O, Khokhlov A, Le Mouël JL, and Léger JM (2001) On the calibration of a vectorial ⁴He pumped magnetometer. *Earth Planets Space* 53: 949–958.

Green AW, Coles RL, Kerridge DJ, and LeMouël J-L (1998) INTERMAGNET, today and tomorrow. In: *Proceedings of the VIIᵗʰ Workshop on Geomagnetic Observatory Instruments, Data Acquisition and Processing '96, Scientific Technical Report STR98/21*, pp. 277–286. Potsdam, Germany: GFZ-Potsdam.

Halls HC (1978) The use of converging remagnetization circles in paleomagnetism. *Physics of the Earth and Planetary Interiors* 16: 1–11.

Henkel H (1991) Petrophysical properties (density and magnetisation) of rocks from the northern part of the Baltic shield. *Tectonophysics* 192: 1–19.

Hillhouse JW (1977) A method for the removal of rotational remanent magnetization acquired during alternating field demagnetization. *Geophysical Journal of the Royal Astronomical Society* 50: 29–34.

Hrvoic I (2001) Standardization of total field magnetometers. *Contributions to Geophysics & Geodesy* 31: 93–97.

Hurwitz L and Nelson JH (1960) Proton vector magnetometer. *Journal of Geophysical Research* 65: 1759–1765.

Irving E, Molyneux L, and Runcorn SK (1966) The analysis of remanent intensities and susceptibilities of rocks. *Geophysical Journal of the Royal Astronomical Society* 137: 451–464.

Jankowski J, Marianiuk J, Ruta A, Sucksdorff C, and Kivinen M (1984) Long-term stability of a torque-balance variometer with photoelectric converters in observatory practice. *Geophysical Surveys* 6: 367–380.

Jankowski J and Sucksdorff C (1996) *IAGA Guide for Magnetic Measurements and Observatory Practice.* International Association of Geomagnetism and Aeronomy.

Josephson BD (1962) Possible new effects in superconducting tunnelling. *Physics Letters* 1: 251–253.

Kastler A (1950) Quelques suggestions concernant la production optique et la détection optique d'une inégalité de population des niveaux de quantification spatiale des atomes. *Journal of Physics, Radium* 11: 255–265.

Kirschvink JL (1980) The least-squares line and plane and the analysis of palaeomagnetic data. *Geophysical Journal of the Royal Astronomical Society* 62: 699–718.

Korepanov V, Berkman R, Rakhlin L, et al. (2001) Advanced field magnetometers comparative study. *Measurement* 29: 137–146.

Langel R, Ousley G, Berbert J, Murphy J, and Settle M (1982) The MAGSAT mission. *Geophysical Research Letters* 9: 243–245.

Larochelle A and Christie KW (1967) An automatic 3-magnet or biastatic magnetometer. *Geological Survey of Canada* Paper 67–26: 28 pp.

Lauridsen KE (1985) Experiences with the DI-fluxgate magnetometer: Inclusive theory of the instrument and comparison with other methods. Danish Meteorological Institute Geophysical Papers, R–71.

Laursen V and Olsen J (1971) Classical methods of geomagnetic observations. In: Flugge S and Rawer K (eds.) *Encyclopedia of Geophysics, Geophysics III/3, Volume XLIX/3*, pp. 276–322. Berlin: Springer.

Lilley FEM and Day AA (1993) D'Entrecasteaux 1792: Celebrating a bicentennial in geomagnetism. *EOS, Transactions of the American Geophysical Union* 74(97): 102–103.

Mackereth FJH (1958) A portable core sampler for lake deposits. *Limnology and Oceanography* 3: 181–191.

Mackereth FJH (1969) A short core sampler for subaqueous deposits. *Limnology and Oceanography* 14: 145–151.

McElhinny (1973) *Palaeomagnetism and Plate Tectonics.* Cambridge, UK: Cambridge University Press.

McElhinny MW, McFadden PL, and Merrill RT (1996) The time-averaged palaeomagnetic field. *Journal of Geophysical Research* 101: 25007–25027.

McFadden PL (1980) The best estimate of Fisher's precision parameter κ. *Geophysical Journal of the Royal Astronomical Society* 60: 397–407.

McFadden PL and Lowes FJ (1981) The discrimination of mean directions drawn from Fisher distributions. *Geophysical Journal of the Royal Astronomical Society* 67: 19–33.

McFadden PL and McElhinny MW (1988) The combined analysis of remagnetization circles and direct observations in paleomagnetism. *Earth and Planetary Science Letters* 87: 161–172.

Melloni M (1853) Du magnétisme des roches. *Comptes Rendus Academie Sciences de Paris* 37: 966–968.

Merayo MG, Brauer R, Primdahl F, Petersen JR, and Nielsen OV (2000) Scalar calibration of vector magnetometer. *Measurement Science and Technology* 11: 120–132.

Meyer O and Voppel D (1954) Ein Theodolit zur Messung des Erdmagnetischen Feldes mit der Förstersonde als Nullindicator. *Deutsche Hydrographische Zeitschrift* 12: 73–77.

Milligan PR, Morse MP, and Rajagopalan S (1991) Pixel map preparation using the HSV colour model. *Exploration Geophysics* 23: 219–224.

Minty BRS (1992) Simple micro-levelling for aeromagnetic data. *Exploration Geophysics* 22: 591–592.

Molyneux LA (1971) A complete result magnetometer for measuring the remanent magnetization of rocks. *Geophysical Journal of the Royal Astronomical Society* 24: 429–433.

Nagata T (1961) *Rock magnetism,* 2nd edn. Tokyo, Japan: Maruzen.

Néel L (1949) Théorie du trainage magnétiique des ferromagnétiques aux grains fins avec applications aux terres cuites. *Annales De Geophysique* 5: 99–136.

Néel L (1955) Some theoretical aspects of rock magnetism. *Advances in Physics* 4: 191–243.

Newitt LR, Barton CE, and Bitterly J (1996) *IAGA Guide for Magnetic Repeat Station Surveys.* International Association of Geomagnetism and Aeronomy.

Nielsen OV, Petersen JR, Primdahl F, *et al.* (1995) Development, construction and analysis of the "Oersted" fluxgate magnetometer. *Measurement Science and Technology* 6: 1099–1115.

O'Sullivan KN (1991) A map for all reasons: The role of image processing in mineral exploration. *Minerals Industry International* 8–14 (January 1991).

Overhauser AW (1953a) Polarization of nuclei in metals. *Physical Review* 91: 476 (abstract only).

Overhauser AW (1953b) Polarization of nuclei in metals. *Physical Review* 92: 411–415.

Packard ME and Varian RH (1954) Free nuclear induction in the Earth's magnetic field. *Physical Review* 93: 941 (abstract only).

Papitshvili VO, Petrov VG, Saxena AB, Clauer CR, and Papitshvili NE (2006) A virtual global magnetic observatory network: VGMO.NET. *Earth Planets Space* 58: 765–774.

Pozzi JP and Thellier E (1963) Sur des perfectionments récents apportés aux magnétometre a très haute sensibilité utilisés en minéralogie magnétique et en paléomagnétisme. *Comptes Rendus Academie Sciences de Paris* 257: 1037–1040.

Primdahl F (2002) Resonance magnetometers. In: Ripka P (ed.) *Magnetic Sensors and Magnetometers*, pp. 267–304. Boston and London: Artech House.

Primdahl F, Merayo JMG, Brauer P, Laursen I, and Risbo T (2005) Internal field of homogeneously magnetized toroid sensor for proton free precession magnetometer. *Measurement Science and Technology* 16: 590–593.

Pulz E, Jäckel KH, and Linthe HJ (1999) A new optically pumped tandem magnetometer: Principles and experiences. *Measurement Science and Technology* 10: 1025–1031.

Rasmussen O (1990) Improvements in fluxgate magnetometers at Danish meteorological Institute's magnetic observatories. In: *Proceedings of the International Workshop on Geomagnetic Observatory Data Acquisition and Processing.* Finnish Meteorological Institute Geophysical Publications No. 15. Helsinki, Finland.

Rasson JL (1996) Progress in the design of an automatic diflux. In: Rasson JL (ed.) *Proceedings of the VIth Workshop on Geomagnetic Observatory Instruments, Data Acquisition and Processing*, Publ. Sci. et Techn. No 003, pp. 190–194. Brussels: Institut Royal Meteorologique de Belgique.

Rasson JL (2005) About absolute geomagnetic measurements in the observatory and in the field. *Publication Scientifique et Technique de l'Institut Royal Météorologique de Belgique* 40: 42.

Reeves CV (2007) The role of airborne geophysical reconnaissance in exploration geoscience. *First Break* 19.9: 501–508.

Reeves CV and Bullock SJ (2006) *Airborne exploration.* Fugro Airborne Surveys Limited (in press).

Reeves CV, Macnab R, and Maschenkov S (1998). Compiling all the world's magnetic anomalies. *EOS American Geophysical Union*, **July 14**, p. 338.

Reeves CV, Reford SW, and Milligan P R (1997) Airborne geophysics: Old methods, new images. In: Gubbins AG (ed.) *Proceedings of Exploration 97, Fourth Decennial International Conference on Mineral Exploration*, pp. 13–30. Toronto, Canada: GEO F/X.

Reford MS and Sumner JS (1964) Aeromagnetics. *Geophysics* 29: 482–516.

Reid AB, Allsop JM, Gransser H, Millett AJ, and Somerton IW (1990) Magnetic interpretation in three dimensions using Euler deconvolution. *Geophysics* 55: 80–61.

Roest WR, Verhoef J, and Pilkington M (1992) Magnetic interpretation using the 3-D analytic signal. *Geophysics* 57: 116–125.

Sapunov V, Denisov A, Denisova O, and Saveliev D (2001) Proton and Overhauser magnetometers metrology. *Contributions to Geophysics & Geodesy* 31: 119–124.

Sapunov V, Rasson J, Denisov A, *et al.* (2006) Theodolite-borne vector Overhauser magnetometer: DIMOVER, In: Yumoto E (ed.) *Proceedings of the XIth Workshop on Geomagnetic Observatory Instruments, Data Acquisition and Processing.* Kakioka.

Schmidt H and Auster V (1971) Neuere Messmethoden de Geomagnetik. In: Flugge S and Rawe K (eds.) *Encyclopedia of Geophysics, Geophysics III/3, Volume XLIX/3*, pp. 323–383. Berlin: Springer.

Schnegg PA and Fischer G (1991) The TURBOMAG vector magnetometer. *Münchner Geophysikalische Mitteilungen* 5: 121–130.

Serson PH and Hannaford WLW (1956) A portable electrical magnetometer. *Canadian Journal of Technology* 34: 232–243.

Shifrin VYa, Alexandrov EB, Chikvadze TI, *et al.* (2000) Magnetic flux density standard for geomagnetometers. *Metrologia* 37: 219–227.

Spector A and Grant FS (1970) Statistical models for interpreting aeromagnetic data. *Geophysics* 35: 293–302.

Stephenson A (1980) Rotational remanent magnetization and the torque exerted on a rotating rock in an alternating magnetic field. *Geophysical Journal of the Royal Astronomical Society* 62: 113–132.

Stephenson A (1981) Gyromagnetic remanence and anisotropy in single domain particles, rocks and magnetic recording tape. *Philosophical Magazine Series B* 44: 635–644.

Tarling DH (1966) The magnetic intensity and susceptibility distributions in some Cenozoic and Jurassic basalts. *Geophysical Journal of the Royal Astronomical Society* 11: 423–432.

Tarlowski C, McEwin AJ, Reeves CV, and Barton CE (1996) Dewarping the composite aeromagnetic anomaly map of Australia using control traverses and base stations. *Geophysics* 61: 696–705.

Tenani M (1941) Nuovo Metodo di misura della declinazione e della inclinazione magnetica. *La Ricerca Scientifica* 20: 1135–1140.

Thellier E (1933) Magnétometre insensible aux champs magnétiques troubles des grandes villes. *Comptes Rendus Academie Sciences de Paris* 197: 224–234.

Trigg DF (1970) A portable D and I magnetometer. *Geomagnetic Laboratory Report n°70-3, Direction de la Physique du Globe, Energy, Mines and Resources*. Ottawa, Canada.

Turner GM, Kamp PJJ, McIntyre AP, Hayton S, McGuire DM, and Wilson GS (2005) A coherent middle Pliocene magnetostratigraphy, Wanganui Basin, New Zealand. *Journal of the Royal Society of New Zealand* 35: 197–227.

van Loo SA and Rasson JL (2006) Development of an automatic declination–inclination magnetometer. In: Rasson JL and Delipetrov T (eds.) *Geomagnetics for Aeronautical Safety: A Case Study in and Around the Balkans*, Nato Science Series, Sub-Series C. Netherlands: Springer.

Watson GS (1983) *Statistics on Spheres*, University of Arkansas Lecture Notes in the Mathematical Sciences, pp. 177–186. New York: Wiley.

Wienert KA (1970) *Notes on Geomagnetic Observatory and Survey Practice*. Brussels: UNESCO.

Wilson RL (1962a) The palaeomagnetic history of a doubly baked rock. *Geophysical Journal of the Royal Astronomical Society* 6: 397–399.

Wilson RL (1962b) The palaeomagnetism of baked contact rocks and reversals of the earth's magnetic field. *Geophysical Journal of the Royal Astronomical Society* 7: 194–202.

Wilson RL and Lomax R (1972) Magnetic remanence related to slow rotation of ferromagnetic material in alternating magnetic fields. *Geophysical Journal of the Royal Astronomical Society* 30: 295–303.

Zijderveld JDA (1967) AC demagnetization of rocks: Analysis of results. In: Collinson DW, Creer KM, and Runcorn SK (eds.) *Methods in Palaeomagnetism*. Amsterdam: Elsevier.

Relevant Websites

http://www.2genterprises.com – 2G Enterprises.

http://www.gemsys.ca – GEM Systems.

http://www.ga.gov.au – Geoscience Australia.

http://www.geosoft.com – Geosoft.

http://www.gfz-potsdam.de – GFZ Potsdam: CHAMP.

http://www.intrepid-geophysics.com – Intrepid Geophysics.

http://www.intermagnet.org – INTERMAGNET.

www.molspin.com – Molspin.

http://www.physics.nist.gov – NIST Physics Laboratory Home Page: Values of physical constants recommended by CODATA, 2002.

http://www-odp.tamu.edu – Ocean Drilling Programme.

http://home.ural.ru – Quantum Magnetometry Laboratory of Ural State Technical University.

http://www.ngdc.noaa.gov – USDOC/NOAA/NESDIS/ National Geophysical Data Center(NGDC) Home. Page: IAGA 2002 Data Exchange Format; IGRF-10; World Data Center Systems.

5 Geomagnetic Secular Variation and Its Applications to the Core

A. Jackson and C. C. Finlay, ETH Zürich, Zürich, Switzerland

5.1   Introduction

The purpose of this chapter is to review the origins of our current knowledge of the secular variation of the Earth's magnetic field, that is, the slow changes that occur on timescales of years to centuries. There is clearly overlap with the description of the present geomagnetic field (Chapters 2 and 9), which treats changes in the field from centuries to millennia.

The source of our knowledge on the timescales we are concerned with is primarily direct historical observations of the field; we review the available data, followed by the treatment of the data to generate mathematical models of the field in space and time. We then discuss interpretations of these models in terms of some of the physical processes occurring at the core surface. We stop short of describing the actual calculation of models of fluid flow at the core surface, but we lay the groundwork by developing an exposition of the governing equations and the approximations that are frequently used.

5.1.1   Historical Background

This section gives a very brief overview of the development of geomagnetism, and does not purport to be comprehensive. Fuller treatments of the history can be found in various places; for example, relevant chapters of Merrill *et al.* (1998) or Chapman and Bartels (1940), Malin (1987), or Stern (2002). A detailed account of geomagnetism up to 1500 can be found in Crichton Mitchell (1932, 1937, 1939); recent articles on nineteenth century geomagnetism are those of Good (1985, 1988). Excellent discussions of geomagnetic instruments can be found in McConnell (1980) and Multhauf and Good (1987). An authoritative source on virtually every aspect of geomagnetic history is the epistle by Jonkers (2000).

It is generally acknowledged that the Chinese were the first to discover the directive property of lodestone, almost certainly by the first century AD. Its development as a primitive navigational device was slow, though the declination had almost certainly been discovered by the ninth century and compasses were certainly in use by the eleventh century; early observations of declination are given by Needham (1962) and Smith and Needham (1967). The first recorded observation of declination in Europe was by George Hartmann in 1510; inclination was discovered by Robert Norman in 1576. The fact that the field underwent slow changes with time (the secular variation)

was not discovered until 1635: by comparing a series of records taken at London previously, Henry Gellibrand showed that secular variation was a real effect. Relative intensities of the field were measured at the end of the eighteenth century by La Perouse, D'Entrecasteaux, and Humboldt, by comparing the periods of oscillation of a magnetic needle at different places. Measurements of the absolute intensity of the field were not made until a method was devised by Gauss in 1832 (see, e.g., Malin, 1982). The method was published in Gauss (1833a); an English translation of the abstract of a paper read in Göttingen in December 1832 can be found in Gauss (1833b).

While early observations of the field are extremely valuable, some problems do exist. For example, before the discovery of secular variation, some observations are undated as the need to record the date was not apparent. The accuracy with which an observer's position was known is also a source of error. Although the measurement of latitude was precise even by late fifteenth century (e.g., an accuracy of 10 min of arc was claimed by 1484 (John II's commission, 1509)), the measurement of longitude at sea remained a problem until approximately 1770 with the introduction of accurate chronometers by Harrison. The result of this poor knowledge of longitude led to the practice of 'running down the parallel', or sailing to the correct line of latitude before sailing due east or west along that parallel to the desired location. Although this practice meant that the ship's company often arrived at their desired destination, it does mean that large navigation errors could occur in the quoted positions of magnetic observations. To a large extent these errors can be alleviated by examination of the original ship's log and plotting the positions on a modern chart. This procedure has been performed for sixteenth, seventeenth, and eighteenth century data by Bloxham (1985, 1986a, 1986b), Hutcheson (1990) (see also Hutcheson and Gubbins, 1990), and Barraclough (1985), and Jackson *et al.* (2000) (hereinafter JJW2000); in addition, the latter authors also developed a statistical theory for accounting for imprecision in longitude.

The Greenwich meridian was adopted as an international longitude standard only in 1884, and some national conventions remained in use later than that date. Consequently, care must be taken as to which of the particular national conventions of Paris, the observatory at Pulkova (Leningrad), Washington, or San Fernando was being used. One example of French marine data measuring longitude from Paris until at least 1895 has been given by

Jackson (1989); this difference of $2°$ $13'$ of longitude between Paris and Greenwich is small, but extremely significant.

5.1.2   Early Theories of the Secular Variation

Beginning with the seminal works *Epistola de Magnete* by Peregrinus (1269) and the better-known *De Magnete* by Gilbert (1600), various authors have sought to explain the Earth's magnetic field by models, some physical, some mathematical. Though Gilbert's model explained a considerable part of the static field, after the discovery of the secular variation a whole new dimension was opened up, requiring explanation. It is not our purpose to adumbrate the numerous models created over time to explain the temporal variation of the field. However, Jonkers (2000) has provided just such a list, comprising a remarkable compilation of theories of the field up to 1800, starting with Peregrinus (1269) and ending with Churchman's (1794) petitions to the English Board of Longitude, requesting acceptance of his theories for use in determining longitude. A shorter description can also be found in Jonkers (2003).

5.2   Data

We refer the reader to Chapter 4 for information on how measurements of the field are taken, and for definitions of the quantities that are typically reported: the declination D, the inclination I, the horizontal and total intensities (H and F, respectively), and the Cartesian components X, Y, and Z in the northerly, easterly, and downward directions. The availability of different data types varies as a function of time, chiefly as a result of the needs for navigation, followed by the drive of scientific curiosity in the eighteenth and nineteenth centuries. It should be noted that until Gauss' invention of a method for the determination of absolute intensity in 1832, only the morphology of the field can be determined from direct measurements.

5.2.1   Catalogs and Compilations of Data

The earliest catalogs, of Stevin (1599), Kircher (1641), and Wright (1657), are deficient in that they contain undated observations. Around 1705, the French hydrographer Guillaume Delisle compiled some 10 000 observations (mostly of declination) in his

notebooks, trying to establish regularity in secular acceleration; these were never published but still exist in the Archives Nationales in Paris. The next important compilation of magnetic data was made by Mountaine and Dodson (1757) who claimed to have based their tables of declination at different points on the Earth on over 50 000 original observations of the field. The original observations of this enormous collection were never recorded and are lost: the work merely printed grids of averaged data with no reference to sources, numbers of data, etc. This claim regarding the number of data involved has attracted some skepticism; however, the work of Jonkers *et al.* (2003) indicates that it is undoubtedly the case that the authors' claim for the number of data is true. The early work of Mountaine and Dodson should almost certainly receive more prominence than it does, representing probably the first large-scale attempt to describe the morphology of the field. Maps based on the data were subsequently produced.

The main era of printed compilations of geomagnetic data was the nineteenth century, featuring the work of Hansteen, Becquerel, Sabine, and Van Bemmelen. In 1819, the Norwegian astronomer and physicist Christopher Hansteen published *Untersuchungen über den Magnetismus der Erde*, which listed data from land surveys and 73 nautical voyages from 1589 to 1816. His collection includes many of the great scientific expeditions during the latter half of the eighteenth century, including Cook's three voyages, contributing over 6500 declination and 1200 inclination observations. Another valuable addition was made by A. C. Becquerel's *Traité Expérimental de l' Électricité et du Magnétisme* (1840), which contains the only comprehensive collection of relative intensities. Several Philosophical Transactions papers by astronomer Edward Sabine span the period 1818-70 with exceptionally good coverage, although, as various authors have noted, they are far from comprehensive. Finally, in the 1890s, Dutch physicist Willem van Bemmelen processed 165 nautical sources prior to 1741 (Bemmelen, 1899).

Finally, one of the largest single compilations, featuring over 28 000 data points of all types, is the *Catalogue of Magnetic Determinations in USSR and in Adjacent Countries from 1556 to 1926* in three volumes. It was compiled and published by Russian physicist B. P. Veinberg in 1929–33 (Veinberg, 1929–33) and contained original data from Russia and neighboring states, mostly obtained in the first decades of the twentieth century. A review of the previous compilations of magnetic data that have been produced over time can be found in Barraclough (1982).

Another category of sources comprise time series for specific locations, normally major cities where investigators have set up permanent instruments, for instance, at national astronomical observatories. Past observers include Graham, London clock maker and the discoverer of diurnal variation, and Gilpin in England, academics Celsius and Hiörter in Sweden (who studied the correlation of needle disturbance with the occurrence of auroras), MacDonald on Sumatra (eighteenth century), and, in France, many scholars and astronomers summarized in Alexandrescu *et al.* (1996). But despite their achievements, a mere handful of cities can boast a series of more or less regular observations spanning over a century prior to 1800. A review of recent efforts to make these data series available to a modern audience is given by Alexandrescu *et al.* (1996), who also list all early geomagnetic observations made in Paris (1541-1883, based in part on earlier work by Raulin (1867) and Rayet (1876)); see **Figure 1**. Other capitals with a sustained tradition of geomagnetic observations include London (Malin and Bullard, 1981, Barraclough *et al.*, 2000; see **Figure 18**), Rome (Cafarella *et al.*, 1992), and Edinburgh (Barraclough, 1995).

It should be noted that only when observations in compilations can be confidently ascribed to individual observations (with well-defined times and locations and not derived by interpolation) have they included

in the database of historical observations of Jonkers *et al.* (2003) (hereinafter JJM2003), and used historical field models such as *gufm1* described in JJW2000.

5.2.2   Surveys, Repeat Stations, and Marine Data

Recent interest in historical secular variation has led to original observations being compiled for other time periods; a comprehensive review of available data has been given recently by JJM2003. It is impossible to detail all characteristics of this data set. Suffice it to say that the largest part of the data set originates in marine observations of the declination, typically taken for the purposes of navigation. It has been possible to characterize the accuracy with which observers measured the declination at sea; it is better than half a degree for the 17th and 18th centuries taken as a whole. When one compares this accuracy with modern measurements, it transpires that the old measurements have a signal/noise ratio which is not too bad compared to modern measurements, for the simple reason that both types of measurements suffer the contaminating effect of the crustal magnetic field. This contributes $\sim 60\sigma/H$ degrees of error in declination where σ is the rms horizontal crustal field, and H the local horizontal field strength; for $\sigma \sim 200$ nT and $H = 20\,000$–$40\,000$ nT the error contribution is 0.3–0.6°, and thus commensurate with the observational error.

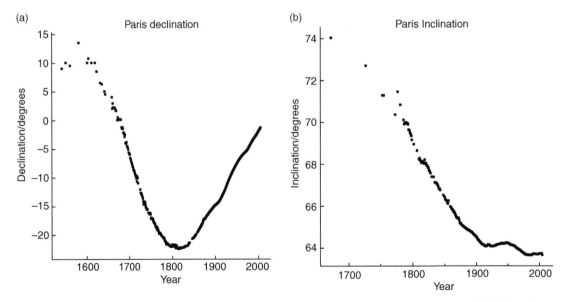

Figure 1   Declination (*D*) and Inclination (*I*) in Paris during historical times, based on Alexandrescu *et al.* (1996) and more recent observatory data. The data have been reduced to the current site of Chambon-la-Forêt; early observations are made by many different observers at several different sites.

An accurate position is of course a crucial part of any magnetic measurement, and thus latitude and longitude need to be known. This poses no problems on land, but at sea the determination of position can be challenging. From the sixteenth century onward, the backstaff provided a method for latitude determination, often said to be accurate to 10 min of arc; the empirical findings of JJW2000 agree with this. For early data, a well-known difficulty is the imprecision in longitude prior to the invention of the marine chronometer. A very detailed study of this was undertaken in JJW2000, who showed that navigational error generally generated a Brownian motion-type effect, such that the errors increased with the square root of voyage duration. Empirically, the data suggest that a typical 25 day voyage might accumulate 2° of error (though voyages often achieved much better than this). The effect can be ameliorated in most voyages by using the fact the voyage arrived at a known location — thus the whole voyage can be retrospectively corrected for the accumulated errors, giving a lower error. The appropriate model for the errors becomes the so-called Brownian bridge, and the error at the mid-point of the journey is reduced to exactly half what it would have been using the simple Brownian motion model — hence typically it is 1° for a 25 day journey. The effect of this imprecision in location is to increase the error budget by an amount depending on the gradient of declination with respect to longitude — a fairly representative figure is less than 1° of declination change per degree of longitude. Hence one can see that even early data have contributions to their error budget (from observational imprecision, crustal magnetic fields, and longitudinal inaccuracy) that are not too dissimilar. Inclination was initially measured on land at London (1576); the next extant measurement was taken shortly after at Uranienborg (1584) by Brahe. The first example of a measurement made on an expedition was by Weymouth in Frobisher Bay, Canada, in 1602. For the first inclination observation south of the equator, one had to wait until 1680 when one Benjamin Harry took observations on board the Berkeley Castle en route to the far east.

The nineteenth century saw burgeoning scientific expeditions on land, which included the measurement of intensity as well as D and I — thus giving the first vector measurements of the field. De Rossel's measurements of the oscillation time of a dip needle in 1791 provides the first measurements of relative intensity between several places on the Earth. Humboldt and Erman also provide well-known relative intensity measurements prior to Gauss' (1833a) invention of a method to determine absolute intensity (see Malin, 1982). The net result of the collation of the known historical observations as described in JJM2003 is a data set which is summarized in **Table 1**. The available data are summarized in geographical plots in **Figures 2–10**.

A vast source of magnetic field data for the twentieth century was prepared by the US Coast and Geodetic Survey for the 1965 world charts (Hendricks and Cain, 1963), and is accessible in machine-readable form from the world data centers. The cutoff date for the collection was arbitrarily set at 1900. This data set has been used by many authors over the years, and has been the basis for many International Geomagnetic Reference Field models; we shall not dwell on a description of this data set, as it has been described numerous times (e.g., Bloxham *et al.*, 1989, Sabaka *et al.*, 1997). Its temporal distribution dominates the temporal distribution of the data used to create *gufm1* during the twentieth century, which is shown in **Figure 11**. Twentieth century data is characterized by a constant improvement in measurement accuracy (see Chapter 2). Marine surveys continued in the twentieth century, the most notable being the voyages of the nonmagnetic surveying ship, the *Carnegie*. A new type of data emerged with the advent of aeromagnetic surveys, most notably Project Magnet; for details, see Langel (1993).

Table 1 Temporal distribution of databased geomagnetic measurements; a single record may contain a land sighting and/or up to three types of measurement (D, I, and intensity (H or F))

Period	Records	D	I	H	F	Total
1510–1589	162	160	2	0	0	162
1590–1699	13 673	12 001	53	0	0	12 054
1700–1799	85 070	68 076	1 747	0	36	69 859
1800–1930	78 162	71 323	17 723	11 404	4 779	105 229
Total	177 067	151 560	19 525	11 404	4 815	187 304

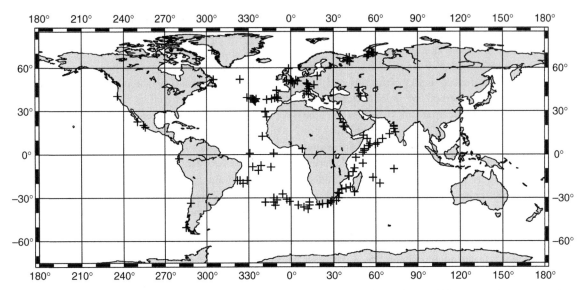

Figure 2   Geographical data distribution of declination observations made before 1590; $n = 160$; some points may overlap; cylindrical equidistant projection.

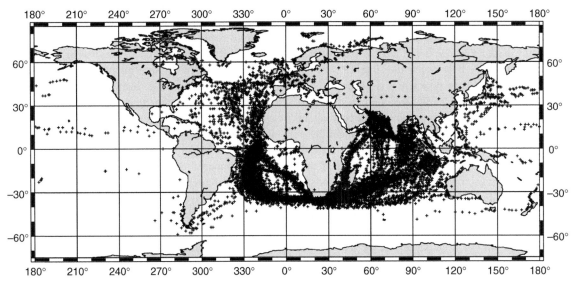

Figure 3   Geographical data distribution of declination observations made in 1590–1699; $n = 12\,001$; some points may overlap; cylindrical equidistant projection.

5.2.3   Observatory Data

The establishment of the Göttingen Magnetic Union (Magnetische Verein) in 1834 by Gauss and Weber heralded the establishment of an observatory network at sites around the world where observations of the magnetic field would be made with regularity. With the adoption of the 'magnetic crusade' of Sabine, Herschel, and Lloyd by the British learned bodies in 1838, Germany and Britain took a lead in driving forward observational geomagnetism

(Cawood, 1977, 1979). The number of observatories gradually grew and their distribution increased toward the distribution of today (**Figure 12**), although some former observatories have closed due to a multitude of factors. Some of the history of the growth of the observatory network can be found in Chapman and Bartels (1940).

From the point of view of studies of the secular variation due to the motions of the liquid outer core, the most important product derived from the

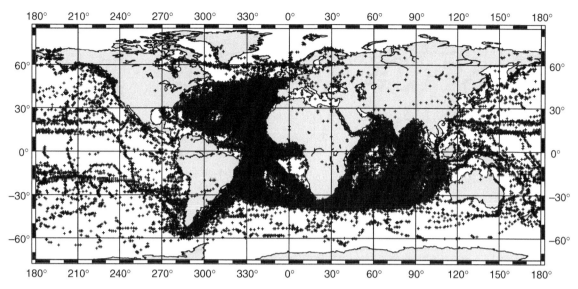

Figure 4 Geographical data distribution of declination observations made in 1700–99; $n = 68\,076$; some points may overlap; cylindrical equidistant projection.

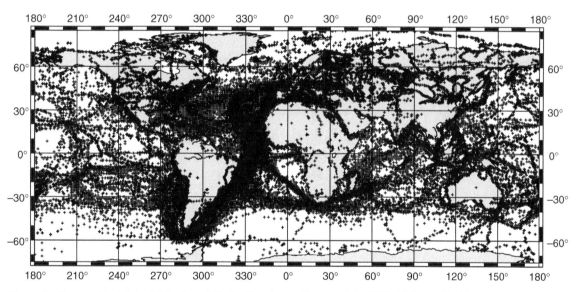

Figure 5 Geographical data distribution of declination observations made in 1800–1930; $n = 71\,323$; some points may overlap; cylindrical equidistant projection.

continuous monitoring of the observatories is the so-called annual mean, representing the yearly averaged value of the geomagnetic elements. Although the current definition of an annual mean is a mean over all data, there has, in the past, been some variability in exactly what is reported as an annual mean. For example, occasionally data reported from the five International Quiet Days every month has been used as an annual mean. **Table 2** shows the frequency of the different types of data that are included in the definitive annual means data file, held by the World

Data Centre for Geomagnetism at the British Geological Survey (Edinburgh). This inconsistency, which cannot be corrected retrospectively (since the original data no longer exist), leads to inevitable difficulties in treating the data, because the data contain different amounts of external magnetic field contribution. Compromises are always required in treating historical magnetic data, and so far these observatory data have been treated as if they were homogeneously recorded; perhaps it will be possible to treat them in a way that recognizes their different characteristics in

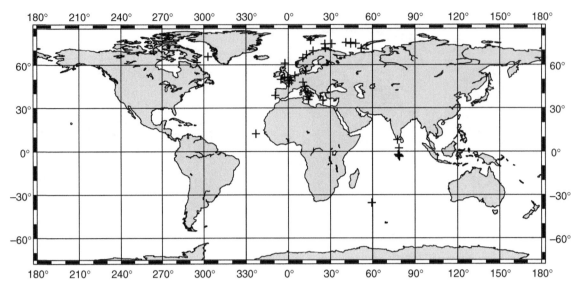

Figure 6   Geographical data distribution of inclination observations made in 1590–1699; $n = 53$; some points may overlap; cylindrical equidistant projection.

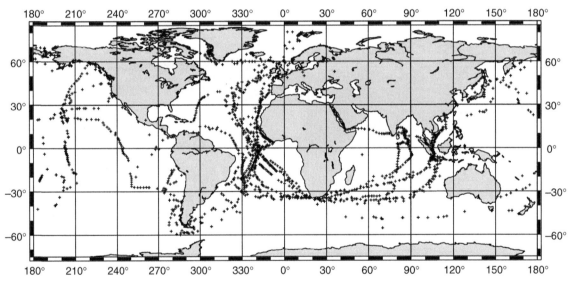

Figure 7   Geographical data distribution of inclination observations made in 1700–99; $n = 1747$; some points may overlap; cylindrical equidistant projection.

the future. **Figure 13** shows the distribution of observatory annual mean data through time, from the first observations originating with the formation of the Göttingen Magnetic Union, to the present day.

It is straightforward to treat single observations of the field (such as those made by surveys or satellites) as being independent measurements that can be fitted simultaneously in a least-squares process. Some words are in order regarding the treatment of observatory data in time-dependent field modeling. Observatories obviously supply critical data on the

secular variation, and indeed the accuracy of many of the modern field models, rests on the observatory time series. A problem that must be recognized, however, is the fact that the observatories are subject to a (quasi-) constant field associated with the magnetization of the crust in the region that they are located. If observatory data are mixed with other types of data (survey, satellite data), this so-called observatory bias must be recognized; otherwise, it will bias the solution for the main field because an observatory time series essentially records it many times. Two approaches have

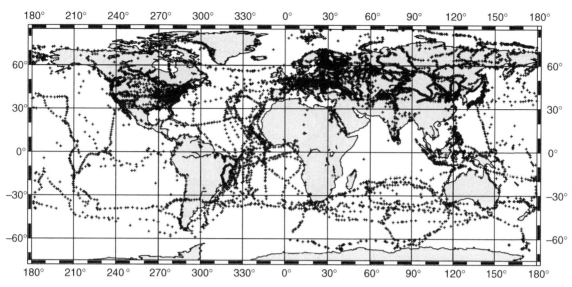

Figure 8 Geographical data distribution of inclination observations made in 1800–1930; $n = 17\,723$; some points may overlap; cylindrical equidistant projection.

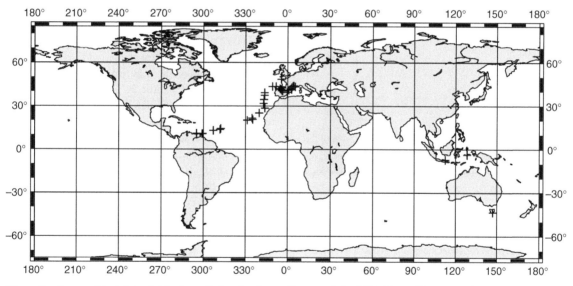

Figure 9 Geographical data distribution of intensity observations made in 1700–99; $n = 36$; some points may overlap; cylindrical equidistant projection.

developed for dealing with this. The first, developed by Langel *et al.* (1982), is to solve for the observatory biases (three per observatory in the X-, Y-, and Z-directions) as unknowns at the same time as solving for the magnetic field. This technique continues to be adopted in the comprehensive series of field models (see below), and works very effectively. The second approach is to desensitize the observatory data to the presence of the bias (see, e.g., Bloxham and Jackson, 1992). An effective way of doing this is to work with the rate of change of the field from the observatory,

and hence first differences of observatory data are used in the *ufm* and *gufm* series of models (see below). There appears to be very little difference in the results of the two approaches.

5.2.4 Satellite Data

Satellite data play a crucial role in determining a detailed global picture of the secular variation. An extensive discussion of the special character of satellite data can be found in Chapter 2, and we shall not

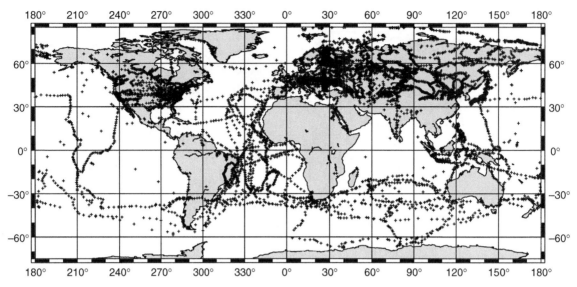

Figure 10 Geographical data distribution of intensity observations made in 1800–1930; $n = 16\,183$; some points may overlap; cylindrical equidistant projection.

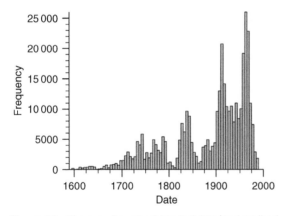

Figure 11 The overall number historical data (as described in JJM2003) together with observatory data, twentieth century survey and repeat station data, and satellite data used in the construction of *gufm1* (Jackson *et al.* 2000). Note that this depicts a subset of data available, as some data selection has taken place, based on criteria designed to avoid the effect of the correlation in errors due to the crust.

duplicate that here. Nevertheless, in **Table 3**, we list some of the satellites that have been used for magnetic field determination over time, and their different characteristics.

5.3 Time-Dependent Models of the Main Field

We now turn to the use that is made of the data sets that have been described in the previous section. The tool that has been most commonly applied has

been spatial spherical harmonic analysis, first applied by Gauss (1839). His analysis demonstrated the predominantly internal origin of the field.

The principles of spherical harmonic analysis are described in Chapter 2. They were applied by many authors in the nineteenth and twentieth centuries, with various amendments in order to deal with the fact that primarily nonlinear functions of the Gauss coefficients were being measured, namely D, I, H, and F; such developments are described fully in Barraclough (1978). As one such example, consider how to treat measurements of declination in the spherical harmonic inverse problem. We have that

$$D = \tan^{-1}\frac{Y}{X} \qquad [1]$$

and the northerly (X) and easterly (Y) components are linearly related to the Gauss coefficients $\{g_l^m; h_l^m\}$ forming the model vector **m**. Let us write these relations as $X = \mathbf{A}_x^{\mathrm{T}}\mathbf{m}$ and $Y = \mathbf{A}_y^{\mathrm{T}}\mathbf{m}$. If we rearrange [1] into the form,

$$X \sin D = Y \cos D \qquad [2]$$

one can form a linear constraint on **m** of the form

$$\left[(\sin D)\mathbf{A}_x - (\cos D)\mathbf{A}_y\right]^{\mathrm{T}}\mathbf{m} = 0 \qquad [3]$$

This can be fit in a least-squares sense, but note that the data enter in defining the linear relation, rather

Geomagnetic observatory locations 1980–2005

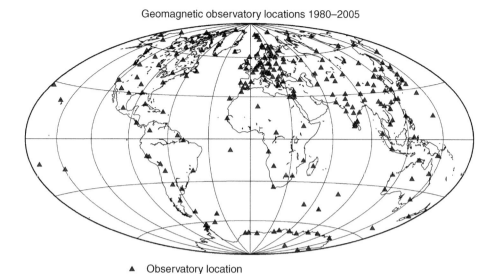

▲ Observatory location

Figure 12 The distribution of observatories operating at some point in the last 25 years.

than as a target for the prediction. Numerous other schemes for dealing with nonlinear data are described in Barraclough (1978). With the advent of significant computer power, the need to deal with nonlinear data in such a way has diminished and iterative schemes, as described in Chapter 2, are more commonplace.

In the years following the early applications of Gauss' method, the technique was applied to the field at different epochs, the interest being primarily in the evolution of global averages such as the dipole moment. Being before the advent of modern computers, it was impossible to deal with true measurements of the field without some preliminary reduction of the data — thus the source for the spherical harmonic analyses were field values at regular intervals read from charts which had been constructed by interpolating the original data by hand. Useful descriptions of these types of model can be found in Barraclough (1978) or Langel (1987).

5.3.1 Methodologies

Chapter 2 discusses the determination of static models of the field. In this section, we will review a selection of the most widely used time-dependent field models, and the techniques used to derive them. We restrict attention to models that have been produced specifically as time-dependent; only passing reference is made to models designed to describe either the static magnetic field or its rate of change at a particular point in time; for models of this type, see Chapter 2.

Our description focuses on models of the magnetic field $\mathbf{B}$ which are simultaneously models of its spatial $((r, \theta, \phi)$ in spherical coordinates) dependence and the temporal dependence (t denotes time). The standard technique which is common to all the analyses we will describe is to employ the spherical harmonic expansion of the field in terms of Gauss coefficients $\{g_l^m; h_l^m\}$ for the internal field; some of the most recent models also incorporate coefficients representing the external field. All the models will

Table 2 Types of observations classed as annual means and used for secular variation modeling

Flag	Annual means derived from	Percentage
1	Data for all days	72.4
2	Data for quiet days	4.0
3	Preliminary data	1.9
4	Absolute observations only	9.0
5	Incomplete data (<12 months, but ≥6 months)	3.4
6	Very incomplete data (<6 months)	1.4
7	Limited absolute control (introduced in 1996)	0.4
0	Unknown	7.6

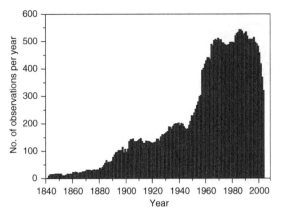

Figure 13 Distribution of observatory annual means through time, reflecting availability as of October 2006. The falloff in recent times is due to the delay in observatories reporting definitive data to the world data centers. Note that an three component observation counts as 3 observations, not 1 observation in this plot.

employ the Schmidt quasi-normalization common in geomagnetism.

A time-dependent model of the field necessarily must be built using a data set spanning a period of time, denoted herein $[t_s, t_e]$. In order that a spherical harmonic analysis can be performed, a parametrization is required for the temporal variation of the field. The unifying idea, common to all analyses, is to use an expansion for the Gauss coefficients of the form

$$g_l^m(t) = \sum_i {}^i g_l^m \, \phi_i(t) \qquad [4]$$

where ϕ_i are a set of basis functions and the ${}^i g_l^m$ are a set of unknown coefficients. (A similar expansion is of

course used for h_l^m.) The different models that have been produced over the last few decades differ in their choice of the $\phi_i(t)$. With an expansion of the form [4], the unknown coefficients $\{ {}^i g_l^m; \; {}^i h_l^m \}$ are denoted as a model vector $\mathbf{m}$, and when linear data such as the elements (X, Y, Z) are required to be synthesized (denoted by vector $\mathbf{d}$) the resulting forward problem is linear and of the form

$$\mathbf{d} = \mathbf{Am} \qquad [5]$$

where $\mathbf{A}$ is often termed the equations of condition or design matrix and describes how model parameters are combined to give predictions that can be compared to the data.

The inverse problem of finding the coefficients $\mathbf{m}$ is generally solved by finding a model minimizing the least-squares difference between the model predictions and the data, sometimes together with a measure of the field complexity to help resolve the issue of nonuniqueness (see Chapter 2). More generally, when I, D, F, and H data are involved so that the relation between the model parameters and the data is a nonlinear function (which we write as $\mathbf{d} = \mathbf{f}(\mathbf{m})$), the model must be found iteratively. If $[\mathbf{A}]_{ij} = \partial f_i / \partial m_j$, and if $\mathbf{C}_e$ is the data covariance matrix, then the model solution is sought by an iterative scheme, such as the quasi-Newton method,

$$\mathbf{m}_{i+1} = \mathbf{m}_i + \left(\mathbf{A}^T \mathbf{C}_e^{-1} \mathbf{A} \right)^{-1} \left[\mathbf{A}^T \mathbf{C}_e^{-1} (\mathbf{d} - \mathbf{f}(\mathbf{m}_i)) \right] \qquad [6]$$

In [6], $\mathbf{m}_i$ stands for the model at the ith iterate, and in principle the matrix $\mathbf{A}$ should be recomputed at every iterate. Such methods converge very rapidly since the effect of the nonlinearity is very mild.

Table 3 Satellite missions of relevance for measurement of the core secular variation. Accuracies refer to the intrinsic accuracies of the instrumentation, combined with the positional and orientation accuracy

Name	Inclination	Dates	Altitude (km)	Accuracy (nT)	Remarks
Cosmos 49	50°	1964	261–488	22	Scalar
OGO-2	87°	1965–67	413–1510	6	Scalar
OGO-4	86°	1967–69	412–908	6	Scalar
OGO-6	82°	1969–71	397–1098	6	Scalar
Magsat	97°	1979–80	325–550	6	Vector
DE-1	90°	1981–1991	568–23 290	?	Vector (spinning)
DE-2	90°	1981–83	309–1012	~30(F)/100	Low accuracy vector
POGS	90°	1990–93	639–769	?	Scalar, timing problems
UARS	57°	1991–94	560	?	Vector (spinning)
Ørsted	97°	1999–	600–850	a	Vector
CHAMP	87°	2000–	350–460	a	Vector
SAC-C	98°	2000–04	702–709	a	Scalar

[a]For accuracies of the present missions, see Chapter 2.
Inclination is measured as the angle at which the satellite crosses the equator while passing from the Southern Hemisphere to the Northern Hemisphere.

5.3.1.1 Taylor series models

The earliest time-dependent models used a Taylor series expansion for the Gauss coefficients of the form

$$g_l^m(t) = g_l^m(t_0) + \dot{g}_m^m(t_0)(t - t_0)$$
$$+ \ddot{g}_l^m(t_0) \frac{(t - t_0)^2}{2!} + \cdots \qquad [7]$$

about some central epoch here denoted t_0. This expansion is of the form [4], with the identification $\phi_n(t) = (t - t_0)^n/n!$ and $^n g_l^m = (\partial_t)^n g_l^m(t_0)$, the nth time derivative at the central epoch.

In the case of the Taylor expansion, **A** is a dense matrix. The first models to be produced this way were those of Cain *et al.* (1965) and Cain *et al.* (1967), who produced models GSFC(4/64) and GSFC(12/66) with temporal expansions truncated at first derivative and second derivative terms, respectively. The truncation level was subsequently raised to third derivative terms in the model GSFC(9/80) of Langel *et al.* (1982). More recently, Taylor series expansion techniques have been used to provide time-dependent models of satellite data, covering the only short intervals of a few years. For example, Olsen (2002) used a first-order expansion and Maus *et al.* (2005) used a second-order expansion; for such models of satellite data covering only a few years, Taylor series expansion models are reasonable.

When one wishes to produce a model of the field spanning a long time period, it is clear that a large number of terms will be required in [7], and it no longer remains an attractive method because of numerical instabilities and lack of flexibility of the parametrization.

5.3.1.2 Two-step models

A variety of models have been made by a two-step process of first making a series of spatial models at particular epochs, followed by some form of interpolation. For example, the International Geomagnetic Reference Fields and Definitive Geomagnetic Reference Fields (DGRFs) are strictly snapshot models of the field for particular epochs, but they can be used to calculate the magnetic field at times intermediate between two epochs by linear interpolation between the models. As a result it is possible to evaluate the DGRFs at any point in time between 1900 and the present day, though from a purist point of view they are not strictly time-dependent models of the magnetic field. The stepping stone between such two-step models and the more sophisticated approach of using a spline representation of temporal behaviour (see Section 5.3.1.3) was the pioneering paper of Langel *et al.* (1986). These authors used a spline temporal basis to interpolate between single epoch secular variation models.

5.3.1.3 Time-dependent models based on cubic B-splines

After the mid-1980s, more flexible representations of the time dependency were introduced. Beginning with Bloxham (1987), who used Legendre polynomials, a variety of functions have been employed. The most commonly used and referenced time-dependent field models along with their time span and modeling approach are summarized in **Table 4**.

The methods employed by different workers have gradually converged toward the use of cubic B-splines as temporal basis functions following the example of Bloxham and Jackson (1992), who were heavily

Table 4 Characteristics of widely used models of the time-varying magnetic field

Model	L	N	Time period	Expansion	Regularized	Author
GSFC(4/64)	5	2	1940–63	Taylor	No	Cain *et al.* (1965)
GSFC(12/66)	10	3	1900–66	Taylor	No	Cain *et al.* (1967)
GSFC(9/80)	13	4	1960–80	Taylor	No	Langel *et al.* (1982)
MFSV/1900/1980/OBS	14	8	1900–80	Legendre	Yes	Bloxham (1987)
	14	10	1820–1900, 1900–80	Chebyshev	Yes	Bloxham and Jackson (1989)
ufm1, ufm2	14	63	1690–1840, 1840–1990	B-spline	Yes	Bloxham and Jackson (1992)
gufm1	14	163	1690–1990	B-spline	Yes	Jackson *et al.* (2000)
CM3	13	14	1960–85	B-spline integrals	Yes	Sabaka *et al.* (2002)
CM4	13	24	1960–2002.5	B-spline integrals	Yes	Sabaka *et al.* (2004)
CHAOS	18	10	1999–2006	B-spline and Taylor	Yes	Olsen *et al.* (2006)

L is the maximum degree of the internal secular variation. *N* is the number of temporal basis functions used for each Gauss coefficient. The CHAOS model uses B-splines up to degree 14 and a first-order Taylor expansion between 15 and 18. For other satellite-derived models, see Chapter 2.

influenced by the approach of Langel *et al.* (1986). There are two reasons for the popularity of the cubic B-spline method. First, when global basis functions such as Legendre or Chebyshev expansions are used (see, e.g., Bloxham (1987) or Bloxham and Jackson (1989)), the design matrix remains dense and requires considerable memory for its storage, whereas a B-spline basis is a 'local' basis, meaning that the basis functions are zero outside a small range (see **Figure 14**). This fact leads to a design matrix which is sparse (in fact it is banded), and storage requirements are minimized. Second, the B-splines provide a flexible basis for smoothly varying descriptions of data. One can show that of all the interpolators passing through a time series of points (say $f(t_i), i = 1, \ldots, N$), an expansion in cubic B-splines of order 4 ($\hat{f}(t)$ say) is the unique interpolator which minimizes a particular measure of roughness $\mathcal{R}$ (see, e.g., De Boor, 2002),

$$\mathcal{R} = \int_{t_s}^{t_e} \left[\frac{\partial^2 \hat{f}(t)}{\partial t^2} \right]^2 dt \qquad [8]$$

The idea of attempting to construct a smooth representation in time is an application of 'Occam's razor', that there should be no extra detail in the representation than that truly demanded by the data. This idea of 'regularization' has been employed in many of the models of **Table 4** from that of Bloxham (1987) onward. Those models that employ regularization typically minimize a combination of norms $\mathcal{N}$ on the core–mantle boundary (CMB) of the form

$$\mathcal{N} = \int_{t_s}^{t_e} \left[\nabla_b^{(n_1)} \partial_b^{(n_2)} B_r \right]^2 d\Omega \, dt \qquad [9]$$

where B_r is the radial field on the CMB. The models produced by Bloxham, Jackson, and co-workers use $n_1 = 0$ and $n_2 = 2$ in one norm, and $n_1 = 1$ and $n_2 = 0$ (approximately, to be precise the ohmic heating norm of Gubbins (1975) is used) in a second norm; this is slightly different to the choices made by Sabaka, Olsen, and co-workers in their comprehensive models and the CHAOS time-dependent model of satellite data (see below). A rather different form of regularization was recently proposed by Jackson (2003) that involves maximizing the entropy of the field model rather than penalising spatial or temporal gradients. This new method has so far been used to produce single epoch models but it could also be applied to provide both spatial and temporal regularization of time-dependent models; such models are currently under development.

Regularized field models are found by minimizing an objective function consisting of a measure (often the L_2 least squares norm) of the misfit of the time-dependent model to the data along with spatial and temporal norms measuring the field complexity. The relative weights of the spatial and temporal norms are scaled by the sizes of so-called damping parameters λ_S and λ_T. The choice of the damping parameters are made by trading off the desire that the data be fit within their estimated errors, the desire that the spatial complexity of the time-dependent model at the core surface be compatible with accurate single-epoch models, and the requirement that no unnecessary temporal oscillations be introduced. The models such as *ufm* and *gufm* satisfy each of these criteria.

5.3.1.3.(i)   The ufm1, ufm2, and gufm1 models

The *ufm1/ufm2* and *gufm1* field models share a common aim, namely to model the long-term secular variation at the core surface as accurately as possible over the past few centuries. They were built using the cubic B-spline basis with knots every 2.5 years, and from the largest data sets possible at the time: *ufm1/ufm2* used over 250 000 data originating from old ships' logs, survey data, observatories and satellite missions; a description of the oldest data can be found in Bloxham (1986b) and Bloxham *et al.* (1989). The *gufm1* model was built from similar data from the twentieth century, but a vastly expanded historical data set, described in Jonkers *et al.* (2003) – the model contains over 365 000 data and consists of 36 512 parameters. **Figure 11** shows the time distribution of the data used in *gufm1*. No account is explicitly taken of external fields in these models.

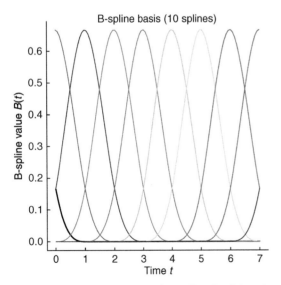

Figure 14   B-splines of order 4 (cubic B-splines). Local temporal basis of cubic B-splines used in the construction of time-dependent geomagnetic field models.

5.3.1.2.(ii) The comprehensive models An effort began in the early 1990s to build a comprehensive series of field models which took account of many effects which are recorded in geomagnetic data in addition to the core secular variation. The first model was reported by Sabaka and Baldwin (1993); Sabaka *et al.* (2002) described the most recent model formulation in detail while Sabaka *et al.* (2004) discuss its extension to include *Ørsted* and *CHAMP* satellite data. We will specifically report on the CM4 model of Sabaka *et al.* (2004). In general terms, the model includes representations of the main field, its secular variation, and both local-time (Sun-synchronous) and seasonal modes of the magnetospheric and ionospheric fields, as well as describing ring-current variations through the Dst index and internal fields induced by time-varying external fields. The data used in creating the model consists of *POGO*, *Magsat*, *Ørsted*, and *CHAMP* satellite data (totalling over 1.6 million observations) and over 500 000 observatory data; the latter consist of either a 1.00 a.m. observation (actually an hourly mean) on the quietest day of the month during the 1960–2002.5 period, plus observations every 2 h on quiet days during the *POGO* and *Magsat* missions.

Comprehensive models take into account not only the time-varying core magnetic field (out to degree 13) but also the static crustal field from degree 14 to degree 65. Because a model of the lithospheric field to this degree captures only a small proportion of the total lithospheric signal, it is necessary to also solve for 1635 observatory biases, generally three components at each observatory. The novel features of CM4 arise in its very sophisticated treatment of the external magnetic fields, and we will discuss these in some detail.

The ionospheric field is modeled as currents flowing in a thin shell at an altitude of 110 km. This leads to magnetic fields which are derived from potentials below and above this layer, which influence the observatory and satellite data, respectively (since all the satellites fly above this layer). In quasi-dipole coordinates, the currents are allowed to vary with 24 h, 12 h, 8 h, and 6 h periods, as well as annually and semi-annually. Induced fields are accounted for by assuming that the conductivity distribution of the Earth varies only in radius, which means that an external spherical harmonic can only excite its corresponding internal spherical harmonic. The magnetospheric field is also parametrized in a similar way, with both daily and seasonal periodicities, but also a modulation is allowed based on the Dst index. In order to take into account the poloidal F-region

currents through which the satellites fly, a parametrization is made in terms of a toroidal magnetic field, which also has periodic time variations.

The model is estimated by a iteratively reweighted least-squares method, using Huber weights, and the core contribution is regularised as in [9] $n_1 = 2$ and $n_2 = 1$ in one norm and $n_1 = 0$ and $n_2 = 2$ in another. This difference from the *ufm/gufm* method simply represents a different approach; the fundamental quantity in the comprehensive models is the secular variation $\partial_t B_r$, which has an expansion in B splines, and the main field B_r is found as the integral of this using the 1980 value as the offset or integration constant. All the other parameters are regularized in a similar way, by smoothing on spheres at different altitudes, representing the physical locations of the sources. In total, CM4 consists of 25 243 free parameters.

5.3.1.2.(iii) CHAOS field model of recent satellite data The satellites *Ørsted*, *CHAMP*, and *SAC-C* are currently providing unprecedented coverage of Earth's magnetic field. Olsen *et al.* (2006) have recently produced a cubic B-spline-based time-dependent field model spanning the interval 1999–2006 covered by data from these satellites. This model is not fully comprehensive in the sense that only a relatively simple external field model is co-estimated and toroidal fields and coupling currents in the ionosphere and magnetosphere are not explicitly considered. For spherical harmonic degrees 1–14 of the internal field, a time-dependent cubic B-spline representation is employed with knot points every year, for degrees 15–18 a static field and the linear secular variation (the first order term in a Taylor series expansion) is solved for, while for degrees 19–50 only the static field is solved for. The near-magnetospheric external fields are represented by a degree 2 spherical harmonic expansion in solar magnetospheric co-ordinates (including Dst-dependence), while a degree 2 zonal expansion in geocentric solar magnetospheric co-ordinates is used to parametrize fields with a far magnetosphere origin. Perhaps the most revolutionary aspect of this model is that the Euler angles required to transform measurements from the satellite magnetometer frame to the geocentric frame are co-estimated along with the model parameters which avoids the inconsistency of alignment using a pre-existing field model (see Olsen *et al.* (2006) for further details). Iteratively reweighted least-squares estimation involving Huber weights in the covariance matrix (see, e.g., Olsen, 2002) is used to find the preferred field model, with regularization in time only by minimizing the mean squared amplitude of the second time derivative of **B** integrated over Earth's surface and averaged over time. The CHAOS model represents the state-of-the-art in terms of determining the secular variation in the twenty-first century; future efforts will aim to incorporate this accurate information into longer time span models of the field evolution.

5.3.1.2.(iv) Comparison between CM3, CM4, and gufm1

To illustrate the fidelity with which the present field models are able to model observatory data, we show in **Figure 15** a comparison of model *gufm1*'s predictions with some observatory annual mean data sets.

To show CM4's performance on very short timescales, **Figure 16** compares the model to hourly mean values for the month of April 1990, data that were not used in deriving the model. It is clear that the model is capable of predicting variations rather well, though with more difficulty at the Antarctic station SBA (Scott Base).

Table 5 compares the performance of models *gufm1* and CM3 against observatory data, showing almost identical performance. This comes about principally because of the large intrinsic variance of the data at some observatories, which neither field model is able to capture.

Figure 17 shows a comparison of the model predictions for the variation in the first six Gauss coefficients over century and decade timescales. Although small differences exist, particularly in estimates of the instantaneous secular variation, it is apparent that modeling has reached a stage where there is considerable consensus between the models.

In the next section, we move on to describe the characteristics of secular variation as observed on Earth's surface and inferred at the CMB. We will ultimately (Section 5.5) describe possible underlying physical mechanisms in terms of core hydromagnetics. Most of the results shown in the next section (unless explicitly stated otherwise) are derived from the *gufm1* field model of Jackson *et al.* (2000), which, as we have shown in this section, provides a good representation of the historical field evolution.

5.4 Historical Field Evolution – Long-Term Secular Variation

5.4.1 Field Evolution at the Earth's Surface

The magnetic field at Earth's surface has changed significantly over the past 400 years. This can clearly be seen, for example, in the long times series of measurements cataloged by Malin and

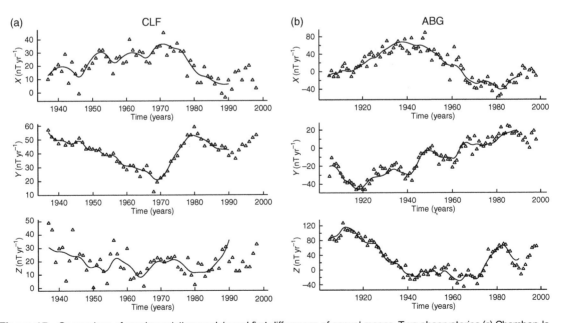

Figure 15 Comparison of secular variation models and first differences of annual means. Two observatories (a) Chambon-la-Foret, France, and b) Alibag, India, are shown comparing observed field rate of change with predictions from the model *gufm1* (solid lines). The symbols show the rate of change of the field, as obtained from first differences of annual means. The *X*-, *Y*-, and *Z*-components are in the northerly, easterly, and downward directions, respectively. Because the post-1990 data were not used in the creation of *gufm1*, there is a small mismatch at the end of the data series — this shows the difficulty in predicting the secular variation.

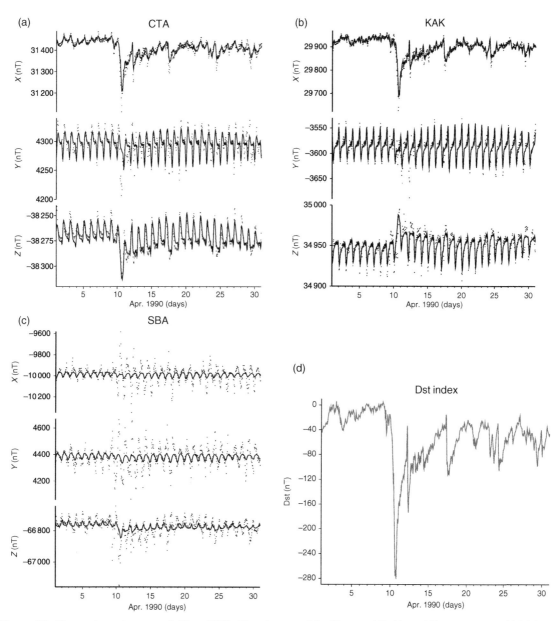

Figure 16 Comparison of one month (Apr. 1990) of hourly mean data. Observed X-, Y-, and Z-, components (dots) from selected observatories and predictions (solid line) from CM4. (a) Charters Towers; (b) Kakioka; (c) Scott Base; (d) the Dst index for April. Note the commencement of a magnetic storm on the 10th day. The Dst index is used in the synthesis of predictions at individual observatories; this is particularly noticeable in the predictions of the X component in (a) and (b).

Bullard (1981) (**Figure 18**). In fact, it was such measurements of changes in declination by Henry Gellibrand in 1634 that first indicated the existence of geomagnetic secular variation. Note that in this figure, since declination and inclination are nonlinearly related to the model **m**, it is not possible to account for observatory crustal biases.

The best way to appreciate global field changes (i.e., secular variation) is for the reader to study contour maps of different field components and compare how they have evolved. In **Figure 19**, the declination (D) at Earth's surface is shown in AD 1590 and 1990, while **Figure 20** shows the inclination anomaly (I_a) (defined as the difference between the observed inclination and that of a geocentric dipole) at the same

Table 5 Comparison of rms differences (in nanotesla) between observatory annual means and predictions from the models *gufm1* and CM3, the latter with or without its external contribution

Component	No. of data	gufm1	CM3 (all)	CM3 (no external)
X	4047	17.71	17.48	18.09
Y	4047	21.27	21.45	21.47
Z	4047	24.55	24.49	24.53

epochs. **Figures 21, 22,** and **23** catalog the evolution of the vertical component of the field (which is much larger than the horizontal components except at low latitudes) at AD 1590, 1690, 1790, 1890 and 1990.

5.4.1.1 The westward drift

Perhaps the most striking aspect of the geomagnetic secular variation over the past 400 years is the westward motion of the field at Earth's surface. This phenomenon has been recognized since the time of Halley (1683, 1692) and was first analyzed in detail by Bullard *et al.* (1950), who concluded that the non-dipole part of the field had moved westward at a rate of 0.18° per year during the first half of the twentieth century. Bullard *et al.* (1950) and later Yukutake (1962) suggested that the westward drift was not globally constant but rather depended on latitude; subsequently, Yukutake and Tachinaka (1969) realised that it could be better explained by separating the field into standing and drifting parts. The latitudinal dependence of the drift rate was conclusively demonstrated by Jault *et al.* (1988).

The westward motion of the field is most easily seen by following the motion of the agonic lines (where $D = 0$) in **Figure 19**. It can be seen that in 1590 one agonic line bisected the African continent, running through the Cape of Good Hope (which at this time was named Cape Agulhas ('needle cape') by sailors due the coincidence of the directions of true and magnetic north there); fast-forwarding 400 years to AD 1990 we find that the same agonic line has now moved westward, so that it now bisects southern America. The maxima and minima of inclination anomalies centered on low latitudes can also be tracked westward, for example, the inclination anomaly high that was present over Africa in 1590 now lies on the western edge of south America. Contour maps of the vertical component of the magnetic field are dominated by the axial dipole component of the field which is unchanged by westward motion due to its axisymmetric nature; however, the westward motion of nonaxial dipole parts of the field can still be

discerned in the maps of **Figures 22** and **23**, especially by following long-lived distortions in the magnetic equator. A southwest-to-northeast trending element of the magnetic equator can be followed from its initial location at the Indian ocean in 1590, through to Africa in 1790 and the Atlantic in 1890, to the eastern edge of southern America in 1990.

5.4.1.2 Hemispherical asymmetry

The description of westward motion of field features in the previous section focused on high-amplitude features moving across the Atlantic hemisphere (longitude 90E to 90W). In contrast, the field evolution in the Pacific hemisphere is characterized by lower-amplitude features and a lack of systematic secular variation. This asymmetry between the hemispheres was first discussed by Fisk (1931), and it has been suggested that this could be a consequence of the influence of lower-mantle inhomogeneities on the dynamo in the core (Doell and Cox, 1971). This interpretation remains controversial, as it is known that asymmetric field morphologies are transiently possible during highly supercritical core convection.

5.4.1.3 Axial dipole decay

The westward drift is only part of the observed secular variation. The largest contribution to present-day secular variation comes from the decay of the axial dipole part of field. The axial dipole has decayed rapidly at an average rate of 5% per century since the first direct measurements of intensity (Barraclough, 1974). Gubbins *et al.* (2006) have recently used paleointensity measurements from the database of Korte *et al.* (2005) along with estimates of field directions from *gufm1* to infer g_1^0 for the interval 1590–1840 and found that field decay rate was much slower (almost constant) during this earlier interval. **Figure 24** shows both the extrapolation of Barraclough (1974) used by Jackson *et al.* (2000) and the result of Gubbins *et al.* (2006). The variability in the rate of change of the axial dipole over the past

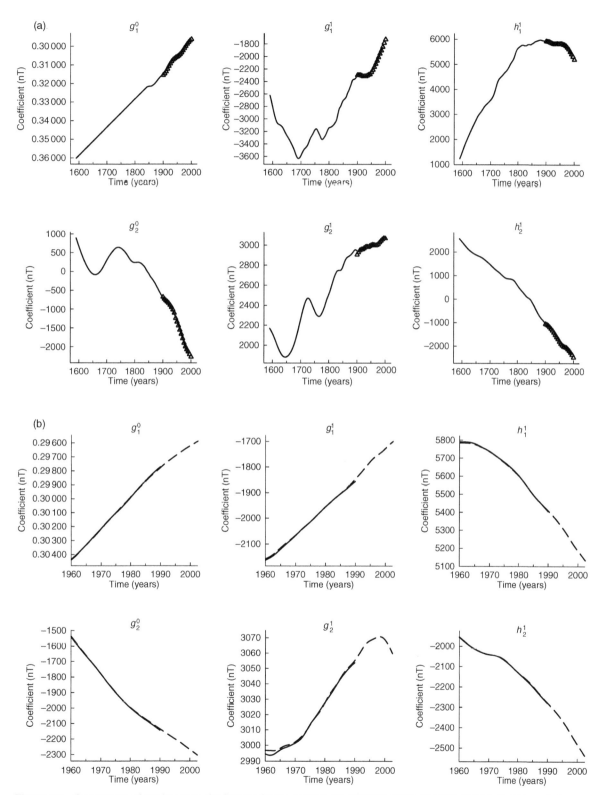

Figure 17 Comparison of model values for first six Gauss coefficients. (a) 1590–1990; (b) 1960–2002.5. Solid is *gufm1*, dashed is CM4, and the triangles are DGRFs. In (a), g_1^0 has been fixed to decrease at a rate of 15nT yr^{-1} prior to 1840; in the absence of intensity data, it is necessary to fix the amplitude of the solutions.

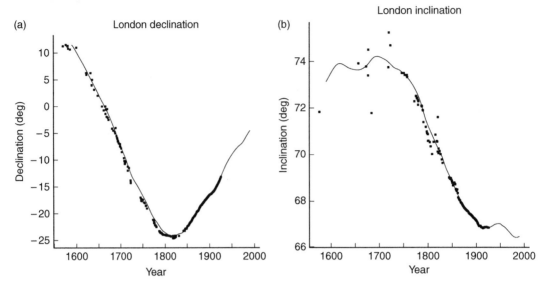

Figure 18   Declination (D) and inclination (I) in London during historical times (Malin and Bullard, 1981), and the fit by *gufm1* of Jackson *et al.* (2000) (line). The data were not used in the construction of *gufm1*, and provide an independent check of its fidelity. Note that the inclination prior to 1700 is very poorly constrained, and the model *gufm1* differs from some archeomagnetic measurements of inclination in Europe (Le Goff, personal communication, 2005).

four centuries suggests that the dynamo process generating the main field is not steady, but is continuously fluctuating in strength.

5.4.1.4   Timescale associated with different wavelengths (spherical harmonic degrees)

A useful statistical estimate of how changes in Earth's magnetic field at the surface depend on the length scale under consideration is the reorganization (or correlation) time $\tau(l)$ introduced by Hulot and Le Mouël (1994),

$$\tau(l) = \sqrt{\frac{\sum_m \left(g_l^m\right)^2 + \left(h_l^m\right)^2}{\sum_m \left(\dot{g}_l^m\right)^2 + \left(\dot{h}_l^m\right)^2}} \qquad [10]$$

This quantity is a measure of how long it takes for power at spherical harmonic degree l to be completely changed (altered by an amount equal to its current value) given its present rate of change. Physically, this corresponds to the time taken to completely reorganize field features of a particular size. In order to calculate $\tau(l)$ one requires only a model of the main field and its time derivative at a given time. In **Figure 25**, $\tau(l)$ derived from the CHAOS model (Olsen *et al.*, 2006) is presented.

The CHAOS model contains global data sets from the *Ørsted*, *CHAMP*, and *SAC-C* satellites (see earlier description) and is the most accurate short-time-span

field model yet derived, thus giving the best picture currently available of τ, to the highest possible degree l. Olsen *et al.* (2006) found that a power law $\tau(l) = 890 l^{-1.35}$ yr provided a good fit to the calculated values of $\tau(l)$ so that for $l = 1$, $\tau \sim 900$ yr while for $l = 16$, $\tau \sim 20$ yr. These results illustrate that secular variation processes span a wide range of timescales but that the small length scales that can now be accurately monitored are evolving on timescales short enough to allow detailed study in the coming decades. Holme and Olsen (2006), analyzing an earlier model (CO2003), found $\tau \sim 1000 l^{-1.45}$; they have argued that the simple power-law functional form of $\tau(l)$ suggests a consistent (l-independent) form of advection–diffusion balance producing the secular variation. For further discussion on the core processes underlying secular variation.

5.4.1.5   Evolution of integrated rate of change of vertical field at Earth's surface

It is also interesting to consider the evolution of a global measure of the amplitude of the instantaneous rate of change of vertical field at Earth's surface. A suitable measure is the root mean square (rms) of $\dot{B}_r$ integrated over Earth's surface. This quantity is plotted in **Figure 26** from 1840 to 1990, for the *ufm1* model.

Dramatic changes in the integrated instantaneous rate of B_r are observed to have occurred. These changes are thought to be robust as *ufm1* is a good

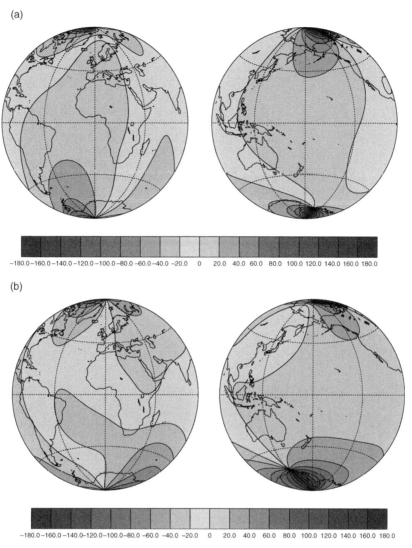

Figure 19   Declination D at Earth's surface in (a) AD 1590 and (b) AD 1990 from the model *gufm1* of Jackson *et al.* (2000). Plots are Lambert equal area projections of the Atlantic and Pacific hemispheres. Color bars are at 20° intervals, red being negative and blue positive. Note the westward displacement of the agonic lines where $D = 0$.

representation of the globally averaged field evolution at the surface, as is evident in comparisons of the model prediction with observed secular variation. Particularly dramatic is the 20% increase in the amplitude of the secular variation at the start of the twentieth century. This is thought to be associated with an increase in the rms core surface flow velocity; Hulot *et al.* (1993) inferred by inversion of the observed secular variation that the rms flow speed increased at this time, and it appears to be the case that zonal (axisymmetric) core flow speeds altered precisely in the required way for observed decadal length-of-day changes to be explained by

geostrophic core motions (Jackson, 1997). It is also remarkable that there are a number of local maxima and minima in **Figure 26** occuring throughout the twentieth century. These extrema seemingly mark reorganizations of the global secular variation, and at least some of them appear coincident with so-called geomagnetic jerks that are discussed in the next section.

5.4.1.6   Geomagnetic jerks

Geomagnetic jerks or secular variation impulses are abrupt changes in the second time derivative of the geomagnetic field at Earth's surface

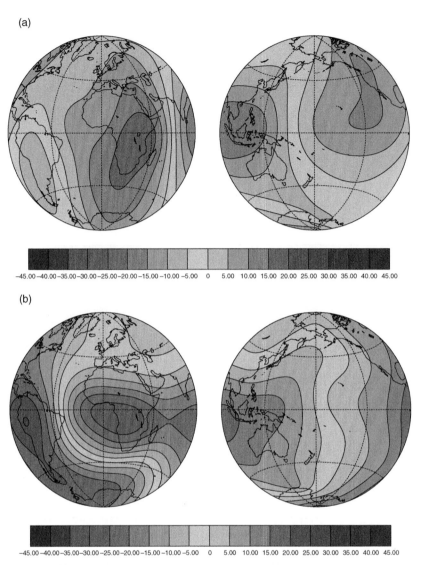

Figure 20   Inclination anomaly I_a at Earth's surface in (a) AD 1590 and (b) AD 1990 from the model *gufm1* of Jackson *et al.* (2000). Plots are Lambert equal area projections of the Atlantic and Pacific hemispheres. Inclination anomaly is the inclination of the field minus that expected for a geocentric dipole. Color bars are at 5° intervals, red being negative and blue positive.

(see, e.g., Courtillot and Le Mouël, 1984). During the twentieth century, they were found to separate intervals of linearly changing secular variation and have been unambiguously identified as having occurred in 1901, 1913, 1925, 1969, 1978, 1991, and 1999 (Alexandrescu *et al.*, 1995; Macmillan, 1996; Mandea *et al.*, 2000). The signature of jerks can be seen particularly clearly at European observatories, for example, in **Figure 27**, which shows the evolution of the linear rate of change of the eastward component of the geomagnetic field ($\dot{Y}$) in Niemegk.

A 12 month running average filter has been applied to the central differences of monthly mean data to produce this time series, following the methodology of Mandea *et al.* (2000). The jerk events are captured (at least in a smoothed manner) by the internal field representation of global models such as CM4 and *gufm1* — this can been seen, for example, in **Figure 15**, which compares model results to observatory annual means. Jerks are not always observed at all locations and those that are observed are not simultaneous; Alexandrescu *et al.* (1996) noted that, for example

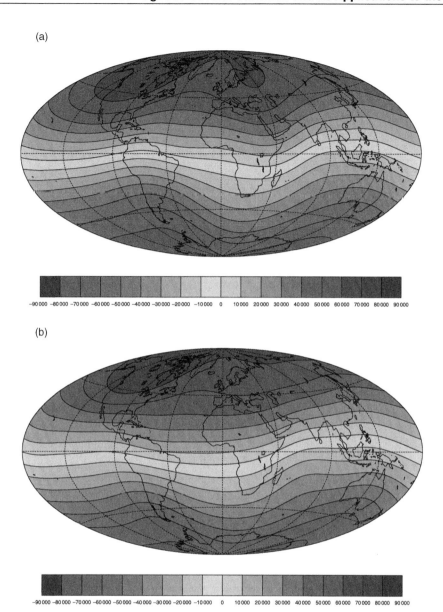

Figure 21 Vertical magnetic field B_Z at Earth's surface in (a) AD 1590 and (b) AD 1690 from the model *gufm1* of Jackson *et al.* (2000). Plots are Molleweide projection; each color interval represents a 10 000 nT increment.

in 1969, the signature of the jerk tended to be observed later in the Southern Hemisphere (see **Figure 28**).

The physical process causing jerks, as well as geographic variations in their detectability and time delays in their observation, are not well understood. Bloxham *et al.* (2002) suggested that jerks might be the surface manifestation of a superposition of torsional oscillations (a special class of axisymmetric, geostrophic, hydromagnetic waves likely to be present in Earth's outer core), and

that variations in their detectability might be the result of variation in the field morphology at the core surface. Alexandrescu *et al.* (1999) and Nagao *et al.* (2003) have suggested that variations in mantle conductivity could explain the observed delays in jerk observations. Much work remains to be carried out in understanding the physical mechanisms involved and in testing the various hypotheses.

Variations in the main geomagnetic field have their origin in Earth's core. Most insight into the

(a)

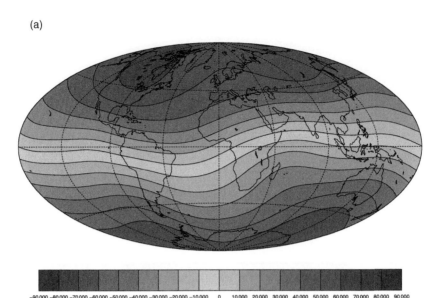

(b)

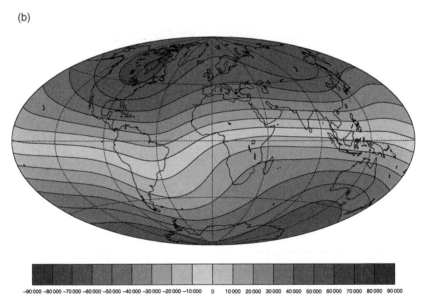

Figure 22 Vertical magnetic field B_z at Earth's surface in (a) AD 1790 and (b) AD 1890 from the model *gufm1* of Jackson *et al.* (2000). Plots are Molleweide projection; each color bar represents a 10 000 nT increment.

physical mechanisms causing the field evolution can therefore be obtained by examining the patterns of field evolution at the core surface. To determine the core field we adopt the approximation of treating the mantle as a perfect insulator. This approximation has been studied by Benton and Whaler (1983), who show that when variations are considered whose periods are longer than annual, the error introduced is small when the mantle has a rather weak electrical conductivity structure as currently believed

(i.e., from 10^{-2} to $10\,\mathrm{S\,m^{-1}}$). In the next section, the patterns of field evolution that result from such an approach are discussed in detail.

5.4.2 Field Evolution at the Core Surface

The evolution of the geomagnetic field at the core surface over the past few centuries was first described in detail by Bloxham and Gubbins (1985) and Bloxham *et al.* (1989) by considering a series of

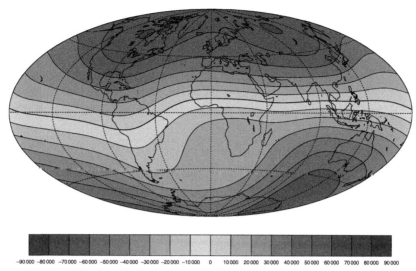

Figure 23 Vertical magnetic field B_Z at Earth's surface in AD 1990 from the model *gufm1* of Jackson *et al.* (2000). Plots are Molleweide projection; each color bar represents a 10 000 nT increment.

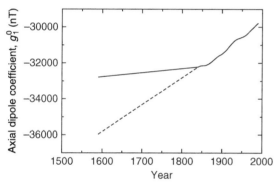

Figure 24 Evolution of g_0^1 in nanotesla since direct measurements began in 1590; absolute intensity measurements began sporadically only in 1832; the two slopes shown before 1840 are (a) the extrapolation based on the average rate of fall since 1840 and (b) derived from indirect paleointensity measurements by Gubbins *et al.* (2006). The solid line indicates g_1^0 from *gufm1* with pre-1840 decay of 2.28 nT and dashed line g_1^0 from *gufm1* with pre-1840 decay of 15 nT. Adapted from Gubbins D, Jones AL, and Finlay CC (2006). Fall in Earth's magnetic field is erratic. *Science* 312: 900–902.

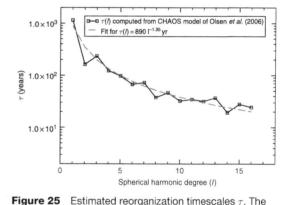

Figure 25 Estimated reorganization timescales τ. The reorganization (or correlation) time $\tau(l) = \sqrt{\sum_m (g_l^m)^2 + (h_l^m)^2 / \sum_m (\dot{g}_l^m)^2 + (\dot{h}_l^m)^2}$ (Hulot and Le Mouël, 1994) is derived from the CHAOS field model (Olsen *et al.*, 2006), giving an instantaneous estimate of the time taken for power at spherical harmonic degree l of the field to be completely changed or renewed. Solid line shows the CHAOS estimates; dashed line shows a two-parameter exponential fit to the data.

single-epoch models. The picture they described has been borne out by the more recent time-dependent field models *ufm1* (Bloxham and Jackson, 1992) and *gufm1* (Jackson *et al.*, 2000), so we shall reiterate their findings here before discussing more recent developments. Contour plots of the historical evolution of the vertical field at the core surface are found in **Figures 29**, **30**, and **31**.

The structure of the vertical field at the core surface is considerably more complicated than at Earth's surface, because higher degree spherical harmonics are amplified more (by a factor $(a/c)^{(l+2)}$ where a is the radius of the Earth, c is the core radius, and l is the spherical harmonic degree) during the downward continuation procedure. This is one reason why it is preferable to downward continue regularized field

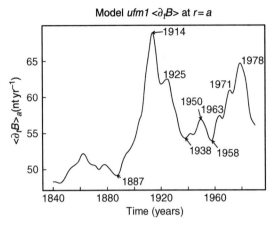

Figure 26 The RMS B_r integrated over Earth's surface from the *ufm1* time-dependent field model.

models rather than those that have been simply truncated, and may contain noise contributions in the higher degree spherical harmonics. Downward continuing truncated field models also unfortunately introduces the possibility of unwanted Gibbs ringing effects due to the sharp cutoff in spectral space (see e.g., Whaler and Gubbins, 1981; Shure *et al.*, 1982; Gubbins, 1983). The *gufm1* model results presented and discussed here have been regularized so the power spectrum for the model has reached negligible values by the time the nominal cutoff at spherical harmonic degree $l = 14$ is reached.

5.4.2.1 *High latitude, approximately stationary flux lobes*

A prominent feature in the maps of the vertical field at the core surface are the high-intensity flux lobes (by which we mean the areas of flux maxima, of either sign) under Arctic Canada, Siberia, and under the eastern and western edges of Antarctica; they can be seen particularly clearly in **Figure 31**. These lobes are responsible for the predominantly axial dipole field structure observed at the surface and have remained approximately stationary (wobbling slightly about a mean position) over the past four centuries. Gubbins and Bloxham (1987) identified these high-latitude flux lobes as the signature of columnar convection rolls in the core (Busse, 1975) which are thought to be a major ingredient in the geodynamo process (Kono and Roberts, 2002). They proposed that flow convergence associated with downwelling in the convection rolls is responsible for producing the observed field concentrations. Bloxham and Gubbins (1987) ascribed the relative stationarity of these flux lobes to the influence of heat flow inhomogeneities at the CMB associated with the structure of mantle convection. More detailed studies using geodynamo simulations (Olson and Christensen, 2002; Bloxham, 2002) have confirmed the feasibility of this mechanism.

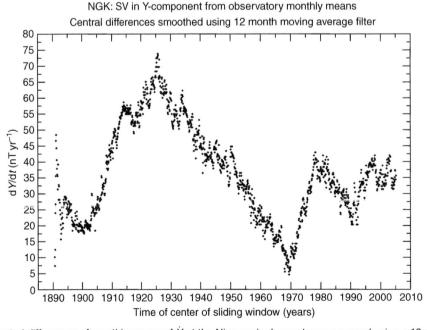

Figure 27 Central differences of monthly means of $\dot{Y}$ at the Niemengk observatory processed using a 12 month moving average filter.

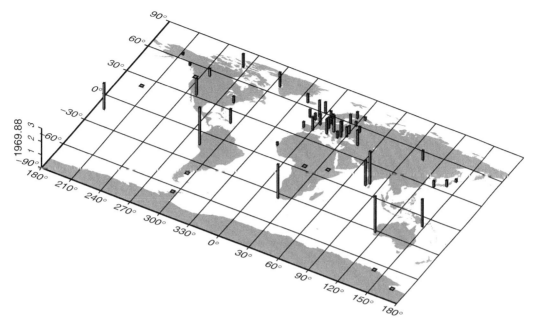

Figure 28 Geographical distribution of the times of occurence of the 1969 jerk measured by Alexandrescu *et al.* (1996). A linear combination of *X*- and *Y*-field components was analyzed using wavelet ridge functions and the jerk onset time estimated. Blue bars represent negative delays relative to the mean occurence time (1969.88) and correspond to earlier jerks, whereas red bars represent positive delays relative to the mean (later jerks). The scale bar varies from 0 to 3 years. Green squares represent locations where jerks were not detected.

5.4.2.2 Reversed flux patches

The presence of reversed flux features at the core surface is a major difference to the field structure observed at Earth's surface. Most prominent of these are the patch that is found close to the geographical North Pole throughout most of the past 400 years, and the large feature that extends from under southern Africa across to under southern America that has been formed by the coalescence of two earlier patches. Gubbins (1987) and Gubbins *et al.* (2006) have linked the growth and migration of the South Atlantic patch to the rapid decay of the axial dipole field observed since 1840. The significance of the changes in the flux through these patches is discussed in Section 5.5.4. If taken at face value, the growth of the South Atlantic patch implies a failure of a particularly attractive approximation for the core, the so-called frozen flux hypothesis, which consequently means that it is very difficult to retrieve fluid motions at the core surface. It is important to recognize that the increase in quality, quantity, and distribution of data throughout time leads to increased complexity in the field models, and it is very difficult to disentangle this effect from true diffusional effects; we refer the reader to the discussion in Section 5.5.4.

5.4.2.3 Low-latitude, westward-drifting field features

Bloxham and Gubbins (1985) pointed out the presence of a number of rapidly westward-moving field concentrations at low and mid-latitudes, especially clear in the Atlantic hemisphere. Bloxham *et al.* (1989) further noted that beneath Europe and the Atlantic Ocean during the twentieth century there was a westward-moving sequence of field highs and lows, and referred to this as a mid-latitude polar wave. They suggested this could be the signature of a wave with azimuthal wave number between $m = 5$ and $m = 9$. Jackson (2003) examined very high resolution images of the field at the core surface in 1980 and 2000, constructed using high-quality satellite data and utilizing a maximum entropy regularization technique. He showed that the wave-like feature identified in the Northern Hemisphere by Bloxham *et al.* (1989) has a counterpart at low latitude on the other side of the geomagnetic equator with amplitude considerably higher than was evident in previous studies.

Since these wave-like features appear to move essentially east–west, their motion can be tracked using plots of field amplitude as a function of time and longitude (TL plots). TL plots of the vertical component of the field at the core surface between

(a)

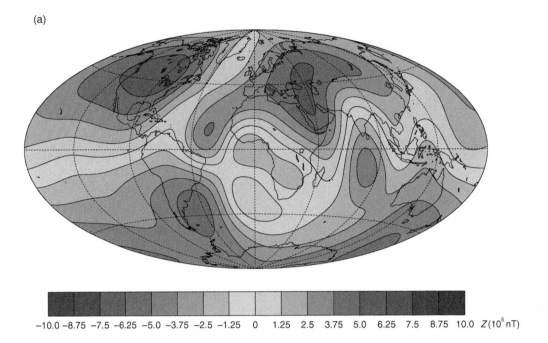

−10.0 −8.75 −7.5 −6.25 −5.0 −3.75 −2.5 −1.25 0 1.25 2.5 3.75 5.0 6.25 7.5 8.75 10.0 $Z(10^5 \text{nT})$

(b)

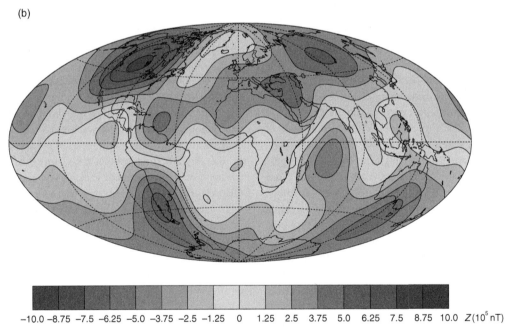

−10.0 −8.75 −7.5 −6.25 −5.0 −3.75 −2.5 −1.25 0 1.25 2.5 3.75 5.0 6.25 7.5 8.75 10.0 $Z(10^5 \text{nT})$

Figure 29 Vertical magnetic field B_Z at the core surface in (a) AD 1590 and (b) AD 1690 from the model *gufm1* of Jackson *et al.* (2000). Plots are Molleweide projection, units are nanotesla.

1590 and 1990 (from the *gufm1* model) at latitudes 60° N, 40° N, 20° N, at the equator, at 20° S, and at 60° S are presented in **Figure 32**.

In TL plots, linear features of high field amplitude running up-down the page indicate stationary flux features such as the high-latitude flux features (see, e.g., at latitudes 60° N between longitudes − 120° and − 90°

as well as between + 90° and + 120°, similarly at 60° S, near longitudes − 90° and + 120°). At lower latitudes, for example at 20° N after 1900, at the equator between longitudes 0 and + 90° and at 20° S there are some hints of diagonal lines of high field intensity that represent azimuthally (east–west) moving field features. Unfortunately, it is rather difficult to analyze these

(a)

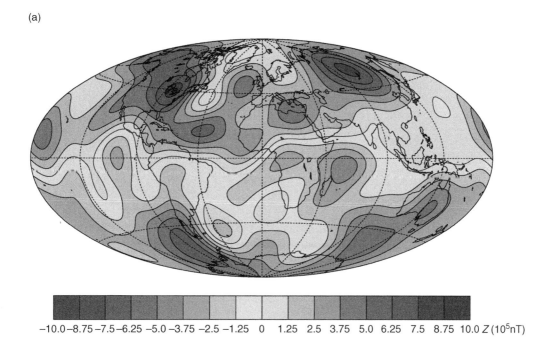

$-10.0\ -8.75\ -7.5\ -6.25\ -5.0\ -3.75\ -2.5\ -1.25\quad 0\quad 1.25\quad 2.5\quad 3.75\quad 5.0\quad 6.25\quad 7.5\quad 8.75\quad 10.0\ Z\ (10^5 nT)$

(b)

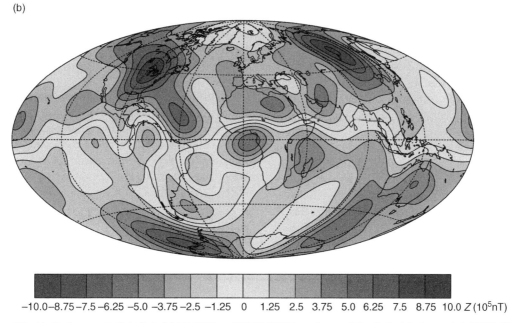

$-10.0\ -8.75\ -7.5\ -6.25\ -5.0\ -3.75\ -2.5\ -1.25\quad 0\quad 1.25\quad 2.5\quad 3.75\quad 5.0\quad 6.25\quad 7.5\quad 8.75\quad 10.0\ Z\ (10^5 nT)$

Figure 30 Vertical magnetic field B_z in (a) AD 1790 and (b) AD 1890 from the model *gufm1* of Jackson *et al.* (2000). Plots are Molleweide projection, units are nanotesla.

features because they are swamped by stationary features that are not of interest in this context. To get round this problem, Finlay and Jackson (2003) high-pass-filtered the vertical field from the *gufm1* model, removing the time-averaged axisymmetric field and all field components varying on timescales longer than 400 years to obtain a field denoted by $\tilde{B}_r$. The result of this processing is shown in TL plots at the same latitudes as before in **Figure 33**. Note that the first and last 40 years of the record have been disregarded to eliminate filter warm-up effects, namely the fact that the edges of the time series affect the filtered output.

The filtering reveals clear westward-moving, wave-like, signals at low latitudes (between 20° N

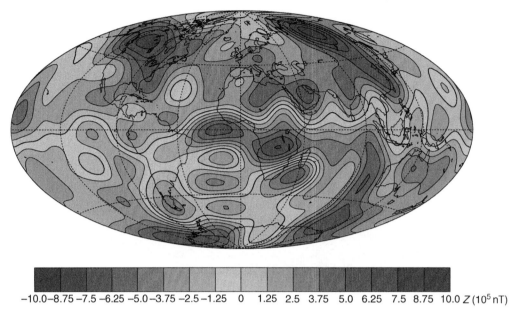

−10.0 −8.75 −7.5 −6.25 −5.0 −3.75 −2.5 −1.25   0   1.25 2.5 3.75 5.0 6.25 7.5 8.75 10.0 Z (10^5 nT)

Figure 31   Vertical magnetic field B_Z at the core surface in AD 1990 from the model *gufm1* of Jackson *et al.* (2000). Plots are Molleweide projection, units are nanotesla.

and S and particularly striking at the equator). No clear wave-like motions were found at higher latitudes, indicating that such patterns of secular variation are confined to low latitudes on timescales shorter than 400 years. By measuring the power traveling at different angles in the TL plots at all latitudes, Finlay (2005) showed that it is possible to construct latitude-azimuthal speed (LAS) plots, which summarize the relative strength, location, and rate of azimuthal secular variation processes. Such plots, constructed when the vertical field at the core surface from *gufm1* is high-pass-filtered with thresholds of 2500, 600, and 200 years, are shown in **Figure 34**.

The LAS power plots suggest that three distinct types of azimuthal secular variation have been operating during the past four centuries. At mid- to high latitudes in the Northern Hemisphere, there are weak signals probably associated with the wobbles of the high-latitude flux lobes. These motions are both eastward and westward and appear most clearly when long timescale field variations are retained in **Figures 34(a)** and **34(b)**. Next, there is the strong equatorially confined signal with speed of approximately 17 km yr^{-1} westward, as described by Finlay and Jackson (2003). This is the dominant signal when only field variations with timescales shorter than 400 years are considered, and looks in TL plots to have the form of a wave-like disturbance. Finally, on all timescales, there is a strong westward signal in the Southern Hemisphere, which is

particularly clear when the filter threshold is much longer than the record length. It seems to be associated with the westward motion of reversed flux features in the South Atlantic and is particularly strong in the twentieth century.

No in-depth study of meridional motions of field features at the core surface has yet been carried out. Such a study would be of interest, especially considering the possible links between meridional motions and proposed reversal mechanisms (Gubbins, 1987; Wicht and Olson, 2004).

5.5   Interpretation in Terms of Core Processes

The observed evolution of the internally generated part of Earth's magnetic field is a consequence of the motions in the liquid metal outer core. In order to understand and model the mechanisms underlying these changes, we must employ the mathematical framework of magnetohydrodynamics – the marriage of Maxwell's laws of electromagnetism and the principles of hydrodynamics or fluid mechanics. In this section, equations describing the evolution of the core magnetic field and the generation of core fluid motions are derived, and useful approximations are discussed; we stop short of describing attempts to invert field observations for core fluid motions at the CMB which in the territory.

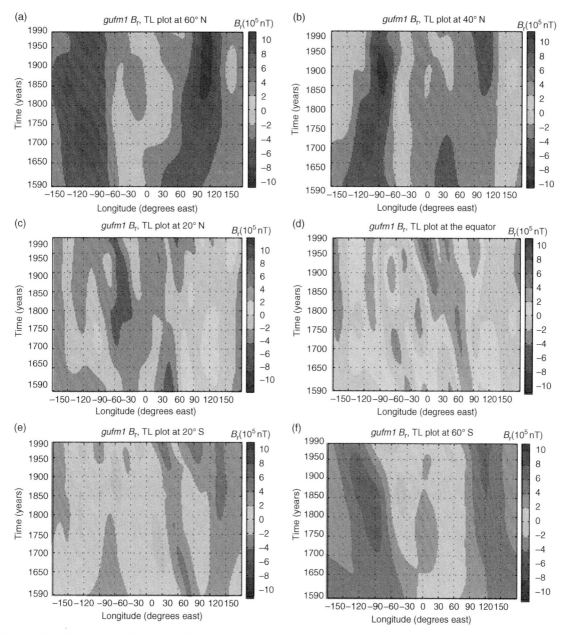

Figure 32 Time–longitude (TL) plots of the unfiltered vertical magnetic field at the core surface from the field model *gufm1* at latitudes 60° N in (a), at 40° N in (b), at 20° N in (c), at the equator in (d), at 20° S in (e), and at 60° S in (f).

5.5.1 Maxwell's Equations and Moving Frames

Maxwell's equations for an electrically conducting fluid moving with a velocity **u** in the presence of a magnetic field **B**, an electric field **E**, and an electric current density **J** are

$$\nabla \cdot \mathbf{B} = 0 \quad \text{Absence of free magnetic monopoles} \quad [11]$$

$$\nabla \times \mathbf{E} = -\partial_t \mathbf{B} \quad \text{Faraday's law of magnetic induction} \quad [12]$$

$$\frac{1}{\mu_0}(\nabla \times \mathbf{B}) = \mathbf{J} \quad \text{Ampere's law of magnetostatics} \quad [13]$$

where μ_0 is the magnetic permeability of free space which is applicable to nonferromagnetic fluids. It should be noticed that these equations are somewhat simpler than the usual, most general form of Maxwell's equations described in, for example,

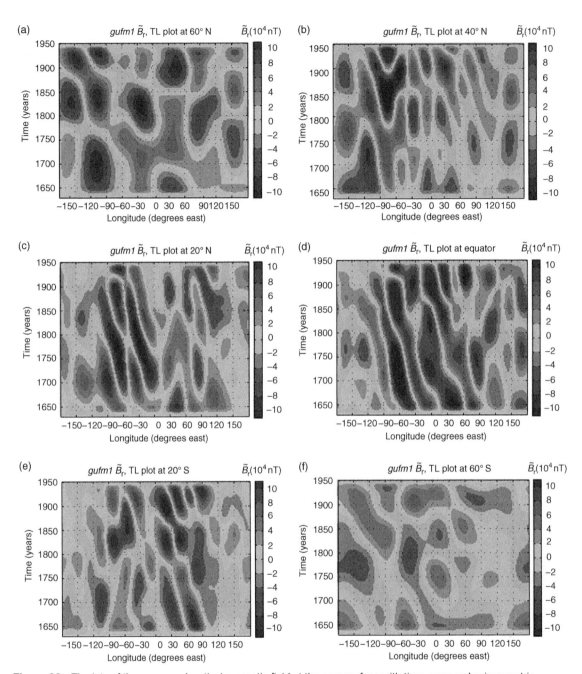

Figure 33 TL plots of the processed vertical magnetic field at the core surface with time-averaged axisymmetric component subtracted and high-pass-filtered with cutoff period 400 years, from the field model *gufm1* at latitudes 60° N in (a), at 40° N in (b), at 20° N in (c), at the equator in (d), at 20° S in (e), and at 60° S in (f).

Jackson (1999) or Backus *et al.* (1996). The fact that the liquid metal flows we are interested in have speeds $|\mathbf{u}| \ll c$ (the speed of light) has enabled the well-known displacement current term in Ampere's law to be neglected and allowed the (decoupled) Gauss law of electrostatics to be dispensed with. This powerful simplification is known as the

magnetohydrodynamic or MHD approximation. In this scenario, Ohm's law for the electrically conducting and moving fluid takes the form,

$$\mathbf{J} = \sigma(\mathbf{E} + \mathbf{u} \times \mathbf{B}) \qquad [14]$$

where σ is the electrical conductivity of the fluid. The mathematical formalism can be further

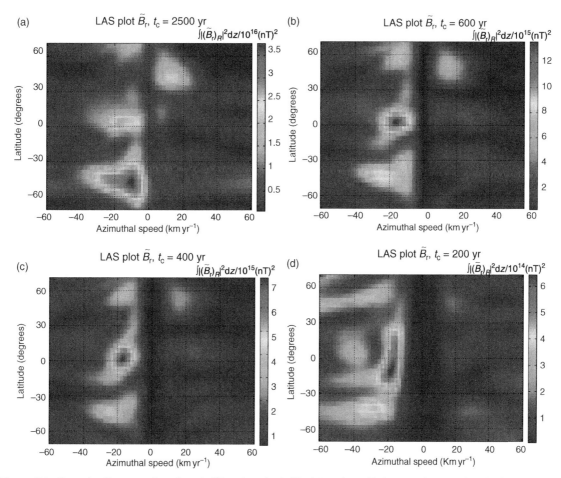

Figure 34 Summing the power traveling at different angles in TL plots using a Radon transform method, LAS power plots are constructed for the high-pass-filtered vertical magnetic field at the core surface from *gufm1*. (a) Result when the high-pass filter threshold is 2500 years, (b) 600 years, (c) 400 years (the case shown for the TL plots in **Figure 33**), and (d) 200 years.

compacted by realising that eqns [11]–[14] can be combined to yield a single prognostic equation governing the evolution of magnetic fields. Substituting from eqn [13] into eqn [14],

$$\nabla \times \mathbf{B} = \mu_0 \sigma (\mathbf{E} + \mathbf{u} \times \mathbf{B}) \quad [15]$$

Taking the curl of this,

$$\nabla \times (\nabla \times \mathbf{B}) = \mu_0 \sigma (\nabla \times \mathbf{E} + \nabla \times (\mathbf{u} \times \mathbf{B})) \quad [16]$$

Substituting from eqn [12] and using the vector identity that $\nabla \times (\nabla \times \mathbf{B}) = \nabla(\nabla \cdot \mathbf{B}) - \nabla^2 \mathbf{B}$, this becomes

$$\nabla(\nabla \cdot \mathbf{B}) - \nabla^2 \mathbf{B} = \mu_0 \sigma \left(-\frac{\partial \mathbf{B}}{\partial t} + \nabla \times (\mathbf{u} \times \mathbf{B}) \right) \quad [17]$$

Using eqn [11], rearranging and defining the magnetic diffusivity $\eta = 1/(\mu_0 \sigma)$ yields

$$\partial_t \mathbf{B} = \nabla \times (\mathbf{u} \times \mathbf{B}) + \eta \nabla^2 \mathbf{B} \quad [18]$$

This is known as the magnetic induction equation. The first term on the right-hand side represents the magnetic induction effects whereby electrical currents are generated when an electrical conductor moves through a magnetic field. The second term represents the changes in magnetic fields due to dissipative Joule heating effects that are the consequence of the flow of electric currents in a material with finite resistivity.

5.5.2 The Induction Equation in a Spherical Earth

Focusing our attention on the MHD of Earth's core, with a view to interpreting secular variation processes, in this section we discuss the induction equation at the core–mantle boundary. In particular, we discuss the approximations commonly made and the consequences of the present uncertainty regarding the electrical conductivity of the lower mantle.

We begin by returning to the magnetic induction equation [18]. We will assume for the moment that the flow **u** is given and examine its effects on the field, the so-called 'kinematic problem;' the question of the dynamics will be examined below. It is useful to introduce the toroidal–poloidal, or Mie (Backus, 1986; Backus et al., 1996), representation:

$$\mathbf{B} = \mathbf{B}_T + \mathbf{B}_P = \nabla \times (T\hat{\mathbf{r}}) + \nabla \times \nabla \times (P\hat{\mathbf{r}}) \quad [19]$$

where $T(r, \theta, \phi)$ and $P(r, \theta, \phi)$ are the toroidal and poloidal scalars, respectively, defining the toroidal $\mathbf{B}_T$ and poloidal $\mathbf{B}_P$ ingredients of **B**. This representation is valid for any solenoidal vector field satisfying

$$\nabla \cdot \mathbf{B} = 0 \quad [20]$$

The sphericity of the core suggests that we continue to work in spherical polar coordinates, though it is the case that the core is ellipsoidal with equatorial radius greater by approximately 1 part in 400 than the polar radius. In terms of approximating the surface for the purposes of plotting fields or fluid motion, the ignorance of the oblate spheroidal nature of the core can be seen to introduce negligible errors; note, however, that the ellipticity might be important in a dynamical context, because of the coupling that it can generate between mantle and core. In spherical polar coordinates (r, θ, ϕ), the toroidal and poloidal ingredients may be written in terms of the toroidal and poloidal scalars in the form,

$$\mathbf{B}_T = \left(0, \; \frac{1}{r\sin\theta}\frac{\partial T}{\partial \phi}, \; -\frac{1}{r}\frac{\partial T}{\partial \theta}\right) \quad [21]$$

and

$$\mathbf{B}_P = \left(\frac{\mathcal{L}^2 P}{r^2}, \; \frac{1}{r}\frac{\partial^2 P}{\partial r \partial \theta}, \; \frac{1}{r\sin\theta}\frac{\partial^2 P}{\partial r \partial \phi}\right) \quad [22]$$

where $\mathcal{L}$ is the angular momentum operator of quantum mechanics, defined by

$$\mathcal{L}^2 = -\left\{\frac{1}{\sin\theta}\frac{\partial}{\partial \theta}\left(\sin\theta\frac{\partial}{\partial \theta}\right) + \frac{1}{\sin^2\theta}\frac{\partial^2}{\partial \phi^2}\right\} \quad [23]$$

We can use the same representation for the fluid flow **u**, if we assume incompressibility. While this is not strictly true (the density increase with pressure across the outer core is approximately 20%), it is a fair approximation.

We now briefly consider the consequences of a mantle with finite conductivity for the spherical induction equation (most studies take no account of the conductivity of the mantle and treat it as a simple insulator). If the mantle has a spherically symmetric distribution of electrical conductivity which is of moderate amplitude, then an insulating mantle is likely to be a reasonable approximation. However, at the present time very little is known about the three-dimensional heterogeneity in conductivity, and in the deep mantle even the spherically symmetric part is poorly known. Thus, the insulating mantle assumption may be in considerable error and it is worth dwelling for a moment on whether this leads to difficulties in interpreting secular variation. In a conducting mantle, we revise [18] (having first set **u** to zero) to account for the fact that η in the mantle may be variable; this gives the diffusion equation for a heterogeneous η:

$$\frac{\partial \mathbf{B}}{\partial t} = -\nabla \times (\eta \nabla \times \mathbf{B}) \quad [24]$$

When we analyze its toroidal and poloidal ingredients, we find that in general there can be poloidal-to-toroidal conversion and vice versa; however, in the case that there is spherical symmetry $(\eta = \eta(r))$, a remarkable simplification occurs: the toroidal and poloidal ingredients of **B** obey

$$\frac{\partial T}{\partial t} = \eta\nabla^2 T + \frac{1}{r}\frac{d\eta}{dr}\frac{\partial}{\partial r}(rT); \; \frac{\partial P}{\partial t} = \eta\nabla^2 P \quad [25]$$

There is thus complete decoupling between the poloidal and toroidal ingredients of the field, even when the mantle conductivity is radially varying. Inasmuch as the spherically symmetric assumption is valid, this simplifies the task of understanding the secular variation.

At the Earth's surface we have electrically insulating boundary conditions and thus the toroidal magnetic field must vanish; this arises because Ampere's law relates fields **B** to currents **J** thus:

$$\begin{aligned}\mu_0\mathbf{J} &= \nabla \times \mathbf{B} = \nabla \times \mathbf{B}_T + \nabla \times \mathbf{B}_P \\ &= \nabla \times \nabla \times (T\hat{\mathbf{r}}) + \nabla \times \nabla \times \nabla \times (P\hat{\mathbf{r}}) \\ &= \nabla \times \nabla \times (T\hat{\mathbf{r}}) + \nabla \times \left[(-\nabla^2 P)\hat{\mathbf{r}}\right]\end{aligned} \quad [26]$$

Comparing this expression for the current with our definition of the toroidal–poloidal decomposition, it is clear that poloidal field results from toroidal currents, and toroidal field from poloidal currents. If **J** = 0, as in an insulator, then $\mathbf{B}_T = 0$, while $\mathbf{B}_P$ does not necessarily vanish, although P must satisfy $\nabla^2 P = 0$ (in which case we call it a potential field).

In principle, we can find the toroidal field within the mantle from measurements of the electric field at the sea bottom (e.g., Runcorn, 1955; Lanzerotti *et al.*, 1993; Shimizu *et al.*, 1998; Shimizu and Utada, 2004), but the electric field associated with the toroidal field from the core is likely to be small compared with the field from other sources, especially that from ocean currents.

We are therefore only able to monitor the poloidal part of the magnetic field at the Earth's surface. Turning now to the CMB, it is extremely unlikely that the toroidal magnetic field vanishes at this location on account of the finite (and possibly large) conductivity there. Several factors come to our rescue to ameliorate what otherwise would seem like a hopeless situation. We temporarily assume that the CMB is a free-slip boundary, so that $\mathbf{u} \cdot \hat{\mathbf{r}} = 0$, but $\mathbf{u} = \mathbf{u}_h \neq 0$). Then, following Bullard and Gellman (1954), we can show that

$$[\nabla \times (\mathbf{u}_h \times \mathbf{B})]_P = [\nabla \times (\mathbf{u}_h \times \mathbf{B}_P)]_P \quad [27]$$

from which we obtain the poloidal induction equation at the CMB,

$$\frac{\partial \mathbf{B}_P}{\partial t} = [\nabla \times (\mathbf{u}_h \times \mathbf{B}_P)]_P + \eta \nabla^2 \mathbf{B}_P \quad [28]$$

In the above, $[\ ]_P$ stands for the poloidal part of the equation. We discover, somewhat counterintuitively, that the poloidal secular variation depends only on the poloidal magnetic field! This result hangs entirely on the fact that radial motions are zero at the CMB. The result would not be true elsewhere in the core, where radial motions are crucial for the production of poloidal field from toroidal magnetic field, in order for the dynamo to operate. Note also that the toroidal part of the induction equation does not separate so easily, and that the rate of change of toroidal field depends on both the toroidal field and the poloidal field. Horizontal flow in this case shears both poloidal and toroidal magnetic field in order to create toroidal secular variation.

It is a fortuitous fact that when one analyzes the different components of the poloidal induction equation, the radial part has a particularly simple form, and gives an equation which will be central to much of our discussion. The radial induction equation reads

$$\partial_t B_r + \mathbf{u}_h \cdot \nabla_h B_r + B_r \nabla_h \cdot \mathbf{u}_h = \frac{\eta}{r} \nabla^2 (r B_r) \quad [29]$$

demonstrating that only radial fields, and their derivatives, need to be known for radial secular variation to be calculable when a particular flow is prescribed.

It is incumbent on us to realise the shortcomings in the above analysis. We applied the condition $\mathbf{u} \cdot \hat{\mathbf{r}} = 0$

rather than the true nonslip condition $\mathbf{u} = 0$ at the CMB. In reality, there is a boundary layer over which the flow adjusts to the nonslip boundary condition, and we really apply the induction equation at the top of the free stream, the bottom of the boundary layer. We need to know the difference in the values of $\mathbf{B}$ across this boundary layer, denoted $[\mathbf{B}]$.

Various analyses of the boundary layer have been carried out, and no consensus has been reached (e.g., Backus, 1968; Hide and Stewartson, 1973; Jault and Le Mouël, 1991). The important issue for our purposes is that the radial derivatives in all three components of $\mathbf{B}$ are expected to be much bigger than the horizontal derivatives. When one uses this fact along with the divergence-free constraint on the field, one finds

$$\frac{\partial B_r}{\partial r} + \nabla_h \cdot \mathbf{B}_h \sim \frac{\partial B_r}{\partial r} = 0 \quad [30]$$

This leads to the conclusion that $[B_r] = 0$, so maps of the radial component of magnetic field immediately above the CMB also represent the radial component of magnetic field at the top of the free stream. The same cannot be said to be true for the horizontal components, but we omit a discussion on the possible jumps in $\mathbf{B}_h$ mainly because the induction equation for $\mathbf{B}_h$ involves the toroidal field, about which we have no direct knowledge.

5.5.3 The Navier–Stokes Equation

Moving on to consider the fluid dynamics of the core, we adopt the hypothesis that on macroscopic length scales core fluid can be well approximated as a continum (see, e.g., Batchelor, 1967), suppose that it is to first approximation incompressible, obeys Newtonian laws of viscosity, and is uniformly rotating. Then, in a frame of reference rotating with the fluid, the conservation of momentum is encapsulated in the Navier–Stokes equation, which under the Boussinesq approximation (e.g., Gubbins and Roberts, 1987) reads

$$\rho_0 \left(\frac{\partial \mathbf{u}}{\partial t} + \mathbf{u} \cdot \nabla \mathbf{u} + 2\mathbf{\Omega} \times \mathbf{u} \right)$$
$$= -\nabla p + \rho' \mathbf{g} + \mathbf{J} \times \mathbf{B} + \rho_0 \nu \nabla^2 \mathbf{u} \quad [31]$$

where ρ_0 and ρ' are the hydrostatic density and departure from hydrostatic density, respectively, $\mathbf{\Omega}$ is the Earth's rotation vector, p is the nonhydrostatic part of the pressure, $\mathbf{g}$ the acceleration due to gravity, ν the kinematic viscosity, and $\mathbf{J}$ the current density. The Boussinesq approximation is a simplification frequently adopted for the core, and ignores variations in density except those that are responsible for thermal buoyancy through the term $\rho' \mathbf{g}$; its applicability to the core remains under scrutiny in the field of numerical simulation of core dynamics, since the compressibility of the core does cause a change of approximately 20% in the core density between the inner- and outer-core boundaries. It suffices as an approximation for our purposes, as we shall focus on the surface of the core.

Considerable simplification can be made if one analyzes the likely sizes of the terms in [31], concentrating on the flow in the main body of the core (outside the boundary layers). First, the Rossby number,

$$R_{\mathrm{o}} = \frac{U}{\Omega L} \simeq 4 \times 10^{-6} \qquad [32]$$

compares the nonlinear advective term on the left-hand side of [31] with the Coriolis term. We take $L \sim 3 \times 10^6$ m as a characteristic length scale for the core. Then the estimate above is based on values for U (roughly half a millimeter per second) gleaned from analysis of the secular variation (e.g., Section 5.4.2.3), and hence (as in much of our analysis) there is a slight sense of circularity. Similarly, the Ekman number (the ratio of viscous forces to the Coriolis force) is given by

$$E = \frac{\nu}{\Omega L^2} \qquad [33]$$

where Ω is the rotation rate. If we take L as before and $\nu \sim 10^{-6}$ m s^{-2}, we find the classic value of $E \sim 10^{-15}$, indicating that viscous effects are negligible in the main body of the core if a laminar value for ν is adopted. The inertial term is somewhat more difficult – there is a mode of oscillation in the core that can occur on decade timescales, the so-called torsional oscillation which may not be negligible. It is easily excited and it is inappropriate to compare it to the Coriolis force because torsional oscillations are unaffected by it. We will neglect the inertial term on the grounds that it is only significant when the period approaches the diurnal period, except in the force balance when averaged over cylinders coaxial with the rotation axis. This leads us to a very useful approximation in core studies, the so-called magnetostrophic approximation,

$$\rho_0 (2\mathbf{\Omega} \times \mathbf{u}) = -\nabla p + \rho' \mathbf{g} + \mathbf{J} \times \mathbf{B} \qquad [34]$$

The difficult term in this equation is the last term on the right-hand-side, the Lorentz force $\mathbf{L}$. We write it in the form,

$$\mathbf{L} = \mathbf{J} \times \mathbf{B} = \frac{1}{\mu_0} (\nabla \times \mathbf{B}) \times \mathbf{B}$$
$$= \frac{1}{\mu_0} \left[-\frac{1}{2} \nabla B^2 + (\mathbf{B} \cdot \nabla) \mathbf{B} \right] \qquad [35]$$

An approximation called the tangential geostrophy approximation, proposed independently by Hills (1979) and Le Mouël (1984), would neglect the horizontal components of this term when compared to all others; to aid our development, we write the horizontal and radial components of $\mathbf{L}$ as $\mathbf{L}_{\mathrm{h}}$ and $\mathbf{L}_{\mathrm{r}}$, respectively. For tangential geostrophy, we require that

$$M = \frac{|\mathbf{L}_{\mathrm{h}}|}{2\rho_0 \Omega U} = \frac{|(\mathbf{B} \cdot \nabla) \mathbf{B}_{\mathrm{h}}|}{2\mu_0 \rho_0 \Omega U} \ll 1 \qquad [36]$$

There are two contributions to $(\mathbf{B} \cdot \nabla) \mathbf{B}_{\mathrm{h}}$:

$$B_{\mathrm{r}} \frac{\partial \mathbf{B}_{\mathrm{h}}}{\partial r} \quad \text{and} \quad (\mathbf{B}_{\mathrm{h}} \cdot \nabla_{\mathrm{h}}) \mathbf{B}_{\mathrm{h}} \qquad [37]$$

We need to estimate $|\mathbf{B}_{\mathrm{h}}|$ and $|\partial \mathbf{B}_{\mathrm{h}} / \partial_r|$. Le Mouël (1984) argues that if the toroidal field is small at the CMB (if the mantle is a perfect insulator it must vanish) and its radial gradient is small, then these terms are of order B_{P}^2 / L, where B_{P} is the size of the poloidal field at the CMB ($\simeq 5 \times 10^{-4}$ T), giving $M \simeq 10^{-3}$.

If we adopt this approximation, we have

$$2\rho_0 (\mathbf{\Omega} \times \mathbf{u}) = -\nabla p + \rho' \mathbf{g} + \mathbf{L}_{\mathrm{r}} \qquad [38]$$

Curling this equation, we obtain

$$2\rho_0 (\mathbf{\Omega} \cdot \nabla) \mathbf{u} = \mathbf{g} \times \nabla \rho' + \nabla \times \mathbf{L}_{\mathrm{r}} \qquad [39]$$

which, in the case that $\mathbf{L}_{\mathrm{r}} = 0$, is the thermal wind equation, of great importance in meteorology. The radial component gives the so-called tangentially geostrophic constraint,

$$\nabla_{\mathrm{h}} \cdot (\mathbf{u}_{\mathrm{h}} \cos \theta) = 0 \qquad [40]$$

which remains true regardless of whether $\mathbf{L}_{\mathrm{r}} \neq 0$. Given these assumptions, we therefore have a strong constraint on the types of allowed fluid motions at the CMB. The interested reader can see that output from self-consistent geodynamo simulations tend to suggest that the tangential geostrophy approximation is reasonably well obeyed. The significance of the constraint is that it vastly reduces the types of allowable fluid motions when the inverse problem for $\mathbf{u}$ is solved.

5.5.4 The Frozen Flux Approximation

5.5.4.1 The neglect of magnetic diffusion and its physical consequences

Interpretation of secular variation using the induction equation [18] is often simplified by neglecting the contribution of magnetic diffusion. In the limit of a perfectly electrically conducting fluid (zero magnetic diffusivity), the induction equation becomes

$$(\partial_t + \mathbf{u} \cdot \nabla)\mathbf{B} = (\mathbf{B} \cdot \nabla)\mathbf{u} \qquad [41]$$

The left-hand side of this equation is the advective derivative describing how the magnetic field changes as one moves along with the fluid, while the right-hand side tells us that such changes occur through the stretching of the magnetic field by fluid motions.

Further intuition follows if we think about how a velocity field $\mathbf{u}$ would advect a material line element $d\mathbf{l}$. We imagine the line element being drawn in the fluid at some instant and subsequently moved along and stretched by the fluid motions. The total rate of change of $d\mathbf{l}$ is then $\mathbf{u}(r + dl) - \mathbf{u}(r)$, where r and $r + dl$ are position vectors at the two ends of $d\mathbf{l}$. The equation describing the evolution of $d\mathbf{l}$ therefore has the form,

$$(\partial_t + \mathbf{u} \cdot \nabla)d\mathbf{l} = \mathbf{u}(r + dl) - \mathbf{u}(r) = (d\mathbf{l} \cdot \nabla)\mathbf{u} \quad [42]$$

Inspection of eqns [41] and [42] reveals they have precisely the same form. This simple example demonstrates that because magnetic fields evolve in an identical manner to material line elements in a fluid if magnetic diffusion is negligible, a field line found on a particular fluid element at some initial instant must continue to lie on that element at all subsequent times. The magnetic field effectively appears to be frozen into the fluid as it moves. This result is known as Alfvén's theorem (part I) after Hannes Alfvén who first derived it; we shall discuss the second part of the theorem in a moment. Neglect of magnetic diffusion in the induction equation and its consequences are most commonly referred to as the frozen flux approximation and we shall use the latter terminology.

Another important property that results from assuming that a fluid is a perfect electrical conductor can be demonstrated by returning to Faraday's law of magnetic induction (eqn [12]). Earlier, this was stated in its differential form. The integral form when applied to material curves of an electrically conducting fluid in motion (see, e.g., Davidson, 2001) takes the form,

$$\oint_C \mathbf{E}' \cdot dl = -\frac{d}{dt}\int_S \mathbf{B} \cdot dS \qquad [43]$$

where $\mathbf{E}' = \mathbf{E} + \mathbf{u} \times \mathbf{B}$ is the total electric field in a reference frame moving along with $d\mathbf{l}$ at velocity $\mathbf{u}$, C is a closed material curve composed of line elements $d\mathbf{l}$, and S is any surface that spans C. Now, from Ohm's law, $\mathbf{J} = \sigma\mathbf{E}'$, so

$$\frac{1}{\sigma}\oint_C \mathbf{J} \cdot dl = -\frac{d}{dt}\int_S \mathbf{B} \cdot dS \qquad [44]$$

but under the frozen flux approximation $\sigma \rightarrow \infty$; therefore,

$$\frac{d}{dt}\int_S \mathbf{B} \cdot dS = 0 \qquad [45]$$

In a perfect electrical conductor, the integrated magnetic field (or magnetic flux) through any material surface is thus always preserved. This is known as Alfvén's theorem (part II).

5.5.4.2 Application of the frozen flux approximation to the generation of secular variation at the core surface

Roberts and Scott (1965) were the first to suggest that the frozen flux approximation could be applied to the problem of modeling secular variation. Many authors (see, e.g., Backus *et al.*, 1996) refer to this as the frozen flux hypothesis. Roberts and Scott argued that there are two distinct timescales associated with the induction equation. Considering a length scale $\mathcal{L}_B$ over which the magnetic field changes, a characteristic flow speed $\mathcal{U}$, and the magnetic diffusivity η, these timescales are defined as

$$\tau_{\text{adv}} = \frac{\mathcal{L}_B}{\mathcal{U}} \quad \text{Advection time scale} \qquad [46]$$

and

$$\tau_{\text{dif}} = \frac{\mathcal{L}_B^2}{\eta} \quad \text{Magnetic diffusion time scale} \qquad [47]$$

The ratio of these timescales $R_m = \tau_{\text{dif}}/\tau_{\text{adv}} = \mathcal{U}\mathcal{L}_B/\eta$ is known as the magnetic Reynolds number and gives a crude measure of the relative strength of advection to magnetic diffusion. Taking estimates of $\mathcal{L}_B = 10^6$ m (the approximate scale of the outer-core container and of large-scale features in Earth's magnetic field), $\mathcal{U} = 5 \times 10^{-4}$ ms^{-1} (speed of observed westward field motions, thought to be caused by core flow), and $\eta = 2$ m^2 s^{-1} (from estimates of liquid iron electrical conductivity at core pressures and temperatures – see Braginsky and Roberts, 1995) gives estimates of $\tau_{\text{adv}} = 65$ yr and $\tau_{\text{dif}} = 1.6 \times 10^4$ yr. In Earth's core (Roberts and Glatzmaier, 2000). On this basis, Roberts and Scott suggested that making the frozen flux assumption was a reasonable (though imperfect) approximation when modeling the action of core motions causing large-scale secular variation. This argument has been the subject of much comment and debate over the past 40 years; we will return later to the question of its validity. First, we will give details on its formal consequences and review attempts to determine whether these are compatible with observed secular variation.

5.5.4.3   Consequences of frozen flux approximation at the core surface

Backus (1968) described the conditions for the main field morphology and secular variation to be consistent with a frozen flux theory of core motions. Neglecting magnetic diffusion, he showed that the radial part of the induction equation reduces to (c.f. eqn [29])

$$\partial_t B_r + \nabla_h \cdot (\mathbf{u}_h B_r) = 0 \qquad [48]$$

He then deduced that this implies a set of conditions on null-flux points and curves (where $B_r = 0$). The most important of these are

$$\int_S \partial_t B_r \mathrm{d}S = 0 \qquad [49]$$

where S is surface bounded by a null-flux curve C and

$$\partial_t B_r = 0 \qquad [50]$$

where two null-flux curves C_1 and C_2 intersect. From the first condition follows a condition closely related to eqn [45], which states that the integrated radial magnetic field through a null flux curve C (an example of a material curve) is preserved. The simple derivation is as follows. We know that

$$\frac{\mathrm{d}}{\mathrm{d}t} \int_S B_r \mathrm{d}S = \int_S \partial_t B_r \mathrm{d}S + \int_C B_r u_n \mathrm{d}l \qquad [51]$$

where $\mathrm{d}l$ is a line element along the null flux curve C and u_n is the normal component of its velocity. Using eqn [49], the first term on the right-hand side is zero and, since C is a null-flux curve, the second term is also zero, giving

$$\frac{\mathrm{d}}{\mathrm{d}t} \int_S B_r \mathrm{d}S = 0 \qquad [52]$$

From eqn [52], it further follows that the sum of the unsigned flux over all null-flux curves must also be conserved,

$$\frac{\mathrm{d}}{\mathrm{d}t} \int_{S'} |B_r| \mathrm{d}S = 0 \qquad [53]$$

with the integration now over the entire core surface S'. It should be noted that eqn [53] is a weaker constraint than eqn [52] on the signed flux through individual flux patches because contributions from small patches will be swamped by those from the larger Northern and Southern Hemisphere patches. The unsigned flux condition will therefore only be violated if there are large amounts of magnetic diffusion occurring on a global scale; it could still be approximately obeyed even if magnetic diffusion was occurring locally. On the other hand, condition [53] is less likely to be adversely affected by errors in the field models; so, it is a reasonable observational test of the global applicability of the frozen flux approximation.

Backus' results additionally require that the topology of the field must be invariant so that null-flux curves cannot split or coalesce. In practice, this is a rather difficult condition to satisfy as it requires only a small amount of diffusion in order to be violated. It therefore seems unlikely that this condition would be satisfied by the magnetic field at the core surface, where frozen flux is at best a useful approximation; we shall therefore not discuss this condition further.

5.5.4.4   Attempts to test the frozen flux approximation using geomagnetic observations

The conditions described in the previous section (which will be referred to collectively as the Backus conditions) have enabled workers to test the validity of the frozen flux approximation using models of the main field and secular variation constructed from geomagnetic observations. These tests fall into two main categories: (1) attempts to estimate whether the Backus conditions are violated in field models constructed without any constraints on field evolution; (2) attempts to build field models that are constrained to obey the Backus conditions but that also satisfactorily fit the observations.

The first category involves investigating whether observations are sufficiently accurate to observe diffusion. Booker (1969) was the first to investigate this issue. He attempted to estimate the integral of $\partial_t B_r$ through the null flux curves represented by the magnetic equator and found that the dipole part of the time derivative was nonzero, possibly violating eqn [49]. However, the field models he used lacked the necessary resolution at high spherical harmonic degrees to definitively calculate the full secular variation integral and he was thus unable to definitively identify the presence of diffusion. Gubbins (1983) repeated the calculation and arrived at a similar conclusion, again being limited by the accuracy of observations and field models.

Hide and Malin (1981) and later Voorhies and Benton (1982) used the criteria for invariance of the unsigned flux at the core surface (eqn [53]) as a method for calculating the outer-core radius and found values that agreed well with the seismologically determined value. These results indicated that the unsigned flux is not changing rapidly at the core surface and was evidence in support in the application of the frozen flux approximation on large length scales. Benton and Voorhies (1987) later extended these analyses to show that there had been little change in the unsigned flux over the interval 1945–85. Bloxham et al. (1989) considered the evolution of the unsigned flux over a much longer interval from 1715 to 1980 and found that it remained approximately constant over that interval. They concluded that this global requirement for frozen flux was satisfied by their field models but pointed out that changes in the flux of smaller null-flux patches would not be seen by this method. Similar results for the unsigned flux were recently obtained by Holme and Olsen (2006) using a high-quality, satellite-derived field model (CO2003).

The second question of possible changes in the flux through individual null flux curves was addressed in a series of papers by Bloxham and Gubbins in the mid- to late 1980s including Bloxham and Gubbins (1985), Gubbins and Bloxham (1985), Bloxham and Gubbins (1986), Bloxham (1986a), and in the landmark study of Bloxham et al. (1989) where the definitive results were reported. In the latter paper, a sequence of single epoch, regularized, field models spanning 1715–1980 were studied and changes of the flux though null-flux curves at the core surface were calculated. Major flux changes through some curves were found; in particular, a patch that moved from the Indian Ocean to southern Africa was found to increase its flux dramatically. Bloxham and Gubbins regarded this as conclusive evidence for the violation of frozen flux; however, others have expressed doubts over the rigor of their arguments. In order for frozen flux to have been demonstrably violated, the changes in the flux must be larger than possible changes in the flux due to errors in the field models. Backus (1988) has argued that the error estimates of Gubbins and Bloxham (1985) and Bloxham and Gubbins (1986) are rather optimistic. He suggests that their models do not fully solve the uniqueness problem because

the damping (regularization) parameter is arbitrarily chosen; their error estimates depend crucially on the value of this parameter and are likely to actually be much larger than those quoted. Furthermore, O'Brien (1996) has demonstrated that even the number of null-flux patches in field models based on excellent satellite data is difficult to definitively determine. It therefore seems that inferences based on the changes in flux through individual null-flux patches in the past (when there were significant variations in the distribution and accuracy of observations) should be viewed with a healthy degree of skepticism.

The compatibility of observations with the Backus conditions for frozen flux has recently become apparent thanks to the construction of field models that are constrained to obey these conditions. Gubbins (1984) used a Lagrangian constraint approach to enforce zero radial secular variation through null-flux curves (eqn [49]) and found that it was possible to do this over the interval 1959–74. Bloxham and Gubbins (1986) and Bloxham et al. (1989) produced field models that satisfied the condition that the flux of B_r through null-flux curves be conserved (eqn [52]). They implemented this by imposing an additional penalty during the inversion process; models with flux integrals differing from a pre-defined reference model were heavily penalized. They reported that it was possible to find models that satisfied these flux constraints, and that their imposition often improved field models where the observations were sparse or of poor quality. However, they also reported that the constrained models had a slightly higher misfit than the unconstrained models, using this to support their contention that frozen flux was in fact violated. This argument is subject to similar caveats regarding errors in field models as the argument concerning changes in the flux integrals in unconstrained models. Benton et al. (1987) constructed field models with constrained flux through null-flux curves covering the interval 1977.5–1982.5 and were able to demonstrate that the secular variation predicted by these models for times outside the span of input data was an improvement on the predictions of unconstrained models. They took this to be evidence in favor of the frozen flux hypothesis.

Most recently, Constable et al. (1993), O'Brien et al. (1997), and Jackson et al. (2007) have attacked the same problem but used a different parametrization of the field based on a spherical triangle tessellation at the core surface. Constable et al. (1993) constructed field models satisfying the flux constraints of eqn [52] for the epochs 1945.5 and 1980.0, O'Brien *et al.* (1997) did the same for the epochs 1915 and 1980, while Jackson *et al.* (2007) managed to do so for the epochs 1882, 1915, 1945, 1980, and 2000. In all cases, it was found that reasonable misfit levels to the observations could be achieved, and it was noted that in order to reject the frozen flux hypothesis it would be necessary to demonstrate that no such models could be found.

We remark here that the published time-dependent field models of Bloxham and Jackson (1992) and Jackson *et al.* (2000) both show an obvious growth in the intensity of reversed flux patches in the Southern Hemisphere in the twentieth century. Gubbins (1987) has pointed out that a simple and physically appealing explanation of this phenomenon is the expulsion of toroidal flux by upwelling core fluid (Allan and Bullard, 1966; Bloxham, 1986a; Drew, 1993). It is therefore unfortunate that the present field observations seem incapable of constraining field models sufficiently to discriminate between this mechanism and a mechanism involving frozen flux advection.

Although our discussion has focused on the radial magnetic field and its secular variation, a limited amount of work has been carried out on whether the horizontal components of the magnetic field are consistent with the frozen flux hypothesis. Assuming continuity of the horizontal magnetic field components across the magnetic field boundary layer close to the core surface (note that this assumption is questionable – see, e.g., Jault and Le Mouël (1991)), a further set of consistency conditions can be constructed (Backus, 1968; Gubbins and Roberts, 1987). An attempt to test these conditions has been made by Barraclough *et al.* (1989); they found no evidence that the constraints are violated, but confessed that the field models used were not yet sufficiently accurate enough to allow a stringent test.

5.5.4.5 *Theoretical issues concerning the frozen flux approximation*

The simple scaling arguments of Roberts and Scott regarding the plausibility of using the frozen flux assumption to model secular variation has been the subject of some debate, especially since the claims in the 1980s that the signature of magnetic diffusion had been observationally detected.

Gubbins and Kelly (1996) have pointed out that, for the special case of steady flows, the frozen flux hypothesis is invalid because the balance in the induction equation must be between secular variation and diffusion, with frozen flux effects being negligible. Although undoubtedly true, this observation appears to be of little relevance when trying to model changes in Earth's magnetic field; the geodynamo is certainly not steady but undergoes fluctuates over a very wide range of frequencies. A striking example of this non-steady behaviour is found in records of geomagnetic excursions over the past 700,000 years (Gubbins, 1999). A more worrisome objection raised by Gubbins and Kelly is that the frozen flux approximation is a singular limit of the induction equation. It lowers the differential order of the system from second to first order, raising the possibility that physically relevant solutions might be filtered out.

Concerns over how the frozen flux approximation arises as a limit of a magnetic advection–diffusion process have been addressed in detail by Gubbins (1996). Gubbins has developed a novel formalism for determining core motions that takes magnetic diffusion into account and shows that the frozen flux approximation can be formally retrieved in the limit when $(\eta/\omega)^{1/2}$ tends to zero, where ω is the frequency of the secular variation. This implies that we should only expect the frozen flux approximation to work well when field variations are rapid and the concomitant fluid motions highly time dependent. Gubbins used his formalism to derive an estimate of the toroidal field gradient close to the core surface necessary to explain the amount of flux change through the South Atlantic reversed flux patch between 1905.5 and 1965.5 as estimated by Bloxham and Gubbins (1985). He arrived at a plausible scenario by considering a 1.2 mT toroidal field at a depth of 60 km in the core (diminishing at a rate of $20\,\text{nT}\,\text{m}^{-1}$ toward the surface) with a horizontal length scale of $10^6\,\text{m}$. When this field was acted on by an upwelling flow with a radial rate of change of $0.02\,\text{yr}^{-1}$ in the presence of a magnetic diffusivity of $1.6\,\text{m}^2\,\text{s}^{-1}$, it was found that it was possible to generate the required flux change of 500 MWb.

This estimate, together with the earlier forward calculations of Allan and Bullard (1966), Bloxham (1986a), and Drew (1993), have established the plausibility of toroidal flux expulsion as a mechanism that could cause localized growth of pairs of reverse and normal flux patches. Such arguments however fail to convince on the need to apply a diffusive formalism globally. Rather, they serve as a caution against interpreting core flow inferred using the frozen flux approximation without first considering whether the specific local patterns of secular variation being explained might be produced by a diffusive mechanism. For example, it might be very reasonable to use the frozen flux approximation to understand the rapid azimuthal motion of flux patches but not to understand rapid growth and decay of field concentrations. Such reasoning was used by Dumberry and Bloxham (2006) in their study of global azimuthal core flows on millennial timescales, derived from archeomagnetic field models.

Love (1999) raised an objection to the frozen flux hypothesis in the case of a nearly steady dynamo. Although clearly correct under the conditions considered, the applicability of his examples to the Earth is questionable since, as noted earlier, the geodynamo does not appear to be steady on any recorded timescales. Recent feasibility tests using more dynamically plausible dynamo models (Roberts and Glatzmaier, 2000; Rau et al., 2000) have, on the other hand, demonstrated that the frozen flux approximation is a useful construct. These important tests will be described in more detail below. The important point to take away from the arguments of Love (1999) is that the original scaling argument of Roberts and Scott was overly simplistic – in real secular variation there is not just one lengthscale and timescale of interest; rather both magnetic and velocity fields will contain power over a range of length and timescales. To address whether the frozen flux hypothesis is a useful approximation really requires experiments studying the induction effects of Earth-like velocity fields acting on Earth-like magnetic fields. Unfortunately, it is still beyond our ability to accurately simulate all aspects of the dynamics of Earth's core (Glatzmaier, 2002), though present geodynamo models already capture many important aspects of Earth's magnetic field and its evolution (Kuang and Bloxham, 1997; Christensen et al., 1998) and are certainly useful tools when considering the validity of the frozen flux hypothesis.

Roberts and Glatzmaier (2000) were the first to use a numerical geodynamo model in an attempt to evaluate the frozen flux approximation. They calculated the variations in the unsigned flux from a high resolution simulation and found that it varied by $\sim$3% over timescales estimated to be equivalent to 150 years in Earth's core. This suggested that for plausible magnetic and velocity fields the frozen flux approximation is reasonable on a global scale. Unfortunately, no attempt was made to investigate how much individual flux integrals varied over the same interval. The authors further remark that for

frozen flux approximation to be useful it is not required to be strictly true; rather, it must only involve errors smaller than those associated with our incomplete knowledge of the field at the core surface. They point out that the errors in the unsigned flux from truncation of the main field at degree 12 in their model (only one aspect of the error present in observationally based field models) are much larger than the amount by which the unsigned flux is varying. Again, a more stringent assessment would involve studying such issues for individual null-flux patches.

Rau et al. (2000) have performed a large number of tests on a different suite of geodynamo models. The main focus of their study was to take the time-dependent magnetic field output from the simulation, carry out inversions for the underlying core surface flow on the basis of the frozen flux approximation, and compare the results with the known simulation flow. We are most interested in their preliminary test that aimed to determine how well the assumption of frozen flux was satisfied in the models. They observed that the known secular variation and the secular variation predicted on the basis of the frozen flux hypothesis using the known flow were similar, finding high correlation coefficients in the range 0.7–0.8. They observed that the diffusive contribution was significant in only a few isolated locations and that the deviations from the frozen flux assumption were not so large as to preclude its use in the determination of large-scale core motions from the magnetic data. It should be noted that their models have R_m in the range 118–320; R_m for the Earth is expected to be rather larger than this (around 500 – see Roberts and Glatzmaier (2000)), and hence the frozen flux approximation might perform even better in reality. Perhaps the most important finding of this study is that even though the frozen flux assumption is not perfectly satisfied (the formal necessary conditions would be violated to an extent), useful information could be extracted concerning fluid motions at the core surface and substantial parts of the secular variation pattern could be explained using the frozen flux approximation. Errors in the field models (due to limits on the range of core field spherical harmonic coefficients that can be determined by observations) and in the oversimplified dynamical assumptions can apparently lead to more serious problems than violation of the frozen flux approximation.

One final remark should be made concerning the inevitable failure of the frozen flux hypothesis (Backus and Le Mouël, 1986). We recollect the radial component of the induction equation in its exact form:

$$\partial_t B_r + \nabla_h \cdot (\mathbf{u}_h B_r) = \eta \frac{1}{r} \nabla^2 (r B_r) \qquad [54]$$

It is clear that the frozen flux hypothesis fails in all locations where $\left| \nabla_h \cdot (\mathbf{u}_h B_r) \right| \leq \left| \eta r^{-1} \nabla^2 (r B_r) \right|$. In particular, the approximation always fails on the curve S where

$$\nabla_h \cdot (\mathbf{u}_h B_r) = 0 \qquad [55]$$

Backus and Le Mouël call these curves 'leaky curves', and the region around them where the approximation fails the 'leaky belt'. The problem, of course, is that one cannot locate these places *a priori*, because one does not know $\mathbf{u}$, and much depends on the relative scales of variation of B_r and $\mathbf{u}$ including the radial variation of B_r.

One thing is certain: any point where B_r and $\nabla_h B_r$ vanish are inevitably on the leaky curve. Backus called such places 'touch points'; it is at these places that null-flux curves can appear or disappear. Of course, it is incredibly difficult, if not impossible, to determine accurately the positions of such points on the CMB (Backus, 1988), and hence very little effort has been expended on such activity.

5.5.4.6 Summary of applicability of the frozen flux approximation in core studies

After 40 years of study using observations, theoretical arguments, and numerical tests, the worth and limitations of using the frozen flux approximation as a tool for explaining of secular variation are now apparent. The approximate invariance of the unsigned flux over intervals of several hundred years in both observation based models and numerical simulations demonstrates that flux is well conserved globally on such timescales. On the other hand, field models spanning over 100 years that satisfy flux conservation conditions and satisfactorily fit observations tell us that geomagnetic data is not yet capable of unambiguously detecting the presence of diffusion. Frozen flux is therefore undoubtedly a useful tool in understanding the advective motions that cause much of the observed secular variation. It should however always be borne in mind that the assumption of frozen flux is only an approximation and that magnetic diffusion will inevitable be present at some level, causing violation of flux constraints over long time intervals or locally where field gradients are very large.

5.5.5 Other Invariants

We have concentrated on the frozen flux approximation, and its observable (and sometimes testable) consequences. Depending on the additional approximations one is willing to make, there are other invariants that can sometimes be testable. When one makes assumptions about the type of fluid flow that may be occurring at the CMB, a number of other invariants arise – see, for example, those reviewed in the cases of tangentially geostrophic or toroidal fluid flow in Bloxham and Jackson (1991), and additional constraints described in Jackson and Hide (1996), Chulliat and Hulot (2001), and Chulliat (2004). The possibility of using deviations from exact satisfaction of the constraints by field models is discussed in Gubbins (1996) and Hulot and Chulliat (2003).

Here we briefly describe an invariant that arises from perhaps the most innocuous of assumptions additional to that of frozen flux, that of a poorly conducting (or to be strict, a perfectly insulating) mantle. The invariant arises from a consideration of eqn [34], namely the Navier–Stokes equation in the magnetostrophic limit. We consider a null-flux curve (on which $B_r = 0$) and examine the form of the Lorentz force $\mathbf{J} \times \mathbf{B}$; we can write the horizontal part of it as

$$[\mathbf{J} \times \mathbf{B}]_h = \mathbf{B}_h J_r - J_h B_r \qquad [56]$$

If the mantle is a sufficiently poor conductor that the radial current J_r is negligible, then all terms on the right-hand side vanish (because $B_r = 0$ also), and we discover that the horizontal Lorentz force vanishes on null-flux curves. Null-flux curves must therefore move as if they are governed by a tangentially geostrophic force balance, even if the true force balance over the core is one of magnetostrophic equilibrium. The repercussion of this is the following: because null-flux curves are material curves (section 5.5.4.1), they must obey Kelvin's celebrated theorem (e.g., Gill, 1982), and conserve their planetary vorticity (we have ignored relative vorticity from the outset by dropping the nonlinear advection term in the Navier–Stokes equation). We therefore conclude that the following integral must hold:

$$\frac{d}{dt} \int_S \cos \theta \, dS = 0 \qquad [57]$$

over any patch S which is a null flux patch, where θ is colatitude. For more details, see Gubbins (1991) and Jackson (1996). The interesting aspect of these constraints is that null flux patches were previously free to shrink *in situ*, while conserving their flux (by concentrating the flux into a smaller area while increasing the amplitude of B_r). This new constraint places demands on their area, so that they cannot shrink *in situ*; they must instead move part of their area to a different latitude, either toward the pole (for shrinking) or away from the pole (for enlargement). These constraints have been imposed on field models by Jackson *et al.* (2007), who find that they can fit 100 years of historical data adequately even when the constraints are imposed.

5.6 Summary

We have described the basis of our knowledge of the secular variation of the magnetic field, starting from the fundamental data sets that are available. We subsequently moved through the mathematical techniques that are used for representing the four-dimensional vector $\mathbf{B}(\mathbf{r}, t)$, and discussed the underlying fluid dynamics and electrodynamics of the core. We ended with a discussion of some of the controversies surrounding the approximations employed, and what can be gleaned about the physical state at the top of the core. We have stopped short of describing the industry of computing models of core velocities, and the interesting conclusions that can be drawn from them. For example, models of flow at the core surface appear to have certain components linked to motions deeper within the core: these are the axisymmetric zonal flows that apparently are linked to certain modes of oscillation in the core, often termed torsional oscillations, that are most easily viewed as motions of nested cylinders (coaxial with the rotation axis) filling the core. These oscillations are almost certainly linked to changes in the rotation rate of the Earth, often coined changes in the length of day. This observation in itself is remarkable, but what is more exciting is the possibility of gleaning something about the state of the magnetic field within the core from exactly these oscillations, since they are strongly affected by the component of the interior magnetic field perpendicular to the rotation axis.

The reader who has consulted Chapter 2 will see that satellites are beginning to provide models of secular variation with unprecedented accuracy; there is a tradeoff between accuracy, and the short observing time span. This means that historical models still have a role to play, since the accumulated effects of slow processes can be seen over long times

even when the signal/noise level is lower. A number of challenges lay ahead, not least the question of how to optimally incorporate the newest satellite, repeat station, and observatory data in a homogeneous way with the older data described herein.

Acknowledgments

We take this opportunity to record our endebtedness to Art Jonkers, for his major contribution to the developments of the historical dataset and model, and for innumerable stimulating conversations. We thank Mioara Mandea for supplying data for some of the figures, and Terry Sabaka and Katia Pinheiro for supplying Figures 16 and 28 respectively. The advice of Susan Macmillan regarding observatory data is gratefully acknowledged. Nils Olsen has constantly given early access to his models, for which we are grateful. We benefitted greatly from reviews by Michael Purucker and Mathieu Dumberry. We acknowledge the crucial role played by observatories and their staff in the long-term monitoring of the field.

References

Alexandrescu M, Courtillot V, and Le Mouël JL (1996) Geomagnetic field direction in Paris since the mid-sixteenth century. *Physics of the Earth and Planetary Interiors* 98: 321–360.

Alexandrescu M, Gibert D, Hulot G, Le Mouël JL, and Saracco G (1995) Detection of geomagnetic jerks using wavelet analysis. *Journal of Geophysical Research* 100(B10): 12557–12572.

Alexandrescu M, Gibert D, Hulot G, Le Mouël JL, and Saracco G (1996) Worldwide wavelet analysis of geomagnetic jerks. *Journal of Geophysical Research* 101(B10): 21975–21994.

Alexandrescu M, Gibert D, Le Mouël JL, Hulot G, and Saracco G (1999) An estimate of average lower mantle conductivity by wavelet analysis of geomagnetic jerks. *Journal of Geophysical Research* 104: 17735–17746.

Allan DW and Bullard EC (1966) The secular variation of the Earth's magnetic field. *Proceedings of the Cambridge Philosophical Society* 62(3): 783–809.

Backus G (1968) Kinematics of geomagnetic secular variation in a perfectly conducting core. *Philosophical Transactions of the Royal Society of London, Series A* 263: 239–266.

Backus GE (1986) Poloidal and toroidal fields in geomagnetic field modeling. *Reviews of Geophysics* 24: 75–109.

Backus G (1988) Bayesian inference in geomagnetism. *Geophysical Journal of the Royal Astronomical Society* 92: 125–142.

Backus GE and Le Mouël JL (1986) The region on the core-mantle boundary where a geostrophic velocity field can be determined from frozen-flux geomagnetic data. *Geophysical Journal of the Royal Astronomical Society* 85: 617–628.

Backus G, Parker R, and Constable C (1996) *Foundations of Geomagnetism.* Cambridge: Cambridge University Press.

Barraclough D (1974) Spherical harmonic analysis of the geomagnetic field for eight epochs between 1600 and 1910. *Geophysical Journal of the Royal Astronomical Society* 36: 497–513.

Barraclough DR (1978) Spherical harmonic analysis of the geomagnetic field, No.8: *Geomagnetic Bulletin of the Institute of Geological Sciences*. London: HMSO.

Barraclough DR (1982) Historical observations of the geomagnetic field. *Philosophical Transactions of the Royal Society of London, Series A* 306: 71–78.

Barraclough DR (1985) Halley's Atlantic magnetic surveys. In: Schröder W (ed.) *Historical Events and People in Geosciences*. New York: Peter Lang.

Barraclough DR (1995) Observations of the Earth's magnetic field in Edinburgh from 1670 to the present day. *Transactions of the Royal Society of Edinburgh: Earth Sciences* 85: 239–252.

Barraclough D, Gubbins D, and Kerridge D (1989) On the use of horizontal components of magnetic field in determining core motions. *Geophysical Journal International* 98: 293–299.

Barraclough DR, Carrigan JG, and Malin SRC (2000) Observed geomagnetic field intensity in London since 1820. *Geophysical Journal International* 141: 83–99.

Batchelor GK (1967) *An Introduction to Fluid Mechanics*. Cambridge: Cambridge University Press.

Becquerel AC (1840) *Traité Expérimental de l'Electricité et du Magnétisme*. Paris: Fermin Didot Frères.

Bemmelen W van (1899) *Die Abweichung der Magnetnadel: Beobachtungen, Säcular-Variation, Wert- und Isogonensysteme bis zur Mitte des XVIIIten Jahrhunderts*, supplement to *Observations Made at the Royal Magnetic and Meteorological Observatory, Batavia*, 21: 1–109.

Benton ER, Estes RH, and Langel RA (1987) Geomagnetic field modeling incorporating constraints from frozen-flux electromagnetism. *Physics of the Earth and Planetary Interiors* 48: 241–264.

Benton ER and Voorhies CV (1987) Testing recent geomagnetic field models via magnetic flux conservation at the core–mantle boundary. *Physics of the Earth and Planetary Interiors* 48: 350–357.

Benton ER and Whaler KA (1983) Rapid diffusion of poloidal geomagnetic field through the weakly conducting mantle: a perturbation solution. *Geophysical Journal of the Royal Astronomical Society* 75: 77–100.

Bloxham J (1985) *Geomagnetic Secular Variation*. Unpublished Ph.D. Thesis, University of Cambridge.

Bloxham J (1986a) The expulsion of magnetic flux from the Earth's core. *Geophysical Journal of the Royal Astronomical Society* 87: 669–678.

Bloxham J (1986b) Models of the magnetic field at the core–mantle boundary for 1715, 1777, and 1842. *Journal of Geophysical Research* 91: 13954–13966.

Bloxham J (1987) Simultaneous stochastic inversion for geomagnetic main field and secular variation 1. A large scale inverse problem. *Journal of Geophysical Research* 92: 11597–11608.

Bloxham J (2002) Time-independent and time-dependent behaviour of high-latitude flux bundles at the core-mantle boundary. *Geophysical Research Letters* 29(18): 1854–1858.

Bloxham J and Gubbins D (1985) The secular variation of Earth's magnetic field. *Nature* 317: 777–781.

Bloxham J and Gubbins D (1986) Geomagnetic field analysis IV – Testing the frozen flux hypothesis. *Geophysical Journal of the Royal Astronomical Society* 84: 139–152.

Bloxham J and Gubbins D (1987) Thermal core–mantle interactions. *Nature* 325: 511–513.

Bloxham J, Gubbins D, and Jackson A (1989) Geomagnetic secular variation. *Philosophical Transactions of the Royal Society of London, Series A* 329(1606): 415–502.

Bloxham J and Jackson A (1989) Simultaneous stochastic inversion for geomagnetic main field and secular variation 2. 1820–1980. *Journal of Geophysical Research* 94: 15753–15769.

Bloxham J and Jackson A (1991) Fluid flow near the surface of the Earth's outer core. *Reviews of Geophysics* 29: 97–120.

Bloxham J and Jackson A (1992) Time-dependent mapping of the magnetic field at the core-mantle boundary. *Journal of Geophysical Research* 97(B13): 19537–19563.

Bloxham J, Zatman S, and Dumberry M (2002) The origin of geomagnetic jerks. *Nature* 420: 685–687.

Booker JR (1969) Geomagnetic data and core motions. *Proceedings of the Royal Society of London Series A - Mathematical Physical and Engineering Sciences* 309: 27–40.

Braginsky SI and Roberts PH (1995) Equations governing convection in Earth's core and the geodynamo. *Geophysical and Astrophysical Fluid Dynamics* 79: 1–97.

Bullard EC, Freedman C, Gellman H, and Nixon J (1950) The westward drift of the Earth's magnetic field. *Philosophical Transactions of the Royal Society of London, Series A* 243: 67–92.

Bullard EC and Gellman H (1954) Homogeneous dynamos and terrestrial magnetism. *Philosophical Transactions of the Royal Society of London, Series A* 247: 213–278.

Busse FH (1975) A necessary condition for the geodynamo. *Journal of Geophysical Research* 80: 278–280.

Cafarella L, De Santis A, and Meloni A (1992) Secular variation in Italy from historical geomagnetic field measurements. *Physics of the Earth and Planetary Interiors* 73: 206–221.

Cain JC, Daniels WE, Hendricks SJ, and Jensen DC (1965) An evaluation of the main geomagnetic field, 1940–1962. *Journal of Geophysical Research* 70: 3647–3674.

Cain JC, Hendricks SJ, Langel RA, and Hudson WV (1967) A proposed model for the International Geomagnetic Reference Field – 1965. *Journal of Geomagnetism and Geoelectricity* 19: 335–355.

Cawood J (1977) Terrestrial magnetism and the development of international collaboration in the early nineteenth century. *Annals of Science* 34: 551–587.

Cawood J (1979) The magnetic crusade: Science and politics in early Victorian Britain. *Isis* 70(no.254): 493–518.

Chapman S and Bartels J (1940) *Geomagnetism*, vol. II, 1049 pp. London: Oxford University Press.

Christensen UR, Olson P, and Glatzmaier G (1998) A dynamo model interpretation of geomagnetic field structures. *Geophysical Research Letters* 25: 1565–1568.

Chulliat A (2004) Geomagnetic secular variation generated by a tangentially geostrophic flow under the frozen-flux assumption – II. Sufficient conditions. *Geophysical Journal International* 157(2): 537–552.

Chulliat A and Hulot G (2001) Geomagnetic secular variation generated by a tangentially geostrophic flow under the frozen-flux assumption – I. Necessary conditions. *Geophysical Journal International* 147(2): 237–246.

Churchman J (1794) *The magnetic atlas or variation charts of the whole terraqueous globe; comprising a system of the variation and dip of the needle, by which, the observations being truly made, the longitude may be ascertained* 2nd edn. London.

Constable CG, Parker RL, and Stark PB (1993) Geomagnetic field models incorporating frozen-flux constraints. *Geophysical Journal International* 113: 419–433.

Courtillot V and Le Mouël JL (1984) Geomagnetic secular variation impulses. *Nature* 311: 709–716.

Crichton Mitchell A (1932) Chapters in the history of terrestrial magnetism I. *Terrestrial Magnetism and Atmospheric Electricity* 37: 105–146.

Crichton Mitchell A (1937) Chapters in the history of terrestrial magnetism II. *Terrestrial Magnetism and Atmospheric Electricity* 42: 241–280.

Crichton Mitchell A (1939) Chapters in the history of terrestrial magnetism III. *Terrestrial Magnetism and Atmospheric Electricity* 44: 77–80.

Davidson P (2001) *An Introduction to Magnetohydrodynamics*. Cambridge: Cambridge University Press.

De Boor C (2002) *A Practical Guide to Splines*. New York: Springer.

Doell RR and Cox A (1971) Pacific Geomagnetic Secular Variation. *Science* 171: 248–254.

Drew S (1993) Magnetic field line expulsion into a conducting mantle. *Geophysical Journal International* 115: 303–312.

Dumberry M and Bloxham J (2006) Azimuthal flows in the Earth's core and changes in the length of day on millennial timescales. *Geophysical Journal International* 165: 32–46.

Finlay CC (2005) *Hydromagnetic Waves in Earth's Core and Their Influence on Geomagnetic Secular Variation*. Unpublished PhD Thesis. University of Leeds.

Finlay CC and Jackson A (2003) Equatorially dominated magnetic field change at the surface of Earth's core. *Science* 300: 2084–2086.

Fisk HW (1931) Isopors and isoporic movements. *Intern. Geodet. Geophys. Union, Section Terrest. Magnet. Elec. Bull. Stockholm* 8: 280–292.

Gauss CF (1833a) *Intensitas vis magneticae terrestris ad mensuram absolutam revocata*. Gottingae: Sumtibus Dieterichianis.

Gauss CF (1833b) Intensitas vis magneticae terrestris ad mensuram absolutam revocata. *Abstracts of the Papers Printed in the Philosophical Transactions of the Royal Society of London* 3: 166–174.

Gauss CF (1839) Allgemeine Theorie des Erdmagnetismus: Resultate aus den Beobachtungen des magnetischen Verein im Jahre 1838. Leipzig.

Gilbert W (1600) *De magnete*. London: P. Short (English translation, Chiswick Press, London, 1900).

Gill AE (1982) *Atmosphere-Ocean dynamics*. Cambridge: Cambridge University Press.

Glatzmaier GA (2002) How realistic and geodynamo simulations? *Annual Review of Earth and Planetary Sciences* 30: 237–257.

Good GA (1985) Geomagnetics and scientific institutions in 19th century America. *Transactions of the American Geophysical Union* 66: 521–526.

Good GA (1988) The study of geomagnetism in the late 19th century. *Transactions of the American Geophysical Union* 69: 218–232.

Gubbins D (1975) Can the Earth's magnetic field be sustained by core oscillations? *Geophysical Research Letters* 2(9): 409–512.

Gubbins D (1983) Geomagnetic field analysis – I. Stochastic inversion. *Geophysical Journal of the Royal Astronomical Society* 73: 641–652.

Gubbins D (1984) Geomagnetic field analysis – II. Secular variation consistent with a perfectly conducting core. *Geophysical Journal of the Royal Astronomical Society* 77: 753–766.

Gubbins D (1987) Mechanism for geomagnetic polarity reversals. *Nature* 326: 167–169.

Gubbins D (1991) Dynamics of the Secular Variation. *Physics of the Earth and Planetary Interiors* 68: 170–182.

Gubbins D (1996) A formalism for the inversion of geomagnetic data for core motions with diffusion. *Physics of the Earth and Planetary Interiors* 98: 193–206.

Gubbins D (1999) The distinction between geomagnetic excursions and reversals. *Geophysical Journal International* 173: F1–F3.

Gubbins D and Bloxham J (1985) Geomagnetic field analysis – III. Magnetic fields on the core-mantle boundary. *Geophysical Journal of the Royal Astronomical Society* 80: 695–713.

Gubbins D and Bloxham J (1987) Morphology of the geomagnetic field and implications for the geodynamo. *Nature* 325: 509–511.

Gubbins D, Jones AL, and Finlay CC (2006) Fall in Earth's magnetic field is erratic. *Science* 312: 900–902.

Gubbins D and Kelly P (1996) A difficulty with using the frozen flux hypothesis to find steady core motions. *Geophysical Research Letters* 23: 1825–1828.

Gubbins D and Roberts PH (1987) Magnetohydrodynamics of the Earth's core. In: Jacobs JA (ed.) *Geomagnetism*, Volume II, pp. 1–183. London: Academic Press.

Halley E (1683) A theory of the variation of the magnetic compass. *Philosophical Transactions of the Royal Society of London* 13: 208–221.

Halley E (1692) On the cause of the change in the variation of the magnetic needle, with a hypothesis of the structure of the internal parts of the Earth. *Philosophical Transactions of the Royal Society of London* 17: 470–478.

Hendricks SJ and Cain JC (1963) World magnetic survey data. *NASA Goddard Space Flight Centre, technical report, August 1963*.

Hide R and Malin SRC (1981) On the determination of the size of the Earth's core from observations of geomagnetic secular variation. *Proceedings of the Royal Society of London Series A - Mathematical Physical and Engineering Sciences* 374: 15–33.

Hide R and Stewartson K (1973) Hydromagnetic oscillations of the Earth's core. *Reviews of Geophysics* 10(2): 579–598.

Hills RG (1979) Convection in the Earth's mantle due to viscous shear at the core–mantle interface and due to large-scale buoyancy, *PhD Thesis*, New Mexico State University, Las Cruces, New Mexico.

Holme R and Olsen N (2006) Core-surface flow modelling from high resolution secular variation. *Geophysical Journal International* 166: 518–528.

Hulot G and Chulliat A (2003) On the possibility of quantifying diffusion and horizontal Lorentz forces at the earth's core surface. *Physics of the Earth and Planetary Interiors* 135(1): 47–54.

Hulot G, Le Huy M, and Le Mouël J-L (1993) Secousses (jerks) de la variation séculaire et mouvements dans le noyau terrestre. *Comples Rendus de l'Académie des Sciences Paris* 317: 333–341.

Hulot G and Le Mouël JL (1994) A statistical approach to earth's main magnetic field. *Physics of the Earth and Planetary Interiors* 82: 167–183.

Hutcheson KA (1990) *Geomagnetic Field Modelling*. Unpublished Ph.D. Thesis, University of Cambridge.

Hutcheson KA and Gubbins D (1990) Earth's magnetic field in the seventeenth century. *Journal of Geophysical Research* 95: 10769–10781.

Jackson A (1989). *The Earth's Magnetic Field at the Core–Mantle Boundary*. Unpublished Ph.D. Thesis, University of Cambridge.

Jackson A (1996) Kelvin's theorem applied to earth's core. *Proceedings of the Royal Society of London* 452: 2195–2201.

Jackson A (1997) Time-dependency of tangentially geostrophic core surface motions. *Physics of the Earth and Planetary Interiors* 103: 293–311.

Jackson A (2003) Intense equatorial flux spots on the surface of Earth's core. *Nature* 424: 760–763.

Jackson A, Constable CG, Walker MR, and Parker R (2007) Models of Earth's main magnetic field incorporating flux and radial vorticity constraints. *Geophysical Journal International* (in press).

Jackson A and Hide R (1996) Invariants in toroidal and tangentially geostrophic frozen-flux velocity fields. *Geophysical Journal International* 125: 925–927.

Jackson A, Jonkers ART, and Walker MR (2000) Four centuries of geomagnetic secular variation from historical records. *Philosophical Transactions of the Royal Society of London, Series A* 358: 957–990.

Jackson JD (1999) *Classical Electrodynamics, 3rd edn.* New York: Wiley.

Jault D, Gire C, and Le Mouël JL (1988) Westward drift, core motions and exchanges of angular momentum between core and mantle. *Nature* 333: 353–356.

Jault D and Le Mouël JL (1991) Physical properties at the top of the core and core surface motions. *Physics of the Earth and Planetary Interiors* 68: 76–84.

John II's commission (1509) *Regimento do astrolabe e do quadrante,* Lisbon.

Jonkers ART (2000) *North by Northwest. Seafaring, Science, and the Earth's Magnetic Field (1600–1800).* Göttingen: Cuvillier Verlag.

Jonkers ART (2003) *Earth's Magnetism in the Age of Sail.* Baltimore, MD: Johns Hopkins University Press.

Jonkers ART, Jackson A, and Murray A (2003) Four centuries of geomagnetic data from historical records. *Reviews of Geophysics* 41: 1006 (doi:10.1029/2002RG000115).

Kircher A (1641) *Magnes sive de Arte Magnetica (. . .),* Rome.

Kono M and Roberts PH (2002) Recent geodynamo simulations and observations of the geomagnetic field. *Reviews of Geophysics* 40: 1013 (doi:10.1029/2000RG00102).

Korte M, Genevey A, Constable C, Frank U, and Schnepp E (2005) Continuous geomagnetic field models for the 7 millennia I: A new global data compilation. *Geochemistry, Geophysics, Geosystems* 6(2): Q02H15 (doi:10.1029/2004GC000800).

Kuang W and Bloxham J (1997) An Earth-like numerical dynamo model. *Nature* 389: 371–374.

Langel RA (1987) The Main Field. In: Jacobs JA (ed.) *Geomagnetism, Volume I,* pp. 249–512. London: Academic Press.

Langel RA (1993) *Types and characteristics of data for geomagnetic field modeling,* NASA Conference Publication 3153. Washington, DC: NASA.

Langel RA, Estes RH, and Mead GD (1982) Some new methods in geomagnetic field modelling applied to the 1960–1980 epoch. *Journal of Geomagnetism and Geoelectricity* 34: 327–349.

Langel RA, Kerridge DJ, Barraclough DR, and Malin SRC (1986) Geomagnetic temporal change: 1903–1982, A spline representation. *Journal of Geomagnetism and Geoelectricity* 38: 573–597.

Lanzerotti LJ, Chave AD, Sayres CH, Medford LV, and Maclennan CG (1993) Large-scale electric-field measurements on the Earth's surface – A review. *Journal of Geophysical Research-Planets* 98(E12): 23525–23534.

Le Mouël JL (1984) Outer core geostrophic flow and secular variation of Earth's magnetic field. *Nature* 311: 734–735.

Love JJ (1999) A critique of frozen-flux inverse modelling of a nearly steady geodynamo. *Geophysical Journal International* 138: 353–365.

Macmillan S (1996) A geomagnetic jerk for the early 1990's. *Earth and Planetary Science Letters* 137: 189–192.

Malin SRC (1982) Sesquicentenary of Gauss's first measurement of the absolute value of magnetic intensity. *Philosophical Transactions of the Royal Society A* 306: 5–8.

Malin SRC (1987) Historical introduction to geomagnetism. In: Jacobs JA (ed.) *Geomagnetism,* Volume I, pp. 1–49. London: Academic Press.

Malin SRC and Bullard EC (1981) The direction of the Earth's magnetic field at London, 1570–1975. *Philosophical Transactions of the Royal Society of London, Series A* 299: 357–423.

Mandea M, Bellanger E, and LeMouel J-L (2000) A geomagnetic jerk for the end of the 20th century. *Earth and Planetary Science Letters* 183: 369–373.

Maus S, Lühr H, Balasis G, Rother M, and Mandea M (2005) Introducing POMME the postdam magnetic model of the Earth. In: Reigber C, Luhr H, and Schwintzer P (eds.) *CHAMP Mission Results II,* pp. 347–352. Berlin: Springer.

McConnell A (1980) *Geomagnetic instruments before 1900; an illustrated account of their construction and use.* London: Harriet Wynter.

Merrill RT, McElhinny MW, and McFadden PL (1998) *The Magnetic Field of the Earth.* London: Academic Press.

Mountaine W and Dodson J (1757) A letter to the Right Honourable the Earl of Macclesfield, President of the Council and Fellow of the Royal Society, concerning the variation of the magnetic needle; with a set of tables annexed, which exhibit the results of upwards of fifty thousand observations, in six periodic reviews, from the year 1700 to the year 1756, both inclusive; and are adapted to every 5 degrees of latitude and longitude in the more frequented oceans. *Philosophical Transactions of the Royal Society of London* 50: 329–350.

Multhauf RP and Good G (1987) *Smithsonian Studies in History and Technology,* Volume 48: *A brief history of geomagnetism and a catalogue of the collections of the National Museum of American History,* 87 pp. Washington: Smithsonian Institution Press.

Nagao H, Iyemori T, Higuchi T, and Araki T (2003) Lower mantle conductivity anomalies estimated from geomagnetic jerks. *Journal of Geophysical Research* 108: (doi 10.1029/2002JB001786).

Needham J (1962) *Science and civilisation in China,* vol. 4, part 1. Cambridge: Cambridge University Press.

O'Brien MS (1996) Resolving magnetic flux patches at the surface of the core. *Geophysical Research Letters* 23: 3071–3074.

O'Brien MS, Constable CG, and Parker RL (1997) Frozen flux modelling for epochs 1915 and 1980. *Geophysical Journal International* 128: 434–450.

Olsen N (2002) A model of the geomagnetic field and its secular variation estimated for epoch 2000. *Geophysical Journal International* 149: 454–462.

Olsen N, Lühr H, Sabaka TJ, *et al.* (2006) CHAOS – A model of Earth's magnetic field derived from CHAMP, Ørsted and SAC-C magnetic satellite data. *Geophysical Journal International* 166: 67–75.

Olson P and Christensen UR (2002) The time-averaged magnetic field in numerical dynamos with nonuniform boundary heat flow. *Geophysical Journal International* 151: 809–823.

Peregrinus P (1269) *Epistola de magnete.*

Rau S, Christensen U, Jackson A, and Wicht J (2000) Core flow inversion tested with numerical dynamo models. *Geophysical Journal International* 141: 484–497.

Raulin V (1867) Sur les variations seculaires du magnetisme terrestre. *Actes de la Société Linnéenne de Bordeaux* 26: 1–92.

Rayet MG (1876) Recherches sur les observations magnétiques faites à l'observatoire de Paris de 1667 à 1872. *Annales de l'Observatoire de Paris Memoires* 13.

Roberts PH and Glatzmaier GA (2000) A test of the frozen flux approximation using a new geodynamo model. *Philosophical Transactions of the Royal Society of London, Series A* 358: 1109–1121.

Roberts PH and Scott S (1965) On the analysis of the secular variation. 1. A hydromagnetic constraint: theory. *Journal of Geomagnetism and Geoelectricity* 17(2): 137–151.

Runcorn SK (1955) The electrical conductivity of the Earth's mantle. *Transactions of the American Geophysical Union* 36: 191–198.

Sabaka TJ and Baldwin RT (1993) Modeling the Sq magnetic field from POGO and Magsat satellite and contemporaneous hourly observatory data. *HSTX/G&G-9302, Hughes STX Corp., 7701 Greenbelt Road, Greenbelt, MD.*

Sabaka TJ, Langel RA, Baldwin RT, et al. (1997) The geomagnetic field 1900–1995, including the large-scale field from magnetospheric sources, and the NASA candidate models for the 1995 revision of the igrf. *Journal of Geomagnetism and Geoelectricity* 49: 157–206.

Sabaka TJ, Olsen N, and Langel RA (2002) A comprehensive model of the quiet-time, near-Earth magnetic field: Phase 3. *Geophysical Journal International* 151: 32–68.

Sabaka TJ, Olsen N, and Purucker ME (2004) Extending comprehensive models of the Earth's magnetic field with Oersted and CHAMP data. *Geophysical Journal International* 159: 521–547.

Shimizu H, Koyama T, and Utada H (1998) An observational constraint on the strength of the toroidal magnetic field at the cmb by time variation of submarine cable voltages. *Geophysical Research Letters* 25(21): 4023–4026.

Shimizu H and Utada H (2004) The feasibility of using decadal changes in the geoelectric field to probe earth's core. *Physics of the Earth and Planetary Interiors* 142(3–4): 297–319.

Shure L, Parker RL, and Backus GE (1982) Harmonic splines for geomagnetic modelling. *Physics of the Earth and Planetary Interiors* 28: 215–229.

Smith PJ and Needham J (1967) Magnetic declination in medieval China. *Nature* 214: 1213–1214.

Stern DP (2002) A millennium of geomagnetism. *Reviews of Geophysics* 40: 1007 (doi:10.1029/2000RG000097).

Stevin S (1599) *De Havenvinding*. Leyden: Plantijn.

Veinberg BP (1929–33) *Catalogue of Magnetic Determinations in USSR and in Adjacent Countries from 1556 to 1926, 3 vols.* Leningrad: *Central Geophysical Observatory.*

Voorhies CV and Benton ER (1982) Pole-strength from MAGSAT and the magnetic determination of the core radius. *Geophysical Research Letters* 9: 258–261.

Whaler K and Gubbins D (1981) Spherical harmonic analysis of the geomagnetic field: An example of a linear inverse problem. *Geophysical Journal of the Royal Astronomical Society* 65: 645–693.

Wicht J and Olson P (2004) A detailed study of the polarity reversal mechanism in a numerical dynamo model. *Geochemistry, Geophysics, Geosystems* 5: Q03H10 (doi:10.1029/2003GC000602).

Wright E (1657) *Certaine errors of navigation detected and corrected, with many additions that were not in the former editions. 3rd edn.* London: Joseph Moxton.

Yukutake T (1962) The westward drift of the magnetic field of the Earth. *Bulletin of the Earthquake Research Institute, University of Tokyo* 40: 1–65.

Yukutake T and Tachinaka H (1969) Separation of the Earth's magnetic field into the drifting and standing parts. *Bulletin of the Earthquake Research Institute, University of Tokyo* 47: 65–97.

6 Crustal Magnetism

M. E. Purucker, Goddard Space Flight Center/NASA, Greenbelt, MD, USA

K. A. Whaler, University of Edinburgh, Edinburgh, UK

Nomenclature

a	radius of Earth
d	magnetic source–observation distance
g_n^m, h_n^m	spherical harmonic coefficients
s_i	structural index or attenuation rate
t	time
v	volume
v_f	volume fraction of magnetite
A	analytic signal
B	magnetic induction
F	non-crustal magnetic field
G	Green's function
H	magnetic field intensity
M	magnetization
M_i	induced magnetization
M_r	remanent magnetization
Ma	million years ago
P	pressure
$P_n^m (\cos \theta)$	Schmidt quasi-normalized associated Legendre functions of degree n and order m
Q	Koenigsberger ratio
T	total field
V	scalar potential
θ	colatitude
μ_0	permeability of free space
τ	temperature
ϕ	longitude
χ	magnetic susceptibility
Γ	Gram matrix
ΔT	total field anomaly

Glossary

analytic signal A transformation formed through the combination of the horizontal and vertical gradients of a magnetic anomaly.

CHAMP A low-earth orbiting satellite launched in 2001 designed to map the vector crustal magnetic field.

crust (lithosphere) The outer shell of the Earth formed by differentiation under the influence of temperature and gravity. The crust overlies the mantle, and the distinction can be expressed in terms of seismic velocities, rock density, rock type, mineralogy, chemical composition, or magnetic properties. The lithosphere, a rheological term, includes the crust, and the uppermost mantle.

Curie depth The depth at which rocks lose their permanent and induced magnetism by virtue of their elevated temperature. The Curie depth is a function of the geothermal gradient within the Earth, and the magnetic mineralogy.

Curie temperature The temperature at which a rock loses its permanent and induced magnetization.

Euler deconvolution method A technique for estimating source positions for magnetic anomalies, which relates the magnetic field to its gradient through the specification of the arrangement of the magnetic sources.

Green's function The relation of magnetization to magnetostatic potential.

harmonic spline A local basis function employed in modeling the magnetic field.

International Geomagnetic Reference Field (IGRF) Values for the spherical harmonic coefficients from which the magnetic field can be calculated at any point in space and time. Each IGRF consists of a set of main field and secular variation coefficients covering a 5-year interval, thereby accounting for the temporal evolution of the main field.

induced magnetization One of two types of magnetization, the other being remanent. Induced magnetizations are proportional in magnitude and generally parallel to the ambient field **H**.

Koenigsberger ratio, Q, or Q-ratio Expresses the relation between the strength of the induced and remanent magnetizations. It is given by $|M_r| \div |M_i|$. Hence, Q's greater than unity indicate dominance by remanent magnetization; Q's less than unity indicate dominance by induced magnetization.

magnetic basement The top of a layer of more strongly magnetized rocks, usually igneous and/or metamorphic rocks, underlying more weakly magnetized sediments.

magnetic susceptibility Expresses the proportionality factor relating ambient field **H** to the induced magnetization.

magnetite Dominant magnetic mineral, often with some Ti, in the Earth's crust.

MAGSAT The first satellite mission to map the Earth's vector magnetic field, including the long-wavelength crustal magnetic field.

matched filter Filter(s) that can be used to decompose observed magnetic anomalies into estimates of the anomalies caused by sources at various depths. These depths are determined by a Fourier domain decomposition of the magnetic anomaly signal.

observatory bias The difference between the magnetic components measured at a magnetic observatory and those predicted by a geomagnetic model truncated at degree 13. This quantity is thought to reflect the higher-degree crustal magnetic field contribution, in part.

pseudogravity transformation Converts a magnetic anomaly into the gravity anomaly that would be caused by a density distribution exactly proportional to the magnetization distribution.

remanent magnetization One of two types of magnetization, the other being induced. The direction and intensity of a remanent magnetization is dependent on the origin and history of a rock.

reduction to pole (RTP) Transforms magnetic anomalies into the anomalies that would be caused by identical magnetic sources but with vertical magnetization and with measurement in a vertical magnetic field.

structural index (s_i) Expresses the rate of attenuation with distance of a magnetic anomaly. This attenuation is a function of the source geometry.

total field anomaly The signed scalar quantity, which is the most common measure of the Earth's crustal magnetic field.

upward (downward) continuation Transforms observed anomalies into the anomalies that would be observed at a higher (lower) altitude.

Werner deconvolution A depth to basement technique that assumes the magnetic body has a specific geometry (dike-like), and solves for the depth to the top of the body based on four or more observations of the magnetic field over the body.

6.1 Introduction

6.1.1 Definition

Crustal magnetism is defined as magnetism originating from rocks below their Curie temperature, in the Earth's crust and uppermost mantle. The dominant magnetism is associated with igneous and metamorphic rocks, whereas sedimentary rocks generally have subordinate, but measurable magnetism. The magnetism of the ferri- and ferromagnetic materials is a function of temperature with a loss of magnetism as the materials approach their Curie temperature (typically 200–700°C). The increase of temperature with depth in the Earth means that rocks below a certain depth, termed the Curie depth, will be non-magnetic. This depth is typically in excess of 20 km in stable continental regions.

6.1.2 Measurement

The measurement of crustal magnetism is done utilizing total and vector field magnetometers, and associated gradiometers. The total field magnetometers exploit fundamental resonances (Primdahl, 2000) to measure the magnitude of the field, whereas the vector instruments typically utilize fluxgate magnetometers (Ripka, 2000). The magnetometers measure these fields from borehole, ground-based, marine, aerial, balloon, or satellite platforms.

6.1.3 Governing Equations

The magnetic induction $B^{(\eta)}(\mathbf{r}_j)$ of the η component of the magnetic field due to a magnetization distribution $\mathbf{M}$ is given by

$$B^{(\eta)}(\mathbf{r}_j) = -\hat{\mathbf{i}}_j^{(\eta)} \cdot \nabla_{\mathbf{r}_j} \int_v \left\{ \frac{\mu_0}{4\pi} \nabla_s \frac{1}{|\mathbf{r}_j - \mathbf{s}|} \right\} \cdot \mathbf{M}(\mathbf{s}) \mathrm{d}v \quad [1]$$

where $\hat{\mathbf{i}}_j^{(\eta)}$ is the unit vector in the η direction, v is the volume of the magnetized crust, and the quantity in brackets is the Green's function relating magnetization $\mathbf{M}$ to magnetostatic potential V. The subscript on the gradient (∇) operator indicates whether derivatives are with respect to observation point coordinates ($\mathbf{r}_j$) or locations within the magnetized crust ($\mathbf{s}$). The magnetization $\mathbf{M}$ is the vector sum of remanent magnetization $\mathbf{M}_r$ and induced magnetization $\mathbf{M}_i$:

$$\mathbf{M} = \mathbf{M}_i + \mathbf{M}_r \quad [2]$$

In the case of terrestrial crustal magnetic field observations, what is often measured is the total field, the magnitude of the total magnetic field without regard to its vector direction. The total field anomaly (ΔT) is then

$$\Delta T = |\mathbf{T}| - |\mathbf{F}| \qquad [3]$$

where $|\mathbf{T}|$ is the magnitude of the magnetic field, and $|\mathbf{F}|$ is the magnitude of the (largely) noncrustal field, determined from a global or regional model. If vector data are available, the total field anomaly is calculated as

$$\Delta T = \hat{\mathbf{F}} \cdot \mathbf{T} \qquad [4]$$

where $\hat{\mathbf{F}}$ is the unit vector in the direction of $\mathbf{F}$. The geometry of a total field anomaly of a magnetic body dominated by induced magnetization is dependent on the geometry of the inducing field. At high latitudes an induced magnetization will give a total field anomaly high (positive) over the source, whereas at low latitudes an induced magnetization will yield a total field anomaly low (negative) over the source.

6.1.4　Previous Reviews

Book-length reviews include those of Langel and Hinze (1998), Blakely (1995), Lindsley (1991), Hahn and Bosum (1986), and Grant and West (1965). Shorter articles, within books or encyclopedias, have included Gubbins and Hererro-Bervera (2007), Shive *et al.* (1992), Frost (1991a, 1991b, 1991c), Reynolds *et al.* (1990c), Blakely and Connard (1989), Harrison (1987), Bosum *et al.* (1985), Haggerty (1976), and Zietz and Andreasen (1967). Reviews in journals include those of Mandea and Purucker (2005), Nabighian *et al.* (2005), Nabighian and Asten (2002), Clark (1997, 1999), Phillips *et al.* (1991), Keller (1988), Mayhew and LaBrecque (1987), Paterson and Reeves (1985), Grant (1985), Mayhew *et al.* (1985), Haggerty (1979), Hinze (1979), and Zietz and Bhattacharyya (1975). Bibliographies include Langel and Benson (1987), Hill (1986), and Reid (in preparation).

6.1.5　Computer Software and Online Applications

Computer software for crustal magnetic field applications includes a collection of Fortran subroutines in Appendix B of Blakely (1995) and online at http://pangea.stanford.edu/~blakely/subroutines.html. Another software resource is the potential field software programs of the US. Geological Survey (Phillips, 1997), which can also be found online at http://

pubs.usgs.gov/fs/fs-0076-95/FS076-95.html. Matlab, C, and Fortran routines for the evaluation of spherical harmonic models are described in Olsen *et al.* (2006a) and can be found online at several sites, including http://www.dnsc.dk/Oested/Field_models, and from the National Geophysical Data Center at http://www.ngdc.noaa.gov/seg/geomag/models.shtml. Programs for high-degree spherical harmonic analysis (SHA) and synthesis (Adam and Swarztrauber, 1997) are associated with Spherepack, available at http://www.cisl.ucar.edu/css/software/spherepack. Occasionally, the journal *Computers and Geosciences* includes articles of relevance. The 'Numerical Recipes' C and Fortran books (Press *et al.*, 1992, 1996, 1997) are another resource for inverse codes, sparse matrix theory, wavelets, interpolation, and Fourier and spectral applications. The publicly available generic mapping package (GMT) is useful for both producing maps, and for analysis of potential field data. It is documented in Wessel and Smith (1998) and available online from http://gmt.soest.hawaii.edu. Commonly used commercial codes include Geosoft, Matlab, and IDL.

Online applications include those for the evaluation of the International Geomagnetic Reference Field (Maus and Macmillan, 2005), available from the NGDC website at the address above. Along these same lines, there is also available an application for the evaluation of the CM4 Comprehensive Model (Sabaka *et al.*, 2004) at http://planetary-mag.net. Finally, the Atlas of Structural Geophysics (Jessel, 2001) can be found online at http://www.mssu.edu/seg-vm/exhibits/structuralatlas.

6.1.6　Structure of the Remainder of the Chapter

The remainder of this chapter begins with a summary of the salient points of magnetic petrology. We then outline the utility of crustal magnetism through a series of case studies, and discuss compilations to produce models at continental or larger scale. This is followed by details of the processing, transformation, and modeling methods that are applied to crustal magnetic data to facilitate interpretation. The issue of the separation of the various contributions to the measured magnetic field is then addressed, and we conclude with one of the key outstanding questions, identifying the induced and remanent components of magnetization.

6.2 Magnetic Petrology

An understanding of the processes that create, alter, and destroy magnetic minerals in rocks is the province of magnetic petrology (Lindsley, 1991; Clark, 1997). This field integrates rock magnetism (*see* Chapter 8) and petrology to address questions such as the effects of metamorphism, hydrothermal alteration, rock composition, and redox state on magnetic properties. Magnetic minerals of major importance to an understanding of crustal magnetism are the Fe–Ti spinel group (magnetite and titanomagnetite), the rhombohedral titanohematites, and monoclinic pyrrhotite. These minerals can possess remanent (permanent) or induced (in response to an inducing field) magnetizations. Induced magnetization is, to first order, proportional to, and parallel to the direction of, the inducing field. The proportionality constant χ is called the magnetic susceptibility, and the governing relationship is of the form $|\mathbf{M}| = \chi|\mathbf{H}|$. Magnetic susceptibility in many rocks is strongly controlled by their magnetite content and the empirically determined relationship (Shive *et al.*, 1992) is

$$\chi \approx 0.2 \times 4\pi v_f \qquad [5]$$

where χ is the susceptibility and v_f is the volume fraction of magnetite. Many authors use k for susceptibility.

Magnetic remanence, on the other hand, while also correlated with titanomagnetite content, is strongly dependent on the grain size, shape, and microstructure of the magnetic minerals. The Koenigsberger ratio (Q) measures the relative strengths of the induced and remanent magnetizations. It is given by $|\mathbf{M}_r|/|\mathbf{M}_i|$. Hence, Q's greater than unity indicate dominance by remanent magnetization; Q-values of less than unity indicate dominance by induced magnetization. Representative tables and values of susceptibility and Q can be found in Clark (1997).

The magnetic properties of igneous and metamorphic rocks are a reflection of the partitioning of iron between oxide and silicate phases, and do not correspond to standard petrologic classifications. This partitioning occurs in the near-surface realm (Clark, 1997) and probably also within the deep lithosphere (cf. Wasilewski and Mayhew, 1992). Standard sedimentary rock classifications, on the other hand, do show a correspondence with magnetic properties. Fe-rich chemical sediments (e.g., banded iron formations) and immature clastic sediments with abundant magnetite are two strongly magnetic sedimentary rock types, for example. Iron sulfide minerals, possibly associated with hydrocarbon migration or abiologic processes, may also produce subtle magnetic anomalies over sedimentary basins (Reynolds *et al.*, 1991, 1994), but the active processes are still controversial and an active area of research (Stone *et al.*, 2004).

Igneous and metamorphic rock types (e.g., granodiorite, rhyolite, and gabbro) often exhibit bimodal susceptibility distributions, a reflection of ferromagnetic and paramagnetic populations (cf. Figure 7 in Clark, 1999). This was first recognized as a consequence of the very large petrophysical sampling program conducted on the Fennoscandian shield (cf. Korhonen, 1993). Iron in the paramagnetic population is incorporated into silicate phases, whereas iron in the ferromagnetic population is typically in magnetite.

The magnetic petrology of granitic rocks provides an example of this bimodal distribution, with the relatively oxidized, magnetite-rich, I-type granitoids contrasting with the relatively reduced, ilmenite-rich, S-type granitoids (Clark, 1999). These granitoid types can often be distinguished by the presence of common minor minerals. Hornblende–biotite granodiorites are usually ferromagnetic, whereas muscovite–biotite granodiorites are not (Clark, 1997). Economic mineralization (Cu, Au, Mo, Sn) also shows patterns (Ishihara, 1981) that are controlled in part by this classification.

Although the rule of thumb that basic rocks are more magnetic than silicic rocks is often violated, rocks from within a single igneous province are more likely to show this tendency than are larger population samples. Hence, interpretation of magnetic surveys should include investigation of the magnetic properties of representative rock samples when possible. Within-province generalizations also find that basalts have slightly higher susceptibilities than related andesites, but phonolites are weakly magnetic. Rhyolites also exhibit a bimodal susceptibility distribution. Rhyolites which are under- or oversaturated with respect to alumina, or which contain iron-rich olivine, are likely to be weakly magnetic (Clark, 1999).

Rapidly chilled basaltic rocks are characterized by high Q-values, and the Q-ratio is strongly correlated to the distance from the chilled margin. As long as the primary remanent magnetization has not been chemically or thermally modified, even relatively thick sills and dikes have high Q-values.

Hydrothermal alteration generally destroys magnetite, and replaces it with paramagnetic phases like zeolites, clays, or more weakly magnetic minerals like titanohematite (Criss and Champion, 1984). One major exception to this generalization is that serpentinization of olivine-rich ultramafic rocks produces abundant magnetite with low Q-values (Saad, 1969). Other notable exceptions include potassic alteration associated with magnetic felsic-intermediate intrusives (Sexton et al., 1995) and potassic and sodic alteration in deeper levels of iron-oxide copper–gold systems (Hitzman et al., 1992). Production of hydrothermal magnetite is enhanced in mafic protoliths.

Metamorphism can produce marked changes in magnetic properties, and these changes are dependent on the composition of the protolith, and the pressure (P), temperature (τ), and time (t) path of the metamorphism. For mafic igneous protoliths undergoing regional metamorphism, primary magnetite remains unchanged during zeolite to prehnite–pumpellyite grade metamorphism in the absence of hydrothermal fluids. Subsequent metamorphism to greenschist grade converts the magnetite to chlorite, epidote, and hematite. In turn, these minerals give way to biotite and amphibole in the amphibolite facies of regional metamorphism. Magnetite is again created during granulite-grade metamorphism. At the highest metamorphic grade (eclogite), the iron returns to silicates such as clinopyroxene and garnet. For sedimentary protoliths, the redox conditions prevailing during sedimentation and diagenesis, and the iron content of the protolith, constrain the mineral assemblage produced during metamorphism.

The magnetic state of the lower crust remains poorly known. Although P and τ can be predicted, the protolith's history and current compositions are the subject of speculation. Because of the lower crust's elevated temperature, induced and viscous remanent magnetizations are expected to be strong (Shive et al., 1992). Inferences from deep drilling and seismic constraints suggest a generally mafic composition. A host of mineralogic and magnetic changes may occur, with maximum magnetizations in the granulite facies zone. Stable large remanence in ilmenite–hematite intergrowths (McEnroe et al., 2001a; McEnroe and Brown, 2000) within granulite-facies rocks provides another mechanism for producing magnetic rocks within the lower crust.

Two mechanisms have been suggested for large-scale magnetizations within the mantle. The conversion of metabasalt to eclogite within subducting oceanic crust releases large amounts of water into the surrounding upper-mantle peridotite, and may produce serpentinite (Hyndman and Peacock, 2003). As long as the mantle wedge in the subduction zone is cooler than the Curie temperature, it is possible that a significant magnetization may form. Blakely et al. (2005) have explained the long-wavelength aeromagnetic and gravity fields above the Cascadia forearc as an example of this process, using matched filters (Section 6.5.5.2) to establish the depth of the source, and a pseudogravity transform (Section 6.5.4.3) to center the magnetic fields over their source. Satellite magnetic anomalies over subduction zones are also common (cf. Clark et al., 1985; Vasicek et al., 1988, Purucker and Ishihara, 2005; Maus et al., 2006), and may have a similar explanation. A second mechanism for magnetizations in the mantle invokes the presence of metallic alloys, which have been detected in xenoliths originating from the upper mantle (Toft and Haggerty, 1988). Significant amounts of metal alloys in the upper mantle could impart magnetic behaviors to depths of almost 100 km. But questions remain about how representative of the upper mantle these metal alloys are (Frost and Shive, 1989; Toft and Haggerty, 1989).

While magnetic petrologic approaches have provided significant insights into the interpretation of crustal magnetism, there still remains the problem of extrapolating from field observations at micron- to hand-sample scale to scales appropriate for aeromagnetic or satellite observation. For example, even within the ferromagnetic population, the distribution of magnetization or susceptibility is usually log normal, and exhibits high variability. Parker (1991) has developed an inverse approach, which incorporates this variability into the creation and testing of a magnetization model. Some of this high variability can also be ascribed to surface processes, such as lightning (Verrier and Rochette, 2002) and weathering, which may not be observable from non-ground-based platforms.

6.3 Continental and Oceanic Magnetic Anomalies

Because magnetic oxide or sulfide-bearing phases are commonly associated with other economic mineral phases, magnetic measurements play a significant role in mineral exploration. Mapping of the crustal magnetic field is a geologic and exploration tool in the terrestrial environment, and provides a third dimension to surface observations of composition and structure. The magnetic method also contributes to plate tectonic theory, oil and gas exploration, structural geology, and geologic mapping. The generation of new seafloor at the ridge crest was established via the magnetic method. The symmetry of the magnetic patterns (Vine and Matthews, 1963) about the ridge crest is often cited as the breakthrough which led to the widespread acceptance of plate tectonics. The magnetic time scale (Heirtzler *et al.*, 1968), suitably calibrated with numerical ages, serves many purposes in the Earth sciences. In particular, readers are referred to Chapter 6 for further details of crustal magnetism within the oceanic realm. Inferences from crustal magnetic fields, interpreted in conjunction with other geological and geophysical information, can locate kimberlite pipes, impact structures, plutons, ophiolites, and other geologic entities which have a magnetic contrast with their surroundings. This permits extrapolation from, or interpolation between, outcrops, drill holes, or regions of localized geophysical measurements into areas where surficial materials may obscure the feature. Magnetic studies can locate faults, folds, and unconformities, and describe their geometrical properties. Magnetic measurements provide constraints on the amount of sediment in a depositional basin by characterizing its depth and dimensions. Magnetic measurements can be used to infer heat flux, and the depth to the bottom of the magnetic crust, because magnetic properties are temperature dependent. Finally, crustal magnetic fields can help delineate suture zones or terrane boundaries, and unravel the history of volcanic terranes.

In the sections that follow, we use a case study approach to illustrate the utility of the magnetic method. We begin with the Chicxulub impact structure, showing how it was first recognized using a combination of aeromagnetic and gravity data, and how these data sets have been used to produce three-dimensional (3-D) models of the structure. We proceed then to review geodynamical interpretations of aeromagnetic data that have been derived from dike swarms, and some of the caveats that must be considered. We then discuss structural and tectonic interpretations of aeromagnetic maps over forearc basins with Cenozoic to Recent faulting, and their role in assessing earthquake risk. We next illustrate how magnetics has been used to infer heat flux under the Antarctic ice cap, and how this may have applications in modeling ice flow, and in identifying undiscovered volcanic regions under the ice. In the

exploration arena, we summarize the role of aero- and ground magnetic surveys in identifying diamond-bearing kimberlites from northern Canada. Finally, we review the structural inferences drawn from magnetic and gravity surveys over the West Siberian Basin, and their relation to the world's largest gas field, the Urengoy. A case study approach such as this might also have included a demonstration of the utility of magnetics in determining the depth to basement in sedimentary basins, and its relevance in petroleum exploration. The proprietary nature of this kind of work means that while there are no shortage of articles discussing depth to basement techniques (e.g., Peters, 1949; Li, 2003; Thompson, 1982; Thurston *et al.*, 2002; Thurston and Smith, 1997; Hsu *et al.*, 1998; Ku and Sharp, 1983; Mushayandebvu *et al.*, 2001; 2004; Naudy 1971; Salem and Ravat, 2003; Silva *et al.*, 2001; Silva and Barbosa, 2003; Nabighian *et al.*, 2005), there is only a single volume (Gibson and Millegan, 1998) which focuses on the role of magnetics in an integrated hydrocarbon exploration program.

6.3.1 Chicxulub

Located below, and straddling the coastline of the northwest Yucatan, Mexico, the Chicxulub impact structure (**Figure 1**) is the world's most widely known impact, and produced major biologic and environmental changes at the end of the Cretaceous Period 65 Ma. The enhanced porosity associated with the collapse of nearby structures (Grieve and Therriault, 2000) from Chicxulub's associated seismic events has been linked to the development of large hydrocarbon deposits in the Campeche Bank region immediately to the NW. The impact site is now covered by up to 1 km of carbonate rock. First recognized by its circular and coincident magnetic (**Figure 1**) and gravity signatures in the aftermath of a 1978 survey (Penfield and Camargo, 1981), the impact was subsequently tied to other diagnostic signatures such as an iridium anomaly and shocked quartz grains by direct drilling into the structure, and dating of the crystallization age of the melt rocks (Hildebrand *et al.*, 1991; Sharpton *et al.*, 1992).

The magnetic signature consists of three concentric zones (Pilkington and Hildebrand, 2000) with radii of 20, 45, and 80 km. The impact occurred in a carbonate sequence several kilometers thick characterized by much longer (hundreds of kilometers) and weaker amplitude magnetic anomalies. The innermost zone is characterized by a single, high-amplitude anomaly

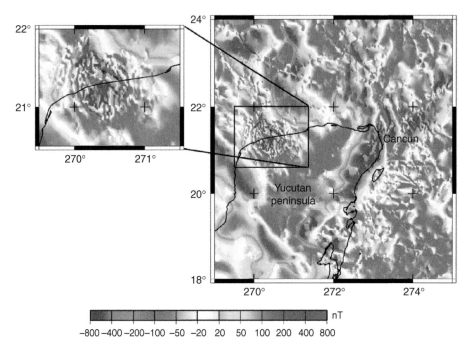

Figure 1 Total field anomaly (ΔT) over Chicxulub impact structure (Pilkington and Hildebrand, 2000), shown in an expanded view in the inset. Coastline is shown as a solid line. Data interpolated to a 1-km grid from digital data grids of the magnetic anomaly map of North America (Bankey *et al.*, 2002). Artificial illumination from the NNE and ESE. Mercator projection.

indicative of a single source. The middle zone consists of numerous, intermediate-amplitude dipolar anomalies. The outermost zone consists of short-wavelength, low-amplitude anomalies. The outermost zone is better defined by its gravity signature, and associated cenotes (freshwater caves), than by its magnetic signature. In their recent interpretation of the aeromagnetic survey data, Pilkington and Hildebrand (2000) perform 3-D modeling of the crater structure by inversion using a two-layer model. The layers, at depths corresponding to the melt sheet and the basement surface, are inverted individually subsequent to separation via a wavelength filter (see Section 6.5.5.2 for the related concept of matched filter). The inner magnetized zones within the melt sheet are interpreted to result from hydrothermal activity at the edge of the central uplift and the collapsed disruption cavity. Although some lines of evidence (Snyder *et al.*, 1999) suggest that Chicxulub may be a multi-ring impact structure, the magnetic data as currently modeled resolve only a single ring with a central peak. Although the magnetic signature of Chicxulub is distinctive, a variety of magnetic signatures are encountered in other terrestrial impact structures (Pilkington and Grieve, 1992; Grieve and Therriault, 2000; Shah *et al.*, 2005; Goussev *et al.*, 2003), dependent on the target rocks,

impact magnetizations, and subsequent evolution of these metastable assemblages. A magnetic low is frequently encountered, due to a reduction in magnetic susceptibility. Large structures such as Chicxulub tend to exhibit a central high-amplitude anomaly. Imaging techniques that emphasize the edges of magnetic bodies via derivatives, or via artificial illumination in one or more directions (Wessel and Smith, 1998) are commonly employed adjuncts to magnetic-survey interpretation of impacts. Specific extensions to impact, and other circular features (e.g., kimberlite pipes) within magnetic data, are circular sunshading as described by Cooper (2003) and Cooper and Cowan (2003), and fractional derivatives (Cowan and Cooper, 2005) for better matching to the available data.

6.3.2 Dike Swarms

The Earth hosts hundreds of radiating, arcuate or linear mafic dike swarms (Ernst *et al.*, 1996) whose mapping has contributed to improved geodynamic models of the Earth. In southern Africa alone, one digital database (Mubu, 1995) has enumerated 14 000 dikes, mapped in large part because of their magnetic expression. While some of these dikes are exposed, most are not, and hence the magnetic method has played a crucial role in their understanding. These magnetically defined dike swarms have been used in global plate reconstructions, and locally to understand the kinematics of rifting. In addition, dikes define fractures and shear zones (**Figure 2**), along which economic mineralization is often found.

Mafic dikes provide evidence for magmatic activity, large igneous provinces, and mantle plumes (Ernst and Buchan, 2001), and are especially useful in older rocks where erosion has removed much of the other evidence for igneous activity. In these older rocks it is frequently only the dikes, representing the igneous plumbing system, that survive. Although dikes are often interpreted as paleo-stress markers, they can also reflect the pre-existing structure of the lithosphere. The Jurassic dikes of southern Africa (Reeves, 2000; Chavez Gomez, 2001; Marsh, 2005), one of the manifestations of the Karoo large igneous province, have been used to enumerate plate motion associated with the breakup of Gondwana. For example, Ernst and Buchan (1997) make the case that the convergence point of these Jurassic dikes defines the location of a paleoplume. The dikes here consist of four distinct swarms: the Okavango, the Save-Limpopo, the Olifants River, and the Lebombo.

Dikes of both Jurassic and Proterozoic age have been identified within the ESE-trending Okavango dike swarm (Jourdan *et al.*, 2006), suggesting that the Jurassic events represent the reactivation of a pre-existing trend, and calling into question Jurassic kinematic reconstructions made using these dikes. Many older dike swarms are now dismembered, as in the well-documented Central Atlantic dike swarm of Africa, North and South America (May, 1971). Magnetic identification of dikes relies on simple pattern matching from contour maps generated from simple source geometries (Vacquier *et al.*, 1951). The depth to the top of dikes can be a valuable indicator of the kinematics of post-dike faulting (Modisi *et al.*, 2000). In Modisi *et al.*'s (2000) study, determination of the depth to the tops of dikes was made using Euler's homogeneity equation (see Section 6.5.4.7). Although the magnetic signature of a dike is usually easy to recognize, little attention has been directed to the important problem of magnetically recognizing dikes of common trend but dissimilar ages from within a single swarm. There are likely to be significant differences in magnetic signature, although the identification and mapping of these differences will require inputs from both field and laboratory studies.

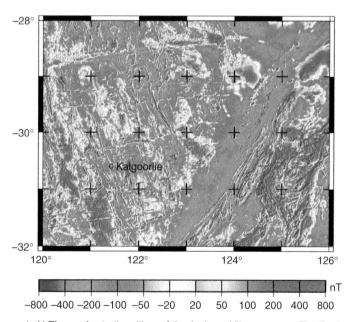

Figure 2 Total field anomaly (Δ*T*) over Australian dikes of the Archean Yilgarn craton. The E-W trending set seen here is part of the Widgiemooltha dike swarm, dated at 2410 Ma. Map based on a 1-km grid rendition of the Magnetic Anomaly Grid Database of Australia (Milligan *et al.*, 2005). Artificial illumination from the East and Southeast. The Kalgoorlie gold and precious metal district is located in the central portion of the figure. Many of the ore deposits in this district are localized along fractures and shear zones (Weinberg *et al.*, 2004). Mercator projection.

6.3.3   Cenozoic–Recent Faulting in Forearc Basins

Forearc basins around the Pacific Rim are the site of devastating earthquakes because of their proximity to large population centers. Three types of earthquakes (mega thrust contact, deep intra-slab, and shallow) are commonly encountered in these basins (Saltus et al., 2005). The faults that host earthquakes occurring along shallow crustal faults in the overriding continental plate can sometimes be located with high-resolution magnetic surveys. The Seattle fault zone, an east-trending zone of reverse faulting extending through Seattle, Washington, was the site of an M7 earthquake about 1100 years ago (Bucknam et al., 1992), and is an example of such a fault. Mapped geologically and with an aeromagnetic survey (Blakely et al., 2002), and studied along several profiles with seismic reflection surveys, this region hosts a tripartite package of rocks in close proximity to the fault zone. The package has a distinct magnetic signature, and allows the fault zone to be traced in areas of poor exposure, or where it is covered. From north to south, the package consists of a magnetic Miocene volcanic conglomerate, a thick sequence of nonmagnetic marine and fluvial rocks, and variably magnetic volcanic and sedimentary rocks of Eocene age. After accounting for remanent magnetization, the magnetic contacts were picked objectively (Blakely and Simpson, 1986). Near-surface features of this magnetic survey have also been enhanced using a matched filter approach (Syberg, 1972; Phillips, 1997; and see Section 6.5.5.2). The deformation front of the Seattle fault zone, as revealed by the seismic reflection data, lies immediately north of, and locally coincident with, the magnetic conglomerate. The aeromagnetic survey can also provide information on individual strands of the fault zone, and whether it is segmented (Blakely et al., 2002). The longer wavelength information within these aeromagnetic surveys (Finn, 1990; Blakely et al., 2005, Wells et al., 1998) can be used to provide a regional context for the tectonics of the Cascadia forearc region that hosts these basins.

6.3.4   Heat Flux beneath the Antarctic Ice Sheet

Using magnetic data to infer heat flux is possible because the magnetic properties of rocks are temperature dependent, and at the Curie temperature rocks lose their magnetization. The geothermal heat flux is an important factor in the dynamics of ice sheets, the occurrence of subglacial lakes and onset of ice streams, and may affect the mass balance. Direct heat flux measurements in ice-covered regions are difficult; thus, Fox Maule et al. (2005) developed a method using first-order features of the satellite magnetic data to estimate the heat flux underneath the Antarctic ice sheet. They found that it varies from 40 to 185 mW m^{-2}, that areas of high heat flux coincide in part with known current volcanism, subglacial lakes, and ice streams, and that some areas landward of the Ronne ice shelf near the shoulder of the West Antarctic rift system may host active, but undiscovered, subice volcanic regions.

Traditional methods for inferring heat flux, or the related magnetic problem of inferring the bottom of the magnetic crust, have relied on the shape of radially averaged spectra from gridded aeromagnetic data sets (Spector and Grant, 1970; Maus et al., 1997). To quote Blakely (1995), "this calculation ranks among the most difficult in potential field inversion." At all wavelengths, the contribution from the bottom of the magnetic source is dominated by contributions from the top. The top of the source must be also be known, in itself a difficult problem. The estimate of the bottom focuses on the lowest wave numbers, which overlap with poorly known regional fields that may be unrelated to the bottom of the magnetic bodies. There also exists a dependence on the characteristic shape of the magnetic bodies, and an assumption about the magnetization distribution. Assuming the magnetization is spatially uncorrelated ('white') is common, although magnetic susceptibility distributions are often correlated (Pilkington and Todoeschuck, 1995).

The method of Fox Maule et al. (2005) uses a self-consistent compositional and thermal model of the mantle and crust (Nataf and Ricard, 1996) as a starting point, and this model is then modified in an iterative fashion with the satellite data until the magnetic field predicted by the model matches the observed magnetic field. At the scale of the surveys used (400+ km wavelength), a unique solution is guaranteed by assuming that induced magnetizations dominate over remanent magnetizations in continental crust, and that vertical crustal thickness variations dominate over lateral susceptibility variations (Purucker and Ishihara, 2005). The resulting magnetic crustal thickness is then used as one boundary condition in a thermal model of the continental crust, assuming one-dimensional (1-D) heat conduction, and using a simple model to account for radioactive heat production in the crust. The largest complications and uncertainties in this approach are (1) uncertainties in determining the magnetic field model in the dynamic, high-latitude auroral, subauroral, and polar cap region, (2) the starting seismic and thermal model, (3) uncertainties in the upper and lower temperature boundary conditions, (4) lateral variations in thermal conductivity, and (5) lateral variations in viscous remanent magnetization.

6.3.5 Northern Canadian Kimberlite Province

Diamond-bearing kimberlites were first recognized in rocks of cratonic North America more than 150 years ago. Exploration interest focused on the Slave Craton in Canada beginning in the 1970s, and the discovery of diamond-bearing kimberlites in the early 1990s set off a mineral staking rush (Krajick, 2001). By 2004 these deposits accounted for 15% of global diamond output by value. The exploration program relied on a complementary suite of geochemical and geophysical techniques, of which the magnetic technique was one. Exploration usually proceeded from a program of indicator mineral sampling, to one of geophysical surveys in favorable regions, and finally to drilling in order to prove the deposits (Power *et al.*, 2004). Airborne total magnetic field and electromagnetic surveys, and follow-up ground surveys, were the most common geophysical surveys performed (Jansen and Witherly, 2004), although sometimes gravity, ground-penetrating radar, and seismic techniques were used. The kimberlite host rock often exhibits a positive magnetic susceptibility contrast, and a strong remanence, compared to the surrounding country rock, commonly a high-grade metamorphic rock, or granite, in the Slave Craton. Kimberlite pipes are often found in geographically localized groups, frequently under lakes because of differential erosion, and the remanence directions within those groups is often similar. Kimberlite pipes are often associated with diabase dikes (see previous section for a discussion of their magnetic signature), and are also commonly intruded along pre-existing zones of weakness (regional faults, geological contacts), many of which will have magnetic signatures. A completely preserved kimberlite pipe may be several hundred meters wide, and is often pipe- or carrot-shaped (Macnae, 1979). The resulting magnetic anomalies are usually circular in form (because the area is near the magnetic pole; see Section 6.5.4.2), and data enhancement techniques

are similar to those used for impact craters (see Section 6.3.1). The use of the analytic signal (see Section 6.5.4.6), and a pattern-recognition technique (Keating and Sailhac, 2004), has been shown to be of some use in identifying possible kimberlite target rocks.

6.3.6 Structural Control of the Urengoy Gas Field

The West Siberian Basin, one of the world's largest sedimentary basins developed on continental crust, hosts a supergiant gas accumulation in the Urengoy field (Littke *et al.*, 1999). The hydrocarbons in the Urengoy are found in an anticlinal trap defined by rejuvenated graben faults (Gibson, 1998; Grace and Hart, 1990). Aeromagnetic (Makarova, 1974) and gravity (Arctic Gravity Project, 2002) mapping (**Figure 3**) over this region reveals north–south-trending positive anomalies that are fundamentally lithologic, originating in Permo-Triassic basalt now found in rift basins. The basalt in these buried grabens is of the same age (Reichow *et al.*, 2002) as the bulk of the Siberian traps exposed further east on the Siberian platform.

The Siberian traps, part of the largest recorded terrestrial flood basalt province, are contemporaneous with the end-Permian extinction (Erwin, 1994), the largest mass extinction of the Phanerozoic, although a causal relation between the two has not yet been established (Elkins-Tanton and Bowring, 2006). The West Siberian rift basins define the base of the sedimentary column, and subsequent post-rift deposition from the Jurassic to the Cretaceous consists of fluviatile and marginal marine sediments. The boundary faults of these basins were reactivated later, and hence the magnetic and gravity anomalies serve to reveal indirectly the faults that define the hydrocarbon trap. The graben faults were rejuvenated in the Early Cretaceous, and created broad arches in the Cretaceous sediments. Maturation of Jurassic source rocks was followed by migration of hydrocarbons into traps located within the Pokur Formation of Cenomanian age.

6.4 Compilations and Models

Because maps of the crustal magnetic field are so useful for regional geologic understanding, and because magnetic surveys are usually acquired over small regions, there is a need to assemble the individual magnetic

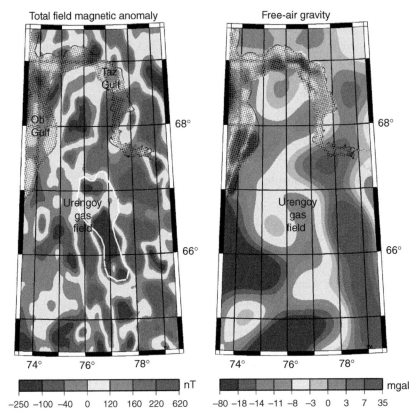

Figure 3   Aeromagnetic total field anomaly (ΔT) and free-air gravity maps of a portion of the West Siberian Basin showing the correspondence of magnetic and gravity lows with the Urengoy gas field. This coincidence is a consequence of both lithologic and structural factors (Gibson, 1998). The magnetic data are extracted from compilations of Makarova (1974), and Geol. Sur. Canada (1995), the gravity data come from the Arctic Gravity Project (2002), and the field boundaries of the Urengoy field are from Grace and Hart (1990). Lambert projection.

surveys into larger compilations. This assembly is often assisted by the addition of longer, higher-altitude magnetic surveys that serve to tie the individual surveys together, and ameliorate the discontinuities that occur at survey boundaries. Perhaps the best known of these higher-altitude surveys are a series of surveys flown in Australia (Tarlowski *et al.*, 1996) for the purpose of 'leveling' the Australian magnetic map, and the Project Magnet surveys (Coleman, 1992) of the US military. In North America, the 1970s saw the first regional compilations, followed by partial compilations of the entire continent in the 1980s, and more complete compilations by 2003. Similar scenarios have played out in Australia, the Former Soviet Union, China, South Asia, Australia, the Arctic and Antarctic, in Europe, and over the world's oceans. In contrast, Africa and South America are less advanced in terms of magnetic compilations, most of which have been led by industrial consortiums. The longest wavelengths of the crustal magnetic field can be measured from

satellites in near-Earth orbits, and beginning in the 1960s, Russian and US satellites began to measure those magnetic fields. This effort continues today as an international effort, with the CHAMP satellite, and the upcoming ESA Swarm mission. Earlier comparisons (Schnetzler *et al.*, 1985) suggested a difference in amplitude between the crustal field measured at or near the surface and from satellites when the data sets were compared at the same altitude, with the satellite amplitude lower, but the two approaches are beginning to converge (e.g., Ravat *et al.*, 2002). Upcoming satellite missions will use a gradiometer configuration to go to spherical harmonic degree 130+, and the wavelength content of near-surface surveys is being enhanced at both ends of the wavelength spectrum. There still remains a gap in our knowledge of magnetic anomalies with wavelengths from about 200 to 400 km. Only in Australia (Milligan *et al.*, 2005) is this gap partially filled. Community efforts are now focused on the development of a World Digital Magnetic Anomaly

Map (WDMAM), planned for release in 2007 (see Section 6.4.2).

In parallel with the development of compilations has been the development of larger and more elaborate models of the magnetic field, built on a deepening understanding of the sources of the magnetic field. These models utilize both forward and inverse approaches, and are frequently tested, and enhanced, using data from the compilations.

6.4.1 Continental-Scale Compilations

The first experimental airborne total field magnetometer was flown in the USSR in 1936 (Gibson, 1998) and in 1974, the Ministry of Geology of the USSR published a mosaic series of 18 sheets at 1:2.5 million scale showing the residual magnetic intensity (Makarova, 1974) over the USSR and surrounding waters. These sheets were digitized in 1982 by the US Naval Oceanographic Office, Stennis Space Center Mississippi in order to produce four regional one-arc-minute grids of magnetic anomaly values covering the entire Former Soviet Union. These digital data were provided to the National Geophysical Data Center (NGDC, 1996) for archival and public dissemination. The digitized data were made available on a 2.5-km grid.

The first continental-scale compilation, of North America (Hinze *et al.*, 1988), was completed in preliminary form as part of the Decade of North American Geology, and released by the Committee for the Magnetic Anomaly Map of North America in 1987. Consisting of the aeromagnetic surveys of Canada and the USA, and surrounding waters, the compilation effort had been preceded by compilations of the USA (Zietz, 1982) and Canada (Hood *et al.*, 1985). The North American compilation was released as a 2-km grid. The addition of aeromagnetic surveys over Mexico, and improved Canadian and US maps, led to a second-generation product (Bankey *et al.*, 2002; Hernandez *et al.*, 2001). The data grids comprising this map have a variety of wavelength content, 1-km grid spacing, and show the total field at 1 km above the terrain. They are projected using a spherical transverse Mercator with a central meridian of 100° W, base latitude of 0°, scale factor of 0.926, and Earth radius of 6 371 204 m. Wavelengths greater than 150 km are poorly represented in this compilation.

Magnetic observations of the North Atlantic and Arctic oceans, and adjacent landmasses, were compiled as part of a Geological Survey of Canada

program (Macnab *et al.*, 1995; Verhoef *et al.*, 1996, and **Figure 4**). The final data set, on a 5-km grid, was merged from three subgrids of (1) digital airborne observations, (2) digital shipborne observations, and (3) pre-existing grids or digitized maps. Only the shipborne observations showed some agreement with the satellite measurements of the crustal magnetic field, and as a consequence, all three subgrids were filtered to remove wavelengths greater than 400 km prior to merging.

European magnetic observations, from northern, western, and eastern Europe, were compiled by Wonik *et al.* (2001) on a 5-km grid at an altitude of 3 km above mean sea level (**Figure 4**). Long wavelengths were retained in this survey, although comparisons with satellite data suggest that wavelengths in excess of 300 km are poorly resolved. The map is projected using a Lambert Conformal Conic with a central meridian of 20° E, and standard parallels at 30° and 60° N.

A compilation of magnetic maps of onshore and offshore regions of China, Mongolia, and Russia with accompanying interpretation was produced by a

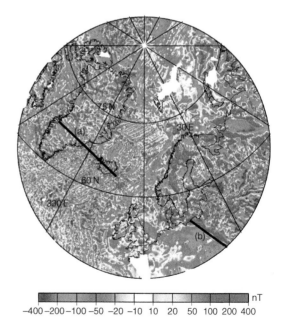

nT
−400 −200 −100 −50 −20 −10 10 20 50 100 200 400

Figure 4 Total field anomaly map (Δ*T*) in a Gnomonic projection showing parts of the merged compilations of Europe (Wonik *et al.*, 2001) and the North Atlantic/Arctic (Verhoef *et al.*, 1996) for use in the ongoing WDMAM project (Purucker and Mandea, 2005). Straight lines on Gnomonic projections are great circle arcs on the sphere. The surface trace of the Iceland hot spot (a) and the Tornquist-Teisseyre line (b) separating Precambrian from Paleozoic Europe are two examples of such features (Purucker and Whaler, 2003).

team from the Geological Survey of Canada (1995). The data were on a 5-km grid, and wavelengths in excess of 400 km have been removed from the map, which is displayed with a transverse Mercator projection.

A digital compilation of marine and aeromagnetic data over South Asia (Geol. Sur. Japan, 2002) was produced on a 2-km grid. A Lambert azimuthal equal-area projection was used with a central point at 15° N 120° E, and a terrestrial radius of 6377 km.

A digital compilation of aeromagnetic data over Australia and the surrounding oceans in now in its fourth edition (Milligan and Franklin, 2004; Milligan et al., 2005). The associated database contains publicly available airborne magnetic grid data for onshore and near-offshore Australia. Flight-line magnetic data for each survey have been optimally gridded and the grids matched in one inverse process. Composite grids at 250- and 400-m grid spacing are available. Aeromagnetic traverses flown around Australia during 1990 and 1994 are used in both quality control of the grids they intersect, and also to constrain grid merging by forcing grid data, where intersected, to the level of the traverse data. The map is displayed with a Lambert Conformal Conic Projection.

A sparse grid of aeromagnetic and marine magnetic data, supplemented by satellite magnetic coverage, is available for the Antarctic (Golynsky et al., 2001). The data set is publicly available as a 5-km grid, referenced to a polar stereographic projection.

The first compilation of onshore and offshore magnetic anomaly maps for China date from the late-1980s (Chinese National Aerogeophysics Survey and Remote Sensing Center, 1989), and has been recently (2004) updated in digital form.

Industry-led consortia have produced magnetic compilations of Africa (Barritt et al., 1993), Arabia, India, and the Middle East (Reeves and Erren, 1994), and South America (Getech, 1996).

Oceanic data sets (GEODAS, 1999) are held by the National Geophysical Data Center of NOAA. The most recent regional compilations are by Ishihara (2004) and Purucker and Ishihara (2005), where the subtraction of non-crustal magnetic field sources was done using the CM4 model of Sabaka et al. (2004) (see Section 6.5.2).

6.4.2 WDMAM Compilation

Although aeromagnetic data have been collected for almost 70 years, no worldwide compilation of them yet exists. An initiative of the International Association of Geomagnetism and Aeronomy has as its goal the production of a 5-km grid of the crustal magnetic field at an altitude of 5 km. The minimum wavelength represented by such a grid will be twice the grid spacing, or 10 km.

The unveiling is planned for the General Assembly of IUGG/IAGA at Perugia, July 2007 (Ravat et al., 2003, Reeves et al., 1998). The map will utilize airborne, marine, and satellite data to capture as many wavelengths as possible between 10 and 2200 km. It will include all freely available major digital national and regional anomaly data sets: Arctic-North Atlantic, North America, Europe, South Asia, North East Asia, Eastern Indian Ocean, Australia, and the Antarctic. It will also include lower-resolution grids extracted from the proprietary coverage of Getech (1996) for Africa and South America. Getech's web page contains maps showing aeromagnetic coverage worldwide in their holdings. Another view of the worldwide coverage can be seen in Reeves et al. (1998).

6.4.3 Satellite Compilations of Crustal Magnetic Fields

Satellite models of crustal magnetic fields are commonly spherical harmonic analyses of data gathered during magnetically quiet times, rather than the field data directly. Two current models of this type are MF-4 (Maus et al., 2006) and CM4 (Sabaka et al., 2004). The two models reflect somewhat different design philosophies, and hence have different strengths: MF-4 is an inversion of data from which estimates of other magnetic field sources have been removed, whilst CM4 solves for all sources, suitably parameterized, simultaneously. Thus MF-4 is a crustal field model only, and extends from degrees 16 to 90. The CHAMP magnetic field satellite input to MF-4 has had removed an internal field model to degree 15, an external field model of degree 2, and the predicted signatures from eight main ocean tidal components. Additional external fields are subsequently removed in a track-by-track scheme. Because of its design philosophy, the MF-4 model can be considered a minimum estimate of the crustal magnetic field, one in which there will be some suppression of along-track magnetic fields. Regularization has been applied to degrees higher than 60 to extract clusters of spherical harmonic coefficients that are well-resolved by the data.

CM4, in contrast, is a comprehensive model, that is, it includes components of internal and external origin, and toroidal fields, in addition to the crustal field (**Figure** 5). It is based on data from all high-precision satellite magnetic field missions, beginning with the POGO missions of the 1960s. It uses an iteratively reweighted least-squares approach to solve for all of the 25000+ parameters using more than 2 million observations. Because of its design

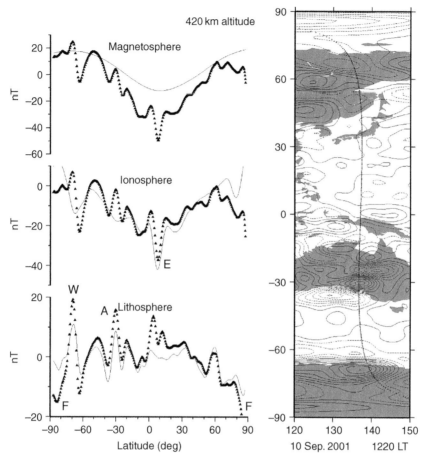

Figure 5 Residual progression versus geographic latitude as magnetic fields from the four main source regions (core, crust, ionosphere, and magnetosphere) are removed with the Comprehensive Model CM4 (Sabaka et al., 2004). This profile shows the total field T in the direction of the main field **F** of a CHAMP descending (South-going) satellite pass on 10 September 2001. The pass is centered at 0300 UT and crosses the Equator at 137° W and 1220 LT. Magnetically quiet conditions prevailed with $Kp = 1^+$ for this period, $Kp = 0^+$ for the previous 3-h period, $Dst = 2$ nT, and $|d(Dst)/dt| < 2$ nT h^{-1}. Siebert and Meyer (1996) discuss magnetic indices, while Mandea and Purucker (2005) discuss their role in data selection (see also Appendix to Chapter 3). For a given panel, the symbols represent residuals with respect to a main field (to spherical harmonic degree 13) plus all fields labeled in the panels above; the line is the prediction from the field source labeled in the current panel. The figure on the right shows the location of the sub satellite point and includes a contour map of the total field anomaly (ΔT) originating in the crust (to spherical harmonic degree 60) from the Comprehensive Model (contour interval = 2 nT, dashed lines indicated negative ΔT). The data from this profile were not included in the construction of the Comprehensive Model. The equatorial electrojet (FF) can be seen (E) because it is most prominent around mid day, following the magnetic dip equator. Although the amplitude of the EE in the model and profile is similar, a slight amplitude offset and latitudinal shift results in a residual anomaly that might be mistaken for a crustal anomaly. While the EE is a robust feature of the low-latitude ionosphere, it does exhibit significant variability on a day-to-day basis (Lühr et al., 2004; Langel et al., 1993), and includes wavelengths shorter than the resolution of CM4 (spherical harmonic degree 45 for the EE). The magnetic field originating in the distant magnetosphere exhibits variations, which are not entirely accounted for by the Dst index, and this may account for some of the mismatch. In contrast, the high-latitude current system (F) exhibits significant variability in time on a minute-to-minute basis, and in space, and CM4 does not attempt to model it. Two significant crustal anomalies, in Wilkes Land, Antarctica (W), and in southern Australia (A) are prominent in the profile. The frequency content of these anomalies again exceeds the cutoff of CM4 (spherical harmonic degree 65 for crustal fields). These two magnetic features (Purucker et al., 1999; Mayhew and Johnson, 1987) were adjacent (Von Frese et al., 1986) in pre-rift reconstructions of Gondwana.

philosophy, the CM4 crustal field component estimate is expected to have more power than MF-4, both because no direct damping is applied to the crustal field coefficients, and because of the along-track approach used by MF-4. No suppression of along-track magnetic fields is expected, and some of them, especially in the vicinity of the dip equator, are of questionable crustal origin.

6.4.4 Global Magnetization Models

Global magnetization models often represent an integration of compositional and thermal models of the crust and mantle with long-wavelength crustal

magnetic field measurements from satellite. Both forward (Hemant and Maus, 2006) and inverse (Fox Maule *et al.*, 2005) approaches are currently under development. For example, one approach (Purucker *et al.*, 2002) has used the 3SMAC (Nataf and Ricard, 1996) compositional and thermal model of the crust and mantle as a starting model, which is then modified in an iterative fashion with the satellite data until the magnetic field predicted by the model (**Figure 6**) matches the observed magnetic field. A unique magnetic crustal thickness solution is obtained by assuming (1) that induced magnetizations dominate in continental crust, (2) the model of Dyment and Arkani-Hamed (1999) describes the

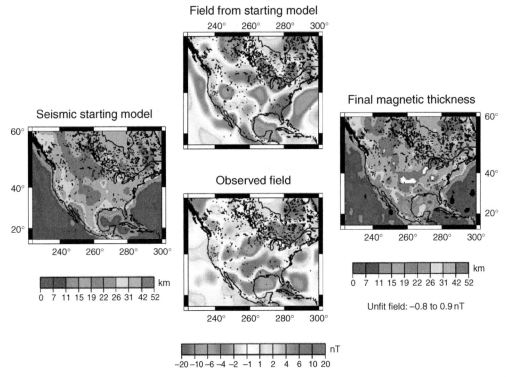

Figure 6 Magnetic crustal thickness map of North America (right), which reproduces satellite observations (bottom) from CHAMP, as represented by MF-4 (Maus *et al.*, 2006). As a starting model the seismic crustal thicknesses (left) from Chulick and Mooney (2002) over North America are used instead of the global 3SMAC crustal and thermal model (Nataf and Ricard, 1996). The magnetic field is calculated from this starting model (top) under the assumption of a constant magnetic susceptibility (χ) of 0.04, and long-wavelength fields (spherical harmonic degree < 15) are removed, simulating a main field subtraction. The observed (bottom) and modeled (top) fields are differenced, and the difference is inverted for a magnetic crustal thickness. The starting model is then updated to reflect this change, and the process continues until convergence is achieved. The process is nonlinear because the total anomaly field (ΔT) is used, and because of the high-pass filter. After three iterations of this technique the residuals to the observations are less than ± 1 nT. Negative magnetic crustal thickness (shown in black, the minimum is −6 km) over a few regions in the ocean could be a consequence of remanent magnetic fields. Large magnetic crustal thicknesses (shown in white; the maximum is 60 km) over parts of the mid-continent region could be a consequence of inaccuracies in the starting model. Purucker *et al.* (2002) applied this approach over North America, and found that if 3SMAC alone was used as a starting model, negative crustal thicknesses were found over the southeastern US landmass. Modification of 3SMAC to place the major crustal thickness change near the Coastal Plain/Piedmont boundary resulted in more realistic crustal thicknesses.

oceanic remanence, and (3) that vertical thickness variations dominate over lateral susceptibility variations. A starting model is necessary for two reasons: (1) to constrain wavelengths obscured by overlap with the core field, and (2) to ensure that most magnetic crustal thicknesses will be non-negative. An approach (**Figure 6**) such as this has been used to define the thickness and thermal properties of cratonic North America (Purucker et al., 2002).

A second approach (Hemant and Maus, 2006; based on earlier work of Hahn and Bosum, 1986) uses the available magnetic petrology, geological age, tectonic and seismic crustal thickness information of the Earth's crust, and assigns magnetization strengths and directions to geological units based on their age and rock compositions. In this way a global magnetization model of the Earth's crust is computed. The model is used to predict the crustal magnetic field at satellite altitude and compared with the observed crustal field measurements. One can match the observed field by varying the boundaries and composition of lower-crustal structures.

6.5 The 'Tools of the Trade'

6.5.1 Survey Design and Resolution

Although magnetic surveys are frequently conducted as 'missions of opportunity', where the mission design is largely dictated by the needs of the primary instrument, or the platform, there are many cases in which the collection of a magnetic survey is the primary goal, and consideration needs to be given to optimizing the return from the survey. A recent example of such a process, and its documentation, has been the planning for the Swarm magnetic field satellite constellation (Olsen et al., 2006b). Survey design of aeromagnetic surveys, including the spacing of flight lines, their altitude, and the inclusion of tie lines, is discussed by Reid (1980). For further details, refer to Chapter 4 on 'observation techniques'.

6.5.2 Removal of Noncrustal Fields

An important part of obtaining crustal anomalies suitable for further processing, modeling, and interpretation is adequate removal of noncrustal fields, primarily that arising from dynamo action in the core, and external fields due to solar–terrestrial interactions. The geodynamo-generated field, often referred to as the main field, has large amplitude but varies slowly, both temporally and spatially.

External fields have much smaller amplitudes but have much more rapid temporal and spatial variations. Time-varying external fields also induce subsurface magnetic fields throughout the crust and mantle, but their amplitudes are generally small compared to those of typical crustal anomalies.

The effect of external fields can be minimized by collecting data at magnetically quiet times, but this is frequently impractical, especially at higher magnetic latitudes. Many surveys are conducted with a continuously recording base station to monitor and correct for external variations. The base station is located at a site where the spatial field gradients are low (i.e., not on a magnetic anomaly), ideally in, roughly, the center of the survey area. It can be used to alert surveyors to magnetic storms, when data acquisition will be suspended, and as a means to judge the quality of the survey data. In periods of normal activity, the temporal variations recorded at the base station will be a reasonable approximation to the external field throughout the survey area. External fields are a minimum at night in low and mid-latitudes. Over several days (or longer) it is usually possible to identify a 'night time quiet value' (NTQV) from the base station record. The difference between the NTQV and the base station value at a given time is an approximation to the external field at that time, and is removed from the survey data. The difficulty of this method lies in the complicated behavior of the external field, combined with the generally unknown conductivity structure of the Earth. A second approach to the correction of external fields is via a least-squares analysis of the misties at intersecting survey lines (Ray, 1985). The two approaches are often used together, with the regression-type analysis used as a refinement, to remove errors not removed by the first approach.

After correcting for external fields, the method for removing the main field depends on the size and scope of the survey. For a small, ground-based survey, it is often sufficient to treat the main field as constant over the survey area. Its amplitude is likely to be well-approximated by the average field (data mean) or the NTQV. Airborne and satellite surveys typically cover much larger areas, over which it may not be reasonable to assume the main field is constant, and therefore more sophisticated main field removal methods are justified. An obvious extension is to remove the best-fitting line (for a 1-D survey) or plane (2-D survey) through the data. More commonly, the predictions of a main field model such as the International Geomagnetic Reference Field

(IGRF) are subtracted. IGRFs consist of values for the spherical harmonic coefficients from which the field can be calculated at any point in space and time (see Chapter 2 for further details). Each IGRF consists of a set of main field and secular variation coefficients covering a 5-year interval, thereby accounting for the temporal evolution of the main field. However, mismatches can occur between crustal anomaly fields in overlapping or abutting areas obtained from surveys at different times because the IGRF representations are not perfect. This is a particular problem when trying to merge surveys to form larger compilations.

The most important data for IGRF modeling have been permanent magnetic observatory night-time values on magnetically quiet days, usually selected on the basis of geomagnetic activity indices (see the Appendix to Chapter 3, and Siebert and Meyer, 1996). Observatories are located in areas of low spatial magnetic field gradients, in areas where crustal fields are a minimum. With the incorporation of large numbers of data from orbiting satellites, and the availability of more powerful computational resources, different methods of analysis become appropriate. As described earlier, rather than attempting to remove certain field sources from the data on a point by point basis prior to main field modeling, they can now be solved for simultaneously. This approach has led to the series of 'comprehensive models' (e.g., Sabaka et al., 2004) in which large numbers of parameters expressing the main, crustal, external, and induced fields are co-estimated (see Chapter 2). These models (**Figure 7**) should enable better estimates of the crustal field to be obtained from survey data (Nabighian et al., 2005). The current version of the 'comprehensive model' is CM4 (Sabaka et al., 2004) and the data envelope extends from 1960 through July of 2002. Usage of the model outside of this time range entails two steps. First, the user must update the values of Dst and F10.7, the indices used for characterizing the state of the ionopshere and magnetosphere. Second, the internal, time-varying low-degree part of the model must be replaced by a model that is valid over the time span considered. This would mean the IGRF for data collected prior to 1960, and a model such as the CHAOS model (Olsen et al., 2006a) for data collected after July 2002. The end result of this data-reduction process should be point representations of our best estimates of the crustal magnetic field, often referred to as the crustal anomaly field.

6.5.3 Representations

Taking the cue from the way seismic trace information is displayed, crustal magnetic data collected along a profile, or even a series of profiles, can be represented as a set of 'wiggles', with areas above the zero line filled, or the areas above and below the line colored differently. More often, data collected over an area are interpolated onto a regular grid for display as a color image, and for further processing and modeling. Various algorithms are suitable for gridding crustal magnetic data, and which is employed in a particular instance will depend on how the data have been collected, and the form of the crustal anomalies encountered. For example, aeromagnetic data usually have a much smaller interval between data points along flight lines than between them (see Chapter 4), making bidirectional spline gridding (Bhattacharyya, 1969) appropriate. For more evenly spaced data, minimum curvature methods (Smith and Wessel, 1990) are often applied. Widely spaced tie lines are often flown perpendicular to the survey direction, and this facilitates the 'leveling' of the survey (Ray, 1985). If the anomalies have a particular directionality to them (e.g., they arise from a series of parallel dykes), interpolations of the crustal anomaly field can be improved by incorporating measured horizontal gradients (Reford, 2006). Besides offering a visual image of the data through imaging, these regular grids form the basis for all the transformations (mostly using wavenumber domain manipulation) discussed below that can be applied to the data. As a consequence this is an area of continuing research (O'Connell et al., 2005; Hansen, 1993; Cordell, 1992; Keating, 1993; Ridsdill-Smith and Dentith, 1999).

There are many methods of modeling the data that can be used to interpolate between data points and extrapolate beyond the survey area. Some of these are only useful for local modeling, others are only applicable to data sets covering all, or at least a large part of, the Earth's surface, and some can be used for both local and global modeling. Global methods are well-summarized in Langel and Hinze's 1998 book, and they present the methods outlined below in a uniform notation.

The most commonly employed global method is SHA, which has been described in Chapter 2 in the context of main field modeling. The potential is expressed as

$$V = a \sum_{n=1}^{N_{max}} \left(\frac{a}{r}\right)^{n+1} \sum_{m=0}^{n} \left(g_n^m \cos(m\phi) + b_n^m \sin(m\phi)\right) P_n^m(\cos\theta)$$

[6]

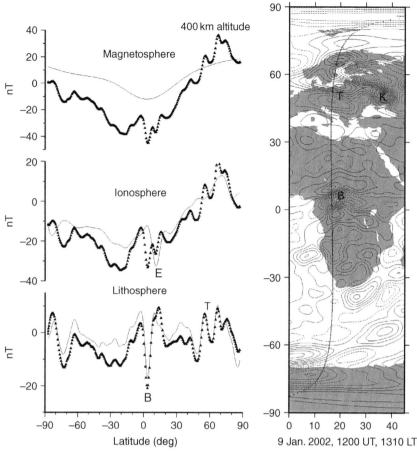

Figure 7 Residual progression versus geographic latitude as magnetic fields from the four main source regions (core, crust, ionosphere, and magnetosphere) are removed with the Comprehensive Model CM4 (Sabaka *et al.*, 2004). This profile shows the total field *T* in the direction of the main field **F** of a CHAMP satellite pass on 9 January 2002, when the magnetic field was in a quiet state. The data from this profile were not included in the construction of the Comprehensive Model. For a given panel, the symbols represent residuals with respect to a main field (to spherical harmonic degree 13) plus all fields labeled in the panels above; the line is the prediction from the field source labeled in the current panel. The figure on the right shows the location of the sub satellite point and includes a contour map of the total field anomaly (ΔT) originating in the crust (to spherical harmonic degree 60) from the Comprehensive Model (contour interval = 2 nT; dashed lines indicate negative ΔT). This figure illustrates the ionospheric/lithospheric separation that is possible with a Comprehensive Model approach with the equatorial electrojet (E) in close proximity to the Bangui (B) crustal anomaly (Girdler *et al.*, 1992). Note also the Tornquist-Teisseyre zone (Taylor and Ravat, 1995) (T), a major litho-tectonic structure in Central Europe, and the Kursk (Taylor and Frawley, 1986) anomaly (K), associated with a substantial Banded Iron Formation.

where *a* is the radius of the Earth, *r* is the radial distance of the observation from the center of the Earth, ϕ denotes longitude, and θ co-latitude, $P_n^m(\cos\theta)$ are the Schmidt quasi-normalized associated Legendre functions of degree *n* and order *m*, and the g_n^m and h_n^m are the spherical harmonic coefficients to be estimated. The difference for crustal field modeling is that the series needs to include much higher harmonic degree terms (N_{max}) to represent the anomaly field adequately. Spherical harmonic models of the crustal field from satellite data (Sabaka *et al.*, 2004; Maus *et al.*, 2006) now go to spherical harmonic

degrees as high as $N_{max} = 90$, corresponding to $N_{max}(N_{max} + 2)$ coefficients. This creates computational difficulties (Cain *et al.*, 1989; Lesur and Gubbins, 1999), and high-resolution data sets require enormous numbers of spherical harmonic coefficients to represent them adequately. SHA is not particularly well-suited to global (or near-global) data sets of varying spatial density: since the basis functions are themselves global, the spherical harmonic series must extend up to the degree representing the shortest spatial wavelength in the data set (approximately c/n, where *c* is circumference, and *n* is spherical harmonic degree). The coefficients multiplying these degree terms will not be well-constrained if only a limited area of the globe has coverage at the spatial sampling rate appropriate to determine them and, unless regularization is applied, their numerical values may generate ringing over other parts of the globe. Basis functions with more local support, such as harmonic splines (HS) (Shure *et al.*, 1982), or wavelet-like functions (Lesur and Maus, 2006) are better suited to data sets with variable resolution over the globe.

Spectral analysis can be applied to data collected along profiles or on a plane, with the usual techniques to avoid ringing, edge effects, and spectral leakage (Parker and O'Brien, 1997; Lowe *et al.*, 2001). This allows high-resolution data sets to be represented by small numbers of model parameters. Alldredge (1981) introduced rectangular harmonic analysis, suitable when the area covered is small enough for the flat Earth approximation to be appropriate. The method is based on the solution to Laplace's equation in a Cartesian geometry. The data are first transformed into the local Cartesian coordinate system with origin at the center of the region, and then the coefficients are determined. They can be used to predict the field at any altitude, and can also be transformed back to a spherical Earth coordinate system. Malin *et al.* (1996) introduced extra coefficients to remove trends. In an analysis of main field data, they concluded the method was only suitable for interpolation, and not for extrapolation. Nakagawa and Yukutake (1985) used a cosine function weight near the edges of the region subject to rectangular harmonic analysis to reduce edge effects.

Alldredge (1982) introduced the related concept of cylindrical harmonic analysis, where the equation to be solved is Laplace's equation in cylindrical polar coordinates. He advocated this when the observations displayed cylindrical symmetry. Again, the size of the area to be modeled and single-valuedness of the potential imposes constraints on the arguments.

A representation useful for both local and global modeling (typically in a Cartesian and spherical coordinate system, respectively), is equivalent source (ES) dipoles, where again the basis support is local. The magnetized crust is divided into blocks, each of which is assumed to have a magnetic dipole at its center. In this case the potential can be expressed as

$$V = -\mathbf{M} \cdot \nabla \frac{1}{d} \qquad [7]$$

where d is the distance between the dipole and the observation location, and $\mathbf{M}$ is the dipole moment. The conversion factor, $\mu_0/4\pi$, between SI and CGS units (cf. Blakely, 1995, p. 67) is implicitly assumed to be included within eqn [7] and subsequent equations relating the potential V to the magnetization $\mathbf{M}$. The model parameters of [7] are the direction and magnetization strength of the dipoles. However, magnetization is often assumed to be purely induced, meaning that the dipole directions are known (parallel to the main field); the problem of inferring strength from vector component anomaly data is then linear. The dipoles can be arranged with variable density according to the data distribution, retaining the resolution of the original data set. This can be a far more efficient (i.e., fewer parameter) modeling method than SHA when the spatial resolution of the data set is uneven. Although the distribution of magnetization in the crust reproducing the anomaly data is highly nonunique (Runcorn, 1975; see Section 6.5.6), it can be interpreted geologically, especially if *a priori* information has been incorporated in the modeling, whether forward or inverse (see Section 6.5.5). ES dipole models (Dyment and Arkani-Hamed, 1998) are widely used for forward modeling since they are intuitively accessible. They can be used straightforwardly to predict the magnetic field at any altitude on or above the Earth's surface, so also provide an excellent tool for upward and downward (analytic) continuation (see Section 6.5.4).

The crustal anomaly field at or above the Earth's surface, even as high as typical orbiting satellite altitudes of a few hundred kilometers, depends on the magnetization of only a small volume of the crust directly beneath the observation point – the footprint of an anomaly measurement is small. Thus when a local basis is used to represent the anomaly field, the matrix relating observations to model parameters is sparse. Using numerical methods for solving sparse matrix systems then allows a large number of basis functions to be included, meaning that the resolution of the original data set can be retained. An application of this to crustal anomaly modeling was by Purucker *et al.* (1996), who applied the iterative conjugate gradient algorithm to ES dipole modeling of a satellite crustal anomaly data set.

Although nonphysical and therefore not suitable as an interpretation tool, crustal anomaly data can be represented by a subsurface distribution of magnetic monopole sources (O'Brien and Parker, 1994). The number and positions of the monopoles on the source sphere are chosen to provide a good representation of the data (again, a spatially variable monopole density can be used to represent spatially variable resolution of the original data); the model parameters are then simply the monopole amplitudes (no assumptions concerning directionality are required).

The potential is expressed as a sum of potential sources $\varphi_k(\mathbf{r})$, $k = 1, \ldots, K$

$$V(\mathbf{r}) = \sum_{k=1}^{K} \alpha_k \varphi_k(\mathbf{r}) \qquad [8]$$

where α_k are the monopole amplitudes to be determined. Monopoles at locations $\mathbf{s}_k$ are represented by functions

$$\varphi_k(\mathbf{r}) = \frac{1}{|\mathbf{r} - \mathbf{s}_k|} \qquad [9]$$

The solution is calculated by minimizing

$$U = \left\| \mathbf{C}^{-1}(\mathbf{d} - \mathbf{G}\alpha) \right\|^2 + \lambda \boldsymbol{\alpha}^T \Gamma \boldsymbol{\alpha} \qquad [10]$$

where $\mathbf{C}$ is the data covariance matrix, $\mathbf{d}$ is the data vector, $\mathbf{G}$ is the matrix of Green's functions relating the monopoles to the measurements, λ is a Lagrange multiplier, and $\boldsymbol{\alpha}^T \Gamma \boldsymbol{\alpha}$ is a quadratic form expressing the field complexity. $\|\cdot\|$ denotes the Euclidean norm or length. Γ is known as the Gram matrix; its (j, k)th element is the inner product of φ_j and φ_k. Thus the first term measures the fit to the data, and the second, the amount of structure in the resulting field model. This is an example of a regularized, or minimum norm, solution; by an appropriate choice of λ, we can relax slightly the fit to the model such that it does not attempt to model noise in the data. The quadratic form (and definition of the inner product for the calculation of the Gram matrix) is chosen to measure some global property of the field such as its mean strength or lateral variability; useful measures lead to closed form, or at least easily calculable, Gram matrix elements. The concept was introduced with main field modeling in mind (see Chapter 2), where some quantities that are expressible as quadratic norms can be bounded theoretically or empirically. It is now widely used as a regularizing tool. For crustal modeling, it ensures that the models have minimum structure for a given fit to the data; if the fit is acceptable, we can then argue that the real Earth has at least as much structure as the model. Since [10] minimizes a global measure of complexity, it does not matter if the monopole sources are distributed unevenly over the surface to reflect the data coverage.

HS are local basis functions introduced by Shure *et al.* (1982) for global main field modeling. They were the first to apply minimum norm modeling to geomagnetism. Using the Green's function for the magnetostatic potential, the solution is constructed as a linear combination of the HS associated with each data point, leading to the solution of a linear system of dimension: the number of data points. This is impractically large even for main field modeling, but naturally preserves the resolution of the original data set. To make the inversion of large data sets computationally tractable, Parker and Shure (1982) expanded the solution in terms of HS at only a subset of the data points, known as the depleted basis. The system then reduces to one of the dimensions of the number of data points. Tests based on small data sets demonstrated that the depleted basis solution had only a slightly larger norm than the minimum value obtained by HS. The resolution of the depleted basis solution depends on the spacing between the basis points. The field can be constructed at any radius beyond which the solution converges, making analytic continuation straightforward.

HS is not used in crustal anomaly modeling because the number of points in practical data sets is too large, but Whaler (1994) inverted a $2° \times 2°$ grid of MAGSAT satellite crustal anomalies using depleted basis HS, using it to downward continue the field from satellite altitude to just above the Earth's surface. With the computational resources available then, she was only able to retain every other basis point in latitude and longitude even over a continental-sized area of the globe (an $80° \times 80°$ area centered on Africa), with consequent loss of resolution. She also found that a sparse distribution of points was required over the remainder of the globe to avoid ringing. Another disadvantage of depleted basis HS is that the arrangement of depleted basis points is subjective. However, it simplifies the inversion of total field anomaly data, since the basis functions for their expansion can be chosen to be those for the vertical component (at a limited subset of the actual data points), simplifying the matrix element calculations (Langel and Whaler, 1996).

A more satisfactory application of HS uses sparse matrix techniques, allowing the full basis to be retained. Unpublished models using the conjugate gradient algorithm based on satellite anomaly data are very similar to those obtained using other global methods, and also compare favorably with Whaler's (1994) depleted basis models over Africa. HS (and depleted basis HS) coefficients can be converted into an infinite series of spherical harmonic coefficients that give power spectra similar to those obtained from SHA at low degree, but typically have less power at higher degrees.

Achache *et al.* (1987) used the same basis functions to model satellite data, but reduced the size of the linear system by recognizing that they fall to negligibly small values quickly with lateral distance from the point at which the solution is being calculated. They thus included only those related to data points close to the point of interest, reducing the dimensions of the matrix to be inverted from the number of data to the number of 'nearby' data points. Their recommendation is that points within a horizontal distance $3h$, where h is satellite altitude, be included. In addition, they used principal component analysis to stabilize the inversion of the resulting (smaller) matrix by including only those eigenvectors associated with the largest eigenvalues. The decision as to how many eigenvectors to include is subjective, but the eigenvalue spectrum shows a rapid fall-off for satellite data acquired above 200-km altitude, making the choice relatively clear-cut. Previously, Langel *et al.* (1984) used principal component analysis to stabilize the calculation of ES solutions.

Based on methods originally developed for modeling seamount magnetism (Parker *et al.*, 1987), then adapted to account for crustal magnetization when modeling the main field (Jackson, 1990, 1994), Whaler and Langel (1996) used a depleted basis minimum norm method to model crustal magnetization from satellite anomaly field data sets. Data are related to magnetization varying continuously in a crust of assumed constant thickness through [1], and hence the solution is expressed as a linear combination of the Green's functions

$$-\hat{\mathbf{i}}_j^{(\eta)} \cdot \nabla_{r_j} \left\{ \frac{\mu_0}{4\pi} \nabla_s \frac{1}{|\mathbf{r}_j - \mathbf{s}|} \right\} \qquad [11]$$

The resulting Gram matrix elements have closed-form expressions involving elliptic integrals, but these can be approximated very accurately by expressions involving only elementary functions since the thickness of the magnetized layer is small in comparison to the Earth's radius. A similar simplification applies if depth-independent magnetization is assumed; this is more appropriate for satellite data modeling, since the thickness of the magnetized layer is very much smaller than satellite altitude, so it is

indistinguishable from a thin sheet. The norm minimized was

$$||\mathbf{M}|| = \sqrt{\int_v \mathbf{M} \cdot \mathbf{M} \, \mathrm{d}v} \qquad [12]$$

that is, the root-mean-square (rms) magnetization amplitude of the crust. The method makes no assumption about the magnetization direction, so allows both remanent and induced magnetization. Whaler and Langel (1996) chose the same data set and distribution of depleted basis points Whaler (1994) used to downward continue the magnetic field to produce a magnetization model for Africa. For ease of comparison with ES dipole models assuming purely induced magnetization, they displayed the model as components in the direction of the main field (consistent with induced magnetization, or remanent magnetization acquired in today's main field), perpendicular to the main field in the meridian plane, and perpendicular to the meridian plane. Whaler and Langel (1996) note that by damping least-squares ES inversion, the solution minimizes the same norm as they employed, that is, minimum rms magnetization. The largest component of magnetization was in the direction of the current main field (or antiparallel to it), but the component of magnetization perpendicular to the main field in the meridian plane was also significant in many areas. The smallest component (perpendicular to the meridian plane) requires rotation and translation of the magnetization vector from that which would be recorded by rocks acquiring a contemporary remanent magnetization.

A similar difficulty of loss of resolution of the solution and subjectivity of the choice of depleted basis points can be overcome in the same fashion as for HS: by employing the iterative conjugate gradient technique to solve the full data-by-data system of equations, taking advantage of the sparseness of the matrix relating data to model parameters. Again, comparisons between Whaler and Langel's (1996) depleted basis magnetization model for Africa and surrounds and the equivalent part of the global conjugate gradient model of Whaler *et al.* (1996) are favorable. Whaler and Purucker (2005) have applied this technique to Martian orbiting satellite data, and compared the model to Langlais *et al.*,'s (2004) ES dipole model. Mars no longer has an active dynamo, so the magnetization direction is unknown. Langlais *et al.* (2004) developed an iterative technique that allowed them to solve for both the amplitude and direction of ES dipoles. Convergence was difficult to achieve, particularly as the dipole spacing was reduced, so their final model had a coarser spacing than the separation between the data points. Nonetheless, there was good agreement between their model and Purucker and Whaler's model, which was also based on a slightly different data set.

Spherical cap harmonic analysis (SCHA) (Haines, 1985), and the related translated origin spherical cap harmonic analysis (TOSCA) (de Santis, 1991), have been developed to model the field over small patches of the globe. As for global SHA, the potential is expressed as a finite sum of spherical harmonics, but including harmonics of non-integer degree. Assuming the cap is centered on $\theta = 0$ and subtends an angle θ_0 at the center of the Earth, the possible values of degree, n_k, which is a function of order, m, are those for which θ_0 is a zero of either $P_{n_k}^m(\theta)$ or $\partial P_{n_k}^m(\theta)/\partial\theta$. SCHA maps harmonics on the sphere to the spherical cap, so their effective wavelength is reduced accordingly. Thus, smaller-scale features can be represented over the cap with fewer coefficients than are required for global SHA. TOSCA moves the origin from the center of the Earth toward the surface along a line joining the Earth's center to the center of the cap. This adjusts the wavelength represented by a given harmonics to be smaller at the center of the cap than at the edge, an advantage if the data distribution is concentrated toward the center of the region. Korte and Holme (2003) present a method for regularizing SCHA, pointing out that, unlike SHA, the basis functions are not orthogonal. This means that it is not possible simultaneously to represent the potential for the vertical and horizontal field components exactly. Analytic continuation is prone to errors that increase with the upward or downward continuation distance, although Thebault *et al.* (2004) have re-posed SCHA as a boundary value problem within a cone extending above the reference surface, thereby allowing satellite data to be downward continued to the Earth's surface.

6.5.4 Transformations

The transformations of the next few subsections are applicable to 2-D data sets expressible in a rectangular geometry, such as those recorded in regional aeromagnetic surveys. They are most easily considered and performed in the wave number domain. The development here follows that of Gunn (1975)

closely, but beware of typographical errors in his manuscript. We begin with the expression for the magnetic scalar potential V resulting from a distribution of magnetization **M** within an infinite half-space. Assuming a uniform direction of magnetization, the potential is (in Cartesian coordinates, with z positive downwards)

$$V(x,y,z) = \frac{\partial}{\partial k_0} \int_0^\infty \int_{-\infty}^\infty \int_{-\infty}^\infty \frac{M(\alpha,\beta,\gamma)}{d}\, d\alpha\, d\beta\, d\gamma$$

[13]

where $\partial/\partial k_0$ is the derivative in the direction of **M**, and d is the source–observation distance:

$$d^2 = (x-\alpha)^2 + (y-\beta)^2 + (z-\gamma)^2$$

[14]

The α, β integral is a convolution, that is,

$$\int_{-\infty}^\infty \int_{-\infty}^\infty \frac{M(\alpha,\beta,\gamma)}{d}\, d\alpha\, d\beta$$
$$\equiv M(x,y,\gamma)^* R(x,y,z-\gamma)$$

[15]

where '*' denotes convolution and

$$R(x,y,z-\gamma) = \frac{1}{\sqrt{x^2+y^2+(z-\gamma)^2}}$$

[16]

Hence, the Fourier transform of V is

$$\tilde{V}(u,v,z) = \frac{\partial}{\partial k_0} \int_0^\infty \tilde{M}(u,v,\gamma) \cdot \tilde{R}(u,v,z-\gamma)\, d\gamma$$

[17]

where

$$\tilde{M}(u,v,\gamma) = \int_{-\infty}^\infty \int_{-\infty}^\infty M(x,y,\gamma)e^{-i(ux+vy)}\, dx\, dy$$

[18]

is the Fourier transform of **M**, using the tilde symbol to denote Fourier-transformed quantities. The Fourier transform of R is

$$\int_{-\infty}^\infty \int_{-\infty}^\infty \frac{1}{\sqrt{x^2+y^2+(z-\gamma)^2}} e^{-i(ux+vy)}\, dx\, dy$$
$$= 2\pi \frac{e^{(z-\gamma)\sqrt{u^2+v^2}}}{\sqrt{u^2+v^2}}$$

[19]

and so

$$\tilde{V}(u,v,z) = 2\pi \frac{\partial}{\partial k_0} \int_0^\infty \tilde{M}(u,v,\gamma) \frac{e^{(z-\gamma)\sqrt{u^2+v^2}}}{\sqrt{u^2+v^2}}\, d\gamma$$

[20]

Let (l,m,n) be the direction cosines of **M**. Then

$$\frac{\partial}{\partial k_0} = l\frac{\partial}{\partial x} + m\frac{\partial}{\partial y} + n\frac{\partial}{\partial z}$$

[21]

But $\partial f/\partial x = iu\tilde{f}$ for any Fourier transform pair f, $\tilde{f}$ (and similarly for differentiation with respect to y). Hence

$$\tilde{V}(u, v, z) = 2\pi \frac{ilu + imv + n\sqrt{u^2 + v^2}}{\sqrt{u^2 + v^2}}$$
$$\times \int_0^\infty \tilde{M}(u, v, \gamma)e^{(z-\gamma)\sqrt{u^2+v^2}}\,d\gamma \quad [22]$$

We now need to relate the (Fourier transformed) potential V to the (Fourier transformed) component of the magnetic field being measured, usually that in the direction of $\mathbf{F}$, which we take to have direction cosines (l', m', n'):

$$\Delta T(u, v, z)$$
$$= 2\pi \frac{\left[ilu + imv + n\sqrt{u^2 + v^2}\right]\left[il'u + im'v + n'\sqrt{u^2 + v^2}\right]}{\sqrt{u^2 + v^2}}$$
$$\times \int_0^\infty M(u, v, \gamma)e^{(z-\gamma)\sqrt{u^2+v^2}}\,d\gamma \quad [23]$$

It is easily shown that $l = \cos I_F \cos D_F$, $m = \cos I_F \sin D_F$, and $n = \sin I_F$, where I_F and D_F denote the inclination and declination of the main field, respectively, and similarly for (l', m', n').

6.5.4.1 Analytic continuation

From [23] it follows straightforwardly that analytic (upward or downward) continuation involves convolving with a filter whose frequency response is e^{Hk}, that is, multiplying by e^{Hk} in the wave number domain, where $k^2 = u^2 + v^2$ is the square of the wave number, and H is the continuation height or depth (also measured positive downwards). e^{Hk} is sometimes referred to as the analytic continuation operator. Thus, downward continuation is an amplifying and roughening operation (conversely for upward continuation). This is analogous to the spherical case where the amplification factor is $(a/r)^{n+2}$, where n is spherical harmonic degree, a is the reference radius of the spherical harmonic expansion, and r the analytic continuation radius. Examples of the uses of analytic continuation are to suppress or enhance short wavelength features, reduce data collected at a variety of altitudes (e.g., different flying heights for airborne surveys) to constant height, either above terrain or relative to mean sea level, compare ground and airborne observations, and estimate depth to sources (see Section 6.5.5).

6.5.4.2 Reduction to the pole

Reduction to the pole (RTP) is achieved by convolving with a filter whose frequency response is

$$\frac{\sqrt{u^2 + v^2}}{ilu + imv + n\sqrt{u^2 + v^2}} \cdot \frac{\sqrt{u^2 + v^2}}{il'u + im'v + n'\sqrt{u^2 + v^2}} \quad [24]$$

(the factor $\sqrt{u^2 + v^2}$ in each term preserves dimensions), that is, [24] is the RTP operator (or filter). It removes the factors in [23] associated with the direction of the main field and the direction of remanent magnetization. Its effect is to reduce the anomalies to those that would be observed at the magnetic North pole with a vertical remanent magnetization direction. If the remanent magnetization direction is unknown, or magnetization can be assumed to be purely induced, the direction cosines of magnetization are replaced by those of the main field. RTP becomes unstable as the magnetic equator is approached since the numerator of [24] approaches zero (recall $n = \sin I_F$). At low magnetic latitudes, it is possible to perform reduction to the equator instead, but the form of the resulting anomalies is not as simple as for RTP. Silva (1986) has treated RTP as an inverse problem, using methods designed for stabilizing inversion to stabilize RTP. Besides aiding interpretation through simplifying the form of the anomalies and centering them over their causative structures, RTP eases the comparison of oceanic magnetic anomalies at different latitudes.

6.5.4.3 Pseudogravity

The pseudogravity transformation follows from Poisson's relation between the magnetic potential and the gravitational field. Consider a body with uniform magnetization (both strength and direction) and density occupying a volume Ω. Then the magnetic scalar potential is

$$V(P) = -\mathbf{M} \cdot \nabla_P \int_\Omega \frac{1}{d}d\Omega \quad [25]$$

where P is the observation point, and d is distance from P, and the gravitational potential is

$$U(P) = G\rho \int_\Omega \frac{1}{d}d\Omega \quad [26]$$

where G is the gravitational constant and ρ is density. Combining the two,

$$V(P) = -\frac{1}{G\rho}\mathbf{M} \cdot \nabla_P U = -\frac{1}{G\rho}Mg_M \quad [27]$$

where g_M is the component of gravity in the direction of $\mathbf{M}$; [27] is Poisson's relation. In fact, it is not necessary for the potential and magnetization to be constant. We can consider a body to be composed of arbitrarily small volumes in which density and magnetization can be regarded as constant. Since potentials add, [27] applies to a body in which density and magnetization vary in proportion. However, pseudogravity is defined as the gravity anomaly that would be observed if the magnetization distribution were replaced by an identical density distribution, that is, M/ρ is a constant. In the wave-number domain, this gives

$$\tilde{g}_{\mathrm{M}} = C\tilde{V} \qquad [28]$$

where $C = -(G\rho/M)$ is a constant. Converting from the magnetic potential to total field anomaly,

$$\tilde{g}_{\mathrm{M}} = C\frac{1}{(il'u + im'v + n'\sqrt{u^2 + v^2})}\Delta\tilde{T} \qquad [29]$$

Finally, converting from the component of gravity in the direction of **M** to the vertical component gives the Fourier-transformed pseudogravity, $\tilde{g}_{\mathrm{ps}}$:

$$\tilde{g}_{\mathrm{ps}} = C\frac{\sqrt{u^2 + v^2}}{(ilu + imv + n\sqrt{u^2 + v^2})(il'u + im'v + n'\sqrt{u^2 + v^2})}$$
$$\times \Delta\tilde{T} \qquad [30]$$

Thus, the pseudogravity operator is

$$\frac{\sqrt{u^2 + v^2}}{(ilu + imv + n\sqrt{u^2 + v^2})(il'u + im'v + n'\sqrt{u^2 + v^2})} \qquad [31]$$

Note that, unlike the expression for the RTP operator, there is only one factor $\sqrt{u^2 + v^2}$ in the numerator of [31] because Poisson's relation [27] relates the magnetic scalar *potential* to the component of the gravitational *field* in the direction of magnetization. This means that the pseudogravity operator alters the frequency content of the signal, preferentially amplifying the longer-wavelength components (in contrast to the spatial derivatives discussed in Section 6.5.4.4, which preferentially amplify the shorter-wavelength components). Like the RTP operator, it can run into problems at low magnetic field or magnetization inclinations. The constant C in [30] means we can predict the pattern (but not the amplitude) of the gravity anomalies that would be obtained over the same structure. The pseudogravity transformation aids the comparison of magnetic and gravity anomalies, allows gravity methods to be used to interpret magnetic anomalies, and can be used in conjunction with gravity data to determine the direction of magnetization and the ratio of magnetization

to density (Cordell and Taylor, 1971). Note that Poisson's relationship provides an example of the ambiguity in magnetization modeling – there is no total field anomaly over a uniformly magnetized sheet (regardless of the direction of magnetization) since its gravity anomaly is constant. The result extends to a spherical geometry, where more complicated magnetization distributions can be shown to produce no external magnetic field (Runcorn, 1975). These magnetic annihilators are discussed more fully in Section 6.5.6.

6.5.4.4 Spatial derivatives

Derivatives are useful for enhancing smaller-scale features of a data set, and anomalies caused by shallow bodies, and directional derivatives for enhancing or suppressing features in a given direction. The second vertical derivative is valuable because it relates to second horizontal derivatives through Laplace's equation, which is satisfied by the total field anomaly as well as the scalar potential if the main field direction is constant. This involves multiplying $\Delta\tilde{T}$ by k^2 and is thus a significantly roughening operation. It is used to suppress regional gradients, and to aid in the determination of source depth and the attitude of interfaces. As noted above, differentiation with respect to x or y multiplies $\Delta\tilde{T}$ by iu or iv, respectively. The horizontal derivative, that is, the derivative in the direction of maximum change, therefore has a Fourier transform $k\Delta\tilde{T}$. This follows because

$$\nabla\Delta T = \frac{\partial\Delta T}{\partial x}\mathbf{i} + \frac{\partial\Delta T}{\partial y}\mathbf{j} \qquad [32]$$

where **i**, **j** are unit vectors in the x-, y-directions, respectively. Thus in the wave-number domain

$$\nabla\Delta\tilde{T} = iu\Delta\tilde{T}\mathbf{i} + iv\Delta\tilde{T}\mathbf{j} \qquad [33]$$

and hence

$$|\nabla\Delta\tilde{T}| = \sqrt{\nabla\Delta\tilde{T}^*\nabla\Delta\tilde{T}} = k\Delta\tilde{T} \qquad [34]$$

where here * denotes complex conjugate. First derivatives are also required in some methods of estimating depth to sources (see, e.g., Section 6.5.4.7). Directional derivatives, for example, in the direction defined by an angle φ_0 with the x-axis, are obtained by taking the magnitude of $\cos\varphi_0$ times the x-derivative, and $\sin\varphi_0$ times the y-derivative. Directions are preserved on Fourier transformation: since the wave vector **k** has components u and v in the x- and y-directions respectively, its magnitude is $\sqrt{u^2 + v^2}$ and its phase is

$$\tan \varphi_0 = \frac{v}{u} = \frac{y}{x} \qquad [35]$$

the latter by definitions of φ_0. Thus, the directional derivative enhances features making an angle φ_0 with the x-axis; to suppress them, but still improve the definition of smaller-scale features, one could take the directional derivative in the orthogonal direction.

6.5.4.5 Pie-crust filter

An alternative method of eliminating anomalies in a particular direction is to use the pie-crust filter or operator, given by

$$W(\varphi) = \begin{cases} 0 & (\varphi_0 - \Delta\varphi < \varphi < \varphi_0 + \Delta\varphi) \\ 0 & (\varphi_0 - \Delta\varphi < \varphi + \pi < \varphi_0 + \Delta\varphi) \quad [36] \\ 1 & \text{otherwise} \end{cases}$$

This removes anomalies within an angle $\Delta\varphi$ either side of φ_0, and preserves those at all other angles, without altering the frequency content. Anomalies within this wedge can be preserved by swapping the values 0 and 1 in [36]. A more sophisticated filter (with a taper between 0 and 1) would reduce ringing when the filtered anomalies are transformed back to the spatial domain.

6.5.4.6 Analytic signal

Another useful interpretational tool is the analytic signal, defined as

$$\mathbf{A}(x,y) = \frac{\partial \Delta T}{\partial x}\mathbf{i} + \frac{\partial T}{\partial y}\mathbf{j} + i\frac{\partial T}{\partial z}\mathbf{k} \qquad [37]$$

where $\mathbf{i}$, $\mathbf{j}$, and $\mathbf{k}$ are unit vectors in the x-, y-, and z-directions respectively. The real and imaginary parts of its Fourier transform are the horizontal and vertical derivatives of ΔT, respectively. They form a Hilbert transform pair, as required for $\mathbf{A}$ to be an analytical signal, a property most easily demonstrated in the wave-number domain (Roest et al., 1992). The amplitude of the analytic signal

$$|\mathbf{A}(x,y)|^2 = \sqrt{\left(\frac{\partial \Delta T}{\partial x}\right)^2 + \left(\frac{\partial \Delta T}{\partial y}\right)^2 + \left(\frac{\partial \Delta T}{\partial z}\right)^2} \qquad [38]$$

is most often used for interpretation. It has maxima at magnetization contrasts, independent of the direction of the ambient magnetic field for 2-D sources and only weakly dependent on these directions for 3-D sources, offering a method for locating the edges of magnetized bodies. If the edges are assumed vertical, it can be used to determine depth to sources. Over a 2-D vertical contrast in magnetization at depth d, the amplitude of the analytic signal is proportional to $1/(x^2 + d^2)$ where x is the horizontal distance from the contact (Nabighian, 1972). Thus the half-width of the curve gives d, the depth to the contrast. Automatic methods for finding the positions of maxima of quantities such as the analytic signal are given by Blakely and Simpson (1986). Other boundary edge finding methods include the terracing operator of Cordell and McCafferty (1989) and the 'amplitude of horizontal gradient method' of Grauch and Cordell (1987).

6.5.4.7 Euler deconvolution

Euler's homogeneity equation can be written (e.g., Reid et al., 1990)

$$(x - x_0)\frac{\partial \Delta T}{\partial x} + (y - y_0)\frac{\partial \Delta T}{\partial y} + (z - z_0)\frac{\partial \Delta T}{\partial z}$$
$$= -s_i \Delta T \qquad [39]$$

Here, (x, y, z) is the position at which the total field anomaly is ΔT, arising from a source at position (x_0, y_0, z_0). s_i is the 'structural index' (SI), a measure of the rate of fall-off of the field with distance, which therefore reflects the source geometry (Thompson, 1982). An SI of 3 corresponds to a point source (dipole), 2 is appropriate for extended line sources, such as pipes and cylinders, a value of 1 for a step, thin dike or sill edge, and values of 0.5 and 0 have been chosen for faults and other contacts. If N is set to zero, the right-hand side of [39] should be replaced by a constant, which depends on the amplitude of the contrast in magnetization, and the strike and dip of the contact (Reid et al., 1990).

Euler deconvolution (which is not strictly a deconvolution) has proved itself particularly useful in a regional context, when grids or profiles of data can be inverted systematically for source parameters and the background field (or the constant on the right-hand side of [39] when $N = 0$). As [39] shows, the method requires derivatives of ΔT, which are usually calculated in the wave-number domain. These are treated as data (along with the SI) in an inversion for the source positions and background field (or right-hand side constant). Obtaining useful solutions depends on careful choice of the window size (i.e., how many adjacent points are included in a single least-squares-type inversion). Each window produces a set of source parameters, and only those with standard deviations below a specified threshold are retained. Even these often plot as quasi-linear features ('strings of pearls'), that is, the solutions tend to be defocused. This may be partly because a region often includes sources represented by more than one SI, and attempts have been made to develop a multiple source approach. Other approaches include methods of determining automatically, or also solving for, SI, for example, wavelet-based methods for estimating degree of homogeneity (Sailhac *et al.*, 2000). The usual methods of damping and generalized inverses are useful in improving the performance of the method and interpretation of the results (e.g., Neil *et al.*, 1991; Mushayandebvu *et al.*, 2004), but damping tends to bias depth estimates.

Several authors have pointed out that Euler deconvolution can be applied to any homogeneous field or function. This includes the horizontal gradient or analytic signal of a field that is itself homogeneous, for example, the magnetic field, or its Hilbert transform; the appropriate SI for the horizontal gradient or analytic signal is then one larger than that of the original field source. The advantage of deconvolving the analytic signal rather than the magnetic field itself is that its calculation effectively removes the background field.

6.5.5 Forward and Inverse Methods

Transformations assist in the characterization of certain features of the magnetic source, thereby facilitating interpretation. Forward and inverse methods take this characterization a step further, determining attributes of the magnetic source. Forward methods begin with one or more magnetic bodies whose salient features are selected *a priori*, on the basis of geologic or geophysical knowledge. Magnetic fields are then predicted for these bodies at the survey location, and model parameters are adjusted on the basis of the closeness of the fit to the observation. This process continues until a sufficiently close fit to the observations is achieved. Inverse methods, in contrast, allow for the direct determination of one or more attributes of the magnetic source, usually through least-squares or Fourier-transform techniques.

6.5.5.1 Forward models
Procedures for calculating magnetic forward models involve simplification of complex bodies into simpler ones, either as collections of rectangular prisms (Bhattacharyya, 1964), magnetic dipoles (Dyment and

Arkani-Hamed, 1998), polygonal laminae (Talwani, 1965; Plouff, 1976), or polyhedrons (Bott, 1963). The calculation can be made either in the space or wave-number domains. Parker (1973) gives a wave-number domain-based algorithm for the rapid calculation of the crustal magnetic field over sources defined by a magnetization contrast over a topographic surface.

6.5.5.2 Inverse approaches
Quantitative interpretations sought using an inverse approach aim to estimate the causative body's depth, dimension, and magnetization contrast. In many applications, depth to the magnetic source is the most important of these properties. Depth to source determinations are of two types: (1) based on the shape of individual anomalies (beginning with Peters, 1949), and (2) based on the statistical properties of ensembles of anomalies (beginning with Spector and Grant, 1970), and implemented in the spectral domain.

The first analytic approximation to determining the depth to source was by Werner (1955), who solved the problem under the assumption that the source was a thin dike. Subsequent work has relaxed that limitation, and allowed for other source geometries (Ku and Sharp, 1983). The exploitation of Euler's homogeneity relation [39] led to a second class of analytic approximations (Reid *et al.*, 1990). This approximation allows for a variety of sources to be treated successfully, as outlined in Section 6.5.4.7.

Wave-number-domain approaches to individual anomalies include the methods of Naudy (1971), applicable to a vertical dike or thin plate, and CompuDepth (O'Brien, 1972).

6.5.5.2.(i) Fourier domain approaches to groups of anomalies
Matched Filters (Syberg, 1972; Phillips, 1997) use the Fourier-domain properties (Spector, 1968; Spector and Grant, 1970) of the magnetic field to estimate the depths of the principal sources. These depths are then used to design wavelength filters, which are in turn used to decompose observed magnetic anomalies into estimates of the anomalies caused by sources at those principal depths. The original Spector and Grant (1970) method estimated the depth from the slope of the radially averaged power spectrum.

6.5.6 Resolving Interpretational Ambiguity

Some of the ways for resolving or better understanding interpretational ambiguity include annihilators

(Runcorn, 1975; Maus and Haak, 2003), ideal body analysis (Parker, 2003), Monte Carlo simulations (Sambridge and Mosegaard, 2002), and, of course, through the use of prior information.

It has long been known that an infinite sheet with constant magnetization produces no magnetic field outside of the sheet, although second-order effects usually ensure that some magnetic fields escape. Only magnetization contrasts produce magnetic fields. Runcorn (1975) demonstrated, in the case of the moon, that a spherical shell of constant susceptibility linearly magnetized by an arbitrary internal field also produces no field outside of the shell. More recently, Maus and Haak (2003) illustrated another class of magnetization solutions that produce no external fields. Their example, based on reasoning from spherical harmonics, is one defined by a magnetic susceptibility profile in a dipolar field that is symmetric about the magnetic equator. In the case of the Earth, South America and Africa are approximately bisected by the magnetic dipole equator, and their shape can be approximated fairly well by an annihilator. Thus, if these continents possess a large-scale magnetic contrast with the surrounding ocean, much of that contrast may be invisible. It is expected that further inquiry will reveal additional classes of annihilators.

The theory of ideal bodies (Parker, 1974, 1975) systematizes the process of placing bounds on the parameters of the source region, such as the depth of burial (Grant and West, 1965). The process for doing this involves the minimization of the infinity norm of the magnetic intensity $|\mathbf{M}|$ within the source region. Parker (2003) showed how such a process could be used to determine the distribution of magnetization that has the smallest possible intensity, without any assumptions about its direction.

The Metropolis algorithm and the Gibbs sampler (Mosegaard and Sambridge, 2002) are Monte Carlo techniques for the exploration of the space of feasible solutions. They also provide measures of resolution and uncertainty. Although not widely used in the field of crustal magnetism, the Metropolis algorithm has been used by Rygaard-Hjalsted et al. (2000) to conduct resolution studies on fluid flow in the Earth's outer core from geomagnetic field observations. Monte Carlo techniques can also be used to find globally optimal solutions, and Dittmer and Szymanski (1995) have applied simulated annealing to magnetic and resistivity data.

6.6 Spectral Overlap with Other Fields

The transition from core-dominated to crust-dominated processes occurs as a relatively sharp break centered at spherical harmonic degree 14 (Alldredge et al., 1963; Cain et al., 1974; Langel and Estes, 1982) corresponding to wavelengths of $40\,000/14 = 2860$ km. This transition can be seen in a power spectrum of the static field (**Figure 8**). The crustal field is much stronger over the continents than over the oceans (Arkani-Hamed and Strangway, 1986; Hinze et al., 1991) and so we expect that the spectral overlap with the core field will be different for continental than for oceanic crust. The crustal field is expected to have power at wavelengths longer than degree 14 because of the markedly different characteristics of continental and oceanic crust (Meissner, 1986) and because of long-wavelength oceanic magnetic anomalies (Dyment and Arkani-Hamed, 1999). These longest wavelengths are masked by the dominant core fields (Hahn and Bosum, 1986), and various forward (Cohen, 1989) and inverse (Purucker et al., 1998, 2002) approaches have been developed in an attempt to include at least some notional idea of these fields. The longest-wavelength crustal magnetic fields remain inaccessible to direct observation, although Mayhew and Estes (1983) suggest that simultaneous modeling of core and crustal fields using ES dipole arrays located both within the core and crust might make possible the separation of the sources. The task they outlined is now computationally feasible (Purucker et al., 1996), but there are reasons for suspecting that a full separation may not be possible. For example, it is possible to represent magnetic fields of long wavelength (say the dimension of a continent) by ES dipoles placed either at the surface, or at the core–mantle boundary. A separation based solely on the radial position of the dipoles is thus likely to be ill-posed, and depend on details of the parameterization, such as the tessellation used, its spacing, and the distance over which the observations are expected to influence the crustal dipoles. The debate over the existence of field-aligned current systems (Dessler, 1986) shows the difficulty of interpreting physically an equivalent current/dipole system deduced solely from observations of the magnetic field.

On the other hand, co-estimation of magnetic fields of internal and external origin through the 'comprehensive model' approach (Sabaka et al.,

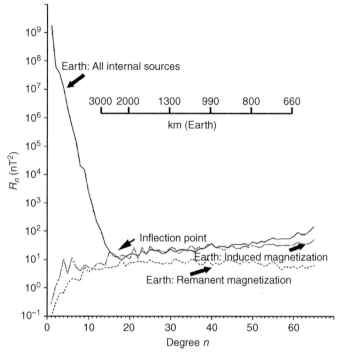

Figure 8 Comparison of the Lowes-Mauersberger (R_n) spectra at the surface of the Earth for a variety of internal fields. The inflection point in the terrestrial power spectra represents the sharp transition from core processes at low n to lithospheric processes at higher n. R_n is the mean square amplitude of the magnetic field over a sphere produced by harmonics of degree n. The spectrum of all internal sources comes from Sabaka et al. (2004), the induced spectrum is derived from Fox Maule et al. (2005), and the remanent magnetization spectrum (of the oceans, and hence a minimum value) was derived from Dyment and Arkani-Hamed (1998).

2002, 2004; Olsen et al., 2006a) has been successful because of the different decay characteristics of the internal and external fields (Langel et al., 1996). This approach utilizes magnetic field measurements from satellites (ionosphere, crust, and core are internal) and the surface (crust and core are internal), and does not separate the internal fields of core and crustal origin, lumping them instead into a spherical harmonic series from spherical harmonic degrees 1–60. These 'comprehensive' models are now widely touted (Nabighian et al., 2005) to replace the existing International Geomagnetic Reference Field Models in many applications.

Another, qualitative, approach to separation is to characterize visually the field, and magnetization solutions deduced from that field, in the hope that features at small scales will provide clues into what is happening at the largest scales. Because of the wider spectral content of recent satellite crustal field models, Purucker and Whaler (2004) were able to recognize two patterns in the vertical component map (**Figure 9**) of the crustal field of the North American region.

The first pattern, which they refer to as 'C', encompasses the North American land mass, the Caribbean and Gulf of Mexico, and northernmost South America. The peak-to-trough magnitude of anomalies in 'C' typically exceeds 50 nT, and the anomalies are either equidimensional or oriented in a direction subparallel to the nearest coastline or tectonic element. The second pattern, which they refer to as 'O', encompasses the Eastern Pacific, the Cocos plate, and the western Atlantic away from continental North America. The peak-to-trough magnitude of anomalies in 'O' is typically less than 30 nT, and the anomalies are commonly narrow and elongate in the direction of the nearest spreading or subduction zone. The 'C' pattern can also be discerned on global maps of the field, when account is taken of the higher altitude. The 'C' pattern is characteristic of much of the Asian landmass, a region centered on but more extensive than Australia, and two broad regions within the African landmass. The 'O' pattern is seen in the eastern Pacific, the North and South Atlantic, and the Indian oceans. The 'C' and 'O' patterns are also evident in a magnetization

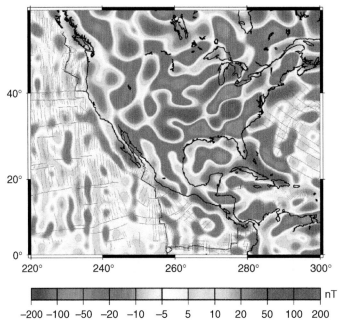

Figure 9   MF-4 vertical field map of the crustal magnetic field (Maus *et al.*, 2006) over North America, evaluated at 50 km altitude.

model based on these observations (Purucker and Whaler, 2004). A map of |**M**| shows these patterns best (**Figure 10**). Regions with magnetizations greater than $0.1 \, A \, m^{-1}$ (red regions on **Figure 10** map) correspond to the 'C' pattern, and regions with magnetization less than $0.06 \, A \, m^{-1}$ (gray regions) correspond to the 'O' pattern.

Intermediate values of magnetization (between 0.06 and $0.1 \, A \, m^{-1}$, pink on above map) generally envelop regions displaying the 'C' pattern. In a general way, the 'C' and 'O' patterns correspond to regions of thick and thin magnetic crustal thickness,

as defined by temperature and seismology in the 3SMAC model (Nataf and Ricard, 1996). There are conspicuous exceptions to this generalization. Most of the South American landmass south of the Equator is characterized by the 'O' pattern, yet crustal thicknesses are typical of continental crust. The other major exception is the Sahara desert, again characterized by the 'O' pattern but with typical continental crustal thicknesses. In both of these regions, seismic crustal thickness and heat flow are poorly known.

Electrical conductivity contrasts (Grammatica and Tarits, 2002) and motional induction by ocean

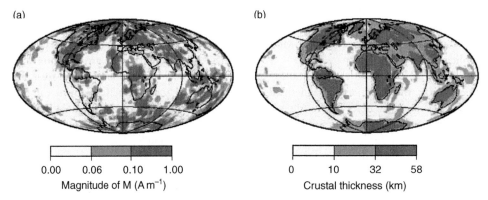

Figure 10   Map showing the magnitude (a) of the magnetization inferred from a model based on MF-3 (Maus *et al.*, 2006), compared with a map of the magnetic crustal thickness (b) from Nataf and Ricard (1996).

currents (Vivier *et al.*, 2004) can also produce quasi-static magnetic fields that overlap with the crustal field. Specific time-variable features of these fields makes separation possible, but their low amplitude makes separation difficult in practice, and no separation has yet been achieved.

6.7 Separation of Induced and Remanent Magnetization

Separation of induced and remanent magnetization is a major outstanding issue. Induced magnetizations in the crust represent one of the largest time-variable geomagnetic fields of internal, non-core origin, and the separation of induced from remanent magnetization remains an area of active research. McLeod (1996) predicted that crustal-source secular variation should dominate core-source secular variation for spherical harmonic degrees in excess of 22. The separation of induced from remanent magnetization can be done without ambiguity only if the source material is available for rock magnetic tests (cf. Chapter 8), and if the time scale is specified. Although magnetizations are often considered either remanent or induced, the distribution of magnetic coercivities in rocks is a continuous function, and viscous remanent magnetizations exist at all time scales.

Both the spatial variability of the static field as a function of inducing field, and an approach using time-variable geomagnetic fields as a probe, have been used to estimate induced magnetizations. Maus and Haak (2002) investigated the long-wavelength power of the crustal field as a function of magnetic dip latitude. Two of the crustal magnetic field models (Sabaka *et al.*, 2002; Cain *et al.*, 1989) they examined showed the strong trend with latitude expected for induced magnetization, while the third model (Arkani-Hamed *et al.*, 1994) they examined was consistent with a purely remanent magnetization. It may be relevant that the models that were consistent with induced magnetization were models of both internal (core and crust) and external fields, while the model that was consistent with remanence was a crustal-only model that employed along-track filtering to remove non-crustal(?) long-wavelength trends. Maus and Haak (2002) also showed that, even in the case of a purely remanent magnetization, ΔT is expected to increase by a factor of two between the equator and the pole, while for a purely induced magnetization, ΔT is expected to increase by a much larger factor. This is due to the preferential

sampling of the weaker tangential part of the crustal field at the equator and the stronger radial part at the poles.

Approaches using time-variable geomagnetic fields can be either direct, using the change of the main geomagnetic field, or indirect, using an EM induction response (Goldstein and Ward, 1966; Yanovsky, 1938; Clark *et al.*, 1998) to time-variable fields like Sq or micropulsations. Lesur and Gubbins (2000) take a direct approach, with a regional study of 20 long-running European observatories. They calculate the difference between the observatory annual mean and the core field model of Bloxham and Jackson (1992). The variability with time of that difference, or observatory bias, was then examined to see if it was consistent with induced or remanent magnetization. They found that for nine observatories a time-dependent induced field fit the data better than a steady remanent field at the 99% confidence level. The other observatories yielded ambiguous results. Some external field signatures remain in observatory annual mean data, and only local distributions of induced and remanent magnetization were considered in the analysis. A global approach to observatory biases by Mandea and Langlais (2002), for observatories operating while the Magsat (1980) and Oersted (2000) missions were in operation, while not attempting to isolate the remanent and induced components, illustrates some of the other difficulties that the direct approach entails. In addition to contamination by external fields, changes in observatory location or measurement practice add to the uncertainty. One can estimate the magnitude of the change in the observatory bias between the Magsat and Oersted epochs, assuming that it is caused solely by changes in the induced magnetization. Values can be as large as 77 nT, with 14 of the 62 observatories having predicted changes in excess of 5 nT. Such a large signal over only 20 years should be measurable, assuming that it can be isolated. Although Mandea and Langlais (2002) determine the biases *a posteriori* the model generation, they could also be solved for in a model, as for example in the Comprehensive model of Sabaka *et al.* (2002).

The separation of induced and remanent magnetization can also be attempted from the spatial distribution of ΔT over isolated magnetic bodies (Zietz and Andreasen, 1967). Zietz and Andreasen (1967) used the position and relative intensity of the maximum and minimum anomaly to infer the declination and inclination of the magnetization vector in the causative body. Schnetzler and Taylor (1984) evaluated this technique globally, and found that the method was more sensitive at higher magnetic latitudes, and that due to zero-level uncertainties, inclination was very difficult to estimate.

Acknowledgments

Purucker is supported under NASA Contract NNG06HX08C. All graphs and maps have been plotted using the Generic Mapping Tools (GMT) software (Wessel and Smith, 1998).

References

Achache JA, Abtout A, and Le Mouël JL (1987) The downward continuation of MAGSAT crustal anomaly field over Southeast Asia. *Journal of Geophysical Research* 92: 11584–11596.

Adam JC and Swarztrauber PN (1997) Spherepack 2.0: A model development facility. *NCAR Technical Note NCAR/TN-436-STR*.

Alldredge LR (1981) Rectangular harmonic analysis applied to the geomagnetic field. *Journal of Geophysical Research* 86: 3021–3026.

Alldredge LR (1982) Geomagnetic local and regional harmonic analysis. *Journal of Geophysical Research* 87: 1921–1926.

Alldredge LR, Vanvoorhis GD, and Davis TM (1963) A magnetic profile around the world. *Journal of Geophysical Research* 68: 3679–3692.

Andreasen G E and Zietz I (1969) Magnetic fields for a 4 × 6 prismatic model. *US Geological Survey Professional Paper 666.* Washington, DC: US Geological Survey.

Arctic Gravity Project (2002) 5′ × 5′; grids compiled by the International Association of Geodesy from airborne, survey, and submarine surveys. [http://earth-info.nga.mil/GandG/wgs84/agp/readme.html]. (accessed Nov 2006).

Arkani-Hamed J, Langel R, and Purucker M (1994) Scalar magnetic anomaly maps of Earth derived from POGO and Magsat data. *Journal of Geophysical Research* 99: 24075–24090.

Arkani-Hamed J and Strangway D (1986) Band-limited global scalar magnetic anomaly map of the Earth derived from Magsat data. *Journal of Geophysical Research* 91: 8193–8203.

Aspler LB, Pilkington M, and Miles WF (2003) Interpretations of Precambrian basement based on recent aeromagnetic data, Mackenzie Valley, Northwest Territories. *Geological Survey of Canada Current Research* 2003-C2: 11.

Aydin I, Karat HI, and Kocak A (2005) Curie-point depth map of Turkey. *Geophysical Journal International* 162: 633–640.

Bankey V, Cuevas A, Daniels D, *et al.* (2002) Digital data grids for the magnetic anomaly map of North America, *USGS Open File Report 02-414.* http://pubs.usgs.gov/of/2002/ofr-02-414 (http://crustal.usgs.gov/geophysics). (accessed Nov 2006).

Baranov V (1957) A new method for interpretation of aeromagnetic maps; pseudo-gravimetric anomalies. *Geophysics* 22: 359–383.

Baranov V and Naudy H (1964) Numerical calculation of the formula of reduction to the magnetic pole. *Geophysics* 29: 67–79.

Barbosa VCF, Silva JBC, and Medeiros WE (1999) Stability analysis and improvement of structural index estimation in Euler deconvolution. *Geophysics* 64: 48–60.

Barritt SA, Fairhead JD, and Misener DJ (1993) The African Magnetic Mapping Project. *ITC Journal* 122–131.

Bean RJ (1966) A rapid graphing solution for the aeromagnetic anomaly of a two-dimensional tabular body. *Geophysics* 31: 963–970.

Bhattacharyya BK (1964) Magnetic anomalies due to prism-shaped bodies with arbitrary polarization. *Geophysics* 29: 517–531.

Bhattacharyya BK (1965) Two-dimensional harmonic analysis as a tool for magnetic interpretation. *Geophysics* 30: 829–857.

Bhattacharyya BK (1966) Continuous spectrum of the total-magnetic-field anomaly due to a rectangular prismatic body. *Geophysics* 31: 97–121.

Bhattacharyya BK (1969) Bicubic spline interpolation as a method for treatment of potential field data. *Geophysics* 34: 402–423.

Bhattacharyya BK (1980) A generalized multibody model for inversion of magnetic anomalies. *Geophysics* 45: 255–270.

Bhattacharyya BK and Chan KC (1977a) Reduction of magnetic and gravity data on an arbitrary surface acquired in a region of high topographic relief. *Geophysics* 42: 1411–1430.

Bhattacharyya BK and Chan KC (1977b) Computation of gravity and magnetic anomalies due to inhomogeneous distribution of magnetization and density in a localized region. *Geophysics* 42: 602–609.

Blakely RJ (1981) A program for rapidly computing the magnetic anomaly over digital topography. *US Geological Survey Open File Report 81–298.*

Blakely RJ (1988) Curie-temperature isotherm analysis and tectonic implications of aeromagnetic data from Nevada. *Journal of Geophysical Research* 93: 11,817–11,832.

Blakely RJ (1995) *Potential Theory in Gravity and Magnetic Application.* New York: Cambridge University Press.

Blakely RJ, Brocher TM, and Wells RE (2005) Subduction zone magnetic anomalies and implications for hydrated forearc mantle. *Geological Society of America Bulletin* 33: 445–448.

Blakely RJ and Connard GG (1989) Crustal studies using magnetic data. In: Pakiser LC and Mooney WD (eds.) *Geophysical Framework of the Continental United States, Geological Society of America Memoir,* 172, Ch. 4, pp. 45–60. Boulder: Geological Society of America.

Blakely RJ and Grauch VJS (1983) Magnetic models of crystalline terrane: Accounting for the effect of topography. *Geophysics* 48: 1551–1557.

Blakely RJ, Jachens RC, Simpson RW, and Couch RW (1985) Tectonic setting of the southern Cascade Range as interpreted from its magnetic and gravity fields. *Geological Society of America Bulletin* 96: 43–48.

Blakely RJ and Simpson RW (1986) Approximating edges of source bodies from magnetic or gravity anomalies. *Geophysics* 51: 1494–1498.

Blakely RJ, Wells RE, Tolan TL, Beeson MH, Trehu AM, and Liberty LM (2000) New aeromagnetic data reveal large strike-slip faults in the northern Willamette Valley, Oregon. *Geological Society of America Bulletin* 112: 1225–1233.

Blakely RJ, Wells RE, Weaver CS, and Johnson SY (2002) Location, structure, and seismicity of the Seattle fault zone, Washington: Evidence from aeromagnetic anomalies, geologic mapping, and seismic-reflection data. *Geological Society of America Bulletin* 114: 169–177.

Blakely RJ, Wells RE, Yelin TS, Madin IP, and Beeson MH (1995) Tectonic setting of the Portland-Vancouver area, Oregon and Washington: Constraints from low-altitude aeromagnetic data. *Geological Society of America Bulletin* 107: 1051–1062.

Bloxham J and Jackson A (1992) Time-dependent mapping of the magnetic field at the core-mantle boundary. *Journal of Geophysical Research* 97: 19537–19563.

Bosum W, Pucher R, and Roeser H (1985) Crustal anomalies and their causes. In: Fuchs K and Soffel H (eds.) *Geophysics of Solid Earth, the Moon and the Planets*, vol V2, subvolume B, Ch 4.2.1, pp. 74–99. Berlin: Springer.

Bott MHP (1963) Two methods applicable to computers for evaluating magnetic anomalies due to finite three dimensional bodies. *Geophysical Prospecting* 11: 292–299.

Bucknam RC, Hemphill-Haley E, and Leopold EB (1992) Abrupt uplift within the past 1700 yr at southern Puget Sound. *Science* 258: 1611–1614.

Cain JC, Davis WM, and Regan RD (1974) An $N = 22$ model of the geomagnetic feld. *EOS, Transaction of the American Geophysical Union* 56: 1108.

Cain JC, Wang Z, Kluth C, and Schmitz DR (1989) Derivation of a geomagnetic model to $n = 63$. *Geophysical Journal* 97: 431–441.

Cady JW (1980) Calculation of gravity and magnetic anomalies of finite-length right polygonal prisms. *Geophysics* 45: 1507–1512.

Campos-Enriquez JO, Dlaz-Navarro R, Espindola JM, and Mena M (1996) Chicxulub – subsurface structure of impact crater inferred from gravity and magnetic data. *The Leading Edge* 15: 357–359.

Caress DW and Parker RL (1989) Spectral interpolation and downward continuation of marine magnetic anomaly data. *Journal of Geophysical Research* 94: 17393–17407.

Chavez Gomez S (2001) A catalogue of dykes from aeromagnetic surveys in eastern and southern Africa. *ITC Publication # 80*.

Chenot D and Debeglia N (1990) Three-dimensional gravity and magnetic constrained depth inversion with lateral and vertical variation of contrast. *Geophysics* 55: 327–335.

Chinese National Aerogeophysics Survey and Remote Sensing Center (1989) *Aeromagnetic Anomaly Map of China and Adjacent Sea, 2nd digital edition in 2004*. Beijing: Geological Publishing House.

Chulick GS and Mooney WD (2002) Seismic structure of the crust and uppermost mantle of North America and adjacent oceanic basins: A synthesis. *Bulletin of the Seismological Society of America* 92: 2478–2492.

Clark DA (1997) Magnetic petrophysics and magnetic petrology; aids to geological interpretation of magnetic surveys. *AGSO Journal of Australian Geology and Geophysics* 17: 83–103.

Clark DA (1999) Magnetic petrology of igneous intrusions: Implications for exploration and magnetic interpretation. *Exploration Geophysics* 30: 5–26.

Clark DA, Schmidt PW, Coward DA, and Huddleston MP (1998) Remote determination of magnetic properties and improved drill targeting of magnetic anomaly sources by differential vector magnetometry. *Exploration Geophysics* 29: 312–312.

Clark SC, Frey H, and Thomas HH (1985) Satellite magnetic anomalies over subduction zones. *Geophysical Research Letters* 12: 41–44.

Cohen Y (1989) Traitements et interpretations de donnees spatiales en geomagnetisme: etude des variations laterales d' aimantation de la lithosphere terrestre, *Docteur es sciences physiques thèse*, Paris, 95pp.

Coleman R (1992) Project magnet high-level vector survey data reduction. In: Langel R and Baldwin R (eds.) *NASA Conference Publication 3153, Types and Characteristics of Data for Geomagnetic field modelling,*, pp. 215–248. Washington, DC: NASA.

Constable CG and Parker RL (1988) Smoothing, splines and smoothing splines. *Journal of computational Physics* 78: 493–508.

Cooper GRJ (2003) Feature detection using sun shading. *Computers and Geosciences* 29: 941–948.

Cooper GRJ and Cowan DR (2003) Sunshading geophysical data using fractional order horizontal gradients. *The Leading Edge* 22: 204.

Cordell L (1992) A scattered equivalent-source method for the interpolation and gridding of potential-field data in three dimensions. *Geophysics* 57: 629–636.

Cordell L and McCafferty AE (1989) A terracing operator for physical property mapping with potential field data. *Geophysics* 54: 621–634.

Cordell L and Taylor PT (1971) Investigation of magnetization and density of North Atlantic Seamount using Poisson's Theorem. *Geophysics* 36: 919–937.

Cowan DR and Cooper GRJ (2005) Enhancement of magnetic signatures of impact structures. In: Kenkmann T, Horz F, and Deutsch A (eds.) *Large Meteorite Impacts,* GSA Special Paper 384, pp. 51–65. Boulder, CO: Geological Society of America.

Cowan DR and Cowan S (1993) Separation filtering applied to aeromagnetic data. *Exploration Geophysics* 24: 429–436.

Craig M (1996) Analytic signals for multivariate data. *Mathematical Geology* 28: 315–329.

Criss RE and Champion DE (1984) Magnetic properties of granitic rocks from the southern half of the Idaho Batholith: Influences of hydrothermal alteration and implications for aeromagnetic interpretations. *Journal of Geophysical Research* 89: 7061–7076.

Davies J, Mushayandebvu MF, and Smith R (2004) Magnetic detection and characterization of tertiary and quaternary buried channels. *SEG Expanded Abstracts* 23: 734.

Dentith M, Cowan D, and Tompkins L (2000) Enhancement of subtle features in aeromagnetic data. *Exploration Geophsics* 31: 104–108.

De Santis A (1991) Translated origin spherical cap harmonic analysis. *Geophysical Journal International*.

Dessler AJ (1986) The evolution of arguments regarding the existence of field-aligned currents. In: Potemra T (ed.) *Magnetospheric Currents*, Geophysical Monograph 28, pp. 22–28. Washington, DC: American Geophysical Union.

Dimri VP (1998) Fractal behaviour and detectability limits of geophysical surveys. *Geophysics* 63: 1943–1946.

Dittmer JK and Szymanski JE (1995) The stochastic inversion of magnetics and resistivity data using the simulated annealing algorithm. *Geophysical Prospecting* 43(3): 397–416.

Dyment J and Arkani-Hamed J (1998) Equivalent sources magnetic dipoles revisited. *Geophysical Research Letters* 25: 2003–2006.

Dyment J and Arkani-Hamed J (1999) Contibution of lithospheric remanent magnetization to satellite magnetic anomalies over the world's oceans. *Journal of Geophysical Research* 103: 15423–15441.

Eaton D and Vasudevan K (2004) Skeletonization of aeromagnetic data. *Geophysics* 69: 478–488.

Elkins-Tanton L and Bowring S (2006) Report on the workshop on the Siberian flood basalts and the end-Permian extinction, NSF-sponsored workshop, Sep 2005.

Ernst RE, Buchan TD, West TD, and Palmer HC (1996) Diabase (dolerite) dyke swarms of the world: First edition, 1:35,000,000 map, 104pp. *Geological Survey of Canada Open File 3241*.

Ernst RE and Buchan KL (1997) Giant radiating dyke swarms: Their use in identifying Pre-Mesozoic large igneous provinces and mantle plumes. In: Mahoney JJ and Coffin MF (eds.) *Large Igneous Provinces: Continental, Oceanic, and Planetary Flood Volcanism,* American Geophysical Union Monograph 100, pp. 297–333. Washington, DC: American Geophysical Union.

Ernst RE and Buchan KL (2001) The use of mafic dike swarms in identifying and locating mantle plumes. In: Ernst RE and

Buchan KL (eds.). *Mantle Plumes: Their Identification Through Time, Geological Society of America Special Paper 352*, pp. 247–265. Boulder CO: Geological Society of America.

Erwin DE (1994) The Permo-Triassic extinction. *Nature* 367: 231–236.

Finn C (1990) Geophysical constraints on Washington convergent margin structure. *Journal of Geophysical Research* 95: 19533–19546.

Fox Maule C, Purucker M, Olsen N, and Mosegaard K (2005) Heat flux anomalies in Antarctica revealed by satellite magnetic data. *Science* 309: 464–467.

Frost BR (1991a) Magnetic petrology: Factors that control the occurrence of magnetite in crustal rocks. In: Lindsley DH (ed.) *Oxide Minerals: Petrologic and Magnetic Significance*, vol. 25, Ch. 14, pp. 489–509. Blacksburg, VA: Mineralogical Society of America.

Frost BR (1991b) Stability of oxide minerals in metamorphic rocks. In: Lindsley DH (ed.) *Oxide Minerals: Petrologic and Magnetic Significance*, vol. 25, Ch. 13, pp. 469–488. Blacksburg, VA: Mineralogical Society of America.

Frost BR (1991c) Introduction to oxygen fugacity and its petrologic importance. In: Lindsley DH (ed.) *Oxide Minerals: Petrologic and Magnetic Significance*, vol. 25, Ch. 1, pp. 1–8. Blacksburg, VA: Mineralogical Society of America.

Frost BR and Shive PN (1989) Comment on limiting depth of magnetization in Cratonic Lithosphere. *Geophysical Research Letters* 16: 477–479.

Geological Survey of Canada (1995) Magnetic anomalies and tectonic elements of Northeast Eurasia. GSC Open-File Report 2574.

Geological Survey of Japan (2002) Magnetic anomaly map of east Asia. *Geological Survey of Japan Digital Geoscience Map P-3* 2nd edn. Tsukuba, Japan: Geological Survey of Japan.

Gerovska D and Araúzo-Bravo MJ (2003) Automatic interpretation of magnetic data based on Euler deconvolution with unprescribed structural index. *Computers and Geosciences* 29: 949–960.

Getech (1996) http://www.getech.com/data.

Gibson RI (1998) Gravity and magnetics in oil exploration: A historical perspective. In: *Geologic Applications of Gravity and Magnetics: Case Histories, Society of Exploration Geophysics. Geophysical Reference Series 8*. Tulsa, OK: Society of Exploration Geophysicists.

Gibson RI (1998) Interpretation of Rift-stage Faulting in the West Siberian Basin from Magnetic Data. In: Gibson RI and Millegan PS (eds.) *Geologic Applications of Gravity and Magnetics*. Exploration Geophysics Geophysical Reference Series 8, pp. 79–81. Tulsa, OK: Society of Exploration Geophysicists.

Gibson RI and Millegan PS (eds.) (1998) *Geologic Applications of Gravity and Magnetics: Case Histories, Society of Exploration Geophysics Geophysical Reference Series 8:* Tulsa, OK: Society of Exploration Geophysicists.

Girdler RW, Taylor PT, and Frawley JJ (1992) A possible impact origin for the Bangui magnetic anomaly (Central Africa). *Tectonophysics* 212: 45–58.

Goldhaber MB and Reynolds RL (1991) Relations among hydrocarbon reservoirs, epigenetic sulfidization, and rock magnetization: Examples from the south Texas costal plain. *Geophysics* 56: 748–757.

Goldstein NE and Ward SH (1966) The separation of remanent from induced magnetism. *in situ. Geophysics* 31: 779–796.

Golynsky AM, Chiappini M, Damaske D, *et al.* (2001) ADMAP – Magnetic anomaly map of the Antarctic, 1:10 000 000 scale map. In: Morris PR and von Frese R (eds.) *BAS (Misc.) 10*, Cambridge: British Antarctic Survey.

Goussev SA, Charters RA, Peirce JW, and Glenn WE (2003) The Meter Reader – Jackpine magnetic anomaly: Identification of a buried meteorite impact structure. *The Leading Edge* 22: 740–741.

Grace JD and Hart GF (1990) Urengoy gas field – USSR West Siberian basin, Tyumen District. In: Beaumont EA and Foster NH (eds.) *Structural Traps III, Tectonic Old and Fault Traps, AAPG Treatise of Petroleum Geology, Atlas of Oil and Gas fields*, A-109, pp. 309–335. Tulsa, OK: Society of Exploration Geophysicists.

Grant FS (1985) Aeromagnetics, geology and ore environments: In: Magnetite in igneous, sedimentary and metamorphic rocks: An overview. *Geoexploration* 23: 303–333.

Grant FS and West GF (1965) *Interpretation Theory in Applied Geophysics*. New York: McGraw-Hill.

Grammatica N and Tarits P (2002) Contribution at satellite altitude of electromagnetically induced anomalies from a three-dimensional heterogeneously conducting Earth, using SQ as an inducing source field. *Geophysical Journal International* 151: 913–923.

Grauch VJS (1987) A new variable-magnetization terrain correction method for aeromagnetic data. *Geophysics* 52: 94–107.

Grauch VJS and Cordell L (1987) Limitations of determining density or magnetic boundaries from the horizontal gradient of gravity or pseudogravity data. *Geophysics* 52: 118–121.

Grieve R and Therriault A (2000) Vredefort, Sudbury, Chicxulub: Three of a kind? *Annual Review of Earth and Planetary Science* 28: 305–338.

Gubbins D and Herrero-Bervera E (eds.) (2007) *Encyclopedia of Geomagnetism and Paleomagnetism*. Berlin: Springer.

Gunn PJ (1975) Linear transformations of gravity and magnetic fields. *Geophysical Prospecting* 23: 300–312.

Gunn PJ (ed.) (1997) Airborne magnetic and radiometric surveys. *AGSO Journal* 17: 2.

Haggerty SE (1976) Opaque mineral oxides in terrestrial igneous rocks. In: Rumble D (ed.) *Oxide Minerals*, vol. 3., Ch. 4, pp. 1–98. Blacksburg: Mineralogical Society of America.

Haggerty SE (1979) The aeromagnetic mineralogy of igneous rocks. *Canadian Journal of Earth Sciences* 16: 1281–1293.

Hahn A, Ahrendt H, Meyer J, and Hufen JH (1984) A model of magnetic sources within the Earth's crust compatible with the field measured by the satellite Magsat. *Geologisches Jahrbuch* A75: 125–156.

Hahn A and Bosum W (1986) *Geomagnetics: Selected Examples and Case Histories*. Berlin: Geopublication Associates.

Haines GV (1985) Spherical cap harmonic analysis of geomagnetic secular variation over Canada 1960–1983. *Journal of Geophysical Research* 90: 12563–12574.

Hanna WF (ed.) (1987) *Geologic applications* of modern aeromagnetic surveys. USGS Bulletin, 1924, 106 pp.

Hansen RO (1993) Interpretative gridding by anisotropic kriging. *Geophysics* 58: 1491–1497.

Hansen RO (2005) 3D multiple-source Werner deconvolution for magnetic data. *Geophysics* 70: L45–L51.

Hansen RO and Miyazaki Y (1984) Continuation of potential fields between arbitrary surfaces. *Geophysics* 49: 787–795.

Hansen RO and Pawlowski RS (1989) Reduction to the pole at low latitudes by Wiener filtering. *Geophysics* 54: 1607–1613.

Hansen RO and Simmonds M (1993) Multiple-source Werner deconvolution. *Geophysics* 58: 1792–1800.

Hansen RO and Suciu L (2002) Multiple-source Euler deconvolution. *Geophysics* 67: 525–535.

Hansen RO and Wang X (1988) Simplified frequency-domain expressions for potential fields of arbitrary three-dimensional bodies. *Geophysics* 53: 365–374.

Harrison C (1987) The crustal field. In: Jacobs J (ed.) *Geomagnetism*, vol 1, Ch. 5, pp. 513–610. London: Academic Press.

Heirtzler JR, Dickson GO, Herron EM, Pitman WC, and Le Pichon X (1968) Marine magnetic anomalies, geomagnetic field reversals, and motions of the ocean floor and continents. *Journal of Geophysical Research* 73: 2119–2136.

Hemant K and Maus S (2006) Geological modeling of the new CHAMP magnetic anomaly maps using a Geophysical Information System (GIS) technique. *Journal of Geophysical Research.* 110: doi: 10.1029/2005JB003837.

Hemant K and Maus S (2005) Why no anomaly is visible over most of the continent-ocean boundary in the global crustal magnetic field. *Physics of the Earth and Planetary Interiors* 149: 321–333.

Hernandez I, Cuevas-Covarrubias A, Campos-Enriquez JO, and Urrutia-Fucugauchi J (2001) Aeromagnetic map of Mexico: An exploration approach for the new millennium – A progress report. *Revista Geofisica Instituto Panamericano de Geografia e Historia* 55, 33pp.

Hildebrand AR, Penfield GT, Kring DA, et al. (1991) Chicxulub crater: A possible Cretaceous–Tertiary boundary impact crater on the Yucatan peninsula. *Geology* 19: 867–871.

Hill PL (1986) Bibliographies and location maps of aeromagnetic and aeroradio metric publications for the United States. *US Geological Survey Open-File Report 86-525*, Reston, Virginia.

Hinze W (1979) Continental magnetic anomalies. *Reviews of Geophysics and Space Physics* 17: 257–273.

Hinze W, et al. (1988) Magnetic anomaly map of North America. *The Leading Edge* 7: 19–21.

Hinze W, Von Frese R, and Ravat D (1991) Mean magnetic contrast between oceans and continents. *Tectonophysics* 192: 117–127.

Hinze WJ (ed.) (1985) *The Utility of Regional Gravity and Magnetic Anomaly Maps*. Tulsa, OK: Society of Exploration Geophysicists.

Hinze WJ and Zietz I (1985) The composite magnetic-anomaly map of the conterminous United States. In: Hinze WJ (ed.) *The Utility of Regional Gravity and Magnetic Anomaly Maps*, pp. 1–24. Tulsa, OK: Society of Exploration Geophysicists.

Hitzman MW, Oreskes N, and Einaudi MT (1992) Geological characteristics and tectonic setting of Proterozoic iron oxide (Cu–U–Au–REE) deposits. *Precambrian Research* 58: 241–287.

Hood PJ, McGrath PH, and Teskey DJ (1985) Evolution of Geological Survey of Canada magnetic-anomaly maps. A Canadian perspective. In: Hinze WJ (ed.) *The Utility of Regional Gravity and Magnetic Anomaly Maps*, pp. 62–87. Tulsa, OK: Society of Exploration Geophysicists.

Hornby P, Boschetti F, and Horowitz FG (1999) Analysis of potential field data in the wavelet domain. *Geophysical Journal International* 137: 175–196.

Hsu S-K, Coppens D, and Shyu C-T (1998) Depth to magnetic source using the generalized analytic signal. *Geophysics* 63: 1947–1957.

Hsu S-K, Sibuet J-C, and Shyu C-T (1996) High-resolution detection of geologic boundaries from potential-field anomalies: An enhanced analytic signal technique. *Geophysics* 61: 373–386.

Hyndman RD and Peacock SM (2003) Serpentinization of the forearc mantle. *Earth and Planetary Science Letters* 212: 417–432.

Ishihara S (1981) The granitoid series and mineralization. In: Skinner BJ *Economic Geology 75th Anniversary Issue* 458–484.

Ishihara T (2004) Application of CM3 model in compilation of marine magnetic anomaly data of North Pacific. *Transactions of the American Geophysical Union* 85(47): Fall Meeting Supplement Abstract GP11D-0882.

Jackson A (1990) Accounting for crustal magnetization in models of the core magnetic field. *Geophysical Journal International* 103: 657–673.

Jackson A (1994) Statistical treatment of crustal magnetization. *Geophysical Journal International* 119: 991–998.

Jackson A (1996) Bounding the long wavelength crustal magnetic field. *Physics of the Earth and Planetary Interiors* 98: 283–302.

Jackson DD (1972) Interpretation of inaccurate, insufficient and inconsistent data. *Geophysical Journal of Royal Astronomical Society* 28: 97–109.

Jansen J and Witherly K (2004) The Tli Kwi Cho Kimberlite Complex, Northwest Territories, Canada: A Geophysical Case Study. *Society of Exploration of Geophysics 74st Annual Mtg Tech Prog* (Ext. Abstr), Denver, CO. 23: 1147–1149.

Jessel MW (2001) Three-dimensional modeling of potential-field data. *Computers and Geosciences* 27: 455–465.

Jessel MW and Fractal Geophysics Pty Ltd (2002) An atlas of structural geophysics II. *Journal of the Virtual Explorer* 5 available online at http://www.mssu.edu/seg-vm/exhibits/structuralatlas.

Jonson SY, et al. (1996) The southern Whidbey Island fault: An active structure in the Puget Lowland. *Geological Society of America Bulletin* 108: 334–354.

Johnson SY, Dadisman SV, Mosher DC, Blakely RJ, and Childs JR (2001) Active tectonics of the Devils Mountain fault and related structures, northern Puget Lowland and eastern Strait of Juan de Fuca region, Pacific Northwest. *US Geological Survey Professional Paper 1643*, 45pp, 2 plates.

Jourdan F, Fraud G, Bertrand H, Watkeys MK, Kampunzu AB, and Le Gall B (2006) Basement control on dyke distribution in Large Igneous Provinces: Case study of the Karoo triple junction. *Earth and Planetary Science Letters* 241: 307–322.

Kanasewich ER and Agarwal ER (1970) Analysis of combined gravity and magnetic fields in wavenumber domain. *Journal of Geophysical Research* 75: 5702–5712.

Keating P (1993) The fractal dimension of gravity data sets and its implication for gridding. *Geophysical Prospecting* 41: 983–994.

Keating P (1995) A simple technique to identify magnetic anomalies due to kimberlite pipes. *Exploration Mining Geology* 4: 121–125.

Keating P and Sailhac P (2004) Use of the analytic signal to identify magnetic anomalies due to kimberlite pipes. *Geophysics* 69: 180–190.

Keating P and Zerbo L (1996) An improved technique for reduction-to-the pole at low latitudes. *Geophysics* 61: 131–137.

Keller GR (1988) The development of gravity and magnetic studies, emphasizing articles published in the GSA Bulletin. *Geological Society of America Bulletin* 100: 469–478.

Kellogg OD (1953) *Foundations of Potential Theory*. New York: Dover.

Korhonen JV (1993) One hundred seventy eight thousand petrophysical parameter determinations from the regional Petrophysical Programme. In: Autio S (ed.) *Geological Survey of Finland, Current Research 1991–1992* Geological Survey of Finland. Special Paper 18, pp. 137–141. Espoo: Geological Survey of Finland.

Koulomzine T, Lamontagne Y, and Nadeau A (1970) New methods for the direct interpretation of magnetic anomalies caused by inclined dikes of infinite length. *Geophysics* 35: 812–830.

Krajick K (2001) *Barren Lands: An Epic Search for Diamonds in the North American Arctic*. New York: Henry Hold.

Ku CC and Sharp JA (1983) Werner deconvolution for automated magnetic interpretation and its refinement using Marquardt inverse modeling. *Geophysics* 48: 754–774.

LaBrecque JL and Ghidella ME (1997) Depth of magnetic basement, and sediment thickness estimates from aerogeophysical data over the western Weddell Basin. *Journal of Geophysical Research* 102: 7929–7945.

LaBrecque JL and Raymond CA (1985) Seafloor spreading anomalies in the MAGSAT field of the North Atlantic. *Journal of Geophycisal Research* 90: 2565–2575.

Langel R and Benson BV (1987) The Magsat bibliography. *NASA Technical Memorandum 87822*. Washington, DC: NASA.

Langel R and Hinze W (1998) *The Magnetic Field of the Earth's Lithosphere*. Cambridge: Cambridge University Press.

Langel RA and Estes RH (1982) A geomagnetic field spectrum. *Geophysical Research Letters* 9: 250–253.

Langel RA and Whaler KA (1996) Maps of the magnetic anomaly field at Earth's surface from scalar satellite data. *Geophysical Research Letters* 23: 41–44.

Langel RA, Purucker ME, and Rajaram M (1993) The equatorial electrojet and associated currents as seen in the Magsat data. *Journal of Atmospheric and Terrestrial Physics* 55: 1233–1269.

Langel RA, Sabaka TJ, Baldwin RT, and Conrad JA (1996) The near-Earth magnetic field from magnetospheric and quiet-day ionospheric sources and how it is modelled. *Physics of the Earth and Planetary Interiors* 98: 235–267.

Langel RA, Slud VE, and Smith PJ (1984) Reduction of satellite magnetic anomaly data. *Journal of Geophysics* 54: 207–212.

Langlais B, Purucker M, and Mandea M (2004) Crustal magnetic field of Mars. *Journal of Geophysical Research* 109: E02008 doi:10.1029/2033JE002048.

Leaman DE (1994) Criteria for evaluation of potential field interpretations. *First Break* 12: 181–191.

LeSchack LA and Van Alstine DR (2002) High-resolution ground-magnetic (HRGM) and radiometric surveys for hydrocarbon exploration: Six case histories in Western Canada. In: Schumacher D and LeSchack LA (eds.) *Surface Exploration Case Histories: Applications of Geochemistry, Magnetics, and Remote Sensing*. AAPG Studies in Geology No. 48 and SEG Geophysical Reference Series No. 11, pp. 67–156. Tulsa, OK: American Association of Petroleum Geologists.

Lesur V and Gubbins D (1999) Evaluation of fast spherical transforms for geophysical applications. *Geophysical Journal Internationl* 139: 547–555.

Lesur V and Gubbins D (2000) Using geomagnetic secular variation to separate remanent and induced sources of the crustal magnetic field. *Geophysical Journal International* 142: 889–897.

Lesur V and Maus S (2006) A global lithospheric magnetic field model with reduced noise in the Ploar regions. *Geophysical Research Letters* 33: doi:10.1029/2006GL025826.

Li X (2003) On the use of different methods for estimating magnetic depth. *The Leading Edge* 22: 1090–1099.

Li Y and Oldenburg DW (1996) 3-D inversion of magnetic data. *Geophysics* 61: 394–408.

Li Y and Oldenburg DW (2003) Fast inversion of large-scale magnetic data using wavelet transforms and logarithmic barrier method. *Geophysical Journal International* 152: 251–265.

Littke R, *et al.* (1999) Gas generation and accumulation in the West Siberian Basin. *AAPG Bulletin* 83: 1642–1665.

Lühr H, Maus S, and Rother M (2004) Noon-time equatorial electrojet: Its spatial features as determined by the CHAMP satellite. *Journal of Geophysical Research* 109: A01306, doi:10.1029/2002JA009656.

Lindsley DH (ed.) (1991) *Oxide Minerals: Petrologic and Magnetic Significance, vol. 25*. Blacksburg, VA: Mineralogical Society of America.

Lowe DAJ, Parker RL, Purucker ME, and Constable CG (2001) Estimating the crustal power spectrum from vector Magsat data. *Journal of Geophysical Research* 106: 8589–8598.

Machel HG and Burton EA (1991) Chemical and microbial processes causing anomalous magnetization in environments affected by hydrocarbon seepage. *Geophysics* 56: 598–605.

Macnae JC (1979) Kimberlites and exploration geophysics. *Geophysics* 44: 1395–1416.

Macnab R, *et al.* (1995) New database documents the magnetic character of the Arctic and North Atlantic. *EOS, Transactions of the American Geophysical Union* 76(45): 449–458.

Makarova ZA (ed.) (1974) Map of the anomalous magnetic field of the Territory of the USSR and adjacent marine areas, scale 1:2,500,000 (18 sheets). *USSR Ministry of Geology*. Leningrad: VSEGEI.

Malin SRC, Duzgit Z, and Baydemir N (1996) Rectangular harmonic analysis revisited. *Journal of Geophysical Research* 101: 28205–28209.

Mandea M and Langlais B (2002) Observatory crustal magnetic biases during MAGSAT and Orsted satellite missions. *Geophysical Research Letters* 29: http://dx.doi.org/10.1029/2001GL013693.

Mandea M and Purucker M (2005) Observing, modeling, and interpreting magnetic fields of the solid Earth. *Surveys in Geophysics* http://dx.doi.org/10.1007/s10712-005-3857-x.

Marsh JS (2005) DISCUSSION: The geophysical mapping of Mesozoic dyke swarms in southern Africa and their origin in the disruption of Gondwana. *Journal of African Earth Sciences* 35: 525–527.

Maus S and Dimri VP (1996) Depth estimation from the scaling power spectrum of potential fields? *Geophysical Journal International* 124: 113–120.

Maus S, Gordon D, and Fairhead JD (1997) Curie temperature depth estimation using a self-similar magnetisation model. *Geophysical Journal International* 129: 163–168.

Maus S and Haak V (2002) Is the long wavelength crustal magnetic field dominated by induced or by remanent magnetisation? *Journal of Indian Geophysical Union* 6: 1–5.

Maus S and Haak V (2003) Magnetic field annihilators: Invisible magnetisation at the magnetic equator. *Geophysical Journal International* 155: 509–513.

Maus S and Macmillan S (2005) 10th generation international geomagnetic reference field. *EOS Transactions of the American Geophysical Union* 86: 159.

Maus S, Rother M, Hemart K, *et al.* (2006) Earth's lithospheric magnetic field determined to spherical harmonic degree 90 from CHAMP satellite measurements. *Geophysical Journal International* 164: 319–330.

May PR (1971) Pattern of Triassic–Jurassic diabase dyks around the North Atlantic in context of predrift position of the continents. *Geological Society of America Bulletin* 82: 1285–1292.

Mayhew M (1982) Application of satellite magnetic anomaly data to Curie isotherm mapping. *Journal of Geophysical Research* 87: 4846–4854.

Mayhew M and Estes R (1983) Equivalent source modeling of the core magnetic field using Magsat data. *Journal of Geomagnetism and Geoelectricity* 35: 119–130.

Mayhew M and Johnson BD (1987) An equivalent layer magnetization model for Australia based on Magsat data. *Earth and Planetary Science Letters* 83: 167–174.

Mayhew M, Johnson BD, and Wasilewski PJ (1985) A review of problems and progress in studies of satellite magnetic anomalies. *Journal of Geophysical Research* 90: 2511–2522.

Mayhew M and LaBrecque J (1987) Crustal geologic studies with Magsat and surface magnetic data. *Reviews of Geophysics* 25: 971–981.

McEnroe SA, Harrison RJ, Robinson P, Golla U, and Jercinovic MJ (2001a) Effect of fine-scale microstructures in titanohematite on the acquisition and stability of natural remanent magnetization in granulite facies metamorphic rocks, southwest Sweden: Implications for crustal magnetism. *Journal of Geophysical Research* 106: 30523–30546.

McEnroe SA, Robinson P, and Panish P (2001b) Aeromagnetic anomalies, magnetic petrology and rock magnetism of hemo-ilmenite-and magnetite-rich cumulates from the Sokndal Region, South Rogaland, Norway. *American Mineralogist* 86: 1447–1468.

McEnroe SA and Brown LL (2000) A closer look at remanence-dominated anomalies: Rock-magnetic properties and magnetic mineralogy of the Russell Belt microcline-sillmanite gneisses, Northwest Adirondacks Mountains, New York. *Journal of Geophysical Research* 105: 16,437–16,456.

McLeod MG (1996) Spatial and temporal power spectra of the geomagnetic field. *Journal of Geophysical Research* 101: 2745–2763.

Meissner R (1986) *The Continental Crust – A Geophysical Approach,* International Geophysics Series 34. San Diego, CA: Academic.

Milligan PR, Franklin R, and Ravat D (2005) Fourth edition Magnetic Anomaly Map of Australia, derived from a new-generation Magnetic Anomaly Grid Database of Australia (MAGDA), International Association of Geomagnetism Aeronomy, IAGA2005-A-01186, Toulouse, France.

Milligan PR and Franklin R (2004) Magnetic anomaly map of Australia, Geoscience Australia, 1:25000000.

Milligan PR, Franklin R, and Ravat D (2004) A new generation of Magnetic Anomaly Grid Database of Australia (MAGDA) – Use of independent data increases the accuracy of long wavelength components of continental-scale merges. *Preview (Australian Society of Exploration Geophysics)* 113: 25–29.

Modisi MP, Atekwana EA, Kampunzu AB, and Ngwisanyi TH (2000) Rift kinematics during the incipient stages of continental extension: Evidence from the nascent Okavango rift basin, northwest Botswana. *Geology* 28: 939–942.

Mooney WD, Laske G, and Masters TG (1998) Crust 5.1: A global crustal model at 5 × 5. *Journal of Geophysical Research* 103: 727–748.

Moreau F, Gibert D, Holschneider M, and Saracco G (1997) Wavelet analysis of potential fields. *Inverse Problems* 13: 165–178.

Mosegaard K and Sambridge M (2002) Monte Carlo analysis of Inverse Problems. *Inverse Problems* 18: R29–R54.

Mubu MS (1995) *Aeromagnetic Mapping and Interpretation of Mafic Dyke Swarms in Southern Africa.* MSc Thesis, ITC Delft, Netherlands, 63pp.

Mushayandebvu MF, Lesur V, Reid AB, and Fairhead JD (2004) Grid Euler deconvolution with constraints for 2D structures. *Geophysics* 69: 489–496.

Mushayandebvu MF, van Driel P, Reid AB, and Fairhead JD (2001) Magnetic source parameters of two-dimensional structures using extended Euler deconvolution. *Geophysics* 66: 814–823.

Nabighian MN (1972) The analytic signal of two-dimensional magnetic bodies with polygonal cross-section: Its properties and use for automated interpretation. *Geophysics* 37: 507–517.

Nabighian MN (1984) Toward a three-dimensional automatic interpretation of potential field data via generalized Hilbert transforms: Fundamental relations. *Geophysics* 49: 780–786.

Nabighian MN and Asten MW (2002) Metalliferous mining geophysics – State of the art in the last decade of the century and the beginning of the new millennium. *Geophysics* 67: 964–978.

Nabighian MN, Grauch VJS, Hansen RO, et al. (2005) The historical development of the magnetic method in exploration. *Geophysics* 70: 33–61.

Nabighian MN and Hansen RO (2001) Unification of Euler and Werner deconvolution in three dimensions via the generalized Hilbert transform. *Geophysics* 66: 1805–1810.

Nakagawa I and Yukutake T (1985) Rectangular harmonic analysis of geomagnetic anomalies derived from Magsat data over the area of the Japanese islands. *Journal of Geomagnetisn and Geolectricity* 37: 957–977.

Naidu PS (1970) Fourier transform of large scale aeromagnetic field using a modified version of Fast Fourier Transform. *Pure and Applied Geophysics* 81: 17–25.

Nataf H and Ricard Y (1996) An a priori tomographic model of the upper mantle based on geophysical modeling. *Physics of the Earth and Planetary Interiors* 95: 101–122.

Naudy H (1971) Automatic determination of depth on aeromagnetic profiles. *Geophysics* 36: 717–722.

Neil C, Whaler KA, and Reid AB (1991) Extension to Euler's method for three-dimensional potential field interpretation. 53rd EAEG (European Assoc. of Exploration Geophysics) Meeting Florence, Italy, Expanded Abstracts.

Nelson JB (1988) Calculation of the magnetic gradient tensor from total field gradient measurements and its application to geophysical interpretation. *Geophysics* 53: 957–966.

NGDC (1996) Magnetic anomaly data of the former USSR, US Naval Oceanographic Office, in collaboration with the Ministry of Geology of the USSR, 1 CD-ROM.

O'Brien DP (1972) CompuDepth, a new method for depth-to-basement computation. *Presented at the 42nd Annual International Society of Exploration Geophysicists.*

O'Brien MS and Parker RL (1994) Regularized geomagnetic field modeling using monoples. *Geophysical Journal International* 118: 566–578.

O'Connell MD, Smith RS, and Vallee MA (2005) Gridding aeromagnetic data using longitudinal and transverse gradients with the minimum curvature operator. *The Leading Edge* 24: 142–145.

Olsen N, Lühr H, Sabaka TJ, et al. (2006a) CHAOS-A model of Earth's magnetic field derived from CHAMP, Orsted, and SAC-C magnetic field data. *Geophysical Journal International* 166: 67–75.

Olsen N, Haagmans R, and Sabaka T (2006b) The Swarm end-to-end mission simulator study: A demonstration of separating the various contributions to Earth's magnetic field using synthetic data. *Earth Planets Space* 58: 359–370.

Parker RL (1972) Inverse theory with grossly inadequate data. *Geophysical Journal of Royal Astronomical Society* 29: 123–138.

Parker RL (1973) The rapid calculation of potential anomalies. *Geophysical Journal of Royal Astronomical Society* 31: 447–455.

Parker RL (1974) Best bounds on density and depth from gravity data. *Geophysics* 39: 644–649.

Parker RL (1975) The theory of ideal bodies for gravity interpretation. *Geophysical Journal of Royal Astronomical Society* 42: 315–334.

Parker RL (1986) Harmonic splines in geomagnetism, in *Function Estimation. American Mathematical Society Proceedings of Contemporary Mathematics* 59: 63–76.

Parker RL (1988) A statistical theory of seamount magnetism. *Journal of Geophysical Research* 93: 3105–3115.

Parker RL (1991) A theory of ideal bodies for seamount magnetism. *Journal of Geophysical Research* 96: 16101–16112.

Parker RL (1997) Coherence of signals from magnetometers on parallel paths. *Journal of Geophysical Research* 102: 5111–5117.

Parker RL (2003) Ideal bodies for Mars magnetics. *Journal of Geophysical Research* 108: E1 5006, doi: 10.1029/2001JE001760.

Parker RL and Huestis SP (1974) Inversion of magnetic anomalies in the presence of topography. *Journal of Geophysical Research* 79: 1587–1593.

Parker RL and Klitgord KD (1972) Magnetic upward continuation from an uneven track. *Geophysics* 37: 662–668.

Parker RL and O'Brien MS (1997) Spectral analysis of vector magnetic field profiles. *Journal of Geophysical Research* 102: 24815–24822.

Parker RL and Shure L (1982) Efficient modeling of the Earth's magnetic field with harmonic splines. *Geophysical Research Letters* 9: 812–815.

Parker RL and Shure L (1985) Gravitational and magnetic fields of some simple solids of revolution. *Geophysical Journal of Royal Astronomical Society* 80: 631–647.

Parker RL, Shure L, and Hildebrand J (1987) The application of inverse theory to seamount magnetism. *Reviews of Geophysics* 25: 17–40.

Paterson NR and Reeves CV (1985) Applications of gravity and magnetic surveys: The state-of-the-art in 1985. *Geophysics* 50: 2558–2594.

Pedersen LB (1977) Interpretation of potential field data- A generalized inverse approach. *Geophysical Prospecting* 25: 199–230.

Pedersen LB (1978) Wavenumber domain expressions for potential fields from arbitrary 2-, $2^{1/2}$-, and 3-dimensional bodies. *Geophysics* 43: 626–630.

Pedersen LB (1979) Constrained inversion of potential field data. *Geophysical Prospecting* 27: 726–748.

Pedersen LB (1991) Relations between potential fields and some equivalent sources. *Geophysics* 56: 961–971.

Pedersen LB and Rasmussen TM (1990) The gradient tensor of potential field anomalies: Some implications on data collection and data processing of maps. *Geophysics* 55: 1558–1566.

Penfield GT and Camargo ZA (1981) Definition of a major igneous zone in the central Yucatan with aeromagnetics and gravity. *Society of Exploration Geophysics 51st Annual Mtg Tech Prog* (Abstr), 37 Los Angeles, CA.

Peters LJ (1949) The direct approach to magnetic interpretation and its practial application. *Geophysics* 14: 290–320.

Phillips JD (1997) Potential-field geophysical software for the PC-version 2.2. *US Geological Surver Open-File Report 97-725*, 34 pp.

Phillips JD, Reynolds RL, and Frey H (1991) Crustal structure interpreted from magnetic anomalies. *Reviews of Geophysics Supplement* 416–427.

Pilkington M (1997) 3-D magnetic imaging using conjugate gradients *Geophysics* 62: 1132–1142.

Pilkington M and Crossley DJ (1986) Determination of crustal interface topography from potential fields. *Geophysics* 51: 1277–1284.

Pilkington M and Grieve RAF (1992) The geophysical signature of terrestrial impact craters. *Reviews of Geophysics* 30: 161–181.

Pilkington M and Hildebrand AR (2000) Three-dimensional magnetic imaging of the Chicxulub crater. *Journal of Geophysical Research* 105: 23,479–23,491.

Pilkington M and Keating P (2004) Contact mapping from gridded magnetic data – A comparison of techniques. *Exploration Geophysics* 35: 306–311.

Pilkington M and Todoeschuck JP (1995) Scaling nature of crustal susceptibilities. *Geophysical Research Letters* 22: 779–782.

Plouff D (1976) Gravity and magnetic fields of polygonal prisms and application to magnetic terrain corrections. *Geophysics* 41: 727–741.

Power M, Belcourt G, and Rockel E (2004) Geophysical methods for kimberlite exploration in northern Canada. *The Leading Edge* 23: 1124–1129.

Press WH, Teukolsky SA, Vetterling WT, and Flannery BP (1992) Numerical Recipes in Fortran. *The Art of Scientific Computing,* 2nd edn. Cambridge: Cambridge Press.

Press WH, Teukolsky SA, Vetterling WT, and Flannery BP (1996) Numerical Recipes in Fortran 90. *The Art of Scientific Computing.* Cambridge: Cambridge Press.

Press WH, Teukolsky SA, Vetterling WT, and Flannery BP (1997) Numerical Recipes in C. *The Art of Scientific Computing,* 2nd edn. Cambridge: Cambridge Press.

Prieto C and Morton G (2003) New insights from a 3D Earth model, deepwater Gulf of Mexico. *The Leading Edge* 22: 356–366.

Primdahl F (2000) Resonance magnetometers. In: Ripka P (ed.), *Magnetic Sensors and Magnetometers*, pp. 267–304. Boston: Artech.

Purucker M (1990) The Computation of vector magnetic anomalies: A comparison of techniques and errors. *Physics of the Earth and Planetary Interiors* 62: 231–245.

Purucker M and Ishihara T (2005) Magnetic images of the Sumatran region crust. *EOS Transactions of the American Geophysical Union* 86(10): 101–102.

Purucker M, Langel R, Rajaram M, and Raymond C (1998) Global magnetization models with a priori information *Journal of Geophysical Research* 103: 2563–2584.

Purucker M, Langlais B, Olsen N, Hulot G, and Mandea M (2002) The southern edge of crationic North America: Evidence from new satellite magnetometer observations. *Geophysical Research Letters* 29(15): 8000 doi:10.1029/2001GL013645.

Purucker M and Mandea M (2005) New research directions based on the World Digital Magnetic Anomaly Map, and Swarm, International Association Geomagnetism of Aeronomy International Meeting, Toulouse, France, 18–29 Jul. 2005.

Purucker M, Sabaka T, and Langel R (1996) Conjugate gradient analysis: A new tool for studying satellite magnetic data sets. *Geophysical Research Letters* 23: 507–510.

Purucker M, Von Frese RRB, and Taylor PT (1999) Mapping and interpretation of satellite magnetic anomalies from POGO data over the Antarctic region. *Annali Di Geofisica* 42: 215–228.

Purucker M and Whaler K (2003) Merging satellite and aeromagnetic data over Europe, the North Atlantic, and the Arctic, 2nd CHAMP Science meeting, 1–4 Sep 2003, Potsdam, Germany.

Purucker M and Whaler K (2004) Recognizing and interpreting the longest wavelength lithospheric magnetic fields obscured by overlap with the core field. *Eos Transactions of the American Geophysical Union* 85(47): Fall Meeting Supplement Abstract GP31A-0821.

Rajaram M, Anand SP, and Balakrishna TS (2006) Composite magnetic anomaly map of India and its contiguous regions. *Journal of Geological Society of India* 68: 569–576.

Rasmussen R and Pedersen LB (1979) End corrections in potential field modeling. *Geophysical Prospecting* 27: 749–760.

Ravat D (1996) Analysis of the Euler method and its applicability in environmental magnetic applications. *Journal of Environmental Engineering and Geophysics* 1: 229–238.

Ravat D, Hildenbrand TG, and Roest W (2003) New way of processing near-sufuce magnetic data: The utillty of the comprehensive model of the magnetic field. *The Leading Edge* 22: 784–785.

Ravat D, Whaler K, Pilkington M, Sabaka T, and Purucker M (2002) Compatibility of high-altitude aeromagnetic and satellite altitude magnetic anomalies over Canada. *Geophysics* 67: 546–554.

Ray RD (1985) Correction of systematic error in magnetic surveys: An application of ridge regression and sparse matrix theory. *Geophysics* 50: 1721–1731.

Reeves CV (2000) The geophysical mapping of Mesozoic dyke swarms in southern Africa and their origin in the disruption of Gondwana. *Journal of African Earth Sciences* 30: 499–513.

Reeves CV and Erren H (1994) AAIME: Aeromagnetics of Arabia, India, and the Middle East. *International Institute for Aerospace Survey and Earth Science* (unpublished).

Reeves CV, Macnab R, and Maschenkov S (1998) Compiling all the world's magnetic anomalies. *Eos Transactions of the American Geophysical Union* 79(28): 338–339.

Reford S (2006) Gradient enhancement of the total magnetic field. *The Leading Edge* 25: 59–66.

Reichow MK, Saunders AD, White RV, *et al.* (2002) Ar–Ar Dates from the West Siberian Basin: Siberian Flood Basalt Province Doubled. *Science* 296: 1846–1849.

Reid AB (1980) Aeromagnetic survery design. *Geophysics* 45: 973–976.

Reid AB, Allsop JM, Granser H, Millett AJ, and Somerton IW (1990) Magnetic interpretation in three dimensions using Euler deconvolution. *Geophysics* 55: 80–91.

Reynolds RL, Fishman NS, Wanty RB, and Goldhaber (1990a) Iron sulphide minerals at Cement oilfield, Oklahoma: Implications for magnetic detection of oilfields. *Geological Society of America Bulletin* 102: 368–380.

Reynolds RL, Webring M, Grauch VJS, and Tuttle M (1990b) Magnetic forward models of Cement oil field, Oklahoma based on rock magnetic, geochemical and petrological constraints. *Geophysics* 55: 344–353.

Reynolds RL, Rosenbaum JG, Hudson MR, and Fishman NS (1990c) Rock magnetism, the distribution of magnetic minerals in the Earth's crust, and aeromagnetic anomalies. In: Hanna WF (ed.) *Geologic Applications of Modern Aeromagnetic Surveys*, pp. 24–45. Denver, CO: US Geological Survery Bulletin 1924.

Reynolds RL, *et al.* (1994) Magnetization and geochemistry of greigite-bearing Cretaceous strata, North Slope Basin, Alaska. *American Journal of Science* 294: 485–528.

Reynolds RL, Fishman NS, and Hudson MR (1991) Sources of aeromagnetic anomalies over Cement oil field (Oklahoma), Simpson oil field (Alaska), and the Wyoming-Idaho-Utah thrust belt. *Geophysics* 56: 606–627.

Ridsdill-Smith TA and Dentith MC (1999) The wavelet transform in aeromagnetic processing. *Geophysics* 64: 1003–1013.

Rigoti A, Padilha AL, Chamalaun FH, and Trivedi NB (2000) Effects of the equatorial electrojet on aeromagnetic data acquisition. *Geophysics* 65: 553–558.

Ripka P (2000) Fluxgate magnetometers. In: Ripka P (ed.) *Magnetic Sensors and Magnetometers*, pp. 75–128. Boston: Artech.

Roest WR and Pilkington M (1993) Identifying remanent magnetization effects in magnetic data. *Geophysics* 58: 653–659.

Roest WR, Verhoef J, and Pilkington M (1992) Magnetic interpretation using the 3-D analytic signal. *Geophysics* 57: 116–125.

Runcorn SK (1975) On the interpretation of lunar magnetism. *Physics of the Earth and Planetary Interiors* 10: 327–335.

Rygaard-Hjalsted C, Mosegaard K, and Olsen N (2000) Resolution studies of fluid flow models near the core–mantle boundary through Bayesian inversion of geomagnetic data. In: Hansen PC, Jacobsen BH, and Mosegaard K (eds.) *Methods and Applications of Inversion: Proceeding of IIC98 conference (Copenhagen, 1998)* pp. 255–275.

Saad AH (1969) Magnetic properties of ultramafic rocks from Red Mountain, California. *Geophysics* 34: 974–987.

Sabaka TJ, Olsen N, and Langel RA (2002) A comprehensive model of the quiet-time, near-the-Earth magnetic field: Phase 3. *Geophysical Journal International* 151: 32–68.

Sabaka TJ, Olsen N, and Purucker ME (2004) Extending comprehensive models of the Earth's magnetic field with Orsted and Champ data. *Geophysical Journal International* 159: 521–547.

Sailhac P, Galdeano A, Gibert D, Moreau F, and Delor C (2000) Identification of sources of potential fields with the continuous wavelet transform: Complex wavelets

and application to aeromagnetic profiles in French Guiana. *Journal of Geophysical Research* 105: 19455–19475.

Salem A and Ravat D (2003) A combined analytic signal and Euler method (AN-EUL) for automatic interpretation of magnetic data. *Geophysics* 68: 1952–1961.

Saltus RW, Blakely RJ, Haeussler PJ, and Wells RW (2005) Utility of aeromagnetic studies for mapping of potentially active faults in two forearc basins: Puget Sound, Washington, and Cook Inlet, Alaska. *Earth Planets Space* 57: 781–793.

Sambridge M and Mosegaard K (2002) Monte Carlo methods in geophysical inverse problems. *Review of Geophysics* 40: 1009 doi: 10.1029/200RG0089.

Schlinger CM (1985) Magnetization of lower crust and interpretation of regional magnetic anomalies-Example from Lofoten and Vesteraen, Norway. *Journal of Geophysical Research* 90: 1484–1504.

Schnetzler CC and Taylor PT (1984) Evaluation of an observational method for estimation of remanent magnetization. *Geophysics* 49: 282–290.

Schnetzler CC, Taylor PT, Langel RA, Hinze WJ, and Phillips JD (1985) Comparison between the recent US composite magnetic anomaly map and Magsat anomaly data. *Journal of Geophysical Research* 90: 2543–2548.

Sexton MA, Morrison GW, Orr TOH, *et al.* (1995) The Mt. Leyshon magnetic anomaly. *Exploration Geophysics* 26: 84–91.

Shah AK, Brozena J, Vogt P, Daniels D, and Plescia J (2005) New surveys of the Chesapeake Bay impact structure suggest melt pockets and target-structure effect. *Geology* 33: 417–420.

Sharpton VL, Dalrymple GB, Marin LE, Ryder G, Schuraftz BC, and Urrutia-Fucugauchi J (1992) New links between the Chicxulub impact structure and the Cretaceous/Tertiary boundary. *Nature* 359: 819–821.

Shive PN, Blakely RJ, Frost BR, and Fountain DM (1992) Magnetic properties of the lower crust. In: Fountain DM, Arculus R, and Kay RW (eds.) *Continental Lower Crust, Developments in Geotectonics 23*, pp. 145–177. Amsterdam: Elsevier.

Shure L, Parker RL, and Backus GE (1982) Harmonic splines for geomagnetic modelling. *Physics of the Earth and Planetary Interiors* 28: 215–229.

Siebert M and Meyer J (1996) Geomagnetic activity indices. In: Dieminger W, Hartmann GK, and Leitinger R (eds.) *The Upper Atmosphere: Data Analysis and Interpretation*, Ch. V.3, pp. 887–911. Berlin: Springer.

Silva JBC (1986) Reduction to the pole as an inverse problem and its application to low-latitude anomalies. *Geophysics* 51: 369–382.

Silva JBC and Barbosa VCF (2003) 3D Euler deconvolution: Theoretical basis for automatically selecting good solutions. *Geophysics* 68: 1962–1968.

Silva JBC, Barbosa VCF, and Medeiros WE (2001) Scattering, symmetry, and bias analysis of source position estimates in Euler deconvolution and its practical implications. *Geophysics* 66: 1149–1156.

Sleep NH (2005) Evolution of the continental lithosphere. *Annual Review of Earth and Planetary Sciences* 33: 369–393.

Smith WHF and Wessel P (1990) Griding with continuous curvature splines in tension. *Geophysics* 55: 293–305.

Snyder DB, Hobbs RW, and Chicxulub Working Group (1999) Ringed structural zones with deep roots formed by the Chicxulub impact. *Journal of Geophysical Research* 104: 10743–10755.

Spector A (1968) *Spectral Analysis of Aeromagnetic Data*. PhD Thesis, University of Toronto.

Spector A and Grant FS (1970) Statistical models for interpreting aeromagnetic data. *Geophysics* 35: 293–302.

Shuey RT and Pasquale AS (1973) End corrections in magnetic profile interpretation. *Geophysics* 38: 507–512.

Shure L, Parker RL, and Backus GE (1982) Harmonic splines for geomagnetic modelling. *Physics of the Earth and Planetary Interiors* 28: 215–229.

Stone VCA, Fairhead JD, and Oterdoom WH (2004) Micromagnetic seep detection in the Sudan. *The Leading Edge* 23: 734–737.

Syberg FGR (1972) A Fourier method for the regional-residual problem of potential fields. *Geophysical Prospecting* 20: 47–75.

Talwani M (1965) Computation with the aid of a digital computer of magnetic anomalies caused by bodies of arbitrary shape. *Geophysics* 30: 797–817.

Tanaka A, Okubo Y, and Matsubayashi O (1999) Curie point depth based on spectrum analysis of the magnetic anomaly data in East and Southeast Asia. *Tectonophysics* 306: 461–470.

Tarlowski C, McEwm AJ, Reeves CV, and Barton CE (1996) Dewarping the composite aeromagnetic anomaly map of Australia using control traverses and base stations. *Geophysics* 61: 696–705.

Taylor PT and Frawley JJ (1986) Magsat anomaly data over the Kursk region, USSR. *Physics of the Earth and Planetary Interiors* 45: 5–15.

Taylor PT and Ravat D (1995) An interpretation of the Magsat anomalies of central Europe. *Applied Geophysics* 34: 83–91.

Teskey DJ, Hood PJ, Morley LW, et al. (1993) The aeromagnetic survey program of the Geological Survey of Canada; contribution to regional geological mapping and mineral exploration. *Canadian Journal of Earth Sciences* 30: 243–260.

Thebault E, Scott JJ, Mandea M, and Hoffbeck JP (2004) A new proposal for spherical cap harmonic analysis (R-SCHA): Validation and properties. *Geophysical Journal International* 159: 83–103.

Thebault E, Schott JJ, and Mandea M (2006) Revised spherical cap harmonic analysis (R-SCHA): Validation and properties. *Journal of Geophysical Research* 111: doi:10.1029/2005JB003836.

Thompson DT (1982) EULDPH-A technique for making computer-assisted depth estimates for magnetic data. *Geophysics* 47: 31–37.

Thurston JB and Smith RS (1997) Automatic conversion of magnetic data to depth, dip, and susceptibility contrast using the SPI^{TM} method. *Geophysics* 62: 807–813.

Thurston JB, Smith RS, and Guillon J-C (2002) A multimodel method for depth estimation from magnetic data. *Geophysics* 67: 555–561.

Toft PB and Haggerty SE (1988) Limiting depth of magnetization in cratonic lithosphere. *Geophysical Research Letters* 15: 530–533.

Toft PB and Haggerty SE (1989) Reply: Limiting depth of magnetization in cratonic lithosphere. *Geophysical Research Letters* 16: 480–482.

Treitel S, Clement WG, and Kaul RK (1971) The spectral determination of depths to buried magnetic basement rocks. *Geophysical Journal of the Royal Astronomical Society* 24: 415–428.

Trumbull RB, Vietor T, Hahne K, Wackerle R, and Ledru P (2004) Aeromagnetic mapping and reconnaissance geochemistry of the Earth Cretaceous Henties Bay-Outjo dike swarm, Etendeka Igneous Province. *Journal of African Earth Sciences* 40: 17–29.

Vacquier V, Steenland NC, Henderson RG, and Zieta I (1951) Interpretation of aeromagnetic maps. *Geological Society of America, Memoir* 47, 190 pp.

Vallée MA, Keating P, Smith RS, and St-Hilaire C (2004) Estimating depth and model type using the continuous wavelet transform of magnetic data. *Geophysics* 69: 191–199.

Vasicek JM, Frey HV, and Thomas HH (1988) Satellite magnetic anomalies and the Middle America trench. *Tectonophysics* 154: 19–24.

Verhoef J, Roest WR, Macnab R, Arkani-Hamed J (1996) Magnetic anomalies of the Arctic and North Atlantic Oceans and adjacent land areas. *Geological Survey of Canada Open file Report 3125*, Dartmouth, Nova Scotia.

Verrier V and Rochette P (2002) Estimating peak currents at ground lightning impacts using remanent magnetization. *Geophysical Research Letters* 29: doi: 10.1029/2002GL015207.

Vine FJ and Matthews DH (1963) Magnetic anomalies over oceanic ridges. *Nature* 199: 947–949.

Vivier F, Maier-Reimer E, and Tyler RH (2004) Simulations of magnetic fields generated by the Antarctic circumpolar current at satellite altitude: Can geomagnetic measurements be used to monitor the flow? *Geophysical Research Letters* L10306: doi: 10.1029/2004GL019804.

Von Frese RRB, Hinze WJ, Olivier R, and Bentley CR (1986) Regional magnetic anomaly constraints on continental breakup. *Geology* 14: 68–71.

Voorhies CV, Sabaka TJ, and Purucker M (2002) On magnetic spectra of Earth and Mars, *Journal of Geophysical Research-Planets,* 107(E6): 5034 doi: 10.1029/2001JE001534.

Wang X and Hansen RO (1990) Inversion for magnetic anomalies of arbitrary three-dimensional bodies. *Geophysics* 55: 1321–1326.

Warner RD and Wasilewski PJ (1997) Magnetic petrology of arc xenoliths from Japan and Aleutian Islands. *Journal of Geophysics Research* 102: 20225–20244.

Wasilewski PJ and Mayhew MA (1992) The Moho as a magnetic boundary revisited. *Journal of Geophysical Research* 19: 2259–2262.

Weinberg RF, Hodkiewicz PF, and Groves DI (2004) What controls gold distribution in Archean cratons? *Geology* 32: 545–548.

Wells RE, Weaver CS, and Blakely RJ (1998) Fore-arc migration in Cascadia and its neotectonic significance. *Geology* 26: 759–762.

Werner S (1955) Interpretation of magnetic anomalies of sheet-like bodies. *Sveriges Geologiska Undersokning, Arsbok* 43: No. 6.

Wessell P and Smith W (1998) New, improved version of Generic Mapping Tools released. *EOS Transactions of the American Geophysical Union* 79: 579.

Whaler KA (1994) Downward continuation of Magsat lithospheric anomalies to the Earth's surface. *Geophysical Journal International* 116: 267–278.

Whaler KA and Langel RA (1996) Minimal crustal magnetizations from satellite data. *Physics of the Earth and Planetary Interiors* 48: 303–319.

Whaler KA, Langel RA, Jackson A, and Purucker ME (1996) Non-uniqueness in magnetization: Crustal models from satellite magnetic data. *EOS Transactions of the American Geophysical Union* 77: F172.

Whaler K and Purucker M (2005) A spatially continuous magnetization model for Mars. *Journal of Geophysical Research* 110: NO. E9, E09001, http://dx.doi.org/10.1029/2004JE002393.

Wonik T, Trippler K, Geipel H, Greinwald S, and Pashkevitch I (2001) Magnetic anomaly map for northern, western, and eastern Europe. *Terra Nova* 13: 203–213.

Yanovsky BM (1938) Variations in elements of terrestrial magnetism in an anomalous field. *Trans (Trudy) Main Geophysical Observatory* 17(3).

Zietz I (1982) Composite magnetic anomaly map of the United States; Part A, conterminous United States: US Geological Survey Geophysical Investigations Map GP-954-A, 59pp, 2 sheets, scale 1:2,500,000

Zietz I and Andreasen GE (1967) Remanent magnetization and aeromagnetic interpretation. In: Ward SH (ed.) *Mining Geophysics, Society of Exploration Geophysics*, vol. 2, pp. 569–590. Tulsa, OK: Society of Exploration Geophysics.

Zietz I and Bhattacharyya BK (1975) Magnetic anomalies over the continents and their analyses. *Reviews of Geophysics and Space Physics* 13: 176–215.

7 Geomagnetic Induction Studies

S. Constable, University of California, San Diego, La Jolla, CA, USA

7.1 Introduction

Interaction of temporal fluctuations in Earth's external magnetic field with the electrically conductive rocks of the planet generates internal secondary electric and magnetic fields. Measurements of the primary and secondary fields may then be used to probe the conductivity structure over depths from a few meters to the lower mantle. Conductivity in turn may be used to infer physical and chemical properties, such as temperature, melt content, water and volatile content, mineral type, and anisotropic fabric. There are three basic steps necessary to obtain useful information from geomagnetic induction studies: estimation of electromagnetic impedances from observations of the electric and magnetic fields; modeling of these impedances in both forward (conductivity to impedance) and inverse (impedance to conductivity) directions; and finally the use of laboratory studies to relate conductivity to the relevant physical and chemical properties. Each of these steps represent significant scientific challenges, and all of them are the subject of current research and recent progress. Clearly, the results of each of these components depend on the quality of the preceding step, and conclusions about the state of Earth's interior depend on the reliability of all three parts of the process.

Electromagnetic induction covers a very broad field which includes applications in exploration, environmental, and mining geophysics, as well as studies of Earth's deep interior. Although many of the underlying principles span the entire range of applications, in the context of this *Treatise on Geophysics* we will concentrate on the 'whole-Earth' aspects of geomagnetic induction. For a good discussion of exploration and crustal applications, the reader is referred to the recent book by Simpson and Bahr (2005).

7.1.1 Historical Beginnings

The history of geophysical electromagnetic induction is entwined with the development of the fundamental physics of electromagnetic energy, closely following Oersted's 1820 observation that an electric current deflected a magnet and Ampere's quantification of this phenomenon in the 1820s. Ampere also noted that an electric current exerts a force on a second electric current and that a coil behaved like a magnetic dipole. In 1831 Faraday observed that moving a magnet through a coil produces an electric current and in 1832 he predicted that water moving through a magnetic field should produce an electric field, an effect that was observed by the British Admiralty much later in 1918.

Induction effects associated with magnetic storms were first seen in telegraph cables: the storm of 1838 being seen on Norwegian telegraph cables; W.H. Barlow (1849) reporting spontaneous currents in telegraph lines in England; and an observation in 1859 by K.T. Clement that the aurora of 29 August disrupted telegraphy. The critical mathematic developments started in 1839 with Gauss' spherical harmonic expansion of the main magnetic field, which allows the separation of internal and external magnetic sources, and later (1864) with Maxwell's equations of electromagnetism. In 1865, the Greenwich Observatory started to observe Earth potentials on 15 km grounded lines. By the early twentieth century, the relationship between magnetic activity and Earth currents were well known. Arthur Schuster (1889) observed the relationship between the diurnal magnetic variation and Earth potentials and Sydney Chapman (1919) inferred that Earth's interior must be more conductive than are crustal rocks, and modeled the daily variation with a conductive sphere of smaller radius than Earth. Although Chapman and Bartels (1940) lamented that a more quantitative relationship between electrical conductivity and magnetic field variation was not yet available, progress in this direction was being made. Lahiri and Price (1939) modeled the internal and external parts of the magnetic field using a radial conductivity profile to a depth of nearly 1000 km. Later, Roger Banks (1969) proposed that the harmonics of the 27-day solar rotation were dominantly of P_1^0 spherical harmonic geometry, associated with the ring of current in the magnetosphere symmetric about the geomagnetic equator, and produced a conductivity profile down to nearly 2000 km. In order to get an estimate of the electrical conductivity of the lowermost mantle, K. L. McDonald (1957) modeled the outward propagation of an inferred geomagnetic secular variation signal originating in the core. **Figure 1** summarizes the results of this early work.

These observations represent the beginnings of the geomagnetic depth sounding (GDS) method, in which magnetic fields alone are used to probe Earth conductivity. Another highly important method, magnetotelluric (MT) sounding, began in the early

1950s when A. N. Tikhonov (1950) and Louis Cagniard (1953) quantified the relationship between induced electric currents and the magnetic field. Both the GDS and MT methods are in extensive use today, and we discuss them in this chapter.

7.1.2 Earth's Geomagnetic Environment

Temporal variations in Earth's magnetic field exist on all timescales from radio frequencies to the reversal record (**Figure 2**). Electromagnetic induction methods can exploit only those fields that are externally generated through interaction of Earth's magnetic field with the solar wind (**Figure 3**), which in principle means periods shorter than about 1 year for GDS methods (with the possible exception of the 11 year sunspot cycle) and, because the induced electric field components become smaller as period increases, periods of a few weeks or shorter for MT methods. As mentioned in the introduction, propagation of internally generated fields of the secular variation upwards through the mantle may be used to estimate conductivity (McDonald, 1957; Alexandrescu *et al.*, 1999), but lack of an independent estimate of the source-field timing limits this approach (Backus, 1983).

7.1.2.1 Earth/ionosphere cavity

Starting at the high-frequency end of the spectrum, the conductive ionosphere and conductive earth are separated by the resistive atmosphere to form a cavity. The electrical conductivity of the ground is of order 10^{-3} S m^{-1}, while the atmosphere near Earth's surface is about 10^{-14} S m^{-1}. What little conductivity there is results from ionization of oxygen and nitrogen by cosmic rays. The mean free path of these ions increases with decreasing atmospheric density, and by an altitude of 100 km, the start of the ionosphere, conductivity is about the same as that of the solid earth. The ionosphere is several hundred kilovolts positive with respect to ground, and a leakage current of about 1000 A flows between the ionosphere and ground through an integrated transverse resistance of about 200 Ω. This leakage current supports a vertical electric field of 100–300 V m^{-1} at the ground's surface. The Earth/atmosphere/ionosphere acts as a large capacitor with a value of about 1 farad and a time constant of a few thousand seconds. Thus there must be a return current to sustain the charge on the ionosphere.

The major contributor to the return current is thunderstorm activity, in which convection carries

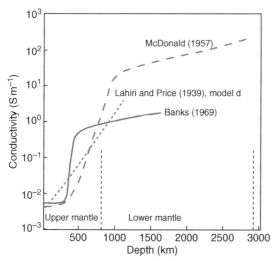

Figure 1 Early electrical conductivity profiles of Earth. These models all feature an increase in conductivity at about the upper-lower-mantle transition, a feature of most modern models.

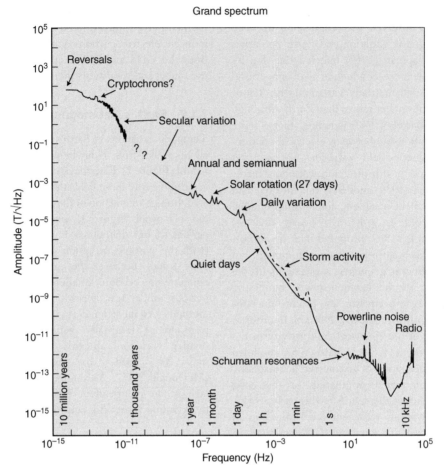

Figure 2 An approximate amplitude spectrum for Earth's magnetic field (to get a power spectrum, double the exponents and change the units to T^2/Hz). While somewhat schematic, this figure illustrates that the source and morphology of the field variations depends on frequency. In particular, at periods shorter than about a year, the field is dominated by sources external to Earth. From Constable CG and Constable SC (2004a) Satellite magnetic field measurements: Applications in studying the deep Earth. In: Sparks RSJ and Hawkesworth CT (eds.) *Geophysical Monograph 150: The State of the Planet: Frontiers and Challenges in Geophysics*, pp. 147–159. Washington, DC: American Geophysical Union.

positively charged particles to the tops of thunderclouds and cloud-to-ground negative charge flow in lightning completes the circuit. Thunderstorms preferentially form in the tropics, in the afternoon, and over land. Since Africa is the dominant equatorial land mass, there is an uneven temporal distribution of lightning activity which peaks as Earth rotates through afternoon in equatorial Africa, or about 16:00 UT. This recharging of the global capacitor can be seen as an increased fair-weather electric field over the oceans anywhere in the world at these times, first observed from the research vessel *Carnegie* and now known as the Carnegie curve.

The atmospheric cavity is excited by lightning strikes, and resonates with a characteristic

frequency of about 8 Hz and harmonics, a phenomenon called the Schumann resonance. Man-made sources of electromagnetic energy such as powerline noise at 50 and 60 Hz and the entire radio frequency spectrum also propagate within the resistive atmosphere.

7.1.2.2 Daily variation and Sq

The dayside ionosphere, heated by the Sun, is more conductive than the nightside and has a pattern of two circulating current systems (one in each hemisphere) that are stationary in solar time, with an intensified, normally eastward-flowing current at the magnetic equator called the equatorial electrojet. Since Earth rotates beneath these current systems,

The Geomagnetic Earth

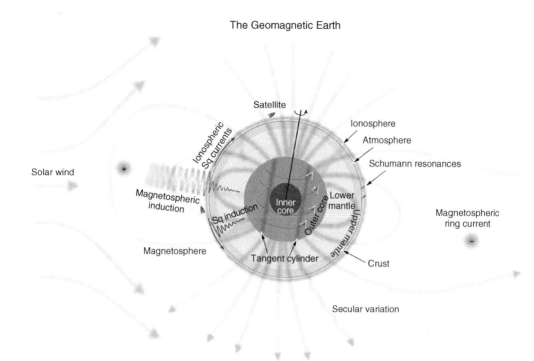

Figure 3   A pictorial representation of the electromagnetic environment of Earth. From Constable CG and Constable SC (2004a) Satellite magnetic field measurements: Applications in studying the deep Earth. In: Sparks RSJ and Hawkesworth CT (eds.) *Geophysical Monograph 150: The State of the Planet: Frontiers and Challenges in Geophysics*, pp. 147–157. Washington, DC: American Geophysical Union.

a daily variation in magnetic field is seen at the surface of Earth, amounting to a few tens of nanotesla.

7.1.2.3   The magnetospheric ring current

The largest component of external field variations comes from a ring of current circulating roughly around the magnetic equator at a distance of 2–9 Earth radii. The solar wind injects charged particles into the ring current, mainly positively charged oxygen ions. Sudden increases in the solar wind associated with coronal mass ejections cause magnetic storms, characterized by a sudden commencement, a small but sharp increase in the magnetic field associated with the sudden increased pressure of the solar wind on Earth's magnetosphere. Following the commencement is a period of fluctuating magnetic field called the initial phase. The main phase of the storm is associated with a large decrease in the magnitude of the fields as a westward circulating ring current is energized – the effect is to cancel Earth's main dipole field slightly. Finally, there is a recovery phase in which the field returns quasi-exponentially back to normal. All this can happen in a couple of hours, or may last days for a large storm.

Although complicated in detail, the ring current behaves somewhat like a large single turn of wire around Earth, creating fields at Earth's surface of predominantly simple P_1^0 spherical harmonic geometry. The frequency content is huge: from minute-by-minute fluctuations during a storm, the hours-long duration of a storm, a large peak at the 27-day rotation period of the Sun (and harmonics), a semiannual line associated with the geometry of the ecliptic and the Sun's equatorial plane, and finally the 11-year solar cycle.

The strength of the ring current is characterized by the 'Dst' (disturbance storm time) index, an index of magnetic activity derived from a network of low- to mid-latitude geomagnetic observatories that measures the intensity of the globally symmetrical part of the equatorial ring current (**Figure 4**). The current method for computing Dst is described in IAGA Bulletin No. 40, a report by Masahisa Sugiura which presents the values of the equatorial Dst index for 1957–1986. Honolulu, Hermanus, San Juan, and Kakioka are the currently contributing observatories. The actual morphology of the ring current fields is asymmetric about the day/night hemispheres, as injection of ions occurs preferentially as a function of solar time (e.g., Balasis and Egbert, 2006).

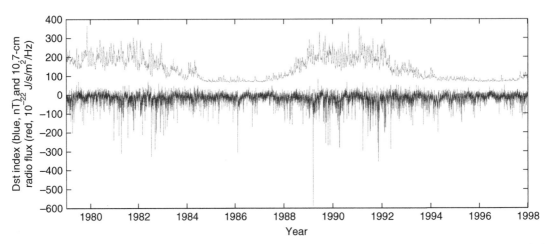

Figure 4 A record of the Dst index (blue) for 19 years between 1979 and 1998. The large negative spikes are individual magnetic storms. To illustrate the relationship between storm activity and the 11-year sunspot cycle, the 10.7-cm solar radio flux is plotted in red.

7.1.2.4 *Pulsations*

Interaction of the solar wind and magnetosphere can generate resonances and waves that have relatively narrow band frequency content, creating quasi-periodic oscillations of the magnetic field called pulsations. The period of oscillation varies between about 1 s and several hundred seconds, and the nomenclature for pulsations distinguishes between various frequency bands and whether the variations are continuous for a time or form a damped oscillation (see Parkinson, 1983).

7.2   Geomagnetic Sounding

7.2.1   Introductory Theory

One of the key concepts in geomagnetic induction is that of the 'skin depth', the characteristic length over which electromagnetic fields attenuate. We can derive the skin depth starting with Faraday's law:

$$\nabla \times \mathbf{E} = -\frac{\partial \mathbf{B}}{\partial t}$$

and Ampere's law:

$$\nabla \times \mathbf{H} = \mathbf{J}$$

where $\mathbf{J}$ is current density ($\mathrm{A\,m^{-2}}$), $\mathbf{E}$ is electric field ($\mathrm{V\,m^{-1}}$), $\mathbf{B}$ is magnetic flux density or induction (T), $\mathbf{H}$ is magnetic field intensity ($\mathrm{A\,m^{-1}}$). We neglect displacement currents, as they are not significant at the frequencies and conductivities relevant to geomagnetic induction (for a rigorous justification of

this, see section 7.2.4 of Backus *et al.* (1996)). We can use the identity $\nabla \cdot \nabla \times \mathbf{A} = 0$ to show that

$$\nabla \cdot \mathbf{B} = 0 \quad \text{and} \quad \nabla \cdot \mathbf{J} = 0$$

in regions free of sources of magnetic fields and currents. $\mathbf{B}$ and $\mathbf{H}$ are related by magnetic permeability μ and $\mathbf{J}$ and $\mathbf{E}$ by conductivity σ:

$$\mathbf{B} = \mu \mathbf{H} \quad \mathbf{J} = \sigma \mathbf{E}$$

(the latter equation is Ohm's law), and so

$$\nabla \times \mathbf{E} = -\mu \frac{\partial \mathbf{H}}{\partial t}$$

$$\nabla \times \mathbf{H} = \sigma \mathbf{E}$$

If we take the curl of these equations and use $\nabla \times (\nabla \times \mathbf{A}) = \nabla(\nabla \cdot \mathbf{A}) - \nabla^2 \mathbf{A}$, for constant σ and μ we have

$$\nabla^2 \mathbf{E} = \mu \frac{\partial}{\partial t}(\nabla \times \mathbf{H}) = \mu\sigma \frac{\partial \mathbf{E}}{\partial t}$$

$$\nabla^2 \mathbf{H} = \sigma(\nabla \times \mathbf{E}) = \mu\sigma \frac{\partial \mathbf{H}}{\partial t}$$

which are diffusion equations. In air and very poor conductors where $\sigma \approx 0$, or if $\omega = 0$, the equations reduce to Laplace's equation. Now if we consider sinusoidally varying fields of angular frequency ω

$$\mathbf{E}(t) = \mathbf{E}_o e^{i\omega t} \quad \frac{\partial \mathbf{E}}{\partial t} = i\omega \mathbf{E}$$

$$\mathbf{H}(t) = \mathbf{H}_o e^{i\omega t} \quad \frac{\partial \mathbf{H}}{\partial t} = i\omega \mathbf{H}$$

and so

$$\nabla^2 \mathbf{E} = i\omega\mu\sigma\mathbf{E}$$

$$\nabla^2 \mathbf{H} = i\omega\mu\sigma\mathbf{H}$$

If we further consider fields that are horizontally polarized in the xy directions and are propagating vertically into a half-space, in Cartesian coordinates these equations decouple to

$$\frac{\partial^2 \mathbf{E}}{\partial z^2} + k^2 \mathbf{E} = 0$$

$$\frac{\partial^2 \mathbf{H}}{\partial z^2} + k^2 \mathbf{H} = 0$$

with solutions

$$\mathbf{E} = \mathbf{E}_0 e^{-ikz} = \mathbf{E}_0 e^{-i\alpha z} e^{-\beta z}$$

$$\mathbf{H} = \mathbf{H}_0 e^{-ikz} = \mathbf{H}_0 e^{-i\alpha z} e^{-\beta z}$$

where we have defined a complex wave number

$$k = \sqrt{i\omega\mu\sigma} = \alpha - i\beta$$

and an attenuation factor, which is called a skin depth

$$z_s = 1/\alpha = 1/\beta = \sqrt{\frac{2}{\sigma\mu_0\omega}}$$

The skin depth is the distance that field amplitudes are reduced to $1/e$, or about 37%, and the phase progresses 1 radian, or about $57°$. In practical units of meters, the skin depth is

$$z_s \approx 500\sqrt{1/\sigma f}$$

where circular frequency f is defined by $\omega = 2\pi f$.

The skin depth concept underlies all of inductive electromagnetism in geophysics. Substituting a few numbers into the equation shows that skin depths cover all geophysically useful length scales from less than a meter for conductive rocks and kilohertz frequencies to thousands of kilometers in mantle

rocks and periods of days (**Table 1**). Skin depth is a reliable indicator of maximum depth of penetration, but one should not think of skin depth as a resolution length. A thin conductive layer can have an effect at periods associated with skin depths much larger than its width, and all external magnetic field variations have to propagate through surface layers. The skin depth describes attenuation associated with induction, but electric and magnetic fields are also modified by variations in conductivity through Ohm's law even in the DC limit, an effect often described as 'galvanic' rather than inductive.

7.2.2   The MT Method

Because the skin depth generates a natural scale length for time-varying fields, the response of Earth to time variations in the externally generated geomagnetic field can be interpreted in terms of electrical conductivity with depth. The estimation of purely geomagnetic response functions and their interpretation in terms of mantle electrical conductivity structure dates from the end of the nineteenth century, but the additional use of electric field measurements only started with Cagniard's work in the 1950s.

Continuing with our half-space formulation, we can define an impedance by taking the ratio of orthogonal field components

$$Z = \frac{E_x}{H_y} = \frac{\mu\omega}{k}$$

where we recall that the complex wave number

$$k = \alpha - i\beta = (\sigma\mu\omega)^{1/2} e^{-i(\pi/4)}$$

and so we can see that E_x leads H_y by $45°$ and that the half-space resistivity is given by

$$\rho = \frac{1}{2\pi f\mu}|Z|^2 = \frac{T}{2\pi\mu}\left|\frac{E_x}{H_y}\right|^2$$

Table 1   Approximate skin depths for a variety of typical Earth environments and a large range of periods

Material	σ (S m^{-1})	1 year	1 month	1 day	1 h	1 s	1 ms
Core	5×10^5	4 km	1 km	209 m	43 m	71 cm	23 mm
Lower mantle	10	883 km	255 km	47 km	9 km	160 m	5 m
Seawater	3	1600 km	470 km	85 km	17 km	290 m	9 m
Sediments	0.1	9000 km	2500 km	460 km	95 km	1.6 km	50 m
Upper mantle	0.01	3×10^4 km	10^4 km	1500 km	300 km	5 km	159 m
Igneous rock	1×10^{-5}	10^6 km	2×10^5 km	5×10^4 km	9500 km	160 km	5 km

One can see that the core is effectively a perfect conductor into which even the longest-period external magnetic field cannot penetrate. However, 1 year variations can penetrate into the lower mantle. The very large range of conductivities found in the crust (sea, sediment, igneous rock) indicates the need for a corresponding large range of frequencies in electromagnetic sounding of that region.

for fields with period T. This is the MT formula made famous by Cagniard (1953). Perhaps more usefully (since measurements tend to be made in **B** rather than **H**),

$$\rho = \frac{\mu}{\omega}\left|\frac{E_x}{B_y}\right|^2$$

Another quantity called the C-response (Weidelt, 1972) is also used, given by

$$c = -\frac{E_x}{i\omega B_y}$$

(note that c is complex) from which it can be seen that

$$\rho = \omega\mu|c|^2$$

If Earth really were a homogeneous half-space, the resistivity obtained at all periods would be the same and equal to the true resistivity of the earth, and the phase between B and E would be constant at $45°$. Of course, this is never the case, but resistivities are computed nevertheless and described as 'apparent resistivities'. Solutions for computing the apparent resistivity response of layered, 2-D, and even 3-D models exist and are used to interpret the MT data. The phase relationship between E and B is also estimated, and although in theory the phase contains no independent information it is useful because it is independent of the amplitudes of E and H.

Complicated near-surface structure can change the amplitude of the E fields very easily, but because these distortions are not inductive, the phase relationships are not altered.

7.2.3   MT in Practice

Figure 5 shows a schematic layout of an MT measurement; the magnetic and electric fields are measured orthogonal to each other. To a first approximation, the magnetic field is the EM source and the electric field is the EM response. For a 1-D earth, this would be the only measurement required; for a 2-D or 3-D earth, measurements are made in both directions, and perhaps even the vertical magnetic field would be measured.

MT measurements can be made from the low end of the radio frequency spectrum, using radio stations and lightning as sources of energy, to periods of several days. The table of skin depths gives some indication of the relationship between conductivity, period, and penetration scale. Structure can profoundly affect the way MT fields propagate, but skin-effect attenuation provides an indication of the limits to deepest sensitivity (see **Figure 5(b)**). The type of instrumentation used to record the E and B fields will depend on the periods required, and the time over which data are collected will determine the

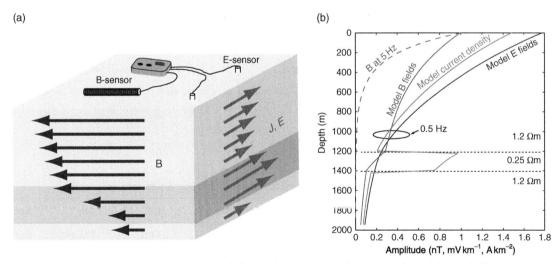

Figure 5 (a) Schematic of an MT measurement. Orthogonal components of the magnetic and electric field are measured at the surface of the earth. In conductive layers (red), electric fields and currents are induced whose secondary magnetic fields attenuate the primary magnetic field more rapidly than in the background material. (b) In a quantitative example of this behavior, magnetic field, electric field, and current density (solid lines) are plotted for 0.5 Hz energy propagating into a model containing a buried conductive layer, in which the magnetic fields can be seen to attenuate rapidly due to induced electric currents. For comparison, magnetic fields for a 5 Hz source field are shown (broken red line), which attenuate before the conductive layer is reached due to the shorter skin depth at the higher frequency.

longest period available. Horizontal spacing of MT sites can vary from continuous electric field dipoles for shallow crustal studies, individual sites spaced tens of kilometers apart for regional profiles, to single long-period soundings for mantle conductivity studies (e.g., Egbert and Booker, 1992; Tarits *et al.*, 2004).

7.2.4 Equipment

7.2.4.1 Electric field measurements

Metal electrodes are too noisy for MT measurements, except at radio frequencies. A nonpolarizing electrode, or porous pot, is almost always used, in which a reversable electrochemical reaction is restricted to that of a single chemistry. Electrical connection is made to a metal rod dipped into an aqueous saturated solution of the metal salt (**Figure 6**). Contact is made with the ground through a semipermeable material such as unglazed porcelain, porous plastic, or plaster. Copper–copper sulphate is a common chemistry for land use, as is lead–lead chloride stabilized in plaster of Paris made with the lead chloride solution (Petiau, 2000). (For marine use, silver–silver chloride provides a low-noise electrode, the silver chloride being compatible with seawater and very low solubility (Webb *et al.*, 1985; Filloux, 1987)) On land, such a system produces a low-noise electrical contact with an impedance of typically a few kΩ. Electrodes are buried at a depth of around 30 cm or more to limit temperature variations, which can generate a time-varying potential on the electrode. Dipole lengths are about 50–200 m. Since signals are of order mV/km, a gain of 10–1000 is applied before digitization. The amplifier needs to be a true differential amplifier because there is no natural concept of 'ground' when making telluric measurements. Often, optical isolation is used to ensure no ground path exists between the recording system and the electrode system.

7.2.4.2 Magnetic field measurements

Two types of magnetic field sensors are in common use, both necessarily vector instruments (which rules out nuclear precession sensors). For periods shorter than a few thousand seconds, induction coils are used, and for periods longer than about 10 s, fluxgate magnetometers provide a lower noise measurement. Induction coils (**Figure 7**) are simply an application of Faraday's law

$$\oint_C \mathbf{E} \cdot \mathrm{dl} = -\frac{\mathrm{d}\Phi}{\mathrm{d}t}$$

so integrating E around a single loop of wire of area A gives a potential difference

$$V = -A\frac{\mathrm{d}B}{\mathrm{d}t}$$

We see that V increases with frequency, but since B is in the pico- to nanotesla range, unless A is huge this does not produce much of a voltage. Two approaches are used to boost V. The first is to simply wind many (say, N) turns in series and the second is to increase the flux through the coil by inserting a core material of high relative permeability μ_r, such as mu-metal or permalloy. The permeable core provides an effective increase in flux through the coil by a factor of a, so

$$V = -aNA\frac{\mathrm{d}B}{\mathrm{d}t}$$

The relative permeability of mu-metal alloys is of order 10^4–10^5, so a can be made quite high. In practice, a long, thin, coil is used because then a is limited by the core geometry, rather than μ_r, thus removing any temperature dependence in μ_r from the sensor. Since most of the flux is trapped in the core material, A becomes the cross-sectional area of the core, rather than the area of each turn of wire. For large N, achieved by making the windings out of very fine wire, the dominant source of noise becomes the thermal resistance noise, or Johnson noise, of the wire:

$$V_J^2 = 4kTR$$

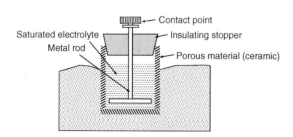

Figure 6 Porous pot, or nonpolarizing, electrode.

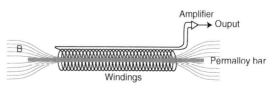

Figure 7 Induction coil magnetometer.

where k is Boltzmann's constant $(1.4 \times 10^{-23}\,\mathrm{J\,K^{-1}})$, T is temperature, and R is resistance. Here we see the second reason for a long, thin, magnetometer. Piling windings upon windings increases the diameter of the turns, and so R starts to increase more rapidly than V.

A good induction coil magnetometer has a noise level of about $10^{-8}\,\mathrm{nT^2\,Hz^{-1}}$ at around 1 Hz (i.e., less than a picotesla), with a red noise spectrum at lower frequencies because of the dB/dt loss of sensitivity. Long-period response is limited to about 5000 s, while high-frequency response is limited by capacitive and inductive losses in the core material. This is addressed by laminating the core material and, ultimately, going to air-cored coils for the highest (low radio) frequencies. Because Earth's main field is around 40 000 nT, minute rotation of an induction coil magnetometer couples this field into the sensor. For a subpicotesla noise floor, this corresponds to a rotation of only a nanoradian, or 1 mm in 1000 km. For these reasons coils are buried 10 cm or so to avoid wind motion, and even so MT sites near trees or coastlines will be observed to have high noise levels.

The fluxgate magnetometer (**Figure 8**) measures the total magnetic field, and so is useful for long-period MT studies as well as magnetic observatory recording. On the other hand, since the main field is so large, dynamic range and sensitivity issues limit the total sensitivity to about 0.01 nT, but this still makes it a better sensor than an induction coil for periods longer than about 10 s. In a fluxgate magnetometer, two primary coils are wound around two identical cores of permeable material, and connected in series so that a current passed through the primary circuit generates magnetic fields in the two cores which oppose each other. A 50–1000 Hz primary current of sufficient strength is used to saturate the cores. In the absence of an external field the magnetization in the cores is equal and opposite, so there is no net magnetic field for the two-core system. In the presence of an external field, the core being magnetized in the direction of the external field saturates sooner than the opposite core. Thus, an asymmetry develops in the magnetization of the cores, leading to a time-varying net magnetization. A secondary coil wound around the entire system measures the rate of change of this magnetization as a series of induced voltage spikes. These spikes are rectified and amplified to produce a voltage signal which is proportional to the magnetic field along the axis of the sensor cores. Note that the fundamental harmonic of the sensing voltage is twice the fundamental harmonic of the primary current.

Because permeability and saturation magnetization will depend on temperature, one common approach to building a fluxgate sensor is to operate it as a null instrument. A solenoid around the fluxgate sensor carries an electric current that creates a magnetic field equal and opposite Earth's field (in that direction). The output of the fluxgate should then be zero (and so need not be calibrated). Any nonzero measurements from the fluxgate are fed back into the

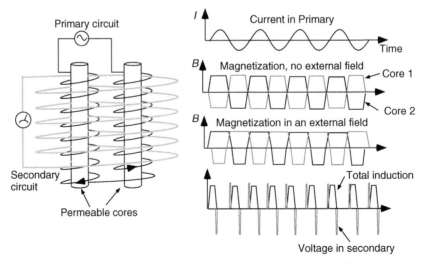

Figure 8   Principle of operation for a fluxgate magnetometer. Two permeable cores are driven into saturation in opposite directions at frequency f. A bias associated with an external field along the axis of the cores produces an asymmetry in the internal fields and a total internal field of frequency $2f$, detected in a secondary winding by a phase-locked amplifier.

solenoid current to return the measurement to zero. The current in the solenoid, rather than the fluxgate output, is then the measurement of the magnetic field. Another common modification is to use a permeable core in the shape of a ring with a toroidal excitation winding. Further details can be found in Nielsen *et al.* (1995).

7.2.5 MT Data Processing

The raw measurements made in the field are time series of magnetic and electric fields, usually in two orthogonal directions. An example is given in **Figure 9**. The objective of MT data processing is to obtain the frequency domain response of the coherent part of these signals. The frequency domain transfer function between the horizontal magnetic and electric field is called the MT impedance tensor Z:

$$\begin{bmatrix} E_x(\omega) \\ E_y(\omega) \end{bmatrix} = \begin{bmatrix} Z_{xx}(\omega) & Z_{xy}(\omega) \\ Z_{yx}(\omega) & Z_{yy}(\omega) \end{bmatrix} \begin{bmatrix} H_x(\omega) \\ H_y(\omega) \end{bmatrix}$$

(we will drop the ω dependence for clarity). The standard treatment of a transfer function would consider H the input and E the output with noise ϵ in the output measurement only:

$$E = ZH + \epsilon$$

which in the classic least-squares approach is estimated by

$$\hat{Z} = \frac{< H^{\dagger} E >}{< H^{\dagger} H >}$$

where $< H^{\dagger} E >$ is the cross spectrum between H and E. The problem is that there is noise in H as well as E, which will bias $< H^{\dagger} H >$ up and bias Z down. The remote reference technique was introduced (Gamble *et al.*, 1979) to reduce bias associated with noise in H by making a second measurement of the magnetic field, H_r, sufficiently far away (remote) that the noise sources are independent, relying on the fact that the magnetic source field is coherent over very large distances. Then

$$\hat{Z} = \frac{<H_r^{\dagger} E>}{<H_r^{\dagger} H>}$$

and with any luck $<H_r^{\dagger}H>$ is now unbiased. The least-squares method also assumes that noise is normally distributed. Non-Gaussian noise can be handled with an iterative robust weight W:

$$\hat{Z} = \frac{<H_r^{\dagger} W E>}{<H_r^{\dagger} W H>}$$

which reduces the effect of outliers. Both the effects of non-Gaussian noise and nonstationary source-field spectra can be handled by various robust estimation techniques (e.g., Egbert and Booker, 1986; Chave *et al.*, 1987). These approaches were generalized into a eigenvalue decomposition for multiple MT sites by Egbert (1997).

The nature of Z depends on the apparent dimensionality of the geology:

For 1-D, $Z_{xx} = Z_{yy} = 0$; $Z_{xy} = - Z_{yx}$

For 2-D, $Z_{xx} = Z_{yy} = 0$; $Z_{xy} \neq - Z_{yx}$

For 3-D, $Z_{xx}, Z_{yy}, Z_{xy}, Z_{yx} \neq 0$

For 2-D structure, $Z_{xx} = Z_{yy} = 0$ only when the coordinate system of the measurement is aligned to the 2-D geometry of the geology. With a prior understanding of structural trends or geological strike, the experimenter will normally arrange for this to be approximately so. Otherwise, the tensor can be

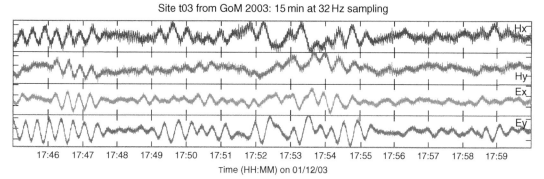

Site t03 from GoM 2003: 15 min at 32 Hz sampling

Hx

Hy

Ex

Ey

17:46 17:47 17:48 17:49 17:50 17:51 17:52 17:53 17:54 17:55 17:56 17:57 17:58 17:59

Time (HH:MM) on 01/12/03

Figure 9 Sample of MT time series measurements, in arbitrary units. Note that H_x and E_y are anticorrelated, and H_y and E_x are correlated.

rotated mathematically to minimize Z_{xx} and Z_{yy}. Conventionally, the impedance with electric field along-strike is called the transverse electric, or TE mode, and the impedance with across-strike electric fields is called the transverse magnetic, or TM mode.

7.2.6   Global Conductivity Studies

If one wants to look deep into Earth using electrical techniques, to study the properties of the deeper mantle, then the MT method is not the ideal tool. Firstly, one needs very long time series of the EM fields, spanning perhaps years, and while magnetic observatories have been collecting records for more than a hundred years, there are few observatories collecting electric field data. Secondly, if one examines the MT relationship

$$\rho_a = \frac{T}{2\pi\mu}\left|\frac{E}{B}\right|^2$$

one observes that for a given magnetic field and resistivity, the amplitude of the induced electric field decreases with period:

$$E = B\sqrt{\frac{2\pi\mu\rho_a}{T}}$$

which is exacerbated by a falling ρ_a with depth and period. At long periods the assumption that the source-field morphology is uniform over scales comparable to the depth of penetration in Earth break down, particularly for the daily variation and harmonics (e.g., Bahr *et al.*, 1993). Finally, there will be an ambiguity in the absolute value of mantle conductivity associated with unknown static effects in the electrical data (**Figure 10**).

7.2.7   Global Response Functions

One can derive other types of electrical response functions, similar in nature to the MT response, using only the three components of the magnetic field recorded by geomagnetic observatories. Much of this work is still based on techniques developed by Banks (1969). The vector magnetic field can be written as a gradient of a scalar potential in the usual fashion:

$$\mathbf{B} = -\mu_0\nabla\Omega$$

and, following Gauss, the scalar potential can be expanded in turn into spherical harmonic coefficients of internal (i_n) and external (e_n) origin:

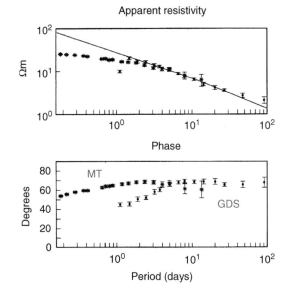

Figure 10 A combination of MT and GDS response functions from Egbert and Booker (1992). Magnetotelluric data are available to a maximum of 10-days period (about 10^6 s), while GDS data can extend this to periods of many months.

$$\boldsymbol{\Omega} = a_0\sum_n\left\{i_n\left(\frac{a_0}{r}\right)^{n+1} + e_n\left(\frac{r}{a_0}\right)^n\right\}P_n(\cos\theta)$$

where n is the degree of spherical harmonic, a_0 is the radius of Earth, θ is geomagnetic colatitude, P_n are spherical harmonics of degree n, and r is the radius of observation. If one chooses geomagnetic coordinates then the analysis is zonal in terms of the main dipole field. The i_n and e_n can be functions of time or frequency, although, like MT, this kind of analysis is usually done in the frequency domain.

One can thus define a geomagnetic response for a radially symmetric Earth as simply the ratio of induced (internal) to external fields:

$$Q_n(\omega) = \frac{i_n(\omega)}{e_n(\omega)}$$

where the frequency domain is now made explicit, and will be implicit in the following. In certain circumstances, such as satellite observations, one has enough data to fit the e_n and i_n directly. However, most of the time one just has horizontal (H) and vertical (Z) components of $\mathbf{B}$ as recorded by a single observatory. We can obtain H and Z from the appropriate partial derivatives of $\boldsymbol{\Omega}$ to obtain

$$H = \mu_0\left(\frac{1}{r}\frac{\partial\boldsymbol{\Omega}}{\partial\theta}\right)_{r=a_0}$$
$$= \mu_0\sum_n A_{Hn}\frac{\partial P_n(\cos\theta)}{\partial\theta}$$

and

$$Z = \mu_0 \left(\frac{\partial \mathbf{\Omega}}{\partial r} \right)_{r=a_0}$$

$$= \mu_0 \sum_n A_{Zn} P_n(\cos \theta)$$

where we have defined new expansion coefficients

$$A_{Hn} = i_n + e_n$$

$$A_{Zn} = ne_n + (n+1)i_n$$

Thus we can define a new electromagnetic response

$$W_n = \frac{A_{Zn}}{A_{Hn}}$$

related to Q by

$$Q_n = \frac{i_n}{e_n} = \frac{n - A_{Zn}/A_{Hn}}{n+1+A_{Zn}/A_{Hn}}$$

The inductive scale length or C-response is, in turn,

$$c_n = \frac{a_0 W_n}{n(n+1)}$$

where the MT apparent resistivity and phase are

$$\rho_a = \omega \mu_0 |ic_n|^2 \quad \phi = \arg(\omega \mu_0 i c_n)$$

Note that for causal systems, the real part of c is positive, and the imaginary part is always negative.

Banks showed that the above analysis could be simplified because, except for the daily and annual variations, the magnetic field was dominantly of P_1^0 geometry as a result of the nature of the ring current.

If one can assume a simple degree-one field geometry, the above discussion of response functions simplifies greatly. W becomes simply the ratio of vertical to horizontal fields with a colatitudinal trigonometric term. The C-response becomes simply $c = a_0 W/2$. Banks looked at the geometry of simultaneous observatory records as a function of geomagnetic colatitude, to demonstrate the P_1^0 geometry. Having shown this, one can take records from a single observatory and, assuming P_1^0, compute a geomagnetic response function. Since some observatories have been recording for over a hundred years, this allows response functions to be computed out to periods of at least 6 months. This approach is usually called 'geomagnetic depth sounding', or GDS.

The daily variation Sq and harmonics can also be used to infer geomagnetic response functions, but this is proportionately more difficult because of the more complicated geometry of the source fields. The nature of the source field cannot be assumed *a priori*, but rather must be estimated using an array of observatory sites. For example, Olsen (1992) used nearly 100 observatories and a spherical harmonic representation of degree 10 to estimate geomagnetic responses for the first six daily harmonics.

7.2.7.1 Using magnetic satellites

The use of magnetic satellites to examine global induction potentially allows the continents and oceans to be examined together. Because satellites sample the entire Earth one can transcend the limitations of the sparse and uneven observatory distribution, and also sample source-field morphology better than for a single-site GDS measurement (although simultaneous processing of data from the global observatory network would sample the source field better than a single satellite could (e.g., Balasis and Egbert, 2006), in practice the observatories are often processed individually). However, because satellites move through the static parts of Earth's magnetic field, folding spatial variations into temporal variations within the satellite frame of reference, considerable effort is required to remove Earth's main field, secular variation, crustal field, and daily variation/equatorial electrojet (if one wishes to process only ring current induction). One tool that attempts to model all these phenomena simultaneously is the Comprehensive Model (Sabaka *et al.*, 2004). Even so, it is difficult to remove the effects of field-aligned currents in the auroral zones, and so data are typically processed between ±50° geomagnetic latitude, and residual effects of the daily variation are usually avoided by using only nighttime data.

To obtain response functions from satellite data (**Figure 11**) after removal of nonmagnetospheric effects, we return to our spherical harmonic expansion of associated Legendre polynomials P_l^m, with Schmidt quasi-normalized spherical harmonic coefficients representing the internal $i_l^m(t)$ and external $e_l^m(t)$ magnetic fields as a function of time t:

$$\Phi(r, \theta, \phi) = a_0 \sum_{l=1}^{\infty} \sum_{m=-l}^{l} \left\{ i_l^m(t) \left(\frac{a_0}{r} \right)^{l+1} + e_l^m(t) \left(\frac{r}{a_0} \right)^l \right\}$$
$$\times P_l^m(\cos \theta) e^{im\phi}$$

Keeping only the P_1^0 contribution and with r, θ, ϕ in geomagnetic coordinates

$$\Phi_1^0(r, \theta) = a_0 \left\{ i_1^0(t) \left(\frac{a_0}{r} \right)^2 + e_1^0(t) \left(\frac{r}{a_0} \right) \right\} P_1^0(\cos \theta)$$

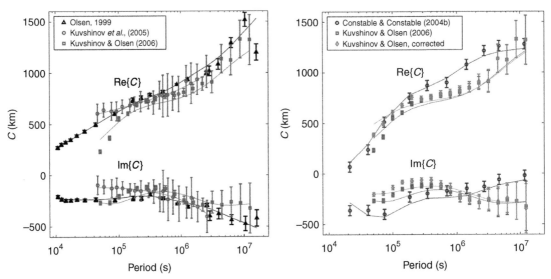

Figure 11 Long-period geomagnetic response functions computed from magnetic satellite records. Kuvshinov and Olsen attribute the difference in the long-period imaginary response between their data and those of Constable and Constable (2004b) as an artifact of Constables' time-series analysis. Their correction removes the 3-D effects of the oceans on the 1-D response functions. Modified from Kuvshinov A and Olsen N (2006) A global model of mantle conductivity derived from 5 years of CHAMP, Orsted, and SAC-C magnetic data. *Geophysical Research Letters* 33: L18301.

and so the magnetic induction **B** is derived from the negative of the gradient in the usual manner

$$\mathbf{B}(r,\,\theta,\,\phi) = -\nabla\Phi_1^0(r,\,\theta,\,\phi)$$

or, expressed as components B_r, B_θ, B_ϕ of a spherical coordinate system:

$$B_r = \left[-e_1^0 + 2i_1^0\left(\frac{a}{r}\right)^3 \right]\cos(\theta)$$

$$B_\theta = \left[e_1^0 + i_1^0\left(\frac{a}{r}\right)^3 \right]\sin(\theta)$$

$$B_\phi = 0$$

For an analysis using a single satellite, these equations can be fit to data from each equatorial pass having different values of altitude (r) and geomagnetic cola-titude (θ), as was done by Didwall (1984) in the first analysis of *POGO* data, and subsequent studies by Olsen *et al.* (2003) and Constable and Constable (2004b), providing estimates of $i_1^0(t)$ and $e_1^0(t)$ at about 100 minute intervals. For studies using multiple satellites (e.g., Olsen *et al.*, 2003; Kuvshinov and Olsen, 2006), a parametrized fit to higher-order coefficients can be made. These time series may then be Fourier transformed into the frequency domain to derive a complex geomagnetic response function of frequency

$$Q_1^0(\omega) = i_1^0(\omega)/e_1^0(\omega)$$

7.3 Interpretation of GDS and MT Data

The subjects of this section are the mechanisms by which one can relate MT and GDS response functions to electrical conductivity distributions in Earth. Predicting the response functions that would be observed over a given conductivity model is the 'forward problem', and estimating conductivity from response function data is the 'inverse problem'.

Forward model studies. If the forward modeling computation is fast enough, and the models simple enough, then it is perfectly reasonable to guess a conductivity model, compute and compare the predicted response to the data visually, and estimate ways in which the model needs to be modified to improve the fit. Forward modeling requires some understanding of how variations in the model affect the predicted data, but allows geological prejudices to be incorporated directly into the process.

Another role of forward model studies is hypothesis testing. If the number of possible geological models is limited, and each model is likely to have a distinct geophysical signature, then geophysical models can be built for each hypothesis and computed to see if the predictions are compatible with the observed data.

Parametrized inversion. If it is sensible to approximate the true complexity of geology to a model having only a few parameters, then there is a whole class of methods that will automatically estimate the value of these parameters (see, for example, Gill *et al.* (1981)). In electrical and electromagnetic methods the problems are almost always nonlinear, so fully nonlinear or linearized approaches need to be chosen.

Regularized inversion. If one does not have an *a priori* justification for parametrizing the subsurface conductivity model, or such a parametrization is impractical, then some other constraint on the model space must be imposed to make the problem well posed. A popular, effective, and useful approach is to generate smooth models using some sort of regularization (see, for example, Constable *et al.* (1987)).

Analytical least-squares solutions. A true least-squares solution for a given model dimensionality is a rare thing; usually the forward models must either be trivially simple, or the nonlinear forward solution must be recast as a linear functional. Such a solution exists for the 1-D resistivity sounding problem (Parker, 1984) and for the 1-D MT problem (Parker and Whaler, 1981).

7.3.1 MT over a Layered Earth

The simplest extension of the half-space apparent resistivity is a horizontally layered Earth. For an N-layered structure numbered downward, we can define a complex wave number for each layer

$$k_j = \sqrt{i\omega\mu_0\sigma_j}, \quad i = 1 \ldots N$$

with layer impedance

$$G_i = \frac{\mu_0\omega}{k_i}$$

Then the total impedance of the layers $i, \ldots, N$ is given by Z_i and defined for the top of each layer by

$$Z_i = G_i \frac{Z_{i+1} + G_i \tanh(ik_i h_i)}{G_i + Z_{i+1} \tanh(ik_i h_i)}$$

where h_i are the layer thicknesses (Schmucker, 1970). Thus Z_1 is the impedance observed at the surface and is easily computed using a recurrence relationship starting with $G_N = Z_N$.

7.3.1.1 *Weidelt's transformation*

An analytical solution exists for a layered, spherically symmetric Earth conductivity model (Srivastava, 1966; see also Parkinson, 1983, p. 313). However,

given the greater simplicity of the layered MT solution (above), and the various inverse solutions available for the flat-earth approximation (below), it is often desirable to analyze global data using flat-earth solutions. Weidelt (1972) provides a compact transformation between the two geometries. If we interpret a GDS C-response $c(\omega)$ obtained for spherical harmonic degree n using a layered model to get a conductivity–depth relationship $\bar{\sigma}(z)$, this can be transformed into a conductivity profile in a spherical Earth $\sigma(r)$ of radius R using

$$\sigma(r) = f^{-4}(r/R)\bar{\sigma}\left(R\frac{(r/R)^{-n} - (r/R)^{n+1}}{(2n+1)f(r/R)}\right)$$

where

$$f(r/R) = \frac{(n+1)(r/R)^{-n} + n(r/R)^{n+1}}{2n+1}$$

which for $n = 1$ reduces to

$$\sigma(r) = f^{-4}(r/R)\bar{\sigma}\left(R\frac{(R/r) - (r/R)^2}{3f(r/R)}\right)$$

and

$$f(r/R) = \frac{2R/r + (r/R)^2}{3}$$

We see from **Figure 12** that the correction is only significant below depths of about 2000 km or, for typical mantle resistivities, periods of about one day or longer.

7.3.2 Forward Modeling in Higher Dimensions

Once the realm of 1-D or radially symmetric models is left, analytical solutions are no longer available for general conductivity structure, and one is forced

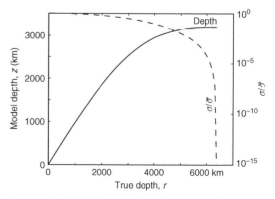

Figure 12 Weidelt's transformation for depth and conductivity, for a P_1^0 source field geometry.

to solve the electromagnetic governing equations numerically. For interpretation of MT data on a local scale, a 2-D model is useful, especially if the data are collected along a single profile (as is often the case), and avoids the complexity of model construction and computation in three dimensions. However, for global induction problems we are forced to take the step from 1-D directly to 3-D. A recent review of 3-D modeling is provided by Avdeev (2005).

Rewriting Maxwell's equations for frequency ω and considering an applied source of current $\mathbf{J}'$ or magnetic field $\mathbf{H}'$ we have

$$\nabla \times \mathbf{E} = -\mathrm{i}\omega\mu(\mathbf{H} + \mathbf{H}')$$

$$\nabla \times \mathbf{H} = \sigma\mathbf{E} + \mathbf{J}'$$

which can be solved if the differential operators are approximated numerically. The most straightforward approach is the finite difference scheme in which the $\mathbf{E}$ and $\mathbf{H}$ are discretized over a rectilinear grid of nodes, resulting in a linear system of equations $\mathbf{A}x = b$, where the b represents the boundary conditions and applied source fields, x is the vector of field values on the nodes, and $\mathbf{A}$ is a matrix of difference operators. $\mathbf{A}$ is large, but sparse, making the inversion of the matrix tractable for even large problems if appropriate sparse matrix solutions are used. Alternatively, one can use algorithms that do not require the storage of the entire matrix $\mathbf{A}$ (e.g. Weiss, 2001).

A disadvantage of the finite-difference approach is that the grid of nodes must be uniformly rectilinear, making it difficult to mesh complex surfaces or structure having a large range of length scales. A solution to this is the finite-element approach, in which $\mathbf{E}$ and $\mathbf{H}$ are defined on the edges of triangular (2-D) or tetrahedral (3-D) elements and interpolated across the elements using basis functions. A linear system of equations again results in a sparse matrix to be solved. Model construction is perhaps more difficult using finite elements, but automated meshing schemes are becoming available (e.g., Key and Weiss, 2006).

Usually the equations are solved for $\mathbf{E}$ and then a discrete approximation to Faraday's law is used to recover $\mathbf{H}$. A common approach to numerical modeling of electromagnetic fields is to break the problem into primary ($\mathbf{E}_0$, $\mathbf{H}_0$, and σ_0) and secondary ($\mathbf{E}_s$, $\mathbf{H}_s$, and σ_s) components,

$$\mathbf{E} = \mathbf{E}_0 + \mathbf{E}_s, \quad \mathbf{H} = \mathbf{H}_0 + \mathbf{H}_s, \quad \sigma = \sigma_0 + \sigma_s$$

where the primary fields have an analytical solution, such as a whole-space, half-space, or layered model. This primary/secondary separation has many advantages, particularly if there is a singularity associated with a man-made source field, but also provides the basis for the third approach to high-dimensional modeling, the 'integral equation' method. If the region over which σ_s is nonzero is a finite volume V, then

$$\mathbf{E}_s = \int_V \mathbf{G}^E(\mathbf{r}, \mathbf{r}')\sigma_s\mathbf{E}(\mathbf{r}')\mathrm{d}v'$$

and

$$\mathbf{H}_s = \int_V \mathbf{G}^H(\mathbf{r}, \mathbf{r}')\sigma_s\mathbf{E}(\mathbf{r}')\mathrm{d}v'$$

where the $\mathbf{G}$ are the 3×3 tensor Green's functions relating the fields at $\mathbf{r}$ to current elements at $\mathbf{r}'$. The discretization of these equations leads to a linear system where $\mathbf{A}$ is now dense, but much smaller than for the finite-element and finite-difference methods. All of these approaches can be projected onto spherical geometries in order to model global electromagnetic induction (e.g., Everett and Schultz, 1996; Weiss and Everett, 1998; Martinec, 1999; Uyeshima and Schultz, 2000; Yoshimura and Oshiman, 2002; Grammatica and Tarits, 2002; Kuvshinov et al., 2002).

A frequency-domain approach, which assumes a stationary excitation of the source field at a given frequency ω, may not be appropriate if one wants to consider the effect of a discrete magnetic storm. In this case, a time-domain approach to modeling is required (Hamano, 2002; Velimsky et al., 2003; Kuvshinov and Olsen, 2004).

7.3.3 Numerical Inversion of Geomagnetic Data

The first thing one needs for an inversion scheme is a measure of how well a given model fits the data. For practical and theoretical reasons, the sum-squared misfit is favored:

$$\chi^2 = \sum_{i=1}^{M} \frac{1}{\sigma_i^2}[d_i - f(x_i, \mathbf{m})]^2$$

where

$$\mathbf{d} = (d_1, d_2, d_3, \ldots\ldots, d_M)$$

are M observed data values,

$$\mathbf{x} = (x_1, x_2, x_3, \ldots\ldots, x_M)$$

are independent data variables (electrode spacing, frequency, etc.), and

$$\mathbf{m} = (m_1, m_2, \ldots\ldots, m_N)$$

are model parameters (layer or block conductivities, layer or block sizes, etc.). Equivalently, the misfit may be written as

$$\chi^2 = \left\|\mathbf{Wd} - \mathbf{W\hat{d}}\right\|^2$$

where $\hat{\mathbf{d}}$ is the predicted response from the model

$$\hat{\mathbf{d}} = f(\mathbf{x}, \mathbf{m})$$

and $\mathbf{W}$ is a diagonal matrix of reciprocal data errors

$$\mathbf{W} = \mathrm{diag}(1/\sigma_1, 1/\sigma_2, \ldots 1/\sigma_M)$$

The least-squares approach attempts to minimize χ^2 with respect to all the model parameters simultaneously. If the data errors σ_i are Gaussian and independent, then least squares provides a maximum likelihood and unbiased estimate of $\mathbf{m}$, and χ^2 is Chi-squared distributed with $M - N$ degrees of freedom. To apply linear inverse theory to nonlinear electromagnetic data we need to linearize the problem by expanding f around an initial model guess $\mathbf{m}_0$

$$\hat{\mathbf{d}} = f(\mathbf{m}_1) = f(\mathbf{m}_0 + \Delta\mathbf{m}) \approx f(\mathbf{m}_0) + \mathbf{J}\Delta\mathbf{m}$$

where $\mathbf{J}$ is a matrix of derivatives of data with respect to the model

$$\mathcal{J}_{ij} = \frac{\partial f(x_i, \mathbf{m}_0)}{\partial m_j}$$

(often called the Jacobian matrix) and

$$\Delta\mathbf{m} = (\delta m_1, \delta m_2, \ldots, \delta m_N)$$

is a model parameter perturbation about $\mathbf{m}_0$. Now our expression for χ^2 is

$$\chi^2 \approx ||\mathbf{Wd} - \mathbf{W}f(\mathbf{m}_0) + \mathbf{WJ}\Delta\mathbf{m}||^2$$

which we minimize in the usual way by setting the derivatives of χ^2 with respect to $\Delta\mathbf{m}$ equal to zero to get N simultaneous equations:

$$\beta = \alpha\Delta\mathbf{m}$$

where

$$\beta = (\mathbf{WJ})^{\mathrm{T}}\mathbf{W}(\mathbf{d} - f(\mathbf{m}_0))$$

$$\alpha = (\mathbf{WJ})^{\mathrm{T}}\mathbf{WJ}$$

The matrix α is sometimes called the curvature matrix. This system can be solved for $\Delta\mathbf{m}$ by

inverting α numerically, and a second model $\mathbf{m}_1 = \mathbf{m}_0 + \Delta\mathbf{m}$ found. Because $\mathbf{J}$ depends on $\mathbf{m}$, one needs to solve this repeatedly and hope for convergence to a solution. Near the least-squares solution, this method will work, but any significant nonlinearity will result in likely failure. An algorithm to compensate for this behavior was suggested by Marquardt (1963). The diagonal terms of the curvature matrix are increased by a factor λ:

$$\alpha_{jk} = \alpha_{jk}(1 + \lambda), \quad \text{for } j = k$$

$$\alpha_{jk} = \alpha_{jk}, \quad \text{for } j \neq k$$

For small λ this method obviously reduces to the linearized inversion. For large λ the diagonal terms dominate and the method reduces to a gradient algorithm that chooses a path in the direction of maximum reduction in χ^2 (the method of steepest descent). The gradient algorithm is robust to nonlinearity but is very inefficient in the parabolic region near the solution (because the gradient gets small; ultimately zero). By adjusting λ to be large far from the solution, and small as we approach the minimum, we have a fairly stable method of determining a model composed of a small number of parameters.

For nonlinear problems that are truly parametrized (e.g., finding the concentration and mobility of a charge carrier in a mineral), the Marquardt method is pretty hard to beat. It also works fairly well for problems where the number of degrees of freedom is large, given by $M - N$ when the M data are truly independent, and the starting model provides a rough fit to the data. In practice, this means 1-D models which consist of a small number of layers. Any attempt to increase the number of layers significantly, to approximate the realistic case of a continuously varying conductivity, or to invert a 2-D or 3-D model with thousands or millions of parameters, will result in wildly fluctuating parameters followed by failure of the algorithm. This is because there will be columns of $\mathbf{J}$ that are small, reflecting the fact that large variations in a model parameter may relate to negligible effects on the combined data.

One approach, suggested by Backus and Gilbert (1967), is to minimize $\Delta\mathbf{m}$, but this and related algorithms converge extremely slowly and are called by Parker (1994) 'creeping methods'. In any case, the true least-squares solutions can be pathologically rough – for the 1-D MT case it is known that the least-squares solution is composed of delta functions in conductivity. Almost all high-dimensional inversion today incorporates some type of regularization, an approach suggested by Tikhonov and Arsenin (1977), which explicitly penalizes bad behavior in the model. For example, instead of minimizing χ^2, we minimize an unconstrained functional

$$U = ||\mathbf{Rm}_1||^2 + \mu^{-1}\left(||\mathbf{Wd} - \mathbf{W}f(\mathbf{m}_1)||^2 - \chi_*^2\right)$$

where χ_*^2 is a target misfit that is greater than the minimum possible, but statistically acceptable, and $\mathbf{Rm}$ is some measure of roughness in the model, often taken to be first differences between adjacent model parameters and easily generated by a matrix $\mathbf{R}$ consisting of $(-1, 1)$ entries on the diagonal. Minimizing U has the effect of minimizing model roughness as well as how far the data misfit is from being acceptable. We substitute our linearization around $f(\mathbf{m}_0)$ and differentiate U with respect to $\mathbf{m}_1$ to accomplish this, rearranging the result to get $\mathbf{m}_1$ directly:

$$\mathbf{m}_1 = \left(\mu\mathbf{R}^{\mathrm{T}}\mathbf{R} + (\mathbf{WJ})^{\mathrm{T}}\mathbf{WJ}\right)^{-1}(\mathbf{WJ})^{\mathrm{T}}\mathbf{W}\left(\mathbf{d} - f(\mathbf{m}_0) + \mathbf{Jm}_0\right)$$

We need only to choose the tradeoff (Lagrange) multiplier μ. The approach of Constable *et al.* (1987) was to note that for each iteration χ^2 is a function of μ, and to use 1-D optimization (simply a line search) to minimize χ^2 when $\chi^2 > \chi_*^2$ and to find μ such that $\chi^2 = \chi_*^2$ otherwise. Constable *et al.* called this approach 'Occam's inversion'. Although the Occam algorithm is reliable and has good convergence behavior, the computation and storage of $\mathbf{J}$ for large models can be limiting, and numerous other regularization approaches using conjugate gradient algorithms and approximate derivatives exist (e.g., Smith and Booker, 1991; Mackie and Madden, 1993; Newman and Alumbaugh, 2000).

7.3.4 Analytical Inversion and Estimation of Bounds

The 1-D MT problem is one of the few problems for which an analytical least-squares solution exists. Parker and Whaler (1981; see also Parker, 1994) showed that the C-response (also called admittance by Parker, since the electric field is a response to the forcing magnetic field) can be cast as a linear functional of a spectral function that is in turn nonlinearly related to conductivity. Linear inverse theory can be used to find a least-squares solution for a given data set in terms of the spectral function, which can then be used to recover a conductivity distribution. The math is nontrivial, and interested readers are referred to the original works, but the result is that least-squares solutions to the 1-D MT problem are delta

functions of conductivity in an infinitely resistive half-space, a model space called D^+ by Parker. This is one of the reasons why layered least-squares solutions are not stable as the layer number increases past the point of restricting variations in conductivity to be few and far between. It may appear that such a pathological solution is not useful, but being able to put a lower bound on the possible 1-D misfit is very useful when choosing desired misfits for regularized inversions, and the inversion machinery of D^+ also allows the estimation of a maximum depth of resolution available for a given data set.

Another approach to interpreting response function data is to put bounds on average conductivity over a given region of the earth. The earliest attempt at this by Backus and Gilbert (1968) suffers from being a linearized theory; this may not be too much of a restriction for seismology, but in geomagnetic induction the range of possible conductivity is so large as to make this approach essentially useless. A nonlinear approach using electromagnetic data was first presented via the funnel functions of Oldenburg (1983). More recently, Medin *et al.* (2006) generalized this by using quadratic programming to solve for electrical conductivity and the forward responses simultaneously, by making both the data and the forward solution equations constraints in the optimization. The flexibility of the optimization approach allows many other constraints, such as monotonicity of conductivity with depth, and bounds on average conductivity, to be applied. By systematically solving for various averages of conductivity over depth, the bounds of acceptable averages can be discovered.

To date, attempts to put bounds on average conductivity has been restricted to the 1-D problem, but in principle the machinery could be extended to higher dimensions.

7.4 Electrical Conductivity of Earth Materials

There are three basic conduction regimes for crustal and mantle rocks (**Table 2**):

1. *Ionic conduction* of crustal rocks containing water. The mineral grains making up most rock-forming minerals are essentially insulators (e.g., quartz conductivity at surface temperatures is about $10^{-12}\ \mathrm{S\,m^{-1}}$), and so the higher conductivity of water in pores and cracks determines the conductivity of the rock.

Table 2   Approximate conductivities and resistivities of various Earth materials

Ionic conduction through water		
Seawater at 20°C	$5\,\text{S m}^{-1}$	$0.2\,\Omega\text{m}$
Seawater at 0°C	$3\,\text{S m}^{-1}$	$0.3\,\Omega\text{m}$
Marine sediments	$1\text{–}0.1\,\text{S m}^{-1}$	$1\text{–}10\,\Omega\text{m}$
Land sediments	$0.001\text{–}0.1\,\text{S m}^{-1}$	$10\text{–}1000\,\Omega\text{m}$
Igneous rocks	$10^{-5}\text{–}10^{-2}\,\text{S m}^{-1}$	$100\text{–}100\,000\,\Omega\text{m}$
Good semiconductors		
Graphite	$10\,000\text{–}1\,000\,000\,\text{S m}^{-1}$	$10^{-4}\text{–}10^{-6}\,\Omega\text{m}$
Galena	$10\text{–}100\,000\,\text{S m}^{-1}$	$10^{-1}\text{–}10^{-5}\,\Omega\text{m}$
Pyrite	$1\text{–}1000\,\text{S m}^{-1}$	$10^{-3}\text{–}1\,\Omega\text{m}$
Magnotito	$20\,000\,\text{S m}^{-1}$	$5\times10^{-5}\,\Omega\text{m}$
Thermally activated conduction		
Olivine at 1000°C	$0.0001\,\text{S m}^{-1}$	$10\,000\,\Omega\text{m}$
Olivine at 1400°C	$0.01\,\text{S m}^{-1}$	$100\,\Omega\text{m}$
Theoleite melt	$3\,\text{S m}^{-1}$	$0.3\,\Omega\text{m}$
Silicate perovskite	$1\,\text{S m}^{-1}$	$1\,\Omega\text{m}$

2. Conduction dominated by minerals that are *good semiconductors or metals*, such as graphite, magnetite, and pyrite, or native metals. These minerals may occur in trace amounts that nevertheless form a well-connected pathway, or in extreme cases they may account for the bulk of the rock, in the case of ore bodies.

3. *Thermally activated conduction* in subsolidus or molten silicates at elevated temperature, either in the deep mantle or crustal volcanic systems.

7.4.1 Moving an Electric Charge

Electrical conduction in a material is accomplished by a movement of charge, which in practice means the movement of electrons, holes (the absence of an electron), ions, or ion vacancies. Materials may be classified on the basis of the nature of the charge carriers and the type of movement. Generally, conductivity in a material is given by

$$\sigma = nq\mu$$

where q is the charge of the mobile species, n is the concentration of the charged species, and μ is the mobility, or drift velocity per applied electric field, $|v/E|$.

7.4.1.1 Conduction in native metals

Conduction in native metals is by means of valence, or conduction, electrons which do not take part in crystal bonding and are loosely bound to atoms. This form of conduction is very efficient, resulting in room temperature conductivities for metals of about $10^7\,\text{S m}^{-1}$. The conductivity of metals decreases with increasing temperature because of thermal agitation of the conduction electrons, which impedes their movement in response to an electric field and therefore lowers the drift velocity. Native metals are fairly rare in surface rocks, but metallic iron is responsible for high conductivity in the core.

Metallic conduction electrons have long mean free paths because they do not get deflected by the periodic arrangement of ion cores, and only rarely get scattered off other conduction electrons. For metals

$$v = -eE\tau/m$$

where τ is the mean free time between collisions and m is the mass of the electron. For metals n is of order $10^{28}\text{–}10^{29}/\text{m}^3$. The mean free path ($v\tau$) is of order 1 cm at 4 K, 10^{-8} m at room temperature. Thus conductivity from metals is thus given by

$$\sigma = \frac{ne^2\tau}{m}$$

which can be thought of as an acceleration under an electric field (e/m) of charge of density ne for a time period τ.

Scattering (which determines mobility μ and mean free time τ) is by means of phonons (lattice vibrations) and impurities. Phonons go to zero at 0 K and are proportional to temperature above some threshold, so

$$\rho \propto T$$

(That threshold is the Debye temperature, below which quantization of energy causes some vibration modes to be 'frozen' out. It is between 100 and 400 K for most metals.)

7.4.1.2 Semiconduction

Semiconduction is the result of behavior in between metallic conduction and conduction in insulators. It is typical of ionically bonded binary compounds, such as PbS, ZnS, MgO, etc. There are no free electrons, as such, in a semiconductor, but an applied electric field can supply enough energy to move electrons from the valence band, across the band gap E_g, into the conduction band at higher energy levels. As a result of these few electrons, and the holes they leave behind in the lower energy levels, moderate conductivities are achieved, between about 10^{-5} and 10^5 S m^{-1}. Sulfide minerals are often semiconductors with quite low resistivities.

For semiconductors,

$$\sigma = ne\mu_e + pe\mu_h$$

where n and p are the carrier densities of electrons and holes (typically around $10^{19} - 10^{23}/m^3$), and μ_e, μ_h are the respective mobilities. Carrier concentration is thermally activated and is

$$n \propto T^{3/2} e^{-E_g/2kT}$$

where k is Boltzmann's constant (1.380×10^{-23} J K^{-1} or 8.617×10^{-5} eV). Mobility is also only proportional to T, so the exponential term dominates. Such thermally activated processes are often termed Arrhenius relationships, and produce linear plots when $\log(\sigma)$ is plotted against $1/T$.

Charge carriers can also be generated by impurities or doping; for example, a divalent cation substituting on a trivalent site will have an excess electron which is available for the conduction band. These two conduction regimes, one determined by statistical mechanics and one determined by crystal impurities, is typical of solid-state conduction (**Figure 13**). The first, purely thermally activated, variation in conductivity is the same for a given material and temperature, and is called 'intrinsic' conduction. The second regime, which also has some temperature dependence but will vary from sample to sample with the amount of impurity, is called 'extrinsic' conduction. In the intrinsic regime, holes must equal electrons because elevation of every electron into the conduction band will leave behind a

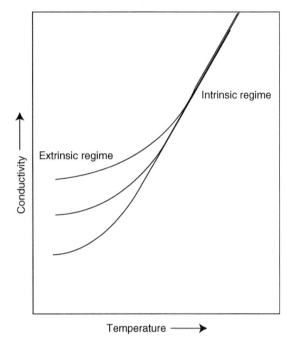

Figure 13 Extrinsic and intrinsic regimes in semiconductors. Extrinsic conduction occurs at lower temperature and is a result of electrons or holes generated by impurities (or doping). Intrinsic conduction is a result only of thermally activated charge carries associated with the pure crystal.

hole. Impurities, however, can donate either holes or electrons to conduction (**Figure 13**).

7.4.1.3 Ionic, or point defect, conduction

Conduction by point defects dominates when the bandgap is too large for semiconduction, and is the result of moving ions and ion vacancies through a crystal structure. It is very difficult to move ionically bonded ions through a crystal lattice, so the resistivities of silicate minerals are very high; quartz has resistivities between 10^{10} and 10^{14} Ω m. Rock salt has a resistivity of 10^{12} Ω m or so. It would, in fact, be close to impossible to move ions through a perfect lattice, but defects in the lattice allow the motion of ions and vacancies. Schottky defects are lattice vacancies associated with ions moving to the surface of the crystal and Frenkel defects are ions which move to interstitial positions. Impurities also constitute defects in the lattice, both in themselves and because they can promote vacancies to maintain charge balance. All these are called point defects, compared with dislocations, line and plane defects, which tend not to contribute substantially to conduction (**Figure 14**).

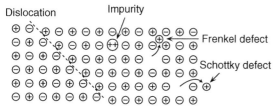

Figure 14 Defects in crystals.

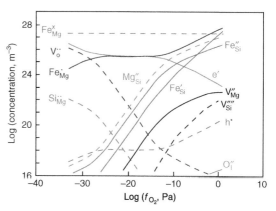

Figure 15 A defect model for silicate perovskite, showing concentrations of various defects as a function of oxygen activity (f_{O_2}). Modeled temperature is 2500 K (2227°C) and pressure is 40 GPa. Redrawn from Hirsch LM and Shankland TJ (1991) Point-defects in (Mg, Fe)SiO$_3$ Perovskite. *Geophysical Research Letters* 18: 1305–1308.

Schottky and Frenkel defects are thermally induced, so the number of defects, and hence conductivity, increases with increasing temperature:

$$n_s = Ne^{-E_s/kT} \quad \text{(Schottky defects or vacancies)}$$

$$n_f = (NN')^{1/2}e^{-E_i/2kT} \quad \text{(Frenkel defects or interstitials)}$$

where N is the total number of atoms and N' is the total number of interstitial sites. Again, however, defects can be generated by impurities. Thus a divalent cation on a monovalent site induces a cation vacancy to maintain charge neutrality. The mobility of point defects can be related to diffusion, the flux $\mathbf{J}_N$ of atoms crossing unit area in unit time under the influence of a concentration gradient ∇N. The diffusivity or diffusion constant D is defined by Fick's law:

$$\mathbf{J}_N = -D\nabla N$$

Diffusivities are found to vary with temperature, again, as a thermally activated Boltzmann process:

$$D = D_0 e^{-E_d/kT}$$

where E_d is another activation energy for diffusivity. Mobilities are related to diffusivities by the Nernst–Einstein equation:

$$kT\mu = qD$$

so $\mu = qD/kT$ and the conductivity of, for example, Schottky defects becomes

$$\sigma = n_s q \mu$$
$$= Ne^{-E_v/kT} q q D/kT$$
$$= Nq^2 D_0/kT e^{-E_v/kT} e^{-E_d/kT}$$

(much of the above material can be found in Kittel (1986)). The exponential terms dominate, and so to a first approximatation conductivity is given by

$$\sigma = \sigma_0 e^{-E/kT}$$

where E is an apparent activation energy associated with both defect concentration and mobility. Often there are several point defects in the mineral;

Figure 15 presents a model of 11 different charge-carrying defects in silicate perovskite, and olivine would have a similar suite of defects. Description of the defects is handled by the Kröger–Vink notation, in which the main character is the actual species involved (a 'V' denotes a vacancy), the subscript is the crystal site for a normal lattice (in interstitial site is denoted by 'I'), and the superscript is negative, positive, or neutral charge shown as ', ·, or x respectively. Thus Fe$_{Mg}$ is a Fe^{3+} ion occupying an Mg^{2+} site resulting in a net charge for the defect of $+e$.

In spite of the variety in defect types, the dominance of the exponential term in the expression for conductivity means that at a given temperature one mechanism is likely to dominate over another. As temperature changes, different activation energies will cause the dominant conduction mechanism to change, so conductivity is often expressed as

$$\sigma = \sum_i \sigma_i e^{-E_i/kT}$$

This Boltzmann relationship between temperature and conductivity can be linearized by taking $\log(\sigma)$ as a function of $1/T$, to form an Arrhenius plot. For a single conduction mechanism, data plotted in this way will be a straight line. For mixed conduction, a number of lines may be observed as a function of temperature. **Figure 16** shows an example of two conductivity mechanisms being expressed as a sample is heated from 500°C to 1200°C.

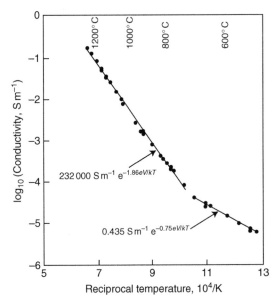

Figure 16 Electrical conductivity data for cobalt olivine (Co_2SiO_4), showing two conduction mechanisms expresses as Arrhenius relationships of two different activation energies. Redrawn from Hirsch LM (1990) Electrical conduction of Co_2SiO_4. *Physics and chemistry of Minerals* 17: 187–190.

7.4.1.4 Pressure and Oxygen

Pressure can also have an effect on conductivity, which may be modeled as

$$\sigma = \sigma_0 e^{-(E + P\Delta V)/kT}$$

where P is pressure and ΔV is called the activation volume. Pressure usually has a minor effect compared with temperature, unless it changes the mineral phase and causes a more mobile or numerous defect to dominate. One of the biggest effects on defect chemistry is oxygen. Oxidation and reduction is associated with an exchange of charge, so oxidation state changes the number of charged defects in a mineral. The model for perovskite in **Figure 15** gives some idea of how defects in a mantle mineral depend greatly on oxidation state.

7.4.1.5 Aqueous electrolytic conduction

Aqueous conduction is associated with ions dissociated in water, and is arguably the most important conduction mechanism in crustal rocks. The dominant dissolved salt is sodium chloride, and it is clear that the number of charge carriers depends simply on the concentration of salt. Mobility increases with temperature because the viscosity of water decreases

with temperature, and for seawater concentration ($35\,\mathrm{g\,l^{-1}}$) conductivity is approximately

$$\sigma = 3 + T/10\,\mathrm{S\,m^{-1}}$$

for temperature T in Celsius. However, for water under pressure, this linear relationship breaks down at around 200°C, and saltwater conductivity peaks at between 300°C and 400°C (Quist and Marshall, 1968).

7.4.1.6 Clay minerals

Clays have an important effect on aqueous conduction in crustal rocks because they have a surface charge, usually negative, associated with their crystal structure. This creates a so-called 'double layer' of charge, the negative charge on the crystal creating a diffuse layer of positive charge in the electrolyte that is mobile and can contribute to electrical conduction. This effect is greatest for low-salinity groundwater, and in practice limits the effective resistivity of pore water. A good description of this surface electrochemistry is found in Morgan *et al.* (1989), and a comprehensive treatment of surface conductance by Revil and Glover (1997).

7.4.1.7 Magma

Magma conductivity is approximately that of seawater, and is associated with diffusion of ions in a highly polymerized structure. So, it is not surprising that again one observes a thermally activated Boltzmann-type conductivity, which for tholeiitic melt above 10 kbar pressure is approximately

$$\sigma = 10^{5.332} e^{-1.533 eV/kT}$$

(Tyburczy and Waff, 1983). How well a mixture of melt and rock conducts electricity depends on melt fraction, temperature, composition, and the geometry of the partial melt (Roberts and Tyburczy, 1999). The relationship between geometry and conductivity of two-component systems is discussed in the next section.

7.4.2 Binary Mixing Laws

We have seen that rocks are often made up of two components, one very much more conductive than the other. For example, a gabbro could be mostly nonconducting silicate minerals with minor amounts of conductive magnetite. A sandstone can be nonconducting quartz saturated with water. A mantle rock can be subsolidus silicates of several hundred

Ohm-m with conductive melt between the grains. Rather than make measurements on every possible combination, binary mixing laws can be used to predict the conductivity of the bulk rock as a function of the conductive fraction. As one might imagine, the physical geometry of the conductive material very much determines how conductive the bulk rock ends up.

The classic binary mixing relationship is Archie's law, which (in spite of the name), is an entirely empirical relationship developed to explain the conductivity of sedimentary rocks containing water:

$$\sigma = \sigma_s + (\sigma_f - \sigma_s)\beta^m$$

Here σ_f is the fluid, or water conductivity, and β is the fluid fraction. In the case of a saturated rock, β is the porosity, but it is easy enough to express fluid content in terms of saturation and porosity. Here we have included the conductivity of the solid, σ_s, although for silicate minerals at crustal temperatures this is usually taken to be zero. The exponent m can be derived experimentally from a suite of samples for a given rock type, but is usually taken to be 2 for sediments. A smaller m corresponds to more efficiently connected pore space.

A model of fluid-filled tubes might be a reasonable representation of well-connected melt in a volcanic system:

$$\sigma = \frac{1}{3}\beta\sigma_f + (1-\beta)\sigma_s$$

Models for binary systems can become quite complicated; various geometries are presented by Schmeling (1986). However, all possible models must lie between the Hashin–Shtrikman (HS) bounds:

$$HS^- = \sigma_s + \beta\left(\frac{1}{\sigma_f - \sigma_s} + \frac{1-\beta}{3\sigma_s}\right)^{-1}$$

$$HS^+ = \sigma_f + (1-\beta)\left(\frac{1}{\sigma_s - \sigma_f} + \frac{\beta}{3\sigma_f}\right)^{-1}$$

Normally, when one measures the electrical conductivity of fluid-filled crustal rocks as a function of pressure, increasing pressure decreases pore volume, and so conductivity decreases. However, if the dominant conduction is by conductive minerals, such as graphite or magnetite, increasing pressure can press these minor minerals into better electrical contact, in which case conductivity will increase with pressure. This is an important diagnostic of conduction mechanism (Shankland *et al.*, 1997).

7.4.3 Polarization

We have discussed various conduction mechanisms; metallic, semiconduction, ionic conduction, and electrolytic conduction, all of which can be found in rocks and minerals and within the apparatus we use to make measurements. When two different types of conductors meet, and one attempts to pass electric current between them, there will be a polarization at the boundary. The classic example of this will be a metallic electrode making contact with the ground or a laboratory sample. Electrons flowing through the circuit set up an electric field in the electrolyte. Ions drift through the fluid in response to this field until they encounter the electrodes, where they have to stop. The initial effect of this is to increase the impedance of the fluid (and change the phase of the conduction current) by depleting the ions available for transport and the distance over which they can migrate. The second effect is that the ions will then participate in reversible and irreversible electrochemical reactions with the electrodes.

7.4.4 Anisotropy

So far we have only considered the case in which conductivity is isotropic. In general conductivity may depend on direction:

$$\sigma = \begin{bmatrix} \sigma_x & & \\ & \sigma_y & \\ & & \sigma_z \end{bmatrix}$$

where most commonly x, y, z correspond to two horizontal and the vertical direction, although any general case can be rotated into three principal axes.

As with porosity, the nature of anisotropy depends on scale, which can loosely be classified as crystallographic anisotropy, textural anisotropy, and structural anisotropy as the scale length gets progressively larger.

7.4.4.1 *Crystallographic anisotropy*
Crystallographic anisotropy is associated with variations in conductivity of the mineral itself. The principal components of crystallographic properties must lie along the crystal axes, and so it is sufficient to measure and specify conductivity along these directions.

7.4.4.2 *Textural anisotropy*
Textural anisotropy is associated with the fabric of the rock. A common example of this type of anisotropy is found in sedimentary rocks that have

well-developed bedding, either as a consequence of platy or tabular mineral grains that preferentially fall flat during deposition, or by fine-scale interbedding. In this case conductivity will always be higher in what were originally the subhorizontal directions (i.e., foliation direction), and lower in the direction across the beds. Anisotropy ratios (σ_y/σ_z) of up to 2–10 are possible in this way.

7.4.4.3 Structural anisotropy

Structural anisotropy is caused by macroscopic features that would normally be resolved by geological mapping but that are too small in relation to the electrical measurements to be resolved. For example, interbedded, meter-thick sedimentary horizons of different resistivity that are buried 10–100 m deep will appear anisotropic to a surface-based electrical method but not to borehole logging.

Although in principle all three principal directions of conductivity can vary, if rocks are anisotropic it is usually through uniaxial anisotropy – that is, only one of the three directions differs from the other two. It is interesting to consider the relationship between the mathematical description and the physical interpretation of the six possibilities associated with uniaxial anisotropy, as shown in **Figure 17**.

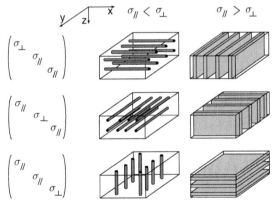

Figure 17 Relationship between geological and mathematical descriptions of anisotropy. Here the more conductive material is shown in grey. Rod-like conductive structure (left) corresponds to the background material ($\sigma_{//}$) being less conductive, and sheet-like conductive structure (right) corresponds to the background material being more conductive. Modified from Everett ME and Constable S (1999) Electric dipole fields over an anisotropic seafloor: Theory and application to the structure of 40 Ma Pacific Ocean lithosphere. *Geophysical Journal International* 136: 41–56.

7.4.5 Laboratory Measurement of Conductivity

While the fundamental physical property of rocks and minerals is conductivity/resistivity, our laboratory equipment is designed to measure circuit resistance. Resistivity ρ and resistance R for a regular prism are related by

$$R = \rho L/A$$

where L is the length of the sample and A is the cross-sectional area of the electrode-covered faces. It can be seen that if one takes a cubic meter of material ($L = 1$; $A = 1$), the series resistance in ohms across two opposite faces of the cube will be the same number as the resistivity in Ωm.

Such a measurement can be made in practice on a regular prism (such as a rock core or mineral sample) when electric current is passed through the sample by means of two metal plates pressed to the end faces. This is called the two-electrode method, but it suffers from the limitation that the contact resistance of the electrodes must be much lower than the resistance of the sample. If L is small compared with A and the sample is of relatively high ρ this approximation is acceptable, because the contact resistance of the electrodes will be less than the resistance through the sample. Otherwise, the solution is the four-electrode method (**Figure 18**). This solves the problem of contact resistance at the electrodes by measuring a potential difference across the center of the rock rather than the ends. Because modern voltmeters have very high input impedances, the contact resistance on the potential measurement circuit does not have a significant impact on the potential across the sample, and the resistivity can be computed from $\rho = \Delta V A/(Is)$ where s is the potential electrode spacing. Practical considerations often force the issue in favor of the two-electrode method (e.g., measurements made on small samples at high temperature and pressure).

Polarization at the sample–electrode interface is controlled by using an AC current, usually of around 1 kHz. Since polarization creates a capacitive response, it is wise to measure the phase, as well as the magnitude, of the conduction current versus applied potential. A technique called impedance spectroscopy exploits this capacitive effect to examine the sample more closely, by varying the frequency of the applied voltage and monitoring both the in-phase (real) and out-of-phase (quadrature or imaginary) current. An impedance diagram is

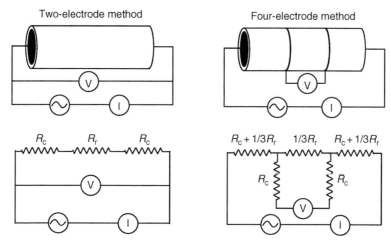

Two-electrode method

Four-electrode method

R_c R_r R_c

$R_c + 1/3R_r$ $1/3R_r$ $R_c + 1/3R_r$

R_c R_c

Figure 18 Two- and four-electrode methods of measuring rock resistivity, showing equivalent circuit elements. R_c is contact or electrode resistance; R_r is the rock or sample resistance. In the two-electrode method, the potential drop across the current electrodes is measured along with the potential drop across the sample. In the four-electrode method, the potential drop across the current electrodes is not part of the voltage-measuring circuit, and the potential electrode impedance simply adds to the already high input impedance of the voltmeter.

constructed, plotting imaginary versus real impedance for various frequencies (such a diagram is also called a Cole–Cole plot). An equivalent circuit of a resistor and capacitor in parallel will create a semicircular arc on such a plot (**Figure 19**). If several conduction mechanisms are present, several arcs will be generated, and in particular, electrode contact impedance in two-electrode measurements will separate from the impedance of the sample. Grain-surface impedance can also be separated from grain-interior impedance this way (e.g., Tyburczy and Roberts, 1990; **Figure 20**).

When measurements are made at high temperature, to predict the behavior in the deep Earth and to establish an activation energy for the conducting species, care must be taken to control the chemical environment of the sample so that it is not altered during measurement and that the measurements are applicable to the natural environment of the sample. The most obvious problem is oxidation during heating; even laboratory-grade inert gases have enough contaminant oxygen to place some minerals outside their stability fields.

To control oxygen in low-pressure experiments, a controlled mixture of $CO:CO_2$ or $H:CO_2$ is passed over the sample. For a given temperature and gas mix ratio, there is a known equilibrium concentration of oxygen developed; any excess is consumed by the reducing agent (CO or H), and any deficiency is liberated by the oxidizing agent (CO_2). For example

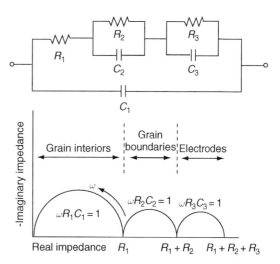

Figure 19 Impedance spectroscopy can identify the different series conduction paths in the sample because they can be associated with different capacitive effects. The top diagram shows an equivalent series circuit, and the lower diagram shows the ideal measurement that would be made on a rock sample. Frequency increases towards the origin. From Tyburczy JA and Roberts JJ (1990) Low frequency electrical response of polycrystalline olivine compacts: Grain boundary transport. *Geophysical Research Letters* 17: 1985–1988.

$$2CO + O_2 \rightleftharpoons 2CO_2$$

so even if the starting gasses contained contaminant oxygen, it will be consumed until only the equilibrium amount remains. Other problems associated

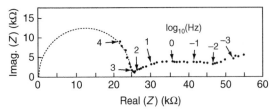

Figure 20 Example of impedance spectroscopy on a sample of San Carlos Olivine at 1200°C and an f_{O_2} of 10^{-15} Pa. Data are from Tyburczy and Roberts (1990). We have highlighted the highest frequency (10^3–10^4 Hz) mechanism, interpreted as grain-interior conduction, with a 12 kΩ radius arc centered on the real axis. Intermediate frequency (10^{-2}–10^2 Hz) impedance is associated with grain boundary conduction, and lowest frequency impedance as polarization on the electrode surfaces.

with the chemical state of the sample are iron loss to platinum electrodes, and migration of chemical species such as silicon between constituents of a multiphase sample.

Since the two-electrode method is usually used for relatively resistive samples, and often at high pressure, care has to be taken to ensure that the dominant conduction path is through the sample, and not through the sample holder and surrounding apparatus. **Figure 21** shows an example of a furnace designed to control the oxygen activity surrounding the sample and to minimize conduction leakage around the sample. The electrodes in contact with the sample are made of iridium, rather than platinum, to minimize iron loss. The thermocouple leads, which allow the temperature to be measured on both sides of the sample (important for thermopower measurements), also serve as electrode leads for a two-electrode measurement. For thermopower measurements, the entire assembly can be moved up or down in the furnace/cooling system to create a temperature gradient across the sample.

7.4.6   High-Pressure Measurements

The surface pressure gas mixing furnace is unsurpassed in terms of making measurements under finely controlled oxygen activity f_{O_2}, but if one wishes to examine electrical conductivity of lower-mantle phases (minerals), these need to be made within a high-pressure apparatus using solid chemical buffers instead of gas mixes.

There are three basic types of high-pressure apparatus. The first is the piston and cylinder apparatus, which is exactly what it says and is suitable only for relatively low pressures, although it does have the advantage that gas mixes could be used to control f_{O_2}. The second is the multianvil cell (**Figure 22**), which uses a set of carefully shaped steel, tungsten carbide, and/or sintered diamond anvils to turn the force of a 1000 + ton uniaxial press into a triaxial confining pressure of up to several tens of GPa (40 GPa is about 1000 km deep). **Figure 23** shows a sample capsule used for conductivity measurements of high-pressure olivine phases in a multianvil cell. Finally, the diamond-anvil cell allows similar or greater pressures to the multianvil cell with a much smaller experimental setup and the possibility of optical interaction with the specimen (**Figure 24**). X-ray experiments can be made through the diamond, and lasers used to locally heat the sample.

7.4.7   Thermopower Measurements

In a gas-mixing furnace such as shown in **Figure 21**, the assembly can be moved up or down to produce a temperature gradient across the sample which can then be measured using the two thermocouples. The resulting thermoelectric effect, also known as the Seebeck effect, can be measured as an EMF between the two electrode wires. The thermopower Q is defined as

$$Q = -\lim_{\Delta T \to 0} \frac{\Delta V}{\Delta T}$$

Mobility of defects is lower on the cold side of the sample, and so an excess of defects 'condenses' there. For a charge-carrying defect, this results in an electric field across the sample that can be measured with a suitable high-impedance voltmeter. The sign convention for Q is such that the sign of the thermopower agrees with charge-carrying defect. Thermoelectric power is inversely proportional to the concentration of charge carriers, and is equivalent to the entropy per charge carrier. Expressions for this are of the form

$$Q = \frac{k}{q}\left[\ln\beta\frac{(1-c)}{c} + \frac{S}{k}\right]$$

Tuller and Nowick (1977). Here c is the fraction of sites which contain a charge carrier, q is the charge of the defect, β is a degeneracy factor (for electrons β is usually taken to be equal to 2 to account for the two possible spin states), S is the vibrational entropy associated with the ions (usually considered negligible), and k is Boltzmann's constant. Because thermopower is dependent only on concentration, and not on mobility, it provides an important tool for the study of charge transport in minerals.

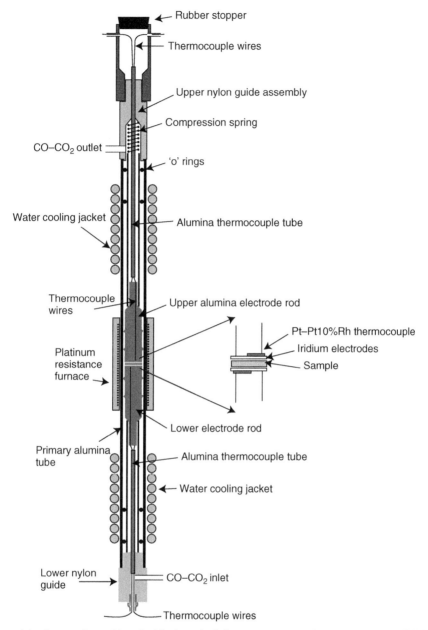

Figure 21 Gas-mixing furnace for making electrical conductivity measurements at room pressure and high temperature. The mixture of CO and CO_2 is controlled by separate mass flow controllers for each gas. The vertical size is about 60 cm. Modified from Duba AG, Schock RN, Arnold E, and Shankland TJ (1990) An apparatus for measurement of electrical conductivity to 1500°C at known oxygen fugacity. In: Duba AG, Durham WB, Handin JW, and Wang HF (eds.) *Geophysical Monograph Series*, vol. 56: The Brittle – Ductile Transitions in Rocks: The Heard Volume, pp. 207–210. Washington, DC: AGU.

7.4.8 Conductivity of Mantle Minerals

Mantle mineralogy depends to a large extent on the increase in pressure with depth, and is inferred from a combination of mantle xenoliths, models of seismic velocity, estimates of bulk chemistry, and high-pressure laboratory studies. **Figure 25** shows such a model of mantle mineralogy in pictorial form.

Figure 26 shows laboratory studies for all the major minerals in the mantle, represented as Arrhenius fits to the original data. Most of these data were collected in a multianvil press, where large (by comparison with diamond anvil cells) samples can be made and solid state buffers used to control f_{O_2}. Most minerals have activation energies

(a)   (b)   (c)

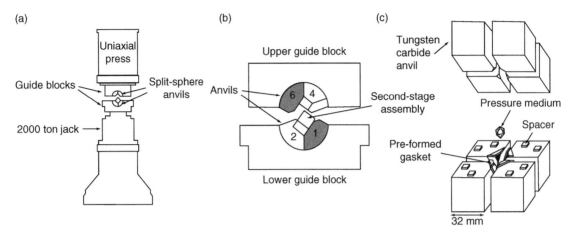

Figure 22   The multianvil cell. From left to right, (a) the uniaxial press and the split-sphere anvils, (b) close up of the split-sphere anvils showing the location of the second-stage assembly, and (c) the second-stage assembly of eight tungsten carbide cubic anvils and an octahedral pressure medium. From Libermann RC, and Wang Y (1992) Characterization of sample environment in a uniaxial split-sphere apparatus. In: Syono Y and Manghnani MH (eds.) *High Pressure Research*: *Applications to Earth and Planetary Sciences*, pp. 19–31. Washington, DC: American Geophysical Union.

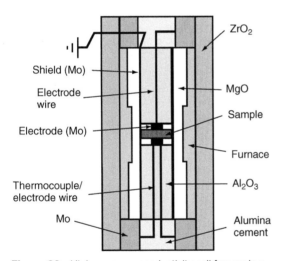

Figure 23   High-pressure conductivity cell for use in a multianvil press. From Xu YS, Poe BT, Shankland TJ, and Rubie DC (1998) Electrical conductivity of olivine, wadsleyite, and ringwoodite under upper-mantle conditions. *Science* 280: 1415–1418.

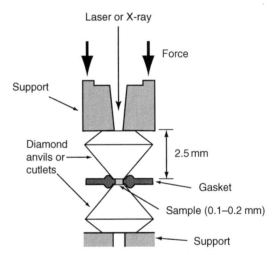

Figure 24   A diamond-anvil setup designed to measure precise strains using an optical technique. The size of the sample area is less than 1 mm in diameter. From Mead C and Jeanloz R (1987) High precision strain measurements at high pressures. In: Manghani MH and Syono Y (eds.) *High-Pressure Research in Mineral Physics*, pp. 41–51. Washington DC: American Geophysical Union.

between 0.5 and 2 eV ($1 \, eV = 1.6 \times 10^{-19}$ J). It is interesting to note that upper-mantle minerals have activation energies of around 1.5 eV, while lower-mantle minerals (magnesiowüstite and silicate perovskite) have activation energies of 0.7 eV. Whether there is a reason or not for the high-pressure phases to have lower activation energies, the implication is that the lower mantle will be more uniform in conductivity given similar variations in temperature.

7.4.8.1   Olivine conductivity

Olivine attracts special attention from the geophysics community because it dominates the mineralogy of the upper mantle, a region that is easiest to study and closest to the crust. Because the next most abundant upper-mantle mineral, clinopyroxene, has a conductivity similar to olivine, it is fairly safe to build models of mantle conductivity on studies of olivine.

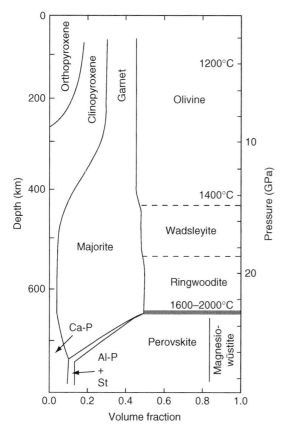

Figure 25 Mantle mineralogy. Olivine is the major component of the upper mantle. It is a orthorhombic magnesium-iron silicate, $(Mg_{1-x}Fe_x)_2SiO_4$, where x is typically 0.1 in the mantle. Wadsleyite and ringwoodite are high-pressure phases of olivine. Garnets are cubic aluminum silicates of the general composition $X_3Al_2Si_3O_{12}$, where X can be magnesium (pyrope), divalent iron (almandine), manganese (spessartine), calcium (grossular) (neglecting garnets where the Al is replaced by other trivalent cations). Majorite is a high-pressure tetragonal phase of $(Mg,Fe)SiO_3$ garnet, Orthopyroxene is typified by enstatite, $(Mg_{1-x}Fe_x)SiO_3$, a chain silicate of orthorhombic structure. Clinopyroxene is typified by the monoclinic diopside series $Ca(Mg,Fe, Mn)Si_2O_6$, another chain silicate. Magnesiowüstite is magnesium-iron oxide $(Mg_{1-x}Fe_x)O$. Stishovite (St) is a high-pressure phase of silica (quartz), or SiO_2. Silicate perovskite is a high-pressure distorted cubic form of magnesium-iron silicate, $Mg_{1-x}Fe_xSiO_3$. Ca-P and Al-P are CaO and Al_2O_3 rich phases, respectively. Modified from Ito E and Takahashi E (1987) Ultrahigh-pressure phase transformations and the constitution of the deep mantle. In: Manghnani MH and Syono Y (eds.) *High-Pressure Research in Mineral Physics*, pp. 221–229. Washington, DC: American Geophysical Union.

Olivine, an orthorhombic mineral, is mildly anisotropic. **Figure 27** shows a recent measurement of olivine anisotropy, made on samples from mantle xenoliths found in San Carlos indian reservation, Arizona. The [001], or c-axis, is most conductive, but still barely

more than a factor of 2 greater than other axes. The [100] (a) axis is of intermediate conductivity. This is, perhaps, unfortunate, because seismic and deformational models of the upper mantle suggest that this axis will be aligned in the direction of plate motion.

SO2 (Constable *et al.*, 1992) is a quantitative model of olivine conductivity based on a geometric average of high-temperature single-crystal measurements, and lower-temperature measurements on polycrystalline olivine rocks (which would melt above 1250°C, fluxed by the pyroxene content). The SO2 model is

$$\sigma = 10^{2.40}e^{-1.6\,eV/kT} + 10^{9.17}e^{-4.25\,eV/kT} \, S\,m^{-1}$$

where the first term corresponds to low-temperature behavior and the second term corresponds to a conduction mechanism that becomes dominant above 1300°C. SO2 served as a useful benchmark for some time, but was often found to underpredict mantle conductivity. An new model, SEO3, was created based on our understanding of the conduction mechanisms in olivine (Constable, 2006). The relationship between the major defect population in olivine and f_{O_2} is thought to be described by

$$8Fe_{Mg}^{\times} + 2O_2 \rightleftharpoons 2V_{Mg}'' + V_{Si}'''' + 4O_O^{\times} + 8Fe_{Mg}^{\cdot}$$

Many studies indicate that the dominant charge carriers are $Fe_{Mg}^{\cdot}$ and V_{Mg}''. For samples buffered by pyroxene, this model predicts that $[Fe_{Mg}^{\cdot}]$ and $[V_{Mg}'']$ will each have a dependence on f_{O_2} of the form

$$\left[Fe_{Mg}^{\cdot}\right] = a_{Fe}f_{O_2}^{1/6} \quad ; \quad \left[V_{Mg}''\right] = a_{Mg}\,f_{O_2}^{1/6}$$

where the pre-exponential terms a are constant and [ ] denotes concentration. For mixed conduction, total conductivity is given by

$$\sigma = \sigma_{Fe} + \sigma_{Mg}$$
$$= \left[Fe_{Mg}^{\cdot}\right]\mu_{Fe}e + 2\left[V_{Mg}''\right]\mu_{Mg}e$$

where the μ are the respective mobilities for the charge carriers and e is the charge on the electron or hole.

Thermopower measurements provide a mechanism for estimating concentration independently of mobility, and Constable and Roberts (1997) created a model of conduction in olivine based on a combined thermopower/conductivity data set. **Figure 28** shows the data set used for this – a dunite from San Quintin in Baja California, Mexico. Natural rocks are useful samples because the crystallographic orientations are usually randomized, and there are enough minor phases to buffer silica. It can be seen that the effect of f_{O_2} has the

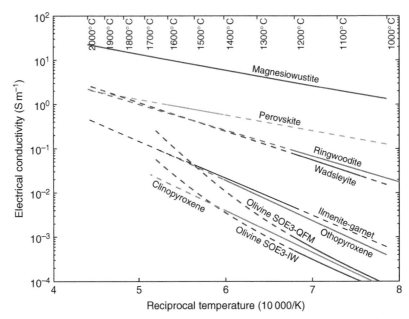

Figure 26 Laboratory studies of conductivity in mantle minerals. Perovskite results are from Xu *et al.* (1998). Olivine SEO3 from Constable (2006) for iron-wüstite (IW) and quartz-fayalite-magnetite (QFM) models of mantle oxygen fugacity. Magnesiouwüstite from Dobson and Brodholt (2000). Other results are from Xu and Shankland (1999). Solid lines represent the temperatures over which measurements were made; the broken lines are extrapolations of the Arrhenius fits to higher and lower temperatures.

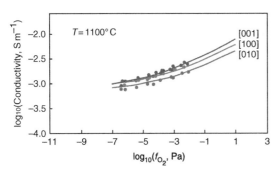

Figure 27 Data (symbols) and model fits (lines) for electrical conductivity of San Carlos Olivine along the three crystallo-graphic axes as a function of f_{O_2} at a temperature of 1200°C. Anisotropy is not strong, and the crystal axis most likely to be aligned in the mantle, [100], is of intermediate conductivity. Data from Du Frane WL, Roberts JJ, Toffelmier DA, and Tyburczy JA (2005) Anisotropy of electrical conductivity in dry olivine. *Geophysical Research Letters* 32: L24315.

same magnitude as a 100 K change in temperature. Iron content also has an effect on conduction, but Fe content in mantle olivines is quite constant at about 10%. This model supports earlier predictions that below 1300°C, small polaron hopping ($Fe_{Mg}^{\cdot}$) dominates the conduction mechanism, and above 1300°C magnesium vacancies (V_{Mg}'') account for conductivity.

Constable (2006) used the mobilities and concentrations as a function of temperature and f_{O_2} from the joint

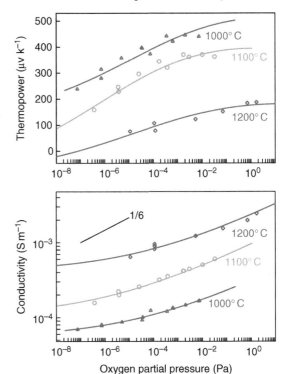

Figure 28 Data (symbols) and model fits (lines) for thermoelectric power and conductivity data for San Quintin dunite, as a function of f_{O_2} for three temperatures. From Constable S and Roberts JJ (1997) Simultaneous modeling of thermopower and electrical conduction in olivine. *Physics and Chemistry of Minerals* 24: 319–325.

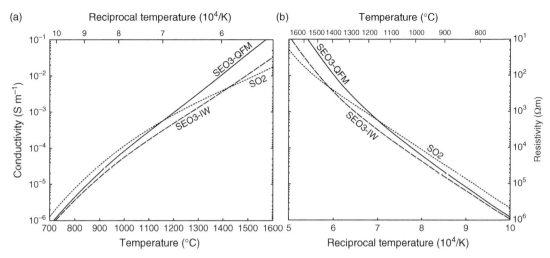

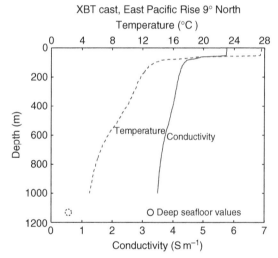

Figure 29 The SEO3 olivine conductivity–temperature model for two different f_{O_2} conditions, quartz-fayalite-magnetite (SEO3-QFM) and iron-wüstite (SEO3-IW) (Constable, 2006). For comparison the SO2 model of Constable *et al.* (1992) is also shown. The models are represented in both 1/T space (a) and temperature space (b).

conductivity/thermopower model to create a quantitative model of conductivity that is perhaps more reliable at high temperature than SO2 (**Figure 29**), because it is based on Boltzmann relationships rather than parametric fits to data. Furthermore, it allows explicit designation of the f_{O_2} conditions; although the quartz-fayalite-magnetite buffer is often used for the upper mantle, it is possible that at depth f_{O_2} may be lower (McCammon, 2005). The effect of pressure is neglected in these models, but Xu *et al.* (2000) show that it is small compared with the sample-to-sample variations associated with high-pressure measurements.

7.5 Global Conductivity Structure

In this section, we examine what is known about the electrical conductivity of Earth's interior, both from field measurements and laboratory studies. Field studies include long-period MT soundings using temporary instrument installations, GDS studies using magnetic observatory data, and, most recently, induction studies using magnetic satellite data.

7.5.1 The Oceans and Crust

The most significant conductivity feature on the surface of Earth is the world ocean. Fortunately, the ocean is also the most easy feature to characterize. Global maps of bathymetry are available, and seawater conductivity is well defined as a function of

Figure 30 Expendable bathy-thermograph (XBT) taken in the Pacific Ocean near the East Pacific Rise at 9° N. The temperature profile has been converted to electrical conductivity. The 'o' symbols show the values near the seafloor in the deeper ocean (4000–5000 m).

salinity (nearly constant) and temperature. **Figure 30** shows an example of the electrical conductivity structure of the upper 1000 m of a site in the Pacific Ocean. The surface water is warm and more than $5\,\mathrm{S\,m^{-1}}$, but both temperature and conductivity drop rapidly below the thermocline, in this case at a depth of about 70 m. By a depth of 1000 m, the conductivity is close to that observed at full ocean depth, and a value of $3.3–3.5\,\mathrm{S\,m^{-1}}$ is usually used to represent ocean conductivity.

The rocks of the oceanic crust are, by and large, also fairly easy to characterize because for the most part the genesis of oceanic crust is uniform and the metamorphic history fairly simple. Boreholes to a depth of a kilometer or so, along with marine-controlled source EM studies, provide *in situ* measurements of conductivity, which, combined with our knowledge of oceanic geology gleaned from dredging and ophiolite studies, allows us to paint a fairly good picture of electrical structure in the oceanic crust (**Figure 31**). Away from ridges, the conductivity is basically a measure of the porosity of the volcanic and igneous rocks that make up the crust. Near ridges conductivity is increased by the presence of magma and hydrothermal fluids in the crust.

In stark contrast, the electrical structure of the continental crust (which includes the continental shelves) is very much more complicated, undoubtedly because it is better studied than the oceanic crust, but mostly because the geological history is much more complicated and the variation in water content (and salinity) is much larger. Data for crustal conductivities on the continents comes mainly from broadband and long-period MT soundings, with some data from large-scale controlled-source EM sounding and resistivity studies. However, neglecting important details and ignoring unusually conductive, volumetrically small, and often economically significant minerals, we can make some gross generalizations:

$$\text{Continental shields}: 10^3 - 10^6 \, \Omega\text{m}$$

$$\text{Continental sediments}: 10 - 100 \, \Omega\text{m}$$

$$\text{Marine sediments}: 1 - 100 \, \Omega\text{m}$$

Everett *et al.* (2003) formalized this approach to generate a global crustal conductance map, using the global sediment thickness map of resolution of $1° \times 1°$ compiled by Laske and Masters (1997). This map contains sediments discretized in up to three layers; a surface layer up to 2 km thick, an underlying layer up to 5 km thick, and, where necessary, a third

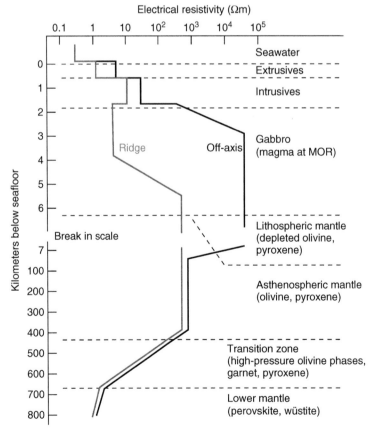

Figure 31 Electrical conductivity structure of the oceanic crust and mantle, based on marine controlled-source EM sounding to depths of about 50 km, and marine MT below that.

Global surface conductance map, logarithmic scale [log10 S]

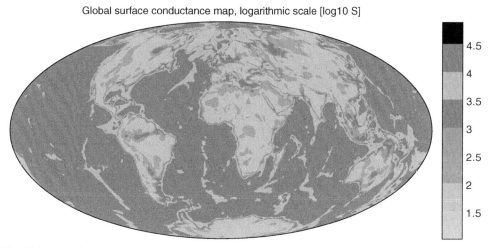

Figure 32 Global electrical conductance generated by the crustal algorithm of Everett *et al.* (2003).

layer to make up total sediment thickness, which often exceeds 10 km. The sediment map was augmented by a $1° \times 1°$ topographic/bathymetric map. Continental and oceanic sediments were mapped heuristically to conductivity, and summed over depth to create a surface conductance map (**Figure 32**). The above algorithm is undoubtedly arbitrary and could be improved upon, but improves on simple ocean depth as a proxy for conductance. Overall, the sedimentary sections defined in this way contribute only 10% to the total surface conductance, but in areas such as the Gulf of Mexico, Arctic Ocean, and Mediterranean/Caspian/Black Seas, accumulated sediment has a conductance comparable to the oceans and sediments drastically alter the shape of the continent/ocean function.

For frequencies where the skin depth is large compared with the thickness of a layer, a surface layer can be characterized by its conductance only, because its thickness cannot be resolved and attenuation of EM energy is determined entirely by the conductivity-thickness product, or conductance. Computations in three dimensions often exploit this phenomenon using a thin-sheet algorithm (e.g., Kuvshinov *et al.*, 1999).

7.5.2 The Mantle

As can be seen from **Figure 11**, global response functions from satellite and observatory measurements are in broad general agreement. This is further illustrated in **Figure 33**, which shows the data set used in an inversion study by Medin *et al.* (2006). These data consist of a combination of

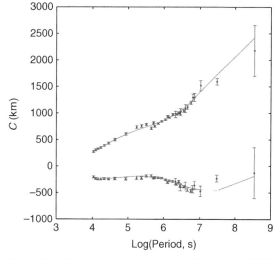

Figure 33 Response function used by Medin *et al.* (2006), consisting of the slightly edited data sets from Olsen (1999, open symbols) and Constable (1993, solid symbols). Real component is plotted as circles, the imaginary as triangles. The error bars are as reported by the authors. The red line is the response of the smooth model shown in **Figure 34**, fitting the data to a conservative RMS 1.2.

stacked global observatory GDS responses (Constable, 1993) which relied on assumptions of P_1^0 geometry, and European observatory responses of Olsen (1999) which allowed more complicated source-field geometry and thus a broader bandwidth. Medin *et al.* excluded four admittances which failed to satisfy the assumption of one-dimensionality, three of which they had reason to believe did not meet the source-field requirements.

Figure 34 shows various inversions of this data set. The green lines represent the D^+ model of Parker and

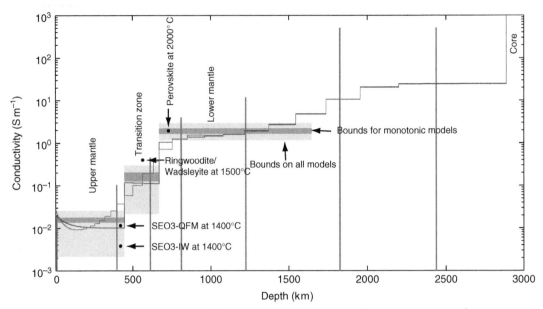

Figure 34 Models fitting the data set shown in **Figure 33**. Green lines are the D$^+$ model divided by 10^5, red line is a first-derivative maximally smooth model, and the blue line is a smooth model that is allowed to jump at 440 and 670 km. The yellow boxes represent bounds on average conductivity from Medin *et al.* (2006). Values of selected mineral conductivities from **Figure 26** are also shown.

Whaler (1981), scaled by 10^5 and fitting the data to RMS 1.044. The red line is a smooth, first-derivative regularized inversion fitting the data to RMS 1.2, and the blue line is a similar inversion but allowing unpenalized jumps at the 440 km and 670 km seismic discontinuities. Finally, the yellow boxes are bounds on average conductivity at the 90% confidence level over the depth intervals 0–418, 418–672, and 672–1666 km. The broader bounds put no restrictions on the models, the tighter bounds shown in the darker color restrict the models to be monotonically increasing in conductivity with depth. These models are representative of the many mantle conductivity profiles published in the literature. One of the most well-resolved features is the jump in conductivity to 2–3 S m^{-1} in the lower mantle, entirely consistent with the more recent measurements on silicate perovskite (see **Figure 26**). The uppermost mantle above the transition zone is likely to be quite heterogeneous and dependent on local tectonic setting (e.g., ridges, subduction zones, plumes could all be expected to be locally much more conductive), but the global average conductivity of 50–100 Ωm is a good agreement with dry olivine conductivity. The order-of-magnitude increase of conductivity in the transition zone is in general agreement with laboratory measurements on the high-pressure phases wadsleyite and

ringwoodite, although biased towards somewhat lower conductivity. This is easily explained in terms of mixing with more resistive phases such as garnet and pyroxene. The order-of-magnitude increase in conductivity in the lowermost mantle is just within the range of thermally activated conduction in perovskite, or could be the influence of other mineral phases. The increase is poorly constrained, relying on the longest-period measurements having largest uncertainty, and probably represents a lower bound on conductivity because the smoothness penalty minimizes slope.

While 1-D global response functions are in general agreement with each other and with laboratory studies, individual response functions vary considerably (e.g., Egbert and Booker, 1992; Schultz *et al.*, 1993; Lizarralde *et al.*, 1995; Tarits *et al.*, 2004), with nearly two-orders-of-magnitude variation in conductivity inferred for the upper mantle. This, of course, begs the question of the reliability of the response functions, particularly for single-site MT soundings where galvanic distortions of the electric field and effects of 2-D and 3-D structure are not all that unusual. Certainly, early single-site MT measurements carried out in the deep ocean basins were distorted by the effect of coastlines (Heinson and Constable, 1992). However, there are several factors which could modify the conductivity of the upper mantle from the averages presented above, and which may become the dominant conduction mechanism locally:

7.5.2.1 Melt

Since magma is very much more conductive than subsolidus olivine, one mechanism for increasing conductivity is to invoke a mantle permeated by partial melt. Tholeitic melt conductivity is more than $1\,S\,m^{-1}$. For small melt fractions to be an effective mechanism for increased conduction, the melt must be well connected. However, well-connected melt is gravitationally unstable, and will migrate upwards, and so while partial melt is clearly an important mechanism in volcanic systems and mid-ocean ridges, it must be replenished on a timescale comparable to the residency time in the mantle for it to contribute to permanent enhancement of conductivity.

7.5.2.2 Carbon

Graphite is extremely conductive (10^4–$10^6\,S\,m^{-1}$), and so it would take only a small amount to increase mantle conductivity over that of dry olivine, as long as it is well connected. There is a significant amount of carbon in the mantle (several hundred ppm) that can exist as CO_2 fluids or graphite. Movement between the two phases depends on f_{O_2}, and so it is conceivable that CO_2 fluids permeating grain boundaries could be reduced to form a connected network of graphite. The idea that carbon might contribute to mantle conductivity was first put forward by Duba and Shankland (1982). There is a big limitation on carbon as a conduction mechanism for mantle rocks; below about 160 km graphite turns into diamond, which is a nonconductor. Note also that carbon has a low thermal activation energy, so we would not expect to see an increase in conductivity with temperature and depth for carbon-enhanced conductivity.

7.5.2.3 Strain-induced defects

Since the early marine MT studies inferred a zone of increased conductivity at the depth of the asthenosphere, one possibility that was considered was that straining generated defects that contribute to conduction. This idea was dispelled by careful measurements by Hirsch and Wang (1986), and now it is clear that marine MT measurements must be interpreted in terms of lower conductivities by taking into consideration the effect of the coastlines.

7.5.2.4 Water

Like carbon, water is known to exist in the mantle – estimates are that several Earth oceans are still resident somewhere in the deep Earth. Where, and in what state, is less well understood.

There are three ways to store water in the mantle: free water, hydrous minerals, and as defects in minerals. Free water could exist for a period of time in the coolest parts of the mantle, but at even modest temperatures will eventually react with olivine to create serpentine minerals. This mechanism is interesting in that magnetite is excluded during the reaction, potentially creating halos of fine magnetite around relic olivine grains that can be electrically connected and lower the resistivity of the rock considerably (Stesky and Brace, 1973). However, hydrous minerals are not otherwise conductive (consider mica). At some temperature/pressure, depending on the mineral but generally at depths shallower than 125 km, hydrous minerals will decompose to produce free water, which is not stable and will usually lower the melting point sufficiently to generate a partial melt into which it dissolves, which will in turn be gravitationally unstable and migrate upwards. This mechanism is largely responsible for volcanism at subduction zones.

The idea that water dissolved as point defects in olivine could enhance electrical conductivity was proposed by Karato (1990), based on observations that chemical diffusion of hydrogen in olivine was very high, particularly along the a-axis. The activation energy is 1.3 eV; not dissimilar to other upper-mantle conduction mechanisms. Although initially put forward as an explanation for early marine MT measurements, the idea has found an enthusiastic following in the EM community. Since the olivine a-axis is considered to be aligned with mantle flow, conduction by hydrogen would create significant anisotropy (only about 100 ppm hydrogen is required to increase a axis conductivity by an order of magnitude if Karato's diffusion model is correct), and evidence of electrical anisotropy in the mantle has been interpreted as support of hydrogen conduction (e.g., Simpson and Tommasi, 2005).

However, the effect of hydrogen on upper-mantle conductivity is a matter for debate. Karato (1990) only considered the fast diffusion along the a-axis, but the effect is not as dramatic when the other two axes are taken into account; adding about 1000 ppm hydrogen per silicon atom predicts a conductivity based on diffusivity only slightly above that of dry olivine. Direct laboratory evidence for such a conduction process has been slow to appear and initial reports are in poor agreement with each other and the diffusion data (e.g., Shankland and Duba, 1997; Poe et al., 2005; Wang et al., 2006; Yoshino et al., 2006), suggesting that hydrogen does not behave as a simple charged defect in olivine. Finally, while observation of anisotropy in MT data has quickly led to support for the idea of hydrogen conduction, many things can result in anisotropic MT responses; for example, structural anisotropy outside the study area could be difficult to identify without adequate data (e.g., Heinson and White, 2005).

Recent high-pressure measurements on ringwoodite and wadeslyite suggest that water solubility in these minerals is very much higher than in upper- or lower-mantle minerals, leading Bercovici and Karato (2003) to suggest a water-saturated transition zone (410–670 km deep) and pooling of water and melt at the 410 km discontinuity. Furthermore, Huang et al. (2005) show that in these minerals conductivity is enhanced by the addition of water. This presents the exciting possibility that electromagnetic methods can provide important constraints on deep Earth processes. Based on conductivity bounds obtained from a rigorous analysis of global response function data, Medin et al. (2006) estimated that the global average water content in the transition zone is <0.27%, less than the 0.4% required for pooling at the 410 km discontinuity. However, these inferences depend critically on relatively new laboratory data, and there remains the possibility that the effect of water is heterogeneous and more significant locally.

7.5.3 The Core

The state of the art with regards conductivity of the core is nicely summarized in the recent paper by Stacey and Anderson (2001). Their estimates are $2.1 \times 10^{-6}\,\Omega$m in the outer core and $1.6 \times 10^{-6}\,\Omega$m in the inner core. Although this refinement on the previously accepted values (1–$3 \times 10^{-6}\,\Omega$m) is not terribly significant in terms of geoelectromagnetism, it is important with respect to thermal conductivity, which is linearly related to electrical conductivity.

Resistivity of metals increases with temperature because lattice vibrations (phonons) can be considered to scatter electrons. Similarly, adding impurity atoms also increases scattering and resistivity. This is important because we know the outer core is alloyed with 10–25% light elements for the alloyed iron to match density estimates. Stacey and Anderson argue that the effects of impurities are independent of

temperature, and indeed somewhat independent of the actual impurity chemistry; the effect of Si and Ni (both probably present in the core) are considered similar when measured in weight %.

Iron behaves as an ideal metal in that the effect of pressure acts to decrease lattice vibrations in a way that affects melting point and resistivity similarly, and so iron at the melting point T_M is expected to have the same resistivity ($1.35 \times 10^{-6}\,\Omega$m) whether at zero pressure or at core pressure. At other temperatures T for a given pressure, resistivity is assumed to be proportional to T/T_M. Thus at the core–mantle boundary (CMB), temperature is estimated to be 3750 K and iron solidus as 4147 K, so the component of pure iron resistivity is $\rho = 3750/4147 \times 1.35 \times 10^{-6}\,\Omega$m or $1.22 \times 10^{-6}\,\Omega$m. At the inner-core boundary, temperature is estimated to be 4971 K and iron solidus as 6000 K, so the component of pure iron resistivity is $1.12 \times 10^{-6}\,\Omega$m.

Because iron converts from body-centered cubic (b.c.c.) to hexagonal close packed (h.c.p.) under pressure, measurements of iron with impurities have been carried out using shock-wave experiments, in which a gas/explosive gun or laser is fired at a sample. The shock front, moving at velocity u_s, separates material at the starting density, pressure, and temperature (ρ_o, P_o, T_o) with material at the shocked conditions (ρ, P, T). The shock front is supersonic and so moves faster than the actual material, or sound, speed u_p. Conservation of mass across the shock front gives

$$u_s \rho_o = \rho\left(u_s - u_p\right)$$

so the density increase produced by the shock is

$$\frac{\rho}{\rho_o} = \left(1 - \frac{u_p}{u_s}\right)^{-1}$$

and pressure is given by the rate of change of momentum per unit area of shock front, which for material starting at rest and room pressure is

$$P = \rho_o u_s \times u_p$$

So, knowing the ratio and product of the two velocities for various-sized shocks gives density as a function of pressure for the material. The shock compression is not adiabatic because the material gains kinetic energy. The temperature curve which the material follows during shocking is called a Hugoniot.

Stacey and Anderson (2001) fit linear models to shock-wave conductivity data to conclude that a silicon impurity of volume fraction X would contribute $3.0X \times 10^{-6}\,\Omega\text{m}$ to resistivity. They expected the silicon to exsolve in a similar way to nickel, and so took averages of the two components to finally come up with an impurity contribution of $0.9 \times 10^{-6}\,\Omega\text{m}$ for a total resistivity of $2.1 \times 10^{-6}\,\Omega\text{m}$. They estimated a lower impurity content for the inner core to derive a total resistivity of $1.6 \times 10^{-6}\,\Omega\text{m}$.

7.6 Conclusions

During the last decade, or even less, the combination of high-quality laboratory measurements of mantle minerals at high pressure and good estimates of global response functions, often assisted by satellite data, has produced a convergence in our understanding of radial electrical conductivity structure, represented to a large extent by **Figure 34**. The challenges for the future lie in extending this work to higher dimensions. An attempt to catalog areas for these next steps is made here.

7.6.1 Anisotropy Measurements

A measurement of bulk conductivity tells us only a small amount about the interior of Earth; the trade-offs between composition and temperature are large and often poorly constrained. A measurement of electrical anisotropy, however, unfolds another dimension in our understanding. For example, strain induced by plate motion could generate anisotropy associated either with mineral texture and fabric or by aligning crystallographic axes. Some conduction mechanisms, such as the diffusion of hydrogen in olivine, predict significant anisotropies in the upper mantle. Combining the interpretation of electrical measurements with other geophysical data, such as seismic velocity, has always been a good idea, and the combination of electrical and seismic anisotropy could prove to be a powerful tool for understanding Earth processes. The reader is referred to a recent review by Eaton and Jones (2006), and the papers in that volume. However, the challenge here is to make a meaningful measurement of mineralogical or fabric anisotropy, and not become confused with anisotropy associated with larger-scale structures.

7.6.2 3-D Conductivity from Satellite Data

While there has been a convergence of globally averaged response functions, the ultimate goal of using satellite data is to obtain information about 3-D structure in the mantle. This presents challenges both for response function estimation and inversion of data. The induction signal in satellite data is buried deeply in signals from the main field, lithosphere, etc., and the data themselves are not always nicely behaved, having significant gaps and sometimes not having the best control on sensor orientation. One can average through these problems when assumptions of P_1^0 geometry are made and a global 1-D response is sought, but the generation of 3-D response functions will require a finer control over data quality and response function estimation, as well as dealing with significant complications of source-field morphology, such as the effect of the ionosphere and non-P_1^0 behavior of the magnetosphere. The multisatellite SWARM experiment should assist greatly in this regard. Also, one needs inversion schemes that are not only fully 3-D, but can handle the surface conductivity structure accurately. The problem is that the average conductance of the upper mantle $(0.01\,\text{S m}^{-1} \times 400\,\text{km} = 4000\,\text{S})$ is less than that of the average ocean $(3\,\text{S m}^{-1} \times 4\,\text{km} = 12\,000\,\text{S})$. The more resistive continents provide windows into the mantle, and regions such as mid-ocean ridges, subduction zones, and plumes are likely to be more conductive than the average, but the existence of a surface conductance that varies by nearly four orders of magnitude puts a great demand on the combination of response function estimation and modeling before mantle structure can be determined.

7.6.3 Water in the Mantle

The current proposals for water pooling in the transition zone provide a great opportunity for geomagnetic induction methods to make fundamental contributions to our understanding of the mantle. Although, based on average conductivity, a case can be made that water pooling does not occur; there is every reason to believe that the distribution of water is not homogeneous, and that locally water does reach the critical threshold. Long-period MT soundings, if they can be made free of distortions due to large-scale crustal structures, could provide important constraints on this hypothesis. Satellite and observatory response functions, if generated in a way that does not reject 3-D structure, could also be bought to bear on this problem.

Acknowledgment

This contribution was supported in part by funding from NASA grant NAG5-13747.

References

Alexandrescu MM, Gibert D, Le Mouel JL, Hulot G, and Saracco G (1999) An estimate of average lower mantle conductivity by wavelet analysis of geomagnetic jerks. *Journal of Geophysical Research* 104: 17735–17745.

Avdeev DB (2005) Three-dimensional electromagnetic modelling and inversion from theory to application. *Surveys in Geophysics* 26: 767–799.

Backus GE and Gilbert JF (1967) Numerical applications of a formalism. *Geophysical Journal of the Royal Astronomical Society* 13: 247–276.

Backus GE and Gilbert JF (1968) The resolving power of gross Earth data. *Geophysical Journal of the Royal Astronomical Society* 16: 169–205.

Backus GE (1983) Application of mantle filter theory to the magnetic jerk of 1969. *Geophysical Journal of the Royal Astronomical Society* 74: 713–746.

Backus G, Parker R, and Constable C (1996) *Foundations of Geomagnetism*. Cambridge, UK: Cambridge University Press.

Bahr K, Olsen N, and Shankland TJ (1993) On the combination of the magnetotelluric and the geomagnetic depthsounding method for resolving an electrical conductivity increase at 400 km depth. *Geophysical Research Letters* 20: 2937–2940.

Balasis G and Egbert GD (2006) Empirical orthogonal function analysis of magnetic observatory data: Further evidence for non-axisymmetric magnetospheric sources for satellite induction studies. *Geophysical Research Letters* 33: L11311.

Banks RJ (1969) Geomagnetic variations and the electrical conductivity of the upper mantle. *Geophysical Journal of the Royal Astronomical Society* 17: 457–487.

Barlow WH (1849) On the spontaneous electrical currents observed in wires of the electric telegraph. *Philosophical Transactions of the Royal Society of London* 139: 61–72.

Bercovici D and Karato SI (2003) Whole-mantle convection and the transition-zone water filter. *Nature* 425: 39–44.

Cagniard L (1953) Basic theory of the magneto-telluric method of geophysical prospecting. *Geophysics* 18: 605–635.

Chapman S (1919) The solar and lunar diurnal variations of terrestrial magnetism. *Philosophical Transactions of the Royal Society of London (A)* 218: 1–118.

Chapman S and Bartels J (1940) *Geomagnetism*. Oxford, UK: Oxford University Press.

Chave AD, Thomson DJ, and Ander ME (1987) On the robust estimation of power spectra, coherences, and transfer-functions. *Journal of Geophysical Research* 92: 633–648.

Constable SC (1993) Constraints on mantle electrical conductivity from field and laboratory measurements. *Journal of Geomagnetism and Geoelectricity* 45: 707–728.

Constable S (2006) SEO3: A new model of olivine electrical conductivity. *Geophysical Journal International* 166: 435–437.

Constable CG and Constable SC (2004a) Satellite magnetic field measurements: Applications in studying the deep Earth. In: Sparks RSJ and Hawkesworth CT (eds.) *Geophysical Monograph 150: The State of the Planet: Frontiers and Challenges in Geophysics*, pp. 147–159. Washington, DC: American Geophysical Union.

Constable S and Constable C (2004b) Observing geomagnetic induction in magnetic satellite measurements and associated implications for mantle conductivity. *Geochemistry Geophysics Geosystems* 5: Q01006 (doi:10.1029/2003GC000634).

Constable SC, Parker RL, and Constable CG (1987) Occam's Inversion: A practical algorithm for generating smooth models from EM sounding data. *Geophysics* 52: 289–300.

Constable S and Roberts JJ (1997) Simultaneous modeling of thermopower and electrical conduction in olivine. *Physics and Chemistry of Minerals* 24: 319–325.

Constable SC, Shankland TJ, and Duba A (1992) The electrical conductivity of an isotropic olivine mantle. *Journal of Geophysical Research* 97: 3397–3404.

Didwall EM (1984) The electrical conductivity of the upper mantle as estimated from satellite magnetic field data. *Journal of Geophysical Research* 89: 537–542.

Dobson DP and Brodholt JP (2000) The electrical conductivity of the lower mantle phase magnesiowustite at high temperatures and pressures. *Journal of Geophysical Research* 105: 531–538.

Duba AG, Schock RN, Arnold E, and Shankland TJ (1990) An apparatus for measurement of electrical conductivity to 1500°C at known oxygen fugacity. In: Duba AG, Durham WB, Handin JW, and Wang HF (eds.) *Geophysical Monograph Series*, vol. 56: The Brittle – Ductile Transitions in Rocks: The Heard Volume, pp. 207–210. Washington, DC: AGU.

Duba AG and Shankland TJ (1982) Free carbon and electrical conductivity in the Earth's mantle. *Geophysical Research Letters* 9: 1271–1274.

Du Frane WL, Roberts JJ, Toffelmier DA, and Tyburczy JA (2005) Anisotropy of electrical conductivity in dry olivine. *Geophysical Research Letters* 32: L24315.

Eaton DW and Jones A (2006) Tectonic fabric of the subcontinental lithosphere: Evidence from seismic, magnetotelluric and mechanical anisotropy. *Physics of the Earth and Planetary Interiors* 158: 85–91.

Egbert GD (1997) Robust multiple-station magnetotelluric data processing. *Geophysical Journal International* 130: 475–496.

Egbert GD and Booker JR (1986) Robust estimation of geomagnetic transfer-functions. *Geophysical Journal of the Royal Astronomical Society* 87: 173–194.

Egbert GD and Booker JR (1992) Very long period magnetotellurics at Tucson-observatory – implications for mantle conductivity. *Journal of Geophysical Research* 97: 15099–15112.

Everett ME and Constable S (1999) Electric dipole fields over an anisotropic seafloor: Theory and application to the structure of 40 Ma Pacific Ocean lithosphere. *Geophysical Journal International* 136: 41–56.

Everett M and Schultz A (1996) Geomagnetic iinduction in a heterogeneous sphere, azimuthally symmetric test computations and the response of an u 660-km discontinuity. *Journal of Geophysical Research* 101: 2765–2783.

Everett ME, Constable S, and Constable CG (2003) Effects of near-surface conductance on global satellite induction responses. *Geophysical Journal International* 153: 277–286.

Filloux JH (1987) Instrumentation and experimental methods for oceanic studies. In: Jacobs JA (ed.) *Geomagnetism*, pp. 143–248. New York: Academic Press.

Gamble TD, Clarke J, and Goubau WM (1979) Magnetotellurics with a remote magnetic reference. *Geophysics* 44: 53–68.

Gill PE, Murray W, and Wright MH (1981) *Practical Optimization*. New York: Academic Press.

Grammatica N and Tarits P (2002) Contribution at satellite altitude of electromagnetically induced anomalies arising from a three-dimensional heterogeneously conducting

Earth, using Sq as an inducing source field. *Geophysical Journal International* 151: 913–923.

Hamano Y (2002) A new time-domain approach for the electromagnetic induction problem in a three-dimensional heterogeneous Earth. *Geophysical Journal International* 150: 753–769.

Heinson G and Constable SC (1992) The electrical conductivity of the oceanic upper mantle. *Geophysical Journal International* 110: 159–179.

Heinson GS and White A (2005) Electrical resistivity of the northern Australian lithosphere: Crustal anisotropy or mantle heterogeneity? *Earth and Planetary Science Letters* 232: 157–170.

Hirsch LM (1990) Electrical conduction of Co_2SiO_4. *Physics and Chemistry of Minerals* 17. 187–190.

Hirsch LM and Shankland TJ (1991) Point-defects in (Mg,Fe)SIO3 Perovskite. *Geophysical Research Letters* 18: 1305–1308.

Hirsch LM and Wang CY (1986) Electrical conductivity of olivine during high-temperature creep. *Journal of Geophysical Research* 91: 429–441.

Huang XG, Xu YS, and Karato SI (2005) Water content in the transition zone from electrical conductivity of wadsleyite and ringwoodite. *Nature* 434: 746–749.

Ito E and Takahashi E (1987) Ultrahigh-pressure phase transformations and the constitution of the deep mantle. In: Manghnani MH and Syono Y (eds.) *High-Pressure Research in Mineral Physics*, pp. 221–229. Washington, DC: American Geophysical Union.

Karato S (1990) The role of hydrogen in the electrical conductivity of the upper mantle. *Nature* 347: 272–273.

Key K and Weiss C (2006) Adaptive finite-element modeling using unstructured grids: The 2D magnetotelluric example. *Geophysics* 71: G291–G299.

Kittel C (1986) *Introduction to Solid State Physics 6th edn.* New York: Wiley.

Kuvshinov AV, Avdeev DB, and Pankratov OV (1999) Global induction by Sq and Dst sources in the presence of oceans: Bimodal solutions for non-uniform spherical surface shells above radially symmetric earth models in comparison to observations. *Geophysical Journal International* 137: 630–650.

Kuvshinov AV, Avdeev DB, Pankratov OV, Golyshev SA, and Olsen N (2002) Modelling electromagnetic fields in 3-D spherical Earth using fast integral equation approach. In: Zhdanov MS and Wannamaker PE (eds.) *Three Dimensional Electromagnetics, Methods in Geochemistry and Geophysics*, vol. 35, pp. 43–54. Amsterdam: Elsevier.

Kuvshinov AV and Olsen N (2004) Modelling the coast effect of geomagnetic storms at ground and satellite altitude. In: Reigber C, Luhr H, Schwintzer, and Wickert J (eds.) *Earth Observation with CHAMP. Results from Three Years in Orbit*, pp. 353–359. Berlin: Springer-Verlag.

Kuvshinov A and Olsen N (2006) A global model of mantle conductivity derived from 5 years of CHAMP, Orsted, and SAC-C magnetic data. *Geophysical Research Letters* 33: L18301.

Lahiri BN and Price AT (1939) Electromagnetic induction in non-uniform conductors, and the determination of the conductivity of the Earth from terrestrial magnetic variations. *Philosophical Transactions of the Royal Society of London (A)* 237: 509–540.

Laske G and Masters G (1997) A global digital map of sediment thickness Eos Transactions AGU, 78(46), Fall Meeting Supplement, Abstract S41E-01, p. F483

Libermann RC and Wang Y (1992) Characterization of sample environment in a uniaxial split-sphere apparatus. In: Syono Y and Manghnani MH (eds.) *High Pressure Research:*

Applications to Earth and Planetary Sciences, pp. 19–31. Washington, DC: American Geophysical Union.

Lizarralde D, Chave A, Hirth G, and Schultz A (1995) Northeastern Pacific mantle conductivity profile from long-period magnetotelluric sounding using Hawaii to California submarine cable data. *Journal of Geophysical Research* 100: 17837–17854.

Mackie RL and Madden TR (1993) 3-Dimensional magnetotelluric inversion using conjugate gradients. *Geophysical Journal International* 115: 215–229.

Marquardt DW (1963) An algorithm for least-squares estimation of non-linear parameters. *Journal of the Society for Industrial and Applied Mathematics* 11: 431–441.

Martinec Z (1999) Spectral-finite element approach to three-dimensional electromagnetic induction in a spherical Earth. *Geophysical Journal International* 136: 229–250.

McCammon C (2005) The paradox of mantle redox. *Science* 308: 807–808.

Mead C and Jeanloz R (1987) High precision strain measurements at high pressures. In: Manghani MH and Syono Y (eds.) *High-Pressure Research in Mineral Physics*, pp. 41–51. Washington DC: American Geophysical Union.

Medin AE, Parker RL, and Constable S (2006) Making sound inferences from geomagnetic sounding. *Physics of the Earth and Planetary Interiors* 160: 51–59.

McDonald KL (1957) Penetration of the geomagnetic secular field through a mantle with variable conductivity. *Journal of Geophysical Research* 62: 117–141.

Morgan FD, Williams ER, and Madden TR (1989) Streaming potential properties of Westerly Granite with applications. *Journal of Geophysical Research* 94: 12449–12461.

Newman GA and Alumbaugh DL (2000) Three-dimensional magnetotelluric inversion using non-linear conjugate gradients. *Geophysical Journal International* 140: 410–424.

Nielsen OV, Petersen JR, Primdahl F, et al. (1995) Development, construction and analysis of the Orsted fluxgate magnetometer. *Measurement Science and Technology* 6: 1099–1115.

Oldenburg DW (1983) Funnel functions in linear and non-linear appraisal. *Journal of Geophysical Research* 88: 7387–7398.

Olsen N (1992) Day-to-day c-repsonse estimation for Sq from 1-cpd to 6-cpd using the Z-Y-method. *Journal of Geomagnetism and Geoelectricity* 44: 433–447.

Olsen N (1999) Long-period (30 days-1 year) electromagnetic sounding and the electrical conductivity of the lower mantle beneath Europe. *Geophysical Journal International* 138: 179–187.

Olsen N, Vennerstrøm S, and Friis-Christensen E (2003) Monitoring magnetospheric contributions using ground-based and satellite magnetic data. In: Reigber CH, Luehr H, and Schwintzer P (eds.) *First CHAMP Mission Results for Gravity, Magnetic and Atmospheric Studies*, pp. 245–250. Berlin: Springer-Verlag.

Parker RL (1984) The inverse problem of resistivity sounding. *Geophysics* 49: 2143–2158.

Parker RL (1994) *Geophysical Inverse Theory*. Princeton, NJ: Princeton University Press.

Parker RL and Whaler K (1981) Numerical methods for establishing solutions to the inverse problem of electromagnetic induction. *Journal of Geophysical Research* 86: 9574–9584.

Parkinson WD (1983) *Introduction to Geomagnetism*. Amsterdam: Elsevier.

Petiau G (2000) Second generation of lead-lead chloride electrodes for geophysical applications. *Pure and Applied Geophysics* 157: 357–382.

Poe B, Romano C, Nestola F, and Rubie D (2005) Electrical conductivity of hydrous single crystal San Carlos olivine, Eos Transactions AGU, 86(52), Fall Meeting Supplement, Abstract5 MR41A-0895

Quist AS and Marshall WL (1968) Electrical conductances of aqueous sodium chloride solutions from 0 to 800° and at pressures to 4000 Bars. *Journal of Physical Chemistry* 71: 684–703.

Revil A and Glover PW (1997) Theory of ionic-surface electrical conduction in porous media. *Physical Review B* 55: 1757–1773.

Roberts JJ and Tyburczy JA (1999) Partial-melt electrical conductivity: Influence of melt composition. *Journal of Geophysical Research* 104: 7055–7065.

Sabaka TJ, Olsen N, and Purucker ME (2004) Extending comprehensive models of the Earth's magnetic field with Orsted and CHAMP data. *Geophysical Journal International* 159: 521–547.

Shankland TJ, Duba AG, Mathez EA, and Peach CL (1997) Increase of electrical conductivity with pressure as an indicator of conduction through a solid phase in midcrustal rocks. *Journal of Geophysical Research* 102: 14741–14750.

Schmeling H (1986) Numerical models on the influence of partial melt on elastic, anelastic, and electrical properties of rocks. Part 2: Electrical conductivity. *Physics of the Earth and Planatery Interiors* 43: 123–136.

Schultz A, Kurtz RD, Chave AD, and Jones AG (1993) Conductivity discontinuities in the upper mantle beneath a stable craton. *Geophysical Research Letters* 20: 2941–2944.

Simpson F and Bahr K (2005) *Practical Magnetotellurics*. Cambridge, UK: Cambridge University Press.

Simpson F and Tommasi A (2005) Hydrogen diffusivity and electrical anisotropy of a peridotite mantle. *Geophysical Journal International* 160: 1092–1102.

Schmucker U (1970) Anomalies of geomagnetic variations in the Southwestern United States. *Bulletin Scripps Institute of Oceanography* 13: 1–165.

Schuster A and Lamb H (1889) The diurnal variation of terrestrial magnetism. *Philosophical Transactions of the Royal Society of London* 45: 481–486.

Shankland TJ and Duba AG (1997) Correlation of olivine electrical conductivity change with water activity, Eos Transactions AGU, 78(46), Fall Meeting Supplement, Abstract T52D-10

Smith JT and Booker JR (1991) Rapid inversion of 2-dimensional and 3-dimensional magnetotelluric data. *Journal of Geophysical Research* 96: 3905–3922.

Srivastava SP (1966) Theory of the magnetotelluric method for a spherical conductor. *Geophysical Journal of the Royal Astronomical Society* 11: 373–387.

Stacey FD and Anderson OL (2001) Electrical and thermal conductivities of Fe–Ni–Si alloy under core conditions. *Physics of the Earth and Planetary Interiors* 124: 153–162.

Stesky RM and Brace WF (1973) Electrical conductivity of serpentinized rocks to 6 kilbars. *Journal of Geophysical Research* 78: 7614–7621.

Tarits P, Hautot S, and Perrier F (2004) Water in the mantle: Results from electrical conductivity beneath the French Alps. *Geophysical Research Letters* 31: L06612.

Tikhonov AN and Arsenin VY (1977) *Solutions of Ill-Posed Problems*. New York: John Wiley and Sons.

Tikhonov AN (1950) Determination of the electrical characteristics of the deep strata of the Earth's crust. *Docklady Academii Nauk SSSR* 73: 295–297.

Tuller HL and Nowick AS (1977) Small polaron electron transport in reduced CeO_2 single crystals. *Journal of Physics and Chemistry of Solids* 38: 859–867.

Tyburczy JA and Roberts JJ (1990) Low frequency electrical response of polycrystalline olivine compacts: Grain boundary transport. *Geophysical Research Letters* 17: 1985–1988.

Tyburczy JA and Waff HS (1983) Electrical conductivity of molten basalt ad andesite to 25 kilobars pressure: Geophysical significance and implications for charge transport and melt structure. *Journal of Geophysical Research* 88: 2413–2430.

Uyeshima M and Schultz A (2000) Geoelectromagnetic induction in a heterogeneous sphere, a new 3-D forward solver using a staggered-grid integral formulation. *Geophysical Journal International* 140: 636–650.

Velimsky J, Everett ME, and Martinec Z (2003) The transient Dst electromagnetic induction signal at satellite altitudes for a realistic 3-D electrical conductivity in the crust and mantle. *Geophysical Research Letters* 30 (doi:10.1029/2002GL016671).

Wang D, Mookherjee M, Xu Y, and Karato S (2006) The effect of water on the electrical conductivity of olivine. *Nature* 443: 977–980.

Webb SC, Constable SC, Cox CS, and Deaton T (1985) A seafloor electric field instrument. *Journal of Geomagnetism and Geoelectricity* 37: 1115–1130.

Weidelt P (1972) The inverse problem in geomagnetic induction. *Zeitschrift für Geophysik* 38: 257–289.

Weiss CJ (2001) A matrix-free approach to solving the fully 3D electromagnetic induction problem. Contributed paper at *SEG 71st Annual Meeting*, San Antonio, USA.

Weiss CJ and Everett ME (1998) Geomagnetic induction in a heterogeneous sphere, fully three-dimensional test computation and the response of a realistic distribution of oceans and continents. *Geophysical Journal International* 135: 650–662.

Yoshimura R and Oshiman N (2002) Edge-based finite element approach to the simulation of geoelectromagnetic induction in a 3-D sphere. *Geophysical Research Letters* 29: (doi:10.1029/2001GL014121).

Yoshino T, Matsuzaki T, Yamashita S, and Katsura T (2006) Hydrous olivine unable to account for conductivity anomaly at the top of the asthensophere. *Nature* 443: 973–976.

Xu YS, Poe BT, Shankland TJ, and Rubie DC (1998) Electrical conductivity of olivine, wadsleyite, and ringwoodite under upper-mantle conditions. *Science* 280: 1415–1418.

Xu YS and Shankland TJ (1999) Electrical conductivity of orthopyroxene and its high pressure phases. *Geophysical Research Letters* 26: 2645–2648.

Xu YS, Shankland TJ, and Duba AG (2000) Pressure effect on electrical conductivity of mantle olivine. *Physics of the Earth and Planatery Interiors* 118: 149–161.

8 Magnetizations in Rocks and Minerals

D. J. Dunlop and Ö. Özdemir, University of Toronto, Toronto, ON, Canada

8.1   Introduction

All matter is magnetic but in geophysics, we are usually concerned with minerals that are either 'ferromagnetic' like iron–nickel in meteorites, 'ferrimagnetic' like magnetite, or 'antiferromagnetic' like hematite. These minerals are orders of magnitude more magnetic than diamagnetic minerals like quartz or paramagnetic minerals like micas. They possess both a reversible 'induced magnetization' (which changes when the inducing field changes, for example, reverses its polarity) and a permanent 'remanent magnetization' (which preserves a memory of paleomagnetic fields).

8.1.1   Diamagnetism and Paramagnetism

'Diamagnetism' is a property of all atoms. It originates in the speeding up or slowing down of electrons in orbital motion when the magnetic Lorentz force is added to the electrostatic Coulomb force. The direction of the resulting magnetic moment per unit volume, or magnetization, M is opposite to that of H, the magnetic field applied, giving a negative susceptibility

$$k_d = dM/dH < 0 \qquad [1]$$

The diamagnetic susceptibility is extremely weak and is overshadowed if any other forms of magnetism are present. However, it is temperature independent and establishes a baseline to the magnetic signal at high T, for example, above the Curie points of ferromagnetic minerals.

'Paramagnetism' is a more powerful effect arising from the partial alignment by H of permanent magnetic dipole moments of atoms, which in turn arise from unpaired electron spins. Every odd-numbered element has at least one unpaired spin and therefore a permanent magnetic moment. The transition elements, in particular, Fe, Co and Ni, have multiple unpaired 3d electrons and correspondingly stronger moments.

Paramagnetic susceptibility is temperature dependent because the aligning effect of H is opposed by thermal disordering. The Langevin theory predicts that M is parallel to H and varies as H/T (Curie's law), giving a positive susceptibility

$$k_p(T) = \partial M(H, T)/\partial H \sim 1/T \qquad [2]$$

At ordinary temperatures, paramagnetic susceptibilities are 1–2 orders of magnitude larger than typical diamagnetic susceptibilities (**Table 1**), but 2–3 orders of magnitude smaller than the susceptibilities of ferrimagnetic minerals, for example, magnetite with a (per unit mass) susceptibility $\chi \approx 5.8 \times 10^{-4} \, \mathrm{m^3 \, kg^{-1}}$. For more extensive tabulations of volume and mass susceptibilities of rocks and minerals, see Hunt *et al.* (1995) or Carmichael (1982).

Paramagnetism comes into prominence in three situations. The first is at very low temperatures, where alignment of atomic moments becomes substantial. Low-T superconducting quantum interference device (SQUID) magnetometry is becoming common in rock magnetism and paramagnetic minerals are sometimes detectable by their $1/T$ signals below $\sim$20 K. The second situation where paramagnetism becomes visible is above the Curie temperature T_C where ferromagnetic and ferrimagnetic minerals lose their spin coupling (see the next section). Third, in determining rock fabric by the anisotropy of magnetic susceptibility, rocks which are virtually barren of ferrimagnetic minerals may still have a measurable anisotropy because of preferential alignment of paramagnetic mineral axes.

8.1.2   Ferromagnetism

Iron, nickel, cobalt, and a few other elements, differ from diamagnets and paramagnets in retaining a 'remanent' magnetization, or 'remanence',

Table 1 Susceptibilities k and χ of some common minerals

Mineral	$k \, (10^{-6} \, \text{SI})$	$\chi \, (10^{-8} \, \text{m}^3 \, \text{kg}^{-1})$
Diamagnetic		
Quartz (SiO_2)	−16.4	−0.62
Orthoclase ($KAlSi_3O_8$)	−14.9	−0.58
Calcite ($CaCO_3$)	−13.6	−0.48
Forsterite (Mg_2SiO_4)	−12.5	−0.39
Paramagnetic		
Troilite (FeS)	$0.6–1.7 \times 10^3$	13–35
Pyrite (FeS_2)	1.5×10^3	30
Siderite ($FeCO_3$)	1.9×10^3	123
Ilmenite ($FeTiO_3$)	$4.7–5.2 \times 10^3$	100–110
Ulvöspinel ($FeTi_2O_4$)	4.8×10^3	100
Orthopyroxene ((Fe,Mg)SiO_3)	$1.55–1.8 \times 10^3$	43–50
Fayalite (Fe_2SiO_4)	5.53×10^3	126
Olivine, intermediate ((Fe,Mg)$_2SiO_4$)	1.56×10^3	36
Serpentinite ($Mg_3Si_2O_5(OH)_4$)	$\geq 3 \times 10^{3a}$	$\geq 120^a$
Amphiboles	$0.5–2.7 \times 10^{3a}$	$16–94^a$
Biotites	$0.5–1.15 \times 10^3$	17–38
	$1.5–2.9 \times 10^{3a}$	$50–97^a$
Clay minerals (illite, montmorillonite)	$0.33–0.41 \times 10^3$	13–15
Ferromagnetic, ferrimagnetic, antiferromagnetic		
Iron, multidomain (αFe)	3.9×10^6	5.0×10^4
Pyrrhotite (Fe_7S_8)	3.2×10^6	6.9×10^4
(Fe_9S_{10})	0.17×10^6	0.38×10^4
Hematite (αFe$_2O_3$)	$0.5–40 \times 10^3$	10–760
Maghemite, multidomain (γFe$_2O_3$)	$2.0–2.5 \times 10^6$	$4.0–5.0 \times 10^4$
Magnetite, multidomain (Fe_3O_4)	3.0×10^6	5.8×10^4
Titanomagnetite, $x = 0.6$ (TM60)	$0.13–0.62 \times 10^6$	$0.25–1.2 \times 10^4$
Titanomaghemite	2.8×10^6	5.7×10^4
Goethite (αFeOOH)	$1.1–12 \times 10^3$	25–280

^{a}Higher values are due to magnetite inclusions.
For ferromagnetic minerals, weak-field initial susceptibilities k_0 and χ_0 are listed.

M_r, in the absence of a field H. In addition, even if the remanence is minor, the 'induced' magnetization kH is strong and readily measurable for relatively weak fields like the Earth's. Magnetic anomalies and the paleomagnetic method both depend on these two basic properties, which in turn reflect coupling of electron spins of neighboring atoms.

Figure 1 illustrates a generic 'hysteresis' or M versus H curve. (In industrial magnetism, flux density $B = \mu_0(H + M)$ is plotted instead of M. The curves are indistinguishable because $M \gg H$ for ferromagnets, making B essentially equal to M, apart from the rescaling required by the permeability of free space, $\mu_0 = 4\pi \times 10^{-7} \, \text{H} \, \text{m}^{-1}$.) Hysteresis, the irreversibility of the magnetization process, ensures that M is not a single-valued function of H. However, provided that maximum fields of equal strength H are applied in opposite directions, a field cycle results in a closed M–H loop.

Hysteresis originates in the energy expended in either increasing the size of one set of 'domains' of coupled spins at the expense of another set or else rotating spins past unfavorably oriented axes in the crystal. Because these processes expend similar amounts of energy for fields in opposite directions, a complete cycle returns the system to its original state of magnetization but with the expenditure of an amount of energy proportional to the area enclosed by the loop. This hysteresis loss is important in industrial applications but not in nature, where field changes are quite modest.

A 'minor' loop measures changes in M in small fields, for example, in the course of two successive reversals of the Earth's field. Note that the ferromagnetic susceptibility $k_f = \mathrm{d}M/\mathrm{d}H$ is not constant but generally increases with increasing H (until H reverses). The isothermal remanence or IRM, M_r, in such a weak-field process is very small, and furthermore is paleomagnetically useless because it

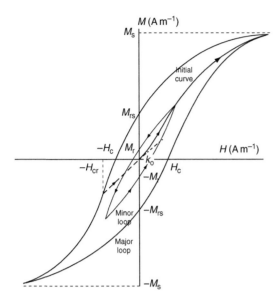

Figure 1 Initial magnetization curve and minor and major hysteresis loops. Standard parameters are k_0 (initial susceptibility), M_s (saturation magnetization), M_{rs} (saturation remanence), M_r (isothermal remanence, minor loop), H_c (coercive force – for either major or minor loop), and H_{cr} (remanent coercive force). Reversing the field at $-H_{cr}$ brings the sample to zero net magnetization (dashed line).

will change with each and every field reversal as well as with secular changes in field intensity between reversals.

The 'saturation remanence' or SIRM, M_{rs}, produced by departing from a field sufficient to saturate all domain enlargements or rotations, is much more substantial and changes little with subsequent small changes (positive or negative) in field. However, it is paleomagnetically irrelevant because the only common fields in nature strong enough to produce SIRM result from lightning currents, which produce their own fields in directions unrelated to the Earth's field. Nevertheless, SIRM is frequently measured as part of the characterization of a sample, as is the 'coercive force' (more properly, 'coercive field') H_c, which is the axis-crossing field in a hysteresis loop, usually the saturation loop. Another field commonly reported is the 'remanent coercive force' H_{cr}, which is the (opposed) field whose application and removal reduce M_r to zero.

Both B and H are called the magnetic field in common parlance. Outside a magnetic material, they can be used interchangeably. H is preferred as the measure of the output of a field-producing device like a solenoid and B (or M) as the magnetic effect of that field, hence B–H loops. In recent years, B_c and B_{cr}

have begun to be quoted in the rock magnetic literature because B converts more conveniently than H between cgs and SI units (see Section 8.1.6). This is incorrect. H is the abscissa in all hysteresis loops; B_c and B_{cr} imply measurements of an internal field, essentially M. If convenience of units is an issue, $\mu_0 H_c$ and $\mu_0 H_{cr}$ can be used. Other situations where we must be careful to use H and not B are in calculating induced magnetization, $M_{in} = kH$, and in dealing with internal fields like the self-demagnetizing field H_d (see Section 8.2.1).

Ferromagnetic ordering results from Heisenberg exchange or superexchange coupling between the spins of 3d electrons in neighboring or next-nearest neighboring atoms, a quantum mechanical manifestation of electrostatic interaction between the electrons. The same mechanism is responsible for covalent bonding. For a discussion, see Dunlop and Özdemir (1997, chapter 2). Before the mechanism of spin coupling was known, Weiss (1907) proposed that a powerful internal field, which he termed the molecular field, must exist in iron and magnetite to produce alignment of their atomic moments. This approach is very fruitful and generates an immediate understanding of the 'spontaneous magnetization' M_s of ferromagnets and its temperature dependence and ultimate disappearance at the Curie point T_C. 'Thermomagnetic' or $M_s(T)$ curves are one of the most basic methods of determining the magnetic mineralogy of a rock (see Section 8.3.1).

8.1.3 Antiferromagnetism and Ferrimagnetism

In ferromagnets, the moments of neighboring atoms are coupled parallel and the origin of spontaneous magnetization is obvious. But of the common magnetic minerals, only iron and iron–nickel are ferromagnetic and they are found mainly on other planets (Moon and meteorites). Magnetite and hematite, Earth's commonest magnetic minerals, are respectively 'ferrimagnetic' and 'antiferromagnetic', terms invented by Néel (1948) to describe new states of matter embodying 'negative' exchange coupling. A schematic picture of these states appears in **Figure 2**.

Negative coupling of atomic moments does not obviously lead to any net magnetization, however strong the coupling may be. This is not actually the case, however, because the 'up' and 'down' spins may not belong to the same type of atom or ion and therefore may have different magnetic moments, represented in **Figure 2** by different lengths of the

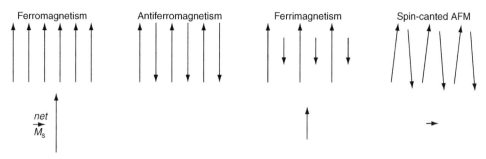

Figure 2 Schematic representation of spin coupling in the ferromagnetic, antiferromagnetic, and ferrimagnetic states of matter. Spin canting in the antiferromagnetic state imparts a weak transverse ferromagnetic moment to hematite. M_s is the net spontaneous magnetization per spin lattice site (or A + B spin sublattice sites, in the case of ferrimagnetism).

arrows. This is the situation in magnetite ($Fe^{2+}Fe^{3+}_2O_4$), for example, where the antiparallel magnetic sublattices have different numbers of Fe^{2+} and Fe^{3+} ions at individual sites. The ferrimagnetism of magnetite, although only a net imbalance between two competing sublattices, creates an $M_s \approx 1/3$ that of ferromagnetic iron, where all atomic moments add.

In antiferromagnetism, as exemplified by hematite ($Fe^{3+}_2O_3$) and its hydrated precursor goethite ($Fe^{3+}OOH$), the up- and down-sublattices have identical numbers and species of ions. Each sublattice has a spontaneous magnetization, but in zero field these cancel because they are oppositely directed. An antiferromagnet in theory has no remanence but a field applied perpendicular to the sublattices will deflect both sets of spins, giving rise to an antiferromagnetic susceptibility k_a. A strong field is required to produce any substantial angular deflection, but on the other hand, the moments being deflected are collectively strong, with the result that k_a is comparable to k_p at ordinary temperatures, although the exchange coupling and thermal disordering mechanisms involved are entirely different. The temperature dependences reveal the different mechanisms, k_a varying only weakly with T. Indeed, before Néel elucidated its origin, antiferromagnetism was known as temperature-independent paramagnetism.

Hematite actually has a weak permanent magnetization, enabling it to carry a remanence. The fundamental mechanism is spin canting, a zero-field deflection of the sublattices out of antiparallelism by a fraction of a degree. As explained by Dzyaloshinsky (1958), spin canting occurs only for certain crystal symmetries and disappears when the symmetry changes, as it does when the sublattices are parallel to the rhombohedral c-axis instead of perpendicular to it. Hematite has two phases, truly antiferromagnetic at temperatures below the Morin

transition T_M (at $-15°C$) and weakly ferromagnetic above T_M. Even in its ferromagnetic phase, M_s is only 0.5% that of magnetite.

8.1.4 Magnetocrystalline Anisotropy

Magnetocrystalline anisotropy is a manifestation of atomic spin–orbit coupling. Electron orbitals tend to lie along preferred or 'easy axes' in a crystal and spin magnetic moments are linked to these axes via the orbital coupling. Macroscopically magnetocrystalline anisotropy energy E_K is created when M_s deviates from an easy axis. In a cubic crystal,

$$E_K = K_1 V(\alpha_1{}^2\alpha_2{}^2 + \alpha_2{}^2\alpha_3{}^2 + \alpha_3{}^2\alpha_1{}^2) + K_2 V\alpha_1{}^2\alpha_2{}^2\alpha_3{}^2 \qquad [3]$$

where K_1 and K_2 are anisotropy constants, V is volume, and the α_i are direction cosines of M_s with respect to the <100> crystal axes. In iron, $K_1 > 0$ and <100> are the easy axes. For magnetite, $K_1 < 0$ and <111> are the easy axes. K_1 and K_2 are measured from torque curves of a magnetically saturated single crystal or from ferromagnetic resonance of a polycrystalline sample.

Minerals with a single crystalline easy axis of magnetization are not common in nature (hematite at low temperatures is one example) but 'uniaxial anisotropy' is often used as a simplified model in magnetic structure calculations. The anisotropy energy is then

$$E_K = K_{u1}V\sin^2\theta + K_{u2}V\sin^4\theta \qquad [4]$$

$\theta = 0°$ or $180°$ being the two preferred orientations of M_s that define the easy axis. Uniaxial anisotropy can arise from 'magnetoelastic anisotropy' due to magnetostriction (Section 8.1.5) and 'shape anisotropy' of the magnetostatic energy of elongated grains (Section 8.2.1).

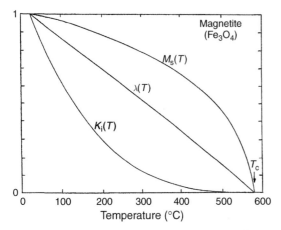

Figure 3 Normalized temperature dependences of spontaneous magnetization M_s, direction-averaged magnetostriction constant λ, and first-order magnetocrystalline anisotropy constant K_1 of magnetite. All three decrease to zero at the Curie point T_C, but at very different rates.

Magnetocrystalline anisotropy decreases with temperature T much more rapidly than does M_s (**Figure 3**). It is generally negligible in the T range near the Curie point where cooling igneous rocks acquire most of their remanence.

8.1.5 Magnetostriction and Magnetoelastic Anisotropy

When a crystal is magnetized, its dimensions change spontaneously. A positive 'magnetostriction' corresponds to an expansion in the direction of the magnetization M. Magnetite expands in the easy direction <111> with maximum strain $\lambda_{111} = +72.6 \times 10^{-6}$ and contracts in the hard direction <100> with maximum strain $\lambda_{100} = -19.5 \times 10^{-6}$.

The magnetostriction averaged over all directions, appropriate for a polycrystalline material, is $\lambda_s = 0.4\lambda_{100} + 0.6\lambda_{111}$.

If a uniform strain is imposed through an external or internal stress σ, and isotropic magnetostriction is assumed, the crystalline anisotropy energy changes by an amount

$$E_\sigma = \lambda \sigma V \approx 1.5\lambda_s \sigma V \sin^2\theta \qquad [5]$$

which has the form of a uniaxial 'magnetoelastic anisotropy'; $\theta = 0$ is the axis of tension. In practice, individual magnetic grains in rocks experience highly variable local strains whose magnetic effects are difficult to describe in a simple way.

Magnetostriction constants decrease with increasing temperature less rapidly than K_1 or K_2 (**Figure 3**). Magnetoelastic anisotropy may thus survive as an important factor in remanence acquisition to higher T than magnetocrystalline anisotropy.

8.1.6 Magnetic Quantities, Units, and Conversions

The basic relation between B, H, and M is $B = \mu_0(H + M)$ in SI, with $\mu_0 = 4\pi \times 10^{-7}\,H\,m^{-1}$, and $B = H + 4\pi M$ in cgs emu. Sometimes the symbol J is used for magnetization in cgs; this is not recommended because J in electromagnetism is current density. Traditionally, cgs was the system of choice in magnetism, and it remains widely used today, largely for pragmatic reasons. B, H, and M have the same dimensions, although their units have different names (**Table 2**), and conversions between B and H are effortless. In SI, B and H, both usually called the magnetic field, differ numerically by a factor of $\sim 10^6$. For geomagnetists, cgs units were particularly natural because the Earth's magnetic field is ~ 1 G or Oe.

Table 2 Units and conversions for magnetic quantities

Quantity	SI unit	cgs unit	Conversion
Magnetic moment, m	$A\,m^2$	emu	$1\,A\,m^2 = 10^3\,emu$
Magnetization, $M\,(=m/V)$	$A\,m^{-1}$	'emu cm^{-3}'	$1\,A\,m^{-1} = 10^{-3}\,emu\,cm^{-3}$
Magnetization per unit mass	$A\,m^2\,kg^{-1}$	'emu g^{-1}'	$1\,A\,m^2\,kg^{-1} = 1\,emu\,g^{-1}$
Magnetic field, H	$A\,m^{-1}$	Oersted	$1\,A\,m^{-1} = 4\pi \times 10^{-3}\,Oe$
Magnetic field or induction, B	Tesla	Gauss	$1\,T = 10^4\,G$
$\quad B = \mu_0(H + M)$ in SI			
$\quad B = H + 4\pi M$ in cgs			
Susceptibility, $k\,(=dM/dH)$ (dimensionless)	'SI unit'	'cgs unit'	$1\,SI\,unit = (1/4\pi)\,cgs$
Susceptibility per unit mass, χ	$m^3\,kg^{-1}$	'emu g^{-1} Oe^{-1}'	$1\,m^3\,kg^{-1} = (10^3/4\pi)\,cgs$
Demagnetizing factor, N (dimensionless)			$1\,(in\,SI) = 4\pi\,(in\,cgs)$

After Moskowitz (1995), table 1, and Dunlop and Özdemir (1997), table 2.1.

The corresponding figures in SI are awkwardly large or small: $\approx 80\,\mathrm{A\,m^{-1}}$ for **H** and $100\,\mu\mathrm{T}$ for **B**.

Because **B** converts from cgs to SI with a simple factor of 10^{-4}, while the conversion for **H** is more cumbersome, fields in SI are seldom quoted in $\mathrm{A\,m^{-1}}$, even when they should be. One example was cited in Section 8.1.2, where H-axis-crossing parameters in hysteresis loops are increasingly being labeled B_c or B_{cr}. They should be called $\mu_0 H_c$ or $\mu_0 H_{cr}$ if the convenience of using T rather than $\mathrm{A\,m^{-1}}$ is compelling.

More serious are quantities determined 'inside' a magnetic material, such as the internal field $\mathbf{H}_i$ of a magnetized grain. One reason for using **H** rather than **B** for internal fields is that **H** is defined even at an atomic level (even though seldom calculated at this scale because of its extreme spatial fluctuation), while **M** and **B** are mesoscopic averages over thousands of atoms. **B** should always be treated as a derivative field, a combination of **H** and **M**, unless one is in free space far from all magnetized bodies. For example, in a **B**–**H** loop, although measurements are made outside the magnetized body, by the nature of the measurement procedure it is **B** inside the body that is determined, not $\mu_0 \mathbf{H}$.

Another situation where H must be used, not B, is in determining susceptibility, k. By definition, $k = \mathrm{d}M/\mathrm{d}H$. Although k is dimensionless, it has different values in SI and cgs: a k value in SI is numerically larger by 4π than the cgs k value. By the same token, demagnetizing factors N (Section 8.2.1) which relate $\mathbf{H}_i$ to **M** are numerically smaller in SI than in cgs by a factor 4π.

8.2   Domains and the Magnetization Process
8.2.1   Demagnetizing Energy and Domain Structure

In exact analogy to a polarized dielectric, at the surface of a magnetized body, bound magnetic poles appear with a density $\sigma = \mathbf{M}\cdot\mathbf{n}$, where **n** is the outward surface normal (**Figure 4**). In turn, these poles act as sources and sinks for lines of the internal 'self-demagnetizing field' $\mathbf{H}_d$, so called because its direction is opposite to **M**. If **M** is a remanent magnetization, $\mathbf{H}_d$ is the total internal field, but if an external applied field $\mathbf{H}_o$ is necessary to maintain an induced magnetization **M**, the internal field is

$$\mathbf{H}_i = \mathbf{H}_o + \mathbf{H}_d = \mathbf{H}_o - N\mathbf{M} \qquad [6]$$

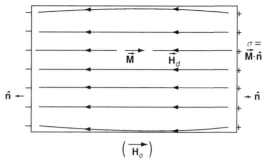

Figure 4   The (almost uniform) self-demagnetizing field $\mathbf{H}_d$ of a material with uniform magnetization **M**. **M** may be induced by an externally applied field $\mathbf{H}_o$ or it may be a permanent (remanent) magnetization. Magnetic poles with surface density $\sigma = \mathbf{M}\cdot\mathbf{n}$, the exact analogs of bound charges associated with electric dipoles in dielectrics, appear wherever **M** cuts the surface. These poles can be viewed as the sources of field lines $\mathbf{H}_d$ within the crystal and of the leakage or stray fields outside the crystal by which domain boundaries are imaged in the Bitter colloid method.

N is the demagnetizing factor and usually has different values for different directions in a crystal, dependent on its shape. For a cube, for example, iron, $N = 1/3$ ($4\pi/3$ in cgs) for **M** parallel to a <100> axis. For an octahedral crystal, for example, magnetite, $N = 1/3$ for **M** along a <111> axis. For platy crystals, for example, hematite, $N \approx 0$ for **M** in the plane (ordinary temperatures) or ≈ 1 transverse to the plane (low temperatures). For elongated grains, $N \approx 0$ for **M** parallel to the long axis (the magnetostatically preferred orientation) and $1/2$ for transverse **M** (hard direction).

The shape dependence of $\mathbf{H}_d$ and the 'demagnetizing energy' $E_d = (1/2)\mu_0 VNM^2$ is large enough in strongly magnetic minerals like magnetite and iron that a slight deviation from perfectly cubic shapes produces a 'shape anisotropy'

$$E_d = \frac{1}{2}\mu_0 V \Delta N M^2 \sin^2\theta \qquad [7]$$

which outweighs the rather small magnetocrystalline anisotropy. Equation [7] applies only to ellipsoidal grains, which have uniform internal fields, but is a useful first-order model for real crystals in nature. ΔN is the difference between N for **M** along the axis of intermediate hardness (usually near $\theta = 90°$) and N when the grain is magnetized along its length (the easy axis, $\theta = 0$).

Measured remanent and induced magnetizations are quite small compared to $\mathbf{M}_s$ for weak fields $\mathbf{H}_o$. How is it that shape anisotropy is so powerful? The

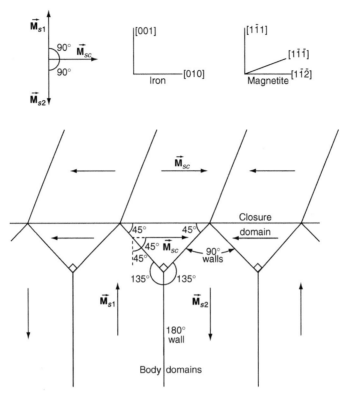

Figure 5 Spontaneous magnetization vectors within body domains ($\mathbf{M}_{s1}$, $\mathbf{M}_{s2}$) and closure domains ($\mathbf{M}_{sc}$). For materials like iron or high-Ti titanomagnetites, with <100> easy axes, $\mathbf{M}_{sc}$ is at 90° to $\mathbf{M}_{s1}$ and $\mathbf{M}_{s2}$, as sketched. For materials like magnetite and low-Ti titanomagnetites, with <111> easy axes, $\mathbf{M}_{sc}$ is at 70.5° or 109.5° to $\mathbf{M}_{s1}$ and $\mathbf{M}_{s2}$. Domain walls bisect the angle between $\mathbf{M}_s$ vectors in adjacent domains, in order to eliminate magnetic poles on the boundary ($\mathbf{M}_{s1} \cdot \mathbf{n} + \mathbf{M}_{sc} \cdot \mathbf{n} = 0$, for example). Modified from Özdemir Ö, Xu S, and Dunlop DJ (1995) Closure domains in magnetite. *Journal of Geophysical Research* 100: 2193–2209.

answer lies in domain structure (**Figure 5**). In very small particles, $\leq$100 nm for magnetite and even smaller for iron, exchange coupling successfully maintains a uniform magnetization $\mathbf{M}_s$ throughout the crystal. Such particles contain a 'single magnetic domain'. In larger particles, E_d for $\mathbf{M} = \mathbf{M}_s$ becomes so large that there are energy savings to be had by subdividing the crystal into two or more domains with their $\mathbf{M}_s$ vectors antiparallel (or at other angles dictated by magnetocrystalline easy axes, as in **Figure 5**). The domains are sheet-like, with $\mathbf{M}_s$ in the plane of the sheet, so that N is reduced for each domain compared to N for a single domain. However, $\mathbf{M} = \mathbf{M}_s$ within each domain and ΔN when the domains are pulled away from the long axis is merely scaled down relative to ΔN for rotation of a single domain. Shape anisotropy remains important.

If the interior of a ferromagnet is magnetized to saturation, globally (single-domain or SD grain) or locally (multidomain or MD grain), how does

it come about that the measured induced or remanent magnetization $\mathbf{M}$ of a rock is proportional to the applied field $\mathbf{H}_o$ that caused it? An individual SD grain has no demagnetized or low-M state. The only possible response to $\mathbf{H}_o$ is rotation of its total magnetic moment $V\mathbf{M}_s$ toward $\mathbf{H}_o$. This added component of magnetization in the direction of $\mathbf{H}_o$ explains induced magnetization – but not remanence. When $\mathbf{H}_o$ disappears, so does $\mathbf{M}$: all the grain moments return to their original easy (long) axes, oriented at random. A remanence can only be produced if some moments are pulled by $\mathbf{H}_o$ past the hard axis, ultimately reversing their moments when $\mathbf{H}_o$ is removed. Because this takes considerable energy, SD grains are hard to magnetize using field alone. On the other hand, once SD moments have been rotated past a shape anisotropy barrier, by whatever means, they are hard to 'demagnetize' under ordinary conditions of temperature and field. They are ideal paleofield recorders.

8.2.2 Domain Walls and Multidomain Magnetization

The magnetization process is entirely different in MD grains. Instead of rotating domain moments, a small field H_o will cause domains with VM_s close ($<90°$) to H_o to enlarge at the expense of unfavorably oriented domains. This process is almost reversible and should result in a large susceptibility. There is, however, a price to be paid in increased E_d because the enlarged domains have increased their N. (The domains that shrink decrease their N but the net effect is still an increase in E_d, as Section 8.4.2 shows.) Self-demagnetization applies a severe brake on increased **M**. Observed susceptibilities of magnetite-bearing rocks are essentially controlled by N, not by the ease of enlarging domains.

The mechanism by which domains enlarge or shrink is resembles the migration of a dislocation when a crystal deforms. The boundaries between domains are not marked by abrupt changes in the direction of M_s because exchange energy favors a gradual rotation from one domain to its neighbor. The regions of rotated M_s are called 'domain walls' (**Figure 6**). They have a definite thickness, dictated by the balance between reducing E_{ex} and E_d and increasing anisotropy energy (next section). They migrate easily because each spin in the wall need only rotate by a small angle in response to H_o to cause 'domain wall displacement'. Spins at one edge of the wall are added to the enlarged domain. New spins from the less-favored domain are added to the opposite edge of the wall. The wall itself has the same structure it had before moving; its energy is unchanged. Only the changing shapes of the domains tilt the energy balance.

The only reason walls are hindered at all in their progress is that they tend to become anchored at crystal imperfections such as surface pits or interior voids and cracks, which reduce the wall's energy by the magnetostatic effect of poles that appear on the crack or void boundaries, or at pileups of dislocations, which interact magnetoelastically with the wall. For a detailed treatment, see Özdemir and Dunlop (1997). The same pinning centers hinder the wall's return motion when H_o is reduced or removed. However, the pinning is not symmetric. The net H_d in any grain favors a return to an overall demagnetized state. This is not necessarily exactly the same as the initial state, but on average, self-demagnetization opposes any remanence and helps walls jump past pins on their way to a low-magnetization state.

By analogy with the macroscopic hysteresis loop, an individual wall can be thought of as having one or several coercive forces H_c that describe how much field is required to impel it past a barrier or barriers to motion. These are often called microcoercivities. The H_c values are generally small compared to fields necessary to rotate SD moments past shape anisotropy barriers, both because pinning by defects is inherently weak and because H_d aids so effectively in demagnetizing any remanence. For these reasons, MD grains are usually poor paleofield recorders. Their remanence is a small fraction of that of an equal volume of SD grains, although this is offset by the high proportion of MD grains in the ferromagnetic fraction of most rocks. More importantly, their coercivities are low and do not stabilize the remanence against changing fields later in the rock's life.

8.2.3 Domain Wall Width and Energy

The width of a wall is an important parameter because it affects the ease of observing domain structures. Along a line of m spins normal to a wall, M_s

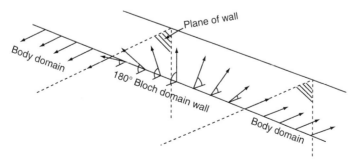

Figure 6 Progressive rotation of spins across a 180° wall between two body domains. In the Bloch wall shown, spins rotate in the plane of the wall so as to eliminate interior magnetic poles. A real domain wall contains a hundred or more spins, not the few shown here.

rotates through small angles $\Delta\theta/m$ from one domain to the other (**Figure 6**). Exchange energy is reduced by making the rotation as gradual as possible: $E_{ex} = \text{const.} + \mathcal{J}_e S^2 \, \Delta\theta^2/m$, where $\mathcal{J}_e$ is the exchange intergral and S atomic spin. However, E_K increases when spins do not lie along crystalline easy axes: $\Delta E_K = Ka^3 m$, where a is interatomic spacing and K ($=1.64 \times 10^3 \, \text{J m}^{-3}$ for magnetite) averages over the line of spins (see Dunlop and Özdemir (1997, chapter 5) for a discussion). Minimizing the sum of the two energies gives a wall width

$$\delta_w = m_{eq} a = \pi \sin(\Delta\theta/2)(A/K)^{1/2} \qquad [8]$$

in which the exchange constant A ($=1.33 \times 10^{-11} \, \text{J m}^{-1}$ for magnetite at 20°C) is $\mathcal{J}_e S^2/a$.

From [8], δ_w is predicted to be 0.28 μm for 180° walls in magnetite, compared to an experimentally measured δ_w of 0.18 μm (Moskowitz *et al.*, 1988). The rather crude theoretical approach is therefore acceptable if only a rough estimate is needed.

The wall energy per unit area is needed to calculate equilibrium domain widths. It is

$$\gamma_w = 2\pi \sin(\Delta\theta/2)(AK)^{1/2} \qquad [9]$$

giving a 180° wall energy in magnetite of $\gamma_w = 0.93 \, \text{J m}^{-2}$. The experimental value is $\gamma_w = 0.91 \, \text{J m}^{-2}$ (Özdemir and Dunlop, 1993a). The close agreement may be deceptive. In the calculation, the demagnetizing field and energy of the wall were ignored. While it is true that no poles appear on the interfaces between a wall and its bounding domains, because spins rotate in the plane of the wall, poles do appear at the edges of the wall where it meets the crystal surface. One way of eliminating this energy is through surface closure domains (see **Figure 5**), discussed below.

8.2.4   Equilibrium Domain Structures

A lattice of spins spontaneously subdivides itself into domains to reduce demagnetizing energy E_d. To a good approximation, both the effective demagnetizing factor N and E_d are inversely proportional to the number of domains n (Dunlop and Özdemir, 1997, figure 5.5). Increasing the number of domains thus reduces E_d to $(1/n)$ of the single-domain energy $(E_d)_{SD}$. However, the process of subdivision cannot continue indefinitely, because each new domain wall adds its energy to the total. In addition, there are other ways of reducing E_d besides making individual

domains longer and narrower. The most effective way is by forming flux-closure domains either in the body of the crystal or at its surface (**Figure 5**).

The simplest domain structure is that of uniaxial materials, which have no closure domains. They form only 180° walls and the domains terminate at the crystal surface. The energy is then the sum of E_d of the n domains and E_w of the $n-1$ walls:

$$E = 1/2\mu_0 LWDN_{SD}M_s^2 (1/n) + \gamma_w LW(n-1) \qquad [10]$$

where L and W are the length and width of the grain and of the plate-like walls, and $D = nd$ is the third dimension of the grain, subdivided into n domains of width d. Minimizing E by setting $dE/dn = 0$ yields

$$d_{eq} = (\mu_0 N_{SD} M_s^2 / 2\gamma_w)^{1/2} D^{1/2} \qquad [11]$$

In multiaxial materials, there is a choice of several crystalline easy axes and closure domains can form. E_d is eliminated or greatly reduced because $\mathbf{M}_s \cdot \mathbf{n} = 0$ on all internal boundaries (180° and 90° walls in iron; 180°, 109.5°, and 70.5° walls in magnetite) and is small if not zero on surface boundaries of body and closure domains (**Figure 5**). A simple calculation for magnetite that considers two main energies, the energy of 180° walls and the magnetoelastic energy E_σ due to incompatible magnetostrictions across 109.5° and 70.5° walls, gives (Özdemir *et al.*, 1995)

$$d_{eq} = (4\gamma_w/9\lambda_{111}c_{44})^{1/2} D^{1/2} \qquad [12]$$

in which c_{44} is an elastic constant.

The domain width predicted by eqn [12] is several times larger than that according to [11]. Closure domains effectively eliminate E_d and so long, narrow body domains are no longer imperative.

Experimental data testing the prediction that $d_{eq} \sim D^{1/2}$ appear in **Figure 7**. For magnetite at least, the prediction is verified most convincingly for grain sizes of 10 μm and less. The proportionality constant is appropriate to eqn [12], not [11], for reasonable values of γ_w, yet relatively few of these small grains exhibited closure domains (**Figure 8**). Large magnetite crystals do have well-developed surface and interior closure domains (**Figure 9**), yet their domain widths are even larger and more discrepant with equilibrium theory (**Figure 7**). That is, they contain fewer domains than equilibrium theory predicts.

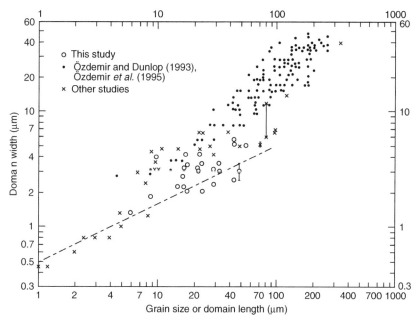

Figure 7 Average width of domains in magnetite as a function of grain size (or in large grains, domain length). The dot-dash line is based on equilibrium domain theory (Eqns [11] and [12]). Most data fall above this line, indicating fewer than the predicted number of domains. Modified from Özdemir Ö and Dunlop DJ (2006) Magnetic domain observations on magnetite crystals in biotite and hornblende grains. *Journal of Geophysical Research* 111: B06103 (doi: 10.1029/2005JB004090).

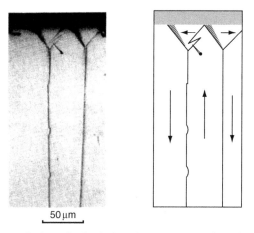

Figure 8 Lamellar body domains and wedge-shaped surface closure domains imaged by the Bitter colloid method on a {110} surface of a single crystal of magnetite. Modified from Özdemir Ö, Xu S, and Dunlop DJ (1995) Closure domains in magnetite. *Journal of Geophysical Research* 100: 2193–2209.

8.2.5 Observations of Domains

The domains in **Figures 8** and **9** were imaged by the Bitter technique. A colloidal suspension of magnetite nanoparticles gathers over places in a polished section through a crystal where magnetic flux leaks out of the surface. The colloidal particles are in essence tracers of magnetic field lines, similar in principle to iron filings used to map the field lines of a bar magnet, but on a greatly reduced size scale. Field lines leave a crystal wherever M_s cuts the surface. Sectioning the crystal in a plane that contains two sets of easy axes, for example, {110} in magnetite, containing two sets of <111> axes, cuts across both body and closure domains. The M_s vectors in the domains are surface parallel, so $M_s \cdot n = 0$ and there are no poles to act as sources for leakage fields. However, M_s vectors in the walls bounding the domains rotate in the plane of the wall and do cut the surface. The poles thus formed attract colloid particles, making the walls visible. The 180° walls are more easily imaged than 109.5° and 70.5° walls because of their greater width (Williams *et al.*, 1992; cf. eqn [8]).

Magnetic force microscopy (MFM), like its sister technique, atomic force microscopy (AFM), measures the force exerted on the tip of a nanometer-scale cantilever as it scans over the surface of a magnetic substance (Pokhil and Moskowitz, 1997; Foss *et al.*, 1998). Companion MFM and AFM images are needed to separate forces due to magnetic leakage fields from atomic Coulomb forces. Although resolutions ≤10 nm are in principle achievable, clear magnetic images like those in **Figure 10** (Pan *et al.*, 2002) are generally seen

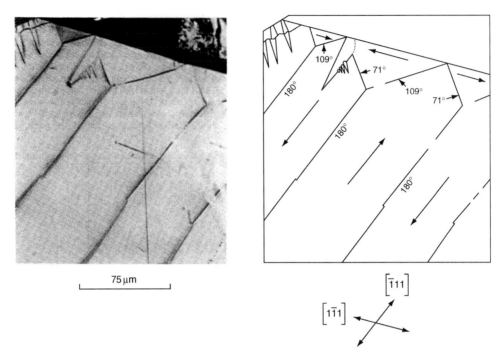

Figure 9 Closure domains observed on a {110} surface in a magnetite single crystal. The boundary (top) is a deep crack. The body and closure domains are magnetized along [$\bar{1}$11] and [1$\bar{1}$1] easy directions, respectively. The 70.5° and 109.5° walls bisect the angles between easy axes, as predicted. Modified from Özdemir Ö, Xu S, and Dunlop DJ (1995) Closure domains in magnetite. *Journal of Geophysical Research* 100: 2193–2209.

only on micrometer scales and for scan heights of ~0.1 μm. The images are of poorer quality when viewed 0.4 μm above the crystals because of divergence of the field lines (Frandsen *et al.*, 2004).

A technique that images domains themselves rather than their boundaries is the magneto-optical Kerr effect (MOKE). Light reflected from a magnetic medium has its plane of polarization rotated a small amount by the surface magnetic field. **Figure 11** illustrates the domain wall displacement process described in Section 8.2.2. Domains with favorably or unfavorably oriented M_s vectors enlarge or shrink with changes in the applied magnetic field. The angle of rotation of the light polarization is extremely small in magnetite; image processing was needed to achieve the contrast seen in **Figure 11** (Heider and Hoffmann, 1992).

Another technique that images the surface magnetization directly rather than through leakage fields is spin-polarized scanning electron microscopy (SEM). It has submicron resolution, and so can detect fine structure that Bitter or MOKE imaging cannot image. Secondary electrons ejected from the surface by the primary electron beam are analyzed at detectors for their spin polarization, which is proportional to the magnetization in the uppermost ≈1 nm of the section. Magnetizations parallel and perpendicular to

the surface can be determined in the same scan, and no corrections for surface topography are needed, unlike MFM. In examples of spin-SEM images given by Haag and Allenspach (1993), titanomagnetite has clear lamellar domain patterns, while titanohematite has rather vague and confused patterns.

A third technique that directly images the in-plane component of **B**, which inside a ferromagnet is essentially **M** (Section 8.1.2), is electron holography transmission electron microscopy (TEM). The resolution is a few tens of nanometers, as **Figure 12** shows (Harrison *et al.*, 2002). Off-axis electron holography measures the amplitude and phase shift of the electron wave that passes through a sample in TEM. From the magnetic phase image, a map of the surface-parallel magnetization is generated. **Figure 12** illustrates a finely exsolved titanomagnetite grain with closely spaced magnetite blocks separated by paramagnetic ulvospinel. The contours trace out magnetic flux lines, which close after linking together a number of neighboring magnetite crystals. The individual grains may be essentially single domains but they interact so strongly that they collectively form superdomains of correlated M_s vectors. Overall flux patterns resemble those of a vortex or a two-domain grain with two closure domains.

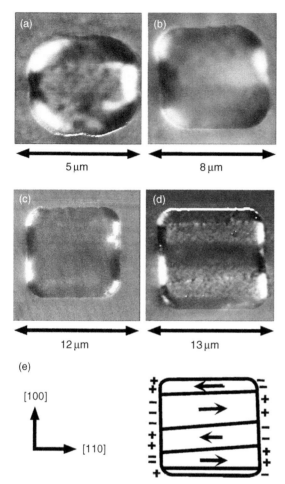

[100]

[110]

(e)

Figure 10 Magnetic force microscope (MFM) images of domains in 250 nm thick epitaxially grown magnetite crystals. The smallest crystal (5 μm) contains three body domains and the largest (13 μm) five domains. Growth on a substrate promotes [110] uniaxial anisotropy and may be the reason for the lack of closure domains. Modified from Pan Q, Pokhil TG, and Moskowitz BM (2002) Domain structures in epitaxial Fe$_3$O$_4$ particles studied by magnetic force microscopy. *Journal of Applied Physics* 91: 5945–5950.

8.2.6 Micromagnetic Modeling

Interacting arrays of magnetic crystals like those in **Figure 12** can be modeled theoretically (Muxworthy *et al.*, 2004) and so can structures within noninteracting crystals. These 'micromagnetic models' are more sophisticated versions of the energy minimization process used in Sections 8.2.3 and 8.2.4 to find domain and wall widths. Instead of imposing a structure of uniformly magnetized domains with straight boundaries, these *ab initio* calculations begin from randomly perturbed spin structures and evolve through

numerous iterations to a minimum energy state. The convergence may be slow but can be speeded by using fast Fourier transforms. If the dynamic behavior of the evolving spin system is included in addition to simple energy minimization, computation time is considerably increased but the solutions are more stable. All relevant energies are considered, including E_{ex}, E_K, E_σ, and E_d, but of these E_d is by far the most time consuming to compute because it requires evaluating the magnetostatic interaction of every pair of spins in the model.

Metastable magnetic states are often found as solutions. These local energy minimum (LEM) states are an impediment if they stall the calculations and prevent the lowest energy or global energy minimum (GEM) state from being found. On the other hand, LEM states have their own interest. They are possible states of the system, albeit less stable than the GEM state. A major change in thinking in magnetism came with the realization that in a system of many particles many different LEM states are occupied as well as the GEM state, and furthermore that each particle can transform its structure from one LEM state to another with the aid of thermal energy. Thus, the equilibrium state of the system is a Boltzmann partition among all LEM states, not the lowest energy state.

8.2.7 Single-Domain Grains

SD grains are of special interest because they have a very strong remanence compared to MD grains and their remanence is extremely stable because to change it requires rotating the grain's moment past a large anisotropy barrier. MD grains by contrast change remanence rather easily by low-energy motion of domain walls. **Figure 10** shows that only very small grains can be SD; these 5–13 μm magnetite crystals contain two or more walls. Grain shape is another factor: an elongated grain may maintain an SD state because of its needle-like shape, which reduces E_d, while a grain of similar size but of more regular shape will nucleate a wall. The maximum size for crystals of a particular mineral to resist subdividing into domains is called the critical SD size d_0.

In reality, this definition of d_0 is ambiguous. **Figure 13** illustrates the various remanent LEM states in a cube (Rave *et al.*, 1998). The structures are of three types: (1) SD, in which $\mathbf{M}_s$ is uniform, except for a 'flowering' of the vector at the corners and edges; (2) lamellar, conventional slab-like domains, appropriate if anisotropy is uniaxial and strong; and (3) vortex, with a continuous curling of the $\mathbf{M}_s$ vector,

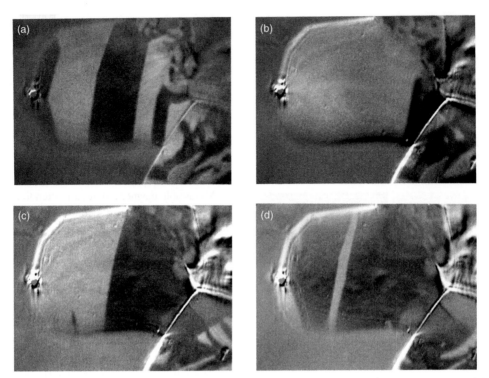

Figure 11 Magneto-optical Kerr effect images of body domains in a ≈70 μm magnetite crystal. A maximum of 6 domains were observed (top left). At saturation remanence, one large domain fills most of the crystal (top right). In reverse fields, a domain wall propagates across the crystal (lower left) and ultimately the grain saturates in the opposite polarity (lower right). Modified from Heider F and Hoffmann V (1992) Magneto-optical Kerr effect on magnetite crystals with externally applied magnetic fields. *Earth and Planetary Science Letters* 108: 131–138.

appropriate if anisotropy is multiaxial and relatively weak. A two-domain structure with two closure domains is a compromise between (2) and (3). But it is possible for an SD flower structure to reverse to a similar state with an intermediate structure of curling spins that resembles a vortex state. Such a reversal is called 'incoherent' in contrast to a 'coherent reversal' in which the structure remains uniform throughout. Thus, d_0 for a uniform remanent state is larger than d_0 for coherent reversal (Enkin and Williams, 1994).

The remanent critical size $(d_0)_r$ is usually quoted but $(d_0)_c$ for coherent reversal is more important paleomagnetically because incoherent reversals greatly reduce the coercive force and the long-term stability of remanence. **Figure 14** illustrates how different $(d_0)_r$ and $(d_0)_c$ are for magnetite grains of different elongations (Newell and Merrill, 1999). Butler and Banerjee (1975a) calculated that $(d_0)_r$ increases by almost an order of magnitude in grains with 5:1 elongation, and **Figure 14** confirms this. Elongated grains can

remain SD in the sense of having a strong and uniform remanence to much larger sizes than equidimensional grains. However, the shape dependence is much weaker for $(d_0)_c$, which changes very little up elongations of 5:1. SD stability – the property of reversing coherently with high coercivity – is hardly dependent on shape at all for geologically reasonable elongations.

8.2.8 Metastable SD Grains

Even $(d_0)_r$ is not as straightforward as it seems. **Figure 15** reveals that starting around 70 nm, energy minimizations with an initial vortex state yield an ultimate low-remanence state (Fabian *et al.*, 1996). The GEM state is vortex. On the other hand, an initial SD (flower) state does not transform during minimization to the GEM state until a much larger size, around 140 nm. In the 70–140 nm size range, the remanent state of a real grain depends on prior history. If the grain began in an SD state, it will remain metastably SD unless

(a) (b)

(c) (d)

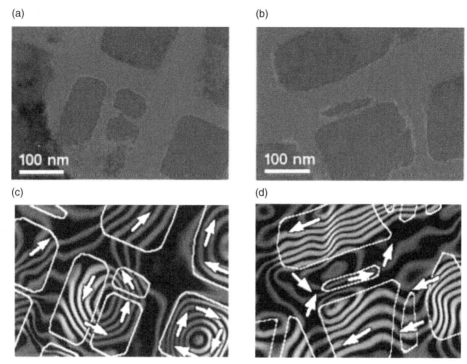

Figure 12 Chemical maps (blue: Fe; red: Ti) of a magnetite-ulvöspinel intergrowth and the magnetic microstructure of the same regions imaged by off-axis electron holography in the transmission electron microscope (TEM). Magnetic field lines link neighboring magnetite crystals across intervening nonferromagnetic ulvöspinel. Isolated crystals have vortex (c) and SD (d) structures but interactions are so strong that groups of particles form 'superdomains'. In (c), the spins of three crystals link in a super-vortex, while in (d), the three SD grains in the middle are flux-coupled with their moments mutually antiparallel. Modified from Harrison RJ, Dunin-Borkowski RE, and Putnis A (2002) Direct imaging of nanoscale magnetic interactions in minerals. *Proceedings of the National Academy of Sciences USA* 99: 16556–16561.

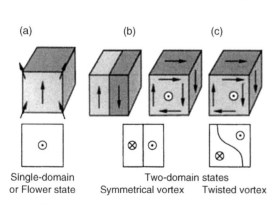

(a)　　(b)　　(c)

Single-domain　　Two-domain states
or Flower state　Symmetrical vortex　Twisted vortex

Figure 13 Schematic drawings of the energetically favoured micromagnetic states in small cubic particles with uniaxial anisotropy. Vortex states evolve from structures with two body and two closure domains, forming a closed internal flux linkage. Modified from Rave W, Fabian K, and Hubert A (1998) Magnetic states of small cubic particles with uniaxial anisotropy. *Journal of Magnetism and Magnetic Materials* 190: 332–348.

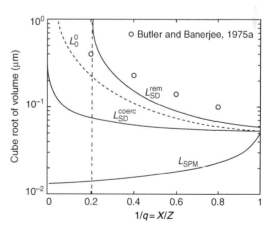

Figure 14 Theoretical stable single-domain and superparamagnetic threshold sizes for magnetites of different sizes and elongations. The usual SD threshold is the upper limit for uniform magnetization in the remanent state, but this is often much larger than the limit for uniform magnetization throughout reversal ('coherent rotation'), which determines coercivity. Modified from Newell AJ and Merrill RT (1999) Single-domain critical sizes for coercivity and remanence. *Journal of Geophysical Research* 104: 617–628.

driven over the energy barrier between LEM states by some external influence (field, long exposure to thermal fluctuations, etc.). The region of possible

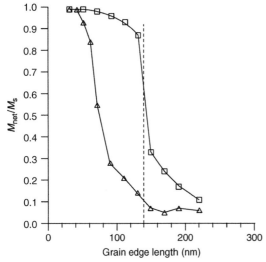

Figure 15 Theoretical micromagnetic states of a uniaxial magnetite grain as a function of grain size. An SD/flower structure with high M/M_s is the minimum energy state at very small sizes and a vortex structure with low M/M_s at large sizes. Between ≈ 70 and ≈ 140 nm, there is a choice between SD and vortex LEM states. The state adopted by the model grain then depends on the initial structure chosen: vortex (triangles) or SD (squares). Modified from Fabian K, Kirchner A, Williams W, Heider F, Leibl T, and Hubert A (1996) Three-dimensional micromagnetic calculations for magnetite using FFT. *Geophysical Journal International* 124: 89–104.

metastable SD behavior is theoretically quite broad in magnetite for all elongations (**Figure 16**). It is noteworthy that the magnetosomes manufactured by magnetotactic bacteria for navigational use have sizes and shapes well outside the 'true' SD region, covering almost exactly the predicted metastable SD region.

Because SD sizes for magnetite are <1 μm, independent experimental proof of metastable SD behavior has been difficult (see, however, Boyd *et al.*, 1984). The same is not true for high-Ti titanomagnetite and pyrrhotite, both of which have $(d_0)_r$ values around 1 μm (**Table 3**). The TM60 (titanomagnetite containing 60 mol.% Ti) grain in **Figure 17** is $>>(d_0)_r$ but has failed to nucleate a domain wall after exposure to a saturating field (Halgedahl and Fuller, 1983). It is metastably SD. However, quite a small reverse field of 1.2 mT causes a wall to nucleate and then rapidly spread across the grain. Thus, as argued in the last section for somewhat different reasons, a grain can have SD-like remanence but relatively low stability.

8.2.9 Pseudo-Single-Domain Grains

The possible choice of LEM states should be fairly narrow for sizes close to $(d_0)_r$ (**Figure 13**). Experimentally, however, grains have a wider choice of remanent LEM states than theory would suggest.

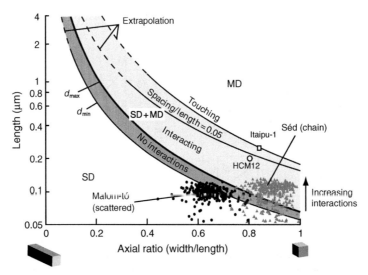

Figure 16 Minimum and maximum critical SD particle lengths, d_{min} and d_{max}, as a function of particle elongation. Strong particle interactions move d_{max} to larger sizes, broadening the metastable SD region. Some interacting magnetosomes (Séd chain) fall within this predicted broader SD region. Modified from Muxworthy AR and Williams W (2006) Critical single-domain/multidomain grain sizes in noninteracting and interacting elongated magnetite particles: Implications for magnetosomes. *Journal of Geophysical Research* 111: B12S12 (doi: 10.1029/2006JB004588).

Table 3   Critical size limits d_0 and d_s for thermally stable single-domain behavior (room temperature, equidimensional grains)

Mineral	Superparamagnetic size d_s (nm)	Single-domain critical size, d_0 (nm)
Iron (αFe)	8–26	17–23
Magnetite (Fe_3O_4)	25–30	50–84
Maghemite (γFe_2O_3)		60
Titanomagnetite, $x = 0.6$ (TM60)	80	200–600
Titanomaghemite		
$\quad x = 0.6, z = 0.4$	50	750
$\quad x = 0.6, z = 0.7$	90	2400
Hematite (αFe_2O_3)	25–30	15 000
Pyrrhotite (Fe_7S_8)		1600

After Dunlop and Özdemir (1997), table 5.1.

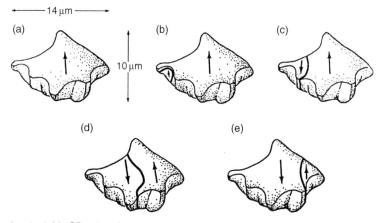

Figure 17   Observed metastable SD saturation remanence state in a TM60 grain (a) and subsequent back-field nucleation (b) and propagation of a domain wall (c–e) to a reversed near-saturation state. Halgedahl SL and Fuller M (1983) The dependence of magnetic domain structure on magnetization state with emphasis on nucleation as a mechanism for pseudo-single-domain behavior. *Journal of Geophysical Research* 88: 6505–6522.

The TM60 grain in **Figure 18** is an example. In the SIRM state, it contains three domains, one of them much larger than the others, but by judicious field cycling, a five-domain state appears. In other TM60 grains, cycled repeatedly to T_C and back in a weak field to produce 'thermoremanent magnetization (TRM)', Halgedahl (1991) observed one to nine domains in the remanent state. These grains were between 20 and 40 μm in size; theoretically, $(d_0)_r$ is 1 μm or less. It is remarkable that such large grains could possess an SD LEM state at all, much less exhibit such an unfavorable state with fair frequency. Such states were never seen after alternating fields (AFs) were applied.

The experimental evidence is clear that the SD threshold is not at all sharp in terms of the variation of properties like coercive force H_c and SIRM M_{rs} with grain size (**Figure 19**). Unlike the precipitous decrease

predicted theoretically for saturation remanence (**Figure 15**), experimental magnetic properties, for magnetite at least, decrease gradually with increasing grain size. The survival above $(d_0)_r$ of remanence intensities and stabilities much higher than those of MD grains is called pseudo-single-domain (PSD) behavior.

Metastable SD grains are one possible source of PSD behavior. Disproportionately large domains in MD grains are another (Halgedahl, 1991; Fabian and Hubert, 1999). The right-hand domain in the grain of **Figure 18** accounts for more than half the area viewed and presumably for more than half the SIRM. Grains like these with domain walls far from a uniform distribution could certainly explain why the remanence does not plummet at the SD threshold $(d_0)_r$. The continuity of coercive force and other measures of stability is not so obvious. As **Figure 17** demonstrates, even a grain with a

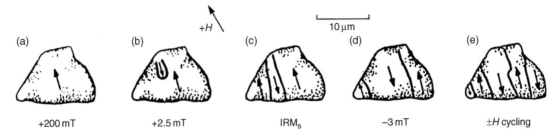

Figure 18 Nucleation of a domain wall in a TM60 grain in a small forward field (b) and alternative remanence states (c, e). The magnetization is dominated by one large domain in (c) and (d); a single wall displacement transforms one state into the other, and reverses the grain's moment. Modified from Halgedahl SL and Fuller M (1983) The dependence of magnetic domain structure on magnetization state with emphasis on nucleation as a mechanism for pseudo-single-domain behavior. *Journal of Geophysical Research* 88: 6505–6522.

metastable SD remanent state has fairly low coercivity. Either a new domain wall (or walls) will nucleate and spread through the grain or existing walls will be driven by the self-demagnetizing field H_d to equalize the size of domains.

8.2.10 Superparamagnetism

Any magnetic structure can become unstable and transform if there is sufficient thermal energy available. The resulting thermally activated changes in magnetization are at the heart of TRM acquisition and the thermal demagnetization of remanences of all types (Sections 8.4.3 and 8.5.3). If a system contains relatively few coupled spins, the probability of correlated thermal impulses (spin waves) sufficient to rotate SD moments or effect other changes in structure (e.g., SD → vortex or two-domain) may become significant even at room temperature. The size below which the anisotropy barrier to magnetization changes can easily be surmounted by thermal energy on a specified (usually laboratory) timescale is called the superparamagnetic (SP) threshold d_s. 'Superparamagnetism' refers to the thermal equilibrium orientation of grain moments in an applied magnetic field, which is analogous to paramagnetism except that grain moments contain hundreds or thousands of coupled atomic moments, hence 'super'.

In practical terms, d_s represents the lower limit to stable SD behavior. Superparamagnetism is not a new state of matter but a limiting case of ferromagnetism or ferrimagnetism. Exchange coupling is not overcome (that occurs at T_C) but remanence and coercivity drop to zero on any reasonable timescale of observation, causing a drop-off in H_c and M_{tr} at grain sizes smaller than those in **Figure 19**. To make a rough estimate of

d_s for equidimensional SD grains, we can equate the thermal energy of $\approx 25kT$ available to the system over times of seconds to minutes to the anisotropy barrier VK that must be surmounted for reversal (normally coherent reversal for very small SD grains):

$$d_s = (25kT/K)^{1/3} \qquad [13]$$

K can be due to any combination of crystalline, magnetoelastic, and shape anisotropies.

Figure 14 indicates that d_s is essentially equal to $(d_0)_r$ or $(d_0)_c$ for equidimensional magnetite grains. Elongated grains have a significant but still narrow stable SD range. It is this very restricted SD range that makes PSD remanence so important for magnetite. In minerals less magnetic than magnetite, there is a larger difference between d_s and d_0 (**Table 3**).

8.3 Magnetic Minerals and Their Properties
8.3.1 Thermomagnetic Curves, Curie Temperatures, and SD Ranges

Thermomagnetic curves of saturation magnetization M_s versus temperature T (**Figure 20**) are often used in determining magnetic mineralogy. The $M_s(T)$ function itself plays a major role in TRM acquisition and thermal demagnetization of a particular mineral (Sections 8.5.3 and 8.6.1). The Curie temperature T_C marks the ferromagnetic → paramagnetic transition, at which $M_s → 0$.

Iron, hematite, and magnetite have T_C values in the range 580–765°C (**Table 4**). Natural remanent magnetizations (NRMs) of these minerals have high thermal stabilities, sometimes remaining essentially unchanged in heating to ∼10–20°C

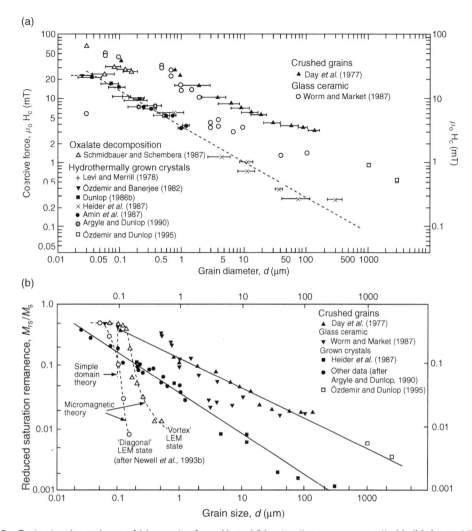

Figure 19 Grain-size dependence of (a) coercive force H_c and (b) saturation remanence ratio M_{rs}/M_s for equidimensional magnetite grains of various origins. Rather than the rather abrupt decrease at the SD threshold predicted theoretically, gradual decreases in properties are observed over several decades of grain diameter. Modified from Dunlop DJ (1986) Hysteresis properties of magnetite and their dependence on particle size: A test of pseudo-single-domain remanence models. *Journal of Geophysical Research* 91: 9569–9584, Dunlop DJ (1995) Magnetism in rocks. *Journal of Geophysical Research* 100: 2161–2174 and Argyle KS and Dunlop DJ (1990) Low-temperature and high-temperature hysteresis of small multidomain magnetites (215-540 nm). *Journal of Geophysical Research* 95: 7069–7083.

below T_C except for reversible decreases dictated by $M_s(T)$. Among other things, this means that they remain sources of magnetic anomalies even when they occur at depths of tens of kilometers in planetary lithospheres. Pyrrhotite has an intermediate T_C of 320°C and intermediate thermal stability. The 300°C isotherm in the crust marks the base of a magnetic layer due to pyrrhotite. High-Ti titanomagnetite ($Fe_{2.4}Ti_{0.6}O_4$ or TM60) and goethite (FeOOH) have $T_C \approx 120$–150°C and are effective magnetic sources only in the uppermost crust.

Relative magnitudes of M_s are also important. For one thing, they determine the critical SD size d_0 (**Figure 21**). Because demagnetizing energy E_d is proportional to M^2 (eqn [7]), grains of weakly magnetic minerals like hematite and goethite ($M_s \approx 2$ kA m^{-1}) have much lower E_d than similar size grains of strongly magnetic minerals like magnetite ($M_s = 480$ kA m^{-1}) or iron ($M_s = 1715$ kA m^{-1}). They remain SD, that is, magnetized to saturation ($M = M_s$), at large sizes (up to 15 μm for hematite), while magnetite subdivides into domains above ~ 0.07 μm and iron above ~ 0.02 μm. All three minerals have similar SP

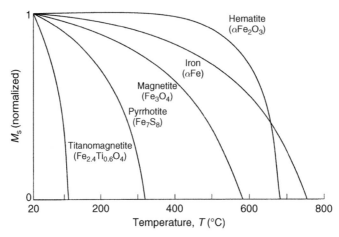

Figure 20   Normalized $M_s(T)$ dependences for five common magnetic minerals. Hematite is unique in having a 'blocky' $M_s(T)$ curve with a steep descent just below the Curie temperature.

Table 4   Some magnetic properties of the more important magnetic minerals

Mineral	M_s (kA m^{-1})	T_C (°C)	K_1 (J m^{-3})	$\lambda \times 10^{-6}$
Iron (αFe)	1715	765	4.8×10^4	-21 (λ_{111})
				21 (λ_{100})
Magnetite (Fe$_3$O$_4$)	480	580	-1.35×10^4	72.6 (λ_{111})
				-19.6 (λ_{100})
Maghemite (γFe$_2$O$_3$)	380	590–675	-4.65×10^3	-8.9
Titanomagnetite, $x = 0.6$	125	150	2.02×10^3	142.5 (λ_{100})
				95.4 (λ_{111})
Hematite (αFe$_2$O$_3$)	$\approx$2.5	675	1.2×10^6 (c-axis)	8
Pyrrhotite (Fe$_7$S$_8$)	$\approx$90	320	$\approx$10^4 (c-plane)	<10

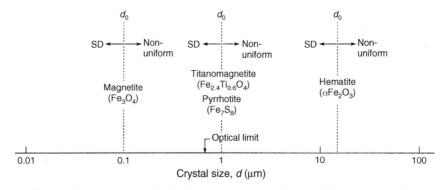

Figure 21   Approximate SD remanence threshold sizes for magnetite, TM60/pyrrhotite, and hematite.

thresholds, ~0.025 μm. Therefore much of the broad size spectrum of hematite in nature falls in the SD range and has large M_{rs}/M_s (Section 8.4.1). Natural examples of SD magnetite are not so common and require a mechanism for producing needle-like grains or subgrains with a very restricted size distribution (see **Figures 14** and **16** and Sections 8.3.2, 8.3.13, and 8.3.14). Naturally occurring SD iron is rare. On the Moon, only impact breccias with a specific range of annealing temperatures develop SD iron (Section 8.3.12). With this one exception, lunar rocks and meteorites contain MD iron with very mobile domain walls – the worst imaginable material for paleomagnetic recording.

Magnetic signal, whether NRM or induced magnetization, depends on more than domain state. The high M_{rs}/M_s and broad SD range of minerals like hematite and goethite are offset by their weak M_s. A small fraction of strongly magnetized magnetite of SD/PSD size will usually dominate the NRM even if a rock also contains large amounts of hematite. On the other hand, thermal demagnetization will ultimately erase the magnetite remanence, leaving a substantial part of the hematite NRM intact above 600 °C.

8.3.2 Magnetite

The most important terrestrial magnetic minerals are oxides of iron and titanium. There are three series: (1) stoichiometric titanomagnetites with spinel structure, solid solutions between end-members magnetite ($Fe^{2+}Fe_2^{3+}O_4$) and ulvöspinel ($Fe_2^{2+}Ti^{4+}O_4$); (2) nonstoichiometric (oxidized) titanomagnetites or titanomaghemites, in which some Fe^{2+} ions migrate to the surface where they are converted to Fe^{3+}, leaving ordered vacancies in the spinel lattice; (3) titanohematites (or hemoilmenites) with rhombohedral structure, solid solutions between hematite ($\alpha Fe_2^{3+}O_3$) and ilmenite ($Fe^{2+}Ti^{4+}O_3$). The compositions of all three series are conveniently represented in a ternary $FeO–Fe_2O_3–TiO_2$ diagram (**Figure 22**).

Magnetite is the single commonest magnetic mineral on Earth. It occurs as a primary magmatic mineral in plutonic rocks and as an end product of either deuteric oxidation or maghemitization followed by inversion (see below) in mafic lavas, dikes, and sills. It is equally common as a primary or secondary mineral in sedimentary and metamorphic rocks.

Magnetite is cubic with inverse spinel structure (Dunlop and Özdemir, 1997, figure 3.4). The O^{2-} ions form a slightly distorted cubic-close-packed lattice with Fe^{2+} and Fe^{3+} cations in interstitial sites. Each cation is surrounded by either four O^{2-} ions in tetrahedral coordination (A sites) or six O^{2-} ions in octahedral coordination (B sites). A unit cell contains 32 O^{2-} ions, 8 Fe^{2+} on B sites, and 16 Fe^{3+} shared equally between A and B sites.

The exchange coupling between Fe cations is indirect (superexchange), with two O^{2-} 2p orbitals providing the linkage between the Fe 3d orbitals. The tetrahedral bonds are in <111> directions and the octahedral bonds are in <100> directions.

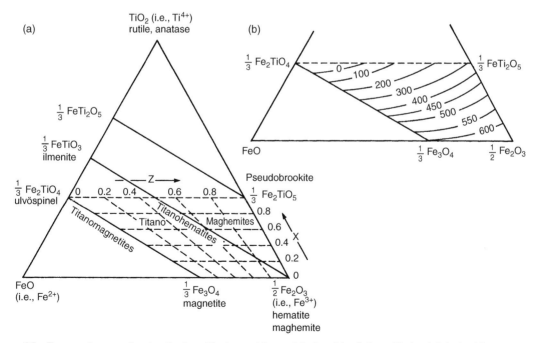

Figure 22 Ternary diagram showing the iron–titanium oxides and their solid solutions. Horizontal dashed lines are titanomaghemite oxidation trends at constant Ti content x; cross-cutting dashed lines are contours of oxidation parameter z. Titanomaghemite Curie temperatures are contoured over the entire oxidation field (upper right). Modified from O'Reilly W (1976) Magnetic minerals in the crust of the Earth. *Reports on Progress in Physics* 39: 857–908.

Exchange coupling is relatively ineffectual for bond angles near 90°. For this reason, A–A and B–B exchange coupling is weak, and the dominant exchange interaction is the negative A–B coupling, which gives magnetite its ferrimagnetic character. The A and B sites can be thought of as two different magnetic sublattices, with oppositely directed cation moments.

The inverse spinel structure, with trivalent cations equally partitioned between A and B sites and the Fe^{2+} ions confined to the B sublattice, produces a spontaneous magnetic moment at low temperatures of $4.1\mu_B$ (Bohr magnetons) per formula weight of Fe_3O_4, close to the theoretical moment of Fe^{2+} of $4\mu_B$. This translates into a spontaneous magnetization $M_s \approx 480\,\text{kA m}^{-1}$ at ordinary temperatures and about half this amount at 500°C.

Below the 'Verwey transition' at $T_V \approx 120\,\text{K}$, the magnetite lattice distorts slightly from cubic to monoclinic. The main magnetic effect is a greatly increased magnetocrystalline anisotropy below T_V. In addition, at a temperature T_K slightly above T_V, the easy axes switch from <111> to <100>. This transition is called the 'isotropic point' because the first magnetocrystalline anisotropy constant K_1 momentarily passes through zero in both heating and cooling. The effect on SD remanence is usually minor (although not always) because the anisotropy barrier is mainly due to shape anisotropy and M_s is continuous at both T_V and T_K. Unpinning of domain walls can be considerable, however. This effect forms the basis of 'low-temperature demagnetization (LTD)', described in Section 8.5.2.

Magnetite, being a cubic mineral, prefers equant crystal forms (cubes, octahedra, dodecahedra). Crystals of these forms have little or no shape anisotropy, and their magnetocrystalline anisotropy leads to modest H_c values of 10–15 mT. These values are not much higher than typical coercivities for MD magnetite. Other paleomagnetically undesirable traits of equidimensional crystals are their narrow SD ranges (**Figure 14**) and low resistance to LTD when cooled through T_K and T_V.

Fortunately, there are several natural mechanisms that produce at least limited quantities of elongated magnetite crystals or chains of crystallites. First is the process of high-temperature oxidation, which generates a lath-like intergrowth of ilmenite lamellae on conjugate magnetite {111} planes. Ilmenite extracts Ti from the titanomagnetite, leaving a Ti-poor phase ('near-magnetite' with T_C usually $\geq 500°C$) in the form of elongated prisms sandwiched between the

ilmenite lamellae. The second process is precipitation of Fe at high temperatures from the lattices of silicate minerals like plagioclase, pyroxenes, amphiboles, micas, and olivine (Section 8.3.13). In some cases, crystallographically oriented nanometer-size needles of magnetite are formed, most commonly in clinopyroxene and plagioclase (Feinberg *et al.*, 2005). A third mechanism is biogenic production of crystallographically aligned nanocrystals of magnetite in elongated chains, for example, by magnetotactic bacteria (Section 8.3.14). The individual SD-size crystals interact strongly, so that the chains act as collective SD particles, although reversing noncoherently.

8.3.3 Titanomagnetites

There is a complete solid solution between magnetite and ulvöspinel at temperatures well above the magnetite Curie point. In principle, any intermediate composition $Ti_xFe_{3-x}O_4$ could be preserved metastably at ordinary temperatures by quenching in an oxygen-poor environment from the melt. In massive flows, stoichiometric titanomagnetites may survive essentially unoxidized for millions of years. In submarine pillow lavas, however, maghemitization begins almost immediately after extrusion, probably aided by iron-leaching bacteria in the seawater (Carlut *et al.*, 2007). For this reason, linear magnetic anomalies over the oceans are strongest over spreading ridges and rapidly decrease in amplitude away from the ridges.

Titanomagnetites have the same inverse spinel structure as magnetite but the A–B exchange interaction weakens and the Curie temperature therefore falls, more or less linearly, with increasing Ti content x. For compositions beyond $x \approx 0.8$, T_C is below room temperature and the oxides are paramagnetic. The replacement of Fe on octahedral sites by nonmagnetic Ti^{4+} causes the spontaneous magnetization to decrease as x increases. The low-temperature spontaneous moment drops almost linearly from $4\mu_B$ in magnetite to 0 in ulvöspinel. These variations of T_C and M_s are useful in determining x. However, they are not diagnostic because cations like Al, Mg, and Mn often substitute for Fe in the lattice.

On Earth, two compositions are commonly found as stoichiometric titanomagnetites. $Fe_{2.4}Ti_{0.6}O_4$ (called TM60 because it contains 60 mol.% Ti) is the primary oxide in mid-ocean ridge basalts (MORBs) and in subaerial basalts (plumes, large igneous provinces). Actual compositions range from TM50 to TM70. The Curie point is $\approx 150°C$ and

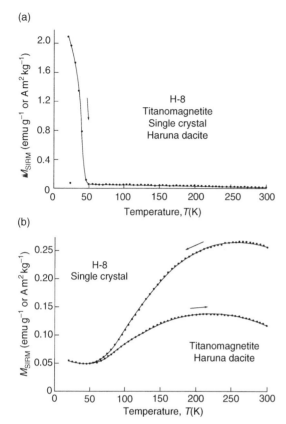

Figure 23 Zero-field temperature cycling of 20 K SIRM (a) and 300 K SIRM (b) of a crystal of iron-rich titanomagnetite ($x \approx 0.15$) from Mt. Haruna, Japan. The isotropic point around 45 K is well marked in (a) but less sharply defined in (b). Modified from Özdemir Ö and Dunlop DJ (2003) Low-temperature behavior and memory of iron-rich titanomagnetites (Mt. Haruna, Japan and Mt. Pinatubo, The Philippines). *Earth and Planetary Science Letters* 216: 193–200.

$M_s \approx 125$ kA m^{-1}, about one-fourth that of magnetite (**Table 4**). Shape anisotropy is correspondingly weaker than in magnetite and magnetocrystalline anisotropy (with positive K_1 and <100> easy axes) is also less than that of magnetite. The magnetostriction constants λ_{100} and λ_{111} are 2–3 times those of magnetite, however. The coercivity of TM60 is in large part magnetoelastic. This stress sensitivity of TM60 makes domain observations difficult.

TM10–TM30 occur in andesitic and dacitic volcanic rocks. The iron-rich titanomagnetites studied by Özdemir and Dunlop (2003) had $T_C = 460$–$490°C$, indicating $x \approx 0.10$–0.15. Even 10 mol.% Ti was enough to depress T_K to 40–55 K (**Figure 23**), making LTD an inconvenient treatment (since T_K is well below the liquid nitrogen temperature, 77 K). Sixty mol.% Ti suppresses the transition, so that remanences produced at either T_0 or low T

change continuously between 50 and 200 K (Moskowitz *et al.*, 1998).

8.3.4 Maghemite and Titanomaghemites

Partially oxidized titanomagnetites are commoner in nature than stoichiometric oxides. They are called 'titanomaghemites' by association with 'maghemite' (γFe_2O_3), the fully oxidized product of magnetite low-temperature oxidation. Maghemitization is an iron-leaching process, usually occurring in the presence of water, that leaves the spinel lattice intact but one-sixth of the octahedral sites vacant. M_s of maghemite is therefore somewhat lower than that of magnetite although T_C is higher, intermediate between that of magnetite and rhombohedral hematite (αFe_2O_3).

Lattice mismatch between the oxidized surface and unoxidized core of partially oxidized (or cation-deficient) magnetite produces a strained transition region between them. The resulting internal stress leads to coercivities higher than those of either γFe_2O_3 or Fe_3O_4 separately. The Verwey transition is also broadened (**Figure 24**), a diagnostic fingerprint of partial oxidation of magnetite (Özdemir *et al.*, 1993). Maghemite itself has no Verwey transition. In nature, maghemitization often accompanies soil formation. In rocks, it suggests a nonprimary NRM resulting from chemical processes (Section 8.6.3).

Although even a secondary NRM can be of interest paleomagnetically, maghemite NRM is difficult to work with. At temperatures of 250–550°C, maghemite inverts to αFe_2O_3 by restacking the lattice and loses most of its magnetism. Thermal demagnetization of a maghemitized rock is usually a losing proposition. The mineralogy changes irreversibly before the NRM components can be 'cleaned' and isolated (Sections 8.5.3–8.5.5), except when aluminum is incorporated into the lattice (Wilson, 1961), leading to inversion temperatures as high as 700°C. Aluminum-substituted maghemites are common in soils.

Titanomaghemites formed by oxidation of TM60 have M_s values that decrease but T_C values that increase with increasing oxidation parameter z (**Figure 25**). Coercive force H_c at first rises, then falls sharply as synthesized SD TM60 is oxidized (**Figure 26**). Thus low-temperature oxidation of TM60 in MORB may ultimately destabilize the NRM. This, rather than decreasing M_s, is probably

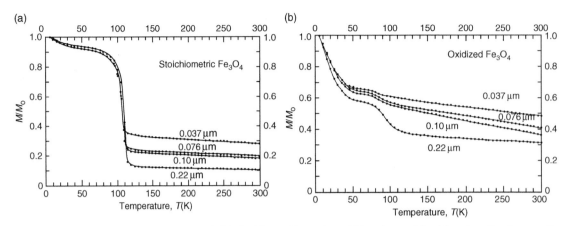

Figure 24 Zero-field warming curves of 5 K SIRM for (a) stoichiometric and (b) surface oxidized (maghemitized) magnetites. The Verwey transition is blurred and ultimately suppressed by oxidation. Modified from Özdemir Ö, Dunlop DJ, and Moskowitz BM (1993) The effect of oxidation on the Verwey transition in magnetite. *Geophysical Research Letters* 20: 1671–1674.

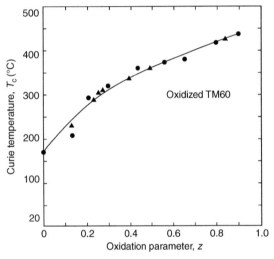

Figure 25 Curie temperature as a function of maghemitization for TM60. Data: circles, Özdemir Ö and O'Reilly W (1982a) Magnetic hysteresis properties of synthetic monodomain titanomaghemites. *Earth and Planetary Science Letters* 57: 437–447.; triangles, Brown K and O'Reilly W (1988) The effect of low-temperature oxidation on the remanence of TRM-carrying titanomagnetite $Fe_{2.4}Ti_{0.6}O_4$. *Physics of the Earth and Planetary Interiors* 52: 147–154.

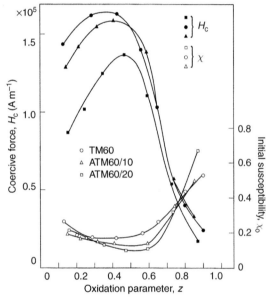

Figure 26 Coercive force and initial susceptibility of SD pure and Al-substituted TM60 as a function of maghemitization. Modified from Özdemir Ö and O'Reilly W (1982a) Magnetic hysteresis properties of synthetic monodomain titanomaghemites. *Earth and Planetary Science Letters* 57: 437–447.

why magnetic anomaly amplitudes decay within at most 1 Ma after the formation of MORB.

If titanomaghemites are heated, by burial beneath later flows on the seafloor for example, they become unstable and 'invert'. This is the analog process to inversion of $\gamma Fe_2O_3 \rightarrow \alpha Fe_2O_3$. However, the inversion products are more varied and so are their intergrowth textures. The usual phase assemblage is a Ti-poor titanomagnetite ('near-magnetite') and a Ti-rich rhombohedral phase ('near-ilmenite'). The rhombohedral phase grows in {111} planes of the spinel, eventually producing a trellis-like texture of intergrown phases very like that resulting from 'high-temperature oxidation' (next section).

The signature of inversion is an irreversible thermomagnetic curve, in air or in vacuum (**Figure 27**).

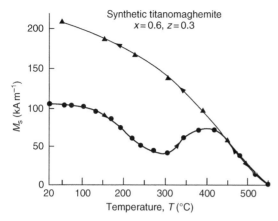

Figure 27 Characteristic irreversible thermomagnetic curve of a partially oxidized TM60. Modified from Özdemir Ö and O'Reilly W (1982a) Magnetic hysteresis properties of synthetic monodomain titanomaghemites. *Earth and Planetary Science Letters* 57: 437–447.

Above 250°C but before the titanomaghemite reaches its T_C, M_s begins to increase and only drops to zero above 500°C at the Curie temperature of a considerably more Fe-rich spinel. During cooling, M_s is much higher than during heating, sometimes by a factor of 4–5. This process is capable of producing a magnetic layer in the seafloor extending to greater depth (the 500°C isotherm) than fresh MORB.

8.3.5 High-Temperature Oxidation

Subaerial basalts exhibit a wide range of oxidation states in the interior of thick flows (**Figure 28**). Oxidation occurs quite rapidly in the melt, producing a phase assemblage similar (for moderate oxidation) to that accompanying inversion of titanomaghemite. However, cation-deficient phases are not part of the assemblage, except possibly during initial single-phase oxidation. The observed phases represent equilibrium under high-temperature oxidizing conditions if cooling is not too rapid.

The 'oxyexsolution' intergrowth texture (**Figure 29**), if sufficiently finely divided, can generate magnetite rods of SD size. Although individual magnetite crystals are separated by nonferromagnetic ilmenite, they do not behave independently. The strong interactions among crystals are similar to those shown in **Figure 12** for unoxidized titanomagnetite exsolved into magnetite + ulvöspinel.

For sufficiently high oxygen fugacities and/or long-term heating, the ultimate phase assemblage is hematite + pseudobrookite (Fe_2TiO_5) and rutile (TiO_2). The hematite (next section) imparts long-term

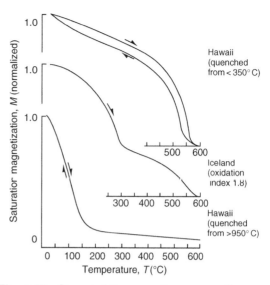

Figure 28 Characteristic vacuum thermomagnetic curves of natural stoichiometric and high-temperature oxidized titanomagnetites, showing a single TM60 Curie point (lower), a single near-magnetite Curie point (upper, high oxidation state), and both Curie points (middle, intermediate oxidation state). Modified from Grommé CS, Wright TL, and Peck DL (1969) Magnetic properties and oxidation of iron-titanium oxide minerals in Alae and Makaopuhi lava lakes, Hawaii. *Journal of Geophysical Research* 74: 5277–5293.

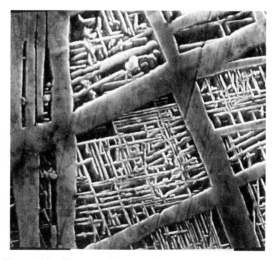

Figure 29 Electron micrograph (magnification ×10 400) of the oxyexsolution intergrowth texture of high-temperature oxidized titanomagnetite. Modified from Gapeyev AK and Tsel'movich VA (1983) Microstructure of synthetic titanomagnetite oxidized at high partial pressures of oxygen (translated from Russian). Izvestiya. *Physics of the Solid Earth* 19: 983–986.

and high-temperature stability. Pseudobrookite and especially rutile testify to terminal oxidation. The presence of pseudobrookite, which cannot form below 585°C, also guarantees that oxidation did not continue below T_C of magnetite. Any magnetite present carries a primary TRM.

8.3.6 Hematite

Hematite (αFe_2O_3), with $T_C = 675°C$, is the fully oxidized end product of magnetite. It is both thermally and chemically stable. Its rhombohedral crystals nucleate and grow topotactically in magnetite {111} planes. The quantity of hematite on the surface of the Earth and other planets with oxidizing atmospheres, notably Mars, far exceeds that of magnetite but hematite is seldom dominant magnetically because of its small M_s.

At ordinary temperatures, hematite is a slightly imperfect antiferromagnet with $M_s \approx 2.5\,kA\,m^{-1}$. The spin sublattices lie in the c-plane, also called the basal plane because of the tabular form of single crystals. Within the c-plane, the sublattices are slightly tilted or canted with respect to each other, giving rise to a tiny but persistent magnetic moment perpendicular to the spins (**Figure 2**). This moment rotates relatively freely in the c-plane but not at all out of the plane until the crystal is cooled below the Morin transition $T_M \approx 260\,K$. At this point, the spin sublattices rotate to the c-axis, and with this new crystal symmetry, spin-canting in principle disappears.

Curiously enough, a fraction (usually but not always small) of hematite's ferromagnetic moment survives below T_M. This low-temperature moment is usually ascribed to either chemical impurities or structural imperfections, hence the name 'defect' moment. Whatever its origin, the defect moment serves to renucleate a fractional memory of any remanence that existed above T_M when the crystal is reheated to room temperature. It has only recently been discovered that cycling through T_M is incidental to memory: a remanence produced 'below' T_M generates a much larger remanence, in the same direction, above T_M (Özdemir and Dunlop, 2005). The importance of these observations is that hematite cycled repeatedly through T_M in nature, for example on Mars, will retain a fraction of its original NRM.

More vital in most applications is the stability of hematite NRM at and above room temperature. Magnetocrystalline anisotropy in the c-plane is small but magnetoelastic anisotropy leads to high coercivities in SD crystals (Özdemir and Dunlop, 2002). MD (>15 μm) hematites are less frequent in

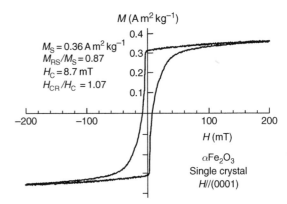

Figure 30 Typical MD hysteresis curve of a large hematite crystal. The intrinsic magnetic properties are revealed because hematite is weakly magnetic and lacks the self-demagnetizing fields that control the hysteresis response of strongly magnetic minerals like magnetite (see **Figure 38(a)**). Modified from Özdemir Ö and Dunlop DJ (2005) Thermoremanent magnetization of multidomain hematite. *Journal of Geophysical Research* 110: B09104 (doi:10.1029/2005JB003820).

most rocks and sediments but they too have adequate coercivities to preserve long-term NRM (**Figure 30**). The $M_s(T)$ curve of hematite is unusually 'blocky' in form (**Figure 20**), resulting in TRM blocking temperatures clustered within 20–30°C of T_C for all but the finest nanometer-size grains. Thus hematite has unparalleled stability against changes in both field and temperature.

8.3.7 Titanohematites (Hemoilmenites)

Titanohematite $Ti_yFe_{2-y}O_3$ is a rhombohedral mineral with bulk composition intermediate between hematite and ilmenite. As with titanomagnetites, single-phase titanohematites of intermediate composition can only be preserved metastably at room temperature by quenching from the melt. The cation distribution ideally is ordered (Fe^{2+} and Ti^{4+} ions segregated on alternate c-planes) for $y \geq 0.5$ and disordered (Fe^{2+} and Ti^{4+} ions in equal numbers on all c-planes) for $y \leq 0.5$.

The ordered distribution gives rise to ferrimagnetism, but T_C drops below room temperature if $y > 0.7$ (**Figure 31**), while for $y < 0.7$ cation ordering is only partial. Ferrimagnetic hemoilmenites are uncommon in nature. Even andesites and dacites with intermediate Fe/Ti ratios usually contain more titanomagnetites than titanohematites.

Disordered titanohematites are weakly ferromagnetic antiferromagnets like hematite. The unique

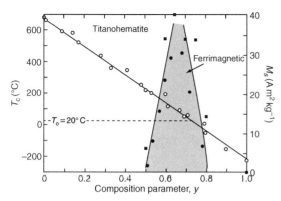

Figure 31 Curie temperature and spontaneous magnetization of titanohematites as a function of Ti content y. T_C drops below room temperature around $y = 0.7$. The ferrimagnetic region (shaded) is therefore rather restricted. Modified from Hunt CP, Moskowitz BM, and Banerjee SK (1995) Magnetic properties of rocks and minerals. In: Ahrens TJ (ed.) *Rock Physics and Phase Relations: A Handbook of Physical Constants*, vol. 3, ch. 14, pp. 189–204. Washington, DC: American Geophysical Union.

phenomenon of 'self-reversing TRM' can occur if the bulk hemoilmenite composition is in the ferrimagnetic range ($y \geq 0.5$) and the oxides are exsolved on a microscale into intergrown ordered and disordered phases (Harrison *et al.*, 2000). The phases are negatively exchange coupled. When cooled in a small magnetic field, the weakly magnetic disordered phase is the first to acquire TRM (in the field direction) because of its higher T_C. With further cooling through T_C of the ordered phase, this strongly ferrimagnetic phase acquires a much larger remanence negatively coupled to the TRM of the first phase, essentially ignoring the external field. The net result is a self-reversed TRM (Prévot *et al.*, 2001; Lagroix *et al.*, 2004).

Since its discovery by Uyeda (1958), self-reversal has fascinated rock magnetists. The fact is that it is a rarity in nature. Samples from Uyeda's type locality, the Haruna dacite, seldom show full self-reversals (Ozima and Funaki, 2003). This may be because the compositions or interlayering of the hemoilmenites are not optimal for strong coupling or because titanomagnetites dominate the whole-rock TRM.

A recently discovered phenomenon in slowly cooled rocks where exsolution of titanohematite has gone to completion is 'lamellar magnetism' (Robinson *et al.*, 2002). The bulk composition of these oxides is in the antiferromagnetic range ($y \leq 0.5$). The finer the scale of the exsolution lamellae, the stronger is the magnetism, which is

the outcome of imbalance between oppositely directed Fe^{2+} spins at interfaces between lamellae (**Figure 32**). According to Robinson *et al.*, M_s values as large as 55 kA m^{-1} are possible by this mechanism. What is certain is that titanohematites with compositions around $y = 0.25$ can impart NRMs as large as 30 A m^{-1} to anorthositic and noritic rocks (McEnroe *et al.*, 2004), on the same order as the TRM of fresh MORB.

8.3.8 Iron Oxyhydroxides

The most important natural hydrous iron oxide is orthorhombic goethite (αFeOOH), a common weathering product and precursor to hematite in sediments and soils. Goethite is antiferromagnetic with a weak superimposed ferromagnetism but its Curie temperature is only $120°C$. Unlike hematite, the sublattice magnetizations lie along the c-axis and so does the weak ferromagnetism. TRM produced parallel to the c-axis is 20 times larger than TRM perpendicular to the c-axis (Özdemir and Dunlop, 1996). Even the c-axis TRM is weaker than TRM of hematite but it is extremely hard. Typical coercivities are several teslas, higher even than those of hematite and far beyond the capabilities of AF demagnetizers (Rochette *et al.*, 2005a). Blocking temperatures of TRM are usually above $90°C$, enough to preserve goethite NRM in nature but very low compared to those of other minerals, making goethite the first mineral to be cleaned in thermal demagnetization.

Goethite may dehydrate under natural conditions, usually with mild heating, to form hematite. Goethite's needle-like crystal form is preserved but each original crystal consists of many hematite crystallites. The effective grain size is often below 30 nm, the superparamagnetic threshold of hematite. Hematite formed in this way carries no NRM. Magnetic parameters useful for discriminating between hematites and goethites of different grain sizes have been developed by Maher *et al.* (2004).

Ferrihydrite is a poorly crystalline iron hydroxide or hydrous ferric oxide which lacks long-range order. It is widespread in aquatic sediments, where it generally forms with the participation of Fe-oxidizing bacteria, for example, near hydrothermal vents on mid-ocean ridges. In terrestrial soils such as tropical laterites, ferrihydrite is a common intermediate weathering product of iron oxides or sulfides, later recrystallizing to goethite or hematite. At room temperature, ferrihydrite has an extremely weak

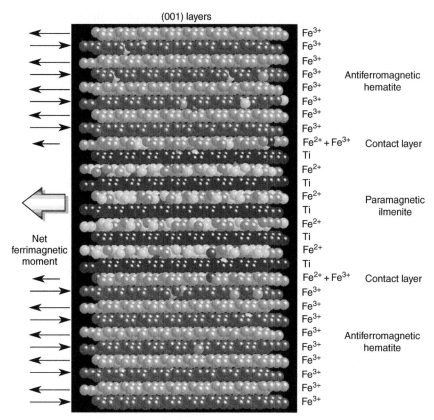

(001) layers

Fe³⁺
Fe³⁺
Fe³⁺
Fe³⁺   Antiferromagnetic
Fe³⁺   hematite
Fe³⁺
Fe³⁺
Fe³⁺
Fe²⁺ + Fe³⁺   Contact layer
Ti
Fe²⁺
Ti
Fe²⁺
Ti   Paramagnetic
Fe²⁺   ilmenite
Ti
Fe²⁺
Ti
Fe²⁺ + Fe³⁺   Contact layer
Fe³⁺
Fe³⁺
Fe³⁺
Fe³⁺   Antiferromagnetic
Fe³⁺   hematite
Fe³⁺
Fe³⁺
Fe³⁺

Net
ferrimagnetic
moment

Figure 32   Monte Carlo simulation of cation ordering and resulting magnetic signal of hematite-rich and ilmenite-rich phases in exsolved titanohematite. 'Lamellar magnetism' is associated with boundary layers between the exsolved phases. Modified from Robinson P, Harrison RJ, McEnroe SA, and Hargraves RB (2002) Lamellar magnetism in the haematite-ilmenite series as an explanation for strong remanent magnetization. *Nature* 418: 517–520.

ferromagnetism, with a saturation remanence $M_{rs} \approx 10^{-4} \, \text{A m}^2 \, \text{kg}^{-1}$ (Pannalal *et al.*, 2005). This is three orders of magnitude less than that of hematite or goethite. However, the grain-size dependence of ferrihydrite's magnetic behavior is dramatic (Guyodo *et al.*, 2006), and even 1–2% silica substitution produces highly ordered ferrihydrite with a Curie point above room temperature (Berquó *et al.*, 2007). This is of importance because ferrihydrite is the ubiquitous first product of silicate diagenesis and silica substitution is common in soils. Ferrihydrite's main interest is as a tracer of biogeochemical cycling of Fe, not as a major remanence carrier.

Also found in soils and sediments is lepidocrocite (γFeOOH), which is antiferromagnetic at low temperatures and paramagnetic at room temperature. Its main interest comes from the fact that it dehydrates when heated above $\approx 200\,°\text{C}$ to strongly magnetic maghemite (Özdemir and Dunlop, 1993b; Gendler *et al.*, 2005). This poses no problem in zero-field heating but could lead to a substantial chemical remanent magnetization (CRM) in experiments like Thellier paleointensity determination (Section 8.5.5) which involve in-field heating–cooling cycles. However, materials containing lepidocrocite are unlikely to be used in paleointensity work.

8.3.9   Iron Sulfides

Pyrrhotite ($Fe_{1-x}S$) is common in terrestrial rocks as an accessory phase but it seldom dominates the magnetic properties. The situation is different in Martian rocks, for example, the SNC meteorites (Rochette *et al.*, 2005b). Pyrrhotite is the stable iron sulfide for oxygen fugacities $fO_2 <$ FMQ (fayalite ($FeSiO4$) $\leftrightarrow$ magnetite + quartz), whereas paramagnetic pyrite (FeS_2) is dominant under the more oxidizing conditions in the Earth's crust (Lorand *et al.*, 2005). Martian meteorites contain about as much pyrrhotite as magnetite.

Natural terrestrial pyrrhotite is a mixture of monoclinic Fe_7S_8 and hexagonal Fe_9S_{10} and $Fe_{11}S_{12}$. The deficiency of Fe relative to S in these minerals results in lattice vacancies. In Fe_7S_8, these vacancies are confined to one of the two magnetic sublattices, giving rise to ferrimagnetism with $M_s \approx 80$–$90\,kA\,m^{-1}$ and $T_C \approx 320°C$ (**Figure 33**) and a monoclinic distortion of the basically hexagonal lattice. Magnetic properties as a function of grain size are reported by Dekkers (1988, 1989). A distinctive low-temperature transition around 34 K is diagnostic of the presence of monoclinic pyrrhotite in a rock or sediment (Dekkers *et al.*, 1989; Rochette *et al.*, 1990).

The *c*-axis is hard in pyrrhotite, as in hematite above T_M, with sublattice magnetizations confined to the *c*-plane. Within the *c*-plane, triaxial magnetocrystalline anisotropy is much stronger than in hematite and overshadows any stress-induced anisotropy. For this reason, pyrrhotite is the only mineral whose domain patterns can easily be viewed on a {0001} crystal plane without removing the stressed layer due to polishing.

Pyrrhotite generally becomes unstable on heating, for example, during thermal demagnetization, and oxidizes or decomposes to magnetite (Bina and Daly, 1994). The transformation occurs well above 320°C, so that no information about the pyrrhotite NRM is lost, but below $T_C = 585°C$ of magnetite. If the field is not perfectly zeroed, a CRM will be produced, corrupting the magnetite NRM.

Hexagonal pyrrhotite (Fe_9S_{10}) is basically antiferromagnetic but has a restricted ferrimagnetic range due to thermally activated vacancy ordering above the λ-transition, so called because of the shape of the $M_s(T)$ curve between the onset of ferrimagnetism around 200°C and its loss at $T_C \approx 265°C$ (Schwarz and Vaughan, 1972). In rapid cooling, the vacancy ordering and ferrimagnetism can be preserved metastably below 200°C.

Both pyrrhotite and greigite (Fe_3S_4) are frequent diagenetic minerals in anoxic, for example, sulfate-reducing, sedimentary environments (Roberts and Turner, 1993). Greigite has the same inverse spinel structure as magnetite but is less magnetic and less anisotropic than its oxide cousin: $M_s \approx 125\,kA\,m^{-1}$ and $T_C \approx 330°C$, close to T_C of troilite (FeS) and monoclinic pyrrhotite. Despite a crystalline anisotropy constant of only $\sim10^3\,J\,m^{-3}$, the coercivities of SD greigite are large enough to preserve a stable NRM (Diaz Ricci and Kirschvink, 1992). A hallmark of natural SD greigite is its high ratio of M_s/k_f compared to pyrrhotite and magnetite of similar sizes (Roberts, 1995).

8.3.10 Iron-Chrome Spinels (Chromites)

Chromites, members of the solid-solution series $FeCr_2O_4$–Fe_3O_4 (also containing variable amounts of Mg, Al, and Ti), are cubic spinel minerals found in nature in such diverse occurrences as meteorites (Weiss *et al.*, 2002), lunar igneous rocks, terrestrial andesitic pumice (Yu *et al.*, 2001), and submarine gabbro and peridotite (Dunlop and Prévot, 1982); $T_C = -185°C$ for $FeCr_2O_4$ and increases with decreasing Cr content. A few naturally occurring chromites carry a significant NRM. Examples are the Kurokami pumices of Mt. Sakujima, Japan (Yu *et al.*, 2001), where the NRM is shared with titanomagnetite, and Martian meteorite SaU 008 (Yu and Gee, 2005), where the chromite dominates the NRM. SaU 008 is an unusual shergottite of plutonic origin, perhaps akin to the gabbro/peridotite affinities of terrestrial chromites.

8.3.11 Iron Carbonates

The commonest iron carbonate is siderite ($FeCO_3$), a frequent constituent of carbonate sediments and rocks on Earth. Siderite has a low-temperature transition at 30–35 K (Housen *et al.*, 1996), distinctively different

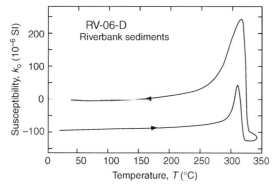

Figure 33 Hopkinson peak (Section 8.4.3) in initial susceptibility k_0 of riverbank sediments from Taiwan just below the 320°C Curie point of pyrrhotite, Fe_7S_8. The overall negative susceptibility is due to diamagnetism of the sample holder (see Section 8.1.1). Modified from Horng C-S and Roberts AP (2006) Authigenic or detrital origin of pyrrhotite in sediments? Resolving a paleomagnetic conundrum. *Earth and Planetary Science Letters* 241: 750–762.

from that of pyrrhotite (Section 8.3.9). Siderite is para-magnetic at ordinary temperatures and carries no NRM. Its interest is as a source of diagenetic magnetite. On Earth, the oxidation of siderite at room temperature and accompanying CRM acquisition is well documented (Hus, 1990). The process becomes rapid and even explosive at higher temperatures and the magnetite CRM effectively masks any remaining NRM of the carbonate rock being thermally demagnetized. Decomposition of siderite in the crust of Mars is more hypothetical but is certainly possible given adequate heat sources (Scott and Fuller, 2004). It could provide a potent source of magnetite crustal magnetization.

8.3.12 Iron and Iron–Nickel

Iron and iron–nickel are the principal NRM carriers in lunar rocks and most meteorites. Chondritic meteorites also contain varying amounts of magnetite. Body-centered-cubic kamacite (αFe) is ferromagnetic with $M_s = 1715\,\mathrm{kA\,m^{-1}}$ and $T_C = 765°C$, making it the epitome of intense and thermally stable magnetization. There are two drawbacks to iron as a remanence carrier, however. First is its extremely narrow SD range: d_0 and d_s are theoretically and experimentally very close (23 and 8 nm, respectively: Kneller and Luborsky (1963) and Butler and Banerjee (1975b); **Table 3**). Only elongated particles are likely to be SD, and iron, being cubic, does not readily form such particles.

The second problem is the transformation during heating from kamacite to face-centered-cubic taenite (γFe). The transformation occurs below 765°C if more than 5–10% Ni is alloyed with Fe. In this case, the NRM, a phase transformation CRM rather than TRM, is destroyed during thermal demagnetization before the kamacite Curie point is reached. The transformation is reversible but sluggish. During cooling, taenite may survive 150–200°C below the $\alpha \rightarrow \gamma$ transition temperature during heating (**Figure 34**).

When the Ni content passes 30%, kamacite is no longer stable above room temperature. However, taenite becomes ferromagnetic in this range. The iron in lunar rocks is seldom this Ni-rich unless it is of meteoritic origin. Magnetic properties of Ni–Fe in different classes of meteorites and synthetic analogs are reviewed by Nagata (1979) and Wasilewski (1981).

Néel *et al.* (1964) discovered that the random (or disordered) arrangement of Fe and Ni atoms in taenite can become ordered in the presence of a strong magnetic field along a <100> axis for compositions of 50–55% Ni. Ordering occurs spontaneously in nature below $\approx 300°C$ over very long times, generating a natural mineral called tetrataenite (Clarke and Scott, 1980; Nagata, 1983), which is common in chrondritic meteorites and is also known from lunar rocks. Ordered tetrataenite has rather different magnetic properties from disordered taenite of the same composition, including high coercivities (30–600 mT) and a 'blocky' $M_s(T)$ curve like that of hematite (cf. **Figure 20**) with $T_C \approx 550°C$.

Because of the complications introduced by phase transformations in Ni–Fe and uncertainty about the nature of NRM, paleointensity work on meteorites is

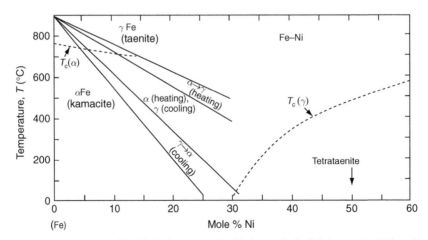

Figure 34 Part of the phase diagram of Fe–Ni. Body-centered cubic kamacite (α-Fe) does not exist for >30 mol.% Ni. Face-centered cubic taenite (γ-Fe) is the stable phase at high temperature and at all T for >30 mol.% Ni. The ordered phase tetrataenite can exist for 50–55 mol.% Ni.

best confined to chondrites and achondrites containing only kamacite or carbonaceous chondrites containing magnetite (Nagata, 1983).

Iron in lunar rocks has four distinct origins: primary igneous Ni-poor iron, often in kamacite–taenite intergrowths; coarse-grained Ni-rich meteoritic iron; secondary iron generated by the impact melting of troilite (FeS); and nanometer-sized pure Fe spheres in impact melted glasses. Of these forms of iron, only the last contains any SD fraction. The other forms are paleomagnetically unappealing, having weak and unstable NRMs. Viscous remagnetization is so pronounced in some of the nearly superparamagnetic fine grains that the advantages of an SD fraction are largely lost. Lunar rocks on the whole are discouraging to work with, and any primary paleomagnetic record is tenuous.

8.3.13 'Magnetic' Silicates

Silicate minerals have no inherent ferromagnetism but they can precipitate ferrimagnetic particles during cooling from the melt (producing primary TRM) or as a result of later oxidation (producing secondary CRM). The most common magnetic precipitate or inclusion in silicates is magnetite (see Section 8.3.2). The intergrowth texture between the precipitated crystals and the host silicate(s) is the best guide to the nature of the NRM but this is not easily observed when the precipitates are submicroscopic.

Figure 35 illustrates crystallographic control of the orientations of needle-like magnetite laths, which lie along two directions in the host clinopyroxene crystal (Renne *et al.*, 2002). The magnetite laths do not continue into an adjacent amphibole grain. These magnetites must have precipitated at high temperature during crystallization and cooling of the pyroxene and the NRM is almost certainly a TRM. The remanence, confined to two crystallographic directions, is markedly anisotropic. TRM directions can be in error by tens of degrees in individual crystals although it is not likely in paleomagnetic samples.

Figure 36 is an AFM/MFM (Section 8.2.5) image of a single inclusion (Feinberg *et al.*, 2005). In this case, the inclusion is not pure magnetite but a high-temperature exsolution intergrowth of magnetite and ulvöspinel (Section 8.3.3). Magnetic domains within the magnetite are truncated by the cross-cutting ulvöspinel bands (cf. textures in **Figure 29**). It is interesting that despite this disruption, the domains

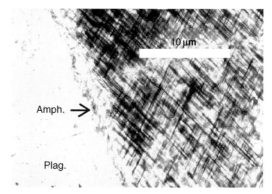

Figure 35 Photomicrograph of two sets of crystallographically oriented magnetite inclusions in a clinopyroxene crystal in gabbro. The magnetite laths do not continue into the amphibole rim, nor are they present in the adjacent plagioclase crystal. Modified from Renne PR, Scott GR, Glen JMG, and Feinberg JM (2002) Oriented inclusions of magnetite in clinopyroxene: Source of stable remanent magnetization in gabbros of the Messum Complex, Namibia. *Geochemistry, Geophysics, Geosystems* 3(12): 1079 (doi:10.1029/2002GC000319).

prefer to align with the short dimension of individual magnetite bands, lining up across the gaps to form 'superwalls' continuing across adjacent grains parallel to the long dimension of the entire inclusion. A similar example of superdomains crossing several grains appears in **Figure 12**.

The earliest work on crystallographically oriented inclusions in pyroxene is that of Evans and Wayman (1970). They used TEM of an etched surface to reveal the arrays of inclusions. These were assumed to be magnetite on the basis of their coercivities, which ranged up to the limit expected for shape anisotropy in magnetite (Evans and McElhinny, 1969). Murthy *et al.* (1971) and Davis (1981) subsequently demonstrated that arrays of crystallographically oriented magnetite needles occur in plagioclase in anorthosites and oceanic gabbros and inferred that they carry primary TRM which dominates the NRM in the particular rocks they studied. Plagioclase crystals extracted from volcanic and plutonic rocks are now being used to refine paleointensity studies (Cottrell and Tarduno, 2000). The magnetites they contain are equidimensional, not needle-like, and record paleomagnetic directions faithfully.

Biotite and amphiboles like hornblende and actinolite can also be quite magnetic. In biotite, magnetite generally occurs in cracks or between the mica sheets but is not so clearly cogenetic with the silicate host. Relatively large magnetite crystals with

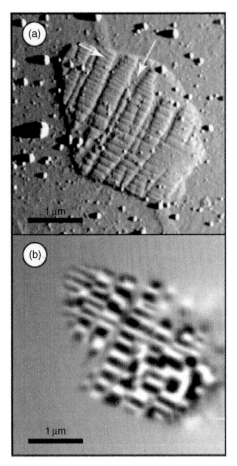

Figure 36 Atomic force microscope (AFM, upper) and magnetic force microscope (MFM, lower) images of a titanomagnetite inclusion in clinopyroxene. The titanomagnetite is exsolved into a boxwork structure (AFM image; see also **Figures 12** and **29**) of magnetite prisms (high relief) separated by ulvöspinel (two sets of cross-cutting channels, white arrows). The individual magnetite prisms appear to be SD (MFM image), with M_s vectors parallel to the long axis of each prism and of the inclusion but transverse to the short axis of bands of magnetite prisms. Because of strong magnetostatic interactions, the M_s vectors line up across the gaps between bands but alternate in polarity along individual bands, similar to the patterns seen in **Figure 12**. Modified from Feinberg JM, Scott GR, Renne PR, and Wenk H-R (2005) Exsolved magnetite inclusions in silicates: Features determining their remanence behavior. *Geology* 33: 513–516.

MD properties tend to dominate the magnetic properties although a population of finer magnetites seems to be present as well, perhaps extending down to SD sizes (Dunlop *et al.*, 2006). Amphibole NRMs are generally MD-like, with rather unstable NRM directions and nonlinear paleointensity behavior (Wu *et al.*, 1974; Dunlop *et al.*, 2005).

8.3.14 Biogenic Magnetic Minerals

Biomagnetism is the science of magnetic fields and magnetic materials in living organisms (Kirschvink *et al.*, 1985). Its main intersection with geophysics is the presence of fossil magnetotactic bacteria and their magnetosomes (chains, intact or disrupted, of magnetic crystals) in sediments, where they carry an NRM of distinctively high stability. The reason for this stability is clear from **Figure 37**. Bacteria, as well as higher organisms, have evolved the ability to produce magnetite (less commonly greigite) whose crystals are of strictly SD size. When crystals are strung together in chains, the added shape anisotropy renders the magnetosome even harder magnetically.

Intact chains seldom survive in sediments because cell walls deteriorate rapidly after death. However, Petersen *et al.* (1986) and Vali *et al.* (1987) have detected what appear to be fossil biomagnetic crystals and chains in deep-sea sediments as old as 150 Ma using scanning electron microscopy. Carbonate globules from ~4 Gy old Martian meteorite ALH84001 contain elongated prismatic magnetite crystals whose size and morphology resemble those of modern bacterial magnetites (Thomas-Keprta *et al.*, 2000). Whether or not these are magnetofossils testifying to the presence of early life on Mars is an open question.

8.4 Induced and Remanent Magnetization
8.4.1 SD Hysteresis and Susceptibility

Self-demagnetizing fields and the magnetizing process for SD and MD grains were discussed in a general way in Sections 8.2.1 and 8.2.2. **Figure 38** illustrates medium-field hysteresis loops synthesized by measuring the difference in total area between positive (field-aligned) and negative domains observed under the microscope at different stages of field cycling (Soffel and Appel, 1982). A moderately large grain (30 μm) exhibits distinctively MD hysteresis. The loop is narrow because lattice defects impeding wall motion produce only slight irreversibility in wall positions in increasing and decreasing fields. M saturates when H_o reaches the maximum value of the internal demagnetizing field, $(H_d)_{max} = NM_s$ (eqn [6]). This state corresponds to zero net internal field, $H_i = 0$.

A smaller grain (5 μm × 10 μm), near magnetite in composition, has a broad SD-like loop, with

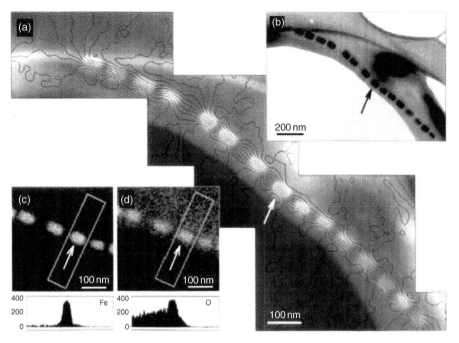

Figure 37 Transmission electron micrograph of part of a single cell of a magnetotactic bacterium, showing a chain of magnetosomes (b) which are imaged by off-axis electron holography in (a). Elemental maps for Fe (c) and O (d) indicate that the magnetosomes are composed of magnetite. Magnetic field lines link the individual magnetosomes along the chain (a), demonstrating strong magnetostatic interactions. As a result, the chain behaves as an elongated SD particle. Modified from Dunin-Borkowski RE, McCartney MR, Frankel RB, Bazylinski DA, Pósfai M, and Buseck PR (1998) Magnetic microstructure of magnetotactic bacteria by electron holography. *Science* 282: 1868–1870.

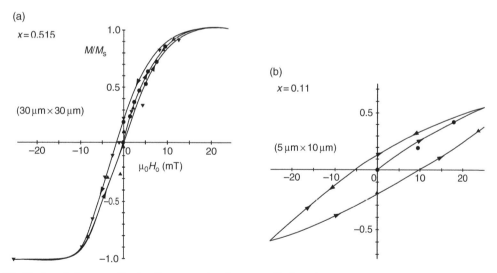

Figure 38 Hysteresis loops synthesized by measuring the areas of '+' and '−' domains imaged by Bitter colloid patterns at different field strengths: (a) MD type saturating loop; (b) SD-type minor loop. Modified from Soffel HC and Appel E (1982) Domain structure of small synthetic titanomagnetite particles and experiments with IRM and TRM. *Physics of the Earth and Planetary Interiors* 30: 348–355.

considerable irreversibility between the magnetization processes in increasing and decreasing H_o. M_r is relatively large and M does not saturate

in the maximum available field $\mu_0 H_o \approx 25$ mT. A broad minor (nonsaturating) loop in intermediate fields is a sign of SD-like behavior, although this grain has a size $>> d_0$ for magnetite (cf. Section 8.2.9).

An SD grain is magnetically hard (high H_c) because it contains no wall. The coupled spins of the entire grain must be bodily rotated away from an easy axis of anisotropy in response to H_o and this takes more energy than moving a wall between domains whose spins remain in easy directions. As explained in Section 8.2.7, complete reversals of M sometimes occur incoherently: for example, 'flower' SD states can reverse by curling, the transitory passing of a wall-like vortex across the grain.

The simplest model of SD magnetization is coherent rotation against a uniaxial anisotropy (Stoner and Wohlfarth, 1948; Néel, 1949). The model is simplistic but surprisingly powerful, and is easily adapted to fit more realistic situations (Egli and Lowrie, 2002; Lanci and Kent, 2003). The model SD grain (**Figure 39**) has its easy axis, $\theta = 0$, at angle ϕ to the applied field H_o, which pulls M_s through an angle θ away from the easy axis. The energies involved are due to (shape) anisotropy and the applied field

(**Figure 39**): $E_d = 1/2\,\mu_0 V\,\Delta N M_s^2 \sin^2\theta$ (eqn [7]) and $E_H = -\mu_0 V \mathbf{M}_s \cdot \mathbf{H}_o = -\mu_0 V M_s H_o \cos(\phi - \theta)$. ΔN is the difference between the demagnetizing factors for M along the hard and easy axes.

When the sum $E = E_d + E_H$ is minimized by setting $dE/d\theta = 0$, the elementary hysteresis loops, shown in **Figure 40**, result. When H_o is applied along the easy axis ($\theta = 0$), the loop is perfectly rectangular. There is no susceptibility because M remains fixed at $\theta = 0$ until H_o reaches a critical value H_K, when M abruptly rotates past the hard axis ($\theta = 90°$) and reverses to the opposite easy axis ($\theta = 180°$). This elementary rectangular loop is sometimes called a hysteron. Its simplicity makes it ideal for further modeling, for example, in the presence of interaction fields (Section 8.4.4).

When H_o is applied along the hard axis, there is in effect no barrier to pass and so no hysteresis. M rotates reversibly at all H_o, reaching $\theta = 90°$ when $H_o = H_K$. At this point, M (measured in the direction of H_o) $= M_s$. This magnetization function is also particularly simple and can be used as a

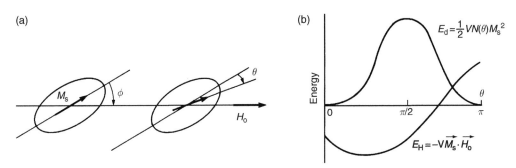

Figure 39   Coherent rotation in a spheroidal SD grain. (a) Orientation of $\mathbf{M}_s$ without and with a field $\mathbf{H}_o$ applied at angle ϕ to the easy axis $\theta = 0$. (b) Contributions of E_d and E_H to the total energy as a function of θ.

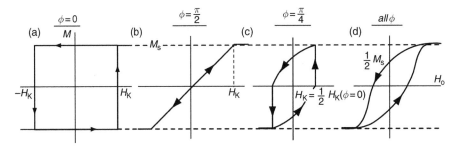

Figure 40   Theoretical hysteresis loops for the SD grains of **Figure 39**. The rectangular loop ((a), $\phi = 0$) and ramp ((b), $\phi = 90°$) are used as hysterons in the Preisach–Néel diagram (**Figure 43**). The angle-averaged loop (d) resembles the loop for $\phi = 45°$ (c) and incorporates both reversible and irreversible changes in magnetization.

hysteron. M varies linearly with H_o, giving a constant susceptibility k_f (or just k) $= M_s/H_K$.

When $\mathbf{H}_o$ is along an intermediate axis, for example, 45°, $\mathbf{M}$ at first rotates reversibly, then abruptly and irreversibly switches at a critical field which is always less than H_K. Averaging over an ensemble of grains with their easy axes randomly oriented relative to $\mathbf{H}_o$ gives a loop that is reminiscent of the experimental SD-like loop in **Figure 38**. The coercive force is $\approx 1/2 H_K = 1/2 \, \Delta N M_s$ and the saturation remanence is exactly $1/2 M_o$.

The saturation hysteresis loop for randomly oriented grains in **Figure 40** appears to have no demagnetized state. That is certainly true for individual SD grains in the ensemble but the ensemble itself possesses a demagnetized state (not accessible by applying continuously increasing or decreasing $\mathbf{H}_o$) in which grain moments are randomly distributed in space when $H_o = 0$. Starting from this state, which can be achieved in the laboratory by a sequence of positive and negative fields (AF demagnetization, Section 8.5.1) or by thermal randomization (thermal demagnetization, Section 8.5.3), the ensemble has a low-field or initial susceptibility $k_o = (2/3) M_s / H_K$ and a saturation remanence $M_{rs} = 1/2 M_s$.

8.4.2 MD Hysteresis and Susceptibility

MD grains respond to a small applied field $\mathbf{H}_o$ by domain wall displacement. Only in large fields do the domains rotate as in SD grains. The amount of wall displacement is determined by how strongly a particular wall is pinned at various locations along its path and by the orientation of $\mathbf{H}_o$ relative to the domain $\mathbf{M}_s$ vectors. The overall magnetostatic energy due to the external field is

$$E_H = -\mu_0 V \mathbf{M} \cdot \mathbf{H}_o = -\mu_0 V M_s(x/d) H_o \cos\psi \quad [14]$$

where x is the average wall displacement in a grain of size d, $\mathbf{M} = (x/d)\mathbf{M}_s$ is the net magnetization of the grain, and ψ is the angle between $\mathbf{M}_s$ and $\mathbf{H}_o$. Since net magnetization governs E_H, the walls can be thought of as interacting with one another: if one wall moves only a small distance, another wall will move a larger distance to produce the desired $\mathbf{M}$.

Ultimately, how far walls can be driven by $\mathbf{H}_o$ depends on the opposition of the internal self-demagnetizing field $\mathbf{H}_d = -N\mathbf{M}$ created by $\mathbf{M}$. The total field inside the grain is the sum of $\mathbf{H}_o$ and $\mathbf{H}_d$. Wall motion will cease when $\mathbf{H}_d$ grows large enough to balance $\mathbf{H}_o$:

$$\mathbf{H}_i = \mathbf{H}_o - N\mathbf{M} \to 0 \quad \text{or} \quad k_o = \mathrm{d}M/\mathrm{d}H_o = 1/N \quad [15]$$

This expression, which ignores both resistance to wall motion by lattice defects and orientation of $\mathbf{H}_o$ relative to $\mathbf{M}_s$, works surprisingly well as a first-order approximation.

Before deriving a more exact expression, we need to consider the meaning of N in an inhomogeneously magnetized grain with a set of domains whose sizes and orientations change with changing $\mathbf{H}_o$. Micromagnetic calculations demonstrate that to a very good approximation, the demagnetizing energy is given by

$$E_d = E_{d0} + 1/2 \mu_0 V N M^2 \quad [16]$$

with $N \approx N_{SD} \, (1 - 1/n)$, where n is the number of domains and N_{SD} is the single-domain demagnetizing factor (see Dunlop and Özdemir, 1997, chapter 5.2). Thus N remains nearly a constant despite the existence of domain structure, so that k_o in eqn [15] is constant, independent of the field H_o and also independent of temperature unless the domain structure alters. Two-domain grains have demagnetizing factors about one-half those of SD grains of the same shape but the difference decreases rapidly with increasing n. Large MD grains with many domains 'look like' SD grains with $M << M_s$.

Equation [15] is exact if there is no resistance to wall displacement, that is, if the intrinsic susceptibility of the material, $k_i = \mathrm{d}M/\mathrm{d}H_i$, is infinite. This is close to being the case in MD iron, but in magnetic oxides and sulfides, there is considerably more resistance to wall motion. Then, solving the equation $M = k_i(H_o - NM)$ for M,

$$k_o = M/H_o = k_i/(1 + Nk_i) \quad [17]$$

For a large equidimensional MD grain of magnetite, containing many domain walls, an estimate of k_i is ≈ 12 (SI units), giving $k_o \approx 12/5 = 2.4$ compared to $k_o = 3$ from eqn [15]. Note that the so-called screening factor $1/(1 + Nk_i)$ is independent of the system of units; for large MD grains of magnetite, it is about $1/5$.

8.4.3 Temperature Effects: Hopkinson Peak and Koenigsberger Ratio Q

MD susceptibility k_o increases with rising temperature T because wall pinning decreases, causing k_i to increase. Walls may be pinned magnetoclastically (by the stress fields of dislocations), or may become anchored on vacancies or actual voids in the crystal, in which case the wall energy is lowered by a combination of magnetocrystalline and demagnetizing effects. Since λ, K, and M_s all decrease as T rises (**Figure 3**), resistance to wall motion also weakens. For magnetite, the increase in k_o is only $\approx 20\%$, from ≈ 2.4 SI at room temperature T_0 to ≈ 3 near the Curie point T_C, because of the strong braking effect of $\mathbf{H}_d$. In minerals like TM60 and pyrrhotite with stronger wall pinning and weaker M_s and H_d, $k_o(T)$ increases more and reaches a peak (the Hopkinson peak) just below T_C.

The Koenigsberger ratio $Q = M_r/k_o H_o$ between remanent magnetization M_r (normally TRM, or NRM assumed to be of TRM origin) and induced magnetization $M_{in} = k_o H_o$ produced by the same (weak) field H_o is often used in magnetic exploration. Although most magnetic anomaly interpretation is done using room temperature values of M_r and k_o, this is not appropriate for long wavelength anomalies having their source in the deep crust, where ambient temperatures are hundreds of Celsius above T_0.

A crude picture of TRM can be obtained by assuming that TRM is due to walls that are strongly pinned below a blocking temperature T_B (Stacey, 1958). The magnetization at T_B is given by eqn [15]: $M(T_B) = H_o/N$. In cooling to room temperature, this is magnified due to the reversible increase in M_s: $M(T_0) = M(T_B)(M_{s0}/M_{sB})$. However, when $\mathbf{H}_o$ is removed at T_0, weakly pinned walls adjust their positions under the influence of $\mathbf{H}_d$ to partially cancel or screen $\mathbf{M}$. The final TRM is then $M(T_0)$ multiplied by the screening factor $1/(1 + Nk_i)$. For MD magnetite, $M_{s0}/M_{sB} \approx 3$ and $1/(1 + Nk_i) \approx 1/5$. Thus $Q \approx 3/5$ at T_0.

We can treat the case of deep crustal anomaly sources by interrupting cooling at some T below T_B but above T_0 and setting $\mathbf{H}_o \to 0$. The TRM acquired will be less than before by the factor $M_s(T)/M_{s0}$, with a further reduction due to the decrease in the screening factor with increasing k_i. When induced magnetization is produced by a later Earth's field $\mathbf{H}_o$, Q will therefore be <0.6 (**Figure 41**). Very deep-seated rocks may remain above their blocking temperatures and record no TRM from earlier magnetic fields ($Q = 0$). However, their induced magnetizations are still limited by self-demagnetization. Any Hopkinson peak is minor if MD magnetite is the mineral involved.

The same is not true for SD grains. Each grain is magnetized to saturation and the only limit on induced magnetization is the number of grains whose moments remain oriented opposite to $\mathbf{H}_o$. At

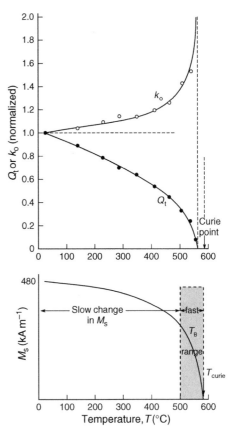

Figure 41 Sketches of the temperature dependences of initial susceptibility k_o and Koenigsberger ratio $Q_t = M_{tr}/k_o H_o$ of SD or SD-like grains of magnetite. When heated to the main blocking temperature range above $500°C$, M_s and therefore M_{tr} and Q_t decrease strongly, while k_o rises to a Hopkinson peak just below the Curie point. Multidomain grains do not display any substantial Hopkinson peak because k_o is limited by self-demagnetization to a value $\leq 1/N$.

sufficiently high T, all barriers to rotation of moments (shape, crystalline, or magneoelastic anisotropy) disappear and only thermal energy prevents perfect alignment of $V\mathbf{M}_s$ of a grain with $\mathbf{H}_o$. This condition is called superparamagnetism (SP) because M_{in} obeys the paramagnetic Langevin and Curie laws:

$$M(H_o, T) = M_s L(\mu_0 V M_s H_o/kT) \quad \text{if} \\ \approx \mu_0 V M_s^2 H_o/3kT \qquad [18] \\ \mu_0 V M_s H_o/kT \ll 1$$

$L(\alpha)$ being the Langevin function $\coth(\alpha) - 1/\alpha$.

In contrast to paramagnetism, SP magnetization approaches saturation for ordinary temperatures and (near T_C) moderate fields. This comes about because the coupled spins in an SD grain are much less perturbed by thermal agitation than are individual atomic moments. The resulting SP susceptibility in weak fields,

$$k_{SP} = \mu_0 V M_s^2 / 3kT \qquad [19]$$

can be orders of magnitude larger than ordinary SD susceptibility due to rotation of moments against an anisotropy barrier (Section 8.4.1) or MD susceptibility. As a result, SD grains have a pronounced Hopkinson peak in k_o at any temperature where anisotropy drops to low values, not only near T_C but also around the magnetite isotropic point $T_K \approx 130$ K (**Figure 42**).

For an SD grain of volume V, the onset of SP occurs quite suddenly in heating above the TRM blocking temperature T_B. At T_B, M is described by an equation similar to [18],

$$M(H_o, T) = M_s \tanh(\mu_0 V M_s H_0 / kT) \qquad [20]$$

which describes thermal equilibrium between populations of grains aligned along two anisotropy-favored easy axes (Section 8.4.1). This distribution appropriate to (H_o, T_B) remains in frozen equilibrium during cooling below T_B and when H_o is removed at T_0. Thus M in eqn [20] is the SD TRM.

Just as SP susceptibility k_{SP} is much higher than anisotropy-limited susceptibility k_o, SD TRM is

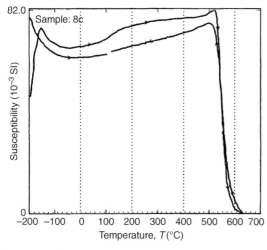

Figure 42 Susceptibility k_o vs. temperature for a monzonite sample containing large grains of magnetite. There is a small Hopkinson peak at 520–540°C and a more prominent peak at the Verwey transition (−153°C). Modified from Dunlop DJ, Schmidt PW, Özdemir Ö, and Clark DA (1997) Paleomagnetism and paleothermometry of the Sydney Basin. 1. Thermoviscous and chemical overprinting of the Milton Monzonite. *Journal of Geophysical Research* 102: 27271–27283.

>>SD induced magnetization $k_o H_o$ at ordinary temperatures. Thus $Q \gg 1$ for SD grains. However, as T increases, TRM decreases reversibly up to T_B, while k_o increases. Therefore, Q drops steadily with increasing T, particularly above T_B (in reality, not a single temperature but a range of T_B's for different values of V; **Figure 41**).

8.4.4 Magnetostatic Interactions

Just as internal self-demagnetizing fields greatly affect the response of MD grains to an external applied field $\mathbf{H}_o$, the $M(H_o)$ response of an SD grain is modified by magnetostatic interaction fields due to surrounding magnetized grains (**Figures 12, 36,** and **37**). Usually neighboring grains are the most influential but because dipole fields drop off as $1/r^3$ while volume increases as r^3, the effect can be over a long range if the grains are fairly uniformly spaced.

A commonly used approximation (Preisach, 1935; Néel, 1954) is the local-field model. The interaction field $\mathbf{H}_i$ acting on an SD grain is taken to be constant in direction and magnitude, independent of the magnetization state of the assemblage of grains in the sample. This is not very realistic but works reasonably well if interactions are due to a few nearby grains that do not switch their moments in the field range of interest; the rest of the grains then provide a background $\mathbf{M}$ which can be accounted for separately by a demagnetizing field. In the special case where both $\mathbf{H}_i$ and $\mathbf{H}_o$ are parallel to the easy axis of a uniaxial SD grain, the rectangular loops of Section 8.4.1 become displaced along the field axis, with modified switching fields $a = H_c - H_i$ and $b = -H_c - H_i$ (**Figure 43**). These provide a set of hysterons for Preisach or FORC (first-order reversal curve) diagrams, which are widely used to characterize domain state, coercivity distribution, and interactions of experimental samples (Roberts *et al.*, 2000; Carvallo *et al.*, 2005).

8.5 NRM and Paleomagnetic Stability
8.5.1 AF Demagnetization

Magnetic anomaly interpretation usually focuses on induced magnetization in the present geomagnetic field. This field has a known direction and intensity at a specified location on Earth. More challenging is determining the direction, and in favorable cases, the intensity of an ancient geomagnetic or planetary

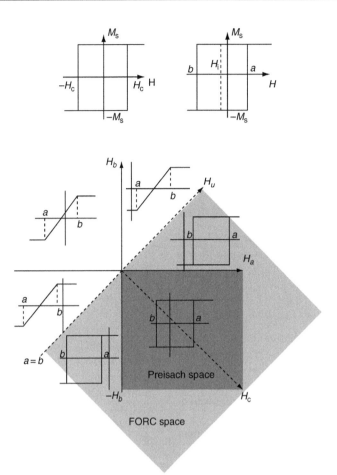

Figure 43 Top: elementary hysteresis loops of an SD grain with and without a constant interaction field H_i acting parallel to the applied field H_o and the easy axis $\theta = 0$ (cf. **Figure 40**). Bottom: elementary reversible and irreversible magnetization cycles, based on the hysterons of **Figure 40** shifted by H_i, for switching fields (a, b) in different sectors of the Preisach/ FORC diagram. Modified from Carvallo C, Dunlop DJ, and Özdemir Ö (2005) Experimental comparison of FORC and remanent Preisach diagrams. *Geophysical Journal International* 162: 747–754.

magnetic field as recorded by the NRM of rocks. To be successful, the paleomagnetic method must be able to separate various generations of NRM acquired by the same rock at different times in its history, including relatively modern 'overprints' that constitute noise. This separation of NRM components is accomplished by one or a combination of 'cleaning' or 'stepwise demagnetization' techniques.

'AF demagnetization', which subjects a sample to fields of alternating polarity whose amplitude gradually diminishes to zero, is univerally used. Raising the initial AF amplitude in a series of steps has the effect of demagnetizing grains with increasing microcoercivities H_c. Individual SD grains do not literally demagnetize because they have asymmetric hysteresis loops (**Figure 43**) and will always be left in the same state (either $+M_s$ or $-M_s$, depending on H_i)

after AF treatment. However, if the distribution of interaction fields $g(H_i)$ is symmetric with mean zero, equal numbers of grains will be magnetized in opposite directions and the net moment will be zero for the population of grains with $H_c - H_i$ less than the peak AF applied in a particular step.

It is not self-evident that higher-microcoercivity grains necessarily have more ancient NRMs. This depends on the mechanism of NRM. Their NRMs are, however, generally more stable or resistant to resetting over long intervals of time (see Section 8.6.2).

If a unidirectional bias field is added to the AF, each ensemble of identical SD grains will acquire a net magnetization rather than being (statistically) demagnetized. This process is called 'anhysteretic remanent magnetization (ARM)' and is sometimes useful as a laboratory analog to TRM. On the other hand, higher-order harmonics leading to asymmetric AFs can produce unwanted ARMs that contaminate rather than clean the NRM. Some older AF demagnetizers produce serious ARMs above ≈60 mT.

8.5.2 Low-Temperature Demagnetization (LTD)

Cooling magnetite grains through the isotropic point ($T_\mathrm{K} \approx 130\,\mathrm{K}$) has the effect of momentarily broadening domain walls, which are then less strongly pinned by defects. The effect on SD grains is less important because shape anisotropy is governed by M_s, which is continuous at T_K and almost continuous at the Verwey transition ($T_\mathrm{V} = 120\,\mathrm{K}$) (Özdemir et al., 2002). As normally applied, 'LTD' is a one-shot technique designed to reduce MD remanence by cooling in zero field to below T_V and back to T_0. Stepwise AF or thermal cleaning generally follows LTD.

It is possible to carry out LTD in stepwise fashion because K_1 of magnetite decreases continuously below T_0, particularly below 200 K, not discontinuously at T_K (Dunlop, 2003a). **Figure 44** demonstrates that saturation IRM of 6 μm magnetite grains is

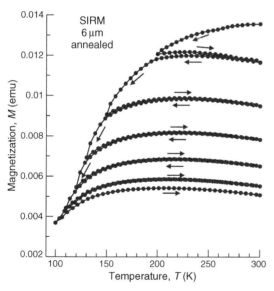

Figure 44 Stepwise low-temperature demagnetization of saturation isothermal remanent magnetization (SIRM) produced at 300 K in 6 μm magnetite grains. Cooling–warming cycles in zero field to 200, 150, 130, 120 and 110 K produced similar decrements in remanence. Modified from Dunlop DJ (2003a) Stepwise and continuous low-temperature demagnetization. *Geophysical Research Letters* 30(11): 1582 (doi:10.1029/2003GL017268).

reduced to ≈25% of its initial value in five approximately equal decrements using stepwise LTD.

8.5.3 Thermal Demagnetization of SD Grains

The simplest NRM mechanism in igneous and metamorphic rocks is a TRM. Even when factors like time and chemical change enter, cooling through a blocking temperature T_B is often a useful first-order model of the NRM process. Reciprocal to the TRM blocking process is 'thermal demagnetization', a zero-field heating–cooling cycle through T_B. Since there is a range of T_B's in any sample, in practice thermal demagnetization occurs continuously over some range of heating temperatures.

For SD grains, Néel (1949) produced a satisfying theory that links the roles of time t and temperature T in magnetization changes. Any out-of-equilibrium magnetization $\Delta M = M(t) - M_\mathrm{eq}$ decays exponentially with time:

$$\Delta M(t) = \Delta M(0)\mathrm{e}^{-t/\tau} \qquad [21]$$

M_eq being thermal equilibrium magnetization and τ relaxation time. In the case of SD grains, following eqn [20] and assuming $\mu_0 V M_\mathrm{s}/kT \ll 1$,

$$M_\mathrm{eq}(H_\mathrm{o}, T) \approx \mu_0 V M_\mathrm{s0}{}^2 m_\mathrm{s}^2(T) H_\mathrm{o}/kT \qquad [22]$$

in which $m_\mathrm{s}(T) = M_\mathrm{s}(T)/M_\mathrm{s0}$. The relaxation time in seconds is given by

$$\tau \approx 10^{-10}\mathrm{e}^{\Delta E/kT} = 10^{-10}\exp(V H_\mathrm{c0} M_\mathrm{s0} m_\mathrm{s}^2(T)/2kT) \qquad [23]$$

where ΔE is the activation energy or barrier to be crossed, here by reversals of SD moments (Section 8.4.1), and $H_\mathrm{c0} = H_\mathrm{c}(T_0)$ is the room-temperature microcoercivity for this process. Shape anisotropy is assumed, in which case $H_\mathrm{c}(T)/H_\mathrm{c0} = m_\mathrm{s}(T)$. For magnetocrystalline anisotropy, for example, in hematite, $H_\mathrm{c}(T)$ varies as a higher power of $m_\mathrm{s}(T)$.

The equilibrium magnetization M_eq is proportional to $m_\mathrm{s}^2(T)$ and $M_\mathrm{s}(T)$ decreases strongly between 500°C and $T_\mathrm{C} = 580$°C in magnetite. However, the relaxation time τ also decreases with increasing T. These two effects oppose each other, the second promoting a closer approach of induced magnetization M to M_eq while the first causes M_eq to decrease with heating. The exponential dependence in eqn [23] makes the second effect dominant. M is thus enhanced over the blocking temperature range and only drops off within 5–10°C of T_C. This results in a strong SD Hopkinson peak in k_o (**Figure 41**).

Blocking temperature implies a sudden rather than a gradual change in M. This is indeed the case for individual grains. Because of the exponential dependence of τ on T, if $T < T_B$, $\tau \gg t$ where t is time measured on a laboratory or a geological scale, depending on the situation. Then according to [21], M remains 'blocked' at its initial value $M(0)$. This is the origin of TRM. Grains acquire a remanence M_{tr} when they are cooled through T_B in a field H_o because M no longer responds to changes in H_o on the timescale of observation.

On the other hand, when grains are above their blocking temperatures, $\tau \ll t$ and $M \to M_{eq}$ during time t. This is the 'unblocked' or superparamagnetic (SP) state, 'super' alluding to the giant susceptibility that results from equilibrium magnetization M_{eq} being achieved in a weak field H_o (cf. eqn [19]). SP is responsible for the Hopkinson peak in $k_o = M_{in}/H_o$ sketched in **Figure 41**, as explained in Section 8.4.3.

When grains are heated through T_B in zero field, they pass from the blocked to the unblocked state and lose all their remanent magnetization. Because no magnetization is induced above T_B in zero field, they do not acquire any new remanence when cooled back through T_B. This process of thermal demagnetization occurs only in the laboratory; zero-field conditions are never encountered in nature, at least on Earth.

For individual SD grains or ensembles of identical grains, the demagnetization is abrupt and occurs over an interval of a few Celsius around T_B. From eqn [23], setting $\tau = 1.78t$ at blocking (the factor 1.78 ensures agreement with more exact theories), we have that

$$VH_{c0} = [2k(23 + \ln 1.78t)/M_{s0}][T_B/m_s^2(T_B)] \quad [24]$$

an implicit equation for T_B.

The magnetic grains in any rock have a spectrum of volumes V and critical fields H_{c0}. Grains with $T_B > T$ will retain their remanence, while those with $T_B < T$ will be demagnetized. For increasing values of T, the rectangular hyperbola described by eqn [24] sweeps through a set of fixed points (V, H_{c0}), each representing a grain or population of similar grains (**Figure 45**). Because of the nonlinear $T/m_s^2(T)$ dependence, the blocking curve sweeps out small areas up to $\sim$500°C (for magnetite) but increasingly large areas above 500°C. Grains below and to

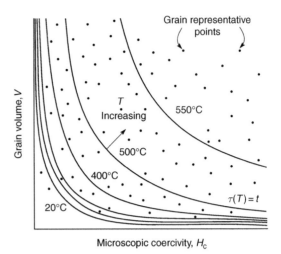

Figure 45 Néel diagram showing representative points (V, H_{c0}) of SD grains in a hypothetical sample. With increasing T, the blocking curve (hyperbolas, eqn [24]) sweeps ever more rapidly through the points, producing the strongly nonlinear variations of k_o and Q_t portrayed in **Figure 41**.

the left of a particular curve are SP and are in thermal equilibrium with H_o at the T indicated. Grains above and to the right of a curve are blocked and unresponsive to H_o at T. Grains intersecting the curve are at their blocking temperature T_B.

The connection between AF and thermal demagnetization is provided by eqn [24]. For grains of a particular volume V, coercivity H_{c0} increases in proportion to $T_B/m_s^2(T_B)$. All other things being equal, SD grains with high blocking temperatures will also have high AF coercivities. However, if there is a wide spectrum of grain sizes or a mixture of remanence-carrying minerals, this correspondence is not clear-cut. Note also that for grains of fixed size and shape (i.e., $VH_{c0} =$ const.), $\ln t$ plays a role similar to T in magnetization and demagnetization. This equivalence is sometimes exploited in zero-field 'storage tests', but only very low-T_B remanence can be demagnetized in practical storage times of days to weeks. The $\ln t - T$ equivalence is also at the root of thermoviscous magnetization (Section 8.6.2).

8.5.4 Thermal Demagnetization of MD Grains

Although one often speaks of T_B of an MD grain (e.g., in Section 8.4.3), the reality is that individual walls have different blocking temperatures and the set of T_B's changes each time one of the walls moves

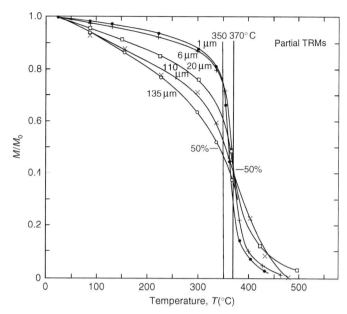

Figure 46 Stepwise thermal demagnetization of partial TRM produced over a narrow T_B range (370–350°C) in magnetites of different grain sizes. In the finest grains, the partial TRM demagnetizes mainly over the original blocking range, but 135 μm MD grains demagnetize over a broad range from room temperature to the Curie point. Open and filled dots, squares, and crosses are arbitary and used to mark the data for the 1 μm (solid dots), 6 μm (upright crosses), 20 μm (open squares), 110 μm (x's) and 135 μm (open dots) samples, respectively. Modified from Dunlop DJ and Özdemir Ö (2001) Beyond Néel's theories: Thermal demagnetization of narrow-band partial thermoremanent magnetizations. *Physics of the Earth and Planetary Interiors* 126: 43–57.

because of the change in H_d. This distribution of T_B values is particularly marked during thermal demagnetization. **Figure 46** illustrates how the spectrum of thermal demagnetization temperatures broadens with increasing grain size. For the larger sizes, demagnetization begins much below the TRM blocking temperature, even in the first heating step above T_0, and continues essentially up to T_C. Thermal demagnetization of MD remanence is not the clean efficient procedure it is for SD grains.

8.5.5 Resolving Multivectorial NRM

A frequent problem in paleomagnetism is to separate NRMs acquired over different T intervals at different times in the rock's history. In the simplest case, these are 'partial TRMs' acquired over nonoverlapping ranges of T_B. If two such components of NRM have sufficiently different ages, they will also have different directions because of plate motion and polarity reversals. The composite NRM is said to be 'multivectorial', and how efficiently the components are resolved can be judged by measuring the resultant NRM vector direction in the course of stepwise demagnetization.

SD grains have 'sharp' TRM blocking and thermal demagnetization temperatures. Their partial TRMs obey the Thellier (1938) laws:

1. Partial TRMs produced in nonoverlapping T intervals covering (T_0, T_C) sum to give the total TRM (T_C, T_0) (additivity law).
2. Partial TRMs produced in nonoverlapping T intervals are independent of one another in direction and magnitude (independence law).
3. A partial TRM produced by cooling in a field from T_1 to T_2 is unaffected by reheating in zero field to T_2 but is completely destroyed by reheating to T_1 (reciprocity law).

The clean thermal demagnetization of small grains evident in **Figure 46** (1 μm grains) is evidence of reciprocity: unblocking is exactly reciprocal to blocking in SD grains. In thermal cleaning of multivectorial NRM, there is a sharp junction between the demagnetization trajectories of two adjacent partial TRMs (**Figure 47**).

MD grains have smeared or distributed unblocking temperatures and do not give sharp junctions in demagnetization trajectories. In extreme cases, the result is that neither NRM component can be correctly resolved. In **Figure 48**, the estimated direction

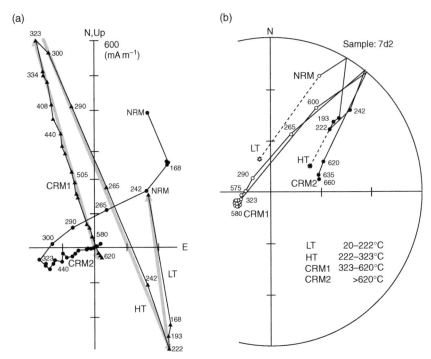

Figure 47   Multivectorial NRM in a monzonite sample, consisting of two partial TRMs (LT and HT) in adjacent T_B ranges (20–222°C and 222–323°C: pyrrhotite) and two CRMs (magnetite and hematite). Stepwise thermal demagnetization results are displayed as orthogonal projections of the full magnetization vector (left) and as unit vectors on a stereographic projection (right). Modified from Dunlop DJ, Schmidt PW, Özdemir Ö, and Clark DA (1997) Paleomagnetism and paleothermometry of the Sydney Basin. 1. Thermoviscous and chemical overprinting of the Milton Monzonite. *Journal of Geophysical Research* 102: 27271–27283.

of the lower-T partial TRM is within $\approx 10°$ of the correct direction but the direction of the higher-T partial TRM is $\approx 30°$ in error. Errors of this magnitude are unacceptable and render MD grains generally unusable as paleomagnetic directional recorders.

8.5.6   Thellier Paleointensity Determination

The Thellier laws are the basis of the standard method for determining paleofield intensity. The reciprocity law is again the key. If a TRM carried by SD grains is heated twice to the same temperature and cooled to room temperature, the first time in zero field and the second time in a field equal to $\mathbf{H_o}$ used to impart the TRM, the remanence lost in the first heating–cooling will be exactly restored in the second. The usual way of presenting such double-heating data is in an Arai plot (Nagata *et al.*, 1963) of NRM remaining after the first heating versus partial TRM gained in the second. For the experiment just described, the Arai plot for ascending T steps is a descending straight line of slope -1.

To make this experiment the basis of a practical technique for determining *a priori* unknown field intensities, H_o requires two additional properties of SD TRM. First is linearity: TRM intensity must be proportional to 'H_o' for weak fields. Second is constancy of blocking/unblocking temperatures: T_B must be independent of H_o, again for weak fields. In practice, these conditions are met if $H_o \leq 1$ Oe or $80\,\text{A m}^{-1}$ approximately.

Paleointensities are much more difficult to determine if MD grains are the TRM carriers. Thermal demagnetization begins immediately above T_0, as explained in the last section, but partial TRM tends to be acquired in higher T ranges (Xu and Dunlop, 2004). The result is a 'sagging' Arai plot in which more TRM is lost than regained in the lower T steps but acquisition outstrips loss of TRM in the higher T steps (**Figure 49**). Although such a plot is perfectly reproducible if caused solely by MD material, it generally leads to immediate rejection on grounds of nonlinearity because most Arai plots of this form result from chemical alteration of the sample during the Thellier experiment For practical details, *see* Chapter 13..

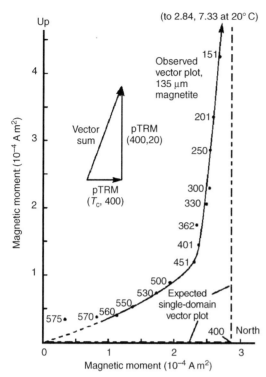

Figure 48 Projection of the full NRM vector in the course of stepwise thermal demagnetization for a sum of orthogonal partial TRMs in 135 μm magnetite. Instead of reproducing the orthogonal pattern of the actual partial TRM sum, the thermal data approximate to a sum of linear demagnetization trajectories with angular errors of about 10° and 30°, respectively. Modified from Dunlop DJ (2003b) Partial thermoremanent magnetization: Louis Néel's legacy in rock magnetism. *Journal of Applied Physics* 93: 8236–8240.

8.5.7 Stability and Domain State Tests

There are many tests that use magnetic parameters to classify samples according to their domain state and implied paleomagnetic stability. Among these are the stepwise demagnetization techniques described in Sections 8.5.1–8.5.4 which deal with NRM directly. The two tests described in this section use instead hysteresis data, which may or may not give a fair representation of the mixture of minerals and grain sizes carrying NRM.

Very popular is the Day diagram (Day *et al.*, 1977), a plot of the saturation hysteresis parameters M_{rs}/M_s versus H_{cr}/H_c (see Section 8.1.2 for definitions of these quantities). The Day diagram is well documented and easily interpretable only for magnetite and TM60 (Dunlop, 2002a, 2002b). For magnetite, it gives a fairly decisive separation into SD + SP, PSD or SD + MD, and MD regions (**Figure 50**). The

underlying theory is too complicated to review here but it is based on results like $M_{rs}/M_s = 0.5$ for uniaxial SD grains (Section 8.4.1) and self-demagnetization of MD grains (Section 8.4.2).

The venerable Preisach diagram, which has been used in many forms by the magnetic recording community over the past four decades, has in recent years been popularized in geophysics as the FORC diagram (Roberts *et al.*, 2000). FORC stands for first-order reversal curve and describes the method used to derive the distribution plotted in the diagram. Starting from saturation, a series of descending partial hysteresis loops are interrupted at different fields and minor loops (the first-order reversal curves) are traced in increasing fields. Rather than simply differencing results at various fields, a quadratic function is fitted to the data in the immediate surroundings of each point in order to reduce noise.

The FORC diagram plots H_u (equivalent to H_i) versus H_c, thus using the Preisach–Néel hysterons of **Figure 43** to interpret the data. Pragmatically, FORC diagrams for different domain and interaction states are distinctive (**Figure 51**). SD grains have a single peak in their distribution, well separated from the origin. If interactions are minimal, the distribution is confined to the H_c axis. Increasing interactions cause spreading parallel to the H_u axis. In PSD grains, the distribution peak moves toward the origin and ultimately merges with the H_u axis. MD grains have distribution contours almost paralleling the H_u axis at low H_c.

It has recently been verified that the Preisach–Néel interpretation does make sense for SD FORC distributions (Muxworthy and Dunlop, 2002; Carvallo *et al.*, 2004), by showing that the distribution parallel to the H_u axis contracts with increasing T as $M_s(T)$, as expected for magnetostatic interactions, and the distribution parallel to the H_c axis contracts in the same way as the bulk coercive force, $H_c(T)$. These results place the FORC method on a firmer physical foundation. H_u in the MD case presumably reflects the effect of self-demagnetizing fields ('domain interactions') rather than interparticle interaction.

The decomposition of magnetic mixtures into their constituent mineral and domain state fractions is a developing field with much promise. Day diagrams are ambiguous in some regions, magnetite + TM60 mixtures and mixtures of SP, SD, and PSD magnetites being indistinguishable (Dunlop, 2002b). For simple mixtures of magnetites with two different domain states, Carter-Stiglitz *et al.*

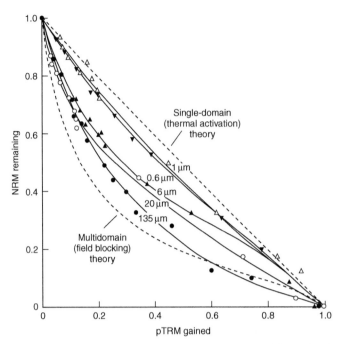

Figure 49 Simulated Thellier paleointensity experiments carried out on magnetite grains of various sizes, compared to the predictions of SD and MD theories. With increasing grain size, the data 'sag' increasingly below the ideal SD line of slope −1. Thy symbols are arbitrary and a different set is used to mark the data for 0.6, 1, 6, 20, and 135 μm samples, respectively. Modified from Xu S and Dunlop DJ (2004) Thellier paleointensity theory and experiments for multidomain grains. *Journal of Geophysical Research* 109: B07103 (doi:10.1029/2004JB003024).

(2001) have devised a decomposition algorithm for hysteresis data that uniquely determines the proportions of the phases. FORC diagrams have the potential to unravel bimodal mixtures of two soft phases or hard and soft phases (Muxworthy *et al.*, 2005). Preisach models without the FORC formalism have also proved useful in analyzing isothermal remanence acquisition curves, themselves widely used to infer components of mixtures (Fabian and von Dobeneck, 1997; Egli, 2003; Heslop *et al.*, 2002, 2004). Finally, Fabian (2003) has suggested several useful parameters that can be obtained with little additional effort from hysteresis and backfield measurements, and Leonhardt (2006) has devised software for analysis of parameters from a broad variety of rock magnetic measurements.

8.6   Remanent Magnetization Processes in Nature
8.6.1   Thermoremanent Magnetization (TRM)

We have touched on many of the properties of SD and MD TRM in Sections 8.4.3 and 8.5.3–8.5.6.

An exhaustive review of TRM has recently been given by Özdemir (2007). Only a limited number of aspects are dealt with here, mostly those of practical paleomagnetic importance.

First is the question of how TRM intensity depends on the strength H_o of the magnetic field applied during cooling from T_C. Only if TRM is proportional to H_o will paleointensity determination (Section 8.5.6) be feasible. If it were possible to choose the laboratory field H_L to exactly match the paleofield H_A, the Arai plot would have a slope of −1 and the TRM field dependence would be irrelevant. In practice, it is difficult or impossible to guess H_A for a particular sample with sufficient accuracy before doing the experiment and the slope NRM/TRM must be assumed to be $-H_A/H_L$. Fortunately, linearity seems to be a good assumption for almost all Earth materials for fields $\leq 100\,\mu$T.

One would anticipate such linearity on the basis of the SD equation for TRM:

$$M_{tr}(H_o, T_0) = M_{s0}\,\tanh(\mu_0 V M_s(T_B)H_0/kT_B)$$
$$\approx \mu_0 V M_{s0}{}^2 m_s(T_B)H_0/kT_B \qquad [25]$$
$$\text{for}\quad \mu_0 V M_{sB}H_0/kT_B \ll 1$$

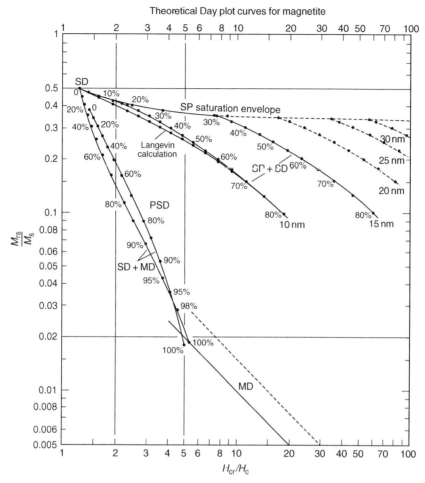

Figure 50 Theoretical Day plot curves for magnetites in various domain states (SD, PSD, MD) and mixtures of different states (SP + SD, SD + MD). Modified from Dunlop DJ (2002a) Theory and application of the Day plot (M_{rs}/M_s versus H_{cr}/H_c), 1. Theoretical curves and tests using titanomagnetite data. *Journal of Geophysical Research* 107(B3): 2056 (doi:10.1029/2001JB000486).

which is an extension of eqns [20] and [22]. For purposes of testing, since V can be difficult to estimate, it is useful to recast [25] in terms of H_{c0} and T_B by substituting from eqn [24] and setting $23 + \ln 1.78t \approx 25$, giving (Dunlop and Kletetschka, 2000)

$$M_{tr}(H_o, T_0) = M_{s0} \tanh[50H_o/H_{c0} \, m_s(T_B)] \quad [26]$$

The TRM intensity predicted by [25] or [26] is often much higher than experimental TRMs of SD magnetites, a discrepancy explained by Dunlop and West (1969) in terms of particle interactions. Magnetostatic interaction is much less important for hematite because of its low M_s, and SD hematites do indeed have TRMs in close agreement with theory (Dunlop and Kletetschka, 2000; Özdemir and Dunlop, 2002).

It is usually assumed that any discrepancy between predicted and measured TRMs in the weak-field region is irrelevant to paleointensity determination because NRM and TRM are affected identically. This is certainly true if the comparison is between total NRM and total TRM: $H_A = (NRM/TRM)H_L$. The theoretical field dependence is not used in determining H_A. All that is required is proportionality between M_{tr} and H_A. However, matters are not so clear for a Thellier-type comparison between sets of partial TRMs and their corresponding NRM fractions because the theoretical proportionality factor, $\mu_0 V M_{s0}^2 m_s(T_B)/kT_B$, is a strong function of V and T_B and will be different for each partial TRM. If the actual proportionality factor depends on additional parameters (H_i, for example), different Thellier steps could give different results. The main effect is likely to be a stretching of portions of the Arai graph relative to others, rather than nonlinearity.

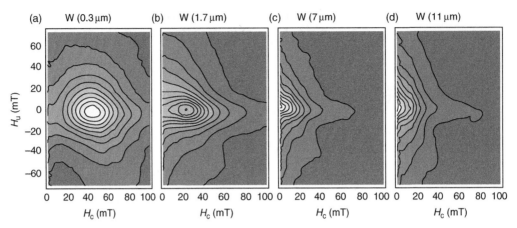

Figure 51 Experimental first-order reversal curve (FORC) distributions for magnetites of various grain sizes. As size increases, the central distribution peak moves to smaller coercivities H_c and ultimately merges with the axis. At the same time, the contours increasingly spread parallel to the H_u axis, a result of self-demagnetizing fields. Modified from Muxworthy AR and Dunlop DJ (2002) First-order reversal curve (FORC) diagrams for pseudo-single-domain magnetites at high temperature. *Earth and Planetary Science Letters* 203: 369–382.

The fundamental basis of TRM linearity in weak fields is less obvious for MD grains. The Stacey (1958) theory (Section 8.4.3) predicts a TRM

$$M_{tr}(H_o, T_0) = [m_s(T_B)(1 + Nk_i)]^{-1}(H_o/N) \quad [27]$$

The proportionality between M_{tr} and H_o in [27] is an illusion, however, because T_B is itself a function of H_o. The Néel (1955) theory, on the other hand, gives an explicit field dependence. $T_B(H_o)$ is accounted for by introducing the temperature dependence of microcoercivity H_c, which through wall pinning or nucleation, controls T_B: $H_c(T)/H_{c0} = h_c(T) = m_s{}^n(T)$. For magnetite, $n = 1–2$, while for hematite, $n = 3$ is usually assumed. The ultimate result is

$$M_{tr}(H_o, T_0) = [n/(n-1)^{1-1/n}][H_{c0}{}^{1/n}/(1 + Nk_i)] \\ \times (H_o{}^{1-1/n}/N) \quad [28]$$

which is more intricate than Stacey's result and highly nonlinear for small values of n. Equation (28) is quite successful in predicting TRM intensities for MD hematite (Kletetschka *et al.*, 2000a, 2000b; Dunlop and Kletetschka, 2000; Özdemir and Dunlop, 2005; **Figure 52**), which are nonlinear below 100 μT, and TRM intensities of PSD magnetites (Dunlop and Argyle, 1997), which are quasi-linear in the same field region.

Experimental TRMs for MD magnetite and TM60 are linear or close to it for fields below about 1 mT and vary approximately as $H_o{}^{1/2}$ (appropriate for $n = 2$) at higher fields (Dunlop and Waddington,

1975; Tucker and O'Reilly, 1980). The generally observed weak-field linearity of TRM inspired Néel (1955) to modify his field-blocking theory (eqn [28]) by postulating a different blocking condition when $H_o < H_f(T_B)$, H_f representing the unpinning effect of thermal fluctuations on domain walls. Then, if $n = 2$,

$$M_{tr}(H_o, T_0) = [H_{c0}{}^{1/2}H_f(T_B)^{1/2}/(1 + Nk_i)](H_o/N) \quad [29]$$

Although the physical reasoning is sound, Néel's result is unsatisfying because $H_f(T_B)$ involves T_B and furthermore is experimentally inaccessible, so that [29] is untestable.

Many other attempts have been made to improve our understanding of MD TRM. Dunlop and Xu (1994) and Xu and Dunlop (1994, 2004) extended the Néel field-blocking theory to partial TRMs and were able to predict quite closely the curvature of MD Arai plots. Fabian (2000, 2001) proposed a phenomenological model closely paralleling the Preisach–Néel model (**Figure 43**), which has since been extended by Leonhardt *et al.* (2004). Biggin and Böhnel (2003) have used Fabian's model to devise an improved experimental protocol that substantially reduces the curvature of Arai plots.

Finally, Kletetschka *et al.* (2004) have reported a linear relationship between $\log M_{tr}/M_{rs}$ and $\log H_o$ which is 'universal', that is, has the same slope for all minerals and domain states. In view of the complexities of TRM and the vastly different blocking mechanisms of SD and MD grains, it is unlikely

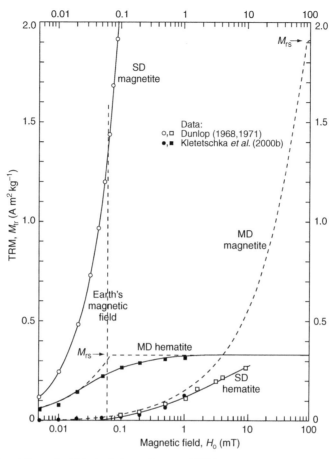

Figure 52 Experimental and theoretical (dashed) dependences of TRM intensity on applied magnetic field H_o for SD and MD magnetite and hematite. Self-demagnetization is the reason why TRM saturates for MD magnetite in fields three orders of magnitude larger than for MD hematite. Modified from Dunlop DJ and Kletetschka G (2000) Multidomain hematite: A source of planetary magnetic anomalies? *Geophysical Research Letters* 28: 3345–3348.

that their parallel data trends reveal an aspect of some universal truth, delightful though that would be. The log–log scale expands and highlights the weak-field region, the crucial field range for paleointensities. The observed parallelism implies that for the materials tested, TRM was approximately linear with field in this range.

Offsets between Kletetschka *et al.*'s lines are in some cases substantial. The offsets reflect the very different fields at which TRM rises toward saturation ($M_{tr} \rightarrow M_{rs}$) for different minerals and domain states. **Figure 52** shows that for hematite, MD TRM is almost saturated in a weak field ($100\,\mu T$) while SD TRM is far from saturation even at $1000\,\mu T$. Just the reverse is true for magnetite, SD TRM rising rapidly but MD TRM theoretically reaching saturation only in fields 200 times larger than the TRM saturation

field for MD hematite. These startling differences are due entirely to the 200-fold contrast between the M_s values of the two minerals. The self-demagnetizing fields that strongly oppose TRM in MD magnetite are negligible in MD hematite. The contrast between the SD TRMs for the two minerals (**Figure 52**) is also due to their contrast in M_s, the ultimate M_{rs} toward which TRM is rising being 200 times larger in magnetite.

These diametrically opposite grain-size dependences of TRM in magnetite and hematite are illustrated further in **Figure 53**. The TRM of magnetite drops more than 2 orders of magnitude between the SD threshold around $0.1\,\mu m$ and true MD sizes of $50{-}100\,\mu m$. Furthermore it decreases continuously, not abruptly. This is one of the defining features of PSD behavior. The TRM of hematite

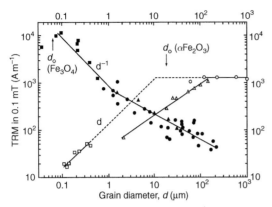

Figure 53 The contrasting grain-size dependences of TRM intensity in magnetite and hematite. TRM of magnetite decreases steadily with increasing grain size above the critical SD size d_0, initially as d^{-1}. TRM of hematite increases with grain size up to a constant plateau (saturation remanence) above the SD–MD transition size d_0 (αFe_2O_3). Modified from Dunlop DJ and Arkani-Hamed J (2005) Magnetic minerals in the Martian crust. *Journal of Geophysical Research* 110: E12S04 (doi:10.1029/2005JE002404).

increases over the same size range, reaching a constant and unexpectedly high (for such a weakly magnetic mineral) level above its SD threshold around 15 μm.

The grain-size dependence of TRM in titanomagnetites with various Ti contents and in pyrrhotite has been reported by Day (1977), O'Donovan *et al.* (1986), and Menyeh and O'Reilly (1998). Over the 1–30 μm range, there is a general decrease in TRM intensity, but not as marked a decrease as in magnetite. There are also measurements of TRM in SD TM60 which link with those for the larger sizes (Özdemir and O'Reilly, 1982b).

8.6.2 Viscous and Thermoviscous Magnetization

The equivalence between the effect of T in thermal demagnetization and TRM acquisition and the effect of an equivalent change in $\ln t$, which produces the so-called 'viscous magnetization', was pointed out in Section 8.5.3. If T and $\ln t$ both change by comparable amounts, for example, during very slow cooling in nature, 'thermoviscous magnetization' results. Viscous and thermoviscous magnetization have recently been reviewed by Dunlop (2007).

Equation [24] implies that a change in T by a fixed amount ΔT has a disproportionate effect on the motion of (V, H_{c0}) contours on the Néel diagram (**Figure 45**) at high T where $m_s(T)$ is changing

rapidly (>500°C in the case of magnetite). Hence, blocking temperatures tend to be concentrated close to T_C. The same is true of a fixed change $\Delta(\ln t) = \Delta t/t$. Viscous magnetization is usually minor near T_0 but large at high T, especially near T_C (**Figure 54**). This effect amounts to a thermoviscous Hopkinson peak which augments the peak due to SP magnetization. In **Figure 55**, for temperatures above 100°C, the extrapolated viscous magnetization M_v due to exposure to the Earth's field for 700 ka (the length of the Brunhes polarity chron) equals M_{in} due to short-term exposure to the same field, doubling the overall induced magnetization. In the Hopkinson peak region above 500°C, the increase is by more than a factor 2.5.

The rocks illustrated in **Figures 54** and **55** are not exceptional. Almost any paleomagnetic sample has a measurable viscous remanent magnetization (VRM)

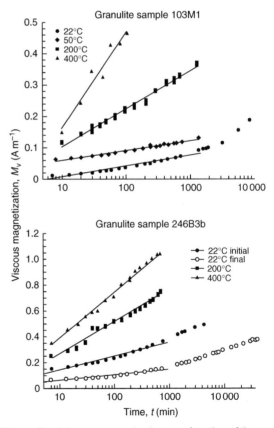

Figure 54 Viscous magnetization as a function of time measured at temperatures up to 400°C for two continental mafic granulites. Samples are much more viscous at 400°C than at room temperature. Modified from Kelso PR and Banerjee SK (1994) Elevated temperature viscous remanent magnetization of natural and synthetic multidomain magnetite. *Earth and Planetary Science Letters* 122: 43–56.

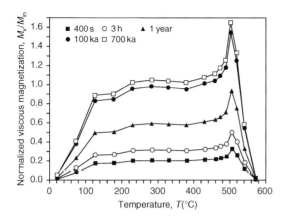

Figure 55 Viscous magnetization measured after 400 s and 3 h and extrapolated assuming a ln t dependence to 1 yr, 100 ka, and 700 ka for an oceanic serpentinized peridotite. M_v is much larger at 120°C and above than at room temperature, and increases to a Hopkinson peak in the T_B range around 500°C. Modified from Pozzi J-P and Dubuisson G (1992) High temperature viscous magnetization of oceanic deep crustal- and mantle-rocks as a partial source for Magsat magnetic anomalies. *Geophysical Research Letters* 19: 21–24.

produced by exposure to the geomagnetic field during the Brunhes at ambient temperatures. The question that then arises is the most efficient means of erasing this contaminant NRM component. In the case of hematite especially, AF demagnetization is not the answer (see Dunlop and Özdemir, 1997 chapter 10.3). Thermal cleaning, on the other hand, is very efficient because of the reciprocal roles of time and temperature in SD magnetization relaxation. VRM carried by MD grains is less amenable to thermal cleaning because of the spreading of the blocking temperature spectrum by self-demagnetizing fields, but fortunately MD grains tend to be less viscous and the VRMs to be cleaned are usually minor.

A useful graphical depiction of SD thermoviscous effects is the Pullaiah diagram (Pullaiah *et al.*, 1975). Equation [24] can be used to find when blocking occurs for the same SD ensemble, specified by VH_{c0}, under different combinations (T_{B1}, t_1) and (T_{B2}, t_2) of temperature and time. The result is

$$\begin{aligned}(23.6 + \ln t_1)[T_{B1}/m_s^2(T_{B1})] \\ = (23.6 + \ln t_2) \times [T_{B2}/m_s^2(T_{B2})]\end{aligned} \quad [30]$$

for magnetite. For hematite, the appropriate factor is $m_s^4(T_B)$. When plotted on a time–temperature diagram, the contours of **Figure 56** result. Each curve is the locus of all combinations (t, T) that produce blocking for a specified ensemble VH_{c0}. For example,

if $t_1 = 700$ ka, $T_{B1} = T_0$ (Brunhes VRM), it is easy to calculate the thermal demagnetization temperature T_{B2} that will erase the VRM on a laboratory heating timescale ($t_2 \sim 15$ min, say) by following one of the blocking curves. SD magnetites follow the Pullaiah contours closely, while larger grains require systematically higher temperatures to demagnetize their VRMs (Dunlop and Özdemir, 2000).

Another way of studying viscous magnetization is to measure how initial susceptibility depends on the frequency ω of a weak applied magnetic field H_0 sin ωt. Néel (1949, equation 73) showed that

$$\partial k_0/\partial \ln \omega = (1/H_0)(\partial M_r/\partial \ln t) \quad [31]$$

Frequency-dependent susceptibility focuses on grains at their blocking condition, passing from stably magnetized to SP. By varying T, the Néel diagram can be scanned, a method of 'magnetic granulometry', that is, of determining (V, H_{c0}). Usually, ultrafine grains are targeted; these are SP at T_0 but become stably magnetized at low T (Jackson *et al.*, 2004).

In discussing TRM and thermal demagnetization, we treated T_B as a sharp temperature. In reality, both processes are thermoviscous rather than truly thermal because natural heating or cooling times even for a rapidly chilled lava flow are >>atomic reorganization times ($\sim 10^{-10}$ s). Blocking occurs over a range of temperatures, indicated on **Figure 56** as 5% and 95% blocking contours flanking the average contour specifying T_B.

The bounds on blocking are calculated in the following way. York (1978a, 1978b) derived an expression for blocking temperature T_B during slow cooling at a rate $-dT/dt$. This same T_B can be obtained from an isothermal magnetization equation, for example, [24], if t is identified with a characteristic cooling time t^*:

$$\begin{aligned}t^* &= [kT_B^2/\Delta E(T_B)](-dT/dt)^{-1} \\ &= [T_B/(23 + \ln 1.78t^*)](-dT/dt)^{-1}\end{aligned} \quad [32]$$

(Dunlop and Özdemir, 1997, p. 480), the second expression being obtained by substituting for $\Delta E(T_B)$ from eqn [23] and substituting $\tau = 1.78t$. An alternative approach that makes no approximations is given by Dodson and McClelland-Brown (1980).

To find 5% and 95% limits on blocking, note that according to eqn [21],

$$\begin{aligned}\Delta M(t)/\Delta M(0) &= \exp[-t/\tau(T_n)] = n \\ \text{if} \quad \tau(T_n) &= t/\ln(1/n)\end{aligned} \quad [33]$$

(a)

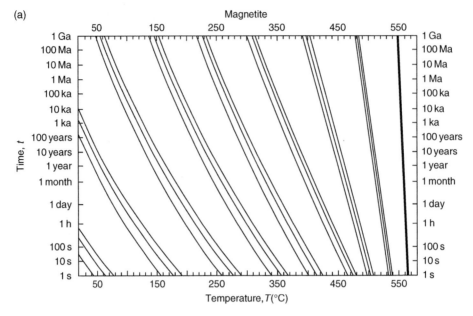

(b)

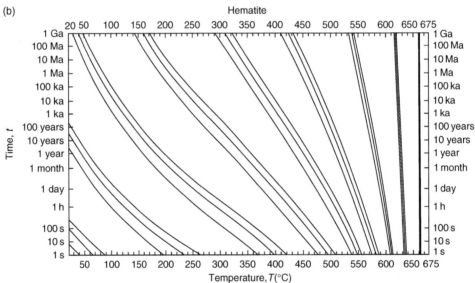

Figure 56   Time–temperature blocking contours for SD grains of magnetite (a) and hematite (b), calculated from Néel theory (eqn [30] or eqns [31]–[35]). Along each contour, the product VH_{c0} is constant and each combination (t, T) produces 5%, 57%, or 95% blocking (upper, middle, lower curves, respectively, in each set). The blocking temperature is strongly t dependent at low T but almost independent of t near the Curie point where $M_s(T)$ changes rapidly. For details of the calculations, see the text.

where T_n is the temperature at which ΔM is reduced to $n = 0.05$ or 0.95. Specifically we have

$$n = 0.05, \tau(T_{0.05}) = 0.334t$$
$$n = 0.57, \tau(T_B) = 1.78t \qquad [34]$$
$$n = 0.95, \tau(T_{0.95}) = 19.5t$$

The middle line of [34] shows that T_B as defined in Section 8.5.3 corresponds to 57% blocking. The

blocking equation corresponding to eqn [24] for a general n is

$$VH_{c0} = [2k/M_{s0}]\{23 + \ln[t/\ln(1/n)]\}[T_n/m_s^2(T_n)] \quad [35]$$

Pullaiah diagrams (**Figure 56**) are computed as follows:

1. Choose values of V, H_{c0}, and T_B. Determine the value of $m_s(T_B)$ for the chosen T_B. Calculate the matching value of t from [24].

2. For the same t, determine the values of $T_{0.05}$ and $T_{0.95}$ that satisfy [35]. This is a trial-and-error process. Even if $m_s(T)$ can be represented analytically (e.g., Dunlop *et al.*, 2000), the resulting transcendental equation is best solved by iteration.

3. Find the corresponding cooling rate by substituting T_B and $t^* = t$ in [32].

For outside limits $10\,s \leq t^* \leq 1\,Ga$ and $300\,K \leq T_B \leq 950\,K$, $5 \leq t^*(-dT/dt) \leq 35$. Thus, t^* can be interpreted as 5–35 times the time to cool through $1°C$. For example, for a typical magnetite blocking temperature of $540°C = 813\,K$ and a cooling timescale $t^* = 1$ Ma, $t^* \approx 15(-dt/dT)$ or the time to cool through about $15°C$ if $(-dT/dt)$ is constant.

We must not confuse t^* with the time to pass from 5% to 95% blocking. The blocking interval $\Delta t = t_{0.05} - t_{0.95}$ is of constant width on a logarithmic scale for short or long t (**Figure 56**). At lower T, as the $t-T$ contours become more oblique, the blocking range $\Delta T_n = T_{0.05} - T_{0.95}$ widens. At high T, where $m_s^2(T)$ controls blocking in [24] or [35], T_n or T_B are almost independent of t, so the contours are steep and ΔT_n is narrow. To be specific, a magnetite partial TRM with $T_B \approx 550°C$ blocks over about a $3°C$ interval, whatever the cooling rate, while for $T_B \approx 300°C$ blocking requires cooling through $\approx 15–25°C$, depending on $-dT/dt$.

As an example of the use of the Pullaiah diagram in assessing magnetization or remagnetization during slow cooling, consider the case of Mars (Dunlop and Arkani-Hamed, 2005). Mars' core dynamo is thought to have ceased by 4.0 Ga. Did this early and short-lived field allow enough time for Mars' crust to become stably magnetized by TRM? For $T_B \approx 500–550°C$, TRM blocking requires about $5°C$ of cooling or about 10 Ma at an assumed cooling rate $-dT/dt \approx 0.5°C/Ma$. To cover the entire $500–550°C$ T_B range would require about 100 Ma, which is well within the lifetime of Mars' dynamo. A second question is whether the primary TRM could have resisted thermoviscous demagnetization over 4 Gy since the field shut off. This question can be answered by following one of the average blocking contours. The grains most susceptible to demagnetization are those with the lowest T_B values, namely $500°C$, measured on a laboratory timescale of a few minutes. The corresponding temperature for a timescale of 4 Ga is $350–400°C$. Provided cooling to this temperature level occurred before the field ceased, all TRM would in principle have survived to the present day. This point is crucial in interpreting the origin of the intense magnetic anomalies on Mars.

Cooling rate affects the intensity of TRM as well as its blocking, a vital point in determining paleointensities in slowly cooled orogens. Dodson and McClelland-Brown (1980) and Halgedahl *et al.* (1980) predicted using Néel SD theory that the intensity of TRM in SD grains should increase for longer cooling times by as much as 40% between laboratory and geological settings. Their predictions are based on the change in T_B with t in eqn [25], but this equation often fails to predict accurately the proportionality factor between TRM and H_o. Experimentally, Fox and Aitken (1980) found for baked clay samples containing SD magnetite about a 7% increase in the intensity of TRM when the cooling time changed from 3 min to 2.5 h. The only serious attempt to apply a cooling-rate correction in paleointensity work is by Selkin *et al.* (2000). In their study of the Archean Stillwater Complex, the uncorrected paleointensity was estimated to be 45% higher than the true paleofield recorded during 0.8 Ma cooling. Bowles *et al.* (2005) found that submarine basaltic glasses cool in nature at about the same rate as in the laboratory, so that minimal or no cooling rate corrections were necessary.

8.6.3 Chemical Remanent Magnetization (CRM)

VRM is not the only source of secondary NRM in rocks. Equally common, and serious because of its often high coercivities and T_B's, is CRM, produced by magnetization blocking in a growing magnetic phase. In **Figure 47**, two CRMs are present in the rock and both have higher blocking temperatures (since they reside in magnetite and hematite) than the paleomagnetically interesting pyrrhotite partial TRMs. In this particular example, the CRM directions mimic those of the partial TRMs, effectively dating the CRMs. Generally, however, CRMs are difficult to date with any precision and are merely noise obscuring the primary paleomagnetic signal.

One setting in which CRM is useful, indeed vital, is the seafloor. Primary TM60 in pillow lavas produced at mid-ocean ridges is soon oxidized in seawater to titanomaghemite (Sections 8.3.3 and 8.3.4). Oxidation proceeds inwards from the surface of grains, probably aided by iron removal by sulfate-reducing bacteria (Carlut *et al.*, 2007). Exchange coupling seems to remain unbroken across the moving phase boundary between the two spinel phases. Coupling is aided by the diffuse nature of the boundary region, which has a gradient of lattice vacancies extending inward from the surface. Nevertheless the lattice mismatch between the two minerals ultimately causes cracking of the surface titanomaghemite and a considerable reduction in effective grain size. This in turn makes the titanomaghemite CRM considerable more coercive than the parent TM60 TRM, and ensures stability of the remanence over the lifetime of an ocean basin. Linear magnetic anomalies over the oceans, the hallmark of seafloor spreading, are usually thought of in terms of TRM but their actual source is CRM. Geophysically speaking, it is fortunate indeed that the CRM direction 'remembers' that of the parent TRM. The oxidation cannot be pushed beyond about $z = 0.65$, where multiphase oxidation products begin to appear (Section 8.3.5). The CRM direction then increasingly deviates from the TRM direction (Özdemir and Dunlop, 1985; **Figure 57**). The fidelity of magnetic stripes over the oceans suggests that oxidation of the seafloor typically stays below this level.

Controlled CRM experiments are difficult. Conditions in nature are hard to reproduce in the laboratory and if heating is used to accelerate the kinetics, undesired additional changes may occur. One recent study in which additional oxidation and unmixing was intentionally produced during experiments (Draeger *et al.*, 2006) concerns paleointensity

determination on basalts, which very frequently reequilibrate chemically in the course of a Thellier experiment. Draeger *et al.* find that CRM (produced near ambient temperature) and TCRM produced at higher T are virtually indistinguishable in their Arai plots from pure TRM. However, there is no reason to suppose that the proportionality between remanence and field is the same for them as it is for pure TRM. Since basalts are by far the most frequently studied rock in paleointensity work, this finding is sobering. Paleofield estimates could be seriously in error despite apparently high-quality Arai plots.

CRMs are particularly common in chemically active environments such as soils and recently deposited sediments. Their detection and elimination is a whole science in itself (e.g., Rowan and Roberts, 2006; see also Verosub and Roberts, 1995).

8.6.4 Detrital and Post-Depositional Remanent Magnetizations (DRM and PDRM)

'Detrital remanent magnetization (DRM)' is produced when previously magnetized mineral grains settle through a water column in the geomagnetic field. The partial alignment of the magnetic moments of the grains may subsequently be altered by changes in the water-rich sediment as it is buried beneath later sediments, compressed, and reworked by burrowing organisms. The altered remanence is called

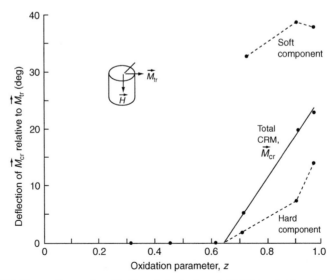

Figure 57 Direction of CRM formed by maghemitization of single-domain TM60 in a field applied perpendicular to the initial TRM direction. CRM forms parallel to initial TRM until multiphase oxidation products begin to appear around $z = 0.6$. Modified from Özdemir Ö and Dunlop DJ (1985) An experimental study of chemical remanent magnetizations of synthetic monodomain titanomaghemites with initial thermoremanent magnetizations. *Journal of Geophysical Research* 90: 11513–11523.

post-depositional remanent magnetization (PDRM). For a review, see Kodama (1992).

The key question for paleomagnetic directional studies is whether or not DRM and PDRM faithfully record the geomagnetic field at the time of deposition or shortly thereafter. In order for sediments and the sedimentary rocks that eventually form from them to be useful as recorders of the paleofield intensity, DRM/PDRM must in addition be proportional to the field acting during deposition. Both questions are the subject of lively debate. Since DRM is strongly dependent on the grain size, composition and concentration of the magnetic phase(s) as well as the properties of the non-magnetic matrix, there is no simple answer to either question.

Almost any theory indicates that the alignment of magnetic moments during settling in a magnetic field occurs rapidly (within seconds for $\leq 100\,\mu m$ magnetite grains) or not at all (for mm-size grains) (see Dunlop and Özdemir, 1997, chapter 15). However, at the bottom of the water column, elongated grains will tend to rotate until the long axis is horizontal. Since magnetic moments of elongated grains prefer to be parallel to the long axis to reduce self-demagnetizing fields, this rotation produces a shallowing of the DRM vector, called the inclination error. DRM with no substantial post-depositional reworking does exhibit an inclination error in both natural and laboratory redepositional settings (Tauxe and Kent, 1984). So do the remanences of deep-sea sediments redeposited in the laboratory and compacted under hydrostatic pressure. The origin of the inclination shallowing in this case seems to be locking of magnetite grains to clay particles during formation of clay fabric (Sun and Kodama, 1992). It may be possible to "undo" inclination shallowing using a correction based on measured magnetic fabric anisotropy (Kodama and Sun, 1992; Hodych and Bijaksana, 1993).

Post-depositional reworking of a sediment may reduce or remove the inclination error. Although demonstrated in laboratory experiments (e.g., Kent, 1973), it is less clear that this is always the case under natural conditions. Within a fluid layer at the top of the sediment column, magnetic grains continue to rotate in water-filled voids until either lack of pore space or adhesion to clay particles causes lock-in. Generally, slow deposition in the deep sea means that there could be a time lag of several thousand years between initial deposition and lock-in. This estimate is based on continuous deposition

experiments in the laboratory (e.g., Løvlie, 1976), which suggest not only a time lag but also some blurring or loss of resolution of rapid geomagnetic field changes. The complex array of factors that potentially affect PDRM has led to increasingly intricate and sophisticated modeling (e.g., Heslop et al., 2004; Witt et al., 2004).

The question of linearity of DRM/PDRM intensity with applied field H_o is far from clear despite decades of experimentation. In classic studies, Johnson et al. (1948) redeposited varved clays and found linear DRM intensities for $H_o \leq 1$ Oe or $80\,A\,m^{-1}$, while there was essentially no linear region for PDRM in sediments redeposited by Tucker (1980). Tauxe et al. (2006) conclude from their experiments that DRM intensity is not linearly related to field strength under many natural conditions.

Apart from the uncertainty about linearity of sedimentary paleointensities, there is another difficulty. Because the DRM/PDRM process cannot really be reproduced in the laboratory, sediments and sedimentary rocks yield only relative paleointensities. Tauxe et al. (1995) developed a pseudo-Thellier method that compares different coercivity fractions of the NRM with corresponding fractions of a laboratory-produced anhysteretic remanent magnetization (ARM). ARM avoids heating and alteration of the material, and for this reason is sometimes used as an assumed analog of TRM in determining (absolute) paleointensities of igneous rocks see chapter 13. In the pseudo-Thellier method, ARM is taken to be analogous in its coercivity spectrum to DRM/PDRM. Although there is no a priori reason to expect such equivalence, the success of the pseudo-Thellier method (e.g., Brachfeld and Banerjee, 2000) justifies the assumption.

Yu et al. (2002a, 2002b, 2003) have investigated whether or not partial ARMs in magnetite obey laws of additivity, reciprocity and independence analogous to Thellier's laws for partial TRMs (Section 8.5.5). This is a necessary condition for success of the pseudo-Thellier method. They found that:

1. For SD, PSD, and MD grains, partial ARMs produced in non-overlapping H_c intervals covering the AF interval $(0, H_{max})$ sum to give the total ARM $(0, H_{max})$ (additivity law).
2. For SD and PSD grains, a partial ARM produced by applying a steady field while ramping down the AF from H_{c2} to H_{c1} is AF demagnetized mainly over the interval (H_{c1}, H_{c2}) (reciprocity law). In the

case of MD grains, however, >50% of the partial ARM is demagnetized below H_{c1} (non-reciprocity).

3. As a result of 2, for SD and PSD grains partial ARMs produced in nonoverlapping H_c intervals are independent of one another in direction and magnitude (independence law). For MD grains, non-reciprocity results in overlap between AF demagnetization ranges such that the direction of the higher-coercivity partial ARM is recoverable but the direction of the lower-coercivity partial ARM is spurious. On the other hand, the intensity of the lower-coercivity partial ARM is relatively well estimated by pseudo-Thellier analysis of the multivectorial total ARM but the intensity of the higher-coercivity partial ARM is always underestimated.

As a result of 2 and 3, pseudo-Thellier plots are increasingly nonlinear for larger grain sizes (**Figure 58**).

Finally although magnetite is typically the primary magnetic mineral in clastic sediments, it is perfectly possible for hematite to acquire DRM and PDRM. Hematite-rich redbeds are of considerable importance in the paleomagnetic record; indeed they dominated early studies because of their typically strong NRMs. The ultrafine grain hematite pigment that colors redbeds is secondary and carries CRM (or no remanence at all if the grains are SP). AF treatment generally fails to demagnetize the CRM completely because of SD hematite's high coercivities, but thermal demagnetization may succeed because the very fine grain sizes lead to relatively low T_B's (eqn [24]). The remaining primary DRM/PDRM signal in coarser detrital hematite grains, if any, is then revealed. Most paleomagnetists would feel uncomfortable in accepting any hematite NRM as primary unless supported by field evidence, for example, correlation between NRM and bedding tilt.

8.7 Summary

The induced and remanent magnetizations of rocks are the result of a complex interplay among factors like mineralogy, grain size, domain structure, temperature, time, and ambient geomagnetic field. For a given mineral, material properties like M_s, K, and λ are slowly varying functions of T except in the vicinity of the Curie point or other phase transitions. It is relatively sharp changes in properties that are mainly responsible for the recording and preservation of a stable paleomagnetic remanence, however. One such transformation is from single-domain to nonuniform domain structures with changing grain size. Although this transition is not as abrupt as imagined in classical domain theory, the first appearance of a spin vortex or other embryonic wall nucleus signals the end of high-coercivity coherent rotation of the entire coupled spin structure of a grain. Reversal by spin curling or, in larger grains, by wall displacement compromises paleomagnetic stability. A single-domain–multidomain dichotomy may be unrealistic but SD or SD-like coercivity is still the paleomagnetic ideal.

The other crucial change in properties occurs during NRM acquisition and is truly sharp on ordinary timescales. This is blocking, the passage from a thermal equilibrium state to a frozen out-of-equilibrium state that may survive unchanged for geological lengths of time. Blocking is not the result of any change in domain structure. It is due simply to the exponential dependence of relaxation time on grain properties like volume and coercivity and on T (eqn [23]). Just below T_C, where SD blocking temperatures are concentrated, TRM blocking typically occurs over an interval of a few °C. CRM blocking requires only a few percent increase in V. Of course blocking is only sharp in a time sense for

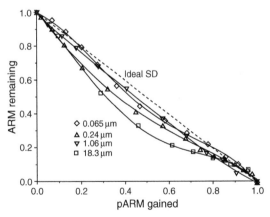

Figure 58 Simulated pseudo-Thellier paleointensity experiments comparing the remaining fraction of an initial ARM after AF demagnetization with partial ARM gained in the same AF for magnetites of various grain sizes. As with the simulated Thellier experiments of **Figure 49**, the results increasingly 'sag' below the ideal SD line of slope −1 as grain size increases. Modified from Yu Y, Dunlop DJ, and Özdemir Ö (2002b) Partial anhysteretic remanent magnetization in magnetite. 2. Reciprocity. *Journal of Geophysical Research* 107(B10): 2245 (doi:10.1029/2001JB001269).

SD grains with reasonably fast rates of cooling or grain growth.

Any MD grain has a multiplicity of magnetization states and consequently a set of both blocking and demagnetization temperatures. The whole concept of blocking becomes blurred. When one wall in a grain moves, all the other walls seek new equilibrium positions so as to minimize the overall self-demagnetizing field. Therefore, blocking is conditional: a domain wall is only locked in place as long as none of its neighbors moves. During cooling, blocking $\rightarrow$ remobilization $\rightarrow$ blocking will occur repeatedly. This complexity has so far defeated efforts to model MD grains containing more than a single wall (Néel, 1955) in any direct way. The set of unblocking temperatures during zero-field heating is easier to model and is well studied experimentally when a narrow range of T_B has been isolated as a partial TRM (e.g., **Figure 46**).

Rock magnetism has mainly concentrated on serving the needs of paleomagnetism, explaining how NRM is produced and stabilized over geological history and inventing techniques for characterizing and selectively enhancing the signal of the most desirable of the remanence carriers. What of induced magnetization? Most continental magnetic anomalies probably originate in the induced magnetization of MD magnetite, although there are notable exceptions (e.g., McEnroe *et al.*, 2004). Oceanic magnetic anomalies are more likely to have an SD or SD-like source because of the oxidation/granulation processes the minerals have undergone on the seafloor or in the shallow crust. Even at high T in the deep crust, MD grains exhibit only slight susceptibility enhancement but SD grains can have a sizeable $k_o(T)$ (Hopkinson) peak which is further increased, possibly even doubled, by thermoviscous effects over the Brunhes epoch (**Figure 55**).

Extremely fine grains are often of interest in paleoclimate studies using loess-paleosol successions or in other environmental applications (e.g., Evans and Heller, 2003). Here susceptibility is generally used simply as a measure of the quantity of magnetic material and its grain-size dependence is ignored. It is true that k_o is an insensitive probe of size variation in the SD $\rightarrow$ MD range (Heider *et al.*, 1996) but this is not the case for ultrafine grains which are SP or nearly SP (viscous) at ordinary temperatures. These grains have strongly size-dependent k_o which may be orders of magnitude greater than k_o for SD or MD grains. Scanning $k_o(T)$ at low T is an effective

granulometric probe since k_{SP} is proportional to V for small fields (eqn [19]).

The classic interpretation (the 'Chinese model', based on observations from the rainy east-central Chinese loess plateau) is that superparamagnetic magnetite forms primarily in response to high rainfall. However, the magnetic enhancement of modern topsoil is due rather to 3–30 nm cuboidal magnetite, likely produced by iron-reducing bacteria (Banerjee, 2006). Furthermore, detailed work on loess/paleosol sequences from northern and western China, Alaska and Argentina is also calling into question the generality of the old Chinese model for magnetic enhancement (e.g., Carter-Stiglitz *et al.*, 2006).

In this chapter we have attempted to summarize briefly the most useful aspects of rock magnetism, emphasizing recent developments. A more comprehensive coverage with separate treatments of igneous, sedimentary and metamorphic rocks appears in Dunlop and Özdemir (1997). Magnetic mineralogy is covered thoroughly by O'Reilly (1984) and the physics of rock magnetism by Stacey and Banerjee (1974). These sources, although older and less linked to applications, are often invaluable. A recent brief but useful primer on rock magnetism is found in Tauxe (1998).

References

Argyle KS and Dunlop DJ (1990) Low-temperature and high-temperature hysteresis of small multidomain magnetites (215–540 nm). *Journal of Geophysical Research* 95: 7069–7083.

Banerjee SK (2006) Environmental magnetism of nanophase iron minerals: Testing the biomineralization pathway. *Physics of the Earth and Planetary Interiors* 154: 210–221.

Berquó TS, Banerjee SK, Ford RG, and Lee Penn R (2007) High crystallinity Si-ferrihydrite: As insight into its Néel temperature and size dependence of magnetic properties. *Journal of Geophysical Research* 112: B02102 (doi:10.1029/2006JB004583).

Biggin AJ and Böhnel HN (2003) A method to reduce the curvature of Arai plots produced during Thellier palaeointensity experiments performed on multidomain grains. *Geophysical Journal International* 155: F13–F19.

Bina M and Daly L (1994) Mineralogical change and self-reversed magnetizations in pyrrhotite resulting from partial oxidation; geophysical implications. *Physics of the Earth and Planetary Interiors* 85: 83–99.

Bowles J, Gee JS, Kent DV, Bergmanis E, and Sinton J (2005) Cooling rate effects on paleointensity estimates in submarine basaltic glass and implications for dating young flows. *Geochemistry, Geophysics, Geosystems* 6: Q07002 (doi:10.1029/2004GC000900).

Boyd JR, Fuller M, and Halgedahl SL (1984) Domain wall nucleation as a controlling factor in the behaviour of fine magnetic particles in rocks. *Geophysical Research Letters* 11: 193–196.

Brachfeld SA and Banerjee SK (2000) A new high-resolution geomagnetic relative paleointensity record for the North

American Holocene: A comparison of sedimentary and absolute intensity data. *Journal of Geophysical Research* 105: 821–834.

Brown K and O'Reilly W (1988) The effect of low-temperature oxidation on the remanence of TRM-carrying titanomagnetite $Fe_{2.4}Ti_{0.6}O_4$. *Physics of the Earth and Planetary Interiors* 52: 147–154.

Butler RF and Banerjee SK (1975a) Theoretical single-domain grain-size range in magnetite and titanomagnetite. *Journal of Geophysical Research* 80: 4049–4058.

Butler RF and Banerjee SK (1975b) Single-domain grain-size limits for metallic iron. *Journal of Geophysical Research* 80: 252–259.

Carlut J, Horen H, and Janots D (2007) Impact of microorganisms activity on the natural remanent magnetization of the young oceanic crust. *Earth and Planetary Science Letters* 253: 497–506.

Carmichael RS (1982) Magnetic properties of minerals and rocks. In: Carmichael RS (ed.) *Handbook of Physical Properties of Rocks,* vol. 2, ch. 2, pp. 229–287. Boca Raton, FL: CRC Press.

Carter-Stiglitz B, Banerjee SK, Gourlan A, and Oches E (2006) A multi-proxy study of Argentina loess: Marine oxygen isotope stage 4 and 5 environmental record from pedogenic hematite. *Palaeogeography Palaeoclimatology Palaeoecology* 239: 45–62.

Carter-Stiglitz B, Moskowitz B, and Jackson M (2001) Unmixing magnetic assemblage and the magnetic behavior of bimodal mixtures. *Journal of Geophysical Research* 106: 26397–26411.

Carvallo C, Dunlop DJ, and Özdemir Ö (2005) Experimental comparison of FORC and remanent Preisach diagrams. *Geophysical Journal International* 162: 747–754.

Carvallo C, Özdemir Ö, and Dunlop DJ (2004) First-order reversal curve (FORC) diagrams of elongated single-domain grains at high and low temperatures. *Journal of Geophysical Research* 109: B04105 (doi:10.1029/2003JB002539).

Clarke RS and Scott ERD (1980) Tetrataenite – ordered FeNi, a new mineral in meteorites. *American Mineralogist* 65: 624–630.

Cottrell RD and Tarduno JA (2000) In search of high-fidelity geomagnetic paleointensities: A comparison of single plagioclase crystal and whole rock Thellier–Thellier analyses. *Journal of Geophysical Research* 105: 23579–23594.

Davis KE (1981) Magnetite rods in plagioclase as the primary carrier of stable NRM in ocean floor gabbros. *Earth and Planetary Science Letters* 55: 190–198.

Day R (1977) TRM and its variation with grain size. *Journal of Geomagnetism and Geoelectricity* 29: 233–265.

Day R, Fuller M, and Schmidt VA (1977) Hysteresis parameters of titanomagnetites: Grain size and composition dependence. *Physics of the Earth and Planetary Interiors* 13: 260–267.

Dekkers MJ (1988) Magnetic properties of natural pyrrhotite. I. Behaviour of initial susceptibility and saturation-magnetization-related rock-magnetic parameters in a grain-size dependent framework. *Physics of the Earth and Planetary Interiors* 52: 376–393.

Dekkers MJ (1989) Magnetic properties of natural pyrrhotite. II. High- and low-temperature behaviour of J_{rs} and TRM as a function of grain size. *Physics of the Earth and Planetary Interiors* 57: 266–283.

Dekkers MJ, Mattéi J-L, Fillion G, and Rochette P (1989) Grain-size dependence of the magnetic behaviour of pyrrhotite during its low-temperature transition at 34 K. *Geophysical Research Letters* 16: 855–858.

Diaz Ricci JC and Kirschvink JL (1992) Magnetic domain state and coercivity preduitions for biogenic greigite (Fe_3S_4): A

comparison of theory with magnetosome observations. *Journal of Geophysical Research* 97: 17309–17315.

Dodson MH and McClelland-Brown E (1980) Magnetic blocking temperature of single domain grains during slow cooling. *Journal of Geophysical Research* 85: 2625–2637.

Draeger U, Prévot M, Poidras T, and Riisager J (2006) Single-domain chemical, thermochemical and thermal remanences in a basaltic rock. *Geophysical Journal International* 166: 12–32.

Dunin-Borkowski RE, McCartney MR, Frankel RB, Bazylinski DA, Pósfai M, and Buseck PR (1998) Magnetic microstructure of magnetotactic bacteria by electron holography. *Science* 282: 1868–1870.

Dunlop DJ (1986) Hysteresis properties of magnetite and their dependence on particle size: A test of pseudo-single-domain remanence models. *Journal of Geophysical Research* 91: 9569–9584.

Dunlop DJ (1995) Magnetism in rocks. *Journal of Geophysical Research* 100: 2161–2174.

Dunlop DJ (2002a) Theory and application of the Day plot (M_{rs}/M_s versus H_{cr}/H_c), 1. Theoretical curves and tests using titanomagnetite data. *Journal of Geophysical Research* 107(B3): 2056 (doi:10.1029/2001JB000486).

Dunlop DJ (2002b) Theory and application of the Diay plot (M_{rs}/M_s versus H_{cr}/H_c), 2. Application to data for rocks, sediments, and soils. *Journal of Geophysical Research* 107(B3): 2057 (doi:10.1029/2001JB000487).

Dunlop DJ (2003a) Stepwise and continuous low-temperature demagnetization. *Geophysical Research Letters* 30(11): 1582 (doi:10.1029/2003GL017268).

Dunlop DJ (2003b) Partial thermoremanent magnetization: Louis Néel's legacy in rock magnetism. *Journal of Applied Physics* 93: 8236–8240.

Dunlop DJ (2007) Magnetization, viscous remanent. In: Gubbins D and Herrero-Bervera E (eds.) *Encyclopedia of Geomagnetism and Paleomagnetism*. Dordrecht, The Netherlands: Springer.

Dunlop DJ and Argyle KS (1997) Thermoremanence, anhysteretic remanence and susceptibility of submicron magnetites: Nonlinear field dependence and variation with grain size. *Journal of Geophysical Research* 102: 20199–20210.

Dunlop DJ and Arkani-Hamed J (2005) Magnetic minerals in the Martian crust. *Journal of Geophysical Research* 110: E12S04 (doi:10.1029/2005JE002404).

Dunlop DJ and Kletetschka G (2000) Multidomain hematite: A source of planetary magnetic anomalies? *Geophysical Research Letters* 28: 3345–3348.

Dunlop DJ and Özdemir Ö (1997) *Rock Magnetism: Fundamentals and Frontiers*. New York: Cambridge University Press.

Dunlop DJ and Özdemir Ö (2000) Effect of grain size and domain state on thermal demagnetization tails. *Geophysical Research Letters* 27: 1311–1314.

Dunlop DJ and Özdemir Ö (2001) Beyond Néel's theories: Thermal demagnetization of narrow-band partial thermoremanent magnetizations. *Physics of the Earth and Planetary Interiors* 126: 43–57.

Dunlop DJ, Özdemir Ö, Clark DA, and Schmidt PW (2000) Time-temperature relations for the remagnetization of pyrrhotite (Fe_7S_8) and their use in estimating paleotemperature. *Earth and Planetary Science Letters* 176: 107–116.

Dunlop DJ, Özdemir Ö, and Rancourt DG (2006) Magnetism of biotite crystals. *Earth and Planetary Science Letters* 243: 805–819.

Dunlop DJ and Prévot M (1982) Magnetic properties and opaque mineralogy of drilled submarine intrusive rocks. *Geophysical Journal of the Royal Astronomical Society* 69: 763–802.

Dunlop DJ, Schmidt PW, Özdemir Ö, and Clark DA (1997) Paleomagnetism and paleothermometry of the Sydney Basin. 1. Thermoviscous and chemical overprinting of the Milton Monzonite. *Journal of Geophysical Research* 102: 27271–27283.

Dunlop DJ and Waddington ED (1975) The field dependence of thermoremanent magnetization in igneous rocks. *Earth and Planetary Science Letters* 25: 11–25.

Dunlop DJ and West GF (1969) An experimental evaluation of single domain theories. *Reviews of Geophysics* 7: 709–757.

Dunlop DJ and Xu S (1994) Theory of partial thermoremanent magnetization in multidomain grains. 1. Repeated identical barriers to wall motion (single microcoercivity). *Journal of Geophysical Research* 99: 9005–9023.

Dunlop DJ, Zhang B, and Özdemir Ö (2005) Linear and nonlinear Thellier paleointensity behavior of natural minerals. *Journal of Geophysical Research* 110: B01103 (doi:10.1029/2004JB003095).

Dzyaloshinsky I (1958) A thermodynamic theory of 'weak' ferromagnetism of antiferromagnetics. *Journal of Physics and Chemistry of Solids* 4: 241–255.

Egli R (2003) Analysis of the field dependence of remanent magnetization curves. *Journal of Geophysical Research* 108(B2): 2081 (doi:10.1029/2001JB002023).

Egli R and Lowrie W (2002) Anhysteretic remanent magnetization of fine magnetic particles. *Journal of Geophysical Research* 107(B10): 2209 (doi:10.1029/2001JB000671).

Enkin RJ and Williams W (1994) Three-dimensional micromagnetic analysis of stability in fine magnetic grains. *Journal of Geophysical Research* 99: 611–618.

Evans ME and Heller F (2003) *Environmental Magnetism: Principles and Applications of Enviromagnetics*. London: Academic Press.

Evans ME and McElhinny MW (1969) An investigation of the origin of stable remanence in magnetite-bearing rocks. *Journal of Geomagnetism and Geoelectricity* 21: 757–773.

Evans ME and Wayman ML (1970) An investigation of small magnetic particles by means of electron microscopy. *Earth and Planetary Science Letters* 9: 365–370.

Fabian K (2000) Acquisition of thermoremanent magnetization in weak magnetic fields. *Geophysical Journal International* 142: 478–486.

Fabian K (2001) A theoretical treatment of paleointensity determination experiments on rocks containing pseudo-single or multidomain magnetic particles. *Earth and Planetary Science Letters* 188: 45–58.

Fabian K (2003) Some additional parameters to estimate domain state from isothermal magnetization measurements. *Earth and Planetary Science Letters* 213: 337–345.

Fabian K and Hubert A (1999) Shape-induced pseudo-single-domain remanence. *Geophysical Journal International* 138: 717–726.

Fabian K, Kirchner A, Williams W, Heider F, Leibl T, and Hubert A (1996) Three-dimensional micromagnetic calculations for magnetite using FFT. *Geophysical Journal International* 124: 89–104.

Fabian K and von Dobeneck T (1997) Isothermal magnetization of samples with stable Preisach function: A survey of hysteresis, remanence, and rock magnetic parameters. *Journal of Geophysical Research* 102: 17659–17677.

Feinberg JM, Scott GR, Renne PR, and Wenk H-R (2005) Exsolved magnetite inclusions in silicates: Features determining their remanence behavior. *Geology* 33: 513–516.

Foss S, Moskowitz BM, Proksch R, and Dahlberg ED (1998) Domain wall structures in single-crystal magnetite investigated by magnetic force microscopy. *Journal of Geophysical Research* 103: 30551–30560.

Fox JMW and Aitken MJ (1980) Cooling-rate dependence of thermoremanent magnetisation. *Nature* 283: 462–463.

Frandsen C, Stipp SLS, McEnroe SA, Madsen MB, and Knudsen JM (2004) Magnetic domain structures and stray fields of individual elongated magnetite grains revealed by magnetic force microscopy (MFM). *Physics of the Earth and Planetary Interiors* 141: 121–129.

Gapeyev AK and Tsel'movich VA (1983) Microstructure of synthetic titanomagnetite oxidized at high partial pressures of oxygen (translated from Russian). Izvestiya. *Physics of the Solid Earth* 19: 983–986.

Gendler TS, Shcherbakov VP, Dekkers MJ, Gapeev AK, Gribov SK, and McClelland E (2005) The lepidocrocite–maghemite–haematite reaction chain – I. Acquisition of chemical remanent magnetization by maghemite, its magnetic properties and thermal stability. *Geophysical Journal International* 160: 815–832.

Grommé CS, Wright TL, and Peck DL (1969) Magnetic properties and oxidation of iron-titanium oxide minerals in Alae and Makaopuhi lava lakes, Hawaii. *Journal of Geophysical Research* 74: 5277–5293.

Guyodo Y, Banerjee SK, Lee Penn R, *et al.* (2006) Magnetic properties of synthetic six-line ferrihydrite nanoparticles. *Physics of the Earth and Planetary Interiors* 154: 222–233.

Haag M and Allenspach R (1993) A novel approach to domain imaging in natural Fe/Ti oxides by spin-polarized scanning electron microscopy. *Geophysical Research Letters* 20: 1943–1946.

Halgedahl SL (1991) Magnetic domain patterns observed on synthetic Ti-rich titanomagnetites as a function of temperature and in states of thermoremanent magnetization. *Journal of Geophysical Research* 96: 3943–3972.

Halgedahl SL, Day R, and Fuller M (1980) The effect of cooling rate on the intensity of weak-field TRM in single domain magnetite. *Journal of Geophysical Research* 85: 3690–3698.

Halgedahl SL and Fuller M (1983) The dependence of magnetic domain structure on magnetization state with emphasis on nucleation as a mechanism for pseudo-single-domain behavior. *Journal of Geophysical Research* 88: 6505–6522.

Harrison RJ, Becker U, and Redfern SAT (2000) Thermodynamics of the R3 to R3c phase transition in the ilmenite-hematite solid solution. *American Mineralogist* 85: 1694–1705.

Harrison RJ, Dunin-Borkowski RE, and Putnis A (2002) Direct imaging of nanoscale magnetic interactions in minerals. *Proceedings of the National Academy of Sciences USA* 99: 16556–16561.

Heider F and Hoffmann V (1992) Magneto-optical Kerr effect on magnetite crystals with externally applied magnetic fields. *Earth and Planetary Science Letters* 108: 131–138.

Heider F, Zitzelsberger A, and Fabian K (1996) Magnetic susceptibility and remanent coercive force in grown magnetite crystals from 0.1 μm to 6 mm. *Physics of the Earth and Planetary Interiors* 93: 239–256.

Heslop D, Dekkers MJ, Kruiver PP, and van Oorschot IHM (2002) Analysis of isothermal remanent magnetisation acquisition curves using the expectation–maximisation algorithm. *Geophysical Journal International* 148: 58–64.

Heslop D, McIntosh G, and Dekkers MJ (2004) Using time- and temperature-dependent Preisach models to investigate the limitations of modeling isothermal remanent magnetisation acquisition curves with cumulative log Gaussian functions. *Geophysical Journal International* 157: 55–63.

Heslop D, Witt A, von Dobeneck T, Huhn K, Fabian K, Bleil U (2004) Discrete element modeling of postdepositional remanent magnetization acquisition: First results. *EOS Transactions of the American Geophysical Union* 85(47), Fall Meeting suppl., Abstract GP21C-08 (available online).

Hodych JP and Bijaksana S (1993) Can remanence anisotropy detect paleomagnetic inclination shallowing due to compaction? A case study using Cretaceous deep-sea limestones. *Journal of Geophysical Research* 98: 22429–22441.

Horng C-S and Roberts AP (2006) Authigenic or detrital origin of pyrrhotite in sediments? Resolving a paleomagnetic conundrum. *Earth and Planetary Science Letters* 241: 750–762.

Housen BA, Banerjee SK, and Moskowitz BM (1996) Low-temperature properties of siderite and magnetite in marine sediments. *Geophysical Research Letters* 23: 2843–2846.

Hunt CP, Moskowitz BM, and Banerjee SK (1995) Magnetic properties of rocks and minerals. In: Ahrens TJ (ed.) *Rock Physics and Phase Relations: A Handbook of Physical Constants,* vol. 3, ch. 14, pp. 189–204. Washington, DC: American Geophysical Union.

Hus JJ (1990) The magnetic properties of siderite concretions and the CRM of their oxidation products. *Physics of the Earth and Planetary Interiors* 63: 41–57.

Jackson M, Sølheid P, Carter-Stiglitz B, Rosenbaum J, and Till J (2004) Tiva Canyon Tuff: Superparamagnetic samples available. *Institute of Rock Magnetism Quarterly (University of Minnesota)* 14(3): 1 and 9–11.

Johnson EA, Murphy T, and Torreson OW (1948) Pre-history of the Earth's magnetic field. *Terrestrial Magnetism and Atmosphric Electricity* 53: 349–372.

Kelso PR and Banerjee SK (1994) Elevated temperature viscous remanent magnetization of natural and synthetic multidomain magnetite. *Earth and Planetary Science Letters* 122: 43–56.

Kent DV (1973) Post-depositional remanent magnetization in deep-sea sediment. *Nature* 246: 32–34.

Kirschvink JL, Jones DS, and MacFadden BJ (eds.) (1985) *Magnetite Biomineralization and Magnetoreception in Organisms: A New Biomagnetism.* New York: Plenum.

Kletetschka G, Acuña MH, Kohout T, Wasilewski PJ, and Connerney JEP (2004) An empirical scaling law for acquisition of thermoremanent magnetization. *Earth and Planetary Science Letters* 226: 521–528.

Kletetschka G, Wasilewski PJ, and Taylor PT (2000a) Unique thermoremanent magnetization of multidomain hematite: Implications for magnetic anomalies. *Earth and Planetary Science Letters* 176: 469–479.

Kletetschka G, Wasilewski PJ, and Taylor PT (2000b) Hematite vs. magnetite as the signature for planetary magnetic anomalies. *Physics of the Earth and Planetary Interiors* 119: 259–267.

Kneller EF and Luborsky FE (1963) Particle size dependence of coercivity and remanence of single-domain particles. *Journal of Applied Physics* 34: 656–658.

Kodama KP (1992) Depositional remanent magnetization. In: Nierenberg WA (ed.) *Encyclopedia of Earth System Science,* vol. 2, pp. 47–54. San Diego, CA: Academic Press.

Kodama KP and Sun WW (1992) Magnetic anisotropy as a correction for compaction-caused palaeomagnetic inclination shallowing. *Geophysical Journal International* 111: 465–469.

Lagroix F, Banerjee SK, and Moskowitz BM (2004) Revisiting the mechanism of self reversed thermoremanent magnetization based on observations from synthetic ferrian ilmenite ($y = 0.7$). *Journal of Geophysical Research* 109: B12108 (doi:10.1029/2004JB003076).

Lanci L and Kent DV (2003) Introduction of thermal activation in forward modeling of hysteresis loops for single-domain magnetic particles and implications for the interpretation of the Day diagram. *Journal of Geophysical Research* 108(B3): 2142 (doi:10.1029/2001JB000944).

Leonhardt R (2006) Analyzing rock magnetic measurements: The RockMagAnalyzer 1.0 software. *Computers and Geosciences* 32: 1420–1431.

Leonhardt R, Krása D, and Coe RS (2004) Multidomain behavior during Thellier paleointensity experiments: A

phenomenological model. *Physics of the Earth and Planetary Interiors* 147: 127–140.

Lorand J-P, Chevrier V, and Sautter V (2005) Sulphide mineralogy and redox conditions in some shergottites. *Meteoritics and Planetary Science* 40: 1257–1272.

Løvlie R (1976) The intensity pattern of post-depositional remanence acquired in some marine sediments deposited during a reversal of the external magnetic field. *Earth and Planetary Science Letters* 30: 209–214.

Maher BA, Karloukovski VV, and Mutch TJ (2004) High-field remanence properties of synthetic and natural submicrometre haematites and goethites: Significance for environmental contexts. *Earth and Planetary Science Letters* 226: 491–505.

McEnroe SA, Langenhorst F, Robinson P, Bromiley GD, and Shaw CSJ (2004) What is magnetic in the lower crust? *Earth and Planetary Science Letters* 226: 175–192.

Menyeh A and O'Reilly W (1998) Thermoremanence in monoclinic pyrrhotite particles containing few domains. *Geophysical Research Letters* 25: 3461–3464.

Moskowitz BM (1995) Fundamental physical constants and conversion factors. In: Ahrens TJ (ed.) *Rock Physics and Phase Relations: A Handbook of Physical Constants,* vol. 1, ch. 23, pp. 346–55. Washington, DC: American Geophysical Union.

Moskowitz BM, Halgedahl SL, and Lawson CA (1988) Magnetic domains on unpolished and polished surfaces of titanium-rich titanomagnetite. *Journal of Geophysical Research* 93: 3372–3386.

Moskowitz BM, Jackson M, and Kissel C (1998) Low-temperature behavior of titanomagnetites. *Earth and Planetary Science Letters* 157: 141–149.

Murthy GS, Evans ME, and Gough DI (1971) Evidence of single-domain magnetite in the Michikamau anorthosite. *Canadian Journal of Earth Sciences* 8: 361–370.

Muxworthy A, Heslop D, and Williams W (2004) Influence of magnetostatic interactions on first-order-reversal-curve (FORC) diagrams: A micromagnetic approach. *Geophysical Journal International* 158: 888–897.

Muxworthy AR and Dunlop DJ (2002) First-order reversal curve (FORC) diagrams for pseudo-single-domain magnetites at high temperature. *Earth and Planetary Science Letters* 203: 369–382.

Muxworthy AR, King JG, and Heslop D (2005) Assessing the ability of first-order reversal curve (FORC) diagrams to unravel complex magnetic signals. *Journal of Geophysical Research* 110: B01105 (doi:10.1029/2004JB003195).

Nagata T (1979) Meteorite magnetism and the early solar system magnetic field. *Physics of the Earth and Planetary Interiors* 20: 324–341.

Nagata T (1983) Meteorite magnetization and paleointensity. *Advances in Space Research* 2: 55–63.

Nagata T, Arai Y, and Momose K (1963) Secular variation of the geomagnetic total force during the last 5000 years. *Journal of Geophysical Research* 68: 5277–5281.

Néel L (1948) Propriétés magnétiques des ferrites: Ferrimagétisme et antiferromagnétisme. *Annales de Physique* 3: 137–198.

Néel L (1949) Théorie du traînage magnétique des ferromagnétiques en grains fins avec applications aux terres cuites. *Annales de Géophysique* 5: 99–136.

Néel L (1954) Remarques sur la théorie des propriétés magnétiques des substances dures. *Applied Scientific Research (The Hague) B* 4: 13–24.

Néel L (1955) Some theoretical aspects of rock magnetism. *Advances in Physics* 4: 191–243.

Néel L, Pauleve J, Pauthenet R, Laugier J, and Dautreppe D (1964) Magnetic properties of an iron-nickel single crystal ordered by neutron bombardment. *Journal of Applied Physics* 35: 873–876.

Newell AJ and Merrill RT (1999) Single-domain critical sizes for coercivity and remanence. *Journal of Geophysical Research* 104: 617–628.

O'Donovan JB, Facey D, and O'Reilly W (1986) The magnetization process in titanomagnetite ($Fe_{2.4}Ti_{0.6}O_4$) in the 1–30 μm particle size range. *Geophysical Journal of the Royal Astronomical Society* 87: 897–916.

O'Reilly W (1976) Magnetic minerals in the crust of the Earth. *Reports on Progress in Physics* 39: 857–908.

O'Reilly W (1984) *Rock and Mineral Magnetism.* New York: Blackie, Glasgow and Chapman & Hall.

Özdemir Ö (2007) Magnetization, thermoremanent. In: Gubbins D and Herrero-Bervera E (eds.) *Encyclopedia of Geomagnetism and Paleomagnetism.* Dordrecht, The Netherlands: Kluwer (in press).

Özdemir Ö and Dunlop DJ (1985) An experimental study of chemical remanent magnetizations of synthetic monodomain titanomaghemites with initial thermoremanent magnetizations. *Journal of Geophysical Research* 90: 11513–11523.

Özdemir Ö and Dunlop DJ (1993a) Magnetic domain structures on a natural single crystal of magnetite. *Geophysical Research Letters* 20: 1835–1838.

Özdemir Ö and Dunlop DJ (1993b) Chemical remanent magnetization during γFeOOH phase transformations. *Journal of Geophysical Research* 98: 4191–4198.

Özdemir Ö and Dunlop DJ (1996) Thermoremanence and Néel temperature of goethite. *Geophysical Research Letters* 23: 921–924.

Özdemir Ö and Dunlop DJ (1997) Effect of crystal defects and internal stress on the domain structure and magnetic properties of magnetite. *Journal of Geophysical Research* 102: 20211–20224.

Özdemir Ö and Dunlop DJ (2002) Thermoremanence and stable memory of single-domain hematites. *Geophysical Research Letters* 29(18): 1877 (doi:10.1029/2002GL015597).

Özdemir Ö and Dunlop DJ (2003) Low-temperature behavior and memory of iron-rich titanomagnetites (Mt. Haruna, Japan and Mt. Pinatubo, The Philippines). *Earth and Planetary Science Letters* 216: 193–200.

Özdemir Ö and Dunlop DJ (2005) Thermoremanent magnetization of multidomain hematite. *Journal of Geophysical Research* 110: B09104 (doi:10.1029/2005JB003820).

Özdemir Ö and Dunlop DJ (2006) Magnetic domain observations on magnetite crystals in biotite and hornblende grains. *Journal of Geophysical Research* 111: B06103 (doi: 10.1029/2005JB004090).

Özdemir Ö, Dunlop DJ, and Moskowitz BM (1993) The effect of oxidation on the Verwey transition in magnetite. *Geophysical Research Letters* 20: 1671–1674.

Özdemir Ö, Dunlop DJ, and Moskowitz BM (2002) Changes in remanence, coercivity and domain state at low temperature in magnetite. *Earth and Planetary Science Letters* 194: 343–358.

Özdemir Ö and O'Reilly W (1982a) Magnetic hysteresis properties of synthetic monodomain titanomaghemites. *Earth and Planetary Science Letters* 57: 437–447.

Özdemir Ö and O'Reilly W (1982b) An experimental study of the intensity and stability of thermoremanent magnetization acquired by synthetic monodomain titanomagnetite substituted by aluminium. *Geophysics Journal of the Royal Astronomical Society* 70: 141–154.

Özdemir Ö, Xu S, and Dunlop DJ (1995) Closure domains in magnetite. *Journal of Geophysical Research* 100: 2193–2209.

Ozima M and Funaki M (2003) Hemoilmenite as a carrier of SRTRM in dacitic pumice from Akagi, Ontake and Sambe volcanoes, Japan. *Earth and Planetary Science Letters* 213: 311–320.

Pan Q, Pokhil TG, and Moskowitz BM (2002) Domain structures in epitaxial Fe_3O_4 particles studies by magnetic force microscopy. *Journal of Applied Physics* 91: 5945–5950.

Pannalal SJ, Crowe SA, Cioppa MT, Symons DTA, Sturm A, and Fowle DA (2005) Room-temperature properties of ferrihydrite: A potential magnetic remanence carrier? *Earth and Planetary Science Letters* 236: 856–870.

Petersen N, von Dobeneck T, and Vali H (1986) Fossil bacterial magnetite in deep-sea sediments from the South Atlantic Ocean. *Nature* 320: 611–615.

Pokhil TG and Moskowitz BM (1997) Magnetic domains and domain walls in pseudo-single-domain magnetite studied with magnetic force microscopy. *Journal of Geophysical Research* 102: 22681–22694.

Pozzi J-P and Dubuisson G (1992) High temperature viscous magnetization of oceanic deep crustal- and mantle-rocks as a partial source for Magsat magnetic anomalies. *Geophysical Research Letters* 19: 21–24.

Preisach F (1935) Über die magnetische Nachwirkung. *Zeitschrift fur Physik* 94: 277–302.

Prévot M, Hoffman KA, Goguitchaichvili A, Doukhan J-C, Shcherbakov V, and Bina M (2001) The mechanism of self-reversal of thermoremanence in natural hemoilmenite crystals: New experimental data and model. *Physics of the Earth and Planetary Interiors* 126: 75–92.

Pullaiah G, Irving E, Buchan KL, and Dunlop DJ (1975) Magnetization changes caused by burial and uplift. *Earth and Planetary Science Letters* 28: 133–143.

Rave W, Fabian K, and Hubert A (1998) Magnetic states of small cubic particles with uniaxial anisotropy. *Journal of Magnetism and Magnetic Materials* 190: 332–348.

Renne PR, Scott GR, Glen JMG, and Feinberg JM (2002) Oriented inclusions of magnetite in clinopyroxene: Source of stable remanent magnetization in gabbros of the Messum Complex, Namibia. *Geochemistry, Geophysics, Geosystems* 3(12): 1079 (doi:10.1029/2002GC000319).

Roberts AP (1995) Magnetic properties of sedimentary greigite (Fe_3S_4). *Earth and Planetary Science Letters* 134: 227–236.

Roberts AP, Pike CP, and Verosub KL (2000) First-order reversal curve diagrams: A new tool for characterizing the magnetic properties of natural samples. *Journal of Geophysical Research* 105: 28461–28475.

Roberts AP and Turner GM (1993) Diagenetic formation of ferrimagnetic iron sulphide minerals in rapidly deposited marine sediments, South Island, New Zealand. *Earth and Planetary Science Letters* 115: 257–273.

Robinson P, Harrison RJ, McEnroe SA, and Hargraves RB (2002) Lamellar magnetism in the haematite–ilmenite series as an explanation for strong remanent magnetization. *Nature* 418: 517–520.

Rochette P, Fillion G, Mattéi J-L, and Dekkers MJ (1990) Magnetic transition at 30–34 kelvin in pyrrhotite: Insight into a widespread occurrence of this mineral in rocks. *Earth and Planetary Science Letters* 98: 319–328.

Rochette P, Gattacceca J, *et al.* (2005b) Matching Martian crustal magnetization and magnetic properties of Martian meteorites. *Meteoritics and Planetary Science* 40: 529–540.

Rochette P, Mathé P-E, *et al.* (2005a) Non-saturation of the defect moment of goethite and fine-grained hematite up to 57 teslas. *Geophysical Research Letters* 32: L0 (doi:10.1029/2005GL024196).

Rowan CJ and Roberts AP (2006) Magnetite dissolution, diachronous greigite formation, and secondary magnetizations from pyrite oxidation: Unravelling complex magnetizations in Neogene marine sediments from New Zealand. *Earth and Planetary Science Letters* 241: 119–137.

Schwarz EJ and Vaughan DJ (1972) Magnetic phase relations of pyrrhotite. *Journal of Geomagnetism and Geoelectricity* 24: 441–458.

Scott ERD and Fuller M (2004) A possible source for the Martian crustal magnetic field. *Earth and Planetary Science Letters* 220: 83–90.

Selkin PA, Gee JS, Tauxe L, Meurer WP, and Newell AJ (2000) The effect of remanence anisotropy on paleointensity estimates: A case study from the Archean Stillwater Complex. *Earth and Planetary Science Letters* 183: 403–416.

Soffel HC and Appel E (1982) Domain structure of small synthetic titanomagnetite particles and experiments with IRM and TRM. *Physics of the Earth and Planetary Interiors* 30: 348–355.

Stacey FD (1958) Thermoremanent magnetization (TRM) of multidomain grains in igneous rocks. *Philosophical Magazine* 3: 1391–1401.

Stacey FD and Banerjee SK (1974) *The Physical Principles of Rock Magnetism*. Amsterdam: Elsevier.

Stoner EC and Wohlfarth EP (1948) A mechanism of magnetic hysteresis in heterogeneous alloys. *Philosophical Transactions of the Royal Society of London A* 240: 599–642.

Sun WW and Kodama KP (1992) Magnetic anisotropy, scanning electron microscopy, and X-ray pole figure goniometry study of inclination shallowing in a compacting clay-rich sediment. *Journal of Geophysical Research* 97: 19599–19615.

Tauxe L (1998) *Paleomagnetic Principles and Practice*. Dordrecht: Kluwer.

Tauxe L and Kent DV (1984) Properties of a detrital remanence carried by haematite from study of modern river deposits and laboratory redeposition experiments. *Geophysical Journal of the Royal Astronomical Society* 76: 543–561.

Tauxe L, Pick T, and Kok YS (1995) Relative paleointensity in sediments: A pseudo-Thellier approach. *Geophysical Research Letters* 22: 2885–2888.

Tauxe L, Steindorf JL, and Harris A (2006) Depositional remanent magnetization: Toward an improved theoretical and experimental foundation. *Earth and Planetary Science Letters* 244: 515–529.

Thellier E (1938) Sur l'aimantation des terres cuites et ses applications géophysiques. *Annales de l'Institut de Physique du Globe de l'Université de Paris* 16: 157–302.

Thomas-Keprta KL, Bazylinski DA, *et al.* (2000) Elongated prismatic magnetite crystals in ALH84001 carbonate globules. *Geochimica et Cosmochimica Acta* 64: 4049–4081.

Tucker P (1980) A grain mobility model of post-depositional realignment. *Geophysical Journal of the Royal Astronomical Society* 63: 149–163.

Tucker P and O'Reilly W (1980) The acquisition of thermoremanent magnetization by multidomain single-crystal titanomagnetite. *Geophysical Journal of the Royal Astronomical Society* 60: 21–36.

Uyeda S (1958) Thermo-remanent magnetism as a medium of paleomagnetism, with special reference to reverse thermoremanent magnetism. *Japanese Journal of Geophysics* 2: 1–123.

Vali H, Förster O, Amarantidis G, and Petersen N (1987) Magnetotactic bacteria and their magnetofossils in sediments. *Earth and Planetary Science Letters* 86: 389–400.

Verosub K and Roberts AP (1995) Environmental magnetism: Past, present, and future. *Journal of Geophysical Research* 100: 2175–2192.

Wasilewski P (1981) Magnetization of small iron-nickel spheres. *Physics of the Earth and Planetary Interiors* 26: 355–377.

Weiss P (1907) L'hypothèse du champ moleculaire et la propriété ferromagnétique. *Journal de Physique* 6: 661–690.

Weiss BP, Vali H, Baudenbacher FJ, Kirschvink JL, Stewart DL, and Shuster DL (2002) Records of an ancient Martian magnetic field in ALH84001. *Earth and Planetary Science Letters* 201: 449–463.

Williams W, Enkin RJ, and Milne G (1992) Magnetic domain wall visibility in Bitter pattern imaging. *Journal of Geophysical Research* 97: 17443–17448.

Wilson RL (1961) Palaeomagnetism in Northern Ireland. 1. The thermal demagnetization of natural magnetic moments in rocks. *Geophysical Journal of the Royal Astronomical Society* 5: 45–58.

Witt A, Heslop D, von Dobeneck T, Huhn K, Fabian K, and Bleil U (2004) Discrete element modelling of post-depositional remanent magnetization acquisition: Numerical principles. *EOS Transactions of the American Geophysical Union* 85(47), Fall Meeting suppl. Abstract GP23A-0172 (available online).

Wu YT, Fuller M, and Schmidt VA (1974) Microanalysis of N.R.M. in a granodiorite intrusion. *Earth and Planetary Science Letters* 23: 275–285.

Xu S and Dunlop DJ (1994) Theory of partial thermoremanent magnetization in multidomain grains. 2. Effect of microcoercivity distribution and comparison with experiment. *Journal of Geophysical Research* 99: 9025–9033.

Xu S and Dunlop DJ (2004) Thellier paleointensity theory and experiments for multidomain grains. *Journal of Geophysical Research* 109: B07103 (doi:10.1029/2004JB003024).

York D (1978a) A formula describing both magnetic and isotopic blocking temperatures. *Earth and Planetary Science Letters* 39: 89–93.

York D (1978b) Magnetic blocking temperature. *Earth and Planetary Science Letters* 39: 94–97.

Yu Y, Dunlop DJ, and Özdemir Ö (2002a) Partial anhysteretic remanent magnetization in magnetite. 1. Additivity. *Journal of Geophysical Research* 107(B10): 2244 (doi:10.1029/2001JB001249).

Yu Y, Dunlop DJ, and Özdemir Ö (2002b) Partial anhysteretic remanent magnetization in magnetite. 2. Reciprocity. *Journal of Geophysical Research* 107(B10): 2245 (doi:10.1029/2001JB001269).

Yu Y, Dunlop DJ, and Özdemir Ö (2003) Testing the independence law of partial ARMs: Implications for paleointensity determination. *Earth and Planetary Science Letters* 208: 27–39.

Yu Y, Dunlop DJ, Özdemir Ö, and Ueno H (2001) Magnetic properties of Kurokami pumices from Mt. Sakurajima, Japan. *Earth and Planetary Science Letters* 192: 439–446.

Yu Y and Gee JS (2005) Spinel in Martian meteorite SaU 008: Implications for Martian magnetism. *Earth and Planetary Science Letters* 232: 287–294.

9 Centennial- to Millennial-Scale Geomagnetic Field Variations

C. Constable, University of California at San Diego, La Jolla, CA, USA

9.1 Introduction

The historical record contains comparatively few direct observations of the geomagnetic field prior to the sixteenth century, when the expanding number of global marine expeditions led to a great increase in the use of the magnetic compass. Although ideas about relative variations in field strength were used much earlier, absolute measurements of magnetic field strength were not routinely available until the middle of the nineteenth century when Gauss initiated systematic measurements at a number of geomagnetic observatories. Earlier observations of the geomagnetic field must be supplemented by indirect measurements, made using a range of paleomagnetic techniques on geological materials or archeological artifacts that preserve a record of the field either as a thermal or (post) depositional magnetic remanence. The primary focus

of this chapter is on records for the time interval ranging from about 0-10 ka, and it includes a discussion of archeomagnetic records preserved in man-made objects, paleomagnetic work on young volcanic materials, and rapidly deposited lake and marine sediments. To be useful as a geomagnetic record the age of magnetization must be obtained by a method that is independent of the geomagnetic field, but in some studies the magnetic direction or field intensity is used as a mechanism for dating materials for archeological, sedimentary, or other geological applications.

The idea that archeological artifacts could be used as a means of extending the geomagnetic record was known by the last decade of the nineteenth century. Folgerhaiter (1899) summarizes the method and his earlier results on magnetic inclinations, and subsequently Chevallier (1925) applied the same ideas to oriented historic lavas from Mount Etna and published a directional record of secular variation. Thellier (1938) used archeological materials in the development of his now well-known technique for recovering magnetic field strength (Thellier and Thellier, 1959; *see* Chapter 13 for more on paleointensities), and went on to produce a record of intensity and directional changes in France that extended over 2000 years (Thellier, 1981). The first varved sedimentary records of secular variations were attempted in the middle of the twentieth century (Johnson *et al.*, 1948; Griffiths, 1953). Subsequently, there have been numerous studies of sediments, greatly facilitated by the use of portable coring technology that has allowed routine sampling of unconsolidated lake sediments, and of high accumulation rate marine sediments.

In early studies paleo- and geomagnetic secular variation were often described in terms of motion of the geomagnetic dipole axis, westward drift of geomagnetic sources and variations in the geomagnetic dipole moment. These results are well summarized in chapter 4 of the monograph by Merrill *et al.* (1996). This was a reasonable approach given the predominantly dipolar structure in the geomagnetic field combined with the sparse spatial and temporal data coverage and lower accuracy generally available for millennial-scale paleomagnetic data compared with direct observations. Changes in dipole moment are of interest for geomagnetic studies, but are also important in other areas of Earth science: the geomagnetic field effectively screens Earth's surface from cosmic radiation, and the magnitude of the dipole moment influences radiogenic nuclide production (Elsasser *et al.*, 1956; Lal and Peters, 1967), with corresponding implications for climate studies that model solar activity and/or ocean circulation in the past (e.g., Bard *et al.*, 2000; Muscheler *et al.*, 2004a, 2000b) and surface exposure dating (e.g., Lal, 1991; Dunai, 2001; Gosse and Phillips, 2001; Lifton *et al.*, 2005). Long-term secular variation causes temporal changes in the location of the dipole axis, and associated changes in auroral sightings that can be tracked through the historical record (Siscoe and Siebert, 2002; Siscoe *et al.*, 2002; Willis and Stephenson, 2001).

Historical field behavior provides a fairly detailed view of geomagnetic field changes for the past 400 years. The current understanding, supported by a wealth of survey, observatory, and satellite data, is that the radial magnetic field at high latitudes shows relatively stable structures often characterized as persistent flux lobes. The westward drift observed by Halley and often considered a ubiquitous part of secular variation is pronounced in the Atlantic

hemisphere, and has recently been interpreted in terms of dynamo waves in the core (Finlay and Jackson, 2003), but the secular variation in the Pacific hemisphere is different in style, and appears to be dominated by large-scale longer-term variations. Differences between the Atlantic and Pacific hemisphere secular variation may well be controlled by thermal structure at the core–mantle boundary (CMB) (Bloxham, 2002; Olson and Christensen, 2002; Bloxham and Gubbins, 1987). The Atlantic hemisphere also shows a large region of low field strength known as the South Atlantic Anomaly, that presents a radiation hazard for low-Earth-orbiting satellites (Heirtzler *et al.*, 2002) and has been tied to ongoing decreases in geomagnetic dipole moment. The longevity of this feature remains unknown, but it has certainly contributed to North–South Hemispheric asymmetries in dipole moment changes for the past few hundred years (Gubbins *et al.*, 2006).

The acquisition of enough reliable paleomagnetic data to represent the global magnetic field is a painstaking business, but for the time interval 0–10 ka enormous progress has been made in recent years. Long-term regional and global secular variation studies, combined with archeomagnetic dating and high-resolution magnetostratigraphy contribute to global interpretations about the magnetic field. It is beginning to be feasible to use these data to look at issues such as the evolution and longevity of the South Atlantic Anomaly, persistence of flux lobes seen in the historical field, dynamo waves, and symmetries in the millennial-scale geomagnetic secular variation. The remainder of this chapter discusses the status of these recent magnetic field data and how they are currently being used.

9.2 Data Types and Methods

Fundamental observations in archeomagnetic and paleomagnetic work involve the recovery of one or more geomagnetic elements (declination, D, inclination, I, or geomagnetic field strength, B or F, see **Figure 1**) from the time of acquisition of magnetization and/or an age estimate (perhaps with inferred stratigraphic relationships to other observations). These observations usually relate to an archeological structure or artifact or some geological material that is the object of study, and knowledge of the context in which this object is found should play an important role in assessing the uncertainties tied to the observations. The geographic location to be associated with

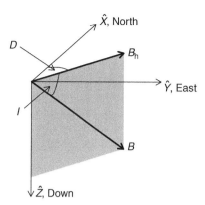

Figure 1 Geomagnetic elements in a local coordinate system. D, declination; I, inclination; B, magnetic field strength; B_h, horizontal component of magnetic field.

the field elements is in many cases known extremely well: in others it is not recorded with the desired accuracy, and in some cases there may be confusion about the place where specific artifacts originated. To acquire a paleomagnetic direction requires knowledge of the relationship between the coordinate system in which measurements are made and the geographic system in which the magnetization was originally acquired. If a structure, sediment, or lava flow has been disturbed after the magnetization was acquired or its original orientation was unknown, this can add a systematic bias to the uncertainty already present in translating directions from sample to geographic coordinates. Age constraints come in an enormous range of flavors: some give only broad stratigraphic relationships, while in the best case scenario they can be very accurate, for example, when there is a correctly recorded eyewitness account of some historical spectacle like a volcanic eruption, or for a specific style of pottery that can be closely tied to a specific time and place. General paleomagnetic techniques are discussed in Chapter 4, but some consideration needs to be given here to the influence of specific sampling strategies and kinds of materials in the context of assessing uncertainties.

9.2.1 Geomagnetic Directional Information

Sampling of young lava flows usually follows standard techniques described in Chapter 4, with multiple samples drilled from a single flow using a hand-held portable drill or occasionally hand samples taken. Orientation data are collected using Sun

or magnetic compass to allow transfer of directions from the sample coordinate system into local coordinates as in **Figure 1**. Each sample may be cut into multiple specimens. Averaging of directions from independently oriented samples helps reduce uncertainty in the paleomagnetic directional results. Archeomagnetic sampling methods are more heterogeneous, and the hierarchy is described by Lanos *et al.* (2005). In some cases very large samples are collected (e.g., Tanguy *et al.*, 1999). It is quite common to stabilize fragile material and provide a suitable surface for orientation by forming some kind of plaster or other cap over the structure to be sampled. The large sample size allows for high precision in the initial orientation process. Later a core can be drilled from the oriented sample or sometimes it is treated whole and measured in a large inductometer (e.g., Le Goff, 1975). Other researchers collect standard paleomagnetic drill samples of 1 inch diameter, when this is acceptable for the structure in question. The large sample method has also been applied to volcanic rocks and its proponents (Tanguy *et al.*, 2003; Arrighi *et al.*, 2004) argue that with this method it is possible to achieve uncertainties of less than 2° in averaged results for a single site. This is lower than what is usually achievable using drill cores and it has been suggested by Arrighi *et al.* (2004) that the large sample method is preferable for high-resolution work. In contrast Speranza *et al.* (2005) argue that laboratory and data processing techniques also play an important role and that work by Lanza *et al.*, (2005) demonstrates that the drill samples estimates provide a more realistic assessment of the uncertainty for both the large samples and drill core sites when compared with the historical geomagnetic field directions. Arrighi *et al.* (2005) note that there are few direct field observations, from the relevant time intervals – Lanza *et al.*'s (2005) comparisons in large part depend on predictions from the *GUFM* field model of Jackson *et al.* (2000), and suggest that a conservative estimate of the error is 2.5–3.0° for Italian lavas.

A wide range of laboratory protocols continue to be used in archeomagnetic work, and there is no overall consensus on quality criteria. Gallet *et al.* (2002) note in a recent compilation of western European results, that laboratory cleaning efforts range from none to the kind of complete demagnetization that allows principal component analysis. It should be noted that the latter is now widely accepted as the standard expected for high-quality paleomagnetic work. The same remark could equally be applied to sedimentary directional data. Where available it is probably reasonable to suppose that the Fisherian α_{95}, the estimate of precision parameter k and number of samples treated can provide some idea of quality control. Several studies have attempted to ground truth results from historical lava flows with somewhat mixed results (Castro and Brown, 1987; Rolph and Shaw, 1986; Baag et al., 1995; Valet and Soler, 1999; Tanguy and Le Goff, 2004). Some have inferred large distortions of the field at individual sites from local magnetic effects, but when large number of samples have been taken distributed along the volcanic unit as in the Mt. Etna study by Tanguy and Le Goff (2004) and studies of the 1960 Hawaiian flow (discussed further below) it seems that any large-scale influences are undetectable.

A variety of coring devices are used in acquiring sedimentary samples ranging from box cores (Klovan, 1964) designed to sample the topmost 0.5–1 m of sediment to the hydraulic piston cores used for large-scale projects like the Ocean Drilling Program (ODP) and drilling lakes like Lake Baikal. Most unconsolidated lake sediments are collected with some kind of portable piston coring device, for example, the Mackereth corer (Mackereth, 1958, Barton and Burden, 1979) or Livingstone corer (Livingstone, 1955). Freeze coring (Rymer and Neale, 1981) can be used to recover material near the sediment–water interface, which is frequently disturbed in the piston coring process, but in practice a multiple corer is the most widely used apparatus for recovering undisturbed material near the sediment–water interface. Freeze cores are not generally used as paleomagnetic samples, but they can provide important age constraints.

Once a sediment core has been acquired, the material is generally subsampled in some way so that a sequence of magnetization measurements and magnetic cleaning can be preformed in the laboratory. U-channel samples (Tauxe et al., 1983) taken from the center of a split core can be measured in a pass-through magnetometer, or closely spaced individual specimens (typically a few cubic centimeter in volume) can be used to generate a stratigraphic record of magnetization within the core. Each core must then be correlated with others from nearby, using suitable stratigraphic markers (which may involve the identification of characteristic sediment, volcanic, or rock magnetic properties), and eventually placed on an appropriate chronology. With appropriate correlation tools in place magnetic records from multiple cores are often stacked in attempts to improve the signal-to-noise ratio. In general, one might expect paleomagnetic directions acquired from sediment cores to have lower accuracy than volcanic sites because orientation errors (from core twisting and nonvertical core penetration, as well as from subsequent subsampling) can be significant for individual cores, and stacked records tend to be temporally blurred by lack of resolution in correlating multiple cores within a given lake or region.

9.2.2 Paleointensity Data and Uncertainties

The theoretical and practical basis for recovering paleomagnetic intensity estimates is described in Chapter 13, and the details will not be duplicated here. However, there are some important considerations in assessing the reliability of paleointensity data for geomagnetic field studies. We make a major distinction between absolute and relative peleointensity measurements and consider them as distinct entities requiring different evaluation techniques.

Absolute paleointensity observations are based on the thermal remanent magnetization (TRM) preserved in a sample after it has been cooled through some range of magnetic blocking temperatures in a magnetic field, and it is widely believed on the basis of Néel (1955) theory that, for single domain and pseudo single domain magnetic grains this TRM acquisition can be at least approximately replicated in the laboratory. One of the greatest hazards is that the magnetic minerals present can alter during the paleointensity experiment in the laboratory, and numerous strategies have been developed for attempting to detect such alteration. Nevertheless, it is to be expected that some changes will go undetected during the experiment. An important question for the purpose of recovering geomagnetic field behavior in the past (recently summarized by Lanos et al., 2005) is whether the experiment can be regarded as unbiased in the statistical sense – if it is unbiased, then even observations with large uncertainties are in principle useful for studying geomagnetic behavior, but if systematic effects result in persistently high or low records from specific regions this will clearly pose a problem.

A number of studies have been conducted on historical lava flows with a view to assessing the accuracy and reproducibility of the paleomagnetic method (e.g., Castro and Brown, 1987; Rolph and Shaw, 1986; Böhnel et al., 1997; Mochizuki et al., 2004; Oishi et al., 2005; Yamamoto et al., 2003).

Recent studies by Mochizuki *et al.* (2004), Oishi *et al.* (2005), and Yamamoto *et al.* (2003) suggest that in some cases high mean paleointensities can be obtained with the modified Thellier method depending on the degree of deuteric oxidation in the sample. They suggest using a modified Shaw method with double heating and low-temperature demagnetization instead. Other checks on the Thellier method are discussed in Chapter 13. Detailed and specific information of this kind is often not available for prehistoric flows, but one can assess the bias expected overall by combining historic data obtained using a broad variety of methods. Love and Constable (2003) combined 86 directional and 95 intensity data available from the 1960 Kilauea flow on the island of Hawaii and used maximum likelihood to estimate the mean and variance of the resulting field vector. Comparisons with the field values predicted from Bloxham and Jackson's (1992) field model (a predecessor to GUFM) indicate reasonable agreement with the field values, but the analysis of the distribution of observations reveals that the uncertainties are quite large – the standard deviation in F is between 16% and 20% of the average value for a single flow that should provide a spot reading of the field. In principle this large uncertainty can be reduced by averaging results from multiple specimens, but the time-consuming nature of Thellier type measurements combined with the usually rather low success rate means that in the past the number of specimens treated has generally been rather small. Recent studies seem to suggest that the variance is lower for more homogeneous materials like submarine basaltic glass (Pick and Tauxe, 1993) and archeomagnetic artifacts (Ziegler *et al.*, 2006; Donadini *et al.*, 2006) which are often clay or ceramic, but currently even in the best cases it seems optimistic to expect to recover the field using a Thellier paleointensity experiment to better than 10% standard error. These results are broadly in agreement with those presented by Korte *et al.* (2005) who compared nineteenth and twentieth century results with predictions from the GUFM field model (Jackson *et al.*, 2000).

The results of the Kilauea study did not suggest any systematic bias in the mean, but two important effects can result in systematic errors to individual results, namely anisotropy in the material being studied and the influence of cooling rate on the TRM acquired (Halgedahl *et al.*, 1980, Dodson and McClelland-Brown, 1980; Fox and Aitken, 1980; Aitken *et al.*, 1981). Samples generally cool more rapidly in the laboratory than in firing pots, bricks,

or kilns, or in cooling of a lava flow (submarine basaltic glass being an exception) and this leads to an overestimate of the field in which the original remanence was acquired. In principle, this can be corrected, but one needs an estimate of the original cooling rate. A number of careful studies of glass give independent measure of cooling rate from relaxation geospeedometry, and these can be used for corrections (Bowles *et al.*, 2006a; Leonhardt *et al.*, 2006). Genevey *et al.* (2003) conducted some empirical studies for French pottery by experimental cooling of pots at different rates in a special kiln designed to mimic the original conditions. But such attention to detail is rare. In many cases crude estimates are used to correct for cooling rate, while in others no correction is performed or it is not recorded. For thin lavas and archeomagnetic artefacts this might impart a 10% upward bias to the results. Very thick flows or intrusions may suffer worse effects, and this might be one contributing factor to the wide range of results found in different positions for the extensively studied Xitle flow in Mexico (Böhnel *et al.*, 1997).

Magnetic anisotropy is also known to impart bias to individual results (Rogers *et al.*, 1979). The need for anisotropy corrections has often been avoided in archeomagnetic studies, by applying the laboratory field so the TRM is acquired parallel to natural remanent magnetization (NRM) (Aitken *et al.*, 1981). However, Veitch and Mehringer (1984) initiated the idea of correcting for anisotropy using the TRM anisotropy tensor (e.g., Chauvin *et al.*, 2000; Selkin and Tauxe, 2000). Sometimes anisotropy of magnetic susceptibility (AMS) has been used instead, but the fact that different magnetic mineral fractions generally contribute to AMS and NRM makes this a questionable approach. Cooling rate corrections would be expected to lower the overall intensities, but one might hope that the effects of anisotropy would average out, so that when corrections have not been applied, this would just result in an overall increase in scatter when data from different sources are combined.

Time series of relative paleointensity variations are produced using a variety of normalization techniques discussed in Chapter 13. Such estimates range from crude ratios of NRM to bulk anhysteretic remanent magnetization (ARM), isothermal remanent magnetization (IRM) or susceptibility to methods that mimic the Thellier method in a so-called pseudo-Thellier method devised by Tauxe *et al.* (1995) for marine sediments, but applied, for example, by Snowball and Sandgren (2004) to varved lake sediments in Sweden. Many factors apart from geomagnetic paleointensity variations can affect the normalized records, including changes in magnetic mineralogy, hiatuses, or other variation in sedimentation leading to incomplete records, inclination errors or other anisotropy in the record, and temporal smoothing of the geomagnetic field record. There is no currently accepted general theory of how depositional or post-depositional remanence is acquired; consequently there is no analog for the quality checks used in the Thellier-type absolute paleointensity experiments. Although data selection criteria based on uniform rock magnetic properties have been in place for some time (see King *et al.*, 1983; Tauxe, 1993; Chapter 13) these remain ad hoc approaches to the problem that offer no guarantees of reliability. The best quality assessment in many cases comes from regional compatibility among a broad range of different records.

9.2.3 Age Controls

A wide variety of dating techniques are applied to volcanics, archeological, and sedimentary materials, including radiometric ages, varve counting, archeological ages, thermoluminescence, various stratigraphic and other relative methods. In general, varve counting and archeological methods are able to provide the strictest constraints, nevertheless the quality can vary considerably depending on the specific study and material available. Radiocarbon dating of organic material is widely used to provide age constraints for both archeological sites and lake sediments. For uniformity uncalibrated radiocarbon years before present (BP) are conventionally measured relative to AD 1950. The ^{14}C ages must be corrected to account for the fact that the production rate in the atmosphere varies with time. The relationship between ^{14}C radiometric age BP and calender age is plotted in **Figure 2**. The data are from dendrochronologically dated tree-ring samples and form the basis for the standard IntCal04 terrestrial radiocarbon age calibration (Reimer *et al.*, 2004). It is readily seen that there are substantial differences in age prior to about 3000 years ago, and that the curve is not monotonic. Such calibrations have been steadily improving since an early curve proposed by Clark (1975; see also Stuiver and Reimer (1993)), and it is important to know details of the calibration scheme in order to assess reliability of the age estimates. A separate calibration is available for marine environments (Hughen *et al.*, 2004) that takes account

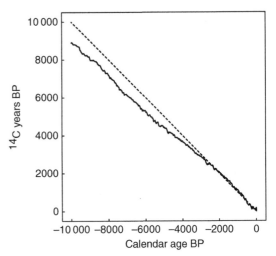

Figure 2 Plot of ^{14}C age vs calendar age from 0–10 ky BP. Dashed line indicates one–one relationship.

of the variable mixing times for oceans and atmosphere.

One concern in using radiocarbon dating of organic material in lake sediments is the age of material being incorporated into the sediment. Mixing of old organic carbon into the sediment can cause systematic age biases; such influences can sometimes be identified by dating of the surface interface or independent age comparisons. Another possibility for introducing both age bias and temporal smoothing of the paleomagnetic signal lies in the lock-in depth for the remanence – this may vary over time and it is not unusual for it to generate a smoothing effect over 100–200 years (e.g., Sagnotti *et al.*, 2005), with corresponding uncertainties in the ages assigned to the magnetization.

9.3 Local and Regional Secular Variation Studies

As was noted in the introduction, Thellier was a pioneer of developing the archeomagnetic record in France and established techniques that are acknowledged as fundamental today. Europe, Scandinavia, and surrounding regions extending eastward and southward to the Far East are natural regions in which to conduct archeomagnetic studies, because of the long history of human structures and associated archeological artifacts. Extensive historical records can make dating more accurate, and consequently there are a number of well established regional reference curves for secular variation, some with multiple revisions and improvements for example in France (Thellier, 1981; Bucur, 1994; Chauvin et al., 2000, Genevey and Gallet, 2002; Gallet et al., 2002), Germany (Schnepp et al., 2004), Hungary (Márton, 2003), Japan (Yoshihara et al., 2003). There are also significant records from the Americas but in some other parts of the world the archeomagnetic record is not so well established because the human record is sparse and less well documented. Places where fired archeological structures and artifacts are sparse can sometimes provide numerous if sporadic volcanic records instead, with Hawaii and the Western United States providing two of the best studied examples. Japan and Italy are well represented by mixed archeological and volcanic records. There remains considerable scope for acquiring new data that could fill some holes in geographical and temporal coverage.

Available data were reviewed recently by Korte et al. (2005) when they attempted a comprehensive compilation of archeomagnetic, volcanic, and sediment data available for the past 7 ky. General characteristics of their data set are discussed in the next section. It draws heavily on both local and regional efforts, but a number of important sediment records are missing, for example, studies on varved Swedish lakes (Snowball and Sandgren, 2002, 2004) and Holocene records from St. Lawrence Estuary in eastern Canada (St-Onge et al., 2003). Some other missing records are listed by Korte and Constable (2006a) in a recent assessment of relative geomagnetic paleointensity variations. Many of the lake records show excellent internal consistency at a local level, and an example from Snowball and Sandgren (2002) is shown in **Figure 3**. Sarsjön and Frängsjon are two closely

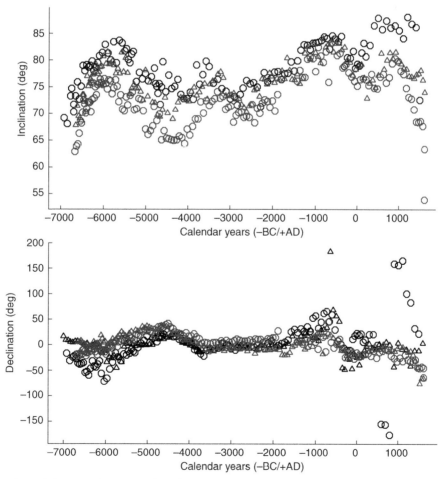

Figure 3 Secular variation records obtained from Swedish varved lakes: black triangles and circles are from two Sarsjön cores, blue from Frängsjon.

spaced varved lakes situated in northern Sweden (locations at 64° 02′ N, 19° 36′ E, and 64° 01′ N, 19° 42′ E, respectively). In **Figure 3** black triangles and circles correspond to data from two separate cores from Sarsjön and blue from Frängsjon. In general the agreement within a lake is to better than a few degrees. The occasional intervals with high scatter highlight the need for replicate records with good internal consistency.

New archeomagnetic and volcanic data sets have also appeared, including Spanish archaeomagnetic directional and intensity data (Gómez-Paccard *et al.*, 2006a, 2006b), directions from Syria (Speranza *et al.*, 2006), and from Greece (Evans, 2006). Several studies have been conducted on volcanics from Southern Italy (Arrighi *et al.*, 2004; Tanguy and Le Goff, 2004; Incoronato *et al.*, 2002; Principe *et al.*, 2004) and on archeomagnetic materials (Evans and Hoye, 2005). Some of this activity has been directed toward providing age constraints for eruptions or archeological sites (see also the preliminary Austrian reference curve presented by Schnepp and Lanos, 2006), but there are also many new independently dated magnetic observations.

Building a regional secular variation curve often requires the reduction of widely distributed data to a central location, and there are established strategies for accomplishing this. Directional data are commonly adjusted for the gross geographic variations by mapping through the virtual geomagnetic pole (VGP), using a method described by Shuey *et al.* (1970), while intensity is equivalently adjusted by mapping through the virtual dipole moment (VDM) (see, e.g., Daly and Le Goff, 1996). **Figure 4** shows an example from Gallet *et al.* (2002, their figure 4) where archeomagnetic directional data from around western Europe are reduced to Paris, then combined and smoothed with a sliding window to produce a reference curve for *D* and *I* that is compared with the British lake sediment curve of Turner and Thompson (1982), also relocated to Paris. It is noted that the curves are not in agreement throughout the time interval, and the authors comment on the discrepancies, and note the difficulties associated with merging sedimentary and archeomagnetic data. Similar discrepancies are readily found elsewhere in the world both at the individual and reference curve level. **Figure 5** shows directional results from individual Hawaiian lavas compared with the Lake Waiau record of Peng and King (1992).

A range of smoothing techniques has been used in constructing reference curves for both sediment and

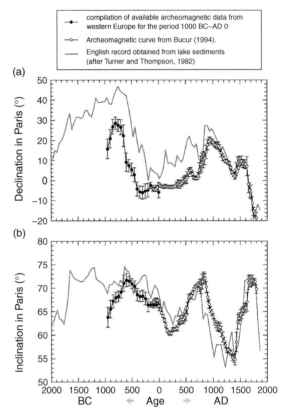

Figure 4 Reference curves for *D* and *I* for western Europe from archeomagnetic and lake sediment data. Reproduced from Gallet Y, Genevey A, and Le Goff M (2002) Three millennia of directional variation of the Earth's magnetic field in western Europe as revealed by archeological artefacts. *Physics of the Earth and Planetary Interiors* 131(1): 81–89, with permission from Elsevier.

archeomagnetic data. These range from scalar averaging of individual geomagnetic elements to more sophisticated bivariate moving average approaches (Daly and Le Goff, 1996). The topic of consistent and optimal construction of reference curves for secular variation has been addressed in some detail in a paper by Lanos (2004) laying out and applying a Bayesian approach to incorporating all the chronological and archeomagnetic information and their uncertainties. The method produces a penalized spherical spline smoothing of the vector observations that takes account of data uncertainties and the uneven temporal distribution of observations. A functional envelope for the 95% confidence limit is defined. Undated or poorly constrained archeomagnetic directions can then be assigned a probabilistic archeomagnetic age using the reference curve. The method has been successfully applied to a number of European studies, including German and Austrian

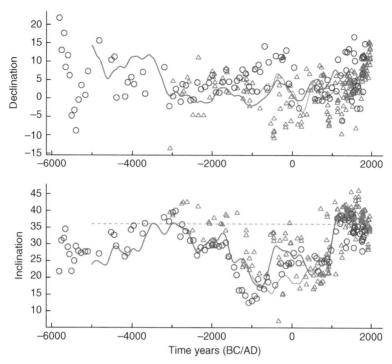

Figure 5 *D* and *I* from Hawaiian lavas, compiled by Korte *et al.* (2005) in red, and Lake Waiau sediment data from Peng and King (1992) in blue. Brown curve is prediction from model *CALS7K.2* described in Section 9.5, green is from *CALS3K.1*. Dashed green line gives inclination expected for an axial dipole field.

data (Schnepp *et al.*, 2003; Schnepp and Lanos, 2005, 2006), and is sufficiently flexible that as the number of archeomagnetic data available continues to grow it should be possible to compute custom reference curves for specific localities rather than relying on existing national curves for the UK, France, or other specific countries. **Figure 6** shows the recently derived reference curve for Austria.

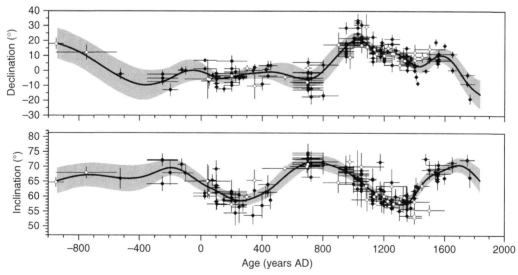

Figure 6 *D* and *I* plotted vs timescale (2σ or archeological estimate for age; 95% confidence limit for *D*, *I*) for data used to construct the Austrian reference curve. Open symbols are from France, Germany, and Switzerland, closed from Austria, Bosnia, and Hungary. Bold black lines give the reference curves, gray area is the 95% confidence band. Reproduced from Schnepp E and Lanos P (2006) A preliminary secular variation reference curve for archaeomagnetic dating in Austria. *Geophysical Journal International* 166: 91–96 (doi:10.1111/j.1365-246X.2006.03012.x), with permission from Blackwell.

The Bayesian approach has recently been contrasted with a hierarchical extension of the moving average bivariate analysis used by Daly and Le Goff (1996). Lanos *et al.* (2005) have conducted a detailed analysis of a heirarchichal bivariate moving average modeling procedure for producing reference curves, and conclude that it is appropriate for very well dated data that are evenly distributed in time, but that such data sets are only rarely achievable.

Lake sediment data from multiple cores are often stacked together and smoothed in an attempt to improve signal-to-noise ratio in the record. This requires identifying equivalent stratigraphic markers, and interpolation onto common timescales. When the chronology is poorly constrained stacking often seems to result in an apparent loss of resolution. In principle the Bayesian approach of Lanos (2004) can be applied to stacks of lake sediment data, but this has not yet been done.

Regional reference curves in the European region confirm that the structure of secular variation is broadly similar across length scales of several thousand kilometers. However, regional correlations break down on large scales, and Asian records cannot easily be reconciled with European versions. Similarly there are regional similarities across North America, but not between Europe and North America. This breakdown in correlation on large scales can be attributed to systematic non-dipole contributions to the geomagnetic field.

9.4   Global Data Compilations – Geographical and Temporal Sampling

In addition to the regional efforts described in the previous section, there have been a number of initiatives directed at acquiring global collections of observations for further synthesis. In the late 1980s, the International Association for Geomagnetism and Aeronomy (IAGA) encouraged the development of a series of paleomagnetic databases including the ARCHEO and SECVR databases, respectively archeomagnetic directional compiled by Donald Tarling of Southampton University, and lake sediment records by McElhinny and Lock (1996). These data sets are archived at the National Geophysical Data Center, but have been largely superseded by others described below.

Daly and Le Goff (1996) compiled available archeomagnetic intensity and directional data for the past 2000 years, sorted them into nine geographical regions,

corrected the data to a standard reference location within each region, and produced smoothed curves presenting *D*, *I*, and *F* at 25-year intervals where data were available. These smoothed curve predictions are referred to as the DLG96 data set. Five sites had more or less complete directional data for the past 2000 years, only three also had good coverage with paleointensity. All were in the Northern Hemisphere, and most confined to mid-latitude regions. Hongre *et al.* (1998) extended the data collection (here designated DLG96+) to include archeomagnetic data from Peru, sediment data from Argentina and New Zealand, and lava data from Hawaii and Sicily, giving a slightly better global coverage with these 14 localities. Constable *et al.* (2000) extended the temporal span to 1000 BC and produced 24 data series based on 100-year averages of *D* and *I* using the previously listed directional sites, but grouped slightly differently and augmented with globally distributed lake sediments and further archeomagnetic data including new sites from China, Mongolia, Australia, and New Zealand. This data set, known as PSVMOD1.0, provided the first reasonable global distribution of directional observations suitable for global field modeling.

The most comprehensive global data sets currently available are to be found in Geomagia50 an online absolute paleointensity database developed by Fabio Donadini of the University of Helsinki geophysics group containing data for the past 50 ky in a web-based searchable format (see Appendix 1), and in a collection of archeomagnetic and sediment data, here designated KGCFS, and made by Korte *et al.* (2005). Geomagia50 and KGCFS are described in the following subsections.

9.4.1   The KGCFS Data Compilation

The global KGCFS data compilation spans the past 7 ky and was made for the explicit purpose of developing the time-varying field model *CALS7K.2* described in Section 9.6. Details are given by Korte *et al.* (2005) and summarized here. The data were assembled from a variety of published sources, using the paleomagnetic databases Archeo00 and Secvr00 archived at the World Data Center in Boulder to acquire digital directional observations where possible, and supplementing these with information acquired directly from the investigators involved in data collection. All the archeomagnetic intensity data were obtained from the original literature or directly from the authors. Three kinds of data are distinguished (1) directional records from lake and rapidly deposited marine sediments, (2) archeomagnetic directional data, and (3) archeomagnetic intensity data. No intensity data from sediments were used, because at the time no systematic evaluation of their validity and internal consistency had been undertaken. All data used must have a chronology that is independent from the magnetic measurements. The numbers of each data type are given in **Table 1**, while the locations and relative concentrations are shown in **Figure 7**. The compiled data files can be found in the Earth Ref Digital Archive at http://earthref.org/cgi-bin/erda.cgi?n=331. Sediments provide a time series of

Table 1 KGCFS data compiled for the interval 5000 BC–AD 1950 and used in *CALS7K.2*

Component	All	Sediment	Archeomag. regions	Archeomag. data
Inclination				
N. Hemisphere	12182	8444	23	3738
S. Hemisphere	3903	3872	1	31
Declination				
N. Hemisphere	10728	8316	23	2412
S. Hemisphere	2352	2321	1	31
Absolute intensity				
N. Hemisphere	2960		17	2960
S. Hemisphere	228		2	228
Total	32353	22953		9400

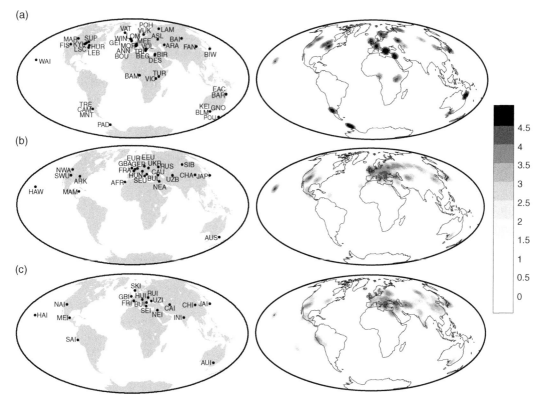

Figure 7 Locations represented in the current KGCFS global compilation of archeomagnetic and paleomagnetic data. Sites of (a) lakes, (b) archeomagnetic directional data, and (c) archeomagnetic intensity data. Left side gives the locations of sediment records, and average locations for archeomagnetic regions, while the right side gives contours of the data concentration.

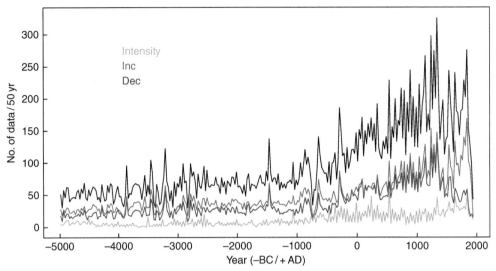

Figure 8 Numbers of *D* (blue), *I* (red), *F* (green) individual elements, and all combined (black) available in KGCFS data set as function of time.

observations from a single location, and each lake or marine site is represented separately in **Figure 7**. Archeomagnetic data rarely form a time series at a single site so they are grouped into average locations for plotting in **Figure 7**, although when used for modeling they are of course treated at their correct locations. This approach contrasts with that used for PSVMOD1.0 and DLG96, where the data have been corrected to a common location. The temporal distributions of the various data elements are illustrated in **Figure 8** from which it is obvious that the observations are concentrated toward more recent times, and that intensity data constitute a rather small fraction of the total. The KGCFS data set is far from complete, and restricted to the time interval 0–7000 BP. There are ongoing attempts to update the directional data, extend the existing data set to longer time intervals, and include relative geomagnetic paleointensity estimates from high-accumulation-rate lake and marine sediments.

9.4.2 Geomagia50 Intensity Data

Currently the most comprehensive collection of absolute paleointensity data for the time period 0–50 ka is available from the searchable Geomagia50 online database (see Appendix 1). Over 3700 results (from about 7300 specimens) are present, drawn from 160 publications and grouped into 38 distinct locations. Various additional information is provided,

including paleointensity method, type of material, specimen type, and dating method. **Figure 9** shows the 3719 virtual axial dipole moments (VADMs) plotted as a function of age back to 50 ka. There is substantial overlap between the KGCFS and Geomagia50 absolute paleointensity data. **Figure 9** confirms that the average VADMs have been higher over the most recent few millennia than for most of the past 50 ky.

9.5 The Global Geomagnetic Field and Its Secular Variation on Millennial Timescales

Attempts to build global geomagnetic field models on archeological timescales have been going on since the 1970s (e.g., Márton, 1970; Braginskiy and Burlatskaya, 1979; Sakai, 1980; Ohno and Hamano, 1993), but a recent resurgence of interest in such activities (Hongre *et al.* 1998; Constable *et al.* 2000; Korte and Constable, 2003, 2005a) has followed from the much-improved data sets that have been compiled over the past decade. These investigations began with smoothed time series of archeomagnetic directional data (Daly and Le Goff, 1996), then extended to include lake sediment directional data for the past 3 ky (Constable *et al.*, 2000), and most recently have used the directional and paleointensity data compiled by Korte *et al.* (2005) discussed in the previous

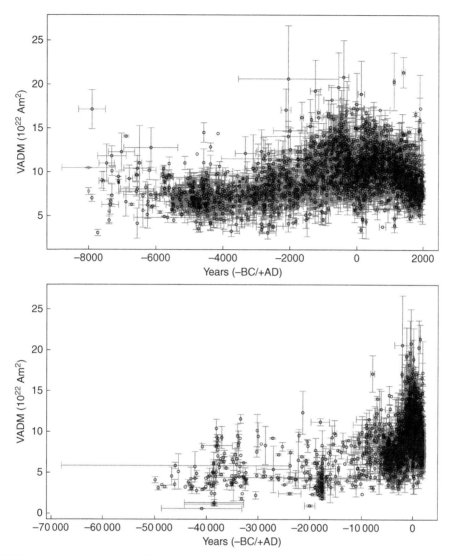

Figure 9 VADMs for the time intervals 0–10 ka (upper) and 0–50 ka (lower) available from the Geomagia50 database on 3 February 2007. Error bars in gray are 1 standard error in age and VADM when available.

section. The basic method for building global models of millennial-scale geomagnetic field variations has followed that used for the present and historical field. The parametrization is in terms of a truncated spherical harmonic expansion of a scalar potential representing the magnetic field originating in Earth's core.

$$\psi(r,\ \theta,\ \phi) = a \sum_{l=1}^{\infty} \sum_{m=0}^{l} \left(\frac{a}{r}\right)^{l+1}$$
$$\times \left(g_l^m(t)\cos m\phi + b_l^m(t)\sin m\phi\right) P_l^m(\cos\theta)$$

P_l^m are partially normalized Schmidt functions of degree l and order m, and position is specified in a geocentric spherical coordinate system with radius r, θ as colatitude, and ϕ as longitude. A magnetic field model is a listing of the Gauss coefficients g_l^m, b_l^m up to some maximum degree L: the coefficients may or may not have an explicit parametrization in time, t, that expresses the temporal dependence arising from secular variation. When no explicit temporal parametrization is specified, the model is usually either an average over some time interval or can be regarded as a snapshot model of a specific moment in time. The gradient of the potential provides the magnetic field $\mathbf{B}(r,\ \theta,\ \phi,\ t) = -\nabla\psi(r,\ \theta,\ \phi,\ t)$ and as appropriate the secular variation can be evaluated from $\partial\mathbf{B}/\partial t$. The approximation that there are no magnetic sources in

Earth's (more-or-less) insulating mantle means that field models built from observations at Earth's surface can be downward continued to map the radial magnetic field at the CMB. However, the accompanying amplification of small-scale features means that such activities require appropriate caution when conducted on already uncertain models.

As with historical field modeling, activities on millennial-scale data have gradually evolved from least-squares fits of the data with low-degree spherical harmonic representations at fixed intervals (Márton, 1970; Hongre et al., 1998) to regularized snapshot inversions (Constable et al., 2000) and most recently to regularized time-varying representations with the temporal variations parametrized by cubic splines (Korte and Constable 2003, 2005a). The hazards of making spherical harmonic models with least squares and poor data coverage are well known and the results must be treated with appropriate caution. The usual strategy is to truncate the expansion at sufficiently low degree that the results are stable for the given data distribution. The maximum degree for least-squares modeling is usually much less than in regularized inversions, which typically extend to spherical harmonic degree and order 10 and may have spline representations with knots every 50 years. The actual spatial and temporal resolutions for these regularized millennial-scale models are much lower than this. The rationale for regularized field modeling has been discussed elsewhere (see Parker, 1994), but loosely speaking one can say that it recognizes that there is some acceptable measure of misfit to the observations, and seeks a model that trades the misfit for finding a model with minimal complexity. Unnecessary structure, of the form specified by the particular complexity or roughness penalty in use, will be excluded from the model. A hypothesis testing approach using several different kinds of roughness penalty (e.g., size of the field, or its spatial gradients over a specified spherical surface) can be useful in evaluating whether particular features in a given model are robust, but has not so far been reported for millennial-scale field modeling. An alternative strategy is to explore the sensitivity of the model to the specific data used via a bootstrap technique (e.g., Korte and Constable, 2006b, 2007). Such evaluations tend to highlight the fact that there are large regions with few or no data, and others may have mutually inconsistent observations, so that the reliability of the models can be quite variable for different times and places. It is worth noting that a flaw in this approach is that regions with no data will tend to produce very consistent minimum complexity results, even though the absence of data means that little is known about the field in this region. Although regularized inversions will minimize the complexity in poorly sampled regions, it should be kept in mind that even under the best circumstances the data uncertainties and poor geographic coverage discussed in earlier sections of this chapter make these models considerable less reliable than their historical or recent counterparts.

Table 2 catalogs some basic information about five efforts at millennial-scale field modeling, derived from data compilations and reductions discussed in Section 9.4. The temporal spans range from 1700 to almost 7000 years, but we can group the various models according to inversion strategies and some similarities among the data sets. *HHK* as an earlier effort has the least number of data and shortest time span, with only 14 distinct locations somewhat incompletely sampled every 25 years. The modeling was done by a nonlinear least-squares Levenberg-Marquardt algorithm to recover Gauss coefficients up to degree and order 2 and also g_3^3, h_3^3. This choice was based on stability tests for the parameters

Table 2   Holocene magnetic field models

Model	Time (AD)	Time parm'n	Max l	Data	Type	Reference
HHK	0–1700	25 yr MA	<3[a]	DLG96+	LS	Hongre et al., 1998
ALS3K	−1000–1800	100 yr BA	10	PSVMOD1.0	SR	Constable et al., 2000
CALS3K.1	−1000–1800	60 yr Spline	10	PSVMOD1.0	TR, SR	Korte and Constable, 2003
CALS3K.2	−1000–1950	60 yr Spline	10	KGCFS	TR, SR	Korte et al., 2005
CALS7K.2	−5000–1950	60 yr Spline	10	KGCFS	TR, SR	Korte et al., 2005

[a]Model *HHK* only estimates g_3^3 and h_3^3 for $l = 3$.
MA means that coefficients were filtered by a moving-average-type temporal filter after estimation at 25-year intervals. BA means that data were averaged in 100-year intervals and snapshot models constructed for each one. Negative times in column 2 mean years BC.
LS, least squares; SR, spatial regularization during inversion; TR, temporal regularization.
Data sets are described in Section 9.4.

recovered (see Hongre *et al.*, 1998, for details). *ALS3K* is like *HHK* in that it is a series of snapshot models. Differences are that the individual models extend to degree and order 10, they are spatially regularized, the time span extends an additional 1000 years, and *ALS3K* is based on the larger PSVMOD1.0 data set with 24 distinct locations, incompletely sampled every 100 years. In *ALS3K* only variations relative to g_1^0 were recovered, and the results were arbitrarily scaled using a fixed value of $g_1^0 = 30\mu T$. *ALS3K* and *CALS3K.1* are the most directly comparable in that they span the same time interval and are both derived from PSVMOD1.0, differing only in the time parametrization of the models, and the fact that in *CAL3K.1* the axial dipole is constrained to follow the VADM evolution of McElhinny and Senanayake (1982). All of the *CALSxK.n* models are constructed using the same inversion strategy (described in Korte and Constable, 2003) with spatial regularization under the diffusion norm and temporal regularization of the second derivative of the field. The acronym stands for Continuous Archeomagnetic and Lake Sediment models spanning x ky and model version n. *CALS3K.1* and *CALS3K.2* differ hugely in the data sets used. In both *CALS3K.2* and *CALS7K.2* each magnetic element (*D*, *I*, or *B*) is treated at its appropriate place in time and space, rather than reducing to a common epoch or reference location. The KGCFS data set is much more comprehensive than earlier ones, but also more heterogeneous in quality.

These properties help to illustrate the differences among the various model predictions shown in **Figures 10–12**. **Figure 10** shows the temporal evolution of the Gauss coefficients up to degree and order 2 for each of the models in **Table 2**, along with predictions from the historical model *GUFM* for the period AD 1590–1990. In general the millennial models converge to something close to the same value as for *GUFM*, although the overlap is minimal for *HHK* and *ALS3K*. Notable exceptions are for g_1^1, h_1^1, and g_2^1 for *CALS7K.2* and *CALS3K.2*, and h_2^1 for all models. Korte and Constable (2007) attribute the poor agreement with *GUFM* for *CALS3K.2* and *CALS7K.2* to end effects in the spline temporal parametrization combined with a drop-off in the number of directional observations for the most recent times. The evolution of *CALS3K.2* and *CALS7K.2* is very similar, and we will not distinguish them further except on the basis of time interval covered. The greatest difference among the models is in the magnitude of the axial dipole g_1^0, which is substantially lower for *CALS7K.2* than for *HHK* and *CALS3K.1*. The

VADM constraints used on g_1^0 for *CALS3K.1* are heavily biased to the European region, and the direct conversion of intensity to VADM takes no account of the non-axial-dipole field contributions. This is believed to result in a bias of g_1^0 to high values, and would certainly influence the other coefficients in the early part of the model where the magnitude of g_1^0 seems to be severely overestimated. *ALS3K* has a rather jittery temporal evolution resulting from the separate inversion for each time slot. In *HHK* the snapshot Gauss coefficients have been temporally smoothed with a bandpass filter with a cutoff period of 600 years for $l = 1$ and about 250 years for higher degrees. *HHK* generally has the largest fluctuations and differs most from the other models. This can be attributed to the sparse data set, and spatial aliasing expected from least-squares estimation. One should, however, keep in mind that the heavy regularization imposed in the *CALSxK.n* models will tend to produce models with considerably less structure than the real geomagnetic field.

Of the models listed in **Table 2** *HHK* and *CALS7K.2* have been subjected to the most systematic assessment about resolution and accuracy. With the available data *HHK* was unable to recover stable estimates of Gauss coefficients above degree 2, except for g_3^3 and h_3^3 (Hongre *et al.*, 1998). Synthetic experiments with *GUFM* and data distributions for the various CASLSxK.n demonstrated broad-scale agreement between *GUFM* and the recovered models. The actual resolution of *CALS7K.2* is not exactly known, but Korte and Constable (2007) infer from the rapid fall-off in the spatial power spectrum that it can be no better than spherical harmonic degree 4. In the time domain the averaging is expected to be over several centuries, and this is supported by the absence of power in the secular variation at periods shorter than a few hundred years (Korte and Constable, 2006b). Korte and Constable (2007) used bootstrap methods to evaluate expected uncertainty in model predictions at Earth's surface, and found directional uncertainties that were generally less than 2° and intensities less than $1.5\mu T$, but these can be 3–4 times as large for some times and places.

9.6 The Average Field

Figure 11 maps inclination anomalies (left) and declination (right) at Earth's surface for temporal averages of the field models *HHK*, *ALS3K*, *CALS3K.1*, *CALS3K.2*, *CALS7K.2*, and *GUFM* over their

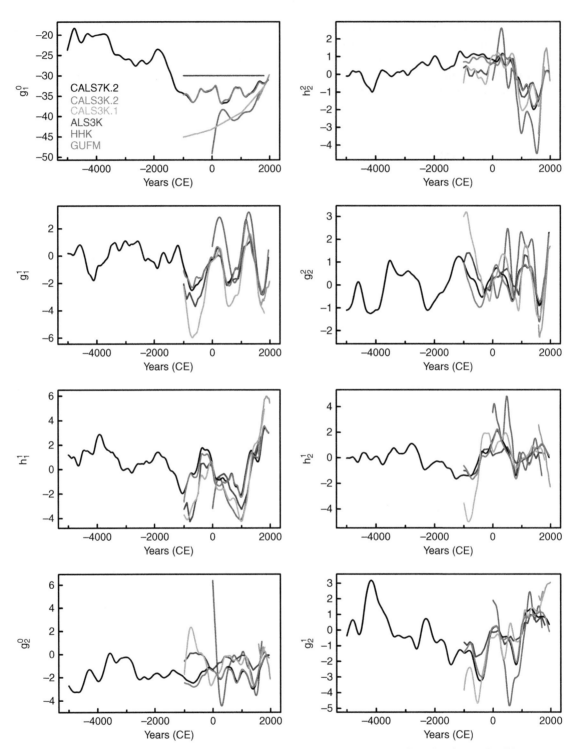

Figure 10 Comparison of degree $l = 1, 2$ Gauss coefficients for field models *HHK, ALS3K, CALS3K.1, CALS3K.2, CALS7K.2,* and *GUFM*. All coefficients are given in microtesla at Earth's surface.

respective time spans. These averages reflect the differences among the models discussed above, as well as the range of averaging from 400–7000 years.

Of the millennial-scale model *HHK* is the most distinct, and one might safely suppose that the data distribution is too sparse and the parametrization

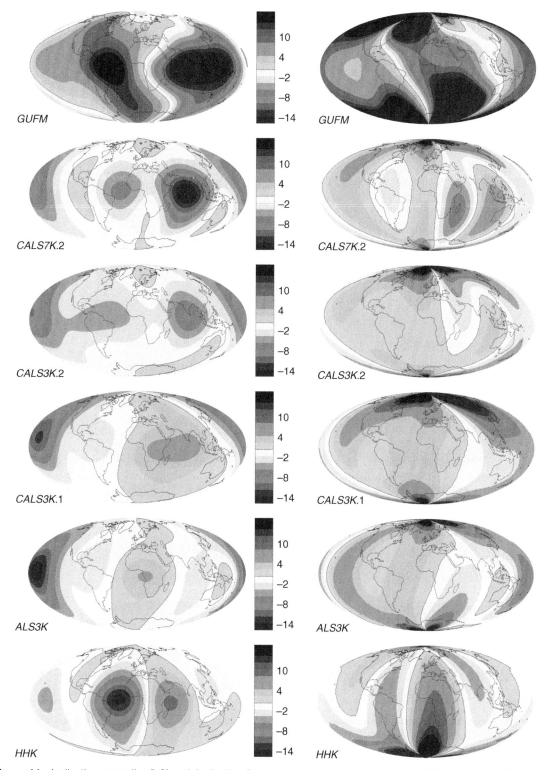

Figure 11 Inclination anomalies (left) and declination for temporally averaged field models *HHK, ALS3K, CALS3K.1, CALS3K.2, CALS7K.2,* and *GUFM.*

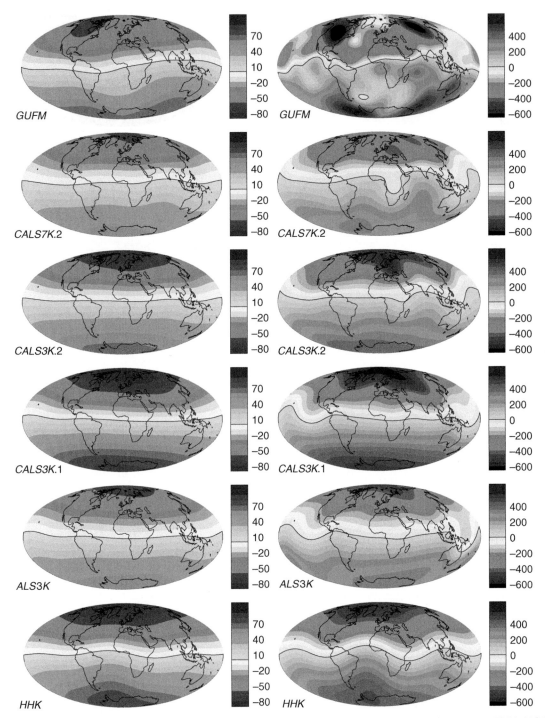

Figure 12 Radial magnetic field at Earth's surface (left) and CMB (right) for temporally averaged field models *HHK*, *ALS3K*, *CALS3K.1*, *CALS3K.2*, *CALS7K.2*, and *GUFM*.

too limited to capture all the field structure present. The large-scale directional variation is broadly similar for *ALS3K* and *CALS3K.1*, with longitudinal asymmetry in both ΔI and *D*. However, the larger

data set used in *CALS3K.2* imposes some substantial changes on the model, with generally negative inclination anomalies emerging in equatorial regions, as one might expect for a field with a persistent axial quadrupole contribution. This signal grows in amplitude in some regions for the longer *CALS7K.2*, but over the Americas the inclination anomaly diminishes and actually reverses sign. In general *CALS7K.2* has larger directional anomalies than the other millennial models, probably because of the diminished magnitude of g_1^0 in the early part of the record. The millennial-scale model directional anomalies are distinct from those found in *GUFM* – the temporal averaging has a powerful effect, not just in reducing their scale but also altering the distribution.

Figure 11 provides a useful means of assessing the size of directional anomalies expected in various regions, but a more direct view of the average geodynamo including poloidal field strength is given by mapping the radial component of the field, B_r. **Figure 12** shows B_r at Earth's surface (left column) and after downward continuation to the CMB (right). Note that both *ALS3K* and *CALS7K.2* are substantially weaker than the others, in the first case because of the artificially imposed constraint that $g_1^0 = 30 \mu T$, and in the latter because the axial dipole is much weaker during the interval 5000–1500 BCE. The millennial-scale models suggest the presence of quasi-stationary high-latitude magnetic flux patches in the Northern Hemisphere, as seen in *GUFM* and other historical models. Experiments with numerical dynamo simulations produce similar time-dependent and time-averaged field morphology when a heat flow related to lower mantle seismic tomography is imposed at

the CMB (Bloxham, 2002; Olson and Christensen, 2002) lending support to the idea that these patches are controlled by thermal interactions between core and mantle. In the Southern Hemisphere the models have poor resolution, and supposing that such patches are present it is unlikely that they could be resolved. Linearized data kernels for changes to an axial dipole field structure shown in **Figure 13** give some guidance on how geomagnetic data elements sampled at various locations are sampling structures at the CMB. From these it can be seen that even though the data coverage is spartan large-scale features at the CMB will contribute to surface observations at distant locations, rather than just to those directly above. The relative sampling of the CMB achieved with the whole KGCFS data set used in *CALS7K.2* is illustrated in **Figure 14** which shows the geographical bias to mid-northern latitudes.

The *CALS7K.2* model and source code for predicting elements of the field are available at Earthref.org, along with animations showing its evolution over time. **Figure 15** shows the average field intensity at Earth's surface for the time interval 0–7 ka next to the contemporary version derived by Olsen (2002) from modern satellite observations. Although the global average field strength is about the same, the structure is markedly different. The South Atlantic magnetic anomaly is notably absent, and the flux lobes are highly attenuated. However, the field in the Pacific remains lower than in the Atlantic hemisphere, and

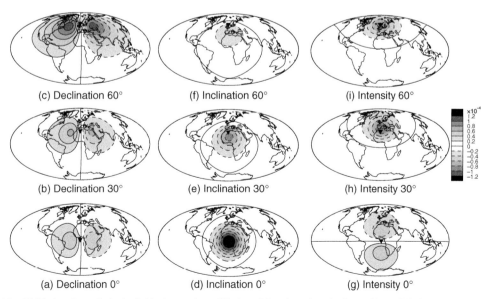

Figure 13 CMB data kernels for individual samples of *D*, *I*, and *B* at locations indicated by solid triangles.

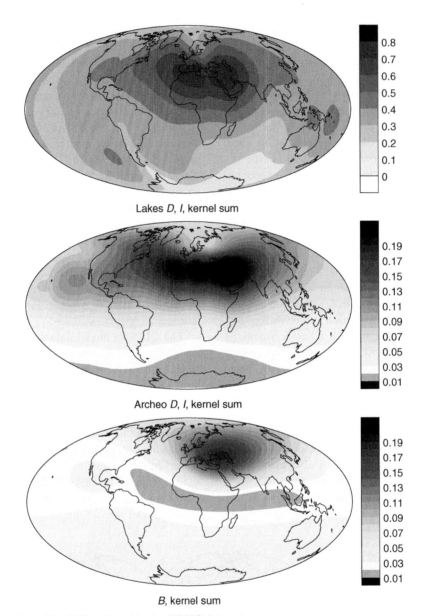

Lakes *D*, *I*, kernel sum

Archeo *D*, *I*, kernel sum

B, kernel sum

Figure 14   Sampling of the CMB achieved by the KGCFS data set.

for the 7 ky average has its lowest value somewhat to the northeast of Australia.

The absence of the South Atlantic Anomaly might be taken as an indication that this is a transient feature, but we should keep in mind limitations imposed by the temporal and spatial distribution of the available data, especially in the Southern Hemisphere. There are no archeomagnetic directional or intensity data to constrain the field in this region – the SAI region (**Figure 7(c)**) is too distant and includes only 147 data from Bolivia, Ecuador, Peru, and the Pacific. Sediment directional data in

general have relatively poorly constrained age scales compared with the best archeomagnetic records, and the age constraints for the African data are the weakest. In many cases the largest contribution to the data uncertainty comes from poor age constraints (and associated secular variation in the magnetic field), and the assigned uncertainties in the observations average 4.0° for inclination, 6.7° for declination, and 11 μT for intensity. The regularized inversion results in a tradeoff between misfit to the data and minimum complexity in the model, and with these large uncertainties we expect the model to lack both temporal

(a)

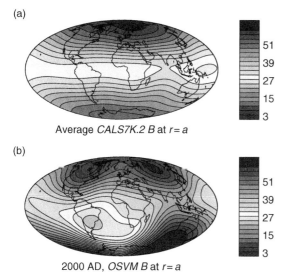

Average *CALS7K.2 B* at *r = a*

51
39
27
15
3

(b)

2000 AD, *OSVM B* at *r = a*

51
39
27
15
3

Figure 15 (a) Scalar magnetic field at Earth's surface in μT for the average of the *CALS7K.2* model over the 0–7 ka time interval, and (b) from Olsen's (2002) Ørsted secular variation model, *OSVM*.

and spatial resolution. The experimental procedures used for deriving intensity data lead to uncertainties in the form of percentage errors and are usually larger for high field values. This leads to a small systematic bias to positive values in the distributions

of intensity residuals from regularized models. In the case of *CALS7K.2* the bias is about 3.6 μT. For the directional data there are no significant biases in the residual distributions.

The low average intensity structure in the western Pacific region in **Figure 15**(a) should also be viewed with some caution, although if it is not required by the observations it should have been eliminated by the inversion strategy using quadratic regularization. It is possible that it might be real geomagnetic behavior, and this view is supported by comparisons made in **Figure 16**, where the radial component of the magnetic field at Earth's surface is plotted after removing the contribution due to the axial-dipole part of the field. This emphasizes geographic variations in structure that represent departures from the geocentric axial dipole field. The different maps in **Figure 16** represent averages over vastly different timescales. The two panels on the left side are for AD 2000 and the average over the past 400 years, while on the right we have the 7 ky average model and a model for the past 5 Ma. There are broad similarities in structure for the two left hand models. The lower panel looks like an attenuated version of the upper left, as might be expected when small-scale rapidly changing features are averaged over time. The same holds true when the two right hand panels

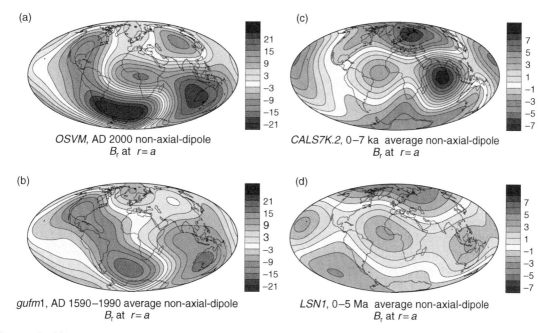

(a)

OSVM, AD 2000 non-axial-dipole
B_r at *r = a*

21
15
9
3
−3
−9
−15
−21

(c)

CALS7K.2, 0–7 ka average non-axial-dipole
B_r at *r = a*

7
5
3
1
−1
−3
−5
−7

(b)

*gufm*1, AD 1590–1990 average non-axial-dipole
B_r at *r = a*

21
15
9
3
−3
−9
−15
−21

(d)

LSN1, 0–5 Ma average non-axial-dipole
B_r at *r = a*

7
5
3
1
−1
−3
−5
−7

Figure 16 (a) Vertical component of the non-axial-dipole field in μT evaluated at Earth's surface (*r* = 6371.2 km) using the geomagnetic field models *OSVM* for AD 2000, (b) *GUFM* averaged over 400 years, (c) *CALS7K.2* averaged over 7 ky, and (d) LSN1 (Johnson and Constable, 1997) averaged over 5 My. Note scales for (c) and (d) differ by factor of 3 from (a) and (b).

are compared with one another, but these longer-term averages have features that are quite distinct from the historical record. On the longer time intervals the largest contributions appear to be latitudinal variations, but there is also a systematic nonzonal contribution. It seems unlikely that this reflects spatial sampling biases, since the data distribution is quite different for the 0–5 Ma data used by Johnson and Constable (1997) from that for 0–7 ka.

9.6.1 Jerks, Drifts, and Waves

The westward drift of geomagnetic features noted by Halley during the seventeenth century is widely interpreted to reflect aspects of the fluid motion at the surface of Earth's core (Bullard *et al.*, 1950; Yukutake and Tachinaka, 1969). Two possible dynamical sources for these effects have received attention – the first discussed by Braginskiy (1972, 1974) is the superposition of propagating MAC (magnetic, Archimedean, Coriolis) waves at the core surface resulting in phase propagation of the field, and reflecting a balance among Lorentz, Coriolis, and buoyancy forces; the second interpretation is that the drifts of the field reflect mean azimuthal flow at the core surface. Core surface flows have been mapped under various constraints about the field evolution in addition to the frozen flux approximation, which specifically requires that the magnetic field is advected with the material flow. Conservation of angular momentum also provides a firm theoretical foundation linking decadal geomagnetic secular variation to changes in the length of day (LOD). These changes are linked to torsional oscillations in the core fluid as it adjusts to small departures from the Taylor state, and have been tied to the sharp accelerations in geomagnetic field records known as geomagnetic jerks (Bloxham *et al.*, 2002). No such firm theoretical footing exists to link longer-term changes in LOD to millennial-scale core flows, as there is little justification for supposing that the core motions are without axial shear over these longer timescales. Nevertheless, Dumberry and Bloxham (2006) have made some substantial preliminary efforts to assess the contributions of millennial-scale geomagnetic field changes to the longer-term changes in LOD now documented back to 2700 BP (Stephenson and Morrison, 1984, 1995; Morrison and Stephenson, 2001).

Studies of regional westward drift in distinct paleomagnetic records are generally limited by inaccuracies in the assigned timescales for records that are separated by a few thousand kilometers or less

(e.g., Lund, 1996). In more distant records the identification of the same magnetic features at different times has proved challenging, probably reflecting the temporal and spatial scales associated with westward drift. An assessment of drift in individual paleomagnetic records has often been made on the basis of the interpretation of Bauer (1895) plots, generally of declination versus inclination centered on the axial dipole or mean field direction at a site. Occasionally the path traced by the VGP is used. Under Runcorn's (1959) rule clockwise motion is taken to represent westward drift of an underlying magnetic source, although it is well known that this interpretation is nonunique (e.g., Skiles, 1970; Dodson, 1979). Nevertheless the general idea can be supported by an analysis of the *GUFM* historical field model shown in **Figure 17**. Regional VGP trajectories mainly show clockwise motion, and the motion is largest in amplitude in the regions where westward drift of the magnetic field is most pronounced. Local application of Runcorn's Rule has led to numerous reports of eastward drift in millennial-scale magnetic records (e.g., Constable and McElhinny, 1985; Snowball and Sandgren, 2002).

Gallet *et al.* (2003) have recently drawn attention to coincident features in the western European archeomagnetic directional secular variation curve and archeomagnetic intensity changes in France and the Middle East. Changes in curvature in the directional variations on Bauer plots are inferred to occur at the same time as local maxima in intensity variations. They call these coincident features archeomagnetic jerks, and raise questions about their regional versus global significance and whether such relationships are a general characteristic of short-term geomagnetic field variations. It should be noted that despite the similarity in terminology they should not be thought of as similar to the geomagnetic jerks observed by direct observations for the most recent century. Instead, as outlined below these archeomagnetic jerks seem to be associated with broad regional changes in directions of drift and wave motions as recently shown in a global assessment by Dumberry and Finlay (2007).

The advent of time-varying millennial-scale geomagnetic models enables analyses of drift to move from a regional to a global context, and there have been two recent attempts using *CALS7K.2* to map the predominant direction of drift (Dumberry and Finlay, 2007), and to determine whether oscillating mean azimuthal flows are the predominant cause of the observed drift (Dumberry and Bloxham, 2006).

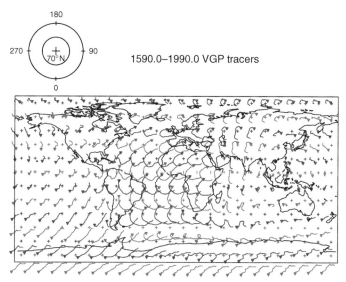

1590.0–1990.0 VGP tracers

Figure 17 VGP trajectories calculated on a 15° geographic grid using field predictions from *GUFM*. Pluses indicate the geographic location for which the VGP is calculated, and form the center for each local VGP projection with scale as at upper left, where 70° N is the inner circle, and 50° N the outer one. Paths are for the time interval AD 1590–1990 with arrowheads indicating the position in 1990. Note the broad clockwise looping in the Atlantic/African region.

Dumberry and Finlay (2007) have used the global time-varying *CALS7K.2* model to conduct a global study of episodes of eastward and westward drift during the past 3 ky. They find both eastward and westward motion at mid-to-high latitudes in the Northern Hemisphere, corresponding to displacements and changes in the two major quasi-stationary high-latitude magnetic flux patches. Poor resolution in the model may prevent similar motions from being discovered in the Southern Hemisphere. They note that the direction changes are associated with the times of the archeomagnetic jerks identified by Gallet *et al.* (2003) – these direction changes are moderately rapid (centennial timescales) but there is no evidence for a sharp change in the geomagnetic field. Dumberry and Bloxham (2006) showed that the observed drift is consistent with the motions being caused by advection of magnetic field features by azimuthal flows, although it is worth noting that their arguments clearly spell out that this is a plausible rather than a unique interpretation of the observations. The quality of the *CALS7K.2* model makes it difficult to determine whether westward drift is a persistent feature of the field for 0–3 ka, but it does seem that the longitudinal region in which it is seen has remained fixed, lending support to the idea that thermal core–mantle interactions may produce both the high-latitude flux lobes, and a regime that produces westward drift in the Atlantic hemisphere. However, the westward drift seen in the

millennial-scale models should be distinguished from that seen in the modern field which tends to be dominated by rapid motions in equatorial regions – these cannot be resolved with the limited temporal and geographical resolution achieved in the current time-varying global paleomagnetic models.

9.6.2 Dipole Moment Variations

Global paleofield models derived from both directional and intensity data allow the recovery of a direct estimate of the geomagnetic dipole moment rather than relying on the proxy obtained by averaging VADMs. **Figure 18(a)** (from Constable and Korte, 2006) shows the dipole moment variation for *CALS7K.2* and proxy estimates derived from block averaging (solid symbols) or spline curve fitting (blue) to VADMs. Also shown in gray is the dipole moment recovered from both fitting all the data (direction and intensity elements) to a time-varying dipolar field. The dipole moment for *CALS7K.2* is lower than the proxy estimates by about 19%. Korte and Constable (2005b) consider the VADM proxies to be biased high and attribute the differences to non-dipole field effects arising from the limited geographic distribution of the intensity data, and data quality influences. It is possible that inadequate cooling rate corrections play a role here. Despite the offset in amplitude there is general agreement in

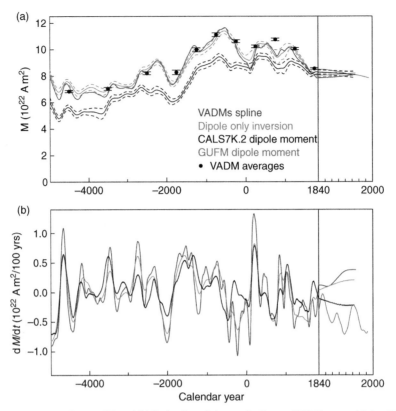

Figure 18 (a) Dipole moment estimates (*M*) and (b) their rates of change for the past 7000 years obtained from the same data by different methods: 500 or 1000 year average VADMs (black dots) and weighted spline fit (blue line) to individual VADMs, spherical harmonic inversion of intensity and directional data for a dipole only (gray line), dipole moment estimate from *CALS7K.2* (black line) and in the expanded right part also from the historical model *GUFM* (red, Jackson *et al.*, 2000). Uncertainty estimates for the spherical harmonic models were determined by a bootstrap method (dashed lines).

the temporal structure of the various rates of change per century plotted in **Figure 18(b)**. Comparisons with *GUFM* in the expanded timescale from AD 1840 to 2000 where there are historical measurements of field intensity show *CALS7K.2* in good agreement with changes in the historical dipole moment when the much lower temporal resolution of *CALS7K.2* is taken into account. Recently Gubbins *et al.* (2006) used the paleointensity data from KGCFS combined with historical directional data used for *GUFM* to confirm that earlier extrapolations of historical dipole moment trend back to 1590 were inaccurate. These results reinforce those already acquired with CALS7K.2. **Figure 18** clearly shows that the current decrease in dipole moment is not anomalous compared with values attained over the past 7 ky at the temporal resolution of the *CALS7K.2* model. Constable and Korte (2006) use this and other data to argue that a geomagnetic reversal cannot be considered imminent. Gubbins *et al.* (2006) have tied the historical decrease in dipole moment to growth of

the low field South Atlantic Anomaly, and associated north/south hemispherical asymmetry. Any extension of such analyses for the paleofield will need to take careful account of geographic variations in resolution and accuracy for CALS7K.2.

9.7 The Geomagnetic Spectrum

Two important statistical properties of the geomagnetic field that should be reproducible by adequate numerical simulations of the geodynamo are the temporal and spatial power spectra (Kono *et al.*, 2000; McMillan *et al.*, 2001; Kono and Roberts, 2002). Korte and Constable (2005a) note that both the spatial power spectra $R_l(t)$ and the spectrum of the secular variation $S_l(t)$ for *CALS7K.2* have greatly diminished power above spherical harmonic degree 4 or 5 compared with the present field. This reflects the limited spatial and temporal resolution in the model, imposed by the distribution and quality of the data.

The long-term temporal variation of R_l with time (here designated P_l) can be measured by the variance of the Gauss coefficients about their mean value, yielding information about the power in the secular variation for comparison with long-term (million-year) statistical models of paleosecular variation (PSV). For *CALS7K.2* we estimate P_l directly from the standard deviation in the estimated Gauss coefficients,

$$P_l = <\vec{\sigma}_{B_l} \cdot \vec{\sigma}_{B_l}>_{r_a} = (l+1) \sum_{m=0}^{l} \left[\left(\sigma_{gl}^m\right)^2 + \left(\sigma_{bl}^m\right)^2 \right]$$

P_l is shown in **Figure 19(a)** along with the equivalent spectra for two candidate PSV models for the time interval 0–5 Ma, *CJ98* and *TK03.gad*, of Constable and Johnson (1999) and Tauxe and Kent (2004), respectively. Both *TK03.gad* and *CJ98* are based on the premise that the spectrum of geomagnetic variations is essentially flat at the CMB for $l > 2$, and they are quite similar in this range. But otherwise there is rather poor agreement among the three models. The differences in *CALS7K.2* for $l > 5$ can be readily explained by the limited resolution in *CALS7K.2*. But both 0–5 Ma models have more power in quadrupole variations than *CALS7K.2*, a feature which is worth further investigation. Also plotted on **Figure 19(a)** is S_l for the *OSVM* model for epoch AD 2000 *OSVM* has most

power in the secular variation at $l = 2$, and diminished power in the dipole. The low dipole power is to be expected from the longer time constant associated with dipole variations. But in *CALS7K.2* P_2 drops below the general trend, contrasting strongly with *CJ98* and *OSVM*. The disagreement among these models highlights our lack of knowledge about long-term (large-scale) field variations. If the low power at $l = 2$ in *CALS7K.2* proves to be robust in future modeling efforts this may be an important feature of long-term geomagnetic secular variations.

The best-known part of the temporal spectrum of variations is that due to dipole moment variations, and on very long timescales even that is rather poorly known. **Figure 19(b)** (after Constable and Johnson, 2005) shows a composite spectrum derived from Cande and Kent's (1995) 0–160 Ma reversal record, (the CK95 (black), and 0–83 Ma reversal record including cryptochrons, CK95cc (gray) portion), various marine sediments records from Site 522 (blue, Constable *et al.*, 1998), VM93 (red, Valet and Meynadier, 1993), Site 983 (green, Channell *et al.*, 1998), Site 984 (brown, Channell, 1999), Sint-800 (orange, Guyodo and Valet, 1999), and *CALS7K.2* (pink). *CALS7K.2*, which is the highest frequency contributor to the paleomagnetic power spectrum, fits well in the frequency overlap with the spectrum derived

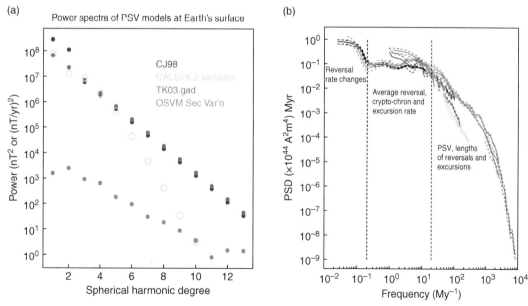

Figure 19 (a) Spatial power spectrum showing the variance as a function of spherical harmonic degree for PSV models *CJ98*, *TK03*, and for *CALS7K.2*. Spectrum of secular variation from *OSVM* is also shown. (b) Composite paleomagnetic spectrum for dipole moment variations (after Constable and Johnson, 2005): the spectrum is derived from the 0–160 Ma reversal record, CK95 (black), 0–83 Ma reversal record including cryptochrons, CK95cc (gray), various marine sediments records from Site 522 (blue), VM93 (red), Site 983 (green), Site 984 (brown), Sint-800 (orange), and *CALS7K.2* (pink).

from the high sedimentation rate marine sediments at the North Atlantic Sites 983 and 984. The individual 983 and 984 records show substantially more power than the global average Sint-800 which is to be expected from dating uncertainties among the cores contributing to the stack. Further comparisons are need with records from a broader range of geographical areas for the older part of the record, as well as with historical field models like *GUFM*. Korte and Constable (2006b) have recently undertaken a more detailed study of the frequency content of higher-degree spherical harmonic terms in *CALS7K.2*. The average temporal resolution of *CALS7K.2* is no better than a few centuries, and it would be of considerable interest to be able to study higher-frequency dipole moment variations, as well as to extend the global dipole moment estimates further back in time.

9.8   Applications

9.8.1   Calibrating Relative Geomagnetic Intensity Variations

For several decades researchers have been acquiring records of relative variations in geomagnetic field strength from lacustrine and marine sediments (e.g., Levi and Banerjee, 1976; Constable, 1985; Tauxe, 1993). Calibrating these data to absolute field

strengths remains a thorny problem (Constable and Tauxe, 1996: Valet, 2003), and most current strategies rely on the idea that average global VADMs provide a suitable dipole moment for scaling individual records. *CALS7K.2* can provide a direct prediction of intensity variations at any geographic location for the past 7 ky, and it can be seen from **Figure 20** that there are substantial geographic variations. The gray curve shows the variation expected from the axial dipole field as $-g_1^0$ and corresponds to the intensity expected at the equator from that term alone. Comparisons of these predictions with relative intensity time series then allow a calibration that can take account of non-dipole field contributions that have not been averaged out over millennial timescales. Korte and Constable (2006a) carried out this calibration for a series of 22 distinct records and concluded that many of them contain significant information about geomagnetic paleointensity variations.

9.8.2   Cosmogenic Isotope Production Rates – the Global View Linking Geomagnetic and Climate Studies

It is well known that the geomagnetic field acts as a shield against cosmic radiation, preventing particles with insufficient energy from penetrating Earth's atmosphere and producing cosmogenic nuclides such as ^{14}C,

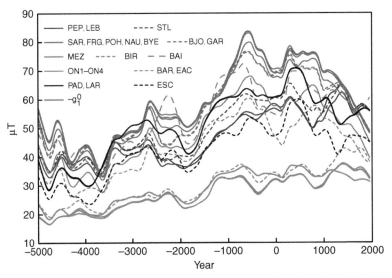

Figure 20   Predictions from *CALS7K.2* of field intensity variations at a variety of locations for which relative intensity records are available. Records are grouped in color codes by region and latitude from South to North, Palmer Deep (PAD), Larsen Ice Shelt (LAR), Lake Escondido, Argentina (ESC), Lakes Barrine (BAR) and Eacham (EAC), Australia, the Ontong-Java Plateau (ON1–ON4), Birkat Ram (BIR) Israel, Greece (MEZ), Lake Baikal (BAI), Lakes Pepin (PEP) and LeBoeuf (LEB), North America, St Lawrence Estuary (STL), Scandinavian lakes (SAR, FRG, POH, NAU, BYE), Bjorn and Garder Drift (BJO GAR) ODP sites 983 and 984 N. Atlantic. See Korte and Constable (2006a) for details.

^{10}Be, and ^{36}Cl. The shielding effect is latitude dependent (diminishing with increasing latitude) as shown in early work by Elsasser *et al.* (1956) who used neutron monitor data, and archeomagnetic observations made by Thellier to initially infer an approximately inverse square root relationship between magnetic dipole moment and nuclide production rate in the upper atmosphere. This led to the idea that reconstructions of past atmospheric radiocarbon concentrations might be used to derive a so-called 'radiocarbon dipole moment' as a means of studying past geomagnetic field behavior (Merrill *et al.*, 1996). Changes in solar activity (that is modulation of the so-called solar constant) affect the incident cosmic ray flux and are also significant in affecting production rates in the upper atmosphere – such changes are well known for the 11 year solar cycle variations, and have been inferred at longer periods using proxies such as sunspot numbers.

In situ production of ^{14}C, ^{10}Be, and ^{36}Cl can be measured in ice cores (e.g., Muscheler *et al.*, 2004b; Lal *et al.*, 2005), and at high latitudes the variation in cosmic ray flux is minimally influenced by the geomagnetic field making these records most sensitive to solar modulation. Lower latitude proxy records drawn from tree rings, corals, speleothems, and sediment cores (e.g., Laj *et al.*, 2002; Snowball and Sandgren, 2002) are influenced by the geomagnetic field and are also affected by Earth's climate, making the reconstruction of the paleoproduction rates a nontrivial task. The individual isotopes have distinct chemistries with differing residence times and mixing effects in the atmosphere, ocean, or sediment sources so that the measured series may reflect quite different response to climate variations and recording environment.

Reconstructions of past atmospheric radiocarbon concentration are usually presented as $\Delta^{14}C$ defined as per mil deviations from the National Institute of Standards and Technology ^{14}C standard, after correction for decay and fractionation. **Figure 21** (upper part) shows the variation with time for the past 10 ky based on the terrestrial dendrochronological data set compiled for IntCal98. One of the major influences on this concentration curve is the change in Earth's

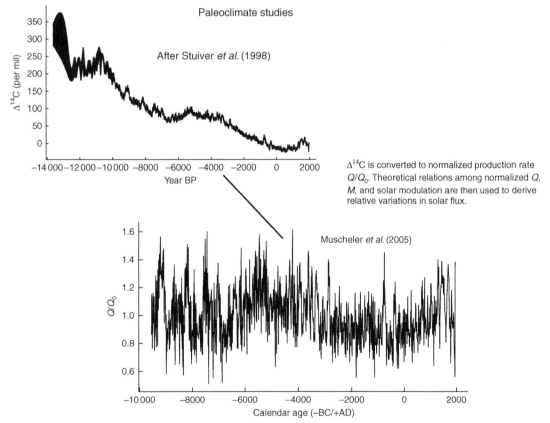

$\Delta^{14}C$ is converted to normalized production rate Q/Q_0. Theoretical relations among normalized Q, M, and solar modulation are then used to derive relative variations in solar flux.

Figure 21 Reconstruction of atmospheric concentration of ^{14}C for the past 10 ky (top). This is converted to a normalized production rate like that given by Muscheler *et al.* (2005) in the lower part of the figure.

dipole moment over the past 10 ky, but this is far from the only one. Changes in solar activity also significant in affecting production, are clearly apparent in the highest frequency variations in **Figure 21**. ^{14}C produced in the atmosphere also plays an important role in the global carbon cycle as it exchanges between the atmosphere, biosphere, and ocean. The ocean is the largest reservoir in the ^{14}C cycle, with a variable climate driven residence time of the same order as the half-life for ^{14}C. The conversion of atmospheric concentration to the normalized production rate in the lower panel is achieved using a box diffusion carbon cycle model (Siegenthaler *et al.*, 1980) that is essentially constant over time – considered a reasonable approximation for much of the Holocene (see Muscheler *et al.* (2005) for details).

The interplay among climate, geomagnetic intensity, and solar modulation is complicated, and it is unlikely that cosmogenic isotopes alone can be used to unravel details of past geomagnetic field behavior. However, paleomagnetic studies provide an important independent record of field strength variations that can be used to model the geomagnetic influence on production rate by studying relationships among production rate, solar flux, and dipole moment (e.g., Masarik and Beer, 1999; Solanki *et al.*, 2004; Usoskin *et al.*, 2006). To date corrections have been largely based on 500 or 1000 year average of the VADMs, like those published by Yang *et al.* (2000), but see also Usoskin *et al.* (2006). The *CALS7K.2* dipole moment shows substantial changes over 1000 years, and it is likely that this is an underestimate of the true variability, because of poor temporal resolution in the model. The dipole moment results presented in **Figure 18** point to the need for continued improvements in temporal resolution for millennial-scale geomagnetic field models like *CALS7K.2*, and these are especially important for paleoclimate studies that make use of production rates of cosmogenic nuclides (^{14}C, ^{10}Be, and ^{36}Cl) in the upper atmosphere to infer the solar flux and associated changes in insolation (e.g., Damon and Peristykh, 2004; Muscheler *et al.*, 2004a).

One way to improve this situation might be to build a high resolution model for 0–7 ka using only the most accurate data with the best age control. It is of some interest to note that relative paleointensity variations derived in several Holocene studies have been linked to millennial- and even centennial-scale variability within cosmogenic isotope records (St-Onge *et al.*, 2003; Snowball and Sandgren, 2002),

raising the possibility of links between climate change and geomagnetic variability, with either one possibly serving as the driving force. On a regional level Gallet *et al.* (2005) have also proposed that the so-called archeomagnetic jerks in the European archeomagnetic record can be further associated with cooling events in western Europe, and that such events may be related to the mechanism for centennial-scale climate change; in particular they seem to view geomagnetic variations as a potential trigger for climate change, suggesting that the geomagnetic field may have a smaller axial dipole component during archeomagnetic jerks. The connection to centennial climate change would come from changing interactions between the internal geomagnetic field and the solar wind. This view that the geomagnetic field might be a partial driving force for climate change is opposite to one suggested for longer timescales (Yokoyama and Yamazaki, 2000), where the possibility has been raised that long term (around 100 ky) orbital changes influencing climate may provide a partial energy source for the geodynamo. A clear need exists for improved understanding and physical modeling that could support the perceived correlations in this area.

9.8.3 Surface Exposure Dating and Local Variations

Calculation of cosmogenic nuclide production rates is also important in surface exposure studies that focus on processes operating on timescales from 10^3 to 10^5 years (Gosse and Phillips, 2001), usually involving the production of radioisotopes on or near the surface of rocks, that have either been exposed since their formation or recently uncovered. The exposure time can be inferred from careful study of an appropriate radioisotope. A major component of these studies involves determining the geographic distribution of cosmic radiation and how to account for scaling effects on the production due to losses in the atmosphere. First scaling models were proposed by Lal (1991), and allowed the calculation of production rates as a function of geomagnetic latitude and altitude. Subsequently attention has focused more on geographic variability and the effects of changes in geomagnetic field strength and non-dipolar magnetic field structure on local production rates (Dunai, 2001; Pigati and Lifton, 2004; Lifton *et al.*, 2005). The most recent of these studies include the idea of using cosmic ray tracing (Shea *et al.*, 1965) and geomagnetic field models to determine cumulative surface exposure rates. It remains to be seen whether millennial-scale magnetic field models can significantly improve the accuracy in Holocene surface exposure dating.

9.8.4 Historical Applications – Navigation and Aurorae

The past motion of the geomagnetic dipole axis can be mapped from Holocene field models, and this is of some interest to scholars who find references to auroral sightings in medieval and ancient texts (Siscoe and Verosub, 1983; Silverman, 1998; Willis and Stephenson, 2001; Siscoe et al., 2002). Historians interested in mapping the paths taken by early maritime explorers can compare the records from early ships' logs with the declination values predicted by the model.

9.8.5 High-Resolution Magnetostratigraphy

From the discussion of local and regional secular variation studies it should already be clear that paleointensity and directional changes in high accumulation rate sediments are considered useful for high-resolution magnetostratigraphy, and that regional secular variation curves based on archeomagnetic and volcanic data find widespread application in providing age constraints, for both lake and archeological studies. A number of recent studies have continued to exploit the paleosecular variation record to unravel the eruptive history of southern Italian volcanoes. Although the timing of eruptions are well documented in the historical record, in some cases their locations are poorly known, leading to misidentification of specific flows and resulting in timing errors in the secular variations record. Recent work by Le Goff et al. (2002) (see also Tanguy et al., 2003) is specifically directed at assessing the limits of archeomagnetic dating in this region. Others (including Arrighi et al., 2004; Speranza et al., 2005; Arrighi et al., 2005; Principe et al., 2004; Incoronato et al., 2002) have used recent paleomagnetic studies to assess the eruptive history of various volcanoes.

Two other magnetostratigraphic applications exploit recent improvements in knowledge about Holocene secular variations. Hagstrum et al. (2004) and St-Onge et al. (2004) were able to recover paleomagnetic directions from sediments associated with paleoseismic excavation in northwestern USA and

Quebec respectively, and use these to make inferences about the timing of the earthquakes studied. Bowles et al. (2006a, 2006b) have used the paleointensity recorded by submarine basaltic glass and comparison with trends in the record from VADMs to constrain the ages of young flows on the East Pacific Rise, with corresponding implications for timing and extent of eruptive activity.

To date most of these high-resolution stratigraphic applications have invoked regional records (but see Blackfeld et al., 2003), and it is mainly on longer timescales that global or interhemispheric comparisons have been made with corresponding lack of resolution because of lower sedimentation rates (see e.g., Laj et al., 2000, 2004; Stoner et al., 2000, 2002; McMillan et al., 2002, 2004; McMillan and Constable, 2006). However, millennial-scale field models provide a basis for evaluating the field everywhere and with appropriate assessment of the resolution attainable (see Korte and Constable, 2007) the *CALS7K* model or future versions of it may develop into a useful magnetostratigraphic tool.

9.9 Outstanding Problems and Scope for Future Progress

Over the past decade there has been considerable progress in integrating millennial-scale geomagnetic field measurements to provide improved understanding of geomagnetic field behavior on timescales longer than the historical record. The current state of global paleomagnetic field modeling on millennial scales can be summarized as follows:

- Time-varying geomagnetic field models back to 5000 BC are feasible.
- Data are unevenly distributed in time and space and of mixed quality.
- Resolution in *CALS7K.2* is poor beyond degree 4 or 5 resulting in steep drop-offs in the spatial power spectra.
- Power in the secular variation is underestimated because of limited temporal resolution.
- Dipole moments derived from average VADMs for the past millennia appear to have been overestimated.
- The secular variation at degree 2 has lower power in *CALS7K.2* than might be anticipated from both longer term and present field secular variation models.

- The 7 Ky average for the non-axial-dipole field differs substantially from both the present field and the average for AD 1590–1990, and looks much more like the structure inferred for the 0–5 Ma field.
- In addition to the geomagnetic interest, numerous applications for the model exist in other fields, including historical, archeological, stratigraphic, and paleoclimate studies.

The above list focuses on results for the past 7 Ky. The number of archeomagnetic data drops off rapidly with increasing age, and the spatial distribution is in any case limited to areas where human development has provided suitable materials. Recent lava flows help augment the data distribution, but in general have less tightly controlled age constraints. The episodic nature of both developments of civilizations and volcanic eruptions mean that the temporal coverage is uneven. However, there have been substantial efforts to acquire new high-quality data with tight age controls, and well-documented laboratory procedures. Public archive of these results in databases like Geomagia50 and MagIC will allow an assessment of systematic influences related to material types, data corrections, and experimental methods. Realistic assessments of age uncertainties are particularly important for making further progress in field modeling.

There is substantial potential for acquiring more sedimentary data, which can augment geographic coverage and also offer the possibility of extending the global view further back in time. Relative paleointensity records can be calibrated by inclusion in the modeling procedure, and it is apparent that a 10 ky time-varying field model is within reach. A higher resolution model is also feasible for a time period like 0–3 ka that only uses the highest quality data with tight age constraints. One goal here would be to detect variations in dipole moment on the shortest possible timescales, interesting in its own right, but also a critical step toward separating solar from geomagnetic variations in cosmogenic nuclide production. In new modeling efforts it will be important to conduct a careful assessment of how to implement more robust data processing tools, and assess the quality, temporal and spatial resolution of the resulting models. Limited high-quality data coverage may make it desirable to consider developing regional models using harmonic splines (Shure et al., 1982), or monopole models (O'Brien and Parker, 1993).

Future models could be used to address the following series of outstanding geomagnetic questions:

1. How dominant is the dipole on average? Most paleomagnetic data point to the present field lying on the high side of the average (Tauxe and Staudigel, 2004; Tauxe, 2006), suggesting that features like the South Atlantic Magnetic Anomaly are probably a normal part of the field. New high-resolution models might be used to assess the longevity of the current feature, and its temporal evolution. In the early part of *CALS7K.2* the dipole moment is a little over half its current value, but the rate of change with time does not seem to be anomalous. Improved temporal resolution will help clarify this, while improved spatial resolution will allow a better assessment of the dominant long wavelength structure in the secular variation.

2. Is the magnetic field and its secular variation persistently lower in the Pacific than in the Atlantic hemisphere, as Walker and Backus (1996) found for the present field? The question of persistent hemispheric asymmetries in the millennial-scale field and its secular variation could be straightforwardly addressed given an adequate time-varying spherical harmonic model.

3. Is there observational evidence for wave motion in the geomagnetic field? On what timescales? What is its physical origin? As improved time-varying geomagnetic models have become available there has recently been substantial interest in understanding the role that might be played by torsional oscillations and other kinds of waves in Earth's core (Zatman and Bloxham, 1997; Bloxham et al., 2002; Finlay and Jackson, 2003). The torsional oscillations detected in historical models are too short period to be detected in the *CALS* models, but the work by Dumberry and Bloxham (2006) suggest that millennial-scale field may be of use in developing new ideas about the observational evidence and physical understanding for longer period waves.

Acknowledgment

This work was partly supported by NSF Grant EAR05-37986.

Appendix 1: Electronic Links

A number of standard and overlapping data sets have been used to produce landmark millennial-scale geomagnetic field models. Links to electronic archives of some standard data sets, models, and animations of field variations are given below:

Data Compilations

IAGA SECVR: http://www.ngdc.noaa.gov/seg/geomag/paleo.shtml

IAGA ARCHEO: http://www.ngdc.noaa.gov/seg/geomag/paleo.shtml

PSVMOD1.0: http://igpphome.ucsd.edu/cathy/Projects/Holocene/Psvmod1.0/

Geomagia50: http://data.geophysics.helsinki.fi/archeo/form.php

CALS7K.2: http://earthref.org/cgi-bin/erda-s2-list.cgi?database_name=erda&search_start=main&arch_id=331

MagIC: http://www.earthref.org/MAGIC/

Field Models

ALS3K http://igpphome.ucsd.edu/cathy/Projects/Holocene/Models/als3k

CALS3K.1 http://earthref.org/cgi-bin/erda.cgi?n=333

CALS3K.2 http://earthref.org/cgi-bin/erda-s2-list.cgi?database_name=erda&search_start=main&arch_id=336

CALS7K.1 http://earthref.org/cgi-bin/erda-s2-list.cgi?database_name=erda&search_start=main&arch_id=334

CALS7K.2 http://earthref.org/cgi-bin/erda-s2-list.cgi?database_name=erda&search_start=main&arch_id=413

Dipole Moments

CALS7K.2 http://earthref.org/cgi-bin/erda-s2-list.cgi?database_name=erda&search_start=main&arch_id=653

References

Aitken MJ, Alcock PA, Bussel GD, and Shaw CJ (1981) Archaeomagnetic determination of the past geomagnetic intensity using ancient ceramics: Allowance for anisotropy. *Archaeometry* 23(1): 53–64.

Arrighi S, Rosi M, Tanguy J-C, and Courtillot V (2004) Recent eruptive history of Stromboli (Aeolian Islands, Italy) determined from high-accuracy archeomagnetic dating. *Geophysical Research Letters* 31: L19603 (doi:10.1029/2004GL020627).

Arrighi S, Tanguy J-C, Courtillot V, and Le Goff M (2005) Reply to comment by F. Sperenza *et al.* on 'Recent eruptive history

of Stromboli (Aeolian Islands, Italy) determined from high-accuracy archeomagnetic dating'. *Geophysical Research Letters* 32: L23305 (doi:10.1029/2005GL023768).

Baag C, Helsley CE, Xu SZ, and Lienert BR (1995) Deflection of paleomagnetic directions due to magnetization of underlying terrain. *Journal of Geophysical Research* 100: 10013–10027.

Bard E, Raisbeck G, Yiou F, and Jouzel J (2000) Solar irradiance during the last 1200 years based on cosmogenic nuclides. *Tellus* 52B: 985–992.

Barton CE and Burden FR (1979) Modification to the Mackereth corer. *Limnology and Oceanography* 24: 977–983.

Bauer LA (1895) On the secular motion of a free magnetic needle II. *Review of Physics* 3: 34–48.

Bloxham J (2002) Time-dependent and time-independent behavior of high-latitude flux bundles at the core-mantle boundary. *Geophysical Research Letters* 29: 1854 (doi:10.1029/2001GL014543).

Bloxham J and Gubbins D (1987) Thermal core-mantle interactions. *Nature* 325: 511–513.

Bloxham J and Jackson A (1992) Time-dependent mapping of the magnetic field at the core–mantle boundary. *Journal of Geophysical Research* 97: 19537–19563.

Bloxham J, Zatman S, and Dumberry M (2002) The origin of geomagnetic jerks. *Nature* 420: 65–68 (doi:10.1038/nature01134).

Böhnel H, Morales J, Caballero C, *et al.* (1997) Variation of rock magnetic parameters and paleointensities over a single Holocene lava flow. *Journal of Geomagnetism and Geoelectricity* 49: 523–542.

Bowles J, Gee JS, Kent DV, Bergmanis E, and Sinton J (2006a) Cooling rate effects on paleointensity estimates in submarine basaltic glass and implications for dating young flows. *Geochemistry Geophysics Geosystems* 6: Q07002 (doi:10.1029/2004GC000900).

Bowles J, Gee JS, Kent DV, Perfit MR, Soule SA, and Fornari DJ (2006b) Paleointensity applications to timing and extent of eruptive activity, 9–10 N East Pacific Rise. *Geochemistry Geophysics Geosystems* 7: Q06006 (doi:10.1029/2005GC001141).

Brachfeld S, Domack E, Kissel C, *et al.* (2003) Holocene history of the Larsen Ice Shelf constrained by paleointensity dating. *Geology* 31: 749–752.

Braginskiy SI (1972) Analytical description of geomagnetic field of the past and spectral analysis of magnetic waves in the Earth's core. *Geomagnetism and Aeronomy* 12: 1092–1105.

Braginskiy SI (1974) Analytical description of geomagnetic field of the past and spectral analysis of magnetic waves in the Earth's core II. *Geomagnetism and Aeronomy* 14: 522–529.

Braginskiy SI and Burlatskaya SP (1979) Spherical analysis of the geomagnetic field based on archeomagnetic data. *Physics of the Solid Earth* 15: 891–895.

Bucur I (1994) The direction of the terrestrial magnetic field in France, during the last 21 centuries, Recent Progress. *Physics of the Earth and Planetary Interiors* 87: 95–109.

Bullard ED, Freeman C, Gellman H, and Nixon J (1950) The westward drift of the Earth's magnetic field. *Philosophical Transactions of the Royal Astronomical Society Series A* A243: 67–92.

Cande SC and Kent DV (1995) Revised calibration of the geomagnetic polarity timescale for the late Cretaceous and Cenozoic. *Journal of Geophysical Research* 100: 6093–6095.

Castro J and Brown L (1987) Shallow paleomagnetic directions from historic lava flows, Hawaii. *Geophysical Research Letters* 14: 1203–1206.

Channell JET (1999) Geomagnetic paleointensity and directional secular variation at Ocean Drilling Program (ODP) Site 984 (Bjorn Drift) since 500 ka; comparisons with ODP

Site 983 (Gardar Drift). *Journal of Geophysical Research* 104(B10): 22937–22951.

Channell JET, Hodell DA, McManus J, and Lehman B (1998) Orbital modulation of geomagnetic paleointensity. *Nature* 394: 464–468.

Chauvin A, Garcia Y, Lanos Ph, and Laubenheimer F (2000) Paleointensity of the geomagnetic field recovered on archaeomagnetic sites from France. *Philosophical Transactions of the Royal Astronomical Society* 120(1-2): 111–136.

Chevallier R (1925) L'aimantation des laves de l'Etna et l'orientation du champ terrestre en Sicile du XIIe et XVIIe siècle. *Annales de Physique* 4: 5–161.

Clark RM (1975) A calibration curve for radiocarbon dates. *Antiquity* 49: 251–266.

Constable CG (1985) Eastern Australian geomagnetic field intensity over the past 14 000 yr. *Geophysical Journal of the Royal Astronomical Society* 81: 121–130.

Constable CG and Johnson CL (1999) Anisotropic paleosecular variation models: Implications for geomagnetic observables. *Physics of the Earth and Planetary Interiors* 115: 35–51.

Constable CG and Johnson CL (2005) A Paleomagnetic power spectrum. *Physics of the Earth and Planetary Interiors* 153: 61–73 (doi:10.1016/j.pepi.2005.03.015).

Constable CG, Johnson CL, and Lund SP (2000) Global geomagnetic field models for the past 3000 years: Transient or permanent flux lobes? *Philosophical Transactions of the Royal Astronomical Society A* 358: 991–1008.

Constable CG and Korte M (2006) Is Earth's magnetic field reversing? *Earth and Planetary Science Letters* 246: 1–16 (doi:10.1016/j.epsl.2006.03.038).

Constable CG and McElhinny MW (1985) Holocene geomagnetic secular variation records from north-eastern Australian lake sediments. *Geophysical Journal of the Royal Astronomical Society* 81: 103–120.

Constable CG and Tauxe L (1996) Towards absolute calibration of sedimentary paleointensity records. *Earth and Planetary Science Letters* 143: 269–274.

Constable CG, Tauxe L, and Parker RL (1998) Analysis of 11 Myr of geomagnetic intensity variation. *Journal of Geophysical Research* 103: 17735–17748.

Daly L and Le Goff M (1996) An updated homogeneous world secular variation database. Part 1: Smoothing of the archeomagnetic results. *Physics of the Earth and Planetary Interiors* 93: 159–190.

Damon PE and Peristykh AN (2004) Solar and climatic implications of the centennial and millennial periodicities in atmospheric $\Delta^{14}C$ variations. In: Pap JM and Fox P (eds.) *AGU Geophysical Monograph, 141: Solar Variability and its Effects on Climate*, pp. 237–249. Washington DC: American Geophysical Union.

Dodson RE (1979) Counterclockwise precession of the geomagnetic field vector and westward drift of the non-dipole field. *Journal of Geophysical Research* 84: 637–644.

Dodson MH and McClelland-Brown E (1980) Magnetic blocking temperatures of single-domain grains during slow cooling. *Journal of Geophysical Research* 85(B5): 2625–2637.

Donadini F, Korhonen K, Riisager P, and Pesonen LJ (2006) Database for holocene geomagnetic intensity information. *EOS Transactions, American Geophysical Union* 87: 137–143 (See also http://data.geophysics.helsinki.fi/).

Dumberry M and Bloxham J (2006) Azimuthal flows in the Earth's core and changes in length of day at millennial timescales. *Geophysical Journal International* 165: 1–399.

Dumberry M and Finlay CC (2007) Eastward and westward drift of the Earth's magnetic field for the last three millennia. *Earth and Planetary Science Letters* 254: 146–157 (doi:10.1016/j.epsl.2006.11.026).

Dunai T (2001) Influence of secular variation of the geomagnetic field on production rates of *in situ* produced cosmogenic nuclides. *Earth and Planetary Science Letters* 193: 197–212.

Elsasser WE, Ney EP, and Winckler JR (1956) Cosmic ray intensity and geomagnetism. *Nature* 178: 1226–1227.

Evans ME (2006) Archaeomagnetic investigations in Greece and their bearing on geomagnetic secular variation. *Physics of the Earth and Planetary Interiors* 159: 90–95 (doi:10.1016/j.pepi.2006.06.005).

Evans ME and Hoye GS (2005) Archaeomagnetic results form southern Italy and their bearing on geomagnetic secular variation. *Physics of the Earth and Planetary Interiors* 151: 155–162 (doi:10.1016/j.pepi.2005.02.002).

Finlay CC and Jackson A (2003) Equatorial dominated magnetic field change at Earth's core surface. *Science* 300: 2084–2086.

Folgerhaiter G (1899) Sur les variations séculaire de l'inclinaison magnétique dans l'antiquité. *Journal of Physics* 8(3): 5–16.

Fox JMW and Aitken MJ (1980) Cooling-rate dependence of thermoremanent magnetisation. *Nature* 283: 462–463.

Gallet Y, Genevey A, and Fluteau F (2005) Does Earth's magnetic field secular variation control centennial climate change? *Earth and Planetary Science Letters* 236: 339–347.

Gallet Y, Genevey A, and Courtillot V (2003) On the possible occurrence of 'archaeomagnetic jerks' in the geomagnetic field over the past three millennia. *Earth and Planetary Science Letters* 214: 237–242.

Gallet Y, Genevey A, and Le Goff M (2002) Three millennia of directional variation of the Earth's magnetic field in western Europe as revealed by archeological artefacts. *Physics of the Earth and Planetary Interiors* 131(1): 81–89.

Genevey A and Gallet Y (2002) Eight thousand years of geomagnetic field intensity variations in the eastern Mediterranean. *Journal of Geophysical Research-Solid Earth* 108(B5): 2228 (doi:10.1029/2001JB001612).

Genevey A, Gallet Y, and Margueron JC (2003) Intensity of the geomagnetic field in western Europe over the past 2000 years. *Journal of Geophysical Research-Solid Earth* 107(B11): 2285 (doi:10.1029/2001JB000701).

Gómez-Paccard M, Chauvin A, Laanos P, Thiriot J, and Jiminez-Castillo P (2006b) Archeomagnetic study of seven contemporaneous kilns from Murcia (Spain). *Physics of the Earth and Planetary Interiors* 157: 16–32 (doi:10.1016/j.pepi.2006.03.001).

Gómez-Paccard M, Catanzariti G, Ruiz VC, *et al.* (2006a) A catalogue of Spanish archaeomagnetic data. *Geophysical Journal International* 166: 1125–1143 (doi:10.1111/j.1365-246X.2006.03020.x).

Gosse JC and Phillips FM (2001) Terrestrial *in situ* cosmogenic nuclides: Theory and application. *Quaternary Science Reviews* 20: 1475–1560.

Griffiths DH (1953) Remanent magentism of varved clays from Sweden. *Nature* 172: 539.

Gubbins D, Jones AL, and Finlay CC (2006) Fall in Earth's magnetic field is erratic. *Science* 312: 900–903.

Guyodo Y and Valet JP (1999) Global changes in geomagnetic intensity during the past 800 thousand years. *Nature* 399: 249–252.

Hagstrum JT, Atwater BF, and Sherrod BL (2004) Paleomagnetic correlation of late Holocene earthquakes among estuaries in Washington and Oregon. *Geochemistry Geophysics Geosystems* 5: Q10001 (doi:10.1029/2004GC000736).

Halgedahl SL, Day R, Fuller M, *et al.* (1980) The effect of cooling rate on the intensity of weak-field TRM in single domain magnetite. *Journal of Geophysical Research* 85(B7): 3690–3698.

Heirtzler JR, Allen JH, and Wilkinson DC (2002) Ever-present south Atlantic anomaly damages spacecraft. *EOS Transaction of the American Geophysical Union* 83: 165.

Hongre L, Hulot G, and Khokhlov A (1998) An analysis of the geomagnetic field over the past 2000 years. *Physics of the Earth and Planetary Interiors* 106: 311–335.

Hughen KA, Baillie MGL, Bard E, *et al.* (2004) Marine04 Marine radiocarbon age calibration, 0–26 Cal Kyr BP. *Radiocarbon* 46: 1059–1086.

Incoronato A, Angelino A, Romano R, *et al.* (2002) Retrieving geomagnetic secular variations from lava flows: Evidence from Mounts Arso, Etna and Vesuvius (southern Italy). *Geophysical Journal International* 149: 724–730.

Jackson A, Jonkers ART, and Walker MR (2000) Four centuries of geomagnetic secular variation from historical records. *Philosophical Transactions of the Royal Astronomical Society, Series A* 358: 957–990.

Johnson CL and Constable CG (1997) The time-averaged geomagnetic field: Global and regional biases for 0–5 Ma. *Geophysical Journal International* 131: 643–666.

Johnson EA, Murphy T, and Torreson OW (1948) Pre-history of the Earth's magnetic field. *Terrestrial Magnetism and Atmospheric Electricity* 53: 349–372.

King JW, Banerjee SK, and Marvin J (1983) A new rock magnetic approach to selecting sediments fro geomagnetic paleointensity studies: Application to paleointensity for the last 4000 years. *Journal of Geophysical Research* 88: 5911–5921.

Klovan JE (1964) Box-type sediment-coring device. *Journal of Sedimentary Research* 34(1): 185–189.

Kono M and Roberts PH (2002) Recent geodynamo simulations and observations of the geomagnetic field. *Reviews of Geophysics* 40: 1013 (doi:10.1029/2000RG000102).

Kono M, Sakuraba A, and Ishida M (2000) Dynamo simulation and palaeosecular variation models. *Philosophical Transactions of the Royal Astronomical Society, Series A* 358: 1123–1139.

Korte M and Constable CG (2003) Continuous global geomagnetic field models for the past 3000 years. *Physics of the Earth and Planetary Interiors* 140: 73–89.

Korte M and Constable CG (2005a) Continuous geomagnetic models for the past 7 millennia II: CALS7K. *Geochemistry Geophysics Geosystems* 6(2): Q02H16 (doi:10.1029/2004GC000801).

Korte M and Constable CG (2005b) The geomagnetic dipole moment over the last 7000 years – new results from a global model. *Earth and Planetary Science Letters* 236: 348–358.

Korte M and Constable CG (2006a) On the use of calibrated relative paleointensity records to improve millennial-scale geomagnetic field models. *Geochemistry Geophysics Geosystems* 7: Q09004 (doi:10.1029/2006GC001368).

Korte M and Constable CG (2006b) Centennial to millennial geomagnetic secular variation. *Geophysical Journal International* 167: 43–52 (doi:10.1111/j.1365-246X.2006.03088.x).

Korte, M and Constable CG (2007) Spatial and temporal resolution of millennial scale geomagnetic field models. *Advances in Space Research* doi:10.1016/j.asr.2007.03.094.

Korte M, Genevey A, Constable CG, Frank U, and Schnepp E (2005) Continuous geomagnetic models for the past 7 millennia I: A new global data compilation. *Geochemistry Geophysics Geosystems* 6(2): Q02H15 (doi:10.1029/2004GC000800).

Laj C, Kissel C, and Beer J (2004) High resolution global paleointensity stack since 75 kyr (GLOPIS-75) calibrated to absolute values. In: Channell JET, Kent DV, Lowrie W, and Meet JG (eds.) *AGU Geophysical Monograph 145: Timescale of Paleomagnetic Field*, pp. 255–265. Washington DC: AGU.

Laj C, Kissel C, Mazaud A, Michel E, Muscheler R, and Beer J (2002) Geomagnetic field intensity, north Atlantic deep water circulation and atmospheric $\Delta^{14}C$ during the last 50 kyr. *Earth and Planetary Science Letters* 200: 177–190.

Laj C, Kissel C, Mazaud A, Channel JET, and Beer J (2000) North Atlantic Paleointensity Stack since 75ka (NAPIS-75) and the duration of the Laschamp event. *Philosophical Transactions of the Royal Astronomical Society, Series A* 358: 1027–1047.

Lal D (1991) Cosmic ray labeling of erosion surfaces: *In situ* nuclide production rates and erosion models. *Earth and Planetary Science Letters* 104: 424–439.

Lal D, Jull AJT, Pollard D, and Vacher L (2005) Evidence for large century time-scale changes in solar activity in the past 32 Kyr, based on *in-situ* cosmogenic ^{14}C in ice at Summit, Greenland. *Earth and Planetary Science Letters* 234: 335–349 (doi:10.1016/j.epsl.2005.02.011).

Lal D and Peters B (1967) Cosmic ray produced radioactivity on the Earth. In: Flugge S (ed.) *Handbuch der Physik*, pp. 551–612. Berlin: Springer.

Lanos P (2004) Bayesian inference of calibration curves: Application to Archeomagnetism. In: Buck CE and Millard AR (eds.) *Lecture Notes in Statistics: Tools for Constructing Chronologies: Crossing Disciplinary Boundaries*, pp. 43–82. London: Springe Velarg.

Lanos P, Le Goff M, Kovacheva M, and Schnepp E (2005) Hierarchical modelling of archaeomagnetic data and curve estimation by moving average technique. *Geophysical Journal International* 160(2): 440–476.

Lanza R, Meloni A, and Tema E (2005) Historical measurements of the Earth's magnetic field compared with remanence directions from lava flows in Italy over the past four centuries. *Physics of the Earth and Planetary Interiors* 148: 97–107.

Le Goff M (1975) Inductomètre à rotation continue pour la mesure des faibles aimantations rémanentes et induites en magnétisme des roches. Mém, Diplôme d'Ingénieur. CNAM, Paris.

Le Goff M, Gallet Y, Genevey A, and Warmé N (2002) On archaeomagnetic secular variation curves and archaeomagnetic dating. *Physics of the Earth and Planetary Interiors* 134(3–4): 203–211 (doi:10.1016/S0031-9201(02)00161-9).

Leonhardt R, Matzka J, Nichols A, and Dingwell DB (2006) Cooling rate correction of paleointensity determination for volcanic glasses by relaxation geospeedometry. *Earth and Planetary Science Letters* 243(1–2): 282.

Levi S and Banerjee SK (1976) On the possibility of obtaining relative palaeointensities from lake sediments. *Earth and Planetary Science Letters* 29: 219–226.

Lifton N, Bieber JW, Clem JM, *et al.* (2005) Addressing solar modulation and long-term uncertainties in scaling secondary cosmic rays for *in situ* cosmogenic nuclide applications. *Earth and Planetary Science Letters* 239: 140–161.

Livingstone DA (1955) A lightweight piston sampler for lake deposits. *Ecology* 36: 137–139.

Love JJ and Constable CG (2003) Gaussian Statistics for Paleomagnetic Vectors. *Geophysical Journal International* 152: 515–565.

Lund SP (1996) A comparison of Holocene paleomagnetic secular variation records from North America. *Journal of Geophysical Research* 101: 8007–8024.

Mackereth FJH (1958) A portable core sampler for lake deposits. *Limnology and Oceanography* 3: 181–191.

Márton P (1970) Secular variation of the geomagnetic virtual dipole field during the last 2000 yr as inferred from the spherical harmonic analysis of the available archaeomagnetic data. *Pure and Applied Geophysics* 81: 163–176.

Márton P (2003) Recent achievements in archaeomagnetism in Hungary. *Geophysical Journal International* 153(3): 675–690.

Masarik J and Beer J (1999) Simulation of particle fluxes and cosmogenic nuclide production in the Earth's atmosphere. *Journal of Geophysical Research* 104D: 12099–12111.

McElhinny MW and Lock J (1996) IAGA paleomagnetic databases with Access. *Surveys in Geophysics* 17: 575–591 (doi:10.1007/BF01888979).

McElhinny MW and Senanayake WE (1982) Variations in the geomagnetic dipole 1: The past 50,000 years. *Journal of Geomagnetism and Geoelectricity* 34: 39–51.

McMillan DG and Constable CG (2006) Limitations in correlation of regional relative geomagnetic paleointensity. *Geochemistry Geophysics Geosystems* 7: Q09009 (doi:10.1029/2006GC001350).

McMillan DG, Constable CG, and Parker RL (2002) Limitations on stratigraphic analyses due to incomplete age control and their relevance to sedimentary paleomagnetism. *Earth and Planetary Science Letters* 201: 509–523.

McMillan DG, Constable CG, and Parker RL (2004) Assessing the dipolar signal in stacked paleointensity records using a statistical error model and geodynamo simulations. *Physics of the Earth and Planetary Interiors* 145: 37–54.

McMillan DG, Constable CG, Parker RL, and Glatzmaier GA (2001) A statistical appraisal of magnetic fields of geodynamo models. *Geochemistry Geophysics Geosystems* 2: (doi:10.129/2000GC000130,2001).

Merrill RT, McElhinny MW, and McFadden PL (1996) *The Magnetic Field of the Earth: Paleomagnetism, the Core and the Deep Mantle*. San Diego, CA: Academic Press.

Mochizuki N, Tsunakawa H, Oishi Y, Wakai S, Wakabayashi K, and Yamamoto Y (2004) Palaeointensity study of the Oshima 1986 lava in Japan: Implications for the reliability of the Thellier and LTD-DHT Shaw methods. *Physics of the Earth and Planetary Interiors* 146: 395–416.

Morrison LV and Stephenson FR (2001) Historical eclipses and the variability of Earth's rotation. *Journal of Geodynamics* 32: 247–265.

Muscheler R, Beer J, and Kubik PW (2004a) Long-term solar variability and climate change based on radionuclide data from ice cores. In: Pap JM and Fox P (eds.) *AGU, GeophysicalMonograph, 141: Solar Variability and its Effects on Climate*, pp. 221–235. Washington, DC: American Geophysical Union.

Muscheler R, Beer J, Kubik PW, and Synal HA (2005) Geomagnetic field intensity during the last 60,000 years based on ^{10}Be and ^{36}Cl from the Summit ice cores and ^{14}C. *Quaternary Science Reviews* 24: 1849–1860 (doi:10.1016/j.quascirev.2005.01.012).

Muscheler R, Beer J, Wagner G, *et al.* (2004b) Changes in the carbon cycle during the last deglaciation as indicated by the comparison of ^{10}Be and ^{14}C records. *Earth and Planetary Science Letters* 219: 325–340.

Néel L (1955) Some theoretical aspects of rock-magnetism. *Advances in Physics* 4: 191–243.

O'Brien MS and Parker RL (1993) Regularized geomagnetic field modeling using monopoles. *Geophysical Journal International* 118: 566–578.

Ohno M and Hamano Y (1993) Spherical harmonic analysis of paleomagnetic secular variation curves. *Central Core of the Earth* 3: 205–212.

Oishi Y, Tsunakawa H, Mochizuki N, Yamamoto Y, Wakabayashi K, and Shibuya H (2005) Validity of the LTD-DHT Shaw and Thellier palaeointensity methods: A case study of the Kilauea 1970 lava. *Physics of the Earth and Planetary Interiors* 149: 243–257.

Olsen N (2002) A Model of the geomagnetic field and its secular variation for Epoch 2000. *Geophysical Journal International* 149: 454–462.

Olson P and Christensen U (2002) The time-averaged magnetic field in numerical dynamos with noon-uniform boundary heat flow. *Geophysical Journal International* 151: 809–823.

Parker RL (1994) *Geophysical Inverse Theory*. Princeton, NJ: Princeton University Press.

Peng L and King JW (1992) A late Quaternary geomagnetic secular variation record from Lake Waiau, Hawaii, and the question of the Pacific nondipole low. *Journal of Geophysical Research* 97: 4407–4424.

Pick T and Tauxe L (1993) Holocene paleointensities: Thellier experiments on submarine basaltic glass from the East Pacific Rise. *Journal of Geophysical Research* 98(B10): 17949–17964.

Pigati JS and Lifton NA (2004) Geomagnetic effects on time-integrted cosmogenic nuclide production with emphasis on *in situ* ^{14}C and ^{10}Be. *Earth and Planetary Science Letters* 226: 193–205.

Principe C, Tanguy J-C, Arrighi S, Paiotti A, Le Goff M, and Zoppi U (2004) Chronology of Vesuvius' activity from A.D. 79 to 1631 based on archeomagnetism of lavas and historical sources. *Bulletin of Volcanology* 66(8): 703–724 (doi:10.1007/s00445-004-0348-8).

Reimer PJ, Baillie MGL, Bard E, *et al.* (2004) IntCal04 Terrestrial Radiocarbon Age Calibration, 0–26 Cal Kyr BP. *Radiocarbon* 46: 1029–1058.

Rogers J, Fox JMW, and Aitken MJ (1979) Magnetic anisotropy in ancient pottery. *Nature* 277: 644–646.

Rolph TC and Shaw J (1986) Variations of the Geomagnetic field in Sicily. *Journal of Geomagnetism and Geoelectricity* 38: 1269–1277.

Runcorn SK (1959) On the theory of geomagnetic secular variation. *Annales de Géophysique* 15: 87–92.

Rymer L and Neale J (1981) Freeze coring as a method of collecting unconsolidated lake sediments. *Australian Journal of Ecology* 6: 123–126.

Sagnotti L, Budillon F, Dinarès-Turell J, Iorio M, and Macri P (2005) Evidence for a variable paleomagnetic lock-in depth in the Holocene sequence from the Salerno Gulf (Italy): Implications for 'high-resolution' paleomagnetic dating. *Geochemistry Geophysics Geosystems* 6: Q11013 (doi:10.129/2000GC001043).

Sakai H (1980) Spherical harmonic analysis of the geomagnetic field during the last 2000 years. *Rock Magnetism and Paleogeophysics* 7: 8–15.

Schnepp E and Lanos P (2005) Archaeomagnetic secular variation in Germany during the past 2500 years. *Geophysical Journal International* 163:2: 479–490.

Schnepp E and Lanos P (2006) A preliminary secular variation reference curve for archaeomagnetic dating in Austria. *Geophysical Journal International* 166: 91–96 (doi:10.1111/j.1365-246X.2006.03012.x).

Schnepp E, Pucher R, Goedicke C, Manzano A, Muller U, and Lanos P (2003) Paleomagnetic directions and thermoluminescence dating from a bread oven-floor sequence in Lubeck (Germany): A record of 450 years of geomagnetic secular variation. *Journal of Geophysical Research* 108(B2): 2078 (doi:10.1029/2002JB001975).

Schnepp E, Pucher R, Reinders J, Hambach U, Soffel H, and Hedley I (2004) A German catalogue of archaeomagnetic data. *Geophysical Journal International* 157(1): 64–78.

Selkin P and Tauxe L (2000) Long term variations in palaeointensity. *Philosophical Transactions of the Royal Astronomical Society* A358: 1065–1088.

Shea MA, Smart DF, and McCracken KG (1965) A study of vertical cutoff rigidities using sixth degree simulation of the geomagnetic field. *Journal of Geophysical Research* 70: 4117–4130.

Shuey R, Cole E, and Mikulich M (1970) Geographic correction of archaeomagnetic data. *Journal of Geomagnetism and Geoelectricity* 41: 485–489.

Shure L, Parker RL, and Backus GE (1982) Harmonic splines for geomagnetic modelling. *Physics of the Earth and Planetary Interiors* 28: 215–229.

Siegenthaler U, Heimann M, and Oeschger H (1980) ^{14}C variations caused by changes in the global carbon cycle. *Radiocarbon* 22(2): 177–191.

Silverman S (1998) Early auroral observations. *Journal of Atmospheric and Solar-Terrestrial Physics* 60: 997–1006.

Siscoe GL and Siebert KD (2002) Solar-terrestrial effects possibly stronger in biblical times. *Journal of Atmospheric and Solar-Terrestrial Physics* 64: 1905–1909.

Siscoe GL, Silverman SM, and Siebert KD (2002) Ezekiel and the Northern lights: Biblical aurora seems plausible. *EOS, Transactions, American Geophysical Union* 83: 16.

Siscoe GL and Verosub KL (1983) High medieval auroral incidence over China and Japan: Implications for the medieval aitc of tho goomagnotic polo. *Geophysical Research Letters* 10: 345–348 (doi:10.1029/0GPRLA000010000004000345000001).

Skiles DD (1970) A methjod of inferring the direction of drift of the geomagnetic field from paleomagnetic data. *Journal of Geomagnetism and Geoelectricity* 22: 441–462.

Snowball I and Sandgren P (2002) Geomagnetic field variations in northern Sweden during the Holocene quantified from varved lake sediments and their implications for cosmogenic nuclide production rates. *The Holocene* 12(5): 517–530.

Snowball I and Sandgren P (2004) Geomagnetic field intensity changes in Sweden between 9000 and 450 cal BP: Extending the record of 'archaeomagnetic jerks' by means of lake sediments and the pseudo-Thellier technique. *Earth and Planetary Science Letters* 227(3–4): 361–376.

Solanki SK, Usoskin LG, Kromer B, Schüssler M, and Beer J (2004) Unusual activity of the Sun during recent decades compared tot he previous 11,000 years. *Nature* 431: 1084–1087.

Speranza F, Maritan L, Mazzoli C, Morandi Bonacossi D, and D'Ajello Caracciolo F (2006) First directional archaeomagnetic results from Syria: Evidence from Tell Mishrifeh/Qatna. *Geophysical Journal International* 165: 47–52.

Speranza F, Sagnotti L, and Meloni A (2005) Comment on 'Recent eruptive history of Stromboli (Aeolian Islands, Italy) determined from high-accuracy archeomagnetic dating' by S. Arrighi et al.. *Geophysical Research Letters* 32: L23306 (doi:10.1029/2005GL022590).

Stephenson FR and Morrison LV (1984) Long-term changes in the rotation of the Earth: 700Bc to AD 1980. *Philosophical Transactions of the Royal Astronomical Society* A313: 47–70.

Stephenson FR and Morrison LV (1995) Long-term fluctuations in the Earth's rotation: 700Bc to AD 1980. *Philosophical Transactions of the Royal Astronomical Society* A351: 165–202.

St-Onge G, Mulder T, Piper DJW, Hillaire-Marcel C, and Stoner JS (2004) Earthquake and flood-induced turbidites in the Saguenay Fjord (Qubec): A Holocene paleoseismicity record. *Quaternary Science Reviews* 23: 283–294 (doi:10.1016/j.quascirev.2003.03.001).

St-Onge G, Stoner JS, Stoner JS, and Hillaire-Marcel C (2003) Holocene paleomagnetic records from the St. Lawrence Estuary, eastern Canada: Centennial- to millennial-scale geomagnetic modulation of cosmogenic isotopes. *Earth and Planetary Science Letters* 209(1–2): 113–130.

Stoner JS, Channel JET, Hillaire-Marcel C, and Kissel C (2000) Geomagnetic paleointensity and environmental record from Labrador sea core MD95-2024: Global marine sediment and ice core chronostratigraphy for the last 110 kyr. *Earth and Planetary Science Letters* 183: 161–177.

Stoner JS, Laj C, Channel JET, and Kissel C (2002) South Atlantic and North Atlantic geomagnetic paleointensity stacks (0–80 ka): Implications for inter-hemispheric correlation. *Quaternary Science Reviews* 21: 1141–1151.

Stuiver M and Reimer P (1993) Extended ^{14}C data base and revised CALIB 3.0 ^{14}C age calibration program. *Radiocarbon* 35: 215–230.

Tanguy J-C and Le Goff M (2004) Distortion of the geomagnetic field in volcanic terrains: An experimental study of the Mount Etna stratovolcano. *Physics of the Earth and Planetary Interiors* 141(1): 59–70.

Tanguy J-C, Le Goff, Chillemi M, et al. (1999) Secular variation of the geomagnetic field direction recorded in lavas from Etna and Vesuvius during the last two millennia. *Comptes Rendus de l'Academie de Sciences Serie Ia: Sciences de la Terre et des Planetes* 329: 557–564.

Tanguy J-C, Le Goff M, Principe C, et al. (2003) Archeomagnetic dating of Mediterranean volcanics of the last 2100 years: Validity and limits. *Earth and Planetary Science Letters* 211(1–2): 111–124.

Tauxe L (1993) Sedimentary records of relative paleointensity of the geomagnetic field: Theory and practice. *Reviews of Geophysics* 31: 319–354.

Tauxe L (2006) Long term trends in paleointensity. *Physics of the Earth and Planetary Interiors* 156: 223–241 (doi:10.1016/j.pepi.2005.03.022).

Tauxe L and Kent DV (2004) A simplified statistical model for the geomagnetic field and the detection of shallow bias in paleomagnetic inclinations. In: Channell JET, Kent DV, Lowrie W, and Meert JG (eds.) AGU Geophysical Monograph, 145: *Timescales of the Paleomagnetic Field*, pp. 101–116. Washington, DC: American Geophysical Union.

Tauxe L, LaBrecque J, Dodson R, and Fuller M (1983) U-channels: A new technique for paleomagnetic analysis of hydraulic piston cores. *EOS Transaction of the American Geophysical Union* 64: 219.

Tauxe L, Pick T, and Kok YS (1995) Relative paleointensity in sediments: A pseudo-Thellier approach. *Geophysical Research Letters* 22(21): (doi:10.1029/95GL03166).

Tauxe L and Staudigel H (2004) Strength of the geomagnetic field in the Cretaceous Normal Superchron: New data from submarine basaltic glass of the Troodos Ophiolite. *Geochemistry Geophysics Geosystems* 5: Q02H06 (doi:10.1029/2003GC000635).

Thellier E (1938) Sur l'aimantation des terres cuites et ses applications geophysique. *Annales Institut Physique du Globe de Paris* 16: 157–302.

Thellier E (1981) Sur la direction du champ magnétique terrestre, en France, durant les deux dernières millenaires. *Physics of the Earth and Planetary Interiors* 24: 89–132.

Thellier E and Thellier O (1959) Sur l'intensité du champ magntique terrestre dans le pass historique et géologique. *Annales de Geophysique* 15: 285–375.

Turner G and Thompson R (1982) Detransformation of the British geomagnetic variation record for Holocene times. *Geophysical Journal of the Royal Astronomical Society* 70: 789–792.

Usoskin IG, Solanki SK, and Korte M (2006) Solar activity reconstructed over the last 7000 years: The influence of geomagnetic field changes. *Geophysical Research Letters* 33(8): L08103 (doi:10.1029/2006GL025921).

Valet J-P (2003) Time variations in geomagnetic intensity. *Reviews of Geophysics* 41: 1 (doi:10.1029/2001RG000104).

Valet J-P and Meynadier L (1993) Geomagnetic field intensity and reversals during the last four million years. *Nature* 366: 234–238.

Valet JP and Soler V (1999) Magnetic anomalies of lava fields in the Canary islands. Possible consequences for paleomagnetic records. *Physics of the Earth and Planetary Interiors* 115: 109–118.

Veitch RJ and Mehringer PJ (1984) An investigation of the intensity of the geomagnetic field during Roman times using magnetically anisotropic bricks and tiles. *Archives des Sciences (Geneva)* 37: 359–373.

Walker AD and Backus GE (1996) On the difference between the average values of B_r^2 in the Atlantic and Pacific hemispheres. *Geophysical Research Letters* 23: 1965–1968.

Willis DM and Stephenson FR (2001) Solar and auroral evidence for an intense recurrent geomagnetic storm during December in AD 1128. *Annales Geophysicae* 19: 289–302.

Yang S, Odah H, and Shaw J (2000) Variations in the geomagnetic dipole moment over the last 12000 years. *Geophysical Journal International* 140: 158–162.

Yamamoto, Tsunakawa YH, and Shibuya H (2003) Palaeointensity study of the Hawaiian 1960 lava: Implications for possible causes of erroneously high intensities. *Geophysical Journal International* 153: 263–276.

Yokoyama Y and Yamazaki T (2000) Geomagnetic paleointensity variation with a 100 kyr quasi-period. *Earth and Planetary Science Letters* 181: 7–14 (doi:10.1016/ S0012-821X(00)00199-0).

Yoshihara A, Kondo A, Ohno M, and Hamano Y (2003) Secular variation of the geomagnetic field intensity during the past 2000 years in Japan. *Earth and Planetary Science Letters* 210(1–2): 219–231.

Yukutake T and Tachinaka H (1969) Separation of the Earth's magnetic field into drifting and standing parts. *Bulletin of the Earthquake Research Institute Tokyo University* 47: 65.

Zatman S and Bloxham J (1997) Torsional oscillations and the magnetic field within the Earth's core. *Nature* 388: 760–763 (doi:10.1038/41987).

Ziegler L, Constable CG, and Johnson CL (2006) A Statistical Assessment of 0–1 Ma Absolute Paleointensity Data. *10th Symposium on Study of the Earth's Deep Interior*, S7.40, p19. Prague, Czech Republic, 9–14 July.

Relevant Website

http://www.columbusnavigation.com – The Columbus Navigation.

10 Geomagnetic Excursions

C. Laj, Laboratoire des Sciences du Climat et de l'Environment, Unité Mixte CEA-CNRS-UVSQ, Gif-sur-Yvette, France

J. E. T. Channell, University of Florida, Gainesville, FL, USA

10.1 Introduction

10.1.1 History of the Polarity Timescale and Excursions

David (1904) and Brunhes (1906) were the first to measure magnetization directions in rocks that were approximately antiparallel to the present Earth's field. Brunhes (1906) recorded magnetizations in baked sedimentary rocks that were aligned with reverse magnetization directions in overlying Miocene lavas from central France (Puy de Dome). In so doing, Brunhes (1906) made first use of a field test for primary thermal remanent magnetization (TRM) that is now referred to as the 'baked contact' test (see Laj *et al.* (2002) for an account of Brunhes' work). Matuyama (1929) was the first to attribute reverse magnetizations in (volcanic) rocks from Japan and China to reversal of geomagnetic polarity, and to differentiate mainly Pleistocene lavas from mainly Pliocene lavas based on the polarity of the magnetization. In this respect, Matuyama (1929) was the first person to use magnetic stratigraphy as a means of ordering rock sequences.

The modern era of paleomagnetic studies began with the studies of Hospers (1951, 1953–54) in Iceland, and Roche (1950, 1951, 1956) in the Massif Central of France. The work of Hospers on Icelandic lavas was augmented by Rutten and Wensink (1960) and Wensink (1964) who subdivided Pliocene–Pleistocene lavas on Iceland into three polarity zones from young to old: N–R–N. Magnetic remanence measurements on basaltic lavas combined with K/Ar dating, pioneered by Cox *et al.* (1963) and McDougall and Tarling (1963a, 1963b, 1964), resulted in the beginning of development of the modern geomagnetic polarity timescale (GPTS). These studies, and those that followed in the mid-1960s, established that rocks of the same age carry the same magnetization polarity, at least for the last few million years. The basalt sampling sites were scattered over the globe. Polarity zones were linked by their K/Ar ages, and were usually not in stratigraphic superposition. Doell and Dalrymple (1966) designated the long intervals of geomagnetic polarity of the last 5 My as magnetic epochs, and named them after pioneers of geomagnetism (Brunhes, Matuyama, Gauss, and Gilbert). The shorter polarity intervals (events) were named after localities: e.g., Jaramillo (Doell and Dalrymple, 1966), Olduvai (Grommé and Hay, 1963, 1971), Kaena and Reunion (McDougall and Chamalaun, 1966), Mammoth (Cox *et al.*, 1963), and Nunivak (Hoare *et al.*, 1968). The nomenclature for excursions has continued this trend by naming excursions after the localities from where the earliest records were derived (e.g., Laschamps in France).

The fit of the land-derived polarity timescale, from paleomagnetic and K/Ar studies of exposed basalts, with the polarity record emerging from marine magnetic anomalies (MMAs) (Vine and Matthews, 1963; Vine, 1966; Pitman and Heirtzler, 1966; Heirtzler *et al.*, 1968) resulted in a convincing argument for synchronous global geomagnetic polarity reversals, thereby attributing them to the main axial dipole. The results relegated self-reversal, detected in the 1950s in the Haruna dacite from the Kwa district of Japan (Nagata, 1952; Nagata *et al.*, 1957; Uyeda, 1958), to a rock-magnetic curiosity rather than to a process generally applicable to volcanic rocks.

The first magnetic stratigraphies in sedimentary rocks may be attributed to Creer *et al.* (1954) and Irving and Runcorn (1957) who documented normal and reverse polarities at 13 locations of the Proterozoic Torridonian Sandstones in Scotland, and in rocks of Devonian and Triassic age. For the Torridonian Sandstones, normal and reverse magnetizations were observed from multiple outcrops, and an attempt was made to arrange the observed polarity zones in stratigraphic sequence. Meanwhile, Khramov (1958) published magnetic polarity stratigraphies in Pliocene–Pleistocene sediments from western Turkmenia (central Asia), and made chronostratigraphic interpretations based on equal duration of polarity intervals. Early magnetostratigraphic studies were carried out on Triassic red sandstones of the Chugwater Formation (Picard, 1964), on the European Triassic Bundsandstein (Burek, 1967, 1970), and on the Lower Triassic Moenkopi Formation (Helsley, 1969). The above studies were conducted on poorly fossiliferous mainly continental (red) sandstones and siltstones; therefore, the correlations of polarity zones did not have support from biostratigraphic correlations. It was the early magnetostratigraphic studies of Plio-Pleistocene marine sediments, recovered by piston coring in high southern latitudes (Opdyke *et al.*, 1966; Ninkovitch *et al.*, 1966; Hays and Opdyke, 1967) and in the equatorial Pacific Ocean (Hays *et al.*, 1969), that mark the beginning of modern magnetic stratigraphy. These studies combined magnetic polarity stratigraphy and biostratigraphy and in so doing refined and extended the GPTS derived from paleomagnetic studies of basaltic outcrops (e.g., Doell *et al.*, 1966).

Heirtzler *et al.* (1968) produced a polarity timescale for the last 80 My using South Atlantic MMA record (V-20) as the polarity template. They assumed constant spreading rate, and extrapolated ages of polarity chrons using an age of 3.35 Ma for the Gilbert/Gauss polarity chron boundary. This was a dramatic step forward that extended the GPTS to ∼ 80 Ma, from ∼5 Ma based on magnetostratigraphic records available at the time (e.g., Hays and Opdyke, 1967). Heirtzler *et al.* (1968) made the inspired choice of a particular South Atlantic MMA record (V-20). Several iterations of the polarity timescale (e.g., Labrecque *et al.*, 1977) culminated with the work of Cande and Kent (1992a) that re-evaluated the MMA record, and used the South Atlantic as the fundamental template with inserts from faster spreading centers in the Indian and Pacific Oceans. They then interpolated between nine numerical age estimates that could be linked to polarity chrons over the last 84 My. Eight of these ages were based on 'high-temperature' radiometric ages (no glauconite ages were included), and the youngest age (2.60 Ma) was the astrochronological age estimate for the Gauss/

Matuyama boundary from Shackleton *et al.* (1990). In the subsequent version of their timescale, hereafter referred to as CK95, Cande and Kent (1995) adopted astrochronological age estimates for all Pliocene–Pleistocene polarity reversals (Shackleton *et al.*, 1990; Hilgen, 1991a, 1991b) and modified the age tie-point at the Cretaceous–Tertiary boundary from 66 to 65 Ma.

The astrochronological estimates for Pliocene–Pleistocene polarity chrons (Shackleton *et al.*, 1990; Hilgen, 1991a, 1991b), incorporated in CK95, have not undergone major modification in the last 15 years, and indicate the way forward for timescale calibration further back in time. Subsequent to CK95, astrochronological estimates of polarity chron ages have been extended into the Miocene, Paleogene, and Cretaceous (Krijgsman *et al.*, 1994, 1995; Hilgen *et al.*, 1995, 2000; Abdul-Aziz, 2000, 2003, 2004). Many of these advances are included in the timescale of Lourens *et al.* (2004). At present, more or less continuous astrochronologies tied to polarity chrons are available back to the Oligocene (e.g., Billups *et al.*, 2004).

Recognition of brief polarity excursions as an integral part of the Earth's paleomagnetic field behavior has developed alongside astrochronological calibration of the polarity timescale in the last 20 years. Although the first recognized polarity excursions (the Laschamp and Blake excursions) were documented in the late 1960s (Bonhommet and Babkine, 1967; Smith and Foster, 1969), excursions were widely considered to represent either spurious recording artifacts or, at best, local anomalies of the geomagnetic field and of doubtful utility in stratigraphy. It was not until the late 1980s that the tide began to turn. As high-resolution sedimentary records from the deep sea became available, it became accepted that excursions are frequent (with perhaps more than seven in the Brunhes Chron) and that they appear to be globally recorded and not local anomalies of the geomagnetic field. Geomagnetic excursions were not recorded in the early days of

magnetic stratigraphy because the high sedimentation rate sequences, required to record brief (few kiloyears duration) magnetic excursions, were generally not targeted during conventional piston coring expeditions. Indeed, at the time, low sedimentation rate sequences were often preferred as they allowed the record to be pushed further back in time. The development of hydraulic piston coring (HPC) techniques, first used in early 1979 during the Deep Sea Drilling Project (DSDP) Leg 64, brought high sedimentation rate sequences within reach by increasing penetration by a factor of ~15, from ~20 m for conventional piston coring to ~300 m for the HPC.

10.1.2 Nomenclature for Excursions and Polarity Intervals

The International Stratigraphic Commission (ISC) has guided the use of magnetostratigraphic units (polarity zones), their time equivalents (polarity chrons), and chronostratigraphic units (polarity chronozones) (see Anonymous, 1977; Opdyke and Channell, 1996). The revelation in the last 20 years of numerous short-lived excursions within the Brunhes and Matuyama chrons, and probably throughout the history of the Earth's magnetic field, requires an extension of terminology to include excursions and brief polarity microchrons (**Table 1**). Terminology based on duration of polarity intervals is obviously problematic at the low end of the duration spectrum. Records are always compromised by limitations of the recording medium and, in addition, estimates of duration are limited by the availability of chronological tools of adequate precision. In the realm of MMAs, La Brecque *et al.* (1977) applied the term 'tiny wiggles' to minor, but lineated, MMAs. The time equivalent of the 'tiny wiggle' was labeled 'cryptochron' (Cande and Kent, 1992a, 1992b), expressing the uncertain origin of 'tiny

Table 1 Nomenclature for polarity intervals and excursions

Magneto-stratigraphic polarity zone	Geochronologic (time) equivalent	Chronostratigraphic equivalent	Duration (yr)
Polarity megazone	Megachron	Megachronozone	10^8–10^9
Polarity superzone	Superchron	Superchronozone	10^7–10^8
Polarity zone	Chron	Chronozone	10^6–10^7
Polarity subzone	Subchron	Subchronozone	10^5–10^6
Polarity microzone	Microchron	Microchronozone	$<10^5$
Excursion zone	Excursion		Brief departure from normal secular variation
Polarity cryptochron	Cryptochron	Cryptochronozone	Uncertain existance

wiggles' as either short-lived polarity intervals or paleointensity fluctuations. These authors placed the duration separating 'polarity chrons' and 'cryptochrons' at 30 ky, representing an estimate of the minimum duration of polarity intervals that can be resolved in MMA records. For magnetostratigraphy, Krijgsman and Kent (2004) advocated a duration cutoff for separating 'subchrons' and 'excursions', at 9–15 ky . The drawback of such a scheme is that, in most cases, the chronostratigraphic precision will be inadequate to establish the distinction. An alternative is that an excursion be defined in terms of a 'brief' ($<10^4$ years) deviation of virtual geomagnetic poles (VGPs) from the geocentric axial dipole (GAD) that lies outside the range of secular variation for a particular population of VGPs. Some authors have adopted an arbitrary VGP cut-off (say a co-latitude of 45°) to define an excursion, although Vandamme (1994) has advocated a method of calculating the cut-off for a specific VGP population based on VGPs lying outside 'normal' secular variation. As higher fidelity records of excursions have become available, it appears that the majority of 'excursions' is manifest as directional changes through ∼180°, followed by a return to the pre-excursional directions within a few thousand years. For example, the Laschamp and Iceland Basin excursions, although short-lived with durations of a few kiloyears, represent paired reversals of the geomagnetic field as VGPs reach high southerly latitudes for both excursions (Channell, 1999; Laj et al., 2006). 'Excursions' displaying low or mid-latitude VGPs, rather than high latitude reverse VGPs, probably very often reflect inadequacy of the recording medium (Roberts and Winklhofer, 2004, Channell and Guyodo, 2004), inadequate rates of sediment accumulation, and/or inadequate sampling methods, rather than geomagnetic characteristics. We therefore advocate the use of the term Microchron for brief polarity chrons with established duration less than 10^5 years (**Table 1**). The term 'excursion' would then be used only for features that represent departures from normal secular variation, for which full polarity reversal has not been established. As these features become better documented, they could then be elevated to the status of Microchron, a term that denotes a brief polarity chron. Under this nomenclature, the Laschamp and Iceland Basin 'excursions' would be elevated to Polarity Microchron status as it has been established that these represent paired full polarity reversals defining a distinct polarity interval.

10.2 Geomagnetic Excursions in the Brunhes Chron

10.2.1 Introduction

In the last few decades numerous geomagnetic excursions have been discovered in the previously believed stable Brunhes Chron (**Figure 1** and **Table 2**).

The Laschamp excursion, which is now known to have an age of ∼40 ka, was the first geomagnetic excursion to be recognized, in lavas from the French Massif Central (Bonhommet and Babkine, 1967). It is the most thoroughly studied excursion, and its existence is proved beyond doubt. This is not the case for other reported excursions in the late Brunhes Chron, and some of these can be attributed to sedimentological and/or sampling artifacts: for example, the Starno event (Noel and Tarling, 1975), the Gothenburg 'flip' (Mörner and Lanser, 1974), and the Lake Mungo excursion (Barbetti and McElhinny, 1972). Other early papers that documented excursions have been ratified by later work: for example, the reported excursion (Blake Event) in marine isotope stage (MIS) 5 from the Blake Outer Ridge (Smith and Foster, 1969). Wollin et al. (1971) documented three short intervals of reverse paleomagnetic inclination in cores from the Caribbean and the Eastern Mediterranean. An approximate age model suggested that the most recent of these was coeval with the Blake Event, and the ages of the other two were estimated to be around 180–210 and 270 ka. Kawai et al. (1972) and Yaskawa et al. (1973) reported a paleomagnetic study of a 197 m core from Lake Biwa (Japan) that documented five short episodes of reverse polarity in the Brunhes Chron. On the basis of a tentative correlation with the results of Wollin et al. (1971), the youngest of these episodes was correlated to the Blake Event and the other two were labeled Biwa I at about 176–186 ka and Biwa II at about 292–298 ka. Later, Kawai (1984), on the basis of fission track ages on zircons from tephra layers interbedded in the sediments, obtained ages of 100, 160, and 310 ka for the first three episodes, and ∼380 ka for a fourth episode which he called Biwa III. As with much of the evidence for directional excursions published in the 1960s and 1970s, it was based on magnetization directions that were poorly defined (by modern standards) and on age control that provided plenty of latitude in their correlation to other supposed excursions. Apart from the Blake and the Laschamp excursions that have stood the test of time, we advocate abandoning the labels such as Biwa, that refer to excursions that are poorly defined with ages that are poorly constrained.

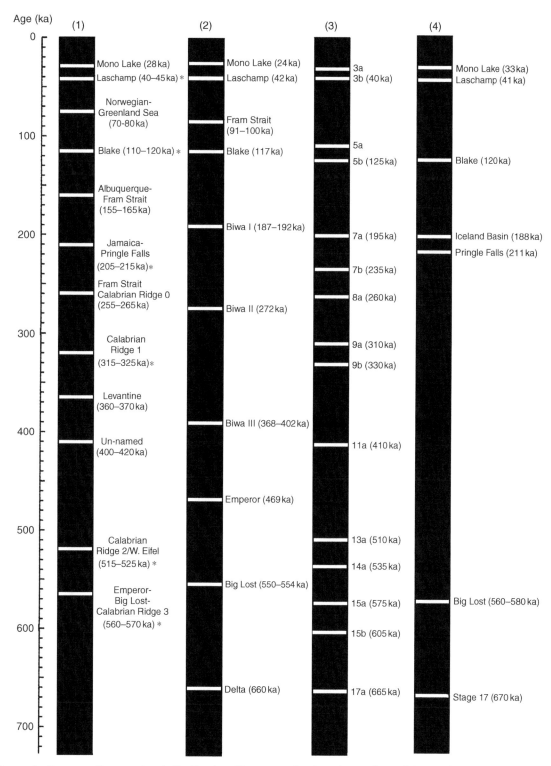

Figure 1 Geomagnetic excursions in the Brunhes Chron according to various authors. Column 1: Langereis *et al.* (1997): asterisks mark 'well-dated, global' excursions, others were deemed 'restricted or not as certain'; column 2: Worm (1997); column 3: Lund *et al.* (2001a); column 4: well-documented excursions with acceptable age control (this study).

Table 2 Excursions within the Brunhes Chron

Excursion name	MIS	Estimated age (ka)	Estimated duration (kyr)	Location or ODP Leg	Principal references
Mono Lake	3	33	1	Mono Lake/Arctic/152	(1)
Laschamp	3	41	1.5	Atlantic realm/Arctic/177/172/152	(2)
Norwegian-Greenland Sea		*60*		*Arctic*	*(3)*
Blake	5d/5e	120	5	North Atlantic/Med/172	(4)
Iceland Basin	6/7	188	3	Atlantic/Pacific/Baikal/ 152/162/172/177/184	(5)
Pringle Falls	7	211		Western US/New Zealand/172/152	(6)
Calabrian Ridge 0	*8*	*260*		*Med/172*	*(7)*
Calabrian Ridge I	*9*	*325*		*Med/172*	*(8)*
Un-named	*11*	*400*		*172*	*(9)*
Calabrian Ridge II	*13*	*525*		*Med/172*	*(10)*
Big Lost	14/15	560–580		Yellowstone/172/162	(11)
Stage 17	17	670		Osaka Bay/172/162	(12)

Italics indicate poorly documented excursions, MIS: marine isotope stage.
Reference: (1) Denham and Cox (1971); Liddicoat and Coe (1979); Negrini *et al*. (1984, 2000); Lund *et al*. (2001a); Nowaczyk and Antonow (1997); Nowaczyk and Knies (2000); Channell (2006). (2) Bonhommet and Babkine, (1967); Bonhommet and Zahringer (1969); Levi *et al*. (1990); Laj *et al*. (2000); Lund *et al*. (2001a); Mazaud *et al*. (2002); Laj *et al*. (2006); Channell (2006). (3) Nowaczyk and Frederichs (1994). (4) Smith and Foster (1969); Tric *et al*. (1991); Zhu *et al*. (1994); Thouveny *et al*. (2004). (5) Channell *et al*. (1997); Weeks *et al*. (1995); Roberts *et al*. (1997); Channell (1999); Oda *et al*. (2002); Stoner *et al*. (2003); Laj *et al*. (2006); Channell (2006). (6) Herrero-Bervera *et al*. (1989, 1994); McWilliams (2001); Singer *et al*. (2005); Channell (2006). (7) Kawai (1984); Langereis *et al*. (1997); Lund *et al*. (2001b). (8) Lund *et al*. (2001b). (9) Lund *et al*. (2001b). (10) Langereis *et al*. (1997), Lund *et al*. (2001b). (11) Champion *et al*. (1988); Lund *et al*. (2001b); Quidelleur *et al*. (1999); Singer *et al*. (2002); Channell *et al*. (2004). (12) Biswas *et al*. (1999); Lund *et al*. (2001b); Channell and Raymo (2003); Channell *et al*. (2004).

During a search for the Laschamp excursion in exposed sediments at Mono Lake (California), an excursion apparently younger than the Laschamp excursion was documented by Denham and Cox (1971) and subsequently by Liddicoat and Coe (1979). This excursion was named the Mono Lake excursion after its type locality. Although the excursion has apparently been observed elsewhere, the age of the excursion at the type locality remains controversial.

Creer *et al*. (1980) studied a drill-core section at Gioia Tauro in southern Italy, and suggested that an episode of reverse polarity found in the upper part of the section could be the Blake Event, at about 105–114 ka. Four additional episodes of low inclination were named α, β, γ, and δ. Kawai (1984) suggested that the youngest of the Biwa episodes could correspond to the Blake Event, the α episode to Biwa I, the β excursion to Biwa II, and the γ episode to Biwa III. Poor age control and poor definition of magnetization directions at Gioia Tauro (and Lake Biwa) make these correlations equivocal.

Champion *et al*. (1981) found evidence for a brief reversal in a sequence of basalt flows in the Snake River Plain (Idaho) to which they assigned an age of 465 ± 50 ka, using K–Ar ages of bracketing normally magnetized flows. They correlated this event with the Emperor Event of Ryan (1972). Later, however,

Champion *et al*. (1988) revised the age of these reverse polarity lavas to 565 ± 10 ka. The new age implies a new reverse episode, which they named the Big Lost excursion. Based on literature available at the time, Champion *et al*. (1988) proposed the existence of eight reverse polarity microchrons in the Brunhes Chron (**Figure 1**).

Langereis *et al*. (1997) reviewed evidence for excursions in the Brunhes Chron, made the case for seven well-dated 'global' excursions and five 'restricted' less-certain Brunhes excursions, for a total of 12 excursions in the Brunhes Chron (**Figure 1**). From a central Mediterranean core, these authors added a series of excursions (Calabrian Ridge 0, 1, 2, 3) to the excursion vocabulary, the oldest (CR3) being correlative to the Big Lost excursion. Worm (1997) suggested a link between geomagnetic reversals/excursions and glaciations, and listed 10 geomagnetic excursions in the Brunhes Chron that he considered to be adequately documented (**Figure 1**). In more recent papers attempting a synthesis, Lund *et al*. (2001a, 2001b, 2006) proposed 17 excursions during the Brunhes Chron. The database for geomagnetic excursions has clearly evolved rapidly in recent years, but with little consensus on the existence or age of Brunhes-aged excursions (**Figure 1**).

In this section, we first describe evidence for the five best-documented excursions in the Brunhes

Chron: the Laschamp, Mono Lake, Blake, Iceland Basin, and Pringle Falls excursions. We then describe other less well-documented excursions recorded in the early Brunhes Chron (**Table 2**).

10.2.2 The Laschamp Excursion

The Laschamp excursion was the first reported geomagnetic excursion, and is certainly the best known excursion in the Brunhes Chron. Bonhommet and Babkine (1967) discovered this excursion and named it after the Puy de Laschamp, in the French Chaîne des Puys (Massif Central, France). The flow carries an anomalous characteristic paleomagnetic direction, up to 160° away from the expected dipole field direction. The excursion was initially discovered in the flow and scoria in the Puy de Laschamp, but the same anomalous direction has also been found in the nearby Olby flow (**Figure 2**).

Considerable effort has been dedicated to determination of the age of the Laschamp excursion. The first determination, using K/Ar methods on whole rock by Bonhommet and Zahringer (1969), yielded an age between 8 and 20 ka. Subsequent determinations did not confirm such a young age: Hall and York (1978) obtained whole rock K–Ar and ^{40}Ar/^{39}Ar dates of 47.4 ± 1.9 and 45.4 ± 2.5 ka. Condomines (1978),

using the ^{230}Th/^{238}U radioactive disequilibrium method, obtained an age of 39 ± 6 ka, which is more than double the age of Bonhommet and Zahringer (1969). Gillot et al. (1979), using the unspiked (Cassignol) K–Ar method, obtained an age of 43 ± 5 ka for the Laschamp flow and 50 ± 7.5 ka for the Olby flow. An early attempt to date the Laschamp excursion using thermoluminescence (Wintle, 1973) failed because of anomalous fading of the signal. Huxtable et al. (1978) mitigated the problem by using sediments baked by the Olby flow and obtained an age of 25.8 ± 1.7 ka, which is somewhat younger than the 35 ± 5 ka age determined by Gillot et al. (1979) from quartz contained in a granitic inclusion found in the Laschamp flow. Gillot et al. (1979) also obtained an age of 38 ± 6 ka from five quartz pebbles found in a baked paleosol beneath the Olby flow.

In the most recent determination of the age of the excursion, Guillou et al. (2004) combined K–Ar and ^{40}Ar/^{39}Ar results from the two basaltic flows at Laschamp and Olby to better resolve the age of the Laschamp excursion. This was possible in part due to recent advances in ^{40}Ar/^{39}Ar incremental heating methods using a resistance furnace which can yield ages for basaltic lava flows between 100 and 20 ka with a precision better than 5% at the 95% confidence level (Heizler et al., 1999; Singer et al., 2000). Similarly,

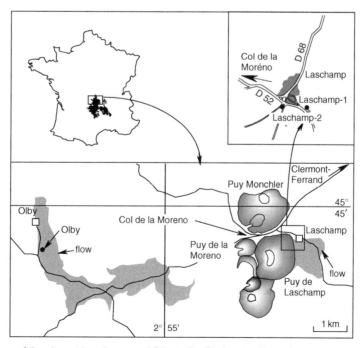

Figure 2 Location map of the sites at Laschamp and Olby in the Chaîne des Puys, Central France. From Guillou H, Singer BS, Laj C, Kissel C, Scaillet S, and Jicha B (2004) On the age of the Laschamp geomagnetic event. *Earth and Planetary Science Letters* 227: 331–343.

the unspiked K–Ar dating method, with the new calibration procedure of Charbit *et al.* (1998), allows accurate and precise measurements of minute quantities of radiogenic argon (Guillou *et al.* (1998). The results of six new unspiked K–Ar and 13 ^{40}Ar/^{39}Ar incremental heating experiments on subsamples from three sites in the Laschamp and Olby flows are concordant and yield a weighted mean age of 40.4 ± 1.1 ka. Consideration of the uncertainties in the ^{40}K $\rightarrow$ ^{40}Ar decay constant led to 40.4 ± 2.0 ka (2σ analytical error plus decay constant uncertainties) as the most probable age for the Laschamp excursion. The incremental progress in determining a precise age for the Laschamp excursion is depicted in **Figure 3**.

Over 10 years after the Laschamp excursion was first reported in the literature, doubts were expressed about its geomagnetic origin. Heller (1980) and Heller and Petersen (1982) noted partial or complete self-reversal of the NRM of many Olby samples, and to a lesser extent of Laschamp samples, during thermal laboratory experiments. They suggested that the almost reverse magnetization in the two flows could be attributed to self-reversal properties, rather than to

the geomagnetic field. This suggestion was at odds with earlier work of Whitney *et al.* (1971) who concluded that the excursional magnetization was carried by single-domain magnetite grains, a magnetic mineralogy inconsistent with suggested mechanisms for self-reversal. Roperch *et al.* (1988) pointed out that, when more than one magnetic phase is present, the species with the lowest blocking temperature usually experiences self-reversal by magnetostatic interactions, while the species with higher (>200 °C) blocking temperature invariably carries the primary reverse direction. Furthermore, Roperch *et al.* (1988) found a reverse direction in a sample of clay which had been baked by the overlying Olby flow. Although limited to one sample (due to the difficulty of finding clays baked by the flow), this result provided evidence for a geomagnetic origin of the Laschamp excursion. Roperch *et al.* (1988) also used the Thellier–Thellier paleointensity method to obtain a value of 7.7 μT (i.e., less than one-sixth of the present field) for the reverse polarity flows at Laschamp and Olby. They argued that this low value is more characteristic of transitional geomagnetic field behavior, and, therefore, that the paleomagnetic directions of the Laschamp and Olby flows were not acquired during a stable period of reverse polarity. In the Chaîne des Puys, Barbetti and Flude (1979) obtained low paleointensity values from sediments baked by the lava flow at Royat. Here, paleointensity data imply a field strength about 30% of its present value, and an ^{40}Ar/^{39}Ar age of around 40 ka (Hall *et al.*, 1978), which is close to that of the Laschamp and Olby flows. In the Chaîne des Puys, Chauvin *et al.* (1989) obtained paleointensity results from different flows on the Louchadière volcano, where K–Ar ages indicated approximate synchroneity with the Laschamp flow. Chauvin *et al.* (1989) documented anomalous paleomagnetic directions (although different from the Laschamp directions) and determined a paleointensity of 12.9 ± 3.3 μT, which is about one-third the present field value, which confirms the transitional character of the Louchadière flow.

In Iceland, Kristjansson and Gudmundsson (1980) found evidence for an anomalous paleomagnetic direction from three different localities in the Reykjanes Peninsula, which they called the Skalamaelifell excursion. Levi *et al.* (1990) subsequently identified the same excursion at four additional localities in the same region, where the K–Ar age is 42.9 ± 7.8 ka, thereby associating the Skalamaelifell excursion with the Laschamp excursion. Levi *et al.* (1990) also obtained a paleointensity determination of 4.2 ± 0.2 μT, which is consistent with an earlier result by Marshall *et al.* (1988) and similar to

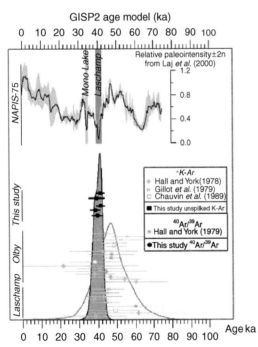

Figure 3 Radioisotopic ages of the Laschamp and Olby flows. *N* is the number of individual age determinations. The broad probability–density curve with a maximum at 46 ka represents previously published K–Ar and ^{40}Ar/^{39}Ar ages, whereas the narrow envelope peaking at 40.5 ka is from Guillou *et al.* (2004).

that determined by Roperch *et al.* (1988) on the Laschamp and Olby flows, and almost an order of magnitude less than the present field strength in Iceland.

In central France, the search for the Laschamp excursion in sediments from Lac du Bouchet, a maar lake only about 100 km from Laschamps (Creer *et al.*, 1990), did not yield evidence of a departure of the geomagnetic field vector from the normal direction, but the interval corresponding to the excursion was marked by low paleomagnetic field intensity. In a later paper, Thouveny and Creer (1992) discussed in more detail possible reasons for the absence of directional changes in the Lac du Bouchet record and concluded that, given the sediment accumulation rate, the duration of the Laschamp excursion could not have exceeded some 200 years, a duration that has subsequently proved to be an underestimate (see Laj *et al.*, 2000). This negative result is probably due to the inability of the Lac du Bouchet sediments to record the brief directional changes associated with the Laschamp excursion (see Roberts and Winklhofer, 2004).

Although sedimentary records often document the characteristic paleointensity minimum associated with the Laschamp excursion, the majority of records fail to document excursional directions associated with the minimum. This is likely due to low sedimentation rates (low resolution of the sedimentary records) combined with the fact that the directional anomaly associated with the Laschamp excursion may be brief (less than a few kiloyears) compared to the duration of the associated paleointensity minimum. Any delay in remanence acquisition due to bioturbation and a progressive magnetization lock-in may lead to smearing of the NRM directions, thereby inhibiting the recording of brief directional events (Roberts and Winklhofer, 2004; Channell and Guyodo, 2004).

Nowaczyk and Antonow (1997) found evidence for the Laschamp excursion in four sediment cores from the Greenland Sea. Subsequently, Nowaczyk and Knies (2000) reported directional and relative paleointensity records of the Laschamp and Mono Lake excursions from the Arctic Ocean. One of the cores documented in this study (PS 2212-3 KAL) illustrates an anomaly common to these high latitude cores. The cumulative percentage thickness of reverse polarity zones in sediment accumulated during the last few 100 ky is far greater than expected ($\sim$50% for the top 4 m of the section representing the last 120 ky in PS2212-3KAL). Fortuitous fluctuations in sedimentation rates, that are not resolvable in the age models, appear to have 'amplified' these excursional records.

In the North Atlantic Paleointensity Stack (NAPIS-75) of Laj *et al.* (2000), all but one of the six cores display abrupt directional changes at around 41 ka. In three of the cores, the component inclinations reach negative values in excess of $-30°$. The chronology of NAPIS-75 is based on correlation of the planktic $\delta^{18}O$ record from core PS2644-5 (Voelker *et al.*, 1998) to the $\delta^{18}O$ record from the GISP2 ice core (Grootes and Stuiver, 1997). Correlation among the NAPIS-75 cores was determined using isotopic stage boundaries with a second step matching cycles observed in the downcore profiles of the anhysteretic remanent magnetization (ARM). The cores were then aligned to the Greenland ice core chronology by correlation of each ARM record to that of PS2644-5 (Kissel *et al.*, 1999). The Laschamp excursion occurred at the end of interstadial 10 (Baumgartner *et al.*, 1997; Laj *et al.*, 2000; Wagner *et al.*, 2000), at about 41 ky, consistent with the most recent radiometric dating of the excursional lava flow at the Laschamp locality (Guillou *et al.*, 2004).

The Laschamp excursion has also been recognized in several piston cores from the sub-Antarctic South Atlantic Ocean (Channell *et al.*, 2000) where the age models indicate an excursional age of $\sim$40 ka and an excursion duration of <2 ky. In the cores that best display the directional excursion, mean sedimentation rates in the excursional interval exceeded 20 cm ky^{-1}. Age models were determined by $\delta^{18}O$ data from the same cores, and by utilization of accelerator mass spectrometry (AMS) ^{14}C ages from a nearby core (RC11-83) (Charles *et al.*, 1996) that could be correlated to the magnetically studied cores using both $\delta^{18}O$ and $\delta^{13}C$ data.

Mazaud *et al.* (2002) resolved the Laschamp excursion in core MD94-103 from the southern Indian Ocean (east of Kerguelen Plateau). Sedimentation rates approached 20 cm ky^{-1} in the excursional interval, the paleointensity record can be unambiguously correlated to NAPIS-75, and the age and duration of the excursion can be estimated as $\sim$41 ka and approximately 2 ky, respectively.

Lund *et al.* (2005) presented two records of the Laschamp excursion from sediment cores from the Bermuda Rise and the Blake Outer Ridge in the Western North Atlantic Ocean. The high sediment accumulation rate (estimated to be 20–28 cm ky^{-1} in the interval recording the Laschamp excursion) and the uncomplicated magnetic properties resulted in well-defined characteristic remanent magnetizations. The excursional directions are present during times of apparently low paleomagnetic field intensity. The declinations display almost a full reversal (120° change in direction), while the inclinations change from 49° to −49° with intermediate values reaching −80°. The directions have been interpreted to display oscillatory behavior, growing in amplitude during the decrease in paleointensity, with an estimated time constant of 1200 years based on a constant sedimentation rate of 28 cm ky^{-1} during the excursion. The duration of the excursional interval was estimated to be ∼2 ky, in agreement with previous estimates (Laj *et al.*, 2000; Channell *et al.*, 2000), but much shorter than the duration estimate from the Arctic (Nowaczyk and Knies, 2000). Lund *et al.* (2005) calculated the rates of intensity and directional changes during the excursion, based on the oxygen isotope stratigraphy. In paleointensity, these rates are typically less than 50 nT yr^{-1} and never more than 150 nT yr^{-1}. Although these values are averaged over some 100–130 years, and are strongly dependent on the age model, they are not significantly different from the mean annual rate of change of the historic field (Peddie and Zunde, 1988). Similarly, the rates of directional change are of the order of 30–70 arc min yr^{-1}, which compares with the maximum rate of 40 arcmin yr^{-1} observed historically.

Laj *et al.* (2006) reported five records of the Laschamp excursion obtained from rapidly deposited sediments at widely separated sites (the Bermuda Rise (MD95-2034), the Greenland Sea (PS2644-5), the Orca Basin of the Gulf of Mexico (MD02-2551 and MD02-2552), and the southern Indian Ocean (MD94-103). The records of the Laschamp excursion from cores JPC-14 (Blake Outer Ridge) and 89-9 (Bermuda Rise)

(Lund *et al.*, 2005), as well as from cores 4-PC03 and 5-PC01 from the sub-Antarctic South Atlantic (Channell *et al.*, 2000), provide evidence for global manifestation of the Laschamp excursion (**Figure 4**).

A coherent picture of excursional field behavior during Laschamp excursion has emerged (Laj *et al.*, 2006): the excursional VGPs trace a clockwise loop (**Figure 5**), moving southward over east Asian/western Pacific longitudes, reaching high southern latitudes, followed by a northward-directed VGP path over Africa and western Europe. The turning point where the VGPs change from being southward- to northward-directed coincides with the minimum in relative paleointensity. A recently published record of the Laschamp Event at ODP Site 919 in the Irminger Basin (off east Greenland) (Channell, 2006) yields a VGP path that also describes a large clockwise loop, albeit not exactly on the same longitudes as those described above. Here, mean sedimentation rates in the vicinity of the excursion exceed 20 cm ky^{-1} and, according to the oxygen isotope age model, the age and duration of the excursion is 40 ka and <2 ky, respectively.

It is now firmly established that the strength of the geomagnetic field is the most important factor controlling cosmogenic radionuclide production (Lal and Peters, 1967; Masarik and Beer, 1999): the smaller the field intensity, the larger the production rate. As shielding of cosmic rays by the geomagnetic field occurs at distances of several Earth radii, only changes in the dipole field (i.e., global changes) are relevant to the regulation

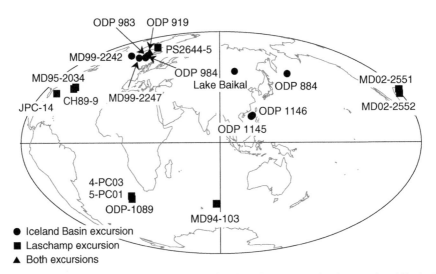

Figure 4 Map of core locations that have yielded records of the Laschamp excursion (squares) and the Iceland Basin excursion (circles).

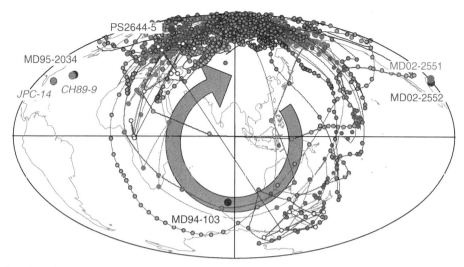

Figure 5 Transitional magnetization directions for the Laschamp excursion represented as virtual geomagnetic polar (VGP) paths. The large arrow illustrates the sense of looping which is consistently clockwise.

of cosmogenic radionuclide production. Cosmogenic nuclide flux therefore provides an independent way of establishing the global character of the intensity minimum associated with the Laschamp excursion. Initial work along these lines demonstrated that ^{10}Be flux and geomagnetic field intensity anticorrelate in a North Atlantic core (Robinson *et al.*, 1995). Subsequently, Frank *et al.* (1997) correlated a reconstruction of paleofield intensity, obtained from ^{10}Be flux records in marine sediments, to long-term trends in the SINT-200 paleointensity stack (Guyodo and Valet, 1996). The 96–25 ka record of ^{36}Cl flux from the GRIP ice core (Baumgartner *et al.*, 1998) agrees reasonably well with a production rate calculation from a paleointensity stack from the Somali basin (Meynadier *et al.*, 1992). More recently, Wagner *et al.* (2000) compared the NAPIS-75 paleointensity stack (Laj *et al.* 2000) to the field intensity record estimated from an improved, higher-resolution ^{36}Cl record, assuming that the variations in the ^{36}Cl flux are entirely due to modulation by the geomagnetic field. The ^{36}Cl-derived profile has been smoothed out using a 3000-year window in order to filter the solar modulation. The coincidence of a prominent maximum in ^{36}Cl flux with the paleointensity profile through the Laschamp excursion (**Figure 6**) leaves little doubt concerning the geomagnetic nature of the increase in cosmogenic production at around 40 ka, and provides strong evidence for the global nature of the intensity minimum associated with the Laschamp excursion.

10.2.3 The Mono Lake Excursion

About 35 years ago, the age of the Laschamp excursion was thought to be ~20 ka from studies in central France (Bonhommet and Zahringer, 1969). Denham and Cox (1971) sampled the Wilson Creek Formation at Mono Lake (California) in search of the same excursion. Based on radiocarbon ages available at the time, this section was thought to span the 13-30 ka time interval. Denham and Cox (1971) detected an excursion at Mono Lake, at a horizon with an estimated age of 24 ka. This excursion has since been named the Mono Lake excursion. Denham (1974) interpreted the excursion as arising from eastward drift of the nondipole field, and showed that the pattern of observed paleomagnetic directions could be reproduced assuming an eccentric radial dipole source first increasing and then decreasing during its eastward movement.

Liddicoat and Coe (1979) confirmed the apparent existence of a geomagnetic excursion at Mono Lake, documented the directional record in more detail, and extended the record further down section. There was initially widespread skepticism about the validity of this excursion. Verosub (1977) did not find evidence for it in sediments from Clear Lake (California) and Palmer *et al.* (1979) failed to find it in gravity cores from Lake Tahoe (California). Turner *et al.* (1982) obtained a distinctive pattern of 'normal' paleosecular variation, but no excursion, from an 18 m sedimentary sequence at Bessette Creek (British Columbia), that was radiocarbon

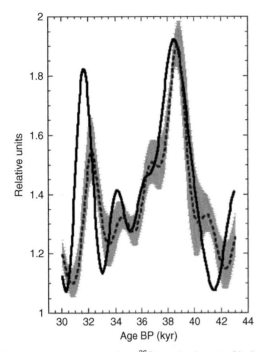

Figure 6 Comparison of the ^{36}Cl production rate (dashed line) calculated from paleomagnetic field data (modified NAPIS-75 stack) and the decay-corrected ^{36}Cl flux (solid line) from the GRIP ice core. The shaded area corresponds to the $\pm 2\sigma$ uncertainties in the ^{36}Cl production rate due to uncertainties in the relative paleointensity data. Both records are low-pass filtered and normalized to the present field values. From Wagner G, Beer J, Laj C, (2000) Chlorine-36 evidence for the Mono lake event in the Summit GRIP ice core. *Earth and Planetary Science Letters* 181: 1–6.

dated to 31.2–19.5 ka. Verosub *et al.* (1980) did not find this excursion in sediments exposed on the shores of Pyramid Lake (Nevada), which covers the time interval 25–36 ka and is only about 230 km from Mono Lake. Absence of the excursion at both Pyramid Lake and Clear Lake led Verosub *et al.* (1980) to conclude that the estimated age of the Mono Lake excursion was in error.

The interpretation of the excursion at Mono Lake has progressively evolved. Liddicoat *et al.* (1982) discovered the excursion in sediments from Lake Lahontan exposed in Carson Sink (Nevada), some 60 km east of Mono Lake. Negrini *et al.* (1984, 2000) discovered the excursion in sediments from Summer Lake, Oregon, 550 km north of Mono Lake. The excursion was apparently detected in sediments recovered by hydraulic piston core from the DSDP Site 480 in the Gulf of California (Levi and Karlin, 1989). Liddicoat (1992) resampled at Carson Sink and Pyramid Lake (Nevada), at the same locality studied

by Verosub *et al.* (1980), and showed that, while the older half of the excursion is not recorded, the younger half can be recognized. Liddicoat (1996) also reported the excursion from the Lahontal Basin, Nevada.

As was the case for the Laschamp excursion, proof of the global character of the Mono Lake excursion has come from the study of marine cores. Nowaczyk and Knies (2000) obtained a record from cores in the Arctic Ocean, in which they identified directional and relative paleointensity changes that they associated to the Laschamp and to the Mono Lake excursions (dated at 35 and 27–25 ka, respectively). In the Irminger Basin (ODP Site 919), off east Greenland, two excursions with duration <2 ky are centered at 34 and 40.5 ka according to the oxygen isotope age model (Channell, 2006). These examples represent the rare occurrence of both the Mono Lake and Laschamp directional excursions in the same section, and there is now accumulating evidence for the existence of two excursions (Mono Lake and Laschamp) separated by 7–8 ky.

As for the Laschamp excursion, although the Mono Lake directional excursion is rarely recorded due to its brevity, a relative paleointensity minimum at about 32–34 ka is often recorded in relative paleointensity records. This Mono Lake paleointensity minimum is manifest as an increased cosmogenic isotope flux in ice core records. For example, in the ^{36}Cl record obtained from the GRIP ice core (Wagner *et al.*, 2000), in addition to the peak in production associated with the Laschamp excursion described above, a second distinct peak is present between Dansgaard–Oeschger (D–O) events 6 and 7, at approximately 33 ka in the GISP2 age model (**Figure 6**). This production peak can therefore be attributed to the geomagnetic field intensity minimum associated with the Mono Lake excursion, providing evidence for the global character of this paleointensity minimum.

Kent *et al.* (2002) have obtained new radiocarbon data from lacustrine carbonates (ostracodes and tufa nodules) from 11 stratigraphic horizons from the lower part of the Wilson Creek (Mono Lake) section. These authors argued that previous age estimates for the excursion recorded in the Mono Lake section, based on a series of 27 published radiocarbon measurements on tufa or ostracodes (Benson *et al.*, 1990, 1998; Lund *et al.*, 1988), do not take into account radiocarbon reservoir effects, modern carbon contamination effects, or radiocarbon production variations. As it proved virtually impossible to find clean shells, Kent *et al.* (2002) made measurements on pairs of uncoated and variably tufa-encrusted ostracodes and tufa nodules. Invariably, the more encrusted ostracode or tufa nodule samples yielded ages that were younger than the uncoated samples (by 700–2100 years). This could be attributed to modern carbon contamination, as supported by the younging ages from sample aliquots subjected to progressive acid leaching. Because (maximum) age plateaus were not always recorded during the leaching process, Kent *et al.* (2002) concluded that the radiocarbon ages should be viewed as minimum age constraints. The difference between these new determinations and those of Benson *et al.* (1990) increases with age, and reaches ~4 ka in the interval of the Mono Lake excursion.

Kent *et al.* (2002) also obtained $^{40}Ar/^{39}Ar$ age estimates from sanidine crystals imbedded in the different ash layers in the Wilson Creek section. These ages are at the younger limit of the method, and are further complicated by residence time in the magma chamber, and other sources of inherited (old) argon. As a consequence, the ash layers are usually characterized by a wide range of sanidine ages. In the middle of the Mono Lake excursion, 34 individual sanidine samples from Ash #15 yielded ages between 49 and 108 ka, a range that far exceeds analytical precision. In such cases, Chen *et al.* (1996) suggested that the youngest $^{40}Ar/^{39}Ar$ ages provide realistic estimates of eruption age, as they are generally closer to accompanying radiocarbon ages. Using similar logic, Kent *et al.* (2002) considered 49 ka as the maximum depositional age for Ash #15.

Using radiocarbon dates as minima and $^{40}Ar/^{39}Ar$ dates as maxima, Kent *et al.* (2002) derived a new age model for the Wilson Creek section, and estimated the age of the lower part of the Wilson Creek section as >46 ka. On this basis, Kent *et al.* (2002) proposed that the excursion recorded at Mono Lake is, in reality, the Laschamp excursion. Indirect evidence for this interpretation is provided by the absence of a second excursion that might otherwise be identified with the Laschamp excursion in the lower part of the section.

Benson *et al.* (2003) argued against the conclusions of Kent *et al.* (2002). First, they noted that Kent *et al.* (2002) assumed a reservoir age of 1000 years, while previous work by Benson *et al.* (1990) had concluded that the carbonates deposited in the Mono Basin during the past 650 calendar years exhibit a reservoir effect ranging from 1100 to 5300 years. The radiocarbon dates obtained by Kent *et al.* (2002) may

therefore have provided overestimates of the age of deposition. In addition, Benson *et al.* (2003) took advantage of the unique chemical composition of Ash #15, which allows it to be distinguished from other tephra layers outside the Mono Basin, and dated this layer in the Pyramid Lake Basin. The advantage of this strategy is that the total organic carbon (TOC) fraction in the Pyramid Lake Basin is mostly composed of algae that obtain their carbon from dissolved CO_3^{2-} and HCO_3^{-}. Therefore, while the radiocarbon ages from the Pyramid Lake Basin still incorporate reservoir effects, they are not likely to be seriously contaminated by modern carbon. Based on a previous estimate of the reservoir effect of around 600 years in the last 3000 years, Benson *et al.* (2003) concluded that the age of Ash #15 is 28 620 ± 300 years (not corrected for the reservoir age). Benson *et al.* (2003) therefore concluded that the Mono Lake excursion recorded at Wilson Creek occurred in the 31.5–33.3 ka interval, which is consistent with the estimated age of the Mono Lake excursion in the Irminger Basin (Channell, 2006), with the paleointensity minimum of this age in NAPIS-75 (Laj *et al.*, 2000), and with the peak in ^{36}Cl flux observed in the GRIP ice core record (Wagner *et al.*, 2000). The age of the excursion reported at the Mono Lake type locality remains controversial. The uncertainty in age makes this excursion difficult to distinguish from the Laschamp excursion (~41 ka).

10.2.4 The Blake Excursion

Smith and Foster (1969) first defined the Blake Event from a paleomagnetic study of four deep-sea cores recovered from the Blake Outer Ridge. Their study established the existence of a short interval of reverse polarity in the later part of the Brunhes Chron. The paleomagnetic declinations could not be reliably determined in these piston cores, but the event was clearly revealed by anomalous inclinations, reaching $-10°$ to $-70°$, compared to an expected inclination at this location of $+40°$. On the basis of the position of the episode within the biostratigraphy of Ericson *et al.* (1961), and the age estimate by Broecker *et al.* (1968), the boundaries of the Blake Event were placed at 108 and 114 ka. Subsequently, the oxygen isotope data of Broecker and Van Donk (1970) indicated that the Blake excursion occurred within marine isotope stage (MIS) 5.

Denham and Cox (1971) and Denham (1976) provided corroborating evidence for the Blake Event from the Greater Antilles Outer Ridge and the Blake-Bahama Outer Ridge. These studies also provided the first evidence that the Blake Event might consist of two short intervals of almost reverse polarity separated by a short interval of almost normal polarity. The evidence for the Blake Event in these early studies came from a limited geographic area and the reality of the Blake excursion as a global phenomenon remained questionable. Denham (1976) considered the possibility of a 'local' reversal. He designed a model involving a small dipole source antiparallel to the main dipole and situated near the core–mantle boundary, directly beneath the 'Blake region'. This model accounted for a reversal in the vicinity of the zone where the Blake excursion had been documented, but affecting only a small portion of the globe.

Subsequently, Creer et al. (1980) reported evidence for the Blake Event in clays from Gioia Tauro in southern Italy, where the double structure of the episode was also apparent. Verosub (1982) pointed out that the record of the Blake excursion is characterized by two reverse intervals separated by a short normal interval in records of the Blake excursion from Italy, Japan (Lake Biwa record), and the North Atlantic Ocean, thereby providing evidence for the global character of this complex directional characteristic.

Tucholka et al. (1987) reported a paleomagnetic and oxygen isotope study of five cores from the eastern and central Mediterranean that could be intercorrelated using sapropel occurrences. The quality of the results from cores from the central Mediterranean (sampled several years after coring) was not high, but the eastern Mediterranean cores yielded unambiguous results. In all three eastern Mediterranean cores, a change to negative inclinations was observed, accompanied by reverse declinations in two cores and a smaller declination change in the third. The excursion was recorded in sediment directly overlying sapropel S5. In two of the cores, the record is probably incomplete, as no underlying normal polarity directions were found above the sapropel layer. One record, on the other hand, shows the bipartite structure that is characteristic of the Blake Event in the North Atlantic, Italy, and Japan. Detailed oxygen isotope analyses, made on four of the cores, established the stratigraphic position of the Blake Event between MIS 5e and MIS 5d. Estimates of the duration of the event, based on constant sediment accumulation rate

between tie points, varied in the 2.8–8.6 ky range, with a mean of 5.3 ± 2.7 ky.

Tric et al. (1991) reported a detailed record of the Blake Event from two Mediterranean cores, one from the Tyrrhenian Sea and the other being one of the cores already studied by Tucholka et al. (1987) that was resampled at much higher resolution. In the Tyrrhenian Sea core, only five excursional (intermediate) directions were recorded possibly due to a lower sediment accumulation rate synchronous with the Blake excursion. On the other hand, the eastern Mediterranean core yielded much more detailed results, with 70 intermediate and reverse polarity directions, making this record of the Blake Event the most detailed obtained so far, and the only one for which intermediate VGPs were obtained (**Figure 7**). Although no record of field intensity was obtained, the directional variations provide insight into the complex dynamics of the Blake Event. First, a sudden jump of inclination from positive to negative values was observed, accompanied by intermediate declinations and ending with a recovery of normal polarity. The VGPs in this first phase, which lasted about 1100 years according to the age model, lie over North and South America followed by an abrupt jump to Australia. The excursion in this core is characterized by two intervals of reverse polarity directions with an intervening normal polarity period. The total duration of the excursion, based on the isotope stratigraphy, appears to be close to 5 ky.

Thouveny et al. (2004) observed two intervals of negative magnetization inclination in the 115–122 ka

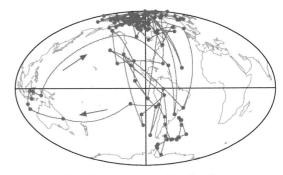

Figure 7 Transitional magnetization directions represented as virtual geomagnetic polar (VGP) paths for the Blake excursion recorded in Core MD84-627 from the Eastern Mediterranean. Data from Tric E, Laj C, Valet J-P, Tucholka P, Paterne M, and Guichard F (1991) The Blake geomagnetic event: Transition geometry, dynamical characteristics and geomagnetic significance. *Earth and Planetary Science* Letters 102: 1–13.

interval of core MD95-2042 from the Portuguese Margin. Oxygen isotope stratigraphy in this core places this record of the Blake excursion in MIS 5c according to table 2 of Thouveny *et al.* (2004), although it appears to span the MIS 5d/5e boundary in their figure 12. The latter designation would be consistent with the findings of Tucholka *et al.* (1987).

Zhu *et al.* (1994) obtained a record of the Blake excursion from a loess section at Xining (western China). The sequence of directional changes is contained in paleosol S1, which corresponds to MIS 5 (An *et al.*, 1991; Li *et al.*, 1992). Zhu *et al.* (1994) estimated the top and base of the studied section to be at 72 and 130 ka, and estimated ages of 117 and 111 ka for the onset and demise of the Blake excursion, respectively, yielding a duration estimate of 5.6 ky. The directional record of the excursion is characterized by stop-and-go behavior, reminiscent of the (Miocene) volcanic record at Steen's Mountain (Mankinen *et al.*, 1985). The Xining record contains three clearly defined periods of reverse polarity separated by two short intervals of normal polarity. The structure of this excursional record is more complex than has been previously reported for this excursion. The documented nonuniform timing of acquisition of magnetization for Chinese loess/paleosol sequences (e.g., Heslop *et al.*, 2000; Spassov *et al.*, 2003) may contribute to this apparent complexity, and to the fact that searches for the Blake Event in Chinese loess have not always been successful (Parès *et al.*, 2004).

Although the Blake excursion has also been documented in the Junzhoutai loess of the excursion section at Lanzhou (Fang *et al.* 1997), the structure is not the same as at Xining. At Lanzhou, it comprises two short reverse polarity intervals separated by a short normal polarity interval, which is similar to some records of the excursion in marine sediments. Based on thermoluminescence and astronomically tuned cycles of magnetic susceptibility, the age of the Blake Event was bracketed between Paleosol S1-c (equivalent to MIS 5e) and Loess 2-2 (MIS 5d) (Fang *et al.*, 1997). This corresponds to an age range of 120–115 ka, and a total duration of 5.5 ky, which is consistent with the estimate of Tucholka *et al.* (1987) from the Mediterranean Sea.

Observations from the highest resolution marine cores and from loess sections indicate that the Blake excursion is a global geomagnetic feature, with a characteristic structure comprising two short periods of almost reverse polarity separated by a short period of almost normal polarity.

10.2.5 The Iceland Basin Excursion

Over the last 10–15 years, many studies have provided evidence for geomagnetic excursions in the 180–220 ka interval. Labels Jamaica, Pringle Falls, and Biwa I have been used somewhat arbitrarily by different authors in referring to excursions with estimated ages in the 180–220 ka interval. As the fidelity and age control of available records has improved, evidence has accumulated for two excursions in the 180–220 ka interval. The younger one is labeled the Iceland Basin excursion (~185–190 ka), and the older one takes its name from Pringle Falls (~211 ka).

As mentioned above, paleomagnetic analyses of cores from Lake Biwa provided early evidence of several excursions during the late Brunhes Chron (Kawai *et al.*, 1972; Yaskawa *et al.*, 1973). The authors suggested that the youngest could be the Blake event, and two other episodes of reverse inclination were tentatively correlated to intervals of anomalous inclination in the record of Wollin *et al.* (1971) from the North Pacific Ocean. From this correlation, an age between 176 and 186 ka was assigned to an event that they called the Biwa I event. These early records have remained problematic due to inadequate paleomagnetic analysis by modern standards, and poor age control.

In the marine realm, records corresponding to this time interval obtained from Norwegian–Greenland Sea revealed several intervals of negative inclination (Bleil and Gard, 1989). These intervals have also been observed in a series of piston cores from further North in Fram Strait (Nowaczyk and Baumann, 1992) and on the Yarmak Plateau (Nowaczyk *et al.*, 1994). These early studies, however, lacked a well-defined $\delta^{18}O$ record, leading to uncertainties in assessing the precise age of the intervals of negative inclination.

Weeks *et al.* (1995) reported a paleomagnetic study of four piston cores from the North Atlantic Ocean. The age model was based on oxygen isotope stratigraphy from planktic foraminifera. The paleomagnetic record indicates that a large swing in inclination to negative values and a marked decrease in relative paleointensity occurred around 180–200 ka. Lehman *et al.* (1996) reported a paleomagnetic study of three marine cores in the Açores area in the North Atlantic Ocean. For one of the cores, a detailed $\delta^{18}O$ record was obtained, and the two others were correlated to it using sediment gray-scale reflectance data. The paleomagnetic record contains fluctuations in declination and inclination and a marked drop in paleointensity at ~190 ka, but no clear shift in the inclination. Nowaczyk and Antonow (1997) produced $\delta^{18}O$ and magnetostratigraphic records from the Greenland Basin that indicate the presence of negative inclinations with ages corresponding to the Mono Lake (27–28 ka) and Laschamp (~40 ka) excursions, with an additional excursion at around 188 ka. Nowaczyk and Antonow (1997) refer to this latter excursion as the 'Biwa I/Jamaica' excursion, thereby associating it with the excursions documented in early papers by Wollin *et al.* (1971), Ryan (1972), and Kawai *et al.* (1972). Roberts *et al.* (1997) documented an excursion in the North Pacific Ocean at ODP Site 884 (**Figure 4**). Preservation of foraminifera was poor at this site, so the age model was obtained by transferring the oxygen isotope stratigraphy from ODP Site 883 (Keigwin, 1995), by correlation of magnetic susceptibility records. However, the $\delta^{18}O$ record was not well defined at Site 883 in the vicinity of the excursion (MIS 6/7 boundary), so the age of the excursion is relatively poorly constrained.

Channell *et al.* (1997) reported $\delta^{18}O$ and paleomagnetic (directions and paleointensity) data from rapidly deposited ($\geq$10 cm ky^{-1}) sediments recovered at ODP Site 983 on the Gardar Drift in the Iceland Basin (**Figure 4**). Paleomagnetic analyses revealed, in the 186–189 ka interval, a short-lived excursion in which the VGPs move to high southern latitudes. The age of this excursion is constrained close to the MIS 6/7 boundary (~188 ka) and appears to be distinct from the Pringle Falls excursion (see

below). A paleointensity minimum with an onset age of 218 ka at ODP Site 983 may be coeval with the Pringle Falls excursion, although no directional excursion is recognized in association with this paleointensity minimum at this site. Channell *et al.* (1997) named this excursion the Iceland Basin Event (now Iceland Basin excursion) and this labeling is adopted here when referring to the geomagnetic excursion at the MIS 6/7 boundary. The Iceland Basin excursion appears to be coeval with the geomagnetic excursions recognized in the central North Atlantic Ocean by Weeks *et al.* (1995), as well as in the western equatorial Pacific Ocean (Yamazaki and Yoka, 1994).

The record of the Iceland Basin excursion from ODP Sites 984 (Channell, 1999) and 980 (Channell and Raymo, 2003), where mean sedimentation rates in the Brunhes Chron exceeded 11 cm ky^{-1}, is similar to that obtained at Site 983. At all three sites, inclinations are negative in the 180–195 ka interval, and the estimated duration of the excursion is ~3 ky. The apparent discrepancy in age at Sites 980, 983, and 984 can be attributed to uncertainties in the isotopic age model, particularly at Site 984 where the age model is not well defined. The VGP paths for the Iceland Basin excursion from Sites 983 and 984 are similar (**Figure 8**) and feature a large-scale counterclockwise loop that is located over Europe and Africa in the first N → S part of the loop, and returns to northern latitudes over the western Pacific Ocean. The directional changes coincide with a prominent paleointensity minimum that is a characteristic feature of sedimentary relative paleointensity records.

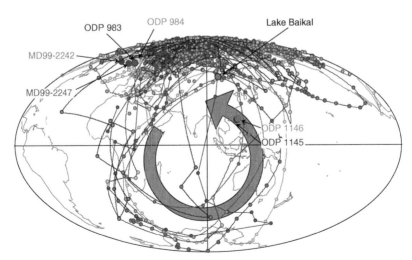

Figure 8 Transitional magnetization directions represented as virtual geomagnetic polar (VGP) paths for the Iceland Basin excursion. The large arrow illustrates the sense of looping of the VGPs which is consistently counterclockwise.

Oda *et al.* (2002) obtained a paleomagnetic record, including a paleointensity record, from core Ver 98-1 from Academician Ridge, Lake Baikal. The age model was derived using sediment bulk density as a proxy for biogenic silica content. Interglacial intervals were then correlated to maxima in silica (density). An age-depth plot was then produced based on SPECMAP ages (Martinson *et al.*, 1987) for isotope substages back to MIS 7.3. The average sedimentation rate was estimated to be around 4.5 cm ky^{-1}. A directional excursion in the paleomagnetic record between 6.70 and 6.96 m below lake floor was estimated to have an age of 177–183 ka. The excursion coincides with a marked minimum in the relative paleointensity proxy that can be correlated with the record from ODP Site 983. Oda *et al.* (2002) also redated an excursion from another core from Academician Ridge previously inferred to record the Blake event (Sakai *et al.*, 1997). The new age is 223 ka, which provides evidence for an excursion (possibly Pringle Falls) preceeding the Iceland Basin excursion in sediments from Lake Baikal.

The general pattern of the VGP paths for the Lake Baikal record is similar, albeit much more scattered than those observed at ODP Sites 983 and 984. After an initial swing over the western Atlantic Ocean (which does not appear in the ODP records), the VGPs move southward over Africa reaching high southern latitudes. The S → N return path is defined by only three points, all of which are in the Southern Hemisphere, which consequently makes the return path ill-defined. The overall counterclockwise loop is, however, similar to that observed from the two North Atlantic ODP cores.

Stoner *et al.* (2003) recorded the Iceland Basin excursion at ODP Site 1089 in the sub-Antarctic South Atlantic. Mean sedimentation rates are in the 15–20 cm ky^{-1} range for a record that extends back to ~580 ka. The relative paleointensity record from this site can be correlated to those from the North Atlantic and Lake Baikal that record the same excursion; the excursion occurs within the same prominent paleointensity low. At this site, the oxygen isotope age model yielded an age of 189–191 ka and a duration of ~2 ky for this excursion.

Laj *et al.* (2006) reported four records (directions and relative paleointensities) of the Iceland Basin excursion. Two of the records are from the North Atlantic, core MD99-2242 was taken on the Eirik Drift (south of Greenland), and core MD99-2247 was from the western flank of the Reykyanes Ridge

(**Figure 4**). The two other records were obtained from the South China Sea, at ODP Sites 1145 and 1146 (ODP Leg 184). Average sedimentation rates ranged from 7 to 10 cm ky^{-1} for cores in the Atlantic Ocean, and ~15 cm ky^{-1} for cores from the South China Sea. The excursion can be correlated to the MIS 6/7 boundary. The VGP paths are uniform for each of the dispersed site locations, and are consistent with the paths obtained from ODP Sites 983 and 984 (**Figure 8**). During the first part of the excursion, the VGPs move southward over Africa along a narrow band of longitudes before crossing the equator and reaching high southern latitudes (fully reverse polarity directions). The VGP path for core MD99-2242 is less well resolved, reaching mid-latitudes in the Indian Ocean. For all four cores, the VGP return path to the Northern Hemisphere lies within a longitudinal band over east Asia.

Channell (2006) recorded the Iceland Basin excursion at ODP Site 919 in the Irminger Basin (off east Greenland). According to the oxygen isotope age model, the excursion occurs in the 180–188 ka interval. Its manifestation at this site is considerably more complex than other records of the same excursion. Although the sediments have all the attributes for high fidelity recording, and the excursion is recorded by both discrete samples and u-channel samples, the excursion is characterized by a first VGP loop to high southern latitudes characterized by an outward and return path over Africa, followed by a second VGP loop with an outward path over east Asia and a return path over Africa. These two VGP loops are followed by a complex group of VGPs at low latitudes over western Africa. In view of the uniformity in magnetic hysteresis properties through the excursion, the complexity of the VGP path cannot be readily attributed to lithological variability. There are two feasible explanations for the discrepancy between this record and those compiled by Laj *et al.* (2006): either unrecognized sediment deformation has affected parts of this record or excursional records are more complex when fully recorded at some (high latitude) locations.

10.2.6 The Pringle Falls Excursion

A detailed record of a geomagnetic excursion obtained from a sedimentary lacustrine sequence near Pringle Falls (Oregon) was initially thought to represent the Blake Event (Herrero-Bervera *et al.*, 1989). Subsequently, however, the age of this episode was revised on the basis of ^{40}Ar/^{39}Ar dating of plagioclase feldspars from an ash layer (Ash D) located close to the base of the excursion (Herrero-Bervera *et al.*, 1994). The isochron age and the plateau age obtained from step heating were not significantly different, but the isochron age had a larger relative uncertainty, resulting from low radiogenic yield. The authors, therefore, took the plateau age of 218 ± 10 ka to be the best estimate of the age of the Ash D layer, which corresponds to the onset of the excursion. Herrero-Bervera *et al.* (1994) associated this excursion with the Jamaica excursion of Wollin *et al.* (1971) and Ryan (1972). By modern standards, the Jamaica excursion was not adequately resolved either paleomagnetically or stratigraphically in these early publications. Following the practice of naming geomagnetic excursions after the location where they have been unambiguosly recorded, we advocate use of Pringle Falls as the name for this excursion.

Herrero-Bervera *et al.* (1994) used the chemical and petrographic characteristics, stratigraphic position, and available age data to correlate the tephra from Pringle Falls to tephra layers present in other sequences in western North America, including Summer Lake, Mono Lake, and Long Valley. Coeval paleomagnetic records of excursions from each of these localities had been previously published (Negrini *et al.*, 1998, 1994; Liddicoat and Bailey, 1989; Liddicoat, 1990; Herrero-Bervera and Helsley, 1993), but only after this correlation of tephras were they recognized as records of the same excursion. The paleomagnetic records from two

sites at Pringle Falls (the second one being at a distance of 1.5 km from the original one) and from Long Valley, some 700 km away, are strikingly similar (**Figure 9**). The VGP paths lie over the Americas in the first N → S part of the excursion and then move to the northwest Pacific for the S → N return path (Herrero-Bervera *et al.*, 1994).

Subsequently, a series of transitional paleomagnetic directions were obtained from the Mamaku ignimbrite in the North Island of New Zealand (Tanaka *et al.*, 1996). Three new $^{40}Ar/^{39}Ar$ ages from plagioclase from the Mamaku ignimbrite (Houghton *et al.*, 1995) yielded a weighted mean of 223 ± 3 ka, which is statistically indistinguishable from the age of Ash D at Pringle Falls according to Herrero-Bervera *et al.* (1994). McWilliams (2001) plotted 29 VGPs for the Mamaku ingimbrite together with those from the two sites at Pringle Falls and at Long Valley. The agreement in time and space is remarkable (**Figure 9**). The VGPs from volcanic rocks in New Zealand record only a fraction of the total excursion and are not nearly as complete as the records from Pringle Falls or Long Valley, as expected due to the episodic nature of volcanic eruption. Nonetheless, this remarkable agreement suggests that the excursion was manifest globally, and that the transitional field was dominated by a dipolar field component (McWilliams, 2001).

Singer *et al.* (2005) recently reported 16 laser incremental heating ages from plagioclase crystals derived

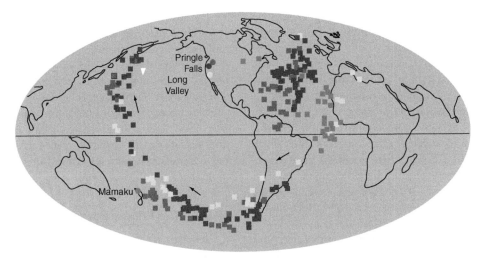

Figure 9 Transitional magnetization directions represented as virtual geomagnetic polar (VGP) paths for the Pingle Falls excursion. Squares are individual VGPs for the event recorded at Pringle Falls (two records), Long Valley, and New Zealand. Circles are sampling sites, color-coded to match respective points on the VGP paths. Arrows indicate the progression of the VGP paths from older to younger. From McWilliams M (2001). Global correlation of the 223 ka Pringle falls event. *International Geological Reviews* 43: 191–195.

from Ash D (deposited during the onset of the excursion) at Pringle Falls, that define an ^{40}Ar/^{39}Ar isochron of 211 ka ±13 ka. At ODP Site 919, off east Greenland, Channell (2006) recorded an excursion that predates the Iceland Basin excursion and, according to the stable isotope age model, correlates to the 206–224 ka age interval. The VGP path has a complex series of loops to equatorial latitudes over the Pacific and South America, followed by a clockwise loop to high southern latitudes. This last loop is in the same sense (clockwise) and follows a similar path as those recorded at Pringle Falls, Long Valley, and New Zealand for supposedly the same excursion. On the other hand, there is no trace in the Pringle Falls and Long Valley records of the initial complex structure seen at ODP Site 919, despite the high sediment accumulation rates at these two localities (the excursion is recorded over stratigraphic intervals of 100 cm at Long Valley and over 700 cm at Pringle Falls).

Well-grouped excursional site mean magnetization directions (mean declination = 101°, mean inclination = −36°) were obtained from 65 sites collected from the six major flows that represent the Albuquerque Volcanoes Field (Geissman et al., 1990). Age determinations using the ^{238}U–^{230}Th method yielded an isochron age of 156 ka ± 29 ka (Peate et al., 1996). More recent ^{40}Ar/^{39}Ar analyses in the Albuquerque volcanics yield an isochron age of 211 ± 22 ka (Singer et al., 2005) implying that the excursional magnetization directions in the Albuquerque Volcanoes Field are coeval with the Pringle Falls excursion in Oregon.

10.2.7 Excursions in the Early Brunhes Chron

Although fewer studies have been conducted for the Brunhes Chron prior to 250 ka, there is evidence for geomagnetic excursions in the early part of the Brunhes Chron. The use of different names, combined with imprecise age control, has led to confusion in labeling and correlation.

Ryan (1972) defined the so-called Emperor Event as a short reverse polarity zone at the base of Caribbean core V12-22. Based on the work of Ericsson et al. (1961) and Broecker and Van Donk (1969), the age of this interval was estimated to be about 460-480 ka. The excursion was, however, defined by only one sample that was only demagnetized in peak alternating fields of 5–15 mT. Nevertheless, some support for the excursion was found in the study of axial MMA records from the Galapagos spreading center that suggested a brief reverse polarity event at 490 ± 50 ka (Wilson and Hey, 1981). This age depends on the assumption of linear spreading rate of the Galapagos Ridge and is therefore not robust. Interestingly, the apparent occurrence of the Emperor Event on the Galapagos Ridge MMA data provides the basis for the one/only 'cryptochron' within the Brunhes Chron in the timescale of Cande and Kent (1992a).

As mentioned above, initial dating of a reverse polarity flow in Idaho at 490 ± 50 ka (Champion et al., 1981) provided land-based support for the existence of the Emperor Event. Later, however, Champion et al. (1988) revised the age of these reverse polarity lavas to 565 ± 10 ka which therefore corresponds to a different (older) reverse episode, which they named the Big Lost excursion. In their seminal paper, Champion et al. (1988) reviewed evidence for excursions from both volcanic and sedimentary sequences and concluded that at least eight excursions existed during the Brunhes Chron. In addition to the newly discovered Big Lost excursion and the well-known excursions discussed above, they presented evidence for: (1) the δ excursion, first documented by Creer et al. (1980) at Gioia Tauro in Southern Italy, dated at 640 ka; (2) the Emperor excursion (~470 ka) (Ryan, 1972); (3) the Biwa III excursion (~390 ka); and the (4) Levantine excursion (or Biwa II) (~290 ka) (Ryan, 1972; Yaskawa et al., 1973; Kawai et al., 1972; Kawai, 1984).

Langereis et al. (1997) documented four short excursions in piston core KC-01B collected from the Calabrian Ridge in the Ionian Sea (Mediterranean Sea). Astronomical calibration of the age sapropels, recognized on the basis of rock-magnetic and geochemical properties, allowed development of an age model, according to which the ages of the four excursions are 255–265 ka, 318 ± 3 ka, 515 ± 3 ka, and 560–570 ka. Langereis et al. (1997) labeled these excursions Calabrian Ridge 0 (CR0), CR1, CR2, and CR3. The authors proposed that CR0 could correspond to the Fram Strait excursion (Nowaczyk et al., 1994), while the oldest one could be the Big Lost excursion (which they associated with the Emperor excursion). The complex redox conditions in the multicolored sediments of core KC-01B (see Langereis et al., 1997), and the fact that the CR excursions are defined by single samples, indicates that they require independent corroboration prior to their incorporation into the library of geomagnetic excursions.

Biswas *et al.* (1999) reported a magnetostratigraphy of a 1700 m core from the Osaka Bay, southwestern Japan, spanning the last 3.2 My. Although the mean sedimentation rate in the Brunhes Chron (50 cm ky^{-1}) is far higher in these coastal sediments than in deep marine sequences, only one excursion is recorded in the Brunhes Chron. This Brunhes Chron excursion is recorded in lacustrine silty clays that are immediately overlain by marine clays, and is characterized by intensity fluctuations with steep inclinations (up to +72° and −82°). It occurred during marine isotopic stage (MIS) 17, probably during substages 17.4–17.3, based on a sea-level interpretation of the marine/terrestrial sequence. This would correspond to an excursion age of 690 ka. The same age is obtained from the assumption of uniform sedimentation rate between the Matuyama/Brunhes (M/B) boundary and a characteristic tuff layer (Aira Tuff dated at 24.5 ka). Biswas *et al.* (1999) named this excursion the 'Stage 17 Event'. They correlated it to the δ event reported by Creer *et al.* (1980) from Gioia Tauro. The age of the δ event can be revised to 680 ka, using the modern value of 780 ka for the M/B reversal. Interestingly, the Stage 17 excursion is manifest in Osaka Bay as a double excursion with steep negative inclinations and an intervening interval with positive inclinations. An excursion within MIS 17 has also been recognized at ODP Site 980 collected from the Feni drift, North Atlantic (Channell and Raymo, 2003). The oxygen isotope age model yields an excursional age of 687–696 ka within MIS 17, and an excursional duration of 9 ky. Paleomagnetic inclinations reach low negative values accompanied by a ∼180° swing in declination.

Lund *et al.* (2001a, 2001b) documented many apparent excursions in the Brunhes Chron in marine cores taken from sediment drifts in the Western North Atlantic Ocean (Blake Outer Ridge, Bahama Outer Ridge and Bermuda Rise) during ODP Leg 172. From initial shipboard measurements, made on half-cores with a low-resolution pass-through cryogenic magnetometer, 14 geomagnetic episodes were identified as 'plausible' Brunhes Chron magnetic field excursions (Lund *et al.*, 1998). Laboratory measurements conducted on u-channels and discrete samples documented narrow intervals of relatively low or high inclination or westerly and easterly declinations, which could be traced among the records from independent holes separated by distances <1 km and were correlated using variations in magnetic susceptibility. Shore-based measurements

led Lund *et al.* (2001a, 2001b) to confirm 12 of the 14 originally defined excursions, one of which appears to be synchronous with the Stage 17 excursion. The large number of geomagnetic excursions in the Brunhes Chron at ODP Sites 1060–1063 (ODP Leg 172) indicates that they are not rare, episodic disturbances of an otherwise stable geomagnetic field, but an integral component of the field. In addition, Lund *et al.* (2001b) proposed that most of the excursions tend to occur in bundles of two or three close together separated by intervals of 'regular' secular variation. This observation, in addition to highlighting an important characteristic of the geomagnetic field, may also explain why there has been considerable difficulty in identifying and distinguishing among near-coeval excursions (i.e., Iceland Basin – Pringle Falls) prior to the availability of high-resolution stratigraphy.

Lund *et al.* (2001b) stated, "for almost any previously identified excursion anywhere in the world, we can find an excursion record in Sites 1060–1063 that is not significantly different in age". Here lies the problem: the ODP Leg 172 sites lack continuous oxygen isotope data that would aid precise age model construction. The most recent age models for ODP Leg 172 sediments were determined by tuning filtered records of carbonate percentage, derived from gray-scale reflectance records calibrated with shipboard and postcruise carbonate measurements, to the astronomical solution for precession and obliquity (Grützner *et al.*, 2002). The age models are more robust for the shallow water sites (Sites 1055–1059) than for the deeper water sites (Sites 1060–1063). Precession-related cycles are weak, particularly in the MIS 6-7 interval and prior to 0.5 Ma for the deeper water sites (Grützner *et al.*, 2002). The tuning was performed on one 'reference' site from each group (Sites 1058 and 1062), and the other sites in the group were correlated to these reference sites using the filtered and unfiltered carbonate records. The correlation and labeling of ODP Leg 172 excursions would be substantially aided by age models based on oxygen isotope data.

Some claims for new excursions in the early Brunhes Chron remain controversial, particularly those from lava flows where precise radiometric dating is often lacking and critical to adequate correlation among excursions. For instance, Quidelleur and Valet (1996) conducted a paleomagnetic study in the Barranco de los Tilos, La Palma (Canary Islands) at a location previously studied by Abdel-Monem *et al.* (1972). Initially, they proposed that transitional magnetization directions from the southern side of the Barranco represented post-transitional rebound associated with the M/B boundary reversal. Later, on the basis of three transitionally magnetized lavas that gave a mean unspiked K/Ar age of 602 ± 24 ka, Quidelleur *et al.* (1999) recognized a new, significantly younger episode and proposed the name La Palma excursion. In support of this claim, Quidelleur *et al.* (1999) noted that the proposed excursion coincides with a marked minimum in the SINT-800 paleointensity stack of Guyodo and Valet (1999). Subsequently, Singer *et al.* (2002) obtained ^{40}Ar/^{39}Ar ages of 580.7 ± 7.8 ka for the same transitional flows studied by Quidelleur *et al.* (1999). This age is indistinguishable at the 95% confidence level from the ^{40}Ar/^{39}Ar age (558 ± 20 ka) obtained by Lanphere *et al.* (2000) for the Big Lost excursion in the Snake River Plain.

Table 2 provides a summary of excursions within the Brunhes Chron. Those entries in normal type are considered (by us) as adequately defined, whereas the entries in italics await further ratification either in definition of the magnetic data or in refinement of the age models.

10.3 Geomagnetic Excursions in the Matuyama Chron

10.3.1 Background

The age and structure of the predominantly reverse polarity Matuyama Chron has been progressively refined since the coupled K/Ar and paleomagnetic studies on basaltic lavas that began with the work of Cox *et al.* (1963) and McDougall and Tarling (1963a, 1963b, 1964). The Olduvai Subchron takes its name from normal polarity lavas dated at 1.72 Ma from the Olduvai Gorge, Tanzania (Grommé and Hay, 1963). The Jaramillo Subchron was first recognized by Doell and Dalrymple (1966) and takes its name from Jaramillo Creek (New Mexico). The Réunion Subchron originates from the work of Chamalaun and McDougall (1966) who found both normal and reverse magnetizations in basaltic rocks yielding K/Ar ages close to 2.0 Ma from the island of La Réunion. These normal polarity directions were, at that time, considered to be coeval with those from Olduvai Gorge as documented by Grommé and Hay (1963). McDougall and Watkins (1973) sampled two basaltic sections on La Réunion and documented a normal polarity zone dated by K/Ar methods to the 1.95–2.04 Ma interval, which corresponds to

~2.07 Ma using more modern decay constants (Steiger and Jager, 1977). By the early 1970s, it was realized that the Réunion Event is significantly older than the Olduvai Subchron. Grommé and Hay (1971) considered that a bimodal distribution of K/Ar ages for normally magnetized lavas with ages of ~2.00–2.14 Ma from a variety of locations indicated the existence of two Réunion Events, although there was, and continues to be, no evidence for two events within any single stratigraphic section (see review in Channell *et al.*, 2003a). In their compilation of the GPTS for the last 5 My, Mankinen and Dalrymple (1979) adopted the two Réunion Events proposed by Grommé and Hay (1971) and estimated their ages as 2.01–2.04 and 2.12–2.14 Ma, respectively.

Mankinen *et al.* (1978) detected normal polarities in 1.1 Ma volcanics from Cobb Mountain (Coso Range, California); however, Mankinen and Dalrymple (1979) were not sufficiently confident in a normal polarity subchron of this age to include it in their timescale. Evidence for this normal polarity subchron was strengthened by further study of volcanics in the Coso Range by Mankinen and Grommé (1982). Subsequent observation in high sedimentation rate sediment cores from the Carribbean (Kent and Spariosu, 1983) and North Atlantic (Clement and Kent, 1987; Clement and Martinson, 1992) cemented the Cobb Mountain Subchron as a feature of the Matuyama Chron.

The GPTS of Mankinen and Dalrymple (1979), based on coupled K/Ar and paleomagnetic studies of basaltic lavas (**Figure 10** and **Table 3**), remained the reference for the 0–5 Ma GPTS for over 10 years. Beginning in the 1980s, astrochronologies from sedimentary sequences demonstrated that the K/Ar ages for polarity reversals in volcanic rocks compiled by Mankinen and Dalrymple (1979) were young by an average of about 7% (due to argon loss). The first astrochonological evidence that the generally accepted K–Ar age for the M/B boundary (0.73 Ma) was too young can be attributed to Johnson (1982) who gave an age of 0.79 Ma for the M/B reversal based on matching the orbital insolation curve to oxygen isotope records from two cores (V28-238/9) that record the M/B reversal. It was, however, the study of ODP Site 677 in the equatorial Pacific Ocean (Shackleton *et al.*, 1990) that opened the door to astrochronological revision of the polarity timescale. Benthic oxygen isotope data at ODP Site 677 are dominated by orbital obliquity, and the planktic record is controlled by orbital precession. Because the precession signal is modulated by eccentricity, the planktic oxygen isotope data at

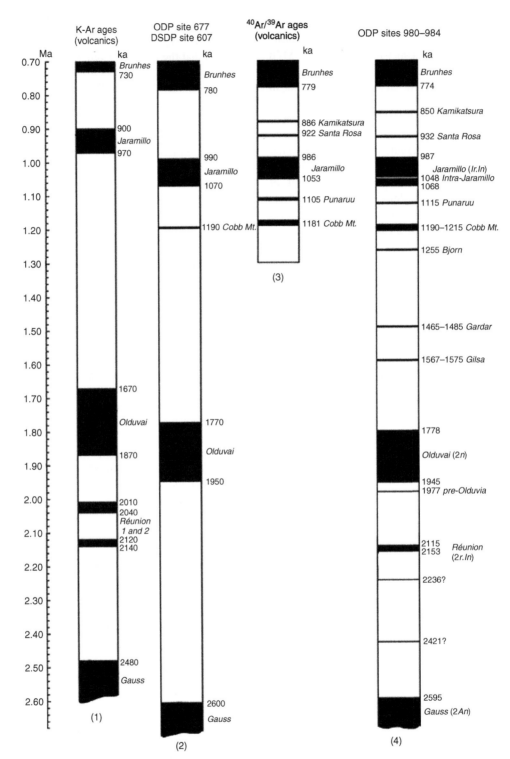

Figure 10 The geomagnetic polarity timescale for the Matuyama Chron. From left: (1) coupled K–Ar and paleomagnetic studies summarized by Mankinen and Dalrymple (1979); (2) ODP Site 677 and DSDP Site 607 (Shackleton *et al.*, 1990) as adopted by Cande and Kent (1995); (3) $^{40}Ar/^{39}Ar$ and paleomagnetic studies of volcanics (summarized by Singer *et al.*, 2004); (4) ODP Sites 980–984 (Channell and Kleiven, 2000; Channell *et al.*, 2002, Channell and Guyodo, 2004).

Table 3   Excursions within the Matuyama Chron

Excursion or subchron	Label	MIS Sites 607, 609, 677 Ref. 1	MIS Site 659 Ref. 2	MIS Italy Ref. 3	MIS (Sites 980–984) Ref. 4	Age (ka) (Sites 980–984) Ref. 4	Duration (kyr) (Sites 980–984) Ref. 4	Age (ka) ^{40}Ar/^{39}A Ref. 5
base Brunhes	base 1n	base 19			19	774		791
Kamakitsura					21	850	?	899
Santa Rosa					top 25	932	3	936
top Jaramillo	top 1r.1n	mid 27	27		base 27	985	8	1001
	1r.1n.1r				base 30	1050	3	
base Jaramillo	base 1r.1n	mid 31	31		base 31	1070	5	1069
Punaruu					mid 34	1115	5	1122
top Cobb. Mt.		base 35			base 35	1190		
Cobb Mt.							35	1194
base Cobb Mt.		base 35			top 37	1125		
Bjorn					top 38	1255	3	
Gardar					49	1465–1485	20	
Gilsa		53			54	1567–1575	8	
top Olduvai	top 2n	base 63	64	64	63	1778	4	1775
base Olduvai	base 2n	base 71	72	71	base 71	1945	5	1922
pre-Olduvai					top 73	1977	3	
Huckleberry Ridge					75	2040		2086
top Reunion	top 2r.1n	79		81	79	2115		
Reunion							38	2137
base Reunion	base 2r.1n			81	81	2153		
pre-Reunion 1					85/86	2236	?	
pre-Reunion 2					95	2421	?	
top Gauss		104			base 103	2595		

MIS: marine isotope stage

Ref. 1: Ruddiman et al. (1989); Raymo et al. (1989); Shackleton et al. (1990). Ref. 2: Tiedemann et al. (1994). Ref. 3: Lourens et al. (1996). Ref. 4: Channell and Kleiven (2000); Channell et al. (2002); Channell et al. (2003a); Channell and Guyodo (2004). Ref. 5: summarized in Singer et al. (2004).

ODP Site 677 can be matched to astronomical data with more confidence than the obliquity signal at this site or at DSDP Site 607 (Ruddiman et al., 1989) where the number of obliquity cycles within the Brunhes Chron was underestimated by a single cycle, thereby inadvertently supporting the K–Ar M/B boundary age (0.73 Ma). Although ODP Site 677 played a pivotal role in revamping the GPTS for the Matuyama Chron, the site itself did not yield a polarity stratigraphy. The Cobb Mt. Subchron, the M/B and Gauss–Matuyama (G/M) boundaries, and the boundaries of the Jaramillo and Olduvai subchronozones, were recorded at DSDP Site 607 (Clement and Robinson, 1987), and their ages were determined by transferring astrochronological ages from ODP Site 677 to DSDP Site 607 using oxygen isotope correlations between the two sites (Shackleton et al., 1990) (**Figure 10** and **Table 3**).

The sequence of reversals of the Matuyama Chron in the CK95 differs from that in the timescale of Mankinen and Dalrymple (1979). In CK95, one (as opposed to two) normal polarity subchron comprises the Réunion Subchron. CK95 adopted the astrochonological age estimates for Pliocene–Pleistocene polarity reversals (Shackleton et al., 1990; Hilgen, 1991a, 1991b). These original astrochronological age estimates have generally stood the test of further astrochronological dating of sediments during the last 15 years (see **Table 3** and **Figure 10**). The exception is the age of the Réunion Subchron in CK95 (2.14–2.15 Ma with 10 ky duration, derived from Hilgen (1991a, 1991b)), which should now be amended to 2.115–2.153 Ma (38 ky duration) based on data from North Atlantic high-sedimentation-rate drift sites (Channell et al., 2003a).

Soon after the K–Ar ages for Pliocene–Pleistocene polarity reversals (Mankinen and Dalrymple, 1979) were superceeded by astrochronological determinations, a large number of $^{40}Ar/^{39}Ar$ ages confirmed the astrochronological ages of polarity chrons. For example, $^{40}Ar/^{39}Ar$ age determinations for the M/B boundary and boundaries of the Jaramillo Subchron (Izett and Obradovich, 1991; Spell and McDougall, 1992; McDougall *et al.*, 1992; Tauxe *et al.*, 1992; Baksi *et al.*, 1993), for the Cobb Mountain Subchron (Turrin *et al.*, 1994), for the Réunion Subchron (Baksi *et al.*, 1993) and for the boundaries of the Olduvai Subchron (Walter *et al.*, 1991; Baksi, 1994) were all close to the astrochronological estimates. The exercise becomes somewhat academic in view of the suggestion of Renne *et al.* (1994) that the $^{40}Ar/^{39}Ar$ standard (Fish Canyon sanadine, Mmhb-1) should be calibrated using the astrochronological ages of polarity reversals.

In CK92/95, 'tiny wiggles' in MMA data, interpreted either as brief polarity chrons (with duration <30 ky) or paleointensity fluctuations, were labeled 'cryptochrons'. Fifty-four 'cryptochrons' were recognized over the last 83 My, since the middle of the Late Cretaceous. In CK92/95, two 'cryptochons' are listed within the Matuyama Chron at 1.201–1.212 Ma (Cobb Mt. Subchron) and 2.420–2.441 Ma (Anomaly X in Heirtzler *et al.* (1968)). As outlined above, the Cobb Mt. Subchron is now a well-established feature within the Matayama Chron, whereas Anomaly X has yet to be well established in magnetostratigraphic records.

The Jaramillo, Réunion, and Olduvai normal polarity subchrons were identified almost 40 years ago. Since then, starting with the unequivocal identification of the Cobb Mountain Subchron in marine sediments (e.g., Clement and Kent, 1987), up to nine additional normal polarity excursions, and one reverse polarity excursion within the Jaramillo Subchron, have now been identified within the Matuyama Chron (**Figure 10** and **Table 3**).

10.3.2 Excursions between the Gauss–Matuyama Boundary and the Réunion Subchron

At ODP Site 982, two intervals with anomalous magnetization directions are observed in MIS 85/86 and MIS 95 (**Figure 10** and **Table 3**) (Channell and Guyodo, 2004). The older of the two coincides in age

(2420 ka) with Anomaly X from the MMA data of Heirtzler *et al.* (1968) and hence with the 'cryptochron' of this age featured in CK92/95.

10.3.3 Huckleberry Ridge

Reynolds (1977) recorded anomalously shallow inclinations and southwestward-directed declinations in paleomagnetic data from 57 sites representing 23 separate localities in the Huckleberry Ridge Tuff (Yellowstone group). The anomalous directions were considered by Reynolds (1977) to record a polarity transition or an excursion, possibly associated with the Réunion Subchron. The tuff has recently been dated at 2.06 Ma using $^{40}Ar/^{39}Ar$ methods (Lanphere *et al.*, 2002), indicating that the transitional directions are younger than the Réunion Subchron for which the mean $^{40}Ar/^{39}Ar$ age is 2.14 Ma (Baksi and Hoffman, 2000; Baksi *et al.*, 1993; Roger *et al.*, 2000; Singer *et al.*, 2004). At ODP Site 981, low inclinations are recorded in MIS 75/76, stratigraphically above the Réunion Subchronozone which occurs in MIS 79–81 (Channell *et al.*, 2003a). It remains to be confirmed whether the anomalous directions within MIS75/76 correlate to the Huckleberry Ridge excursion as recorded in the Huckleberry Ridge Tuff.

10.3.4 Gilsa

The name Gilsa, in a geomagnetic context, owes its origin to McDougall and Wensink (1966) who detected two normal polarity flows separated by a reverse polarity flow within the lower part of the Matuyama Chronozone at Jokuldalur (Iceland). They assigned the older normal polarity flow to the Olduvai Subchronozone and the younger normal polarity interval, dated at 1.60 Ma, was named the Gilsa Event. Watkins *et al.* (1975) subsequently sampled the same sections and measured normal polarity directions in superposed lava flows that yielded ages of 1.58 and 1.67 Ma, but with no intervening reverse polarity interval. More recently, Udagawa *et al.* (1999), from work on five separate sections in the Jokuldalur region, confirmed the existence of a reverse polarity flow below the 1.60 Ma normal polarity flow and above the normal polarity flows associated with the Olduvai Chronozone. This work supports the conclusion of McDougall and Wensink (1966) that the Gilsa excursion (at

~1.60 Ma) is an interval of normal polarity distinct from the underlying Olduvai Chronozone.

In view of the Watkins *et al.* (1975) study, Mankinen and Dalrymple (1979) were not sufficiently confident to include the Gilsa Event in their timescale, nor was it included in CK95 as a 'cryptochron' as it is not evident in MMA records. The presence of a normal polarity interval at ~1.55 Ma in two holes at DSDP Site 609, however, confirmed its existence (Clement and Kent, 1987). This was corroborated at ODP Sites 983 and 984, where the correlative normal polarity excursion correlates to MIS 54 at 1567–1575 ka (**Table 3**), implying duration of 8 ky (Channell *et al.*, 2002).

10.3.5 Gardar and Bjorn

At ODP Sites 983 and 984, two normal polarity excursions are present between the Gilsa excursion zone and the Cobb Mountain subchronozone (Channell *et al.*, 2002). These two normal polarity intervals which occur in MIS 49 (Gardar excursion) and MIS 38 (Bjorn excursion), with estimated durations of 8 ky (Gardar) and 3 ky (Bjorn), have not been detected elsewhere, presumably due to their brief duration and paucity of deep-sea cores of this age with sufficiently high sedimentation rates. The names for these excursions are derived from the locations of Sites 984 (Bjorn Drift) and 983 (Gardar Drift); however, both excursions are present at both sites.

10.3.6 Cobb Mountain

Mankinen *et al.* (1978) documented a normal polarity site in the Alder Creek rhyolite at Cobb Mountain (California) which has yielded an $^{40}Ar/^{39}Ar$ age of 1186 ka (Turrin *et al.*, 1994). Further work by Mankinen and Grommé (1982) in the Cosa Range (California) supported the existence of the Cobb Mountain Event. DSDP Site 609 in the central North Atlantic Ocean provided the first unequivocal documentation of a normal polarity zone of similar age in deep-sea sediments (Clement and Kent, 1987). At Site 609, the so-called Cobb Mountain Subchron can be correlated to MIS 35/36 (Ruddiman *et al.*, 1989) (**Table 3**). This subchron has also been recognized at ODP Hole 647B in the southern Labrador Sea (Clement and Martinson, 1992), in the Celebes and Sulu Seas (Hsu *et al.*, 1990; Clement, 1992), in the Lau Basin (Abrahamsen and Sager, 1994), in New Zealand (Pillans *et al.*, 1994), off the California

Margin (Guyodo *et al.*, 1999; Hayashida *et al.*, 1999), in the western Philippine Sea (Horng *et al.*, 2002, 2003), on the Bermuda Rise (ODP Leg 172) (Yang *et al.*, 2001), and at ODP Sites 980 and 983/984 where it correlates to MIS 35 at ~1.2 Ma and has an estimated duration of 35 ky (Channell *et al.*, 2002; Channell and Raymo, 2003).

10.3.7 Punaruu

The Punaruu excursion originates from the Punaruu Valley (Tahiti) where normal polarity magnetizations were recorded in basaltic lava flows stratigraphically below, and distinct from, the Jaramillo Subchronozone (Chauvin *et al.*, 1990). The Punaruu excursion in the type section on Tahiti has yielded an age of 1105 ka using $^{40}Ar/^{39}Ar$ methods (Singer *et al.*, 1999). This excursion appears in the sediment record from ODP Site 1021 (California Margin) where an age of 1.1 Ma is based on assumed uniform sedimentation rate between the M/B boundary and the top of the Jaramillo Subchronozone (Guyodo *et al.*, 1999). At both sites 983 and 984, the same excursion lies within MIS 34, which corresponds to an age of 1115 ka (Channell *et al.*, 2002).

10.3.8 Intra-Jaramillo Excursion (1r.1n.1r)

An intra-Jaramillo excursion has been detected in the Jingbian loess sequence from Northern China within loess 10 (Guo *et al.*, 2002), in marine cores from New Zealand (Pillans *et al.*, 1994), and at ODP sites 983 and 984 where it is correlated to MIS 30 at 1048 ka (Channell and Kleiven, 2000; Channell *et al.*, 2002).

10.3.9 Santa Rosa

Transitional magnetization directions from volcanic rocks at Cerro Santa Rosa I dome in New Mexico (Doell and Dalrymple, 1966; Doell *et al.*, 1968) were originally interpreted as recording the polarity transition at the end of the Jaramillo Subchron. $^{40}Ar/^{39}Ar$ ages (Spell and McDougall, 1992; Izett and Obradovich, 1994) from this dome are significantly younger than the end of the Jaramillo Subchron. More recently, anomalous (transitional) magnetization directions (mean: dec/inc: 103°/−63°, $\alpha_{95} = 9.5°$) and a mean $^{40}Ar/^{39}Ar$ age of 936 ka (Singer and Brown, 2002) support the existence of a Santa Rosa excursion. An excursion of similar age (932 ka) has been observed at ODP Sites 983 and 984 at the top of MIS 25

(Channell *et al.*, 2002) and at the same position in the western Philippine Sea (Horng *et al.*, 2002, 2003).

10.3.10 Kamikatsura

The name 'Kamikatsura excursion' originates from the work of Maenaka (1983) who documented excursional magnetization directions in the Kamikatsura Tuff of the Osaka group (SW Japan). The existence of this excursion is supported by a 0.83 Ma normal-polarity flow from near Clear Lake (Mankinen *et al.*, 1981), and was promoted by Champion *et al.* (1988) in their review of Matuyama–Brunhes excursions. More recently, Takatsugi and Hyodo (1995) documented another excursion in marine clays about 10 m above the Kamikatsura Tuff, and 1 m above another tuff (Azuki Tuff) that has yielded a K–Ar age of 0.85 Ma. The Kamikatsura and Azuki Tuffs are separated by about 10 m of pebbly sand in the Osaka group (see Takasugi and Hyodo, 1995) and rapid deposition of this facies may mean that the two excursional intervals record a single geomagnetic excursion. On the other hand, two excursions with age estimates of 0.89 and 0.92 Ma have been detected in loess 9 (L9), which corresponds to MIS 22 according to the loess chronology of Heslop *et al.* (2000), in the Baoji loess section in southern China (Yang *et al.*, 2004). This observation follows earlier work in the Lishi and Luochuan regions (China) where an apparent excursion, also within L9, has been recorded (Wang *et al.*, 1980; Liu *et al.*, 1985). Although excursional directions in loess deposits have often been dismissed as remagnetizations in view of the probable delay of remanence acquisition of Chinese loess (see Spassov *et al.*, 2003), an interval of anomalously low VGP latitudes also occurs at ODP Site 983 in MIS 21 at about 850 ka (Channell and Kleiven, 2000). An interval of negative (reverse) inclination in equatorial Pacific core KK78O30 between the M/B boundary and the Jaramillo Subchronozone (Laj *et al.*, 1996) was assigned to the Kamikatsura excursion. Its age (based on uniform sedimentation rate between the M/B boundary and the top Jaramillo Subchronozone) is closer to that of the Santa Rosa excursion, which was not recognized at the time. Low VGP latitudes from volcanics on Maui and Tahiti associated with the Kamikatsura excursion yield mean ages of 866 ka (Singer *et al.*, 1999). Coe *et al.* (2004) associated a 25 m thick interval of anomalous declinations on Maui with

the Kamikatsura excursion, and gave a weighted mean $^{40}Ar/^{39}Ar$ age of 900.3 ± 4.7 ka for this interval. To add to the confusion in this interval, a transitionally magnetized flow from La Palma (Canary Islands) has yielded a K–Ar age of 821 ka (Quidelleur *et al.*, 2002). There is clearly much to be done to resolve the behavior of the geomagnetic field in this interval immediately prior to the M/B boundary. The relationship of the Kamikatsura excursion and paleointensity minimum immediately prior to the M/B boundary (Kent and Schneider, 1995; Hartl and Tauxe, 1996) remains to be determined.

10.4 Geomagnetic Excursions in Pre-Matuyama Time

In the Gauss and Gilbert chrons, there are no well-documented excursions in magnetostratigraphic records, nor are there 'tiny wiggles' in MMA data denoting 'cryptochrons' of this age. Based on 'tiny wiggles' in MMA data, CK92/95 included four 'cryptochrons' in the Late Miocene, 18 in the Oligocene, 3 in the Late Eocene, and 23 in the Paleocene and Early Eocene. These 'tiny wiggles' can often be correlated between ship's tracks, and they have been thought to represent fluctuations in geomagnetic field intensity (Cande and LaBrecque, 1974; Cande and Kent, 1992b) and/or short polarity intervals (e.g., Blakely and Cox, 1972; Blakely, 1974).

From magnetostratigraphic studies of sedimentary sequences, there is evidence for ~14 polarity excursions in the Miocene (**Figure 11**), four of which correlate with cryptochrons in CK92/95. The Miocene magnetostratigraphic record, like that in the Brunhes Chron, implies that there are far more polarity excursions than are represented in the cryptochron record from MMA data. On the other hand, whereas 44 Paleocene to Oligocene (Paleogene) cryptochrons are given in CK92/95, only three excursions have been documented in the Paleogene, and only one of these appears to correlate with a cryptochron in CK92/95.

In the Middle Cretaceous, at the base of the Cretaceous Long Normal interval, the existence of reverse polarity subchrons younger than CM0 has been advocated (e.g., Tarduno *et al.*, 1992) although their existence remains controversial. Here, the uncertainty is due to the lack of continuous sedimentary sections recording both CM0 (at the base of the

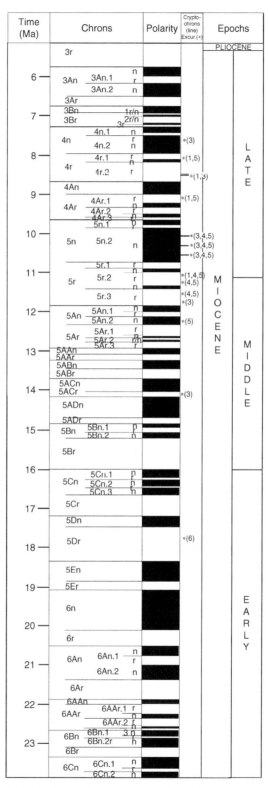

Figure 11 Miocene geomagnetic polarity timescale of Cande and Kent (1995) with cryptochrons from Cande and Kent (1992a) and additional polarity subchrons/excursions according to various authors: 1 = Schneider (1995), 2 = Acton *et al.* (2006), 3 = Evans and Channell (2003), Evans *et al.* (2004), 4 = Abdul Aziz and Langereis (2004), 5 = Krijgsman and Kent (2004), 6 = Channell *et al.* (2003b).

Cretaceous Long Normal) and the younger proposed subchrons, and the inability of $^{40}Ar/^{39}Ar$ age dating to clearly distinguish these subchrons in discontinuous volcanic sequences.

10.4.1 C5n.2n (Late Miocene)

Late Miocene polarity subchron C5n.2n in the CK92/95 GPTS includes three 'cryptochrons' first recorded as 'tiny wiggles' in MMA profiles from the North Pacific Ocean (Blakely, 1974). Over the last 25 years, 'tiny wiggles' in MMA records of C5n.2n have been represented as polarity subchrons in several versions of the GPTS notably those of Ness et al. (1980) and Harland et al. (1982, 1990). They were not included in the GPTS by Lowrie and Alvarez (1981) or Berggren et al. (1985) because of the lack of confirmation for polarity subchrons of this age from magnetostratigraphy.

A single track of deep-tow magnetic anomaly data, covering polarity subchron C5n.2n, was collected in 1998 at 19°S on the flanks of the East Pacific Rise (Bowers et al., 2001). The half-spreading rate at this site was estimated to have been 9 cm yr^{-1} for the Late Miocene. This profile revealed more than twice as many 'tiny wiggles' as seen in profiles from the North Pacific region, where Blakely (1974) first identified the four short wavelength anomalies within C5. Bowers et al. (2001) identified a number of short wavelength magnetic anomalies within C5n.2n, three of which were interpreted as correlative to the three cryptochrons in CK92/95. Bowles et al. (2003) addressed the question as to whether these tiny wiggles within C5n.2n represent polarity subchrons/excursions or paleointensity fluctuations. These authors compared the directional and paleointensity record from C5n.2n in the sedimentary sequence at ODP Site 887 with the deep-tow record of Bowers et al. (2001). The absence of directional excursions in the magnetostratigraphic record of C5n.2n led Bowles et al. (2003) to conclude that the tiny wiggles in MMA records of C5n.2n were produced by paleointensity fluctuations rather than directional excursions or brief polarity subchrons.

The magnetic polarity stratigraphy at ODP Site 884, on the slopes of the Detroit Seamount (NW Pacific), was initially interpreted from the shipboard pass-through magnetometer measurements of split half-cores (Weeks et al., 1995). The C5n part of the section was reinterpreted by Roberts and Lewin-Harris (2000) who recognized two short polarity subchrons and one excursion within a polarity zone correlative to C5n.2n (see figure 3 in Roberts and Lewin-Harris, 2000). The excursion and subchrons were estimated to have durations of 6, 23, and 28 ky, respectively, assuming constant sedimentation rates in C5n.2n.

At ODP Site 1092, in the sub-Antarctic South Atlantic, the stratigraphic interval correlative to C5n.2n is 34 m thick with a mean sedimentation rate of 3.3 cm ky^{-1}. At this site, three polarity excursions are recognized within C5n.2n, all of which have estimated durations <10 ky (Evans and Channell, 2003; Evans et al., 2004). The presence of directional excursions within C5n.2n has recently been supported by measurements of new discrete samples collected at DSDP Site 608 (Krijgsman and Kent, 2004), although at this site the excursions correlated to C5n.2n are represented by single samples with shallow negative inclinations.

Continental records of short polarity events in C5n.2n have been reported from NE Spain (Garces et al., 1996), Bolivia (Roperch et al., 1999), and western China (Li et al., 1997). The section from the Pyrenees in NE Spain consists of a composite section from two terrestrial sequences that cover the interval from 8.7 to 11.1 Ma (Garces et al., 1996). The magnetic stratigraphy can be interpreted on the basis of polarity zone pattern fit to the GPTS aided by the first occurrence datum of Hipparion. The polarity zone correlative to C5n.2n has a thickness of 175 m and includes a ~5 m polarity zone in its upper part that was correlated with cryptochron C5n.2n-1 in CK92/95. A 4.5 km thick composite section of red beds from the Bolivian Altiplano produced a magnetic stratigraphy for the 9–14 Ma interval (Roperch et al., 1999). The polarity zone correlative to C5n.2n is 1 km thick in this section and contains a single 5 m thick polarity zone that was correlated with cryptochron C5n.2n-1. The interpretation of the polarity stratigraphy is supported by $^{40}Ar/^{39}Ar$ age determinations on two tuff layers within the section, one of which falls within C5n.2n. An upper Cenozoic section in western China recorded three short reverse polarity zones within C5n.2n (Li et al., 1997). The polarity interpretation at the site is fairly unambiguous, and is corroborated by the presence of vertebrate fossils. The events are, however, only represented by single discrete samples. The authors correlate these events with the three cryptochrons in CK92/95.

10.4.2 Other Miocene Excursions/Subchrons

In sediments from the equatorial Pacific Ocean recovered during ODP Leg 138, a normal polarity subchron was recognized within C4r.2r (Schneider, 1995) that correlates to polarity subchrons recorded in sediments from ODP Site 1092 in the South Atlantic (Evans and Channell, 2003; Evans *et al.*, 2004) and from ODP Site 1095 off the Antarctic Peninsula (Acton *et al.*, 2006), as well as to a cryptochron in CK92/95 (**Figure 1**). Independent verification from multiple sites argues strongly for an excursion or brief subchron within C4r.2r (**Figure 11**).

Three other subchrons or excursions within C4r.1r, C4Ar.1r, and C5r.2r (**Figure 11**) were apparent from the study of ODP Leg 138 sediments (Schneider, 1995). All three of these subchrons (or excursions) are apparently supported by recent data from a resampling of DSDP Site 608 (Krijgsman and Kent, 2004). One of these subchrons/excursions, that within C5r.2r, is also supported by the studies of Abdul-Aziz and Langereis (2004) in continental sediments from NE Spain. Abdul-Aziz and Langereis (2004) documented a total of three subchrons within C5r (**Figure 11**). The oldest of the C5r excursions (that within C5r.3r) may correlate with a subchron found within the same polarity chron at ODP Site 1092 (Evans and Channell, 2003). At ODP Site 1092, yet another newly recognized subchron has been observed within chron C5ACr (reinterpreted from the initial placement within C5AAr by Evans and Channell, 2003).

The plethora of Late and Middle Miocene subchrons recognized in stratigraphic sections clearly out-numbers the cryptochrons identified in CK92/95 derived from MMA data (**Figure 11**). At least 13 subchrons that were not included in the CK92/95 polarity chron template have been documented in stratigraphic section since the publication of CK95. Four of 13 newly recognized subchrons correlate with cryptochrons in CK92/95.

10.4.3 Oligocene and Eocene

Three 'cryptochrons' appear in C13r in the CK92/95 GPTS (**Figure 12**) based on 'tiny wiggles' in MMA data. Bice and Montanari (1988) found one normal polarity sample in the top third of C13r at Massignano, Italy, in the Eocene–Oligocene boundary stratotype section. Lowrie and Lanci (1994)

resampled the Massignano section and concluded that there were no normal polarity intervals within C13r. This conclusion appears to be corroborated by the study of cores recovered from a borehole drilled at Massignano (Lanci *et al.*, 1996). At ODP Site 1090, in the sub-Antarctic South Atlantic Ocean, the high sedimentation rates in C13r appear to have facilitated the recording of a short subchron represented by a 3 m thick normal polarity zone, implying a duration for the C13r.1n subchron of 79 ky, which is

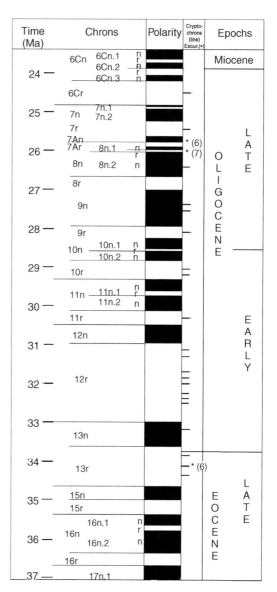

Figure 12 Late Eocene–Oligocene geomagnetic polarity timescale of Cande and Kent (1995) with cryptochrons from Cande and Kent (1992a) and additional polarity subchrons/excursions according to: 6 = Channell *et al.* (2003b) and 7 = Lanci *et al.* (2005).

comparable to the duration of the Jaramillo Subchron (Channell *et al.*, 2003b).

Apart from the apparent subchron within C13r, two additional polarity subchrons not included in the CK92/95 GPTS are observed at ODP Site 1090 (Channell *et al.*, 2003b). One of these (within C5Dr) is listed as a cryptochron in table 7 of CK92, and was included as a subchron in versions of the GPTS prior to CK92/95. This normal polarity subchron was originally identified in the North Pacific MMA stack (Blakely, 1974). Two other normal polarity subchrons have been documented in the Late Oligocene (**Figure 12**). One of these occurs within C7Ar and is recorded at ODP Site 1090 (Channell *et al.*, 2003b) and another occurs within C8n.1n at ODP Site 1218 in the equatorial Pacific Ocean (Lanci *et al.*, 2005). Neither of these subchrons has been recognized in other magnetostratigraphic records, or in MMA records.

10.4.4 Middle Cretaceous

The record of polarity subchrons in the Middle Cretaceous begins with the so-called ISEA reversal that was originally recognized in a bed of reddish Aptian limestone along a road outcrop in the Umbrian Apennines (Italy) by VandenBerg *et al.* (1978). Lowrie *et al.* (1980) resampled this anomalous limestone bed, and confirmed the reverse polarity magnetization but were unable to substantiate the existence of this reverse polarity zone in other coeval Umbrian sections. Some 10 years later, Tarduno (1990) documented reverse polarity magnetizations in two samples within the *G. algerianus* foraminiferal zone of the middle Aptian at DSDP Site 463. Documentation of this short polarity subchron is strengthened by two samples with reverse magnetization spanning a 43 cm interval at ODP Site 765 (Ogg *et al.*, 1992). Reverse magnetizations in basalts from the Tarim Basin (China) yielding $^{40}Ar/^{39}Ar$ whole rock and plagioclase fraction ages of 113 and 119 Ma, respectively (Sobel and Arnaud, 2000), have been associated with ISEA (Gilder *et al.*, 2003). According to some timescales (e.g., Channell *et al.*, 1995a), an age of 119 Ma is less than 2 My younger than CM0 (the youngest reverse polarity chron of the M-sequence) at the Aptian/Barremian boundary. On the other hand, in the timescale of Gradstein *et al.* (2004), an age span of 113–119 Ma would correspond to the Late Aptian, with 119 Ma corresponding closely to the supposed age of ISEA. An andesitic lava sequence from Liaoning Province (China) also yields

reverse magnetization directions and $^{40}Ar/^{39}Ar\pm$ ages of 116.8 Ma $\pm$ 3.0 Ma (Zhu *et al.*, 2004) and these authors associate the reverse magnetizations either to ISEA or to CM0.

Tarduno *et al.* (1992) documented seven reverse polarity zones in the middle Albian of the Contessa section (Umbria, Italy). The cyclostratigraphy of Herbert *et al.* (1995) applied to the Contessa section implies a duration of ~800 ky for the thickest of these polarity zones. As this duration is greater than the estimated duration of CM0, the polarity chron at the base of the Aptian that is easily recognized in MMA records, it is unlikely the Albian polarity zones documented in the Contessa section represent reverse polarity chrons. As the reversely magnetized limestones in the Albian at Contessa are reddish, and the NRM is partly carried by hematite, the hematite magnetizations may be recording a Late Cretaceous or younger magnetization rather than an Albian one. On the other hand, reverse polarity magnetizations have been retrieved from sediments of Albian age recovered during ODP Leg 171 although these varicolored sediments indicate extensive iron mobilization and probably indicate a (Late Cretaceous) remagnetization (Ogg and Bardot, 2001).

The template for the Late Jurassic and Early Cretaceous polarity sequence (CM0–CM25) stems from the MMA records published by Larson and Hilde (1975). Correlation of this sequence of polarity chrons to biozonations/stage boundaries and numerical ages has been reviewed by Channell *et al.* (1995a). The only modification of this template in the intervening 30 years since publication of Larson and Hilde (1975) has been the identification of a second reverse polarity subchron between CM11 and CM12. The two reverse polarity subchrons within CM12n have been recognized in MMA records (Tamaki and Larson, 1988) and in magnetostratigraphic sections in Italy (Channell *et al.*, 1987; Channell *et al.*, 1995b).

10.5 Duration of Geomagnetic Excursions

Precise evaluation of the duration of geomagnetic excursions has become a point of interest since the proposal of Gubbins (1999) that, during excursions, the geomagnetic field reverses polarity in the Earth's liquid outer core but that this outer-core field does not persist for long enough in a reversed polarity state for diffusion of the field into the solid inner core. Diffusion times of ~3 ky for the inner core, therefore, provide a prediction for excursion duration. It is not trivial to determine excursion durations of a few kiloyears in the geologic record. Only in cases where astronomical tuning is possible to the level of orbital precession can we expect durations of the order of thousands of years to be adequately resolved. Interpolation, assuming constant sedimentation rates, among tie points matching an oxygen isotope record to an (astronomically tuned) isotope target curve is unlikely to provide duration estimates of sufficient precision. Radiometric (^{10}Ar/^{39}Ar or K/Ar) ages are also unlikely to have realistic uncertainties within a few kiloyears, limiting their utility for estimating the duration of excursions.

Nonetheless, for the Laschamp excursion, Laj *et al.* (2000) presented a duration estimate of 1500 years based on correlation of marine cores to the GISP2 (Greenland) layer-counted ice-core record. Here, correlation among the NAPIS-75 cores was achieved through correlation of the susceptibility records (Kissel *et al.*, 1999), and the oxygen isotope record from one of the cores provided a correlation to GISP2 (Voelker *et al.*, 1998). Approximately, the same duration estimate was obtained from each of the five NAPIS-75 cores that record the excursion. As these cores are spread over a distance of about 5000 km, it is unlikely that the duration estimate is affected by local changes in the sediment deposition rate at the time of the excursion. Further evidence for the duration of the Laschamp excursion is obtained from the record of flux of ^{36}Cl in the GRIP ice core. Wagner *et al.* (2000) showed that variations in ^{36}Cl flux (assumed to be entirely due to modulation by the geomagnetic field) are similar (in inverse sense) to changes in geomagnetic field intensity (e.g., **Figure 6**). A duration estimate of 1500 years for the ^{36}Cl anomaly, corresponding to the paleointensity minimum associated with the Laschamp excursion, supports the excursion duration determined from paleomagnetic records. The ^{36}Cl flux ice-core record also indicates that a second peak, associated with the Mono Lake excursion, has approximately the same duration, about 1500–2000 years (on the GRIP–GISP age model).

Although none of the sedimentary records containing the Laschamp excursion are accompanied by primary astrochronologic tuning to orbital solutions, there is good consistency in estimates of excursion duration. For example, Lund *et al.* (2001a, 2005) estimated a duration of 2 ky for the Laschamp excursion using age models based on AMS radiocarbon ages from Keigwin and Jones (1994). A similar duration is

apparent for records of the Laschamp excursion in the South Atlantic Ocean (Channell *et al.*, 2000) and the southern Indian Ocean (Mazaud *et al.*, 2002). In the Irminger Basin, both the Mono Lake and Laschamp excursions have apparent durations of ~1 ky (Channell, 2006). Similarly, duration estimates for the Iceland Basin excursion are 2–3 ky for the records that have oxygen isotope age control (e.g., Channell *et al.*, 1997; Channell, 1999; Stoner *et al.*, 2003; Laj *et al.*, 2006). A longer duration (~8 ky) for the Iceland Basin excursion recorded at ODP Site 919 (Channell, 2006) may be attributable to inadequacy of the oxygen isotope age model. A somewhat longer duration has been documented for the Blake Event, with values of 5–8 ky (Tric *et al.*, 1991). These duration estimates are about 2–3 times larger than most of those observed for the Laschamp, Mono Lake, and Iceland Basin excursions, although the duration estimates are heavily dependent on the quality of the individual age models.

In general, longer duration estimates for geomagnetic excursions have come from high-latitudes cores from the Atlantic and Arctic Oceans. In the work of Nowaczyk and Antonow (1997) and Nowaczyk and Knies (2000), excursional durations of >10 ky are implied by the age models. The structure of the excursions is also different, with rapid transitions, and distinct periods of time with fully reversed polarity VGPs. As mentioned above, the results from the Arctic cores are anomalous in that, for the last 100 ky, the reverse polarity intervals occupy almost 50% of the recovered sedimentary sequence, which is obviously much greater than expected. It has been suggested that these anomalously long duration estimates for excursions in this region is due to large increases in sedimentation rate during the excursions, which would be particularly noticeable at high latitudes, since, as noted by Worm (1997), many excursions seem to occur preferentially during cold or cooling climatic stages. According to Clement's (2004) review, reversal duration at the Matuyama–Brunhes boundary, and at the boundaries of the Jaramillo and Olduvai subchrons, varies with latitude and falls in the 2–10 ky range. This distribution of reversal durations with latitude is broadly consistent with a model in which nondipole fields are allowed to persist while the axial dipole decays to zero and then builds in the opposite direction. The apparent increase in excursion duration in high latitude cores (Nowaczyk and Antonow, 1997; Nowaczyk and Knies, 2000) appears to be greater than can be accommodated by such a model. King *et al.* (2005) reported preliminary results from Pleistocene sediments from the Lomonosov Ridge, Central Arctic Ocean, collected during IODP Leg 302. Excursional magnetization directions are apparently present during the early Brunhes Chron. However, these authors observe a strong correlation between rock magnetic variations, color changes and physical properties, and the observed excursions. They therefore attribute the observed Arctic paleomagnetic behavior to environmental controls rather than to anomalous geomagnetic behavior.

In summary, the duration estimates for the Laschamp, Mono Lake, and Iceland Basin excursions (**Table 2**) and for most Matuyama Chron excursions (**Table 3**) provide support for the excursion mechanism of Gubbins (1999) in that the excursion durations are estimated to be no more than a few kiloyears, which is comparable with the magnetic diffusion time of the inner core. This result is consistent with the concept that excursions can lead to a full polarity reversal and a subsequent prolonged polarity interval only if the excursion (with directions approximately antiparallel to the pre-excursional field) is maintained for times exceeding the magnetic diffusion time of the inner core.

10.6 Excursional Field Geometry

Determination of the field geometry during an excursion requires availability of multiple records from different and widely separated geographical locations. Only an incomplete picture can be obtained, because only two excursions, the Laschamp and Iceland Basin excursions, have been studied at distant localities from rapidly deposited sediments that yield detailed VGP paths (e.g., Laj *et al.*, 2006). For the Iceland Basin excursion (**Figure 8**), a consistent picture of VGP paths is obtained for widely distributed site locations. VGPs move first southward over Africa along a rather narrow band of longitudes before crossing the equator and reaching high southern latitudes. The return paths to the Northern Hemisphere are contained in a longitudinal band over eastern Asia. The overall picture is that of a large counterclockwise loop of the VGPs. A coherent picture is also observed for VGP paths for the Laschamp excursion (**Figure 5**). The first southward part of the paths passes over east Asian–western Pacific longitudes and then reaches high southern latitudes. The northward-directed part of the VGP paths proceeds over Africa and

Western Europe. For the Laschamp as well as for the Iceland Basin excursion, the turning point, where the VGPs change from being southward to northward, coincides with the minimum in relative paleointensity. In other words, we see no clear evidence for recovery of paleointensity within the excursion interval, although a recovery in paleointensity may be filtered by the DRM acquisition process.

There are similarities among the records obtained for the Laschamp and Iceland Basin excursions. For instance, although the sense of movement of the VGP path is opposite for the Iceland Basin (counterclockwise) and Laschamp (clockwise) excursions (**Figures 5** and **8**), the two sets of VGP paths pass over similar longitudinal bands. This suggests that a similar core–mantle boundary structure may have prevailed during the two excursions. In addition, the repetitive structure of the VGP paths for dispersed site locations can be taken as an indication of a simple, possibly dipolar, geometry for the transitional geomagnetic field for both excursions. The VGP paths and the intensity records for the two excursions are consistent with a decrease in strength of the axial dipole, a substantial transitional equatorial dipole, and a reduced nondipole field relative to the axial dipole. During the first, N → S part of the Iceland Basin VGP path, over Europe and Africa, the g_1^1 term of the equatorial dipole appears to be preponderant, while h_1^1 appears to be dominant during the S → N part of the excursion. For the Laschamp excursion, the opposite is true, with the first part dominated by h_1^1 and the second part dominated by g_1^1 (Laj *et al.*, 2006).

These results do not support the view that the geomagnetic intensity minima that coincide with directional excursions reflect the emergence of nondipole geomagnetic components (e.g., Merrill and McFadden, 1994; Guyodo and Valet, 1999). If nondipole fields were dominant, one should observe widely different VGP paths at the different sites, which is not the case here, nor was it the case in Clement's (1992) analysis of the Cobb Mountain subchron or in McWilliams' (2001) analysis of the Pringle Falls excursion. As originally pointed out by Valet and Meynadier (1993), the intensity of the geomagnetic field is substantially reduced during excursions. As an important corollary, the dominance of nonaxial dipolar fields during excursions implies that the amplitudes of the nondipole components may have been relatively reduced during excursions. Remarkably, the transitional VGPs appear to follow the same paths for the Laschamp and Icelandic Basin excursions which are separated in time by ~140 ky. This is obviously longer than the time constant associated with fluid motions in the outer core, and therefore might suggest a deep Earth (lower mantle) control on the excursional field geometry (Laj *et al.*, 2006). The Blake excursion, however, situated in time between the Laschamp and Iceland Basin excursions, does not appear to have the same transitional field geometry. For the Blake excursion, there is only a single record (Tric *et al.*, 1991) that incorporates enough transitional magnetization directions for the transitional VGP path to be mapped. The VGP path for this record lies over the Americas during the N → S part reaching high southern latitudes (**Figure 7**). The return path, although much less detailed, is situated over Australia and South Eastern Asia. The VGP path, albeit based on a single record, is thus different from that of the Laschamp and Iceland Basin excursions. For the Pringle Falls excursion, on the other hand, several records from different sites yield consistent results. As noted by McWilliams (2001), this is an indication that the transitional field had a large dipolar component during the Pringle Falls excursion. Curiously, the path for the Pringle Falls excursion is similar to that observed for the Blake excursion from the Mediterranean record of Tric *et al.* (1991).

10.7 Concluding Remarks

The paleomagnetic records for the two best-documented excursions (Laschamp and Iceland Basin excursions) imply that the field during excursions was characterized by a simple transitional geometry. The axial dipole underwent a substantial decrease in strength, while equatorial dipoles were apparently relatively enhanced during the excursions. Contrary to a commonly held view, the nondipole field may have undergone a decrease during excursions. Although the sense of looping of the VGP paths is opposite for the Iceland Basin and Laschamp excursions, the VGPs for the two excursions (separated in time by ~140 ky) followed the same path, which suggests a repetitive lower mantle control on excursional field geometry. Some excursions, notably the Blake excursion (Tric *et al.*, 1991; Zhu *et al.*, 1994) and some records of the Iceland Basin and Pringle Falls excursions (Channell, 2006), appear to be characterized by multiple VGP swings to high latitudes. Such field instability may be explained by the observation

that the critical Reynolds number for the onset of core convection is very sensitive to the poloidal field, and the strength of core convection varies wildly in response changes in magnetic field strength particularly during intensity minima (Zhang and Gubbins, 2000).

The duration of excursions in the Brunhes Chron (**Table 2**), as well as excursions with the Matuyama Chron (**Table 3**), based mainly on constraints from oxygen isotope stratigraphy, is usually estimated to be <5 ky. This duration is comparable with the ~3 ky timescale for diffusive field changes in the Earth's solid inner core, which must reverse polarity in order for a full geomagnetic reversal to be sustained. The fact that our estimate of excursion duration is comparable with the ~3 ky time constant for inner core magnetic diffusion provides support for the suggestion by Gubbins (1999) that there is a mechanistic distinction between polarity reversals, that define polarity chrons, and excursions. Nevertheless, the similarity of some VGP paths of excursions and reversal transitions bounding polarity chrons suggests an inherent link between the mechanisms that give rise to geomagnetic excursions and reversals. For instance, the East Asian longitudinal bands identified in the excursional VGP paths coincides with one of the preferred longitudinal bands for transitional VGP paths during reversals (Laj *et al.*, 1991; Clement, 1991), and are featured in VGP clusters that appear in volcanic and high-resolution sedimentary records of reversal transitions (e.g., Hoffman, 1992; Channell and Lehman, 1997; Channell *et al.*, 2004). The longitudinal band over western Europe and Africa, on the other hand, is less evident in compilations of sedimentary VGP reversal paths; however, this region does feature in a recent compilation of transitional VGP paths from volcanic rocks (**Figure 13**; Valet and Herrero-Bervera, 2003).

An increasing number of relative paleointensity records, based on normalized remanence data from marine sediments, are now available for the Matuyama Chron (Valet and Meynadier, 1993; Meynadier *et al.*, 1995; Kok and Tauxe, 1999; Hayashida *et al.*, 1999; Guyodo *et al.*, 1999, 2001; Channell and Kleiven, 2000; Dinares-Turell *et al.*, 2002; Channell *et al.*, 2002; Horng *et al.*, 2003). The sedimentary sequences that have yielded high-quality relative paleointensity records only rarely capture the directional changes associated with excursions. Presumably, paleointensity minima are captured more readily because the paleointensity features are longer lasting than the accompanying directional

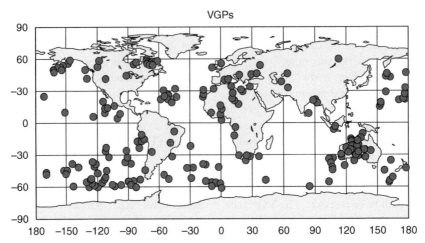

Figure 13 Polarity transition VGPs with latitudes lower than 60° obtained from volcanic sites with large global distribution. Adapted from Valet J-P and Herero-Bervera E (2003) Some characteristics of geomagnetic reversals inferred from detailed volcanic records. *Comptes Rendus Geoscience* 335: 79–90.

excursions, and are perhaps less susceptible to magnetic overprinting and the filtering affect of a finite magnetization lock-in zone during DRM acquisition. The short duration of directional excursions (**Tables 2 and 3**) result in them being rarely recorded in sediments with mean sedimentation rates of less than several cm ky^{-1}. It has often been noted that

the ages of paleointensity minima in paleointensity records correspond to the excursions found elsewhere (e.g., Valet and Meynadier, 1993). As an example of this correspondence, we show the ODP Site 983 paleointensity record for the 750–1800 ka interval (**Figure 14**) with the position of polarity reversals and excursions as recorded in the same

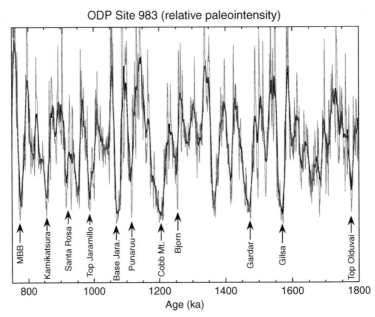

Figure 14 ODP Site 983 relative paleointensity record (blue) with a 10 kyr running mean (black) of the same record. Ages of principal polarity reversals and excursions are indicated, and correspond to paleointensity minima. Data from Channell JET and Kleiven HF (2000) and Channell *et al.* (2002).

sediment sequence (Channell and Kleiven, 2000; Channell et al., 2002).

In the next few years, paleomagnetic studies of rapidly deposited sediments will be combined with astrochronological age models, and coupled with $^{40}Ar/^{39}Ar$ dating and paleomagnetic studies from volcanic rocks. These will converge toward a concensus on the short-term (excursional) behavior of the geomagnetic field. At present, the majority of documented excursions is in the Brunhes Chron (**Figure 1**), the Matuyama Chron (**Figure 10**), and in the Late and Middle Miocene (7–14 Ma) interval (**Figure 11**). It remains to be seen whether these time intervals are representative of excursion frequency in general, or whether they remain exceptional, as the catalog of excursions expands. It is interesting to note that the high frequency (and short duration) of excursions in these intervals, if applied to the entire Cenozoic GPTS, would lead to a distribution of chron/excursion durations inconsistent with a Poisson distribution (see Lowrie and Kent, 2004), implying that polarity excursions are mechanistically distinct from polarity chrons.

The study of polarity excursions, today, is analogous to the study of polarity chrons 40 years ago when the principal polarity chrons of the last 5 My were in the process of being resolved by coupled K–Ar and paleomagnetic studies in volcanic rocks, and by studies of conventional piston cores from the oceans. The challenge of resolving brief excursions, that have remained obscure and poorly documented until very recently, has only just begun.

Acknowledgments

The authors thank M. McWilliams and J.-P. Valet for providing **Figures 9** and **13** respectively, and N. Nowaczyk and H. Oda for providing numerical data, respectively.

References

Abdel-Monem A, Watkins ND, and Gast PW (1972) Potassium–argon ages, volcanic stratigraphy, and geomagnetic polarity history of the Canary Islands: Tenerife, La Palma and Hierro. *American Journal of Science* 272: 805–825.

Abdul-Aziz H, Hilgen F, Krijgsman W, Sanz E, and Calvo JP (2000) Astronomical forcing of sedimentary cycles in the middle to late Miocene continental Calatayud Basin (NE Spain). *Earth and Planetary Science Letters* 177: 9–22.

Abdul-Aziz H, Krijgsman W, Hilgen FJ, Wilson DS, and Calvo JP (2003) An astronomical polarity timescale for the late Middle Miocene based on cyclic continental sequences. *Journal of Geophysical Research* 108: 2159 (doi:10.1029/2002JB001818).

Abdul-Aziz H and Langereis CG (2004) Astronomical tuning and duration of three new subchrons (C5r.2r-1n, C5r.2r-2n and C5r.3r-1n) recorded in a middle Miocene continental sequence in SE Spain. In: Channell JET, et al. (eds.) *AGU Geophysical Monograph, 145: Timescales of the Paleomagnetic Field*, pp. 141–160. Washington DC: AGU.

Abrahamsen N and Sager W (1994) Cobb mountain geomagnetic polarity event and transitions in three deep-sea sediment cores from the Lau Basin. In: Hawkins J, Parson L, Allan J, et al. (eds.) *Proceedings of ODP Science Results*, 135, pp. 737–762. College Station, TX: Ocean Drilling Program.

Acton G, Guyodo Y, and Brachfield S (2006) The nature of a cryptochron from a paleomagnetic study of chron C4r.2r recorded in sediments off the Antarctic Peninsula. *Physics of the Earth and Planetary Interiors* 156: 213–222.

An ZS, Kukla GJ, Porter SC, and Xiao J (1991) Magnetic suceptibility evidence of monsoon variation on the loess plateau of central China during the last 130,000 years. *Quaternary Research* 36: 29–36.

Anonymous (1977) Magnetostratigraphic polarity units. A supplementary chapter of the international subcommission on stratigraphic classification international stratigraphic guide. *Geology* 7: 578–583.

Baksi AK (1994) Concordant sea-floor spreading rates obtained from geochronology, astrochronology and space geodesy. *Geophysical Research Letters* 21: 133–136.

Baksi AK and Hoffman KA (2000) On the age and morphology of the Reunion event. *Geophysical Research Letters* 27: 2297–3000.

Baksi AK, Hoffman KA, and McWilliams M (1993) Testing the accuracy of the Geomagnetic polarity time-scale (GPTS) at 0-5 Ma utilizing $^{40}Ar/^{39}Ar$ incremental heating data on whole-rock basalts. *Earth and Planetary Science Letters* 118: 135–144.

Barbetti M and Flude K (1979) Paleomagnetic field strengths from sediments baked by lava flows of the Chaîne des Puys, France. *Nature* 278: 153–156.

Barbetti M and McElhinny MW (1972) Evidence of a geomagnetic excursion 30,000 yr BP. *Nature* 239: 327–330.

Baumgartner S, Beer J, Masarik J, Wagner G, Meynadier L, and Synal H-A (1998) Geomagnetic modulation of the ^{36}Cl flux in the GRIP Ice Core, Greenland. *Science* 279: 1330–1332.

Baumgartner S, Beer J, Suter M, et al. (1997) ^{36}Cl fallout in the Summit Greenland Ice Core Project ice core. *Journal of Geophysical Research* 102: 26659–26662.

Benson L, Liddicoat J, Smoot J, Sarna-Wojcicki A, Negrini R, and Lund S (2003) Age of the Mono Lake excursion and associated tephra. *Quaternary Science Reviews* 22: 135–140.

Benson LV, Currey DR, Dorn RI, et al. (1990) Chronology of expansion and contraction of four Great Basin lake systems during the past 35,000 years. *Palaeogeography, Palaeoclimatology, Palaeoecology* 78: 241–286.

Benson LV, Lund SP, Burdett JW, et al. (1998) Correlation of late Pleistocene lake-level oscillations in Mono Lake, California, with North Atlantic climate events. *Quaternary Research* 49: 1–10.

Bice DM and Montanari A (1988) Magnetic stratigraphy of the Massignano section across the Eocene/Oligocene boundary in the Marche–Umbria basin (Italy). In: Premoli-Silva I, Coccioni I, and Montanari A (eds.) *The Eocene/Oligocene Boundary in the Marche–Umbria Basin (Italy)*. International Subcommission on Paleogene Stratigraphy, Special Publication II, vol. 4, pp. 111–117. Ancona, Italy: International Subcommission on Paleogene Stratigraphy.

Billups K, Pâlike H, Channell JET, Zachos J, and Shackleton NJ (2004) Astronomic calibration of the late Oligocene through Early Miocene geomagnetic polarity time scale. *Earth and Planetary Science Letters* 224: 33–44.

Biswas DK, Hyodo M, Taniguchi Y, *et al.* (1999) Magnetostratigraphy of Plio-Pleistocene sediments in a 1700-m core from Osaka Bay, southwestern Japan and short geomagnetic events in the middle Matuyama and early Brunhes chrons. *Palaeogeography, Palaeoclimatology, Palaeoecology* 148: 233–248.

Blakely RJ (1974) Geomagnetic reversals and crustal spreading rates during the Miocene. *Journal of Geophysical Research* 79: 2979–2985.

Blakely RJ and Cox A (1972) Evidence for short geomagnetic polarity intervals in the Early Cenozoic. *Journal of Geophysical Research* 35: 7065–7072.

Bleil U and Gard G (1989) Chronology and correlation of quaternary magnetostratigraphy and nannofossil biostratigraphy in Norwegian–Greenland sea sediments. *Geology Rundschau* 78: 1173–1187.

Bonhommet N and Babkine J (1967) Sur la presence d'aimentation inverse dans la Chaine des Puys. *Comptes Rendus Hebdomadaires des Seances de l'Academie des Sciences Series B* 264: 92–94.

Bonhommet N and Zahringer J (1969) Paleomagnetism and potassium–argon determinations of the Laschamp geomagnetic polarity event. *Earth and Planetary Science Letters* 6: 43–46.

Bowers NE, Cande S, Gee J, Hildebrand JA, and Parker RL (2001) Fluctuations of the paleomagnetic field during chron C5 as recorded in near bottom marine magnetic anomaly data. *Journal of Geophysical Research* 106: 26379–26396.

Bowles J, Tauxe L, Gee J, McMillan D, and Cande S (2003) Source of tiny wiggles in Chron 5n: A comparison of sedimentary relative paleointensity and marine magnetic anomalies. *Geochemistry, Geophysics, Geosystems* 4: 6 (doi:10.1029/2002GC000489).

Broecker WS, Thurber DL, Goddard J, Ku T, Matthews RK, and Mesolella KJ (1968) Milankovitch hypothesis supported by precise dating of coral reefs and deep-sea sediments. *Science* 159: 297–300.

Broecker WS and Van Donk J (1970) Insolation changes, ice volumes, and the ^{18}O record in deep-sea cores. *Reviews of Geophysics and Space Physics* 8: 169–197.

Brunhes B (1906) Recherches sur la direction de l'aimantation des roches volcaniques. *Journal de Physique* 5: 705–724.

Burek PJ (1967) Korrelation revers magnatisierter Gesteinfolgen in Oberen Bandsandstein S.W. Deutschland. *Geologisches Jahrbuch* 84: 591–616.

Burek PJ (1970) Magnetic reversals: Their application to stratigraphic problems. American. *Association of Petroleum Geologists, Bulletin* 54: 1120–1139.

Cande SC and Kent DV (1992a) A new geomagnetic polarity timescale for the late Cretaceous and Cenozoic. *Journal of Geophysical Research* 97: 13917–13951.

Cande SC and Kent DV (1992b) Ultrahigh resolution marine magnetic anomaly profiles: A record of continuous paleointensity variations? *Journal of Geophysical Research* 97: 15075–15083.

Cande SC and Kent DV (1995) Revised calibration of the geomagnetic polarity timescale for the Late Cretaceous and Cenozoic. *Journal of Geophysical Research* 100: 6093–6095.

Cande SC and LaBrecque JL (1974) Behaviour of the Earth's paleomagnetic field from small scale marine magnetic anomalies. *Nature* 247: 26–28.

Chamalaun FH and McDougall I (1966) Dating geomagnetic polarity episodes in Reunion. *Nature* 210: 1212–1214.

Champion DE, Dalrymple GB, and Kuntz MA (1981) Radiometric and paleomagnetic evidence for the emperor reversed polarity event at 0.46 ± 0.005 Myr in basalt lava flows from the eastern Snake river plain, Idaho. *Geophysical Research Letters* 8: 1055–1058.

Champion DE, Lanphere MA, and Kuntz MA (1988) Evidence for a new geomagnetic reversal from lava flows in Idaho: Discussion of short polarity reversals in the Brunhes and late Matuyama polarity chrons. *Journal of Geophysical Research* 93: 11667–11680.

Channell JET (1999) Geomagnetic paleointensity and directional secular variation at Ocean Drilling Program (ODP) Site 984 (Bjorn Drift) since 500 ka: Comparisons with ODP Site 983 (Gardar Drift). *Journal of Geophysical Research* 104: 22937–22951.

Channell JET (2006) Late Brunhes polarity excursions (Mono Lake, Laschamp, Iceland Basin and Pringle Falls) recorded at ODP Site 919 (Irminger Basin). *Earth and Planetary Science Letters* 244: 378–393.

Channell JET, Bralower TJ, and Grandesso P (1987) Biostratigraphic correlation of Mesozoic polarity chrons (M1 to CM23) at Capriolo and Xausa (southern Alps, Italy). *Earth and Planetary ScienceLetters* 85: 203–321.

Channell JET, Cecca F, and Erba E (1995b) Correlations of Hauterivian and Barremian (early Cretaceous) stage boundaries to polarity chrons. *Earth and Planetary Science Letters* 134: 125–140.

Channell JET, Curtis JH, and Flower BP (2004) The Matuyama–Brunhes boundary interval (500–900 ka) in north Atlantic drift sediments. *Geophysical Journal International* 158: 489–505.

Channell JET, Erba E, Nakanishi M, and Tamaki K (1995a) A late Jurassic–early Cretaceous timescale and oceanic magnetic anomaly block models. In: Berggren W, Kent D, Aubry M, and Hardenbol J (eds.) *Geochronology, Timescales, and Stratigraphic Correlation, SEPM Special Publication* 54, pp. 51–64. Tulsa: Society of Economic Paleontologists and Mineralogists.

Channell JET, Galeotti S, Martin EE, Billups K, Scher H, and Stoner JS (2003b) Eocene to Miocene magnetostratigraphy, biostratigraphy, and chemostratigraphy at ODP Site 1090 (sub-Antarctic south Atlantic). *Geological Society of America Bulletin* 115: 607–623.

Channell JET and Guyodo Y (2004) The Matuyama Chronozone at ODP Site 982 (Rockall Bank): Evidence for decimeter-scale magnetization lock-in depths. In: Channell *et al.* (eds.) *AGU Geophysical Monograph Seminar, 145: Timescales of the Geomagnetic Field*, pp. 205–219. Washington, DC: AGU.

Channell JET, Hodell DA, and Lehman B (1997) Relative geomagnetic paleointensity and δ^{18}O at ODP Site 983 (Gardar Drift, north Atlantic) since 350 ka. *Earth and Planetary Science Letters* 153: 103–118.

Channell JET and Kleiven HF (2000) Geomagnetic palaeointensities and astrochronological ages for the Matuyama–Brunhes boundary and the boundaries of the Jaramillo Subchron: Palaeomagnetic and oxygen isotope records from ODP Site 983. *Philosophical Transactions of the Royal Society of London A* 358: 1027–1047.

Channell JET, Labs J, and Raymo ME (2003a) The Réunion Subchronozone at ODP Site 981 (Feni Drift, North Atlantic). *Earth and Planetary Science Letters* 215: 1–12.

Channell JET and Lehman B (1997) The last two geomagnetic polarity reversals recorded in high-deposition-rate sediment drifts. *Nature* 389: 712–715.

Channell JET, Mazaud A, Sullivan P, Turner S, and Raymo ME (2002) Geomagnetic excursions and paleointensities in the 0.9–2.15 Ma interval of the Matuyama Chron at ODP Site 983 and 984 (Iceland Basin). *Journal of Geophysical Research* 107: (B6) (10.1029/2001JB000491).

Channell JET and Raymo ME (2003) Paleomagnetic record at ODP Site 980 (Feni Drift, Rockall) for the past 1.2 Myrs. *Geochemistry, Geophysics, Geosystems* 4(4): 1033 (doi:10.1029/2002GC000440).

Channell JET, Stoner JS, Hoddell DA, and Charles C (2000) Geomagnetic paleointensity for the last 100 kyr from the subantarctic south Atlantic: A tool for interhemispheric correlation. *Earth and Planetary Science Letters* 175: 145–160.

Charbit S, Guillou H, and Turpin L (1998) Cross calibration of K–Ar standard minerals using an unspiked Ar measurement technique. *Chemical Geology* 150: 147–159.

Charles CD and Lynch-Stieglitz J (1996) U.S. Ninnemann and R.G. Fairbanks, climate connections between the hemisphere revealed by deep sea sediment core ice core correlations. *Earth and Planetary Science Letters* 142: 19–27.

Chauvin A, Duncan RA, Bonhommet N, and Levi S (1989) Paleointensity of the earth's magnetic field and K–Ar dating of the Louchadière volcanic flow (central France): New evidence for the Laschamp Excursion. *Geophysical Research Letters* 16: 1189–1192.

Chauvin A, Roperch P, and Duncan RA (1990) Records of geomagnetic reversals from volcanic islands of French Polynesia, 2-paleomagnetic study of a flow sequence (1.2 to 0.6 Ma) from the island of Tahiti and discussion of reversal models. *Journal of Geophysical Research* 95: 2727–2752.

Chen Y, Smith PE, Evensen NM, York D, and Lajoie KR (1996) The edge of time: Dating young volcanic ash layers with the ^{40}Ar/^{39}Ar laser probe. *Science* 274: 1176–1178.

Clement BM (1991) Geographical distribution of transitional V.G.P.'s: Evidence for non-zonal equatorial symmetry during the Matuyama–Brunhes geomagnetic reversal. *Earth and Planetary Science Letters* 104: 48–58.

Clement BM (1992) Evidence for dipolar fields during the Cobb mountain geomagnetic polarity reversals. *Nature* 358: 405–407.

Clement BM (2004) Dependence of the duration of geomagnetic polarity reversals on site latitude. *Nature* 428: 637–640.

Clement BM and Kent DV (1987) Short polarity intervals within the Matuyama: Transition field records from hydraulic piston cored sediments from the North Atlantic. *Earth and Planetary Science Letters* 81: 253–264.

Clement BM and Martinson DG (1992) A quantitative comparison of two paleomagnetic records of the Cobb Mountain subchron from North Atlantic deep-sea sediments. *Journal of Geophysical Research* 97: 1735–1752.

Clement BM and Robinson R (1987) The magnetostratigraphy of Leg 94 sediments. In: Ruddiman W, Kidd RB, Thomas E, et al. (eds.) *Init. Reports, DSDP*, 94, pp. 635–650. Washington: U.S. Governmentt Printing Office.

Coe RS, Singer BS, Pringle MS, and Zhao X (2004) Matuyama-Brunhes reversal and Kamikatsura event on Maui: Paleomagnetic directions, ^{40}Ar/^{39}Ar ages and implications. *Earth and Planetary Science Letters* 222: 667–684.

Condomines M (1980) Age of the Olby-Laschamp geomagnetic polarity event. *Nature* 286: 697–699.

Cox A, Doell RR, and Dalrymple GB (1963) Geomagnetic polarity epochs and Pleistocene geochonometry. *Nature* 198: 1049–1051.

Creer KM, Irving E, and Runcorn SK (1954) The direction of the geomagnetic field in remote epochs in Great Britain. *Journal of Geomagnetism and Geoelectricity* 6: 164–168.

Creer KM, Readman PW, and Jacobs AM (1980) Paleomagnetic and paleontological dating of a section at Gioa Tauro, Italy: Identification of the Blake Event. *Earth and Planetary Science Letters* 50: 289–300.

Creer KM, Thouveny N, and Blunk I (1990) Climatic and geomagnetic influences on the Lac du Bouchet palaeomagnetic SV record through the last 110 000 years. *Physics of the Earth and Planetary Interiors* 64: 314–341.

David P (1904) Sur la stabilité de la direction d'aimantation dans quelques roches volcaniques. *Comptes Rendus Hebdomadaires des Seances de l'Academie des Sciences Paris* 138: 41–42.

Denham C, Anderson R, and Bacon M (1997) Paleomagnetism and radiochemical estimates for late Brunhes polarity episodes. *Earth and Planetary Science Letters* 35: 384–397.

Denham CR (1974) Counterclockwise motion of paleomagnetic directions 24,000 years ago at Mono Lake, California. *Journal of Geomagnetism and Geoelectricity* 26: 487.

Denham CR (1976) Blake polarity episode in two cores from the Greater Antilles outer ridge. *Earth and Planetary Science Letters* 29: 422–443.

Denham CR and Cox A (1971) Evidence that the Laschamp polarity event did not occur 13,300–34,000 years ago. *Earth and Planetary Science Letters* 13: 181–190.

Dinarès-Turell J, Sagnotti L, and Roberts AP (2002) Relative geomagnetic paleointensity from the Jaramillo Subchron to the Matuyama/Brunhes boundary as recorded in a Mediterranean piston core. *Earth and Planetary Science Letters* 194: 327–341.

Doell RR and Dalrymple GB (1966) Geomagnetic polarity epochs: A new polarity event and the age of the Brunhes-Matuyama boundary. *Science* 152: 1060–1061.

Doell RR, Dalrymple GB, Smith RL, and Bailey RA (1968) Paleomagnetism, potassium–argon ages and geology of rhyolites and associated rocks of the Valles caldera, New Mexico. *Geological Society of America Memoris* 116: 211–248.

Ericson DB, Ewing M, Wollin G, and Heezen BC (1961) Atlantic deep-sea sediment cores. *Geological Society of America Bulletin* 72: 193–206.

Evans HF and Channell JET (2003) Late Miocene magnetic stratigraphy from ODP Site 1092 (sub-Antarctic south Atlantic): Recogniton of 'cryptochrons' in C5n.2n. *Geophysical Journal International* 153: 483–496.

Evans HF, Westerhold T, and Channell JET (2004) ODP Site 1092: Revised composite depth section has implications for Upper Miocene 'cryptochrons'. *Geophysical Journal International* 156: 195–199.

Fang X, Li J, Van der Voo R, et al. (1997) A record of the Blake event during the last interglacial paleosol in the western loess plateau of China. *Earth and Planetary Science Letters* 146: 73–82.

Frank M, Schwarz B, Baumann S, Kubik P, Suter M, and Mangini A (1997) A 200 kyr record of cosmogenic radionuclide production rate and geomagnetic field intensity from ^{10}Be in globally stacked deep-sea sediments. *Earth and Planetary Science Letters* 149: 121–129.

Garces M, Agusti J, Cabera L, and Pares JM (1996) Magnetostratigraphy of the Vallesian (late Miocene) in the Valles-Penedes Basin (northeast Spain). *Earth and Planetary Science Letters* 142: 381–396.

Geissman JW, Brown L, Turrin BD, McFadden LD, and Harlan SS (1990) Brunhes chron excursion/polarity-episode recorded during the late Pleistocene, Albuquerque volcanoes, New Mexico, USA. *Geophysical Journal International* 102: 73–88.

Gilder S, Chen Y, Cogné J-P, et al. (2003) Paleomagnetism of upper Jurassic to lower Cretaceous volcanic and sedimentary rocks from the western Tarim Basin and implications for inclination shallowing and absolute dating of the M-0 (ISEA) chron. *Earth and Planetary Science Letters* 206: 587–600.

Gillot PG, Labeyrie J, Laj C, et al. (1979) Age of the Laschamp paleomagnetic excursion revisited. *Earth and Planetary Science Letters* 42: 444–450.

Gradstein F, Ogg JG, and Smith A (2004) *A Geologic Time Scale*, 589 pp. Cambridge, New York: Cambridge University Press.

Grommé CS and Hay RL (1963) Magnetization of basalt of Bed I, Olduvai Gorge. *Nature* 200: 560–561.

Grommé CS and Hay RL (1971) Geomagnetic polarity epochs: Age and duration of the Olduvai normal polarity event. *Earth and Planetary Science Letters* 18: 179–185.

Grootes PM and Stuiver M (1997) Oxygen 18/16 variability in Greenland snow and ice with 10^{-3} to 10^5 year time resolution. *Journal of Geophysical Research* 102: 26455–26470.

Grützner J, Giosan L, Franz SO, et al. (2002) Astronomical age models for Pleistocene drift sediments from the western north Atlantic (ODP Sites 1055-1063). *Marine Geology* 189: 5–23.

Gubbins D (1999) The distinction between geomagnetic excursions and reversals. *Geophysical Journal International* 137: F1–F3.

Guillou H, Carracedo JC, and Day S (1998) Dating of the upper Pleistocene–Holocene volcanic activity of La Palma using the unspiked K–Ar Technique. *Journal of Volcanology and Geothermal Research* 86: 137–149.

Guillou H, Singer BS, Laj C, Kissel C, Scaillet S, and Jicha B (2004) On the age of the Laschamp geomagnetic event. *Earth and Planetary Science Letters* 227: 331–343.

Guo B, Zhu R, Florindo F, Ding Z, and Sun JA (2002) A short, reverse polarity interval within the Jaramillo subchron: Evidence from the Jingbian section, northern China Loess plateau. *Journal of Geophysical Research* 107: 10029–10040.

Guyodo Y, Acton GD, Brachfeld S, and Channell JET (2001) A sedimentary paleomagnetic record of the Matuyama chron from the western Antarctic margin. *Earth and Planetary Science Letters* 191: 61–74.

Guyodo Y, Richter C, and Valet J-P (1999) Paleointensity record from Pleistocene sediments (1.4–0 Ma) off the California margin. *Journal of Geophysical Research* 104: 22953–22964.

Guyodo Y and Valet J-P (1996) Relative variations in geomagnetic intensity from sedimentary records: The past 200.000 years. *Earth and Planetary Science Letters* 143: 23–36.

Guyodo Y and Valet J-P (1999) Global changes in intensity of the earth's magnetic field during the past 800 kyr. *Nature* 399: 249–252.

Hall CM and York D (1978) K–Ar and ^{40}Ar/^{39}Ar age of the Laschamp geomagnetic polarity reversal. *Nature* 274: 462–464.

Hall CM, York D, and Bonhommet N (1979) ^{40}Ar/^{39}Ar age of the Laschamp event and associated volcanism in the Chaîne des Puys. *EOS, Transactions, American Geophysical Union* 60: 244.

Harland WB, Armstrong R, Cox AV, Craig L, Smith AG, and Smith D (1990) *A Geologic Time Scale 1989*, 263 pp. Cambridge, New York: Cambridge University Press.

Harland WB, Cox AV, Llewellyn PG, Pickton CAG, Smith AG, and Walters RA (1982) *Geologic Time Scale*, 131 pp. Cambridge, New York: Cambridge Univ. Press.

Hartl P and Tauxe L (1996) A precursor to the Matuyama/Brunhes transition-field instability as recorded in pelagic sediments. *Earth and Planetary Science Letters* 138: 121–135.

Hayashida A, Verosub KL, Heider F, and Leonhardt R (1999) Magnetostratigraphy and relative paleointensity of late Neogene sediments at ODP Leg 167 Site 1010 off Baja California. *Geophysical Journal International* 139: 829–840.

Hays JD and Opdyke ND (1967) Antarctic radiolaria, magnetic reversals, and climatic change. *Science* 158: 1001–1011.

Hays JD, Saito T, Opdyke ND, and Burckle LH (1969) Pliocene and Pleistocene sediments of the equatorial Pacific: Their paleomagnetic, biostratigraphic and climatic record. *Geological Society of America Bulletin* 80: 1481–1514.

Heirtzler JR, Dickson GO, Herron Em, Pittman WC, III, and LePichon X (1968) Marine magnetic anomalies, geomagnetic field reversal and motions of the ocean floor and continents. *Journal of Geophysical Research* 73: 2119–2136.

Heizler MT, Perry FV, Crowe BM, Peters L, and Appelt R (1999) The age of Lathrop Wells Volcanic Center: An ^{40}Ar/^{39}Ar dating investigation. *Journal of Geophysical Research* 104: 767–804.

Heller F (1980) Self-reversal of natural remanent magnetization in the Olby–Laschamp lavas. *Nature* 284: 334–335.

Heller F and Evans ME (1995) Loess magnetism. *Reviews of Geophysics* 33: 211–240.

Heller F and Petersen N (1982) The Laschamp excursion. *Philosophocal Transaction of the Royal Society A* 306: 169–177.

Helsley CE (1969) Magnetic reversal stratigraphy of the lower Triassic Moenkope formation of western Colorado. *Geological Society of America Bulletin* 80: 2431–2450.

Herrero-Bervera E, Helsley CE, Hammond SR, and Chitwood LA (1989) A possible lacustrine paleomagnetic record of the Blake episode from Pringle falls, Oregon, USA. *Physics of the Earth and Planetary Interiors* 56: 112–123.

Herrero-Bervera E and Helsley CE (1993) Global paleomagnetic correlation of the Blake geomagnetic polarity episode. In: Aissaoui DM, McNeill DF, and Hurley NF (eds.) *Applications of Paleomagnetism to Sedimentary Geology, SEPM Special Publication*, 49, pp. 71–82. Tulsa: Society of Economic Paleontologists and Mineralogists.

Herrero-Bervera E, Helsley CE, Sarna-Wojcicki AM, et al. (1994) Age and correlation of a paleomagnetic episode in the western United States by ^{40}Ar/^{39}Ar dating and tephrochronology: The Jamaica, Blake, or a new polarity episode? *Journal of Geophysical Research* 99: 24091–24103.

Heslop D, Langereis CG, and Dekkers MJ (2000) A new astronomical timescale for the loess deposits of Northern China. *Earth and Planetary Science Letters* 184: 125–139.

Hilgen FJ (1991a) Extension of the astronomically calibrated (polarity) time scale to the Miocene/Pliocene boundary. *Earth and Planetary Science Letters* 107: 349–368.

Hilgen FJ (1991b) Astronomical calibration of Gauss to Matuyama sapropels in the Mediterranean and implications for the Geomagnetic Polarity Time Scale. *Earth and Planetary Science Letters* 104: 226–244.

Hilgen FJ, Abdul-Aziz H, Krijgsman WJ, Raffi I, and Turco E (2003) Integrated stratigraphy and astronomical tuning of the Serravallian and lower Tortonian at Monti dei Corvi (Middle–Upper Miocene, northern Italy). *Palaeogeography, Palaeoclimatology, Palaeoecology* 199: 229–264.

Hilgen FJ, Krijgsman W, Langereis CG, Lourens LJ, Santarelli A, and Zachariasse WJ (1995) Extending the astronomical (polarity) time scale into the Miocene. *Earth and Planetary Science Letters* 136: 495–510.

Hilgen FJ, Krijgsman W, Raffi I, Turco E, and Zachariasse WJ (2000) Integrated stratigraphy and astronomical calibration of the Serravallian/Tortonian boundary section at Monte Gibliscemi (Sicily, Italy). *Marine Micropaleontology* 38: 181–211.

Hoare JM, Condon WH, Cox A, and Dalrymple GB (1968) Geology, paleomagnetism, and potassium-argon ages of basalts from Nunivak Island, Alaska. *Geological Society of America Memoris* 116: 377–413.

Hoffman KA (1992) Dipolar reversal states of the geomagnetic field and core–mantle dynamics. *Nature* 359: 789–794.

Horng CS, Lee MY, Pälike H, et al. (2002) Astronomically calibrated ages for geomagnetic reversals within the Matuyama Chron. *Earth Planets Space* 54: 679–690.

Horng CS, Roberts AP, and Liang WT (2003) Astronomically tuned record of relative geomagnetic paleointensity from the western Phillipine Sea. *Journal of Geophysical Research* 108: 2059 (doi:10.1029/2001JB001698).

Hospers J (1951) Remanent magnetization of rocks and the history of the geomagnetic field. *Nature* 168: 1111–1112.

Hospers J (1953–1954) Reversals of the main geomagnetic field I, II, and III. *Proceedings of the Koninklijke Nederlandse Akademie van Wetenschappen - Series B* 56: 467–491; 57: 112–121.

Houghton BF, Wilson CJN, McWilliams MO et al. (1995) Chronology and dynamics of a large silicic magmatic system: Central Taupo volcanic zone, New Zealand. *Geology* 23: 13–16.

Hsu V, Merrill DL, and Shibuya H (1990) Paleomagnetic transition records of the Cobb mountain event from sediment of the Celebes and Sulu seas. *Geophysical Research Letters* 17: 2069–2072.

Huxtable J, Aitken MJ, and Bonhommet N (1978) Thermoluminescence dating of sediments baked by lava flows of the Chaîne des Puys. *Nature* 275: 207–209.

Irving E and Runcorn SK (1957) Analysis of the palaeomagnetism of the Torridonian sandstone series of north-west Scotland. *Philosophical Transactions of the Royal Society of London A* 250: 83–99.

Izett GA and Obradovich JD (1994) ^{40}Ar/^{39}Ar age constraints for the Jaramillo normal subchron and the Matuyama–Brunhes geomagnetic boundary. *Journal of Geophysical Research* 99: 2925–2934.

Johnson RG (1982) Brunhes–Matuyama magnetic reversal dated at 790,000 yr B.P. by marine–astronomical correlations. *Quaternary Research* 17: 135–147.

Kawai N (1984) Paleomagnetic study of the Lake Biwa sediments. In: Horie S and Junk W (eds.) *Lake Biwa*. Dordrecht: Dr. W Junk Publisher.

Kawai N, Yaskawa K, Nakajima T, Torii M, and Horie S (1972) Oscillating geomagnetic field with a recurring reversal discovered from Lake Biwa. *Proceedings of Japan Academy* 48: 186–190.

Keigwin L and Jones G (1994) Western North Atlantic evidence for millennial-scale changes in ocean circulation and climate. *Journal of Geophysical Research* 99: 12397–12410.

Keigwin LD (1995) Oxygen isotope stratigraphy from Hole 883D, Leg 145. *Proceedings of ODP Science Research* 145: 257–264.

Kent DV, Hemming SR, and Turrin BD (2002) Laschamp excursion at Mono Lake? *Earth and Planetary Science Letters* 197: 151–164.

Kent DV and Olsen PE (1999) Astronomically tuned geomagnetic timescale for the late Triassic. *Journal of Geophysical Research* 104: 12831–12841.

Kent DV and Schneider DA (1995) Correlation of the paleointensity variation records in the Brunhes/Matuyama polarity transition interval. *Earth and Planetary Science Letters* 129: 135–144.

Kent DV and Spariosu DJ (1983) High resolution magnetostratigraphy of Caribbean Plio-Pleistocene sediments. *Paleogeography Paleoclimatology Paleoecology* 42: 47–64.

Khramov AN (1960) *Palaeomagnetism and stratigraphic correlation, Lojkine AJ(trans.)*. Canberra: Geophysics Department, A.N.U.

King JW, Heil C, O'Regan M et al. (2005) Paleomagnetic results from Pleistocene sediments of Lomonosov ridge, central Arctic ocean, IODP Leg 302. *EOS Transactions AGU*, 85 (47). *Fall Meeting Suppliment*, Abstract GP44A-04.

Kissel C, Laj C, Labeyrie L, Dokken T, Voelker A, and Blamart D (1999) Rapid climatic variations during marine isotopic stage 3: Magnetic analysis of sediments from Nordic seas and north Atlantic. *Earth and Planetary Science Letters* 171: 489–502.

Kok YS and Tauxe L (1999) A relative geomagnetic paleointensity stack from Ontong–Java plateau sediments for the Matuyama. *Journal of Geophysical Research* 104: 25401–25413.

Krijgsman W, Hilgen FJ, Langereis CG, Santarelli A, and Zachariasse WJ (1995) Late Miocene magnetostratigraphy, biostratigraphy and cyclostratigraphy in the Mediterranean. *Earth and Planetary Science Letters* 136: 475–494.

Krijgsman W, Hilgen FJ, Langereis CG, and Zachariasse WJ (1994) The age of the Tortonian/Messinian boundary. *Earth and Planetary Science Letters* 121: 533–547.

Krijgsman W and Kent DV (2004) Non-uniform occurrence of short-term polarity fluctuation in the geomagnetic field? New results from middle to late Miocene sediments of the north Atlantic (DSDP Site 608). In: Channell JET, *et al.* (eds.) AGU Geophysical Monograph Series, 145: Timescales of the Geomagnetic Field, pp. 161–174. Washington, DC: AGU.

Kristjansson L and Gudmundsson A (1980) Geomagnetic excursions in late-glacial basalt outcrop in south-western Iceland. *Geophysical Research Letters* 7: 337–340.

LaBrecque JL, Kent DV, and Cande SC (1977) Revised magnetic polarity time-scale for the late Cretaceous and Cenozoic time. *Geology* 5: 330–335.

Laj C, Kissel C, Garnier F, and Herrero-Bervera E (1996) Relative geomagnetic field intensity and reversals for the last 1.8 My from a central Pacific core. *Geophysical Research Letters* 23: 3393–3396.

Laj C, Kissel C, and Guillou H (2002) Brunhes' research revisited: Magnetization of volcanic flows and baked clays. *Eos, Transactions American Geophysical Union* 83(35): 381 and 386–387.

Laj C, Kissel C, Mazaud A, Channell JET, and Beer J (2000) North Atlantic paleointensity stack since 75 ka (NAPIS-75) and the duration of the Laschamp event. *Philosophical Transactions of the Royal Society of London* 358: 1009–1025.

Laj C, Kissel C, and Roberts AP (2006) Geomagnetic field behavior during the Icelandic basin and Laschamp geomagnetic excursions: A simple transitional field geometry? *Geochemistry, Geophysics, Geosystems* 7: Q03004 (doi:10,1029/2005GC001122).

Laj C, Mazaud A, Fuller M, and Herrero-Bervera E (1991) Geomagnetic reversal paths. *Nature* 351: 11–26.

Lal D and Peters B (1967) Cosmic ray produced radioactivity on the Earth. In: Flügge S (ed.) *Handbuch für Physik*, pp. 551–612. Berlin: Springer.

Lanci L, Lowrie W, and Montanari A (1996) Magnetostratigraphy of the Eocene/Oligocene boundary in a short drill-core. *Earth and Planetary Science Letters* 143: 37–48.

Lanci L, Pares JM, Channell JET, and Kent DV (2005) Oligocene magnetostratigraphy from equatorial Pacific sediments (ODP Sites 1218 and 1219, Leg 199). *Earth and Planetary Science Letters* 237: 617–634.

Langereis CG, Dekkers MJ, de Lange GJ, Paterne M, and Van Santvoort PJM (1997) Magnetostratigraphy and astronomical calibration of the last 1.1 Myr from an eastern Mediterranean piston core and dating of short events in the Brunhes. *Geophysical Journal International* 129: 75–94.

Lanphere MA (2000) Comparison of conventional K-Ar and ^{40}Ar/^{39}Ar dating of young mafic rocks. *Quaternary Research* 53: 294–301.

Lanphere MA, Champion DE, Christiansen RL, Izett GA, and Obradovich JD (2002) Revised ages for tuffs of the Yellowstone plateau volcanic field: Assignment of the Huckleberry ridge tuff to a new geomagnetic polarity event. *Geological Society of America Bulletin* 114: 559–568.

Larson RL and Hilde TWC (1975) A revised time scale of magnetic reversals for the early Cretaceous and late Jurassic. *Journal of Geophysical Research* 80: 2586–2594.

Lehman B, Laj C, Kissel C, Mazaud A, Paterne M, and Labeyrie L (1996) Relative changes of the geomagnetic field intensity during the last 280 kyear from piston cores in the Azores area. *Physics of the Earth and Planetary Interiors* 93: 269–284.

Levi S, Gudmunsson H, Duncan RA, Kristjansson L, Gillot PV, and Jacobsson SP (1990) Late Pleistocene geomagnetic excursion in Icelandic lavas: Confirmation of the Laschamp excursion. *Earth and Planetary Science Letters* 96: 443–457.

Levi S and Karlin R (1989) A sixty thousand year paleomagnetic record from Gulf of California sediments: Secular variation, late Quaternary excursions and geomagnetic implicationq. *Earth and Planetary Science Letters* 92: 443–447.

Li J-J, Fang X-M, Van der Voo R, *et al.* (1997) Late Cenozoic magnetostratigraphy (11-0 Ma) of the Dongshanding and Wangjiashan sections in the Longzhong basin, western China. *Geology en Mijnbouw* 76: 121–134.

Li J-J, Zhu J-J, Kang J-C, *et al.* (1992) The comparison of Lanzhou loess profile with Vostok ice core in Antarctica over the last glaciation cycle. *Science in China* 35(B): 476–487.

Liddicoat JC (1990) Aborted reversal of the paleomagnetic field in Brunhes normal Chron in east-central California. *Geophysical Journal International* 102: 747–752.

Liddicoat JC (1992) Mono Lake excursion in Mono basin, California, and at Carson sink and Pyramid lake, Nevada. *Geophysical Journal International* 108: 442–452.

Liddicoat JC (1996) Mono lake excursion in the Lahontan basin, Nevada. *Geophysical Journal International* 125: 630–635.

Liddicoat JC and Bailey RA (1989) Short reversal of the paleomagnetic field about 280,000 years ago at Long Valley, California. In: Lowes FJ, Collinson DW, Parry JH, Runcorn SK, Tozer DC, and Soward AM (eds.) *Proceedings of NATO Advanced Study Institute on Geomagnetism and Paleomagnetism, NATO Advanced Study Institutes Series; Mathematical and Physical Sciences* pp. 137–153.

Liddicoat JC and Coe RS (1979) Mono lake geomagnetic excursion. *Journal of Geophysical Research* 84: 261–271.

Liddicoat JC, Lajoie KR, and Sarna-Wojcicki AM (1982) Detection and dating of the Mono lake excursion in the lake Lohontan Sehoo formation, Carson sink, Nevada. *Transactions, American Geophysical Union* 63: 920.

Liu TS, An ZS, Yuan BY, and Han JM (1985) The loess-paleosol sequence in China and climatic history. *Episodes* 8: 21–28.

Liu X, Liu T, Xu T, Liu C, and Chen M (1988) The Chinese loess in Xifeng. 1. The primary study on magnetostratigraphy of a loess profile in Xifeng area, Gansu province. *Geophysical Journal of the Royal Astronomical Society* 92: 345–348.

Lourens LJ, Antonarakou A, Hilgen FJ, Van Hoof AAM, Vergnaud-Grazzini C, and Zachariasse WJ (1996) Evaluation of the Plio-Pleistocene astronomical timescale. *Paleoceanography* 11: 391–413.

Lourens LJ, Hilgen FJ, Laskar J, Shackleton NJ, and Wilson D (2004) The Neogene Period. In: Gradstein FM, Ogg JG, and Smith AG (eds.) *Geologic Time Scale 2004*, pp. 409–440. Cambridge, New York: Cambridge University Press.

Lowrie W and Alvarez W (1981) One hundred million years of geomagnetic polarity history. *Geology* 9: 392–397.

Lowrie W, Alvarez W, Premoli Silva I, and Monechi W (1980) Lower Cretaceous magnetic stratigraphy in Umbrian pelagic carbonate rocks. *Geophysical Journal of Royal Astronomical Society* 60: 263–281.

Lowrie W and Kent DV (2004) Geomagnetic polarity time scales and reversal frequency regimes. In: Channell JET, *et al.* (ed.) *AGU Geophysical Monograph Series, 145: Timescales of the Geomagnetic Field*, pp. 117–129. Washington, DC: AGU.

Lowrie W and Lanci L (1994) Magnetostratigraphy of Eocene–Oligocene boundary sections in Italy: No evidence for short subchrons within chrons 12R and 13R. *Earth and Planetary Science Letters* 126: 247–258.

Lund SP, Acton G, Clement B, *et al.* (1998) Geomagnetic field excursions occurred often during the last million years. *Transactions, American Geophysical Union (EOS)* 79(14): 178–179.

Lund SP, Acton GD, Clement B, Okada M, and Williams T (2001a) Brunhes chron magnetic field excursions recovered from Leg 172 sediments. In: Keigwin LD, Rio D, Acton GD, and Arnold E (eds.) *Proceedings of the ODP Science Research*, 172, pp. 1–18 (Online).

Lund SP, Acton GD, Clement B, Okada M, and Williams T (2001b) Paleomagnetic records of Stage 3 excursions, Leg 172. In: Keigwin LD, Rio D, Acton GD, and Arnold E (eds.) *Proceedings of the ODP Science Research*, 172, pp. 1–20 (Online).

Lund SP, Liddicoat JC, Lajoie KL, Henyey TL, and Robinson S (1988) Paleomagnetic evidence for long-term 104 year memory and periodic behavior in the earth's core dynamo process. *Geophysical Research Letters* 15: 1101–1104.

Lund SP, Schwartz M, Keigwin L, and Johnson T (2005) Deep-sea sediment records of the Laschamp geomagnetic field excursion (<41,000 calendar years before present). *Journal of Geophysical Research* 110: Q12006 (doi :10,1029, 2005GC001036).

Lund SP, Stoner JS, Channell JET, and Acton G (2006) A summary of Brunhes paleomagnetic field variability recorded in ODP Cores. *Physics of the Earth and Planetary Interiors* 156: 194–204.

Maenaka K (1983) Magnetostratigraphic study of the Osaka group, with special reference to the existence of pre and past-Jaramillo episodes in the late Matuyama polarity Epoch. Mem. Hanazono Univ. 14: 1–65.

Mankinen EA and Dalrymple GB (1979) Revised geomagnetic polarity time scale for the interval 0–5 m.y.b.p. *Journal of Geophysical Research* 84: 615–626.

Mankinen EA, Donnelly JM, and Grommé CS (1978) Geomagnetic polarity event recorded at 1.1 m.y.b.p. on Cobb mountain, Clear Lake volcanic field, California. *Geology* 6: 653–656.

Mankinen EA, Donnelly-Nolan JM, Grommé CS, and Hearn BC, Jr. (1981) Paleomagnetism of the Clear lake volcanics and new limits on the age of the Jaramillo normal-polarity event. *US Geological Survey Professional Paper* 1141: 67–82.

Mankinen EA and Grommé CS (1982) Paleomagnetic data from the Cosa Range, California, and current status of the Cobb mountain normal geomagnetic polarity event. *Geophysical Research Letters* 9: 1239–1282.

Mankinen EA, Prévot M, Grommé CS, and Coe RS (1985) The Steens mountain (Oregon) geomagnetic polarity transition. 1. Directional history, duration of the episodes, and rock magnetism. *Journal of Geophysical Research* 90: 393–416.

Marshall M, Chauvin A, and Bonhommet N (1988) Preliminary paleointensity measurements and detailed magnetic analyses of basalts from the Skalamaelifell excursion, south-west Iceland. *Journal of Geophysical Research* 93: 11681–11698.

Martinson DG, Pisias NG, Hays JD, Imbrie J, Moore TC, Jr., and Shackleton NJ (1987) Age dating and the orbital theory of the Ice Ages: Development of a high-resolution 0 to 300,000-year chronostratigraphy. *Quaternary Research* 27: 1–29.

Masarik J and Beer J (1999) Simulation of particle fluxes and cosmogenic nuclide production in the Earth's atmosphere. *Journal of Geophysical Research* 12000 12111.

Matuyama M (1929) On the direction of magnetization of basalts in Japan, Tyosen and Manchuria. *Proceedings of the Imperial Academy (Tokyo)* 5: 203–205.

Mazaud A, Sicre MA, and Ezat U (2002) Geomagnetic-assisted stratigraphy and sea surface temperature changes in core MD94-103 (southern Indian Ocean): Possible implications for north–south climatic relationships around H4. *Earth and Planetary Science Letters* 201: 159–170.

McDougall F, Brown FH, Cerling TE, and Hillhouse JW (1992) A reappraisal of the geomagnetic polarity time scale to 4 Ma using data from Turkana basin, east Africa. *Geophysical Research Letters* 19: 2349–2352.

McDougall I and Chamalaun FH (1966) Geomagnetic polarity scale of time. *Nature* 212: 1415–1418.

McDougall I and Tarling DH (1963a) Dating of reversals of the earth's magnetic field. *Nature* 198: 1012–1013.

McDougall I and Tarling DH (1963b) Dating of polarity zones in the Hawaiian Islands. *Nature* 200: 54–56.

McDougall I and Tarling DH (1964) Dating geomagnetic polarity zones. *Nature* 202: 171–172.

McDougall I and Watkins ND (1973) Age and duration of the Réunion geomagnetic polarity event. *Earth and Planetary Science Letters* 19: 443–452.

McDougall I and Wensink J (1966) Paleomagnetism and geochronology of the Pliocene–Pleistocene lavas in Iceland. *Earth and Planetary Science Letters* 1: 232–236.

McWilliams M (2001) Global correlation of the 223 ka Pringle falls event. *International Geological Reviews* 43: 191–195.

Merrill RT and McFadden PL (1994) Geomagnetic field stability : Reversal events and excursions. *Earth and Planetary Science Letters* 121: 57–69.

Meynadier L Valet J-P and Shackleton NJ (1995) Relative geomagnetic intensity during the last 4 MY from the equatorial Pacific. *Proceedings of ODP Science Research* 138: 779–793.

Meynadier L, Valet J-P, Weeks R, Shackleton NJ, and Hagee VL (1992) Relative geomagnetic intensity of the field during the last 140 ka. *Earth and Planetary Science Letters* 114: 39–57.

Mörner NA and Larsen JP (1952) Gothenburg magnetic "flip". *Nature* 251: 408–409.

Nagata T (1952) Reverse thermal-remanent magnetism. *Nature* 169: 704.

Nagata T, Uyeda S, and Ozima M (1957) Magnetic interaction between ferromagnetic minerals contained in rocks. *Philosophical Magazine, Supplement Advanced Physics* 6: 264–287.

Negrini RM, Davis JO, and Verosub KL (1984) Mono lake geomagnetic excursion found at Summer lake, Oregon. *Geology* 12: 643–646.

Negrini RM, Erbes DB, Roberts AP, Verosub KL, Sarna-Wojcicki AM, and Meyer CE (1994) Repeating waveform initiated by a 180–190 ka geomagnetic excursion in eastern north America: Implications for field behavior during polarity transitions and susequent secular variation. *Journal of Geophysical Research* 99: 24105–24119.

Negrini RM, Erbes DB, Faber K, *et al.* (2000) A paleoclimate record for the past 250,000 years from Summer lake, Oregon, USA: I. Chronology and magnetic proxies for lake level. *Journal of Paleolimnology* 24: 125–149.

Negrini RM, Verosub KL, and Davis JO (1987) Long-term non-geocentric axial dipole directions and a geomagnetic excursion from the middle Pleistocene sediments of the Humboldt river Canyon, pershing county, Nevada. *Journal of Geophysical Research* 92: 10617–10627.

Negrini RM, Verosub KL, and Davis JO (1988) The middle to late Pleistocene geomagnetic field recorded in fine-grained sediments from Summer lake, Oregon, and double hot springs, Nevada, U.S.A. *Earth and Planetary Science Letters* 87: 173–192.

Ness G, Levi S, and Couch R (1980) Marine magnetic anomaly timescales for the Cenozoic and Late Cretaceous: A precis, critique and synthesis. *Reviews of Geophysics and Space Physics* 18: 753–770.

Ninkovitch D, Opdyke ND, Heezen BC, and Foster JH (1966) Paleomagnetic stratigraphy, rates of deposition and tephrachronology in north Pacific deep-sea sediments. *Earth and Planetary Science Letters* 1: 476–492.

Noel M and Tarling DH (1992) The Laschamp 'event'. *Nature* 253: 705–707.

Nowaczyk NR and Antonow M (1997) High resolution magnetostratigraphy of four sediment cores from the Greenland Sea - I. Identification of the mono lake excursion, Laschamp and Biwa I/Jamaica geomagnetic polarity events. *Geophysical Journal International* 131: 310–324.

Nowaczyk NR and Baumann M (1992) Combined high-resolution magnetostratigraphy and nannofossil biostratigraphy for late Quaternary Artic ocean sediments. *Deep Sea Research* 39: 567–601.

Nowaczyk NR and Frederichs TW (1994) A. Eisenhauer and G. Gard, Magnetostratigraphic data from late Quaterny sediments from the Yermak plateau, Arctic Ocean: Evidence for four geomagnetic polarity events within the last 170 ka of the Brunhes Chron. *Geophysical Journal International* 117: 453–471.

Nowaczyk NR and Knies J (2000) Magnetostratigraphic results from the eastern Arctic ocean: AMS ^{14}C ages and relative paleointensity data of the mono lake and Laschamp geomagnetic reversal excursions. *Geophysical Journal International* 140: 185–197.

Oda H, Nakamura K, Ikehara K, Nakano T, Nishimura M, and Khlystov O (2002) Paleomagnetic record from Academician ridge, Lake Baikal: A reversal excursion at the base of marine oxygen isotope stage 6. *Earth and Planetary Science Letters* 202: 117–132.

Ogg JG and Bardot L (2001) Aptian through Eocene magnetostratigraphic correlation of the BlakeNose Transect (Leg 171B), Florida continental margin. In: Kroon D, Norris RD and Klaus A (eds.) *Proceedings of the ODP, Science Results*, 171B, College Station, TX (available online).

Ogg JG, Karl SM, and Behl RJ (1992) Jurassic through early Cretaceous sedimentation history of the central equatorial Pacific and of Sites 800 and 801. In: *Proceedings of ODP, Science Research*, 129, pp. 571–613 pp. College Station, TX: Ocean Drilling Program.

Opdyke ND (1972) Paleomagnetism of deep sea cores. *Reviews of Geophysics and Space Physics* 10: 213–249.

Opdyke ND, Burkle LH, and Todd A (1974) The extension of the magnetic time scale in sediments of the central Pacific ocean. *Earth and Planetary Science Letters* 22: 300–306.

Opdyke ND and Channell JET (1996) *Magnetic Stratigraphy*, 346 pp. San Diego, CA: Academic Press.

Opdyke ND, Glass B, Hays JP, and Foster J (1966) Paleomagnetic study of Antarctic deep-sea cores. *Science* 154: 349–357.

Palmer DF, Henyey TL, and Dodson RE (1979) Paleomagnetic and sedimentological studies at lake Tahoe, California, Nevada. *Earth and Planetary Science Letters* 46: 125–137.

Parés JM, Van der Voo R, Yan M, and Fang X (2004) After the dust settles: Why is the Blake event imperfectly recorded in Chinese Loess? In: Channell JET, Kent DV, Lowrie W, and Meert J (eds.) *AGU Geophysical Monogrograph Series, 145: Timescales of the Paleomagnetic Field*, pp. 191–204. Washington, DC: AGU.

Peate DW, Chen JH, Wasserburg GJ, Papanastassiou DA, and Geissman JW (1996) ^{238}U-^{230}Th dating of a geomagnetic excursion in Quaternary basalts of the Albuquerque volcanoes field, New Mexico (USA). *Geophysical Research Letters* 23: 2271–2274.

Peddie N and Zunde A (1988) The magnetic field of the Earth: 1985., U.S. Geol. Surv. Geophys. Invest. Map, GP-987-1.

Picard MD (1964) Paleomagnetic correlation of units within Chugwater (Triassic) formation, west-central Wyoming. *American Association of Petroleum Geologists Bulletin* 48: 269–291.

Pillans BJ, Roberts AP, Wilson GS, Abbott ST, and Alloway BV (1994) Magnetostratigraphic, lithostratigraphic and tephrostratigraphic constraints on Lower and Middle Pleistocene sea-level changes, Wanganui Basin, New Zealand. *Earth and Planetary Science Letters* 121: 81–98.

Pitman III, WC and Heirtzler JR (1966) Magnetic anomalies over the Pacific–Antarctic ridge. *Science* 154: 1164–1171.

Quidelleur X, Carlut J, Gillot PY, and Soler V (2002) Evolution of the geomagnetic field prior to the Matuyama–Brunhes transition: Radiometric dating of a 820 ka excursion at La Palma. *Geophysical Journal International* 151: F6–F10.

Quidelleur X, Gillot PY, Carlut J, and Courtillot V (1999) Link between excursions and paleointensity inferred from abnormal field directions recorded at La Palma around 600 ka. *Earth and Planetary Science Letters* 168: 233–242.

Quidelleur X and Valet J-P (1994) Paleomagnetic records of excursions and reversals: Possible biases caused by magnetization artefacts. *Physics of the Earth and Planetary Interiors* 82: 27–48.

Quidelleur X and Valet J-P (1996) Geomagnetic changes across the last reversal record in lava flows from La Palma, Canary Islands. *Journal of Geophysical Research* 101: 13755–13773.

Renné PR, Deino PL, Walter RC, et al. (1994) Intercalibration of astronomical and radioisotopic time. *Geology* 22: 783–786.

Reynolds R (1977) Paleomagnetism of welded tuffs of the Yellowstone group. *Journal of Geophysical Research* 82: 3677–3693.

Roberts AP, Lehman B, Weeks RJ, Verosub KL, and Laj C (1997) Relative paleointensity of the geomagnetic field over the last 200,000 years from ODP Sites 883 and 884, north Pacific ocean. *Earth and Planetary Science Letters* 152: 11–23.

Roberts AP and Lewin-Harris JC (2000) Marine magnetic anomalies: Evidence that 'tiny wiggles' represent short-period geomagnetic polarity intervals. *Earth and Planetary Science Letters* 183: 375–388.

Roberts AP and Winklhofer M (2004) Why are geomagnetic excursions not always recorded in sediments? Constraints from post-depositional remanent magnetization lock-in modeling. *Earth and Planetary Science Letters* 227: 345–359.

Robinson C, Raisbeck GM, Yiou F, Lehman B, and Laj C (1995) The relationship between [10]Be and geomagnetic field strength records in central North Atlantic sediments during the last 80 ka. *Earth and Planetary Science Letters* 136: 551–557.

Roche A (1950) Sur les caracteres magnétiques du système éruptif de Gergovie. *Comptes Rendus de l'Académia des Sciences Paris* 230: 113–115.

Roche A (1951) Sur les inversions de l'aimentation remanente des roches volcaniques dans les monts d'Auvergne. *ComptesRendus del' Académie des Sciences, Paris* 223: 1132–1134.

Roche A (1956) Sur la date de la derniere inversion du champ magnetique terrestre. *ComptesRendus de l'Académie des Sciences, Paris* 243: 812–814.

Roger S, Coulon C, Thouveny N, et al. (2000) ^{40}Ar/^{39}Ar dating of a tephra layer in the Pliocene Seneze maar lacustrine sequence (French Massif central): Constraint on the age of the Réunion-Matuyama transition and implications on paleoenvironmental archives. *Earth and Planetary Science Letters* 183: 431–440.

Roperch P, Bonhommet N, and Levi S (1988) Paleointensity of the earth's magnetic field during the Laschamp excursion and its geomagnetic implications. *Earth and Planetary Science Letters* 88: 209–219.

Roperch P, Herail G, and Fornari M (1999) Magnetostratigraphy of the Miocene Corque basin, Bolivia: Implications for the geodynamic evolution of the Altiplano during the late Tertiary. *Journal of Geophysical Research* 104: 20415–20429.

Ruddiman WF, Raymo ME, Martinson DG, Clement BM, and Backman J (1989) Pleistocene evolution: Northern hemisphere ice sheet and north Atlantic ocean. *Paleoceanography* 4: 353–412.

Rutten MG and Wensink H (1960) Paleomagnetic dating, glaciations and chronology of the Plio-Pleistocene in Iceland. *International Geological Congress Session 21* IV: 62–70.

Ryan WBF (1972) Stratigraphy of late Quaternary sediments in the eastern Mediterranean. In: Stanley DJ (ed.) *The Mediterranean Sea*, pp. 149–169. Stroudsberg, PA: Dowden, Hutchinson and Ross.

Sakai H, Nakamura T, Horii M, et al. (1997) Paleomagnetic study with ^{14}C dating analysis on three short cores from Lake Baikal. *Bulletin of Nagoya University Furukawa Museum* 13: 11–22.

Schneider DA (1995) Paleomagnetism of some Ocean Drilling Program Leg 138 sediments: Detailing Miocene stratigraphy. In: Pisias NG, Mayer LA, Janecek TR, Palmer-Julson A, and Van Andel TH (eds.) *Proceedings of the ODP Science Research*, 138, pp. 59–72. College Station, TX: ODP.

Shackleton NJ, Berger A, and Peltier WR (1990) An alternative astronomical calibration of the lower Pleistocene timescale based on ODP Site 677. *Transactions of the Royal Society of Edinburgh: Earth Sciences* 81: 251–261.

Singer B and Brown LL (2000) The Santa Rosa event: ^{40}Ar/^{39}Ar and paleomagnetic results from the Valles rhyolite near Jaramillo Creek, Jemez Mountains, New Mexico. *Earth and Planetary Science Letters* 197: 51–64.

Singer B, Hildreth W, and Vincze Y (2000) ^{40}Ar/^{39}Ar evidence for early deglaciation of the central Chilean Andes. *Geophysical Research Letters* 27: 1663–1666.

Singer BS, Brown LL, Rabassa JO, and Guillou H (2004) ^{40}Ar/^{39}Ar chronology of late Pliocene and early Pleistocene geomagnetic and glacial events in southern Argentina. In: Channell JET et al. (ed.) *AGU Geophysical Monograph Series 145: Timescales of the Paleomagnetic Field*, pp. 175–190. Washington, DC: AGU.

Singer BS, Hoffman KA, Chauvin A, Coe RS, and Pringle MS (1999) Dating transitionally magnetized lavas of the late Matuyama Chron: Toward a new ^{40}Ar/^{39}Ar timescale of reversals and events. *Journal of Geophysical Research* 104: 679–693.

Singer BS, Jicha BR, Kirby BT, Zhang X, Geissman JW, and Herrero-Bervera E (2005) An ^{40}Ar/^{39}Ar age for geomagnetic instability recorded at the Albuquerque volcanoes and Pringle Falls, Oregon. *EOS Transaction AGU 86 (53) Fall Meeting Supplement Abstract* GP21A-0019.

Singer BS, Relle MK, Hoffman KA, et al. (2002) ^{40}Ar/^{39}Ar ages from transitionally magnetized lavas on La Palma, Canary Islands, and the geomagnetic instability timescale. *Journal of Geophysical Research* 107: 2307 (doi: 10.1029/2001JB001613).

Smith JD and Foster JH (1969) Geomagnetic reversal in the Brunhes normal polarity epoch. *Science* 163: 565–567.

Sobel ER and Arnaud N (2000) Cretaceous–Paleogene basaltic rocks of the Tuyon Basin, NW China and the Kyrgyz Tian Shan: The trace of a small plume. *Lithos* 50: 191–215.

Spassov S, Heller F, Kretzschmar R, Evans ME, Yue LP, and Nourgaliev DK (2003) Detrital and pedogenic magnetic mineral phases in the loess/palaeosol sequence at Lingtai (Central Chinese Loess Plateau). *Physics of the Earth and Planetary Interiors* 140: 255–275.

Spell TL and McDougall I (1992) Revisions to the age of the Brunhes/Matuyama boundary and the Pleistocene geomagnetic polarity timescale. *Geophysical Research Letters* 19: 1182–1184.

Steiger RH and Jager E (1977) Subcommission on geochronology: Convention on the use of decay constants in geo- and cosmochronology. *Earth and Planetary Science Letters* 36: 359–362.

Stoner JS, Channell JET, Hodell DA, and Charles C (2003) A 580 kyr paleomagnetic record from the sub-Antarctic south Atlantic (ODP Site 1089). *Journal of Geophysical Research* 108: 2244 (doi:10.1029/2001JB001390).

Takatsugi KO and Hyodo (1995) A geomagnetic excursion during the late Matuyama Chron, the Osaka group, southwest Japan. *Earth and Planetary Science Letters* 136: 511–524.

Tamaki K and Larson RL (1988) The Mesozoic tectonic history of the Magellan microplate in the western central Pacific. *Journal of Geophysical Research* 93: 2857–2874.

Tanaka H, Turner GM, Houghton BF, Tachibana T, Kono M, and McWilliams MO (1996) Paleomagnetism and chronology of the central Taupo Volcanic zone, New Zealand. *Geophysical Journal International* 124: 919–934.

Tarduno JA (1990) Brief reversed polarity interval during the Cretaceous normal polarity superchron. *Geology* 18: 683–686.

Tarduno JA, Lowrie W, Sliter WV, Bralower TJ, and Heller F (1992) Reversed polarity characteristics magnetizations in the Albian Contessa section, Umbrian Appenius, Italy: implications for the existence of a mid-Cretaceous mixed polarity interval. *Journal of Geophysical Research* 97: 241–271.

Tarling DH and Mitchell JG (1976) Revised Cenozoic polarity time scale. *Geology* 4: 133–136.

Tauxe L, Deino AL, Behrensmeyer AK, and Potts R (1992) Pinning down the Brunhes/Matuyama and upper Jaramillo boundaries; a reconciliation of orbital and isotopic time scales. *Earth and Planetary Science Letters* 190: 561–572.

Thouveny N, Carcaillet J, Moreno E, Leduc G, and Nérini D (2004) Geomagnetic moment variation and paleomagnetic excursions since 400 kyr BP: A stacked record from sedimentary sequences of the Portugese margin. *Earth and Planetary Science Letters* 219: 377–396.

Thouveny N and Creer KM (1992) On the brievity of the Laschamp excursion. *Bulletin de la Societe Géologique de France* 6: 771–780.

Tiedemann R, Sarnthein M, and Shackleton NJ (1994) Astronomic timescale for the Pliocene Atlantic d18O and dust flux records of Ocean Drilling Program Site 659. *Paleoceanography* 9: 619–638.

Tric E, Laj C, Valet J-P, Tucholka P, Paterne M, and Guichard F (1991) The Blake geomagnetic event: Transition geometry, dynamical characteristics and geomagnetic significance. *Earth and Planetary Science Letters* 102: 1–13.

Tucholka P, Fontugne M, Guichard F, and Paterne M (1987) The Blake polarity episode in cores from the Mediterranean Sea. *Earth and Planetary Science Letters* 86: 320–326.

Turner GM, Evans ME, and Hussin IB (1982) A geomagnetic secular variation study (31,000-19,000 yr BP) in western Canada. *Geophysical Journal of the Royal Astronomical Society* 71: 159–171.

Turrin BD, Donnelly-Nolan JM, and Hearn BC Jr, (1994) ^{40}Ar/^{39}Ar ages from the rhyolite of Alder Creek, California: Age of the Cobb mountain normal polarity subchron revisited. *Geology* 22: 251–254.

Udagawa S, Kitagawa H, Gudmundsson A, *et al.* (1999) Age and magnetism of lavas in Jokuldalur area, eastern Iceland: Gilsa event revisited. *Physics of the Earth and Planetary Interiors* 115: 147–171.

Uyeda S (1958) Thermoremanent magnetism as a medium of palaeomagnetism with special reference to reverse thermoremanent magnetism, Japan. *Journal of Geophysics* 2: 1–123.

Valet J-P and Herero-Bervera E (2003) Some characteristics of geomagnetic reversals inferred from detailed volcanic records. *Comptes Rendus Geoscience* 335: 79–90.

Vandamme D (1994) A new method to determine paleosecular variation. *Physics of the Earth and Planetary Interiors* 85: 131–142.

Valet J-P and Meynadier L (1993) Geomagnetic field intensity and reversals during the last four million years. *Nature* 366: 234–238.

VandenBerg J, Klootwijk CT, and Wonders AAH (1978) Late Mesozoic and Cenozoic movements of the Italian peninsula; further paleomagnetic data from the Umbrian sequence. *Geological Society of America Bulletin* 89: 133–150.

Vine FJ (1966) Spreading of the ocean floor: New evidence. *Science* 154: 1405–1415.

Vine FJ and Matthew DH (1963) Magnetic anomalies over oceanic ridges. *Nature* 199: 947–949.

Verosub KL (1975) Paleomagnetic excursions as magnetostratigraphic horizons : A cautionary note. *Science* 190: 48–50.

Verosub KL (1977) The absence of the Mono Lake geomagnetic excursion from the paleomagnetic record at Clear lake, California. *Earth and Planetary Science Letters* 36: 219–230.

Verosub KL (1982) Geomagnetic excursions : A critical assessment of the evidence as recorded in sediments of the Brunhes epoch. *Philosophical Transaction of the Royal Society London* A306: 161–168.

Verosub KL, Davis JO, and Valastro S (1980) A paleomagnetic record from Pyramid lake, Nevada, and its implications for proposed geomagnetic excursions. *Earth and Planetary Science Letters* 49: 141–148.

Voelker A, Sarnthein M, Grootes PM, *et al.* (1998) Correlation of marine ^{14}C ages from the Nordic sea with GISP2 isotope record: Implication for ^{14}C calibration beyond 25 ka BP. *Radiocarbon* 40: 517–534.

Wagner G, Beer J, and Laj C (2000) Chlorine-36 evidence for the Mono lake event in the Summit GRIP ice core. *Earth and Planetary Science Letters* 181: 1–6.

Walter RC, Manega PC, Hay RL, Drake RE, and Curtis GH (1991) Laser fusion ^{40}Ar/^{39}Ar dating of Bed 1, Olduvai Gorge, Tanzania. *Nature* 354: 145–149.

Wang Y, Evans ME, Rutter N, and Ding ZL (1990) Magnetic susceptibility of Chinese loess and its bearing on paleoclimate. *Geophysical Research Letters* 17: 2449–2451.

Watkins ND, McDougall I, and Kristjansson L (1975) A detailed paleomagnetic survey of the type location for the Gilsa geomagnetic polarity event. *Earth and Planetary Science Letters* 27: 436–444.

Weeks R, Laj C, Endignoux L, *et al.* (1995) Normalized NRM intensity during the last 240,000 years in piston cores from the central north Atlantic ocean: Geomagnetic field intensity or environmental signal? *Physics of the Earth and Planetary Interiors* 87: 213–229.

Weeks RJ, Roberts AP, Verosub KL, Okada M, and Dubuisson GJ (1995) Magnetostratigraphy of upper Cenozoic sediments from Leg 145, north Pacific ocean. In: Rea DK, Basov IA, Scholl DW, and Allan J (eds.) *Proceedings of the ODP, Science Research*, 145, pp. 209–302. College Station, TX: Ocean Drilling Program.

Wensink H (1966) Paleomagnetic stratigraphy of younger basalts and intercalated Plio-Pleistocene tillites in Iceland. *Geology Rundschau* 54: 364–384.

Whitney J, Johnson HP, Levi S, and Evans BW (1971) Investigations of some magnetic and mineraological properties of the Laschamp and Olby flows, France. *Quaternary Research* 1: 511–521.

Wilson DS and Hey RN (1980) The Galapagos axial magnetic anomaly: Evidence for the emperor reversal within the Brunhes and for a two-layer magnetic source, (abs). *EOS* 61: 943.

Wintle AG (1973) Anomalous fading of thermoluminescence in mineral samples. *Nature* 245: 143–144.

Wollin G, Ericson DB, Ryan WBF, and Foster JH (1971) Magnetism of the earth and climatic changes. *Earth and Planetary Science Letters* 12: 175–183.

Worm H-U (1997) A link between geomagnetic reversals and events and glaciations. *Earth Planetary Science Letters* 147: 55–67.

Yamazaki T and Yoka N (1994) Long-term secular variation of the geomagnetic field during the last 200 kyr recorded in sediment cores from the western equatorial Pacific. *Earth and Planetary Science Letters* 128: 527–544.

Yang T, Hyodo M, Yang Z, and Fu J (2004) Evidence for the Kamikatsura and Santa Rosa excursions recorded in eolian deposits from the southern Chinese Loess plateau. *Journal of Geophysical Research* 109: B12105 (doi:10.1029/2004JB002966).

Yang Z, Clement BM, Acton GD, Lund SP, Okada M, and Williams T (2001) Record of the Cobb mountain subchron from the Bermuda Rise (ODP Leg 172). *Earth and Planetary Science Letters* 193: 303–313.

Yaskawa K, Nakajima T, Kawai N, Torii M, Natsuhara N, and Horie S (1973) Paleomagnetism of a core from Lake Biwa (I). *Journal of Geomagnetism and Geoelectricity* 25: 447.

Zhang K and Gubbins D (2000) Is the geodynamo intrinsically unstable?. *Geophysical Journal International* 140: F1–F4.

Zhu R, Hoffman KA, Nomade S, *et al.* (2004) Geomagnetic paleointensity and direct age determination of the ISEA (M0r?) chron. *Earth and Planetary Science Letters* 217: 285–295.

Zhu RX, Zhou LP, Laj C, Mazaud A, and Ding DL (1994) The Blake geomagnetic polarity episode recorded in Chinese loess. *Geophysical Research Letters* 21: 697–700.

11 Time-Averaged Field and Paleosecular Variation

C. L. Johnson, University of British Columbia, Vancouver, BC, Canada

P. McFadden, Geoscience Australia, Canberra, ACT, Australia

11.1 Introduction

The geomagnetic field measured at Earth's surface includes contributions from sources internal and external to the planet. The main field – that generated by magnetohydrodynamic processes in Earth's liquid iron outer core – exhibits spatial and temporal variations that can be observed directly today via surface, aeromagnetic, and satellite measurements, and indirectly over geological timescales via remanent magnetization in crustal rocks. This chapter deals with the geometry of the geomagnetic field and its temporal variability as recorded in volcanic and sedimentary rocks over the past few million years. The focus of the chapter is the information provided by directional records, as historically these have been the main contributing data sets. A complete understanding of paleofield behavior requires measurements of the full vector field – that is, both direction and intensity, and we discuss paleointensity data in this context. (A detailed review of paleointensity data is provided in Chapter 13.) We focus attention on the time period 0–5 Ma. This interval bridges the gap between the past few hundred or few thousand years for which global, continuously time-varying field models can be constructed, and long paleo timescales (tens of million years and longer for which data sets are sparse and geographical information is limited. Paleomagnetic data for the past 5 My provide sufficient temporal and spatial coverage to enable regional and global investigations; although we shall see that most of the debates surrounding field behavior stem from the need for improvements and additions to current data sets.

Characterization of magnetic field behavior on 10^5–10^6 year timescales is required to understand not only the evolution of the geodynamo, but also interactions among crust, mantle, and core processes. From an observational perspective, knowledge of the temporal evolution of the magnetic field globally provides essential context for understanding specific aspects of field behavior. Examples include whether there are conditions (e.g., low or variable intensity) under which magnetic reversals are more likely and how magnetic reversals relate to the overall temporal spectrum of field variations. Paleomagnetic data play an important role in tectonics – from studies of local or regional rotations to global plate reconstructions. Their usefulness depends critically on the validity of a very simple model for the long-term average magnetic field direction.

Electromagnetic coupling between the inner and outer core, together with the very presence of the inner core, affects the geometry of fluid flow in the outer core. The tangent cylinder – a hypothetical cylinder coaxial with Earth's rotation axis and tangent to the inner core at the equator – separates regions in the inner and outer core in which the fluid flow and the resulting magnetic field are expected to be quite different. The inner core has been suggested as the cause of some of the features seen in historical field models (see, e.g., Bloxham *et al* (1989) and Jackson *et al* (2000)), and as providing

needed stability against reversals (Hollerbach and Jones, 1993a, 1993b; Gubbins, 1999), although this is debated (Wicht, 2002). Intersection of the tangent cylinder with the core–mantle boundary (CMB) occurs at latitudes of $\pm 70°$, and an obvious question is whether there are observable differences in the paleomagnetic field at latitudes above and below this at Earth's surface, although sampling locations are clearly restricted. The presence, geometry, and strength of the field preserved in ancient (Archaean) rocks (Hale and Dunlop, 1984) provides a leading order constraint for the early thermal evolution of our planet, including perhaps key insight into the age and growth of the inner core (Labrosse, 2003; Jaupart *et al.*, 2007). Posing testable hypotheses regarding such earliest field behavior requires proper characterization of 'recent' paleomagnetic field behavior (past few million years). For example, several studies have suggested that the onset of inner–core growth might be diagnosed geomagnetically via a low intensity field (Valet, 2003). However, what constitutes a 'weak' field requires an understanding of the mean global field strength over a range of timescales, and this is a topic of intense debate (*see* Chapter 13)

Over timescales of millions of years and longer, geomagnetic field spatial and temporal variability may reflect not just the presence of the inner core but also the nature of CMB thermal (Gubbins and Richards, 1986; Merrill *et al.*, 1990; Gubbins, 1988; Bloxham and Jackson, 1990) and electromagnetic (Runcorn, 1992; Clement, 1991; Laj *et al.*, 1991; Costin and Buffett, 2005) coupling. There is growing evidence that the lowermost mantle is laterally heterogeneous in physical and chemical properties over a large range of spatial scales (see review in, e.g., Jellinek and Manga (2004)). Long wavelength thermal anomalies, evidenced in seismic tomographic images of the lower mantle, have been suggested to account for some of the features observed in the historical magnetic field (e.g., Bloxham *et al.*, 1989), leading to the speculation that such features might persist over timescales consistent with mantle convection. As we shall see, the question of whether spatial variations in CMB conditions result in detectable signals in the 0–5 My paleomagnetic record is an area of considerable discussion.

Over the past decade, enormous advances have been made in numerical simulations of the geodynamo (see reviews in Roberts and Glatzmaier (2000), Busse (2002), Kono and Roberts (2002)). Such simulations have had remarkable success in capturing the main qualitative features of the real field (dipole dominated, occurrence of reversals, some aspects of temporal variability), despite their current inability to operate in the parameter regime applicable to the outer core. This success has, in turn, led to comparisons of dynamo simulations with paleomagnetic observations (e.g., Glatzmaier *et al.*, (1999)). Clearly paleomagnetic data provide important constraints for future simulations; in particular in enabling comparison of statistical aspects of real and simulated global field behavior (McMillan *et al.*, 2001; Bouligand *et al.*, 2005).

The temporal and spatial scales of variability in Earth's magnetic field provide insight into the underlying magnetohydrodynamics of the outer core, and into how the dynamics of that system are affected by its boundaries at the inner core and the mantle. Determining those spatial and temporal variations requires geographically distributed data over a range of timescales. We deal here with paleomagnetic data on timescales of tens of thousands to millions of years; thus we are examining processes that have a signature at Earth's surface over periods much longer than core overturn times. Our interest is in both the time-averaged field (TAF) geometry and the temporal variations about that long-term average. The most basic approximate description of the field over time is that due to a geocentric axial dipole (GAD): we examine temporal variations in the field (paleosecular variation (PSV)) and long-term time-averaged departures from GAD (the TAF).

The chapter is organized as follows. In the remainder of the introduction we provide further background and motivation, specifically for studies of PSV and the TAF. We then outline some of the essential concepts that are needed to navigate the material that follows and the relevant literature (Section 11.2). In Section 11.3 we review global data sets that have been used in global and regional TAF and PSV studies. In Sections 11.4 and 11.5 we summarize global models for PSV and the TAF. The review is not exhaustive, we highlight contributions that exemplify the major viewpoints of the community as they evolved. We analyze the important issues related to data sets and modeling approaches that have led to differing conclusions in the literature (Section 11.6). Several reviews of the TAF have been published previously (Merrill *et al.*, 1996; McElhinny *et al.*, 1996; Merrill and McFadden, 2003), but the concluding remarks in these reviews represent one interpretation of existing models. We discuss the current status of both modeling and

database efforts (Section 11.6) and (Section 11.7) offer some thoughts on future directions.

11.1.1 The Geocentric Axial Dipole (GAD) Approximation

The geomagnetic field at Earth's surface can be approximated by that due to a dipole at Earth's center and aligned with the rotation axis. The instantaneous field shows significant departures from this GAD approximation (e.g., present magnetic north is not co-located with geographic north); however, over geological timescales it might be expected that temporal variability in the field means that such departures will average to zero. Using the statistical methods developed by Fisher (1953), Hospers (1954) showed that, averaged over several thousand years, GAD provides a good approximation to the observed magnetic field. Opdyke and Henry (1969) used observations of paleomagnetic directions from deep-sea piston cores to demonstrate that the GAD approximation has held over the past 2.5 My (**Figure 1**). It is now well known that Earth's field can be described to first order by GAD, oriented in either its current (normal) or reverse configuration. The GAD approximation has been heavily exploited in tectonic studies and is central to models for plate reconstruction.

11.1.2 Paleosecular Variation

The magnetic field generated in the outer core exhibits variability on all timescales. PSV describes the temporal variations in the paleomagnetic field, for which observations are of course obtained at Earth's surface. While the upper bound on timescales associated with PSV is clearly the age of the geomagnetic field itself, the term 'PSV' is often used in the context of characterizing field variations during stable polarity periods. Furthermore, the use of the term PSV usually carries the implicit assumption that only variations in direction will be discussed. These rather arbitrary (and limiting) separations of timescale and components of the field vector arise from the historical availability of specific data types, and a desire to distinguish temporal variations during stable polarity periods from geomagnetic reversals.

PSV is studied via time series of field variations or via statistical descriptions of paleomagnetic records with incomplete age control. These provide quite different measures of temporal variations. Secular variation can be defined as the rate at which the

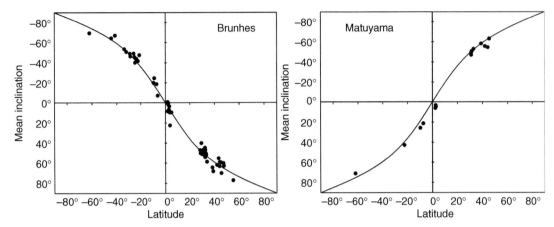

Figure 1   Inclination data from the deep sea sediment cores of Opdyke and Henry (1969) (filled circles) for the Brunhes (left) and Matuyama (right) epochs. Solid line is the inclination predicted by a geocentric axial dipole (GAD) field.

field changes, $\partial_t \vec{B}$ (Courtillot and Valet, 1995; Love, 2000), and time series of paleomagnetic observations allow this to be assessed. Relative paleointensity or direction from long sediment cores are candidates for time series analyses, especially the use of spectral estimation techniques. However, even when such data sets exist, characterization of PSV is far from straightforward due to issues such as age calibration, smoothing of the magnetic field during remanence acquisition, and the comparison of data from different geographical locations. In contrast to sedimentary records, lava flows provide geologically instantaneous recordings of the paleofield, but the data distribution is determined by the occurrence of volcanism and the present-day accessibility of flows. Radiometric ages are typically available for only a small percentage of flows sampled for paleomagnetic purposes. Consequently, statistical methods must be employed to assess PSV, via for example, the variance in directions and/or intensity. These summary statistics provide a measure of PSV, but no rate information. (In cases where sequences of flows are available, the correlations between vectors from stratigraphically adjacent flows can provide some rate information (Love, 2000)). Confusingly, the paleomagnetic literature often refers to analyses based on time series from sedimentary records as 'PSV' studies, and analyses based on spot recordings from volcanic rocks as 'PSVL' studies (paleosecular variation from lavas).

Here we use the term PSV to describe temporal variations in the field manifest by the magnetic field vector preserved in the rock record. We examine directional and intensity records (ideally the full

vector field at any location) from lavas or sediments. Of particular interest are geographical variations in PSV: are differences recorded inside and outside the tangent cylinder, and are there longitudinal, as well as latitudinal variations? Also important to understanding PSV is the issue of how variations in direction are related to variations in intensity (e.g., Love, 2000).

Analyses of the paleomagnetic power spectrum (Barton, 1982; Constable and Johnson, 2005) indicate a continuum of temporal variations, but of interest historically has been the question of whether one can usefully characterize PSV over a specified timescale. Barton (1982) used sediment records to suggest that timescales of at least 10^3 years provided reasonable estimates of PSV, more recent studies suggest at least 10^4–10^5 years (Carlut et al., 1999; Merrill and McFadden, 2003). One of the major difficulties in understanding the temporal and spatial scales of PSV is the limited and uneven distribution of paleomagnetic data.

11.1.3   The Time-Averaged Field (TAF) – Departures from GAD?

While the GAD approximation describes most of the structure in the geomagnetic field, it has been known for some time (Wilson, 1970, 1971) that the paleomagnetic record displays small but persistent departures from GAD. Records of paleodirection from Europe and Asia were found to have time-averaged pole positions that showed an offset from geographic north; specifically poles that were far-sided (on the other side of geographic north relative to the observation location) and right-handed (eastward of the observation location for normal polarities). Far-sided pole positions result from time-averaged values of inclination that have a small negative bias compared with the GAD prediction (**Figure 2**). Wilson (1970, 1971) interpreted the deviations from GAD as due to a dipole offset northward along the rotation axis, although this is of course a nonunique solution.

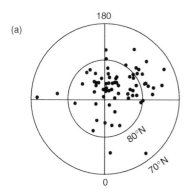

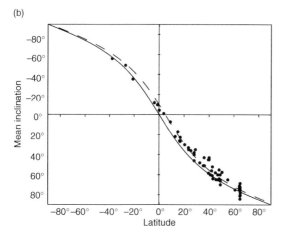

Figure 2 (a) Stable polarity continental sediment and lava flow VGPs from 72 sites for the past 25 My (data are the AF demagnetized data of Wilson, (1971)). Center of the projection is the geographic pole, contours are virtual geomagnetic pole (VGP) latitudes of 80° and 70°. VGPs are plotted as if all the sites were on a common site longitude, taken here as 0°. VGPs are seen to plot on the far-side of geographic north, compared with the common observation site longitude. The so-called right-handed effect is also seen – VGPs appear to plot on average eastward of 180°. (b) Upper Tertiary inclination data from Wilson (1970) (solid circles). Solid line is the predicted inclination as a function of latitude from a GAD model. The dashed line is the predicted inclination for a dipole with an offset r = 306 km (Wilson, 1970). The observed inclinations are systematically shallower than those predicted by GAD, and are better matched by the offset dipole model.

Departures of the time-averaged geomagnetic field (TAF) from GAD are a topic of this chapter. An issue that becomes immediately apparent is whether in fact there is a stable TAF geometry, in other words whether the process by which the magnetic field is generated is statistically stationary. Even if this is not the case, the question remains as to over what intervals the field should be averaged in order to provide useful information.

Over the historical period, AD 1590–1990, satellite, observatory, and survey measurements of the geomagnetic field have permitted the construction of a spatially detailed, temporally varying magnetic field model, GUFM1 (Jackson *et al.* (2000), and see also, e.g., Bloxham *et al.* (1989) and Bloxham and Jackson (1992)). A representation of the time-averaged structure can be obtained by averaging the model over its respective time interval. The radial component of the magnetic field, B_r, is shown in **Figure 3** after downward continuation to the surface of Earth's core under the assumption that the mantle is an insulator. The field due to GAD is also shown and displays only latitudinal structure. GUFM1 has significant non-GAD structure, which has been extensively discussed elsewhere (Jackson *et al.*, 2000), and is thought to be influenced by the presence of the inner core and by lateral heterogeneity in the lowermost mantle. Regions of increased radial flux at high latitudes, commonly referred to as flux lobes, have persisted in much the same locations for 400 years. Low radial field over the North Pole has been interpreted as a manifestation of magnetic thermal winds and polar vortices within the tangent cylinder (Hulot *et al.*, 2002; Olson and Arnou, 1999; Sreenivasan and Jones, 2005, 2006). Equatorial flux patches that are pronounced in the time-varying version of GUFM1 and in modern satellite models (e.g., Hulot *et al.*, 2002), and that appear to propagate westward in the Atlantic hemisphere, are attenuated in the 400 year temporal average. Overturn times for fluid motions in the outer core are on the order of a few hundred years, and thus the persistence of features over 400 years in GUFM1 suggests that magnetic field generation in the outer core may be influenced by the inner core and outer core boundaries. Of particular interest in the context of other areas of deep Earth geophysics have been suggestions of inner core/outer core electromagnetic and core–mantle thermal coupling (Bloxham *et al.*, 1989; Jackson *et al.*, 2000). If conditions at the CMB are important, then the timescales associated with these will be those of mantle convection, and some of the

(a)   1590–1990 time-averaged field

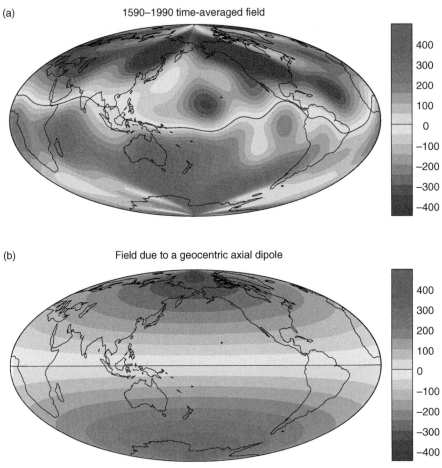

(b)   Field due to a geocentric axial dipole

Figure 3   Radial field at the CMB in μT; Hammer–Aitoff projection. (a) Model GUFM1 (Jackson *et al.*, 2000), (b) Field due to a geocentric axial dipole – note that this possesses only latitudinal structure in contrast to the time-averaged historical field in (a).

steady features in GUFM1 might be expected to persist over the past few million years, albeit damped by PSV.

11.2   Essential Concepts

In this section we cover some of the essential concepts needed to navigate the PSV and TAF literature. Some are soundly based in mathematical formulation; others are (for better or worse) conventions that have been adopted to assist the analysis of paleomagnetic data.

11.2.1   Paleomagnetic Observations

Paleomagnetic observations comprise records of ancient field direction and intensity recorded in

sedimentary and igneous rocks. Paleodirections are specified by declination, D, and inclination, I (**Figure 4**). Declination is the angle between the field vector projected onto the horizontal plane and geographic north. Inclination is the dip of the magnetic field vector from the local horizontal. Paleodirections are related to the local north (X), east (Y), and down (Z) magnetic field elements as follows:

$$D = \tan^{-1}\left(\frac{Y}{X}\right), \quad I = \tan^{-1}\left(\frac{Z}{(X^2 + Y^2)^{1/2}}\right) \quad [1]$$

Ideally one would like a measure of the full vector field; however a consequence of practical difficulties inherent to obtaining intensity measurements (*see* Chapter 13) is that current data sets spanning the past 5 My are dominated by paleodirections. Lava flow data are (D, I) pairs, but lack of azimuthal

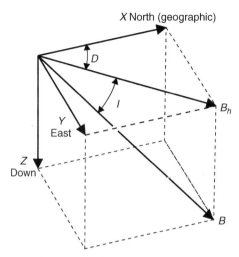

Figure 4 Geomagnetic field elements – local orthogonal components north, east, and down, and paleomagnetic measurements of declination, D and inclination, I.

orientation information means that often only I measurements are available for deep-sea sedimentary cores. The former provide intermittent or spot readings of the paleofield, the latter provide continuous time series of observations. **Figures 5** and **6** show typical representations of each of these data types; specific data sets are discussed in Section 11.3.

Neel theory (1949) suggests that under certain conditions the thermal remanent magnetization (TRM) acquired by, for example, lava flows can be reproduced in the laboratory, allowing measurements of absolute paleointensity. In practice, such data are more difficult to acquire than paleodirections due to alteration of the sample during the laboratory procedure. In sedimentary rocks, remanence is acquired during deposition and compaction, and unlike the case for lava flows no theoretical basis exists for absolute paleointensity measurements. Instead, one measures the natural remanent magnetization (NRM) of the sample and normalizes by a proxy that can account for the amount and type of magnetic carrier present. These relative paleointensity data are easier to acquire, although a whole field of study surrounds the choice of normalizer, and the issue of how best to calibrate relative to absolute paleointensity has not yet been satisfactorily explored.

Details of paleomagnetic sampling techniques are covered in Chapter 4. Typically, samples are drilled in the field from individual lava flow or sedimentary units. Each flow or sedimentary unit allows an estimate to be made of the paleofield at a specific

location at the time of cooling (lavas) or compaction (sediments). This is known as a site. At a given site, multiple samples are collected to allow averaging of directions to reduce the influence of measurement error, in particular orientation error. These samples are demagnetized in the laboratory to remove (unwanted) secondary remanence. Important issues for the use of paleomagnetic data in field modeling studies are whether site-level measurement noise can be adequately assessed, and whether overprints have been removed.

11.2.2 Paleomagnetic Data Analysis

11.2.2.1 Comparing data from different locations

For a GAD field, inclination and intensity of the field will vary with latitude (λ), with inclination predicted by

$$\tan I = 2\tan\lambda \qquad [2]$$

A standard approach in paleomagnetism is to calculate the equivalent geocentric (but not axial) dipole that would give rise to the observed site-level paleodirection. The pole position of this equivalent dipole is known as a 'virtual geomagnetic pole' or VGP, and, under the assumption of a geocentric dipole, is independent of the site location. An alternative approach, proposed by Hoffman (1984), but used less frequently is to compute the direction, D', I', relative to the expected GAD direction at the site. An example of a data set represented by D, I pairs, D', I' pairs, and VGP positions is shown in **Figure 7**. D, I pairs from 175 lavas (115 normal polarity, 60 reverse polarity) from Hawaii are shown (there is no significance to the choice of data) plotted on an equal area projection. The closed (open) circles represent projections onto the lower (upper) hemisphere and for Northern Hemisphere sites represent normal (reverse) polarities. The GAD predicted direction at the mean site location is given for normal (reverse) polarities by the filled (open) red triangles. Azimuth on the figure represents declination, measured clockwise from north; distance from the center of the figure gives inclination, where the center of the figure is $I = 90°$. On average the directions are shallower than GAD (a negative inclination anomaly), most obvious visually in the reverse polarity data. Some directions are quite far from the GAD prediction. The VGP equal area projection gives VGP longitude (azimuth) and latitude (distance from center, with center being 90°).

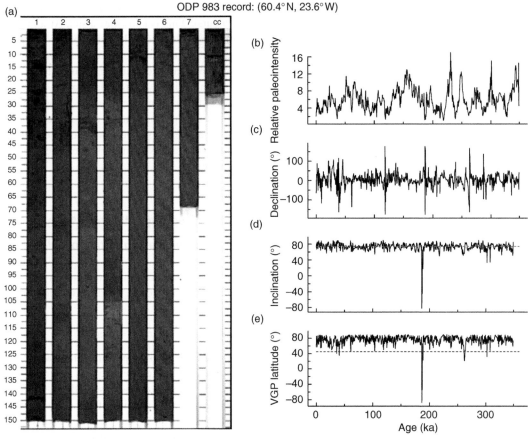

Figure 5 Example of sediment record (a) core photo, (b) –(e) time series of observations. In (d) dashed red line is the inclination predicted by GAD. The large variability in declination compared with inclination is due to the high site latitude. Black dashed line (e) is at a VGP latitude of 45°; several places in the time series indicate directions with VGP latitudes lower than 45° – whether these are excursions or part of typical stable polarity PSV is not immediately obvious. A reversed polarity direction is clearly recorded at ~180 ka. The total VGP dispersion, S_T (eqn 8) calculated for directions with VGP latitudes greater than 45° is $S_T = 16.6°$.

Several directions in this data set would be regarded as transitional (VGP latitudes less than 45°), and are also seen to be quite anomalous in the D', I' representation. Note that the shape of the distributions in all three figures appear different because of the nonlinear mappings involved in converting D, I to VGP or to D', I'.

11.2.2.2 Measures of PSV and the TAF

Global studies of PSV and the TAF have used data sets for which temporal control is incomplete, and so summary statistics are needed to characterize the field. Given a collection of N sites, that span a sufficiently long time period to enable an estimate of PSV and the TAF, the TAF can be described via the mean direction at a given location, and PSV via the variance (dispersion) in the field.

The paleodirection specified by a (D, I) pair can be expressed as direction cosines:

$$x = \cos D \cos I, \quad y = \sin D \cos I, \quad z = \sin I \quad [3]$$

Given a set of N pairs of D and I measurements, the unit vector mean direction has direction cosines

$$X = \frac{1}{R}\sum_{i=1}^{N} x_i, \quad Y = \frac{1}{R}\sum_{i=1}^{N} y_i, \quad Z = \frac{1}{R}\sum_{i=1}^{N} z_i \quad [4]$$

R is the vector sum of the individual unit vectors, given by

$$R = \left[\left(\sum_{i=1}^{N} x_i\right)^2 + \left(\sum_{i=1}^{N} y_i\right)^2 + \left(\sum_{i=1}^{N} z_i\right)^2 \right]^{1/2} \quad [5]$$

X, Y and Z from eqn [4] can be used in eqn [1] to estimate the mean declination, D, and inclination,

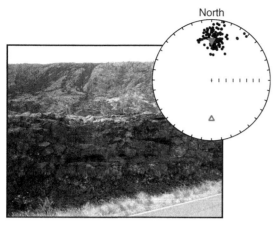

North

Figure 6 Lava sequence from the Southwest Rift zone of Kilauea Volcano, Hawaii. Paleomagnetic directions shown in the equal area plot represent samples taken from sites surrounding Kilauea with ages from the mid-Pleistocene to the Quaternary. (Data are a subset of the compilation in Lawrence *et al.*, 2006.) The direction predicted by GAD is shown in red. Paleomagnetic directions in the equal area plot are specified by declination (azimuth clockwise in plot from North) and inclination (radial distance, where the center of the plot is $I = 90°$ and the circumference is $I = 0°$). Photograph courtesy of Roi Granot.

I, for the set of N directions. Interest in the time-averaged field stems, quite naturally, from the desire to quantify departures from GAD (if any). A TAF direction is usually expressed in terms of its deviation from that predicted at the site location by GAD, that is, in terms of the inclination anomaly (ΔI) and declination anomaly (ΔD), where

$$\Delta I = \bar{I} - I_{GAD}, \quad \Delta D = \bar{D} \qquad [6]$$

I_{GAD} can be calculated from eqn [2] and $D_{GAD} = 0$.

The nonlinear mapping between the field direction (D, I) at a particular site and the latitude, λ, and longitude, ϕ of the VGP position, means that the scatter caused by variations in the field manifests itself differently in the VGP frame of reference from the (D, I) frame of reference (**Figure 7**). Consequently, a decision has to be made as to which frame of reference is to be used. Historically, discussions of PSV have used the VGP frame of reference. A mean VGP position can be calculated as for the mean direction using eqns [3]–[5] and the procedure described above. An angular standard deviation, S, can be defined, as

$$S = \left(\frac{1}{N-1} \sum_{i=1}^{N} \Theta_i^2 \right)^{1/2} \qquad [7]$$

where Θ_i is the angle between the ith direction and the sample mean, and again N is the number of sites. Approaches vary among studies: sometimes the scatter about the mean VGP (computed from the data themselves) is calculated, often the scatter is calculated assuming that the mean VGP coincides with geographic north (the GAD hypothesis). This difference in approach is in fact significant and the choice is critical in the development of models of PSV (McFadden *et al.*, 1988).

The scatter, S, calculated as above, can be assessed at the site level in which case it measures within-site error (S_W). If the site properly represents a single point in time then S_W should represent only our own measurement errors plus errors caused by imperfection in the original recording mechanism of the rock; it should not include any variation due to changes in the geomagnetic field itself. The scatter S_T, again calculated as above, but using the individual site means from several sites in a PSV study as the observations, then gives the total scatter; this is a combination of the within-site scatter and the scatter caused by the variation of the geomagnetic field from site to site, referred to as the between-site dispersion, S_B. This between-site scatter is a measure of PSV and is given by

$$S_B^2 = S_T^2 - \frac{S_W^2}{n} \qquad [8]$$

where n is the average number of samples measured at each site. Eqn [8] can be modified to account explicitly for unequal numbers of samples at each site:

$$S_B = \sqrt{\frac{1}{N-1} \sum_{i=1}^{N} \left(\Theta_i^2 - \frac{S_{W_i}^2}{n_i} \right)} \qquad [9]$$

S_{W_i} is the within-site dispersion determined from the n_i samples at the i'th site.

Alternative statistics used to describe PSV include standard deviation in inclination/declination (σ_I, σ_D, respectively), and the root mean square angular deviation of paleodirections (S_d). Site-level intensity measurements are usually converted to the equivalent virtual axial dipole moment (VADM) (*see* Chapter 13), and variations in intensity are quantified by, for example, the standard deviation in VADMs.

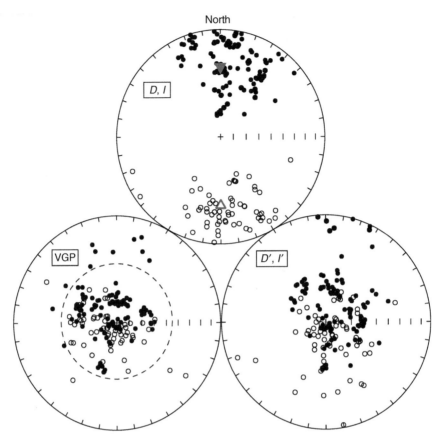

Figure 7 Equal area projections of a data set from Hawaii (see text for description) to show (D, I) projection (top figure) with GAD prediction, representation in terms of VGP positions (lower left), and representation in (D', I') coordinates. Directions that have large angular deviations from GAD can be seen in the top figure, and are manifest as VGPs with latitudes less than 45° (dashed circle) and (D', I') coordinates that plot far from the center of the (D', I') figure. Solid (open) circles represent normal (reverse) directions.

11.2.3 Global Field Models: Spherical Harmonic Representation

Models of the geomagnetic field can be constructed using a spherical harmonic representation. The magnetic scalar potential in a source-free region due to an internal field obeys Laplace's equation, and can be written as

$$\Psi(r, \theta, \phi, t) = a \sum_{l=1}^{\infty} \sum_{m=0}^{l} \left(\frac{a}{r}\right)^{l+1}$$
$$\times \left(g_l^m(t)\cos m\phi + h_l^m(t)\sin m\phi\right) P_l^m(\cos\theta) \quad [10]$$

where $g_l^m(t)$ and $h_l^m(t)$ are the Schmidt partially normalized Gauss coefficients at a time t, a is the radius of the Earth, r, θ, and ϕ are radius, colatitude, and longitude, respectively, and P_l^m are the partially normalized Schmidt functions. l is the spherical harmonic degree and m is the spherical harmonic

order. The magnetic field, $\vec{B}$, is the negative gradient of the potential Ψ, so

$$B_r = -\frac{\partial\Psi}{\partial r}, \quad B_\theta = -\frac{1}{r}\frac{\partial\Psi}{\partial\theta}, \quad B_\phi = -\frac{1}{r\sin\theta}\frac{\partial\Psi}{\partial\phi} \quad [11]$$

For paleomagnetic purposes (where we assume a spherical reference surface) the relationship between the local coordinate system in **Figure 4**, and the spherical coordinate system is $B_r = -Z$, $B_\theta = -X$, and $B_\phi = Y$.

A given field model is specified by the Gauss coefficients $g_l^m(t)$ and $h_l^m(t)$. The $m = 0$ terms correspond to spherical harmonic functions with no azimuthal structure – that is, they are axially symmetric or zonal. Such fields have predicted declinations of zero. g_1^0, g_2^0, and g_3^0 are the Gauss coefficients representing the axial dipole, axial quadrupole, and axial octupole terms, respectively.

Paleomagnetic observations, D, I, and intensity, are all nonlinearly related to the Gauss coefficients and because of limited temporal information it is not currently possible to determine the variation of these coefficients as a function of time. Instead, paleomagnetic field modeling to date has typically involved two approaches: (1) the use of mean field directions along with parameter estimation or inversion to estimate mean values for g_l^m and h_l^m over time intervals of interest – TAF models, and/or, (2) specification of the variance in the Gauss coefficients *a priori* and forward modeling of summary statistics (e.g., S_B) for PSV.

11.3 Data Sets

In this section we review paleodirectional data sets that were designed to be suitable for global and/or regional TAF and PSV studies. We document compilations spanning at least the Brunhes normal polarity chron (past 780 ky) and as long as the past 5 My. As much of the discussion surrounding TAF and PSV modeling is related to issues of data distribution and quality we summarize the selection criteria used in each study. We mention paleointensity data sets briefly – but we note that good quality paleointensity data will provide important constraints in future PSV and TAF research.

For paleomagnetic data to be useful for TAF or PSV studies the following criteria need to be met:

1. The sampling sites should not have been subjected to tectonic effects since the acquisition of magnetic remanence.
2. A sufficient number of temporally independent sites, N, need to have been sampled, covering a time period of at least 10^4 years (and preferably longer), to minimize bias in estimates of PSV and the TAF.
3. Multiple samples per site (n) should have been taken to allow assessment of within-site error.
4. The remanence should have been established as primary via laboratory cleaning (demagnetization).

Because of the historical desire to examine PSV during stable polarity periods as a distinct entity, separate from reversals, transitional data (those that deviate greatly from GAD) have been excluded from PSV and TAF studies. The measure used to assess whether a paleodirection is transitional is the VGP latitude; various choices

for a VGP latitude that discriminates between these two states have been proposed (Vandamme, 1994; McElhinny and McFadden, 1997), but whatever the choice, the distinction is artificial, and purely for convenience.

We review data sets based on paleodirections from lavas and sediments separately. Lava flows offer the advantage of declination and inclination observations, but temporal information is poor in older collections (e.g., often only the polarity chron is known, and there is little information on the time interval spanned by a series of sites at a given location). Deep-sea sediment cores provide sampling in the ocean basins, but the lack of declination information, means that in practice only latitudinal variations in the TAF and PSV can be assessed with these data. Furthermore, the deep-sea sediment cores considered below are piston cores with no independent orientation information. The advantage offered by these data though is a better temporal average due to smoothing during remanence acquisition, and the possibility of estimating the time interval spanned by a core of a given length if sedimentation rates are known.

11.3.1 Global Database – Paleosecular Variation from Lavas (PSVRL) Database

To date the most comprehensive lava flow database assembled for PSV studies is that of McElhinny and McFadden (1997), hereafter *MM97*. It comprises data from 3719 lava flows and thin dikes covering the peroid 0–5 Ma. The restriction to lava flows and dikes ensures that data are from igneous units that cooled quickly and for which the paleodirection is an instantaneous recording of the field.

This database, known as the PSVRL database, grew out of a series of IAGA-sponsored databases (Lock and McElhinny, 1991; McElhinny and Lock, 1993, 1996), and preceeding data sets, the earliest of which was that of McElhinny and Merrill (1975). The goal of this database was to be as inclusive as possible and so only 2 samples per site ($n = 2$) were required, and data subjected to any level of laboratory demagnetization were accepted. A demagnetization code was assigned to all sites, this DMAG code varies between 1 (no cleaning) and 4 (modern lab methods, typically employed in most studies from the early 1990s onward). At a given location, 5 sites ($N = 5$) were considered sufficient to estimate PSV and the TAF, and stable polarity data

Normal polarity data, GPMDB

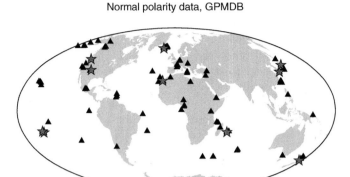

Reverse polarity data, GPMDB

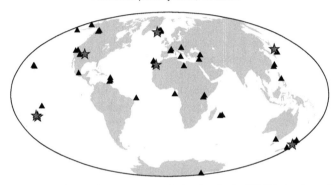

Figure 8 Geographical distribution of normal and reverse polarity sites in the PSVRL database (McElhinny and McFadden, 1997). All data (black triangles), DMAG 4 data only (red stars). Note that there are very few distinct locations (8 normal polarity, 6 reverse polarity) with data that meet current laboratory protocols.

were defined as those with VGP latitudes greater than 45°.

The geographical distributions of normal (0–5 Ma) and reverse (0–5 Ma) data are shown in **Figure 8**. MM97 argue that because modern magnetic cleaning methods are required to remove secondary overprints, only the DMAG = 4 data are suitable for TAF and PSV modeling – these data comprise less than 12% of the database and are from only 8 distinct locations (**Figure 8**). As part of their assessment MM97 listed many previous studies that should be replaced by new data, subjected to current lab protocols for demagnetization; this list has led to many of the recent paleomagnetic sampling efforts.

11.3.2 Other Global Lava Flow Data Sets

Several other lava flow data sets have been assembled for TAF and PSV modeling. We summarize two here;

those of Quidelleur et al. (1994) and Johnson and Constable (1996), hereafter Q94 and JC96, respectively. Together with the PSVRL database, these two data sets have formed the basis for most of the TAF and PSV studies over the past 13 years. Both were built on the compilation by Lee (1983), adding newer references from the published literature.

Q94 and JC96 contain many data in common; over 50% of the data in JC96 are included in Q94. Q94 includes more data than JC96, as criteria associated with the number of samples per site, n, and the cutoff VGP latitude for transitional data, were less stringent. The number of samples per site needed to assess within-site error has been a subject of debate. In the absence of systematic noise, the uncertainty in the site mean direction should decrease as $1/\sqrt{n}$, and the estimation of a variance requires at least 3 samples per site. JC96 required $n = 3$ and this results in the exclusion of many sites, notably large numbers of data from Iceland and other strategic locations in the context of global data coverage. For a site with $n > n_{cutoff}$, choices as to what constitutes large within-site error also differ among studies. Measures of within-site error usually assume that paleofield directions at a given location can be represented by a Fisher distribution. (The Fisher distribution is the analog, for vectors on a unit sphere, of the two-dimensional normal distribution (Fisher, 1953)). Some studies define a within-site error requirement based on k, the best estimate of the Fisher precision parameter, κ (Fisher, 1953):

$$k = \frac{n-1}{n-R} \qquad [12]$$

where n is the number of samples at the site, and R is defined in eqn [5]. The precision parameter is the analog, for vectors on a unit sphere, of the inverse of the variance of a two-dimensional normal distribution. Other studies use the size of the 95% cone of confidence, α_{95}:

$$\alpha_{95} \approx \frac{140°}{\sqrt{kn}} \qquad [13]$$

We do not belabor differences among the Q94 and JC96 data sets as the collective contribution of sampling efforts by the paleomagnetic community over the past decade are already resulting in global data sets that supersede these existing compilations in number and quality (see Section 11.7).

The data set of Q94 comprises 3179 lava flows, that of JC96 2187 records. The number of distinct locations reported in each study are 86 (Q94) and 104 (JC96), although about 50 of those in JC96 differ by less than 1° spatially. Both data sets contain about twice as many normal polarity records as reverse data, and the data sets are dominated by Brunhes age paleodirections. The spatial and temporal distributions of JC96 are shown in **Figure 9** and **10**; those for Q94 are not shown but are similar. Northern Hemisphere data coverage of both JC96 and Q94 is reasonable, especially for the 0–5 Ma normal polarity data combined. Reverse polarity data provide poorer spatial coverage. Coverage of the Southern Hemisphere is limited, in part because the data sets only include land-based observations.

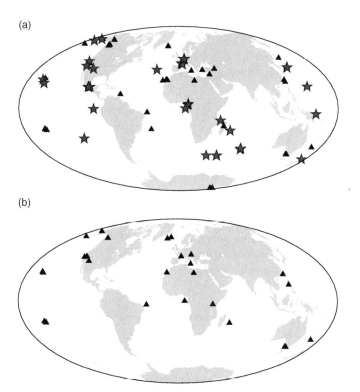

(a)

(b)

Figure 9 Geographical data distribution of 0–5 Ma lava flow data set of Johnson and Constable (1996). (a) Brunhes (blue stars) and 0–5 Ma normal polarity data (black triangles); (b) 0–5 Ma reverse polarity data.

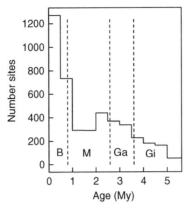
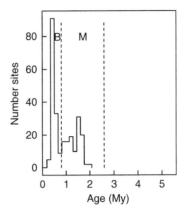

Figure 10 Age distribution of JC96 lava flow data (left) and SK90 sediment data (right). Polarity chrons are delineated by vertical dashed lines: B, Brunhes; M, Matuyama; Ga, Gauss; Gi, Gilbert.

11.3.3 Recent Regional Lava Flow Compilations

Several recent studies have combined new paleo-magnetic data with 'legacy' data (i.e., data sets already included in, e.g., the PSVRL database) to produce large regional data sets. Six compilations published to date are for paleomagnetic directions from the NW and SW USA (Tauxe *et al.*, (2004a, 2004b, 2003), respectively), Mexico (Mejia *et al.*, 2005; Lawrence *et al.*, 2006), Hawaii, the South Pacific and Reunion (Lawrence *et al.*, 2006). These data sets, combined with other studies will provide greatly improved lava flow data sets for TAF and PSV studies, and we return to this in Section 11.7.

11.3.4 Sedimentary Records

Deep-sea sediment cores fill a large gap in the geographical distribution of PSV/TAF data. For deep-sea cores the time interval spanned for a given core can be estimated via mean sedimentation rates, and so the time-averaging at each location is better controlled than for lava flows. The sedimentation rate, along with type of laboratory methods used (in particular whether pass-through magnetometers are used) determines the along-core sampling interval for which adjacent samples can be considered temporally independent.

The first compilation of deep-sea sediment cores was the piston core data set used by Opdyke and Henry (1969) to demonstrate the validity of the GAD approximation over the past ~2.5 My. This was expanded and improved by Schneider and Kent (1988, 1990), with the goal of investigating small, but long-lived departures from GAD. The data set of

Schneider and Kent (1990), hereafter *SK90*, comprises piston-cores from the low-mid latitudes (**Figure 11**) and contains 176 cores with Brunhes data and 125 cores with Matuyama data. The standard error in inclination for each core is typically on the order of 1°, however as shown in Johnson and Constable (1997) this does not reflect the strong regional variability in inclination estimates. For lava flow data this variability might be caused by inadequate temporal sampling at one or more sites; however sediment cores should provide a good time-averaged field direction because the record is almost continuous and a large number of sample directions are averaged to give an inclination anomaly for each core. Regional variability in inclination anomalies recorded by sediments suggests inconsistency among the observations; this can be caused by inadequate demagnetization of some cores, and by nonvertical coring. Cores from the North Pacific region appear to be particularly inconsistent (Johnson and Constable, 1997). These cores were some of the first piston cores to be collected and were demagnetized in low alternating fields (5–15 mT compared with 15–40 mT for some later cores (*SK90*), and may contain viscous overprints.

Inclination records provided by more recent, very long drill cores (such as those of the Ocean Drilling Program (ODP) and Deep Sea Drilling Program) will contribute greatly to PSV and TAF studies in the future (Section 11.7). For example, 15 cores worldwide contain data that span several thousand years or more in the period 0–2 My (Valet *et al.*, 2005) and provide not only relative paleointensity records, but also inclination data. In some cases cores or core sections are sufficiently well oriented to provide absolute declination.

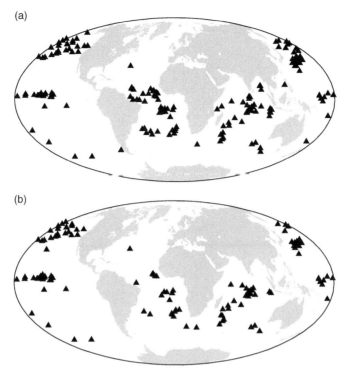

Figure 11 Geographic distribution of Brunhes and Matuyama piston cores in the Schneider and Kent (1990) data set: (a) Brunhes polarity piston cores; (b) Matuyama polarity piston cores.

11.4 Paleosecular Variation

The study of PSV has been a major part of paleomagnetism over the past four decades. In some cases it is possible to examine continuous sedimentary records using traditional time-series analysis approaches (for an early paper see, e.g., Creer *et al.* (1983)). More commonly, analyses rely heavily on statistical approaches where the variance in the field is quantified. The two most commonly used descriptions are the angular dispersion in field directions, S_D, or the angular dispersion in the VGPs, S_B (eqn [8]). Globally, VGP dispersion has been observed to increase with latitude (see red symbols in **Figure 12**) and the form of this increase has motivated many of the PSV models to date. Regionally, one of the earliest-noted signatures of PSV was low S_B at Hawaii, as reported by Doell and Cox (1963, 1965). This observation was interpreted as low variability in the nondipole contributions to the field, and combined with observations from the historical record (Bloxham and Gubbins, 1985) led to the suggestion of persistent low non-dipole fields in the Pacific: the so-called 'Pacific Dipole Window'.

The statistical approach inherent in using S_B, or some other summary statistic for field variability, circumvents the lack of detailed age control and the absence of time series of observations that are inevitable in working with lava flows from disparate locations. An additional view, continued until recently, was the idea of characterizing 'typical' PSV as distinct from reversals or excursions.

Early models for PSV attributed secular variation to three sources: variations in the direction of the dipole (dipole wobble), variations in the intensity of the dipole wobble, and variations in the nondipole field. Later statistical descriptions have used present-day properties of the field to constrain PSV models; for example, McFadden *et al.* (1988) used the present-day field to establish the general form of a VGP dispersion curve, while Constable and Parker (1988) used the present-day power spectrum as a constraint on the paleopower spectrum. It is important to note that none of these models are based on any physical theory: earlier models were based on the empirical observation that most of the power in the paleofield observed at Earth's surface can be accounted for by a geocentric dipolar field structure,

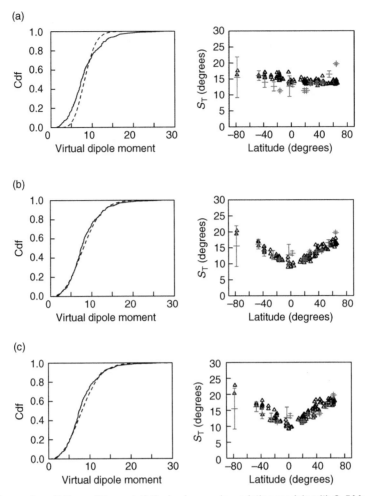

Figure 12 Comparisons of predictions of three statistical paleosecular variation models with 0–5 Ma paleointensity data of Tanaka *et al.* (1995) and the 0–5 Ma directional data JC96. (a) Model CP88, (b) Model CJ98, (c) Model CJ98.nz. The left panels show the cumulative distribution functions for the paleointensity data (solid) and model (dashed), where the 100 simulations of the model are run with the same site distribution and number of data per site as in the intensity data set. The right panel shows VGP dispersion, S_T, as a function of latitude for the JC96 data set (red) and the models (black). The data are averaged in latitude bands, and the mean dispersions along with the 1 standard deviation error bar are shown. For the models 10 simulations at each data site are shown.

while later models are based on statistical properties of the field.

11.4.1 Early PSV Models

The earliest global PSV models considered dipole wobble (models A (Irving and Ward, 1964) and B (Creer *et al.*, 1959; Creer, 1962)), and were followed by a suite of models that considered both dipole wobble and nondipole variations: (models C (Cox, 1962), D (Cox 1970), E (Baag and Helsley, 1974), M (McElhinny and Merrill, 1975) and its modification

(Harrison, 1980), F (McFadden and McElhinny, 1984)). For a thorough review of these studies the reader is referred to the discussion in Merrill *et al.* (1996).

McFadden *et al.* (1988) proposed a different representation of PSV, in which they introduced the idea of separating contributions to the variance in the field into two parts – the dipole and quadrupole families. This provided a direct link to the spherical harmonic description of the field as the dipole family comprises terms for which $l - m$ is odd (i.e., asymmetric about the equator) and the quadrupole family comprises terms for which $l - m$ is even (symmetric about the equator). The VGP dispersion due to the quadrupole family is constant with latitude, whereas that due to the dipole family varies linearly (from zero at the equator) up to a latitude of $70°$. McFadden *et al.* (1988) used this form for VGP dispersion to establish the dipole and quadrupole family contributions to PSV for the period 0–5 Ma. Their model G has survived remarkably well as a reasonable description of one measure (S_B) of PSV. One of the important aspects of model G, was its attempt to tie PSV statistics to dynamo theory, specifically mean-field dynamos in which the magnetic field solutions separate into the dipole and quadrupole families. We return to possible links between PSV and dynamo models in Section 11.7.

11.4.2 Giant Gaussian Process (GGP) Models

In the same year that model G was proposed, an alternative approach was put forward by Constable and Parker (1988). They proposed that the power spectrum of the present field be used as a guide in constructing models for paleosecular variation. In CP88 PSV is described by statistical variability of each Gauss coefficient in a spherical harmonic description of the geomagnetic field, with each coefficient treated as a normally distributed random variable: the Gauss coefficients of the non-dipole part of the field exhibit isotropic variability, and the variances are derived from the present field spatial power spectrum. *CP88* and its descendants are known collectively as GGP models. The dipole terms have a special status in *CP88*, with a nonzero mean for the axial dipole, and lower variance than predicted from the spatial power spectrum. All non-dipole terms have zero mean except the axial quadrupole. Isotropic means that the statistical variations of Gauss coefficients about their mean value do not depend on the orientation of the coordinate system in which the Gauss coefficients are defined. Apart from the mean values the statistical distributions for g_l^m and h_l^m vary only with spherical harmonic degree, l, not with the order m. The isotropic variations in the Gauss coefficients for the geomagnetic potential suggest a secular variation model in which there is no preferred directional dependence, but this isotropy does not extend to the actual physical measurements (X, Y, Z or D, I, F) of the local geomagnetic field.

One advantage of this model over earlier descriptions is that it supplies a complete (albeit oversimplified) description of the model field, even though its only parameters are the means and variances of distributions of Gauss coefficients. It allows direct calculation or simulation of the expected distributions for any geomagnetic observable and these can be tested against the distributions of available observations.

CP88 was found to match distributions of declination and inclination taken from the data set of Lee (1983). The most notable shortcomings of CP88 are its inability to predict the observed form of S_B with latitude and its underprediction of the variance in paleointensity data (**Figure 12**). Kono and Tanaka (1995) showed that an isotropic model like CP88 cannot generate significant variations in VGP dispersion with latitude, but that extra variance in Gauss coefficients of degree two and order one might generate the observed variation in VGP dispersion with latitude. This idea was developed further by Hulot and Gallet (1996) who suggested the dominance of order one terms up to about spherical harmonic degree four, emphasizing the need for anisotropy in the field variations, and/or cross-correlations among the Gauss coefficients. Quidelleur and Courtillot (1996) carried out extensive simulations for available data sites with a variety of different parameters for the means and variances of the Gauss coefficients, and concluded that the simplest modification to CP88 compatible with observed VGP dispersion required a standard deviation for the g_2^1 and h_2^1 terms about three times larger than for the rest of the quadrupole terms.

Four parameters associated with CP88 can be adjusted to fit the observations: α determines the variance $(\sigma_l)^2$ for $l = 2$–12; additionally there are the mean values of the axial dipole and axial quadrupole contributions to the field and the variance $(\sigma_1)^2$ in the dipole contributions to the field. These parameters reflect constraints on the model imposed by the observations; for *CP88* the values of the average axial quadrupole term, and the dipole terms' variance were chosen to fit the paleodirectional data set of Lee (1983). g_1^0 was arbitrarily set to $30\,\mu T$, and the variance of the nondipole terms was derived from the white spectral fit for the nondipole present field.

Constable and Johnson (1999) proposed variations on CP88 in which anisotropy in the variability in the Gauss coefficients was introduced. This was motivated by the desire to examine models in which proxies for homogeneous versus heterogeneous CMB conditions could be introduced. The incorporation of large variance in the axial dipole, and in the nonaxial quadrupole Gauss coefficients, g_2^1 and h_2^1, results in predictions that provide an improved fit (over CP88) to the latitudinal form of VGP dispersion and the distribution of paleointensity data in the Tanaka *et al.* (1995) data compilation (**Figure 12**). The resulting variance in paleomagnetic observables depends only on latitude (zonal models), unless the variance in h_2^1 is different from that in g_2^1 (nonzonal models). Specific candidate models discussed in the paper are CJ98 and CJ98.nz. While the particular model CJ98.nz is an end member of a possible choice of nonzonal models, it demonstrates several aspects of PSV that could be tested with such models, given sufficiently good data sets (**Figure 13**). Nonzonal (longitudinal) variations in PSV such as the high secular variation associated with the flux lobes seen in the historical magnetic field are simulated by such nonzonal models. VGP dispersion is rather insensitive to longitudinal variations in structure of PSV, whereas other measures such as inclination dispersion have the potential to be more informative. Finally, geographic variations in the frequency of occurrence of excursional directions are predicted by nonzonal models, and could eventually be tested against observed distributions.

Tauxe and Kent (2004) suggested an alternative approach to modifying CP88 to fit both intensity and directional data. Rather than treating the dipole and $l = 2$, $m = 1$ variances as special cases, they proposed a different variance for the dipole and quadrupole families (McFadden *et al.*, 1988). The parameter β in their model (TK03) describes the ratio of the dipole/quadrupole family variance. They also set the mean value of the axial dipole term to be 50% of that used in CP88, CJ98, and CJ98.nz, reflecting their view of a lower global average value for paleointensity (Selkin and Tauxe, 2000).

Hulot and Bouligand (2005) and Bouligand *et al.* (2005) extended the symmetry ideas introduced by Constable and Johnson (1999) further, providing the mathematical constraints that a GGP model must obey, if it is to satisfy spherical, axisymmetric, or equatorially symmetric symmetry properties (see Gubbins and Zhang (1993) for a discussion of symmetry properties of the dynamo equations). They use the results of two numerical dynamo simulations – one with homogeneous CMB conditions, and one with heterogeneous CMB conditions (simulations described in Glatzmaier *et al.* (1999)) – to demonstrate that calculation of the mean and the covariance matrix

allowed the symmetry breaking properties introduced by the CMB conditions to be identified correctly. This lays some groundwork for analyses of PSV data, assuming the ability to estimate covariance among the Gauss coefficients.

11.5   The Time-Averaged Field

Following the observation that the GAD hypothesis provides a first-order description of field behavior for 0–2 Ma (Opdyke and Henry, 1969), studies of the TAF geometry have addressed how good this approximation is and the nature of any departures from GAD. Such signals, while second order in magnitude, may be profoundly important for understanding the interplay of deep-Earth processes. Several avenues of investigation are thus required: (1) the detection of non-GAD structure, (2) discrimination among possible sources (real field behavior vs, e.g., sampling bias or rock magnetic effects), and (3) the explanation of any non-GAD field structure. In this section we review studies of the TAF, focussing on the inferred non-GAD structure. We return to the issues of discrimination and explanation in Sections 11.6 and 11.7.

Models of the historical geomagnetic field (**Figure 3**) suggest that the CMB region and inner core affect both the TAF and its temporal fluctuations (Bloxham *et al.*, 1989). In the early 1990s, reports that magnetic records of reversals showed VGP paths that were confined to two longitude bands (Clement, 1991; Laj *et al.*, 1991) caused much controversy in the paleomagnetic community (Valet *et al.*, 1992; McFadden *et al.*, 1993; Prévot and Camps, 1993). The apparent coincidence of these longitude bands with the static flux lobes seen in the historical field models, and with seismically high lower mantle velocities in tomographic models available at the time was interpreted as the influence of lower mantle thermal conditions on core flow and field generation during reversals. The scarcity of records of field reversals and the debate about their fidelity, prompted a series of studies of the paleomagnetic TAF during stable polarities to investigate the persistent non-GAD structure in the field.

Studies of the TAF have focused on the past 5 My due to the availability of global data sets (Section 11.4), with sufficient data to investigate both normal and reverse polarity periods. Departures of the TAF direction from GAD at any location are small (on the

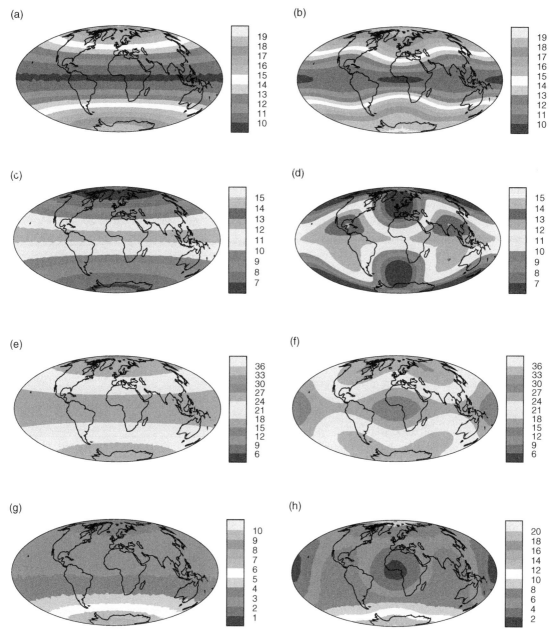

Figure 13 Predictions for various summary statistics for PSV at Earth's surface. Left column shows predictions of a zonal PSV model (CJ98), right column predictions of a nonzonal model (CJ98.nz). Rows from top to bottom show VGP dispersion, standard deviation in inclination, standard deviation in intensity, and percentage of excursional directions. (a) CJ98 vgp dispersion (degrees); (b) CJ98.nz vgp dispersion (degrees); (c) CJ98 sd inclination (degrees); (d) CJ98.nz sd inclination (degrees); (e) CJ98 sd intensity (μT); (f) CJ98.nz sd intensity (μT); (g) CJ98, percentage of VGP latitudes <45°; and (h) CJ98.nz, percentage of VGP latitudes <45°.

order of a few degrees), and so restricting studies to the past 5 My means that plate motion corrections can be performed with sufficient accuracy to distinguish magnetic field behavior from geographic effects.

11.5.1 Early Studies

Studies by Wilson and colleagues (Wilson, 1970, 1971) noted small departures from GAD in the TAF recorded by sediments and lavas. Inclination data appeared shallower than the predictions of

GAD, with correspondingly far-sided VGPs (**Figure 2**). Wilson's interpretation was that these data were best approximated by a dipole, offset northward by ~150–300 km from Earth's center, along the rotation axis. In addition, he noted that the pole positions from volcanic rocks showed a 'right-handed' effect, corresponding to a persistent nonzero declination, with a magnitude on the order of 3°–5°.

In a seminal contribution, Merrill and McElhinny (1977) used time-averaged declination and inclination anomalies (eqn [6]) from their global compilation of paleomagnetic data to estimate time-averaged values of the Gauss coefficients. Their data set comprised D, I pairs from 101 normal and 31 reverse Quaternary (0–2 Ma) and 70 normal and 64 reverse Pliocene (2–5 Ma) records, to which they added I data from 50 normal and 50 reverse polarity piston cores of Opdyke and Henry (1969). Because of the incorporation of sediment data, the geographical distribution was quite good (see their figure 2). An analysis of declination anomalies led the authors to conclude that the right-handed effect is an artifact, resulting from unevenly distributed, and insufficient, data. Based on this conclusion, and on the desire to use as large a number of data as possible in estimating the TAF direction, the authors binned their data into 10° latitude bands (separately for normal and reverse polarities). Mean inclination anomalies (possibly an arithmetic average of contributing sites, as there is no specific mention of a vector mean), along with the standard error were calculated for each bin. Predominantly negative inclination anomalies were seen, with the greatest signal (~5° for normal polarity, larger for reverse) at low latitudes (**Figure 14**). This general observation has held for three decades. A zonal spherical model was visually matched to the observed inclination anomalies. The predicted inclination for a zonal model, truncated at spherical harmonic degree l_{max}, is given by

$$\tan I = \frac{\sum_{l=1}^{l_{max}} (l+1) g_l^0 P_l(\cos\theta)}{\sum_{l=1}^{l_{max}} -g_l^0 P_l'(\cos\theta)} \quad [14]$$

where $P_l'(\cos\theta) = dP_l(\cos\theta)/d\theta$ (see Merrill *et al.* (1996), appendix B5 for details.) Merrill and McElhinny (1977) used $l_{max} = 3$. Values for the axial quadrupole, and the axial octupole, (g_2^0, g_3^0) terms as a percentage of g_1^0 were (5%, −1.7%) for

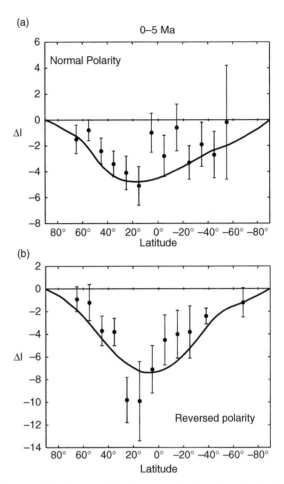

Figure 14 Average inclination anomaly (ΔI) and its standard error within 10° latitude bins for (a) normal and (b) reverse polarity during the past 5 My (data from Merrill and McElhinny, 1977). The solid lines represent models of inclination anomaly calculated from eqn 14 and (a) $g_2^0/g_1^0 = 1/20$, $g_3^0/g_1^0 = -1/60$ (b) $g_2^0/g_1^0 = 1/1$, $g_3^0/g_1^0 = -1/20$. In (b) the sign of the inclination anomaly has been reversed, for direct comparison with the normal polarity data in (a).

normal polarity and (8.3%, −3.4%) for reverse polarity. The difference between the normal and reverse time-averaged signature was shown to be statistically significant. Lateral temperature variations in the lower mantle (Dziewonski *et al.*, 1987) were suggested as a possible cause of the long-term non-GAD TAF and the normal/reverse polarity asymmetry.

Schneider and Kent (1990) investigated the TAF using only sediment cores because of the better time averaging provided by a core as compared with a limited number of lava flows of unknown ages. They compiled and used an updated piston core data set (Section 11.3.4) that was greater in number than that used by Merrill and McElhinny (1977), and in which a larger percentage of cores had been subjected to magnetic cleaning. For each core an average inclination anomaly was calculated using a maximum likelihood technique (McFadden and Reid, 1982). A best-fit zonal TAF model truncated at spherical harmonic degree 4 was calculated for Brunhes and Matuyama separately, by minimizing $\sum_{i=1}^{N} \left[I_{obs}^i - I_{pred}^i \right]^2$. Values for (g_2^0, g_3^0) as a percentage of g_1^0 were (2.6%, −2.9%) for the Brunhes and (4.6%, −2.1%) for the Matuyama, again reflecting the larger reverse polarity inclinations.

In summary, these early studies of the TAF detected a latitudinal signal in inclination anomalies from lavas and sediments, best fit by two-parameter models. Models proposed included an axial quadrupole contribution of 2.5% to 5% of the axial dipole term for normal polarities, and about twice this for reverse polarities. Axial octupole terms were generally on the order of −1.5% to −3.5% of the axial dipole term.

11.5.2 The 1990s: Longitudinal Structure in the TAF?

As motivated earlier, structure seen in historical field models, along with the provocative suggestion of lower mantle control on magnetic records of reversals prompted interest in whether there is persistent longitudinal structure in the field. Two independent groups pursued this using the lava flow compilations of Q94 and JC96. In contrast to the least squares fitting, and/or grid search approaches of previous studies, the tools of inverse theory (Parker, 1994; Gubbins, 2004) were used to construct spherical harmonic models that fit the observations, but that also possess minimum structure. The results of the two groups are documented in a series of papers – Gubbins and Kelly (1993); Johnson and Constable (1995, 1997, 1998); Kelly and Gubbins (1997) (hereafter GK93, JC95, JC97, JC98, KG97, respectively).

GK93 modeled the normal polarity average field structure using the data set of Q94 for the past 2.5 My, and a technique similar to that applied to the historical data (Bloxham *et al.*, 1989). In a parallel study, JC95 used the JC96 data set, and a different inversion algorithm to model the Brunhes, 0–5 Ma normal and 0–5 Ma reverse polarity data. The 0–2.5 Ma normal polarity model of GK93, along with their preferred model in a later paper (KG97) and the 0-5 Ma normal polarity model of JC95 are remarkably similar, both displaying some features seen in historical field,

notably the presence of two Northern Hemisphere flux lobes (**Figure 15**). The JC95 Brunhes polarity model contains less structure, interpreted as resulting from the smaller data set. As in previous zonal models for the TAF, JC95 found larger deviations from GAD during reverse polarities: this is seen in the raw inclination anomaly observations. Both groups followed up these results with studies that include sediment data, and in the case of KG97 also some intensity data (Tanaka *et al.*, 1995). Because only inclination data are available from sediments, models that include these data are smoother than their lava-flow-only counterparts (JC97; KG97). Several tests of the robustness of the results were conducted: for example JC97 showed that purely zonal models could not fit the lava flow data, and they conducted bootstrap estimates of uncertainty to demonstrate resolvable features in their models. As the discussion of structure in these TAF models is helped by comparing maps of the radial field, B_r, at the CMB, JC97 also showed how the geographical distribution of observations of declination and inclination samples B_r at the CMB.

In a series of subsequent papers, longitudinal structure in TAF models was discussed at length. McElhinny *et al.* (1996) noted that the time-averaged declination anomalies were not statistically distinguishable from zero, and so argued that if any nonzonal features exist they are too small to be reliably resolved with the then-available data. The authors advocated zonal model fits to latitudinally averaged inclination anomalies, as in Merrill and MElhinny (1977). Carlut and Courtillot (1998) used the JC95 code to investigate TAF models for both the Q94 and JC96 data sets, truncating their inversions at degree and order 4. They found values for the axial quadrupole, compatible with other studies and in the range of 3.5–5.5% of the axial dipole term, and concluded that model coefficients are not significantly different from zero below a threshold of 300 nT, rendering only these two terms robust. However, some of these conclusions result from the truncation level used in implementing the inversion algorithm, rendering the resulting models more akin to those derived from truncated least squares, than smooth inversions.

Thus, regularized inversions of paleodirections from lava flows, and from lava flow and sediment data combined, result in nonzonal models for the TAF, with longitudinal structure that bears some resemblance to that seen in the historical field. Different views exist as to the robustness of this structure: we discuss some of the underlying issues

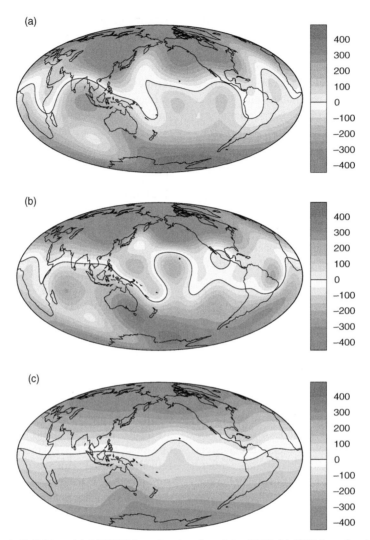

Figure 15 Normal polarity field models (a) GK93, based on lava flow data of Q94, (b) JC95, based on lava flow data of JC96, (c) JC97, based on JC96 lava flow compilation and SK90 sediment data set. The figure shows the radial field at the CMB in μT.

in Section 11.6, and in Section 11.7 offer some suggestions for future work that can address this question.

11.5.3 Recent Studies: Joint Estimation of PSV and the TAF

Recent global time-averaged field studies have begun to assess another issue – that of whether inversions of the TAF direction alone can result in a bias due to the influence on the TAF of PSV. Statistical models of the type described in Section 11.4.2 produce non-Fisherian distributions of directions locally (Khokhlov *et al.*, 2001; Tauxe and Kent, 2004; Lawrence *et al.*, 2006), as has been observed in data sets (Tanaka, 1999; Lawrence *et al.*, 2006).

Hatakeyama and Kono (2001, 2002) invert for both the TAF and PSV using the JC96 data set. They conclude that the nonzonal TAF structure in their resulting models is not robust and find an average axial quadrupole ∼4.3% of the axial dipole term, similar to that found in other studies. They find a PSV model with large variance in the degree 2, order 1 terms, as proposed in several PSV models (see Section 11.4.2). However, the absence of paleointensity data in the inversion results in their PSV model having insufficient variance in the dipole terms. In Section 11.6, we discuss the issue of bias resulting from the absence of paleointensity data and the implications for TAF investigations. We return to joint PSV/TAF models in Section 11.7, as the simultaneous estimation of PSV and the TAF is likely to be an important aspect of future paleomagnetic field models.

11.6 Discussion

In this section we first address technical issues that give rise to many of the discussions surrounding PSV and the TAF. Some are driven by the existing data sets, others by differences in modeling approaches. We then summarize the successes and limitations of current models with respect to understanding paleo-field behavior.

11.6.1 Issues in PSV and TAF Modeling

The number, distribution, and quality of existing global paleomagnetic data sets are the underlying cause of much of the debate surrounding regional variations in PSV and persistent non-GAD structure in the 0–5 Ma field. Specific aspects of the debate (e.g., the merits of one data set vs another) have been discussed elsewhere (e.g., Carlut and Courtillot, 1998; Merrill *et al.*, 1996; McElhinny and McFadden; 1997); here we review the general issues and how they have led to different conclusions in the literature. Some of these issues are being directly addressed in new data sampling and laboratory efforts.

It is important to remember that early (1960s and 1970s) paleomagnetic data were collected to investigate, for example, magnetic polarity and the GAD hypothesis, not to pursue the kinds of studies we wish to conduct now. Hence the kind of data coverage, and quality that we require from data sets today, simply does not exist in some of these earlier, but quite large collections.

Below we address issues that arise from four sources: (1) data distribution – temporal and spatial, (2) data quality, (3) bias due to the lack of intensity data, and (4) modeling approaches.

11.6.1.1 Data sets: temporal sampling
This is perhaps the most thorny issue in studies to date, and can be subdivided into two problems at any location: (1) whether paleodirections span a long-enough time interval to provide a reliable estimate of PSV and the TAF; and (2) whether site mean paleodirections are temporally independent. For deep-sea sediment cores, neither of these problems is acute, since a time series of observations is available for a given core. However, the uneven, and discrete nature of lava flows, means the situation is

quite different. Data sets from different locations globally may all have data of Brunhes polarity age, but may sample different, and quite short, parts of the polarity chron. While 10^4 years has been loosely used as a time interval thought to be sufficient to characterize PSV (Merrill *et al*, 1996; Carlut *et al*, 1999; Johnson and Constable, 1996), other studies (Merrill and McFadden, 2003) suggest that at least 10^5 years are necessary. Studies of the paleomagnetic power spectrum (Constable and Johnson, 2005) indicate that different estimates of PSV might be obtained by averaging over, say, 10^3 versus 10^6 years. Whatever the case, significant improvements could be made by having data that sample the same, known time interval at different locations globally. Obtaining radiometric ages was not previously a routine part of paleomagnetic PSV/TAF data collection efforts; this has changed recently with ~20% of flows being dated on average in a given study (Johnson *et al.*, 2007).

Temporal independence is also problematic for lava flow sites, since thick sequences can be erupted in very short time intervals, and paleodirections from one flow to the next are often serially correlated. Unfortunately the problem is worst at places where there are large amounts of data such as Hawaii and Iceland. In a 'catch-22', the more conservative viewpoint then attributes any non-GAD structure in regions with greatest data sampling to 'too much of the same'. Inadequate temporal sampling will lead to biased mean directions, and artificially low estimates of PSV. This is the root of most of the discussion surrounding the evidence for long-term anomalous TAF and PSV at Hawaii (e.g., McElhinny *et al.*, 1996), and the resulting Pacific/Atlantic hemisphere asymmetries in paleofield models. It is then a matter of philosophy as to how to analyze the data. Johnson and Constable (1997) advocate careful data selection (omitting the lava flow sequences that are the most problematic) and test for bias by 'thinning' data sets from lava flow sequences to reduce the effect of oversampling. A conservative approach is to attribute all the longitudinal variation in the mean field direction, at a given latitude, to poor temporal sampling of the data and to small undetected tectonic movements (e.g., McElhinny *et al.*, 1996), and to advocate only TAF and PSV models that are a function of latitude.

11.6.1.2 Data sets: spatial distribution
The geographical distribution of paleomagnetic data spanning a prescribed time interval affects existing TAF and PSV models differently. Most global PSV

models prescribe statistics that vary only as a function of latitude. For these models, the spatial distribution of data is reasonable, although Southern Hemisphere coverage is poor compared with its Northern Hemisphere counterpart. TAF models, especially those that investigate nonzonal structure discussed in Section 11.6.2, are affected by differences in not only the spatial distribution, but also the type of contributing data sets.

Equation [9] gives the solution to Laplace's equation in terms of spherical harmonic functions, and for a particular set of spherical harmonic coefficients allows calculation of the resulting potential or associated magnetic field at any point outside the assumed source region in Earth's core. An alternative way of writing the magnetic field at Earth's surface is in terms of the Green's function for $B_r(\hat{s})$, the radial magnetic field at the CMB:

$$\vec{B}(\vec{r}) = \int_S \vec{G}(\vec{r}|\hat{s}) B_r(\hat{s}) d^2\hat{s} \qquad [15]$$

The Green's functions for the field elements X, Y, and Z have been published several times in the geomagnetic literature (see, e.g., Gubbins and Roberts (1983) and Constable *et al.* (1993)).

Paleomagnetic observations – declination, inclination and intensity, F, – are nonlinear functionals of $B_r(\hat{s})$, so one can formulate linearized data kernels that describe how these observations respond to changes in $B_r(\hat{s})$ (Johnson and Constable, 1997). These data kernels can be used to indicate the response of D, I, and F to departures from an axial dipole field configuration. They vary with geographic location (**Figure 16**). Importantly, sampling of $B_r(\hat{s})$ at locations other than immediately below the observation site is obtained. Declination provides longitudinal information, preferentially sampling $B_r(\hat{s})$ at locations east and west of the observation location. Inclination observations at the equator preferentially sampling $B_r(\hat{s})$, directly beneath the observation site, but at non-equatorial site locations,

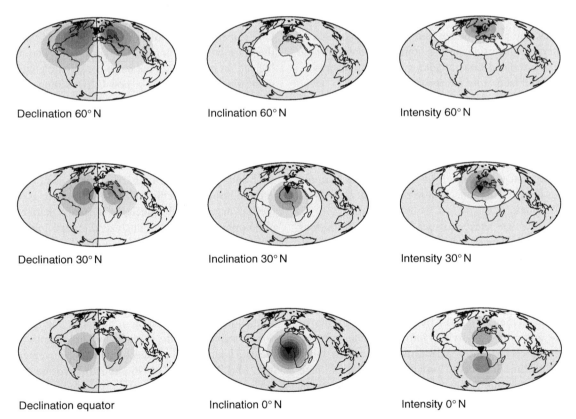

Figure 16 Sampling kernels to show how observations of *D*, *I*, and *F* at the surface sample the radial field at the CMB. Color scale denotes relative sampling: darker regions are sampled more heavily than lighter ones. Positive(negative) Br is shown in red(blue). Declination observations provide the best longitudinal coverage; inclination(intensity) observations preferentially sample Br at the CMB at lower(higher) latitudes than the observation site, except at the equator.

sampling of $B_r(\hat{s})$ is biased toward lower latitudes. In contrast, intensity data at nonequatorial sites provide maximum sampling of $B_r(\hat{s})$ at latitudes higher than the observation latitude.

The CMB sampling offered by a particular data set can be investigated by summing the magnitudes of the contributions from the kernels for each sampling location. This is done by assuming that each location contributes an estimate of the TAF – in other words, one observation of the time-averaged inclination and, for lavas, one of the time-averaged declination. In **Figure 17** we show this sampling for the JC96 normal

polarity lava flow distribution and the JC96 plus SK90 sediment data distribution. The longitudinal coverage provided by the declination information from lavas is apparent. The sediment data sample mainly low latitude regions, and this sampling dominates the combined lava plus sediment data set, explaining the smoother field models, and reduction in high latitude structure in these joint data set inversions. A somewhat different picture of CMB sampling is obtained if one sums the contributions from the sampling kernels for each site – this would be appropriate for inversions where site level data, rather than time-averaged

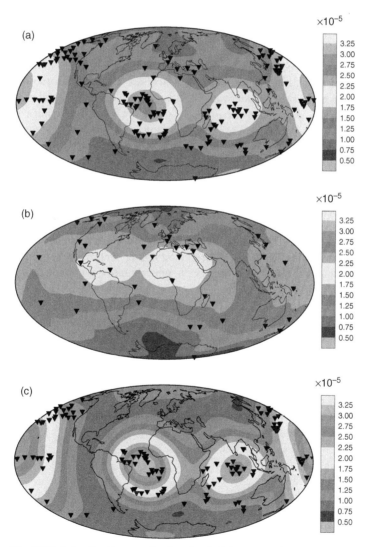

Figure 17 Sampling of the CMB defined by the sampling function in Johnson and Constable (1997), and reported in the text here. Maps show the relative sampling of B_r at the CMB by measurements from (a) sediment cores (*I* only, SK90), (b) lava flows (*I* and *D*, JC96), and (c) sediments and flows. Notice that the large number of sediment cores with only inclination data focusses sampling at low-mid latitudes and accounts for models based on lavas and sediments with less nonzonal and less high latitude structure than models based on lava flow data alone.

directions, are inverted. Clearly, in such a picture, locations with many contributing sites (such as Hawaii) would dominate CMB sampling.

11.6.1.3   Data quality

Reliable PSV and TAF models require that at the site level, mean paleodirections must represent the primary remanence, and within-site error should be a minimum. As discussed earlier, estimates of within–site error require several samples per site. Traditionally two or three samples per site were taken in the field, much of the discussion in the literature has been about whether sites with fewer than 3 samples should be used in field modeling. Today ~10 samples per site are typically taken in the field and, after laboratory cleaning, site mean directions are usually obtained for 5–10 samples. Simulations from statistical PSV models also suggest that at least 5 samples per site are desirable (Tauxe et al., 2003). Similarly, as stepwise demagnetization procedures and the estimation of directions by principal component analysis have become routine, concern about data contaminated by overprints is decreasing.

Assuming sufficient samples per site and removal of overprints, the question then is how and whether data should be excluded on the basis of poor quality. As already noted, two measures of data quality are the cone of 95% confidence about the mean direction, α_{95}, and k (eqns [12] and [13]) and choices for both have been used in assembling data sets. Recent studies suggest that a cutoff of $k = 100$ (i.e., including only data with $k > 100$ assures a robust estimate of the TAF and PSV (Tauxe et al., 2004; Lawrence et al., 2006; Johnson et al., 2007)). Until now, TAF studies have discarded poor quality data; obviously in the future a better approach is a more appropriate choice of data weights.

11.6.1.4   Transitional data

Another selection criterion used in generating PSV and TAF models is to discard transitional data. For PSV models this is done on the basis that a few sites with low latitude VGPs can significantly increase the estimate of dispersion, and that field behavior during such times may be distinct from that during stable polarities. This effect is shown in **Figure 18**, using simulations from a modified version of the statistical model CJ98, in which a TAF of GAD was prescribed (instead of GAD plus an axial quadrupole as in CJ98). Directions were simulated at 5° latitude increments, with 10 000 sites at each latitude. VGP dispersion was calculated using sites with VGPs greater than a

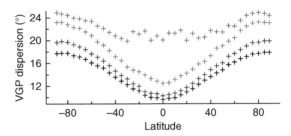

Figure 18   Effect of VGP latitude cutoff on estimate of VGP dispersion. Simulations use statistics prescribed in PSV model CJ98, but with GAD as the TAF field. 10 000 sites at each latitude were simulated. The total VGP dispersion (assuming zero within-site error) was computed for VGP latitude cutoffs of 55° (black), 45° (blue), 0° (green) and −90° (brown).

specified VGP cutoff, where the latter was varied from 55° to −90°. A VGP latitude cutoff of −90° includes all data, even those that may in fact be of opposite polarity, hence the large increase in dispersion. The VGP dispersion curves for VGP latitude cutoffs of 0°, 45°, and 55° show an overall increase in dispersion as lower latitude VGPs are permitted, and also shows that the increase itself is latitude dependent. The use of a VGP latitude cutoff is not ideal, since the identification of two classes of field behavior – transitional versus stable polarity – is arbitrary. With more comprehensive and accurately dated data sets one could explore the influence of temporal sampling and what happens with the appropriate representative sampling of transitions, but this is still not possible with current data sets.

11.6.1.5   Bias from unit vectors

It has long been known (Creer, 1983; see also discussion in Merrill and McFadden (2003)) that biased estimates of direction are obtained when unit vectors are averaged. If full vector measurements of the field are available then the resultant vector, R, of eqn [5] is modified as follows:

$$R = \left[\left(\sum_{i=1}^{N} b_i x_i \right)^2 + \left(\sum_{i=1}^{N} b_i y_i \right)^2 + \left(\sum_{i=1}^{N} b_i z_i \right)^2 \right]^{1/2} \quad [16]$$

where b_i are the contributing individual intensities. PSV results in a distribution of b_i at any given location, so the unit vector mean is not equal to the full vector mean. The bias in the mean direction inferred from unit vectors depends on the statistics of the PSV, and in general there is no closed-form analytical solution for this. Love and Constable (2003) provide an analytical solution for the bias in inclination, obtained by averaging only measurements of inclination, when the underlying statistics of PSV are given by a locally isotropic process (equal variances in B_x, B_y, and B_z).

Here we show the kind of bias that might be expected on the basis of our current understanding of 0–5 Ma PSV. We first use the statistical model CJ98, but with a TAF of GAD, to simulate the apparent inclination anomalies obtained by averaging unit vectors, for a range of VGP cutoffs (**Figure 19(a)**). Inclination anomalies are well approximated by an apparent axial octupole

signature as noted previously (Creer, 1983; Merrill and McFadden, 2003), with a magnitude of 2–3% of g_1^0, depending on the choice of VGP latitude cutoff. For a VGP latitude cutoff of 45°, declination anomalies are close to zero except at high (greater than 75°) latitudes, and they show no longitudinal dependence (**Figure 19(b)**).

We examine the dependence of the bias on the partitioning of variance in the statistical PSV models, by comparing CJ98 with TK03 (**Figure 19(c)**). For a VGP latitude cutoff of 45°, CJ98 results in bias approximated by that due to an axial octupole term,

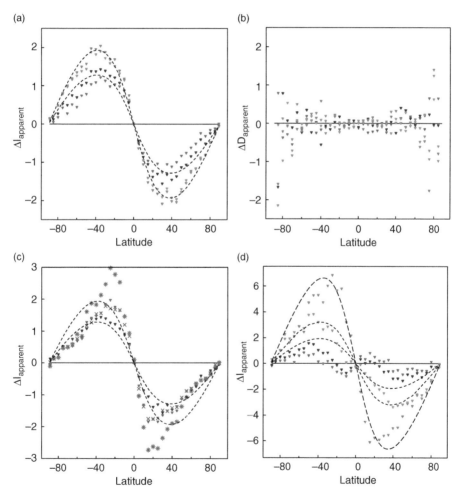

Figure 19 Simulations from PSV statistical models to show bias as a function of latitude incurred by averaging of unit vectors. (a) Apparent inclination anomaly for CJ98, using a TAF of GAD and four choices of VGP latitude cutoff: 55° (red), 45° (blue), 0° (green), and −90° (brown). Predictions for an axial dipole plus axial octupole field are shown by dashed curves, with $g_3^0 = 2\% g_1^0$ (smaller magnitude anomalies) or $g_3^0 = 3\% g_1^0$ (b) Apparent declination anomaly computed along five lines of longitude: 0° (red), 45° (blue), 90° (green), 135° (brown), and 180° (orange). (c) Apparent inclination anomaly using a VGP latitude cutoff of 45° for four prescriptions of PSV – CJ98 (blue triangles, as in Figure 19(a)), TK03 (red triangles), modified TK03, with $\alpha = 17.6$, $\beta = 1$ (red stars – note that this is essentially CP88), modified TK03 with $\alpha = 0.75$, $\beta = 40.5$ (red crosses). (d) Apparent inclination anomaly, colors as in (a), for a modified version of TK03 in which $\alpha = 15$ and $\beta = 7.6$, i.e., twice their values in TK03. Dashed lines as in (a), long dashed line (largest amplitude inclination anomaly) is for $g_3^0 = 10\% g_1^0$.

$g_3^0 = 2\% g_1^0$; TK03 suggests a larger (3%) axial octupole term, although the shape of the apparent inclination anomaly is less well described by only an axial octupole term. We also simulate the bias obtained by two modified versions of TK03. In the first the variance in the dipole family is increased, at the expense of the variance in the quadrupole family. We set the parameter α of TK03 = 0.75 and β of TK03 = 40.5; the choice of α and β is made so as to keep the total variance in the spherical harmonic degree one terms constant, and as predicted by TK03. This results in little change in the bias compared with that obtained from TK03. In the second experiment we set $\beta = 1$, and $\alpha = 17.6$, allowing equal variances in the dipole and quadrupole families. Note that this is essentially model CP88, one that predicts VGP dispersion that is invariant with latitude. This results in a lager magnitude bias at low-mid latitudes, and a latitudinal dependence that is less well modeled by only a g_3^0 signature. Finally, we show how the magnitude of the bias could change if the overall level of PSV were higher. We double values of α and β given in TK03. (This results in values for VGP dispersion that range from 14° at the equator to 22° near the poles, higher but not greatly so, than those predicted by TK03 or CJ98 – see blue crosses in **Figure 18**). The resulting bias in inclination is shown in **Figure 19(d)** for the values of VGP cutoff, used in **Figure 19(a)**. The higher PSV results in similar magnitude (on the order of 1° bias) for VGP cutoffs around 45°, compared with that predicted by TK03, but the structure in the bias is no longer described by an axial octupole. VGP cutoffs of −90° result in a bias that is reasonably approximated by an axial octupole, but one that has a magnitude of 10% of g_1^0. We note that this has interesting implications for paleofield investigations for older times, for which plate motion corrections are not possible. In such cases it may not be possible to distinguish excursional or reverse polarity directions, and large axial octupole terms could erroneously be inferred.

In summary, the bias incurred by averaging of unit vectors is zonal, and thus cannot be invoked to argue against longitudinal structure in TAF models. It is small, but of similar magnitude at some latitudes to the observed zonal non-GAD signature. For the past 5 My it is likely that the bias is reasonably approximated by an axial octupole term, $g_3^0 = 2–3\%$ of g_1^0, when low-latitude VGP sites are excluded from analyses. However, this is not so for other PSV scenarios, in particular cases in which the overall level of PSV is increased.

11.6.1.6   Modeling approaches

Some of the differences among existing studies are related to modeling approaches. Studies that bin data in latitude bands can of course only examine zonal models for the TAF – that is, ones lacking longitudinal structure. Another difference lies in the philosophy as to how one should define 'simple' models. In the parameter estimation approach (least squares or grid searches) models are found that have minimal parameters but that fit the data (McElhinny, 2004). In the inverse theory approach, 'simple' models are described via a smoothness or regularization criterion (e.g., minimizing the power in the non-GAD coefficients). Models are found that fit the data to within a specified tolerance, but that also have minimal structure in the specified sense. The regularization results in smooth models in regions where data coverage is sparse. To ensure that the minimal structure model is found, the spherical harmonic truncation level must not be set too low. This difference in truncation levels is one of the causes of the differing interpretations of the same data set (JC96) by Johnson and Constable (1997) and Carlut and Courtillot (1998). Different philosophies as to how to choose the Lagrange multiplier – the relative importance of the regularization constraint – also exist (see discussions in Johnson and Constable (1997) and Hatakeyama and Kono, 2002, along with Kelly and Gubbins (1997) for three different approaches). Various ways of assessing model uncertainties have also been used: for example, examining the covariance matrix (Kelly and Gubbins, 1997; Hatakeyama and Kono, 2002), choice of a resolvable threshold level for all spherical harmonic model coefficients (Carlut and Courtillot, 1998), and nonparametric approaches Johnson and Constable (1995, 1997).

11.6.2   Successes and Limitations of Current TAF and PSV Models

PSV and TAF models to date have been quite successful in fitting summary statistics (the mean and measures of variance) of observations of paleodirection and intensity. The need for such models has motivated the compilation of several global data sets of directions recorded in lavas and sediments. However several limitations and outstanding questions remain; most result from the issues raised in Section 11.6.1.

Still unanswered satisfactorily is the question of symmetries in the TAF and PSV: do longitudinal and/or polarity asymmetries exist? This is important to resolve since there is increasing evidence from numerical simulations, as well as centennial to millenial scale field models, that lateral heterogeneities in the lowermost mantle influence magnetic field generation in the core (Bloxham and Gubbins, 1987; Bloxham, 2000; Olson and Glatzmaier, 1996). **Figures 20** and **21** show the current status in understanding the time average of the global magnetic field on timescales of hundreds, thousands, and millions of years. **Figure 20** shows the geomagnetic field averaged over three quite different time periods: the past 400 years (model GUFM1 discussed earlier), the past 7 ky (model CALS7K.2, (Korte and Constable, 2005)) and three models for the past 5 My. When averaged over 0–7 ka, CALS7K.2 shows longitudinal structure that suggests the presence of flux lobes seen in the historical field. The radial magnetic field is attenuated in CALS7K.2 compared with GUFM1: the resolution and accuracy is clearly inferior, but averaging of millennial scale secular variation also plays a major role in subduing the structure. Three quite different field models for the past 5 My are shown, spanning the range of

proposed geographical structure in the field. A purely zonal model is shown in **Figure 20(c)**, with a zonal quadrupole contribution, g_2^0, of 5% of g_1^0. **Figures 20(d)** and **20(e)** show model LN1 (lava flows only) of JC95 and model LSN1 (lavas and sediments) of JC97.

Figure 21 shows the signal expected at Earth's surface for the GUFM1, CALS7K.2, and LSN1 average field models, in the form of geographic variations in inclination anomaly and declination anomaly. The structure in the archeo- and paleofield anomalies is rather similar (despite the very different data distributions from which they are derived) and contrasts with that seen in GUFM1. Note that the magnitude of the signal decreases over longer timescales. From **Figure 21(f)** we see that the average inclination anomaly in LSN1 is rather small, and we can expect the largest signal at equatorial latitudes. If this view of the time-averaged field is approximately correct, then at mid-to-high latitudes it will be difficult to detect departures from GAD without large data sets that provide accurate measures of ΔI.

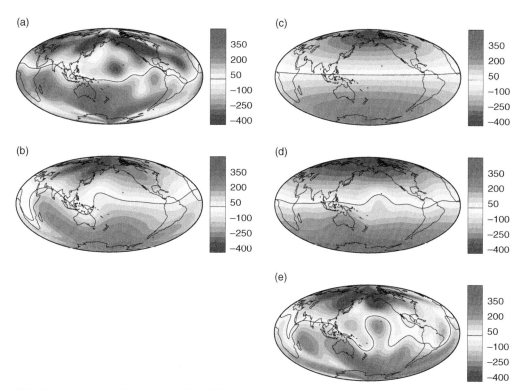

Figure 20 Time-averaged radial magnetic field, (B_r), at the core–mantle boundary (CMB), on different timescales. Units are µT. (a) Historical field: 1590–1990, Model GUFM1 (Jackson *et al.*, 2000) (b) Archeofield: 0 – 7 ka, Model CALS7K.2 (Korte and Constable, 2005), (c) Paleofield: 0–5 Ma, axial dipole plus axial quadrupole field (see text), (d) Model LSN1 (Johnson and Constable, 1997) (e) Model LN1 (Johnson and Constable, 1995).

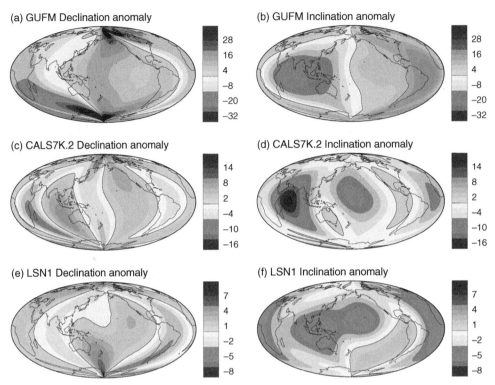

(a) GUFM Declination anomaly

(b) GUFM Inclination anomaly

(c) CALS7K.2 Declination anomaly

(d) CALS7K.2 Inclination anomaly

(e) LSN1 Declination anomaly

(f) LSN1 Inclination anomaly

Figure 21 Declination (a), (c), (e) and inclination (b), (d), (f) anomalies in degrees (deviations from GAD direction) at Earth's surface predicted from models for the three time intervals in **Figure 20**: (a) and (b) Model GUFM1, (c) and (d) CALS7K.2, (e) and (f) Model LSN1. The scale bar for the historical field is twice that for the archeofield , and four times that for paleofield anomalies.

Long-lived longitudinal variations in PSV are equally difficult to assess with current data sets and modeling approaches. Low PSV in the Pacific, as suggested by Doell and Cox (1963), has been refuted by many on the basis of oversampling of short (and hence unrepresentative) time intervals (e.g., McElhinny *et al.*, 1996). On historical timescales hemispheric differences in Atlantic and Pacific PSV exist, with much greater secular variation in the Atlantic (Bloxham and Gubbins, 1985; Bloxham *et al.*, 1989), and this has been suggested to extend to the millennial paleomagnetic record (Johnson and Constable, 1998; Gubbins and Gibbons, 2004; Korte and Constable, 2005). Recent data compilations for four regions at 20° latitude indicate that VGP dispersion at Hawaii is lower, but not significantly so, than at Mexico or Reunion. The paleomagnetic record of Pacific PSV is complicated because data sets come primarily from Hawaii and French Polynesia: the former have been argued to include oversampling of short stable polarity intervals, but the latter include unrepresentatively high sampling of reversal records. The discussion in Section 11.4

along with **Figure 13** indicates that detection of regional variations in PSV requires measures other than VGP dispersion, and clearly resolution of this question requires the development of time-varying field models.

The equations for the dynamo problem are such that if **B** is a solution for the magnetic field then so is −**B**. This means that polarity asymmetries are not explained by dynamo theory, and so it has been suggested that, like longitudinal TAF structure, they may result from long-term spatially heterogenous CMB conditions (Merrill and McElhinny, 1977; Johnson and Constable, 1995, 1977; Kelly and Gubbins, 1997). Others have argued (see review in Merrill *et al.*, 1996) that these result from overprints in the reverse polarity data. Viscous overprints from the current normal polarity period, should have a different blocking temperature spectrum from the primary remanence and modern techniques are capable of removing these. Interestingly, many recent data sets still have reverse polarity directions that are farther from GAD than their normal counterpart (Johnson *et al.*, 1998, Tauxe *et al.*, 2004b), although it is often true that the 95% confidence limits of the two directions overlap.

Bias in existing TAF models can result from the averaging of unit vectors. Simulations using the statistical models of TK03 and CJ98 indicate that for zonal TAF and PSV models this bias is manifest as an inclination anomaly as a function of latitude described by a g_3^0 term with a magnitude of 2% of g_1^0 (Johnson et al., 2007).

Recent work demonstrates that none of the existing GGP models predict local directional distributions that fit large regional data sets (Lawrence et al., 2006). More generally, it is not yet clear whether Gaussian distributions for the spherical harmonic coefficients are appropriate. Comparisons with numerical dynamo simulations (McMillan et al., 2001) and millenial-scale time-varying field models (Korte and Constable, 2006) demonstrate the kinds of tests that could be performed, given a new generation of time-varying spherical harmonic paleo field models.

11.7   Future Directions

11.7.1   Toward New Global Data Sets

Over the past decade significant effort has been expended on the collection of 0–5 Ma paleomagnetic data from lavas. These new data differ in many ways from previous data sets: more samples per site are taken in the field, stepwise alternating field or thermal demagnetization is routine, and radiometric dating of sampled flows is increasingly an integral component of the study. It is not possible to summarize all the studies here, but we refer to one set of studies targeted specifically to address questions of global TAF or PSV behavior. The project known as the 'Time-Averaged Field Investigations' (TAFI) project has involved the collection of paleodirections (and some intensities) from 883 lava flows at 17 locations (Figure 22). The TAFI study locations were chosen to improve the geographical coverage of 0–5 Ma paleomagnetic directions at high latitudes (Spitzbergen, Aleutians, Antarctica), and in the Southern Hemisphere (various South American locales, Easter Island, and Australia). Several of the TAFI studies were conducted to replace previous, inadequately demagnetized data. Samples in four studies – Aleutians (Stone and Layer et al., 2006; Coe et al., 2000), Antarctica (Tauxe et al., 2004a), Easter Island (Brown, 2002) – were collected in the 1960s and 1970s, all other sites required new field work. New radiometric dates, along with 95% uncertainties have been obtained for 203 of the TAFI sites (Johnson et al., 2007).

As mentioned earlier several regional compilations have been published recently (the SW USA (Tauxe et al., 2003), the NW USA (Tauxe et al., 2004b), and 20° latitude (Lawrence et al., 2006)), and

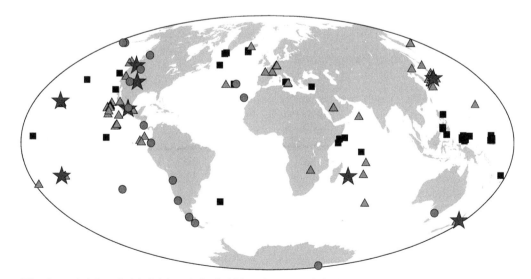

Figure 22   Current status of global data sets for 0–2 Ma paleomagnetic field modeling. The figure does not include all individual PSV studies of lava flows, only those that were part of a recent multidisciplinary sampling effort (red circles, see text). Blue stars indicate published regional compilations (see text) of directional information from lava flows. Black squares are sediment cores included in either Sint800 or Sint2000. Green triangles are absolute paleointensity data sites, where Thellier–Thellier measurements with pTRM checks have been performed.

several more are underway including Japan, New Zealand, Iceland, and Germany. The published regional compilations, combined with the TAFI data set already results in a data compilation with more than eight times the number of data in the 'DMAG 4' category of (McElhinny and McFadden, 1997). The incorporation of data from the regional compilations in preparation will result in a new data set that supersedes and replaces all of the *MM97 DMAG 4* sites, and spans latitudes from 78° S to 65° N. In particular, the TAFI data significantly improve coverage in the Southern Hemisphere. This was previously restricted (using only *MM97* DMAG 4 data) to three locations. The age distribution of even an interim global data set (Johnson *et al.*, 2007) indicates that it should now be possible to examine field behavior during the Brunhes and Matuyama chrons specifically, rather than needing to average all 0–5 Ma normal or reverse polarity data. In addition, sufficient data may exist to examine the period 0–100 ka, distinct from the rest of the Brunhes.

Figure 22 also shows the distribution of two other data types: records of absolute paleointensity from the past 2 My and deep-sea sediment cores that span all or part of the past 2 My for which directional and relative paleointensity records are available. The combination of these new lava flow direction and absolute paleointensity data, and sediment inclination (sometimes declination) and relative paleointensity data provide exciting new opportunities in global and regional field modeling.

11.7.2 A New Generation of Paleomagnetic Field Modeling

Future paleomagnetic field models will be greatly aided not only by the incorporation of the new, larger data sets, but also by the joint modeling of multiple types of data. This is nontrivial since much work remains to be done in for example, how best to calibrate absolute and relative paleointensity data. The sampling kernels described in Section 11.6 demonstrate that the inclusion of absolute paleointensity data will aid sampling of B_r at the CMB, especially at high latitudes. The joint study of both intensity and directional data will allow investigation of whether directional variability is indeed less during periods of higher field strength as has been suggested (e.g., Love, 2000), and is hinted at (via lower PSV and smaller deviations from GAD in the TAF for the Brunhes compared with the rest of the

past 5 My) in the ±20° latitude compilation of Lawrence *et al.* (2006).

While a good start on joint inversions of data for TAF and PSV has been made, studies to date (Hatakeyama and Kono, 2001, 2002) have solved iteratively for the TAF and the PSV. In the algorithm used in these studies an inversion for the time-averaged Gauss coefficients is performed, followed by an inversion for the variance in these coefficients, and the procedure repeated until convergence. Because the full inversion is nonlinear, it is unclear to what extent this approximation is a satisfactory approach. The heavy reliance on the GGP of field modeling may in fact be unwarranted, as is suggested by the inability of current statistical models to match regional empirical directional distributions (Lawrence *et al.*, 2006) and even some global distributions (Khokhlov *et al.*, 2001, 2006). In the case of the 20° latitude data set of Lawrence *et al.*, (2006), the empirical distributions cannot be fit by simply adjusting the relative variance contributions to the PSV. Interestingly, a better fit is obtained to the 20° latitude data set by reducing the ratio of the dipole to quadrupole family variance (β/α of Tauxe and Kent, 2004). Globally this implies less variation in VGP dispersion with latitude, a conclusion that is compatible with the recent (though still incomplete) global data set reported in Johnson *et al.* (2007).

This result suggests that future work should consider inversions of regional distributions of (D, I, and where possible F) for the local mean and variance in the field. Love and Constable (2003) have already demonstrated how to formulate the inverse problem for locally isotropic (Gaussian) fields, given mixed combinations of directions and intensities. A natural, but substantive, extension of this work is to consider local anisotropy. Clearly, at the very least, TAF modeling efforts should include non-Gaussian uncertainties in all the contributing data types and the use of robust statistics. In principle, future PSV models could also include temporal covariance estimated from long time series of paleomagnetic observations, as done by Hulot and Le Mouël (1994) for the historical field.

It is clear that the separation of PSV from the TAF is artificial and our understanding of how to deal with reversal and transitional data is poor. Detailed regional studies of field behavior can examine the question of field behavior over different timescales, via a combination of different kinds of studies. Regional paleomagnetic power spectra, along with estimates of PSV, and percentage of transitional/excursional data may offer better insight than separate global analyses of the TAF, PSV, or reversals. Such regional studies are particularly needed to try to determine longitudinal variations in field behavior that might reflect CMB influences on field behavior, and to characterize field behavior at high latitudes to understand what if signatures of the influence of the inner core are observable in the paleofield.

The statistical models for PSV offer some potentially exciting avenues of investigation because they can link analyses of paleomagnetic data themselves, as well as the output from dynamo simulations. Some such studies have been conducted (Glatzmaier *et al.*, 1999; McMillan *et al.*, 2001), but much remains to be done, subject of course to the caveats inherent in interpreting current numerical dynamo simulations. For example, the partitioning of dipole to quadrupole variance during reversals versus stable polarity periods, as determined from dynamo simulations, might provide an avenue to link or discriminate between reversal statistics and PSV statistics in the paleomagnetic record. One such recent study, confirming a prediction by McFadden *et al.* (1988), is that of Coe and Glatzmaier (2006) in which an inverse correlation between the stability and equatorial symmetry of the simulated field is reported, suggesting that reversal rates may have been low when the inner core was smaller than at present. This kind of study exemplifies the need to consider field behavior over a variety of timescales.

An obvious long-term goal is to incorporate all the disparate data types, along with their temporal information and invert for time-varying paleomagnetic field models, as is now possible on millennial timescales (Korte and Constable, 2005). This goal is still some way off (at least if the results are to be believed!), since much work is required to calibrate and link the information provided by different data sets, and to develop the needed modeling tools. However, **Figure 22** shows that the wealth of recently collected high quality data affords exciting avenues for future field modeling.

11.8 Concluding Remarks

The past two decades have seen an enormous transition in the kind of information we require from paleomagnetic data. Earlier studies needed data of sufficient quality to establish whether the GAD approximation was a good first order description of the paleomagnetic field. Deviations from GAD in the TAF direction are small; typically on the order of a few degrees, and good temporal control of data is required to interpret them correctly. These more stringent data requirements have necessitated the collection of many new data. This is a time-consuming and incremental process, but the ensemble effort of the paleomagnetic community is just beginning to result in global data sets that outstrip their predecessors in number and quality, and that can enable a new generation of field models.

Studies of the time-averaged field have provided tantalizing suggestions for the persistence of longitudinal structure in the field and they continue to affirm early recognition of apparent asymmetries in normal and reverse polarity fields. The current situation is unsatisfying, however, as the only agreed-upon non-GAD signal in the TAF is a contribution that can be described by an axial quadrupole term (g_2^0) in global spherical harmonic models that is 2–5% of the axial dipole term. The new directional data sets available from lava flows will allow better identification and separation of spatial and temporal field variations, and mapping of the paleomagnetic field will be greatly improved, by, at the very least, inclusion of high quality, absolute paleointensity data.

The GGP class of statistical models for PSV have enabled significant steps forward in PSV studies. Their major advantage stems from the ability to predict distributions of any magnetic observable at the Earth's surface, and their success has been to predict summary statistics of intensity and directional data sets. Several models have been proposed that can predict the latitudinal variation in VGP dispersion, and the empirical distribution of virtual axial dipole moments calculated from intensity data globally, but none are able to predict regional distributions of declination and inclination.

The development of new tools is needed to address many of the outstanding questions regarding paleofield behavior. These are critical to understanding the long-term role of boundary conditions on core flow and magnetic field generation, to link dynamo and paleofield models, and to propose sensible hypothesis tests for examining field behavior further back in time. Emerging data compilations indicate that the development of continuously time-varying regional and global field models spanning the past ~ 1 My is a realistic goal for the next decade, and could afford great insight into interactions among deep-Earth processes.

Acknowledgments

The authors thank Kristin Lawrence for help with figures. CLJ acknowledges support from several NSF grants over the past decade.

References

Baag C and Helsley CE (1974) Geomagnetic secular variation model E. *Journal of Geophysical Research* 79: 4918–4922.

Barton CE (1982) Spectral analysis of palaeomagnetic time series and the geomagnetic spectrum. *Philosophical Transactions of the Royal Society of London A* 306: 203–209.

Bloxham J (2000) The effect of thermal core–mantle interactions on the paleomagnetic secular variation. *Philosophical Transactions of the Royal Society of London, Series A* 358: 1171–1179.

Bloxham J and Gubbins D (1985) The secular variation of Earth's magnetic field. *Nature* 317: 777–781.

Bloxham J and Gubbins D (1987) Thermal core–mantle interactions. *Nature* 325: 511–513.

Bloxham J, Gubbins D, and Jackson A (1989) Geomagnetic secular variation. *Philosophical Transactions of the Royal Society of London, Series A* 329: 415–502.

Bloxham J and Jackson A (1990) Lateral temperature variations at the core–mantle boundary deduced from the magnetic field. *Geophysical Research Letters* 17: 1997–2000.

Bloxham J and Jackson A (1992) Time-dependent mapping of the magnetic field at the core-mantle boundary. *Journal of Geophysical Research* 97: 19537–19563.

Bloxham J, Gubbins D, and Jackson A (1989) Geomagnetic secular variation. *Philosophical Transactions of the Royal Society of London A* 329: 415–502.

Bouligand C, Hulot G, Khokhlov A, and Glatzmaier GA (2005) Statistical paleomagnetic field modelling and dynamo numerical simulation. *Geophysical Journal International* 161: 603–626.

Brown L (2002) Paleosecular variation from Easter Island revisited: Modern demagnetization of a 1970s data set. *Physics of the Earth and Planetary Interiors* 133: 73–81.

Busse FH (2002) Convective flows in rapidly rotating spheres and their dynamo action. *Physics of Fluids* 14: 1301–1314.

Carlut J and Courtillot V (1998) How complex is the time-averaged geomagnetic field over the last 5 million years? *Geophysical Journal International* 134: 527–544.

Carlut J, Courtillot V, and Hulot G (1999) Over how much time should the geomagnetic field be averaged to obtain the mean-paleomagentic field? *Terra Nova* 11: 239–243.

Christensen UR and Olson P (2003) Secular variation in numerical geodynamo models with lateral variations of boundary heat flow. *Physics of the Earth and Planetary Interiors* 138: 39–54.

Clement BM (1991) Geographic distribution of transitional VGPs: Evidence for non-zonal equatorial symmetry during the Matuyama–Brunhes geomagnetic reversal. *Earth and Planetary Science Letters* 104: 48–58.

Coe RS and Glatzmaier GA (2006) Symmetry and stability of the geomagnetic field. *Geophysical Research Letters* 33: (doi:10.1029/2006GL027903).

Coe RS, Zhao X, Lyons JJ, Pluhar CJ, and Mankinen EA (2000) Revisiting the 1964 collection of Nunivak lava flows. 81, Fall Meeting Supplement GP62A-06.

Constable CG and Johnson CL (1999) Anisotropic paleosecular variation models: Implications for geomagnetic observables. *Physics of the Earth and Planetary Interiors* 115: 3551.

Constable C and Johnson C (2005) A paleomagnetic power spectrum. *Physics of the Earth and Planetary Interiors* 153: 61–73.

Constable CG and Parker RL (1988) Statistics of the geomagnetic secular variation for the past 5 m.y. *Journal of Geophysical Research* 93: 11569–11581.

Constable CG, Parker RL, and Stark PB (1993) Geomagnetic field models incorporating frozen flux constraints. *Geophysical Journal International* 113: 419–433.

Costin S and Buffett B (2005) Preferred reversal paths caused by a heterogeneous conducting layer at the base of the mantle. *Journal of Geophysical Research* 109 (doi:10.1029/2003JB002853).

Courtillot V and Valet JP (1995) Secular variation of the Earth's magnetic field: From jerks to reversals. *Comptes rendus de l'Académie des Sciences, Paris* 320: 903–922.

Cox A (1962) Analysis of the present geomagnetic field for comparison with paleomagnetic results. *Journal of Geomagnetics and Geoelectricity* 13: 101–112.

Cox A (1970) Latitude dependence of the angular dispersion of the geomagnetic eld. *Geophysical Journal of the Royal Astronomical Society* 20: 253–260.

Creer KM (1962) The dispersion of the geomagnetic field due to secular variation and its determination from remote times from paleomagnetic data. *Journal of Geophysical Research* 67: 3461–3476.

Creer KM (1983) Computer synthesis of geomagnetic paleosecular variations. *Nature* 304: 695–699.

Creer KM, Irving E, and Nairn AEM (1959) Paleomagnetism of the great Whin sill. *Geophysical Journal of the Royal Astronomical Society* 2: 306–323.

Doell RR and Cox AV (1963) The accuracy of the paleomagnetic method as evaluated from historic Hawaiian lava flows. *Journal of Geophysical Research* 68: 1997–2009.

Doell RR and Cox AV (1965) Paleomagnetism of Hawaiian lava flows. *Journal of Geophysical Research* 70: 3377–3405.

Dziewonski AM (1984) Mapping the lower mantle: Determination of lateral heterogeneity in P velocity up to degree and order 6. *Journal of Geophysical Research* 89: 5929–5952.

Dziewonski AM, Hager BH, and O'Connell RJ (1987) Large scale heterogeneities in the lower mantle. *Journal of Geophysical Research* 82: 239–255.

Fisher RA (1953) Dispersion on a sphere. *Proceedings of the Royal Society of London A* 217: 295–305.

Glatzmaier GA, Coe RS, Hongre L, and Roberts PH (1999) The role of the Earth's mantle in controlling the frequency of geomagnetic reversals. *Nature* 401: 885–890.

Gubbins D (1988) Thermal core–mantle interactions and the time-averaged paleomagnetic field. *Journal of Geophysical Research* 93: 3413–3420.

Gubbins D (1994) Geomagnetic polarity reversals: A connection with secular variation and core-mantle interaction? *Reviews of Geophysics* 32: 61–83.

Gubbins D (1999) The distinction between geomagnetic excursions and reversals. *Geophysical Journal International* 137: F1–F3.

Gubbins D (2004) *Time Series Analysis and Inverse Theory for Geophysicists*. New York: Academic Press.

Gubbins D and Bloxham J (1985) Geomagnetic field analysis – III. Magnetic fields on the core-mantle boundary. *Geophysical Journal of the Royal Astronomical Society* 80: 695–713.

Gubbins D and Gibbons SJ (2003) Low Pacific secular variation. In: Channell JETC, Kent DV, Lowrie W, and Meert JG (eds.) *Geophysical Monograph Series*, Vol. 145: *Timescale of the Internal Geomagnetic Field*, pp. 279–286, 10.1029/145GM07. Washington, DC: American Geophysical Union.

Gubbins D and Kelly P (1993) Persistent patterns in the geomagnetic field over the past 2.5 Myr. *Nature* 365: 829–832.

Gubbins D and Richards M (1986) Coupling of the core dynamo and mantle: Thermal or topographic? *Geophysical Research Letters* 13: 1521–1524.

Gubbins D and Roberts N (1983) Use of the frozen-flux approximation in the interpretation of archaeomagnetic and palaeomagnetic data. *Geophysical Journal of the Royal Astronomical Society* 73: 675–687.

Gubbins D and Zhang K (1993) Symmetry properties of the dynamo equations for paleomagnetism and geomagnetism. *Physics of the Earth and Planetary Interiors* 75: 225–241.

Hale CJ and Dunlop DJ (1984) Evidence for an early Archean geomagnetic field: A paleomagnetic study of the Komati Formation, Barberton greenstone belt, South Africa. *Geophysical Research Letters* 11: 97–100.

Harrison CGA (1980) Secular variation and excursions of the Earth's magnetic field. *Journal of Geophysical Research* 85: 3511–3522.

Hatakeyama T and Kono M (2001) Shift of the mean magnetic field values: Effect of scatter due to secular variation and errors. *Earth Planets Space* 53: 3144.

Hatakeyama T and Kono M (2002) Geomagnetic field models for the last 5 Myr: Time-averaged field and secular variation. *Physics of the Earth and Planetary Interiors* 133: 181215.

Hoffman K (1984) A method for the display and analysis of transitional paleomagnetic data. *Journal of Geophysical Research* 89: 6285–6292.

Hollerbach R and Jones C (1993a) A geodynamo model incorporating a finitely conducting inner core. *Physics of the Earth and Planetary Interiors* 75: 317–327.

Hollerbach R and Jones C (1993b) Influence of Earth's inner core on geomagnetic fluctutations and reversals. *Nature* 365: 541–543.

Hospers J (1954) Reversals of the main geomagnetic field III. *Proc. Kon. Kederl. Akad. Wetensch., B* 57: 112–121.

Hulot G and Bouligand C (2005) Statistical paleomagnetic field modelling and symmetry considerations. *Geophysical Journal International* 161: 591–602.

Hulot G, Eymin C, Langlais B, Mandea M, and Olsen N (2002) Small-scale structure of the geodynamo inferred from Oersted and Magsat satellite data. *Nature* 416: 620–623.

Hulot G and Gallet Y (1996) On the interpretation of virtual geomagnetic pole (VGP) scatter curves. *Physics of the Earth and Planetary Interiors* 95: 37–53.

Hulot G and Le Mouël JL (1994) A statistical approach to Earth's main magnetic field. *Physics of the Earth and Planetary Interiors* 82: 167–183.

Irving E and Ward MA (1964) A statistical model of the geomagnetic field. *Pure and Applied Geophysics* 57: 47–52.

Jackson A, Jonkers ART, and Walker MR (2000) Four centuries of geomagnetic secular variation from historical records. *Philosophical Transactions of the Royal Society of London, Series A* 358: 957–990.

Jellinek AM and Manga M (2004) Links between long-lived hotspots, mantle plumes, D″ and plate tectonics. *Reviews of Geophysics* 42: (doi: 10.1029/2003RG000144).

Johnson CL and Constable CG (1995) The time-averaged geomagnetic field as recorded by lava flows over the past 5 Myr. *Geophysical Journal International* 122: 489–519.

Johnson CL and Constable CG (1996) Paleosecular variation recorded by lava flows over the last 5 Myr. *Philosophical Transactions of the Royal Society of London, Series A A* 354: 89–141.

Johnson CL and Constable CG (1997) The time-averaged geomagnetic field: Global and regional biases for 0–5 Ma. *Geophysical Journal International* 131: 643–666.

Johnson CL and Constable CG (1998) Persistently anomalous Pacific geomagnetic fields. *Geophysical Research Letters* 25: 1011–1014.

Johnson CL, Constable C, Tauxe L, *et al.* (in press) Recent investigations of the 0–5 Ma geomagnetic field recorded by lava flows. *Geochemistry Geophysics Geosystems*.

Johnson CL, Wijbrans JR, Constable CG, *et al.* (1998) $^{40}Ar/^{39}Ar$ ages and paleomagnetism of São Miguel lavas, Azores. *Earth and Planetary Science Letters* 160: 637–649.

Kelly P and Gubbins D (1997) The geomagnetic field over the past 5 Myr. *Geophysical Journal International* 128: 315–330.

Kokhlov A, Hulot G, and Carlut J (2001) Towards a self-consistent approach to palaeomagnetic field modelling. *Geophysical Journal International* 145: 157–171.

Khokhlov A, Hulot G, and Bouligand C (2006) Testing statistical palaeomagnetic field models against directional data affected by measurement errors. *Geophysical Journal International* 167: 635–648.

Kono M and Roberts PH (2002) Recent geodynamo simulations and observations of the geomagnetic field. *Reviews of Geophysics* 40: 10.1029/2000RG000102.

Kono M and Tanaka H (1995) Mapping the Gauss coefficients to the pole and the models of paleosecular variation. *Journal of Geomagnetics and Geoelectricity* 47: 115–130.

Korte M and Constable CG (2005) Continuous geomagnetic models for the past 7 millennia II: CALS7K. *Geochemistry Geophysics Geosystems* 6(2): Q02H16 (doi:10.1029/ 2004GC000801).

Korte M and Constable CG (2006) Centenntial to millenial geomagnetic secular variation. *Geophysical Journal International* 167: 43–52.

Labrosse S (2003) Thermal and magnetic evolution of the Earth's core. *Physics of the Earth and Planetary Interiors* 140: 127–143.

Laj C, Mazaud A, Weeks R, Fuller M, and Herrero-Bervera E (1991) Geomagnetic reversal paths. *Nature* 351: 447.

Lawrence K, Constable CG, and Johnson CL (2006) Paleosecular variation and the average geomagnetic field at $\pm 20°$ latitude. *Geochemistry Geophysics Geosystems* 7 (doi:10.1029/2005GC001181).

Lee S (1983) A study of the time-averaged paleomagnetic field for the past 195 million years. *Phd Thesis*, Australian National University.

Lock J and McElhinny M (1991) The global paleomagnetic database: Design, installation, and use with ORACLE. *Surveys in Geophysics* 12: 317–491.

Love JJ (2000) Paleomagnetic secular variation as a function of intensity. *Philosophical Transactions of the Royal Society of London, Series A* 358: 1191–1223.

Love JJ and Constable CG (2003) Gaussian statistics for palcomagnetic vectors. *Geophysical Journal International* 152: 515–565.

Masters G, Johnson S, Laske G, and Bolton H (1996) A shear-wave velocity model of the mantle. *Philosophical Transactions of the Royal Society of London, Series A* 354: 1385–1411.

McElhinny MW (2004) Geocentric axial dipole hypothesis: A least squares perspective. In: Channell JETC, Kent DV, Lowrie W, and Meert JG (eds.) *Geophysical Monograph Series 145: Timescale of the Internal Geomagnetic Field*, pp. 101–115, 10.1029/145GM07. Washington, DC: American Geophysical Union.

McElhinny M and Lock J (1993) Global Paleomagnetic database supplement number one, update to 1992. *Surveys in Geophysics* 14: 303–329.

McElhinny M and Lock J (1996) IAGA paleomagnetic databases with access. *Surveys in Geophysics* 14: 303–329.

McElhinny MW and McFadden PL (1997) Palaeosecular variation over the past 5 Myr based on a new generalized database. *Geophysical Journal International* 131: 240–252.

McElhinny MW, McFadden PL, and Merrill RT (1996) The time averaged paleomagnetic field 0–5 Ma. *Journal of Geophysical Research* 101: 25007–25027.

McElhinny MW and Merrill RT (1975) Geomagnetic secular variation over the past 5 my. *Reviews of Geophysics and Space Physics* 13: 687–708.

McElhinny MW, McFadden PL, and Merrill RT (1996a) The time averaged paleomagnetic field 0 – 5 Ma. *Journal of Geophysical Research* 101: 25007–25027.

McElhinny MW, McFadden PL, and Merrill RT (1996b) The myth of the Pacific dipole window. *Earth and Planetary Science Letters* 143: 13–22.

McFadden PL and Reid AB (1982) Analysis of paleomagnetic inclination data. *Geophysical Journal of the Royal Astronomical Society* 69: 307–319.

McFadden PL and McElhinny MW (1984) A physical model for paleosecular variation. *Earth and Planetary Science Letters* 67: 19–33.

McFadden PL, Merrill RT, and McElhinny MW (1988) Dipole/quadrupole family modelling of paleosecular variation. *Journal of Geophysical Research* 93: 11583–11588.

McFadden PL, Barton CE, and Merrill RT (1993) Do virtual geomagnetic poles follow preferred paths during geomagnetic reversals? *Nature* 361: 344–346.

McMillan DG, Constable CG, Parker RL, and Glatzmaier GA (2001) A statistical appraisal geodynamo models. *Geochemistry Geophysics Geosystems* 2 (doi:10.129/2000GC000130,2001).

Mejia V, Böhnel H, Opdyke ND, Otega-Rivera MA, Lee JKW, and Aranda-Gomez JJ (2005) Paleosecular variation and time-averaged field recorded in late Pliocene–Holocene lava flows from Mexico. *Geochemistry Geophysics Geosystems* 6(7) (doi:10.1029/2004GC000871).

Merrill RT and McElhinny MW (1977) Anomalies in the time averaged paleomagnetic field and their implications for the lower mantle. *Reviews of Geophysics and Space Physics* 15: 309–323.

Merrill R and McFadden P (2003) The geomagnetic axial dipole field assumption. *Physics of the Earth and Planetary Interiors* 139: 171–185.

Merrill RT, McElhinny MW, and Stevenson DJ (1979) Evidence for long-term asymmetries in the Earth's magnetic field and possible implications for dynamo theories. *Physics of the Earth and Planetary Interiors* 20: 75–82.

Merrill RT, McFadden PL, and McElhinny MW (1990) Paleomagnetic tomography of the core–mantle boundary. *Physics of the Earth and Planetary Interiors* 64: 87–101.

Merrill RT, McElhinny MW, and McFadden PL (1996) *The Magnetic Field of the Earth: Paleomagnetism, the Core and the Deep Mantle*. San Diego, CA: Academic Press.

Néel L (1949) Théorie du trainage magnétique des ferromagnétiques aux grains fins avec applications aux terres cuites. *Annals of Geophysics* 5: 99–136.

Olson P and Aurnou J (1999) A polar vortex in the Earth's core. *Nature* 402: 170–173.

Opdyke ND and Henry KW (1969) A test of the dipole hypothesis. *Earth and Planetary Science Letters* 6: 139–151.

Parker RL (1994) *Geophysical Inverse Theory*. Princeton, NJ: Princeton University Press.

Prévot M and Camps P (1993) Absence of preferred longitude sectors for poles from volcanic records of geomagnetic reversals. *Nature* 366: 53–56.

Quidelleur X and Courtillot V (1996) On low degree spherical harmonic models of paleosecular variation. *Physics of the Earth and Planetary Interiors* 95: 55–77.

Quidelleur X, Valet JP, Courtillot V, and Hulot G (1994) Long-term geometry of the geomagnetic field for the last 5 million years; an updated secular variation database from volcanic sequences. *Geophysical Research Letters* 21: 1639–1642.

Roberts PH and Glatzmaier GA (2000) Geodynamo theory and simulations. *Reviews of Modern Physics* 72: 1081–1123.

Runcorn SK (1992) Polar path in geomagnetic reversals. *Nature* 356: 654–656.

Schneider DA and Kent DV (1988) Inclination anomalies from Indian Ocean sediments and the possibility of a standing non-dipole field. *Journal of Geophysical Research* 93: 11621–11630.

Schneider DA and Kent DV (1990) The time-averaged paleomagnetic field. *Reviews of Geophysics* 28: 71–96.

Selkin P and Tauxe L (2000) Long term variations in paleointensity. *Philosophical Transactions of the Royal Society of London, Series A* 358: 1065–1088.

Sreenivasan B and Jones CA (2005) Structure and dynamics of the polar vortex in the Earth's core. *Geophysical Research Letters* 32: L20301.

Sreenivasan B and Jones CA (in press) Azimuthal winds, convection and dynamo action in the polar regions of planetary cores. *Geophysical and Astrophysical Fluid Dynamics* 100(4–5): 319–339.

Stone D and Layer P (2006) Paleosecular variation and GAD studies of 02 Ma flow sequences from the Aleutian Islands, Alaska. *Geochemistry Geophysics Geosystems* 7(4) (doi:10.1029/2005GC001007).

Tanaka H (1999) Circular asymmetry of the paleomagnetic directions observed at low latitude volcanic sites. *Earth Planets Space* 51: 1279–1285.

Tanaka H, Kono M, and Uchimura H (1995) Some global features of paleointensity in geological time. *Geophysical Journal International* 120: 97–102.

Tauxe L and Kent D (2004) A simplified statistical model for the geomagnetic field and the detection of shallow bias in paleomagnetic inclinations. Was the ancient magnetic field dipolar? In: Channell JETC, Kent DV, Lowrie W, and Meert JG (eds.) *Geophysical Monograph Series* 145: *Timescales of the Internal Geomagnetic Field*, pp. 101–115, 10.1029/145GM07. Washington, DC: American Geophysical Union.

Tauxe L, Constable C, Johnson CL, Koppers AAP, Miller WR, and Staudigel H (2003) Paleomagnetism of the Southwestern USA recorded by 0-5 Ma igneous rocks. *Geochemistry Geophysics Geosystems* 4(4): 8802 (doi:10.1029/2002GC000343).

Tauxe L, Gans P, and Mankinen E (2004a) Paleomagnetism and 40Ar/39Ar ages from volcanics extruded during the Matuyama and Brunhes Chrons near McMurdo Sound, Antarctica. *Geochemistry Geophysics Geosystems* 5(6) (doi:10.1029/2003GC000656).

Tauxe L, Luskin C, Selkin P, Gans P, and Calvert A (2004b) Paleomagnetic results from the Snake River Plain: Contribution to the time-averaged field global database. *Geochemistry Geophysics Geosystems* 5(8) (doi:10.1029/2003GC000661).

Tauxe L, Staudigel H, and Wijbrans J (2000) Paleomagnetism and $^{40}Ar/^{39}Ar$ ages from La Palma in the Canary Islands. *Geophysical Journal International* 1, ISSN: 1525–2027.

Valet J-P (2003) Time variations in geomagnetic intensity. *Reviews of Geophysics* 41: (doi:10.1029/2001RG000104).

Valet JP, Tucholka P, Courtillot V, and Meynadier L (1992) Paleomagnetic constraints on the geometry of the geomagnetic field during reversals. *Nature* 356: 400–407.

Vandamme D (1994) A new method to determine paleosecular variation. *Physics of the Earth and Planetary Interiors* 85: 131–142.

Van der Voo R and Torsvik TH (2001) Evidence for late Paleozoic and Mesozoic non-dipole fields provides an explanation for the Pangea reconstruction problems. *Earth and Planetary Science Letters* 187: 71–81.

Wicht J (2002) Inner-core conductivity in numerical dynamo simulations. *Physics of the Earth and Planetary Interiors* 132: 281–302.

Wilson RL (1970) Permanent aspects of the earth's non-dipole magnetic field over upper tertiary times. *Geophysical Journal of the Royal Astronomical Society* 19: 417–439.

Wilson RL (1971) Dipole offset – The time-average paleomagnetic field over the past 25 million years. *Geophysical Journal of the Royal Astronomical Society* 22: 491–504.

12 Source of Oceanic Magnetic Anomalies and the Geomagnetic Polarity Timescale

J. S. Gee, University of California, San Diego, La Jolla, CA, USA

D. V. Kent, Rutgers University, Piscataway, NJ, USA

12.1 Introduction

As new seafloor is created at the ridge crest, it cools and acquires a thermoremanence that captures a record of past geomagnetic field variations. The dominant geomagnetic signal recorded in the ocean crust is the pattern of field reversals; the juxtaposition of normal and reverse polarity crust constitutes a large magnetization contrast that generates readily identifiable variations in the magnetic field measured at the sea surface. The lineated and globally correlatable character of these sea-surface magnetic anomalies has proven remarkably useful in documenting the pattern of geomagnetic reversals since ~160 Ma. This reversal pattern, calibrated with suitable numerical ages, provides the basis for the geomagnetic polarity timescale (GPTS) that has broad reaching applications throughout Earth sciences including mapping the age distribution of the ocean floor constituting more than half of Earth's surface area.

The inferred pattern of geomagnetic reversals has changed relatively little since the pioneering work of Heirtzler *et al.* (1968), who used an assumption of constant seafloor spreading in the South Atlantic to derive the first GPTS. The stability of polarity interval widths in subsequent timescales testifies to the regularity of seafloor spreading, at least for carefully chosen spreading corridors. The anomaly profile in **Figure 1**, calculated from a simple block model with vertical polarity boundaries and for a full spreading rate of $140\,\mathrm{km\,My^{-1}}$, illustrates the optimal resolution that is possible at the fastest spreading known from the present ocean basins. Note that the terms slow-, intermediate-, fast-, and superfast spreading refer to full spreading rates of $<40\,\mathrm{km\,My^{-1}}$, $40–90\,\mathrm{km\,My^{-1}}$, $90–140\,\mathrm{km\,My^{-1}}$, $>140\,\mathrm{km\,My^{-1}}$, respectively. This simple model highlights some important aspects of magnetic anomalies measured at the sea surface. First, the depth of magnetization source (varying from about 2.6 km at the ridge crest to 5.6 km for the oldest seafloor (Stein and Stein, 1994) exerts a fundamental control on the wavelengths that can be resolved at the sea surface. The depth and thickness of the magnetized layer act as a bandpass filter (Earth filter of Schouten and McCamy, 1972; with the least attenuation at

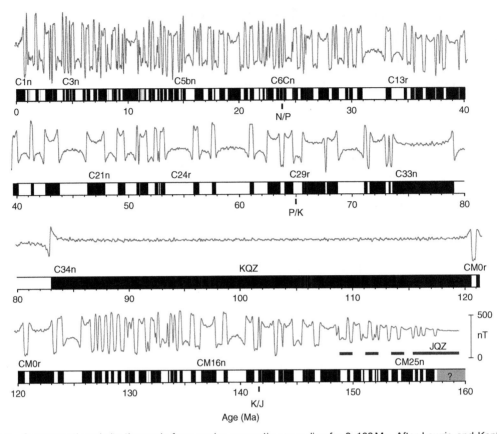

Figure 1 Geomagnetic polarity timescale from marine magnetic anomalies for 0–160 Ma. After Lowrie and Kent (2004); largely based on Cande and Kent (1995) and Channell *et al.* (1995). Filled and open blocks represent intervals of normal and reverse geomagnetic field polarity. (**Table 1**); key chrons that were used as calibration tiepoints are identified above the bar graph (C1n, C3n, etc.) and correlated positions of geologic period boundaries are indicated by ticks below the bar graph (N/P, Neogene/Paleogene; P/K, Paleogene/Cretaceous, K/J, Cretaceous/Jurassic). Idealized magnetic anomaly profile uses this polarity pattern in a 1 km thick source layer that has a mean magnetization of 5 A m^{-1} with gaussian variation ($\sigma = 42\%$ of mean) to mimic paleointensity variations, and vertical magnetization boundaries. The anomaly is calculated at the pole (no skewness, magnetization and ambient field direction vertical) at a full-spreading rate of 140 km My^{-1}, with depth to the upper surface that increases with age$^{0.5}$ (Ma) (Stein and Stein, 1994). Note the amplitude modulation, which primarily reflects the sequence effect since source properties otherwise stay constant other than modest monotonic change in water depth, except for magnetization ramp from JQZ into M-sequence (dashed line; factor of 5 increase from 157 Ma to 143 Ma) that is likely due to a systematic increase in geomagnetic intensity (Cande *et al.*, 1978; McElhinny and Larson, 2003).

wavelengths of 8–15 km for typical seafloor depths). Polarity intervals corresponding to these spatial scales ($\sim$100–200 ky for fast-spreading crust) are particularly well represented in the sea-surface anomaly profile (e.g., Anomaly 2; 1.77–1.95 Ma). Significantly shorter polarity intervals ($\sim$10^4 years) are also recognizable in high-resolution profiles from fast-spreading ridges, although such brief magnetization changes appear as broader, substantially attenuated anomalies at the sea surface. Thus, the combination of rapid spreading and typical seafloor depths of 3–4 km is well suited to capture essentially

all polarity intervals with durations exceeding $\sim$10^4 years in sea-surface anomaly data. Indeed, independent magnetostratigraphic studies corroborate that the GPTS can be viewed as essentially complete at timescales greater than about 30 ky.

Establishing the origin and significance of anomaly fluctuations at scales both longer and shorter than the dominant timescale of reversals (10^4–10^6 years) can potentially provide valuable additional insights into geomagnetic field behavior. To the extent that the magnetization of the ocean crust is a thermoremanence, the magnetization of the ocean crust might be expected to preserve a broad spectrum of geomagnetic field variations, from brief ($<10^3$ years) excursions or intensity fluctuations to longer-term variations on the scale of superchrons (10^7-10^8 years). Documenting the relative importance of geomagnetic field fluctuations in controlling magnetic anomaly amplitudes has proven difficult, however, since these anomalies are the end product of the recording of a paleofield signal (e.g., paleointensity variations, directional excursions, and polarity reversals), modulated by crustal accretionary processes (e.g., variations in geochemistry or the pattern of lava accumulation), and geometry of the source region(s). Despite limitations in accounting for the recording medium, significant progress has been made in documenting short wavelength anomaly fluctuations that may be attributed to geomagnetic field behavior. A key element in many of these studies is the collection of anomaly data nearer the magnetic source layer to preserve higher frequency variations that are attenuated in sea-surface anomaly data. The coherence among near-bottom anomaly profiles and the similarity of these coherent fluctuations with independent records of field intensity from sediments suggest that a substantial geomagnetic intensity signal may be preserved in the ocean crustal magnetization.

Some longer-term variations in anomaly amplitudes may likewise have a geomagnetic origin although there has been less progress in differentiating this from other causes of variation in source properties. For example, the gradual increase in anomaly amplitudes following the Jurassic Quiet Zone (JQZ; crust older than $\sim$154 Ma) has been attributed to an increase in paleofield intensity (Cande et al., 1978) whereas the amplitude envelope observed over some ridge axes has long been interpreted as evidence of low-temperature alteration of the source layer (e.g., Bleil and Petersen, 1983; Raymond and LaBrecque, 1987). The particular pattern of polarity reversals and the relative widths of adjacent polarity intervals (sometimes termed the sequence effect) can modulate anomaly amplitudes, for example, the variations in amplitudes that are apparent over the past 40 My in the simple model shown in **Figure 1**, so recognition of long-term trends is not straightforward. There are also first-order geomagnetic questions about the marked contrast in the character of magnetic anomalies in the Cretaceous Quiet Zone (KQZ; 83–120.6 Ma) and the JQZ, the former often associated with large amplitude anomalies and the latter with small amplitude

anomalies, which may be related to significant differences in mean field strength in the Jurassic and the Cretaceous.

Despite more than 40 years of study, many aspects of the magnetization source responsible for lineated marine magnetic anomalies remain uncertain. Early studies (Atwater and Mudie, 1973; Talwani et al., 1971) indicated that the source layer is thin and dominated by extrusives whereas compilations of lava magnetizations suggest that additional deeper sources may be required and indeed in some cases (e.g., near Ocean Drilling Program (ODP) Hole 735B; Dick et al., 1991) recognizable anomaly lineations are present where only the intrusive portion of the crust is preserved. Since the last review of crustal magnetization (Smith, 1990), there has been significant progress in sampling of the dikes, lower crustal gabbros, and upper mantle material exposed in tectonic windows, and these studies indicate that dikes and gabbros are likely significant contributors to sea-surface magnetic anomalies. In addition to the need for a fundamental characterization of possible magnetization sources within the crust, some understanding of the details of the accretionary process (e.g., width of the neovolcanic zone, off-axis volcanism) is also essential since these provide an intrinsic limit on the fidelity of the crustal recording process. As our understanding of geomagnetic field variations improves (particularly over timescales on the order of 10^4 years over which much of the ocean crust is constructed), near-bottom magnetic anomaly data can be used to constrain aspects of crustal accretion. Such studies can, in turn, be useful in evaluating the timescales at which geomagnetic information can be recovered from magnetic anomaly records.

In this chapter we will focus primarily on the promise and limitations of the ocean crust as a recorder of geomagnetic field variations, emphasizing the record of past geomagnetic field variations recorded in anomalies (and therefore in source magnetization) on timescales of 10^3 years (excursions) to 10^4-10^6 (reversals) and 10^7-10^8 (superchrons). We review the origin of the magnetization in the various crustal source layers responsible for lineated magnetic anomalies and conclude by mentioning some applications to deciphering how oceanic crust formed and by speculating on future directions. The chapter is based mostly on published literature that appeared since the last major review of ocean crust magnetization by Smith (1990).

12.2 Magnetic Anomalies as Records of Geomagnetic Field Behavior

12.2.1 Polarity Reversals and the Geomagnetic Polarity Timescale

Marine magnetic anomalies (Vine, 1966; Vine and Matthews, 1963) provide a complete record of geomagnetic polarity reversals, making the age-calibrated geomagnetic polarity sequence the basis for global correlations and geochronology for the past 160 My (**Figure 1**). The relative widths of the magnetic polarity intervals for the Late Jurassic to Recent were determined from magnetic profiles initially by Heirtzler *et al.* (1968) for the Central Anomaly (Anomaly 1) to Anomaly 32 (C-sequence) and by Larson and Pitman (1972) for Anomalies M1–M22 (M-sequence). Larson and Pitman recognized that the M-sequence anomalies were bracketed by long intervals with apparently few or no reversals: the KQZ between the M-sequence and the ridge crest C-sequence of Heirtzler *et al.* (to which Larson and Pitman added Anomalies 33 and 34) and the JQZ prior to the M-sequence (which was soon extended to M0 at the younger end and to M25 at the older end by Larson and Hilde (1975) and to M29 by Cande *et al.* (1978). Key aspects of the geomagnetic interpretations were verified by magnetostratigraphic studies. For example, Helsley and Steiner (1969) documented an interval of constant normal polarity (the Cretaceous Normal Polarity Superchron, CNPS) corresponding to the KQZ observed in magnetic anomaly profiles. Similarly, Lowrie and Alvarez (1981) verified the overall correspondence between magnetostratigraphic polarity intervals and the ~100 My anomaly record from Anomaly 6C to Anomaly M3 including the CNPS and KQZ, and Ogg and Lowrie (1986) provided magnetostratigraphic confirmation for the central part of the M-sequence. Correlations between anomaly interpretations and magnetostratigraphy are being refined in land sections and marine cores (e.g., Billups *et al.*, 2004; Channell *et al.*, 2003; Lanci *et al.*, 2004, 2005; Speranza *et al.*, 2005) and provide the linkage to numerical ages based on radioisotopic dates or astronomical cyclicity in stratigraphic sections for age calibration of the anomaly sequence in the construction of geomagnetic polarity timescales.

Standard geomagnetic polarity chron nomenclature is based on the long-standing numbering schemes (sometimes with lettered additions) for prominent but irregularly spaced magnetic anomalies (Larson and Pitman, 1972; Pitman and Heirtzler, 1966; Pitman *et al.*, 1968). Polarity subdivisions for the past ~5 My based on compilations of radiometric dating of discrete lavas (Cox *et al.*, 1964) were initially labeled after prominent geomagneticians (Brunhes, Matuyama, Gauss, and Gilbert for chrons, formerly called epochs; Anonymous, 1979) and type localities (Jaramillo, Olduvai, Mammoth, etc., for subchrons, formerly called events) but this nomenclature system was impractical to be extended to the numerous older polarity intervals delineated by magnetic anomalies. By convention, identifiers usually refer to positive anomalies, which correspond to normal polarity for the C-sequence (Anomalies 1–34) but mostly to reverse polarity intervals for the M-sequence (M0–M29) because they formed in the Southern Hemisphere but are now in the Northern Hemisphere in the Pacific where the M-sequence is well developed. A chron corresponds to the interval from the younger boundary of the eponymous anomaly to the younger boundary of the preceding anomaly and has the prefix 'C' (e.g., Chron C3A or Chron CM20). However, each of these chrons is usually subdivided into the two constituent intervals of predominantly normal and reverse polarity, which are designated by adding to the name the suffix 'n' for normal polarity and 'r' for the preceding reverse polarity interval (e.g., Chron C3An and Chron C3Ar, or Chron CM20n and Chron CM20r). When these polarity chrons are further subdivided into shorter polarity intervals they are referred to as subchrons and identified by appending, from youngest to oldest, '.1', '.2', etc., to the polarity chron name, and adding an 'n' for normal polarity or an 'r' for reverse polarity (e.g., Chron C3An.1r, or Chron CM20n.1r). Finally, the designation '–1', '–2', etc., is used following a chron or subchron name to denote apparently brief geomagnetic features corresponding to short wavelength anomalies or 'tiny wiggles', which, upon calibration, convert to durations less than 30 ky. In view of their uncertain origin, these globally mapped geomagnetic features are referred to as cryptochrons (Cande and Kent, 1992a); they can be elevated to subchron status by the addition of 'n' for normal polarity and 'r' for reverse polarity if confirmed as polarity intervals in magnetostratigraphic studies (e.g., Chron C1r.2r-1n).

Anomaly spacings for the Central Anomaly (Anomaly 1) to Anomaly 34 (younger end of KQZ) were comprehensively refined by Cande and Kent (1992a) and for the M-sequence (M0–M29) by

Channell *et al.* (1995). Cande and Kent (1992a) used a combination of nine finite rotation poles to describe seafloor spreading in the South Atlantic and 61 stacked profiles distributed over the finite rotation pole intervals to develop a continuous framework for the anomaly sequence; finer scale information was derived from faster spreading rate ridges in the Pacific and Indian Oceans and inserted into the South Atlantic pattern. For age calibration, Cande and Kent (1992a) assumed that spreading rates in the South Atlantic were smoothly varying and fit a cubic interpolation spline function to a set of nine tiepoints that link radioisotopic ages with distances of correlative anomalies from the zero-age ridge axis to derive a GPTS for 0–83 Ma (Anomaly 34). The revised GPTS of Cande and Kent (1995) (CK95) includes astrochronological estimates for polarity reversals for the past 5.23 My (Hilgen, 1991; Shackleton *et al.*, 1990), which are negligibly different (within 0.03 My) from the most recent retuning by Lourens *et al.* (2004). For the M-sequence, Channell *et al.* (1995) compared profiles from the Phoenix, Japanese, and Hawaiian lineation sets and derived a representative anomaly sequence for M0 to M29 from a new block model for the Hawaiian lineations. There are still relatively few age-diagnostic data available for calibration of M-sequence anomalies – Channell *et al.* (1995) used only three tiepoints that were regarded as reliable and consistent with constant spreading on the new model of the Hawaiian lineations to derive a GPTS (referred to by the authors as CENT94) for 120.6 Ma (M0) to 157.53 Ma (M29).

The listing of polarity intervals given in **Table 1** is the same as used for statistical analysis by Lowrie and Kent (2004) and is basically a combination of CK95 for the C-sequence and CENT94 for the M-sequence. This polarity sequence for the past 160 My is believed to be complete to a resolution of better than 30 ky and includes those relatively few cryptochrons that have been detected as short polarity intervals in high-resolution magnetostratigraphic studies and elevated to subchrons, mostly in the Cenozoic (see summary by Krijgsman and Kent 2004). Small-scale magnetic anomalies have been identified beyond M29 (Handschumacher *et al.*, 1988), most recently to M41 with an apparent age of ~167 Ma (Sager *et al.*, 1998; Tivey *et al.*, 2006). However, the character of the pre-M29 anomalies resembles small-scale magnetic anomalies or tiny wiggles in the Cenozoic, which have generally been attributed to paleointensity variations rather than

geomagnetic reversals (e.g., Cande and Kent, 1992b). Accordingly, the pre-M29 anomalies along with the cryptochrons in the Cenozoic that have not been confirmed as polarity intervals are not included in **Table 1** or shown in **Figure 1**.

A recently published timescale (GTS2004: Gradstein *et al.*, 2004) utilizes the distances to anomalies in the South Atlantic from Cande and Kent (1992a, 1995) but with a somewhat different array of age calibration tiepoints. The largest difference between CK95 and GTS2004 is less than 1 My and occurs at chron C6Cn.2n, which was set at 23.8 ± 1 Ma in CK95 but has been recalibrated using astrochronology to 23.03 ± 0.04 Ma (Lourens *et al.*, 2004; Shackleton *et al.*, 2000). CK95 also maintains age registry with the compilation of Cenozoic cryptochrons by Cande and Kent (1992b), which were not migrated to the chronology of GTS2004. For the M-sequence, age calibration of M0 is a major difference: GTS2004 sets the age of Chron CM0 (base of Aptian) at about 125 Ma, which is 4 My older than in the CENT94 timescale of Channell *et al.* (1995) and needs to be corroborated because of the profound implications for the rate of seafloor spreading in the KQZ (e.g., Cogné and Humler, 2006).

On a plot of the age of each reversal against the order of its occurrence (**Figure 2**), the CENT94 data define a straight line whereas the CK95 C-sequence can be divided into two nearly linear segments that intersect near Chron C12r at about 32 Ma. The 37.6 My-long CNPS is evident as an abrupt discontinuity between the C-sequence and the M-sequence and is the longest polarity interval in the entire ~160 My-long sequence. Polarity intervals C33n and C33r that immediately follow the CNPS are the 2nd and 3rd longest chrons and may have a closer affinity to the CNPS than to the rest of the reversal sequence. Linear segments imply that the reversal process is stationary and allow calculation of representative statistical parameters for each of the reversal frequency regimes. According to Lowrie and Kent (2004), mean polarity interval lengths are 0.25 My for C1n–C12r, 0.75 My for C13n–C32r, and 0.415 My for M0–M29. Despite the large differences in mean polarity interval length and taking into account a finite reversal transition time of a few thousand years, the distributions of chron lengths within each of the regimes are not significantly different from a Poisson or exponential distribution (Lowrie and Kent, 2004), which implies that the reversal process is essentially free of memory (Cox,

Table 1   Geomagnetic polarity chrons from marine magnetic anomalies

Age range (Ma)	Normal polarity Chrons	Age range (Ma)	Reverse polarity chrons
0.000–0.780	C1n	0.780–0.990	C1r.1r
0.990–1.070	C1r.1n	1.070–1.201	C1r.2r·
1.201–1.211	C1r.2r–1n	1.211–1.770	C1r.2r ·
1.770–1.950	C2n	1.950–2.140	C2r.1r
2.140–2.150	C2r.1n	2.150–2.600	C2r.2r
2.581–3.040	C2An.1n	3.040–3.110	C2An.1r
3.110–3.220	C2An.2n	3.220–3.330	C2An.2r
3.330–3.580	C2An.3n	3.580–4.180	C2Ar
4.180–4.290	C3n.1n	4.290–4.480	C3n.1r
4.480–4.620	C3n.2n	4.620–4.800	C3n.2r
4.800–4.890	C3n.3n	4.890–4.980	C3n.3r
4.980–5.230	C3n.4n	5.230–5.894	C3r
5.894–6.137	C3An.1n	6.137–6.269	C3An.1r
6.269–6.567	C3An.2n	6.567–6.935	C3Ar
6.935–7.091	C3Bn	7.091–7.135	C3Br.1r
7.135–7.170	C3Br.1n	7.170–7.341	C3Br.2r
7.341–7.375	C3Br.2n	7.375–7.432	C3Br.3r
7.432–7.562	C4n.1n	7.562–7.650	C4n.1r
7.650–8.072	C4n.2n	8.072–8.225	C4r.1r
8.225–8.257	C4r.1n	8.257–8.606	C4r.2r∗
8.606–8.664	C4r.2r–1n	8.664–8.699	C4r.2r∗
8.699–9.025	C4An	9.025–9.097	C4Ar.1r∗
9.097–9.117	C4Ar.1r–1n	9.117–9.230	C4Ar.1r∗
9.230–9.308	C4Ar.1n	9.308–9.580	C4Ar.2r
9.580–9.642	C4Ar.2n	9.642–9.740	C4Ar.3r
9.740–9.880	C5n.1n	9.880–9.920	C5n.1r
9.920–10.949	C5n.2n	10.949–11.052	C5r.1r
11.052–11.099	C5r.1n	11.099–11.167	C5r.2r∗
11.167–11.193	C5r.2r–1n	11.193–11.352	C5r.2r∗
11.352–11.363	C5r.2r–2n	11.363–11.476	C5r.2r∗
11.476–11.531	C5r.2n	11.531–11.555	C5r.3r∗
11.555–11.584	C5r.3r–1n	11.584–11.935	C5r.3r∗
11.935–12.078	C5An.1n	12.078–12.184	C5An.1r
12.184–12.401	C5An.2n	12.401–12.678	C5Ar.1r
12.678–12.708	C5Ar.1n	12.708–12.775	C5Ar.2r
12.775–12.819	C5Ar.2n	12.819–12.991	C5Ar.3r
12.991–13.139	C5AAn	13.139–13.302	C5AAr
13.302–13.510	C5ABn	13.510–13.703	C5ABr
13.703–14.076	C5ACn	14.076–14.178	C5ACr
14.178–14.612	C5ADn	14.612–14.800	C5ADr
14.800–14.888	C5Bn.1n	14.888–15.034	C5Bn.1r
15.034–15.155	C5Bn.2n	15.155–16.014	C5Br
16.014–16.293	C5Cn.1n	16.293–16.327	C5Cn.1r
16.327–16.488	C5Cn.2n	16.488–16.556	C5Cn.2r
16.556–16.726	C5Cn.3n	16.726–17.277	C5Cr
17.277–17.615	C5Dn	17.615–17.793	C5Dr∗
17.793–17.854	C5Dr–1n	17.854–18.281	C5Dr∗
18.281–18.781	C5En	18.781–19.048	C5Er
19.048–20.131	C6n	20.131–20.518	C6r
20.518–20.725	C6An.1n	20.725–20.996	C6An.1r
20.996–21.320	C6An.2n	21.320–21.768	C6Ar
21.768–21.859	C6AAn	21.859–22.151	C6AAr.1r
22.151–22.248	C6AAr.1n	22.248–22.459	C6AAr.2r
22.459–22.493	C6AAr.2n	22.493–22.588	C6AAr.3r
22.588–22.750	C6Bn.1n	22.750–22.804	C6Bn.1r
22.804–23.069	C6Bn.2n	23.069–23.353	C6Br
23.353–23.535	C6Cn.1n	23.535–23.677	C6Cn.1r
23.677–23.800	C6Cn.2n	23.800–23.999	C6Cn.2r

(Continued)

Table 1 (Continued)

Age range (Ma)	Normal polarity Chrons	Age range (Ma)	Reverse polarity chrons
23.999–24.118	C6Cn.3n	24.118–24.730	C6Cr
24.730–24.781	C7n.1n	24.781–24.835	C7n.1r
24.835–25.183	C7n.2n	25.183–25.496	C7r
25.496–25.648	C7An	25.648–25.678	C7Ar*
25.678–25.705	C7Ar–1n	25.705–25.823	C7Ar*
25.823–25.951	C8n.1n	25.951–25.992	C8n.1r
25.992–26.554	C8n.2n	26.554–27.027	C8r
27.027–27.972	C9n	27.972–28.283	C9r
28.283–28.512	C10n.1n	28.512–28.578	C10n.1r
28.578–28.745	C10n.2n	28.745–29.401	C10r
29.401–29.662	C11n.1n	29.662–29.765	C11n.1r
29.765–30.098	C11n.2n	30.098–30.479	C11r
30.479–30.939	C12n	30.939–33.058	C12r
33.058–33.545	C13n	33.545–34.655	C13r
34.655–34.940	C15n	34.940–35.343	C15r
35.343–35.526	C16n.1n	35.526–35.685	C16n.1r
35.685–36.341	C16n.2n	36.341–36.618	C16r
36.618–37.473	C17n.1n	37.473–37.604	C17n.1r
37.604–37.848	C17n.2n	37.848–37.920	C17n.2r
37.920–38.113	C17n.3n	38.113–38.426	C17r
38.426–39.552	C18n.1n	39.552–39.631	C18n.1r
39.631–40.130	C18n.2n	40.130–41.257	C18r
41.257–41.521	C19n	41.521–42.536	C19r
42.536–43.789	C20n	43.789–46.264	C20r
46.264–47.906	C21n	47.906–49.037	C21r
49.037–49.714	C22n	49.714–50.778	C22r
50.778–50.946	C23n.1n	50.946–51.047	C23n.1r
51.047–51.743	C23n.2n	51.743–52.364	C23r
52.364–52.663	C24n.1n	52.663–52.757	C24n.1r
52.757–52.801	C24n.2n	52.801–52.903	C24n.2r
52.903–53.347	C24n.3n	53.347–55.904	C24r
55.904–56.391	C25n	56.391–57.554	C25r
57.554–57.911	C26n	57.911–60.920	C26r
60.920–61.276	C27n	61.276–62.499	C27r
62.499–63.634	C28n	63.634–63.976	C28r
63.976–64.745	C29n	64.745–65.578	C29r
65.578–67.610	C30n	67.610–67.735	C30r
67.735–68.737	C31n	68.737–71.071	C31r
71.071–71.338	C32n.1n	71.338–71.587	C32n.1r
71.587–73.004	C32n.2n	73.004–73.291	C32r.1r
73.291–73.374	C32r.1n	73.374–73.619	C32r.2r
73.619–79.075	C33n	79.075–83.000	C33r
83.00–120.60	C34n (CNPS)	120.60–121.00	CM0r
121.00–123.19	CM1n	123.19–123.55	CM1r
123.55–124.05	CM2n	124.05–125.67	CM3r
125.67–126.57	CM4n	126.57–126.91	CM5r
126.91–127.11	CM6n	127.11–127.23	CM6r
127.23–127.49	CM7n	127.49–127.79	CM7r
127.79–128.07	CM8n	128.07–128.34	CM8r
128.34–128.62	CM9n	128.62–128.93	CM9r
128.93–129.25	CM10n	129.25–129.63	CM10r
129.63–129.91	CM10Nn.1n	129.91–129.95	CM10Nn.1r
129.95–130.22	CM10Nn.2n	130.22–130.24	CM10Nn.2r
130.24–130.49	CM10Nn.3n	130.49–130.84	CM10Nr
130.84–131.50	CM11n	131.50–131.71	CM11r.1r
131.71–131.73	CM11r.1n	131.73–131.91	CM11r.2r
131.91–132.35	CM11An.1n	132.35–132.40	CM11An.1r
132.40–132.47	CM11An.2n	132.47–132.55	CM11Ar

(Continued)

Table 1 (Continued)

Age range (Ma)	Normal polarity Chrons	Age range (Ma)	Reverse polarity chrons
132.55–132.76	CM12n	132.76–133.51	CM12r.1r
133.51–133.58	CM12r.1n	133.58–133.73	CM12r.2r
133.73–133.99	CM12An	133.99–134.08	CM12Ar
134.08–134.27	CM13n	134.27–134.53	CM13r
134.53–134.81	CM14n	134.81–135.57	CM14r
135.57–135.96	CM15n	135.96–136.49	CM15r
136.49–137.85	CM16n	137.85–138.50	CM16r
138.50–138.89	CM17n	138.89–140.51	CM17r
140.51–141.22	CM18n	141.22–141.63	CM18r
141.63–141.78	CM19n.1n	141.78–141.88	CM19n.1r
141.88–143.07	CM19n	143.07–143.36	CM19r
143.36–143.77	CM20n.1n	143.77–143.84	CM20n.1r
143.84–144.70	CM20n.2n	144.70–145.52	CM20r
145.52–146.56	CM21n	146.56–147.06	CM21r
147.06–148.57	CM22n.1n	148.57–148.62	CM22n.1r
148.62–148.67	CM22n.2n	148.67–148.72	CM22n.2r
148.72–148.79	CM22n.3n	148.79–149.49	CM22r
149.49–149.72	CM22An	149.72–150.04	CM22Ar
150.04–150.69	CM23n.1n	150.69–150.91	CM23n.1r
150.91–150.93	CM23n.2n	150.93–151.40	CM23r
151.40–151.72	CM24n.1n	151.72–151.98	CM24n.1r
151.98–152.00	CM24n.2n	152.00–152.15	CM24r
152.15–152.24	CM24An	152.24–152.43	CM24Ar
152.43–153.13	CM24Bn	153.13–153.43	CM24Br
153.43–154.00	CM25n	154.00–154.31	CM25r
154.31–155.32	CM26n	155.32–155.55	CM26r
155.55–155.80	CM27n	155.80–156.05	CM27r
156.05–156.19	CM28n	156.19–156.51	CM28r
156.51–157.27	CM29n	157.27–157.53	CM29r

Geomagnetic polarity time scale tabulated from Cande and Kent (1995), with additional short subchrons summarized by Lowrie and Kent (2004), for after the Cretaceous Normal Polarity Superchron (CNPS), which corresponds to the Cretaceous Quiet Zone, and from Channell *et al.* (1995) for before the CNPS. The names of chrons marked by asterisks are repeated because the reverse polarity intervals contain one or two cryptochrons that have been elevated to the status of new normal polarity subchrons.

1968). It has been suggested that the CNPS may represent part of a continuous, long-term evolution of reversal rate (McFadden and Merrill, 2000). However, the lack of precursory field behavior that might have heralded this superchron (Gallet and Hulot, 1997; Lowrie and Kent, 2004) suggests that the CNPS may represent either an abrupt perturbation of the reversal process or a separate (non)reversal regime.

12.2.2 Geomagnetic Intensity Fluctuations

Sea-surface magnetic anomaly profiles from fast-spreading ridges allow recognition of polarity intervals as short as $\sim$30 ky. Yet the geomagnetic field is known to fluctuate on significantly shorter time-scales. For example, direct observations indicate that the dipole field is decreasing at 15 nT yr^{-1} so that g_1^0 has decreased about 8% since 1832 when Gauss first estimated the field intensity from a spherical harmonic analysis (Jackson *et al.*, 2000). Absolute paleointensities from archeomagnetic materials suggest that field intensities at 1–3 ka were approximately 40–50% higher than the present field (Yang *et al.*, 2000) although nondipolar field variations undoubtedly result in considerable spatial variability (Korte and Constable, 2005). On somewhat longer timescales, more sparse absolute intensity data from volcanic materials provide evidence for extremely low field values ($\sim$20% or less of the present value) at approximately 40 ka associated with the Laschamp excursion (Levi *et al.*, 1990; Roperch *et al.*, 1988).

Short-term ($\sim$$10^2$–$10^3$ years) fluctuations in geomagnetic intensity, even if quite substantial, will

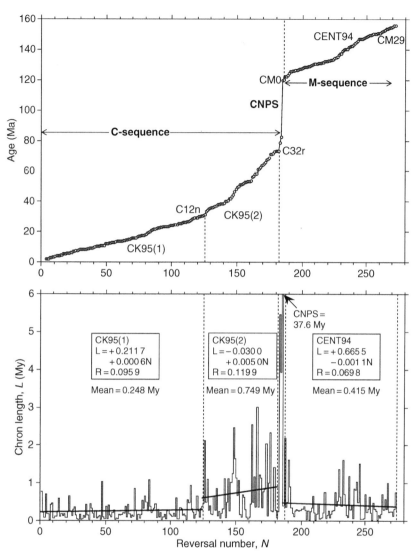

Figure 2 Top, ages of magnetic reversals (Ma), and bottom, polarity chron lengths (My) in the Cenozoic and Mesozoic oceanic anomaly sequences plotted against their age order (from Lowrie and Kent, 2004). Statistical properties are shown for two segments of the CK95 timescale and the CENT94 timescale in which linear fits have slopes not distinct from zero. CNPS, Cretaceous normal polarity superchron.

hardly be evident in sea-surface anomaly data. Intensity variations on timescales of 10^4 years may be detectable in sea-surface profiles from fast-spreading ridges but will have greatly reduced amplitudes and longer wavelengths as dictated by the Earth filter. In order to document these higher-frequency anomaly variations (some of which may reflect geomagnetic field behavior), it is useful to make magnetic field observations nearer the magnetic source (Larson and Spiess, 1969). One might well ask whether much additional information can be gained from these near-bottom records since even polarity

intervals as short as 10^4 may be recorded, albeit in filtered form, in sea-surface profiles from the fastest-spreading ridges in the present ocean basins. The answer is twofold. First, comparison of sea-surface and near-bottom data can sometimes differentiate between short wavelength anomalies due to short reversals and those arising from intensity fluctuations. For example, Bowers *et al.* (2001) have documented that many of the coherent tiny wiggles in sea-surface profiles from Anomaly 5 evolve into more complex, but still lineated, anomalies near the seafloor, suggesting these short wavelength anomalies represent geomagnetic intensity fluctuations rather than discrete short polarity events (which should be manifest by a higher amplitude and simpler anomaly pattern). A second benefit of near-bottom data is that two closely spaced short events (whether intensity variations or reversals) can be resolved as distinct features but must be much more widely separated to be differentiated in sea-surface anomaly data.

Near-bottom magnetic anomaly data from near the axis of fast-spreading ridges are particularly useful for assessing any geomagnetic signal, since they can provide high temporal resolution over a time period where independent estimates of geomagnetic intensity fluctuations are best known. Such anomaly data from fast-spreading ridges often show a short wavelength (2–4 km) axial anomaly high with a superimposed anomaly minimum or notch of even shorter wavelength (Gee *et al.*, 2000; Perram *et al.*, 1990; Shah *et al.*, 2003). Possible mechanisms that have been advanced to explain this notch in the axial anomaly include: (1) variations in the thickness of the magnetized layer, (2) variations in magnetization intensity as a result of alteration, and (3) geomagnetic intensity fluctuations (Gee and Kent, 1994; Perram *et al.*, 1990; Tivey and Johnson, 1987). Variations in the thickness of the extrusive magnetic source layer might arise from an axial keel of less magnetic dikes or from elevated temperature that may exceed the Curie point of the Ti-rich titanomagnetite in the extrusives (Shah *et al.*, 2003). The magnetization of seafloor basalts can be substantially reduced by alteration in localized hydrothermal upflow zones (e.g., Tivey *et al.*, 1996; Tivey and Johnson, 2002). As will be discussed further in Section 12.3.1, the long-standing notion that low-temperature alteration of the magnetic source is the dominant process in controlling crustal magnetization has significantly influenced interpretation of magnetic anomaly data both in the near-ridge environment (e.g., Schouten *et al.*, 1999), as well as on much longer timescales (e.g., Bleil and Petersen, 1983; Zhou *et al.*, 2001).

Although a variety of mechanisms undoubtedly influence magnetic anomaly amplitudes, near-bottom anomaly data and associated absolute paleointensities from the superfast-spreading southern East Pacific Rise (EPR) illustrate that geomagnetic intensity variations are likely sufficient to account for much of the near-ridge anomaly signal (**Figure 3**). Models of crustal alteration predict a monotonic decrease in magnetization intensity away from the ridge axis. The initial increase in anomaly amplitudes on the ridge flanks is thus apparently at odds with low-temperature alteration being the dominant process

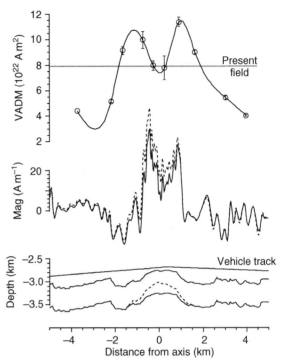

Figure 3 Comparison of glass paleointensities (top) and inversion magnetization solution (middle) for near-bottom magnetic profile across East Pacific Rise at 19.75° S. For source-layer geometries (bottom), both a constant-thickness source (solid line) and a variable-thickness source (dashed line) approximating off-axis doubling of seismic layer 2A thickness produce inversions indicating that a lower magnetization is required on-axis with flanking areas of high magnetization. Modified from Gee *et al.* (2000).

controlling crustal magnetization. Thinning of the extrusive magnetic source layer at the ridge axis may also contribute to the axial anomaly notch (Shah *et al.*, 2003). However, even a factor of 2 thinner extrusive source at the ridge axis, as suggested by seismic data (e.g., Christeson *et al.*, 1992; Vera and Diebold, 1994), is apparently not sufficient to account for the lower axial magnetization (**Figure 3**). Absolute paleointensity data from surface glass samples near the ridge axis reveal a substantial variation in paleointensity (Bowles *et al.*, 2006; Gee *et al.*, 2000; Mejia *et al.*, 1996), with moderate paleointensity values at the ridge axis higher values 1–2 km off-axis and much lower values farther from the ridge. This pattern is remarkably similar to independent estimates of field intensity variations established from archeomagnetic and volcanic materials, which document a substantial dipole moment increase from a low of $\sim 2 \times 10^{22}$ A m^2 at ~ 40 ka to a peak of $\sim 11 \times 10^{22}$ Am^2, followed by a rapid decrease to the present value of 7.9×10^{22} A m^2

(e.g., Merrill *et al.*, 1996). The similarity of these paleointensity data, both in magnitude and pattern, and the near-bottom anomaly data strongly suggests that geomagnetic intensity fluctuations are an important, if not the dominant, factor controlling anomaly amplitudes at many ridges.

Comparison of geomagnetic intensity variations over the past 800 ky from sedimentary records and near-bottom magnetic anomalies from the EPR at 19° S (Gee *et al.*, 2000) (**Figure 4**) also points to the potential importance of geomagnetic intensity variations in producing tiny wiggles. Absolute intensity information from well-dated archeomagnetic and

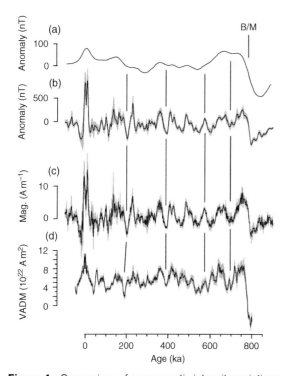

Figure 4 Comparison of geomagnetic intensity variations over the past 800 ky from sedimentary records and in sea-surface and near-bottom magnetic anomalies from the East Pacific Rise at 19° S. (a) Stack of sea-surface anomaly profiles coincident with near-bottom magnetic anomaly (b) and inversion solution (c) stacks (see Gee *et al.*, 2000 for details of inversion). Ages calculated assuming constant spreading rate and an age of 780 ky for Brunhes/Matuyama (B/M) boundary. Lower panel (d), shows Sint800 sedimentary relative paleointensity stack for 10–800 ky (Guyodo and Valet, 1999) combined with global archeomagnetic data for past 10 ky (Merrill *et al.*, 1996), all scaled as virtual axial dipole moment (VADM). Modified from Gee JS, Cande SC, Hildebrand JA, Donnelly K, and Parker RL (2000) Geomagnetic intensity variations over the past 780 kyr obtained from near-seafloor magnetic anomalies. *Nature* 408: 827–832.

volcanic materials are too sparse prior to about 50 ka to allow construction of a time series of field intensity fluctuations. However, marine sediments can provide continuous, globally distributed records of relative paleointensity (*see* Chapter 12 for a discussion of these relative intensity records) over longer timescales. A stack of sedimentary relative paleointensity records spanning the Brunhes (Guyodo and Valet, 1999) shows many similarities with coherent anomaly (and magnetization) fluctuations on eight profiles from the southern EPR. Because these anomaly profiles are separated by up to 60 km, it is unlikely that crustal accretionary variables (variations in the source thickness or geochemistry) would result in coherent anomaly variations. Thus, variations in geomagnetic intensity are the most likely cause of the coherent fluctuations in the near-bottom anomaly profiles.

The pattern of geomagnetic intensity fluctuations in the near-bottom data is also recognizable in many sea-surface anomaly profiles across the Central Anomaly. At the superfast spreading southern EPR, sea-surface anomalies reveal two broad minima at ~300 and ~550 ky (**Figure 4**). A similar pattern is discernible in many sea-surface profiles, particularly from intermediate- and fast-spreading ridges (**Figure 5**). The systematic variation of profiles of the Central Anomaly with spreading rate and their similarity to progressively smoothed records of sediment-derived paleointensity suggest that the short wavelength anomalies within the Central Anomaly are global features representing filtered variations in geomagnetic dipole intensity.

Near-bottom anomaly profiles of Anomaly 5 in the northeast Pacific (Bowers *et al.*, 2001) and comparison with the corresponding sea-surface anomaly data provide additional evidence that many short wavelength anomalies reflect paleointensity variations rather than short polarity reversals (**Figure 6**). Although the tiny wiggles in sea-surface anomaly profiles can conveniently be modeled as due to short reverse polarity intervals (Blakely, 1974), these anomaly minima in the sea-surface data correspond to more complex but still lineated anomalies in the near-bottom data. For example, the sea-surface minima labeled 5.4 corresponds to a local anomaly maximum within a broader low in the near-bottom data (Bowers *et al.*, 2001). Numerous lineated magnetic lows were found in the 12 near-bottom profiles from the northeast Pacific. If modeled as short polarity reversals, the approximately ~1 My duration normal polarity interval would be interrupted by 26 reverse polarity

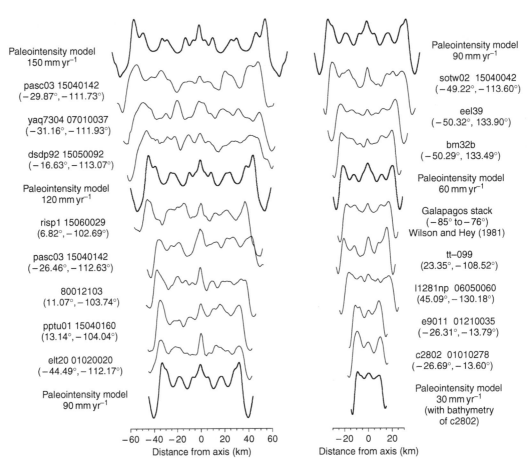

Figure 5 Sea-surface magnetic anomaly profiles illustrating the character of the Central Anomaly at different full spreading rates and the similarity to synthetic profiles (heavier lines) based on sedimentary paleointensity variations. All profiles have been reduced to the pole; actual profiles are identified by the location of the ridge crossing and the NDGC designations where available. Modified from Gee J, Schneider DA, and Kent DV (1996) Marine magnetic anomalies as recorders of geomagnetic intensity variations. *Earth and Planetary Science Letters* 144: 327–335.

intervals ranging from 1.2 to 19 ky in duration (Bowers *et al.*, 2001). More than three-quarters of these would have durations less than 10 ky and so would be comparable to or less than the time needed for two reversals (average reversal duration 7 ky; Clement, 2004). The implausibly large number and short duration of required reversed polarity intervals suggest that most of the near-bottom anomaly fluctuations reflect geomagnetic intensity fluctuations rather than short polarity intervals. This interpretation is also supported by the fact that some correlative sedimentary records show no polarity reversals and yet have relative intensity variations that appear to correlate with the short wavelength anomaly features (Bowles *et al.*, 2003). Other sedimentary records, however, reveal a small number of polarity fluctuations within Chron C5n (e.g., Evans and Channell, 2003; Evans *et al.*, 2004; Roberts and Lewin-Harris, 2000) though

the number, duration, and timing of these features is often conflicting. In some cases, the inferred long duration (e.g., 23 and 28 ky events identified by Roberts and Lewin-Harris, 2000) is incompatible with magnetic anomaly data (such long polarity intervals would be readily apparent in profiles from fast-spreading ridges). In other cases, proposed events are sufficiently brief (5–11 ky) to not conflict with existing anomaly data, though these events may represent excursions rather than polarity subchrons (Evans *et al.*, 2004). While the sedimentary data provide evidence for a small number of polarity fluctuations, the bulk of the variations evident in near-bottom anomaly records are likely to reflect intensity variations rather than short polarity reversals. A linkage between directional excursions and low intensities is also suggested by a statistical model of the geomagnetic field (Tauxe and Kent, 2004).

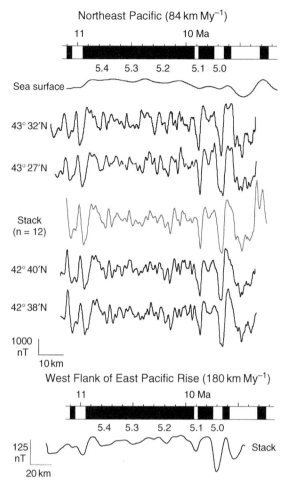

Figure 6 Magnetic anomaly profiles for Anomaly 5. Top shows sea-surface stack compared to representative near-bottom profiles and coincident near-bottom stack from moderate (84 km My^{-1} full-rate) spreading seafloor in Northeast Pacific; bottom shows sea-surface stack from fast (180 km My^{-1} full-rate) spreading seafloor on west flank of the southern East Pacific Rise. Modified from Bowers NE, Cande SC, Gee JS, Hildebrand JA, and Parker RL (2001) Fluctuations of the paleomagnetic field during chron C5 as recorded in near-bottom marine magnetic anomaly data. *Journal of Geophysical Research* 106: 26379–26396.

The geomagnetic intensity signal documented in the Central Anomaly and Anomaly 5 is likely to be a more general feature of marine magnetic anomalies, since the thermoremanence in the ocean crust should record geomagnetic intensity as well as polarity information. Apparently lineated, short-wavelength magnetic anomalies that most plausibly represent geomagnetic intensity fluctuations have indeed been documented within several portions of the C-sequence, for example, in sea-surface anomaly profiles from Anomalies 12–13 (Cande and

LaBrecque, 1974) and Anomalies 24–27 (Cande and Kent, 1992b). They are mostly attributed to intensity fluctuations because attempts to find corresponding short polarity intervals in sedimentary records have generally not been successful (e.g., Bowles *et al.*, 2003; Evans and Channell, 2003; Hartl *et al.*, 1993; Lanci and Lowrie, 1997; Schneider, 1995); these magnetostratigraphic data were summarized by Krijgsman and Kent (2004). Although the detailed pattern of pre-Brunhes intensity variations is not known, the statistical properties of the recent field (with ~40% intensity fluctuations about the mean; Merrill *et al.*, 1996) provide a plausible model for generating short wavelength anomaly variations (Cande and Kent, 1992b). The models shown in **Figure 7** illustrate that coherent short wavelength anomaly variations are likely to be present throughout the marine magnetic anomaly sequence.

The amplitude and character of such short wavelength anomaly fluctuations, particularly in anomalous intervals such as the KQZ and JQZ, may therefore provide important clues about the origin of the geomagnetic field (e.g., by testing proposed links between reversal rate and geomagnetic intensity; Tauxe and Hartl, 1997). Sea-surface magnetic anomalies within the JQZ, arbitrarily delineated as older than anomaly M25, typically have very low amplitudes and there is a gradual increase in amplitude, extending from at least M29 to around M19, which may be field related (Cande *et al.*, 1978; McElhinny and Larson, 2003) (**Figure 8(a)**). Although the lineated nature of even older M-sequence anomalies has been documented over fast spreading crust and used for tectonic reconstructions (Handschumacher *et al.*, 1988; Nakanishi *et al.*, 1989; 1992; Nakanishi and Winterer, 1998), magnetostratigraphic evidence for polarity reversals over the time interval represented by the JQZ, that is, prior to ~M25, is still ambiguous (e.g., Channell *et al.*, 1984; Juarez *et al.*, 1994; Steiner *et al.*, 1985). These older anomalies resemble the kind of short-wavelength, low-amplitude anomalies observed in the Cenozoic that have been attributed to paleointensity variations (Cande and Kent, 1992b). In contrast, magnetic anomalies in the KQZ can often be of high amplitude (e.g., **Figure 8(b)**) even though magnetostratigraphic studies and paleomagnetic compilations provide strong evidence that the KQZ formed during predominantly normal geomagnetic polarity (e.g., Irving and Pullaiah, 1976; Lowrie and Alvarez, 1981). It is not clear whether the anomalies in the KQZ are field related (e.g., Cronin *et al.*, 2001) but their large amplitude does suggest that the field

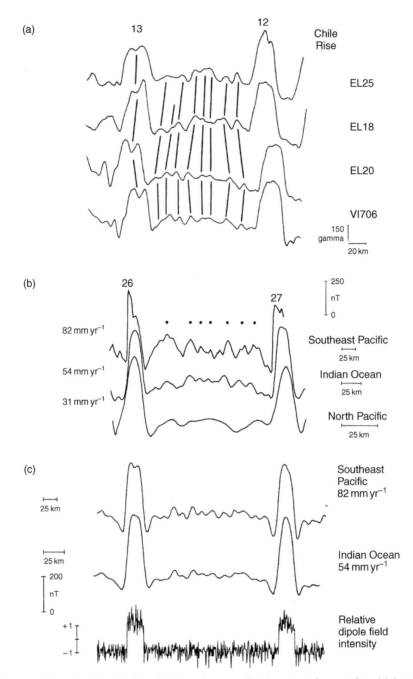

Figure 7 Further examples of short wavelength magnetic anomalies in sea-surface profiles (a) from between Anomalies 12 and 13 on Chile Rise and (b) at various spreading rates from between Anomalies 26 and 27. A model for the source of the short wavelength magnetic anomalies between Anomalies 26 and 27 at different spreading rates is shown in (c) and assumes that the anomalies are due to random fluctuations in the dipole field paleointensity whose statistical properties are compatible with observed paleointensity variations over the past 5 My. (a) Modified from Cande SC and LaBrecque JL (1974) Behaviour of the Earth's paleomagnetic field from small scale marine magnetic anomalies. *Nature* 247: 26–28. (b,c) Modified from Cande SC and Kent DV (1992b) Ultrahigh resolution marine magnetic anomaly profiles: A record of continuous paleointensity variations. *Journal of Geophysical Research* 97: 15075–15083.

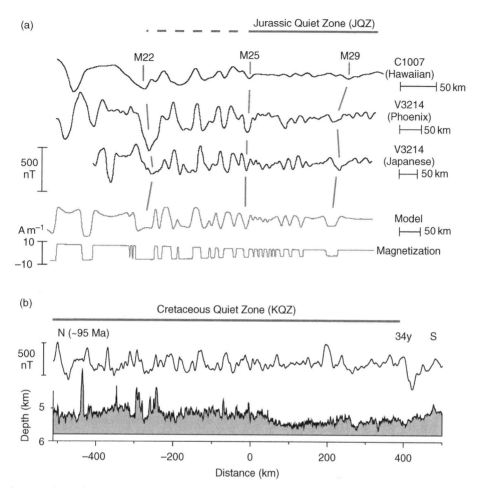

Figure 8 A comparison of sea-surface magnetic profiles from the Jurassic Quiet Zone (JQZ) in the western Pacific (top; after Cande *et al.*, 1978) and the Cretaceous Quite Zone (KQZ) from the southwest Pacific (bottom; data from Pockalny *et al.*, 2002). Representative anomaly profiles from the JQZ from the Hawaiian, Phoenix, and Japanese lineation sets have been reduced to the pole. Note the pronounced decrease in amplitude from anomaly M22 to M29, and the low (<100 nT) amplitudes prior to M29. In contrast, many anomaly profiles within the KQZ show large amplitude variations that in some cases are as large as known polarity reversals (e.g., anomaly 34y – young end of anomaly 34 (as for chrons in **Table 1**)).

intensity during the KQZ may have been high (Tarduno *et al.*, 2001; Tauxe and Staudigel, 2004), especially compared to the JQZ, whose low-amplitude anomalies might be indicative of low average field intensity in the Jurassic (McElhinny and Larson, 2003; Prévot *et al.*, 1990).

12.2.3 Anomaly Skewness and Nondipole Field

In addition to the wealth of information on geomagnetic intensity and polarity reversals, the shape (skewness) of magnetic anomalies can also provide an estimate of remanent inclination that can, in turn, be used for tectonic or geomagnetic studies. The skewness of a magnetic anomaly is a function of the effective inclinations (i.e., those projected in a plane perpendicular to the azimuth of the magnetized blocks) of the ambient field and of the remanence of the source blocks (Schouten and Cande, 1976). Skewness amounts to a uniform phase shift as a measure of the deviation in shape between the observed magnetic anomalies and the ideal symmetric profile that would be produced in a vertical field by vertically oriented magnetizations separated by vertical source boundaries. Determination of the skewness from anomaly profiles, together with the known geometry of the spreading lineation and ambient field, allows the remanent inclination to be estimated. For the Central Anomaly, Schneider (1988) applied this type of skewness analysis to 14 magnetic anomaly profiles from the Galapagos Ridge to derive a precise estimate of the mean remanent inclination, which was found to differ by several degrees from the expected field inclination and indicated that there is a small but significant (~5%) long-term nondipole (mainly axial quadrupole) contribution to the geocentric axial dipole field (**Figure 9**). This result was confirmed by an analysis of skewness in 203 profiles of the

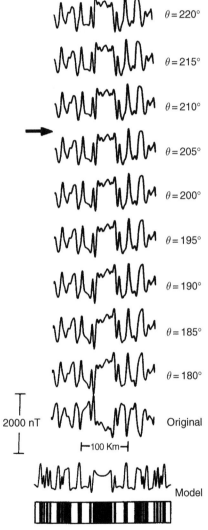

Figure 9 Representative sea-surface magnetic anomaly profile from Galapagos spreading center illustrating technique used to determine skewness of Central Anomaly. Arrow points to approximate phase shift ($\theta = 207.5°$) required to match the shape of the Central Anomaly to the model profile formed at the pole (bottom). Modified from Schneider DA (1988) An estimate of the long-term non-dipole field from marine magnetic anomalies. *Geophysical Research Letters* 15: 1105–1108.

Central Anomaly from the global ridge system (Acton *et al.*, 1996).

In applications of the skewness method outside of the Central Anomaly (or Brunhes), where both iso-chronous boundaries of the source block contribute and thus simplify interpretation of the anomaly shape, other aspects of the source that may not be symmetrically distributed to cancel out (e.g., sloping polarity boundaries, tectonic rotations) need to be considered for interpretation of field behavior (Cande and Kent, 1976). The skewness of magnetic anomalies that is not accounted for by a simple block model is termed anomalous skewness (Cande, 1976). While such anomalous skewness is a complicating factor in extracting geomagnetic or tectonic information from magnetic anomalies (e.g., Petronotis *et al.*, 1992), the details of the anomaly shape provide strong constraints on the geometry and tectonic deformation of the magnetization source that will be discussed further in Section 12.4. Assessing the fidelity of the geomagnetic field record in anomalies requires an understanding of both these nonfield related variations in source properties (e.g., transition zone width, geometry of polarity boundaries), as well as the intrinsic magnetic properties of the various source layers. The nature and origin of the remanent magnetization of these layers are the subject of the following section.

12.3 Magnetic Source Regions

Analysis of sea-surface and near-bottom magnetic anomalies suggest that most of the signal comes from the extrusive layer (**Figure 10**). For example, Talwani *et al.* (1971) determined the mean magnetization of basement topography from sea-surface magnetic surveys made along strike, which allowed an estimate of the thickness of the layer responsible for the across-strike (seafloor spreading) magnetic anomalies. The analysis was made on the slow-spreading Reykjanes Ridge and yielded a thickness of about 500 m, approximately the thickness of the volcanic layer. Atwater and Mudie (1973) did a comparable analysis using near-bottom magnetic profiles on the Gorda Rise in the Pacific. They also found that a 500 m thick source layer could account for the seafloor spreading magnetic anomalies. Dredged and drilled samples of oceanic basalts have remanent magnetizations that are more or less compatible with a relatively thin source layer but firm conclusions are inhibited by the several orders of magnitude range in the magnetization values (Lowrie, 1977).

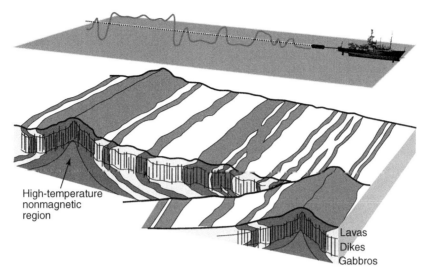

Figure 10 3-D perspective cartoon of ocean crust (after Johnson *et al.*, 1997), showing spreading centers separated by a transform fault. Crust generated during normal/reverse geomagnetic polarity (shown as shaded/unshaded regions) can be inferred from magnetic anomalies, for example, as measured from ship-towed magnetometers. Magnetization polarity boundaries are expected to be vertical in sheeted dikes and are likely to slope but in opposite directions in extrusive basalt layer (due to successive emplacement of lavas) and in intrusive gabbroic layer (due to progressive cooling with depth of gabbros).

Thicker and even deeper sources have also been invoked, usually as a counterbalance to supposedly altered and diminished shallower sources to explain the apparent loss of fidelity in earlier Cenozoic anomaly sequences (e.g., Blakely, 1976) or to produce anomalous skewness from sloping blocking temperature isotherms (Cande and Kent, 1976). Although broader transition zone widths (and consequently some loss of short polarity events) might be expected if the slowly cooled gabbroic layer is a substantial contributor, the still older (M-sequence) anomalies do not have markedly broader transition zone widths (e.g., Cande *et al.*, 1978) and magnetostratigraphic data confirm that the reversal frequency in the earlier Cenozoic was in fact lower (Lowrie and Alvarez, 1981; Lowrie and Kent, 2004). Similarly, the general absence of anomalous skewness for Anomalies M0–M4 (Cande, 1978; Larson, 1977; Larson and Chase, 1972) places limits on the contribution from the gabbroic layer. These observations suggest that contributions from deeper source layers are not necessarily required, although they are also not precluded (Harrison, 1987). Magnetic data from samples of these deeper layers suggest that they do indeed contribute significantly to magnetic anomalies.

Below we summarize the magnetization of the oceanic crust according to the main subdivisions:

extrusives, sheeted dikes, and gabbros, plus serpentinized mantle peridotites, and show that sample remanence and anomaly data can be reconciled by a three-layer model, with relatively narrow transition zones in lavas and dikes and broader polarity boundaries in the gabbros. Oceanic basement rocks acquire a thermal remanent magnetization (TRM) during initial cooling. The magnitude of the TRM is proportional to the strength of ambient geomagnetic field but also depends on the geochemistry (e.g., iron content) of the rocks and magnetic grain size. The rocks may alter with age, which may also modify the TRM. The resultant magnetization measured on rock samples in the laboratory is referred to as the natural remanent magnetization (NRM). In order to facilitate comparison of magnetization values from different locations (dipole intensity varies by a factor of 2 from equator to pole), all magnetization values are reported as equatorial values. Although the arithmetic mean is most pertinent for comparison with magnetic anomalies, most large magnetization data sets are characterized by approximately lognormal distributions. Geometric mean values are used where possible (many authors do not specify how mean values were calculated) for comparisons since these are less influenced by extreme values in smaller data sets that are typical of many studies of oceanic rocks.

12.3.1 Magnetization of Lavas

Mid-ocean ridge basalts that constitute the oceanic extrusive layer have a magnetization that is characteristically dominated by a strong stable remanence compared to an induced component, making these shallow seated rocks a traditional prime source of seafloor spreading magnetic anomalies. Titanium-rich titanomagnetite ($Fe_{3-x}Ti_xO_4$, where $x \sim 0.6$) that has often undergone variable degrees of low-temperature oxidation (maghemitization) is the principal magnetic mineral. However, it was recognized early on that the very wide range of observed NRM intensities could not be explained by a mere variation in low-temperature oxidation; instead, it is necessary to invoke substantial variations in titanomagnetite composition, concentration, and grain size (Lowrie, 1977).

12.3.1.1 Initial grain size and composition

The rapidly chilled margins of submarine lavas produce a large gradient in cooling rate-dependent grain size and compositional effects (Marshall and Cox, 1971). For example, detailed sampling perpendicular to the chilled margin shows that the NRM and magnetic hysteresis parameters can vary systematically by an order of magnitude on centimeter scales (Gee and Kent, 1997, 1998, 1999; Kent and Gee, 1996; Marshall and Cox, 1971) (**Figure 11**). Near the chilled margin, hysteresis parameters indicate grain sizes spanning the superparamagnetic/single domain boundary (~ 30 nm; Özdemir and O'Reilly, 1981). The presence of stable single domain grains, associated with a peak in NRM intensity, presumably accounts for much of the strong, stable remanence associated with the extrusive layer. The pronounced grain-size variation, however, presents a practical problem in characterizing the magnetic properties of oceanic basalts because random sampling of such large within-flow variations can seriously bias estimates of between-flow and between-site variations.

Unblocking temperatures of NRM as well as Curie temperatures can also show a large systematic variation with depth in a flow (Grommé et al., 1979; Ryall and Ade-Hall, 1975), even in a zero-age pillow such as is illustrated in **Figure 11**. The within-flow variation in Curie temperatures in these and other studies has usually been interpreted as reflecting progressive low-temperature oxidation of an originally homogeneous titanomagnetite host (e.g., Kent and Gee, 1996). However, transmission electron microscopic

(a)

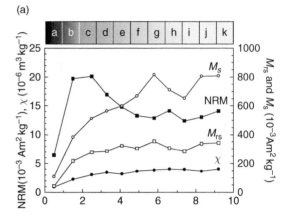

(b)

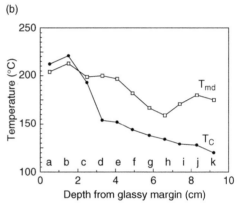

Figure 11 Within-lava flow variability in magnetic properties that can be ascribed to cooling rate-dependent grain size effects. (a) Variation in NRM, susceptibility (χ), saturation remanence (M_{rs}) and saturation magnetization (M_s) as a function of distance from glassy chilled margin in New Flow (erupted 1993) pillow fragment R-1 from the Juan de Fuca Ridge. (b) Variation in Curie temperature (T_c) and median demagnetizing temperature of NRM (T_{md}) in the same samples. Similar cooling-rate dependent variations in magnetic properties in oceanic basalts have been documented by Marshall and Cox (1971), Gee and Kent (1997), and others. Modified from Kent DV and Gee J (1996) Magnetic alteration of zero-age oceanic basalt. *Geology* 24: 703–706.

observations (Zhou *et al.*, 2000) on a zero-age pillow fragment from the Juan de Fuca Ridge show no oxidation and that the titanium content of titanomagnetite grains varies as a function of distance from the chilled margin (**Figure 12**). The larger grains in the interior have a more uniform composition of around $x = 0.6$ whereas the smaller grains toward the glassy margin have more variable compositions with a lower average x value of ~ 0.45. Ultrafine magnetite ($x \sim 0$) has been documented both in interstitial glass, as well as in the chilled glassy

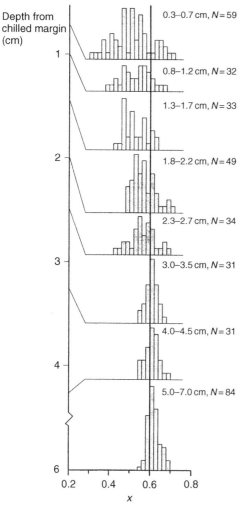

Figure 12 Variation of composition parameter, x, of titanomagnetite as a function of depth from the glassy chilled margin in a New Flow (erupted 1993) pillow fragment from the Juan de Fuca Ridge. Modified from Zhou W, Van der Voo R, Peacor DR, and Zhang Y (2000) Variable Ti-content and grain size of titanomagnetite as a function of cooling rate in very young MORB. *Earth and Planetary Science Letters* 179: 9–20.

margin (Pick and Tauxe, 1994; Zhou *et al.*, 1999b) and these grains may remain unaltered for tens of millions of years (Zhou *et al.*, 1999b). Kent and Gee (1996) also documented trace amounts of low-Ti magnetite in the crystalline interiors of very young flows.

A more general geochemical factor, iron and titanium enrichment of basaltic melts, is expected to exert a fundamental control on the magnetization of oceanic basalts and has been conveniently cast in terms of the magnetic telechemistry hypothesis (Vogt and Johnson, 1973). The iron content of basaltic melts increases during low-pressure fractionation

(e.g., Juster *et al.*, 1989) and this iron enrichment should be accompanied by increased abundance of titanomagnetite, resulting in higher NRM and thus in enhanced magnetic anomaly amplitudes. Several attempts to test the magnetic telechemistry hypothesis produced uncertain results, largely because of large scatter in the magnetization data. For example, Johnson and Tivey (1995) found a poor correlation ($R \sim 0.2$) between FeO^* (total iron expressed as FeO) and NRM from the Juan de Fuca Ridge. However, good linear correlations (R up to ~ 0.8) between NRM and FeO^* were obtained from data from the southern East Pacific Rise (EPR) (Gee and Kent, 1997, 1998) (**Figure 13(a)**). The greatly improved correlations can be attributed to detailed sampling that spanned the full range of cooling-related magnetization changes within a flow, as well as to the young age of the axial samples, which effectively minimized age-dependent magnetization changes. An inversion of the axial magnetic anomaly profile shows a close correspondence between the magnitude of the magnetization solution and the range of intensities observed in the axial samples (**Figure 13(b)**). The equivalent NRM values calculated from the mean FeO^* content of each dredge using the linear regression also provide confirmation of the predicted link between geochemistry and anomaly amplitude.

The geochemical dependence demonstrated on the southern EPR indicates that the magnetization of oceanic basalts can vary by up to a factor of ~ 4 as a function of their iron contents alone, from around $12 \, A \, m^{-1}$ for 9% FeO^* to more than $50 \, A \, m^{-1}$ for the highest FeO^* values ($\sim 15\%$; separation of early crystallizing phases cause FeO^* to increase but at higher degrees of fractionation FeO^* decreases as FeTi oxide phenocrysts begin crystallizing). Geochemical variation is thus expected to exert a fundamental control on the magnetization of basalt and the source of magnetic anomalies, which should vary more or less proportionately in amplitude as has been observed on the southern EPR. Ridge crest discontinuities, where enhanced fractionation is expected, are often accompanied by higher amplitude magnetic anomalies (Bazin *et al.*, 2001; Wilson and Hey, 1995). In addition, the average degree of fractionation of lavas appears to vary with spreading rate (Sinton and Detrick, 1992). Slow-spreading ridges, which lack steady magma chambers, have systematically less-evolved magmas with lower FeO^* than faster-spreading ridges, differences that might be reflected in overall values of basalt magnetizations and anomaly amplitudes.

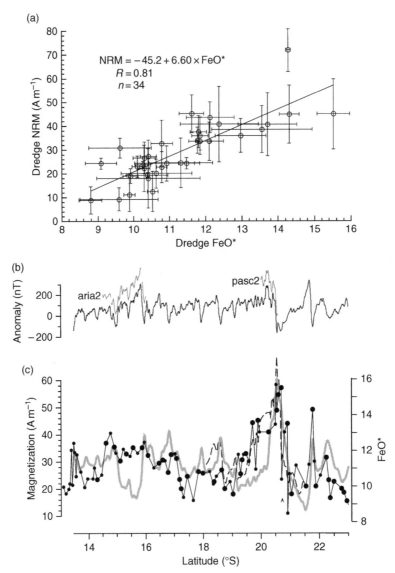

Figure 13 Magnetic and geochemical variations along the axis of the southern East Pacific Rise. (a) Variation in NRM as a function of FeO* (total Fe as FeO) for dredges from the southern East Pacific Rise. NRM values represent the arithmetic mean intensity (±s.d.) of three or more flows/pillows, each with multiple subspecimens (Gee and Kent, 1997). Dredge mean FeO* content (±s.d.) calculated from microprobe analysis of glass chips (Sinton *et al.*, 1991) from samples used for magnetic study. (b) Composite axial magnetic anomaly (prior to reduction to pole). (c) Magnetization inferred from profile inversion (heavy grey line) and grid inversion (dashed, both left scale) is compared to FeO* content (filled circles, with larger circles indicating dredges for which sample magnetizations are available), with the scaling and offset between the axes determined by the regression result from panel (a). Prior to inversion, the axial anomaly was reduced to pole using an effective source thickness of 375 m, corresponding to an extrusive layer that thickens from 250 to 500 m within approximately 2.5 km of the ridge crest. A cosine-tapered band-pass filter was applied where wavelengths < 10 km and > 600 km were cut and wavelengths >20 km and <300 km were passed unattenuated. Modified from Gee J and Kent DV (1998) Magnetic telechemistry and magmatic segmentation on the southern East Pacific Rise. *Earth and Planetary Science Letters* 164: 379–385.

12.3.1.2 Low-temperature alteration

Low-temperature oxidation of stoichiometric titanium-rich titanomagnetite is widely considered to be the dominant process of magnetic alteration of oceanic basalts. The magnetic consequences of such low-temperature alteration include a fourfold reduction in the saturation magnetization (for complete oxidation; O'Reilly, 1984), as well as a significant increase in the Curie temperature (Xu *et al.*, 1996), with the possible acquisition of a chemical remanent magnetization (CRM) that might replace the initial TRM (Raymond and LaBrecque, 1987). Substantial (up to an order of magnitude) changes in the magnetization of the extrusive layer and the amplitude of the associated magnetic anomalies have been attributed to low-temperature alteration (Bleil and Petersen, 1983; Irving, 1970). As documented below, zero-age lavas do have a remanence that is higher, by a factor of 4, than that of older lava samples. However, magnetic anomaly data and direct determination of the degree of low-temperature oxidation suggest that this process is responsible for only about half of this decrease in magnetization, with the remaining discrepancy plausibly attributed to higher paleofield intensity in the axial lavas. Moreover, the heterogeneity of alteration makes definitive resolution of the relevant timescales difficult.

Sea-surface magnetic profiles and inversion solutions from fast-spreading ridges like the Pacific-Antarctic Ridge (**Figure 14(a)**) record a short (~10 km) wavelength axial Central Anomaly magnetic high (CAMH) (Klitgord, 1976). Near-bottom anomaly data from fast-spreading ridges indicate that the width of this high magnetization zone is ~2–3 km (**Figure 3**). At slow-spreading ridges, the CAMH occupies a larger proportion of the Central Anomaly and consequently may not be readily distinguishable in sea-surface profiles from the Central Anomaly (**Figure 14(b)**). In concert with a dominant Central Anomaly (e.g., Vine, 1966), the evidence for a decay in basalt magnetizations has traditionally come from the slow-spreading ridges in the Atlantic Ocean where oceanic basalts have been thought to suffer decay related to aging and alteration with a time constant on the order of 0.5 My (Irving, 1970; Johnson and Atwater, 1977). Alteration-induced magnetization decay has also often been linked to the CAMH (Klitgord, 1976), although this would require very different time constants for fast- and slow-spreading ridges. For example, profiles from the fast spreading EPR at 12° N suggest that the magnetization contrast must occur over a much shorter timescale to balance a presumed negative magnetic anomaly that would result from the thickening of Layer 2A deduced from seismic imaging (**Figure 15**).

Although originally interpreted as reflecting low-temperature alteration on rapid timescales (Gee and Kent, 1994), remanence data from near-ridge basalts at the EPR at 12° N do not require a substantial decay in magnetization related to alteration. This is because low NRM intensities and magnetic

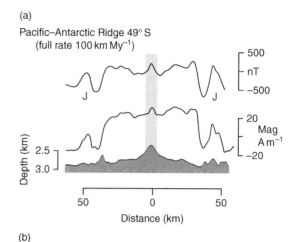

(a)

Pacific–Antarctic Ridge 49° S
(full rate 100 km My⁻¹)

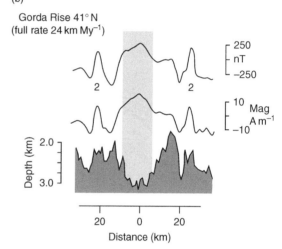

(b)

Gorda Rise 41° N
(full rate 24 km My⁻¹)

Figure 14 Sea-surface magnetic profiles, inversion solutions, and bathymetric profiles from (a) the fast-spreading Pacific–Antarctic Ridge and (b) the slow-spreading Gorda Rise illustrating the short (~10 km) wavelength axial Central Anomaly magnetic high (CAMH, vertical shading). The CAMH occupies a larger proportion of the Central Anomaly at slow-spreading ridges where it is not readily distinguishable in sea-surface profiles from the Central Anomaly because the wavelengths become similar. Modified from Klitgord KD (1976) Sea-floor spreading: The central anomaly magnetization high. *Earth and Planetary Science Letters* 29: 201–209.

susceptibilities and high Curie temperatures, which might be an indication of alteration, occur at the ridge axis (zero age) and as well as off-axis (**Figure 16**). Indeed, efforts to induce alteration of young oceanic basalts in the laboratory have generally failed (e.g., Kent and Gee, 1994). Instead, it now appears that the high magnetization values on-axis producing the contrast that accounts for the CAMH on fast-spreading ridges is mostly due to a paleointensity signal, which is also suggested by the central

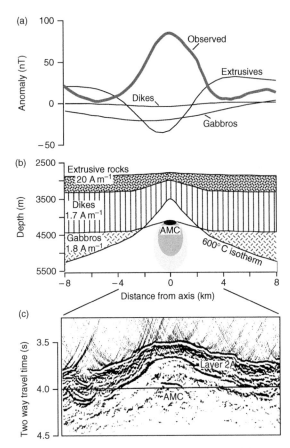

Figure 15 (a) Forward models illustrating the axial magnetic anomaly minimum generated by off-axis thickening of a uniformly magnetized extrusive layer or cooling of deeper layers as shown in (b), compared to the observed sea-surface profile at 19.5° S on the East Pacific Rise. (b) Source models based on seismic data from near 19.5° S. (c) Wide aperture seismic profile from East Pacific Rise near 14° S (Detrick *et al.*, 1993) that illustrates rapid thickening of Layer 2A and presence of axial magma chamber (AMC). Modified from Gee J and Kent DV (1994) Variations in layer 2A thickness and the origin of the central anomaly magnetic high. *Geophysical Research Letters* 21: 297–300.

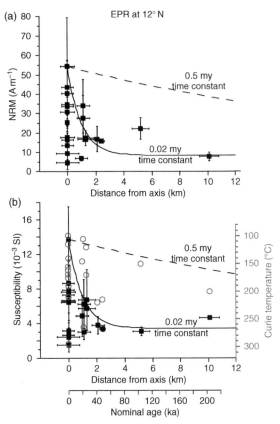

Figure 16 (a) Natural remanent magnetization (NRM), and (b) magnetic susceptibility for samples of dredged oceanic basalts from 12° N on the East Pacific Rise as a function of distance from the ridge axis (or nominal age based on full spreading rate of 98 km My^{-1}). Each data point ($\pm$ s.d.) is the average of 3–6 specimens collected within 5–8 cm of the glassy margin. Exponential magnetization decay curves with time constants of 0.5 My and 0.02 My are shown for reference but the large variability in the magnetization data, as well as in the previously unpublished Curie temperature data shown in (b) for these samples do not make a compelling case for any simple decay scheme. Modified from Gee J and Kent DV (1994) Variations in layer 2A thickness and the origin of the central anomaly magnetic high. *Geophysical Research Letters* 21: 297–300.

notch in the CAMH in near-bottom profiles over faster-spreading rate ridges (Gee *et al.*, 2000; Perram *et al.*, 1990) (**Figure 3**).

Evidence has also been sought for longer-term changes in oceanic basalt magnetic properties that could be attributed to the effects of alteration to explain an apparent envelope of decreasing anomaly amplitudes with distance from the ridge axis. Several magnetic anomaly inversion studies have suggested a decrease in magnetization by a factor of 2–3 within about 10 My of the ridge axis (Sayanagi and Tamaki, 1992; Wittpenn *et al.*, 1989). However, it is difficult to

evaluate the effects of various potential artifacts associated with data processing (filtering, gridding) and the sequence effect in biasing the results, for example, in producing the apparent substantial (factor of 3–5) increase in magnetization with increasing age beyond 10 Ma and into the KQZ. An influential and more direct analysis in this regard was the compilation of oceanic basalt magnetizations from Deep Sea Drilling Project (DSDP) sites by Bleil and Petersen (1983), who proposed that the data showed an initial sharp decrease in NRM to minimum values at around 20 Ma followed by a gradual increase to ages of around 120 Ma. Subsequent compilations (Furuta, 1993; Johnson and Pariso, 1993) (**Figure 17**) show a similar pattern of variation with age although the data quality is highly variable with more than one-third of the sites represented by 10 or fewer samples (e.g., six sites have only one sample!). The overall arithmetic mean NRM intensity for all 64 site means reduced to the equator is $4.4 \, A \, m^{-1}$; high values of around $10 \, A \, m^{-1}$ occur at both very young ages as well as at around 120 Ma, whereas values of $1 \, A \, m^{-1}$ or less occur at around 40 Ma. There is considerable scatter throughout with only a handful of sites older than 40 Ma penetrating more than 50 m of basalt or represented by more than a nominal number (20–30) of samples. Because of the typically small number of samples, as well as the lack of control on geochemical variations, within-flow grain size variations, or geomagnetic intensity fluctuations that strongly influence remanent intensity, suggestions of any temporal trends in magnetization intensity of seafloor lavas from such plots should be regarded as tentative.

A more robust estimate of age-dependent changes can be obtained by comparing the distribution of NRM values for a large collection of near-axis (zero-age) samples from the southern EPR with histograms of NRM values for the best characterized DSDP/ODP drill sites that have the deepest penetration and are represented by about 100 or more samples (**Figure 18**). The NRM distributions are reasonably approximated by log-normal distributions. The zero-age southern EPR samples have a geometric mean NRM of $18.9 \, A \, m^{-1}$. This is virtually identical to the geometric mean NRM intensity of $18 \, A \, m^{-1}$ for the axial samples from $12° N$ on the EPR, which were mostly taken within about 3 km of the axis and thus are less than $\sim 50 \, ky$ old (Gee and Kent, 1994). Published data from the youngest drill site with significant penetration (DSDP Site 482, which was sited on ~ 0.4 Ma crust; Lewis et al., 1983) indicate NRM intensities of about $5 \, A \, m^{-1}$, almost a factor of 4 less than the axial samples. Still older drill sites have geometric mean NRM intensities that range from $2.3 \, A \, m^{-1}$ for Site 395 (7.8 Ma) to $14.1 \, A \, m^{-1}$ for Site 1256 (15.4 Ma) although the data at this latter site may be biased to higher NRM values by a particularly strong drilling remanence (Wilson et al., 2003). Nevertheless, there

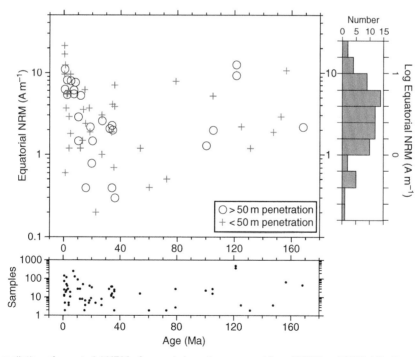

Figure 17 Compilation of equatorial NRM of oceanic basalts recovered from DSDP and ODP drill sites as a function of crustal age and sorted by depth of basement penetration. Lower plot shows number of sample measurements available for calculation of site mean (from references listed in Johnson and Pariso, 1993); histogram at right is for all site mean NRM values. NRM data are from compilation of Johnson and Pariso (1993) with some corrections (e.g., deletion of data for samples from sills). Ages are based on updated anomaly or biostratigraphic assignments and the timescales of Berggren et al. (1995) and Channell et al. (1995).

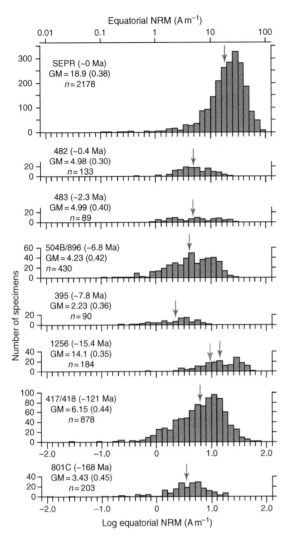

Figure 18 Histograms of equatorial NRM for oceanic basalts for sites with largest datasets, including from the axial zone of the southern EPR versus off-axis drill sites – Sites 482, 483, 504B/896, 395, 1256, 417/418, and 801C (see **Table 2** for data sources). Geometric mean (GM) is indicated by red arrow for each distribution. The green arrow for Hole 1256D indicates average NRM for samples with initial inclinations <30° and >–30° that should be least affected by drilling remanence.

does not appear to be an obvious age-dependent trend in the drillsite-mean NRM values since the oldest sites (Site 417/418 at 121 Ma and Hole 801C at 169 Ma) have about the same average NRM intensities as some of the youngest drill sites (e.g., Sites 482, 483 at 2.3 Ma, and Hole 504B at ~6.8 Ma), which are all between about 4–6 A m^{-1}. What does seem significant is the factor of almost 4 decrease in mean NRM intensity from the axial zone to older sites; if

the entire reduction is ascribed to alteration, the time constant must be considerably shorter than about 0.4 My, the age of the youngest off-axis site (Site 482) with reduced magnetization. A similar pattern of magnetization change was observed in the FAMOUS area (37° N on the slow-spreading Mid-Atlantic Ridge) based on the magnetization of topography derived from near-bottom observations (Macdonald, 1977).

Direct observations of the degree of low-temperature alteration suggest that the fourfold difference in intensity of axial lavas and older drill sites cannot be attributed solely to maghemitization. The oxidation parameter ($0 < z < 1$, where 1 indicates complete oxidation to titanomaghemite; O'Reilly, 1984) has now been estimated from lattice spacing and composition using transmission and analytical electron microscopy on small titanomagnetite grains in fine-grained oceanic basalts over a broad age range (Wang *et al.*, 2005; Wang *et al.*, 2006; Zhou *et al.*, 2001). The development and application of microanalytical techniques (Zhou *et al.*, 1999a) represent a major advance in the study of oceanic basalts because they allow more direct characterization of the important fine-grained magnetic carriers that previously had to be inferred from larger magnetic grains or bulk properties. Although there is considerable scatter (**Figure 19**), the oxidation parameter data are consistent with an initial, very rapid increase in alteration such that z values up to 0.8 occur in basalts less than 1 My old, as suggested by Bleil and Petersen (1983). Much less convincing is any systematic long-term trend because although values of z greater than 0.8 do tend to occur in samples older than about 30 Ma, the oldest sample measured (Hole 801C, 169 Ma) has a z value of only ~0.2. Thus, while low-temperature alteration evidently can occur rapidly, the extrusive layer is not uniformly oxidized even for the oldest seafloor.

Since complete oxidation would be required to account for a factor of 4 reduction in saturation magnetization (O'Reilly, 1984), some additional parameter is apparently required to explain the factor of 4 difference in magnetization of axial lavas and older drilled basalts. Paleofield changes might explain the difference if the present field is unusually high and it is worth noting that magnetization variations over long timescales have been attributed by some authors to paleofield variations (e.g., Wang *et al.*, 2005; Juarez *et al.*, 1998). Even the traditional model of alteration-related variations in saturation magnetization (Ms) is still

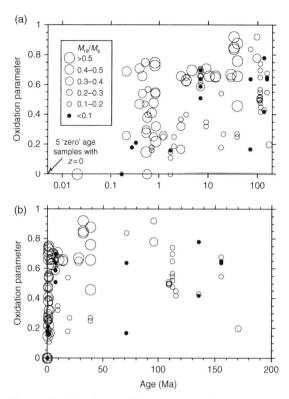

Figure 19 Variation in oxidation parameter (and hysteresis ratio, M_{rs}/M_s) of oceanic basalts samples as a function of crustal age (log scale in (a) and linear scale in (b)). Data are from Wang *et al.* (2005, 2006) and Zhou *et al.* (2001). The plot shows no obvious evidence of any systematic evolution of magnetic properties that could be ascribed to low-temperature alteration, for example, the basalt samples from Site 801C, which are the oldest at 169 Ma, have an oxidation parameter value of only 0.2.

possible, for example, Matzka *et al.* (2003) argue that the low NRM values for 10–40 Ma samples are primarily due to the reduction in saturation magnetization accompanying maghemitization, similar to the model of Bleil and Petersen (1983). Some other alternative explanations are described below.

Xu *et al.* (1997c) have suggested a model of preferential dissolution with time of the finest-grained titanomagnetites as the major process contributing to long-term temporal changes in remanent intensity of mid-ocean ridge basalts. If widespread, such dissolution could enhance the effect of the reduction in saturation magnetization accompanying low-temperature oxidation. However, the data in **Figure 19** show no obvious coarsening in magnetic grain size with age or alteration, for example, some of the highest M_{rs}/M_s values (i.e., high ratios of saturation remanence to saturation magnetization) indicate

the finest magnetic grain sizes) are associated with the highest z values (greatest low-temperature oxidation) in ∼30-My-old samples. Instead, the results in **Figure 19** are consistent with the more voluminous hysteresis data shown in **Figure 20** that illustrate that substantial intra-sample variability in oceanic basalts precludes recognition of any systematic trend in magnetic grain size with age from DSDP/ODP drill sites (Gee and Kent, 1999).

Directional changes associated with low-temperature alteration might also contribute to reduction in seafloor basalt magnetization with time if a CRM is acquired in a direction that opposes the initial thermoremanence. Several early studies (e.g. Ryall and Ade-Hall, 1975; Marshall and Cox, 1972) concluded that the remanence direction was unaffected by low-temperature alteration since the oxidized titanomaghemite would inherit the original TRM direction. These results are not conclusive, however, since these studies were conducted on young Brunhes-age samples where no large directional difference would be expected. More extensive studies on altered lavas from a range of ages (Beske-Diehl, 1990) were also interpreted as reflecting no alteration-related directional changes since the observed directional shifts could also be attributed to viscous remanence acquisition. Although the low Curie point of TM60 makes direct experimental investigation of seafloor oxidation difficult, Kelso *et al.* (1991) found that CRM in synthetic TM40 grains was acquired parallel to the applied field direction, with secondary magnetizations particularly enhanced at lower pH values. If broadly applicable, such field-parallel CRM acquisition coupled with the pattern of geomagnetic polarity reversals could provide a mechanism for long-term variations in NRM intensity as suggested by Raymond and Labrecque (1987).

In addition, low-temperature oxidation has been suggested as a possible mechanism by which partial self-reversal might occur, leading to the acquisition of an antipodal overprint that could significantly reduce the NRM intensity (Doubrovine and Tarduno, 2004). These authors found that the majority of thermally demagnetized samples from Detroit Seamount had two nearly antipodal remanence components. Some samples showed a maximum in the $Ms(T)$ curves above room temperature that might be indicative of N-type ferromagnetic behavior and partial self-reversal (Doubrovine and Tarduno, 2004). A similar peak in $Ms(T)$ and importantly a corresponding reversible peak in NRM above room temperature were reported for 10–40 Ma oceanic basalts (Matzka

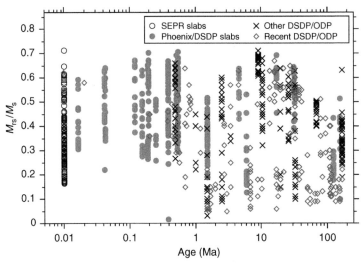

Figure 20 Variation of M_{rs}/M_s with age in oceanic basalt samples. Data acquired from systematic sampling relative to the chilled flow margin (labeled slabs) include samples from the southern East Pacific Rise (SEPR), Phoenix dredge collection, and DSDP sites (see Gee and Kent, 1999 and references therein). Other DSDP/ODP designates additional hysteresis data from isolated samples used in this same compilation. More recent hysteresis data (Recent DSDP/ODP) includes data from Wang *et al.* (2005, 2006) and Zhou *et al.* (2001).

et al., 2003). Krasa *et al.* (2005) provided compelling documentation for partial self-reversal associated with titanomaghemite in subaerial lavas, although they argue that the 'low-temperature' oxidation occurred at temperatures above the blocking temperatures of the original titanomagnetite phase during initial cooling. Thus, while such partial self-reversals may occur, the importance of this process in generating long-term changes in seafloor magnetization remains to be demonstrated.

A common motivation for seeking evidence of a time-dependent change in the magnetization of oceanic basalts is the widespread perception of a decreasing amplitude envelope for ridge crest anomalies. This concept stems largely from long-standing observations that the Central Anomaly in the slow-spreading Atlantic and Indian Oceans tends to be anomalously high (e.g., Vine, 1966). For example, Bleil and Petersen (1983) show a magnetic profile across the North Atlantic as evidence of this phenomenon, which had also been described to occur over some sectors of the Australia–Antarctic Ridge by Weissel and Hayes (1972). However, the Central Anomaly is not such a pronounced feature in magnetic profiles from the faster-spreading sectors of the East Pacific, which are also typically described as showing only a gradual decrease in amplitude on the ridge flanks (e.g., Pitman and Heirtzler, 1966; Pitman *et al.*, 1968). Indeed, the prominence of the

Central Anomaly (and the associated amplitude envelope) appears to be closely tied to spreading rate (**Figure 21**; Cande and Gee, 2001).

The fact that neither a prominent Central Anomaly nor a pronounced anomaly envelope are present on all ridge systems places limits on systematic time-dependent processes that are expected to affect the magnetization of all oceanic basalts, such as long-term changes in paleointensity and alteration. Suggested time constants of magnetic alteration, assuming an exponential decay, have ranged over several orders of magnitude, from ~0.05 My (Gee and Kent, 1994) to ~0.5 My (Johnson and Atwater, 1977; Macdonald, 1977), and to ~5 My or longer (Bleil and Petersen, 1983; Raymond and LaBrecque, 1987; Xu *et al.*, 1997b; Zhou *et al.*, 2001). We use these order of magnitude increments in the estimated time constants to model the expected effects on magnetic anomalies in **Figure 22**. For fast-spreading ridges (**Figure 22(a)**), models using a factor of 4 decay in magnetization, corresponding to the nominal decrease in NRM from the ridge axis to off-axis drill sites (**Figure 18**), are virtually precluded over the various timescales. These models produce either a very prominent axial high at a short (0.05 My) decay constant or an obvious amplitude envelope at a long (5 My) decay constant, but neither of these features is compatible with the representative Pacific–Antarctic Ridge profile shown in **Figure 21**. This comparison suggests that the effective

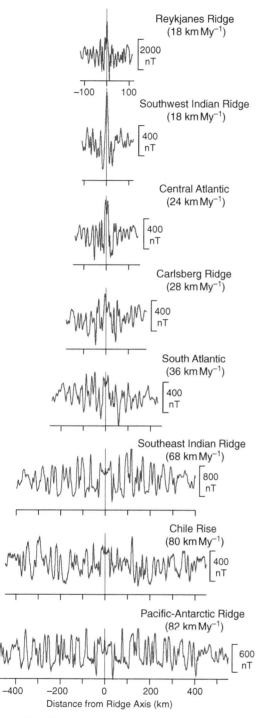

Figure 21 Representative sea-surface anomaly profiles illustrating the spreading rate dependence of the amplitude of the Central Anomaly. All profiles extend approximately to Anomaly 5 and have been reduced to the pole. The Central Anomaly has notably higher amplitude only at full spreading rates less than $\sim$30 km My^{-1}.

magnetization reduction from alteration is likely much smaller than a factor of 4, perhaps reflecting the heterogeneous nature of low-temperature oxidation in the crust. Indeed, models using a 2x factor of decay in magnetization at a variety of decay constants for fast-spreading ridges (**Figure 22(a)**) cannot be excluded as resembling the observed Pacific–Antarctic Ridge profile. Models for different decay factors and time constants for a slow-spreading ridge (**Figure 22(b)**), on the other hand, seem to favor a larger decay (e.g., compare 4x models to observed Central Atlantic or Southwest Indian Ridge profiles in **Figure 21**).

The independent evidence for rapid changes in oxidation parameter (**Figure 19**) favors a fast decay time constant, although the entire observed decrease in NRM intensity may not be entirely due to alteration. For example, the NRM of the near-axis samples may be biased by high paleointensities over the past few thousand years that were up to a factor of 2 higher than today's field and perhaps a factor of 4 higher than the average long-term paleofield intensity (Selkin and Tauxe, 2000). Absolute paleointensity data, which show a 3x change near the ridge axis (Bowles *et al.*, 2006; Gee *et al.*, 2000), are a powerful indication that a large part of that near-ridge signal is field-related. A factor of 2 decrease in NRM intensity due to alteration might be a reasonable estimate for oceanic basalts, with a possibly larger contrast in the slow-spreading ridges due to more intense near-axis tectonics, for example.

12.3.2 Magnetization of Dikes

One of the most significant changes in our understanding of the magnetization of oceanic crust since the review by Smith (1990) has been in the role of dikes, which some earlier compilations based on ophiolites and dredge metabasalt samples from fracture zones suggested were characterized by weak, rather unstable remanences (e.g., Kent *et al.*, 1978). The penetration, albeit with limited recovery, of $\sim$1000 m section of dike rocks at Site 504B (**Figure 23**) has allowed the documentation of the physical and magnetic properties of this important constituent of oceanic crust in a more typical setting. An $\sim$100 m-thick transition between pillow lavas and dikes (at $\sim$1000 mbsf) does have weak magnetizations, 0.1 A m^{-1} and less, although they have very high coercivities. Deeper levels in the sheeted dike complex have moderate NRM ($\sim$2 A m^{-1}) and moderate median demagnetizing fields. Moreover, analysis of logging data suggests that, due to void

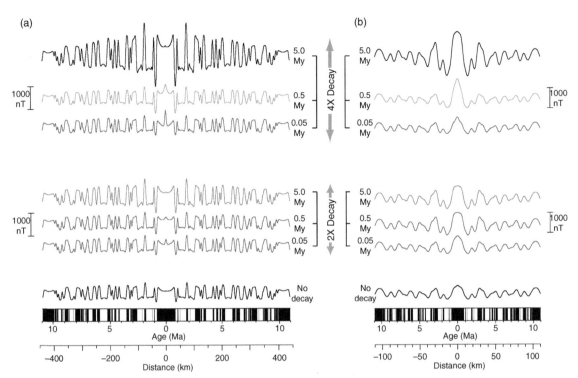

Figure 22 Magnetic anomaly models illustrating the effect of alteration-related exponential decay in magnetization for (a) intermediate-spreading crust (full rate of 80 km My^{-1}) and (b) slow-spreading crust (full rate of 20 km My^{-1}). For each spreading rate, anomaly profiles are shown for a fourfold decrease in magnetization (20 A m^{-1} to 5 A m^{-1}) and a twofold decrease in magnetization (10 A m^{-1} to 5 A m^{-1}) for a range of decay time constants. The magnetization model (block diagram) and model with constant remanence intensity are shown for comparison. In all models, the anomaly is calculated for a 1 km thick source layer, with the depth to the upper surface that increases with age$^{0.5}$ (Ma) (Stein and Stein, 1994). A gaussian filter has been applied to the magnetization ($\sigma = 0.5$ km and $\sigma = 1.0$ km for the intermediate- and slow-spreading ridges, respectively) to simulate the transition zone width in the extrusive layer.

spaces, the effective NRM of the lavas from Hole 504B may be only 3 A m^{-1} rather than the 5 A m^{-1} based on sample measurements (Worm *et al.*, 1996), which would further reduce the magnetization contrast with the dike complex.

Dike samples obtained by both dredges and submersible sampling also highlight the importance of dikes to crustal magnetization. Submersible sampling of ~1.2 my-old fast-spread crust exposed at Hess Deep provides a direct comparison of magnetic properties of extrusives, dikes, and gabbros (**Figure 24**). Here, lavas and dikes have comparable NRM values (~4 A m^{-1}). Dredged dike samples from ~2 my crust exposed along the Blanco Fracture Zone also have moderately strong magnetizations (~1.7 A m^{-1} when reduced to equatorial value; Johnson and Salem, 1994), although this is substantially less than the magnetization of extrusives (~6.4 A m^{-1}) inferred from near-bottom magnetic anomaly surveys in this area (Tivey, 1996; Tivey *et al.*, 1998a). Although samples of the dike complex are still limited, it

appears that they may have a remanence that is significant relative to that of the extrusive layer.

A major difference from the pillows, however, is the magnetic mineralogy, which in pillows is variably maghemitized titanomagnetite (Curie temperatures typically 200–400°C) whereas in the dikes it is dominated by nearly pure magnetite (Curie temperatures >550°C; **Figures 23** and **24**). The origin of the magnetite is critical to assessing role of dikes as a source of seafloor spreading magnetic anomalies. If magnetite is primarily produced by alteration at temperatures below the Curie point, as some have suggested, then the remanence may be largely chemical in origin and may be acquired significantly later than that of the extrusive layer (e.g., Hall and Muzzatti, 1999). Based on a transmission electron microscopy study of dike samples from Hole 504B, Shau *et al.* (2000) documented multiple pathways for formation of magnetite (**Figure 25**). Initial oxyexsolution, at temperatures > 600°C, would have produced only slightly Ti-depleted titanomagnetite, with a Curie temperature too low to acquire a

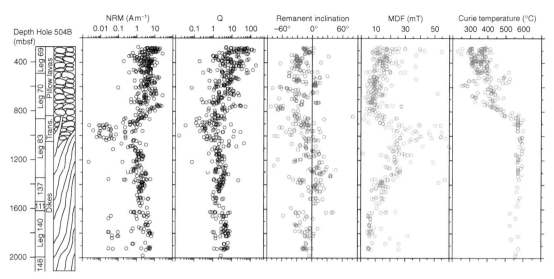

Figure 23 Depth variation in magnetic properties (NRM, Q calculated for 40 A m⁻¹ field, stable remanent inclination, median demagnetization field (MDT), and Curie temperature) of samples of Miocene (~6.8 Ma) oceanic basement at Site 504B on the Galapagos spreading center. Schematic lithology and penetration during seven drilling legs shown at left. Data compiled from Facey *et al.* (1985); Furuta (1983); Furuta and Levi (1983); Kinoshita *et al.* (1985); Pariso and Johnson (1989); Pariso *et al.* (1995); Pechersky *et al.* (1983a); Smith and Banerjee (1985, 1986).

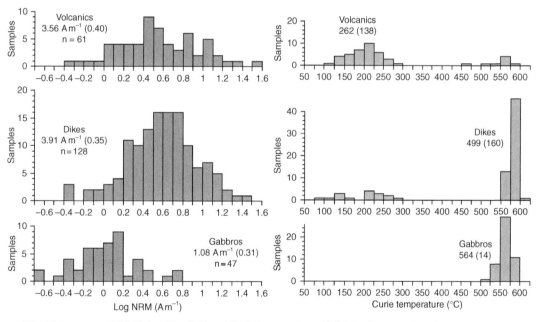

Figure 24 Histograms of NRM intensities (left) and Curie temperatures (right) for dive samples of volcanics, dikes and gabbros from ~1.2 Ma fast-spread oceanic crust at Hess Deep (data from Gee *et al.*, 1992; Varga *et al.*, 2004). Note that the mean NRM intensity for dikes is higher than reported at Hole 504B and is more similar to the volcanics, suggesting that the dikes are an important source of magnetic anomalies at least in the Hess Deep area.

remanence at these elevated temperatures. Further oxidation or true exsolution and reduction, likely at temperatures of ~500–400°C, then produces the fine Ti-poor magnetite that acquires a thermal and/or chemical remanence (Shau *et al.*, 2000). While the remanence in dikes maybe, at least in part, chemical in origin, the temperatures at which the alteration occurs imply that remanence would be acquired very

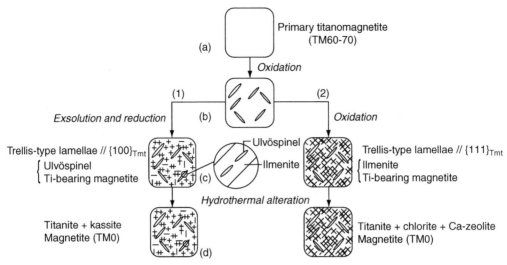

Figure 25 Alteration pathways for primary titanomagnetite in sheeted dikes. From Shau *et al.*, 2000. Either oxyexsolution or true exsolution and reduction may produce fine-grained magnetite (*x* ~ 0) that is distinct from primary titanomagnetite in the extrusives.

near the ridge axis. This result is confirmed by the consistent (reverse) polarity in the lavas and dikes inferred from magnetic logging data at Hole 504B (Worm and Bach, 1996).

12.3.3   Magnetization of Gabbros

Early estimates of the gabbroic contribution to marine magnetic anomalies relied on dredged samples and studies of ophiolites, with the results from ophiolites yielding variable and sometimes contradictory results (Banerjee, 1980). For example, Swift and Johnson (1984) concluded from the low remanent intensities and Königsberger ratios (the ratio of remanent to induced magnetization) of unaltered gabbros from the Bay of Islands ophiolite that only the altered intrusive portion of the crust (characterized by higher magnetization values) would contribute to magnetic anomalies. In contrast, results from dredges and other ophiolites (Kent *et al.*, 1978; Vine and Moores, 1972) indicated that gabbros had significant magnetization (~1 A m^{-1}) and high Königsberger ratios and likely would contribute to magnetic anomalies.

At the time of the last review of ocean crustal magnetization (Smith, 1990), relatively little direct information was available on the magnetic properties of oceanic gabbros. Initial results from the first significant drillcore penetration of gabbros (500 m at Hole 735B; Robinson *et al.*, 1989) indicated highly variable NRM intensities, with some FeTi oxide gabbros

having extremely high values. With only a single drill-core, however, it was difficult to evaluate whether these magnetic (and petrologic) results were generally applicable. In the intervening time, lower crustal gabbros have been sampled by drilling at three portions of the slow-spreading Mid-Atlantic Ridge (Mid-Atlantic Ridge near the Kane Fracture Zone or MARK area at 23 °N (Cannat *et al.*, 1995), near the 15° 20′ Fracture Zone (Kelemen *et al.*, 2004), Atlantis Massif (Blackman *et al.*, 2006)), at fast-spreading crust at Hess Deep (Gillis *et al.*, 1993), and Hole 735B has been deepened by an additional kilometer (Dick *et al.*, 1999) and supplemented by additional drilling on the Atlantis Bank (Allerton and Tivey, 2001; Pettigrew *et al.*, 1999). Although these sites all made use of tectonic exposures, recent drilling in the Guatemala Basin (Wilson *et al.*, 2006) penetrated lavas, dikes, and a small amount of gabbro; continued drilling at this site could provide the first *in situ* gabbroic section from the ocean crust.

These various drill sites into lower crustal gabbros provide a remarkably consistent average magnetization of ~1–2 A m^{-1} (**Table 2**), suggesting that these rocks should constitute an important magnetization source for magnetic anomalies. Although average magnetic properties are similar, gabbroic sections (particularly at slow-spreading ridges) exhibit substantial variability both in lithology and magnetic properties as exemplified by results from Hole 735B (**Figure 26**). FeTi oxide gabbros with >2% (and occasionally >10%) oxide minerals constitute a significant fraction of the core from Hole 735B. Magnetic susceptibility

Table 2 Equatorial natural remanent magnetization values for oceanic crustal rocks

Site	Sample type	Age (Ma)	n	A.M. (A m^{-1})	std. (A m^{-1})	G.M. (A m^{-1})	std (log)	Ref.
Lavas								
EPR	Dredge	0	2178	24.68	15.54	18.94	0.38	1
Hess Deep	Dive	1.2	61[a]	5.37	5.70	3.56	0.40	2, 3
Site 482	Drill	0.4	133	6.20	4.19	4.98	0.30	4, 5
Site 483	Drill	2.3	89	7.42	6.49	4.99	0.40	4, 5
Hole 504B/896[b]	Drill	6.8	430	6.15	4.95	4.23	0.42	6–12
Site 395[c]	Drill	7.8	90	2.95	2.01	2.23	0.35	15
Site 1250[d]	Drill	15.4	184	18.36	12.22	11.06	0.35	16
Sites 417/418	Drill	121	878	9.22	8.03	6.15	0.44	17
Hole 801C[e]	Drill	168	203	4.92	3.89	3.43	0.45	18, 19
Deep Drillholes	Drill	–	7	7.89	5.01	4.92	0.25	
Dikes								
Hess Deep	Dive	1.2	128[a]	5.27	4.50	3.91	0.35	2, 3
Hole 504B[f]	Drill	6.8	171	2.11	1.67	1.32	0.53	9–14
All	–	–	2	3.69	2.23	2.27	0.33	
Gabbros								
Hess Deep	Dive	1.2	47[a]	1.40	1.17	1.08	0.31	2, 3
Hole 735B	Drill	12	600	3.34	8.42	1.19	0.57	20, 21
Leg 209[g]	Drill	0.5–2.0	61	6.15	6.23	2.36	0.78	22
Leg 153[h]	Drill	~1	252	1.19	1.74	0.62	0.53	23, 24
Hole 894G[i]	Drill	~1	87	2.25	1.56	1.65	0.41	25
Exp. 304/305[j]	Drill	~1	472	3.11	6.28	0.56	0.97	26
All Drillholes	Drill	–	5	2.91	1.81	1.09	0.24	
Peridotites								
Oufi compilation[k]	Drill	–	234	5.84	5.05	3.15	0.62	27
Leg 209[l]	Drill	0.5–2.0	99	2.75	3.57	0.79	0.99	22
All	Drill	–	333	4.92	4.86	2.09	0.79	

[a]Each sample value based on three or more specimens.
[b]Average based on Hole 504B lava samples above 898 mbsf (corresponding to low-temperature altered zone of Pariso and Johnson, 1991) and all lava samples from Site 896.
[c]Excludes gabbro and peridotites recovered at this site.
[d]Average based on shipboard data from 252–749 mbsf. Wilson *et al.* (2003) note particularly large drilling induced remanence.
[e]Average based on samples from 551–933 mbsf (excluding younger alkalic lavas and hydrothermal deposits).
[f]Average based on samples deeper than 1055 mbsf (dike/lava transition zone [878–1055 mbsf] excluded).
[g]Average based on all gabbroic samples from Sites 1268–1275 (near 15° 20′ N Fracture Zone).
[h]Average based on all gabbroic samples from Sites 921–924 (MARK Area).
[i]Minor recovery of gabbroic material in other drillsites at Hess Deep not included.
[j]Includes gabbroic samples from Holes 1309B and 1309D (excluding diabase and ultramafic samples).
[k]Compilation of peridotite samples from DSDP/ODP drilling up to and including Leg 153.
[l]Average based on all peridotite samples from Sites 1268–1274 (near 15° 20′ N Fracture Zone).
A.M., arithmetic mean; G.M., geometric mean.
References: 1, Gee and Kent (1997); 2, Varga *et al.* (2004); 3, Gee *et al.* (1992); 4, Day *et al.* (1983); 5, Pechersky *et al.* (1983b); 6, Allerton *et al.* (1996); 7, Furuta and Levi (1983); 8, Pechersky *et al.* (1983a); 9, Kinoshita *et al.* (1985); 10, Facey *et al.* (1985); 11, Smith and Banerjee (1985); 12, Smith and Banerjee (1986); 13, Pariso and Johnson (1989); 14, Pariso *et al.* (1995); 15, Johnson (1978); 16, JANUS database for ODP Leg 206 (Wilson *et al.*, 2003); 17, Levi (1980); 18, JANUS database for ODP Legs 129 (Lancelot *et al.*, 1990) and 185 (Plank *et al.*, 2000); 19, Wallick and Steiner (1992); 20, Kikawa and Pariso (1991); 21, Dick *et al.* (1999); 22, Kelemen *et al.* (2004); 23, Gee *et al.* (1997); 24, Cannat *et al.* (1995); 25, Pariso *et al.* (1996a); 26, Blackman *et al.* (2006); 27, Oufi *et al.* (2002).

data indicate the presence of several hundred oxide-rich zones, ranging from centimeter to several meters in thickness, throughout the core (Natland, 2002). These oxide-rich intervals may be associated with extremely high NRM values (in some cases >10^3 A m^{-1}; Robinson *et al.*, 1989) though these high values likely reflect a substantial drilling induced overprint and thus are not representative of the *in situ* magnetization. Borehole magnetic data from the upper 500 m of Hole 735B indicate an *in situ* equatorial remanence of 4.4 A m^{-1} for oxide-rich gabbros and 1.8 A m^{-1} for other gabbroic rocks (Pariso and Johnson, 1993a). Although NRM

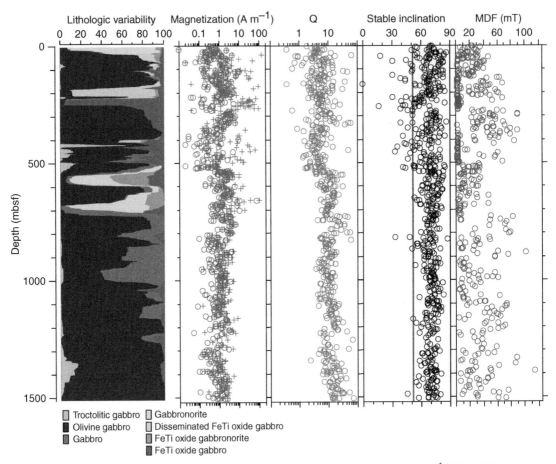

Figure 26 Depth variation in lithology and magnetic properties (NRM, Q calculated for 30 A m^{-1} field, stable remanent inclination and MDF, median demagnetization field) for samples of gabbroic section in oceanic crust at ODP Hole 735B. Magnetic data compiled from Kikawa and Pariso (1991) and Dick *et al.* (1999); lithology column from Dick *et al.* (2000). Heavy line in inclination plot is expected dipole value. Magnetization data have not been reduced to equatorial values (open circles and + symbols represent the intensity after removal of the drilling-induced remanence and NRM, respectively).

intensities show significant variability, results from both drillcore and submersible samples (**Table 2**) suggest that an average intensity of $\sim$1 A m^{-1} is typical for oceanic gabbroic rocks.

Although less compelling than for lavas, there appears to be a weak correlation between more fractionated (Fe-rich) gabbros and higher NRM intensities (Gee *et al.*, 1997; Kikawa and Ozawa, 1992). For a subset of gabbroic rocks from the MARK area, Hess Deep, and Hole 735B (**Figure 27(a)**), the lowest NRM values are exclusively associated with the least evolved gabbros (with high $Mg\# = Mg^{2+}/(Fe^{2+} + Mg^{2+})$). Kikawa and Ozawa (1992) have noted a comparable broad increase in remanent intensity from Hole 735B samples as fractionation proceeds from troctolite to olivine gabbro. These same authors have suggested

that gabbro magnetization values also may be reduced by up to an order of magnitude as a result of alteration (as measured by the percentage of secondary mafic minerals). This trend is less obvious in a more qualitative comparison of NRM and shipboard estimates of alteration from MARK area gabbros (**Figure 27(b)**). Nonetheless, it appears that earlier suggestions of increasing remanent intensity with degree of alteration (Swift and Johnson, 1984) are not generally applicable to oceanic gabbros.

Curie temperatures of gabbroic samples are uniformly near 580°C (e.g., Gee *et al.*, 1997; Pariso and Johnson, 1993b; Worm, 2001), indicating that Ti-poor magnetite is the dominant remanence carrying phase. Primary discrete magnetite (associated with more abundant ilmenite) is abundant in FeTi oxide gabbros

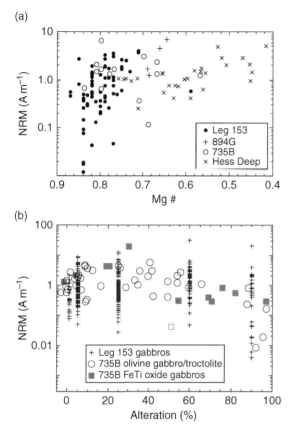

(a)

(b)

Figure 27 Variation of NRM of oceanic gabbros as a function of Mg# (whole rock $Mg^{2+}/(Mg^{2+}+Fe^{2+})$) and alteration (after Gee et al., 1997, with additional data from Kikawa and Ozawa, 1992). Qualitative alteration data for Leg 153 gabbros are from shipboard descriptions; percent alteration for Hole 735B gabbros (Kikawa and Ozawa, 1992) is based on pyroxene alteration. NRM data have not been reduced to equatorial values.

but rare in the less-evolved rocks that constitute the bulk of the gabbroic material. In these latter samples, much of the stable remanence may be carried by fine-grained magnetite enclosed in silicate grains. Fine-grained oxides are present in olivine, pyroxene, and plagioclase in nearly all troctolites and olivine gabbros from Hole 735B (Natland, 2002). Magnetite along cracks in olivine is attributed to amphibolite–granulite facies alteration at temperatures of 500–700°C (Pariso and Johnson, 1993b). Crystallographically oriented elongate magnetite within pyroxene is demonstrably a high temperature (~800°C) 'exsolution' product in some cases (Renne et al., 2002). A similar exsolution origin for crystallographically controlled magnetite in plagioclase has also been suggested (Natland, 2002; Selkin et al., 2000; Xu et al., 1997a). Silicate-hosted magnetite, at least in plagioclase and pyroxene, appears to be a common feature in oceanic gabbros (Gee and Meurer, 2002) and may be responsible for much of the stable remanence in these rocks. Hysteresis and Curie temperature data from single plagioclase crystals from gabbros at the MARK area and from Leg 209 (**Figure 28**) illustrate that these fine-grained magnetite grains may have substantial coercivities.

With the exception of FeTi oxide gabbros, most oceanic gabbros have high stability magnetizations and Königsberger ratios (Q) > 1, indicating that they may constitute an ideal magnetization source. For example, Königsberger ratios for gabbros from Hole 735B are essentially all >1 and exceed 10 for much of the lower kilometer of the section (**Figure 26**). The increase in Q downhole is paralleled by a general increase in coercivity, as measured by the median

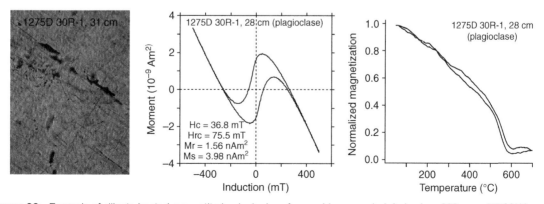

Figure 28 Example of silicate hosted magnetite in plagioclase from gabbro sampled during Leg 209, near 15° 20′ N Fracture Zone Left, photomicrograph (25X, oil immersion) of plagioclase with elongate magnetite; center, hysteresis loop; and right, Curie temperature curve for a single plagioclase crystal.

destructive field (MDF). Oxide-rich gabbros are characterized by lower stabilities, as might be expected from the abundance of coarse-grained magnetite grains that are particularly susceptible to acquiring a drilling-induced remanence. The olivine gabbros and gabbros that constitute the bulk of the material from Hole 735B are characterized by significantly higher MDF values (several 10's mT). Comparable MDF values (average ~30 mT) have been reported for gabbros from Hess Deep (Pariso *et al.*, 1996b) and for the ~1400 m of gabbro sampled from the Atlantis Massif (Blackman *et al.*, 2006).

The variability of inclination values in gabbroic rocks (**Figure 26**) is considerably less than for extrusive sections and presumably reflects significant averaging of secular variation during slow cooling. For the lowermost 1000 m of Hole 735B sampled during Leg 176, the directional dispersion parameter (kappa $= 59$, n $= 339$) is approximately 1.6 times the value expected from the paleosecular variation model of McFadden *et al.* (1988). The higher value of kappa indicates tighter grouping of the inclination data and the implied time averaging during slow cooling is a necessary prerequisite for using the average inclination data to infer tectonic tilts. Although the inclinations from Hole 735B are offset from the expected dipole value due to tilting, the uniform polarity and relatively high magnetization of this 1500 m gabbroic section (where lavas and dikes have been tectonically removed) are sufficient to account for the sea-surface anomaly amplitudes at the site (Dick *et al.*, 2000). While uniform polarity is likely the norm, it should be noted that complex multicomponent remanences and dual polarities within a site have been reported from gabbros from the MARK area and the Atlantis Massif (Cannat *et al.*, 1995; Gee and Meurer, 2002; Blackman *et al.*, 2006). These complex remanence characteristics have been attributed to crustal accretion and cooling spanning multiple polarity intervals (Gee and Meurer, 2002) and, if more generally applicable, would serve to reduce the gabbroic contribution to the magnetic anomalies.

12.3.4   Mantle-Derived Peridotites

Although pristine ultramafic rocks are nonmagnetic (containing only paramagnetic silicates and spinels), serpentinization results in the production of magnetite (e.g., Toft *et al.*, 1990) and magnetizations that are sufficient to constitute a potential significant source for marine magnetic anomalies (e.g., Nazarova, 1994).

Well-documented exposures of serpentinized peridotite at slow-spreading ridges (Cannat, 1993; Cannat *et al.*, 1997) indicate that such upper mantle material undoubtedly contributes, at least locally, to magnetic anomalies. Serpentinization of peridotites and lower crustal gabbros has also been suggested as the cause of higher anomalous skewness of anomalies at slow-spreading ridges (Dyment and Arkani-Hamed, 1995; Dyment *et al.*, 1997). In addition, the enhanced positive magnetization (both for normal and reverse polarity crust) at the ends of slow-spreading ridge segments has been attributed to induced magnetization of serpentinized peridotites and gabbros (Pariso *et al.*, 1996b; Pockalny *et al.*, 1995; Tivey and Tucholke, 1998).

As with sampling of the gabbroic portion of the ocean crust, drilling at tectonic exposures (Site 895 at Hess Deep (Gillis *et al.*, 1993); Sites 670 and 920 in the MARK area (Cannat *et al.*, 1995; Detrick *et al.*, 1988); and near the 15° 20' Fracture Zone (Kelemen *et al.*, 2004)) has significantly enhanced our understanding of the magnetic properties of oceanic ultramafic rocks. Minor amounts of serpentinized peridotite were also (unexpectedly) recovered from a number of Atlantic drillsites (Sites 334, 395, 556, 558, 560) designed to sample oceanic lavas. Oufi *et al.* (2002) provide a thorough analysis and review of magnetic studies on serpentinized peridotites up to and including Leg 153 drilling in the MARK area. These results and the more recent remanence data from Leg 209 (Kelemen *et al.*, 2004) are summarized in **Table 2**.

Magnetizations of serpentinized peridotites, whether obtained by dredging or drilling, vary from $<0.1\,A\,m^{-1}$ to values higher than $30\,A\,m^{-1}$ that are comparable to typical values for older oceanic basalts (**Figure 29**; **Table 2**). The magnetite content, susceptibility, and magnetization of ultramafic rocks increases with the degree of serpentinization, though this increase is not linear and depends on the particular alteration phases produced (Toft *et al.*, 1990). The degree of serpentinization may be estimated from the sample density (Miller and Christensen (1997), with a correction for the amount of substantially denser magnetite (Oufi *et al.*, 2002)), and **Figure 30** shows that relatively little magnetite is produced until the serpentinization exceeds 50%. This nonlinear behavior may be attributed to the FeO content of the silicate alteration phases. For $<75\%$ serpentinization, early FeO-rich lizardite is formed and with continued alteration this lizardite is replaced by more FeO-poor chrysotile, with a concomitant release of iron and production of

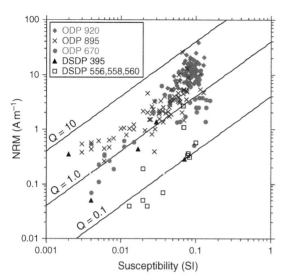

Figure 29 Log–log plot of NRM and susceptibility of oceanic serpentinized peridotite samples from various DSDP and ODP drill sites (data compiled by Oufi et al., 2002). Distribution of values compared to lines of constant Q (Königsberger ratio of remanent to induced magnetization; calculated for field of $40\,A\,m^{-1}$) shows that remanence is greater than induced magnetization in most samples. NRM data have not been reduced to equatorial values.

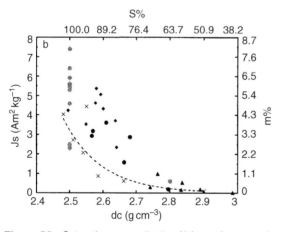

Figure 30 Saturation magnetization (Js) as a function of corrected density (dc) of serpentinized peridotite samples from various DSDP and ODP sites, showing how percentage of magnetite (m%) increases with degree of serpentinization (S%). Dashed curve is best-fitting relationship between Js and dc for a subset of samples. After Oufi O, Cannat M, and Horen H (2002) Magnetic properties of variably serpentinized abyssal peridotites. *Journal of Geophysical Research* 107: (B5): (doi:10.1029/2001JB000549).

magnetite (Oufi *et al.*, 2002). Alteration assemblages that include even more FeO-rich brucite may further suppress magnetite production, as evident in some of the lower magnetization samples from Sites 895 and

670 (**Figure 29**). Most dredged peridotites and some drillcore samples (particularly from breccias or fault zones) have experienced post-serpentinization low-temperature oxidation, as evidenced by the presence of maghemite (e.g., Nazarova *et al.*, 2000). Although magnetic properties may vary considerably, the geometric mean intensity ($2.1\,A\,m^{-1}$ equatorial value including results from Leg 209; **Table 2**) and average Königsberger ratio of ~2 suggest that both remanent and induced magnetization from oceanic serpentinized peridotites may contribute significantly to magnetic anomalies. The influence of serpentinized peridotites on the anomaly pattern, however, may largely be limited to fracture zones and slow-spreading ridges where mantle-derived peridotite exposures are relatively common.

Although many serpentinites may predominantly have a viscous remanence, the origin of the remanence in oceanic serpentinized peridotites is likely to include a chemical remanence and perhaps also a partial thermoremanence as demonstrated by remanence directions that deviate substantially from the present field direction (e.g., Garces and Gee, 2007). Temperature estimates for serpentinization based on oxygen isotope fractionation between magnetite and serpentine vary widely, though most serpentinites from Hess Deep and the MARK area yield temperatures conservatively estimated as >350 °C (Agrinier and Cannat, 1997; Früh-Green *et al.*, 1996). Based on these relatively high temperatures and the simple demagnetization behavior of MARK area peridotites, Lawrence *et al.* (2002) suggested that much of remanence might be thermal in origin. While coarse-grained magnetite is abundant, values of M_{rs}/M_s as high as 0.4, as well as direct grain size estimates from image analysis indicate that relatively fine-grain magnetite is also present (Lawrence *et al.*, 2002; Oufi *et al.*, 2002). These finer particles are presumably responsible for the relatively stable remanence in the MARK area peridotites, which was used to infer that little tilting occurred during uplift and exposure (Lawrence *et al.*, 2002).

12.4 Crustal Accretion and Structure of the Magnetic Source

Crustal accretionary processes exert a fundamental control on the spatial distribution of magnetization sources, limiting the temporal resolution of the geomagnetic signal resolvable from marine magnetic anomalies and, in some cases, modulating their shape (skewness). In addition, the remanent magnetization vector may be rotated during accretion (e.g., tilt due to lava loading) or by subsequent tectonic deformation. Any tectonic rotation of the magnetic source layer about a ridge-parallel rotation axis will result in an equivalent phase shift of the magnetic anomaly (Cande and Kent, 1985). Anomalous skewness, the phase shift not accounted for by a standard thin layer block model (Cande, 1976), therefore provides a strong constraint on any tectonic tilt of the magnetization source, as well as nonvertical polarity boundaries. In the discussion below, we address the consequences of the (generally accepted) pattern of nonvertical magnetic boundaries separately from the effects related to (more speculative) rotation of the magnetic source.

12.4.1 Nonvertical Magnetic Boundaries

The pattern of magnetization boundaries (isochrons) in the ocean crust is shown schematically in **Figure 31**. Isochrons within the extrusive layer are expected to dip toward the ridge (e.g., Kidd, 1977) as new lavas progressively cover pre-existing flows. The resulting sloping (sigmoidal) magnetization boundary in the extrusives can be conveniently modeled by a gaussian filter applied to the standard magnetization block model. The transition zone width, the width over which 90% of the magnetization change occurs, is approximately four times the standard deviation (σ) of the gaussian filter (Atwater and Mudie, 1973). In a classic survey at 21° N on the EPR, Macdonald *et al.* (1983) documented both the validity of this general model, as well as the spatial extent of lava spillover onto pre-existing flows by comparing the location of the reversal boundary inferred from near-bottom magnetic anomaly data and the polarity of individual volcanic features at the surface from submersible observations. The surface polarity boundary was displaced (away from the ridge) by 250–500 m relative to the vertically averaged polarity boundary determined from the near-bottom anomaly data, suggesting a transition zone width of 1.4–1.8 km at this intermediate (60 mm yr^{-1} full rate) spreading rate (Macdonald *et al.*, 1983). Comparable transition zone widths ($\sim$1.5–3.0 km) have been determined from near-bottom anomaly data at intermediate to superfast (up to 150 mm yr^{-1} full rate) spreading ridges (Bowers *et al.*, 2001; Sempere *et al.*, 1987). Less focused accretion at slow-spreading ridges likely results in broader

transition zones (Sempere *et al.*, 1987), though the spreading rate dependence has not been well established.

The narrow (1–3 km) transition zone widths (which represent an upper bound due to the effects of minor extension and the $\sim$7 ky (Clement, 2004) necessary for the field to reverse) imply a relatively high-fidelity recording process at intermediate- and fast-spreading ridges, compatible with the recognition of short polarity events (e.g., the $\sim$10 ky Cobb Mountain or Reunion events; **Table 1**) in anomaly profiles at these spreading rates. These estimates are also consistent with independent estimates of the width over which lavas accumulate. Submersible observations at intermediate- and fast-spreading ridges suggest that most volcanic activity is concentrated within 1 km of the axis (Karson *et al.*, 2002; Perfit and Chadwick, 1998). Flows erupted off-axis or transported downslope may extend for a few kilometers from the ridge (Fornari *et al.*, 1998; Perfit *et al.*, 1994) and off-axis accumulation of lavas can significantly affect the pattern of magnetic anomalies (Gee *et al.*, 2000; Schouten *et al.*, 1999). The off-axis thickening of seismic Layer 2A (commonly interpreted as corresponding to the extrusive layer) within 1–3 km of the axis (e.g., Harding *et al.*, 1993; Vera and Diebold, 1994) also suggests that lava accumulation occurs over a narrow zone comparable to the transition zone widths inferred from anomaly data.

The pattern of magnetization contrasts in the sheeted dikes and gabbros is less well constrained. The zone of dike intrusion, where mixed polarity dikes would occur (**Figure 31**), is likely much narrower than for the lava flows. In order to match seismic constraints and the sharp extrusive/dike boundary observed in ophiolites, Hooft *et al.* (1996) found that the zone of dike intrusion at the EPR must be less than a few hundred meters wide (1σ of 10–50 m). To the extent that the lower crust is conductively cooled, polarity boundaries in the gabbroic portion of the crust should approximately follow the shape of the 580°C isotherm (Cande and Kent, 1976) (**Figure 32**). Half-space conductive cooling models (Oxburgh and Turcotte, 1969; Sclater and Francheteau, 1970) predict that magnetization isochrons in the gabbroic layer should dip away from the ridge at 4–30° (for full spreading rates from 150 to 16 km My^{-1}, respectively). Near-bottom magnetic anomaly data from tectonic exposures of lower crustal gabbro at Atlantis Bank, together with inclination data from wireline rock drill samples and from the

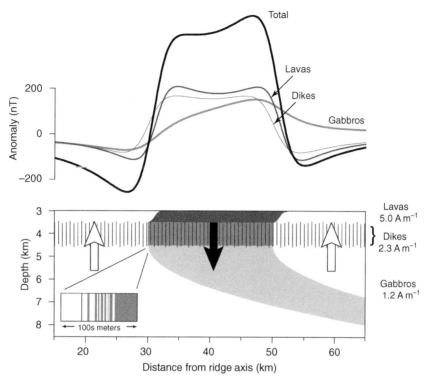

Figure 31 Schematic illustration of magnetization boundaries in oceanic crust (bottom) and magnetic anomaly models for these magnetic source bodies (top). The magnetization values of the three layers are based on the overall geometric mean values in **Table 2**. Anomalies are calculated for vertical magnetization and ambient field for a full spreading rate of 80 km My^{-1}. No gaussian filter has been applied to the magnetization.

1.5 km penetration at Hole 735B, have been interpreted as indicating the presence of a sloping isochron with a present day dip of ~25° (Allerton and Tivey, 2001). However, the well-determined remanent inclination (−71°; **Figure 26**) from Hole 735B suggests an outward tilt of about 20°, so that the original dip of this isochron was approximately 5°. While the dip direction is away from the ridge, as expected, the very shallow dip angle is difficult to reconcile with that expected (~30°) from conductive cooling (John *et al.*, 2004) at this slow-spreading ridge.

The effect of nonvertical magnetization boundaries on the shape and amplitude of anomalies should be minor for lavas and dikes but is significant for the gabbroic portion of the crust (**Figure 31**). For 1–3 km wide transitions in the extrusive layer, the corresponding anomaly is shifted slightly away from the ridge but there is little detectable effect on the anomaly shape or amplitude. Based on vertical magnetic profiling data from intermediate-spreading crust exposed at the Blanco escarpment, substantially broader (~10 km) transition zone widths have been

suggested (Tivey, 1996; Tivey *et al.*, 1998a). While this novel profiling technique has considerable promise in mapping the distribution of polarities, the very broad transitions at the Blanco escarpment predict substantial changes in anomaly amplitude and shape that are not generally apparent. The narrow polarity transition in the sheeted dikes ($\sigma_{dikes} \ll \sigma_{flows}$) is likewise not expected to significantly affect the anomaly shape or amplitude.

In contrast, the gently sloping boundaries in the gabbroic lower crust should give rise to significant anomalous skewness (Cande and Kent, 1976). Anomalies generated by such gently dipping polarity boundaries are illustrated in **Figure 32** for a range of polarity interval lengths and spreading rates. The resulting anomalous skewness is most easily recognized for broad polarity intervals and amounts to a phase shift of ~45°, in the same sense as global observations of anomalous skewness. The distortion is not a pure phase shift, however, as evidenced by the asymmetric shoulder at the older edge that coincides with the magnetization contrast at the base of the gabbroic layer. Although the anomaly shapes for

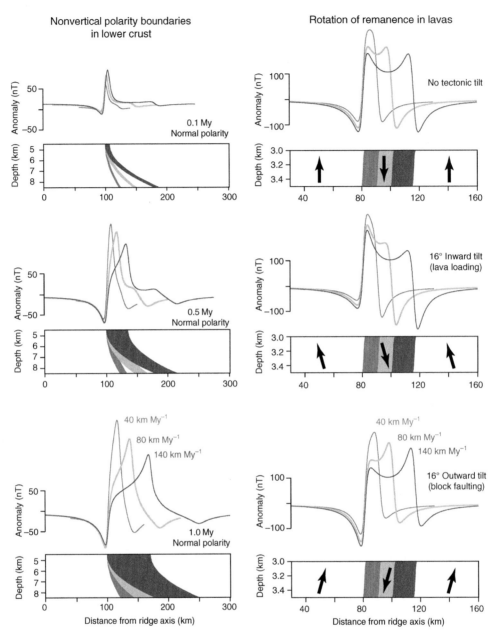

Figure 32 Magnetic anomaly models illustrating the effect of sloping cooling isotherms in gabbroic layer and various lava tilts. Left, conductively cooled magnetization boundaries in the gabbroic layer are illustrated for three different length polarity intervals (0.1, 0.5, and 1.0 My) at three spreading rates. This normal polarity interval is indicated by the three shaded regions (green, red, blue for 40, 80, 140 km/My^{-1} full spreading rates). The corresponding magnetic anomalies have an anomalous skewness of ~45°. Anomalies are calculated for vertical magnetization and ambient field. Right, effect of tilt of the extrusive layer on anomalies from a 0.5 My normal polarity interval at the same three spreading rates. The sigmoidal magnetization boundary reflects lava accumulation over ~2 km and introduces no detectable change in the anomaly shape (upper plot). Ridgeward or outward tilting of 16° (the average tilt implied by lava accumulation over ~2 km) results in an equivalent amount of anomalous skewness.

narrow intervals or at different spreading rates appear quite different, the anomalous skewness introduced by a conductively cooled gabbroic layer is essentially independent of spreading rate. A three-layer source model (with subequal contributions from the lavas, dikes, and gabbros as indicated by the average remanence of these layers and their thickness and depth; **Table 2**) illustrates that the anomalous skewness imparted by the gabbroic layer results in an overall anomalous skewness of ~10° (**Figure 31**). This value is well within the range (5–25° for a full spreading rate of 80 km My^{-1}) determined from global compilations of anomalous skewness data (e.g., Roest *et al.*, 1992; Petronotis *et al.*, 1992). Although rotations of the magnetic remanence direction (which are not included in the model shown in **Figure 31**) may result in additional anomalous skewness, such rotations are apparently not required to explain the modest anomalous skewness at intermediate and faster spreading rates.

12.4.2 Rotations of the Magnetic Source Layer

Normal faulting is an integral part of extensional seafloor spreading and therefore rotations, either toward or away from the ridge, about ridge parallel axes are not unexpected. Inward (toward ridge) dipping faults are dominant at all but the fastest spreading rates (Carbotte and Macdonald, 1990; Kriner *et al.*, 2006), and even at superfast spreading rates these faults accommodate most of the strain (Bohnenstiehl and Carbotte, 2001). These inward dipping faults should result in outward tilt of the magnetic source layer, which will produce a phase shift (1° for each 1° of tilt) of the anomalies that is compatible with the observed sense of anomalous skewness (**Figure 32**; Cande, 1978).

Tilting in response to lava loading has recently been suggested as a potential way of rotating the remanence vector of seafloor lavas (Schouten and Denham, 2000; Schouten *et al.*, 1999; Tivey *et al.*, 2005). In these models, the sloping isochrons within the extrusive layer are taken as also reflecting the inward tilt of individual lava flows. The resulting downhole distribution of inclinations (possibly modified by later block rotation) can then be used to infer the average pattern of lava accumulation. This type of model requires that the remanence in relatively small intervals within the extrusive layer represent the time-averaged field direction, so that downhole trends in the average inclination can be detected. If prevalent, such lava loading would produce systematic inward tilt of the lavas and anomalous skewness in the opposite sense of that observed. The relatively thin extrusive layer and the generally narrow region over which flows accumulate limit the amount of tilting expected from lava loading. For accumulation of a 0.5 km thick extrusive layer over a typical transition zone width of ~2 km, the average dip of the

isochrons (and therefore inward tilt of the remanence) is ~16°, which is accompanied by a readily observable anomalous skewness of 16° (**Figure 32**).

A significant number of drill sites sampling the extrusive layer have inclinations that deviate from expected values, and block rotations and/or lava loading undoubtedly contribute to some of these deviations. In the most recent (although 30 years ago) compilation of inclination results from drillcore data, Lowrie (1977) found that observed inclination values were well correlated with expected values, though the results exhibit appreciable scatter about the ideal 1:1 line. Although no significant bias toward either shallower or steeper inclinations was found, two aspects of the drill core data make it unlikely that any systematic bias (whether from block rotations or lava loading) would be evident. First, paleosecular variation can result in significant scatter even when the entire lava section is sampled (e.g., 504B; **Figure 23**) and so average inclination results from shallow penetration holes should be viewed with caution. Second, the lack of azimuthal orientation of the drillcore samples precludes recognition of a systematic sense for ridges that strike nearly N–S, as does much of the Pacific and Atlantic ridge system. In principle, E–W striking ridges might provide information on systematic tilt in the extrusives though it is unclear whether sufficient data exist to conduct such an analysis.

Magnetic anomaly shapes provide the most robust estimate of the average tilt of the magnetization source. Comparison of drillcore data from three relatively deep penetration holes in Cretaceous (M0 age) crust (**Figure 33**) and correlative anomaly shapes (**Figure 34**) illustrate how models of block tilting and lava loading can be independently evaluated. The mean inclinations in Holes 417A and 418A (where there is a polarity reversal at 500 mbsf) are eminently compatible with expected inclinations from Early Cretaceous paleopoles for the North America plate (Bosum and Scott, 1988; Levi, 1980). The inclinations from the upper portion of Hole 417D are steeper and have been attributed to tectonic tilting by some authors (Verosub and Moores, 1981), although Levi (1980) argues that the inclination data are compatible with paleosecular variation and do not demand any tectonic rotation. Schouten (2002) interprets the Hole 417A data (with reverse polarity inclination very similar to the expected value) as reflecting two oppositely directed tectonic tilts: a large initial tilt toward the ridge axis as function of lava loading (requiring the bulk of the lavas to

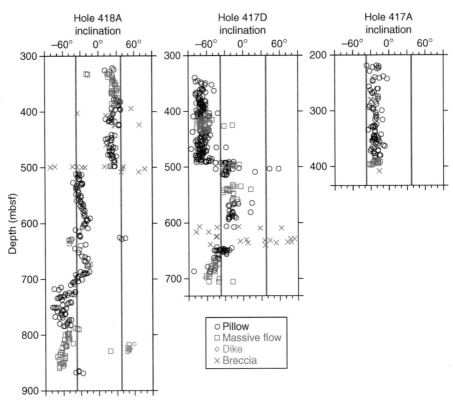

Figure 33 Stable remanent inclinations (i.e., after partial demagnetization to remove spurious components) from discrete samples from seafloor lavas at DSDP Holes 417A, 417D, and 418A. The heavy lines indicate the expected Cretaceous inclination values at the site. Data from Levi (1980).

accumulate within 300 m of the axis), followed by 41° outward block tilting (**Figure 34**). Magnetic anomaly data across Anomaly M0, including from the immediate vicinity of Sites 417/418 (**Figure 34**; Cande and Kent, 1985), indicate the M0-age crust from both ridge flanks is characterized by little or no anomalous skewness and therefore little or no net tilt of the magnetic source. While such fortuitous offsetting tilts could, of course, explain the lack of anomalous skewness near Sites 417 and 418, this is unlikely to be generally applicable. The magnitude and sense of anomalous skewness indicate that substantial tilting (whether from lava loading or block rotation) is not generally applicable, at least for fast- and intermediate-spreading ridges where anomalous skewness magnitudes are generally small.

Relatively few observations are available that constrain possible rotations of the intrusive portion of the crust. Based on submersible studies of tectonic exposures of fast-spread crust at Hess Deep, much of the sheeted dike complex appears to have a systematic dip (~60°) away from the ridge (Karson et al., 2002) although areas with approximately vertical dikes are

also present (Francheteau et al., 1992). As with the lavas, anomalous skewness constraints suggest that such inward tilting of the sheeted dikes is unlikely to be a general feature, provided that the dike remanence is a significant contributor to the magnetic anomalies as available data seem to now indicate. Substantial rotations of the gabbroic layer have been documented at slow-spreading ridges. For example, tectonic tilts of ~20° have been well documented for gabbros at the Atlantis Bank (Dick et al., 1999) and even larger rotations (40–80°) have been reported from gabbros (and associated serpentinized peridotites) from the vicinity of the 15° 20' N Fracture Zone in the Atlantic (Garces and Gee, 2007; Kelemen et al., 2004). The outward rotation of the gabbroic sections, presumably from flexural/isostatic adjustment in response to unloading, is compatible with the sense of anomalous skewness data. Although the magnitude of anomalous skewness appears to increase at slower spreading rates (Roest et al., 1992), the very large rotations noted from the 15° 20' N area exceed the largest values of anomalous skewness and thus are unlikely to be representative of more typical slow spread crust.

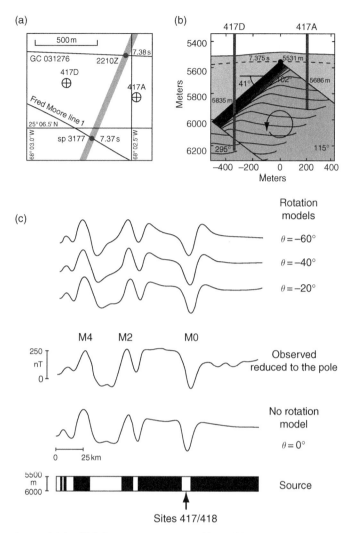

Figure 34 Tectonic rotation model for ODP Sites 417A and 417D (a) and (b) location map; (c) diagrammatic cross-section) from Schouten (2002), compared to observed and modeled magnetic anomaly data for a profile nearby Site 417A and 417D (from Cande and Kent, 1985). Note that the observed profile is matched best by a model with no tectonic rotation and is poorly represented by a model with 40° of rotation suggested from inclination data at Sites 417A and 417D by Schouten (2002).

12.5 Future Directions

There is increasing evidence that marine magnetic anomalies are capable of recording a broad spectrum of geomagnetic field behavior, ranging from millennial-scale paleointensity variations to polarity reversals to apparent polar wander to, more speculatively, long-term changes in average field strength as suggested, for example, by the ramp in anomaly amplitudes at the older end of the M-sequence. However, because of the inherent ambiguity in determining source properties from potential field data, independent geomagnetic field estimates will be needed to guide and calibrate inferences

made from magnetic anomalies. Here we highlight some general approaches – the use of autonomous vehicles, oriented samples, absolute paleointensity of near-ridge lavas, and measurements of the vector anomalous field – that are likely to significantly advance our understanding of the geomagnetic signal recorded in the oceanic crust, as well as our ability to utilize this information in addressing outstanding problems in crustal accretion processes.

One general area that is ripe for further development and applications is near-bottom observations, which are required to obtain high spatiotemporal resolution. Such near-bottom observations from towed systems (often incorporating side scan sonar or other instrument packages as well) have already provided seminal contributions in our understanding of the fine structure of anomalies and source geometries (e.g., Atwater and Mudie, 1973; Larson and Spiess, 1969). More recently, near-bottom data have been used to demonstrate the lineated (i.e., geomagnetic) nature of small-scale magnetic anomalies, which have been shown to mainly reflect paleointensity variations (Bowers *et al.*, 2001; Bowles *et al.*, 2003; Gee *et al.*, 2000), and to examine crustal accretionary processes (e.g., Hussenoeder *et al.*, 1996; Smith *et al.*, 1999; Tivey *et al.*, 2003). These towed packages are robust, proven marine geophysics tools but tend to be used sparingly because such wireline deployments require a dedicated vessel and slow towing speeds. An exciting recently developed alternative is ABE (Autonomous Benthic Explorer), which can be operated simultaneously with other shipboard programs. ABE is capable of following preprogrammed, closely spaced track lines at low altitudes (5–40 m above the seafloor) to yield fine-scale bathymetry and high-resolution magnetic data. An example of an ABE survey at 17° S on the EPR (Shah *et al.*, 2003) is shown in **Figure 35**. In this study, a magnetic field low was found extending several kilometers along the axial trough, which was interpreted as reflecting the presence of a few-hundred-meter-wide region of weakly or nonmagnetic shallow dikes (Shah *et al.*, 2003) but might also delineate lavas erupted during the recent (relative) geomagnetic intensity low (**Figure 3**). When combined with the submeter resolution bathymetry, ABE magnetic anomaly data also provide a powerful technique to examine the relationship between individual volcanic features and their corresponding anomaly signatures (e.g., Tivey *et al.*, 1998b). Similar high-resolution anomaly data can also be obtained with the new generation of tethered vehicles, that allow far more motion control than towed systems and also may incorporate high-resolution swath mapping and sampling capabilities. Such high-resolution mapping on the Endeavor segment of the Juan de Fuca ridge has revealed circular (~100 m diameter) anomaly lows that correlate with present-day and fossil hydrothermal upflow zones (**Figure 36**; Tivey and Johnson, 2002).

The more detailed magnetic anomaly observations that are possible with tethered vehicles, ABE, and manned submersibles are nicely complemented by the availability of high-resolution sampling techniques that allow collection of fully oriented samples. Block samples oriented with the Geocompass (which uses a

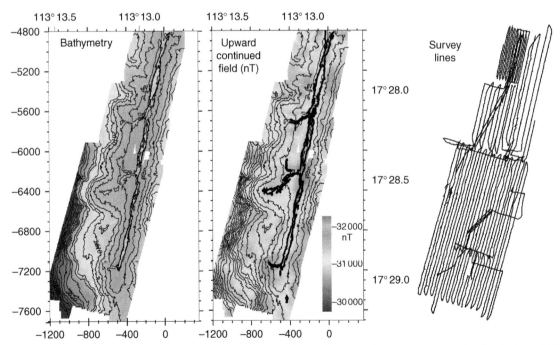

Figure 35 High resolution bathymetric (left) and magnetic data (center) obtained by an Autonomous Benthic Explorer (ABE) survey (tracks on right) on the East Pacific Rise at 17° S. Note the presence of lower anomaly amplitudes along the axial summit trough (heavy lines in center figure). Modified from Shah AK, Cormier M-H, and Ryan WBF *et al.* (2003) Episodic dike swarms inferred from near-bottom magnetic anomaly maps at the southern East Pacific Rise. *Journal of Geophysical Research* 108(B2): 2097 (doi:10.1029/2001JB000564).

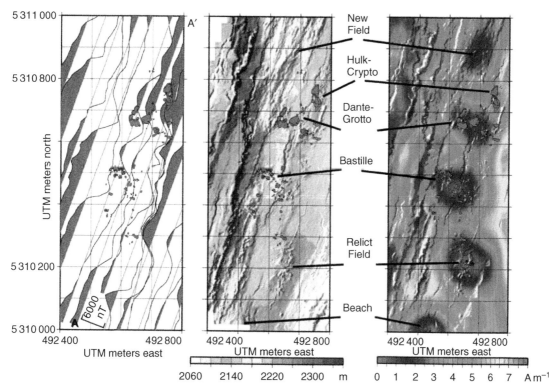

Figure 36 Near-bottom bathymetric and magnetic observations on the Endeavor segment of the Juan de Fuca Ridge showing hydrothermal burn-holes. Red areas indicate locations of active hydrothermal vents. Modified from Tivey MA and Johnson HP (2002) Crustal magnetization reveals subsurface structure of Juan de Fuca Ridge hydrothermal vent fields. *Geology* 30: 979–982.

compass and digital tilt meters to measure attitude) have been collected from submersible (Cogne *et al.*, 1995; Hurst *et al.*, 1994; Varga *et al.*, 2004) and this technique has also been successfully used with tethered remotely operated vehicles (ROVs). Oriented samples from a wireline rockdrill (Allerton and Tivey, 2001) further expand the range of seafloor outcrops that can be sampled to address tectonic problems. Indirect information on the remanence direction can also be obtained from borehole magnetometer measurements with gyro-oriented logging (Bosum and Scott, 1988). Such borehole measurements have substantial advantages in characterizing the magnetic source region, since they can provide representative magnetization values by avoiding drilling-induced remanence and directly account for void and porosity effects on assessments of discrete sample values.

Our understanding of the neovolcanic zone at fast-spreading ridges, especially the EPR, has evolved and become more sophisticated in recent years as higher-resolution observations, such as with ABE, became available. Placing accurate age constraints to determine eruptive recurrence intervals on near-axis lava flows have become increasingly important given the structural and volcanic complexity of the neovolcanic zone. In this regard, geomagnetic paleointensity of submarine basaltic glass (Pick and Tauxe, 1993) holds particular promise for placing quantitative age constraints on near-axis flows. For example, multi-specimen Thellier paleointensity results from four independent samples distributed over several kilometers from a single (Animal Farm) flow on the EPR at ~18° S were found to be in excellent mutual agreement (35.6 ± 1.0 μT), and by comparison to a geomagnetic reference curve projected to the site location, the Animal Farm paleointensity value indicated a fairly recent time of eruption (AD 1910 ± 20) in general agreement with other observations (Carlut and Kent, 2000) (**Figure 37**). An integrated bathymetric, geochemical, and paleointensity study of adjacent and contrasting ridge segments at ~15° N on the EPR (Carlut *et al.*, 2004) and one at 9–10° N that involved the analysis of what may be the largest published dataset of absolute (Thellier) paleointensity

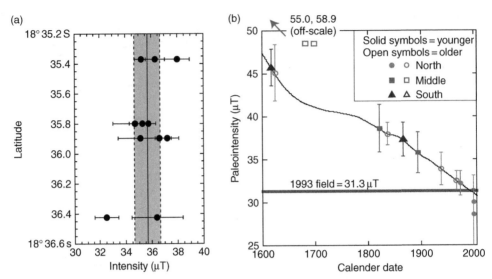

Figure 37 Absolute paleointensity results from submarine basaltic glass illustrating the potential as a dating tool for young mid-ocean ridge lavas. (a) Paleointensity results from multiple samples within a single flow (mapped by submersible) highlight the degree of reproducibility (~1 μT) possible. (b) Glass paleointensity results from the southern East Pacific Rise compared to global field model of Jackson et al. (2000) that illustrate how paleointensity data can be used to infer eruption ages. The ages inferred from this comparison are consistent with all geological data and radiometric dates from these flows. (a) Modified from Carlut J, Cormier M-H, Kent DV, Donnelly KE, and Langmuir CH (2004) Timing of volcanism along the northern East Pacific Rise based on paleointensity experiments on basaltic glasses. *Journal of Geophysical Research* 109: B04104 doi:10.1029/ 2003JB002672. (b) Modified from Bowles J, Gee JS, Kent DV, Bergmanis E, and Sinton J (2005) Cooling rate effects on paleointensity estimates in submarine basaltic glass and implications for dating young flows. *Geochemistry Geophysics Geosystems* 6: Q07002 doi:1029/2004GC000900.

determinations (551 accepted of 991 basalt glass samples analyzed; Bowles *et al.*, 2006), further illustrate the potential for dating lavas and more generally, for discriminating whether eruptions were synchronous. A cooling rate bias in Thellier paleointensities was shown not to be as important as initially suspected in rapidly cooled submarine basaltic glasses (Bowles *et al.*, 2005). However, terrain effects may be an important source of uncertainty in absolute plaeointensity values in some settings, such as the Juan de Fuca Ridge where there are large amplitude magnetic anomalies (Carlut and Kent, 2000). Ultimately, the reliability of paleointensities for dating depends on the reference field model, which could be improved considerably with higher quality (and better dated) Thellier data on archeological and geologic materials. This would also allow better calibration of sedimentary relative paleointensity records, which are used for making comparisons over time intervals beyond the radiocarbon-dated record (e.g., Gee *et al.*, 2000).

Marine magnetic surveys have traditionally used total field sensors that rely on well-established physical constants (e.g. the proton gyromagnetic ratio for proton precession magnetometer) and thus provide accurate (~1nT) field intensity data with negligible

drift. The utility of such total field measurements, however, is limited in certain geometries (N\S spreading ridges near the magnetic equator generate very low amplitude total field anomalies) and external field variations, particularly at low magnetic latitudes, may further inhibit the identification of spreading-related anomalies. The use of total field magnetic gradiometers, two sensors typically deployed with a horizontal separation of >100 m, allow recognition and removal of a significant portion of the external field variations (e.g., Roeser *et al.*, 2002). With the development of high resolution (~0.01 nT) total field sensors, this technique shows considerable promise for facilitating the recognition of low-amplitude anomalies at equatorial latitudes. Measurement of the vector components of the anomalous field also provides two distinct advantages that may aid in recognition of lineated (geomagnetic) anomalies. Perhaps the most useful attribute of vector anomaly data is the ability to characterize, with a single profile, the degree to which the magnetic source is two dimensional (Blakely *et al.*, 1973; Parker and O'Brien, 1997). Aeromagnetic vector profiles, where the sensor is typically mounted on a gyro-stabilized platform, have been used to estimate the spreading

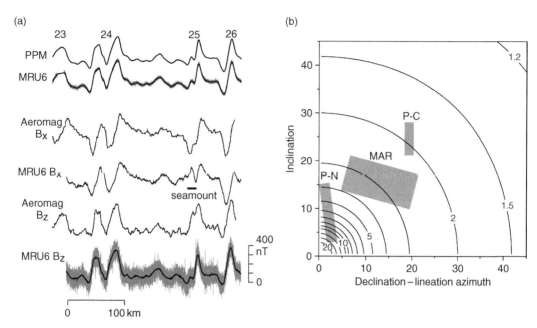

Figure 38 Anomaly data from a surface towed vector magnetometer and plot of amplitude ratio of vector components relative to total field anomalies (Gee and Cande, 2002). (a) Comparison of vector anomalies measured by aircraft and with towed vector magnetometer. Upper panel shows comparison of total field anomaly from proton precession magnetometer (PPM) and vector magnetometer (MRU6). Bz and Bx are the vertical and along track vector anomalies. (b) Ratio of amplitudes of vector components and total field anomaly (contours) for a perfectly two-dimensional source. This ratio depends on the angular difference between the spreading lineation and the ambient field declination, as well as the remanent inclination. The boxes delineate the range of these parameters for the equatorial Pacific-Nazca (P-N), Pacific-Cocos (P-C) and Mid-Atlantic Ridge (MAR) ridge systems.

lineation direction and to map low-amplitude anomalies in the equatorial Pacific (Horner-Johnson and Gordon, 2003; Parker and O'Brien, 1997). Shipboard three axis magnetometers have also been used to determine the location and azimuth of magnetization contrasts in several areas (e.g., Korenaga, 1995; Seama et al., 1993; Yamamoto et al., 2005). Towed vector magnetometer systems can effectively eliminate the ship effect and resolve vector anomalies on the order of 30–50 nT (**Figure 38**; Gee and Cande, 2002).

Although many high-resolution magnetic studies are conducted near the ridge crest, emerging technologies may also allow documentation of the seafloor geomagnetic signal throughout the ocean basins. A relevant new platform for vector and total field magnetic instruments would be on unmanned aerial vehicles (UAVs) for surveying in the marine environment. Deployments of UAVs from oceanographic research vessels would allow multifold increases in magnetic coverage at spatial resolutions comparable to sea-surface data. High-density magnetic data are necessary to quantitatively evaluate the relative contributions from a lineated or coherent geomagnetic source and nonlineated heterogeneities

in the recording medium originating from crustal accretionary processes. Of particular interest are regions like the ~25% of the ocean floor that is encompassed by the KQZ (~84–122 Ma), which remain virtually uncharted in terms of fundamental properties such as spreading history and the possible presence of lineations that might be related to geomagnetic variations. There are also virtually no deep crustal drill holes in the KQZ to document the magnetization of oceanic crust that formed in this unusual time interval of no geomagnetic reversals. Tools are thus available that should allow us to significantly increase our understanding of the source of oceanic magnetic anomalies and geomagnetic variations over timescales ranging from 10^3 to 10^8 years.

Acknowledgments

The authors gratefully acknowledge the support of various NSF grants over more than a decade that allowed much of this work to be completed. Lamont-Doherty Earth Observatory contribution 6992.

References

Acton GD, Petronotis KE, Cape CD, Ilg SR, Gordon RG, and Bryan PC (1996) A test of the geocentric axial dipole hypothesis from an analysis of the skewness of the central marine magnetic anomaly. *Earth and Planetary Science Letters* 144: 337–346.

Agrinier P and Cannat M (1997) Oxygen-isotope constraints on serpentinization processes in ultramafic rocks from the Mid-Atlantic Ridge (23° N). *Proceedings of the Ocean Drilling Program, Scientific Results* 153: 381–388.

Allerton S and Tivey MA (2001) Magnetic polarity structure of the lower oceanic crust. *Geophysical Research Letters* 28: 423–426.

Allerton S, Worm HU, and Stokking LB (1996) Paleomagnetic and rock magnetic properties of Hole 896A. *Proceedings of the Ocean Drilling Program, Scientific Results* 148: 217–226.

Anonymous (1979) Magnetostratigraphic polarity units–A supplementary chapter of the ISSC International Stratigraphic Code. *Geology* 7: 578–583.

Atwater T and Mudie JD (1973) Detailed near-bottom geophysical study of the Gorda Rise. *Journal of Geophysical Research* 78: 8665–8686.

Banerjee SK (1980) Magnetism of the oceanic crust: Evidence from ophiolite complexes. *Journal of Geophysical Research* 85: 3557–3566.

Bazin S, Harding AJ, Kent GM, *et al.* (2001) Three-dimensional shallow crustal emplacement at the 9 degrees 03 N overlapping spreading center on the East Pacific Rise: Correlations between magnetization and tomographic images. *Journal of Geophysical Research* 106: 16101–16117.

Berggren WA, Kent DV, Swisher CC, and Aubry MP (1995) A revised Cenozoic geochronology and chronostratigraphy. In: Berggren WA, Kent DV, Aubry MP, and Hardenbol J (eds.) *Geochronology Time Scales and Global Stratigraphic Correlations*, SEPM Special publication 54, pp. 129–212. Tulsa (OK): SEPM.

Beske-Diehl SJ (1990) Magnetization during low-temperature oxidation of sea-floor basalts – no largescale chemical remagnetization. *Journal of Geophysical Research* 95: 21413–21432.

Billups K, Palike H, Channell JET, Zachos JC, and Shackleton NJ (2004) Astronomic calibration of the Late Oligocene through Early Miocene geomagnetic polarity time scales. *Earth and Planetary Science Letters* 224: 33–44.

Blackman DK, Ildefonse B, John BE, Ohara Y, Miller DJ, and MacLeod CJ (2006) Oceanic core complex formation, Atlantis Massif, *Proceedings of the IODP-MI*, Vol. 304/305. Integrated Ocean Drilling Program Management International, Inc., for the Integrated Ocean Drilling Program. http://iodp.tamu.edu/publications/exp304_305/30405title.htm (accessed on Jan 2007).

Blakely RJ (1974) Geomagnetic reversals and crustal spreading rates during the Miocene. *Journal of Geophysical Research* 79: 2979–2985.

Blakely RJ (1976) An age dependent, two-layer model for marine magnetic anomalies. American *Geophysical Union Monographs* 19: 227–234.

Blakely RJ, Cox A, and Lufer EJ (1973) Vector magnetic data for detecting short polarity intervals in marine magnetic profiles. *Journal of Geophysical Research* 78: 6977–6983.

Bleil U and Petersen N (1983) Variations in magnetization intensity and low-temperature titanomagnetite oxidation of ocean floor basalts. *Nature* 301: 384–388.

Bohnenstiehl DR and Carbotte SM (2001) Faulting patterns near 19° 30' S on the East Pacific Rise: Fault formation and growth at a superfast spreading center. *Geochemistry Geophysics Geosystems* 2(9): doi:10.1029/2001GC000156.

Bosum W and Scott JH (1988) Interpretation of magnetic logs in basalt, Hole 418A: *Proceedings of the Ocean Drilling Program, Scientific Results* 102: 77–95.

Bowers NE, Cande SC, Gee JS, Hildebrand JA, and Parker RL (2001) Fluctuations of the paleomagnetic field during chron C5 as recorded in near-bottom marine magnetic anomaly data. *Journal of Geophysical Research* 106: 26379–26396.

Bowles J, Gee JS, Kent DV, Bergmanis E, and Sinton J (2005) Cooling rate effects on paleointensity estimates in submarine basaltic glass and implications for dating young flows. *Geochemistry Geophysics Geosystems* 6: Q07002 (doi:1029/2004GC000900).

Bowles J, Gee JS, Kent DV, Perfit M, Soule A, and Fornari D (2006) Paleointensity results from 9°–10°N on the East Pacific Rise: Implications for timing and extent of eruptive activity. *Geochemistry Geophysics Geosystems* 7: Q06006 (doi:1029/2005GC001141).

Bowles J, Tauxe L, Gee J, McMillan D, and Cande SC (2003) The source of tiny wiggles in chron C5: A comparison of sedimentary relative intensity and marine magnetic anomalies. *Geochemistry Geophysics Geosystems* 4(6): 1049 (doi:10.1029/2002GC000489).

Cande SC (1976) A paleomagnetic pole from Late Cretaceous marine magnetic anomalies in the Pacific. *Geophysical Journal of the Royal Astronomical Society* 44: 547–566.

Cande SC (1978) Anomalous behavior of the paleomagnetic field inferred from the skewness of Anomalies 33 and 34. *Earth and Planetary Science Letters* 40: 275–286.

Cande SC and Gee JS (2001) Evidence for the rapid low-temperature alteration of oeceanic crust on the Pacific–Antarctic Ridge from the amplitude envelope of magnetic anomalies. *Eos, Transactions American Geophysical Union* 82: F348.

Cande SC and Kent DV (1976) Constraints imposed by the shape of marine magnetic anomalies on the magnetic source. *Journal of Geophysical Research* 81: 4157–4162.

Cande SC and Kent DV (1985) Comment on "Tectonic rotations in extensional regimes and their paleomagnetic consequences for ocean basalts" by K.L. Verosub and E.M. Moores. *Journal of Geophysical Research* 90(B6): 4647–4651.

Cande SC and Kent DV (1992a) A new geomagnetic polarity time scale for the Late Cretaceous and Cenozoic. *Journal of Geophysical Research* 97: 13917–13951.

Cande SC and Kent DV (1992b) Ultrahigh resolution marine magnetic anomaly profiles: A record of continuous paleointensity variations. *Journal of Geophysical Research* 97: 15075–15083.

Cande SC and Kent DV (1995) Revised calibration of the geomagnetic polarity time scale for the Late Cretaceous and Cenozoic. *Journal of Geophysical Research* 100: 6093–6095.

Cande SC and LaBrecque JL (1974) Behaviour of the Earth's paleomagnetic field from small scale marine magnetic anomalies. *Nature* 247: 26–28.

Cande SC, Larson RL, and LaBrecque JL (1978) Magnetic lineations in the Pacific Jurassic Quiet Zone. *Earth and Planetary Science Letters* 41: 434–440.

Cannat M (1993) Emplacement of mantle rocks in the seafloor at mid-ocean ridges. *Journal of Geophysical Research* 98: 4163–4172.

Cannat M, Karson JA, and Miller DJ (1995) *Proceedings of the Ocean Drilling Program, Initial Reports* 153: 798 (College Station, TX, Ocean Drilling Program).

Cannat M, Lagabrielle Y, Bougault H, *et al.* (1997) Ultramafic and gabbroic exposures at the Mid-Atlantic Ridge: Geological mapping in the 15° N region. *Tectonophysics* 279: 193–213.

Carbotte SM and Macdonald KC (1990) Causes of variation in fault-facing direction on the ocean floor. *Geology* 18: 749–752.

Carlut J, Cormier M-H, Kent DV, Donnelly KE, and Langmuir CH (2004) Timing of volcanism along the northern East Pacific Rise based on paleointensity experiments on basaltic glasses. *Journal of Geophysical Research* 109: B04104 (doi:10.1029/2003JB002672).

Carlut J and Kent DV (2000) Paleointensity record in zero-age submarine basalt glasses: Testing a new dating technique for recent MORBs. *Earth and Planetary Science Letters* 183: 389–401.

Channell JET, Erba E, Nakanishi M, and Tamaki K (1995) Late Jurassic–Early Cretaceous time scales and oceanic magnetic anomaly block models. In: Berggren WA, Kent DV, Aubry M-P, and Hardenbol J (eds.) *Geochronology, Time Scales and Global Stratigraphic Correlations*, SEPM Special publication 54, pp. 51–63. Tulsa (OK): SEPM.

Channell JET, Galeotti S, Martin EE, Billups K, Scher HD, and Stoner JS (2003) Eocene to Miocene magnetostratigraphy, biostratigraphy, and chemostratigraphy at ODP Site 1090 (sub-Antarctic South Atlantic). *Geological Society of America Bulletin* 115: 607–623.

Channell JET, Lowrie W, Pialli P, and Venturi F (1984) Jurassic magnetic stratigraphy from Umbrian (Italian) land sections. *Earth and Planetary Science Letters* 68: 309–325.

Christeson GL, Purdy GM, and Fryer GJ (1992) Structure of young upper crust at the East Pacific Rise near 9° 30' N. *Geophysical Research Letters* 19: 1045–1048.

Clement BM (2004) Dependence of the duration of geomagnetic polarity reversals on site latitude. *Nature* 428: 637–640.

Cogné JP and Humler E (2006) Trends and rhythms in global seafloor generation rate. *Geochemistry Geophysics Geosystems* 7: 17.

Cogné JP, Francheteau J, Courtillot V, *et al.* (1995) Large rotation of the Easter microplate as evidenced by oriented paleomagnetic samples from the ocean floor. *Earth and Planetary Science Letters* 136: 213–222.

Cox A (1968) Lengths of geomagnetic polarity intervals. *Journal of Geophysical Research* 73: 3247–3260.

Cox A, Doell RR, and Dalrymple GB (1964) Reversals of the Earth's magnetic field. *Science* 144: 1537–1543.

Cronin M, Tauxe L, Constable C, Selkin P, and Pick T (2001) Noise in the quiet zone. *Earth and Planetary Science Letters* 190: 13–30.

Day R, Osterhoudt M, and Bleil U (1983) Rock magnetism of igneous rocks from Deep Sea Drilling Project Sites 482, 483, and 485. *Initial Reports of the Deep Sea Drilling Project* 65: 727–736.

Detrick R, Honnorez J, Bryan WB, and Juteau T (1988) *Proceedings of the Ocean Drilling Program, Initial Reports* vol. 106/109: 249 (College Station, TX: Ocean Drilling Program).

Detrick RS, Harding AJ, Kent GM, Orcutt JA, Mutter JC, and Buhl P (1993) Seismic structure of the southern East Pacific Rise. *Science* 259: 499–503.

Dick HJB, Natland JH, Alt JC, *et al.* (2000) A long *in situ* section of the lower ocean crust: Results of ODP Leg 176 drilling at the Southwest Indian Ridge. *Earth and Planetary Science Letters* 179: 31–51.

Dick HJB, Natland JH, and Miller DJ (1999) *Proceedings of the Ocean Drilling Program, Initial Reports* 176, College Station, TX: Ocean Drilling Program.

Dick HJB, Schouten H, Meyer PS, *et al.* (1991) Tectonic evolution of the Atlantis II Fracture Zone. *Proceedings of the Ocean Drilling Program, Scientific Results* 118: 359–398.

Doubrovine PV and Tarduno JA (2004) Self-reversed magnetization carried by titanomaghemite in oceanic basalts. *Earth and Planetary Science Letters* 222: 959–969.

Dyment J and Arkani-Hamed J (1995) Spreading-rate-dependent magnetization of the oceanic lithosphere inferred from the anomalous skewness of marine magnetic anomalies. *Geophysical Journal International* 121: 789–804.

Dyment J, Arkani-Hamed J, and Ghods A (1997) Contribution of serpentinized ultramafics to marine magnetic anomalies at slow and intermediate spreading centers: Insights from the shape of the anomalies. *Geophysical Journal International* 129: 691–701.

Evans HF and Channell JET (2003) Upper Miocene magnetic stratigraphy at ODP Site 1092 (sub-Antarctic South Atlantic): Recognition of 'cryptochrons' in C5n.2n. *Geophysical Journal International* 153: 483–406.

Evans HF, Westerhold T, and Channell JET (2004) ODP Site 1092: Revised composite depth section has implications for Upper Miocene 'cryptochrons'. *Geophysical Journal International* 156: 195–199.

Facey D, Housden J, and O'Reilly W (1985) A magneto-petrological study of rocks from Leg 83, Hole 504B, Deep Sea Drilling Project. *Initial Reports of the Deep Sea Drilling Project* 83: 339–346.

Fornari DJ, Haymon RM, Perfit MR, Gregg TKP, and Edwards MH (1998) Axial summit trough of the East Pacific Rise 9°–10° N: Geological characteristics and evolution of the axial zone on fast spreading mid-ocean ridges. *Journal of Geophysical Research* 103: 9827–9855.

Francheteau J, Armijo R, Cheminee JL, Hekinian R, Lonsdale P, and Blum N (1992) Dyke Complex of the East Pacific Rise Exposed in the Walls of Hess Deep and the Structure of the Upper Oceanic-Crust. *Earth and Planetary Science Letters* 111: 109–121.

Früh-Green GL, Plas A, and Lecuyer C (1996) Petrologic and stable isotope constraints on hydrothermal alteration and serpentinization of the EPR shallow mantle at Hess Deep (Site 895). *Proceedings of the Ocean Drilling Program, Scientific Results* 147: 255–288.

Furuta T (1983) Magnetic properties of basalt samples from Holes 504B and 505B on the Costa Rica Rift, Deep Sea Drilling Project Legs 69 and 70. *Initial Reports of the Deep Sea Drilling Project* 69: 711–720.

Furuta T (1993) Magnetic properties and ferromagnetic mineralogy of oceanic basalts. *Geophysical Journal International* 113: 95–114.

Furuta T and Levi S (1983) Basement paleomagnetism of Hole 504B. *Initial Reports of the Deep Sea Drilling Project* 69: 697–703.

Gallet Y and Hulot G (1997) Stationary and nonstationary behaviour within the geomagnetic polarity time scale. *Geophysical Research Letters* 24: 1875–1878.

Garces M and Gee JS (2007) Paleomagnetic evidence of large footwall rotations associated with low-angle faults at the Mid-Atlantic Ridge. *Geology* 35: 279–282.

Gee J and Kent DV (1994) Variations in layer 2A thickness and the origin of the central anomaly magnetic high. *Geophysical Research Letters* 21: 297–300.

Gee J and Kent DV (1997) Magnetization of axial lavas from the southern East Pacific Rise (14°–23° S): Geochemical controls on magnetic properties. *Journal of Geophysical Research* 102: 24,873–24,886.

Gee J and Kent DV (1998) Magnetic telechemistry and magmatic segmentation on the southern East Pacific Rise. *Earth and Planetary Science Letters* 164: 379–385.

Gee J and Kent DV (1999) Calibration of magnetic granulometric trends in oceanic basalts. *Earth and Planetary Science Letters* 170: 377–390.

Gee J and Meurer WP (2002) Slow cooling of middle and lower oceanic crust inferred from multicomponent magnetizations of gabbroic rocks from the Mid-Atlantic Ridge south of the

Kane fracture zone (MARK) area. *Journal of Geophysical Research* 107: 3-1/3–18.

Gee J, Natland JH, Hurst SD, and Nilsson K (1992) Magnetic properties of ocean crust samples from Hess Deep: Implications for marine magnetic anomalies. *EOS, Transactions American Geophysical Union* 73: 490.

Gee J, Schneider DA, and Kent DV (1996) Marine magnetic anomalies as recorders of geomagnetic intensity variations. *Earth and Planetary Science Letters* 144: 327–335.

Gee JS and Cande SC (2002) A surface-towed vector magnetometer. *Geophysical Research Letters* 29(14): doi: 10.1029/2002GL015245.

Gee JS, Cande SC, Hildebrand JA, Donnelly K, and Parker RL (2000) Geomagnetic intensity variations over the past 780 kyr obtained from near-seafloor magnetic anomalies. *Nature* 408: 827–832.

Gee JS, Lawrence RM, and Hurst SD (1997) Remanence characteristics of gabbros from the MARK area: Implications for crustal magnetization. *Proceedings of the Ocean Drilling Program, Scientific Results* 153: 429–436.

Gillis K, Mevel C, and Allan J (1993) *Proceedings of the Ocean Drilling Program, Initial Reports* 147: 366 (College Station, TX, Ocean Drilling Program).

Gradstein FM, Ogg JG, and Smith AG (2004) *A Geologic Time Scale 2004*, 589 pp. Cambridge: Cambridge University Press.

Grommé CS, Mankinen EA, Marshall M, and Coe RS (1979) Geomagnetic paleointensities by the Thelliers' method from submarine pillow basalts: Effects of seafloor weathering. *Journal of Geophysical Research* 84: 3553–3575.

Guyodo Y and Valet J-P (1999) Global changes in intensity of the Earth's magnetic field during the past 800 kyr. *Nature* 399: 249–252.

Hall JM and Muzzatti A (1999) Delayed magnetization of the deeper kilometer of oceanic crust at Ocean Drilling Project Site 504. *Journal of Geophysical Research* 104: 12843–12851.

Handschumacher DW, Sager WW, Hilde TWC, and Bracey DR (1988) Pre-Cretaceous tectonic evolution of the Pacific plate and extension of the geomagnetic polarity reversal time scale with implications for the origin of the Jurassic 'Quiet Zone'. *Tectonophysics* 155: 365–380.

Harding AJ, Kent GM, and Orcutt JA (1993) A multichannel seismic investigation of upper crustal structure at 9° N on the East Pacific Rise: Implications for crustal accretion. *Journal of Geophysical Research* 98: 13925–13944.

Harrison CGA (1987) Marine magnetic anomalies–the origin of the stripes. *Annual Review of Earth and Planetary Science Letters* 15: 505–543.

Hartl P, Tauxe L, and Constable C (1993) Early Oligocene geomagnetic field behavior from Deep Sea Drilling Project Site 522. *Journal of Geophysical Research* 98: 19649–19665.

Heirtzler JR, Dickson GO, Herron EM, Pitman WC, and Le Pichon X (1968) Marine magnetic anomalies, geomagnetic field reversals, and motions of the ocean floor and continents. *Journal of Geophysical Research* 73: 2119–2136.

Helsley CE and Steiner MC (1969) Evidence for long intervals of normal polarity during the Cretaceous period. *Earth and Planetary Science Letters* 5: 325–332.

Hilgen FJ (1991) Extension of the astronomically calibrated (polarity) time scale to the Miocene/Pliocene boundary. *Earth and Planetary Science Letters* 107: 349–368.

Hooft EEE, Schouten H, and Detrick RS (1996) Constraining crustal emplacement processes from the variation in seismic layer 2A thickness at the East Pacific Rise. *Earth and Planetary Science Letters* 142: 289–309.

Horner-Johnson BC and Gordon RG (2003) Equatorial Pacific magnetic anomalies identified from vector aeromagnetic data. *Geophysical Journal International* 155: 547–556.

Hurst SD, Karson JA, and Verosub KL (1994) Paleomagnetism of tilted dikes in fast spread oceanic crust exposed in the Hess Deep Rift: Implications for spreading and rift propagation. *Tectonics* 13: 789–802.

Hussenoeder SA, Tivey MA, Schouten H, and Searle RC (1996) Near-bottom magnetic survey of the Mid-Atlantic Ridge axis, 24°–24° 40′ N: Implications for crustal accretion at slow spreading ridges. *Journal of Geophysical Research* 10: 22051–22069.

Irving E (1970) The mid-Atlantic Ridge at 45° N. XIV. Oxidation and magnetic properties of basalt; review and discussion. *Canadian Journal of Earth Sciences* 7: 1528–1538.

Irving E and Pullaiah G (1976) Reversals of the geomagnetic field, magnetostratigraphy, and relative magnitude of paleosecular variation in the Phanerozoic. *Earth Science Reviews* 12: 35–64.

Jackson A, Jonkers ART, and Walker MR (2000) Four centuries of geomagnetic secular variation from historical records. *Philosophical Transactions of the Royal Society London A* 358: 957–990.

John BE, Foster DA, Murphy JM, *et al.* (2004) Determining the cooling history of insitu lower oceanic crust – Atlantis Bank, SW Indian Ridge. *Earth and Planetary Science Letters* 222: 145–160.

Johnson HP (1978) Rock magnetic properties of igneous rock samples – Leg 45. *Initial Reports of the Deep Sea Drilling Project* 45: 397–406.

Johnson HP and Atwater T (1977) Magnetic study of basalts from the Mid-Atlantic Ridge, lat 37° N. *Geological Society of America Bulletin* 88: 637–647.

Johnson HP, Kent DV, Tivey MA, Gee JS, Larson RL, and Embley RW (1997) Conference on the Magnetization of the Oceanic Crust Steers Future Research. *EOS, Transactions, American Geophysical Union* 78: 199–200.

Johnson HP and Pariso JE (1993) Variations in oceanic crustal magnetization: Systematic changes in the last 160 million years. *Journal of Geophysical Research* 98: 435–445.

Johnson HP and Salem BL (1994) Magnetic properties of dikes from the oceanic upper crustal section. *Journal of Geophysical Research* 99: 21733–21740.

Johnson HP and Tivey MA (1995) Magnetic properties of zero-age oceanic crust; a new submarine lava flow on the Juan de Fuca ridge. *Geophysical Research Letters* 22: 175–178.

Juarez MT, Osete ML, Melendez G, Langereis CG, and Zijderveld JDA (1994) Oxfordian magnetostratigraphy of the Aguilon and Tosos sections (Iberian Range, Spain) and evidence of a pre-Oligocene overprint. *Earth and Planetary Science Letters* 85: 195–211.

Juarez MT, Tauxe L, Gee JS, and Pick T (1998) The intensity of the Earth's magnetic field over the past 160 million years. *Nature* 394: 878–881.

Juster TC, Grove TL, and Perfit MR (1989) Experimental constraints on the generation of FeTi basalts, andesites, and rhyodacites at the Galapagos spreading center, 85° W and 95° W. *Journal of Geophysical Research* 94: 9251–9274.

Karson JA, Klein EM, Hurst SD, *et al.* (2002) Structure of uppermost fast-spread oceanic crust exposed at the Hess Deep Rift: Implications for subaxial processes at the East Pacific Rise. *Geochemistry Geophysics Geosystems* 3(1): 4 (doi:10.129/2001GC000155).

Kelemen PB, Kikawa E, and Miller DJ (2004) *Proceedings of the Ocean Drilling Program, Initial Reports* 209: (College Station, TX, Ocean Drilling Program).

Kelso PR, Banerjee SK, and Worm HU (1991) The effect of low-temperature hydrothermal alteration on the remanent magnetization of synthetic titanomagnetites; a case for acquisition of chemical remanent magnetization. *Journal of Geophysical Research* 96: 19545–19553.

Kent DV and Gee J (1994) Grain size-dependent alteration and the magnetization of oceanic basalts. *Science* 265: 1561–1563.

Kent DV and Gee J (1996) Magnetic alteration of zero-age oceanic basalt. *Geology* 24: 703–706.

Kent DV, Honnorez BM, Opdyke ND, and Fox PJ (1978) Magnetic properties of dredged oceanic gabbros and the source of marine magnetic anomalies. *Geophysical Journal of the Royal Astronomical Society* 55: 513–537.

Kidd RGW (1977) The nature and shape of the sources of marine magnetic anomalies. *Earth and Planetary Science Letters* 33: 310–320.

Kikawa E and Ozawa K (1992) Contribution of oceanic gabbros to sea-floor spreading magnetic anomalies. *Science* 258: 796–799.

Kikawa E and Pariso JE (1991) Magnetic properties of gabbros from Hole 735B, Southwest Indian Ridge. *Proceedings of the Ocean Drilling Program, Scientific Results* 118: 285–307.

Kinoshita H, Furuta T, and Kawahata H (1985) Magnetic properties and alteration in basalt, Hole 504B, Deep Sea Drilling Project Leg 83. *Initial Reports of the Deep Sea Drilling Project* 83: 331–338.

Klitgord KD (1976) Sea-floor spreading: The central anomaly magnetization high. *Earth and Planetary Science Letters* 29: 201–209.

Korenaga J (1995) Comprehensive analysis of marine magnetic vector anomalies. *Journal of Geophysical Research* 100: 365–378.

Korte M and Constable CG (2005) The geomagnetic dipole moment over the last 7000 years – new results from a global model. *Earth and Planetary Science Letters* 236: 348–358.

Krasa D, Shcherbakov VP, Kunzmann T, and Petersen N (2005) Self-reversal of remanent magnetization in basalts due to partially oxidized titanomagnetites. *Geophysical Journal International* 162: 115–136.

Krijgsman W and Kent DV (2004) Non-uniform occurrence of short-term polarity fluctuations in the geomagnetic field? New results from Middle to Late Miocene sediments of the North Atlantic (DSDP Site 608). In: Channell JET, Kent DV, Lowrie W, and Meert J (eds.) *AGU Geophysical Monograph 145: Timescales of the Paleomagnetic Field*, pp. 161–174. Washington, DC: American Geophysical Union.

Kriner KA, Pockalny RA, and Larson RL (2006) Bathymetric gradients of lineated abyssal hills: Inferring seafloor spreading vectors and a new model for hills formed at ultra-fast rates. *Earth and Planetary Science Letters* 242: 98–110.

Lancelot YP, Larson RL, and Fisher A (1990) *Proceedings of the Ocean Drilling Program, Initial Reports* 129: 488 (College Station, TX, Texas A & M University, Ocean Drilling Program).

Lanci L and Lowrie W (1997) Magnetostratigraphic evidence that 'tiny wiggles' in the oceanic magnetic anomaliy record represent geomagnetic paleointensity variations. *Earth and Planetary Science Letters* 148: 581–592.

Lanci L, Pares JP, Channell JET, and Kent DV (2004) Miocene Magnetostratigraphy from Equatorial Pacific sediments (ODP Site 1218, Leg 199). *Earth and Planetary Science Letters* 226: 207–224.

Lanci L, Pares JP, Channell JET, and Kent DV (2005) Oligocene magnetostratigraphy from equatorial Pacific sediments (ODP Sites 1218 and 1219, Leg 199). *Earth and Planetary Science Letters* 237: 617–634.

Larson RL (1977) Early Cretaceous breakup of Gondwanaland off western Australia. *Geology* 5: 57–60.

Larson RL and Chase CG (1972) Late Mesozoic evolution of the western Pacific Ocean. *Geological Society of America Bulletin* 83: 3627–3644.

Larson RL and Hilde TWC (1975) A revised time scale of magnetic reversals for the Early Cretaceous and Late Jurassic. *Journal of Geophysical Research* 80: 2586–2594.

Larson RL and Pitman WC (1972) Worldwide correlation of Mesozoic magnetic anomalies, and its implications. *Geological Society of America Bulletin* 83: 3645–3662.

Larson RL and Spiess FN (1969) The East Pacific Rise crest: A near-bottom geophysical profile. *Science* 163: 68–71.

Lawrence RM, Gee JS, and Karson JA (2002) Magnetic aniso-tropy of serpentinized peridotites from the MARK area: Implications for the orientation of mesoscopic structures and major fault zones. *Journal of Geophysical Research* 107: 14.

Levi S (1980) Paleomagnetism and some magnetic properties of basalts from the Bermuda Triangle. *Initial Reports of the Deep Sea Drilling Project* 51,52,53: 1363–1378.

Levi S, Audunsson H, Duncan RA, Kristjansson L, Gillot P-Y, and Jakobsson SP (1990) Late Pleistocene geomagnetic excursion in Icelandic lavas: Confirmation of the Laschamp excursion. *Earth and Planetary Science Letters* 96: 443–457.

Lewis BTR, Snydsman WE, McClain JS, Holmes ML, and Lister CRB (1983) Site survey results at the mouth of the Gulf of California, Leg 65, Deep Sea Drilling Project: *Initial Reports of the Deep Sea Drilling Project* 65: 309–317.

Lourens LJ, Hilgen FJ, Shackleton NJ, Laskar J, and Wilson D (2004) The Neogene period. In: Gradstein FM, Ogg JG, and Smith AG (eds.) *A Geologic Time Scale 2004*, pp. 409–440. Cambridge: Cambridge University Press.

Lowrie W (1977) Intensity and direction of magnetization in oceanic basalts. *Journal of the Geological Society of London* 133: 61–82.

Lowrie W and Alvarez W (1981) One hundred million years of geomagnetic polarity history. *Geology* 9: 392–397.

Lowrie W and Kent DV (2004) Geomagnetic polarity timescales and reversal frequency regimes. In: Channell JET, Kent DV, Lowrie W, and Meert J (eds.) *AGU Geophysical Monograph 145: Timescales of the Paleomagnetic Field*, pp. 117–129. Washington, DC: American Geophysical Union.

Macdonald KC (1977) Near-bottom magnetic anomalies, asymmetric spreading, oblique spreading, and tectonics of the Mid-Atlantic Ridge near lat 37° N. *Geological Society of America Bulletin* 88: 541–555.

Macdonald KC, Miller SP, Luyendyk BP, Atwater T, and Shure L (1983) Investigation of a Vine–Matthews magnetic lineation from a submersible: The source and character of marine magnetic anomalies. *Journal of Geophysical Research* 88: 3403–3418.

Marshall M and Cox A (1971) Magnetism of pillow basalts and their petrology. *Geological Society of America Bulletin* 82: 537–552.

Marshall M and Cox A (1972) Magnetic changes in pillow basalt due to sea-floor weathering. *Journal of Geophysical Research* 77: 6459–6469.

Matzka J, Krasa D, Kunzmann T, Schult A, and Petersen N (2003) Magnetic state of 10–40 Ma old ocean basalts and its implications for natural remanent magnetization. *Earth and Planetary Science Letters* 206: 541–553.

McElhinny MW and Larson RL (2003) Jurassic dipole low defined from land and sea data. *EOS, Transactions of the American Geophysical Union* 84: 362–366.

McFadden PL and Merrill RT (2000) Evolution of the geomag-netic reversal rate since 160 Ma: Is the process continuous?. *Journal of Geophysical Research* 105: 28455–28460.

McFadden PL, Merrill RT, and McElhinny MW (1988) Dipole/quadrupole family modeling of paleosecular variation. *Journal of Geophysical Research* 93: 11583–11588.

Mejia V, Opdyke ND, and Perfit MR (1996) Paleomagnetic field intensity recorded in submarine basaltic glass from the East Pacific Rise, the last 69 KA. *Geophysical Research Letters* 23: 475–478.

Merrill RT, McElhinny MW, and McFadden PL (1996) The Magnetic Field of the Earth: Paleomagnetism, the Core, and the Deep Mantle, 531 pp. San Diego: Academic Press.

Miller DJ and Christensen NI (1997) Seismic velocities of lower crustal and upper mantle rocks from the slow-spreading Mid-Atlantic Ridge, south of the Kane Transform Zone (MARK). *Proceedings of the Ocean Drilling Program, Scientific Results* 153: 437–454.

Nakanishi M, Tamaki K, and Kobayashi K (1989) Mesozoic magnetic anomaly lineations and seafloor spreading history of the Northwestern Pacific. *Journal of Geophysical Research* 94: 15437–15462.

Nakanishi M, Tamaki K, and Kobayashi K (1992) A new Mesozoic isochron chart of the Northwestern Pacific Ocean: Paleomagnetic and tectonic implications. *Geophysical Research Letters* 19: 693–696.

Nakanishi M and Winterer EL (1998) Tectonic history of the Pacific–Farallon–Phoenix triple junction from Late Jurassic to Early Cretaceous: An abandoned Mesozoic spreading system in the Central Pacific Basin. *Journal of Geophysical Research* 103: 12453–12468.

Natland JH (2002) Magnetic susceptibility as an index of the lithology and composition of gabbros, ODP Leg 176, Hole 735B, Southwest Indian Ridge. *Proceedings of the Ocean Drilling Program, Scientific Results* 176.

Nazarova KA (1994) Serpentinized peridotites as a possible source for oceanic magnetic anomalies. *Marine Geophysical Researches* 16: 455–462.

Nazarova KA, Wasilewski PJ, and Dick HJB (2000) Magnetic study of serpentinized harsburgites from the Islas Orcadas Fracture Zone. *Marine Geophysical Researches* 21: 475–488.

O'Reilly W (1984) *Rock and Mineral Magnetism*, 220 pp. New York: Chapman and Hall.

Ogg JG and Lowrie W (1986) Magnetostratigraphy of the Jurassic/Cretaceous boundary. *Geology* 14: 547–550.

Oufi O, Cannat M, and Horen H (2002) Magnetic properties of variably serpentinized abyssal peridotites. *Journal of Geophysical Research* 107(B5): doi:10.1029/2001JB000549.

Oxburgh ER and Turcotte DL (1969) Increased estimate for heat flow at oceanic ridges. *Nature* 223: 1354–1355.

Özdemir O and O'Reilly W (1981) High-temperature hysteresis and other magnetic properties of synthetic monodomain titanomagnetites. *Physics of the Earth and Planetary Interiors* 25: 406–418.

Pariso JE and Johnson HP (1989) Magnetic properties and oxide petrography of the sheeted dike complex in Hole 504B. *Proceedings of the Ocean Drilling Program, Scientific Results* 111: 159–167.

Pariso JE and Johnson HP (1991) Alteration processes at Deep Sea Drilling Project Ocean Drilling Program Hole 504B at the Costa Rica Rift – Implications for magnetization of oceanic crust. *Journal of Geophysical Research* 96: 11703–11722.

Pariso JE and Johnson HP (1993a) Do Layer 3 rocks make a significant contribution to marine magnetic anomalies? In situ magnetization of gabbros at Ocean Drilling Program Hole 735B. *Journal of Geophysical Research* 98: 16033–16052.

Pariso JE and Johnson HP (1993b) Do lower crustal rocks record reversals of the Earth's magnetic field? Magnetic petrology of oceanic gabbros from Ocean Drilling Program Hole 735B. *Journal of Geophysical Research* 98: 16013–16032.

Pariso JE, Kelso P, and Richter C (1996a) Paleomagnetism and rock magnetic properties of gabbro from Hole 894G, Hess Deep. *Proceedings of the Ocean Drilling Program, Scientific Results* 147: 373–381.

Pariso JE, Rommevaux C, and Sempere J-C (1996b) Three-dimensional inversion of marine magnetic anomalies: Implications for crustal accretion along the Mid-Atlantic Ridge (28°–31° 30′ N). *Marine Geophysical Researches* 18: 85–101.

Pariso JE, Stokking LB, and Allerton S (1995) Rock magnetism and magnetic mineralogy of a 1-km section of sheeted dikes, Hole 504B. *Proceedings of the Ocean Drilling Program. Scientific Results* 137/140: 253–262.

Parker RL and O'Brien MS (1997) Spectral analysis of vector magnetic field profiles. *Journal of Geophysical Research* 102: 24815–24824.

Pechersky DM, Tikhonov LV, and Pertsev NN (1983a) Magnetic properties of basalts, Deep Sea Drilling Project Legs 69 and 70. *Initial Reports of the Deep Sea Drilling Project* 69: 705–710.

Pechersky DM, Tikhonov LV, and Zolotarev BP (1983b) Rock magnetism and paleomagnetism of basalts drilled during Deep Sea Drilling Project Leg 65. *Initial Reports of the Deep Sea Drilling Project* 65: 717–726.

Perfit MR and Chadwick WW (1998) Magmatism at mid-ocean ridges: Constraints from volcanological and geochemical investigations. In: Buck WR (ed.) *Faulting and Magmatism at Mid-Ocean Ridges*, Geophysical Monograph Series 106, pp. 59–115. Washington, DC: AGU.

Perfit MR, Fornari DJ, Smith MC, Bender JF, Langmuir CH, and Haymon RM (1994) Small-scale spatial and temporal variations in mid-ocean ridge crest magmatic processes. *Geology* 22: 375–379.

Perram LJ, Macdonald KC, and Miller SP (1990) Deep-tow magnetics near 20° S on the East Pacific Rise: A study of short wavelength anomalies at a very fast spreading center. *Marine Geophysical Researches* 12: 235–245.

Petronotis KE, Gordon RG, and Acton GD (1992) Determining palaeomagnetic poles and anomalous skewness from marine magnetic anomaly skewness data from a single plate. *Geophysical Journal International* 109: 209–224.

Pettigrew TJ, Casey JF, and Miller DJ (1999) *Proceedings of the Ocean Drilling Program, Initial Reports* 179. (College Station, TX: Ocean Drilling Program).

Pick T and Tauxe L (1993) Holocene paleointensities: Thellier experiments on submarine basaltic glass from the East Pacific Rise. *Journal of Geophysical Research* 98: 17949–17964.

Pick T and Tauxe L (1994) Characteristics of magnetite in submarine basaltic glass. *Geophysical Journal International* 119: 116–128.

Pitman WC and Heirtzler JR (1966) Magnetic anomalies over the Pacific–Antarctic ridge. *Science* 154: 1164–1171.

Pitman WC, Herron EM, and Heirtzler JR (1968) Magnetic anomalies in the Pacific and sea floor spreading. *Journal of Geophysical Research* 73: 2069–2085.

Plank T, Ludden JN, and Escutia C (2000) *Proceedings of the Ocean Drilling Program, Initial Reports* 185: (accessed Jan 2007) http://www-odp.tamu.edu/publications/185 IR/185TOC.HTM. (College Station, TX: Texas A&M University, Ocean Drilling Program).

Pockalny RA, Larson RL, Viso RF, and Abrams LJ (2002) Bathymetry and gravity data across a mid-Cretaceous triple junction in the southwest Pacific basin. *Geophysical Research Letters* 29: 1007.

Pockalny RA, Smith A, and Gente P (1995) Spatial and temporal variability of crustal magnetization of a slowly spreading ridge: Mid-Atlantic Ridge (20°–24° N). *Marine Geophysical Research* 17: 301–320.

Prévot M, Derder ME-M, McWilliams M, and Thompson J (1990) Intensity of the Earth's magnetic field: Evidence for a Mesozoic dipole low. *Earth and Planetary Science Letters* 97: 129–139.

Raymond CA and LaBrecque JL (1987) Magnetization of the oceanic crust: Thermoremanent magnetization or chemical remanent magnetization? *Journal of Geophysical Research* 92: 8077–8088.

Renne PR, Scott GR, Glen JMG, and Feinberg JM (2002) Oriented inclusions of magnetite in clinopyroxene: Source of stable remanent magnetization in gabbros of the Messum Complex, Namibia. *Geochemistry Geophysics Geosystems* 3(12): 1079 (doi:10.1029/2002GC000319).

Roberts AP and Lewin-Harris JC (2000) Marine magnetic anomalies: Evidence that 'tiny wiggles' represent short-period geomagnetic polarity intervals. *Earth and Planetary Science Letters* 183: 375–388.

Robinson PT, Von Herzen RP, and Adamson AC (1989) *Proceedings of the Ocean Drilling Program, Initial Reports* 118: 826 (College Station, TX, Ocean Drilling Program).

Roeser HA, Steiner C, Schreckenberger B, and Block M (2002) Structural development of the Jurassic Magnetic Quiet Zone off Morocco and identification of Middle Jurassic magnetic lineations. *Journal of Geophysical Research* 107(10): doi:10.1029/2000JB000094.

Roest WR, Arkani-Hamed J, and Verhoef J (1992) The seafloor spreading rate dependence of the anomalous skewness of marine magnetic anomalies. *Geophysical Journal International* 109: 653–669.

Roperch P, Bonhommet N, and Levi S (1988) Paleointensity of the earth's magnetic field during the Laschamp excursion and its geomagnetic implications. *Earth and Planetary Science Letters* 88: 209–219.

Ryall PJ and Ade-Hall JM (1975) Radial variation of magnetic properties in submarine pillow basalt. *Canadian Journal of Earth Sciences* 12: 1959–1969.

Sager WW, Weiss MA, Tivey MA, and Johnson HP (1998) Geomagnetic polarity reversal model of deep-tow profiles from the Pacific Jurassic 'Quiet Zone'. *Journal of Geophysical Research* 103: 5269–5286.

Sayanagi K and Tamaki K (1992) Long-term variations in magnetization intensity with crustal age in the Northeast Pacific, Atlantic, and Southeast Indian Oceans. *Geophysical Research Letters* 12: 2369–2372.

Schneider DA (1988) An estimate of the long-term non-dipole field from marine magnetic anomalies. *Geophysical Research Letters* 15: 1105–1108.

Schneider DA (1995) Paleomagnetism of some Leg 138 sediments: Detailing Miocene magnetostratigraphy. *Proceedings of the Ocean Drilling Program, Scientific Results* 138: 59–72.

Schouten H (2002) Paleomagnetic inclinations in DSDP Hole 417D reconsidered: Secular variation or variable tilting? *Geophysical Research Letters* 29(7): doi:10.1029/2001GL013581.

Schouten H and Cande SC (1976) Paleomagnetic poles from marine magnetic anomalies. *Geophysical Journal of the Royal Astronomical Society* 44: 567–575.

Schouten H and Denham CR (2000) Comparison of volcanic construction in the Troodos ophiolite and oceanic crust using paleomagnetic inclinations from Cyprus Crustal Study Project (CCSP) CY-1 and CY-1A and Ocean Drilling Program (ODP) 504B drill cores. In: Dilek Y, Moores EM, Elthon D, and Nicolas A (eds.) *Ophiolites and Oceanic Crust: New Insights from Field Studies and the Ocean Drilling Program*, Geological Society of America Special Paper 349, pp. 181–194. Boulder, CO: Geological Society of America.

Schouten H and McCamy K (1972) Filtering marine magnetic anomalies. *Journal of Geophysical Research* 77: 7089–7099.

Schouten H, Tivey MA, Fornari DJ, and Cochran JR (1999) Central anomaly magnetization high: Constraints on the volcanic construction and architecture of seismic layer 2A at a fast-spreading mid-ocean ridge, the EPR at 9° 30–50 N. *Earth and Planetary Science Letters* 169: 37–50.

Sclater JG and Francheteau J (1970) The implications of terrestrial heat flow observations on current tectonic and geochemical models of the crust and upper mantle of the Earth. *Geophysical Journal of the Royal Astronomical Society* 20: 509–542.

Seama N, Nogi Y, and Isezaki N (1993) A new method for precise determination of the position and strike of magnetic boundaries using vector data of the geomagnetic anomaly field. *Geophysical Journal International* 113: 155–164.

Selkin PA, Gee JS, Tauxe L, Meurer WP, and Newell AJ (2000) The effect of remanence anisotropy on paleointensity estimates: A case study from the Archean Stillwater Complex. *Earth and Planetary Science Letters* 183: 403–416.

Selkin PA and Tauxe L (2000) Long-term variations in paleointensity. *Philosophical Transactions of the Royal Society of London A* 358: 1065–1088.

Sempere J-C, Macdonald KC, Miller SP, and Shure L (1987) Detailed study of the Brunhes/Matuyama reversal boundary on the East Pacific Rise at 19° 30′ S: Implications for crustal emplacement processes at an ultra fast spreading center. *Marine Geophysical Researches* 9: 1–23.

Shackleton NJ, Berger A, and Peltier WR (1990) An alternative astronomical calibration of the lower Pleistocene timescale based on ODP Site 677. *Transactions of the Royal Society of Edinburgh, Earth Sciences* 81: 251–261.

Shackleton NJ, Hall MA, Raffi I, Tauxe L, and Zachos J (2000) Astronomical calibration age for the Oligocene–Miocene boundary. *Geology* 28: 447–450.

Shah AK, Cormier M-H, Ryan WBF, et al. (2003) Episodic dike swarms inferred from near-bottom magnetic anomaly maps at the southern East Pacific Rise. *Journal of Geophysical Research* 108(B2): 2097 (doi:10.1029/2001JB000564).

Shau Y-H, Torii M, Horng C-S, and Peacor DR (2000) Subsolidus evolution and alteration of titanomagnetite in ocean ridge basalts from Deep Sea Drilling Project/Ocean Drilling Program Hole 504B, Leg 83: Implications for the timing of magnetization. *Journal of Geophysical Research* 105: 23635–23649.

Sinton JM and Detrick RS (1992) Mid-ocean ridge magma chambers. *Journal of Geophysical Research* 97: 197–216.

Sinton JM, Smaglik SM, Mahoney JJ, and Macdonald KC (1991) Magmatic processes at superfast spreading mid-ocean ridges: Glass compositional variations along the East Pacific Rise 13°–23° S. *Journal of Geophysical Research* 96: 6133–6156.

Smith DK, Tivey MA, Schouten H, and Cann JR (1999) Locating the spreading axis along 80 km of the Mid-Atlantic Ridge south of the Atlantis Transform. *Journal of Geophysical Research* 104: 7599–7612.

Smith GM (1990) The magnetic structure of the marine basement. *Reviews in Aquatic Sciences* 2: 205–227.

Smith GM and Banerjee SK (1985) Magnetic properties of basalts from Deep Sea Drilling Project Leg 83: The origin of remanence and its relation to tectonic and chemical evolution. *Initial Reports of the Deep Sea Drilling Project* 83: 347–357.

Smith GM and Banerjee SK (1986) Magnetic structure of the upper kilometer of the marine crust at Deep Sea Drilling Project Hole 504B, Eastern Pacific Ocean. *Journal of Geophysical Research* 91: 10337–10354.

Speranza F, Satolli S, Mattioli E, and Calamita F (2005) Magnetic stratigraphy of Kimmeridgian–Aptian sections from Umbria–March (Italy): New details on the M polarity sequence. *Journal of Geophysical Research* 110: B12109.

Stein CA and Stein S (1994) Comparison of plate and asthenospheric flow models for the thermal evolution of oceanic lithosphere. *Geophysical Research Letters* 21: 709–712.

Steiner MB, Ogg JG, Melendez G, and Sequeiros L (1985) Jurassic magnetostratigraphy, 2. Middle–Late Oxfordian of Aguilon, Iberian cordillera, northern Spain. *Earth and Planetary Science Letters* 76: 151–166.

Swift BA and Johnson HP (1984) Magnetic properties of the Bay of Islands ophiolite suite and implications for the magnetization of oceanic crust. *Journal of Geophysical Research* 89: 3291–3308.

Talwani M, Windisch CC, and Langseth MG (1971) Rekjanes Ridge crest: A detailed geophysical study. *Journal of Geophysical Research* 76: 473–517.

Tarduno JA, Cottrell RD, and Smirnov AV (2001) High geomagnetic intensity during the Mid-Cretaceous from Thellier analyses of single plagioclase crystals. *Science* 291: 1779–1782.

Tauxe L and Hartl P (1997) 11 million years of Oligocene geomagnetic field behaviour. *Geophysical Journal International* 128: 217–229.

Tauxe L and Kent DV (2004) A simplified statistical model for the geomagnetic field and the detection of shallow bias in paleomagnetic inclinations: Was the ancient magnetic field dipolar?. In: Channell JET, Kent DV, Lowrie W, and Meert J (eds.) *Geophysical Monograph 145: Timescales of the Paleomagnetic Field*, pp. 101–115. Washington, DC: AGU.

Tauxe L and Staudigel H (2004) Strength of the geomagnetic field in the Cretaceous Normal Superchron: New data from submarine basaltic glass of the Troodos Ophiolite. *Geochemistry, Geophysics, Geosystems* 5(2): doi:10.1029/2003GC000635.

Tivey M (1996) Vertical magnetic structure of ocean crust determined from near-bottom magnetic field measurements. *Journal of Geophysical Research* 101: 20275–20296.

Tivey M, Johnson HP, Fleutelot C, *et al.* (1998a) Direct measurement of magnetic reversal boundaries in a cross-section of oceanic crust. *Geophysical Research Letters* 25: 3631–3634.

Tivey M, Kona PA, and Kleinrock MC (1996) Reduced crustal magnetization beneath relict hydrothermal mounds: TAG hydrothermal field, Mid-Atlantic Ridge, 26° N. *Geophysical Research Letters* 23: 3511–3514.

Tivey M, Larson R, Schouten H, and Pockalny R (2005) Downhole magnetic measurements of ODP Hole 801C: Implications for Pacific oceanic crust and magnetic field behavior in the Middle Jurassic. *Geochemistry Geophysics Geosystems* 6: Q04008 doi:10.1029/2004GC000754.

Tivey M, Schouten H, and Kleinrock MC (2003) A near-bottom magnetic survey of the Mid-Atlantic Ridge axis at 26N: Implications for the tectonic evolution of the TAG segment. *Journal of Geophysical Research* 108(B5): 2277 (doi:10.1029/2002JB001967).

Tivey MA and Johnson HP (1987) The Central Anomaly Magnetic High: Implications for ocean crust construction and evolution. *Journal of Geophysical Research* 92: 12685–12694.

Tivey MA and Johnson HP (2002) Crustal magnetization reveals subsurface structure of Juan de Fuca Ridge hydrothermal vent fields. *Geology* 30: 979–982.

Tivey MA, Johnson HP, Bradley A, and Yoerger D (1998b) Thickness of a submarine lava flow determined from near-bottom magnetic field mapping by autonomous underwater vehicle. *Geophysical Research Letters* 25: 805–808.

Tivey MA, Sager WW, Lee S-M, and Tominaga M (2006) Origin of the Pacific Jurassic quiet zone. *Geology* 34: 789–792.

Tivey MA and Tucholke BE (1998) Magnetization of 0–29 Ma ocean crust on the Mid-Atlantic Ridge, 25° 30' to 27° 10' N. *Journal of Geophysical Research* 103: 17807–17826.

Toft PD, Arkani-I lamed J, and I laggerty SE (1000) Tho offoots of serpentinization on density and magnetic susceptibility: A petrophysical model. *Physics of the Earth and Planetary Interiors* 65: 137–157.

Varga RJ, Karson JA, and Gee JS (2004) Paleomagnetic constraints on deformation models for uppermost oceanic crust exposed at the Hess Deep Rift: Implications for axial processes at the East Pacific Rise. *Journal of Geophysical Research* 109: B02104 (doi:10.1029/2003JB002486).

Vera EE and Diebold JB (1994) Seismic imaging of oceanic layer 2A between 9°30'N and 10° N on the East Pacific Rise from two-ship wide-aperture profiles. *Journal of Geophysical Research* 99: 3031–3041.

Verosub KL and Moores EM (1981) Tectonic rotations in extensional regimes and their paleomagnetic consequences for oceanic basalts. *Journal of Geophysical Research* 86: 6335–6349.

Vine FJ (1966) Spreading of the ocean floor: New evidence. *Science* 154: 1405–1415.

Vine FJ and Matthews DH (1963) Magnetic anomalies over oceanic ridges. *Nature* 199: 947–949.

Vine FJ and Moores EM (1972) A model for the gross structure, petrology and magnetic properties of oceanic crust. *Geological Society of America Memoir* 132: 195–205.

Vogt PR and Johnson GL (1973) Magnetic telechemistry of oceanic crust? *Nature* 245: 373–375.

Wallick BP and Steiner MB (1992) Paleomagnetic and rock magnetic properties of Jurassic Quiet Zone basalts, Hole 801C. *Proceedings of the Ocean Drilling Program, Scientific Results* 129: 455–470.

Wang D, Van der Voo R, and Peacor DR (2005) Why is the remanent magnetic intensity of Cretaceous MORB so much higher than that of Mid- to Late Cenozoic MORB? *Geosphere* 1(3): 138–146 (doi: 10.1130/GES00024.1).

Wang D, Van der Voo R, and Peacor DR (2006) Low-temperature alteration and magnetic changes of variably altered pillow basalts. *Geophysical Journal International* 164: 25–35.

Weissel JK and Hayes DE (1972) Magnetic anomalies in the Southeast Indian Ocean. In: Hayes DE (ed.) *Antarctic Research Series, 19: Antarctic Oceanology II: The Australian–New Zealand Sector*, pp. 165–196. Washington, D.C: American Geophysical Union.

Wilson DS and Hey RN (1981) The Galapagos axial magnetic anomaly: Evidence for the Emperor event within the Brunhes and for a two-layer magnetic source. *Geophysical Research Letters* 8: 167–188.

Wilson DS and Hey RN (1995) History of rift propagation and magnetization intensity for the Cocos–Nazca spreading center. *Journal of Geophysical Research* 100: 10041–10056.

Wilson DS, Teagle DAH, and Acton GD (2003) Proceedings of the Ocean Drilling Program, Initial Reports, 2006. College Station, TX: Ocean Drilling Program.

Wilson DS, Teagle DAH, Alt JC, *et al.* (2006) Drilling to gabbro in intact ocean crust. *Science* 312: 1016–1020.

Wittpenn NA, Harrison CGA, and Handschumacher DW (1989) Crustal magnetization in the South Atlantic from inversion of magnetic anomalies. *Journal of Geophysical Research* 94: 15463–15480.

Worm H-U (2001) Magnetic stability of oceanic gabbros from ODP Hole 735B. *Earth and Planetary Science Letters* 193: 287–302.

Worm H-U and Bach W (1996) Chemical remanent magnetization in oceanic sheeted dikes. *Geophysical Research Letters* 23: 1123–1126.

Worm H-U, Bohm V, and Bosum W (1996) Implications for the sources of marine magnetic anomalies derived from magnetic logging in Holes 504B and 896A. *Proceedings of the Ocean Drilling Program, Scientific Results* 148: 331–338.

Xu W, Geissman JW, Van der Voo R, and Peacor DR (1997a) Electron microscopy of Fe oxides and implications for the origin of magnetizations and rock magnetic properties of Banded Series rocks of the Stillwater Complex, Montana. *Journal of Geophysical Research* 102: 12139–12157.

Xu W, Peacor D, Van der Voo R, Dollase W, and Beaubouef R (1996) Modified lattice parameter/Curie temperature diagrams for titanomagnetite/titanomaghemite within the quadrilateral Fe_3O_4–Fe_2TiO_4–Fe_2O_3–Fe_2TiO_5. *Geophysical Research Letters* 23(20): 2811–2814 (doi:10.1029/96GL01117).

Xu W, Peacor DR, Dollase W, Van der Voo R, and Beaubouef R (1997b) Transformation of titanomagnetite to titanomaghemite:

A slow, two-step, oxidation-ordering process in nature. *American Mineralogist* 82: 1101–1110.

Xu W, Van der Voo R, Peacor DR, and Beaubouef RT (1997c) Alteration and dissolution of fine-grained magnetite and its effects on magnetization of the ocean floor. *Earth and Planetary Science Letters* 151: 279–288.

Yamamoto M, Seama N, and Isezaki N (2005) Geomagnetic paleointensity over 1.2 Ma from deep-tow vector magnetic data across the East Pacific Rise. *Earth Planets and Space* 57: 465–470.

Yang S, Odah H, and Shaw J (2000) Variations in the geomagnetic dipole moment over the last 12 000 years. *Geophysical Journal International* 140: 158–162.

Zhou W, Peacor DR, Van der Voo R, and Mansfield JF (1999a) Determination of lattice parameter, oxidation state, and composition of individual titanomagnetite/titanomaghemite grains by transmission electron microscopy. *Journal of Geophysical Research* 104: 17689–17702.

Zhou W, Van der Voo R, and Peacor DR (1999b) Preservation of pristine titanomagnetite in older ocean-floor basalts and its significance for paleointensity studies. *Geology* 27: 1043–1046.

Zhou W, Van der Voo R, Peacor DR, Wang D, and Zhang Y (2001) Low-temperature oxidation in MORB of titanomagnetite to titanomaghemite: A gradual process with implications for marine magnetic anomaly amplitudes. *Journal of Geophysical Research* 106: 6409–6421.

Zhou W, Van der Voo R, Peacor DR, and Zhang Y (2000) Variable Ti-content and grain size of titanomagnetite as a function of cooling rate in very young MORB. *Earth and Planetary Science Letters* 179: 9–20.

Relevant Website

http://www.fugroairborne.com – Fugro Airborne Surveys is a versatile and technologically advanced airborne survey company.

13 Paleointensities

L. Tauxe, University of California San Diego, La Jolla, CA, USA

T. Yamazaki, Geological Survey of Japan, AIST, Tsukuba, Ibaraki, Japan

13.1   Introduction

The geomagnetic field acts both as an umbrella, shielding us from cosmic radiation and as a window, offering one of the few glimpses of the inner workings of the Earth. Ancient records of the geomagnetic field can inform us about geodynamics of the early Earth and changes in boundary conditions through time. Thanks to its essentially dipolar nature, the geomagnetic field has acted as a guide, pointing to the axis of rotation thereby providing latitudinal information for both explorers and geologists. A complete understanding of the geomagnetic field requires not only a description of the direction of field lines over the surface of the Earth, but information about its strength as well. While directional information is relatively straightforward to obtain, intensity variations are much more difficult and are the subject of this chapter.

In his treatise *De Magnete*, published in 1600, William Gilbert described variations in field strength with latitude based on the sluggishness or rapidity with which a compass settled on the magnetic direction. Magnetic intensity was first measured quantitatively in the late 1700s by French scientist Robert de Paul, although all records were lost in a shipwreck. Systematic measurement of the geomagnetic field intensity began in 1830 (see, e.g., Stern (2003) for a review). Despite studies of the geomagnetic field that included some mention of its strength, stretching back to at least the time of Gilbert, basic questions such as what is the average field strength and whether there are any predictable trends remain subject to debate. To study field intensity in the past requires us to use 'accidental' records; we rely on geological or archeological materials which can reveal much about the behavior of the Earth's magnetic field in ancient times.

There have been several fine reviews of the field of paleointensity recently (see, e.g., Valet, 2003) and the subject is developing very rapidly. Paleointensity data derived from archeological materials will be considered elsewhere (*see* Chapter 9). This chapter reviews the theoretical basis for paleointensity

experiments in igneous and sedimentary environments especially with regard to experimental design. We then turn to new and updated existing databases. Finally, we highlight current topics of interest.

13.2   Theory of Paleointensity

In principle, it is possible to determine the intensity of ancient magnetic fields because the primary mechanisms by which rocks become magnetized (e.g., thermal, chemical, and detrital remanent magnetizations or TRM, CRM, and DRM, respectively) can be approximately linearly related to the ambient field for low fields such as the Earth's. Thus we have by assumption

$$M_{NRM} \simeq \alpha_{anc} H_{anc}$$

and

$$M_{lab} \simeq \alpha_{lab} H_{lab} \qquad [1]$$

where α_{lab} and α_{anc} are dimensionless constants of proportionality; M_{NRM} and M_{lab} are natural and laboratory remanent magnetizations, respectively; and H_{anc} and H_{lab} are the magnitudes of the ancient and laboratory fields, respectively. If α_{lab} and α_{anc} are the same, we can divide the two equations and rearrange terms to get

$$H_{anc} = \frac{M_{NRM}}{M_{lab}} H_{lab}$$

In other words, if the laboratory remanence has the same proportionality constant with respect to the applied field as the ancient one, the remanences are linearly related to the applied field, and the natural remanence (NRM) is composed solely of a single component, all one needs to do to get the ancient field is measure M_{NRM}, give the specimen a laboratory proxy remanence M_{lab} and multiply the ratio by H_{lab}.

In practice, paleointensity is not so simple. The remanence acquired in the laboratory may not have the same proportionality constant as the original remanence (e.g., the specimen has altered its capacity to acquire remanence or was acquired by a mechanism not reproduced in the laboratory). The assumption of linearity between the remanence and the applied field may not hold true. Or, the natural remanence may have multiple components acquired at different times with different constants of proportionality.

In Sections 13.3 and 13.4 we will discuss the assumptions behind paleointensity estimates and outline various approaches for getting paleointensity data. We will start by considering thermal remanences and then address depositional ones. (To our knowledge, no one has deliberately attempted paleointensity using other remanence types such as chemical or viscous remanences.) In Section 13.5 we will briefly consider ways in which these remanences can be compromised by remagnetization processes. Section 13.6 considers how paleointensity data can be evaluated as to their reliability, and Section 13.7 reviews the published data and database initiatives. We concentrate here on data prior to the Holocene as the Holocene is discussed in Chapter 9. Finally, Section 13.8 highlights some of the major issues posed by the paleointensity data.

13.3 Paleointensity with Thermal Remanence

It appears that Folgheraiter (1899) was the first to propose that normalized thermal remanences of pottery be used to study the ancient magnetic field, although Königsberger and/or Thellier are most often given credit. Königsberger (1936) described an experimental protocol for estimating the ratio of natural remanence to a laboratory-acquired TRM (**Figure 1(a)**) and assembled data from igneous and metamorphic rocks that spanned from the Pre-Cambrian to the recent (**Figure 1(b)**). He noted that with few exceptions, the ratio $M_{\mathrm{NRM}}/M_{\mathrm{lab}}$ decreased with increasing age, and discussed various possible explanations for the trend, including changing geomagnetic field strength and shaking by earthquakes. His preferred reason for the trend in normalized remanence, however, was that magnetized bodies lose their magnetism over time, a phenomenon we now recognize as magnetic viscosity. In fact Königsberger (1938a, 1938b) believed that the trend in normalized remanence could be used to date rocks. It was Thellier (1938) who argued

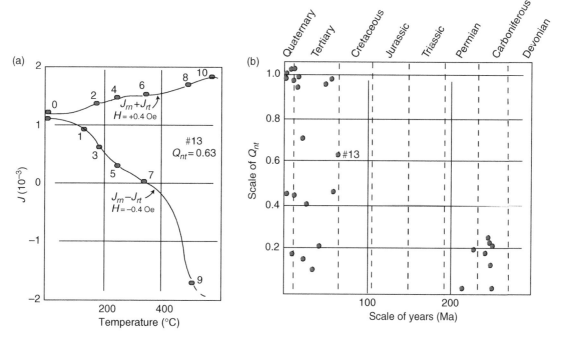

Figure 1 (a) Example of thermal normalization experiment of Königsberger (1938a). A specimen is heated to given temperature and cooled in a field of $+0.4$ Oe ($40\,\mu$T) (e.g., step labeled # 1). Then the specimen is heated to same temperature and cooled in field of -0.4 Oe (e.g., step # 2). The two curves can be decomposed to give M_{NRM} and M_{lab}, the ratio of which was termed Q_{nt} by Königsberger. (b) Q_{nt} data for a number of specimens compiled by Königsberger (1938b). The specimen from (a) is labelled # 13. These data were interpreted by Königsberger to reflect the decay of magnetic remanence with time.

strongly for the use of the thermal remanences of archeological artifacts normalized by laboratory TRMs for studying the past magnetic fields.

Königsberger's approach was largely empirical; he knew that TRMs were proportional to the magnetic fields in which they cooled and that remanences tended to decay over time, and he was well aware of the relationship between coercivity and thermal blocking. Nonetheless, he had very few tools at his disposal to discriminate among the myriad possible explanations for his observed trend that the NRM/TRM ratio appeared to decay with increasing age; for example, he did not call on apparent polar wander to explain deviant directions, relying instead on the idea that parts of lava flows tend to cool below their Curie temperatures before they stop moving.

The theoretical basis for how ancient magnetic fields might be preserved was clarified with the Nobel Prize-winning work of Néel (1949, 1955). Modern theory of TRM is discussed in detail in Chapter 8 (see also Tauxe (2005) and a recent review by Valet, 2003) but we review only the essential ideas here.

Briefly, a magnetized rod in the absence of a magnetic field will tend to be magnetized in one of several (often two) 'easy' directions. In order to overcome the intervening energy barrier and get from one easy direction to another, a magnetic particle must have energy sufficient to leap through some intervening 'hard' direction. According to the Boltzmann distribution law, the probability of a given particle having an energy ϵ is proportional to $e^{-\epsilon/kT}$ where k is Boltzmann's constant and T is the temperature in kelvin (yielding thermal energy for the product kT). Therefore, it may be that at a certain time, the magnetic moment may have enough thermal energy to flip the sense of magnetization from one easy axis to another.

If we had a collection of magnetized particles with some initial statistical alignment of moments giving a net remanence M_0, the random flipping of magnetic moments from one easy direction to another over time will eventually lead to the case where there is no preferred direction and the net remanence will have decayed to zero. The rate of approach to magnetic equilibrium is determined by the 'relaxation time' which describes the frequency of moments flipping from one easy axis to another.

Relaxation time in zero field according to Néel theory is given by

$$\tau = \frac{1}{C} \exp \frac{[\text{anisotropy energy}]}{[\text{thermal energy}]} = \frac{1}{C} \exp \frac{[Kv]}{[kT]} \quad [2]$$

where C is a frequency factor with a value of something like $10^{10}\,\text{s}^{-1}$, v is volume, and K is an 'anisotropy constant'. Equation [2] is sometimes called the 'Néel equation'.

The energy barrier for magnetic particles to flip through a 'hard direction' into the direction of the applied field H (the anisotropy energy) requires less energy than to flip the other way, so relaxation time must be a function of the applied field. The more general equation for relaxation time is given by

$$\tau = \frac{1}{C} \exp \frac{[Kv]}{[kT]} \left[1 - \frac{H}{H_c} \right]^2 \quad [3]$$

where H and H_c are the applied field and the field required to overcome the anisotropy energy and change the moment of the particle (known as the 'coercivity').

From eqn [2] we know that τ is a strong function of temperature. As described by Néel (1955), there is a very sharply defined range of temperatures over which τ increases from geologically short to geologically long timescales (see Dunlop and Özdemir (1997) and Tauxe (2005) for more details). Taking reasonable values for magnetite, the most common magnetic mineral, we can calculate the variation of relaxation time as a function of temperature for a cubic grain of width = 25 nm as shown in **Figure 2**. At room temperature, such a particle has a relaxation time longer than the age of the Earth, while at a few hundred degrees centigrade, the grain has a relaxation time that allows the magnetization to flip frequently between easy axes and can maintain an

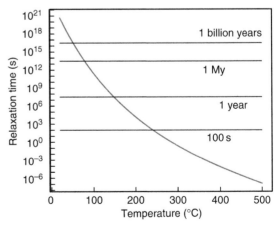

Figure 2 Variation of relaxation time versus temperature for a 25-nm-width cube of magnetite. Data from Tauxe L (2005) *Lectures in Paleomagnetism*, earthref.org/Magic/Books/Tauxe/2005, San Diego.

equilibrium with the external field. Such populations will have a slight statistical preference for the direction of the applied field because of the small difference in relaxation time between directions closer to the applied field direction from eqn [3].

The temperature at which τ is equal to about 10^2–10^3 s is defined as the 'blocking temperature', T_b. At or above the blocking temperature, but below the Curie temperature (the temperature at which all spontaneous magnetization is lost), a population of these grains is in equilibrium with the applied field and are called 'superparamagnetic.' Further cooling increases the relaxation time such that the magnetization is effectively blocked and the rock acquires a thermal remanence.

Consider a lava flow which has just been extruded (see **Figure 3**). First, the molten lava solidifies into rock. While the rock is above the Curie temperature, there is no remanent magnetization; thermal energy dominates the system. As the rock cools through the Curie temperature of its magnetic phase(s), exchange energy (the energy that encourages electronic spins to align with each other) becomes more important and the rock acquires a magnetization. The magnetization, however, is free to track the prevailing magnetic field because anisotropy energy is still less important than the energy encouraging alignment with the magnetic field (the magnetostatic energy). At this high temperature, the magnetic moments in the lava flow are superparamagnetic and tend to flop from one easy direction to another, with a slight statistical bias toward the direction with the minimum angle to the applied field (**Figure 3(c)**). The equilibrium magnetization of superparamagnetic grains is only slightly aligned, and the degree of alignment is a quasi-linear function of the applied field for low fields like the Earth's. The magnetization approaches saturation at higher fields, depending on the details of the controls on anisotropy energy like shape, size, mineralogy, etc.

13.3.1 Linearity Assumption

From theory we expect thermal remanences of small single-domain particles to be approximately linearly related to the applied field for low fields like the Earth's. However, as particle size increases, TRMs can become quite nonlinear even at relatively low fields (see **Figure 4**). Predicted TRM curves with respect to the applied field for randomly oriented populations of single-domain particles ranging in size from 20 to 100 nm widths are plotted in **Figure 4(a)**. We calculated these curves assuming quasi-equidimensional grains (1.5:1) and highly elongate grains (10:1). For the elongate grains, the TRM is predicted to be distinctly nonlinear even for the 80 nm particles. (The approximate range of the present Earth's field is shown as the shaded box.) Particles of magnetite larger than about 90 nm will have more complicated remanent states (flower, vortex, multidomain) and will not necessarily follow the predicted curves which are based on single-domain theory.

We note in passing that Kletetschka *et al.* (2006) have postulated that multidomain particles have TRMs that are highly nonlinear at fields below some threshold value with linear behavior at higher field values. This behavior was observed using a Schonstedt oven, which has very poor field control and we were unable to reproduce the observations in the SIO laboratory which has excellent field control; linear behavior was observed in fields as low as 10 nT (Yongjae Yu, personal communication, 2006).

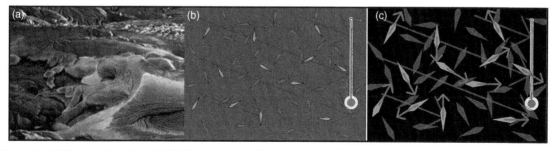

Figure 3 (a) Picture of lava flow. (b) While the lava is still well above the Curie temperature, crystals start to form, but are nonmagnetic. (c) Below the Curie temperature but above the blocking temperature, certain minerals become magnetic, but their moments continually flip among the easy axes with a statistical preference for the applied magnetic field. As the lava cools down, the moments become fixed, preserving a thermal remanence. (a) Courtesy of Daniel Standigel. (b) and (c) Modified from animation of Genevieve Tauxe available at http://magician.ucsd.edu/Lab_tour/movs/TRM.mov. Data from Tauxe L (2005) *Lectures in Paleomagnetism*, earthref.org/Magic/Books/Tauxe/2005, San Diego.

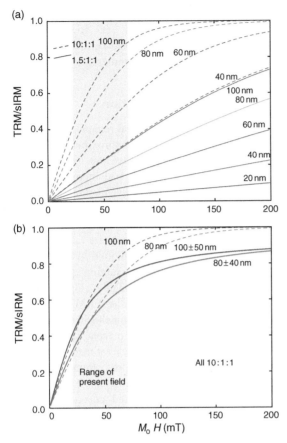

Figure 4 Predicted TRM expressed as a fraction of saturation for various particle sizes and distributions of magnetite. Note the nick point for which the linearity assumption fails is a strong function of particle size, but linearity holds true for equant particles in fields less than a 100 µT. Strongly elongate particles will behave in a more nonlinear fashion.

Dunlop and Argyle (1997) discovered strongly nonlinear TRM acquisition behavior in synthetic specimens with mean grain sizes in the single-domain grain range. Although their laboratory fields were mostly much higher than those of the Earth's field (up to 9 mT!), the results should give practitioners of paleointensity pause. Moreover, Selkin *et al.* (2007) have found nonlinear TRM behavior in natural specimens with single-domain behavior and the nonlinearity is distinct in fields as low as 50 µT.

The modeling exercises shown in **Figure 4** and the experimental results of Dunlop and Argyle (1997) and Selkin *et al.* (2007) suggest that it would be a wise practice to incorporate a test of TRM linearity into paleointensity experiments as a matter of routine. If the relationship between TRM and applied field is known empirically, then biased results can be corrected to the true ancient intensity.

13.3.2 Alteration during Heating

The second assumption for absolute paleointensity determinations is that the laboratory and ancient proportionality constants are the same (i.e., $\alpha_{lab} = \alpha_{anc}$ in eqn [1]). Simply measuring the NRM and giving the specimen a total TRM does nothing to test this assumption. For example, alteration of the specimen during heating could change the capacity to acquire thermal remanence and give erroneous results with no way of assessing their validity.

There are several ways of checking the ability of the specimen to acquire thermal remanence in paleointensity experiments. The most commonly used are experiments that employ stepwise replacement of the natural remanence with a laboratory thermal remanence (Königsberg/Thellier-Thellier or KTT family of experiments) and those that compare anhysteretic remanenence before and after heating ('Shaw' family of experiments). Other approaches attempt to prevent the alteration from occuring, for example, by heating in controlled atmospheres or vacuum, or by using microwaves to heat just the magnetic phases, leaving the rest of the specimen cool. Another approach is to find materials that are particularly resistant to alteration (e.g., submarine basaltic glass or single plagioclase crystals). Finally, some methods attempt to normalize the remanence with saturation isothermal remanent magnetization (sIRM) and avoid heating altogether. We will briefly describe each of these in turn, beginning with the KTT family of experiments.

Figure 4(b) shows the effect of having a distribution of grain sizes. We calculated curves for populations with normally distributed particle widths (all with 10:1 elongation) with mean widths of 80 and 100 nm, respectively. The effect of the distribution of particle sizes is to depress the TRM below that for a uniform distribution because smaller particles have much lower TRMs at a given field strength and the effect is asymmetric (the difference between 80 and 100 nm width at say 100 µT is much less than the difference between 60 and 80 nm at the same field). (Note also, grains smaller than about 20 nm are superparamagnetic at room temperature and do not contribute to the TRM at all, so distributions that include small particles will have suppressed TRMs relative to their theoretical saturation remanences.)

13.3.2.1 KTT family of experiments

Detection of changes in the proportionality constant caused by alteration of the magnetic phases in the rock during heating has been a goal in paleointensity experiments since the earliest days. As we have already seen (**Figure 1(a)**), Königsberger (1936, 1938a, 1938b) heated up specimens in stages, progressively replacing the natural remanence with partial thermal remanences (pTRMs), an experiment that was elaborated on by Thellier (1938) and Thellier and Thellier (1959). The so-called 'KTT' approach is particularly powerful when lower-temperature steps are repeated, to verify directly that the ability to acquire a thermal remanence has not changed.

The stepwise approach relies on the assumption that pTRMs acquired by cooling between any two temperature steps (e.g., $500°$C and $400°$C in **Figure 5**) are independent of those acquired between any other two temperature steps. This assumption is called the 'Law of independence' of pTRMs. The approach also assumes that the total TRM is the sum of all the independent pTRMs (see **Figure 5**), an assumption called the 'Law of additivity.'

There are several possible ways to progressively replace the NRM with a pTRM in the laboratory. In the original KTT method (see, e.g., **Figure 1(a)**), the

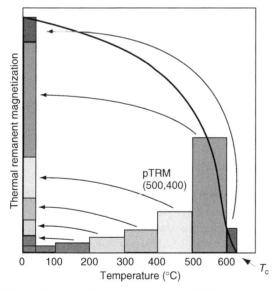

Figure 5 Laws of independence and additivity. Partial thermal remanences (pTRMs) acquired by cooling between two temperature steps are independent from one another and sum together to form the total TRM. Modified from Yu Y and Tauxe L (2004) On the use of magnetic transient hysteresis in paleomagnetism for granulometry. *Geochemistry, Geophysics, Geosystems* 6: Q01H14 (doi: 10.1029/2004GC000839).

specimen is heated to some temperature (T_1) and cooled in the laboratory field B_{lab}. After measurement the combined remanence (what is left of the natural remanence plus the new laboratory pTRM) is

$$M_1 = M_{NRM} + M_{pTRM}$$

Then the specimen is heated a second time and cooled upside down (in field $- B_{lab}$). The second remanence is therefore

$$M_2 = M_{NRM} - M_{pTRM}$$

Simple vector subtraction allows the determination of the NRM remaining at each temperature step and the pTRM gained (see **Figure 6(a)**). These are nowadays plotted against each other in what is usually called an 'Arai plot' (Nagata *et al.*, 1963) as in **Figure 6(b)**. The KTT method implicitly assumes that a magnetization acquired by cooling from a given temperature is entirely replaced by reheating to the same temperature (i.e., $T_b = T_{ub}$), an assumption known as the 'Law of reciprocity.'

As magnetic shielding improved, more sophisticated approaches were developed. In the most popular paleointensity technique (usually attributed to Coe, 1967), we substitute cooling in zero field for the first heating step. This allows the direct measurement of the NRM remaining at each step. The two equations now are

$$M_1 = M_{NRM}$$

and

$$M_2 = M_{NRM} + M_{pTRM}$$

The laboratory M_{pTRM} in this 'zero-field/in-field' (or ZI) method is calculated by vector subtraction. Alternatively, the first heating and cooling can be done in the laboratory field and the second in zero field (Aitken *et al.*, 1998; see also Valet *et al.*, 1998), here called the 'in-field/zero-field' or (IZ) method.

In all three of these protocols, lower temperature in field cooling steps can be repeated to determine whether the remanence-carrying capacity of the specimen has changed. These steps are called 'pTRM checks.' Differences between the first and second M_{pTRM}s at a given temperature indicate a change in capacity for acquiring thermal remanences and are grounds for suspicion or rejection of the data after the onset of such a change. Some have proposed that paleointensity data can be 'fixed' even if the pTRM checks show significant alteration (e.g., Valet *et al.*, 1996 and McClelland and Briden, 1996). The argument is that if pTRM checks can be brought back

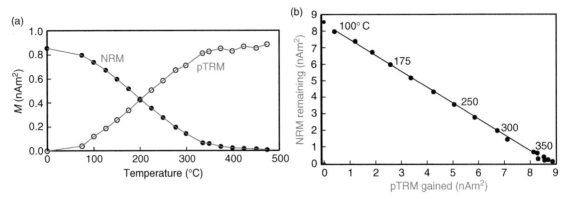

Figure 6 Illustration of the KTT method for determining absolute paleointensity. (a) Thermal demagnetization of NRM shown as filled circles and the laboratory-acquired pTRM shown as open symbols. (b) Plot of NRM component remaining versus pTRM acquired for each temperature step. Data from Tauxe L (2005) *Lectures in Paleomagnetism*, earthref.org/Magic/Books/Tauxe/2005, San Diego.

into accordance with the original pTRM measurements using a correction factor, then if that same correction factor is applied to all subsequent pTRM measurements, the effect of the alteration has been accounted for and the data can be considered 'reliable.' We consider this correction to carry some risk and 'corrected' data should be clearly marked as such.

Despite its huge popularity and widespread use, the approach of progressively replacing the natural remanence with a thermal remanence has several drawbacks. Alteration of the ability to acquire a pTRM is not the only cause for failure of the assumption of equality of α_{lab} and α_{anc}. Both experiment (Bol'shakov and Shcherbakova, 1979; Shcherbakova *et al.*, 2000) and theory (e.g., Dunlop and Xu (1994) and Xu and Dunlop (1994) suggest that the essential assumption of equivalence of blocking and unblocking temperatures may break down for larger particles.

Micromagnetic modeling of hysteresis behavior can shed some light on what might be going on. In simulated hysteresis experiments, particles can be subjected to a large DC applied magnetic field, sufficient to completely saturate them. As the field is lowered, certain particles form vortex structures at some applied field strength (see **Figure 7**). These vortex structures are destroyed again as the field is ramped back up to saturation. However, the field at which the vortex is destroyed is higher than the field at which it formed. This is the phenomenon responsible for 'transient hysteresis' (see Fabian, 2003; Yu and Tauxe, 2004).

One can imagine that something similar to transient hysteresis could occur if we cooled a particle from its Curie temperature, and then heated it back up again. Just below the Curie temperature, the particle would be in a saturated state (because the magnetization is quite low and the vortex structure is just an attempt by the particle to reduce its external field). As the specimen cools down, the magnetization grows. At some temperature a vortex structure may form. As the specimen is heated back up again, the vortex may well remain stable to higher temperatures than its formation temperature by analogy to the behavior in the simulated hysteresis experiment.

If the particle is large enough to have domain walls in its remanent state, then the scenario is somewhat different and not easily tractable by theory (*see* Chapter 8). At just below its Curie temperature the particle is at saturation. As the particle cools, domain walls will begin to form at some temperature. The remanent state will have some net moment because the domain walls are distributed such that there is incomplete cancellation leaving a small net remanence, proportional to the applied field for moderate field strengths. As the temperature ramps up again, the walls 'walk around' within the particle seeking to minimize the magnetostatic energy and are not destroyed until temperatures very near the Curie temperature.

The fact that blocking and unblocking of remanence occur at different temperatures in some particles means that a pTRM blocked at a given temperature will remain stable to higher temperatures; the unblocking temperature is not equal to the blocking temperature. This means that $\alpha_{lab} \neq \alpha_{anc}$ and the key assumptions of the KTT-type methods are not met. The Arai plots may be curved (see Dunlop and Özdemir (1997) for a more complete discussion) and if any portion of the NRM/TRM

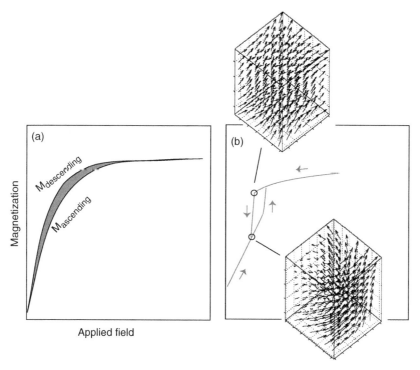

Figure 7 (a) Illustration of a zero FORC (ZFORC) whereby the descending loop from saturation is terminated at zero field and the field is then ramped back up to saturation. The transient hysteresis (TH) of Fabin (2003) is the shaded area between the two curves. (b) Micromagnetic model of a ZFORC for a 100 nm cube of magnetite. Two snap shots of the internal magnetization on the descending and ascending loops are shown in the insets. Modified from Yu Y Tauxe L (2004) On the use of magnetic transient hysteresis in paleomagnetism for granulometry. *Geochemistry, Geophysics, Geosystems* 6: Q01H14. (doi: 10.1029/2004GC000839).

data are used instead of the entire temperature spectrum, the result could be biased. For example, the lower-temperature portion might be selected on the grounds that the higher-temperature portion is affected by alteration, or the higher-temperature portion might be selected on the grounds that the lower-temperature portion is affected by viscous remanence. Both of these interpretations would be wrong.

In order to detect inequality of blocking and unblocking and the presence of unremoved portions of the pTRM known as 'high-temperature pTRM tails,' several embellishments to the KTT-type experiment have been proposed and more are on the way. In one modification, a second zero-field step is inserted after the in-field step in the ZI method. This so-called 'pTRM-tail check' (e.g., Riisager and Riisager, 2001) assesses whether the pTRM gained in the laboratory at a given temperature is completely removed by reheating to the same temperature. If not, the specimen is said to have a 'pTRM tail,' a consequence of an inequality of the unblocking temperature T_{ub} and the original blocking temperature T_b in violation of the law of reciprocity and grounds for rejection. A second

modification is to alternate between the IZ and ZI procedures (the so-called 'IZZI' method first conceived with Agnès Genevey and described by Yu *et al.*, 2004.) The IZZI method is also extremely sensitive to the presence of pTRM tails and may obviate the need for the pTRM-tail check step. An example of a complete IZZI experiment is shown in **Figure 8**.

There are several other violations of the fundamental assumptions that require additional tests and/or corrections in the paleointensity experiment besides alteration or failure of the law of reciprocity. For example, if the specimen is anisotropic with respect to the acquisition of thermal remanence, the anisotropy tensor must be determined and the intensity corrected (e.g, Fox and Aitken, 1980). The detection and correction for anisotropy can be very important in certain paleomagnetic (and archeomagnetic) materials. The correction involves determining the TRM (or the ARM proxy) anisotropy tensor and matrix multiplication to recover the original magnetic vector (see Selkin *et al.*, 2000, for a more complete discussion). Moreover, because the approach to equilibrium is a function of time,

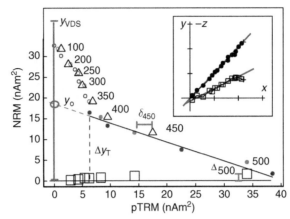

Figure 8 Data from an IZZI experiment. Circles are the pTRM gained at a particular temperature step versus the NRM remaining. Solid symbols are those included in the slope calculation. Blue (darker) symbols are the infield–zerofield steps (IZ) and the brown (lighter) symbols are the zerofield–infield steps (ZI). The triangles are the pTRM checks and the squares are the pTRM tail checks. The difference between the pTRM check and the original measurement is δ_i as shown by the horizontal bar labeled δ_{450}. The difference between the first NRM measurement and the repeated one (the pTRM tail check) is shown by the vertical bar labeled Δ_{500}. The vector difference sum (VDS) is the sum of all the NRM components (tall vertical bar labeled VDS). The NRM fraction is shown by the vertical dashed bar. The insets are the vector components (x, y, z) of the zero-field steps. The solid symbols are (x, y) pairs and the open symbols are (x, z) pairs. The specimen was unoriented with respect to geographic coordinates. The laboratory field was applied along the z-axis in the in-field steps. Redrawn from Tauxe L and Staudigel H (2004) Strength of the geomagnetic field in the Cretaceous Normal Superchron: New data from submarine basaltic glass of the Troodos Ophiolite. *Geochemistry, Geophysics, Geosystems* 5(2): Q02H06 (doi:10.1029/2003GC000635).

slower cooling results in a larger TRM, hence differences in cooling rate between the original remanence acquisition and that acquired in the laboratory will lead to erroneous results (e.g., Halgedahl *et al.*, 1980). Compensating for differences in cooling rate is relatively straightforward if the original cooling rate is known and the specimens behave according to single-domain theory. Alternatively, one could take an empirical approach in which the rock is allowed to acquire a pTRM under varying cooling rates (e.g., Genevey and Gallet, 2003), an approach useful for cooling rates of up to a day or two.

13.3.2.2 Shaw family of experiments

The previous section was devoted to experiments in which detection of nonideal behavior is done by repeating various temperature steps. In this section we will consider an alternative approach, long in use in paleointensity studies, which employs the laboratory proxy anhysteretic remanence (ARM). The so-called 'Shaw method' (e.g., Shaw, 1974) is based on ideas first explored by van Zijl *et al.* (1962a, 1962b). In its simplest form, we measure the NRM, then progressively demagnetize the NRM with alternating fields (AF) to establish the coercivity spectrum of the specimen prior to heating. The specimen is then given an anhysteretic remanence (M_{ARM_1}) by

subjecting the specimen to progressively higher peak AF which decay in the presence of a small bias field. The use of ARM has been justified because it is in many ways analogous to the original TRM (*see* Chapter 8). M_{ARM_1} is then progressively demagnetized to establish the original relationship between the coercivity spectrum of the M_{NRM} (presumed to be a thermal remanence) and ARM prior to any laboratory heating.

As with the KTT-type methods, M_{NRM} is normalized by a laboratory thermal remanence. But in the case of the Shaw-type methods, the specimen is given a total TRM, (M_{TRM_1}) which is AF demagnetized as well. Finally, the specimen is given a second ARM (M_{ARM_2}) and demagnetized for the last time.

The general experiment is shown in **Figures 9(a)** and **9(b)**. If the first and second ARMs do not have the same coercivity spectrum as in **Figure 9(b)**, the coercivity of the specimen has changed and the NRM/TRM ratio is suspect.

Rolph and Shaw (1985) suggested that the ratio M_{ARM_1}/M_{ARM_2} at each demagnetizing step be used to 'correct' for the alteration bias of M_{TRM_1} by

$$M_{TRM_1}* = M_{TRM_1}\frac{M_{ARM_1}}{M_{ARM_2}}$$

So doing can, in some cases, restore linearity between NRM and TRM as shown in **Figure 9(c)**.

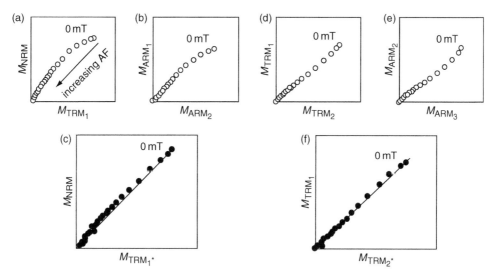

Figure 9 Shaw family of methods (see text). (a) Plot of pairs of NRM and the first TRM for each AF demagnetization step. (b) Plot of pairs of the first ARM and the second ARM for each AF demagnetization step. (c) Plot of pairs of NRM and TRM adjusted by the ratio of ARM_1/ARM_2 for that AF step from (b) ($TRM_1{}^*$). (d) Same as (a) but for the first and second TRMs. (e) Same as (a) but for the second and third ARMs. (f) Same as (c) but for first and second TRM where $TRM_2{}^*$ is adjusted using ARM_2/ARM_3 ratio from (e). Data from Yamamoto Y, Tsunakawa H, and Shibuya H (2003) Palaeointensity study of the Hawaiian 1960 lava: Implications for possible causes of erroneously high intensities. *Geophysical Journal International* 153(1): 263–276.

Valet and Herrero-Bervera (2000) argued that only data requiring no correction and utilizing the entire coercivity spectrum should be used. They further pointed out that many specimens are required to lend credibility to a paleointensity experiment. As the former requirement generally leaves very few specimens, Valet and Herrero-Bervera reasoned that a quicker experimental procedure would ultimately result in more acceptable data, hence a better overall outcome, even if the results from many experiments are discarded. To speed up the measurement process, they employed a truncated Shaw method in which no ARMs are imparted, but both the NRM and the laboratory TRM are completely demagnetized using AF. Linearity of the two when plotted as in **Figure 9(a)** is taken as the sole criterion for acceptance.

Tsunakawa and Shaw (1994) suggested that a tendency for chemical alteration could also be detected if the specimen is heated to above the Curie temperature twice, each followed by AF demagnetization (see **Figures 9(d)–9(f)**). During the second heating step, the specimen is left at high temperature for a longer period of time than the first heating step to encourage alteration to continue so that it may be detected by the method. If the slope $M_{TRM_1}/M_{TRM_2}{}^*$ differs by more than the experimental error, the experimental results are rejected.

The issue of contamination of the remanence by multidomain particles has also been considered in the Shaw-type methods. It has long been known (Ozima *et al.*, 1964) that specimens can lose much of their remanence by cooling to temperatures below about $-160°$ C and warming in zero field. This behavior is generally attributed to magnetocrystalline-dominated remanences cycling through the so-called 'Verwey transition' at which the axis of magnetocrystalline anisotropy changes, erasing the magnetic memory of these particles (see, e.g., Dunlop and Özdemir, 1997). This behavior is frequently assumed to occur most readily in multidomain particles, hence their contribution could be minimized if specimens are pretreated to low temperatures (low-temperature demagnetization, or LTD) prior to measurement. Yamamoto *et al.* (2003) and Yamamoto and Tsunakawa (2005) argued that one of the major causes of failure in paleointensity experiments is the effect of multidomain particles, which violate the essential assumption that the original blocking temperature is the same as the laboratory unblocking temperature. They therefore treat specimens to LTD prior to AF demagnetization of each remanence. This 'LTD–DHT' method gave improved results for the otherwise disappointingly difficult Hawaiian 1960 lava flow (see, e.g., Tanaka and Kono (1991), Valet and Herrero-Bervera (2000), and Valet (2003)).

The LTD–DHT experiment assumes that mainly the multidomain particles are affected by the LTD step. However, Carter-Stiglitz *et al.* (2002, 2003) found that single-domain magnetites can and do lose substantial remanence by LTD as well. This behavior means that LTD treatment may demagnetize part of the desired as well as the undesired NRM. It is possible that the SD remanence removed by LTD may be the low-coercivity contribution and unimportant to the paleointensity.

One other note of caution for all paleointensity experiments using specimens with multidomain grains was raised by Dunlop and Argyle (1997). They showed experimentally that the acquisition of TRM in grains with domain walls was very non-linear, even in low fields like the Earth's, a problem that no one has addressed yet for multidomain specimens.

The primary reasons stated for using the Shaw method are that (1) it is faster and (2) alteration is minimized, as the specimen is only heated once (albeit to a high temperature). The first rationale is no longer persuasive because modern thermal ovens have high capacities and the KTT method is certainly not slower than the Shaw method on a per specimen basis. This is particularly true for the LTD–DHT Shaw method as this experiment takes approximately 8 h to complete per specimen. The second rationale may have some validity. The key features of any good experiment are the built-in tests of the important assumptions.

13.3.3 Methods that Minimize Alteration

Several alternative approaches have been proposed which instead of detecting nonideal behavior, such as alteration, attempt to minimize it. These methods include reducing the number of heating steps required (as in the Shaw methods), heating specimens in controlled atmospheres, reducing the time at temperature by, for example, measuring the specimens at elevated temperature, and using microwaves to excite spin moments as opposed to direct thermal heating. Finally, there has been some effort put into finding materials that resist alteration during the heating experiments.

13.3.3.1 *Reduced number of heating steps*

Kono and Ueno (1977) describe in detail a single heating per temperature step method suggested by Kono (1974) whereby the specimen is heated in a laboratory field applied perpendicular to the NRM. M_{pTRM} is gotten by vector subtraction. Reducing the number of heatings can reduce the alteration to some extent. However, this method has only rarely been applied because it can only be used for strictly uni-vectorial NRMs (an assumption that is difficult to test with the data generated by this method) and requires rather delicate positioning of specimens in the furnace or fancy coil systems that generally have a limited region of uniform field, reducing the number of specimens that can be analyzed in a single batch. While pTRM checks are possible with this method, they necessitate additional heating steps and are not generally performed.

A second strategy for reducing the number of heating steps was proposed by Hoffman *et al.* (1989), and modified by Hoffman and Biggin (2005; see also Dekkers and Böhnel, 2006). In the Hoffman–Biggin version, at least five specimens from a given cooling unit are sliced into four specimens each, one of which is dedicated to rock-magnetic analysis. The remaining specimens (at least 15) are heated a total of five times giving remanence measurements M_1–M_5. (Please note that bold face parameters are vectors, while normal text variables are scalars, in this case the magnitudes.) In the first three heating steps, the specimens are treated to increasingly high temperatures (T_0, T_1, and T_2) and cooled in zero field. The first heating step ostensibly removes any secondary overprint (e.g., a viscous remanent magnetization (VRM)) and M_1 serves as the baseline for normalizing all subsequent steps so that data from different specimens can be combined. After the three zero-field heating steps, the specimens are heated again to T_2 and cooled with the laboratory field switched on between T_2 and T_0 after which it is switched off. This treatment step gives the pTRM acquired between T_2 and T_0 by vector subtraction of $\mathbf{M_4}-\mathbf{M_3}$. The fifth heating step is to T_1 followed by a zero-field cooling. This final step serves both to supply the pTRM acquired between T_2 and T_1 by vector subtraction of $\mathbf{M_4}-\mathbf{M_5}$ and a kind of 'pseudo pTRM check' step as explained later.

In interpreting results, there are two data points from each specimen with estimates for NRM remaining versus pTRM gained, denoted T_1 and T_2. The NRM remaining part of T_1 and T_2 are ratios M_2/M_1 and M_3/M_1, respectively. The pTRMs gained at T_1 and T_2 are $|\mathbf{M_5}-\mathbf{M_4}|/M_1$ and $|\mathbf{M_4}-\mathbf{M_3}|/M_1$. Because all remanences are normalized by the NRM remaining after zero-field cooling from T_0 (M_1) measured for each specimen, we can combine data from the different specimens together on a single Arai-like plot (see **Figure 10**).

Hoffman and Biggin (2005) have several criteria that help screen out 'unreliable' data. First, they require that the directions of the zero-field steps trend to the origin on an orthogonal plot and have low scatter. This helps eliminate data for which the characteristic remanence has not been isolated (although three zero-field steps is not generally considered sufficient for this purpose). Second, they require the y-intercept to be between 0.97 and 1.03 and that the correlation coefficient must be ≥ 0.97. If the T_1 data are displaced from the line connecting the T_2 point and a y-intercept of 1.0, then the specimen may have altered during laboratory heating (e.g., open symbols in **Figure 10**) and can be rejected.

Noting that the results of the multispecimen procedure when applied to the 1971 Hawaiian flow (shown in **Figure 10**) were significantly different than the known field ($37\,\mu T$), Hoffman and Biggin (2005) suggested that the data, which are heavily influenced by data from a single, low blocking temperature specimen (green symbols in **Figure 10**), could be reweighted to remove the bias. Furthermore, they proposed averaging all the data by accepted specimen, and including the y-intercept in the calculation. These modifications yielded a concordant result with the known field within error. Finally, they redefined many of the parameters typically used in paleointensity experiments (see Section 13.6) for use with the multispecimen method.

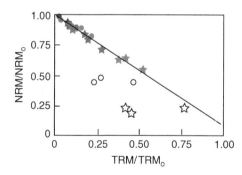

Figure 10 Illustration of multispecimen approach for specimens from the 1971 Hawaiian lava flow. Different colors represent data from different specimens. Different symbols are different heating steps. Open symbols are from specimen that failed initial selection criteria. Best-fit line represents a slope that predicts a 'paleo' field of $33\,\mu T$, whereas the actual field was $37\,\mu T$. Data from Hoffman KA and Biggin AJ (2005) A rapid multi-sample approach to the determination of absolute paleointensity. *Journal of Geophysical Research* 110: B12108 (doi:10.1029/2005JB003646).

The primary advantage of the multispecimen approach put forward by Hoffman and Biggin (2005) is the speed with which measurements can be made, allowing many more specimens to be analyzed. While the method may be fast, it loses multiple pTRM checks and any ability to assess the equivalence of blocking and unblocking. Moreover, the method strongly emphasizes the lower-blocking-temperature portion of the blocking temperature spectrum (especially in the moment-corrected version). This means that the remanence is contaminated by viscous or multidomain remanences leading to a concave downward curve in the Arai plot, the multispecimen result will overestimate the true value of the paleointensity. Finally, it is experimentally very difficult to turn the laboratory field off precisely when the specimen's internal temperature is T_0 because only the oven temperature is known and the specimen temperature lags behind that of the oven by variable and unknown amounts, depending on the exact disposition of the specimens in the oven. This bias will lead to scatter and contribute to a systematic bias (the field will always be turned off at too high a temperature, thereby underestimating the pTRM gained).

On the positive side, the presentation of all specimen data on a single Arai diagram (also proposed by Chauvin *et al.*, 2005) is an interesting modification of the traditional Arai diagram. Plotting all the KTT data from specimens from a given cooling unit on a single Arai diagram allows instant assessment of the reproducibility of data and of course can be done with traditional experimental results.

Dekkers and Böhnel (2006) argue that their multispecimen procedure, which employs a single heating/cooling step with the laboratory field oriented parallel to the NRM, can be used on specimens of any domain state. The fundamental assumption of this method is the assumed linearity of pTRM with applied field which the authors claim is independent of domain state. As already discussed, this may not be true, particularly for multidomain grains, which may explain the high degree of scatter in their experimental results.

13.3.3.2 Use of controlled atmospheres to reduce alteration

Alteration during heating is caused by oxidation (or reduction) of the magnetic minerals in the specimen. There have been several strategies to reduce this effect with varying degrees of success. Thellier (1938) tried using vacuum and nitrogen atmospheres.

Taylor (1979) developed a technique for paleointensity determination that encapsulated specimens in silica glass. By placing an oxygen 'getter' such as titanium in the evacuated glass tube along with the specimen, he suggested that oxygen fugacity could be maintained and alteration would be reduced. This technique was tested by Sugiura *et al.* (1979) who claimed some improvement in experimental results on lunar glass specimens. More recently, Valet *et al.* (1998) performed paleointensity experiments by heating in argon atmospheres and cooling in nitrogen atmospheres. They reported a significant improvement in their argon results over those performed in air.

The difficulty of heating and cooling in vacuum and controlled atmospheres are (1) difficulty in achieving a uniform and reproducible temperature in the oven and (2) unintended oxidation or reduction reactions. It appears that reduction in alteration can be achieved using these techniques, although the overwhelming majority of paleointensity experiments are done in air.

13.3.3.3 *Measurement at elevated temperature*

Boyd (1986) suggested that measurements could be made more rapidly if they were measured at elevated temperatures instead of cooling back to room temperature for measurement. The idea was that alteration could be detected immediately and the experiment aborted, before wasting time finishing the entire measurement sequence. This idea was recently warmed up by Le Goff and Gallet (2004) who developed a vibrating sample magnetometer equipped with magnetic field coils which allow the specimen to be measured at temperature and in controlled fields, greatly speeding up the measurement process and, one hopes, reducing the effects of cooling rate and specimen alteration. Preliminary data are promising.

13.3.3.4 *Use of microwaves for thermal excitation*

Until now we have not concerned ourselves with HOW the magnetic moment of a particular grain flips its moment. Earlier, we mentioned 'thermal energy' and left it at that. But how does thermal energy do the trick?

An external magnetic field generates a torque on the electronic spins, and in isolation, a magnetic moment will respond to the torque in a manner similar in some respects to the way a spinning top

responds to gravity: the magnetic moment will precess about the applied field direction, spiraling in and come to a rest parallel to it. Because of the strong exchange or superexchange coupling in magnetic phases, spins tend to be aligned parallel (or antiparallel) to one another and the spiraling is done in a coordinated fashion, with neighboring spins as parallel as possible to one another. This phenomenon is known as a 'spin wave.'

Raising the temperature of a body transmits energy (via 'phonons') to the electronic spins, increasing the amplitude of the spin waves. This magnetic energy is quantized in 'magnons.' In the traditional KTT experiment, the entire specimen is heated and the spin waves are excited to the point that some may flip their moments as described in Section 13.3.

As in most kitchens, there are two ways of heating things up: the conventional oven and the microwave oven. In the microwave oven, molecules with certain vibrational frequencies (e.g., water) are excited by microwaves. These heat up, passing their heat on to the rest of the pizza (or whatever). If the right microwave frequency is chosen, ferromagnetic particles can also be excited directly, inviting the possibility of heating only the magnetic phases, leaving the matrix alone (e.g., Walton *et al.*, 1993). The rationale for developing this method is to reduce the degree of alteration experienced by the specimen because the matrix often remains relatively cool, while the ferromagnetic particles themselves get hot. (The magnons get converted to phonons, thereby transferring the heat from the magnetic particle to the matrix encouraging alteration, but there may be ways of reducing this tendency (see Walton 2004).)

The same issues of nonlinearity, alteration, reciprocity, anisotropy, and cooling rate differences, etc. arise in the microwave approach as in the thermal approach. Ideally, the same experimental protocol could be carried out with microwave ovens as with thermal ovens. In practice, however, it has proved quite difficult to repeat the same internal temperature, making double (or even quadruple) heatings problematic although progress toward this end may have been made recently (e.g., Böhnel *et al.*, 2003.) It is likely that the issues of reciprocity of blocking and unblocking in the original (thermally blocked) and the laboratory (microwave unblocked) and differences in the rate of blocking and unblocking will remain a problem for some time as they have for thermally blocked remanences. It is also worth echoing the concerns raised by Valet (2003) and LeGoff and Gallet

(2004) that the theoretical equivalence between thermal unblocking and microwave unblocking has not yet been explained. In fact, Walton (2005) pointed out that resonance within the magnetic particles is wavelength dependent. This raises the possibility that unblocking may occur in an entirely different manner in microwave processes than in thermal ones (by chords instead of scales to use a musical metaphor) leading to serious questions about the applicability of the method for recovery of paleointensity estimates. Nonetheless, if alteration can be prevented by this method, and the theoretical underpinnings can be worked out, it is worth pursuing.

13.3.3.5 Using materials resistant to alteration

Another very important approach to the paleointensity problem has been to find and exploit materials that are themselves resistant to alteration. There is an increasing variety of promising materials, ranging from quenched materials, to single crystals extracted from otherwise alteration-prone rocks, to very slowly cooled plutonic rocks (e.g., layered intrusions). Quenched materials include volcanic glasses (e.g., Pick and Tauxe, 1993), metallurgical slag (e.g., Ben Yosef et al., 2005), and welded tuffs (unpublished results). Single crystals of plagioclase extracted from lava flows (see review by Tarduno et al., 2006) can yield excellent results while the lava flows themselves may be prone to alteration or other nonideal behavior. Parts of layered intrusions (e.g., Selkin et al., 2000b) can also perform extremely well during the paleointensity experiment.

While some articles have called the reliability of submarine basaltic glass results into question (e.g., Heller et al., 2002), Tauxe and Staudigel (2004) and Bowles et al. (2005) addressed these concerns in great detail and the reader is referred to those papers and the references therein for a thorough treatment of the subject. In any case, results from alteration-resistant materials are quite promising and more results will be available in the near future.

13.3.4 Use of IRM Normalization

Sometimes it is difficult or impossible to heat specimens because they will alter in the atmosphere of the lab, or the material is too precious to be subjected to heating experiments (e.g., lunar samples and some archeological artifacts). Looking again at **Figure 4** suggests an alternative for order of magnitude guesstimates for paleointensity without heating at all.

TRM normalized by a saturation remanence (IRM) is quasi-linearly related to the applied field up to some value depending on mineralogy and grain-size population.

Cisowski and Fuller (1986; see also, e.g., Kletetschka et al., 2004) advocated the use of IRM normalization of the NRMs of lunar samples to estimate paleointensity. They argued that, especially when both remanences were partially demagnetized using alternating field demagnetization, the NRM:IRM ratio gave order-of-magnitude constraints on absolute paleointensity and reasonable relative paleointensity estimates. Their argument is based on monomineralic suites of rocks with uniform grain size. They further argue optimistically that multidomain contributions can be eliminated by the AF demagnetization.

As can be seen by examining **Figure 4**, at best only order-of-magnitude estimates for absolute paleointensity are possible. The monomineralic and uniform grain size constraints make even this unlikely. Finally, the behavior of multidomain TRMs and IRMs is not similar under AF demagnetization, the former being much more stable than the latter. Nonetheless, if magnetic uniformity can be established, it may in fact be useful for establishing relative paleointensity estimates as is done routinely in sedimentary paleointensity studies (see Section 13.6). The caveats concerning single-component remanences are still valid and perhaps complete AF demagnetization of the NRM would be better than a single 'blanket' demagnetization step. Moreover, we should bear in mind that for larger particles, TRM can be strongly nonlinear with applied field at even relatively low fields (30 μT) according to the experimental results of Dunlop and Argyle (1997; see also figure 1(a) of Kletetschka et al. 2006). The problem with the IRM normalization approach is that domain state, linearity of TRM, and nature of the NRM cannot be assessed. The results are therefore difficult to interpret in terms of ancient fields.

13.4 Paleointensity with Depositional Remanences

Sediments become magnetized in quite a different manner than igneous bodies. Detrital grains are already magnetized, unlike igneous rocks which crystallize above their Curie temperatures. Magnetic particles that can rotate freely will turn into the direction of the applied field which can result in a

DRM. Sediments are also subject to post-depositional modification through the action of organisms, compaction, diagenesis, and the aquisition of VRM all of which will affect the magnetization and our ability to tease out the geomagnetic signal. In the following, we will consider the syn-depositional processes of physical alignment of magnetic particles in viscous fluids (giving rise to the primary DRM), then touch on the post-depositional processes important to paleointensity in sedimentary systems.

13.4.1 Physical Alignment of Magnetic Moments in Viscous Fluids

The theoretical and experimental foundation for using DRM for paleointensities is far less complete than for TRM. Tauxe (1993) reviewed the literature available through 1992 thoroughly and the reader is referred to that paper for background (see also Valet, 2003). In the last decade there have been important contributions to both theory and experiment and we will outline our current understanding here.

Placing a magnetic moment **m** in an applied field **B** results in a torque Γ on the particle $\mathbf{\Gamma} = \mathbf{m} \times \mathbf{B}$. The magnitude of the torque is given by $\Gamma = mB \sin \theta$, where θ is the angle between the moment and the magnetic field vector. This torque is what causes compasses to align themselves with the magnetic field. The torque is opposed by the viscous drag and inertia and the equation of motion governing the approach to alignment is

$$I \frac{d^2 \theta}{dt^2} = -\lambda \frac{d\theta}{dt} - mB \sin \theta \qquad [4]$$

where λ is the viscosity coefficient opposing the motion of the particle through the fluid and I is the moment of inertia. Nagata (1961) solved this equation by neglecting the inertial term (which is orders of magnitude less important that the other terms) as

$$\tan \frac{\theta}{2} = \tan \frac{\theta_0}{2} e^{(-mBt/\lambda)} \qquad [5]$$

where θ_0 is the initial angle between **m** and **B**. He further showed that by setting $\lambda = 8\pi r^3 \eta$ where r is the particle radius and η to the viscosity of water ($\sim 10^{-3} \, \mathrm{kg \, m^{-1} \, s^{-1}}$), the time constant γ of eqn [5] over which an initial θ_0 is reduced to $1/e$ of its value is

$$\gamma = \frac{\lambda}{mB} = \frac{6\eta}{MB} \qquad [6]$$

where M is the volume-normalized magnetization.

Now we must choose values η, M, and B. As noted by many authors since Nagata himself (see recent discussion by Tauxe, 2006), plugging in reasonable values for η, M, and B and assuming isolated magnetic particles, the time constant is extremely short (microseconds). The simple theory of unconstrained rotation of magnetic particles in water, therefore, predicts that sediments with isolated magnetic particles should have magnetic moments that are fully aligned and insensitive to changes in magnetic field strength. Yet even from the earliest days of laboratory redeposition experiments (e.g., Johnson *et al.*, 1948; see **Figure 11(a)**) we have known that depositional remanence (DRM) can have a strong field dependence and that DRMs are generally far less than saturation magnetizations ($\sim 0.1\%$). Much of the research on DRM has focussed on explaining the strong field dependence observed for laboratory redepositional DRM.

The observation that DRM is usually orders of magnitude less than saturation and that it appears to be sensitive to changing geomagnetic field strengths implies that the time constant of alignment is much longer than predicted by eqn [6]. To increase γ, one can either increase viscosity or decrease magnetization.

One can increase γ by using the viscosity in the sediment column (e.g., Denham and Chave, 1982) instead of the water column. However, something must act to first disrupt the alignment of particles prior to burial, so calling on changes in viscosity is at best an incomplete explanation.

There are several ways of increasing γ by reducing the value of M hence inhibiting the alignment in the first place. For example, one could use values for M much lower than the saturation magnetizations of common magnetic minerals (e.g., Collinson, 1965; Stacey, 1972). However, even using the magnetization of hematite, which is two orders of magnitude lower than magnetite, results in a time constant of alignment that is still less than a second.

There are two mechanisms by which the time constant of alignment can be reduced which account for experimental results of laboratory redeposition experiments: Brownian motion and flocculation. Collinson (1965) called on Brownian motion to disrupt the magnetic moments by analogy to paramagnetic gases. Reasonable parameter assumptions suggest that particles smaller than about 100 nm will be affected by Brownian motion suggesting a possible role in DRM of isolated magnetite grains free to rotate in water. Furthermore, Yoshida and Katsura (1985) presented

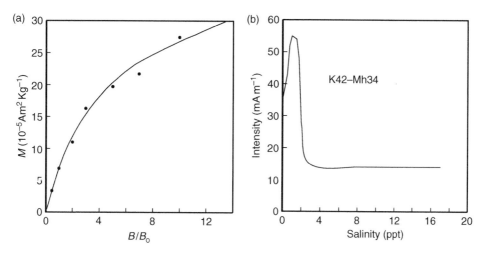

Figure 11 (a) Depositional remanence verus applied field for redeposited glacial varves. B_o was the field in the lab. Data from Johnson *et al.* (1948). (b) Relationship of DRM intensity and salinity for synthetic sediment composed of a mixture of kaolinte and maghemite. Data of Van Vreumingen, 1993b. Data from Tauxe L (2005) *Lectures in Paleomagnetism*, earthref.org/Magic/Books/Tauxe/2005, San Diego.

experiments on the magnetization of suspensions in response to applied fields that were entirely consistent with a Brownian motion model. Flocculation was fingered by Shcherbakov and Shcherbakova (1983) (see also Katari and Bloxham, 2001) who noted that in saline environments, sedimentary particles tend to flocculate and that isolated magnetic particles would be highly unlikely. When magnetic moments are attached to nonmagnetic 'fluff' it is the net magnetization of the floc that must be used in eqn [6], that is, much smaller than the magnetization of the magnetic mineral alone.

The role of water chemistry (e.g., pH and salinity) has been investigated by several authors since the early 1990s (Lu *et al.*, 1990; van Vreumingen, 1993a, 1993b; Katari and Tauxe, 2000; and Tauxe, 2006). In **Figure 11(b)** we replot data from one of the van Vreumingen experiments. The data were obtained by depositing a synthetic mixture of kaolinite, illite, and maghemite under various conditions of salinity. There is an intriguing increase in intensity with small amounts of NaCl followed by a dramatic decrease in intensity which stabilizes for salinities greater than about 4 ppt.

Both the increase and the decrease can be explained in terms of Brownian motion and flocculation, which is encouraged by increasing salinity. The initial increase in intensity with small amounts of NaCl could be the result of the maghemite particles adhering to the clay particles, increasing viscous drag, hence reducing the effect of Brownian motion. The subsequent decrease in intensity with higher salinities could be caused by building composite flocs with decreased net moments, hence

lowering the time constant of alignment. The decrease in net moment with increasing flocculation was also supported by the redeposition experiments of Lu *et al.* (1990), Katari and Tauxe (2000), and Tauxe (2006).

There are therefore two completely different systems when discussing DRM: ones in which magnetic particles remain isolated (e.g, freshwater lakes; see **Figure 12(a)**) and ones in which flocculation plays a role (e.g., marine environments; see **Figure 12(b)**). For the case of magnetite in fresh water, Brownian motion may well be the dominant control on DRM efficiency. In saline waters, the most important control on DRM is the size of the flocs in which the magnetic particles are embedded. In the following we briefly explore these two very different environments.

13.4.1.1 Nonflocculating environments

In freshwater we expect to have isolated magnetic particles whose magnetic moments would presumably be a saturation remanence. The overwhelming majority of laboratory redeposition experiments have been done in deionized water (e.g., Kent, 1973; Lovlie, 1974), hence are in the nonflocculating regime. However, only a few studies have attempted to model DRM using a quantitative theory based on Brownian motion (e.g., Collinson, 1965; King and Rees, 1966; Stacey, 1972; Yoshida and Katsura, 1985). Here we outline the theory to investigate the behavior of DRM that would be expected from a Brownian motion mechanism (henceforth a Brownian remanent magnetization or BRM).

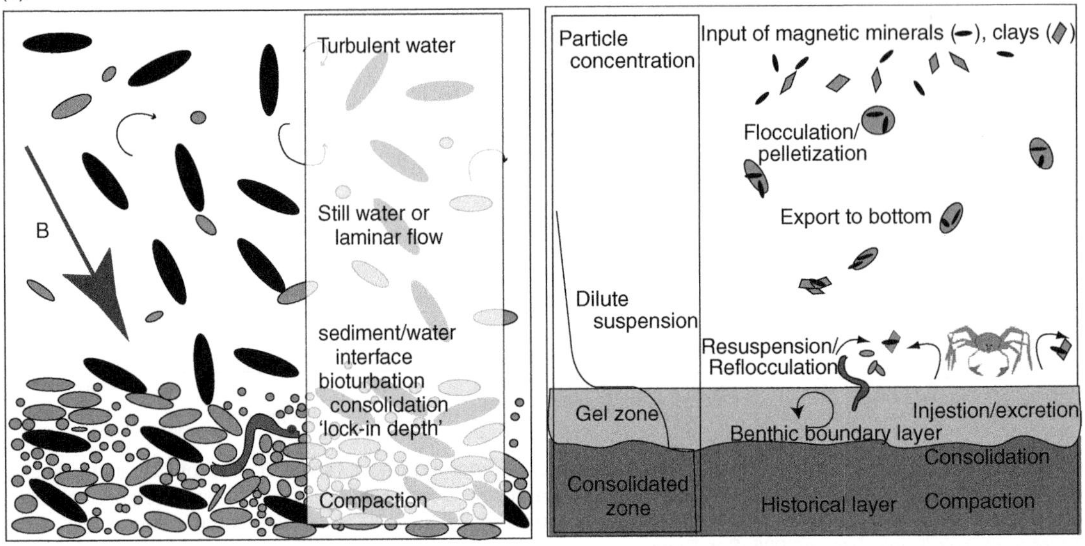

(a)

(b)

Figure 12 (a) Schematic drawing of traditional view of the journey of magnetic particles from the water column to burial in a nonflocculating (freshwater) environment. Magnetic particles are black. (b) View of depositional remanence in a flocculating (marine) environment. (a) Redrawn from Tauxe L (1993) Sedimentary records of relative paleointensity of the geomagnetic field: Theory and practice. *Reviews of Geophysics* 31: 319–354. (b) Redrawn from Tauxe L (2006) Long term trends in paleointensity: The contribution of DSDP/ODP submarine basaltic glass collections. *Physics of the Earth and Planetary Interiors* 156: 223–241.

To estimate the size of particles effected by Brownian motion, Collinson used the equation

$$\frac{1}{2}mB\phi^2 = \frac{1}{2}kT \qquad [7]$$

where ϕ is the Brownian deflection about the applied field direction (in radians), k is Boltzmann's constant $(1.38 \times 10^{-23} \, \mathrm{J\,K^{-1}})$, and T is the temperature in kelvin. The effect of viscous drag on particles may also be important when the magnetic moments of the particles are low (see Coffey *et al.* (1996) for a complete derivation), for which we have

$$\frac{\phi^2}{\delta} = \frac{kT}{4\pi\eta r^3}$$

where δ is the time span of observation (say, 1 s). According to this relationship, weakly magnetized particles smaller than about a micron will be strongly effected by Brownian motion. Particles that have a substantial magnetic moment, however, will be partially stabilized (according to eqn [7]) and might remain unaffected by Brownian motion to smaller particle sizes (e.g., 0.1 μm). In the case of isolated particles of magnetite, therefore, we should use eqn [7] and BRM should follow the Langevin equation for paramagnetic gases, that is,

$$\frac{\mathrm{BRM}}{\mathrm{sIRM}} = \coth\left(\frac{mB}{kT}\right) - \frac{kT}{mB} \qquad [8]$$

To get an idea of how BRMs would behave, we first find m from $M(r)$ (here we use the results from micromagnetic modeling of Tauxe *et al.* (2002)). Then, we evaluate eqn [8] as a function of B for a given particle size (see **Figure 13(a)**). We can also assume any distribution of particle sizes (e.g., that shown as the inset to **Figure 13(b)**), and predict BRM/sIRM for the distribution (blue line in **Figure 13(b)**). It is interesting to note that BRMs are almost never linear with the applied field unless the particle sizes are very small.

BRMs would be fixed when the particles are no longer free to move. The fixing of this magnetization presumably occurs during consolidation, at a depth (known as the lock-in depth) where the porosity of the sediment reduces to the point that the particles are pinned (see **Figure 12(a)**). Below that, the magnetization may be further affected by compaction (e.g., Deamer and Kodama, 1990) and diagenesis (e.g., Roberts, 1995).

13.4.1.2 Flocculating environments

DRM in flocculating environments (saline waters) has been studied in the laboratory by Lu *et al.*

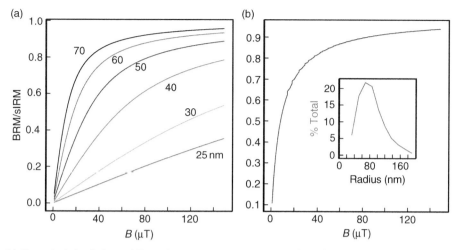

Figure 13 (a) Numerical simulations of Brownian remanent magnetization (BRM) for various sizes of magnetite. (b) BRM simulated for distribution of particle sizes of magnetite shown in inset.

(1990), van Vreumingen (1993a, 1993b), Katari and Tauxe (2000), and Tauxe (2006), and theoretically by Shcherbakov and Shcherbakova (1983), Katari and Bloxham (2001), and Tauxe (2006). We summarize the current state of the theory in the following.

Katari and Bloxham (2001) rearranged eqn [5] by replacing time with settling distance l, a parameter that is more easily measurable in the laboratory using the empirical relationship of settling velocity to radius of Gibbs (1985). They got

$$\tan\frac{\theta}{2} = \tan\frac{\theta_0}{2}\exp\left(-\,mBl/8.8\pi\eta r^{3.78}\right) \qquad [9]$$

As in Nagata (1961), a magnetic moment $\mathbf{m}$ making an initial angle θ_0 with the applied field $\mathbf{B}$ will begin to turn toward the direction of the magnetic field. After time t (or equivalently, settling distance, l), the moment will make an angle θ with the field. Tauxe (2006) showed that the new coordinates of $\mathbf{m}$ (x', y', z') are related to the initial values (x_0, y_0, z_0) by

$$x' = \cos\theta \cdot y' = \sqrt{\frac{1-x_0^2}{1+z_0^2/y_0^2}} \quad \text{and} \quad z' = y'\frac{z_0}{y_0} \quad [10]$$

From the preceding, we can make a simple numerical model to predict the DRM for an initially randomly oriented assemblage of magnetic moments, after settling through l. For an initial set of simulations, Tauxe (2006) followed Katari and Bloxham, using the viscosity of water, m of 5 fAm^2 (where femto (f) $= 10^{-15}$), and a settling length l of 0.2 m. **Figures 14(a)** and **14(b)**, show the predicted DRM curves as a function of magnetic field and radius. We see that particles, in general, are either nearly aligned with the magnetic

field, or nearly random with only a narrow band of radii in between the two states for a given value of B. Increasing B increases the size for which particles can rotate into the field, giving rise to the dependence of DRM intensity on applied field strength. Taking a given particle size and predicting DRM as a function of the applied field (**Figure 14(b)**) predicts the opposite behavior for DRM than the Brownian motion theory (**Figure 13**) in that the larger the floc size, the weaker the DRM and also the more linear with respect to the applied field. The theories of Brownian motion, which predicts low DRM efficiency for the smallest particles increasing to near saturation values for particles around 0.1 μm and composite flocs theory, which predicts decreased DRM efficiency for larger floc sizes can therefore explain the experimental data of van Vreumingen 1993 shown in **Figure 11(b)**.

The flocculation model of DRM makes specific predictions which can in principle be tested if the model parameters can be estimated or controlled. Tauxe (2006) tested the theory by dispersing natural sediments in settling tubes to which varying amounts of NaCl had been introduced. Prior to dispersal, each specimen of mud was given a saturation IRM. They measured DRM as a function of floc size (increasing salinity enhanced floc size) and the applied field (see **Figure 14(c)**). In general their results suggest the following: (1) The higher the NaCl concentration, the lower the net moment (confirming previously published efforts); (2) the higher the salinity, the faster the particles settled (a well-known phenomenon in coastal environments, see, e.g., Winterwerp

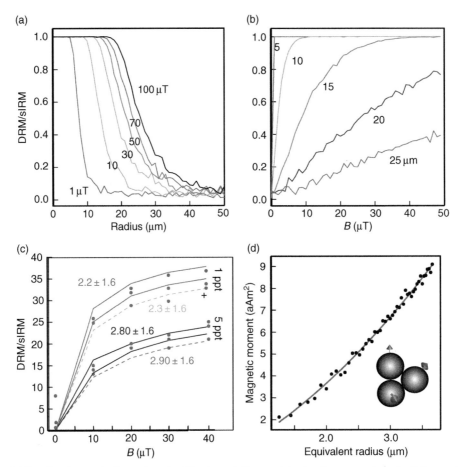

Figure 14 (a) Results of numerical experiments of the flocculation model using the parameters: $\ell = 0.2$ m and the viscosity of water. M/M_o is the DRM expressed as a fraction of saturation, holding $\bar{m}$ constant and varying B. For a given field strength, particles are either at saturation or randomly oriented, except for within a very narrow size range. (b) Same as (a) but plotted versus applied field (B). (c) Results of settling experiments as a function of field (B) in a flocculating environment. The assumed mean and standard deviations of truncated log-normal distributions for floc radii are shown in the legends and are indicated using the different line styles in the figure. (d) m versus equivalent radius for composite flocs as in inset. Line given by polynomial fit $m = ar^2 + br + c$, where $a = 3.61 \times 10^{-7}$, $b = 1.2 \times 10^{-12}$, $c = -2.1 \times 10^{-19}$ is based on a fundamental floc of 1 μm with a measured saturation remanence. Redrawn from Tauxe (2006).

and van Kestern, 2004); (3) the higher the applied field, the higher the DRM, although a saturation DRM appears to be nearly achieved in the 1 ppt NaCl set of tubes by 30 μT (**Figures 14(c)**) and **4**) the relationship of DRM to B was far from linear with applied field in all cases. Moreover, in the Katari and Bloxham (2001) model of DRM, a single magnetic particle is assumed to be embedded in each floc; hence the magnetization of the flocs is independent of floc size. In this view, the saturation DRM (sDRM) should equal the sum of all the individual flocs, that is, sIRM in the case of these experiments. sDRM was well below sIRM in all experiments (see, e.g., **Figure 14(c)**) and no Katari–Bloxham-type model can account for the results.

Tauxe (2006) modified the simple theory of Katari and Bloxham (2001) by incorporating the understanding of flocculation from the extensive literature on the subject. In nature, flocs are formed by coalescing of 'fundamental flocs' into composite flocs. Each fundamental floc would have tiny magnetic particles adhering to them and would have the sIRM imparted prior to settling. As the composite flocs grow by chance encounters with other flocs, the net moment of the composite floc will be the vector sum of the moments of the fundamental flocs (see, e.g., inset to **Figure 14(d)**). They modeled the magnetization of flocs as a function of floc radius (assuming a quasi-spherical shape) through Monte Carlo simulation, an example of which is shown in

Figure 14(d). By choosing reasonable log-normal distributions of flocs for settling tube, their model predicts the curves shown in **Figure 14(c)**, in excellent agreement with the redeposition data.

13.4.2 PostDepositional Processes

It appears that by combining the effects of Brownian motion for nonflocculating environments and a composite floc model for flocculating environments we are on the verge of a quantitative physical theory that can account for the acquisition of depositional remanence near the sediment/water interface. At some point after deposition, this DRM will be fixed because no further physical rotation of the magnetic particles in response to the geomagnetic field is possible. The depth at which moments are pinned is called the lock-in depth. If lock-in depth is selective and some magnetic particles would be fixed while others remain free, there will be some depth (time) interval over which remanence is fixed, resulting in some temporal smoothing of the geomagnetic signal. Physical rotation of particles in response to compaction can also change the magnetic remanence. Other processes not involving postdepositional physical rotation of magnetic particles, including 'viscous' (in the sense of magnetic viscosity) remagnetization and diagenetic alteration resulting in a chemical remanence, may also modify the DRM. All of these processes influence the intensity of remanence and hamper our efforts to decipher the original geomagnetic signal. We will briefly discuss the effects specific to sediments in the following; chemical alteration and viscous remagnetization effect of both TRMs and DRMs will be addressed in Section 13.5.

The 'standard model' of depositional remanence (DRM) acquisition was articulated, for example, by Verosub (1977) and Tauxe (1993). In this view, detrital remanence is acquired by locking in different grains over a range of depths. This phased lock-in leads to both significant smoothing and to an offset between the sediment/water interface and the fixing of the DRM. Many practitioners of paleomagnetism still adhere to this concept of DRM which stems from the early laboratory redeposition experiments which were carried out under nonflocculating conditions (see Section 13.4.1). Several studies on natural marine sediments (e.g., deMenocal et al., 1990; Lund and Keigwin, 1994; and Kent and Schneider, 1995; see also Channell et al., 2004) are frequently cited which suggest a high degree of mobility of magnetic

particles after deposition resulting in sedimentary smoothing and delayed remanence acquisition.

The work of deMenocal et al. (1990) called for a deep lock-in depth of up to ~16 cm for marine sediments based on a compilation of deep sea sediment records with oxygen isotopes and the Matuyama–Brunhes boundary (MBB). However, Tauxe et al. (1996) updated the compilation with twice the number of records and, using the same logic, concluded that, on average, the magnetization is probably recorded within the top few centimeters.

Several papers have revived the deep lock-in debate (e.g., Bleil and von Dobeneck, 1999; and Channell et al., 2004). The former used a complicated lock-in model to explain results not observed anywhere else (substantial reversely magnetized intervals in apparently Late Brunhes Age equatorial sediments). The latter noted that in North Atlantic drift deposits, the mid-point of the MBB is 'younger' isotopically than records with lower sedimentation rates, implying a deep lock-in. However, drift deposits by nature collect sediments from a large catchment area. A particular bit of plankton from the surface waters of the North Atlantic will be transported along the bottom for some time before it finds a permanent home in the drift. The age offset between the isotopic (acquired at the surface) and magnetic ages (acquired at the final point of deposition) obviates the need for a deep lock-in depth.

The most-quoted examples of significant smoothing in natural sediments are those of Lund and Keigwin (1994) and Kent and Schneider (1995). On close examination, the evidence is weak. Lund and Keigwin (1994) postulated that the PSV record of Bermuda Rise, western North Atlantic Ocean was systematically subdued with respect to the PSV recorded in Lake St. Croix stemming from the observed difference in sediment accumulation rate, the Lake St. Croix record having been deposited at a rate several times that of the Bermuda Rise record. They suggested that smoothing the Lake St. Croix data with a 10 or 20 cm moving average window reproduced the Bermuda Rise data with high-frequency features smoothed out, and the amplitude of variation significantly reduced. However, they ignored the age constraints present in the original St. Croix record. Tauxe (2006) showed that a substantially better fit of the Bermuda data could be achieved when the available age constraints are used and no smoothing was required by the data.

The study of Kent and Schneider (1995) showed three records of relative paleointensity across the Matyama/Brunhes boundary (MBB) and interpreted these in terms of sedimentary smoothing. These records came from low and moderate sediment accumulation rates. Hartl and Tauxe (1996) augmented the database of peri-MBB relative paleointensity records using an additional ten records obtained from a wide range of sediment accumulation rates and showed that the single low-sedimentation-rate core of Kent and Schneider (V16–58) most probably had a poorly constrained timescale. Once again, little, if any, smoothing of sedimentary paleointensity records is required.

As sediments lose water and consolidate, compaction can have a strong effect on DRM intensity (e.g., Anson and Kodama, 1987). Consolidation is a continuous process starting from the sediment–water interface when sedimentary particles first gel (see, e.g., **Figure 12(b)**) and continuing until the sediment is completely compacted, perhaps as deep as hundreds of meters. The effect on magnetic remanence depends on volume loss during compaction which depends largely on clay content, so clay-rich sediments will have the largest effect.

13.4.3 Note on Aeolian Deposits

The theoretical and experimental foundations of relative paleointensity studies have all been done on water-borne sedimentary deposits. Nonetheless it is clear that aeolian sediments, in particular, loess, can retain a natural remanence that appears to record the direction of the geomagnetic field (e.g., Heller and Liu, 1982). Details of how the geomagnetic field is impressed on loess deposits are not well known, but mechanisms must include viscous remanence, pedogenic modification (chemical remanence), and perhaps also a remanence acquired at deposition (see e.g., Spassov et al., 2003 for discussion). What controls the intensity of remanence acquired during deposition of wind-blown dust is unknown, yet there have been several attempts to use the normalized remanence in loess as a proxy for geomagnetic intensity variations (e.g., Zhu et al., 1994; Pan et al., 2001; and Liu et al., 2005). These studies rely heavily on the theoretical and experimental work developed for water-lain sediments (see also Spassov et al., 2003); theoretical and experimental efforts must be carried out for the mechanism involved in remanence acquisition in loess.

13.4.4 Normalization

Until now we have considered only how magnetic moments behave when placed in a magnetic field and are allowed to rotate freely. Paleointensity studies in sediments make the *a priori* assumption that DRM is quasi-linear with the applied field (although as we have seen in the section on DRM theory that is only true under certain circumstances). However, we have not yet considered the effect of changing the magnetic content of the sediment which of course will have a profound effect on the intensity of the remanence. Such changes must be compensated for through some sort of normalization process (see e.g., Kent, 1982). Methods of normalization were reviewed thoroughly by King et al. (1983) and Tauxe (1993) but there have been a few contributions to the subject published since. Here we briefly summarize the most commonly used methods of normalization.

Most studies use some easily measured bulk magnetic parameter such as saturation remanence (Johnson et al., 1948), magnetic susceptibility (χ, Harrison, 1966) or anhysteretic remanence (Johnson et al., 1975) which will compensate for changes in concentration of the magnetic minerals in a relatively crude way. Levi and Banerjee (1976) proposed a more sophisticated approach in which the natural remanence is partially demagnetized as is the anhysteretic remanent normalizer to ensure that the same coercivity fraction is used to normalize the remanence as is carrying the natural remanence. Following up on this line of reasoning, Tauxe et al. (1995) suggested that the natural remanence be normalized by anhysteretic remanence in a manner similar to the KTT experiments for thermal remanences using a technique they called 'pseudo-Thellier' normalization. King et al. (1983) reminded us that anhysteretic remanence itself is a strong function of concentration with higher magnetite concentrations being less efficient at ARM acquisition than lower concentrations. As a result, zones with varying concentrations will be normalized differently (in effect, different αs in eqn [1] and violate the fundamental assumptions of the method. More recently, Brachfeld and Banerjee (2000) proposed a secondary correction for normalized intensity that attempted to remove some of the nonlinear effects of the normalization process. Tauxe and Wu (1990) argued that if the power spectrum of the normalizer was coherent with the normalized remanence, the normalization process was insufficient. Constable et al. (1998)

expanded on this idea, suggesting that the normalizer most coherent with the remanence should be used.

One of the important implications of the composite floc model of DRM of Tauxe (2006) described in Section 13.4.1 is that current methods of normalizing sedimentary records for changes in magnetic grain size and concentration do not account for changes in floc size, hence will be only partially effective in isolating the geomagnetic contribution to changes in DRM. This has practical implications in the role of climate in influencing relative paleointensity records. For example, changes in the clay content could well lead to differences in flocculation, which in turn could influence paleointensity with no observable change in the magnetic mineralogy apart from a change in concentration. Other 'stealth' influences could be miniscule changes in salinity of lakes, which could result in profound changes in the paleointensity recorded, with no means of detecting it. However, in stable environments with only small changes in magnetic mineralogy and concentration, we can only hope that the normalization procedures chosen will give records that are reasonably linear with the applied field.

13.5 Remagnetization

Theoretical treatment of how rocks get magnetized and how that magnetization might be used for paleointensity studies assume that the remanence was blocked either thermally (Section 13.3) or depositionally (Section 13.4). Yet almost no natural remanence remains completely unchanged for long. Thermodynamics teaches us that all substances out of equilibrium with their environments will approach equilibrium as the energy available permits. Magnetic particles out of equilibrium with the magnetic field in which they sit are subject to magnetic viscosity. If they are out of chemical equilibrium, they will alter chemically. The former results in the acquisition of a viscous remanence and the latter a chemical one. These are discussed in more detail in Chapter 8. We will briefly describe their importance to paleointensity in the following.

13.5.1 Magnetic Viscosity

Returning to **Figure 2**, we see that magnetic moments can respond to external fields even if the magnetic crystal itself is fixed on timescales determined by the magnetic relaxation time τ. When the

relaxation time is short relative to the time span of observation, the magnetization is in equilibrium with the external field and the particles are called 'superparamagnetic.' This means that magnetic particles have sufficient thermal energy to overcome intervening energy barriers and flip their magnetic moments from one easy direction to another. The energy barrier is in part controlled by the external field, with a lower threshold into the direction of the applied field than out of it. Therefore magnetic moments will tend to 'pool' in the direction of the applied field.

The magnetization that is acquired in this isochemical, isothermal fashion is termed 'viscous remanent magnetization' or VRM. With time, more and more grains will have sufficient thermal energy to overcome anisotropy energy barriers and flip their magnetizations to an angle more in alignment with the external field. The lower the value of τ, the quicker the approach to equilibrium.

According to eqn [2], relaxation time varies with external factors such as temperature (as seen in **Figure 2**) and applied field B and with factors specific to the magnetic particle such as volume and its intrinsic resistance to changing external fields reflected in its anisotropy constant K. In any natural substance, there will be a range of values for τ that could span from seconds (or less) to billions of years. It is interesting to note that a TRM is in effect the equilibrium magnetization (see, e.g., Yu and Tauxe, 2006) and TRMs will only be subject to magnetic viscosity if the field changes. DRMs, however, are typically one or two orders of magnitude less than the TRM that would be acquired in the same field, hence are almost never in equilibrium and therefore will nearly always be subject to viscous remagnetization, depending on the spectrum of τ values (see Kok and Tauxe, 1996a for discussion).

13.5.2 Chemical Alteration

Geological materials form in one environment (e.g., extruding red hot from the mouth of a volcano!) and wind up in quite different environments. Inevitably, they will break down as part of the rock cycle. Magnetic minerals are no exception and growth, alteration and dissolution of magnetic minerals change the original remanence. The magnetization that is fixed by growth or alteration of magnetic minerals is termed chemical remanent magnetization (CRM) and while this too is controlled in part by the external magnetic field, the theory of how to normalize CRM to retrieve the geomagnetic signal has never been properly developed. In general, paleointensity studies strive to recognize CRMs and exclude such remanences from interpretation.

13.6 Evaluating Paleointensity Data

13.6.1 Thermally blocked remanences

A well done paleointensity experiment allows us to test (1) whether the NRM was a single-component magnetization, (2) whether alteration occurred during laboratory reheating, and (3) whether blocking and unblocking were reciprocal, and (4) whether the TRM is a linear function of the applied field. Parameters can be calculated to provide measures of overall quality (scatter about the best-fit line, distribution of temperature steps, fraction of the NRM, etc.) of a given experiment. Some useful parameters are listed for convenience in **Table 1**.

13.6.2 Depositional Remanences

How can sedimentary relative paleointensity data be judged? Here are some thoughts:

1. The natural remanence must be carried by a detrital phase of high magnetic stability. Furthermore, the portion of the natural remanent vector used for paleointensity should be a single, well-defined component of magnetization. The nature of the NRM can be checked with progressive demagnetization using AF and thermal techniques. Supplementary information from hysteresis and rock-magnetic experiments can also be useful.

2. The detrital remanence must be an excellent recorder of the geomagnetic field, exhibit no inclination error, and if both polarities are present the two populations should be antipodal. The associated directional data should therefore be plotted on equal area projections (or at least histograms of inclination) whenever possible.

3. Large changes in concentration (more than about an order of magnitude) and changes in magnetic mineralogy or grain size should be minimized. These changes can be detected with the use of bi-plots of, for example, IRM versus χ. Such bi-plots should be linear, with low scatter.

4. The relative paleointensity estimates that are coherent with bulk rock-magnetic parameters should be treated with caution. Coherence can be assessed using standard spectral techniques.

5. Records from a given region should be coherent within the limits of a common timescale. Whenever possible duplicate records should be obtained and compared.

6. For a relative paleointensity record to have the maximum utility, it should have an independent

Table 1 Parameters

Parameter	Name	Definition/notes	Ref.
$\lvert b \rvert$	Best-fit slope	Slope of pTRM acquired versus NRM remaining	1
B_{anc}	Ancient field estimate	$\lvert b \rvert$ times the laboratory field	1
β	Scatter parameter	Standard error of the slope over $\lvert b \rvert$	1
Q	Quality factor	Combines several parameters	1
VDS	Vector difference sum	Sum of vector differences of sequential demagnetization steps	3
F_{vds}	Fraction of the total NRM	Total NRM is VDS	2
δ_i	pTRM check	Difference between pTRM at pTRM check step at T_i	2
T_{max}	Maximum blocking temperature	Highest step in calculation of $\lvert b \rvert$	2
DRATS	Difference RATio sum	$\sum \delta_i$ normalized by pTRM(T_{max})	2
N_{pTRM}	Number of pTRM checks	Below T_{max}	2
Δ_i	pTRM tail check	Difference between NRM remaining after first and second zero-field steps	2
MD%	Percent maximum difference	$100 \times$ maximum value of Δ_i/VDS	2
T	Orientation matrix	Matrix of sums of squares and products of demagnetization data	3
τ_i	Eigenvalues of **T**	$\tau_1 > \tau_2 > \tau_3$	3
$\mathbf{V_i}$	Eigenvectors of **T**	Best-fit direction is $\mathbf{V_1}$	3
MAD	Maximum angle of deviation	$\tan^{-1}\left(\sqrt{(\tau_2^2 + \tau_3^3)}/\tau_1\right)$	4
DANG	Deviation ANGle	Angle between origin and $\mathbf{V_1}$	2

1, Coe *et al.*, (1978); 2, Tauxe and Staudigel (2004); 3, Tauxe (1998); 4, Kirschvink (1980).
Table from Tauxe L (2006) Long term trends in paleointensity: The contribution of DSDP/ODP submarine basaltic glass collections. *Physics of the Earth and Planetary Interiors* 156: 223–241.

timescale. Many deep-sea sediment records are calibrated using oxygen-isotopic curves or magnetostratigraphic age constraints (or both). Lake sediments are more difficult to date and rely for the most part on radiocarbon ages.

13.7 Current State of the Paleointensity Data

13.7.1 Paleomagnetic Databases

There has been an enormous effort in collecting and preserving paleomagnetic data since the early 1960s (e.g., Irving, 1964) but since the 1987 meeting of the IAGA in Vancouver the effort has been more concerted with seven IAGA-sponsored databases. Absolute paleointensities have been assembled in a series of compilations by Tanaka and Kono (1994), Perrin and Shcherbakov (1997), Perrin *et al.* (1998), and more recently Perrin and Schnepp (2004). This most recent version (here referred to as PINT03) contains data from 3128 cooling units from 215 references through 2003 and is available for downloading along with the other IAGA-sponsored databases at the National Geophysical Data Center (NGDC) website. There are also a number of paleointensity data (both relative and absolute) in the TRANS database also available at the NGDC website. This database contains data associated with polarity transitions.

In their assessment of the most recent release of the absolute paleointensity database, Perrin and Schnepp (2004) stated:

For the future, a harmonization or a combination of all IAGA databases would be desirable. Furthermore, the input of raw data at the specimen level would be useful in order to allow reinterpretation of data with more developed and sophisticated methods based on our increasing understanding of rock magnetism.

In order to address this widely felt sentiment, the MagIC database was created which can be accessed at its website. This database has merged several of the existing IAGA databases and allows for data ranging from original magnetometer output (including magnetometer, hysteresis, thermomagnetic, susceptibility, and other measurements) and their interpretations. Detailed descriptions of the data are possible by using 'method codes.'

The absolute paleointensity database of Perrin and Schnepp (2004) can now be accessed through the MagIC website by searching for that reference.

We have updated it to include data published through late 2006 (Bowles *et al.* 2006; Carvallo *et al.*, 2004, Garcia *et al.*, 2006; Gee *et al.*, 2000; Halls and Davis, 2004; Herrero-Bervera and Valet, 2005; Leonhardt *et al.*, 2003; Macouin *et al.*, 2006; Riisager *et al.*, 2004; Smirnov and Tarduno, 2005; Tarduno and Cottrell, 2005; Tarduno *et al.*, 2002; Tauxe *et al.*, 2004a, 2004b; Tauxe and Staudigel, 2004; Tauxe, 2006; Yamamoto and Tsunakawa, 2005; Yoshihara and Hamano, 2004; Yoshihara *et al.*, 2003; Zhao *et al.*, 2004, Zhu *et al.*, 2001a, 2001b). Data can be retrieved through each individual reference, or by searching for this reference (Tauxe and Yamazaki, 2007) to retrieve the entire absolute paleointensity database. Data from the Tauxe references include everything down to the original measurement data as an example of the power of the MagIC database concept. We refer to our compilation of paleointensity data as PINT06 in the following.

The data in the PINT03 and PINT06 databases include information on geographic location (see map of data locations in **Figure 15(a)**), rock type and age of the sampling sites, type of paleointensity experiment, the remanence vector (including direction if available), and summary statistics such as the standard deviation of replicate specimens from a given cooling unit. If no measurement data are available, we may know that, for example, pTRM checks were performed (e.g, studies with pTRM listed under alteration check), but we do not know whether they 'passed' any particular criterion. The only reliability criteria included is the standard deviation of the replicate measurements from a given cooling unit. Because there are many useful reliability criteria for judging paleointensity data (see Section 13.6), efforts should be made to update the contributions in the MagIC database to include as many of these as are available.

Perrin and Schnepp (2004) ably summarized the characteristics of the PINT03 database and we will not repeat their analysis here. Nonetheless, it is useful to reiterate that most of the data come from the last million years and are from the Northern Hemisphere. The temporal bias is particularly egregious when only the most 'reliable' data are used (i.e., that employed TRM normalization with pTRM checks.)

There is no IAGA database for relative paleointensity data (except those included in the TRANS database). As a step toward rectifying this problem, we summarize the published literature with relative paleointensity data in **Table 2**. Locations of records

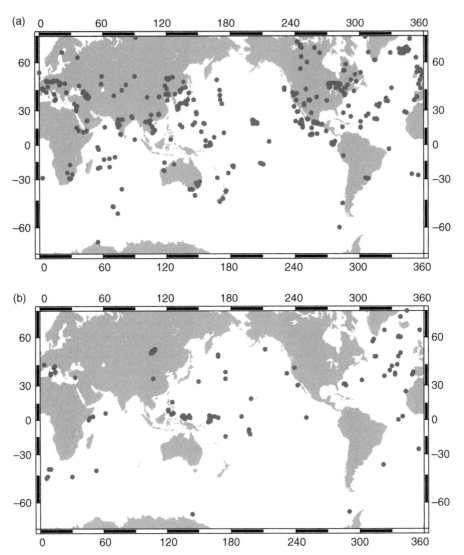

Figure 15 Locations of all paleointensity data in the (a) absolute (PINT06) and (b) relative (SEDPI06) databases compiled for this chapter.

are shown in **Figure 15(b)**. We have obtained data from nearly 100 references and contributed them to the MagIC database (obtainable individually through the original reference or collectively through this reference). We refer to this compilation of relative paleointensity data as SEDPI06 in the following. Authors are encouraged to contribute or augment their own data in the database. In Section 13.8 we will discuss the highlights of the available paleointensity data from both the PINT06 (absolute) and SEDPI06 (relative) compilations. Before we discuss the global dataset, we will first describe methods of converting to virtual dipole moment.

13.7.2 Conversion to Virtual Dipole Moment

13.7.2.1 Absolute paleointensity data

The intensity of the magnetic field varies by a factor of two from equator to pole simply as a result of a dipole source, so data from different latitudes must be normalized to take this inherent dipole variation into account. In the following we discuss methods for converting intensities to 'virtual dipole moments' in both absolute and relative paleointensity data sets.

There are several ways to calculate equivalent dipole moments for paleointensity data. Early studies tended to present a given intensity result as a ratio

Table 2   Summary of relative paleointensity records

Record	Latitude/Longitude	Age range	Dating method	Compilation	TS	Ref.
10-pc03[a]	−47/6	23–115 ka	RPI	69	44	69,7
1010[a]	30/−118	19–2036 ka	POL,MS	22	4,1	40,27
1021[a, b]	39/−128	13–1562 ka	POL	23	4,1	40,24
1089[a]	−41/10	20–578 ka	δ^{18}O	69	44	70
1092	−46/7	5.9–3.5 Ma	POL		4	18
1101[a, b]	−64/−70	706–1105 ka	POL,MS, δ^{18}O	22	4,66	21
21-pc02[a]	−41/8	0–81 ka	δ^{18}O	69	44	69,70,7
305-a5[a]	53/106	0–11 ka			44	54
337-l2[a]	53/106	13–84 ka	RPI		44	54
4-pc03[a]	−41/10	9–44 ka	RPI	69	44	69,7
5-pc01[a]	−41/10	8–64 ka	RPI	69	44	69,7
522[a]	−26/−5	22.8–34.7 Ma	POL		4	76
606a[a]	37/323	773–792 ka	POL	26	4	13
609b[a]	50/336	777–825 ka	POL	26	4	26
664d[a]	0/336	670–807 ka	POL	26	4	81
665a[a]	3/340	770–817 ka	POL	26	4	81
767[a, b]	5/124	601–1518 ka	POL	22	4	22,61
767b[a]	5/124	759–829 ka	POL	22	4	22,61
768a[a]	8/121	5–94 ka	δ^{18}O, [14]C	23	41,79	63
768b[a]	8/121	9–130 ka	δ^{18}O, [14]C	23	41,79	63
769[a]	9/121	5–831 ka	δ^{18}O, [14]C	23	41,79,66	62,63
803a[a, b]	2/161	783–2178 ka	POL	22	4	33
803b[a, b]	2/161	1487–2786 ka	POL	22	4	33
804c[a]	1/161	1448–1470 ka	POL	26	4	26
805b[a]	1/160	770–821 ka	POL	26	4	26
848–851[a]	2/−110	34–4035 ka	POL	23	4	82
877[a]	54/−148	9.4–11.3 Ma	POL		4	3
882b	50/168	0–200 ka	∼		66	52
883	51/168	15–200 ka	δ^{18}O		66	58
884	51/168	15–200 ka	MS		66	58
884[a, b]	51/168	9.9–10.3 Ma	POL		4	56
983[a, b]	60/−24.1	0–1889 ka	POL, δ^{18}O	22,34	4,66	10,11,12,6,8,9
984[a, b]	60.4/−23.6	0–2151 ka	POL, δ^{18}O		4,66	11,8,9
Chewaucan	43/−121	65–102 ka	[14]C, ThL			57
Sed-17aK[a]	25/−17	9–224 ka	δ^{18}O		29	25
ch88-10p[a]	30/−73	10–70 ka			44	64
ch88-11p	31/−74	10–70 ka	∼		44	64
ch89-1p	31/−75	12–71 ka	∼		44	65
con-01-603-2	54/109	10–200 ka	POL, ^{14}C, ∼			16
con-01-604-2	52/106	0–60 ka	ARM			16
con-01-605-3	52/105	0–40 ka	ARM			16
ded8707[a]	40/14	10–60 ka	Ash	23	29	80
ded8708	40/14	40–80 ka	Ash		29	80
e113p[a]	−2/159	4–380 ka	δ^{18}O	22	1	74
hu90-013-012	59/−47	10–110 ka	RPI, δ^{18}O, [14]C			71
hu90-013-013	58/−48	10–110 ka	RPI, [14]C			71
hu91-045-094	50/−45	10–110 ka	RPI, δ^{18}O, [14]C			71
ket8251[a]	40/14	8–95 ka	ash	23	29	80
kh73-4-7	3/165	0–2000 ka	POL			59
kh73-4-8	2/168	0–2000 ka	POL			59
kh90-3-5	4/160	32–1159 ka	POL		4	60
kk78-030[a]	19/−161	601–1785 ka	POL	22	4	35
kr9912-pc2[a]	−11/−163	1003–3000 ka	POL, ARM		4,42	86
kr9912-pc4[a]	−13/−162	2002–2845 ka	POL		4,42	86
kr9912-pc5[a]	−9/−163	1295–2118 ka	POL		4,42	86
ks87-752[a, b]	−38/−38	311–1023 ka	POL, MS	23	4	83
lc07[a]	38/10	754–1033 ka	POL, ∼	22	66	17
ldb[a]	45/4	20–308 ka	MS, [14]C		15	85,78

(Continued)

Table 2   (Continued)

Record	Latitude/Longitude	Age range	Dating method	Compilation	TS	Ref.
massicore	44/14	32–35 Ma	POL		4	37
md01-2440	38/−11	2–400 ka	RPI, MS		44	77
md01-2441	38/−11	30–54 ka	RPI, MS		44	77
md84-528	−42/53	15–80 ka	δ^{18}O		29	80
md84-629	36/33	15–62 ka	δ^{18}O		29	80
md85-668[a]	0/46	21–187 ka	δ^{18}O	23	44	45
md85-669[a]	2/47	20–138 ka	RPI	23	44	45
md85-674[a]	3/50	18–138 ka	RPI	23	44	45
md90-0940[a]	6/62	108–1954 ka	POL, MS, fossils	23	4	46
md95-2009[a]	63/−4	10–76 ka	ARM	34	20	32,34
md95-2024[a]	50/−46	1–117 ka	MS, δ^{18}O	68	2	68
md95-2034[a]	34/−58	12–76 ka	ARM	34	20	32,34
md95-2039	40/−10	0–320 ka	δ^{18}O, ^{14}C		44	77
md95-2042	40/−10	32–160 ka	δ^{18}O, ^{14}C		44	77
md97-2140[a, b]	2/142	568–1465 ka	POL		4	5
md97-2143[a, b]	16/125	601–2226 ka	POL, δ^{18}O	22	4,38	28
md98-2181[a]	6/126	12–660 ka	δ^{18}O, ^{14}C		67	73
md98-2183[a, b]	2/135	20–1193 ka	POL, MS, ARM	86	4,42	86
md98-2185[a, b]	3/134	9–2256 ka	POL, MS, ARM	22	4,42	86
md98-2187[a, b]	4/135	51–3053 ka	POL, MS, ARM		4,42	86
md99-2334	38/−11	0–38 ka	RPI		44	77
ngc16[a]	2/135	2–191 ka	MS	23	44	87
ngc26[a]	3/135	1–120 ka	MS	23	44	87
ngc29[a]	4/136	2–192 ka	MS	23	44	87
ngc36[a]	−1/161	1–546 ka	δ^{18}O	23	29,1	88
ngc38[a]	−15/175	9–406 ka	δ^{18}O	23	29,1	88
ngc65[a]	35/175	6–635 ka	S		1	89
ngc69[a, b]	40/175	7–881 ka	S		1	89
np35[a]	4/141	127–798 ka	δ^{18}O	23	29,1	88
np5[b]	1/137	8–196 ka	δ^{18}O	23	44	87
np7[a]	2/138	6–199 ka	MS	23	44	87
p012[a]	59/−47	14–177 ka	RPI	23	23	72
p013[a]	58/−48	14–277 ka	RPI	23	23	72
p094[a]	50/−46	2–111 ka	RPI	23	23	72
p226[a]	3/−170	41–780 ka	POL	23	4	88
ps1535-10	79/2	0–100 ka	^{14}C			48
ps1535-6	79/2	0–100 ka	^{14}C			48
ps1535-8	79/2	0–100 ka	^{14}C			48
ps1707-2	73/−14	0–80 ka	MS		44	49
ps1852-2[a]	70/−16	4–283 ka	MS		29,44	50
ps1878-3	73/−9	0–100 ka	^{14}C			48
ps1878-3	73/−10	0–45 ka	δ^{18}O, ^{14}C		44	49
ps2138-1	82/31	10–75 ka	δ^{18}O, ^{14}C			47
ps2644-5[a]	68/−22	12–76 ka	ARM	34	20	32,34
rc10-167[a]	33/150	11–781 ka	POL	26,23	4	30
rndb75p[a]	2/160	124–668 ka	δ^{18}O	23	29,66	75
su90-24[a]	63/−37	11–76 ka	ARM	34	20	32,34
su90-33[a]	60/−22	12–76 ka	ARM	34	20	32,34
su9003	41/−32	10–240 ka	Color		55	84
su9004	41/−32	0–240 ka	Color		55	84
su9008	44/−30	10–180 ka	δ^{18}O		55	84
su9039	52/−22	0–240 ka	δ^{18}O		55	84
su92-17[a]	39/−27	4–280 ka	Color		44	39
su92-18[a]	38/−27	4–280 ka	δ^{18}O	23	44	39
su92-19[a]	38/−27	4–279 ka	Color	23	44	39
v16-58[a]	−46/30	767–770 ka	POL	26	4	31
ver98-1-1	53/108	20–60 ka	ARM			16
ver98-1-14	54/108	0–350 ka	ARM			16

(Continued)

Table 2 (Continued)

Record	Latitude/Longitude	Age range	Dating method	Compilation	TS	Ref.
ver98-1-3	54/108	50–250 ka	ARM			16
ver98-1-6[a]	54/108	65–235 ka	Silica		44	51
Kotsiana	36/24	—				36
Lingtai[a]	35/107	10–73 ka	MS, ^{14}C, ThL			53
Potamida	36/24	—				36
Ir[a, b]	43/13	90–94.9 Ma	POL, fossils		19	14
WEGAstack	−65/144	0–800 ka	RPI, ^{14}C, fossils			43
MBstack[a]	3/162	32–1159 ka	POL	22	4	60
PMstack[a]	30/10	0–402 ka	RPI, δ^{18}O, [14]C		44	77
NAstack[a]	45/−25	10–250 ka	δ^{18}O		55	84

[a]Submitted to the MagIC database.
[b]Ages recalculated.
TS, timescale.
Dating methods: RPI, relative paleointensity; POL, polarity stratigraphy; MS, correlation of magnetic susceptibility; ARM, correlation of ARM; carb., correlation of calcium carbonate; ash, tephrostratigraphy; color, correlation of color; δ^{18}O, oxygen isotopes; ^{14}C, radiocarbon; ∼, correlation of some unspecified wiggle; ThL, thermoluminescence; S, correlation of high to low coercivity IRM; silica, correlation of silica variations; fossil, correlation based on fossils.
References: 1, Bassinot *et al.* (1994); 2, Bender *et al.* (1994); 3, Bowles *et al.* (2003); 4, Cande and Kent (1995); 5, Carcaillet *et al.* (2003); 6, Channell *et al.* (2000); 7, Channell and Kleiven (2000); 8, Channell *et al.* (2002); 9, Channell *et al.* (2004); 10, Channell *et al.* (1997); 11, Channell *et al.* (1998); 12, Channell (1999); 13, Clement and Kent (1986); 14, Cronin *et al.* (2001); 15, Dansgaard *et al.* (1993); 16, Demory *et al.* (2005); 17, Dinares-Turell *et al.* (2002); 18, Evans and Channell (2003); 19, Stratigraphy (2004); 20, Grootes and Stuiver (1997); 21, Guyodo *et al.* (2001); 22, Guyodo and Valet (2006); 23, Guyodo and Valet (1999); 24, Guyodo *et al.* (1999); 25, Haag (2000); 26, Hartl and Tauxe (1996); 27, Hayashida *et al.* (1999); 28, Horng *et al.* (2003); 29, Imbrie *et al.* (1984); 30, Kent and Opdyke (1977); 31, Kent and Schneider (1995); 32, Kissel *et al.* (1999); 33, Kok and Tauxe (1999); 34, Laj *et al.* (2000); 35, Laj *et al.* (1996); 36, Laj *et al.* (1996b); 37, Lanci and Lowrie (1997); 38, Laskar *et al.* (1993); 39, Lehman *et al.* (1996); 40, Leonhardt *et al.* (1999); 41, Linsley and Thunnell (1990); 42, Lisiecki and Raymo (2005); 43, Macri *et al.* (2005); 44, Martinson *et al.* (1987); 45, Meynadier *et al.*, (1992); 46, Meynadier *et al.* (1994); 47, Nowaczyk and Knies (2000); 48, Nowaczyk *et al.* (2003); 49, Nowaczyk and Antonow (1997); 50, Nowaczyk and Frederichs (1999); 51, Oda *et al.* (2002); 52, Okada (1995); 53, Pan *et al.* (2001); 54, Peck *et al.* (1996); 55, Pisias *et al.* (1984); 56, Roberts and Lewin-Harris (2000); 57, Roberts *et al.* (1994); 58, Roberts *et al.* (1997); 59, Sato and Kobayashi (1989); 60, Sato *et al.* (1998); 61, Schneider *et al.* (1992); 62, Schneider (1993); 63, Schneider and Mello (1996); 64, Schwartz *et al.* (1996); 65, Schwartz *et al.* (1998); 66, Shackleton *et al.* (1990); 67, Sowers *et al.* (1993); 68, Stoner *et al.* (2000); 69, Stoner *et al.* (2002); 70, Stoner *et al.* (2003); 71, Stoner *et al.* (1995); 72, Stoner *et al.* (1998); 73, Stott *et al.* (2002); 74, Tauxe and Wu (1990); 75, Tauxe and Shackleton (1994); 76, Tauxe and Hartl (1997); 77, Thouveny *et al.* (2004); 78, Thouveny *et al.* (1994); 79, Tiedemann *et al.* (1994); 80, Tric *et al.* (1992); 81, Valet *et al.* (1989); 82, Valet and Meynadier (1993); 83, Valet *et al.* (1994); 84, Weeks *et al.* (1995); 85, Williams *et al.* (1998); 86, Yamazaki and Oda (2005); 87, Yamazaki and Ioka (1994); 88, Yamazaki *et al.* (1995); 89, Yamazaki (1999).

with some expected field. For example, Thellier and Thellier (1959) normalized intensity data to a reference inclination of 65°, using the paleomagnetically determined inclination and the relationship between inclination and field strength expected from a magnetic dipole. Most studies published over the last few decades however express paleointensity in terms of the equivalent geocentric dipole moment which would have produced the observed intensity at that (paleo) latitude. There are two ways in which this is done, the virtual dipole moment (VDM; **Figure 16(a)**) and the virtual axial dipole moment (VADM; **Figure 16(b)**). The VDM (Smith, 1967) is the moment of an geocentric dipole that would give rise to the observed magnetic field vector at location *P*. (The piercing point on the surface of the globe of this moment is the virtual geomagnetic pole or VGP.) To get the VDM, we first calculate the magnetic (paleo) colatitude θ_m from the observed inclination I and the so-called dipole formula ($\tan I = 2 \cot \theta_m$).

Then, assuming a centered (but not axial) magnetic dipole with moment VDM we have

$$\text{VDM} = \frac{4\pi r^3}{\mu_0} B_{\text{anc}} \left(1 + 3\cos^2\theta_m\right)^{-1/2} \quad [11]$$

The VDM calculation requires a good estimate of the inclination, which is not always available, especially when unoriented specimens are used. In such cases, it may be possible to use either the site (co) latitude or a paleo (co) latitude estimated by a plate reconstruction in the place of magnetic colatitude in eqn [11]. This moment is known as the VADM (Barbetti *et al.*, 1977).

In order to compare the two forms of normalization, we selected data from the PINT06 database for the last 200 My that (1) were obtained with thermal normalization and used pTRM checks, (2) had multiple specimens that had standard deviations less than 15% of the mean or were less than 5 µT. We estimated paleolatitudes for the sampling sites using the

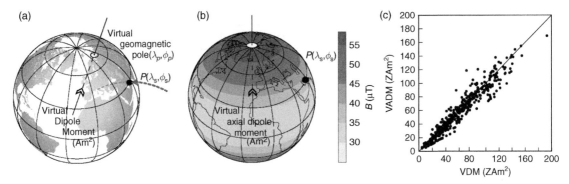

Figure 16   (a) The virtual dipole moment VDM is the geocentric dipole that would give rise to the observed geomagnetic field vector at the location P. λ_s, ϕ_s are the site latitude and longitude, respectively. (b) The virtual axial dipole moment is the geocentric axial dipole that would give rise to the observed intensity at P. (c) Comparison of VDM and VADM for paleointensity data (see text).

global apparent polar wander paths of Besse and Courtillot (2002) and used these to calculate VADMs for many sites. We show the two estimates of dipole moment in **Figure 16(c)**; the two are essentially equivalent representations.

It is important to note here that neither VDMs nor VADMs actually represent the true dipole moment (see Korte and Constable, 2005). They do not take into account the rather substantial effect of the nondipole field contributions and in fact overestimate the true dipole moment based on an evaluation of data for the last 7000 years.

13.7.2.2   Relative paleointensity data

Sedimentary paleointensity data are at best 'relative' paleointensity. Nonetheless several studies have attempted to calibrate relative paleointensity data into a quasi-absolute form and cast them as 'VADMs' in order to compare them with the igneous data sets. There are different strategies for accomplishing this conversion (see **Figure 17**): setting the 'floor' to some minimum value expected for the field (Constable and Tauxe, 1996), setting parts of the sedimentary record to be equal to coeval igneous records (e.g., Guyodo and Valet, 1999; Valet *et al.*, 2005) or setting the mean value to be some assumed value.

In the Constable and Tauxe (1996) method, the nonaxial dipole field is assumed to be on average 7.5 µT as it is for the present field. Reasoning that because the axial dipole must go through zero in a polarity transition, the average transitional field should be about 7.5 µT. **Figure 17(a)** illustrates an application of this method to calibrate the Oligocene relative paleointensity data from DSDP Site 522 of Tauxe and Hartl (1997; see **Table 2**) into VADM values. Setting the average value of the intensity in

transitional records (red square in **Figure 17(a)**) to a value of 7.5 µT calibrates the entire record to B* in µT. Assuming a paleolatitude of 32° S allows these B* values to be converted to VADM values using eqn [11]. The problem with this method is that it is extremely sensitive to the choice of the nonaxial dipole field 'floor' value, which is not known for ancient times. Small changes in the choice of floor result in large changes for the calibrated record.

A different approach was take by Guyodo and Valet (1999) who collected together many relative paleointensity records spanning the last 800 ky (see compilation reference #23 in **Table 2**). The 'SINT-800' stack (see **Figure 17(b)**) overlapped a sequence of absolute paleointensity data whose ages were well known (red dots). These absolute data were used to calibrate the SINT-800 stack into VADM. Valet *et al.* (2005) extended the relative paleointensity stack to span the last 2 My (see **Figure 17(c)**). In this latest version, known as the SINT-2000 stack (a subset of the records compiled in reference #22), they took the global paleointensity data in the PINT03 database (with no selection criteria), and averaged them into 100 000 year bins (red dots). These were used to convert the SINT-2000 stack to VADM values. The two calibrations are somewhat different, with the latter version being higher on average.

Because amplitudes of relative paleointensity records must be related to latitude it is preferable to convert individual records to VADM prior to stacking, instead of stacking first and then converting to VADM. However, the 'floor setting' method of Constable and Tauxe (1996) required transitional data, which are not always available and has severe drawbacks of its own as mentioned before.

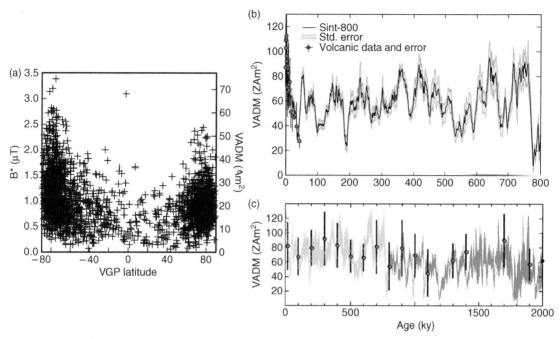

Figure 17 Calibration of sedimentary relative paleointensity data to quasi-absolute values. (a) Approach of Constable and Tauxe (1996) on relative paleointensity data from the Oligocene record at DSDP Site 522. Data associated with transitional fields (low VGP latitudes) are assumed to be on average 7.5 μT, the value of the present average non-GAD field. This converts the relative intensity data to μT. Taking a paleolatitude of 32°, Tauxe and Staudigel (2004) converted these data to VADM using eqn [12]. (b) Stacked record of relative paleointensity data spanning the last 800 ky (SINT-800) of Guyodo and Valet (1999). They used the VADM data from contemporaneous lava flows to calibrate the record to VADM. (c) Similar to (b) but data span the last 2 My (SINT-2000) from Valet *et al.* (2005). They used averages of the PINT03 database (red dots) to calibrate the SINT record to VADM.

13.8 Discussion

In the following, we will discuss some of the 'hot topics' in paleointensity. The issues for many of these are still under debate and conclusions are still tentative. Nonetheless, the spirit of this volume is to present the 'state of the field' and we will endeavor to do so.

13.8.1 Selection Criteria from the PINT06 Database

For the purpose of this discussion, we selected data from the PINT06 database that had either standard deviations ≤5 μT or 15% or the mean. These were divided into those that used the so-called KTT experimental protocol with pTRM checks (here called 'strict') and those that did not (mostly KTT without pTRM checks and 'Shaw'-type experiments; here called 'loose'). Those that did not meet the consistency standard are labeled 'rejected.' As noted before, there is very little else to go on in the database as it currently stands. We plot these data versus age in **Figure 18**.

There are a total of 1504 cooling unit averages from the 'strict' group with a mean and standard deviation of 63.6 ± 34 ZAm^2 ('Z' stands for Zetta (10^{21})) and 1133 from the 'loose' experimental group with a mean of 62.8 ± 37 ZAm^2. A total of 1115 results were rejected on the grounds of poor within-cooling-unit reproducibility (or single-specimen results). These had a mean and standard deviation of 68.0 ± 35.5 ZAm^2. The standard errors of the means of these three data sets are: 0.9, 1.1, and 1.1, respectively, so while the 'strict' and 'loose' data sets have indistinguishable means, the rejected mean is significantly higher than the other two. The differences between the three data sets are illustrated using cumulative distributions in **Figure 19**. The strict and loose data sets have indistinguishable means, with the 'loose' data set having more scatter (more of both high and low values), while the mean of the 'rejected' group is significantly higher.

In this very large data set, there is no statistical difference between means of the 'strict' and 'loose' data sets (although the 'loose' data set has significantly more scatter) and the mean is significantly lower

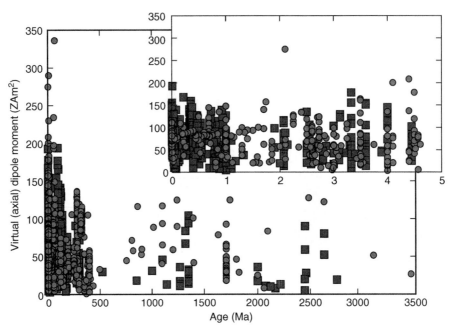

Figure 18 Summary of published data meeting the criteria of standard deviations of the mean being either less than 15% of the mean or less than 5 μT. Blue squares are the 'strict' data selection (KTT with pTRM checks) from the PINT06 database and the red dots are the 'loose' (mostly KTT without pTRM checks and 'Shaw'-type experiments); see text. The inset shows only data from the last 5 million years.

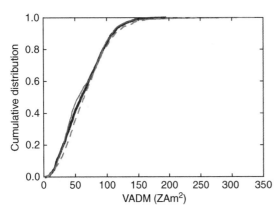

Figure 19 Cumulative distributions of 'strict' (heavy blue line; K T T-type experiment with pTRM checks), 'loose' (light red line; other experimental designs), and 'rejected' (dashed green line); all data points with standard deviations >15% of the mean and > 5 μT. The strict and loose data sets are similar while the rejected data are significantly different with a higher mean.

'loose' criteria data set suggests that some of the scatter is not solely geomagnetic in origin. Furthermore, using data with poor within-site consistency results in a significantly different and higher estimate for the mean value. We will not consider these 'rejected' data in the following discussion.

Although the data set is by now quite large, the age distribution of paleointensity data is still quite patchy with 39% of the data being younger than 1 Ma. By far the most data has come from the Northern Hemisphere (see **Figure 15(a)**). So one point of agreement among all papers on the subject and that is that more and better data would be helpful for defining the average paleofield intensity and its variation.

13.8.2 What is the Average Strength of the Geomagnetic Field?

The present Earth's magnetic field is well approximated by a geocentric magnetic dipole with a moment of about 80 ZAm². But what is the average value of the dipole moment? A great deal of effort has been put into assembling paleointensity databases over more than three decades, yet there remains little consensus on the answer to this most basic question.

Early studies suggested that the average field strength has either been quite a bit lower than the

than the present field. While the highest values are in the 'loose' category, there are few of them and the means are indistinguishable. This does not mean that experimental design makes no difference; the discussion of theory makes it quite clear that many things can give an erroneous result and these things should be tested for. The fact that there is higher scatter in the present (e.g., Smith (1967) and Coe (1967)) or approximately equivalent to today's field (Kono, 1971; Bol'shakov and Solodonikov, 1980; and McFadden and McElhinny, 1982). Some studies found no trend with age in VDMs (e.g., Bol'shakov and Solodonikov, 1980) over the last few hundred million years, while others found a significant increase in dipole moment from the Mesozoic toward the present (e.g., Smith, 1967). Tanaka *et al.* (1995) estimated the average dipole moment for the last 20 My to be approximately $84 \, ZAm^2$ with significantly lower values in the Mesozoic (the so-called 'Mesozoic Dipole Low' of Prévot *et al.*, 1990), a view also held by Perrin and Shcherbakov (1997) and recently reiterated by Biggin *et al.* (2003). But a series of recent papers have argued for a lower average (Juarez *et al.*, 1998; Juarez and Tauxe, 2000; Selkin and Tauxe, 2000; Yamamoto and Tsunakawa, 2005; Tauxe, 2006). In this view, the Mesozoic Dipole Low was not 'low' but was of average paleomagnetic intensity.

The lack of consensus stems in part from differing views on which data to include as well as the explosive growth of paleointensity data available (compare, e.g., Selkin and Tauxe, 2000; Heller *et al.*, 2002; Biggin *et al.*, 2003; and Goguitchaichvili *et al.*, 2004). While Biggin *et al.* (2003) argue that because such procedures, as the so-called 'pTRM check' designed to identify alteration during the paleointensity experiment, cannot guarantee the quality of a particular result, there is no need to reject data that do not have pTRM checks; others (e.g., Riisager and Riisager, (2001) and Tauxe and Staudigel (2004)) have tried to develop more rigorous experimental protocols to detect and reject 'bad' data. For the purposes of discussion, we here take the broad view advocated by Biggin *et al.* (2003), relying strongly on strict consistency tests at the cooling unit level.

As a first look at what the average field might be, we consider the data from **Figure 18** (excluding the 'rejected data') that meet one of the following conditions: (1) they have ages less than 5 Ma and we can use the present latitude as the 'paleolatitude' (λ); (2) they have model latitudes based on plate reconstructions or derived from the apparent polar wander paths of Besse and Courtillot (2002); or (3) they have inclination (I) data associated with intensities whereby we can approximate paleolatitude using the dipole formula ($\tan (I) = 2 \tan \lambda$). We binned the data in $10°$ latitudinal bins (see **Figure 20**). The error bars are the standard errors of the mean. Also shown is the intensity at each latitude expected from the average

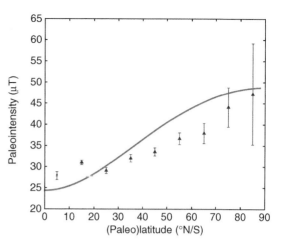

Figure 20 Paleointensity versus (paleo) latitude (see text). The error bars are the standard error of the mean for each paleolatitude bin.

dipole moment of $63 \, ZAm^2$ (the average estimated in the previous section for the whole data set).

Although there is some trend with latitude, the overall fit of the paleointensity data shown in **Figure 20** to a dipole field is poor. The equatorial results are 'too high' and mid-latitude results are 'too low.' Reasons for the failure of the dipole hypothesis in the PINT06 data compilation include: (1) the data may be 'no good'; (2) there may be long-term nonaxial dipole field contributions to the geomagnetic field; (3) the average may be nonstationary and the data in different bins in **Figure 20** are from different ages and average fields. The first two hypotheses cannot be tested with the present data set, but the third can, to some extent by looking at the more recent past for which there are a lot of data.

In order to address the issue of the average field, we look at the data from the last 170 My meeting the consistency test described in the previous section in **Figure 18**. We have split these out into data from submarine basaltic glasses (SBG, blue dots), from single crystals (SX, red diamonds), and 'other' (mostly lava flows, triangles).

The present compilation is sufficient to clarify a few key misconceptions in the literature: (1) that SBG data are generally lower than other (mostly lava flow) data; (2) that single crystal results are 'high'; and (3) there is a stationary average field. The average of the 421 data points from SBG for the last 170 My is 66.9 ± 34 (standard error of the mean of 1.7) while that of all other data is 62.6 ± 35.6 (standard error of the mean of 0.75); that is, the SBG data are in fact higher than the other data over this age interval.

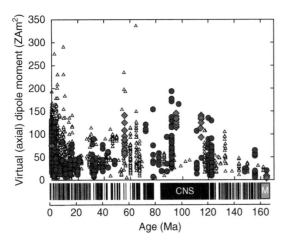

Figure 21 Summary of published data meeting minimum criteria for last 200 My. Blue dots are submarine basaltic glass data. Red diamonds are single-crystal results. Triangles are all other data meeting the same consistency criteria (σ <5% of mean or <5 μT); VADMs calculated as in text.

There are insufficient data in the database from single crystals, but it is clear from **Figure 21** that the single-crystal data are consistent with other data of similar age. Finally, the concept of a stationary average field is difficult to support with the data shown in **Figure 21**. It appears that there are periods of time (e.g., from about 20 Ma to about 55 Ma and older than about 125 Ma) during which the field was relatively low, with an average dipole moment of around 50 ZAm². There are also periods of time during which the field is much higher (e.g., the last 20 My and the period from about 55 to 122 Ma).

Another clear result from the data in **Figure 18** is that the geomagnetic field is highly variable on both short and long timescales. Therefore it is likely that unless the same time period is considered in all latitudinal bins, in figures like **Figure 20** there will be scatter introduced from comparing tmes with different average field strength. Some effort should be put in obtaining data for certain time slices as a function of paleolatitude.

13.8.3 Are There Any Trends?

13.8.3.1 Intensity versus polarity interval length

Although there are no clear long-term trends in the paleointensity data shown in **Figures 18 and 21**, there are times when the field is stronger than others as noted previously. Cox (1968) suggested that strong geomagnetic fields could inhibit reversals of the geomagnetic field and this makes sense with the observation that geomagnetic fields are low when the field is reversing.

Tauxe and Hartl (1997) and Constable *et al.* (1998) demonstrated a weak correlation between the length of a given polarity interval and the average paleointensity in the relative paleointensity data from DSDP Site 522. So one of the primary motivations for initiating the study of the DSDP/ODP submarine basaltic glasses for paleointensity was to test the hypothesis that long intervals of stable polarity (like the Cretaceous Normal Superchron or CNS in **Figure 21**) were associated with unusually strong fields (see, e.g., Pick and Tauxe, 1993). It was therefore puzzling and a bit disappointing when Selkin and Tauxe (2000) compared paleofield strength with reversal rate and found no clear relationship. There were just too few data from the CNS to make a definitive statement.

Now there are many more data from the last 175 My with several data sets available from intervals whose polarity chron are known and the data can be associated with known polarity interval length. In **Figure 22** we show the data from **Figure 21** associated with known polarity intervals. These data are from SBG obtained from holes drilled on clearly identifiable magnetic anomalies compiled by Tauxe (2006) and data from lava flows in magnetostratigraphic sections correlated to the timescale (Riisager *et al.*, 2003; Herrero-Bervera and Valet, 2005). It appears that the correlation suggested by Tauxe and Hartl (1997) and Constable *et al.* (1998) based on relative paleointensity in sediments is supported by the absolute paleointensity data set.

13.8.3.2 Source of scatter in the CNS

Another prediction made from the relative paleointensity data from Site 522 was that the scatter in the data is proportional to the average value and strongly linked to polarity interval length (Constable *et al.*, 1998). This observation also appears to be weakly supported by the absolute paleointensity data set. One question that springs to mind, however, is whether the scatter is geomagnetic in origin. To address this issue, Granot *et al.* (2007) assembled a data set from the Troodos ophiolite which formed during the CNS. Their data set includes new data from gabbros as well as the submarine basaltic glass data of Tauxe and Staudigel (2004). Many of the gabbro data came from a sequence of small plutons with a clear relationship to the ancient spreading axis and their relative age relationships were therefore known. Tauxe and Staudigel (2004) had sampled

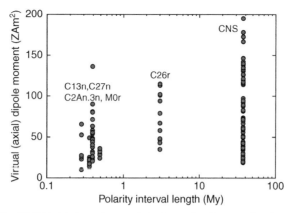

Figure 22   VADM data from the PINT06 database from polarity intervals of known duration. Data from the CNS are from Cottrell and Tarduno (2000); Riisager *et al.* (2003); Sherwood *et al.* (1993); Tarduno *et al.* (2001, 2002); Tauxe (2006); Tauxe and Staudigel (2004); Tanaka and Kono (2002); Zhao *et al.* (2004); Zhu *et al.* (2004b). C26r and C27n are from Riisager and Abrahmsen (2000); C12n, C13n, C32n are from Tauxe (2006); M0r are from Riisager *et al.* (2003), Tauxe (2006) and Zhu *et al.* (2001); and C2An.3n are from Herrero-Bervera and Valet (2005).

two transects through the entire oceanic extrusive layer, separated by some 10 km. Data from these two transects are in stratigraphic order, so their age relationships are also known. In **Figure 23** we show their plot of the three time sequences. The data exhibit remarkable serial correlation, which Granot *et al.* (2007) used to argue that the scatter in the CNS data is largely geomagnetic in origin.

13.8.3.3   The oldest paleointensity records

Under the topic of 'trends in paleointensity,' one of the most interesting questions concerns the earliest records of paleointensity. In **Figure 24** we show published results satisfying minimum consistency constraints obtained from Archean aged rocks. Until recently, there were very few studies that were based

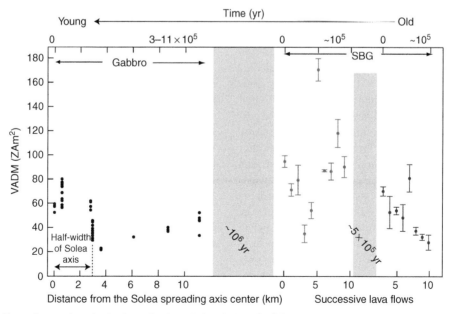

Figure 23   Three time-series of paleointensity data during the late CNS from the gabbros and glasses. Gray areas represent gaps in time estimated from moderate spreading rate (full spreading rate of 20–75 mm y^{-1}). Results for the gabbro specimens are shown as individual points and are corrected for cooling rate and anisotropy. Results for the SBG sites correspond to the average results from successive cooling units. Data from Granot *et al.* (2007) and Tauxe and Staudigel (2004). Modified from Granot *et al.* (2007).

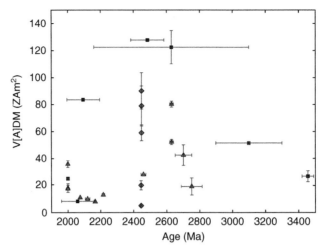

Figure 24 Archean data from the database that have standard deviations of <20% of the mean or are <5 μT. Age uncertainties are indicated by the horizontal bars. Triangles are data from K T T-type experiments with pTRM checks and squares are from other types of experiments. The diamonds are from single-crystal experiments. Data from Morimoto *et al.* (1997); Yoshihara and Hamano (2000); Sumita *et al.* (2001); Macouin *et al.* (2003); Bergh (1970); Hale (1987); McElhinny and Evans (1968); Schwartz and Simons (1969); Smirnov *et al.* (2005); and Selkin *et al.* (2000).

on experiments that used pTRM checks (triangles and diamonds in **Figure 24**). Recent data meet the highest experimental standards and show that the field had a large range in intensity, similar to more recent times, although the highest values come from experiments not done with pTRM checks (squares in **Figure 24**). The inescapable conclusion from these data is that the geomagnetic field was 'alive and well' by ∼3 Ga.

13.8.3.4   The paleointensity 'saw-tooth'

Valet and Meynadier (1993) (detailed data set in Meynadier *et al.*, 1995) presented a relative paleointensity record for the last 4 My using sediment cores of ODP Leg 138 taken from the eastern equatorial Pacific (see **Figure 25**). They postulated an 'asymmetric saw-tooth pattern' of paleointensity variations, that is, a rapid intensity growth just after a polarity transition and a gradual decrease since then towards the next reversal (see **Figure 25**). They also suggested that the length of a polarity zone is proportional to the magnitude of the intensity jump. The latter observation is consistent with the data shown in **Figure 22** whereby long polarity intervals appear to have higher average fields. Nonetheless, the 'saw-tooth' idea became the subject of heated arguments.

The saw-tooth was originally epitomized by the long-term decreasing trends observed in the Matuyama, one toward the Brunhes, one toward the Olduvai, and in the Gauss, one toward the Kaena, and one toward the Matuyama boundaries. The saw-tooth envisioned by Valet *et al.* (2005; see SINT-2000 in **Figure 25**), however, is a shorter trend immediately

preceding the reversal boundary. We will refer to this, more restricted view as the 'short saw-tooth' hypothesis in the following. First, we review the debate about the saw-tooth in the literature.

While all transitional records with paleointensity data have low field intensities associated with transitional directions, not all records display a long-term decreasing trend toward a reversal. Arguments supporting the existence of the saw-tooth pattern were presented in rapid succession. Valet *et al.* (1994) examined paleointensity records near the Brunhes–Matuyama transition from the Atlantic, Indian, and Pacific Oceans, and found the rapid intensity growth after the transition. Meynadier *et al.* (1994) recognized the 'saw-tooth pattern' in a relative paleointensity record spanning the last 4 My obtained from a core in the Indian Ocean. Verosub *et al.* (1996) presented a record focused near the MBB and the Jaramillo Subchron of a sediment core from the central north Pacific which supported the 'saw-tooth pattern.'

Counter arguments to the saw-tooth also began to appear. For example, Laj *et al.* (1996a, 1997 (for correction)) failed to see the 'saw-tooth pattern' utilizing the same core as that of Verosub *et al.* (1996) when examining a longer period of time up to the Olduvai Subchron. Laj *et al.* (1996b) reported that no rapid intensity increase after polarity transitions was observed in relative paleointensity records from two Late Miocene sections in Crete (Kotsiana and Potamida in **Table 2**).

Arguments against the 'saw-tooth pattern' were presented also from paleointensity estimations based on recording mechanisms different from

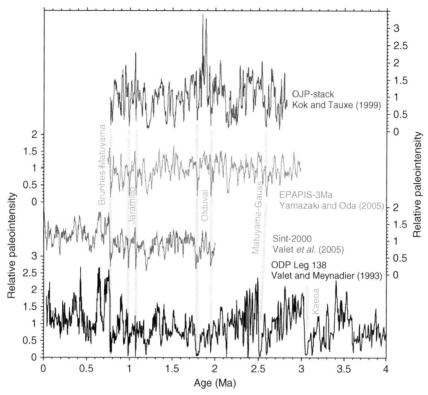

Figure 25 The paleointensity record from ODP Leg 138 cores (Valet and Meynadier, 1993) showed the 'asymmetric sawtooth pattern', whereas other records which reached the Gauss–Matuyama boundary do not show such pattern: Ontong-Java Plateau stack of Kok and Tauxe (1999) and equatorial Pacific Paleointensity Stack of Yamazaki and Oda (2005).

sediments. Records of ^{10}Be/^{9}Be reflect geomagnetic paleointensity through a control on the production rate of the cosmogenic nuclide (^{10}Be). Raisbeck *et al.* (1994) argued that a ^{10}Be/^{9}Be record at Site 851 of ODP Leg 138, which is the same site as of Valet and Meynadier (1993), is inconsistent with the 'saw-tooth pattern.' Westphal and Munschy (1999) showed that the 'saw-tooth pattern' cannot explain the shape of stacked magnetic anomaly profiles over the East Indian Ridge, Juan de Fuca Ridge, and East Pacific Rise. McFadden and Merrill (1997) presented an analysis of the Cenozoic polarity reversal chronology in which it was argued that effective inhibition of a future reversal can only last for about 50 ky at most, which contradicts the 'saw-tooth pattern' requiring much longer inhibition.

The 'saw-tooth pattern' was also questioned from remanent magnetization acquisition processes. Kok and Tauxe (1996a) proposed a cumulative viscous remanence model for remanence acquisition of sediments that can yield intensity variations like the 'saw-tooth' pattern. Then, Kok and Tauxe (1996b) reproduced the 'saw-tooth pattern' of the ODP Leg 138 sediments by the cumulative viscous remanence

model using the values of the equilibrium magnetization constrained by results of a Thellier-type paleointensity experiments applied to the ODP Leg 138 sediments. Furthermore, they resampled the Site 851 sediments near the Gauss–Gilbert boundary and showed that the 'saw-tooth pattern' disappeared by thermal demagnetization to 400°C.

Meynadier *et al.* (1998) made a counterargument to the cumulative viscous remanence model of Kok and Tauxe (1996a, 1996b). They suggested that to produce the saw-tooth pattern and to preserve the magnetostratigraphy with the cumulative viscous remanence model, a very narrow distribution of the relaxation times is required. They also showed that the saw-tooth pattern of the Site 851 sediments did not change by thermal demagnetization and AF demagnetization of stronger fields, which is inconsistent with the result of Kok and Tauxe (1996b) despite using the sediments from the same site. Kok and Tauxe (2000) on the comments to Meynadier *et al.* (1998) stressed the nonuniqueness of relaxation time distribution explaining the 'saw-tooth pattern', and pointed out that the τ distribution of Kok and Tauxe (1996b) is not all relaxation times present in the sediments, but merely a part of them that behave viscously. In the reply, Meynadier and Valet (2000) mentioned that the remanent magnetization with blocking temperatures between 150 and 300° C which may carry cumulative viscous remanence is only minor part of NRM of the sediments. The two groups also argued about the validity of thermal demagnetization on relative paleointensity estimation from sediments (Kok and Tauxe, 1999; Valet and Meynadier, 2001; Kok and Ynsen, 2002).

Mazaud (1996) proposed another model of magnetization acquisition which produces the 'saw-tooth pattern': a large fraction (say, two-thirds) of magnetic grains acquires NRM at deposition time while remaining grains reorientate or acquire magnetization after deposition. Although Meynadier and Valet (1996) considered this model unlikely from the knowledge of pDRM acquisition processes prevailing at that time, recent studies suggest that this model may have difficulties. As discussed in Section 13.4, the depth lag of pDRM acquisition is estimated to be very small ($\sim$2 cm) from the compilation of B/M boundary positions and the oxygen-isotope chronology recorded in marine carbonate cores with various sedimentation rates (Tauxe et al., 1996). An experiment of Katari et al. (2000) using natural undisturbed sediments suggests that pDRM (reorientation of magnetic particles) is a rare phenomenon, probably because of the effects of flocculation: magnetic minerals would be aggregated with clay.

In the 851 record of Valet and Meynadier (1993), the 'saw-tooth pattern' is most apparent in the early Matuyama Chron after the Gauss–Matuyama transition and during the Gauss Chron. The number of paleointensity records reported so far that reached the Gauss–Matuyama transition is still small. However, available records from different groups do not support the 'saw-tooth pattern': neither a stacked record since 2.8 Ma from the Ontong-Java Plateau (Kok and Tauxe, 1999; OJP-stack in **Figure 25**) nor a stacked record since 3.0 Ma from the equatorial Pacific (Yamazaki and Oda, 2005; EPAPIS stack in **Figure 25**) shows variations like the 'saw-tooth pattern' after the Gauss–Matuyama transition.

We do not yet understand well the rock-magnetic processes which produce the 'saw-tooth pattern' only for some sediments. If the DRM acquisition model of Mazaud (1996) works in general, all sedimentary paleointensity records should display 'saw-tooth'-like changes. From the cumulative viscous remanence model of Kok and Tauxe (1996a), the sediments producing the 'saw-tooth pattern' are expected to have a

particular magnetic grain-size distribution favorable for the long-term viscous remanence acquisition, but this has not yet been fully tested.

Recently, Valet et al. (2005) used 10 of the 15 records of relative paleointensity data compiled by Guyodo and Valet (2006) to create the so-called SINT-2000 stack (see **Table 2** and **Figure 25**). Based on this subset of the data, Valet et al. (2000) argue for the 'short saw-tooth' pattern in the 80 ky interval immediately prior to the four reversals included in the stack. Of the four reversals in the SINT-2000 stack (figure 4 of Valet et al., 2000), only the Upper Jaramillo and the lower Olduvai show convincing short saw-tooth patterns.

Relative paleointensity records spanning the last 2 My have steadily been produced (see, e.g., **Table 2**). Because the SINT-2000 is a stack which did not include many of the records in **Table 2**, we have plotted those records that span at least the period from 800 to 900 ky in **Figure 26** for the interval including the Brunhes and Jaramillo. Considering all the records, it appears that even the short saw-tooth is only observed in a small subset of the records, although an intensity peak just after the B/M boundary (Guyodo and Valet, 1999; Valet et al., 2005) does appear in all records. In general, the long-term saw-tooth pattern originally observed by Valet and Meynadier (1993) has not been universally observed, as would be expected from the behavior of a dipole source.

13.8.4 High-Resolution Temporal Correlation

13.8.4.1 Sediments

Oxygen-isotope stratigraphy revolutionized paleoceanography by providing a global signal with a resolution on the order of tens of thousands of years (e.g., Hays et al., 1976). Yet oxygen-isotope stratigraphy has its drawbacks. It cannot be applied in sediments deeper than the carbonate compensation depth or lakes. Oxygen-isotopic data are often difficult to interpret in marginal seas, where records may not reflect global ice-volume changes. Finally, temporal resolution better than 10^4 yr is critical for assessing the global nature of climatic events and their durations. The prospect of using relative paleointensity from sediments as a high-resolution correlation and dating tool has therefore been met with great enthusiasm (e.g., Stott et al., 2002). Here we examine the prospects and problems with so-called 'paleointensity-assisted chronology' or PAC.

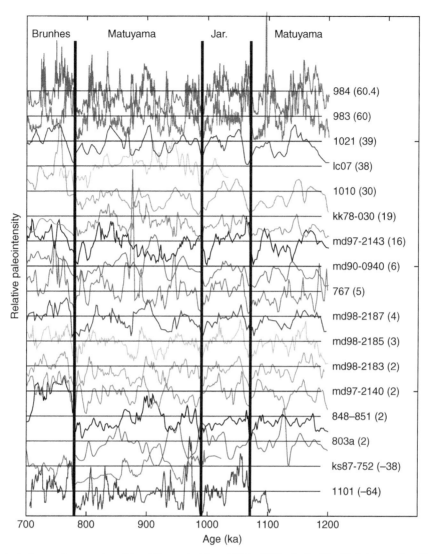

Figure 26 Plot of all records spanning at least the period from 800 to 900 ka in the MagIC database (see **Table 2**) for the range 700–1200 ka. Records are plotted in order of latitude (in parentheses to right of record name. The positions of the Matuyama/Brunhes, and the upper and lower Jaramillo transitions are indicated by heavy black lines.

The importance of PAC is not simply as a substitution for $\delta^{18}O$. It can also be used to examine consistency of other chronologies such as $\delta^{18}O$ and ^{14}C, because it is quasi-independent of them. Regional and global inter-core correlations tied by paleointensity variations revealed discrepancies of up to several thousand years between those based on ^{14}C and $\delta^{18}O$ (Stoner et al., 1995), and between GISP2 and the $\delta^{18}O$ chronologies (Stoner et al., 1995, 2000) during the last c. 100 ky. Moreover, paleointensity stratigraphy can have higher resolution than $\delta^{18}O$ stratigraphy because the variations contain shorter-wavelength components than those of $\delta^{18}O$. Truly dipolar features of geomagnetic field variations have a potential for providing a

time reference for an inter-hemisphere paleoclimatic relationship with unprecedented resolution.

On a more limited scale, there is a possibility that inter-core correlation and age estimation can be performed using paleointensity by correlating patterns among cores with a standard curve such as SINT-800 (Guyodo et al., 2001) and NAPIS-75 (Laj et al., 2000), which is exemplified by Stoner et al. (1995; reference # 71 in **Table 2**) in the Labrador Sea, Demory et al. (2005; reference # 16 in **Table 2**) in the Lake Bikal, and Macri (2005; reference # 43 in **Table 2**) in the Wilkes Land Basin off Antarctica.

We feel that while regional correlations can be achieved, much is lost by using PAC as a primary dating tool. These records can no longer be used to constrain paleointensity models or global stacks, because the age information is not independent and features correlate by assumption. Such records have been clearly labeled in **Table 2** (RPI) and in the MagIC database.

If we desire a global correlation tool, we require a dipolar signal. However, it is as yet unclear at which wavelength the dipole terms give way to nondipole terms. Korte and Constable (2005) caution us that variations in VADMs may not be global and their variations need not be synchronous because they can be strongly influenced by nondipole field effects. Moreover, the duration of a polarity reversal is thought to be dependent on latitudes, ranging from about 2 ky near the equator to 10 ky at ± 60°

(Clement, 2004). Furthermore, the contribution of nondipole effects are probably higher when paleointensity is low, and both the duration and shape of a particular paleointensity low will be site dependent.

Doubts about the global nature of paleointensity features notwithstanding, global inter-core correlations on a millennial scale have been attempted. A series of papers by Channell *et al.* (2000), Stoner *et al.* (2000), and Mazaud *et al.* (2002) correlated records between the high latitudes of the North Atlantic and the South Atlantic and Indian Ocean sectors. As an example of the method, we show the records of Stott *et al.* (2002), who tied cores together between the North Atlantic and the western equatorial Pacific (see **Figure 27**). The large-scale features (labeled H1–H10 and L1–L8 correlate reasonably

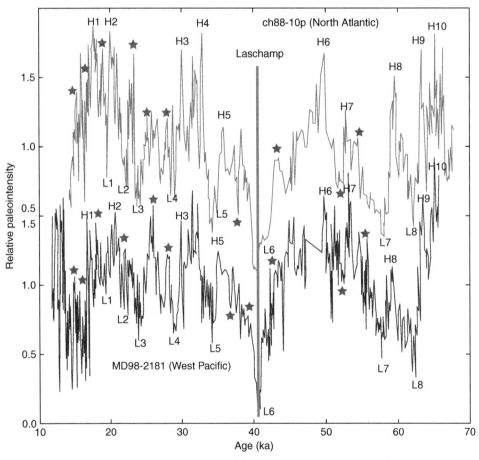

Figure 27 Comparison of deep-sea sediment relative paleointensity records from md81-2181 and the North Atlantic Ocean site ch88-10p. The two records were dated by correlation of their individual oxygen-isotopic stratigraphies to the GISP2 record. Distinctive paleointensity highs (H1–10) and lows (L1–8) are identified in the two records. Less-convincing features (stars) are marked as well. The ages of the named features are synchronous to within an uncertainty of less than ± 500 years. The 'Laschamp Excursion' correlates with a distinctive ^{36}Cl excursion in the GISP2 record and provides an independent tie point to ice cores. Redrawn from supplement to Stott L, Poulsen C, Lund S, and Thunell R (2002) Super ENSO and global climate oscillations at millennial time scales. *Science* 297: 222–226.

well and are consistent with the oxygen-isotopic records from the two cores. These allow correlation with a resolution of 2–3 ky. The difficulty of identifying global features on a submillenial scale is made apparent by the rather unconvincing correlation of features marked by the stars. Although these features may well be synchronous, they do not resemble each other very much in the two hemispheres and without the excellent and very detailed chronological control of the independent oxygen-isotopic records, their identification would not have been possible. How much should these 'millennial' features look like each other and how synchronous they are expected to be requires much more detailed knowledge of the process of secular variation, a topic of active research (*see* Chapter 7).

13.8.4.2 Ridge crest processes

The potential of paleointensity for high-resolution dating is not restricted to sediments. Thellier-type paleointensity data can be used to estimate ages of basalts near mid-ocean ridges, and it is expected that paleointensity will be useful for studying crustal accretion processes at ridges (Gee *et al.*, 2000; Ravilly *et al.*, 2001; Bowles *et al.*, 2006). This work is discussed in more detail by Gee and Kent (*see* Chapter 12).

13.8.5 Atmospheric Interaction

Radioactive forms of carbon, beryllium, and chlorine are produced in the atmosphere by cosmic ray bombardment. The decay of these isotopes is used for dating purposes in a wide variety of disciplines. There are large variations in ages predicted from tree ring, varve, ice layer counting or U/Th dating, and those estimated by radiocarbon dating. An example of such a comparison is shown in **Figure 28(a)** which shows the age based on ^{14}C dating from a core in the Cariaco Basin versus layer counting in the Greenland Ice Sheet Project 2 (GISP2) ice core from a core in the Cariaco Basin (Hughen *et al.*, 2004). The correlation of the marine sediment core to the ice core was based on tying dark sedimentary layers to interstadials in the GISP2 core.

The difference between radiocarbon and other age estimates in **Figure 28(a)** is used to calculate variations in initial radiocarbon in the atmosphere relative to the concentration in the modern atmosphere (see atmospheric Δ^{14}C plotted as dots in **Figure 28(b)**). An excess of radiocarbon (positive Δ^{14}C) results in an underestimation of the age

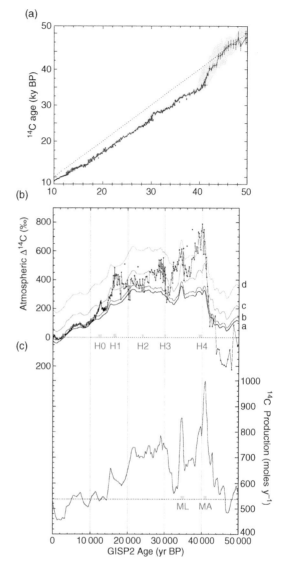

Figure 28 (a) Radiocarbon calibration data from Cariaco ODP Leg 165, Holes 1002D, and 1002E (blue circles), plotted versus calendar age assigned by correlation of detailed paleoclimate records to the Greenland Ice Core GISP2. The thin black line is high-resolution radiocarbon calibration data from tree rings joined at 12 cal. k BP to the varve counting chronology. Red squares are paired ^{14}C-U/Th dates from corals. Light gray shading represents the uncertainties in the Cariaco calibration. The radiocarbon dates are 'too young', falling well below the dashed line of 1:1 correlation. (b) Compilation of data interpreted as production rate changes in radiocarbon (Δ^{14}C) versus calender age (symbols same as in (a)). (c) Predicted variation of Δ^{14}C from the geomagnetic field intensity variations from sediments of the north Atlantic (Laj *et al.*, 2000) using the model of Masarik and Beer (1999). Figure modified from Hughen K, Lehman S, Southon J, *et al.*, (2004) C-14 activity and global carbon cycle changes over the past 50,000 years. *Science* 303(5655): 202–207.

because there is 'too much' radiocarbon in the sample for its age. Changes in $\Delta^{14}C$ have been attributed to differences in the production of radiocarbon in the atmosphere by cosmic ray bombardment and changes in the carbon balance between the atmosphere and the deep ocean, which is a reservoir of old carbon (see, e.g., Bard *et al.*, 1990). If, for example, the transfer of atmospheric carbon into the ocean was less efficient in the past or the release of old carbon from the deep ocean was less efficient ('ventilation' was slower), then there would be an excess of radiocarbon in the atmosphere relative to the modern atmosphere, resulting in 'too young' ^{14}C ages.

Radiocarbon production is thought to be strongly controlled by changes in magnetic field strength because the magnetic field shields the atmosphere from cosmic rays (*see* Chapter 7). Changes in the intensity of the magnetic field should therefore result in changes in radiocarbon production (among other things); hence, the variation in intensity is a key parameter in deriving accurate age information. Hughen *et al.* (2004) used a paleointensity stack from the North Atlantic (the NAPIS stack of Laj *et al.*, 2000) as a proxy for changes in the dipole moment of the Earth's magnetic field over the last 70 ky. By using the Monte Carlo simulations of the relationship between geomagnetic field strength and radiocarbon production of Masarik and Beer (1999), Hughen *et al.* (2004) predicted radiocarbon production for the past 50 ky (**Figure 28(c)**). The different curves in **Figure 28(b)** (a–d) use the predicted radiocarbon production values as input into box models using different ocean/atmosphere boundary conditions relating to different models of deep-sea ventilation, resulting in different estimates for atmospheric $\Delta^{14}C$.

None of the curves in **Figure 28(b)** based on the model predictions using the NAPIS stack provide a satisfactory fit to the observed variations. Hughen *et al.* (2002) suggested that either the model of Masarik and Beer (1999) for translating geomagnetic field intensity to radiocarbon production is incorrect at low field strengths or that our understanding of the global carbon cycle is insufficient. It is of course also possible that the NAPIS model of North Atlantic relative paleointensity is a poor proxy for the global paleomagnetic field intensity variations. We explore the latter in the following.

The NAPIS stack (compilation #34 in **Table 2**) is based on a number of relative paleointensity records in the North Atlantic and as such it is a regional stack. Variations in cosmic ray bombardment are largely controlled by the strength of the dipole term, the estimation of which requires global coverage. In **Figure 29** we plot all the records in the MagIC database that have independent age control based on oxygen isotopes. Hughen *et al.* (2004) labeled two peaks in predicted radiocarbon production 'LA' and 'ML' for the Laschamp and Mono Lake excursions (see Laj *et al.* (2000) for correlations) at 41 and 34 ka, respectively. Although excursions are defined on the basis of anomalous directions, they are thought to be related to low field strengths and decreases in paleointensity (DIPs) are frequently correlated to excursions (e.g., Guyodo and Valet, 1999).

The '41' and '34' ky ages are also marked in **Figure 29(a)** and the nearest DIPs are marked with red and blue bars for the so-called LA and ML features, respectively. Many records in the figure show only one DIP, or a broad low-paleointensity zone. The age agreement is rather poor. In particular, the two records from ODP Site 983 and 984 (see **Table 2**) were taken from very close to one another and have quite different character in the 41 ka DIP with 984 showing two distinct DIPs, but 983 only one. Moreover, Kent *et al.* (2002) argue that the Mono Lake is actually 38–41 ka and that it may well have the same feature as the Laschamp. Finally, as was discussed in Section 13.4, the response of sedimentary relative paleointensity may not be linear with the applied field and the amplitudes could be biased.

The absolute paleointensity data from the PINT06 database (with σs < 15% of the mean or <5 μT) that had reasonable age uncertainties (better than 20%) are shown in **Figure 29(b)**. There is generally high scatter prior to about 15 ka, but after that there is a sharp upward trend toward the present.

While both the relative and absolute paleointensity data support a general period of low intensity spanning from perhaps 20 ka to as much as 50 ka, the details are at best unclear. There are not enough data to establish the existence of both the Mono Lake and the Laschamp as two separate globally observed excursions.

Rough agreement of paleointensity patterns between predictions based on cosmogenic radionuclide production rates and sedimentary relative paleointensity has been reported also for older ages: ^{36}Cl flux in GRIP Ice Core for the past 100 ky (Baumgartner *et al.*, 1998), a global ^{10}Be stack of the last 200 ky (Frank *et al.*, 1997), and $^{10}Be/^9Be$-based paleointensity during the last 300 ky in the North Atlantic and from 500 to 1300 ka in the western equatorial Pacific (e.g., Carcaillet *et al.*, 2004). These results show the potential of ^{10}Be for providing independently geomagnetic field intensity

(a)

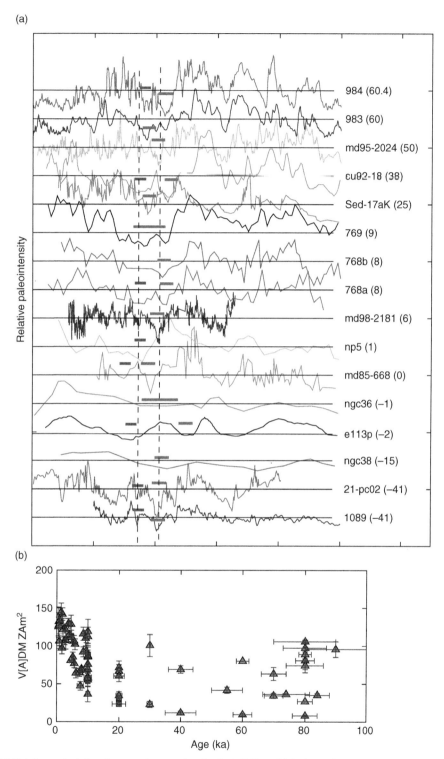

(b)

Figure 29 (a) Relative paleointensity records spanning the last 100 ky with independent age control based on $\delta^{18}O$. The solid red bars indicate intensity lows that are possibly related to the 'Laschamp Excursion' and the blue bars are a later paleointensity low, sometimes referred to as the 'Mono Lake Excursion.' (b) Absolute paleointensity data with age uncertainties of less than 20% meeting the 'strict' selection criteria (see text).

information for the past several million years, although uncertainty originating from transportation and sedimentation processes of the radionuclide still remains (e.g., Christl *et al.*, 2003; McHargue and Donahue, 2005).

Overall, variations in dipole intensity are not as well constrained as we would like. As these variations are key to refining the radiocarbon calibration, more records with independent and accurate age constraints are needed.

13.8.6 Frequecy of Intensity Fluctuations and the Climatic Connection

Arguments regarding the possible relationship between the intensity of the geomagnetic field and climate have a long history of more than 30 years (e.g., Wollin *et al.*, 1971). Discussions in the 1970s were based on the remanent intensity variations which were not corrected for differences in magnetizability of sediments. The relationship was convincingly rejected by Kent (1982), who showed that the remanent intensity variations of the relevant sediment cores were controlled by climatically induced variations in carbonate contents. Kent and Opdyke (1977) were the first to suggest the presence of the obliquity frequency, $\sim$43 ky in a normalized intensity record, but this idea was not seriously discussed at that time.

Rapid progress in relative paleointensity studies revived the orbital modulation issue in the late 1990s, and it has been heatedly argued since then. Tauxe and Shackleton (1997) found significant power in the power spectrum of a record from the Ontong-Java Plateau in the 30–50 ky band, but showed that the intensity fluctuations came in and out of phase of the associated oxygen-isotopic record, arguing against a strong relationship between climate and paleointensity. In contrast, Channell *et al.* (1998) and Channell (1999) proposed a $\sim$40 ky obliquity frequency from a power spectrum analysis of their relative paleointensity records during the Brunhes Chron obtained from ODP Sites 983 and 984 in the North Atlantic. They interpreted it as geomagnetic field behavior from the observations that no power exists at $\sim$40 ky in bulk magnetic properties and there is no coherence between the relative intensity and the normalizer (IRM), percent carbonate, and a magnetic grain-size proxy at $\sim$40 ky. Their paleointensity records showed significant power also at the $\sim$100 ky eccentricity frequency, but they rejected this because this frequency was observed also in bulk magnetic properties. Yamazaki (1999) instead proposed the possible

presence of the $\sim$100 ky frequency in his relative paleointensity records from the North Pacific based on the same logic as that of Channell *et al.* (1998): occurrence of $\sim$100 ky power in relative paleointensity but not in the normalizer on the power spectra. Yokoyama and Yamazaki (2000) applied a wavelet analysis to five paleointensity records from the Pacific Ocean reported in Yamazaki *et al.* (1995) and Yamazaki (1999), and found a quasi-period of $\sim$100 ky. They considered the $\sim$100 ky period inherent to the geomagnetic field, because of the good coincidence of the relative intensity records in this scale despite significant phase differences in magnetic properties. Thouveny *et al.* (2004) reported a $\sim$100 ky period in a paleointensity record during the last 400 ky from Portuguese Margin sediments, North Atlantic. Possible occurrence of 100 ky period in paleointensity would not be limited in the Brunhes Chron: Yamazaki and Oda (2005; see **Figure 30**) found significant power at $\sim$100 ky period in paleointensity records from 0.8 to 3.0 Ma. Kok and Tauxe (1999) found point at $\sim$150 ky in paleointensity records during the Matuyama Chron from the Ontong-Java Plateau.

Whether the orbital frequencies found in sedimentary paleointensity records reflect geomagnetic field behavior or not has been discussed mainly on the following three points: significance and stability of the orbital periodicities, error in age control, and lithological contamination to paleointensity records. Guyodo and Valet (1999) argued that there is no stable periodicity during the Brunhes Chron by a spectrum analysis on the SINT-800 stack using sliding windows. However, the orbital modulation may be a nonstationary process. Sato *et al.* (1998) suggested that there is no constant period but continuous shifts between 50 and 140 ky based on a stacked paleointensity record from three cores in the western equatorial Pacific during the last 1.1 My. Horng *et al.* (2003) argued that the orbital frequencies in paleointensity variations are not statistically significant by applying a wavelet analysis on their relative paleointensity record during the last 2.14 My from the western Philippine Sea. On the other hand, Teanby and Gubbins (2000) proposed that the periodicities of several tens of thousand years observed from sedimentary paleointensity records could be due to aliasing, an artifact of coarse sampling, and simulated using archaeointensity data with a 2 ky period how false longer periods appear by aliasing. However, orbital periodicities have been reported even from sediment cores with significant variations in

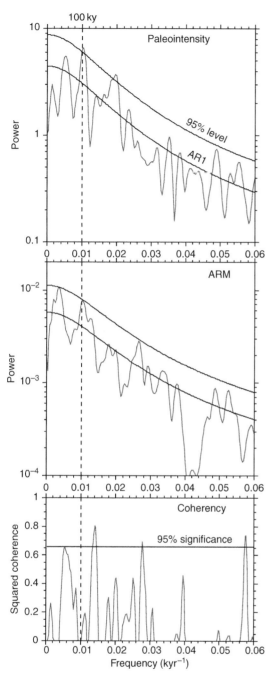

sedimentation rates, which is difficult to be explained by aliasing. Guyodo and Channell (2002) performed numerical simulation of paleointensity records with various sedimentation rates and variable quality of age control, and showed how spectral information is lost with decreasing sedimentation rates: the power spectra are reliable for periods as short as 4 ky in records with a sedimentation rate of 15 cm ky^{-1} with good age control, whereas periods of only *c.* 50 ky or longer are reliable in records with a sedimentation rate of 1 cm ky^{-1}. McMillan *et al.* (2002, 2004) evaluated effects of various sources of errors such as dating errors, misidentified tiepoints, changes in sedimentation rate, and the effect of nondipole components. They simulated coherence of records among various sites, and evaluated the accuracy of a stacked record suggesting that dipole variations with periods of longer than 20 ky can be recovered (but shorter ones would be problematic).

Strong arguments against orbital modulation of the geomagnetic field come from possible lithological contamination to sedimentary paleointensity records. Because paleointensity records during the last 200 ky look coherent with the oxygen-isotope curve, Kok (1999) suspected that sedimentary paleointensity records including those derived from ^{10}Be are controlled by paleoclimate due to inadequate normalization. Guyodo *et al.* (2000) performed a wavelet analysis on the records of paleointensity and magnetic properties from ODP Site 983, which was the same data set as those from which Channell *et al.* (1998), and Channell and Kleiven (2000) proposed the ~41 ky periodicity. They found that relative paleointensity has coherency with the normalizer and a magnetic grain-size proxy (ARM/k), and concluded that orbital frequencies in paleointensity records may be the expression of lithological variations.

At present, we cannot exclude a possibility that relative paleointensity estimation is significantly influenced by an unknown rock-magnetic mechanism which amplifies effects of minor variations in magnetic properties, because the mechanism of remanent-magnetization acquisition of sediments has not yet been fully understood, and no natural sediments are completely free from lithological variations induced by environmental changes. If variations of magnetic properties of sediments such as magnetic grain size and mineralogy contain the orbital periodicities and show coherence with paleointensity changes, this suggests the possible contamination of magnetic property changes to paleointensity records. However, this cannot exclude

Figure 30 Power spectra of relative paleointensity and the normalizer (ARM), and cross-correlation between them of Core MD982187 in the western equatorial Pacific between 0.8 and 3.0 Ma (Yamazaki and Oda, 2005). This core is one of six cores which was used to construct the EPAPIS-3 Ma curve shown in **Figure 25** and covers the longest period of time among them. The relative paleointensity variations contain quasi-priod of ~100 kyr, but no significant correlation between relative paleointensity and the normalizer. The statistical significance of the spectral peaks was tested against the red-noise background from a first-order autoregressive (AR1) process, and the 95% confidence level is indicated.

the possibility of orbital modulation of paleointensity, because the two can also have coherency if the orbital parameters affect both the geomagnetic field and depositional environment. To solve the problem, it is necessary to understand the quantitative relationship between the magnitude of magnetic property (e.g., magnetic grain size and mineralogy) and the magnitude of induced changes in normalized intensity. At present, we cannot even predict whether normalized intensity increases or decreases when magnetic grain size increases in a certain grain-size range. It is also important to examine phase relationships in coherency analyses. Patterns of lithological and magnetic property variations induced by paleoclimatic changes may vary place to place; for example, magnetic grain size would increase in a certain period of time in some areas, but in other areas it would decrease in the same period of time. Paleointensity, on the other hand, should be globally synchronous. Thus, even when paleointensity and magnetic property variations have coherency, lithological contamination is suggested to be minor if the phase angles between the two from various places differ significantly.

On the possibility of orbital modulation of the geomagnetic field, a relationship of excursions and reversals with paleoclimate has also been discussed since 1970s. Rampino (1979, 1981) suggested that excursions may have occurred at about 100 ky intervals, and the ages of the excursions seem to coincide with times of peak eccentricity of the Earth's orbit (but see Rampino and Kent, 1983). Worm (1997) revisited the problem, and suggested that excursions and reversals tend to have occurred during periods of global cooling or during cold stages. He also gave an explanation to the intriguing observation that Arctic sediments seem to have recorded apparently longer duration of excursions (e.g., Nowaczyk et al., 2001): larger sedimentation rates during glacials would cause a higher chance to record excursions if probability of excursions is higher during glacials. On the contrary, Kent and Carlut (2001) rejected the relationship. They concluded that six excursions in the Brunhes Chron and 21 reversals since 5.3 Ma have no tendency to occur at a consistent amplitude or phase of obliquity and eccentricity. Recently, the number of possible excursions during the Brunhes Chron have increased significantly (up to c. 20) (Lund et al., 2001). Hence the problem of the possible connection between excursions and paleoclimate is not independent to the arguments on the orbital frequencies in relative paleointensity. Besides paleointensity and excursions,

a discussion of orbital frequencies in paleomagnetic directions has been revitalized. Yamazaki and Oda (2002) reported a ~100 ky periodicity in an inclination record during the last 2.2 My from the western equatorial Pacific, whereas Roberts et al. (2003) concluded that it is not statistically significant.

As noted in Section 13.4, there can be a 'stealth' link between lithological factors, like clay content, which are controlled by climate and the relative paleointensity records which would be difficult to detect using the standard methods of normalization. To date, the significance and implications of possible climatic controls on paleointensity have not yet been adequately addressed.

13.9 Conclusions

Scientists have dreamed of analyzing the ancient magnetic field intensity for over four centuries. Since the first serious attempts to acquire paleointensity data in the 1930s, experimental design has improved dramatically. The theoretical foundations for interpreting paleointensity data from both thermal and depositional remanences are steadily improving and there has been explosive growth in publications with paleointensity data. The data are slowly being contributed to the communal database whose scope has increased dramatically recently.

We have highlighted some of the major topics involving paleointensity data in this treatise. While these topics are still fresh and arguments abound, we can make the following statements regarding paleointensity:

1. Everyone agrees that more and better data could resolve many of the current debates. Experiments with built-in assessments of the fundamental assumptions of the method and the ability to estimate reliability indices are essential. Data sets are being contributed to the MagIC database, including measurements and full documentation of methods and data processing. This new generation of a database has the potential to go a long way toward settling some of the major issues discussed in this chapter.

2. In general, the ancient magnetic field has been highly variable on both short and long timescales. There have been extended periods of time with intensities lower than the present field but there have also been intervals with field strengths greater than the present field in the past. These

periods of increased field strength may be related to the length of the polarity interval in which they are found.

3. The geomagnetic field in the Archean appears to have been 'alive and well.'

4. Arguments against the 'asymmetric saw-tooth' in paleointensity data associated with polarity reversals appear to be winning while arguments about the coherence of paleointensity with orbital frequencies and the ability to correlate globally paleointensity features on a millenial scale remain unresolved.

Acknowledgments

We thank the individuals who supplied data for the MagIC database compilation: J.E.T. Channell, C. Laj, Y. Guyodo, M. Perrin, L. Schnepp, J. Stoner, N. Thouveny, and J.-P. Valet. This work was in part supported by NSF Grant EAR 033-7399 and the Grant-in-Aid for Scientific Research ((A)(2) No. 16204034) of the Japan Society for the Promotion of Science to TY. Acknowledgment is also made to the donors of the American Chemical Society Petroleum Research Fund for partial support of this research.

References

Aitken MJ, Allsop AL, Bussell GD, and Winter MB (1988) Determination of the intensity of the Earth's magnetic field during archeological times: Reliability of the Thellier technique. *Reviews of Geophysics* 26: 3–12.

Anson GL and Kodama KP (1987) Compaction-induced inclination shallowing of the post-depositional remanent magnetization in a synthetic sediment. *Geophysical Journal of the Royal Astronomical Society* 88: 673–692.

Barbetti M (1977) Measurements of recent geomagnetic secular variation in southeastern Australia and the question of dipole wobble. *Earth and Planetary Science Letters* 36: 207–218.

Bard E, Hamelin B, Fairbanks RG, and Zindler A (1990) Calibration of the 14C timescale over the past 30,000 years using mass spectrometric U–Th ages from Barbados corals. *Nature* 345: 405–410.

Bassinot FC, Labeyrie LD, Vincent E, Quidelleur X, Shackleton NJ, and Lancelot Y (1994) The astronomical theory of climate and the age of the Brunhes–Matuyama magnetic reversal. *Earth and Planetary Science Letters* 126: 91–108.

Baumgartner S, Beer J, Masarik J, Wagner G, Meynadier L, and Synal H-A (1998) Geomagnetic modulation of the ^{36}Cl flux in the GRIP Ice Core, Greenland. *Science* 279: 1330–1331.

Ben Yosef E, Tauxe L, Ron H, *et al.* (2005) The intensity of the geomagnetic field during the last 6 millennia as recorded by slag deposits from archaeological sites in the Southern Levant. *EOS Transactions of the American Geophysical Union, Fall Meeting Supplement* 86(62): GP21A-0007.

Bender M, Sowers T, Dickson ML, *et al.* (1994) Climate correlations between Greenland and Antarctica during the past 100,000 years. *Nature* 372(6507): 663–666.

Bergh HW (1970) Paleomagnetism of the Stillwater Complex. Montana. *Paleogeophysics* 17: 143–158.

Besse J and Courtillot V (2002) Apparent and true polar wander and the geometry of the geomagnetic field over the last 200 Myr. *Journal of Geophysical Research* 107: 2300 (doi:10.1029/2000JB000050).

Biggin AJ, Bohnel HN, and Zuniga FR (2003) How many paleointensity determinations are required from a single lava flow to constitute a reliable average? *Geophysical Research Letters* 30(11): 1575.

Bleil U and von Dobeneck T (1999) Geomagnetic events and relative paleointensity records: Clues to high-resolution paleomagnetic chronostratigraphies of Late Quaternary marine sediments. In: Fischer G and Wefer G (eds.) *Use of Proxies in Paleoceanography: Examples from the South Atlantic*, pp. 635–654. Hiedelberg, Germany: Springer-Verlag.

Böhnel H, Biggin AJ, Walton D, Shaw J, and Share JA (2003) Microwave palaeointensities from a recent Mexican lava flow, baked sediments and reheated pottery. *Earth and Planetary Science Letters* 214(1–2): 221–236.

Bol'shakov A and Shcherbakova V (1979) A thermomagnetic criterion for determining the domain structure of ferrimagnetics. *Izvestiya, Physics of the Solid Earth* 15: 111–117.

Bol'shakov A and Solodonikov GM (1980) Paleomagnetic data on the intensity of the magnetic field of the Earth. *Izvestiya Earth Physics* 16: 602–614.

Bowles J, Gee J, Kent D, Bergmanis E, and Sinton J (2005) Cooling rate effects on paleointensity estimates in submarine basaltic glass and implications for dating young flows. *Geochemistry, Geophysics, Geosystems* 6: Q07002 (doi:10.1029/2004GC000900).

Bowles J, Gee J, Kent DV, Perfit M, Soule A, and Fornari D (2006) Paleointensity applications to timing and extent of eruptive activity, 9°–10°N East Pacific rise. *Geochemistry, Geophysics, Geosystems* 7: Q06006 (doi:10.1029/2005GC001141).

Bowles J, Tauxe L, Gee J, McMillan D, and Cande SC (2003) Source of tiny wiggles in Chron C5: A comparison of sedimentary relative intensity and marine magnetic anomalies. *Geochemistry, Geophysics, Geosystems* 4(6): 1049 (doi:10.1029/2002GC000489).

Boyd M (1986) A new method for measuring paleomagnetic intensities. *Nature* 3196: 208–209.

Brachfeld SA and Banerjee SK (2000) A new high-resolution geomagnetic relative paleointensity record for the North American Holocene: A comparison of sedimentary and absolute intensity data. *Journal of Geophysical Research* 105(B1): 821–834.

Cande SC and Kent DV (1995) Revised calibration of the geomagnetic polarity timescale for the late Cretaceous and Cenozoic. *Journal of Geophysical Research* 100: 6093–6095.

Carcaillet J, Bourles DL, Thouveny N, and Arnold M (2004) A high resolution authigenic ^{10}Be/^{9}Be record of geomagnetic moment variations over the last 300 ka from sedimentary cores of the Portuguese margin. *Earth and Planetary Science Letters* 219(3–4): 397–412.

Carcaillet JT, Thouveny N, and Bourles DL (2003) Geomagnetic moment instability between 0.6 and 1.3 Ma from cosmonuclide evidence. *Geophysical Research Letters* 30(15): 1792.

Carter-Stiglitz B, Jackson M, and Moskowitz B (2002) Low-temperature remanence in stable single domain magnetite. *Geophysical Research Letters* 29: 1129 (doi:10.1029/2001GL014197).

Carter-Stiglitz B, Moskowitz B, and Jackson M (2003) Low-temperature remanence in stable single domain magnetite. *Geophysical Research Letters* 30(21): 2113.

Carvallo C, Ozdemir O, and Dunlop DJ (2004) Palaeointensity determinations, palaeodirections and magnetic properties of basalts from the Emperor seamounts. *Geophysical Journal International* 156(1): 29–38.

Channell JET (1999) Geomagnetic paleointensity and directional secular variation at Ocean Drilling Program (ODP) Site 984 (Bjorn Drift) since 500 ka: Comparisons with ODP Site 983 (Gardar Drift). *Journal of Geophysical Research* 104(B10): 22937–22951.

Channell JET, Curtis JH, and Flower BP (2004) The Matuyama–Brunhes boundary interval (500–900 ka) in North Atlantic drift sediments. *Geophysical Journal International* 158(2): 489–505.

Channell JET, Hodell DA, and Lehman B (1997) Relative geomagnetic paleointensity and ^{18}O at ODP Site 983 (Gardar Drift, North Atlantic) since 350 ka. *Earth and Planetary Science Letters* 153: 103–118.

Channell JET, Hodell DA, McManus J, and Lehman B (1998) Orbital modulation of the Earth's magnetic field intensity. *Nature* 394(6692): 464–468.

Channell JET and Kleiven HF (2000) Geomagnetic palaeointensities and astrochronological ages for the Matuyama–Brunhes boundary and the boundaries of the Jaramillo Subchron: Palaeomagnetic and oxygen isotope records from ODP Site 983. *Philosophical Transactions of the Royal Society of London, Series A* 358(1768): 1027–1047.

Channell JET, Mazaud A, Sullivan P, Turner S, and Raymo ME (2002) Geomagnetic excursions and paleointensities in the Matuyama Chron at Ocean Drilling Program Sites 983 and 984 (Iceland Basin). *Journal of Geophysical Research* 107(B6): 2114.

Channell J, Stoner J, Hodell D, and Charles C (2000) Geomagnetic paleointensity for the last 100 kyr from the sub-antarctic South Atlantic: A tool for inter-hemispheric correlation. *Earth and Planetary Science Letters* 175: 145–160.

Chauvin A, Roperch P, and Levi S (2005) Reliability of geomagnetic paleointensity data: The effects of the NRM fraction and concave-up behavior on paleointensity determinations by the Thellier method. *Physics of the Earth and Planetary Interiors* 150(4): 265–286.

Christl M, Strobl C, and Mangini A (2003) Beryllium-10 in deep-sea sediments: A tracer for the Earth's magnetic field intensity during the last 200,000 years. *Quaternary Science Reviews* 22: 725–739.

Cisowski S and Fuller M (1986) Lunar paleointensities via the IRM(s) normalization method and the early magnetic history of the moon. In: Hartmann W, Phillips R, and Taylor G (eds.) *The Origin of the Moon*, pp. 411–421. Houston, TX: Lunar and Planetary Science Institute.

Clement BM (2004) Dependence of the duration of geomagnetic polarity reversals on site latitude. *Nature* 428(6983): 637–640.

Clement BM and Kent DV (1986) Short polarity intervals within the Matuyama: Transitional field records from hydraulic piston cored sediments from the North Atlantic. *Earth and Planetary Science Letters* 81: 253–264.

Coe RS (1967) The determination of paleo-intensities of the Earth's magnetic field with emphasis on mechanisms which could cause non-ideal behavior in Thellier's method. *Journal of Geomagnetism and Geoelectricity* 19: 157–178.

Coe RS, Grommé S, and Mankinen EA (1978) Geomagnetic paleointensities from radiocarbondated lava flows on Hawaii and the question of the Pacific nondipole low. *Journal of Geophysical Research* 83: 1740–1756.

Coffey W, Kalmykov Y, and Waldron J (1996) *World Scientific Series in Contemporary Chemcical Physics, Vol. 11: The Langevin Equation with Applications in Physics, Chemistry and Electrical Engineering*. Singapore: World Scientific.

Collinson DW (1965) DRM in sediments. *Journal of Geophysical Research* 70: 4663–4668.

Constable CG, Tauxe L, and Parker RL (1998) Analysis of 11 Myr of geomagnetic intensity variation. *Journal of Geophysical Research* 103(B8): 17735–17748.

Constable C and Tauxe L (1996) Towards absolute calibration of sedimentary paleointensity records. *Earth and Planetary Science Letters* 143: 269–274.

Cottrell R and Tarduno J (1999) Geomagnetic paleointensity derived from single plagioclase crystals. *Earth and Planetary Science Letters* 169: 1–5.

Cottrell R and Tarduno J (2000) Late Cretaceous true polar wander: Not so fast. *Science* 288: 2283a.

Cox AV (1968) Lengths of geomagnetic polarity intervals. *Journal of Geophysical Research* 73: 3247–3260.

Cronin M, Tauxe L, Constable C, Selkin P, and Pick T (2001) Noise in the quiet zone. *Earth and Planetary Science Letters* 190: 13–30.

Dansgaard W, Johnsen S, Clausen H, et al. (1993) Evidence for general instability of past climate from a 250-kyr ice core record. *Nature* 364: 218–220.

Deamer GA and Kodama KP (1990) Compaction-induced inclination shallowing in synthetic and natural clay-rich sediments. *Journal of Geophysical Research* 95: 4511–4529.

Dekkers M and Böhnel H (2006) Reliable absolute paleointensities independent of magnetic domain state. *Earth and Planetary Science Letters* 248: 508–517.

deMenocal PB, Ruddiman WF, and Kent DV (1990) Depth of p-DRM acquisition in deep-sea sediments – A case study of the B/M reversal and oxygen isotopic stage 19.1. *Earth and Planetary Science Letters* 99: 1–13.

Demory F, Nowaczyk NR, Witt A, and Oberhansli H (2005) High-resolution magnetostratigraphy of late Quaternary sediments from Lake Baikal, Siberia: Timing of intracontinental paleoclimatic responses. *Global and Planetary Change* 46(1–4): 167–186.

Denham CR and Chave AD (1982) Detrital remanent magnetization: Viscosity theory of the lock-in zone. *Journal of Geophysical Research* 87: 7126–7130.

Dinares-Turell J, Sagnotti L, and Roberts AP (2002) Relative geomagnetic paleointensity from the Jaramillo Subchron to the Matuyama/Brunhes boundary as recorded in a Mediterranean piston core. *Earth and Planetary Science Letters* 194(3–4): 327–341.

Dunlop D and Argyle K (1997) Thermoremanence, anhysteretic remanence and susceptibility of submicron magnetites: Nonlinear field dependence and variation with grain size. *Journal of Geophysical Research* 102: 20199–20210.

Dunlop DJ and Xu S (1994) Theory of partial thermoremanent magnetization in multidomain grains, 1 repeated identical barriers to wall motion (single microcoercivity). *Journal of Geophysical Research* 99: 9005–9023.

Dunlop D and Ozdemir O (1997) *Cambridge Studies in Magnetism: Rock Magnetism: Fundamentals and Frontiers*. Cambridge, UK: Cambridge University Press.

Evans HF and Channell JET (2003) Upper Miocene magnetic stratigraphy at ODP site 1092 (sub-Antarctic South Atlantic): Recognition of 'cryptochrons' in C5n.2n. *Geophysical Journal International* 153(2): 483–496.

Fabian K (2003) Some additional parameters to estimate domain state from isothermal magnetization measurements. *Earth and Planetary Science Letters* 213(3–4): 337–345.

Folgheraiter M (1899) Sur les variations sèculaires de l'inclinaison magnètique dans l'antiquitè. *Journal de Physique* 5: 660–667.

Fox JMW and Aitken MJ (1980) Cooling-rate dependence of thermoremanent magnetization. *Nature* 283: 462–463.

Frank M, Schwarz B, Baumann S, Kubik P, Suter M, and Mangini A (1997) A 200 kyr record of cosmogenic radionuclide production rate and geomagnetic field intensity from ^{10}Be in globally stacked deep-sea sediments. *Earth and Planetary Science Letters* 149: 121–129.

Garcia AS, Thomas DN, Liss D, and Shaw J (2006) Low geomagnetic field intensity during the Kiaman superchron: Thellier and microwave results from the Great Whin Sill intrusive complex, northern United Kingdom. *Geophysical Research Letters* 33(16): L11701.

Gee J, Cande S, Hildebrand J, Donnelly K, and Parker R (2000) Geomagnetic intensity variations over the past 780 kyr obtained from near-seafloor magnetic anomalies. *Nature* 408: 827–832.

Genevey A and Gallet Y (2003) Eight thousand years of geomagnetic field intensity variations in the eastern Mediterranean. *Journal of Geophysical Research* 108: 2228 (doi:10.1029/2001JB001612).

Gibbs R (1985) Estuarine flocs: Their size, settling velocity and density. *Journal of Geophysical Research* 90: 3249–3251.

Goguitchaichvili A, Alva-Valdivia LM, Luis M, Rosas-Elguera J, Urrutia-Fucugauchi J, and Sole J (2004) Absolute geomagnetic paleointensity after the Cretaceous Normal Superchron and just prior to the Cretaceous–Tertiary transition. *Journal of Geophysical Research* 109: B01105 (doi:10.1029/2003JB002477).

Granot R, Tauxe L, Gee JS, and Ron H (2007) A view into the Cretaceous geomagnetic field from analysis of gabbros and submarine glasses. *Earth and Planetary Science Letters* 256: 1–11.

Grootes P and Stuiver M (1997) Oxygen 18/16 variability in Greenland snow and ice with 10^{-3} to 10^5 year time resolution. *Journal of Geophysical Research* 102: 26455–26470.

Guyodo Y, Acton GD, Brachfeld S, and Channell JET (2001) A sedimentary paleomagnetic record of the Matuyama chron from the Western Antarctic margin (ODP Site 1101). *Earth and Planetary Science Letters* 191(1–2): 61–74.

Guyodo Y and Channell JET (2002) Effects of variable sedimentation rates and age errors on the resolution of sedimentary paleointensity records. *Geochemistry, Geophysics, Geosystems* 3: 1048.

Guyodo Y, Gaillot P, and Channell JET (2000) Wavelet analysis of relative geomagnetic paleointensity at ODP Site 983. *Earth and Planetary Science Letters* 184(1): 109–123.

Guyodo Y, Richter C, and Valet JP (1999) Paleointensity record from Pleistocene sediments off the California Margin. *Journal of Geophysics* 104: 22953–22965.

Guyodo Y and Valet JP (1999) Global changes in intensity of the Earth's magnetic field during the past 800 kyr. *Nature* 399(6733): 249–252.

Guyodo Y and Valet JP (2006) A comparison of relative paleointensity records of the Matuyama Chron for the period 0.75–1.25 Ma. *Physics of the Earth and Planetary Interiors* 156: 205–212.

Haag M (2000) Reliability of relative palaeointensities of a sediment core with climatically-triggered strong magnetisation changes. *Earth and Planetary Science Letters* 180(1–2): 49–59.

Hale CJ (1987) Palaeomagnetic data suggest link between the Archean-Proterozoic boundaryand inner-core nucleation. *Nature* 329: 233–237.

Halgedahl S, Day R, and Fuller M (1980) The effect of cooling rate on the intensity of weak-field TRM in single-domain magnetite. *Journal of Geophysical Research* 95: 3690–3698.

Halls H and Davis DW (2004) Paleomagnetism and Ub–P geochronology of the 2.17 Ga Biscotasing dyke swarm, Ontario, Canada: Evidence for vertical-axis crustal rotation across the Kapuskasing Zone. *Canadian Journal of Earth Sciences* 41: 255–269.

Heller R, Merrill RT, and McFadden PL (2002) The variation of intensity of Earth's magnetic field with time. *Physics of the Earth and Planetary Interiors* 131(3–4): 237–249.

Harrison CGA (1966) The paleomagnetism of deep sea sediments. *Journal of Geophysical Research* 71: 3033–3043.

Hartl P and Tauxe L (1996) A precursor to the Matuyama/Brunhes transition- field instability as recorded in pelagic sediments. *Earth and Planetary Science Letters* 138: 121–135.

Hayashida A, Verosub KL, Heider F, and Leonhardt R (1999) Magnetostratigraphy and relative palaeointensity of late Neogene sediments at ODP Leg 167 Site 1010 off Baja California. *Geophysical Journal International* 139(3): 829–840.

Hays JD, Imbrie J, and Shackleton NJ (1976) Variations in the Earth's orbit: Pacemaker of the ice ages. *Science* 194: 1121–1132.

Heller F and Liu T (1982) Magnetostratigraphical dating of loess deposits in China. *Nature* 300: 431–433.

Herrero-Bervera E and Valet JP (2005) Absolute paleointensity and reversal records from the Waianae sequence (Oahu, Hawaii, USA). *Earth and Planetary Science Letters* 234(1–2): 279–296.

Hoffman KA and Biggin AJ (2005) A rapid multi-sample approach to the determination of absolute paleointensity. *Journal of Geophysical Research* 110: B12108 (doi:10.1029/2005JB003646).

Hoffman KA, Constantine VL, and Morse DL (1989) Determination of absolute palaeointensity using a multi-specimen procedure. *Nature* 339: 295–297.

Horng CS, Roberts AP, and Liang WT (2003) A 2.14-Myr astronomically tuned record of relative geomagnetic paleointensity from the western Philippine Sea. *Journal of Geophysical Research* 108(B1): 2059.

Hughen K, Lehman S, Southon J, et al. (2004) C-14 activity and global carbon cycle changes over the past 50,000 years. *Science* 303(5655): 202–207.

Imbrie J, Hays JD, Martinson DG, et al. (eds.) (1984) *Milankovitch and Climate, Part 1: The Orbital Theory of Pleistocene Climate: Support from a Revised Chronology of the Marine δ^{18} O Record*, Hingham, MA Reidel.

Irving E (1964) *Paleomagnetism and Its Application to Geological and Geophysical Problems*. New York: John Wiley.

Johnson EA, Murphy T, and Torreson OW (1948) Pre-history of the Earth's magnetic field. *Terrestrial Magnetism and Atmospheric Electricity* 53: 349–372.

Johnson HP, Lowrie W, and Kent DV (1975) Stability of anhysteretic remanent magnetization in fine and coarse magnetite and maghemite particles. *Geophysical Journal of the Royal Astronomical Society* 41: 1–10.

Juarcz M and Tauxe L (2000) The intensity of the time averaged geomagnetic field: The last 5 m.y. *Earth and Planetary Science Letters* 175: 169–180.

Juarez T, Tauxe L, Gee JS, and Pick T (1998) The intensity of the Earth's magnetic field over the past 160 million years. *Nature* 394: 878–881.

Katari K and Bloxham J (2001) Effects of sediment aggregate size on DRM intensity: A new theory. *Earth and Planetary Science Letters* 186(1): 113–122.

Katari K and Tauxe L (2000) Effects of surface chemistry and flocculation on the intensity of magnetization in redeposited sediments. *Earth and Planetary Science Letters* 181: 489–496.

Katari K, Tauxe L, and King J (2000) A reassessment of post depositional remenent magnetism: Preliminary experiments with natural sediments. *Earth and Planetary Science Letters* 183: 147–160.

Kent D, Hemming S, and Turrin B (2002) Laschamp excursion at Mono Lake? *Earth and Planetary Science Letters* 197: 151–164.

Kent DV (1973) Post-depositional remanent magnetization in deep-sea sediment. *Nature* 246: 32–34.

Kent DV (1982) Apparent correlation of paleomagnetic intensity and climatic records in deep-sea sediments. *Nature* 299: 538–539.

Kent DV and Carlut J (2001) A negative test of orbital control of geomagnetic reversals and excursions. *Geophysical Research Letters* 28(18): 3561–3564.

Kent DV and Opdyke ND (1977) Paleomagnetic field intensity variation recorded in a Brunhes epoch deep-sea sediment core. *Nature* 266: 156–159.

Kent DV and Schneider DA (1995) Correlation of paleointensity variation records in the Brunhes/Matuyama polarity transition interval. *Earth and Planetary Science Letters* 129: 135–144.

King JW, Banerjee SK, and Marvin J (1983) A new rock magnetic approach to selecting sediments for geomagnetic paleointensity studies: Application to paleointensity for the last 4000 years. *Journal of Geophysical Research* 88: 5911–5921.

King RF and Rees AI (1966) Detrital magnetism in sediments: An examination of some theoretical models. *Journal of Geophysical Research* 71: 561–571.

Kirschvink JL (1980) The least-squares line and plane and the analysis of paleomagnetic data. *Geophysical Journal of the Royal Astronomical Society* 62: 699–718.

Kissel C, Laj C, Labeyrie L, Dokken T, Voelker A, and Blamart D (1999) Rapid climatic variations during marine isotopic stage 3: Magnetic analysis of sediments from Nordic Seas and North Atlantic. *Earth and Planetary Science Letters* 171(3): 489–502.

Kletetschka G, Acuna MH, Kohout T, Wasilewski PJ, and Connerney JEP (2004) An empirical scaling law for acquisition of thermoremanent magnetization. *Earth and Planetary Science Letters* 226(3–4): 521–528.

Kletetschka G, Fuller M, Kohout T, et al. (2006) TRM in low magnetic fields: A minimum field that can be recorded by large multidomain grains. *Physics of the Earth and Planetary Interiors* 154: 290–298.

Koenigsberger J (1936) Die abhaengigkeit der natuerlichen remanenten magnetisierung bei eruptivgesteinen von deren alter und zusammensetzung. *Beitriige zur Angewandte Geophsik* 5: 193–246.

Koenigsberger J (1938a) Natural residual magnetism of eruptive rocks, Pt I. *Terrestrial Magnetism and Atmospheric Electricity* 43: 119–127.

Koenigsberger J (1938b) Natural residual magnetism of eruptive rocks, Pt II. *Terrestrial Magnetism and Atmospheric Electricity* 43: 299–320.

Kok YS and Tauxe L (1996a) Saw-toothed pattern of relative paleointensity records and cumulative viscous remanence. *Earth and Planetary Science Letters* 137: 101–108.

Kok YS and Tauxe L (1996b) Saw-toothed pattern of sedimentary paleointensity records explained by cumulative viscous remanence. *Earth and Planetary Science Letters* 144: E9–E14.

Kok YS and Tauxe L (1999) A relative geomagnetic paleointensity stack from Ontong-Java Plateau sediments for the Matuyama. *Journal of Geophysical Research, Solid Earth* 104(B11): 25401–25413.

Kok YS and Ynsen I (2002) Reply to comment by J.-P. Valet and L. Meynadier on 'A relative geomagnetic paleointensity stack from Ontong-Java Plateau sediments for the Matuyama'. *Journal of Geophysical Research* 107(B3): 2051.

Kok Y and Tauxe L (2000) Comment on 'Saw-toothed variations of relative paleointneisty and cumulative viscous remanence:

Testing the records and the model', by L. Meynadier, J.-P. Valet, Y. Guyodo, and C. Richter. *Journal of Geophysical Research* 105: 16609–16612.

Kono M (1971) Intensity of the Earth's magnetic field during the Pliocene and Pleistocene in relation to the amplitude of mid-ocean ridge magnetic anomalies. *Earth and Planetary Science Letters* 11: 10–17.

Kono M (1974) Intensities of the Earth's magnetic field about 60 m.y. ago determined from the Deccan Trap basalts, India. *Journal of Geophysical Research* 79: 1135–1141.

Kono M and Ueno N (1977) Paleointensity determination by a modified Thellier method. *Physics of the Earth and Planetary Interiors* 13: 305–314.

Korte M and Constable C (2005) The geomagnetic dipole moment over the last 7000 years – New results from a global model. *Earth and Planetary Science Letters* 236: 348–358.

Laj C, Kissel C, and Garnier F (1996a) Relative geomagnetic field intensity and reversals for the last 1.8 My from a central equatorial Pacific core. *Geophysical Research Letters* 23: 3393–3396.

Laj C, Kissel C, and Lefevre I (1996b) Relative geomagnetic field intensity and reversals from Upper Miocene sections in Crete. *Earth and Planetary Science Letters* 141(1–4): 67–78.

Laj C, Kissel C, Mazaud A, Channell JET, and Beer J (2000) North Atlantic palaeointensity stack since 75 ka (NAPIS-75) and the duration of the Laschamp event. *Philosophical Transactions of the Royal Society of London* 358(1768): 1009–1025.

Laj C, Rais A, Surmont J, et al. (1997) Changes of the geomagnetic field vector obtained from lava sequences on the island of Vulcano (Aeolian Islands, Sicily). *Physics of the Earth and Planetary Interiors* 99: 161–177.

Lanci L and Lowrie W (1997) Magnetostratigraphic evidence that 'tiny wiggles' in the oceanic magnetic anomaly record represent geomagnetic paleointensity variations. *Earth and Planetary Science Letters* 148(3–4): 581–592.

Laskar J, Joutel F, and Boudin F (1993) Orbital, precessional, and insolation quantities for the Earth from −20 Myr to + 10 Myr. *Astronomy and Astrophysics* 270(1–2): 522–533.

Le Goff M and Gallet Y (2004) A new three-axis vibrating sample magnetometer for continuous high-temperature magnetization measurements: Applications to paleo- and archeo-intensity determinations. *Earth and Planetary Science Letters* 229(1–2): 31–43.

Lehman B, Laj C, Kissel C, Mazaud A, Paterne M, and Labeyrie L (1996) Relative changes of the geomagnetic field intensity during the last 280 kyr from piston cores in the Acores area. *Physics of the Earth and Planetary Interiors* 93: 269–284.

Leonhardt R, Heider F, and Hayashida A (1999) Relative geomagnetic field intensity across the Jaramillo subchron in sediments from the California margin: ODP Leg 167. *Journal of Geophysical Research* 104: 29133–29146.

Leonhardt R, Matzka J, and Menor EA (2003) Absolute paleointensities and paleodirections of miocene and Pliocene lavas from Fernando de Noronha, Brazil. *Physics of the Earth and Planetary Interiors* 139(3–4): 285–303.

Levi S and Banerjee SK (1976) On the possibility of obtaining relative paleointensities from lake sediments. *Earth and Planetary Science Letters* 29: 219–226.

Linsley B and Thunnell R (1990) The record of deglaciation in the Sulu Sea: Evidence for the Younger Dryas event in the western tropical Pacific. *Paleoceanography* 5: 1025–1039.

Lisiecki L and Raymo ME (2005) A Pliocene–Pleistocene stack of 57 globally distributed benthic δ^{18}O records. *Paleoceanography* 20: PA1003 (doi:10.1029/2004PA001071).

Liu Q, Banerjee SK, Jackson MJ, Deng C, Pan Y, and Zhu R (2005) Inter-profile correlation of the Chinese loess/paleosol sequences during Marine Oxygen Isotope Stage 5 and indications of pedogenesis. *Quaternary Science Reviews* 24(1–2): 195–210.

Lovlie R (1974) Post-depositional remanent magnetization in a re-deposited deep-sea sediment. *Earth and Planetary Science Letters* 21: 315–320.

Lu R, Banerjee SK, and Marvin J (1990) Effects of clay mineralogy and the electrical conductivity of water on the acquisition of depositional remanent magnetization in sediments. *Journal of Geophysical Research* 95: 4531–4538.

Lund SP and Keigwin L (1994) Measurement of the degree of smoothing in sediment paleomagnetic secular variation records: An example from late Quaternary deep-sea sediments of the Bermuda Rise, western North Atlantic Ocean. *Earth and Planetary Science Letters* 122: 317–330.

Lund S, Williams T, Acton GD, Clement BM and Okada M (2001) Proceedings of the ocean Drilling Program, Scientific Results 17, Vol. 172: *Brunhes Chron magnetic field excursions recovered from Leg 172 sediments*. Ocean Drilling Program.

Macouin M, Valet J, Besse J, and Ernst R (2006) Absolute paleointensity at 1.27 Ga from the Mackenzie dyke swarm (Canada). *Geochemistry, Geophysics, Geosystems* 7: Q06H21 (doi:10.1029/2005GC000960).

Macouin M, Valet JP, Besse J, Buchan K, Ernst R, LeGoff M, et al. (2003) Low paleointensities recorded in 1 to 2.4 Ga Proterozoic dykes, Superior Province, Canada. *Earth and Planetary Science Letters* 213(1–2): 79–95.

Macri P, Sagnotti L, Dinares-Turell J, and Caburlotto A (2005) A composite record of Late Pleistocene relative geomagnetic paleointensity from the Wilkes Land Basin (Antarctica). *Physics of the Earth and Planetary Interiors* 151(3–4): 223–242.

Martinson DG, Pisias NG, Hays JD, Imbrie J, Moore TC, Jr., and Shackleton NJ (1987) Age dating and the orbital theroy of the Ice ages: Development of a high-resolution 0-300,000 year chronostratigraphy. *Quaternary Research* 27: 1–29.

Masarik J and Beer J (1999) Simulation of particle fluxes and cosmogenic nuclide production in the Earth's atmosphere. *Journal of Geophysical Research* 104(D10): 12099–12111.

Mazaud A (1996) 'Sawtooth' variation in magnetic intensity profiles and delayed acquisition of magnetization in deep sea cores. *Earth and Planetary Science Letters* 139: 379–386.

Mazaud A, Sicre MA, Ezat U, et al. (2002) Geomagnetic-assisted stratigraphy and sea surface temperature changes in core MD94-103 (Southern Indian Ocean): Possible implications for North-South climatic relationships around H4. *Earth and Planetary Science Letters* 201(1): 159–170.

McClelland E and Briden J (1996) An improved methodology for Thellier-type paleointensity determination in igneous rocks and its usefulness for verifying primary thermoremanence. *Journal of Geophysical Research* 101: 21995–22013.

McElhinny MW (1973) *Paleomagnetism and Plate Tectonics*. Cambridge: Cambridge University Press.

McElhinnhy MW and Evans ME (1968) An investigation of the strength of the geomagnetic field in the Early Precambrian. *Physics of the Earth and Planetary Interiors* 1: 485–497.

McFadden PL and McElhinny MW (1982) Variations in the geomagnetic dipole 2: Statistical analysis of VDM's for the past 5 million years. *Journal of Geomagnetism and Geoelectricity* 34: 163.

McFadden P and Merrill R (1997) Sawtooth paleointensity and reversals of the geomagnetic field. *Physics of the Earth and Planetary Interiors* 103: 247–252.

McHargue L and Donahue D (2005) Effects of climate and the cosmic-ray flux on the ^{10}Be content of marine sediments. *Earth and Planetary Science Letters* 232: 193–207.

McMillan DG, Constable CG, and Parker RL (2002) Limitations on stratigraphic analyses due to incomplete age control and their relevance to sedimentary paleomagnetism. *Earth and Planetary Science Letters* 201(3–4): 509–523.

McMillan DG, Constable C, and Parker RL (2004) Assessing the dipolar signal in stacked paleointensity records using a statistical error model and geodynamo simulations. *Physics of the Earth and Planetary Interiors* 145: 37–54.

Meynadier L and Valet JP (1995) Relative geomagnetic intensity during the last 4 m.y. from the equatorial Pacifid. *Proceedings of the Ocean Drilling Program, Scientific Results* 138: 779–795.

Meynadier L and Valet JP (1996) Post-depositional realignment of magnetic grains and asymmetrical saw-tooth patterns o magnetization intensity. *Earth and Planetary Science Letters* 126: 109–127.

Meynadier L, Valet JP, Bassinot F, Shackleton NJ, and Guyodo Y (1994) Asymmetrical saw-tooth pattern of the geomagnetic field intensity from equatorial sediments in the Pacific and Indian Oceans. *Earth and Planetary Science Letters* 126: 109–127.

Meynadier L, Valet JP, Guyodo Y, and Richter C (1998) Saw-toothed variations of relative paleointensity and cumulative viscous remanence: Testing the records and the model. *Journal of Geophysical Research* 103: 7095–7105.

Meynadier L, Valet JP, Weeks R, Shackleton NJ, and Hagee VL (1992) Relative geomagnetic intensity of the field during the last 140 ka. *Earth and Planetary Science Letters* 114: 39–57.

Morimoto C, Otofuji Y, Miki M, Tanaka H, and Itaya T (1997) Preliminary palaeomagnetic results of an archaean dolerite dyke of west Greenland: Geomagnetic field intensity at 2.8 Ga. *Geophysical Journal International* 128: 583–585.

Nagata T (1961) *Rock Magnetism*. Tokyo: Maruzen Co., Ltd.

Nagata T, Arai Y, and Momose K (1963) Secular variation of the geomagnetic total force during the last 5000 years. *Journal of Geophysical Research* 68: 5277–5282.

Néel L (1949) Théorie du trainage magnétique des ferromagneétiques en grains fines avec applications aux terres cuites. *Annales de Geophysique* 5: 99–136.

Néel L (1955) Some theoretical aspects of rock-magnetism. *Advances in Physics* 4: 191–243.

Nowaczyk NR and Antonow M (1997) High-resolution magnetostratigraphy of four sediment cores from the Greenland Sea .1. Identification of the Mono Lake excursion, Laschamp and Biwa I Jamaica geomagnetic polarity events. *Geophysical Journal International* 131(2): 310–324.

Nowaczyk NR, Antonow M, Knies J, and Spielhagen RF (2003) Further rock magnetic and chronostratigraphic results on reversal excursions during the last 50 ka as derived from northern high latitudes and discrepancies in precise AMS C-14 dating. *Geophysical Journal International* 155(3): 1065–1080.

Nowaczyk NR and Frederichs TW (1999) Geomagnetic events and relative palaeointensity variations during the past 300 ka as recorded in Kolbeinsey Ridge sediments, Iceland Sea: Indication for a strongly variable geomagnetic field. *International Journal of Earth Sciences* 88(1): 116–131.

Nowaczyk NR, Harwart S, and Melles M (2001) Impact of early diagenesis and bulk particle grain size distribution on estimates of relative geomagnetic palaeointensity variations in sediments from Lama Lake, northern Central Siberia. *Geophysical Journal International* 145(1): 300–306.

Nowaczyk NR and Knies J (2000) Magnetostratigraphic results from the eastern Arctic Ocean: AMS ^{14}C ages and relative palaeointensity data of the Mono Lake and Laschamp geomagnetic reversal excursions. *Geophysical Journal International* 140(1): 185–197.

Oda H, Nakamura K, Ikehara K, Nakano T, Nishimura M, and Khlystov O (2002) Paleomagnetic record from Academician Ridge, Lake Baikal: A reversal excursion at the base of marine oxygen isotope stage 6. *Earth and Planetary Science Letters* 202(1): 117–132.

Okada M (1995) Detailed variation of geomagnetic field intensity during the late Pleistocene at Site 882. *Proceedings of the Ocean Drilling Program, Scientific Results* 145: 469–474.

Ozima M, Ozima M, and Akimoto S (1964) Low temperature characteristics of remanent magnetization of magnetite. *Journal of Geomagnetism and Geoelectricity* 16: 165–177.

Pan Y, Zhu R, Shaw J, Liu Q, and Guo B (2001) Can relative paleointensities be determined from the normalized magnetization of the wind-blown loess of China? *Journal of Geophysical Research* 106: 19221–19232.

Peck J, King J, Colman S, and Kravchinsky V (1996) An 84-kyr paleomagnetic record from the sediments of Lake Baikal, Siberia. *Journal of Geophysical Research* 101: 1365–1385.

Perrin M and Schnepp E (2004) IAGA paleointensity database: Distribution and quality of the data set. *Physics of the Earth and Planetary Interiors* 147(2–3): 255–267.

Perrin M, Schnepp E, and Shcherbakov V (1998) Paleointensity database updated. *EOS Transactions of the American Geophysical Union* 79: 198.

Perrin M and Shcherbakov V (1997) Paleointensity of the Earth's magnetic field for the past 400 Ma: Evidence for a dipole structure during the Mesozoic low. *Journal of Geomagnetism and Geoelectricity* 49: 601–614.

Pick T and Tauxe L (1993) Geomagnetic paleointensities during the Cretaceous normal superchron measured using submarine basaltic glass. *Nature* 366: 238–242.

Pisias NG, Martinson DG, Moore TC, *et al.* (1984) High-resolution stratigraphic correlation of benthic oxygen isotopic records spanning the last 300,000 years. *Marine Geology* 56(1–4): 119–136.

Prévot M, Derder MEM, McWilliams M, and Thompson J (1990) Intensity of the Earth's magnetic field: Evidence for a Mesozoic dipole low. *Earth and Planetary Science Letters* 97: 129–139.

Raisbeck G, Yiou F, and Zhou S (1994) Paleointensity puzzle. *Nature* 371: 207–208.

Rampino M (1979) Possible relations between changes in global ice volume, geomagnetic excursions, and the eccentricity of the Earth's orbit. *Geology* 7: 584–587.

Rampino MR (1981) Revised age estimates of Brunhis paleomagnetic events: Support for a link between geomagnetism and eccentricity. *Earth and Planetary Science Letters* 8: 1047–1050.

Rampino MR and Kent DV (1983) Geomagnetic excursions and climate change. *Nature* 302: 455.

Ravilly M, Horen H, Perrin M, Dyment J, Gente P, and Guillou H (2001) NRM intensity of altered oceanic basalts across the MAR (21 degrees N, 0-1.5 Ma): A record of geomagnetic palaeointensity variations? *Geophysical Journal International* 145(2): 401–422.

Riisager J, Riisager P, and Pedersen A (2003) The C27n–C26r geomagnetic polarity reversal recorded in the west Greenland flood basalt province: How complex is the transitional field? *Journal of Geophysical Research* 108: 2155 (doi:10.1029/2002JB002124).

Riisager J, Riisager P, Zhao XX, Coe RS, and Pedersen AK (2004) Paleointensity during a chron C26r excursion recorded in west Greenland lava flows. *Journal of Geophysical Research, Solid Earth* 109(B4): B04107.

Riisager P and Abrahamsen N (2000) Palaeointensity of West Greenland Palaeocene basalts: Asymmetric intensity around the C27n–C26r transition. *Physics of the Earth and Planetary Interiors* 118(1–2): 53–64.

Riisager P and Riisager J (2001) Detecting multidomain magnetic grains in Thellier palaeointensity experiments. *Physics of the Earth and Planetary Interiors* 125(1–4): 111–117.

Roberts A, Lehman B, Weeks R, Verosub K, and Laj C (1997) Relative paleointensity of the geomagnetic field over the last 200,000 years from ODP Sites 883 and 884, North Pacific Ocean. *Earth and Planetary Science Letters* 152: 11–23.

Roberts AP (1995) Magnetic properties of sedimentary greigite (Fe_3S_4). *Earth and Planetary Science Letters* 134: 227–236.

Roberts AP and Lewin-Harris JC (2000) Marine magnetic anomalies: Evidence that 'tiny wiggles' represent short-period geomagnetic polarity intervals. *Earth and Planetary Science Letters* 183(3–4): 375–388.

Roberts AP, Verosub KL, and Negrini RM (1994) Middle Late Pleistocene relative paleointensity of the geomagnetic-field from lacustrine sediments, Lake Chewaucan, Western United States. *Geophysical Journal International* 118(1): 101–110.

Roberts AP, Winklhofer M, Liang WT, and Horng CS (2003) Testing the hypothesis of orbital (eccentricity) influence on Earth's magnetic field. *Earth and Planetary Science Letters* 216(1–2): 187–192.

Rolph TC and Shaw J (1985) A new method of palaeofield magnitude correction for thermally altered samples and its application to Lower Carboniferous lavas. *Geophysical Journal of the Royal Astronomical Society* 80: 773–781.

Sato T, Kikuchi H, Nakashizuka M, and Okada M (1998) Quaternary geomagnetic field intensity: Constant periodicity or variable period? *Geophysical Research Letters* 25: 2221–2224.

Sato T and Kobayashi K (1989) Long-period secular variations of the Earth's magnetic field revealed by Pacific deep-sea sediment cores. *Journal of Geomagnetism and Geoelectricity* 41: 147–159.

Schneider DA (1993) An estimate of Late Pleistocene geomagnetic intensity variation from Sulu Sea sediments. *Earth and Planetary Science Letters* 120(3–4): 301–310.

Schneider DA, Kent DV, and Mello GA (1992) A detailed chronology of the Australasian impact event, the Brunhes–Matuyama geomagnetic polarity reversal and global climate change. *Earth and Planetary Science Letters* 111: 395–405.

Schneider D and Mello G (1996) A high-resolution marine sedimentary record of geomagnetic iintensity during the Brunhes chron. *Earth and Planetary Science Letters* 144: 297–314.

Schwartz M, Lund SP, and Johnson TC (1996) Environmental factors as complicating influences in the recovery of quantitative geomagnetic-field paleointensity estimates from sediments. *Geophysical Research Letters* 23: 2693–2696.

Schwartz M, Lund SP, and Johnson TC (1998) Geomagnetic field intensity from 71 to 12 ka as recorded in deep-sea sediments of the Blake Outer Ridge, North Atlantic Ocean. *Journal of Geophysical Research, Solid Earth* 103(B12): 30407–30416.

Schwartz E and Symons D (1969) Geomagnetic intensity between 100 and 2500 million years ago. *Physics of the Earth and Planetary Interiors* 2: 11–18.

Selkin P, Gee JS, and Tauxe L (2007) Nonlinear thermoremanence acquisition and implications for paleointensity data. *Earth and Planetary Science Letters* 256: 81–89.

Selkin P, Gee J, Tauxe L, Meurer W, and Newell A (2000a) The effect of remanence anisotropy on paleointensity estimates: A case study from the Archean Stillwater complex. *Earth and Planetary Science Letters* 182: 403–416.

Selkin P and Tauxe L (2000b) Long-term variations in paleointensity. *Philosophical Transactions of the Royal Society of London* 358: 1065–1088.

Shackleton NJ, Berger A, and Peltier WR (1990) An alternative astronomical calibration of the lower Pleistocene timescale based on ODP Site 677. *Transactions of the Royal Society of Edinburgh, Earth Sciences* 81: 251–261.

Shaw J (1974) A new method of determining the magnitude of the paleomanetic field application to 5 historic lavas and five archeological samples. *Geophysical Journal of the Royal Astronomical Society* 39: 133–141.

Shcherbakov V and Shcherbakova V (1983) On the theory of depositional remanent magnetization in sedimentary rocks. *Geophysical Surveys* 5: 369–380.

Shcherbakova V, Shcherbakov V, and Heider F (2000) Properties of partial thermoremanent magnetization in PSD and MD magnetite grains. *Journal of Geophysical Research* 105: 767–782.

Sherwood G, Shaw J, Baer G, and Mallik S (1993) The strength of the geomagnetic field during the Cretaceous Quiet Zone: Paleointensity results from Israeli and Indian Lavas. *Journal of Geomagnetism and Geoelectricity* 45: 339–360.

Smirnov AV and Tarduno JA (2005) Thermochemical remanent magnetization in precambrian rocks: Are we sure the geomagnetic field was weak? *Journal of Geophysical Research* 110: B06103 (doi:10.1029/2004JB003445).

Smirnov AV, Tarduno JA, and Pisakin BN (2003) Paleointensity of the early geodynamo (2.45 Ga) as recorded in Karelia: A single-crystal approach. *Geology* 31(5): 415–418.

Smith PJ (1967) The intensity of the ancient geomagnetic field: A review and analysis. *Geophysical Journal of the Royal Astronomical Society* 12: 321–362.

Sowers T, Bender M, Labeyrie L, *et al.* (1993) A 135,000 year Vostck–Specmap common temporal framework. *Paleoceanography* 8: 737–766.

Spassov S, Heller F, Evans ME, Yue LP, and Dobeneck Tv (2003) A lock-in model for the complex Matuyama–Brunhes boundary record of the loess/palaeosol sequence at Lingtai (Central Chinese Loess Plateau). *Geophysical Journal International* 155(2): 350–366.

Stacey FD (1972) On the role of Brownian motion in the control of detrital remanent magnetization of sediments. *Pure and Applied Geophysics* 98: 139–145.

Stern DP (2003) A millennium of geomagnetism. *Reviews of Geophysics* 41(2): 1007.

Stoner J, Channell J, Hillaire-Marcel C, and Kissel C (2000) Geomagnetic paleointensity and environmental record from Labrador Sea core MD95-2024: Global marine sediment and ice core chronostratigraphy for the last 110 kyr. *Earth and Planetary Science Letters* 5626: 1–17.

Stoner JS, Channell JET, and Hillaire-Marcel C (1995) Late Pleistocene relative geomagnetic paleointensity from the deep Labrador Sea: Regional and global correlations. *Earth and Planetary Science Letters* 134: 237–252.

Stoner JS, Channell JET, and Hillaire-Marcel C (1998) A 200 ka geomagnetic chronostratigraphy for the Labrador Sea: Indirect correlation of the sediment record to SPECMAP. *Earth and Planetary Science Letters* 159(3–4): 165–181.

Stoner JS, Channell JET, Hodell DA, and Charles CD (2003) A ~580 kyr paleomagnetic record from the sub-Antarctic South Atlantic (Ocean Drilling Program Site 1089). *Journal of Geophysical Research* 108(B5): 2244.

Stoner JS, Laj C, Channell JET, and Kissel C (2002) South Atlantic and North Atlantic geomagnetic paleointensity stacks (0-80 ka): Implications for inter-hemispheric correlation. *Quaternary Science Reviews* 21(10): 1141–1151.

Stott L, Poulsen C, Lund S, and Thunell R (2002) Super ENSO and global climate oscillations at millennial time scales. *Science* 297: 222–226.

Stratigraphy ICo (2004) *Geologic Time Scale 2004*. Cambridge, UK: Cambridge University Press.

Sugiura N, Strangway D, Pearce G, Wu Y, and Taylor L (1979) A new magnetic paleointensity value for a 'young lunar glass'.

Proceedings of the Lunar and Planetary Science Conference X: 2189–2197.

Tanaka H and Kono M (1991) Preliminary results and reliability of palaeointensity studies on historical and C14 dated Hawaiian lavas. *Journal of Geomagnetism and Geoelectricity* 43: 375–388.

Tanaka H and Kono M (2002) Paleointensities from a Cretaceous basalt platform in inner Mongolia, Northeastern China. *Physics of the Earth and Planetary Interiors* 133(1–4): 147–157.

Tanaka H, Kono M, and Uchimura H (1995) Some global features of paleointensity in geological time. *Geophysical Journal International* 120: 97–102.

Tanaka H, Otsuka A, Tachibana T, and Kono M (1994) Paleointensities for 10–22 ka from volcanic rocks in Japan and New Zealand. *Earth and Planetary Science Letters* 122: 29–42.

Tarduno JA and Cottrell RD (2005) Dipole strength and variation of the time-averaged reversing and nonreversing geodynamo based on Thellier analyses of single plagioclase crystals. *Journal of Geophysical Research* 110: B11101 (doi:10.1029JB003970).

Tarduno JA, Cottrell RD, and Smirnov AV (2001) High geomagnetic intensity during the mid-Cretaceous from Thellier analyses of single plagioclase crystals. *Science* 291(5509): 1779–1783.

Tarduno JA, Cottrell RD, and Smirnov AV (2002) The Cretaceous superchron geodynamo: Observations near the tangent cylinder. *Proceedings of the National Academy of Sciences of the United States of America* 99(22): 14020–14025.

Tarduno J, Cottrell R, and Smirnov A (2006) The paleomagnetism of single silicate crystals: Recording geomagnetic field strength during mixed polarity intervals, superchrons, and inner core growth. *Reviews of Geophysics* 44: RG1002 (doi:10.1029/2005RG000189).

Tauxe L (1993) Sedimentary records of relative paleointensity of the geomagnetic field: Theory and practice. *Reviews of Geophysics* 31: 319–354.

Tauxe L (1998) *Paleomagnetic Principles and Practice*. Dordrecht, The Netherlands: Kluwer.

Tauxe L (2005) *Lectures in Paleomagnetism*, earthref.org/ Magic/Books/Tauxe/2005, San Diego.

Tauxe L (2006) Long term trends in paleointensity: The contribution of DSDP/ODP submarine basaltic glass collections. *Physics of the Earth and Planetary Interiors* 156: 223–241.

Tauxe L, Steindorf JL, and Harris AJ (2006) Depositional remanent magnetization: Toward an improved theoretical and experimental foundation. *Earth and Planetary Science Letters* 244: 515–529.

Tauxe L, Bertram H, and Seberino C (2002) Physical interpretation of hysteresis loops: Micromagnetic modelling of fine particle magnetite. *Geochemistry, Geophysics, Geosystems* 3, (doi:10.1029/2001GC000280).

Tauxe L, Gans PB, and Mankinen E (2004a) Paleomagnetic and $^{40}Ar/^{39}Ar$ ages from Matuyama/Brunhes aged volcanics near McMurdo Sound, Antarctica. *Geochemistry, Geophysics, Geosystems* 5: Q06H12 (doi:10.1029/2003GC000656).

Tauxe L and Hartl P (1997) 11 million years of Oligocene geomagnetic field behaviour. *Geophysical Journal International* 128: 217–229.

Tauxe L, Herbert T, Shackleton NJ, and Kok YS (1996) Astronomical calibration of the Matuyama–Brunhes Boundary: Consequences for magnetic remanence acquisition in marine carbonates and the Asian loess sequences. *Earth and Planetary Science Letters* 140: 133–146.

Tauxe L and Love J (2003) Paleointensity in Hawaiian Scientific Drilling project Hole (HSDP2): Results from submarine

basaltic glass. *Geochemistry, Geophysics, Geosystems*
4: 8702 (doi:10.1029/2001GC000276).

Tauxe L, Luskin C, Selkin P, Gans PB, and Calvert A (2004b)
Paleomagnetic results from the Snake River Plain:
Contribution to the global time averaged field database.
Geochemistry, Geophysics, Geosystems 5: Q08H13
(doi:10.1029/2003GC000661).

Tauxe L, Pick T, and Kok YS (1995) Relative paleointensity in
sediments; a pseudo-Thellier approach. *Geophysical
Research Letters* 22: 2885–2888.

Tauxe L and Shackleton NJ (1994) Relative paleointensity
records from the Ontong-java Plateau. *Geophysical Journal
International* 117: 769–782.

Tauxe L and Staudigel H (2004) Strength of the geomagnetic
field in the Cretaceous Normal Superchron: New data from
submarine basaltic glass of the Troodos Ophiolite.
Geochemistry, Geophysics, Geosystems 5(2): Q02H06
(doi:10.1029/2003GC000635).

Tauxe L and Wu G (1990) Normalized remanence in sediments
of the Western Equatorial Pacific: Relative paleointensity of
the geomagnetic field? *Journal of Geophysical Research*
95(B8): 12337–12350.

Taylor L (1979) An effective sample preparation technique for
paleointensity determinations at elevated temperatures.
Proceedings of the Lunar and Planetary Science Conference
X: 1209–1211.

Teanby N and Gubbins D (2000) The effects of aliasing and lock-in
processes on palaeosecular variation records from sediments.
Geophysical Journal International 142(2): 563–570.

Thellier E (1938) Sur l'aimantation des terres cuites et ses
applications géophysique. *Annales de Institut de Physique
du Globe Université, Paris* 16: 157–302.

Thellier E and Thellier O (1959) Sur l'intensité du champ
magnétique terrestre dans le passé historique et géologique.
Annales de Geophysique 15: 285–378.

Thouveny N, Carcaillet J, Moreno E, Leduc G, and Nerini D
(2004) Geomagnetic moment variation and paleomagnetic
excursions since 400 kyr BP: A stacked record from
sedimentary sequences of the Portuguese margin, Pt 1.
Earth and Planetary Science Letters 219(3–4): 377–396.

Thouveny N, Debeaulieu JL, Bonifay E, *et al.* (1994) Climate
variations in Europe over the past 140-Kyr deduced from
rock magnetism. *Nature* 371(6497): 503–506.

Tiedemann R, Sarnthein M, and Shackleton NJ (1994)
Astronomical timescale for the Pliocene Atlantic delta ^{18}0
and dust flux record of Ocean Drilling Program Site 659.
Paleoceanography 9: 619–638.

Tric E, Valet JP, Tucholka P, *et al.* (1992) Paleointensity of the
geomagnetic field during the last 80,000 years. *Journal of
Geophysical Research* 97: 9337–9351.

Tsunakawa H and Shaw J (1994) The Shaw method of
paleointensity determinations and its application to recent
volcanic rocks. *Geophysical Journal International*
118: 781–787.

Valet J-P (2003) Time variations in geomagnetic intensity.
Reviews of Geophysics 41: 1004 (doi:10.1029/
2001RG000104).

Valet JP, Brassart J, Lemeur I, *et al.* (1996) Absolute
paleointensity and magnetomineralogical changes. *Journal
of Geophysical Research* 101(B11): 25029–25044.

Valet J-P and Herrero-Bervera E (2000) Paleointensity
experiments using alternating field demagnetization. *Earth
and Planetary Science Letters* 177: 43–58.

Valety JP and Meynadier L (1993) Geomagnetic field intensity
and reversals during the past four million years. *Nature*
366: 234–238.

Valet JP and Meynadier L (2001) Comment on 'A relative
geomagnetic paleointensity stack from Ontong-Java plateau

sediments for the Matuyama' by Yvo S. Kok and Lisa Tauxe.
Journal of Geophysical Research 106: 11013–11015.

Valet JP, Meynadier L, Bassinot FC, and Garnier F (1994)
Relative paleointensity across the last geomagnetic reversal
form sediments of the Atlantic, Indian and Pacific Oceans.
Geophysical Research Letters 21: 485–488.

Valet JP, Meynadier L, and Guyodo Y (2005) Geomagnetic
dipole strength and reversal rate over the past two million
years. *Nature* 435: 802–805.

Valet JP, Tauxe L, and Clement BM (1989) Equatorial and mid-
latitude records of the last geomagnetic reversal from the
Atlantic Ocean. *Earth and Planetary Science Letters*
94: 371–384.

Valet JP, Tric E, Herrero-Bervera E, Meynadier L, and
Lockwood JP (1998) Absolute paleointensity from Hawaiian
lavas younger than 35 ka. *Earth and Planetary Science
Letters* 161: 19–32.

van Vreumingen M (1993a) The magnetization intensity of some
artificial suspensions while flocculating in a magnetic field.
Geophysical Journal International 114: 601–606.

van Vreumingen M (1993b) The influence of salinity and
flocculation upon the acquisition of remanent magnetization
in some artificial sediments. *Geophysical Journal
International* 114: 607–614.

van Zijl JSU, Graham KWT, and Hales AL (1962a) The
paleomagnetism of the Stormberg Lavas 1. *Geophysical
Journal of the Royal Astronomical Society* 7: 23–39.

van Zijl JSU, Graham KWT, and Hales AL (1962b) The
paleomagnetism of the Stormberg Lavas, 2. The behavior of
the Earth's magnetic field during a reversal. *Geophysical
Journal of the Royal Astronomical Society* 7: 169–182.

Verosub KL (1977) Depositional and postdepositional
processes in the magnetization of sediments. *Reviews in
Geophysics and Space Physics* 15: 129–143.

Verosub KL, Herrerobervera E, and Roberts AP (1996) Relative
geomagnetic paleointensity across the Jaramillo Subchron
and the Matuyama/Brunhes boundary. *Geophysical
Research Letters* 23(5): 467–470.

Walton D (2004) Avoiding mineral alteration during microwave
magnetization. *Geophysical Research Letters* 31: L03606
(doi:10.1029/2003GL019011).

Walton D (2005) Isolating viscous overprints with microwaves.
*EOS Transactions of the American Geophysical Union, Fall
Meeting Supplement* 86: GP12A-05.

Walton D, Share J, Rolph TC, and Shaw J (1993) Microwave
magnetisation. *Geophysical Research Letters* 20: 109–111.

Weeks RJ, Laj C, Endignoux L, *et al.* (1995) Normalised natural
remanent magnetisation intensity during the last 240000
years in piston cores from the central North Atlantic Ocean:
Geomagnetic field intensity or environmental signal? *Physics
of the Earth and Planetary Interiors* 87: 213–229.

Westphal M and Munschy M (1994) Saw-tooth pattern of the
Earth's magnetic field tested by magnetic anomalies over
oceanic spreading ridges. *Comptes Rendus de l' Academie
des Sciences Series IIA Earth and Planetary Science*
329: 565–571.

Westphal M and Munschy M (1999) Un test de la variation en
dents de scie du champ magnétique terrestre par les
anomalies du champ magnétique sur les dorsales. *Earth and
Planetary Sciences* 329: 565–571.

Williams T, Thouveny N, and Creer KM (1998) A normalised
intensity record from Lac du Bouchet: Geomagnetic
palaeointensity for the last 300 kyr? *Earth and Planetary
Science Letters* 156(1–2): 33–46.

Winterwerp J and van kesteren W (2004) *Developments in
Sedimentology, Vol. 56: Introduction to the Physcis of
Cohesive Sediment in the Marine Environment*. Amsterdam,
The Netherlands: Elsevier.

Wollin G, Ericson D, and Ryan WBF (1971) Variations in magnetic intensity and climatic changes. *Nature* 232: 549–550.

Worm H (1997) A link between geomagnetic reversals and events and glaciations. *Earth and Planetary Science Letters* 147: 55–67.

Xu S and Dunlop DJ (1994) Theory of partial thermoremanent magnetization in multidomain grains 2 Effect of microcoercivity distribution and comparison with experiment. *Journal of Geophysical Research* 99: 9025–9033.

Yamamoto Y and Tsunakawa H (2005) Geomagnetic field intensity during the last 5 Myr: LTD-DHT Shaw palaeointensities from volcanic rocks of the Society Islands, French Polynesia. *Geophysical Journal International* 162(1): 79–114.

Yamamoto Y, Tsunakawa H, and Shibuya H (2003) Palaeointensity study of the Hawaiian 1960 lava: Implications for possible causes of erroneously high intensities. *Geophysical Journal International* 153(1): 263–276.

Yamazaki T (1999) Relative paleointensity of the geomagnetic field during Brunhes Chron recorded in North Pacific deep-sea sediment cores: Orbital influence? *Earth and Planetary Science Letters* 169(1–2): 23–35.

Yamazaki T and Ioka N (1994) Long-term secular variation of the geomagnetic field during the last 200 kyr recorded in sediment cores from the western equatorial Pacific. *Earth and Planetary Science Letters* 128: 527–544.

Yamazaki T, Ioka N, and Eguchi N (1995) Relative paleointensity of the geomagnetic field during the Brunhes Chron. *Earth and Planetary Science Letters* 136: 525–540.

Yamazaki T and Oda H (2002) Orbital influence on Earth's magnetic field: 100,000-year periodicity in inclination. *Science* 295(5564): 2435–2438.

Yamazaki T and Oda H (2005) A geomagnetic paleointensity stack between 0.8 and 3.0 Ma from equatorial Pacific sediment cores. *Geochemistry, Geophysics, Geosystems* 6: Q11H20 (doi:10.203/2005GC001001).

Yokoyama Y and Yamazaki T (2000) Geomagnetic paleointensity variation with a 100 kyr quasiperiod. *Earth and Planetary Science Letters* 181(1–2): 7–14.

Yoshihara A and Hamano Y (2000) Intensity of the Earth's magnetic field in Late Archean obtained from diabase dikes of the Slave Province, Canada. *Physics of the Earth and Planetary Interiors* 117: 295–307.

Yoshihara A and Hamano Y (2004) Paleomagnetic constraints on the Archean geomagnetic field intensity obtained from komatiites of the Barberton and Belingwe greenstone belts, South Africa and Zimbabwe. *Precambrian Research* 131(1–2): 111–142.

Yoshida S and Katsura I (1985) Characterization of fine magnetic grains in sediments by the suspension method. *Geophysical Journal of the Royal Astronomical Society* 82: 301–317.

Yoshihara A, Kondo A, Ohno M, and Hamano Y (2003) Secular variation of the geomagnetic field intensity during the past 2000 years in Japan. *Earth and Planetary Science Lettters* 210(1–2): 219–231.

Yu Y and Tauxe L (2004) On the use of magnetic transient hysteresis in paleomagnetism for granulometry. *Geochemistry, Geophysics, Geosystems* 6: Q01H14 (doi: 10,1029/2004GC000839).

Yu Y and Tauxe L (2006) Acquisition of viscous remanent magnetization. *Physics of the Earth and Planetary Interiors* 159: 32–42.

Zhao X, Riisager P, Riisager J, Draeger U, Coe R, and Zheng Z (2004) New paleointensity results from Cretaceous basalt of Inner Mongolia, China. *Physics of the Earth and Planetary Interiors* 141: 131–140.

Zhu R, Laj C, and Mazaud A (1994) The Matuyama–Brunhes and upper Jaramillo transitions recorded in a loess section at Weinan, north-central China. *Earth and Planetary Science Letters* 125: 143–158.

Zhu RX, Lo CH, Shi RP, Pan YX, Shi GH, and Shao J (2004a) Is there a precursor to the Cretaceous normal superchron? New paleointensity and age determination from Liaoning province, northeastern China. *Physics of the Earth and Planetary Interiors* 147(2–3): 117–126.

Zhu RX, Lo CH, Shi RP, Shi GH, Pan YX, and Shao J (2004b) Palaeointensities determined from the middle Cretaceous basalt in Liaoning Province, northeastern China. *Physics of the Earth and Planetary Interiors* 142(1–2): 49–59.

Zhu RX, Pan YX, Shaw J, Li DM, and Li Q (2001) Geomagnetic palaeointensity just prior to the Cretaceous normal superchron. *Physics of the Earth and Planetary Interiors* 128(1–4): 207–222.

Relevant Websites

http://earthref.org/MAGIC/ – Magnetics Information Consortium (MagIC).

http://www.ngdc.noaa.gov/seg – National Geophysical Data Center (NGDC).

14 True Polar Wander: Linking Deep and Shallow Geodynamics to Hydro- and Bio-Spheric Hypotheses

T. D. Raub, Yale University, New Haven, CT, USA

J. L. Kirschvink, California Institute of Technology, Pasadena, CA, USA

D. A. D. Evans, Yale University, New Haven, CT, USA

14.1 Planetary Moment of Inertia and the Spin-Axis

Planets, as quasi-rigid, self-gravitating bodies in free space, must spin about the axis of their principal moment of inertia (Gold, 1955). Net angular momentum (the **L** vector), a conserved quantity in the absence of external torques, is related to the spin vector (ω) by the moment of inertia tensor (**I**) such that $\mathbf{L} = \mathbf{I}\omega$. It is well known that the movement of masses within or on the solid Earth, such as sinking slabs of oceanic lithosphere, rising or erupting plume heads, or growing or waning ice sheets, can alter components of that inertial tensor and induce compensating changes in the ω vector so that **L** is conserved. Observable mass redistributions caused by postglacial isostatic readjustment, by earthquakes, and even by weather systems cause detectible ω changes, manifested as shifts in the geographic location of Earth's daily rotation axis and/or by fluctuations in the spin rate ('length of day' anomalies). Generally, such shift of the geographic rotation pole is called true polar wander (TPW).

Because its mantle and lithosphere are viscoelastic, Earth's daily spin distorts its shape slightly from that of a perfect sphere, producing an equatorial bulge of about 1 part in 300, or an excess equatorial radius of ~ 20 km relative to its polar radius. Variations in Earth's spin vector (ω) will cause this hydrostatic bulge to shift in response, with a characteristic timescale of $\sim 10^5$ years, governed by the relaxation time of the mantle to long-wavelength loads (Ranalli, 1995; Steinberger and Oconnell, 1997).

Goldreich and Toomre (1969) demonstrated that this hydrostatic bulge only exerts a stabilizing influence on the orientation of the planetary spin vector on timescales $< \sim 10^5$ years, with little or no influence on the bulk solid Earth over longer timescales (see also Richards *et al.*, 1997; Steinberger and Oconnell, 1997; Steinberger and O'Connell, 2002). For the purposes of this discussion, only the nonhydrostatic component of Earth's inertial tensor is geologically important, and inertial perturbations of sufficient scale should induce TPW in a sense that moves a new principal inertial axis (I_{max}) back toward the average rotation vector, ω.

14.2 Apparent Polar Wander (APW) = Plate motion + TPW

14.2.1 Different Information in Different Reference Frames

For mid-Mesozoic and younger time, seafloor magnetic lineations permit accurate paleogeographic reconstructions of continents separated by spreading ridges. For older times, paleogeographers must rely upon paleomagnetic data referenced to an assumed geocentric, spin-axial dipole magnetic field. There is ample evidence for dominance of this axial dipole field configuration for most of the past 2 Gy (Evans (2006) and references therein). While many authors have used the unfiltered global paleomagnetic database to estimate the maximum permissible contribution of non-dipole terms in the geomagnetic spherical harmonic expansion (e.g., Evans, 1976; Kent and Smethurst, 1998), the likelihood that continents are not distributed uniformly on the globe over time (e.g., Evans, 2005) introduces a ready alternative explanation for the database-wide trends.

Mesozoic–Cenozoic paleogeographic reconstructions usually rely on the collection of a time series of well-dated paleomagnetic poles (called apparent polar wander paths, or APWPs) from distinct continental blocks. When combined with other geological or paleontological constraints, it is possible to match and rotate similarly shaped portions of those paths into overlapping alignment (Irving, 1956; Runcorn, 1956), thereby providing, to first order, the absolute paleolatitude and relative paleolongitude for constituent plate-tectonic blocks of ancient supercontinents such as Pangea. For times prior to the oldest marine magnetic lineations, matching exceptionally quick APWP segments which might correspond to rapid, sustained TPW provides an alternative paleomagnetic method for constraining paleolongitude, relative to the arbitrary meridian of a presumed-equatorial TPW axis (Kirschvink *et al.* (1997) and see subsequent sections).

As discussed subsequently, accumulating evidence suggests that Neoproterozoic and Early Paleozoic time experienced much larger and more rapid TPW episodes than apparent during the Mesozoic and Cenozoic Eras, suggesting fundamental (and intellectually exciting) temporal changes in basic geophysical parameters that control Earth's moment of inertia tensor.

Following theoretical work on 'polar wandering' in the late nineteenth century, Wegener (1929) (English translation by Biram (1966; pp. 158–163)) recognized that continents might appear to drift not only at the behest of spatially varying internal forces, but also by steady or punctuated whole-scale shifting of the solid-Earth reference frame with respect to the ecliptic plane (i.e., the 'celestial' or 'rotational' reference frame):

> ... the geological driving force acts as before and shifts the principal axis of inertia by the amount *x* in the same direction, and the process repeats itself indefinitely. Instead of a single displacement by the amount *x*, we now have a *progressive* displacement, whose rate is set by the size of the initial displacement *x* on the one hand, and by the viscosity of the earth on the other; it does not come to rest until the geological driving force has lost its effect. For example, if this geological cause arose from the addition of a mass *m* somewhere in the middle latitudes, the axial shift can only cease when this mass increment has reached the equator... p. 158 (italics Wegener's)

Since Earth's geomagnetic field derives directly from its spin influence on convection cells in its liquid outer core, which are sustained by growth of the solid inner core and by plate-tectonics-driven secular mantle cooling (Nimmo, 2002; Stevenson, 1983), TPW causes the Earth's mantle and crust to slip on the solid/liquid interface at core–mantle boundary, while Earth's magnetic field most likely remains geocentric and average spin-axial.

Consequently, TPW was recognized as a possibly confounding signal by early paleomagnetists (Irving, 1957; Runcorn, 1955; Runcorn, 1956), although ultimately TPW was assumed to be negligible relative to rates of plate motion (Irving (1957), DuBois (1957); and most comprehensively Besse and Courtillot (2002)). Short intervals of non-negligible TPW have been hypothesized for various intervals of the Phanerozoic and are permitted by the sometimes-coarse resolution of the global paleomagnetic database (Besse and Courtillot (2002, 2003); and see

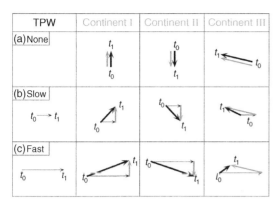

TPW	Continent I	Continent II	Continent III
(a) None			
(b) Slow $t_0 \longrightarrow t_1$			
(c) Fast			

Figure 1 Cartoon showing contribution of TPW to APWP. For zero (a), slow (b), or fast (c) TPW over time increment t_0–t_1 (thin black vectors in top, middle, and bottom rows, respectively), plate-tectonic motion (light gray vectors) may dominate – or be obscured within – the net APW signal. The fidelity by which APW represents plate motion will depend not only on the relative rates of TPW and plate motion, but also on the relative directions in an assumed independent (e.g., geomagnetic) reference frame. Observed coherence of APW between continents with dissimilar plate motion vectors divulges an increased TPW rate. Reprinted from figure 2 in Evans DAD (2003) True polar wander and supercontinents. *Tectonophysics* 362(1–4): 303–320, with permission from Elsevier.

subsequent sections). Irving (1988, 2005) further recounts the historical adoption of paleomagnetism to test hypotheses of continental drift.

In the wake of modern hypotheses suggesting that ancient TPW rates may sometimes match or exceed long-term plate velocities (see subsequent sections), Evans (2003) enunciated the combinatorial possibilities by which TPW may either enhance or mask the plate-tectonic component of APW (**Figure 1**). Most powerfully, TPW must be recognized in the paleomagnetic record as an APW component common to all plates in the celestial or geomagnetic reference frames. Plate-tectonic motion is expected to vary considerably, and even possibly sum to zero (no net rotation) (Gordon, 1987; Jurdy and Van der Voo, 1974, 1975). Lack of kinematic information from ancient areas of oceanic lithosphere that have been lost via subduction precludes rigorous estimation of this no-net-rotation reference frame and thus hinders ancient TPW estimates using that method.

14.2.2 'Type 0' TPW: Mass Redistribution at Clock to Millenial Timescales, of Inconsistent Sense

Although not the principal subject of this review, measurable amounts of small-magnitude TPW have

been observed to follow major earthquakes that redistribute surficial mass nearly instantaneously (e.g., Soldati *et al.*, 2001). The conservation of momentum response of this variety of TPW is often associated with (and more popularly reported as) changes in the length of day, although specific mass redistributions may change Earth's spin rate though not move its inertial axes. On a neotectonic timescale, earthquake-induced TPW may have no net effect on APW, since fault orientation is controlled by subplate scale stress regimes, which vary globally (although the long term effect of earthquake-induced TPW might also be nonzero; see, e.g., Spada (1997)).

Very short timescale TPW is also driven by momentum transfer within and between the circulating ocean and atmosphere systems and Earth's solid surface (e.g., Celaya *et al.*, 1999; Fujita *et al.*, 2002; Seitz and Schmidt, 2005). Such varieties of subannual, itinerant TPW are superimposed upon the larger-magnitude annual and Chandler wobbles (e.g., Stacey, 1992), 12 and *c.* 14 month decametric oscillations of ω (e.g., data of the United States Naval Observatory, International Earth Rotation and Reference Systems Service).

On a somewhat longer timescale, glacial ice loading and postglacial isostatic rebound produces lithospheric mass redistribution that drives TPW at the same rate (centimeters per year) as continental plate motions (Mitrovica *et al.*, 2001a, 2001b, 2005; Nakada, 2002; Nakiboglu and Lambeck, 1980; Sabadini and Peltier, 1981; Sabadini, 2002; Sabadini *et al.*, 2002; Vermeersen and Sabadini, 1999). However, the ~10 ky timescale for dampening of isostatic dynamics, and the cyclic nature of Quaternary ice sheets, probably also relegates this TPW to insignificance when examining APW paths over million-year timescales.

Following Evans (2003) and elaborated subsequently, we term all TPW of fleeting duration relative to plate-tectonic timescales or hydrostatic geoid relaxation as 'type 0'. For more detailed treatment of the considerable literature examining type 0 TPW, we point the reader to Spada *et al.* (2006).

14.2.3 Type I TPW: Slow/Prolonged TPW

For TPW at timescales and magnitudes relevant to plate tectonics, the effects of surficial and internal mass parcels which dominate changes to Earth's net moment of inertia tensor will vary in part with the square of radial distance. Other contributing factors to the relative effects of such anomalies are somewhat more complicated.

For instance, Hager *et al.* (1985) note that while Earth's nonhydrostatic geoid is dominated by the signature of subducting slabs residing in the upper mantle, the residual geoid (nonhydrostatic geoid minus the modeled contribution of those slabs) correlates with presumed lower-mantle thermal anomalies. Positive dynamic topography at viscosity discontinuities (principally the core–mantle boundary, the crust–mantle boundary, and possibly the 660 km discontinuity) could balance lower-mantle density deficiencies such as rising plumes, while negative dynamic topography at the same discontinuities could counteract upper-mantle density excesses due to slabs (Hager *et al.*, 1985).

A buoyant, upper-mantle plume head illustrates the nontrivial compensatory effects of a density anomaly interacting with surrounding mantle of variable viscosity and structure, in different radial positions. While in the lower mantle, this plume head might have created sufficient positive dynamic topography at the core–mantle boundary and 660 km discontinuity to counteract the negative effect of its inherent density deficiency. When rising through the upper mantle, which is ∼10–30 times less viscous than the lower mantle, however, immediate dynamic topographic effects should be diminished. In that case, the density deficiency of the rising mantle plume, plus its greater radial distance, should effectuate a decrease in Earth's inertial moment along the

radial axis of plume ascent, whereas only ∼10 My earlier, the same mantle plume could have produced a positive inertial anomaly along the same axis.

Moreover, entrainment of surrounding material at the front of and beside moving mantle density anomalies may also either exaggerate or mitigate the effects of those parcels (e.g., Sleep, 1988; Zhong and Hager, 2003), depending on anomaly speed and surrounding mantle viscosity, as well as radial position.

Finally, Earth's inertial tensor is also sensitive to the volumes of density anomalies. Considering both size and location of moving mass components, we might expect mantle superswells to exhibit great, consistently positive effects on Earth's net inertial tensor, whereas lesser and variable magnitude effects could be produced by individual mantle plumes or subducting slabs (**Figure 2**).

Evans (2003) relates these putative geodynamic drivers to a TPW-supercontinent cycle. In the first stage of the cycle, individual cratonic components of a future supercontinent amalgamate at any particular latitude on Earth. During amalgamation, plate-tectonic velocities of already assembled fragments will tend to slow. Consequently, mantle beneath the centroid of a supercontinent will tend to become buoyant, as a result of thermal insulation by the supercontinent and its circumferential, subducting slabs, as well as of upward return flow induced by those sinking slabs (this argument adapts Anderson (1982).

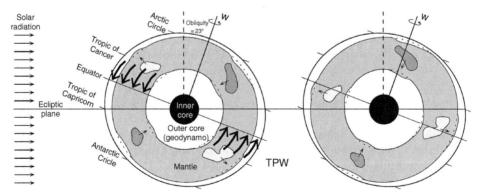

Figure 2 Cartoon of possible mantle phenomena driving TPW. In addition to surface loading, mass redistribution in the mantle also drives changes in Earth's moment of inertia. Entrainment of flowing material along the edges of rising (light) and sinking (dark) mass anomalies will modify their effective contributions. Dynamic topography at viscosity discontinuities leading and trailing the moving anomalies would similarly affect the net inertial change. A dynamic planet (equatorial bulge exaggerated) spins stably and conserves momentum by shifting positive inertial anomalies toward the equator and negative inertial anomalies toward the poles via TPW. Because TPW affects only the solid Earth, whereas Earth's geomagnetic field arises from the vorticity of convection in its liquid outer core, continents will rotate through the geomagnetic and celestial reference frames; while those reference frames remain, on average, fixed with respect to each other. Adapted from figure 1 in Evans DAD (2003) True polar wander and supercontinents. *Tectonophysics* 362(1–4): 303–320.

Elevated dynamic topography from the resulting superswell and the overlying supercontinent that created it will alter Earth's inertial tensor so that slow, continuous TPW places the supercontinent and superswell on Earth's equator (Anderson, 1994; Richards and Engebretson, 1992). Evans (2003) termed this stage of the hypothesized TPW-supercontinent cycle 'type I' TPW.

The Tharsis volcanic region on Mars is plausibly an example of a type I TPW-shifted surface mass (Phillips *et al.*, 2001). In contrast, the south polar water geyser on Saturn's moon Enceladus is an apparent example of the eruption of a hot, buoyant plume associated with migration from mid-latitudes toward a spin-axis (Nimmo and Pappalardo, 2006). Enceladus' mantle is presumably water ice while its core is (mostly nonmolten) silicate and/or iron (Porco *et al.*, 2006). We therefore suggest that the afore-summarized logic of Hager *et al.* (1985) restricts partial silicate melting to Enceladus' uppermost core; or else core diapirism would surely have overwhelmed inertial effects of mantle (ice) upwelling, forcing equator-ward rather than poleward TPW.

14.2.4 Type II TPW: Fast/Multiple/Oscillatory TPW: A Distinct Flavor of Inertial Interchange

Once an equatorial-migrated supercontinent begins to fragment, its superswell might remain essentially fixed in an equator-centered position in the angular momentum reference frame (disregarding orbital oscillations), while the constituent continental fragments disperse away from the superswell.

The centroid of that relict superswell would likely define Earth's minimum moment of inertia axis. If Earth's maximum and intermediate inertial axes were of nearly equal moment – dominated by the first harmonic geoid minimum to the superswell and by the uniformly dispersing fragment plates – then incremental changes in one or the other of those approximately equivalent moments might reverse their relative magnitudes, producing an 'intertial interchange' event (Evans *et al.*, 1998; Evans, 2003; Kirschvink *et al.*, 1997; see also Matsuyama *et al.*, 2006).

For instance, individual mantle plume ascents and/or eruptions, orogenic root delaminations, initiation of new subduction on the trailing edge of one dispersing supercontinent fragment but not another, or even changes in the relative sizes and speeds of those fragments might enhance the

intermediate inertial moment and cause it to transiently equal or slightly exceed that of Earth's previously maximum (average daily spin) axis.

In such a case, Earth's new, equatorial maximum moment of inertia axis would experience poleward drift in the celestial reference frame, until Earth's inertia tensor satisfied simple spin mechanics once more.

If the mass anomaly of the supercontinental superswell were large enough to pin I_{min} stably on the equator, 'normal' plate-tectonic and mantle dynamic events would continue to enhance one of the other two (nonminimum) inertial moments. Repeated, frequent, and sudden switches of the maximum and intermediate inertial moments could produce multiple episodes of TPW adjustment. Thus aforementioned 'type I' TPW might prepare the planet for multiple 'type II' events that could occur much more frequently than would be predicted by considering only randomly moving internal masses (Anderson (1990, p. 252); Fisher (1974)).

In summary, 'type II' TPW might produce back-and-forth rotation or repeated, same-sense tumbling about an equatorial (I_{min}) axis over an interval while $I_{max} \approx I_{int}$. The magnitude of any single event is unpredictable, as it would depend upon the magnitude and duration of the transient changes in the moment of inertia tensor, which in turn are a function of the speed and strength at which the driving anomalies are imposed and compensated.

14.2.5 Hypothesized Rapid or Prolonged TPW: Late Paleozoic–Mesozoic

Marcano *et al.* (1999) note that the Pangean APW path tracks ~35° of equator-ward arc from ~295 to ~205 Ma, indicating a corresponding poleward shift of the lithosphere via plate tectonics and/or TPW. We note that ~25° of arc occurs during the latter half of this interval, at an inferred rate of ~0.45° My^{-1}. Because the leading edge of Pangea in such a plate kinematic reconstruction should have overrun a considerable sector of Panthalassan Ocean lithosphere, those authors suggest the subduction-depauperate geologic record of Pangea's Cordilleran–Arctic Margin favors interpretation of the long APW track as relatively fast, sustained TPW. Irving and Irving (1982) made a similar suggestion of anomalous TPW near the Permian–Triassic boundary interval.

We suggest that the apparently poleward motion of Pangea over this interval might be due to TPW motion caused by upper-mantle ascent of the Siberian Traps plume. For a plume-consistent model incorporating marine accommodation of much early Siberian Trap stratigraphy, see Elkins-Tanton and Hager (2005). Such a plume might have originated in the lower mantle yet avoided triggering equator-ward TPW as discussed previously if I_{max} had significantly exceeded I_{int} during lower-mantle plume ascent, though the two inertial moments converged prior to ~251 Ma.

Perhaps the span of this hypothesized Permian–Triassic Pangea-poleward TPW (≥ 55 My) corresponds to the minimum timescale for buildup of thermal insulation, reflected by later equator-ward APW of the North American Jurassic APW cusps (Beck and Housen, 2003) and possible correlatives on other, less-constrained continents (Besse and Courtillot, 2002), although see Hynes (1990) and subsequent discussion. It is certainly worth testing this scenario with high-resolution paleomagnetic studies in the Permian/Triassic boundary interval, particularly from sedimentary sections that have preserved stable components of that age (e.g., Ward et al., 2005).

For the post-200 Ma global paleomagnetic record, in which a fixed hot-spot reference frame may be hypothesized in order to model true plate-tectonic drift with respect to the mesosphere, Besse and Courtillot (2002, 2003) have presented the most recent thoroughly integrated database and discussion (e.g. Andrews et al., 2006; Besse and Courtillot, 1991; Camps et al., 2002; Cottrell and Tarduno, 2000; Dickman, 1979; Gordon et al., 1984; Gordon, 1987, 1995; Harrison and Lindh, 1982; Prevot et al., 2000; Schult and Gordon, 1984; Tarduno and Smirnov, 2001, 2002; Torsvik et al., 2002; Van Fossen and Kent, 1992). Besse and Courtillot (2002) critically discuss hypothesized Cretaceous fast TPW and provide in-depth consideration of tests and caveats that are only summarized here.

In brief, while many authors have noted that hot spots may form a significantly nonuniform reference frame (e.g., recently Cottrell and Tarduno, 2003; Riisager et al., 2003; Tarduno and Gee, 1995; Tarduno and Cottrell, 1997; Tarduno and Smirnov, 2001; Tarduno et al., 2003, Besse and Courtillot (2002) base their synthesis upon Indo-Atlantic hot spots, which appear more fixed to each other than they are to Pacific hot spots (Muller et al., 1993). Some Pacific hot-spot-motion models, (e.g., DiVenere and Kent, 1999; Petronotis and Gordon, 1999) are similar to Besse and Courtillot's (2002)

Indo-Atlantic hot-spot APWP, although Van Fossen and Kent (1992) argue for incompatibility of the two oceanic reference frames. Ultimately, whether or not hot-spot motion represents coherent or independent drifting in a 'mantle wind', TPW of the whole solid Earth in the geomagnetic/celestial reference frame should still be resolvable if it is rapid or long-lived, by discerning common APW tracks from all continents.

Hynes (1990) invokes a sort of TPW for Early Jurassic time, c. 180–150 Ma, but latter compilations based on better-constrained hot-spot age and position data mark the same period as a standstill (e.g., Besse and Courtillot, 2002). The post-140 Ma interval, which Hynes (1990) marks as a standstill, appears strongest as prolonged, possibly slow TPW in those later models. Since the seafloor hot-spot record before ~150 Ma is poor, and continental hot-spot tracks may be subject to eruptive-tectonic complications, Hynes' (1990) Jurassic-TPW event is of questionable support. Recent studies in lower Jurassic South American strata (Llanos et al., 2006) and Upper Jurassic–Early Cretaceous successions of Adria (Satolli et al., 2007) appear to support possible bursts of fast TPW during this interval.

Both Besse and Courtillot (2002) and Prevot et al. (2000), however, mark the Middle Cretaceous as an interval over which TPW appears to have sustained a net faster, coherent component of motion in the geomagnetic (celestial) reference frame ($\sim <0.5° My^{-1}$) than during TPW-stillstand intervals before and afterward. In both studies, the authors note that even a global synthetic APWP lacks the time resolution (and for some intervals, global pole coverage and/or paleomagnetic pole precision) necessary to discriminate very short, rapid TPW events from somewhat longer ($\sim 5-10$ My) events of slower pace. Thus, while Prevot et al. (2000) favor a short interval of still-quicker TPW at ~115 Ma, the conceptual introduction of oscillatory, type II TPW, further complicates such recognition.

Besse and Courtillot (2002) favor conservative interpretation of the synthetic global APWP dataset and ascribe part or all of Prevot's ~115 Ma event to a small number of data bracketing the interval, and/or to inaccurate dating of supposedly 118–114 Ma, petrologically dissimilar kimberlites in South Africa.

Petronotis and Gordon (1999) and Sager and Koppers (2000a) hypothesize very fast TPW between 80–70 Ma, and near 84 Ma, respectively, possibly as the middle of an oscillatory, triple event. Cottrell and Tarduno (2000) emphasize those authors' enumeration of potential ambiguities and pitfalls of magnetic anomaly modeling and age dating of seamounts (but see Andrews *et al.* (2006) for a promising new treatment), and suggest that hypothesized Campanian–Maastrichtian TPW events fail a global signal test when compared with magnetostratigraphic data from Italy (e.g., Alvarez *et al.*, 1977; Alvarez and Lowrie, 1978). Sager and Koppers (2000b) question whether that Italian magnetostratigraphic data are of sufficient precision to rule out all possible interpretations of the Pacific seamount dataset, within its own uncertainty.

The apparent dispersion of Pacific paleomagnetic data, hypothetically accounted for by TPW, remains unresolved in still more recent studies (Sager, 2006). The younger (<85 Ma) interval of Cretaceous TPW ought to be readily testable by new magnetostratigraphies from various continents to parallel the classic Italian sections. Detailed magnetostratigraphic analysis of Paleocene–Eocene transitional sections exposed on continents led Moreau *et al.* (2007) to hypothesize small-magnitude, fast, back-and-forth TPW during that time, a possibility which may also be supported by paleomagnetic data from the North Atlantic Igneous Province (Riisager *et al.*, 2002).

14.2.6 Hypothesized Rapid or Prolonged TPW: 'Cryogenian'–Ediacaran–Cambrian–Early Paleozoic

Substantial evidence is accumulating that Earth experienced large and rapid bursts of TPW during Neoproterozoic and Early Paleozoic time, of a sort that perhaps has not been experienced since. This 'type II' TPW, invoked as a single event through Early- to Mid-Cambrian time by Kirschvink *et al.* (1997) and debated by Torsvik *et al.* (1998), Evans *et al.* (1998), and Meert (1999), was originally termed inertial interchange true polar wander ('IITPW',). We prefer Evans' (2003) renaming of the process principally because it de-emphasizes the reference frame of the pre-TPW inertial tensor.

Per Kirschvink *et al.* (1997) 'interchange', refers to the beginning of such TPW, when the maximum and intermediate inertial moments of that tensor's reference frame switch identities. One frequent misconception has been that, consequently, the TPW response of the solid Earth must be a before-to-after change of 90°. As noted in, for example, Evans *et al.* (1998) and Matsuyama *et al.* (2006), that change may be any magnitude of rotation at all less than 90°, and continuing until new axes are definable and stable, orthogonal from each other and from the

(effectively stationary) equatorial, I_{min}. The end-member case of a 90° shift would imply that the positive mass anomaly causing the shift was imposed precisely along the spin-axis (as might be the case with an upper-mantle plume head). Off-axis eruption would lead to smaller magnitude events.

We prefer renaming Kirschvink *et al.*'s (1997)'s 'IITPW' as 'type II' TPW also to emphasize its hypothesized proclivity to multiple events and to differentiate it from other patterns, as per Sections 14.2.5 and 14.2.4.

The hypothesized pan-Cambrian type II TPW event of Kirschvink *et al.* (1997) defines an approximate I_{min} axis (the 'TPW axis'). Evans (2003) notes that this TPW axis is essentially identical to both a plausible centroid of the supercontinent of Rodinia, as well as that hypothesized in the reference frame of the largest Early Paleozoic continent, Gondwanaland, by Van der Voo (1994). (Also see Veevers (2004) and later Piper (2006) and **Figure 3** here; and see subsequent sections.)

That coincidence of the Van der Voo (1994) and Kirschvink *et al.* (1997) TPW axes was invoked as an example of the potential long-lived legacy of supercontinental superswell geoid anomalies by Evans (2003), also citing, as a modern analog, Pangea's relict

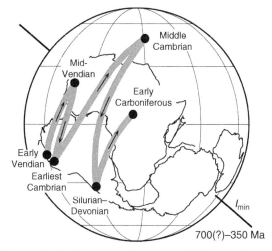

Figure 3 Stability of I_{min}. Van der Voo (1994) notes that Gondwanaland's Cambrian–Carboniferous APWP includes dramatic excursions; and Kirschvink *et al.* (1997) and Evans (1998, 2003) note that Late Neoproterozoic–Cambrian paleopole dispersion marks swings with approximately the same orientation. Evans (2003) suggests that all those swings might represent type II TPW about a paleo-equatorial minimum moment-of-inertia axis defined by a mantle superswell legacy of the Neoproterozoic supercontinent, Rodinia. From figure 3 in Evans DAD (2003) True polar wander and supercontinents. *Tectonophysics* 362(1–4): 303–320.

geoid persisting to the present (Anderson, 1982; Chase and Sprowl, 1983; Davies, 1984; Richards and Engebretson, 1992). Whether relict from a large Ediacaran–Cambrian continent or from the earlier supercontinent Rodinia, Evans and Kirschvink (1999) and Evans (2003) note that, in fact, at least three distinct TPW pulses (of apparently alternating sense) are implied for both Late Neoproterozoic and Early Paleozoic time.

Oscillatory, rapid TPW about an equatorial axis has also been invoked to explain stratigraphically systematic but ≥50° dispersed paleomagnetic poles produced from Svalbard's Akademikerbreen succession deposited on Rodinia's margin *c.* 800 Ma (Maloof *et al.*, 2006). With primary remanence established by a positive syn-sedimentary fold test and by apparently correlatable polarity reversals, this result is highly robust. Although the sequence is undated except by stratigraphic and carbon isotopic correlations to two Australian sub-basins, inferred TPW events appear rapid on a sequence-stratigraphic timescale. Using global carbon isotope correlation and thermal subsidence analysis, the TPW event-recurrence timescale is estimated at <<15 My.

Li *et al.* (2004) invoke type I TPW to explain dramatic paleomagnetic difference between South China's high paleolatitude, 802 ± 10 Ma Xiaofeng dikes and unconformably overlying, gently tilted, low paleolatitude, 748 ± 12 Ma Liantuo glacial deposits (Evans *et al.*, 2000; Piper and Zhang, 1999); and see subsequent sections. Although the age of Li *et al.*'s (2004) hypothesized TPW is imprecisely constrained, and Maloof *et al.*'s (2006) hypothesized events are not dated directly, the two studies appear to invoke different types of TPW for, essentially, a single phase of a supercontinent cycle at *c.* 800 Ma. Large uncertainties in the absolute ages of the Svalbard succession studied by Maloof *et al.* preclude precise correlations to the South China poles; furthermore, uncertainties in Rodinia paleogeography (e.g., Pisarevsky *et al.*, 2003) leave considerable flexibility for reconstructing South China relative to Laurentia (including Svalbard). Identifying the type (I vs II) of TPW at *c.* 800 Ma, and inferring its dynamic cause, must await better constraints in these two topics.

14.2.7 Hypothesized Rapid or Prolonged TPW: Archean to Mesoproterozoic

Evans (2003) suggests that long tracks and loops of Laurentia's Mesoproterozoic APWP might represent oscillatory TPW in the inertial legacy of Earth's Paleoproterozoic supercontinent, Nuna, although global paleomagnetic database synthesis testing of such a hypothesis has yet to be presented robustly.

Considering the oldest volcanic succession with detailed strata-bound paleomagnetic analyses, Strik *et al.* (2003) note that one of several unconformities punctuating the basalt series of Pilbara's late Geon 27 Fortescue Group separates zones of distinct magnetization direction, but with similar petrographic character, marked by a ~27° inclination difference across ~3 My. These authors enumerate rapid TPW as one possible explanation, although they prefer alternative scenarios.

14.3 Geodynamic and Geologic Effects and Inferences

14.3.1 Precision of TPW Magnitude and Rate Estimation

The magnitude and rate of type 0 TPW are directly measured or derived quantities. Precision is often limited, in practice, by technical capabilities of geodetic and satellite instruments. In principle it is ultimately limited by uncertainties in fundamental constants like g, and by observational design, which may not readily quantify special relativistic effects. These uncertainties are probably trivial for the purposes of any geologic application of type 0 TPW data.

The magnitude and rate of type I and II TPW are controlled by the same processes, although we expect type I TPW to be slower than type II. This is because the hypothesized forcing (long-wavelength mantle upwelling in context of initial $I_{min} \approx I_{int}$) is more slowly imposed. Both phenomena should be inherently rate limited by the speed at which Earth's viscoelastic daily-rotational bulge can relax through the incrementally shifting solid Earth. Both phenomena should be resolved in magnitude and rate at the lowest limits of paleomagnetic and geochronologic error and uncertainty.

Quantifying the viscoelastic rate control on type I and II TPW is nontrivial. The three controlling variables are probably forcing magnitude of inertial changes, timescale of inertial change, and effective mantle viscosity (Tsai and Stevenson (2007), expanding on Munk and MacDonald (1960)). Most TPW numerical models assume a single viscosity or two-layered viscosity for the solid Earth; at any rate, there is no consensus on the precise viscosity structure of the modern mantle. We argue that these models might equally well be driven by empirical paleomagnetic data as vice versa.

Simplified mantle modeling confirms that lesser effective mantle viscosity permits faster TPW and that more highly structured mantle viscosity enhances sensitivity to a given forcing (because of the dynamic topography and surrounding entrainment effects). The balance between those influences, however, is unclear. Given a mantle viscosity structure (e.g., Mitrovica and Forte, 2004), simple TPW rate limitation calculations (e.g., Spada *et al.*, 2006; Spada and Boschi, 2006; Steinberger and Oconnell, 1997) may be controlled by average mantle viscosity or perhaps maximum viscosity. Obviously, more experimental and theoretical work is needed on this question.

We project the possible shape of a TPW rate-limitation versus controlling viscosity curve onto Mitrovica and Forte (2004) mantle viscosity structure in such a way that suggests, today, full type I (i.e., pole-to-equator) TPW could occur over an order of magnitude different timescales, within the span of possible controlling mantle viscosity (**Figure 4**).

Besse and Courtillot (2002) synthesize the global paleomagnetic record over the past 200 My, and consider the most conservative solid Earth (e.g., Indo-Atlantic hot spot) reference frame in their analysis. TPW appears to be an episodically measurable contributor to APW. Over 'fast' TPW intervals estimated to last ~20–40(?) My, TPW may occur at $<0.5°\,My^{-1}$. As already discussed, those authors acknowledge that time resolution in the global coverage of the paleomagnetic database limits their analysis to moving 5 or 10 My windows at finest resolution. If faster TPW occurred over a shorter timescale, it would not be readily recognized using a time-averaged approach.

If ancient TPW were estimated – by APWP compilation in older times or by single-location paleomagnetic records of any age – to occur substantially quicker, then it should be possible to use the TPW data to construct inverse models of the controlling mantle viscosity. This approach might allow constraints to be placed on mantle viscosity and related parameters as a function of geological time.

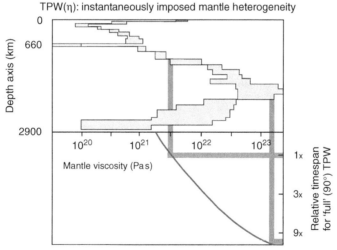

Figure 4 Viscosity structure in Earth's mantle affecting hypothetical TPW dynamics. For a given controlling mantle viscosity, maximum TPW rate may be modeled (red curve). In the cartoon representation shown here, present-day TPW controlled by mantle viscosity of ~10^{23} Pa s would span a given arc of rotation ~10 times slower than TPW controlled by mantle viscosity of ~2 x 10^{21} Pa s. The viscosity structure of Earth's modern mantle is uncertain in sense and magnitude at many levels, particularly in the deep mantle. Mitrovica and Forte (2004), however, suggest a dual inverse-modeled mantle viscosity structure as depicted in yellow (with viscosity uncertainties occupying horizontal space and mantle position and thickness varying vertically). It is not clear to us whether a highly structured mantle, such as is inferred for modern Earth, would be effectively 'controlled' by its maximum-viscosity shell, or whether it will be controlled by some integrated average viscosity, permitting faster net TPW. Projection of ever-more refined TPW rate-response calculations and ever-more sophisticated viscosity structure models is critical to understanding the putative TPW rate limitation on modern Earth. If ancient TPW is demonstrated to occur at significantly faster rates, secular evolution of controlling mantle viscosity (or development of viscosity structure) could be inferred and inversely modeled. Adapted from figure 4(b) in Mitrovica JX and Forte AM (2004) A new inference of mantle viscosity based upon joint inversion of convection and glacial isostatic adjustment data. *Earth and Planetary Science Letters* 225(1–2): 177–189.

14.3.2 Physical Oceanographic Effects: Sea Level and Circulation

George Darwin (1877) was the first to note that changes in Earth's rotation would induce fluctuations in local sea level. In reviewing these early studies that specifically contrasted the delayed response of the solid Earth's viscoelastic bulge compared with the immediate response of the ocean's bulge to changes in the spin vector, we again return to the prophetic words of Alfred Wegener (Biram, 1966; translated from Wegener (1929)):

> ...Since the ocean follows immediately any re-orientation of the equatorial bulge, but the earth does not, then in the quadrant in front of the wandering pole increasing regression or formation of dry land prevails; in the quadrant behind, increasing transgression or inundation. (p. 159)

The physics of this effect have been subsequently modeled with increasing sophistication (Gold, 1955; Mound and Mitrovica, 1998; Mound et al., 1999, 2001, 2003; Sabadini et al., 1990).

Consider only the magnitude of the spin vector, ω. A simple decrease in magnitude of ω (which is an increase in the length of day without any associated TPW) will reduce the size of Earth's equatorial bulge. However, because the solid Earth is viscoeleastic, it requires $\sim 10^4$ to 10^5 years to fully relax, whereas Earth's ocean surface responds instantaneously. Hence, in equatorial regions, mean sea level will experience a transient relative fall

with respect to the land surface, with a compensating rise at higher latitudes. Similarly, simply increasing ω (shortening the day) will produce the opposite effect of equatorial transgressions and high-latitude regressions (Peltier, 1998).

In contrast, altering the orientation of the spin vector relative to the equatorial bulge yields a quadrature pattern in relative sea level. As the ω vector remains constant, dimensions of the overall bulge also remain the same. A point moving toward the equator, for example, will impinge onto the equatorial bulge, and hence tend to increase its distance from the spin-axis.

The solid Earth beneath this point, however, responds with the aforementioned viscoelastic time constant and will lag the ocean's instantaneous shift to the new, itinerant equilibrium shape. That response difference generates a marine transgression. Similarly, points on Earth's surface that move away from the equator during TPW vacate the spin bulge and will relax at the solid Earth's timescale while the adjacent ocean surface 'falls' faster, effecting a regression.

The maximum relative sea-level effect for an instantaneous (and impossible!) shift of exactly 90° would simply be the size of Earth's equatorial bulge, or ~ 10 km. In reality, sea-level excursions are constrained by the effective elastic lithosphere thickness and viscosity structure of the mantle. Mound et al. (1999) estimate that, for reasonable estimates of present viscosity structure, a 90° TPW event acting over 10 and 30 My will generate >200 to ~ 50 m excursions, respectively (**Figure 5**). TPW sea-level effects

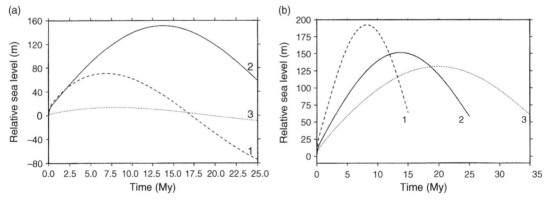

Figure 5 Sea-level fluctuations as a function of TPW rate. (a) Modeled sea-level fluctuations over 25 My of TPW for three locations on the globe: (1), a continent initially at mid-latitude moving through the equatorial bulge to mid-latitudes in the opposite hemisphere, experiences moderate transgression followed by significant regression; (2), an initially polar continent moving toward the equator experiences significant transgression followed by moderate regression; and (3), a continent near the TPW axis experiences little sea-level fluctuation. The period and amplitude of each anomaly in this model are most sensitive to input parameters of lithospheric thickness, upper-mantle viscosity, and TPW rate. (b) The amplitude of any one continent's sea-level anomaly increases nonlinearly with TPW rate, other variables held constant. From figure 2 in Mound JE, et al. (1999) A sea-level test for inertial interchange true polar wander events. *Geophysical Journal International* 136(3): F5–F10.

are certainly of equivalent magnitude, and perhaps larger, than those of standard glacioeustasy and should exert first-order control of global sequence stratigraphy during intervals of time for which TPW was significant, quick, and/or frequent.

Estimates of the sea-level effects of the fastest possible TPW motions tend to flounder on the uncertainties of detailed mantle-viscosity structure, but excursions of approximately several kilometers in scale might be feasible for TPW events on the 1–3 My timescale (Kirschvink *et al.*, 2005; Mound *et al.*, 1999). Of course, sea level will relax to roughly its initial state when a TPW event ends.

Physical oceanography should also show second-order responses to TPW. Earth's spin vector modulates a variety of oceanographic parameters, including temperature, the pole-to-equator atmospheric energy gradient, and the chemical dynamics of the world ocean (e.g., mean and local salinity). It also has a large influence on the tides, which depend

on the orientation and character of seafloor topography and continental geography. The well-known poleward boundary currents on east-facing continental margins and deepwater upwellings on west-facing margins are fundamental outcomes of Earth's sense of spin. Large TPW events will force changes in many of these parameters that could leave fingerprints in the geological record (**Figure 6**).

During TPW events, we expect thermohaline circulation in the ocean to reorganize – possibly several times – at considerably shorter timescales than the TPW events and possibly at shorter timescales than the resulting sea-level fluctuations. As originally west-facing margins near the equator rotate clockwise, upwelling will cease on those margins and may begin anew on the west-rotating, originally southern margins of the continent in consideration. Boundary currents driving basinal sediment advection may reorganize, foundering or winnowing sand drifts, as low-latitude east-facing continental margins migrate

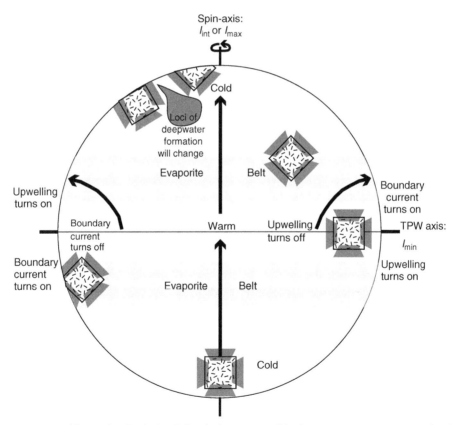

Figure 6 A wide range of first-order physical and chemical oceanographic changes may accompany sea-level fluctuations during TPW, depending on the initial location of a continental margin with respect to the TPW spin-axis. Sediment supply to margins initially girdled by eastern boundary currents may founder; nutrient-rich upwellings on initially west-facing margins may cease and begin anew on initially zonal, later west-facing margins. In general, TPW is expected to leave a legacy of eddy instability and thermohaline reorganization which proceeds episodically during and possibly after an inertial shift.

to polar regions, cross the equator, or rotate into zonality. **Figure 6** illustrates some of the wide range of effects conceivable during a TPW event for distinct continents occupying various original geographic locations. Certainly, hypothesized ancient TPW ought to leave dramatic sedimentological and geochemical signatures in the eu- and miogeoclinal records of most then-extant passive margins (see also Section 14.3.3).

At a dramatic extreme, Li *et al.* (2004) follow Schrag *et al.*'s (2002) geochemical modeling and suggest that *c.* 800–750 Ma TPW may have initiated the 'Snowball Earth' glaciations of the Cryogenian Period. Kirschvink and Raub (2003) invoke changes in eddy circulation and lateral sediment advection, as well as permafrost transgression during multiple Ediacaran–Cambrian TPW to destabilize methane clathrates, increase organic carbon remineralizatoin, and effect 'planetary thermal cycling' on the biosphere. In turn, this stimulated the Cambrian Explosion by alternately stressing species and forcing them to reduce generation time, accumulating genetic mutations which would promote morphological disparity and speciation.

14.3.3 Chemical Oceanographic Effects: Carbon Oxidation and Burial

As noted in Section 14.3.2, TPW is capable of producing a variety of effects on global sea level, ranging from small, meter-scale regional fluctuations associated with isostatic effects from glacial loading and unloading, to possibly larger, even kilometer-scale effects from rapid and 'ultra'-rapid TPW. In turn, sea-level fluctuations can force shoreline migration, alter drainage patterns, and cause geological organic carbon to be remineralized. Potentially, this remineralized carbon could come from bitumens, kerogen, or pressure-destabilized methane clathrates.

As discussed previously, Maloof *et al.* (2006) provide a compelling case for a pair of rapid TPW events in Middle Neoproterozoic platform carbonates of East Svalbard, Norway. Several profiles through the ~650 m thick section record a clear step-function offset in the $\delta^{13}C$ signature of approximately −5‰ identified as the ~800 Ma Bitter Springs Event (Halverson *et al.*, 2005), which lasted ~10 My based on thermal subsidence modeling and younger chronostratigraphic correlations (**Figure 7**).

In Svalbard this isotope shift is bracketed by two sequence boundaries (sea-level erosional horizons). Although magnetization directions before and after

the Bitter Springs Event in the Akademikerbreen Group are similar to each other, precisely at these sequence boundaries the mean magnetic declination rotates >50°. Maloof *et al.* (2006) note that this rotation is present in three separate stratigraphic sections separated by >100 km on a single craton, with congruent carbon isotope signals. A pair of rapid TPW events separated by ~10 My provide a single, unified explanation for two important observations.

First, TPW can account for the Bitter Springs isotope anomaly by producing a sudden shift in the fraction of global carbon burial expressed as organic carbon deposition (f_{org}), plausibly by changing the proportion of riverine sediment (and nutrients) deposited at low-latitude versus at mid-latitudes. Per gram of sediment, low-latitude rivers are more effective at organic carbon burial than those at mid- or high-latitudes (Halverson *et al.*, 2002; Maloof *et al.*, 2006). Although paleomagnetic data argue Svalbard was close to the I_{min} inertial axis, a 50° type II TPW event could swing pan-hemispheric Rodinia, shifting orographic precipitation loci, drainage patterns, and continental sediment delta locations, varying f_{org}.

Second, this pair of type II TPW events would produce transient sea-level variations. Only two sequence boundaries exist in this interval in Svalbard, and they bracket the isotope anomaly. As discussed previously, Li *et al.* (2004) have also argued for rapid TPW in this general interval of time.

On a more speculative note, Kirschvink and Raub (2003) suggested that a hypothesized interval of multiple, type II TPW may have been in part responsible for the production of large and episodic oscillations in inorganic $\delta^{13}C$ during Early Cambrian time, the so-called Cambrian Carbon Cycles (Brasier *et al.*, 1994; Kirschvink *et al.*, 1991; Magaritz *et al.*, 1986, 1991; Maloof *et al.*, 2005). The principal mechanism suggested – remineralization of organic carbon from exposed sediments and/or destabilization of methane clathrate reservoirs along continental margins and in permafrost – was patterned after similar suggestions for methane release associated with a marked inorganic carbon anomaly at the initial Eocene thermal maximum (IETM, formerly called the Paleocene–Eocene thermal maximum, PETM) event (Dickens *et al.*, 1995, 1997).

Subsequent analyses have questioned this interpretation for the IETM event, based on revised estimates of the available methane reservoir stored in Late Mesozoic and Early Tertiary continental shelf and upper-slope sediments, methane clathrate residence times, and the short (decadal) residence time of methane in Earth's present atmosphere

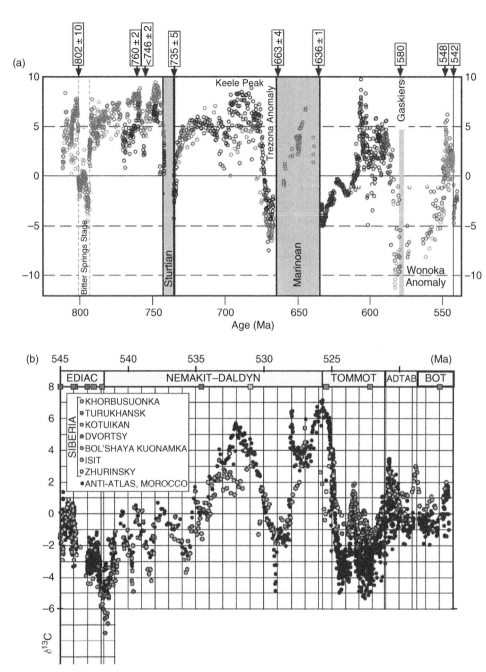

Figure 7 TPW effects on the geological carbon cycle. (a) One possible global inorganic carbon isotopic evolution time series through Mid- and Late-Neoproterozoic time. At least three distinct glacial events punctuate this interval. While successions around the world are ever-better dated by high-precision U–Pb ages on magmatic zircons in interbedded ashes and volcanic flows, ambiguities still exist in sequence-stratigraphic and lithostratigraphic correlations between continents. (As discussed in the text and in captions to **Figures 6** and **7**, global sequence stratigraphic correlations may be severely complicated by TPW.) Maloof *et al.* (2006) consider the long-lived negative carbon isotopic excursion dubbed 'Bitter Springs Stage' a consequence of type II TPW, see text. (b) Cambrian inorganic carbon isotopic evolution time series from well-dated successions in Morocco and undated successions in Siberia. Kirschvink and Raub (2003) note that organic carbon-based isotope curves over parts of the same time interval from Western United States and other areas share many similarly shaped and positioned excursions. Those authors hypothesize that some or all of the negative carbon isotopic excursions in Cambrian time were partly driven by destabilization of seafloor and permafrost methane clathrates during a legacy of eddy instability and thermohaline reorganization associated with Ediacaran–Cambrian type II TPW. Adapted from figure 15(b) in Halverson *et al.* (2005); and part of figure 8 in Maloof AC, *et al.* (2005).

relative to the apparent ∼100 000 + interval of initial Eocene carbon isotopic excursion (Buffett and Archer, 2004; Farley and Eltgroth, 2003). In reviewing the controversy, Higgins and Schrag (2006) argue for the sudden close-off of an epicontinental seaway through tectonic action to provide ∼5000 Gt of organic carbon needed to produce the effect.

We note here that Kirschvink and Raub's (2003) TPW-based 'methane fuse hypothesis' for the Cambrian anticipated many of the arguments against methane summarized by Higgins and Schrag (2005). Several factors argue that the geological context of the multiple Cambrian carbon cycles are markedly different than the singular end-Paleocene event, in our view making at least the partial involvement of a methane-derived light isotopic reservoir more plausible.

First, Neoproterozoic time (including the Ediacaran Period) is characterized by extended intervals during which the inorganic $\delta^{13}C$ remained markedly positive ($+2 - +5‰$), most likely implying relatively high organic carbon burial fractions, plausibly leading to order-of-magnitude larger methane reservoirs than extant. The long-term average inorganic carbon isotopic value for most of the Phanerozoic Eon is 0‰.

Second, Cambrian cycles are individually no longer, though collectively more prolonged, than Phanerozoic negative carbon isotopic excursions, with at least 12 named cycles up to Middle Cambrian time (Brasier *et al.*, 1994; Montanez *et al.*, 2000) spanning an interval of ∼15–20 Ma. When viewed at exceptionally high resolution (e.g., Maloof *et al.*, 2005), many individual Cambrian carbon cycles likely have duration 100 ky–1 My, but with occasional sharp excursions superimposed, reminiscent of the Early Eocene event superimposed on its still-longer Early Eocene climate optimum oscillation. We suspect that the carbon isotopic oscillations discovered recently in the Early Triassic (Payne *et al.*, 2004) are individually of longer average duration, but occur over a shorter overall chronologic interval, than those of the Early Cambrian. The reservoir arguments against methane contributing to the initial Eocene carbon excursion, then, do not apply equally well when applied to Ediacaran–Cambrian time; the geological and physiographic conditions are distinctly different.

TPW-induced changes to the geological carbon cycle are likely to be myriad, with potential positive feedbacks. **Figure 8** elaborates the pervasive, systematic perturbations to the geological carbon cycle expected during TPW-driven sea-level fluctuations

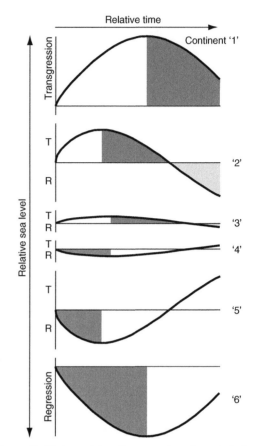

Figure 8 Cartoon elaboration on systematic perturbations to the geological carbon cycle caused by TPW-driven sea-level fluctuations. For schematic sea-level anomalies experienced by continents in six different locations relative to a TPW axis, a different interval of each anomaly will be most susceptible to regression-induced methane clathrate destabilization and oxidation of long-buried organic matter (green); to new burial of organic carbon on TPW-transgressed shelves (white); and subsequent oxidation of part of the same organic-carbon pool during later emergence (blue); and to renewed oxidation of ancient organic matter (yellow). Destabilization of methane clathrate in permafrost, not explicated in this figure, is probably preferred at the very beginning of such a cycle for appropriate continents. Depending on the initial sizes and mean isotopic values of each reservoir, the rate and duration of TPW, and specific parameters including paleogeography and thermohaline circulation, it is possible to imagine (or quantitatively model) a wide range of global inorganic-carbon reservoir isotopic responses. Continent #1 moves from either of Earth's poles to its equator. Continent #2 swings from mid-latitudes on one side of the equator to mid-latitudes in the opposite hemisphere. Continent #3, located near the TPW axis, rotates slightly further away from the equator. Continent #4, located near the TPW axis, rotates slightly closer toward the equator. Continent #5 swings from mid-latitudes across the pole to mid-latitudes in the same hemisphere, but on the opposite side of the globe. Continent #6 moves from near Earth's equator to one of its poles.

for continents in six distinct locations relative to a TPW axis. Some continents will be especially prone to regression-induced methane clathrate destabilization and oxidation of long-buried organic matter; while others will dominantly accommodate new burial of organic carbon on newly TPW-transgressed shelves. If the continental location is appropriate and TPW sufficiently rapid, then TPW transgression may flood continental highlands of exceptionally large area. That same organic carbon pool may oxidize during later emergence related to sea level reequilibration; other continents may only remineralize ancient organic matter for the first time in the TPW event near its end.

Depending on the initial sizes and mean isotopic values of each reservoir, the rate and duration of TPW, and specific parameters including paleogeography and thermohaline circulation, it is possible to imagine (or quantitatively model) a wide range of global inorganic-carbon reservoir isotopic responses. If a global Ediacaran–Cambrian paleogeography were confidently reconstructed, box modeling of reservoirs over relative timescales as indicated in **Figure 8** ought to match the general shape and timescale of Ediacaran–Cambrian carbon isotopic excursions, if the carbon cycles are, in fact, genetically linked to an oceanographic legacy of type II TPW.

14.4 Critical Testing of Cryogenian–Cambrian TPW

14.4.1 Ediacaran–Cambrian TPW: 'Spinner Diagrams' in the TPW Reference Frame

While Kirschvink *et al.* (1997) suggested that a single TPW event in the interval ~530–508 Ma accounted for the anomalous dispersion between Late Neoproterozoic and Late Cambrian poles for most continents, Evans and Kirschvink (1999); adopted by Kirschvink and Raub (2003) noted that, in fact, straightforward interpretation of the timeseries order of dispersed poles for relatively well-constrained continents like Australia demands multiple TPW events.

Although none of the Cryogenian–Cambrian paleomagnetic poles for Australia is 'absolutely' dated, stratigraphic order between those poles is clear; many come from a single sedimentary succession (South Australia's Adelaide 'geosyncline').

We prefer a terminal Cryogenian–Ediacaran–Cambrian APWP for Australia composed of the

poles in **Table 1**. While these poles are of varying quality, none can be demonstrated persuasively to have been remagnetized, or to have dramatically wrong age assignments without basing such an argument on the inexplicability of the magnetization direction itself. Many other paleomagnetic poles for Australia in this time interval have been excluded from our analysis, using data selection and supercedence criteria we believe uncontroversial, but beyond the scope of this chapter.

The oldest group of poles, uppermost 'Cryogenian' in age, are probably ~635 Ma or slightly older, based on 'Snowball Earth' lithological and chemostratigraphic correlation to well-dated successions in Namibia and South China. Four published poles from the Elatina Formation and one from the immediately underlying Yaltipena Formation, presumed genetically related to the onset of Elatina glaciation or deglaciation, cluster well, placing Australia very near the equator. (In today's geomagnetic reference frame, paleopole longitudes for this cluster are near 330, 360 °E and pole latitudes are in 40, 55 °S.) A pole position from immediately overlying, earliest Ediacaran Brachina Formation siliciclastics, is similar though slightly far-sided.

The only Mid-Ediacaran paleomagnetic pole we consider is that of the undated Bunyeroo Formation, ~200–400 m stratigraphically higher, and separated from uppermost Brachina Formation by at least one supersequence boundary in those study areas of South Australia. Although the Bunyeroo Formation is lithologically similar to the Brachina Formation, shares a similar burial and gentle-folding tectonic history, and passes a reliability-by-comparison test with the Acraman Impact meltrock aeromagnetic anomaly on the nearby Gawler Craton (the Bunyeroo Formation hosts ejecta of that impact event), its pole position is significantly different, lying at 16.3° E, −18.1° S (Schmidt and Williams, 1996), and implying mostly counterclockwise rotation of Australia, in the celestial reference frame, during Early Ediacaran time.

Kirschvink's (1978) middle-upper Ediacaran pole from central Australia's Upper Pertatataka Formation and Lower Arumbera Formation returns to the vicinity of Elatina poles, suggesting mid-Ediacaran clockwise rotation for Australia in the celestial reference frame. Terminal Ediacaran magnetization from upper Arumbera Formation is broadly similar.

Lower Cambrian poles from South Australia once more return to the vicinity of 340, 015 °E and ~−45 °S, suggesting terminal Ediacaran or earliest

Table 1   Paleomagnetic poles defining 'Cryogenian'–Ediacaran–Cambrian APW oscillations for Australia

Pole #	Pole ID	Site latitude	Site longitude	Pole latitude	Pole longitude	dp	dm	Age order (1 = oldest)	Reference
14	Hudson	−17	129	18	19	13	13	9	Luck (1972)
13	Hugh River	−23.8	133	11.2	37.2	4.9	9.4	8	Embleton (1972)
12	Billy Creek[a]	−31.1	138.7	−37.4	20.1	7.2	14.4	7	Klootwijk (1980)
11	Kangaroo Isl. A	−35.6	137.6	−33.8	15.1	6.2	12.3	7	Klootwijk (1980)
10	Todd River/ Allua/Eninta	−23.4	133.4	−43.2	339.9	4.5	7.7	7	Kirschvink (1978)
9	Upper Arumbera	−23.4	133.4	−46.6	337.4	2.6	4.6	6	Kirschvink (1978)
8	Lower Arumbera/ upper Pertatataka	−23.4	133.4	−44.3	341.9	7.7	13.6	5	Kirschvink (1978)
7	Bunyeroo	−31.6	138.6	−18.1	16.3	6.5	11.8	4	Schmidt and Williams (1996)
6	Brachina	−30.5	139	−33	328	12	20	3	McWilliams and McElhinny (1980)
5	Elatina SKCB	−31.3	138.7	−39.6	1.7	3.3	6.4	2	Sohl et al. (1999)
4	Elatina SW	−31.2	138.7	−51.5	346.6	7.7	15.3	2	Schmidt and Williams (1995)
3	Elatina SWE	−32.4	138	−54.3	326.9	0.9	1.8	2	Schmidt, Williams, and Embleton (1991)
2	Elatina EW	−32.4	138	−51.1	337.1	1.7	3.4	2	Embleton and Williams (1986)
1	Yaltipena	−31.3	138.7	−44.2	352.8	5.8	11.3	1	Sohl et al. (1999)
	best-fit great circle			114.3	−32.2				

[a]Billy Creek result is excluded from swath-pole statistics by preference for possibly coeval 'Kangaroo Island A' result; but listed here to weigh against oroclinal rotation between distant Flinders Ranges sites as an alternative explanation for pole dispersion.

Cambrian counterclockwise rotation of Australia, and Cambro-Ordovician paleopoles from South, central, and northern Australia all are still further shifted northward in the modern reference frame, permitting one long Ediacaran–Cambrian TPW event (per Kirschvink *et al.*, 1997) or multiple events of unresolved duration and magnitude summing to the same net rotation.

The resulting Australian APWP (**Figure 9(a)**) does not resemble a time-progressive 'path' so much as a pole 'swath' (although in fact, time progression of that path would indicate repeated reversal of rotational motion for Australia in the celestial reference frame).

We suggest that a more insightful way to view such an APW swath is in the hypothesized TPW reference frame itself. We calculate a best-fit great circle to Australia's APW swath using modified least-squares techniques; in so doing, we assume that (multiple) TPW over the time interval spanned by the constituent poles occurred at a rate far greater than Australia's tectonic-plate motion in the deep-mantle reference frame. This best-fitting great circle, and its corresponding pole with 95% confidence, are shown in **Figure 9(b)**.

By rotating Australia and its pole swath and best-fitting great circle so that the pole to that circle is at the center of an equal-area projection, we place the hypothesized TPW axis at the 'pole' of the stereonet. If TPW were sufficiently fast, and since polarity choice of each paleopole is arbitrary, Australia may

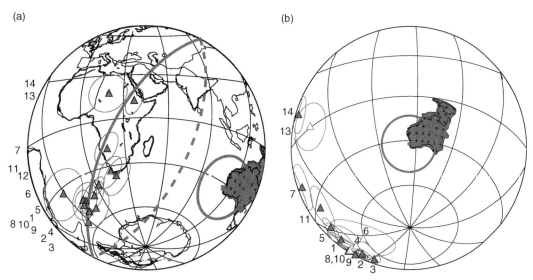

Figure 9   Example of a 'spinner diagram'. (a) Late Cryogenian–Ediacaran–Cambrian APW swath for Australia in the modern geographic reference frame. Fourteen critical poles, dated by biostratigraphy or relative stratigraphic position, appear widely dispersed, and the pole (with 95% confidence trace) to a great circle best-fitting thirteen of those poles lies near the western margin of the Australian Craton. (b) Rotation of Australia, its APW swath, and the swath's best-fitting great circle so that the pole to that swath lies at the center of an equal-area projection creating a 'spinner diagram'. If it is supposed that the back-and-forth time series comprising a continent's APW swath represents multiple (type II) TPW events at rates substantially faster than (or parallel to) plate motion; and if it is assumed, by reasoning outlined in the text and in Kirschvink *et al.* (1997) and Evans (1998, 2003) that the TPW axis lies on Earth's spin-equator, then because of geomagnetic polarity ambiguity in Neoproterozoic time, Australia's actual geographic location at most times during the interval spanning all proposed TPW events may be any rotationally equivalent location that preserves coincidence of its APW swath pole and the TPW axis. For the purposes of paleogeographic reconstruction, Australia's geomagnetic (= rotational) paleolatitude is constrained only for the discrete times of direct paleomagnetic constraints as represented by the individual poles. If rapid type II TPW oscillations were occurring on timescales of a few million years or less, then traditional comparisons of separate plates' paleolatitudes would be meaningless except in the rare instance of extremely precise agreement in pole ages from those plates. Each point on the Earth does, however, maintain a constant angular distance from a type II TPW axis, thus continents can be reconstructed relative to that reference frame. In practical terms from the figured example, Australia may be 'spun' around the (constantly equatorial) TPW axis shown in the center of the diagram. An alternative set of paleogeographic positions occupies the far hemisphere about the antipole to the TPW axis.

effectively occupy any rotationally identical position about that TPW axis, except for precise ages where its paleolatitude is constrained by a corresponding pole. (At those ages, Australia may occupy one of precisely two absolute paleogeographic positions, in the TPW reference frame, recognizing geomagnetic polarity ambiguity).

At all other times, there is rotational paleolatitude ambiguity associated with rapid oscillations about the TPW reference axis. We call this projection a 'spinner' diagram, because the position of a continental block may be freely 'spun' about the TPW axis at the center of the spinner projection. Since all continents must undergo identical TPW, all continental blocks' computed spinner diagrams may be overlapped; and each continent 'spun' to avoid overlaps, to produce a permissible intercontinental paleogeography in the TPW reference frame.

14.4.2   Proof of Concept: Independent Reconstruction of Gondwanaland Using Spinner Diagrams

Although Australia's paleomagnetic APW swath is by far the best constrained of the Gondwanaland-constituent continents, all those Gondwanaland elements show dispersed paleopoles easily fit by great circles. We show those APW swaths for India and Antarctica (Mawsonland), for 'Arabia(-Nubia)', for all of 'Africa' (assuming Congo–Kalahari coherence and ignoring pole-less Sao Francisco), and for all of 'South America' (assuming Amazonia–Rio de la Plata coherence). These swaths are depicted in **Figure 10** using poles in **Table 2**, in both the present geomagnetic reference frame and as spinner diagrams in the inferred TPW reference frame.

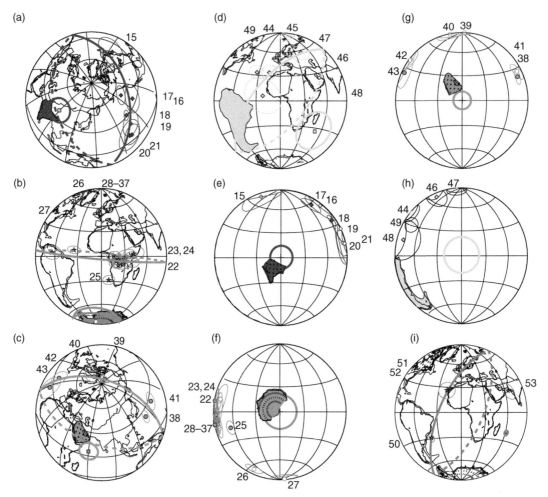

Figure 10 Elaboration of proto-Gondwanaland's APW commonality. Late Cryogenian–Ediacaran–Cambrian APW distributions, in the modern geographic reference frame (but various views) for the Indian and Antarctic Cratons; for 'Arabia', a simplification of the Arabian-Nubian shield and Neoproterozoic basement in modern Iran; for 'South America', a simplification of the Neoproterozoic Rio de la Plata and Amazonia Cratons in late Brasiliano time; and for 'Africa', a simplification of the Neoproterozoic Kalahari and Congo–Sao Francisco Cratons (see text for tectonic reconstruction caveats). All cratons except 'Africa' are also represented in spinner diagrams using poles to best-fitting great circles through critical-pole APW distributions. Adapted from figure 3 in Evans DA, *et al.* (1998) Polar wander and the Cambrian. *Science* 279: 9a–9e.

We are favorably impressed that, by superimposing each continent's APW swath pole at the center of a common spinner projection, continents may be 'spun' so that a nearly Gondwanaland configuration is produced, with relatively minor shifting of each continent's APW swath pole within its 95% confidence limit (**Figure 11**).

Although we recognize that Gondwanaland was not yet entirely formed during Ediacaran–Cambrian time (e.g., Boger *et al.*, 2001; Collins and Pisarevsky, 2005; Valeriano *et al.*, 2004), the post-Ediacaran relative movements between its constituents should have been minor, and even Late-Precambrian stitching of constituent African

and South American Cratons was permissibly non-mobilistic on a tens of degrees spatial scale (Veevers, 2004). We also recognize that our Arabia-Nubia 'craton' includes amalgamated terranes and basement of varying age (Veevers, 2004); therefore, our tectonic definition of 'Gondwanaland', and its implied inheritance from latest Neoproterozoic Rodinia fragmentation, is first-order only.

Nonetheless, this reconstruction restores Gondwanaland to a configuration resembling that calculated by back-rotating post-Pangean seafloor spreading ridges by an entirely independent method, predicated only on the assumption that Late Neoproterozoic–Early Cambrian TPW was

Table 2 Paleomagnetic poles defining 'Cryogenian'–Ediacaran–Cambrian APW oscillations for other Gondwanaland constituents

Pole #	Pole ID	Site Latitude	Site longitude	Pole latitude	Pole longitude	dp	dm	Age order (1 = oldest)	Reference
	India								
21	Salt Range pseudomorph rot. 30°	32.7	73	89.3	38.3	13.8	13.8	6	Wenski, 1972
20	Purple sandstone rot. 30°	32.7	73	7.2	192	9	11.8	5	McElhinny, 1970
19	Khewra A rot. 30°	32.3	71	8.2	189.4	12.5	18.5	4	Klootwijk et al., 1986
18	Bhander sandstone	23.7	79.6	19.3	197	10	14	3	Athavale et al., 1972
17	Upper Vindhyan Rewa-Bhander	27	77.5	31.5	199	3.1	5.9	2.5	McElhinny, 1970
16	Rewa	23.8	78.9	51	217.8	5.6	11.2	2	Athavale et al., 1972
15	Mundwarra Complex	25	73	35	222	9.6	15.9	1	Athavale et al., 1963
	Antarctica (Mawsonland)								
37	Teall Nunatak	−74.9	162.8	234.8	−83.5	3.6	7.2	8	Lanza and Tonarini, 1998
36	Wright Valley granitics	−77.5	161.6	−11	31	4.1	8.2	8	Funaki, 1984
35	Lake Vanda feldspar porphyry dikes	−77.5	161.7	−5.5	18.4	4.3	8.6	7	Grunow, 1995
34	Dry Valley mafic swarm	−77.6	163.4	−4	27	5.5	10.9	7	Manzoni and Nanni, 1977
33	Killer Ridge Mt. Loke diorites	−77.4	162.4	−9.4	26.7	6	12	6	Grunow and Encarnacion, 2000
32	Lake Vanda Bonny pluton	−77.5	161.7	−7.1	21.2	6.6	13.1	6	Grunow, 1995
31	Granite Harbour pink granite	−76.8	162.8	−8.3	27.8	4.4	8.5	6	Grunow, 1995
30	Granite Harbour mafic dikes	−77	162.5	−3.4	16.6	4.7	9	6	Grunow, 1995
29	Granite Harbour grey granite	−77	162.6	0.9	15	3.3	6.5	6	Grunow, 1995
28	Mirnyy Station charnokites	−66.5	93	−3.6	21.7	8	16	5	McQueen et al., 1972
27	Upper Heritage Group	−79.2	274	−1.4	28.6	7.1	12.5	5	Watts and Bramall, 1981
26	Liberty Hills Formation	−80.1	276.9	4	296	4.3	8.6	4	Randall et al., 2000
25	Sor Rondane intrusions	−72	24	7.3	325.4	9.9	5.7	3	Zijderveld, 1968
24	Briggs Hill Bonny pluton	−77.8	163	−28.4	9.5	7.8	15	2	Grunow, 1995
23	Wyatt Ackerman Mt. Paine tonalite	−86.5	170	2.7	41.4	4	8	1	Grunow and Encarnacion, 2000
22	Zanuck granite	−86.5	38.8	1.1	39.3	5.5	10.9	1	Grunow and Encarnacion, 2000
	Arabia								
43	Hornblende diorite porphyry	21.7	43.7	25.8	332.2	9.1	14.7	Unknown	Kellogg and Beckmann, 1982
42	Red sandstone, Jordan	29.7	35.3	36.7	323.4	6.7	10	5	Burek, 1968
41	Jorden dikes	29.5	35.1	26	161	5.1	9.4	4	Sallomy and Krs, 1980
40	Arfan/Jujuq	21.3	43.7	77.4	297.9	6.9	12.2	3	Kellogg and Beckmann, 1982
39	Mt. Timna	29.8	34.9	83.6	223.2	3.3	5.4	2	Marco et al., 1993

(*Continued*)

Table 2 (Continued)

Pole #	Pole ID	Site Latitude	Site longitude	Pole latitude	Pole longitude	dp	dm	Age order (1 = oldest)	Reference
38	Mirbat sandstone	17.1	54.8	23.3	141.8	3.9	7.5	1	Kempf et al., 2000
	'South America' (Rio de la Plata, Amazonia, late Brasiliano)			48.5	−34.3				
49	Mirassol d'Oeste remagnetization	−15	302	33.6	326.9	7.1	10	6	Trindade et al. (2003)
48	Upper Sierra de las Animas	−34.7	304.7	5.9	338.1	19.6	26.7	5	Sanchez-Bettucci and Rapalini (2002)
47	Campo Alegre	−26.5	310.6	57	43	9	9	4	D'Agrella-Filho and Pacca (1988)
46	Playa Hermosa	−34.7	304.7	43	18.4	8.6	16	3	Sanchez-Bettucci and Rapalini (2002)
45	Mirassol d'Oeste characteristic[a]	15	302	82.6	112.6	5.6	9.3	2	Trindade et al. (2003)
44	Urucum	−19	302	34	344	9	9	1	Creer (1965)
	'Africa' (Congo and Kalahari)			61.5	−21.4				
53	Doomport[b]	−23.7	17.2	21.8	64.9	6.8	11	4	Piper (1975)
52	Dedza Mtn. syenite	−14.4	34.3	26	341	10	10	3	McElhinny et al. (1968)
51	Ntonya ring structure	−15.5	35.3	27.8	344.9	1.4	2.3	2	Briden et al. (1993)
50	Sinyai	0.5	37.1	−29	319	3	5	1	Meert and van der Voo (1996)

[a]The Mirassol ChRM pole falls considerably off of the great circle defined by the younger five South America poles. Following Valeriano et al. (2004) we suppose that Mirassol d'Oeste characteristic as an old pole located near a border of the mobile belt, might be rotated. The 'South American' great circle including Mirassol d'Oeste characteristic has a pole at (19.7, 242.9).

[b]Doomport, a likely Cambrian remagnetization, is internal to the Damaride fold belt. It is far from the (admittedly underconstrained) 'African' great circle that it must either indicate inappropriate lumping of date from the constituent cratons, or else it has been rotated by Damaride structures. We acknowledge the relatively unsatisfying character of the 'African' and 'South American' databases, but leave full consideration for a future paper.

Neither 'South America' nor 'Africa' existed in any semblance of the modern, during 'Cryogenian' – Ediacaran – Cambrian time. However, the 'constituent' Precambrian cratons of each – Rio de la Plata and Amazonia for South America; and Kalahari and Congo–Sao Francisco for Africa, might have remained nearly fixed with respect to each other during pan-African and Brasiliano mobilism. Likewise, although the 'Arabian-Nubian' shield incorporates significant juvenile arc material, we suppose it might have occupied a common general area of latest Neoproterozoic real estate, including basement terranes. In any case, it is intriguing that, as a 'block,' Arabia may be treated, essentially, as one of the other cratons, by owning a dispersed APW swat which, when fit to a great circle, reconstructs nearly to Gondwanaland configuration independent of seafloor retro-rotations.

We consider it almost certain that some of these paleomagnetic poles, unconstrained by direct field tests of magnetization fidelity, are in fact remagnetizations or tectonically rotated. However we have attempted to compile a list of possibly primary magnetizations which could not be discounted except by concern for the actual direction obtained.

(a)   (b)

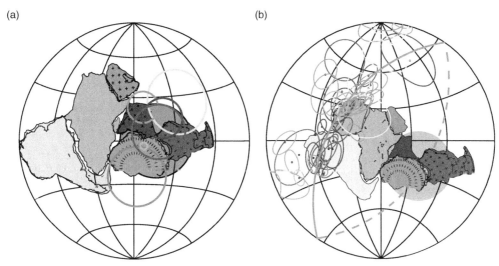

Figure 11   Multiple type II TPW in Late Cryogenian–Ediacaran–Cambrian time, proof of concept. (a) Using only spinner diagrams, spinning each continent about its own APW swath-fitting great-circle pole, and shifting those poles about a common, inferred TPW axis within each 95% confidence limit, it is possible to reconstruct Gondwanaland approximately. Because Gondwanaland formed during the same period of time as these hypothesized multiple TPW events, with possibly minor reshuffling of the constituent Rodinia Cratons, we expect its later-dispersed fragments to share the essentially common APW swath shown in (b). This is equivalent to the earliest APW swings depicted by Evans (2003; **figure 3**) using the same logic, and it expands upon the pole database of Kirschvink *et al.* (1997). Any alternative interpretation to the hypothesis of multiple, rapid type II TPW events for Late Neoproterozoic to Cambrian time must explain not only the individual, dispersed poles and continental APWPs, it must also explain why all of the constituent cratons of Gondwanaland show APWPs which are of differing sense in the modern geographic reference frame, but which restore to near-coincidence when viewed in the Gondwanaland reference frame.

multiple and rapid relative to the rates of plate motion. We consider this proof of concept that the Ediacaran–Cambrian multiple TPW hypothesis is a viable, non-trivially discounted explanation for general dispersion of paleomagnetic poles of that age.

14.5   Summary: Major Unresolved Issues and Future Work

Although little-metamorphosed Precambrian rocks exist and may hold primary magnetization with equal fidelity to unmetamorphosed Phanerozoic rocks, the ancient paleomagnetic database is severely underconstrained, both spatially and temporally. We consider it equally likely that better pan-continental paleomagnetic coverage of TPW intervals already hypothesized will offer non-TPW alternatives for those events, and that more detailed geographic and temporal APW determinations will recognize many new apparent TPW events.

For continental reconstructions of any age, quick and/or repeated, oscillatory type II TPW will be resolved better by detailed magnetostratigraphy of thick, continuous-deposited volcanic or

sedimentary successions. Relatively few such magnetostratigraphies exist, even through Paleozoic and Mesozoic time.

Geodynamic modeling of TPW should address the predicted effects of multiple and/or quick (type II) TPW events, as well as longer-duration, large-magnitude (type I) TPW events. We have not yet been able to satisfactorily fathom the dynamics of viscoelastic relaxation by a highly viscosity-structured mantle, if those dynamics in fact have been addressed in the geophysical literature.

Although mantle-driven TPW seems to predict (and even partly rely upon) dynamic topography, we understand that it is not clear that any modern-day surface physiographic feature is unambiguously isostatically uncompensated. We hope that this paradox will trigger additional work and debate in the near future.

If multiple, fast Ediacaran–Cambrian TPW events are not borne out by future, detailed magnetostratigraphic investigation, we will be fascinated to discover the otherwise incredible misfortune explaining the distribution of the dispersed APW swaths that today rotate those hypothesized TPW events into coincidence with the same parameters which reconstruct Gondwanaland in the Mesozoic Era.

Sequence-stratigraphic compilations for times of hypothesized TPW should show global quadrature, rather than global constructive synchroneity, when paleolatitudes of those deposits are considered in a spinner-diagram-derived global paleogeography.

Physical and chemical oceanographic changes predicted by TPW should be observed in the rock record synchronous with hypothesized or demonstrated TPW of sufficient magnitude; and some of these predictions (e.g., atmospheric temperature anomaly resulting from methane clathrate destabilization) should be testable by independent methods, where the rock record is amenable.

Acknowledgments

The authors are grateful for comments and criticisms from Dennis Kent, Bob Kopp, Adam Maloof, Ross Mitchell, and Will Sager. Adam Maloof developed spinner diagrams in concert with TDR and DADE. Jean-Pascal Cogne's *Paleomac©* program (downloaded with password permission from J-PC) readily produces spinner diagrams. This work is supported in part by a 3-year NSF Graduate Fellowship to TDR, NSF grants 9807741, 9814608, and 9725577 and funds from the NASA National Astrobiology Institute to JLK, and a David and Lucile Packard Foundation Award to DADE.

References

Alvarez W, et al. (1977) Upper Cretaceous–Paleocene magnetic stratigraphy at Gubbio, Italy. Part 5: Type Section for Late Cretaceous–Paleocene geomagnetic reversal time scale. *Geological Society of America Bulletin* 88(3): 383–388.

Alvarez W and Lowrie W (1978) Upper Cretaceous paleomagnetic stratigraphy at Moria (Umbrian Apennines, Italy) – Verification of Gubbio section. *Geophysical Journal of the Royal Astronomical Society* 55(1): 1–17.

Anderson DL (1982) Hotspots, polar wander, Mesozoic convection and the geoid. *Nature* 297: 391–393.

Anderson DL (1990) *Theory of the Earth*, 366 pp. Boston, MA. Blackwell Scientific Publications.

Anderson DL (1994) Superplumes or Supercontinents *Geology* 22(1): 39–42.

Andrews DL, et al. (2006) Uncertainties in plate reconstructions relative to the hotspots; Pacific-hotspot rotations and uncertainties for the past 68 million years. *Geophysical Journal International* 166(2): 939–951.

Athavale RN, et al. (1963) Paleomagnetism of some Indian Rocks. *Geophysical Journal of the Royal Astronomical Society* 7(3): 304–313.

Athavale RN, et al. (1972) Paleomagnetism and age of Bhader and Rewa sandstones from India. *Geophysical Journal of the Royal Astronomical Society* 28(5): 499–509.

Beck ME and Housen BA (2003) Absolute velocity of North America during the Mesozoic from paleomagnetic data. *Tectonophysics* 377(1-2): 33–54.

Besse J and Courtillot V (1991) Revised and synthetic apparent polar wander paths of the African, Eurasian, North-American and Indian Plates, and true polar wander since 200 Ma. *Journal of Geophysical Research-Solid Earth and Planets* 96(B3): 4029–4050.

Besse J and Courtillot V (2002) Apparent and true polar wander and the geometry of the geomagnetic field over the last 200 Myr. *Journal of Geophysical Research-Solid Earth* 107(B11).

Besse J and Courtillot V (2003) Apparent and true polar wander and the geometry of the geomagnetic field over the last 200 Myr (vol 107, art no 2300, 2002). *Journal of Geophysical Research-Solid Earth* 108(B10).

Biram J (1966) *A. Wegener: The Origin of Continents and Oceans 4th edn*. New York, NY: Dover Publications.

Boger SD, et al. (2001) Early Paleozoic tectonism within the east Antarctic craton: The final suture between east and west Gondwana?. *Geology* 29(5): 463–466.

Brasier MD, et al. (1994) Multiple delta-13C excursions spanning the Cambrian explosion to the Botomian crisis in Siberia. *Geology* 22(5): 455–458.

Buffett B and Archer D (2004) Global inventory of methane clathrate: Sensitivity to changes in the deep ocean. *Earth and Planetary Science Letters* 227(3-4): 185–199.

Camps P, et al. (2002) Comment on 'Stability of the Earth with respect to the spin axis for the last 130 million years' by JA. Tarduno and A.Y. Smirnov (2001) [*Earth and Planetary Science Letters* 184: 549–553] *Earth and Planetary Science Letters* 198(3-4): 529–532.

Celaya MA, et al. (1999) Climate-driven polar motion. *Journal of Geophysical Research Solid Earth* 104(B6): 12813–12829.

Chase CG and Sprowl DR (1983) The modern geoid and ancient plate boundaries. *Earth and Planetary Science Letters* 62(3): 314–320.

Collins AS and Pisarevsky SA (2005) Amalgamating eastern Gondwana: The evolution of the Circum-Indian orogens. *Earth-Science Reviews* 71(3-4): 229–270.

Cottrell RD and Tarduno JA (2000) Late Cretaceous true polar wander: Not so fast. *Science* 288: 2283a.

Cottrell RD and Tarduno JA (2003) A Late Cretaceous pole for the Pacific plate: Implications for apparent and true polar wander and the drift of hotspots. *Tectonophysics* 362(1-4): 321–333.

Darwin G (1877) On the influence of geological changes on the Earth's axis of rotation. *Philosophical Transactions of the Royal Society of London* 167: 271–312.

Davies GF (1984) Lagging mantle convection, the geoid and mantle structure. *Earth and Planetary Science Letters* 69(1): 187–194.

Dickens GR, et al. (1995) Dissociation of oceanic methane hydrate as a cause of the carbon-isotope excursion at the end of the Paleocene. *Paleoceanography* 10(6): 965–971.

Dickens GR, et al. (1997) A blast of gas in the latest Paleocene: Simulating first-order effects of massive dissociation of oceanic methane hydrate. *Geology* 25(3): 259–262.

Dickman SR (1979) Continental-drift and true polar wandering. *Geophysical Journal of the Royal Astronomical Society* 57(1): 41–50.

DiVenere V and Kent DV (1999) Are the Pacific and Indo-Atlantic hotspots fixed? Testing the plate circuit through Antarctica. *Earth and Planetary Science Letters* 170(1-2): 105–117.

DuBois PM (1957) Comparison of palaeomagnetic results from selected rocks of Great Britain and North America. *Philosophical Magazine Supplement Advances in Physics* 6: 177.

Elkins-Tanton LT and Hager BH (2005) Giant meteoroid impacts can cause volcanism. *Earth and Planetary Science Letters* 239(3-4): 219–232.

Evans DA, *et al.* (1998) Polar wander and the Cambrian. *Science* 279: 9a–9e.

Evans DAD and Kirschvink JL (1999) Multiple episodes of rapid true polar wander in Neoproterozoic–Cambrian time?, paper presented at Geological Society of America Annual meeting, Geological Society of America (GSA). Boulder, CO.

Evans DAD (2003) True polar wander and supercontinents. *Tectonophysics* 362(1-4): 303–320.

Evans DAD (2006) Proterozoic low orbital obliquity and axial-dipolar geomagnetic field from evaporite palaeolatitudes. *Nature* 444(7115): 51–55.

Evans DAD, *et al.* (2000) A high-quality mid-Neoproterozoic paleomagnetic pole from south China, with implications for ice ages and the breakup configuration of Rodinia. *Precambrian Research* 100(1-3): 313–334.

Evans ME (1976) Test of dipolar nature of geomagnetic field throughout Phanerozoic time. *Nature* 262(5570): 676–677.

Evans ME (2005) Testing the geomagnetic dipole hypothesis: Palaeolatitudes sampled by large continents. *Geophysical Journal International* 161(2): 266–267.

Farley KA and Eltgroth SF (2003) An alternative age model for the Paleocene-Eocene thermal maximum using extraterrestrial He-3. *Earth and Planetary Science Letters* 208(3-4): 135–148.

Fisher D (1974) Some more remarks on polar wandering. *Journal of Geophysical Research* 79(26): 4041–4045.

Fujita M, *et al.* (2002) Oceanic torques on solid Earth and their effects on Earth rotation. *Journal of Geophysical Research-Solid Earth* 107(B8).

Funaki M (1984) Investigation of the Paleomagnetism of the basement-complex of Wright Valley, southern Victoria Land, Antarctica. *Journal of Geomagnetism and Geoelectricity* 36(11): 529–563.

Gold T (1955) Instability of the Earth's axis of rotation. *Nature* 175: 526–529.

Goldreich P and Toomre A (1969) Some remarks on polar wandering. *Journal of Geophysical Research* 74(10): 2555–2567.

Gordon RG, *et al.* (1984) Paleomagnetic Euler Poles and the apparent polar wander and absolute motion of North-America since the Carboniferous. *Tectonics* 3(5): 499–537.

Gordon RG (1987) Polar wandering and paleomagnetism. *Annual Review of Earth and Planetary Sciences* 15: 567–593.

Gordon RG (1995) Plate motions, crustal and lithospheric mobility, and paleomagnetism: Prospective viewpoint. *Journal of Geophysical Research,Solid Earth* 100(B12): 24367–24392.

Grunow AM (1995) Implications for Gondwana of new Ordovician Paleomagnetic data from Igneous rocks in Southern Victoria Land, East Antarctica. *Journal of Geophysical Research,Solid Earth* 100(B7): 12589–12603.

Grunow AM and Encarnacion JP (2000) Cambro-Ordovician palaeomagnetic and geochronologic data from southern Victoria Land, Antarctica: Revision of the Gondwana apparent polar wander path. *Geophysical Journal International* 141(2): 391–400.

Hager BH, *et al.* (1985) Lower mantle heterogenity, dynamic topography and the geoid. *Nature* 313(6003): 541–546.

Halverson GP, *et al.* (2002) A major perturbation of the carbon cycle before the Ghaub glaciation (Neoproterozoic) in Namibia: Prelude to snowball Earth? *Geochemistry, Geophysics and Geosystems* 3.

Halverson GP, *et al.* (2005) Toward a Neoproterozoic composite carbon-isotope record. *Geological Society of America Bulletin* 117(9–10): 1181–1207.

Harrison CGA and Lindh T (1982) Comparison Between the Hot Spot and Geomagnetic-Field Reference Frames. *Nature* 300(5889): 251–252.

Higgins JA and Schrag DP (2006) Beyond methane: Towards a theory for the Paleocene-Eocene Thermal Maximum. *Earth and Planetary Science Letters* 245(3–4): 523–537.

Higgins MD (2005) A new model for the structure of the Sept Iles Intrusive suite, Canada. *Lithos* 83: 199–213.

Hynes A (1990) 2-Stage Rifting of Pangea by 2 different mechanisms. *Geology* 18(4): 323–326.

Irving E (1956) Palaeomagnetic and paleoclimatological aspects of polar wandering. *Geofisica Pura et Applicata* 33: 23–41.

Irving E (1957) Rock magnetism: a new approach to some palaeogeographic problems. *Advances in Physics* 6(22): 194–218.

Irving E (1988) Paleomagnetic confirmation of Continental Drift. *EOS, Transactions of American Geophysical Union* 69(44): 994–1014.

Irving E (2005) The role of latitude in mobilism debates. *Proceedings of the National Academy of Sciences of the United States of America* 102(6): 1821–1828.

Irving E and Irving GA (1982) Apparent polar wander paths carboniferous through cenozoic and the assembly of Gondwana. *Geophysical Surveys* 5(2): 141–188.

Jurdy DM and Van der Voo R (1974) Method for separation of true polar wander and continental-drift, including results for last 55 my. *Journal of Geophysical Research* 79(20): 2945–2952.

Jurdy DM and Van der Voo R (1975) True polar wander since early cretaceous. *Science* 187(4182): 1193–1196.

Kent DV and Smethurst MA (1998) Shallow bias of paleomagnetic inclinations in the Paleozoic and Precambrian. *Earth and Planetary Science Letters* 160(3–4): 391–402.

Kirschvink JL (1978) Precambrian-Cambrian boundary problem - Paleomagnetic directions from Amadeus Basin, Central Australia. *Earth and Planetary Science Letters* 40(1): 91–100.

Kirschvink JL, *et al.* (1991) The Precambrian-Cambrian boundary: Magnetostratigraphy and Carbon Isotopes resolve correlation problems between Siberia, Morocco, and South China. *GSA Today* 1: 69–91.

Kirschvink JL, *et al.* (1997) Evidence for a large-scale Early Cambrian reorganization of continental masses by inertial interchange true polar wander. *Science* 277: 541–545.

Kirschvink JL, et al. (2005) Rapid inertial-interchange true polar wander during ediacaran time induced by intrusion of the sept-îles intrusive suite, Quebec, Canada: A possible trigger for massive sea-level and carbon isotopic excursions, paper presented at Gondwana XII, Mendoza, Argentina.

Kirschvink JL and Raub TD (2003) A methane fuse for the Cambrian explosion: Carbon cycles and true polar wander. *Comptes Rendus Geoscience* 335(1): 65–78.

Klootwijk CT, *et al.* (1986a) Paleomagnetic constraints on formation of the Mianwali Reentrant, Trans-Indus and Western Salt-Range, Pakistan. *Earth and Planetary Science Letters* 80(3–4): 394–414.

Klootwijk CT, *et al.* (1986b) Rotational overthrusting of the northwestern Himalaya – Further paleomagnetic evidence from the Riasi thrust sheet, Jammu foothills, India. *Earth and Planetary Science Letteres* 80(3-4): 375–393.

Lanza R and Tonarini S (1998) Palaeomagnetic and geochronological results from the Cambro-Ordovician granite harbour intrusives inland of Terra Nova Bay (Victoria Land, Antarctica). *Geophysical Journal International* 135(3): 1019–1027.

Li ZX, *et al.* (2004) A 90 degrees spin on Rodinia: Possible causal links between the Neoproterozoic supercontinent, superplurne, true polar wander and low-latitude glaciation. *Earth and Planetary Science Letters* 220(3-4): 409–421.

Llanos MPI, Riccardi AC, and Singer SE (2006) Palaeomagnetic study of Lower Jurassic marine strata from the Neuquen Basin, Argentina: A new Jurassic apparent polar wander path for South America. *Earth and Planetary Science Letters* 252(3–4): 379–397.

Magaritz M, et al. (1986) Carbon-isotope events across the Precambrian–Cambrian boundary on the Siberian platform. *Nature* 320: 258–259.

Magaritz M, et al. (1991) The Precambrian–Cambrian boundary problem: Carbon isotope correlations for Vendian and Tommotian time between Siberia and Morocco. *Geology* 19: 847–850.

Maloof AC, et al. (2005) An expanded record of Early Cambrian carbon cycling from the anti-atlas margin, Morocco. *Canadian Journal of Earth Sciences* 42(12): 2195–2216.

Maloof AC, et al. (2006) Combined paleomagnetic, isotopic, and stratigraphic evidence for true polar wander from the Neoproterozoic Akademikerbreen Group, Svalbard, Norway. *Geological Society of America Bulletin* 118(9-10): 1099–1124.

Manzoni M and Nanni T (1977) Paleomagnetism of Ordovician lamprophyres from Taylor Valley, Victoria Land, Antarctica. *Pure and Applied Geophysics* 115(4): 961–977.

Marcano MC, et al. (1999) True polar wander during the Permo-Triassic. *Journal of Geodynamics* 28(2-3): 75–95.

Matsuyama I, et al. (2006) Rotational stability of dynamic planets with elastic lithospheres. *Journal of Geophysical Research-Planets* 111(E2): E02003.

McQueen DM, et al. (1972) Cambro-Ordovician paleomagnetic pole position and rubidium–strontium total rock isochron for charnockitic rocks from Mirnyy Station, east Antarctica. *Earth and Planetary Science Letters* 16(3): 433–438.

Meert JG (1999) A paleomagnetic analysis of Cambrian true polar wander. *Earth and Planetary Science Letters* 168(1-2): 131–144.

Mitrovica JX, et al. (2001a) Glacial isostatic adjustment on a rotating Earth. *Geophysical Journal International* 147(3): 562–578.

Mitrovica JX, et al. (2001b) Recent mass balance of polar ice sheets inferred from patterns of global sea-level change. *Nature* 409(6823): 1026–1029.

Mitrovica JX, et al. (2005) The rotational stability of an ice-age Earth. *Geophysical Journal International* 161(2): 491–506.

Mitrovica JX and Forte AM (2004) A new inference of mantle viscosity based upon joint inversion of convection and glacial isostatic adjustment data. *Earth and Planetary Science Letters* 225(1-2): 177–189.

Montanez IP, et al. (2000) Evolution of the Sr and C isotope composition of Cambrian oceans. *GSA Today* 10(5): 1–7.

Moreau M-G, et al. (2007) A New global Paleocene–Eocene apparent polar wandering path loop by 'stacking' magnetostratigraphies: Correlations with high latitude climatic data. *Earth and Planetary Science Letters,* doi: 10.1016/j.epsl.2007.05.025.

Mound JE, et al. (1999) A sea-level test for inertial interchange true polar wander events. *Geophysical Journal International* 136(3): F5–F10.

Mound JE, et al. (2001) Sea-level change and true polar wander during the Late Cretaceous. *Geophysical Research Letters* 28(10): 2057–2060.

Mound JE, et al. (2003) The equilibrium form of a rotating Earth with an elastic shell. *Geophysical Journal International* 152(1): 237–241.

Mound JE and Mitrovica JX (1998) True polar wander as a mechanism for second-order sea-level variations. *Science* 279(5350): 534–537.

Muller RD, et al. (1993) Revised plate motions relative to the hotspots from combined Atlantic and Indian-Ocean hotspot tracks. *Geology* 21(3): 275–278.

Munk WH and MacDonald GJF (1960) *The Rotation of the Earth.* Cambridge, UK: Cambridge University Press.

Nakada M (2002) Polar wander caused by the Quaternary glacial cycles and fluid Love number. *Earth and Planetary Science Letters* 200(1-2): 159–166.

Nakiboglu SM and Lambeck K (1980) Deglaciation effects on the rotation of the Earth. *Geophysical Journal of the Royal Astronomical Society* 62(1): 49–58.

Nimmo F (2002) Why does Venus lack a magnetic field? *Geology* 30(11): 987–990.

Nimmo F and Pappalardo RT (2006) Diapir-induced reorientation of Saturn's moon Enceladus. *Nature* 441(7093): 614–616.

Payne JL, et al. (2004) Large perturbations of the carbon cycle during recovery from the end-Permian extinction. *Science* 305(5683): 506–509.

Peltier WR (1998) Postglacial variations in the level of the sea: Implications for climate dynamics and solid-Earth geophysics. *Reviews of Geophysics* 36(4): 603–689.

Petronotis KE and Gordon RG (1999) A Maastrichtian palaeomagnetic pole for the Pacific plate from a skewness analysis of marine magnetic anomaly 32. *Geophysical Journal International* 139(1): 227–247.

Phillips RJ, et al. (2001) Ancient geodynamics and global-scale hydrology on Mars. *Science* 291: 2587–2591.

Piper JDA (2006) A approximate to 90 degrees Late Silurian–Early Devonian apparent polar wander loop: The latest inertial interchange of planet Earth? *Earth and Planetary Science Letters* 250(1-2): 345–357.

Piper JDA and Zhang QR (1999) Palaeomagnetic study of Neoproterozoic glacial rocks of the Yangzi Block: Palaeolatitude and configuration of south China in the Late Proterozoic supercontinent. *Precambrian Research* 94(1-2): 7–10.

Pisarevsky SA, et al. (2003) Models of Rodinia assembly and fragmentation. In: Yoshida M, Windley BF, and Dasgupta S (eds.) *Proterozoic East Gondwana: Supercontinent Assembly and Breakup,* pp. 35–55. London: Geological Society of London.

Porco CC, et al. (2006) Cassini observes the active south pole of Enceladus. *Science* 311(5766): 1393–1401.

Prevot M, et al. (2000) Evidence for a 20 degrees tilting of the Earth's rotation axis 110 million years ago. *Earth and Planetary Science Letters* 179(3-4): 517–528.

Ranalli G (1995) *Rheology of the Earth,* 413 pp. London: Chapman and Hall.

Randall DE, et al. (2000) A new Late Middle Cambrian paleomagnetic pole for the Ellsworth Mountains, Antarctica. *Journal of Geology* 108(4): 403–425.

Richards MA, et al. (1997) An explanation for Earth's long-term rotational stability. *Science* 275(5298): 372–375.

Richards MA and Engebretson DC (1992) Large-scale mantle convection and the history of subduction. *Nature* 355(6359): 437–440.

Riisager P, Riisager J, Abrahamsen N, and Waagskin R (2002) New Paleomagnetic pole and magnetostratigraphy of Faroe Islands flood volcanics, North Atlantic igneous province. *Earth and Planetary Science Letters* 201(2): 261–276.

Riisager P, et al. (2003) Paleomagnetic paleolatitude of Early Cretaceous Ontong Java Plateau basalts: Implications for Pacific apparent and true polar wander. *Earth and Planetary Science Letters* 208(3-4): 235–252.

Runcorn SK (1955) The Earth's magnetism. *Scientific American* 193(3): 152–154.

Runcorn SK (1956) Palaeomagnetic comparisons between Europe and North America. *Proceedings of the Geological Association of Canada* 8(1): 77–85.

Sabadini R (2002) Paleoclimate – Ice sheet collapse and sea level change. *Science* 295(5564): 2376–2377.

Sabadini R, *et al.* (1990) Eustatic sea level fluctuations induced by polar wander. *Nature* 345: 708–710.

Sabadini R, *et al.* (2002) Ice mass loss in Antarctica and stiff lower mantle viscosity inferred from the long wavelength time dependent gravity field. *Geophysical Research Letters* 29(10): 1373.

Sabadini R and Peltier WR (1981) Pleistocene deglaciation and the Earth's rotation – Implications for mantle viscosity. *Geophysical Journal of the Royal Astronomical Society* 66(3): 553–578.

Sager WW (2006) Cretaceous paleomagnetic apparent polar wander path for the pacific plate calculated from Deep Sea Drilling Project and Ocean Drilling Program basalt cores. *Physics of the Earth and Planetary Interiors* 156(3-4): 329–349.

Sager WW and Koppers AAP (2000a) Late Cretaceous polar wander of the Pacific plate: Evidence of a rapid true polar wander event. *Science* 287(5452): 455–459.

Sager WW and Koppers AAP (2000b) Late Cretaceous true polar wander: Not so fast. *Science* 288(5452): 2283a.

Satolli S, Besse J, Speranza F, and Calamita F (2007) The 125–150 Ma high-resolution Apparent Polar Wander Path for Adria from magnetostratigraphic sections in Umbria-Marche (Northern Apennines, Italy): Timing and duration of the global Jurassic–Cretaceous hairpin turn. *Earth and Planetary Science Letters* 257(1–2): 329–342.

Schmidt PW and Williams GE (1996) Palaeomagnetism of the ejecta-bearing Bunyeroo formation, Late Neoproterozoic, Adelaide fold belt, and the age of the Acraman impact. *Earth and Planetary Science Letters* 144(3-4): 347–357.

Schrag DP, *et al.* (2002) On the initiation of a snowball Earth. *Geochemistry Geophysics Geosystems* 3(6): (doi:10.1029/2001GC000219).

Schult FR and Gordon RG (1984) Root mean-square velocities of the continents with respect to the hot spots since the Early Jurassic. *Journal of Geophysical Research* 89(NB3): 1789–1800.

Seitz F and Schmidt M (2005) Atmospheric and oceanic contributions to Chandler wobble excitation determined by wavelet filtering. *Journal of Geophysical Research-Solid Earth* 110: B11406.

Sleep NH (1988) Gradual entrainment of a chemical layer at the base of the mantle by overlying convection. *Geophysical Journal-Oxford* 95(3): 437–447.

Soldati G, *et al.* (2001) The effect of global seismicity on the polar motion of a viscoelastic Earth. *Journal of Geophysical Research Solid Earth* 106(B4): 6761–6767.

Spada G (1997) Why are earthquakes nudging the pole towards 140 degrees E? *Geophysical Research Letters* 24(5): 539–542.

Spada G, *et al.* (2006) Glacial isostatic adjustment and relative sea-level changes: The role of lithospheric and upper mantle heterogeneities in a 3-D spherical Earth. *Geophysical Journal International* 165(2): 692–702.

Spada G and Boschi L (2006) Using the post-Widder formula to compute the Earth's viscoelastic Love numbers. *Geophysical Journal International* 166(1): 309–321.

Stacey F (1992) *Physics of the Earth*. 3rd edn. 513 pp. Brisbane, QLD: Brookfield Press.

Steinberger B and Oconnell RJ (1997) Changes of the Earth's rotation axis owing to advection of mantle density heterogeneities. *Nature* 387(6629): 169–173.

Steinberger B and O'Connell RJ (2002) The convective mantle flow signal in rates of true polar wander. In: Mitrovica JX and Vermeersen LLA (eds.) *Geodynamics Series 29: Ice Sheets, Sea Level and the Dynamic Earth*, pp. 233–256. Washington, DC: American Geophysical Union.

Stevenson DJ (1983) Planetary magnetic fields. *Reports on Progress in Physics* 46(5): 555–620.

Strik G, *et al.* (2003) Palaeomagnetism of flood basalts in the Pilbara Craton, western Australia: Late Archaean continental drift and the oldest known reversal of the geomagnetic field. *Journal of Geophysical Research-Solid Earth* 108(B12): 2551.

Tarduno JA, *et al.* (2003) The Emperor Seamounts: Southward motion of the Hawaiian hotspot plume in Earth's mantle. *Science* 301(5636): 1064–1069.

Tarduno JA and Cottrell RD (1997) Paleomagnetic evidence for motion of the Hawaiian hotspot during formation of the Emperor seamounts. *Earth and Planetary Science Letters* 153(3-4): 171–180.

Tarduno JA and Gee J (1995) Large-scale motion between Pacific and Atlantic hotspots. *Nature* 378(6556): 477–480.

Tarduno JA and Smirnov AV (2001) Stability of the Earth with respect to the spin axis for the last 130 million years. *Earth and Planetary Science Letters* 184(2): 549–553.

Tarduno JA and Smirnov AV (2002) Response to comment on 'Stability of the Earth with respect to the spin axis for the last 130 Million Years' by P. Camps, M. Prevot, M. Daignieres, and P. Machetel. *Earth and Planetary Science Letters* 198(3-4): 533–539.

Torsvik TH, *et al.* (1998) Polar wander and the Cambrian. *Science* 279: 9 (correction p. 307).

Torsvik TH, *et al.* (2002) Relative hotspot motions versus true polar wander. *Earth and Planetary Science Letters* 202(2): 185–200.

Tsai VC and Stevenson DJ (2007) Theoretical constraints on true polar wander. *Journal of Geophysical Research* 112: B05415 (doi:10.1029.2005JB003923).

Valeriano CM, *et al.* (2004) U–Pb geochronology of the southern Brasilia belt (SE-Brazil): Sedimentary provenance, Neoproterozoic orogeny and assembly of west Gondwana. *Precambrian Research* 130: 27–55.

Van der Voo R (1994) True polar wander during the Middle Paleozoic. *Earth and Planetary Science Letters* 122(1-2): 239–243.

Van Fossen MC and Kent DV (1992) Paleomagnetism of 122 Ma plutons in New England and the Mid-Cretaceous paleomagnetic field in North America: True polar wander or large-scale differential mantle motion? *Journal of Geophysical Research* 97: 19651–19661.

Veevers JJ (2004) Gondwanaland from 650–500 Ma assembly through 320 Ma merger in Pangea to 185–100 Ma breakup: Supercontinental tectonics via stratigraphy and radiometric dating. *Earth-Science Reviews* 68(1-2): 1–132.

Vermeersen LLA and Sabadini R (1999) Polar wander, sea-level variations and ice age cycles. *Surveys in Geophysics* 20(5): 415–440.

Ward PD, *et al.* (2005) Abrupt and gradual extinction among land vertebrates in the Karoo basin, South Africa. *Science* 307: 709–714.

Watts DR and Bramall AM (1981) Paleomagnetic evidence from the Ellsworth Mountains supports microplate nature of western Antarctica. *Geophysical Journal of the Royal Astronomical Society* 65(1): 271.

Wegener A (1929) *Die Entstehung der Kontinente und Ozeane. Vierte gaenzlich umgearbeitete Auflage*. Braunschweig, Germany: Druck and Verlag von Fridrich Vieweg und Sohn.

Zhong SJ and Hager BH (2003) Entrainment of a dense layer by thermal plumes. *Geophysical Journal International* 154(3): 666–676.

Zijderveld JD (1968) Natural remanent magnetizations of some intrusive rocks from Sor Rondane Mountains Queen Maud Land Antarctica. *Journal of Geophysical Research* 73(12): 3773–3785.

Printed and bound by CPI Group (UK) Ltd, Croydon, CR0 4YY

03/10/2024

01040325-0016